AF903656

Werner Nagel, Christoph Thäle, Viola Weiß

Iteration Stable Tessellations

De Gruyter Studies in Mathematics

Editor-in-Chief
Guozhen Lu, Storrs, USA

Editorial Board
Carsten Carstensen, Berlin, Germany
Gavril Farkas, Berlin, Germany
Nicola Fusco, Napoli, Italy
Fritz Gesztesy, Waco, Texas, USA
Zenghu Li, Beijing, China
Karl-Hermann Neeb, Erlangen, Germany
René L. Schilling, Dresden, Germany
Volkmar Welker, Marburg, Germany

Volume 105

Werner Nagel, Christoph Thäle, Viola Weiß

Iteration Stable Tessellations

Random Tessellations Generated by Division of Polytopes

DE GRUYTER

Mathematics Subject Classification 2020
05B45, 37A25, 52A22, 60D05, 60F05, 60F17, 60G44, 60G55, 60G57, 60J76

Authors
Werner Nagel
Friedrich Schiller University Jena
Germany
werner.nagel@uni-jena.de

Viola Weiß
Ernst Abbe University of Applied Sciences Jena
Germany
Viola.Weiss@eah-jena.de

Christoph Thäle
Ruhr University Bochum
Germany
christoph.thaele@rub.de

ISSN 0179-0986
ISBN 978-3-11-914264-9
ISBN 978-3-11-222240-9 (PDF)
ISBN 978-3-11-222255-3 (E-PUB)
DOI https://doi.org/10.1515/9783112222409

Library of Congress Control Number: 2026933870

Bibliographic information published by the Deutsche Nationalbibliothek
The Deutsche Nationalbibliothek lists this publication in the Deutsche Nationalbibliografie; detailed bibliographic data are available on the Internet at https://dnb.dnb.de.

© 2026 Walter de Gruyter GmbH, Genthiner Straße 13, 10785 Berlin, Germany

All rights reserved. No part of this publication may be reproduced, translated, stored in a retrieval system, or transmitted in any form or by any means, electronic, mechanical, photocopying, recording or otherwise, without prior written permission from the publisher. Text and datamining for commercial purposes requires the publisher's permission.

De Gruyter and Walter de Gruyter GmbH are part of De Gruyter Brill.
www.degruyterbrill.com

Questions about General Product Safety Regulation: productsafety@degruyterbrill.com

Printing: CPI books GmbH, Leck

MIX
Papier | Fördert gute Waldnutzung
FSC® C083411

Dedicated to the memory of
Joseph Mecke (1938–2014) and Tomasz Schreiber (1975–2010)

Contents

1 Prologue

1.1 Introduction

Tessellations, also known as mosaics or tilings, are patterns created by the repetition of geometric shapes without gaps or overlaps. They are widely used in art and architecture, especially in Islamic design, such as the intricate tile work of the Alhambra in Spain. Artists like M. C. Escher and P. Mondrian also employed tessellations to explore symmetry and visual perception. Mosaics, often associated with ancient Roman and Byzantine art, utilize small tiles – tesserae – to craft elaborate images and patterns. Beyond their artistic allure, tessellations are integral to various scientific disciplines, serving as a foundation in mathematics for studies of geometry, symmetry, and periodic or aperiodic patterns. In physics and materials science, they help to explain crystal structures, atomic arrangements, and quasicrystals, whereas in biology, natural tessellations like honeycombs and cellular structures exemplify efficient space utilization. In computer science, tessellations are indispensable for texture mapping, graphics design, and computational geometry. In supervised learning, tessellations are integral to decision trees and ensemble methods like Random Forests. Another emerging area is in deep learning, where tessellation-inspired structures can be used to interpret the behavior of neural networks, particularly in understanding decision boundaries.

Since the late 1970s, substantial progress in the theory of *random* tessellations has firmly positioned them as a fundamental element of stochastic geometry. This progress has provided a rigorous mathematical framework for analyzing random spatial patterns, proving essential across numerous scientific fields. Random tessellations are now widely used to model and interpret phenomena such as the distribution of cells in biological tissues, the structure of materials in physics, and the design of wireless networks in telecommunication.

One of the most celebrated models in tessellation theory is the Voronoi tessellation, particularly the Poisson–Voronoi tessellation, which plays a pivotal role in stochastic geometry; see the left panel of Figure 1.1. This model was introduced to describe the growth of regions, referred to as zones of influence or cells, originating from a set of points scattered in space. Each point generates a cell encompassing all locations closer to it than to any other point, producing a partition that encapsulates spatial proximity relationships. The Poisson–Voronoi tessellation, in which cells emerge around points from a stationary Poisson point process, is a fundamental model in stochastic geometry. It plays a crucial role in diverse applications, from modeling biological tissue structures to understanding material compositions and spatial distributions in physics.

Another key class of tessellations is the Poisson hyperplane tessellation, where space is divided by randomly placed hyperplanes; see the right panel of Figure 1.1. These hyperplanes are determined by a Poisson point process in the space of all hyperplanes, leading to intricate and versatile spatial structures. These tessellations have been the subject of extensive mathematical study. The comprehensive theory of Poisson

https://doi.org/10.1515/9783112222409-001

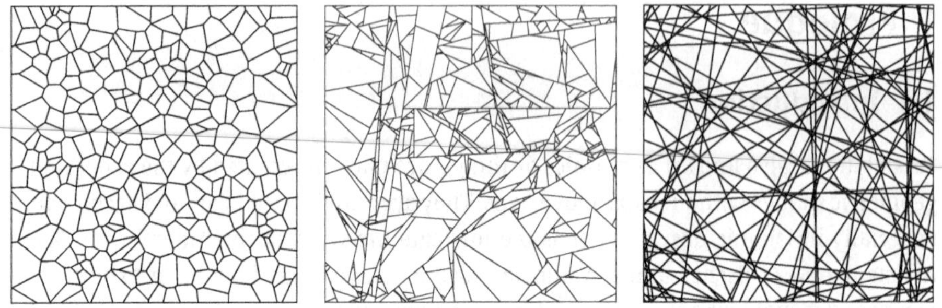

Figure 1.1: Simulations of three tessellation models in the plane. Left: Poisson–Voronoi tessellation. Middle: STIT tessellation. Right: Poisson line tessellation.

hyperplane tessellations is thoroughly detailed in the monograph [82], which serves as a foundational resource for understanding their properties.

STIT – short for **ST**able with respect to **IT**eration – tessellations are a relatively recent discovery in the field of stochastic geometry. Their defining characteristic is the concept of iteration, also known as nesting, which is an operation applied to tessellations. Starting with a single tessellation, referred to as the frame, and a sequence of other tessellations, which are indexed by the cells of the frame, the iteration produces a new tessellation by nesting a cut-out of each sequential tessellation into the corresponding cells of the frame. As a result, different cells in the frame are subdivided by different tessellations, creating a complex pattern in which the cells are not arranged in a so-called face-to-face position; see the middle panel of Figure 1.1.

The origin of STIT tessellations can be traced to a mathematical question posed by R. V. Ambartzumian in the 1980s. Ambartzumian inquired whether the repeated iteration of independent and identically distributed random tessellations, when appropriately rescaled, would converge to a limiting tessellation. If such a tessellation existed, its distribution would need to exhibit a remarkable stability property: starting the iteration process with the limiting tessellation would reproduce in distribution the same tessellation after every iteration step and after rescaling. This intriguing theoretical challenge laid the groundwork for subsequent research. A major breakthrough came in 2005, when the existence and uniqueness in distribution of a STIT tessellation were rigorously established by the first and the last authors of this book in [145]. Their work demonstrated not only that STIT tessellations exist, but also that they possess remarkable mathematical properties, making them a highly tractable model for theoretical investigation.

This book brings together the key advances in STIT tessellation research, offering a comprehensive reference for both theoretical exploration and practical applications. It caters to researchers working on the theoretical development of tessellations and those applying these models across a wide range of scientific fields. By presenting foundational concepts and rigorous mathematical proofs, the book aims to deepen understanding and foster broader applications of STIT tessellations in stochastic geometry and related disciplines.

STIT tessellations belong to a broader class of random tessellations generated by the consecutive division of polytopes, commonly referred to as cells. These processes, known as cell division processes or cell splitting processes, represent a family of stochastic processes where the tessellation evolves over time as cells are successively divided into smaller parts. The conceptual foundation for such processes can be traced back to 1984, when Cowan [37] introduced a model that aimed to represent the consecutive division of biological cells. In his model, he proposed that the life times of cells – the time intervals before they undergo division – should follow independent exponential distributions. This approach was inspired by biological processes where cell divisions occur randomly over time. However, Cowan left open two key questions: how to appropriately choose the parameters governing the exponential life times, and whether the iterative division process would lead to a limiting tessellation. Cowan's pioneering ideas laid the groundwork for subsequent advancements in stochastic geometry and the study of dynamical random tessellations. His questions regarding the choice of life time parameters and the existence of limiting tessellations became central problems in the development of models like the STIT tessellation, which addressed these challenges through rigorous mathematical frameworks.

Cowan [39] further advanced the study of cell division processes by providing a systematic framework for a wide class of such models. This systematization introduced a structured approach for describing and analyzing tessellation-valued stochastic processes. Cowan's framework, adapted to a continuous-time setup, is built around two key characteristics:

- **Life time L**: This refers to the distribution of the random life span of a cell, defined as the time interval between its creation – via the division of its mother cell – and its eventual division. The life time distribution is a fundamental component, governing the temporal dynamics of the process.
- **Division rule D**: This specifies the rule by which a cell divides at the end of its life span. The division rule determines how the space occupied by the cell is subdivided into smaller polytopes, influencing the geometric evolution of the tessellation.

By formalizing these two aspects, Cowan created a unifying framework that applies to a broad spectrum of cell division models. This framework not only supports the analysis of specific cases, such as those with exponentially distributed life times, but also facilitates the exploration of alternative models with varied life time distributions and division rules.

In the case of cell division processes operating in discrete time, the concept of a life time distribution is replaced by a selection rule. This rule determines which cell in the tessellation will be divided at each successive time step. Unlike continuous-time processes, where the timing of cell division is governed by random life spans, discrete-time models focus on sequential decision-making to select the next cell for division.

A nontrivial challenge in the study of cell division processes is determining how to appropriately initiate the process. At least three approaches have been proposed for launching such processes, each with distinct methodologies and implications:

Construction in a window

This approach begins with a d-dimensional polytope W, referred to as the window, in the Euclidean space $\mathbb{R}^d$. At time $t = 0$, the window W serves as the initial cell, and the rules of life time L and division D are applied to it. The process then continues iteratively within W, subdividing the generated cells according to the same rules as illustrated in Figure 1.2. A critical issue with this method is the dependence on the choice of the window W. Specifically, the question of spatial consistency arises: does the distribution of cells within W remain unaffected if the process is instead initiated in a larger window $W' \supseteq W$? This local construction approach is central to the definition of STIT tessellations in windows, as detailed in Definition 4.2, whereas the spatial consistency problem is addressed rigorously in Section 5.3.

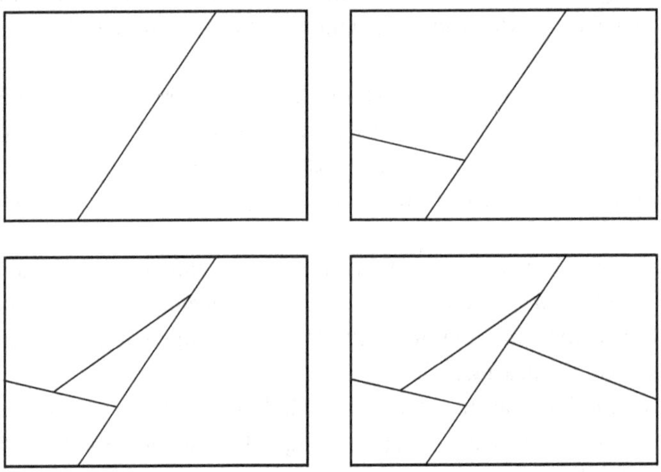

Figure 1.2: Consecutive division of a rectangular window in the plane.

Initial tessellation

In this method, it is assumed that a random tessellation of whole $\mathbb{R}^d$ is already present at time $t = 0$. The rules L and D are then applied to this initial tessellation to evolve the process. Although this approach is conceptually straightforward, it cannot be used directly to construct a STIT tessellation process. However, Chapter 7 demonstrates that STIT tessellations can emerge as a weak limit under rescaled iteration of tessellations, showing their connection to this framework.

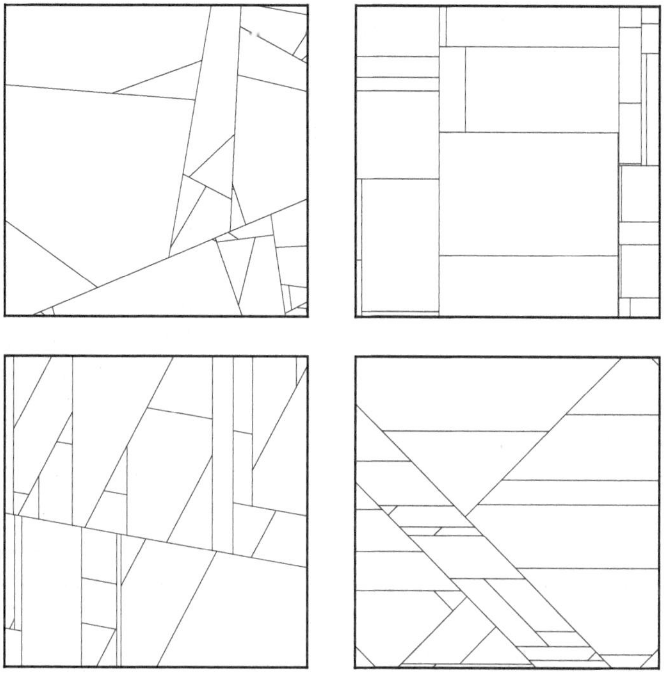

Figure 1.3: Simulations of STIT tessellations for several spherical directional distributions. Upper left: Isotropic. Upper right: Mondrian. Lower panels: Three directions. Figures 4.3 and 4.4 illustrate trajectories of the STIT tessellation processes for the isotropic and Mondrian cases, respectively.

No initial tessellation – use of an auxiliary tessellation process for $t > 0$

In this approach, no explicit initial tessellation of whole $\mathbb{R}^d$ is required. Instead, an auxiliary tessellation process is introduced for $t > 0$ to generate the required structure. An example of this method is provided in the global construction of the STIT tessellation process described in Chapter 8. Here a family of auxiliary Poisson hyperplane tessellation processes is used to define and construct the STIT tessellation globally, bypassing the need for a predefined starting configuration.

A key characteristic of STIT tessellations is that the distribution of their typical cell – the probabilistically representative cell of the tessellation – is identical to that of a Poisson hyperplane tessellation (PHT) with a suitably chosen intensity measure. This remarkable property bridges the STIT tessellations and Poisson hyperplane tessellations through their typical-cell distributions. This feature of STIT tessellations also addresses a problem raised by R. E. Miles and M. S. Mackisack, who sought to construct diverse random tessellations with typical cells distributed identically to those of PHTs. Their work, as detailed in [136], was motivated by the need to understand and expand the variety of tessellations sharing this property. Earlier, in [6], examples of such tessellations in two-dimensional spaces were introduced, providing specific constructions with typical-cell distributions matching those of PHTs. However, STIT tessellations offer a fundamentally different construction compared to these earlier models. Although they share the distribution of the typical cell with PHTs, the geometric and structural properties of STIT tessellations set them apart. For instance, as illustrated in Figure 16.3, certain configurations and principles, such as the mikado principle discussed in other constructions, are satisfied for STIT tessellation but not for the models considered in [6, 136].

When comparing STIT tessellations to PHTs, a fundamental difference lies in the mutual arrangement of their cells. PHTs are face-to-face tessellations, meaning that adjacent cells share entire faces, ensuring a strict geometric alignment. In contrast, STIT tessellations are not face-to-face, leading to configurations where cells may meet at partial faces rather than aligning perfectly. This structural distinction necessitates the development of new concepts and methods for studying and analyzing STIT tessellations, as conventional approaches for face-to-face tessellations are not applicable. Interestingly, the STIT model exhibits characteristics that position it between two classical models, the PHT and the Poisson–Voronoi tessellation. This suggests that STIT tessellations bridge the gap between these two well-known models, offering a versatile framework that captures elements of both while maintaining unique properties. In this sense, STIT tessellations do enrich the landscape of stochastic geometry.

Furthermore, an important feature of the STIT tessellation model is its inherent nature as a tessellation-valued stochastic process in time. Unlike static tessellation models, the STIT model evolves over time, allowing for the study of the dynamic interplay between spatial structures and temporal development. This dynamic nature opens the door to leveraging tools and techniques from the theory of stochastic processes, providing a richer framework for analyzing and understanding the model's behavior.

However, this temporal component also introduces new challenges. Traditional concepts and methodologies for random tessellations, which are primarily spatial, must be augmented to incorporate time as an additional dimension. This augmentation involves adapting definitions and analytical tools to account for the evolution of tessellations over time. The integration of space and time in the STIT model suggests new perspectives and naturally leads to an interplay between spatial and temporal randomness.

This book is dedicated to the memory of Joseph Mecke (1938–2014) and Tomasz Schreiber (1975–2010), whose pioneering work and enthusiastic contributions significantly advanced the theory of STIT tessellation processes. A substantial portion of the results presented in this book are rooted in their groundbreaking contributions. For instance, Joseph Mecke devised the global construction of STIT tessellations, which is comprehensively detailed in Chapter 8. Tomasz Schreiber, on the other hand, made remarkable progress in applying martingale methods to the study of STIT processes. His innovative approaches and results, which have become foundational to the field, are primarily featured in Part II of this book. Their legacy continues to inspire and shape research in stochastic geometry and tessellation theory.

This book also acknowledges with great appreciation the contributions of Richard Cowan to the field of random tessellations, particularly his work on models that are not face-to-face and are generated through the consecutive division of cells. Over the decades, his research, insightful papers, and collaborative discussions have been a profound source of inspiration for our work. His contributions have significantly shaped the development of this field and continue to influence advancements in the study of stochastic tessellation processes.

Furthermore, we gratefully acknowledge the pleasant and fruitful collaboration with our coauthors on various papers related to STIT tessellations: Carina Betken, Nicolas Chenavier, Christian Deuß, Hans-Otto Georgii (†), Julia Hörrmann, Daniel Hug, Tom Kaufmann, Roberto León, Servet Martínez, Kathrin Meier, Ngoc Linh Nguyen, Joachim Ohser, and Claudia Redenbach. Their contributions and expertise have greatly enhanced this work, and we sincerely appreciate the insights and progress made through these collaborations.

Special thanks are due to Servet Martínez and Eliza O'Reilly, who kindly read parts of the manuscript and provided valuable feedback. Their insightful comments and suggestions have been instrumental in improving the quality and clarity of this work.

We extend our heartfelt thanks to our colleagues Anna Gusakova, Lothar Heinrich, Daniel Hug, Zakhar Kabluchko, Günther Last, Matthias Reitzner, Rolf Schneider, Matthias Schulte, Dietrich Stoyan, and Wolfgang Weil (†). Their stimulating discussions and valuable suggestions have encouraged and inspired us over so many years. Their support and intellectual contributions have been indispensable to our work and have profoundly influenced our development.

We would like to express our gratitude to Philipp Tuchel for his excellent work on preparing the figures and to Benedikt Rednoß for his invaluable support with LaTeX formatting and typesetting. Additionally, we are deeply thankful to Roberto León and Claudia Redenbach for providing various simulations, which have significantly enriched this book. Finally, we are deeply indebted to the anonymous referee for his thoughtful and constructive comments, which have led to substantial improvements and helped us to present the results in a clearer and more coherent form.

1.2 Survey of contents

This volume is dedicated to the comprehensive study of STIT tessellation processes, presenting their theoretical foundations and key mathematical properties. The content primarily focuses on the formal development and rigorous analysis of these processes. Although the book is primarily theoretical, an exception is made in Chapter 9, which diverges from the main theoretical discourse. This chapter provides an overview of potential applications of STIT tessellation models, illustrating their relevance and utility in various practical contexts. It also includes references to pertinent modifications of the standard model that may arise in applied scenarios.

The book is divided into four parts, each exploring STIT tessellations from a distinct perspective. Additionally, the Appendix complements the main text by providing background information on topological and measurability issues and an introduction to the theory of point processes.

Part I: Construction and fundamental properties of STIT tessellations

Part I lays the foundation for the theory of STIT tessellations by introducing the concepts and operations central to random tessellations and random tessellation processes. It begins with Chapter 2, which explores the essential concepts for tessellations, with particular attention to those that are not face-to-face. The chapter also introduces the concept of primitives and provides a primer on Poisson hyperplane tessellations. In Chapter 3 the operation of iteration of tessellations is discussed in detail, with a focus on the concept of stability under iteration. This sets the stage for Chapter 4, where the main subject of the book – STIT tessellation processes – is formally introduced. The chapter outlines the construction of a STIT tessellation within a window. The theoretical framework is further supported by Chapter 5, which, though technical in nature, lays the groundwork for key proofs. These include the existence and uniqueness of STIT tessellations in the Euclidean space $\mathbb{R}^d$ and the development of corresponding STIT tessellation processes, detailed in Chapter 6. In Chapter 7 a solution to Ambartzumian's problem is presented, demonstrating that any sequence of rescaled iterations of independent and identically distributed random stationary tessellations converges in distribution to a STIT tessellation. This result also establishes the uniqueness of the STIT model, offering a cornerstone for its theoretical justification. The culmination of Part I is Chapter 8, where the global construction of STIT tessellation processes is presented. Unlike earlier chapters, this construction does not restrict the process to a window, resulting in a more complex formalization. However, this generalization facilitates proofs of several key propositions discussed in Part III. Additionally, the global construction introduces the concept of maximal polytopes – one of the basic building blocks that can be associated with STIT tessellations. These include maximal $(d-1)$-dimensional polytopes that arise when cells are divided by hyperplanes, and their lower-dimensional faces, which are also critical elements of the tessellation. Finally, Part I concludes with Chapter 9, which provides a concise overview of modifications to STIT tessellations, related models, and potential applications.

Part II: Geometry of STIT tessellations via martingale methods

Part II delves into the application of martingale methods to study various functionals associated with STIT tessellation processes. It makes particular use of the fact that the STIT model in a window can be interpreted as a pure jump Markov process taking values in the space of tessellations of that window. This new angle paves the way to powerful tools for analyzing and deriving intricate properties of STIT tessellations. The discussion begins with Chapter 10, a preparatory chapter that serves as a reference for key functionals and their associated martingales. This chapter systematically introduces and organizes the foundational components required for the subsequent in-depth analysis. The most basic device we apply in this context is the classical Dynkin martingale formula. Building on this groundwork, Chapter 11 explores the close relationships between STIT tessellations and Poisson hyperplane tessellations (PHTs). These relationships are examined not only in terms of the cells that compose the tessellations but also in the context of the maximal polytopes. The results presented here highlight the deep connections between these models and their structural elements. Chapter 12 offers a comprehensive study of second-order properties of STIT tessellations. Initially, the analysis is confined to a window, allowing for precise computations. Subsequently, the focus shifts to the asymptotic behavior, extending the results to encompass STIT tessellations in the entire Euclidean space. The application of martingale methods reaches its peak in Chapter 13, which presents a series of limit theorems – even on a functional level. Among the highlights are several central limit theorems, which demonstrate Gaussian fluctuations under specific conditions. A primary example will be the surface content of all $(d-1)$-dimensional maximal polytopes that are born after some fixed time threshold $s_0 > 0$. However, an intriguing departure from Gaussian behavior is observed in dimensions $d \geq 3$, where non-Gaussian limits emerge, when s_0 is sent to zero to capture the total surface content. This striking phenomenon underscores the richness and complexity of the STIT tessellation processes – especially in comparison with the traditional Poisson–Voronoi and Poisson hyperplane tessellations.

Part III: Implications of the global construction

Part III focuses on the global construction of the STIT tessellation process and explores its advanced mathematical properties. This part is built on the concepts introduced earlier, delving into technical but insightful analyses that provide a broader understanding of STIT tessellations in windows. Chapter 14 introduces a Mecke-type equation tailored for STIT tessellations. Inspired by the classical Mecke equation for Poisson point processes on general state spaces, its STIT-adapted formulation presents new challenges. In fact, even in relatively simple specific cases, its structure appears intricate. Despite this complexity, the Mecke-type equation proves to be a powerful tool, enabling the derivation of results concerning maximal polytopes. Chapter 15 is devoted to the study of mixing properties in both spatial and temporal contexts. For spatial dependencies within STIT tessellations, it is shown that they exhibit the property of being β-mixing, also known as absolutely regular. This property ensures a controlled level of dependence between spatial regions, which is essential for various applications and theoretical analyses. In the temporal domain, the STIT tessellation process, viewed as a stochastic process evolving over time, becomes stationary through an exponential transformation of the time axis. For this transformed process, the β-mixing property is also established. Moreover, a remarkable result is proven: the transformed process is isomorphic to a Bernoulli shift or a Bernoulli flow, respectively.

Part IV: STIT tessellations in 2D and 3D

Part IV focuses on STIT tessellations in two and three dimensions, cases that are both intuitive and mathematically rich. These dimensions are particularly well suited for visualization and understanding, offering a more tangible perspective compared to higher-dimensional scenarios. Furthermore, significantly more theoretical results are available for these dimensions, making them a central point of interest for readers. The mathematical tools employed in this part are more elementary, and the text provides detailed explanations to ensure accessibility. This makes Part IV especially appealing to the readers seeking a more intuitive and less abstract understanding of STIT tessellations. For those primarily interested in a direct and comprehensible approach to the model in dimensions $d = 2$ or $d = 3$, it is recommended to focus on Chapters 4, 16, and 17. These chapters not only present the model in a clear and detailed manner but also specify the broader results from other parts of the book for these two dimensions. In particular, Chapter 4 introduces the basic construction of STIT tessellations, serving as a foundational starting point. Chapter 16 focuses on results specific to planar STIT tessellations. Chapter 17 extends this discussion to three-dimensional tessellations, showcasing their complexity and the richness of the model. In Chapters 16 and 17 the basic combinatorial structure of STIT tessellations is analyzed. Initially, the focus is on the primitives, which serve as building blocks of the structure of a random tessellation. Their intensities, mean values, and distributions are derived. In addition, the arrangement of cells and the adjacencies of various components are examined. The structure is successively refined further by classifying certain components. For example, different types of segments are analyzed, and the class of edges is subdivided on the basis of specific characteristics. In this context, the maximal polytopes – especially the maximal 1-polytopes, which are referred to as I-segments – play a decisive role, and their mean values, intensities, and distributions are determined accordingly.

Both chapters conclude with a comparison of STIT tessellations, Poisson–Voronoi tessellations, and Poisson line or plane tessellations. This encompasses first- and second-order properties and ergodic and mixing characteristics and offers valuable insights for selecting the most appropriate model for various applications.

Throughout the text, certain statements and proofs are intentionally presented with a degree of redundancy, derived using different methods. This redundancy is deliberate and serves to provide a deeper understanding of key concepts while ensuring that the readers with varying methodological preferences can engage fully with the material. By approaching results from multiple perspectives, the text highlights the versatility of the underlying techniques. For instance, the identity of the typical cell distributions in STIT and Poisson hyperplane tessellations are explored through complementary approaches. The use of martingale methods introduces a more probabilistic framework for understanding key properties, whereas nonmartingale techniques provide alternative, equally valuable perspectives. Additionally, the global construction and derivation of a

Mecke-type formula offer distinct, yet interconnected, ways to formalize and analyze STIT tessellation processes.

The entire framework of this text is deeply rooted in stochastic geometry, which serves as the underlying mathematical discipline for the study of spatial structures and processes. For a comprehensive and rigorous theoretical foundation, the primary reference is the monograph of Schneider and Weil [172]. For readers with a more application-oriented focus, particularly those interested in geometric statistics, the reference work by Chiu, Stoyan, Kendall, and Mecke [33] offers an excellent alternative. This text bridges the gap between theory and practice, presenting stochastic geometry in a way that is both accessible and rigorous. For background material on Poisson hyperplane tessellations, we direct the reader to the monograph of Hug and Schneider [82].

For the readers interested in specific topics, the scheme in Figure 1.4 indicates suitable paths leading to them, while allowing chapters that may be skipped on a first reading to be avoided. Naturally, all chapters contain numerous cross-references to other parts of the book, pointing to related concepts and results.

1.3 Notation and conventions

In this book, we widely adopt the notation system of [172]. This system has proven effective for a broad range of stochastic geometry and tessellation-related topics. However, our focus on STIT tessellations introduces new challenges and nuances not fully addressed in the original framework. First, STIT tessellations are not face-to-face, meaning that adjacent cells in the tessellation do not necessarily share a full face. This nonstandard structure complicates the application of a notation system originally developed for face-to-face tessellations. Additional modifications were therefore necessary to accurately describe the geometrical and topological relationships in this more complex setting. Second, we extend the system to handle time-dependent STIT tessellation processes, which involve tracking the evolution of the tessellation over time. This requires notation for key dynamic features such as birth times associated with geometric objects of STIT tessellations. These additions demanded compromises in the notation system. While adhering to the foundational structure of [172] wherever possible, we adapted and extended it to ensure that the system remained both functional and intuitive for our specific applications.

The purpose of this section is to introduce and collect some standard notation used throughout the book. This section is intended primarily as a reference. Since several concepts and symbols introduced here will only be defined in later chapters, it is meant to provide an overview rather than to be read sequentially. The readers may return to it whenever specific notation is needed. In addition, we refer to the appendix, Chapters 18 and 19, where essential background material from topology and point process theory is summarized.

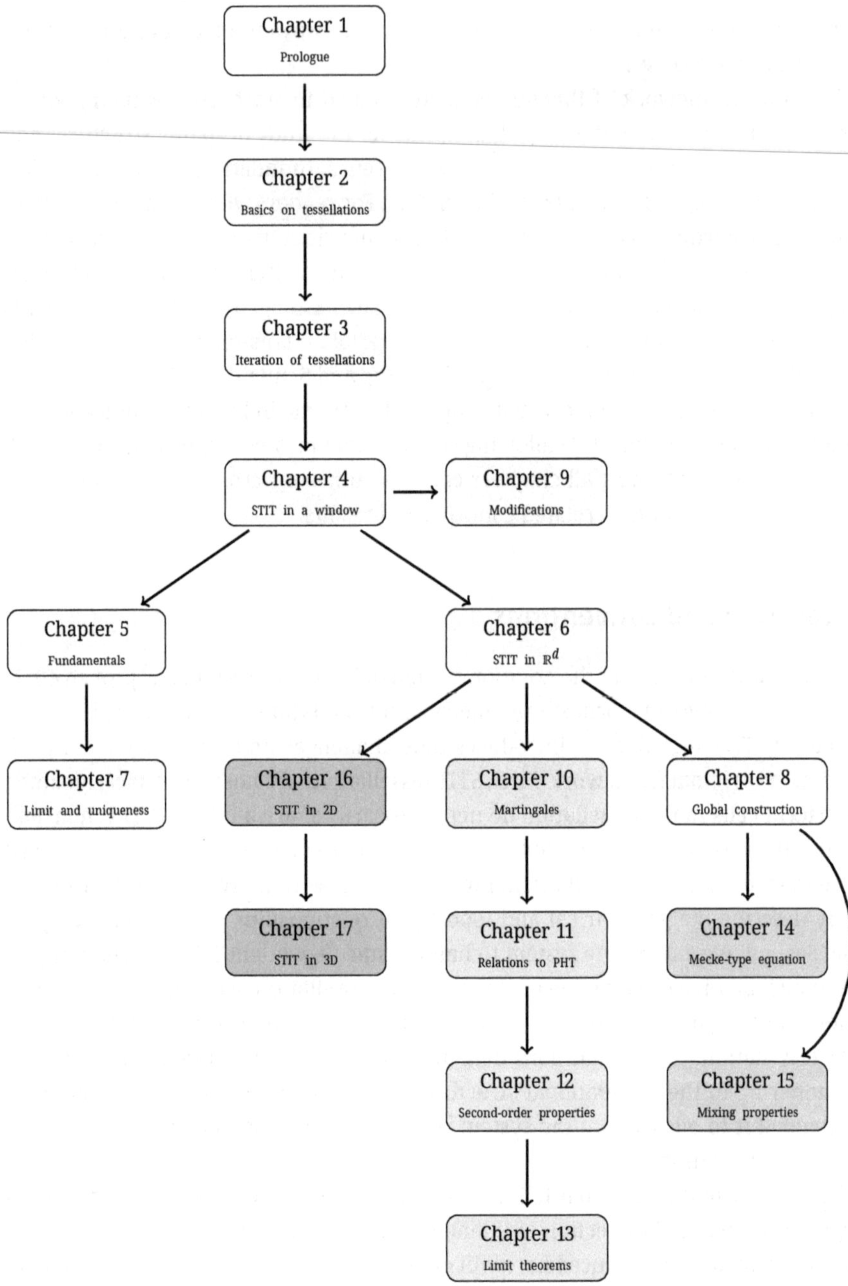

Figure 1.4: Reading paths between the chapters; different grey scales indicate different parts of the book.

Numbers	
$\mathbb{N}$	set of positive integers $\{1, 2, 3, \ldots\}$
$\mathbb{N}_0$	set of nonnegative integers $\{0, 1, 2, 3, \ldots\}$
$\mathbb{Z}$	set of all integers $\{\ldots, -2, -1, 0, 1, 2, \ldots\}$
$\mathbb{Z}^-$	set of nonpositive integers $\mathbb{Z} \setminus \mathbb{N}$
$\mathbb{R}$	set of real numbers
$\mathbb{R}^+$	set of positive real numbers $(0, \infty)$
$\mathbb{R}^-$	set of negative real numbers $(-\infty, 0)$
$\lfloor x \rfloor$	largest integer $n \in \mathbb{Z}$ such that $n \leq x$, $x \in \mathbb{R}$
$\prod_{\ell=m}^{n}(\ldots) := 1$	if $m > n$, $m, n \in \mathbb{Z}$
$\binom{0}{0} := 1$	
$\frac{1}{\log 1} := 0$	

To introduce our notation concerning sets, basic set operations, and spaces of sets, let E be a topological space.

Sets and set spaces	
$\operatorname{int} A$	interior of a set $A \subseteq E$
$\operatorname{relint} A$	relative interior of a set $A \subseteq E$ taken with respect to a subset $S \subseteq E$ in the subspace topology of S
$\operatorname{cl} A$	closure of a set $A \subseteq E$
∂A	boundary of a set $A \subseteq E$
A^c	complement of a set $A \subseteq E$
$A \triangle B$	symmetric difference $A \triangle B := (A \setminus B) \cup (B \setminus A)$
$\lvert A \rvert$	cardinality of a finite or countable set $A \subseteq E$
$A + B$	Minkowski sum $A + B := \{x + y : x \in A, y \in B\}$ of sets $A, B \subseteq \mathbb{R}^d$
$A \ominus B$	Minkowski difference $A \ominus B := (A^c + B)^c$ of sets $A, B \subseteq \mathbb{R}^d$
$\mathbb{1}_A(x)$ or $\mathbb{1}\{x \in A\}$	indicator function of a set $A \subseteq E$ evaluated at $x \in E$
$\mathcal{B}(E)$	σ-algebra of Borel sets in E
$\mathcal{F}$	family of nonempty closed subsets of E
$\mathcal{G}$	family of nonempty open subsets of E
$\mathcal{C}$	family of nonempty compact subsets of E
δ^{Hausd}	Hausdorff metric on $\mathcal{C}$

We remark that our notation for $\mathcal{F}$, $\mathcal{G}$, and $\mathcal{C}$ diverges from that used in [172], which uses the symbols $\mathcal{F}'$, $\mathcal{G}'$, and $\mathcal{C}'$, respectively.

The space $\mathcal{F}$ is supplied with the topology of closed convergence and with the Borel σ-field $\mathcal{B}(\mathcal{F})$ induced by this topology; see Chapter 18. The σ-algebra for $\mathcal{C}$ is the restriction of $\mathcal{B}(\mathcal{F})$ to $\mathcal{C}$. The same will be applied, for example, to the set systems $\mathcal{K}$ and $\mathcal{P}$ we introduce below.

Probability	
$(\Omega, \mathbf{A}, \mathbb{P})$	underlying probability space
$\sigma(\mathcal{S})$	σ-algebra generated by a set system $\mathcal{S}$
$\sigma(X_i, i \in I)$	σ-algebra generated by a family of random variables $(X_i, i \in I)$
$\mathbb{E}\, U$	expectation or mean of a real-valued random variable U
$\operatorname{var} U$	variance of a real-valued random variable U
$\operatorname{cov}(U_1, U_2)$	covariance between real-valued random variables U_1 and U_2
$\overset{D}{=}$	identity of distributions of random variables
$\overset{\text{law}}{\Longrightarrow}$	convergence in law
$\overset{\mathbb{P}}{\Longrightarrow}$	convergence in probability
$\overset{L^p}{\Longrightarrow}$	convergence in L^p, $p \geq 1$
$\operatorname{Norm}(a, \sigma^2)$	normal distribution with mean $a \in \mathbb{R}$ and variance $\sigma^2 \in \mathbb{R}^+$
$\operatorname{Pois}(a)$	Poisson distribution with parameter $a \in \mathbb{R}^+$
$\operatorname{Expo}(a)$	exponential distribution with parameter $a \in \mathbb{R}^+$
$\operatorname{Unif}(a, b)$	uniform distribution on the interval (a, b), $a, b \in \mathbb{R}$, $a < b$
$\operatorname{Geom}(p)$	geometric distribution on $\mathbb{N}_0$ with parameter $p \in (0, 1)$
$\operatorname{Erl}(k, a)$	Erlang distribution with parameters $k \in \mathbb{N}$ and $a \in \mathbb{R}^+$

For families of random variables, we use i. i. d. as the abbreviation of "independent and identically distributed", and we abbreviate "almost surely" by a. s.

In this book, we frequently deal with asymptotic statements. To introduce our corresponding notation, let $f : \mathbb{R} \to \mathbb{R}$ and $g : \mathbb{R} \to \mathbb{R}$ be two real-valued functions. The Landau symbols mean

$$\begin{aligned}
f(x) = \Omega(g(x)) \quad &\text{as } x \to a \quad \text{iff} \quad \limsup_{x \to a} \left| \frac{f(x)}{g(x)} \right| > 0\,,\\
f(x) = O(g(x)) \quad &\text{as } x \to a \quad \text{iff} \quad \limsup_{x \to a} \left| \frac{f(x)}{g(x)} \right| < \infty\,,\\
f(x) = o(g(x)) \quad &\text{as } x \to a \quad \text{iff} \quad \limsup_{x \to a} \left| \frac{f(x)}{g(x)} \right| = 0\,,\\
f(x) = J(g(x)) \quad &\text{as } x \to a \quad \text{iff} \quad 0 < \liminf_{x \to a} \left| \frac{f(x)}{g(x)} \right| \leq \limsup_{x \to a} \left| \frac{f(x)}{g(x)} \right| < \infty\,,
\end{aligned}$$

for $a \in \mathbb{R} \cup \{-\infty, \infty\}$. We remark that in this book, we use the nonstandard notation $f(x) = J(g(x))$ instead of the more traditional $f(x) = \Theta(g(x))$, since the symbol Θ will be reserved for the driving hyperplane measure of a STIT tessellation process. Moreover, we use $\sim$ to describe the asymptotic equivalence of functions, that is,

$$f(x) \sim g(x) \quad \text{as } x \to \infty \quad \text{iff} \quad \lim_{x \to \infty} \frac{f(x)}{g(x)} = 1\,.$$

Next, we introduce some notation related to the geometry of the d-dimensional Euclidean space $\mathbb{R}^d$ with $d \geq 1$.

Geometry	
$\langle\cdot,\cdot\rangle$	Euclidean scalar product on $\mathbb{R}^d$
$\|\cdot\|$	Euclidean norm in $\mathbb{R}^d$
o	origin of $\mathbb{R}^d$
$B(x,r)$	Euclidean ball with radius $r>0$ centered at $x\in\mathbb{R}^d$
$B^d := B(o,1)$	unit ball in $\mathbb{R}^d$
S^{d-1}	unit sphere in $\mathbb{R}^d$
$\kappa_d := \frac{\pi^{d/2}}{\Gamma(1+\frac{d}{2})}$	volume of B^d
$\omega_d := d\,\kappa_d = \frac{2\pi^{d/2}}{\Gamma(\frac{d}{2})}$	total spherical Lebesgue measure of S^{d-1}
$\nabla_k(u_1,\dots,u_k)$	k-volume of the parallelepiped spanned by $u_1,\dots,u_k\in S^{d-1}$
$\operatorname{conv} A$	convex hull of a set $A\subseteq\mathbb{R}^d$
$\operatorname{lin} A$	linear hull of a set $A\subseteq\mathbb{R}^d$
$\operatorname{aff} A$	affine hull of a set $A\subseteq\mathbb{R}^d$
$\operatorname{aff}_o A$	translate of $\operatorname{aff} A$ containing the origin
$\dim A := \dim\operatorname{aff} A$	dimension of a set $A\subseteq\mathbb{R}^d$
$G(d,k)$	Grassmannian of all k-dimensional linear subspaces of $\mathbb{R}^d$, $k\in\{0,\dots,d\}$
$A(d,k)$	Grassmannian of all k-dimensional affine subspaces of $\mathbb{R}^d$, $k\in\{0,\dots,d\}$
$\mathcal{H} := A(d,d-1)$	space of all hyperplanes in $\mathbb{R}^d$
$\mathcal{H}_o := G(d,d-1)$	space of all hyperplanes in $\mathbb{R}^d$ containing the origin
$[A]$	set of hyperplanes intersecting a set $A\subseteq\mathbb{R}^d$, $[A] := \{H\in\mathcal{H} : H\cap A\neq\emptyset\}$
$A^\perp\in G(d,d-k)$	linear subspace orthogonal to $A\in A(d,k)$
$F^\perp$	$(\operatorname{aff} F)^\perp$ for $F\subseteq\mathbb{R}^d$
$A\vert L$	orthogonal projection of $A\subseteq\mathbb{R}^d$ onto $L\in G(d,k)$
$A\vert u := A\vert\operatorname{lin}(u)$	orthogonal projection of a set $A\subseteq\mathbb{R}^d$ onto $\operatorname{lin}(u)$, $u\in S^{d-1}$
$u(L)\in S^{d-1}$	unit vector with $\operatorname{lin}(u(L)) = L$ and nonnegative coordinates, $L\in G(d,1)$
$e(L)$	segment of unit length on $L\in G(d,1)$

We remark that our notation $A^\perp$ is used here for the $(d-k)$-dimensional linear subspace orthogonal to $A\in A(d,k)$. This differs from the common convention, in which $A^\perp = \{x : \langle x,a\rangle = 0 \text{ for all } a\in A\}$.

Each hyperplane in $\mathbb{R}^d$ admits a parametric representation of the form

$$H(u,r) := \{x\in\mathbb{R}^d : \langle x,u\rangle = r\},$$

where $u\in S^{d-1}$ is a unit normal vector, and $r\in\mathbb{R}$ can be interpreted as the signed distance of the hyperplane to the origin. Such a hyperplane generates two closed half-spaces of $\mathbb{R}^d$, namely

$$H^+(u,r) := \{x\in\mathbb{R}^d : \langle x,u\rangle \geq r\} \quad\text{and}\quad H^-(u,r) := \{x\in\mathbb{R}^d : \langle x,u\rangle \leq r\}.$$

Note that $H(u,r) = H(-u,-r)$ and $H^+(u,r) = H^-(-u,-r)$. Still, we will often simply write H^+ and H^-, respectively, if it is not essential which of the half-spaces is the positive or negative one. The normal direction of a hyperplane $H(u,r)$ is given by $u\in S^{d-1}$ as well as by $-u \in S^{d-1}$. It can alternatively be described by its orthogonal complement $H^{\perp}(u,r) := \mathrm{lin}(u) \in G(d,1)$, the linear subspace spanned by u.

Measures	
$\ll$	absolute continuity of measures
$\otimes$	product of measures
$\mu \llcorner A$	restriction of a measure μ to a set A
δ_x	Dirac measure at an element x
λ_d	Lebesgue measure on $\mathbb{R}^d$
λ_+	Lebesgue measure on $\mathbb{R}^+$
λ_-	Lebesgue measure on $\mathbb{R}^-$
λ_A	k-dimensional Lebesgue measure on $A \in A(d,k)$
$\lambda_F := \lambda_A \llcorner F$	for $A \in A(d,k)$ and $F \in \mathcal{B}(A)$
$\mathcal{H}^k$	k-dimensional Hausdorff measure
φ_{iso}	normalized spherical Lebesgue measure on S^{d-1}

We remark that by slight abuse of notation the symbols $\otimes$ and $\llcorner$ will also be applied to σ-algebras to denote the product of σ-algebras or their restrictions, respectively. Moreover, we normalize the Hausdorff measures in such a way that

$$\lambda_A(B) = \mathcal{H}^k(A \cap B) \quad \text{and} \quad \lambda_F(B) = \mathcal{H}^k(F \cap B) \tag{1.1}$$

for $A \in A(d,k)$ and $B, F \in \mathcal{B}(A)$.

Furthermore, for a better readability, we write integrals with respect to a time variable s as $\int_0^t h(s)\,\mathrm{d}s$ instead of $\int_{(0,t)} h(s)\,\lambda_1(\mathrm{d}s)$, even if the function $h : \mathbb{R} \to \mathbb{R}$ is only Lebesgue-integrable.

Next, we recall some key concepts from convex geometry. First, by a convex body in $\mathbb{R}^d$ we understand a nonempty compact and convex subset.

Convexity		
$\mathcal{K}$	space of convex bodies in $\mathbb{R}^d$	
$\mathcal{K}_d$	space of d-dimensional convex bodies in $\mathbb{R}^d$	
$\mathcal{R}$	convex ring consisting of all finite unions of elements of $\mathcal{K}$	
$h(K,\cdot) : S^{d-1} \to \mathbb{R}$	support function of $K \in \mathcal{K}$, $h(K,u) := \sup\{\langle x,u\rangle : x \in K\}, \quad u \in S^{d-1}$	(1.2)
$w(K,\cdot) : S^{d-1} \to \mathbb{R}$	width or breadth function of $K \in \mathcal{K}$, $w(K,u) := \lvert h(K,u) - h(K,-u)\rvert, \quad u \in S^{d-1}$	
$\mathrm{diam}\, K$	diameter of a set $K \in \mathcal{K}$, $\mathrm{diam}\, K := \sup_{u \in S^{d-1}} w(K,u)$	

On $\mathcal{K}$, by a center function $c : \mathcal{K} \to \mathbb{R}^d$ we understand a measurable and translation-covariant function, which means that $c(K+x) = c(K) + x$ for all $K \in \mathcal{K}$ and $x \in \mathbb{R}^d$. With respect to a given center function c, the set of all centered compact convex sets is

$$\mathcal{K}_c := \{K - c(K) : K \in \mathcal{K}\}.$$

The intrinsic volumes $V_j : \mathcal{K} \to [0,\infty)$, $j \in \{0,\dots,d\}$, are defined with the help of the Steiner formula, which ensures the polynomial expansion

$$V_d(K + rB^d) = \sum_{j=0}^{d} r^{d-j}\kappa_{d-j} V_j(K), \quad K \in \mathcal{K},\ r > 0. \tag{1.3}$$

Some of the coefficients $V_j(K)$ have intuitive geometric interpretations:

$$\begin{aligned}
V_d(K) &= \lambda_d(K) && \text{volume of } K,\\
2V_{d-1}(K) &= \mathcal{H}^{d-1}(\partial K) && \text{total surface area of } K,\\
\frac{2\kappa_{d-1}}{d\kappa_d} V_1(K) &= \int_{S^{d-1}} h(K,u)\,\varphi_{\text{iso}}(\mathrm{d}u) && \text{mean width of } K,\\
V_0(K) &= 1 && \text{Euler characteristic.}
\end{aligned}$$

The intrinsic volumes V_j admit additive extensions to the convex ring $\mathcal{R}$, which by slight abuse of notation are denoted by the same symbol $V_j : \mathcal{R} \to \mathbb{R}$, $j \in \{0,\dots,d\}$.

In this monograph, we frequently use two standard integral geometric formulae. The first is the following **translative integral formula**. It says that if $A, B \in \mathcal{B}(\mathbb{R}^d)$ are Borel sets and μ is a σ-finite measure on $\mathbb{R}^d$, then

$$\int_{\mathbb{R}^d} \mu((A + x) \cap B)\,\lambda_d(\mathrm{d}x) = \lambda_d(A)\,\mu(B); \tag{1.4}$$

see Theorem 5.2.1 of [172].

When we are working with the k-dimensional Hausdorff measure on $\mathbb{R}^d$, which is not σ-finite for $k < d$, we use the following variant, which requires finiteness only on the set B.

Lemma 1.1. *Let μ be a Borel measure on $\mathbb{R}^d$, and let $A, B \in \mathcal{B}(\mathbb{R}^d)$ with $\mu(B) < \infty$. Then*

$$\int_{\mathbb{R}^d} \mu((A + x) \cap B)\,\lambda_d(\mathrm{d}x) = \lambda_d(A)\,\mu(B). \tag{1.5}$$

Proof. Define the finite measure μ_B on $\mathbb{R}^d$ by $\mu_B(C) := \mu(C \cap B)$ for $C \in \mathcal{B}(\mathbb{R}^d)$. Then, for all $x \in \mathbb{R}^d$, $\mu_B((A + x) \cap B) = \mu((A + x) \cap B)$. Since μ_B is a finite measure on $\mathbb{R}^d$ and hence is σ-finite, (1.4) can be applied to μ_B. This yields

$$\int_{\mathbb{R}^d} \mu((A+x)\cap B)\,\lambda_d(\mathrm{d}x) = \int_{\mathbb{R}^d} \mu_B((A+x)\cap B)\,\lambda_d(\mathrm{d}x)$$
$$= \lambda_d(A)\,\mu_B(B) = \lambda_d(A)\,\mu(B)\,,$$

and the proof is complete. □

Another integral geometric formula we use is the **Crofton formula**. It provides a relation between the intrinsic volumes of sections of a convex body and the intrinsic volumes of the body itself. Here we only consider the section with hyperplanes. For $K \in \mathcal{K}$ and $j \in \{0,\dots,d-1\}$, it says that

$$\int_{S^{d-1}} \int_{\mathbb{R}} V_j(K \cap H(u,r))\,\lambda_1(\mathrm{d}r)\,\varphi_{\mathrm{iso}}(\mathrm{d}u) = c_{j,d} V_{j+1}(K) \tag{1.6}$$

$$\text{with} \quad c_{j,d} := \frac{\Gamma(\frac{d}{2})\,\Gamma(\frac{j+2}{2})}{\Gamma(\frac{j+1}{2})\,\Gamma(\frac{d+1}{2})}\,.$$

For the general formula, we refer the reader to Theorem 5.1.1 of [172].

Finally, we introduce some notation for polytopes. We start by recalling that a polytope in $\mathbb{R}^d$ is the convex hull of a nonempty finite set of points.

Polytopes	
$\mathcal{P}_k$	set of all k-dimensional polytopes in $\mathbb{R}^d$, $k \in \{0,\dots,d\}$
$\mathcal{P}$	$\mathcal{P} := \bigcup_{k=0}^{d} \mathcal{P}_k$ set of all polytopes in $\mathbb{R}^d$
$\mathcal{F}_j(P)$	set of j-dimensional faces of a polytope $P \in \mathcal{P}_k$, $k \in \{0,\dots,d\}, j \in \{0,\dots,k\}$, $\mathcal{F}_k(P) := \{P\}$
$\mathcal{P}_{k,c}$	set of centered k-dimensional polytopes in $\mathbb{R}^d$, $k \in \{0,\dots,d\}$ and a center function $c : \mathcal{P}_k \to \mathbb{R}^d$

Part I: **Construction and fundamental properties of STIT tessellations**

2 Random tessellations

In this preparatory chapter, we compile definitions and facts for general random tessellations. Special attention is paid to tessellations that are not face-to-face and to their structure, which is more intricate and more varied than in the face-to-face case. Of particular interest are the so-called primitives.

Since the STIT tessellations are in many respects related to Poisson hyperplane tessellations, the last section of this chapter is devoted to that model. In view of the requirements of this book, the tessellation-valued stochastic process of Poisson hyperplane tessellations on the time axis is introduced, and important properties are summarized.

Detailed introductions and surveys on random tessellations can be found in [138, 172] and [33]. The primary reference for Poisson hyperplane tessellations is the book [82].

2.1 Tessellations and their components

A tessellation of the Euclidean space $\mathbb{R}^d$ is defined as a space-filling collection of nonoverlapping polytopes. The objects of interest are these polytopes, the system of their faces, and, in addition, the intersections of adjacent polytopes. These intersections can subdivide the faces of the polytopes and thus generate an internal structure within the faces. This creates a refined system of components that are assigned to a tessellation and reflect essential features of the arrangement of the polytopes. Furthermore, sections of tessellations with affine subspaces of $\mathbb{R}^d$ are analyzed with regard to the qualitative and quantitative properties inherited by the section profiles.

2.1.1 Definition

A **polytope** in $\mathbb{R}^d$ is defined as the convex hull of finitely many points. The set of all k-dimensional polytopes is denoted by $\mathcal{P}_k$, $k \in \{0, \dots, d\}$, and $\mathcal{P} := \bigcup_{k=0}^{d} \mathcal{P}_k$.

Definition 2.1. *A* ***tessellation of*** $\mathbb{R}^d$ *is a countable subset $T \subset \mathcal{P}_d$ of d-dimensional polytopes satisfying the following conditions:*

(a) $\bigcup_{P \in T} P = \mathbb{R}^d$.

(b) *If $P, P' \in T$ and $P \neq P'$, then* $\operatorname{int} P \cap \operatorname{int} P' = \emptyset$.

(c) *The set $\{P \in T : P \cap C \neq \emptyset\}$ is finite for all compact sets $C \subset \mathbb{R}^d$ (local finiteness).*

The polytopes forming a tessellation are called ***cells***.

Let $\mathcal{T}$ denote the set of all tessellations of $\mathbb{R}^d$. The topological spaces and measurability considerations associated with maps related to tessellations are provided in the appendix, Chapter 18. The Borel σ-algebra on $\mathcal{T}$ is $\mathcal{B}(\mathcal{T}) := \mathcal{B}(\overline{\mathcal{F}}(\mathcal{F})) \llcorner \mathcal{T}$, according to (18.3).

https://doi.org/10.1515/9783112222409-003

There are several equivalent ways to describe a tessellation. Instead of the cells, we can take the cell centers and mark them with the centered cells. Furthermore, the set of cells can be characterized by a counting measure on the set of polytopes. Also, the union of the cell boundaries, referred to as the $(d-1)$-skeleton, uniquely determines a tessellation. Accordingly, the representation of the $(d-1)$-skeleton by the set of $(d-1)$-primitives, defined in Section 2.1.2, also describes a tessellation. This is now presented in detail. Moreover, we also consider systems of the lower-dimensional faces of the cells and, more generally, sets of components, which will be introduced in Section 2.1.3. The measurability issues are dealt with in Chapter 18.

For $k \in \{0, \ldots, d\}$, the set of all **k-dimensional faces of all cells of a tessellation** $T \in \mathcal{T}$ is

$$\mathcal{F}_k(T) := \bigcup_{P \in T} \mathcal{F}_k(P) \subset \mathcal{P}_k \tag{2.1}$$

with the convention that $\mathcal{F}_d(P) := \{P\}$ and $\mathcal{F}_d(T) := T$. For simplicity of notation, sometimes, we identify x with $\{x\}$ if $\{x\} \in \mathcal{F}_0(P)$. In Lemma 18.5 the measurability of the map $\mathcal{F}_k(\cdot) : \mathcal{T} \to \overline{\boldsymbol{\mathcal{F}}}_{ef}(\mathcal{F})$ is shown.

It is essential to note that in a tessellation T, an element $F \in \mathcal{F}_k(T)$ can be a k-face of one cell only or of more than one cell of T. Hence, in some propositions the **multiplicity** of a k-face F, that is,

$$m(F) := \left|\{P \in T : F \in \mathcal{F}_k(P)\}\right|,$$

has to be taken into account. A $(d-1)$-face always belongs either to one or two cells. Furthermore, depending on the arrangement of the cells of T, the relative interior of a k-face, $k < d$, can be tessellated by other faces of T. This motivates the classification into face-to-face tessellations and tessellations that are not face-to-face. For the latter ones, besides the k-faces, the k-primitives will be also defined.

For $k \in \{0, \ldots, d-1\}$, the **k-skeleton** of a tessellation T is the union of its k-faces,

$$\operatorname{skel}_k T := \bigcup_{F \in \mathcal{F}_k(T)} F. \tag{2.2}$$

We will also use the abbreviation $\partial T := \operatorname{skel}_{d-1} T$ for the union of cell boundaries of T. In Lemma 18.6, it is shown that the map $\operatorname{skel}_k : \mathcal{T} \to \mathcal{F}$ is measurable.

The skeletons play a central role in the theory of random tessellations and thus also throughout this book. As the $(d-1)$-skeleton ∂T uniquely determines the tessellation T, it is a very useful characteristic of T for many statements and their proofs. Lemma 8.8 provides a representation of the k-skeletons of a STIT tessellation by the so-called maximal k-polytopes. Note that in (2.15) a different concept of a skeleton is introduced, which is based on primitives.

The structure of the k-faces of cells is an essential feature of a tessellation, comprising information about the mutual position and arrangement of the cells. A tessellation T

is called **face-to-face** if

$$P \cap P' \in \left(\bigcup_{k=0}^{d} \mathcal{F}_k(P) \cap \bigcup_{k=0}^{d} \mathcal{F}_k(P') \right) \cup \{\emptyset\} \quad \text{for all } P,\, P' \in T\,. \tag{2.3}$$

This property means that for cells P and P' that have a nonempty intersection, this intersection is a common k-face of *both* cells for some $k \in \{0, \dots, d\}$.

In contrast to hyperplane tessellations and Voronoi tessellations, which are face-to-face, the STIT tessellations are not face-to-face. This is illustrated in Figure 1.1.

2.1.2 Primitives of tessellations

For tessellations that are not face-to-face, the intersection of two cells is not necessarily a face of any of these cells. This motivates the concept of k-primitives of a tessellation $T \in \mathcal{T}$. These are k-dimensional components of T without internal structure.

Definition 2.2. *For $T \in \mathcal{T}$ and $k \in \{0, \dots, d\}$, denote by*

$$A(T) := \bigcup_{n \in \mathbb{N}} \left\{ \bigcap_{P \in B} P : B \subset T, |B| = n \right\}$$

the set of all intersections of finitely many cells of T, and

$$A_k(T) := A(T) \cap \mathcal{P}_k\,.$$

*The set of k-**primitives** of T is defined as*

$$\mathcal{F}_k^{\mathrm{pr}}(T) := \{F \in A_k(T) : \forall F' \in A_k(T) : F' \subseteq F \Rightarrow F' = F\}\,, \tag{2.4}$$

that is, as the set of all minimal elements of $A_k(T)$ with respect to the partial order $\subseteq$.

In Lemma 18.5 the measurability of the map $\mathcal{F}_k^{\mathrm{pr}}(\cdot) : \mathcal{T} \to \overline{\mathcal{F}}_{\ell f}(\mathcal{F})$ is shown.

The d-primitives of T are the cells of this tessellation, that is, $\mathcal{F}_d^{\mathrm{pr}}(T) = \mathcal{F}_d(T)$ for all $T \in \mathcal{T}$. Properties (2.9) and (2.10) below show that a primitive does not have an “internal structure” in its relative interior. All 0-faces of cells are 0-primitives. However, for $d \geq 3$, there can be 0-primitives that are not 0-faces of any of the cells; see the right panel of Figure 2.1. Hence $\mathcal{F}_0(T) \subseteq \mathcal{F}_0^{\mathrm{pr}}(T)$. A 1-primitive is a line segment without any 0-primitive in its relative interior.

Note that in general, for $k \in \{1, \dots, d-1\}$, none of the sets $\mathcal{F}_k(T)$ and $\mathcal{F}_k^{\mathrm{pr}}(T)$ must be a subset of the other one if T is not face-to-face.

Corollary 2.3. *If $T \in \mathcal{T}$ is a face-to-face tessellation, then*

$$\mathcal{F}_k(T) = \mathcal{F}_k^{\mathrm{pr}}(T) \quad \textit{for all } k \in \{0, \dots, d\}\,,$$

that is, the sets of k-primitives and of k-faces coincide.

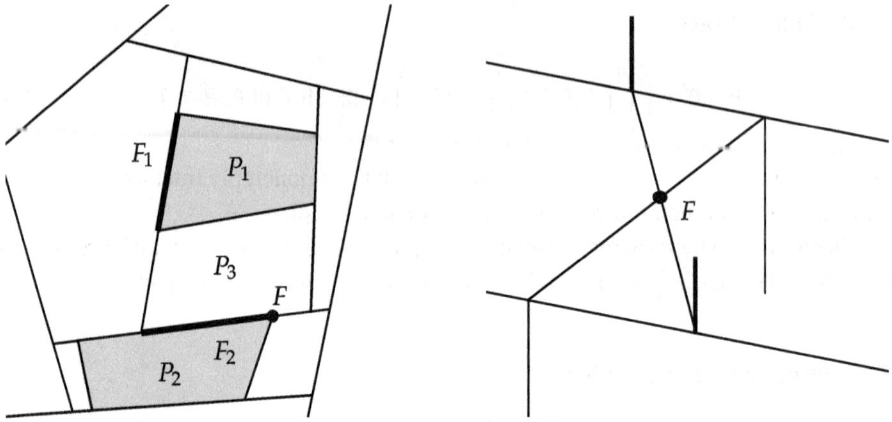

Figure 2.1: Left: In $\mathbb{R}^2$, F is a 0-primitive that is a 0-face of P_2 and is located in the relative interior of a 1-face of P_3. F_1 is a 1-primitive that is also a 1-face of P_1. F_2 is a 1-primitive that is not a 1-face of either P_2 or P_3. Right: In $\mathbb{R}^3$, F is an example of a 0-primitive that is not a 0-face of a cell.

Proof. If T is a face-to-face tessellation, by Definition 2.2 and (2.3) we have that $A_k(T) = \mathcal{F}_k(T)$. This equality also implies that $F' = F$ if $F' \subseteq F$ and $F, F' \in \mathcal{F}_k(T)$. Hence $\mathcal{F}_k^{\mathrm{pr}}(T) = \mathcal{F}_k(T)$. □

Now we provide some alternative representations of primitives. For all $k \in \{0, \dots, d\}$, $T \in \mathcal{T}$, and $F \in \mathcal{F}_k^{\mathrm{pr}}(T)$, we have

$$F = \bigcap_{P \in T,\, F \subseteq P} P\,. \tag{2.5}$$

To see this, note that $F \subseteq \bigcap_{P \in T, F \subseteq P} P$. Because $F \in A_k(T)$ is the section of finitely many cells of T, the assumption that $F \neq \bigcap_{P \in T, F \subseteq P} P$ yields the existence of $P' \in T \setminus \{P \in T : F \subseteq P\}$ such that $F \subseteq P'$, a contradiction.

To give an equivalent characterization of primitives, denote

$$F(x, T) := \bigcap_{P \in T,\, x \in P} P, \quad T \in \mathcal{T}, \quad x \in \mathbb{R}^d,$$

which for a tessellation T and any point $x \in \mathbb{R}^d$, is the intersection of all the cells of T that contain x. First, we show a preliminary result on $F(x, T)$.

Lemma 2.4. *Let $k \in \{1, \dots, d\}$ and $T \in \mathcal{T}$. If $F \in \mathcal{F}_k^{\mathrm{pr}}(T)$ and $x \in \operatorname{relint} F$, then*

$$\{P \in T : x \in P\} = \{P \in T : F \subseteq P\}, \tag{2.6}$$

and, consequently,

$$F(x, T) = F\,. \tag{2.7}$$

Proof. If $x \in F$, then $\{P \in T : F \subseteq P\} \subseteq \{P \in T : x \in P\}$. To show the other inclusion, let $F \in \mathcal{F}_k^{\mathrm{pr}}(T)$ and $x \in \operatorname{relint} F$. Denote by $\operatorname{aff} F \in A(d,k)$ the affine subspace spanned by F, which is the k-dimensional affine subspace containing F. Choose $\varepsilon > 0$ such that:

(a) $B(x,\varepsilon) \cap \operatorname{aff} F \subset F$,

(b) $B(x,\varepsilon) \cap \bigcup_{x\in P, P\in T} P = B(x,\varepsilon)$.

Now assume that there is a cell $P' \in T \setminus \{P \in T : F \subseteq P\}$ with $x \in P'$, and hence $B(x,\varepsilon) \cap \operatorname{int} P' \neq \emptyset$. Recall the notation $A_k(T)$ introduced in Definition 2.2.

Case (i): If $P' \cap F \in \mathcal{P}_k$, then also $P' \cap F \in A_k(T)$, but $F \not\subseteq P'$, which contradicts the definition of $F \in \mathcal{F}_k^{\mathrm{pr}}(T)$ as a minimal element in $A_k(T)$.

Case (ii): If $P' \cap F \in \mathcal{P}_j, j < k \leq d-1$, then choose $u \in S^{d-1}$ such that $(x + \varepsilon u) \in \operatorname{int} P'$ and the affine subspace $A := \operatorname{aff}(F, x+u)$ spanned by F and $x+u$. Denote by $A^+ \subset A$ the half-space of A that contains $x+u$. Then $B(x,\varepsilon) \cap \bigcup_{x\in P, P\in T, P\neq P'} P$ is not convex, and also

$$A^+ \cap B(x,\varepsilon) \cap \bigcup_{x\in P, P\in T, P\neq P'} P$$

is not convex. On the other hand, because the polytopes are closed sets,

$$F \cap A^+ \cap B(x,\varepsilon) \cap \bigcup_{x\in P, P\in T, P\neq P'} P = F \cap B(x,\varepsilon),$$

which is a k-dimensional set. Therefore there is no cell $P \in T$ such that $P \cap A^+ \cap B(x,\varepsilon) = A^+ \cap B(x,\varepsilon)$. Hence there must be a cell $P'' \in T$ with $x \in P''$, $F \not\subseteq P''$, and $P'' \cap F \in \mathcal{P}_k$, which yields a contradiction, because $F \in \mathcal{F}_k^{\mathrm{pr}}(T)$ is a minimal element in $A_k(T)$. Thus (2.6) is proved, and (2.7) follows from (2.5). Figure 2.2 illustrates this case for $d = 2$, $k = 1$, and $j = 0$. □

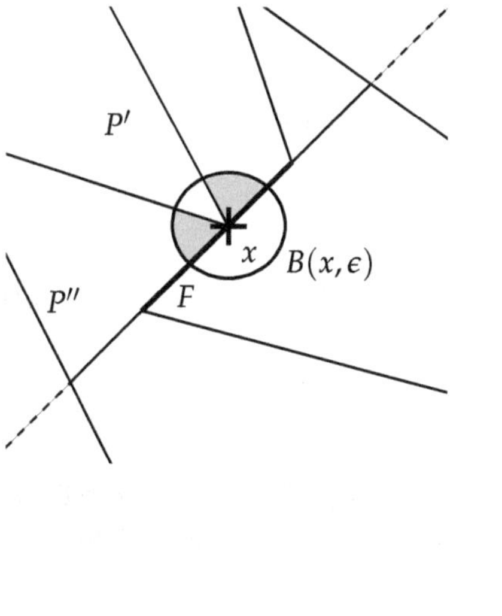

Figure 2.2: Illustration of Case (ii) in the proof of Lemma 2.4 for $d = 2$, $k = 1$, and $j = 0$.

The primitives can also be described as follows.

Corollary 2.5. *For $T \in \mathcal{T}$ and $k \in \{0, \dots, d\}$, we have*

$$\mathcal{F}_k^{\mathrm{pr}}(T) = \{F(x,T) : x \in \mathbb{R}^d\} \cap \mathcal{P}_k . \tag{2.8}$$

Proof. Regarding (2.4) and (2.5), the statement for the cases $k = 0$ and $k = d$ is obvious. Let $k \in \{1, \dots, d-1\}$.

If $F \in \mathcal{F}_k^{\mathrm{pr}}(T)$, then there exists $x \in \operatorname{relint} F$, and hence, according to Lemma 2.4, we obtain that $F(x,T) = F$.

If $F \in \{F(x,T) : x \in \mathbb{R}^d\} \cap \mathcal{P}_k$, then there is $x \in F$ with $F = F(x,T) = \bigcap_{P \in T, x \in P} P$, and thus $F \in A_k(T)$. Keep such $x \in F$ fixed in the rest of the proof. It remains to show that F is a minimal element in $A_k(T)$. For that purpose, let $F' \in A_k(T)$ with minimal $F' \subseteq F$ such that $F' \in \mathcal{F}_k^{\mathrm{pr}}(T)$. If $x \in \operatorname{relint} F'$, then $F' = F(x,T) = F$ by (2.7).

On the other hand, if $x \in F \setminus \operatorname{relint} F'$, then fix $x' \in \operatorname{relint} F'$, and hence $F(x',T) = F'$. Then, either $F = F(x,T) = F(x',T) = F'$, or $F = F(x,T) \supsetneq F(x',T) = F'$. In the second case, there must be a cell $P' \in T$ with $x' \in P'$ and $x \notin P'$. Considering a ball $B(x', \varepsilon)$, analogously as in the proof of Lemma 2.4, the cells intersected by this ball, and taking into account that the cells of a tessellation are convex, we can show that this implies that $x' \notin \operatorname{relint} F$, which contradicts the assumption that $x' \in \operatorname{relint} F'$. This completes the proof of (2.8). □

Essential properties of primitives are the subject of the following proposition: The relative interiors are pairwise disjoint, there is no 0-primitive in the relative interior of a primitive, and all k-faces of a tessellation can be represented as a union of k-primitives.

Corollary 2.6. *Let $T \in \mathcal{T}$. Then*

$$\operatorname{relint} F \cap \operatorname{relint} F' = \emptyset \quad \textit{for all } F, F' \in \bigcup_{k=1}^{d} \mathcal{F}_k^{\mathrm{pr}}(T),\ F \neq F' , \tag{2.9}$$

and

$$F \cap \operatorname{relint} F' = \emptyset \quad \textit{for all } F \in \mathcal{F}_0^{\mathrm{pr}}(T),\ F' \in \bigcup_{k=1}^{d} \mathcal{F}_k^{\mathrm{pr}}(T) . \tag{2.10}$$

Furthermore, for $k \in \{0, \dots, d\}$,

$$F = \bigcup_{F' \subseteq F,\, F' \in \mathcal{F}_k^{\mathrm{pr}}(T)} F' \quad \textit{for all } F \in \mathcal{F}_k(T) . \tag{2.11}$$

Proof. To show (2.9), assume that $F, F' \in \bigcup_{k=1}^{d} \mathcal{F}_k^{\mathrm{pr}}(T)$ and that there is $x \in \operatorname{relint} F \cap \operatorname{relint} F'$. Then (2.7) yields that $F(x,T) = F = F'$.

Assume that there are $\{x\} = F \in \mathcal{F}_0^{\mathrm{pr}}(T)$ and $F' \in \bigcup_{k=1}^{d} \mathcal{F}_k^{\mathrm{pr}}(T)$ with $x \in \operatorname{relint} F'$. Then (2.5) and (2.7) yield that $F' = F(x,T) = F$, a contradiction. This proves (2.10).

To prove (2.11), assume that $F \subset \mathcal{F}_k(T)$ for some $k \in \{0,\dots,d-1\}$. First, we show that $F(x,T) \subseteq F$ for all $x \in F$. Let $P' \in T$ with $F \in \mathcal{F}_k(P')$ and $x \in F$. Because $k \le d-1$, this yields that $x \in \partial P'$ and hence $F(x,T) \subseteq \partial P'$. Because $x \in F(x,T)$ and $F(x,T)$ is convex as an intersection of convex sets, we obtain that $F(x,T) \subseteq F$. This yields that $F = \bigcup_{x\in F} F(x,T)$. By Definition 2.1(c) this union has finitely many different items. All these items are closed sets, and therefore the k-dimensional ones cover F, which means that, with (2.8),

$$F = \bigcup_{x\in F,\, F(x,T)\in\mathcal{P}_k} F(x,T) = \bigcup_{F'\subseteq F,\, F'\in\mathcal{F}_k^{\mathrm{pr}}(T)} F' .$$

This completes the argument. □

As already mentioned, a primitive is not necessarily a face of a cell. However, as shown in the following result, each k-primitive can be embedded into a certain j-face of a cell with $k \le j$.

Corollary 2.7. *Let $T \in \mathcal{T}$, $k \in \{1,\dots,d-1\}$, and $F \in \mathcal{F}_k^{\mathrm{pr}}(T)$, and let $P \in T$ be such that $F \subseteq P$. Then there are a unique $j \ge k$ and a unique $F_j \in \mathcal{F}_j(P)$ with the following two properties:*

$$\operatorname{relint} F \subseteq \operatorname{relint} F_j , \tag{2.12}$$

and

$$F' \cap \operatorname{relint} F = \emptyset \quad \textit{for all } i < j \textit{ and } F' \in \mathcal{F}_i(P) . \tag{2.13}$$

Furthermore, if $F \in \mathcal{F}_0^{\mathrm{pr}}(T)$, $P \in T$, and $F \subseteq P$, then either $F \in \mathcal{F}_0(P)$, or there are a unique $j > 0$ and a unique $F_j \in \mathcal{F}_j(P)$ with

$$F \subseteq \operatorname{relint} F_j . \tag{2.14}$$

Proof. Firstly, we show (2.14). Suppose $T \in \mathcal{T}$, $F \in \mathcal{F}_0^{\mathrm{pr}}(T)$, $P \in T$, $F \subseteq P$, and $F \notin \mathcal{F}_0(P)$. Now consider the sets $\mathcal{F}_j(P)$, starting with $j = 1$ and increasing this dimension step by step. If there is $F_1 \in \mathcal{F}_1(P)$ with $F \subseteq F_1$, then F cannot be on the relative boundary of F_1. Hence $F \subseteq \operatorname{relint} F_1$, F_1 is uniquely determined, and there are no $j > 1$ and $F_j \in \mathcal{F}_j(P)$ such that $F \subseteq \operatorname{relint} F_j$. If there is no $F_1 \in \mathcal{F}_1(P)$ with $F \subseteq F_1$, but there is $F_2 \in \mathcal{F}_2(P)$ with $F \subseteq F_2$, then we can conclude analogously as in the case above for $j = 1$ that $F \subseteq \operatorname{relint} F_2$. If necessary, we consecutively increase the value of j by one. Because $F \subseteq \partial P$, there must be $j \in \{0,\dots,d-1\}$ and $F_j \in \mathcal{F}_j(P)$ such that $F \subseteq F_j$, and j and F_j are uniquely determined.

To prove (2.12), we can proceed analogously as above, beginning with $j = k$. Suppose $T \in \mathcal{T}$, $k \in \{1,\dots,d-1\}$, $F \in \mathcal{F}_k^{\mathrm{pr}}(T)$, $P \in T$, $F \subseteq P$ fixed.

If there is $F_k \in \mathcal{F}_k(P)$ such that $F \subseteq F_k$, then $\operatorname{relint} F \subseteq \operatorname{relint} F_k$. If there is no such $F_k \in \mathcal{F}_k(P)$, then check whether there is $F_{k+1} \in \mathcal{F}_{k+1}(P)$ with $F \subseteq F_{k+1}$ and hence relint $F \subseteq$ relint F_{k+1}. If necessary, we consecutively increase by one the dimension of the faces of P. If for some $j \in \{k,\ldots,d-1\}$, there is $F_j \in \mathcal{F}_j(P)$ such that relint $F \subseteq$ relint F_j, then relint $F \cap$ relint $F'_j = \emptyset$ for all $F'_j \in \mathcal{F}_j(P)$ with $F'_j \neq F_j$, which means that F_j is uniquely determined. Furthermore, for $i > j$ and $F_i \in \mathcal{F}_i(P)$, we have that $F \not\subseteq$ relint F_i. Thus the uniqueness of the value j is also shown.

Now let $i < j$ and $F_i \in \mathcal{F}_i(P)$. Then $F_i \cap$ relint $F_j = \emptyset$ for all $F_j \in \mathcal{F}_j(P)$, and hence (2.12) implies (2.13). □

Since for tessellations that are not face-to-face, the union of k-faces of the cells does not necessarily coincide with the union of k-primitives, a concept of a skeleton is introduced alternatively to (2.2). For $k \in \{0,\ldots,d-1\}$, the **skeleton of k-primitives** of $T \in \mathcal{T}$ is defined as

$$\operatorname{skel}_k^{\mathrm{pr}} T := \bigcup_{F \in \mathcal{F}_k^{\mathrm{pr}}(T)} F . \tag{2.15}$$

In Lemma 18.6, it is shown that the map $\operatorname{skel}_k^{\mathrm{pr}} : \mathcal{T} \to \mathcal{F}$ is measurable.

Now we compare the skeletons with the skeletons of primitives. The right panel in Figure 2.1 illustrates an example in $\mathbb{R}^3$ where the sets $\operatorname{skel}_0 T$ and $\operatorname{skel}_0^{\mathrm{pr}} T$ are different.

Corollary 2.8. *For all $T \in \mathcal{T}$ and $k \in \{0,\ldots,d-1\}$,*

$$\operatorname{skel}_k T \subseteq \operatorname{skel}_k^{\mathrm{pr}} T . \tag{2.16}$$

In particular, for $k = d-1$,

$$\operatorname{skel}_{d-1}^{\mathrm{pr}} T = \operatorname{skel}_{d-1} T = \partial T , \tag{2.17}$$

and if $d \geq 2$, then for $k = d-2$,

$$\operatorname{skel}_{d-2}^{\mathrm{pr}} T = \operatorname{skel}_{d-2} T . \tag{2.18}$$

Proof. Inclusion (2.16) is a direct consequence of (2.11).

Now let $F \in \mathcal{F}_{d-1}^{\mathrm{pr}}(T)$. Then by (2.5) and (2.12) there are $P \in T$ with $F \subseteq P$ and $F_{d-1} \in \mathcal{F}_{d-1}(P)$ such that $F \subseteq F_{d-1}$. Hence $\operatorname{skel}_{d-1}^{\mathrm{pr}} T \subseteq \operatorname{skel}_{d-1} T$, and this proves (2.17).

In the next step, we show (2.18) for $d = 2$, that is, $\operatorname{skel}_0^{\mathrm{pr}} T = \operatorname{skel}_0 T$. Let $T \in \mathcal{T}$ and $F \in \mathcal{F}_0^{\mathrm{pr}}(T)$. If there is $P \in T$ with $F \in \mathcal{F}_0(P)$, then also $F \subset \operatorname{skel}_0 T$. Otherwise, assume that there is no $P \in T$ with $F \in \mathcal{F}_0(P)$. Then by Corollary 2.7 we conclude: For all $P \in T$ with $F \subset P$, there is a unique $F_1(P) \in \mathcal{F}_1(P)$ with $F \subseteq$ relint $F_1(P)$. Hence by the definition of a planar tessellation there are exactly two cells $P, P' \in T$ with $F \subset P$ and $F \subset P'$. Furthermore, relint $F_1(P) \cap$ relint $F_1(P') \supseteq F$, and their intersection has dimension 1. Thus

$$F \neq \bigcap_{P \in T,\, F \subseteq P} P ,$$

and, by (2.5), $F \notin \mathcal{F}_0^{\mathrm{pr}}(T)$, which contradicts the assumption, and (2.18) is proved for $d-2$.

It remains to show (2.18) for $d \geq 3$. Let $T \in \mathcal{T}$ and $F \in \mathcal{F}_{d-2}^{\mathrm{pr}}(T)$. If there exist a cell $P \in T$ and a $(d-2)$-face $F_{d-2} \in \mathcal{F}_{d-2}(P)$ with $F \subseteq F_{d-2}$, then $F \subseteq \operatorname{skel}_{d-2} T$. On the other hand, assume that there is no such F_{d-2}. Then by (2.12) for all cells $P \in T$ with $F \subseteq P$, there is $F_{d-1}(P)$ with $\operatorname{relint} F \subseteq \operatorname{relint} F_{d-1}(P)$. Therefore

$$\bigcap_{P\in T,\, F\subseteq P} P \supseteq \bigcap_{P\in T,\, F\subseteq P} \operatorname{relint} F_{d-1}(P) \supseteq \operatorname{relint} F\,.$$

By the definition of a tessellation there are exactly two cells $P, P' \in T$ such that $\operatorname{relint} F_{d-1}(P) \cap \operatorname{relint} F_{d-1}(P') \supseteq \operatorname{relint} F$, and their intersection has the dimension $d-1$. Because $\dim F = d-2$, we have that $F \neq \bigcap_{P\in T, F\subseteq P} P$. Thus, by (2.5), $F \notin \mathcal{F}_{d-2}^{\mathrm{pr}}(T)$, and the assumption that there are no cell $P \in T$ and face $F_{d-2} \in \mathcal{F}_{d-2}(P)$ with $F \subseteq F_{d-2}$ is falsified. This completes the proof. □

2.1.3 Components of tessellations

Now, as a generic term, we introduce the concept of a set of $\mathsf{A}(T)$-components of a tessellation T. Sets of cells, of k-faces of cells, of primitives, or, for STIT tessellations, sets of maximal polytopes introduced in Section 8.2 will be particular cases. A k-component is a k-dimensional polytope, $k \in \{0, \dots, d\}$.

The group of translations on $\mathbb{R}^d$ also acts on the space of tessellations. For $T \in \mathcal{T}$, $A \subset \mathcal{T}$, and $x \in \mathbb{R}^d$, we write

$$T + x := \{P + x : P \in T\} \quad \text{and} \quad A + x := \{T + x : T \in A\}\,. \tag{2.19}$$

An analogous notation is used for the group SO_d of rotations on $\mathbb{R}^d$. Note that the σ-algebra $\mathcal{B}(\mathcal{T})$ is closed under translations and rotations.

Definition 2.9. *Let $k \in \{0, \dots, d\}$, and let $\mathsf{A} : \mathcal{T} \to \overline{\mathcal{F}}_{\mathit{lf}}(\mathcal{P}_k)$ be a measurable translation-covariant map, that is, $\mathsf{A}(T + x) = \mathsf{A}(T) + x$ for all $T \in \mathcal{T}$ and $x \in \mathbb{R}^d$. Furthermore, assume that $\bigcup_{F\in \mathsf{A}(T)} F \subseteq \partial T$ for $k < d$ and $\mathsf{A}(T) \subseteq T$ for $k = d$. Then, for a tessellation $T \in \mathcal{T}$, the set $\mathsf{A}(T)$ is called the* ***set of A-components of T***.

Several components of interest for random stationary tessellations are dealt with in detail in Section 2.2. A key proposition is the exchange formula in Theorem 2.29.

2.1.4 Restrictions of tessellations to polytopes

In this book, tessellations partitioning a polytope $W \in \mathcal{P}_d$ play an important role. In this context, W is also referred to as a **window**. We adapt Definition 2.1.

Definition 2.10. *Let $W \in \mathcal{P}_d$ be a d-dimensional polytope. A* ***tessellation of*** *W is a finite subset $T \subset \mathcal{P}_d$ of d-dimensional polytopes satisfying*
(a) $\bigcup_{P \in T} P = W$;
(b) *if* $P, P' \in T$ *and* $P \neq P'$, *then* $\operatorname{int} P \cap \operatorname{int} P' = \emptyset$.

For a fixed $W \in \mathcal{P}_d$, let $\mathcal{T}_W$ denote the set of all tessellations of W, endowed with the σ-algebra $\mathcal{B}(\mathcal{T}_W) := \mathcal{B}(\overline{\mathcal{F}}_{ef}(\mathcal{F})) \llcorner \mathcal{T}_W$; see Lemma 18.2. In particular, $\{W\} \in \mathcal{T}_W$.

For $T \in \mathcal{T}_W$ and $k \in \{0, \ldots, d-1\}$, the sets $\mathcal{F}_k(T)$, $\mathcal{F}_k^{\mathrm{pr}}(T)$, $\operatorname{skel}_k T$, $\operatorname{skel}_k^{\mathrm{pr}} T$, and ∂T are defined in exactly the same way as for $T \in \mathcal{T}$. In a window W, for $T \in \mathcal{T}_W$, the skeletons $\operatorname{skel}_k T$ and $\operatorname{skel}_k^{\mathrm{pr}} T$ are elements of the convex ring $\mathcal{R}$.

Furthermore, we define the restriction of a tessellation T of the space $\mathbb{R}^d$ to a window $W \in \mathcal{P}_d$.

Definition 2.11. *Let $W \in \mathcal{P}_d$ be a d-dimensional polytope, and let $T \in \mathcal{T}$ be a tessellation. The* ***restriction of*** *T* ***to*** *W is defined as*

$$T \wedge W := \{P \cap W : P \in T,\ P \cap \operatorname{int} W \neq \emptyset\}. \tag{2.20}$$

This definition yields that $T \wedge W \in \mathcal{T}_W$ for all $T \in \mathcal{T}$ and $W \in \mathcal{P}_d$. Note that $\mathcal{T}_W \not\subseteq \mathcal{T}$. The measurability of the mapping $T \mapsto T \wedge W$ is shown in Lemma 18.10.

Such restrictions of tessellations to polytopes are crucial for the concept of spatial consistency, explained in Section 5.3, and, as we can see in Theorem 5.10, for the existence of STIT tessellations in the space $\mathbb{R}^d$. Indeed, in Chapter 4 the STIT tessellations are defined in windows W only, and the problem arises whether these tessellations can be considered as restrictions of a "global" random tessellation of $\mathbb{R}^d$.

2.1.5 Sections of tessellations with affine subspaces

The section of a tessellation in $\mathbb{R}^d$ with a lower-dimensional flat generates a tessellation within this flat. This is now formalized. To begin with, Definition 2.1 for tessellations of $\mathbb{R}^d$ is straightforwardly transferred to flats.

Definition 2.12. *Let $A \in A(d,k)$ be a k-dimensional affine subspace of $\mathbb{R}^d$, $k \in \{1, \ldots, d-1\}$. A* ***tessellation within*** *A is a countable subset $T \subset \mathcal{P}_k$ of k-dimensional polytopes satisfying*
(a) $\bigcup_{P \in T} P = A$;
(b) *if* $P, P' \in T$ *and* $P \neq P'$, *then* $\operatorname{relint} P \cap \operatorname{relint} P' = \emptyset$;
(c) *the set* $\{P \in T : P \cap C \neq \emptyset\}$ *is finite for all compact sets* $C \in \mathcal{C}$.

Let $\mathcal{T}_A$ denote the set of all tessellations within $A \in A(d,k)$. The Borel σ-algebra $\mathcal{B}(\mathcal{T}_A)$ is the restriction $\mathcal{B}(\overline{\mathcal{F}}(\mathcal{F}(A))) \llcorner \mathcal{T}_A$ of $\mathcal{B}(\overline{\mathcal{F}}(\mathcal{F}(A)))$ to $\mathcal{T}_A$, where $\mathcal{F}(A)$ is the set of all closed subsets of A; see Chapter 18.

For $A \in A(d,k)$, we define the subset $\mathcal{T}(A) \subseteq \mathcal{T}$ of the set of the tessellations of $\mathbb{R}^d$ by

$$\mathcal{T}(A) := \{T \in \mathcal{T} : \forall P \in T : A \cap P \neq \emptyset \Rightarrow A \cap \operatorname{int} P \neq \emptyset\} \tag{2.21}$$

with the Borel σ-algebra $\mathcal{B}(\mathcal{T}(A)) := \mathcal{B}(\mathcal{T}) \llcorner \mathcal{T}(A)$.

Definition 2.13. *Let $A \in A(d,k)$, $k \in \{1, \ldots, d-1\}$, and let $T \in \mathcal{T}(A)$ be a tessellation of $\mathbb{R}^d$ that satisfies the condition formulated in* (2.21). *The* ***section of*** T ***with the*** k***-dimensional flat*** A *is defined as*

$$T \cap A := \{P \cap A : P \in T, A \cap P \neq \emptyset\}. \tag{2.22}$$

The condition that $T \in \mathcal{T}(A)$ implies that $T \cap A \in \mathcal{T}_A$. The measurability of the mapping $T \mapsto T \cap A$ is shown in Lemma 18.12.

Analogously, for $W \in \mathcal{P}_d$ and $A \in A(d,k)$ with $A \cap \operatorname{int} W \neq \emptyset$, we define the set of tessellations $\mathcal{T}_W(A) \subset \mathcal{T}_W$ and the section of $T \in \mathcal{T}_W(A)$ with A as in (2.22).

2.2 Random tessellations

The concepts provided in the previous section are now transferred to a stochastic context. Firstly, random tessellations are defined. Then, to supply a basis for the main part of this book, random tessellation-valued *processes on the time axis* are also considered.

2.2.1 Definition and capacity functional

Definition 2.14. *Let $(\Omega, \mathbf{A}, \mathbb{P})$ be a probability space, and let $I \subseteq \mathbb{R}$.*

(a) *A* ***random tessellation in*** $\mathbb{R}^d$ *is a random variable $X : \Omega \to \mathcal{T}$.*
(b) *A* ***random tessellation in*** $W \in \mathcal{P}_d$ *is a random variable $X_W : \Omega \to \mathcal{T}_W$.*
(c) *A* ***random tessellation-valued process in*** $\mathbb{R}^d$ *is a family $\mathbf{X} = (X_t,\ t \in I)$, where all X_t are random tessellations.*
(d) *A* ***random tessellation-valued process in*** $W \in \mathcal{P}_d$ *is a family $\mathbf{X}_W = (X_{t,W},\ t \in I)$, where all $X_{t,W}$ are random tessellations of W.*

For processes, mostly in this book, $I = \mathbb{R}^+$; for time-stationary tessellation-valued processes, $I = \mathbb{R}$ or $I = \mathbb{Z}$.

A main tool for dealing with the distribution of a random tessellation X and for proving key results will be the capacity functional of the $(d-1)$-skeleton ∂X.

Definition 2.15. *Let B be a random closed set, that is, a measurable map from a probability space $(\Omega, \mathbf{A}, \mathbb{P})$ to $(\mathcal{F}, \mathcal{B}(\mathcal{F}))$. The* ***capacity functional*** $T_B : \mathcal{C} \to [0,1]$ *of B is defined* by

$$T_B(C) := \mathbb{P}(B \cap C \neq \emptyset), \quad C \in \mathcal{C}. \tag{2.23}$$

The following lemma states that the distribution of the skeleton of a random tessellation is already determined by the values of the capacity functional on the convex ring $\mathcal{R}$.

Lemma 2.16. (a) *If X is a random tessellation of $\mathbb{R}^d$, then for all $k \in \{0, \dots, d-1\}$, the k-skeletons $\operatorname{skel}_k X$ and $\operatorname{skel}_k^{\mathrm{pr}} X$ are random closed sets in $\mathbb{R}^d$.*

(b) *If X is a random tessellation of $\mathbb{R}^d$, then its distribution is uniquely determined by the family of probabilities*

$$\mathbb{P}(\partial X \cap C = \emptyset), \quad C \in \mathcal{R}.$$

(c) *If $W \in \mathcal{P}_d$ and X_W is a random tessellation of W, then for all $k \in \{0, \dots, d-1\}$, the k-skeletons $\operatorname{skel}_k X_W$ and $\operatorname{skel}_k^{\mathrm{pr}} X_W$ are random closed sets in $\mathbb{R}^d$. The distribution of X_W is uniquely determined by the family of probabilities*

$$\mathbb{P}(\partial X_W \cap C = \emptyset), \quad C \in \mathcal{R}, \quad C \subseteq \operatorname{int} W.$$

Proof. (a) This statement follows immediately from Lemma 18.6.

(b) If X is a random tessellation of $\mathbb{R}^d$, then according to Lemma 18.7, by Theorem 2.1.3 of [172] its distribution is uniquely determined by the capacity functional of the random closed set ∂X.

The results and comments in Section 1.1.5 of [137] show that the capacity functional of ∂X is already determined by its values on the system of all finite unions of balls in $\mathbb{R}^d$. The convex ring $\mathcal{R}$ is defined as the system of all finite unions of convex bodies.

(c) By Lemma 18.9 the skeletons $\operatorname{skel}_k X_W$ and $\operatorname{skel}_k^{\mathrm{pr}} X_W$, $k \in \{0, \dots, d-1\}$, are random closed sets in $\mathbb{R}^d$. The same arguments as in the proof of (b) yield that the distribution of X_W is uniquely determined by the values of the capacity functional for ∂X_W on the convex ring $\mathcal{R}$. □

The capacity functional for STIT tessellations in windows is determined in Section 5.2. It is the essential tool to show in Theorem 5.10 the spatial consistency of the STIT construction in windows. This, in turn, yields the existence of STIT tessellations of the whole space $\mathbb{R}^d$.

2.2.2 Stationarity and typical components

In this book, we always restrict ourselves to stationary tessellations when considering random tessellations of $\mathbb{R}^d$. Stationarity allows us to define intensities and densities of certain components and to speak of typical components and directional distributions in typical points of a skeleton. Sometimes results are formulated for random tessellations that are stationary and additionally isotropic.

An essential foundation for the study of random tessellations is the theory of point processes. The main tools from this theory are summarized in the appendix, Chapter 19. In particular, we refer to Remark 19.3, which justifies that a simple point process on a space E may also be described by a random subset of E. We will freely switch between these two viewpoints throughout the book. Moreover, the concept of marked point processes, introduced in Definition 19.6, plays an important role in our exposition.

The action of the group of translations on $\mathbb{R}^d$ on the space of tessellations is introduced in the usual way by (2.19).

Definition 2.17. *Let X be a random tessellation of $\mathbb{R}^d$. It is called*

(a) ***stationary**, or **space-stationary**, or **homogeneous** if its distribution $\mathbb{P}_X$ is invariant under the group of translations of $\mathbb{R}^d$, that is,*

$$\mathbb{P}_X(A + x) = \mathbb{P}_X(A) \quad \text{for all } A \in \mathcal{B}(\mathcal{T}),\ x \in \mathbb{R}^d\,;$$

(b) ***isotropic** if its distribution is invariant under the group SO_d of rotations around the origin of $\mathbb{R}^d$, that is,*

$$\mathbb{P}_X(\vartheta A) = \mathbb{P}_X(A) \quad \text{for all } A \in \mathcal{B}(\mathcal{T}),\ \vartheta \in SO_d\,.$$

Consider a certain type of A-components of a tessellation as introduced in Definition 2.9. For random stationary tessellations, we can define the intensity of A-components and the distribution of a typical A-component. In some cases, it also makes sense to introduce a density.

Firstly, we present these concepts for general A-components, and then we specify them for cells, faces, and primitives.

A-components

For some $k \in \{0, \dots, d\}$, let $\mathsf{A} : \mathcal{T} \to \mathcal{P}_k$ be as in Definition 2.9. To define the intensity and distribution of a "typical A-component", let $c : \mathcal{P}_k \to \mathbb{R}^d$ be a center function, which means that c is a measurable translation-covariant function, that is, $c(P + x) = c(P) + x$ for all $P \in \mathcal{P}_k$ and $x \in \mathbb{R}^d$. Then, for a random tessellation X, denote by

$$\mathcal{Z}(\mathsf{A}(X)) := \{(c(\mathsf{a}), \mathsf{a} - c(\mathsf{a})) : \mathsf{a} \in \mathsf{A}(X)\} \tag{2.24}$$

the set of all component centers marked with the centered components.

If X is a random stationary tessellation, then by the translation covariance of A, $\mathsf{A}(X)$ also is a stationary point process on $\mathcal{P}_k$, and $\mathcal{Z}(\mathsf{A}(X))$ is a stationary marked point process on $\mathbb{R}^d$ with mark space $\mathcal{P}_{k,c}$.

For a random stationary tessellation X, define the **intensity of A-components** as

$$\gamma_{\mathsf{A}} := \mathbb{E} \sum_{(x,\mathsf{a})\in\mathcal{Z}(\mathsf{A}(X))} \mathbb{1}\{x \in [0,1]^d\} = \mathbb{E} \sum_{\mathsf{a}\in\mathsf{A}(X)} \mathbb{1}\{c(\mathsf{a}) \in [0,1]^d\}, \tag{2.25}$$

which is the mean number of centers of A-components per unit volume. In the following, we always assume that $0 < \gamma_{\mathsf{A}} < \infty$.

The distribution of the **typical A-component** of X with respect to the center function c is defined as

$$\mathbb{Q}_{\mathsf{A}}(B) := \frac{1}{\gamma_{\mathsf{A}}} \mathbb{E} \sum_{(x,\mathsf{a})\in\mathcal{Z}(\mathsf{A}(X))} \mathbb{1}\{x \in [0,1]^d\}\, \mathbb{1}\{\mathsf{a} \in B\} \tag{2.26}$$

for $B \in \mathcal{B}(\mathcal{P}_{k,c})$. As usual, for a fixed center function, a random polytope with distribution $\mathbb{Q}_{\mathsf{A}}$ is referred to as a typical A-component. Using standard methods of integration theory, equation (2.26) leads to

$$\mathbb{E} \sum_{\mathsf{a}\in\mathsf{A}(X)} f(c(\mathsf{a}), \mathsf{a} - c(\mathsf{a})) = \gamma_{\mathsf{A}} \int_{\mathcal{P}_{k,c}} \int_{\mathbb{R}^d} f(x,\mathsf{a})\, \lambda_d(\mathrm{d}x)\, \mathbb{Q}_{\mathsf{A}}(\mathrm{d}\mathsf{a}) \tag{2.27}$$

for all nonnegative measurable $f : \mathbb{R}^d \times \mathcal{P}_{k,c} \to \mathbb{R}$; see equation (4.3) in [172].

Remark 2.18. *If P_{A} is a typical A-component and $g : \mathcal{P}_k \to \mathbb{R}$ is a translation-invariant measurable function, then the random variable $g(P_{\mathsf{A}})$ does not depend on the choice of the center function c. To see this, let c_1, c_2 be two center functions. Then, for all $P \in \mathcal{P}_k$, there is $x(P) \in \mathbb{R}^d$ such that $c_2(P) = c_1(P) + x(P)$. Hence, by the translation invariance of g, $g(P - c_1(P)) = g(P - c_2(P))$ for all $P \in \mathcal{P}_k$. Thus, when working with translation invariant functionals of typical components, it is not necessary to refer to a specific center function.*

An application of Theorem 4.1.3 of [172] provides the following lemma, which presents some alternative descriptions of the intensity γ_{A} and distribution $\mathbb{Q}_{\mathsf{A}}$.

Lemma 2.19. *Let X be a random stationary tessellation, and let $g : \mathcal{P}_k \to \mathbb{R}$ be a translation-invariant measurable function that is nonnegative or $\mathbb{Q}_{\mathsf{A}}$-integrable. Then for $C \in \mathcal{B}(\mathbb{R}^d)$ with $0 < \lambda_d(C) < \infty$, we have*

$$\gamma_{\mathsf{A}} \int_{\mathcal{P}_{k,c}} g(\mathsf{a})\, \mathbb{Q}_{\mathsf{A}}(\mathrm{d}\mathsf{a}) = \frac{1}{\lambda_d(C)} \mathbb{E} \sum_{\mathsf{a}\in\mathsf{A}(X)} \mathbb{1}\{c(\mathsf{a}) \in C\}\, g(\mathsf{a}). \tag{2.28}$$

Furthermore, let $W \in \mathcal{K}$ with $V_d(W) > 0$. Then

$$\gamma_{\mathsf{A}} \int_{\mathcal{P}_{k,c}} g(\mathsf{a})\, \mathbb{Q}_{\mathsf{A}}(\mathrm{d}\mathsf{a}) = \lim_{r\to\infty} \frac{1}{V_d(rW)} \mathbb{E} \sum_{\mathsf{a}\in\mathsf{A}(X)} \mathbb{1}\{\mathsf{a} \subseteq rW\}\, g(\mathsf{a}). \tag{2.29}$$

If, additionally, $\int_{\mathcal{P}_{k,c}} |g(\mathrm{a})|\, V_d(\mathrm{a} + B^d)\, \mathbb{Q}_{\mathsf{A}}(\mathrm{da}) < \infty$, *then also*

$$\gamma_{\mathsf{A}} \int_{\mathcal{P}_{k,c}} g(\mathrm{a})\, \mathbb{Q}_{\mathsf{A}}(\mathrm{da}) = \lim_{r\to\infty} \frac{1}{V_d(rW)}\, \mathbb{E} \sum_{\mathrm{a}\in \mathsf{A}(X)} \mathbb{1}\{\mathrm{a} \cap rW \neq \emptyset\}\, g(\mathrm{a})\,. \tag{2.30}$$

To deal with typical components together with other adjacent components of a tessellation – for example, the cells containing the typical 0-primitive – an analogy to the Palm method for point processes is applied.

For a random stationary tessellation X and a set of A-components, we define the probability distribution $\mathbb{P}^o_{\mathsf{A}}$ on $\mathcal{T}$ by

$$\mathbb{P}^o_{\mathsf{A}}(B) := \frac{1}{\gamma_{\mathsf{A}}} \mathbb{E} \sum_{\mathrm{a}\in \mathsf{A}(X)} \mathbb{1}\{c(\mathrm{a}) \in [0,1]^d\}\mathbb{1}\{X - c(\mathrm{a}) \in B\} \tag{2.31}$$

for $B \in \mathcal{B}(\mathcal{T})$. Note that $\mathbb{P}^o_{\mathsf{A}}$ is concentrated on

$$\mathcal{T}^{\mathsf{A}} := \{T \in \mathcal{T} :\ \text{exists an a} \in \mathsf{A}(T) \text{ with } c(\mathrm{a}) = o\}\,.$$

With these definitions, we can prove an equation, which reminds the refined Campbell theorem 19.5, but note that it is not the same.

Lemma 2.20. *For a random stationary tessellation X, a set of* A*-components, and a center function c,*

$$\mathbb{E} \sum_{\mathrm{a}\in \mathsf{A}(X)} g(c(\mathrm{a}), X - c(\mathrm{a})) = \gamma_{\mathsf{A}} \int_{\mathcal{T}} \int_{\mathbb{R}^d} g(x,T)\, \lambda_d(\mathrm{d}x)\, \mathbb{P}^o_{\mathsf{A}}(\mathrm{d}T) \tag{2.32}$$

for all nonnegative measurable $g : \mathbb{R}^d \times \mathcal{T} \to \mathbb{R}$.

Proof. Equation (2.31) can be rewritten as

$$\begin{aligned} &\mathbb{E} \sum_{\mathrm{a}\in \mathsf{A}(X)} \mathbb{1}\{c(\mathrm{a}) \in [0,1]^d\}\, \mathbb{1}\{X - c(\mathrm{a}) \in B\} \\ &\quad = \gamma_{\mathsf{A}} \int_{\mathcal{T}} \int_{\mathbb{R}^d} \mathbb{1}\{x \in [0,1]^d\}\, \mathbb{1}\{T \in B\}\, \lambda_d(\mathrm{d}x)\, \mathbb{P}^o_{\mathsf{A}}(\mathrm{d}T)\,. \end{aligned}$$

Then by standard methods of integration theory this identity extends to nonnegative measurable functions $g : \mathbb{R}^d \times \mathcal{T} \to \mathbb{R}$. □

For $T \in \mathcal{T}^{\mathsf{A}}$, we denote by $\mathrm{a}^o(T) \in \mathsf{A}(T)$ the A-component of T with $c(\mathrm{a}^o) = o$. As a particular case of a function g, we consider

$$g(x,T) := \mathbb{1}\{x \in [0,1]^d\} f(a^o(T), T)\,, \tag{2.33}$$

where $f : \mathcal{P}_k \times \mathcal{T} \to \mathbb{R}$ is a nonnegative measurable function. For example, f can be a functional that expresses properties of the A-component $a^o(T)$ jointly with its "surrounding" in T. In Section 2.2.4 a so-called exchange formula for different types of components of a tessellation will be shown, and in Chapters 16 and 17, there will be concrete applications of it.

Furthermore, we will apply **disintegration** of $\mathbb{Q}_A$ with respect to conditional probability measures, conditioning on certain properties of A-components. This can be based on the following result, taken from Theorems 3.4 and 8.5 of [89].

Theorem 2.21. *Let $(U,\mathcal{U})$ and $(V,\mathcal{V})$ be measurable spaces, where V is a Borel space, and $\mathbb{P}$ is a probability measure on the product space $(U \times V, \mathcal{U} \otimes \mathcal{V})$. Denote $\mathbb{P}_U(\cdot) := \mathbb{P}(\cdot \times V)$, which is a probability measure on $(U,\mathcal{U})$. Then there is a probability kernel $\mathbb{P}_{V|u}$ with $u \in U$, from U to V, uniquely determined $\mathbb{P}_U$-almost everywhere, and*

$$\int_{U\times V} g(u,v)\,\mathbb{P}(\mathrm{d}(u,v)) = \int_U \int_V g(u,v)\,\mathbb{P}_{V|u}(\mathrm{d}v)\,\mathbb{P}_U(\mathrm{d}u) \tag{2.34}$$

for all nonnegative measurable $g : U \times V \to \mathbb{R}$.

As an application of this general theorem, consider the case that the distribution $\mathbb{P}_U$ is the image measure of a probability measure on $(V,\mathcal{V})$ under a function $h : V \to U$.

Corollary 2.22. *Let $(U,\mathcal{U})$ and $(V,\mathcal{V})$ be measurable spaces, where V is a Borel space, and $\mathbb{P}$ is a probability measure on $(V,\mathcal{V})$. Furthermore, let $h : V \to U$ be a measurable map and denote by $\mathbb{P}_h := \mathbb{P} \circ h^{-1}$ the image measure. Then there exists a $\mathbb{P}_h$-almost everywhere uniquely determined family $(\mathbb{P}_{V|u},\ u \in U)$ of probability measures on V such that $\mathbb{P}_{V|u}(V \setminus h^{-1}(u)) = 0$ for $\mathbb{P}_h$-almost all $u \in U$ and*

$$\int_V g(v)\,\mathbb{P}(\mathrm{d}v) = \int_U \int_{h^{-1}(u)} g(v)\,\mathbb{P}_{V|u}(\mathrm{d}v)\,\mathbb{P}_h(\mathrm{d}u) \tag{2.35}$$

for all nonnegative measurable $g : V \to \mathbb{R}$.

Proof. For a nonnegative measurable $g : V \to \mathbb{R}$ and measurable $h : V \to U$, define the function $g_1 : U \times V \to \mathbb{R}$ by

$$g_1(u,v) := \mathbb{1}\{u \in h(v)\}\,g(v)\,, \quad (u,v) \in U \times V\,,$$

and a probability measure $\mathbb{P}_1 := \mathbb{P} \circ h_1^{-1}$ on $(U \times V, \mathcal{U} \otimes \mathcal{V})$ with $h_1 : v \mapsto (h(v), v)$ and thus

$$\int_{U\times V} f(u,v)\,\mathbb{P}_1(\mathrm{d}(u,v)) := \int_V f(h(v),v)\,\mathbb{P}(\mathrm{d}v)$$

for nonnegative measurable $f : U \times V \to \mathbb{R}$. For that measure, we have $\mathbb{P}_h(\cdot) = \mathbb{P}_1(\cdot \times V)$. Then by (2.34)

$$\int_V g(v)\,\mathbb{P}(\mathrm{d}v) - \int_V g_1(h(v),v)\,\mathbb{P}(\mathrm{d}v)$$
$$= \int_{U\times V} g_1(u,v)\,\mathbb{P}_1(\mathrm{d}(u,v))$$
$$= \int_U \int_V g_1(u,v)\,\mathbb{P}_{V|u}(\mathrm{d}v)\,\mathbb{P}_h(\mathrm{d}u)$$
$$= \int_U \int_V \mathbb{1}\{v \in h^{-1}(u)\}\, g(v)\,\mathbb{P}_{V|u}(\mathrm{d}v)\,\mathbb{P}_h(\mathrm{d}u)$$
$$= \int_U \int_{h^{-1}(u)} g(v)\,\mathbb{P}_{V|u}(\mathrm{d}v)\,\mathbb{P}_h(\mathrm{d}u)\,.$$

The probability measure $\mathbb{P}_{V|u}$ can be chosen such that $\mathbb{P}_{V|u}(V\setminus h^{-1}(u)) = 0$ for $\mathbb{P}_h$-almost all $u \in U$. □

As an example, consider a random stationary tessellation X and the set of A-components $\mathsf{A}(X) = \mathcal{F}_k(X)$, the set of k-faces of X. Denote by $\mathbb{Q}_k$ the distribution of the typical k-face of X with respect to a fixed center function $c : \mathcal{P}_k \to \mathbb{R}^d$. Note that $\mathcal{P}_k$ is a Borel space by Lemma 18.1 and by the fact that $\mathcal{P}_k \in \mathcal{B}(\mathcal{F})$. Recall the notation $\mathrm{aff}_o F$ for the k-dimensional linear subspace parallel to the affine subspace spanned by $F \in \mathcal{P}_k$. An application of Corollary 2.22 with $V = \mathcal{P}_k$ and $h : \mathcal{P}_k \to G(d,k)$ given by $h(F) := \mathrm{aff}_o F$, $F \in \mathcal{P}_k$, yields the existence of a family $(\mathbb{Q}_{k|L}\,,\ L \in G(d,k))$ of probability measures on $\mathcal{P}_{L,k,c}$ satisfying

$$\int_{\mathcal{P}_{k,c}} g(F)\,\mathbb{Q}_k(\mathrm{d}F) = \int_{G(d,k)} \int_{\mathcal{P}_{L,k,c}} g(F)\,\mathbb{Q}_{k|L}(\mathrm{d}F)\,(\mathbb{Q}_k \circ h^{-1})(\mathrm{d}L) \tag{2.36}$$

for all nonnegative measurable functions $g : \mathcal{P}_k \to \mathbb{R}$, where $\mathcal{P}_{L,k,c} := \{F \in \mathcal{P}_{k,c} : \mathrm{aff}_o F = L\}$ is the set of centered k-polytopes parallel to L. The distribution $\mathbb{Q}_{k|L}$ can be interpreted as the conditional distribution of the typical k-face of X under the condition that this face has direction $L \in G(d,k)$.

2.2.3 Particular components

In this section, cells, k-faces, and k-primitives, $k \in \{0,\dots,d-1\}$, are dealt with in more detail. Later, in Section 4.2, for STIT tessellations in a window, also birth time marked maximal polytopes are introduced; see (4.10) and (4.11). Furthermore, birth time marked maximal polytopes for STIT tessellations in $\mathbb{R}^d$ are defined in Section 8.2, (8.7) and (8.8).

Cells

For random stationary tessellations, it makes sense to speak of the distribution of the typical cell. Furthermore, it is of interest to study the zero-cell and the relationships between the two.

A random tessellation X can be treated as a random particle process, that is, a point process on $\mathcal{P}_d$. As a random measure, it is

$$\delta(X,\cdot) := \sum_{P\in X} \delta_P(\cdot)\,. \tag{2.37}$$

For a given center function $c : \mathcal{P}_d \to \mathbb{R}^d$ and

$$\mathcal{Z}(X) := \{(c(P), P - c(P)) : P \in X\}\,,$$

a random tessellation can be described as a germ-grain process

$$\delta(\mathcal{Z}(X),\cdot) := \sum_{P\in X} \delta_{(c(P),P-c(P))}(\cdot)\,,$$

which is the process of center points $c(P)$ of cells, marked with centered polytopes $P - c(P)$.

If X is stationary, then the intensity of cells, which are d-dimensional polytopes, is defined as in (2.25) and is denoted by γ_d, and the distribution of the typical cell, as a particular case of (2.26), is denoted by $\mathbb{Q}_d$.

The distribution of the typical cell determines the intensity γ_d. By the refined Campbell theorem 19.8 and the translative integral formula (1.4) with $\mu = \lambda_d$,

$$1 = \mathbb{E} \sum_{(x,P)\in\mathcal{Z}(X)} V_d((P+x)\cap[0,1]^d) = \gamma_d\,\mathbb{E}V_d(Z)\,,$$

and thus

$$\gamma_d = \frac{1}{\mathbb{E}V_d(Z)}\,, \tag{2.38}$$

where Z is the typical cell of the random stationary tessellation X, that is, a random polytope with distribution $\mathbb{Q}_d$.

Furthermore, the distribution of the typical cell also determines the **intensity measure** Λ of the particle process assigned to a random tessellation X. It is defined by

$$\Lambda(A) := \mathbb{E}\,\delta(X,A)\,, \quad A \in \mathcal{B}(\mathcal{P}_d)\,.$$

Corollary 2.23. *Let X be a random stationary tessellation. Then the distribution $\mathbb{Q}_d$ of its typical cell with respect to a center function c determines the intensity measure Λ of X as a particle process on $\mathcal{P}_d$ by*

$$\Lambda = \gamma_d\,(\lambda_d \otimes \mathbb{Q}_d) \circ f^{-1},$$

where $f : \mathbb{R}^d \times \mathcal{P}_{d,c} \to \mathcal{P}_d$ *with* $f(x,P) := P + x$.

Proof. By (2.38) the distribution $\mathbb{Q}_d$ determines the intensity γ_d of the marked point process of cells $\mathcal{Z}(X)$. Thus Theorem 4.1.1 of [172] yields the statement. □

The following lemma provides a further characterization of the typical cell. This is used in Section 12.5.4.

Lemma 2.24. *Let X be a random stationary tessellation of $\mathbb{R}^d$, and let $c : \mathcal{P}_d \to \mathbb{R}^d$ be a center function. Denote by Z its typical cell with respect to c. Furthermore, let $K \in \mathcal{K}$ with $0 < \lambda_d(K) < \infty$, and let $g : \mathcal{P}_d \to \mathbb{R}$ be a nonnegative measurable translation-invariant function. Then*

$$\mathbb{E} \sum_{P\in X:\, P\cap K\neq\emptyset} g(P) = \frac{1}{\mathbb{E}V_d(Z)}\,\mathbb{E}\big(g(Z)\,V_d(Z + (-K))\big)\,. \tag{2.39}$$

Proof. By (2.27), the translation invariance of g, and (2.38),

$$\begin{aligned}\mathbb{E} \sum_{P\in X:\, P\cap K\neq\emptyset} g(P) &= \gamma_d \int_{\mathcal{P}_{d,c}} \int_{\mathbb{R}^d} g(P)\,\mathbb{1}\{(P+x)\cap K \neq \emptyset\}\,\lambda_d(\mathrm{d}x)\,\mathbb{Q}_d(\mathrm{d}P)\\ &= \frac{1}{\mathbb{E}V_d(Z)}\,\mathbb{E}\big(g(Z)\,V_d(Z + (-K))\big)\,,\end{aligned}$$

which is (2.39). □

In a random stationary tessellation X, the cell containing the origin o is a. s. uniquely determined, since the d-dimensional volume intensity of the $(d-1)$-skeleton ∂X is zero. Consequently, $\mathbb{P}(o \in \partial X) = 0$; see, for instance, Theorem 2.4.7 of [172]. This cell containing the origin is denoted by $Z_{\langle o\rangle}$ and is called the **zero-cell** of the random tessellation.

Lemma 2.25. *The distribution of the zero-cell $Z_{\langle o\rangle}$ is uniquely determined by the family of probabilities*

$$\mathbb{P}(K \subseteq Z_{\langle o\rangle})\,, \quad K \in \mathcal{P}\,.$$

Proof. By definition, $Z_{\langle o\rangle}$ is a random compact convex set with nonempty interior, and hence, due to Lemma 1.1.67 and Proposition 1.1.69 of [137], its distribution is uniquely determined by the family of probabilities

$$\mathbb{P}(A \subseteq Z_{\langle o\rangle})\,, \quad A \subseteq \mathbb{R}^d, \quad A \text{ finite}\,.$$

Because $Z_{\langle o\rangle}$ is convex, $A \subseteq Z_{\langle o\rangle}$ if and only if $\operatorname{conv} A \subseteq Z_{\langle o\rangle}$. Thus the equality $\{\operatorname{conv} A : A \subseteq \mathbb{R}^d, A \text{ finite}\} = \mathcal{P}$ yields the proposition of the lemma. □

If X is stationary, then its typical cell Z is related to its zero-cell via

$$\mathbb{E}g(Z) = \mathbb{E}V_d(Z)\,\mathbb{E}\Big(\frac{1}{V_d(Z_{\langle o\rangle})}g(Z_{\langle o\rangle})\Big)$$
$$= \Big(\mathbb{E}\frac{1}{V_d(Z_{\langle o\rangle})}\Big)^{-1}\mathbb{E}\Big(\frac{1}{V_d(Z_{\langle o\rangle})}g(Z_{\langle o\rangle})\Big) \tag{2.40}$$

for all nonnegative measurable translation-invariant functions $g : \mathcal{P}_d \to \mathbb{R}$; see [138] or Theorem 10.4.1 of [172]. As g is assumed to be translation invariant, this does not depend on the choice of the center function c.

Faces and primitives

Analogously to the process of cells, for $k \in \{0, \dots, d-1\}$, the point processes $\delta(\mathcal{F}_k(X), \cdot)$ of k-faces, $\delta(\mathcal{F}_k^{\mathrm{pr}}(X), \cdot)$ of k-primitives, and the respective germ-grain processes $\delta(\mathcal{Z}(\mathcal{F}_k(X)), \cdot)$ and $\delta(\mathcal{Z}(\mathcal{F}_k^{\mathrm{pr}}(X)), \cdot)$ are defined.

Now we fix a center function $c : \mathcal{P}_k \to \mathbb{R}^d$ and consider the marked point process of k-primitives

$$\mathcal{Z}(\mathcal{F}_k^{\mathrm{pr}}(X)) := \{(c(F), F - c(F)) : F \in \mathcal{F}_k^{\mathrm{pr}}(X)\} \tag{2.41}$$

as a particular case of (2.24).

For a random stationary tessellation X and $k \in \{0, \dots, d-1\}$, denote the **intensity of k-primitives** by

$$\gamma_k^{\mathrm{pr}} := \mathbb{E} \sum_{(x,F)\in\mathcal{Z}(\mathcal{F}_k^{\mathrm{pr}}(X))} \mathbb{1}\{x \in [0,1]^d\}, \tag{2.42}$$

which is the mean number of centers of k-primitives per unit volume.

The refined Campbell theorem 19.8 for marked point processes yields that there is a family of probability measures $\mathbb{P}^{o,F}$, $F \in \mathcal{P}_{k,c}$, on the space $\mathsf{N}_s(\mathbb{R}^d \times \mathcal{P}_{k,c})$ of simple counting measures, and a probability measure $\mathbb{Q}_k^{\mathrm{pr}}$ on $\mathcal{P}_{k,c}$ such that

$$\mathbb{E} \sum_{(x,F)\in\mathcal{Z}(\mathcal{F}_k^{\mathrm{pr}}(X))} f(x, F, \mathcal{Z}(\mathcal{F}_k^{\mathrm{pr}}(X)))$$
$$= \gamma_k^{\mathrm{pr}} \int_{\mathbb{R}^d}\int_{\mathcal{P}_{k,c}}\int_{\mathcal{T}} f(x, F, \mathcal{Z}(\mathcal{F}_k^{\mathrm{pr}}(T + x)))\,\mathbb{P}^{o,F}(\mathrm{d}T)\,\mathbb{Q}_k^{\mathrm{pr}}(\mathrm{d}F)\,\lambda_d(\mathrm{d}x) \tag{2.43}$$

for all nonnegative measurable $f : \mathbb{R}^d \times \mathcal{P}_{k,c} \times \mathsf{N}_s(\mathbb{R}^d \times \mathcal{P}_{k,c}) \to \mathbb{R}$.

Choosing f as an appropriate indicator function yields an explicit expression for $\mathbb{Q}_k^{\mathrm{pr}}$:

$$\mathbb{Q}_k^{\mathrm{pr}}(A) = \frac{1}{\gamma_k^{\mathrm{pr}}}\,\mathbb{E} \sum_{(x,F)\in\mathcal{Z}(\mathcal{F}_k^{\mathrm{pr}}(X))} \mathbb{1}\{x \in [0,1]^d\}\,\mathbb{1}\{F \in A\} \tag{2.44}$$

for $A \in \mathcal{B}(\mathcal{P}_{k,c})$. A random k-dimensional polytope F_k^{pr} with distribution $\mathbb{Q}_k^{\mathrm{pr}}$ is called the **typical k-primitive** of X with respect to the center function c.

The mean jth intrinsic volume, $j \in \{0, \dots, k\}$, of the typical k-primitive is defined by

$$\overline{V}_{k,j}^{\mathrm{pr}} := \mathbb{E} V_j(F_k^{\mathrm{pr}}) = \int_{\mathcal{P}_{k,c}} V_j(F)\, \mathbb{Q}_k^{\mathrm{pr}}(\mathrm{d}F)$$

$$= \frac{1}{\gamma_k^{\mathrm{pr}}}\, \mathbb{E} \sum_{(x,F) \in \mathcal{Z}(\mathcal{F}_k^{\mathrm{pr}}(X))} V_j(F)\, \mathbb{1}\{x \in [0,1]^d\} \,. \tag{2.45}$$

The distribution $\mathbb{Q}_k^{\mathrm{pr}}$ also determines the **directional distribution $\mathbb{Q}_{k:h}^{\mathrm{pr}}$ of the typical k-primitive** by

$$\mathbb{Q}_{k:h}^{\mathrm{pr}}(A) := \int_{G(d,k)} \mathbb{1}\{\mathrm{aff}_o F \in A\}\, \mathbb{Q}_k^{\mathrm{pr}}(\mathrm{d}F) \tag{2.46}$$

for $A \in \mathcal{B}(G(d,k))$.

Now we deal with the skeletons. For a random tessellation X and $k \in \{0, \dots, d-1\}$, the random closed sets $(\mathrm{skel}_k X) \cap [0,1]^d$ and $(\mathrm{skel}_k^{\mathrm{pr}} X) \cap [0,1]^d$ have values in the convex ring $\mathcal{R}$. Hence their k-volumes are well-defined and can be expressed using the intrinsic volume V_k, the Hausdorff measure $\mathscr{H}^k$, or suitably restricted Lebesgue measures; see also (1.1).

For general $B \in \mathcal{B}(\mathbb{R}^d)$, the respective **intensity measures** are defined by

$$\Lambda_k^{\mathrm{pr}}(B) := \mathbb{E} \mathscr{H}^k(B \cap \mathrm{skel}_k^{\mathrm{pr}} X) = \mathbb{E} \sum_{F \in \mathcal{F}_k^{\mathrm{pr}}(X)} \lambda_F(F \cap B) \,.$$

Analogously, the intensity measure Λ_k for $\mathrm{skel}_k X$ is introduced. If the random tessellation X is stationary, then these intensity measures are translation invariant, and hence they are multiples of the d-dimensional Lebesgue measure λ_d. Thus the **density of the k-skeleton of primitives** of a random stationary tessellation X is defined by

$$\eta_k^{\mathrm{pr}} := \overline{V}_k(\mathrm{skel}_k^{\mathrm{pr}} X) := \lambda_d(B)^{-1}\, \mathbb{E} \mathscr{H}^k(B \cap \mathrm{skel}_k^{\mathrm{pr}} X) \tag{2.47}$$

for $B \in \mathcal{B}(\mathbb{R}^d)$ with $0 < \lambda_d(B) < \infty$. Analogously, the **density of the k-skeleton** is defined as

$$\eta_k := \overline{V}_k(\mathrm{skel}_k X) := \lambda_d(B)^{-1}\, \mathbb{E} \mathscr{H}^k(B \cap \mathrm{skel}_k X) \tag{2.48}$$

for $B \in \mathcal{B}(\mathbb{R}^d)$ with $0 < \lambda_d(B) < \infty$. This is the mean total k-volume of the k-skeleton per unit volume in $\mathbb{R}^d$. By (2.17) and (2.18) we have that $\eta_{d-1}^{\mathrm{pr}} = \eta_{d-1}$ and $\eta_{d-2}^{\mathrm{pr}} = \eta_{d-2}$.

The k-primitives have disjoint relative interiors; see (2.9). Thus the density η_k^{pr} can also be written as

$$\begin{aligned}\eta_k^{\mathrm{pr}} &= \mathbb{E} \sum_{F \in \mathcal{F}_k^{\mathrm{pr}}(X)} \lambda_F(F \cap [0,1]^d) \\ &= \mathbb{E} \sum_{F \in \mathcal{F}_k^{\mathrm{pr}}(X)} V_k(F \cap [0,1]^d) \\ &= \mathbb{E} \sum_{(x,F) \in \mathcal{Z}(\mathcal{F}_k^{\mathrm{pr}}(X))} V_k((F + x) \cap [0,1]^d) \end{aligned} \tag{2.49}$$

with $\mathcal{Z}(\mathcal{F}_k^{\mathrm{pr}}(X))$ as defined in (2.41).

Besides the directional distribution $\mathbb{Q}_{k:h}^{\mathrm{pr}}$ of the typical k-primitive, important attributes of a random stationary tessellation X are directional distributions with respect to the k-skeletons for $k \in \{1, \dots, d-1\}$.

Recall that for $F \in \mathcal{P}_k$, we denote by $\mathrm{aff}_o F \in G(d,k)$ the linear subspace that is a translate of the k-dimensional affine subspace $\mathrm{aff}\, F \in A(d,k)$ spanned by F. For $\mathcal{H}^k$-almost all $x \in \mathrm{skel}_k^{\mathrm{pr}} X$, there is a unique $F(x) \in \mathcal{F}_k^{\mathrm{pr}}(X)$ with $x \in F(x)$. Then the probability measure χ_k^{pr} on $G(d,k)$ is defined by

$$\begin{aligned}\chi_k^{\mathrm{pr}}(A) &:= \mathbb{E} \sum_{F \in \mathcal{F}_k^{\mathrm{pr}}(X)} \mathbb{1}\{\mathrm{aff}_o F \in A\}\, V_k(F \cap [0,1]^d) \\ &= \mathbb{E} \int_{\mathbb{R}^d} \mathbb{1}\{x \in [0,1]^d \cap \mathrm{skel}_k^{\mathrm{pr}} X\}\, \mathbb{1}\{\mathrm{aff}_o F(x) \in A\}\, \mathcal{H}^k(\mathrm{d}x) \end{aligned} \tag{2.50}$$

for $A \in \mathcal{B}(G(d,k))$. It can be understood as the **k-volume-weighted directional distribution** of $\mathrm{skel}_k^{\mathrm{pr}} X$ or as the directional distribution in the "typical point" of the k-skeleton of primitives in the sense of a Palm distribution.

Similarly, the k-volume-weighted directional distribution χ_k of the k-skeleton $\mathrm{skel}_k X$ is defined by simply deleting the upper index "pr" in (2.50). Since by (2.17) and (2.18) we have $\mathrm{skel}_k^{\mathrm{pr}} X = \mathrm{skel}_k X$ for $k \in \{d-2, d-1\}$, it follows that $\chi_{d-2} = \chi_{d-2}^{\mathrm{pr}}$ and $\chi_{d-1} = \chi_{d-1}^{\mathrm{pr}}$.

For $F \in \mathcal{P}_{d-1}$, it is often more convenient to use the orthogonal linear subspace of the hyperplane that contains F, denoted by $F^\perp \in G(d,1)$. Then the respective probability measure ψ_o on $G(d,1)$ is defined by

$$\eta_{d-1}\, \psi_o(A) := \mathbb{E} \sum_{F \in \mathcal{F}_{d-1}^{\mathrm{pr}}(X)} \mathbb{1}\{F^\perp \in A\}\, V_{d-1}(F \cap [0,1]^d) \tag{2.51}$$

for $A \in \mathcal{B}(G(d,1))$. The distribution ψ_o defines the even probability measure ψ on S^{d-1} by

$$\psi(B) := \frac{1}{2}\psi_o(\{\mathrm{lin}(u) : u \in B \setminus (B \cap -B)\}) + \psi_o(\{\mathrm{lin}(u) : u \in B \cap -B\}) \tag{2.52}$$

for $B \in \mathcal{B}(S^{d-1})$, which is also referred to as the **spherical directional distribution in the typical point of ∂X**.

Now we summarize some relations between the characteristics introduced for primitives.

Lemma 2.26. *Let X be a random stationary tessellation in $\mathbb{R}^d$, and let $k \in \{0, \dots, d-1\}$. Then*

$$\eta_k^{\mathrm{pr}} = \gamma_k^{\mathrm{pr}} \overline{V}_{k,k}^{\mathrm{pr}}, \tag{2.53}$$

and for all nonnegative measurable $f : G(d,k) \to \mathbb{R}$,

$$\int_{\mathcal{P}_{k,c}} f(\mathrm{aff}_o F)\, V_k(F)\, \mathbb{Q}_k^{\mathrm{pr}}(\mathrm{d}F) = \overline{V}_{k,k}^{\mathrm{pr}} \int_{G(d,k)} f(L)\, \chi_k^{\mathrm{pr}}(\mathrm{d}L)\,. \tag{2.54}$$

If $k = d-1$, then this can also be written for nonnegative measurable $g : G(d,1) \to \mathbb{R}$ as

$$\int_{\mathcal{P}_{d-1,c}} g(F^{\perp})\, V_{d-1}(F)\, \mathbb{Q}_{d-1}^{\mathrm{pr}}(\mathrm{d}F) = \overline{V}_{d-1,d-1}^{\mathrm{pr}} \int_{G(d,1)} g(L)\, \psi_o(\mathrm{d}L)\,. \tag{2.55}$$

Proof. To prove (2.53), we use for $F \in \mathcal{P}_k$ the translative integral formula (1.5), replacing there the restricted measure μ_B by $\mathcal{H}^k(\cdot \cap F)$. Note that $\mathcal{H}^k(([0,1]^d + x) \cap F) = V_k(([0,1]^d + x) \cap F)$. By (2.49), the refined Campbell theorem 19.8, and the translation invariance of the intrinsic volume V_k we obtain

$$\begin{aligned}
\eta_k^{\mathrm{pr}} &= \mathbb{E} \sum_{(x,F) \in \mathcal{Z}(\mathcal{F}_k^{\mathrm{pr}}(X))} V_k\big((F + x) \cap [0,1]^d\big) \\
&= \gamma_k^{\mathrm{pr}} \int_{\mathbb{R}^d} \int_{\mathcal{P}_{k,c}} V_k\big((F + x) \cap [0,1]^d\big)\, \mathbb{Q}_k^{\mathrm{pr}}(\mathrm{d}F)\, \lambda_d(\mathrm{d}x) \\
&= \gamma_k^{\mathrm{pr}} \overline{V}_{k,k}^{\mathrm{pr}}\,.
\end{aligned}$$

The definitions of the distribution $\mathbb{Q}_k^{\mathrm{pr}}$ in (2.44) and the intensity χ_k^{pr} in (2.50), and (2.49) yield

$$\begin{aligned}
&\gamma_k^{\mathrm{pr}} \int_{\mathcal{P}_{k,c}} f(\mathrm{aff}_o F)\, V_k(F)\, \mathbb{Q}_k^{\mathrm{pr}}(\mathrm{d}F) \\
&\quad = \mathbb{E} \sum_{(x,F) \in \mathcal{Z}(\mathcal{F}_k^{\mathrm{pr}}(X))} V_k\big((F + x) \cap [0,1]^d\big) f(\mathrm{aff}_o F) \\
&\quad = \mathbb{E} \int_{\mathrm{skel}_k^{\mathrm{pr}} X} \mathbb{1}\{x \in [0,1]^d\} f(\mathrm{aff}_o F(x))\, \lambda_k(\mathrm{d}x) \\
&\quad = \eta_k^{\mathrm{pr}} \int_{G(d,k)} f(L)\, \chi_k^{\mathrm{pr}}(\mathrm{d}L)\,,
\end{aligned}$$

and (2.54) follows by (2.53).

By the definitions (2.50) and (2.51) of χ_{d-1}^{pr} and ψ_o, respectively, it follows that

$$\int_{G(d,1)} g(L)\,\psi_o(\mathrm{d}L) = \int_{G(d,d-1)} g(H^\perp)\chi_{d-1}^{\mathrm{pr}}(\mathrm{d}H) \tag{2.56}$$

for nonnegative measurable functions $g : G(d,1) \to \mathbb{R}$, and thus (2.55). □

In (2.38) and Corollary 2.23, it was stated that the distribution $\mathbb{Q}_d$ of the typical cell determines the intensity γ_d and the intensity measure of the particle process of cells. Now we show that also the density η_{d-1} and directional distribution ψ_o of a random stationary tessellation X are specified by $\mathbb{Q}_d$.

Lemma 2.27. *Let X be a random stationary tessellation, and let Z be its typical cell. Then*

$$\eta_{d-1} = \frac{\mathbb{E}V_{d-1}(Z)}{\mathbb{E}V_d(Z)}, \tag{2.57}$$

and for all nonnegative measurable $f : G(d,1) \to \mathbb{R}$,

$$\int_{G(d,1)} f(L)\,\psi_o(\mathrm{d}L) = \frac{1}{2\,\mathbb{E}V_{d-1}(Z)}\mathbb{E}\sum_{F\in\mathcal{F}_{d-1}(Z)} V_{d-1}(F)f(F^\perp). \tag{2.58}$$

Proof. For $\mathcal{H}^{d-1}$-almost all $x \in \partial X$, the orthogonal one-dimensional linear subspace to ∂X in the point x is uniquely determined and is denoted by $L(x) \in G(d,1)$. Furthermore, each of these points x belongs to exactly two cells. Then we use (2.50), (2.51), and relations (2.56) and (2.43). Finally, we apply the translative integral formula (1.5), replacing there the restricted measure μ_B by $\mathcal{H}^k(\cdot \cap F)$ for fixed $F \in \mathcal{P}_k$. Note that $\mathcal{H}^k(([0,1]^d + x)\cap F) = V_k(([0,1]^d + x) \cap F)$. Thus we obtain

$$\begin{aligned}
&\eta_{d-1}\int_{G(d,1)} f(L)\,\psi_o(\mathrm{d}L)\\
&= \mathbb{E}\int_{\mathbb{R}^d} \mathbb{1}(x \in [0,1]^d \cap \partial X) f(L(x))\,\mathcal{H}^{d-1}(\mathrm{d}x)\\
&= \frac{1}{2}\mathbb{E}\sum_{(x,P)\in\mathcal{Z}(X)}\ \sum_{F\in\mathcal{F}_{d-1}(P)} V_{d-1}((F+x)\cap[0,1]^d) f(F^\perp)\\
&= \frac{1}{2}\gamma_d \int_{\mathbb{R}^d}\int_{\mathcal{P}_{d,c}} \sum_{F\in\mathcal{F}_{d-1}(P)} V_{d-1}((F+x)\cap[0,1]^d) f(F^\perp)\,\mathbb{Q}_d(\mathrm{d}P)\,\lambda_d(\mathrm{d}x)\\
&= \frac{1}{2}\gamma_d\,\mathbb{E}\sum_{F\in\mathcal{F}_{d-1}(Z)} f(F^\perp)\int_{\mathbb{R}^d} V_{d-1}((F+x)\cap[0,1]^d)\,\lambda_d(\mathrm{d}x)\\
&= \frac{1}{2}\gamma_d\,\mathbb{E}\sum_{F\in\mathcal{F}_{d-1}(Z)} f(F^\perp)\,V_{d-1}(F)
\end{aligned}$$

for the typical cell Z. For $f = 1$, using the fact that $2\,V_{d-1}(Z)$ is the surface content of Z, by (2.53) and (2.38) this yields

$$\eta_{d-1} = \frac{1}{2}\gamma_d\,\mathbb{E}\sum_{F\in\mathcal{F}_{d-1}(Z)} V_{d-1}(F) = \gamma_d\,\mathbb{E}V_{d-1}(Z) = \frac{\mathbb{E}V_{d-1}(Z)}{\mathbb{E}V_d(Z)}\,.$$

This proves (2.57). Furthermore, it follows that

$$\begin{aligned}\int_{G(d,1)} f(L)\,\psi_o(\mathrm{d}L) &= \frac{\gamma_d}{2\eta_{d-1}}\,\mathbb{E}\sum_{F\in\mathcal{F}_{d-1}(Z)} f(F^\perp)\,V_{d-1}(F)\\ &= \frac{1}{2\,\mathbb{E}V_{d-1}(Z)}\,\mathbb{E}\sum_{F\in\mathcal{F}_{d-1}(Z)} f(F^\perp)\,V_{d-1}(F)\,,\end{aligned}$$

that is, (2.58). □

Throughout this volume, we will make the following assumption, even if it is not necessary for every statement.

Convention 2.28. In this book, we only consider random stationary tessellations where all intensities γ_d, γ_k^{pr}, densities η_k^{pr}, and all mean jth intrinsic volumes of the typical k-primitives $\overline{V}_{k,j}^{\mathrm{pr}}$, defined in (2.45), $j \in \{0,\dots,k\}$, $k \in \{0,\dots,d-1\}$, are finite.

Note that due to inherent relationships between the parameters mentioned in this convention, the finiteness of some parameters may imply the finiteness of some others. For example, by equation (2.53) the finiteness of γ_k^{pr} and η_k^{pr} yields that $\overline{V}_{k,k}^{\mathrm{pr}}$ is finite too.

STIT and Poisson hyperplane tessellations (PHTs) both satisfy our standing convention. For PHT, Theorem 6.1.2 of [82] guarantees that the distribution of the radius of the smallest ball, centered at o and containing the zero-cell, has an exponential tail and hence finite moments of all orders. By (6.3) the zero-cells of STIT tessellations and PHTs coincide in law when their parameters are suitably matched. Monotonicity of intrinsic volumes then implies that all mean jth intrinsic volumes of the typical k-primitives $\overline{V}_{k,j}^{\mathrm{pr}}$ are finite for both models. For a detailed proof concerning STIT tessellations, see Lemma 6.8. Convention 2.28 is crucial in the proof of Theorem 7.1. Moreover, in proving Theorem 13.8, it is used that the radius of the circumball of the typical STIT cell has finite moments of every order.

2.2.4 Exchange formula for random tessellations

For the investigation of tessellations, it is often not sufficient to focus on individual components such as cells or faces of cells, but methods are also required to investigate the mutual arrangement of these components.

For a tessellation $T \in \mathcal{T}$ and two sets of components $\mathsf{A}(T)$, $\mathsf{B}(T)$, we consider relations "$\sim_T$" on $\mathsf{A}(T) \times \mathsf{B}(T)$. It is assumed that the relations are **translation covariant** in the following sense:

$$\text{If } \mathsf{a} \sim_T \mathsf{b}, \text{ then } \mathsf{a} + x \sim_{T+x} \mathsf{b} + x \quad \text{for all } \mathsf{a} \in \mathsf{A}(T),\ \mathsf{b} \in \mathsf{B}(T),\ x \in \mathbb{R}^d. \tag{2.59}$$

In Chapters 16 and 17, several particular relations will be considered. One type of relations are the so-called adjacencies. It means that an A-component a is adjacent to a B-component b if a ⊆ b or b ⊆ a. The concept of adjacent components will be important to describe various linkages in a tessellation structure.

Let f be a real-valued nonnegative measurable function on $\mathcal{P}_k \times \mathcal{T}$. Furthermore, for a set of k-dimensional components $\mathsf{A}(T) \subset \mathcal{P}_k$, the function is assumed to be invariant under translations in $\mathbb{R}^d$ in both arguments, that is,

$$f(\mathsf{a}, T) = f(\mathsf{a} + x, T + x) \quad \text{for all } T \in \mathcal{T},\ \mathsf{a} \in \mathsf{A}(T), \text{ and } x \in \mathbb{R}^d. \tag{2.60}$$

The following theorem describes how for a random stationary tessellation X, the point of view can be exchanged for mean values concerning different components: Either for each component $\mathsf{a} \in \mathsf{A}(X)$, all $\mathsf{b} \in \mathsf{B}(X)$ with $\mathsf{a} \sim_X \mathsf{b}$ are collected, and a certain functional is summed up, or, vice versa, for each component $\mathsf{b} \in \mathsf{B}(X)$, all $\mathsf{a} \in \mathsf{A}(X)$ with $\mathsf{a} \sim_X \mathsf{b}$ are summed up correspondingly.

Theorem 2.29 (Exchange formula). *Let X be a random stationary tessellation with two sets of components $\mathsf{A}(X) \subset \mathcal{P}_k$, $\mathsf{B}(X) \subset \mathcal{P}_j$, $k, j \in \{0, \dots, d\}$, and let $\sim_X$ be a translation-covariant relation on $\mathsf{A}(X) \times \mathsf{B}(X)$.*

For all nonnegative measurable functions $f : \mathcal{P}_k \times \mathcal{T} \to \mathbb{R}$ and $h : \mathcal{P}_j \times \mathcal{T} \to \mathbb{R}$, both satisfying (2.60), and a center function c, we have

$$\begin{aligned} &\mathbb{E} \sum_{\mathsf{a} \in \mathsf{A}(X)} \mathbb{1}\{c(\mathsf{a}) \in [0,1]^d\} f(\mathsf{a}, X) \sum_{\mathsf{b} \in \mathsf{B}(X)} \mathbb{1}\{\mathsf{a} \sim_X \mathsf{b}\}\, h(\mathsf{b}, X) \\ &\quad = \mathbb{E} \sum_{\mathsf{b} \in \mathsf{B}(X)} \mathbb{1}\{c(\mathsf{b}) \in [0,1]^d\}\, h(\mathsf{b}, X) \sum_{\mathsf{a} \in \mathsf{A}(X)} \mathbb{1}\{\mathsf{a} \sim_X \mathsf{b}\} f(\mathsf{a}, X). \end{aligned} \tag{2.61}$$

Proof. The stationarity of the random tessellation X yields

$$\begin{aligned} &\mathbb{E} \sum_{\mathsf{a} \in \mathsf{A}(X)} \mathbb{1}\{c(\mathsf{a}) \in [0,1]^d\} f(\mathsf{a}, X) \sum_{\mathsf{b} \in \mathsf{B}(X)} \mathbb{1}\{\mathsf{a} \sim_X \mathsf{b}\}\, h(\mathsf{b}, X) \\ &\quad = \mathbb{E} \sum_{\mathsf{a} \in \mathsf{A}(X)} \sum_{\mathsf{b} \in \mathsf{B}(X)} \mathbb{1}\{c(\mathsf{a}) \in [0,1]^d\}\, \mathbb{1}\{\mathsf{a} \sim_X \mathsf{b}\} f(\mathsf{a}, X)\, h(\mathsf{b}, X) \\ &\quad \times \int_{\mathbb{R}^d} \mathbb{1}\{c(b) \in [0,1]^d + y\}\, \lambda_d(\mathrm{d}y) \\ &\quad = \int_{\mathbb{R}^d} \mathbb{E} \sum_{\mathsf{a} \in \mathsf{A}(X+y)} \sum_{\mathsf{b} \in \mathsf{B}(X+y)} \mathbb{1}\{\mathsf{a} \sim_{X+y} \mathsf{b}\} f(\mathsf{a}, X + y)\, h(\mathsf{b}, X + y) \\ &\quad \times \mathbb{1}\{c(\mathsf{a}) \in [0,1]^d\}\, \mathbb{1}\{c(\mathsf{b}) \in [0,1]^d + y\}\, \lambda_d(\mathrm{d}y) \end{aligned}$$

$$= \int_{\mathbb{R}^d} \mathbb{E} \sum_{\mathsf{a}\in\mathsf{A}(X)} \sum_{\mathsf{b}\in\mathsf{B}(X)} \mathbb{1}\{\mathsf{a}+y \sim_{X+y} \mathsf{b}+y\} f(\mathsf{a}+y, X+y)\, h(\mathsf{b}+y, X+y)$$
$$\times \mathbb{1}\{c(\mathsf{a}+y) \in [0,1]^d\}\, \mathbb{1}\{c(\mathsf{b}+y) \in [0,1]^d + y\}\, \lambda_d(\mathrm{d}y)\,.$$

The translation covariance $c(\mathsf{a}+y) = c(\mathsf{a}) + y$ of the center function, together with (2.59) and (2.60), implies that the last term equals

$$\int_{\mathbb{R}^d} \mathbb{E} \sum_{\mathsf{a}\in\mathsf{A}(X)} \sum_{\mathsf{b}\in\mathsf{B}(X)} \mathbb{1}\{\mathsf{a} \sim_X \mathsf{b}\} f(\mathsf{a}, X)\, h(\mathsf{b}, X)$$
$$\times \mathbb{1}\{c(\mathsf{a}) \in [0,1]^d - y\}\, \mathbb{1}\{c(\mathsf{b}) \in [0,1]^d\}\, \lambda_d(\mathrm{d}y)$$
$$= \mathbb{E} \sum_{\mathsf{a}\in\mathsf{A}(X)} \sum_{\mathsf{b}\in\mathsf{B}(X)} \mathbb{1}\{\mathsf{a} \sim_X \mathsf{b}\} f(\mathsf{a}, X)\, h(\mathsf{b}, X)\, \mathbb{1}\{c(b) \in [0,1]^d\}$$
$$\times \int_{\mathbb{R}^d} \mathbb{1}\{c(\mathsf{a}) \in [0,1]^d - y\}\, \lambda_d(\mathrm{d}y)\,,$$

which completes the proof. □

Using (2.32), the exchange formula (2.61) can also be written as

$$\gamma_{\mathsf{A}} \int_{\mathcal{T}} f(\mathsf{a}^o(T), T) \sum_{\mathsf{b}\in\mathsf{B}(T)} \mathbb{1}\{\mathsf{a}^o(T) \sim_T \mathsf{b}\}\, h(\mathsf{b}, T)\, \mathbb{P}^o_{\mathsf{A}}(\mathrm{d}T)$$
$$= \gamma_{\mathsf{B}} \int_{\mathcal{T}} h(\mathsf{b}^o(T), T) \sum_{\mathsf{a}\in\mathsf{A}(T)} \mathbb{1}\{\mathsf{a} \sim_T \mathsf{b}^o(T)\} f(\mathsf{a}, T)\, \mathbb{P}^o_{\mathsf{B}}(\mathrm{d}T)\,,$$

where for $T \in \mathcal{T}^{\mathsf{A}}$, the component $\mathsf{a}^o(T) \in \mathsf{A}(T)$ is that with $c(\mathsf{a}^o(T)) = o$. The left-hand side concerns the typical A-component of the tessellation X and the B-components related according to $\sim_X$ to it, whereas the right-hand side regards the typical B-component and the A-components related to it. A special case of the exchange formula is given in equation (16.7), which is an important tool in the mean value theory of random tessellations. It is frequently used in Chapters 16 and 17.

2.2.5 Sections with affine subspaces

The following statement shows that for a random stationary tessellation, a section with a k-dimensional flat A a. s. yields a tessellation in the sense of (2.22).

Lemma 2.30. *Let X be a random stationary tessellation of $\mathbb{R}^d$, and let $A \in A(d,k)$ be a k-dimensional affine subspace, $k \in \{1,\dots,d-1\}$. Then $X \in \mathcal{T}(A)$ a. s., and $X \cap A$ is stationary with respect to A, which means that its distribution is invariant under translations that preserve A.*

Proof. Analogously to the proof on page 130 of [172], where it is shown that the section of a stationary j-flat process in $\mathbb{R}^d$ with a fixed k-dimensional flat A is a. s. a $(k+j-d)$-flat process on A, it follows that for a random stationary tessellation X,

$$\begin{aligned}\mathbb{P}(X \notin \mathcal{T}(A)) &= \mathbb{P}(\exists P \in X : A \cap P \neq \emptyset, A \cap \operatorname{int} P = \emptyset) \\ &= \mathbb{P}(\exists P \in X, j \in \{0,\dots,d-1\}, F \in \mathcal{F}_j(P) : F \subseteq A) = 0 .\end{aligned}$$

Because the distribution of X is assumed to be invariant under all translations of $\mathbb{R}^d$ it is, in particular, invariant under the translations that preserve A, and this is inherited by the tessellation $X \cap A$. □

The section of STIT tessellations with flats is the topic of Sections 6.3 and 6.4. It is an important feature that the STIT property is preserved in lower-dimensional sections.

2.3 The Poisson hyperplane tessellation process

In this book the relationships between the STIT tessellation and the Poisson hyperplane tessellation – abbreviated as PHT – play an important role. The PHTs are dealt with in all detail in the monograph [82]. In this section, we summarize the facts that will be used in the following chapters. For our purposes, it is also necessary to introduce a stochastic PHT process on the time axis $(0, \infty)$.

2.3.1 Definition

To define a stochastic process of hyperplane tessellations, it is appropriate to incorporate birth times of the hyperplanes.

Let φ be an even probability measure on the sphere S^{d-1}, which is a probability measure on S^{d-1} with $\varphi(A) = \varphi(-A)$ for all $A \in \mathcal{B}(S^{d-1})$. Throughout this book, we use a measure Θ on $(\mathcal{H}, \mathcal{B}(\mathcal{H}))$, a hyperplane measure, which satisfies

$$\int_{\mathcal{H}} f(H)\,\Theta(\mathrm{d}H) = \int_{S^{d-1}} \int_{\mathbb{R}} f(H(u,r))\,\lambda_1(\mathrm{d}r)\,\varphi(\mathrm{d}u) \tag{2.62}$$

for all nonnegative measurable functions $f : \mathcal{H} \to \mathbb{R}$, and we assume that

$$\varphi \text{ is not concentrated on a great hypersphere of } S^{d-1} . \tag{2.63}$$

This condition guarantees that the following construction yields a. s. a random tessellation; see Theorem 4.3.1 of [82].

The starting point is a Poisson point process $\widehat{X}^*$ on $\mathcal{H} \times \mathbb{R}^+$ with intensity measure $\Theta \otimes \lambda_+$, where λ_+ is the Lebesgue measure on $\mathbb{R}^+$. Thus $\widehat{X}^*$ can be considered as a Poisson process of hyperplanes marked with birth times. This gives rise to the spatio-temporal process $(\widehat{X}_t^*, t > 0)$ with

$$\widehat{X}_t^* := \{(H, s) \in \widehat{X}^* : s \leq t\} . \tag{2.64}$$

The projection of $\widehat{X}_t^*$ to $\mathcal{H}$, that is, the deletion of the birth times, yields the Poisson hyperplane process

$$\widehat{X}_t := \{H : (H, s) \in \widehat{X}_t^*\} \tag{2.65}$$

with intensity measure $t\,\Theta$.

Note that $\widehat{X}^*$ is not a marked point process in the sense of Definition 19.6, because condition (19.3) is not satisfied. In contrast, $\widehat{X}_t^*$ for $t > 0$ is a marked point process.

For all $t > 0$, the Poisson hyperplane process $\widehat{X}_t$ induces a random tessellation of $\mathbb{R}^d$, referred to as the PHT X_t. Formally, X_t is defined as the set of cells that are the topological closures of the connected components of $\mathbb{R}^d \setminus \bigcup_{H \in \widehat{X}_t} H$.

Because Θ is invariant under translations of $\mathbb{R}^d$, $\widehat{X}_t$ and X_t are stationary. Thus we obtain the **Poisson hyperplane tessellation process driven by** Θ,

$$(X_t, t > 0) . \tag{2.66}$$

The independence property of Poisson point processes (see Definition 19.10) implies that the three processes $(\widehat{X}_t^*, t > 0)$, $(\widehat{X}_t, t > 0)$, and $(X_t, t > 0)$ are continuous-time Markov processes on their respective state spaces.

Lemma 2.31. *The process $(X_t, t > 0)$ of PHTs driven by Θ has the scaling property*

$$t X_t \overset{\mathrm{D}}{=} X_1 \quad \textit{for all } t > 0 . \tag{2.67}$$

Proof. The tessellation $t X_t$ is induced by the hyperplane process $t \widehat{X}_t := \{t H : H \in \widehat{X}_t\}$. Denote its intensity measure by Θ'. Then by definition (2.62) of Θ, for all $A \in \mathcal{B}(\mathcal{H})$,

$$\begin{aligned}
\Theta'(A) &= \mathbb{E}\big|\{H' \in t\widehat{X}_t : H' \in A\}\big| \\
&= \mathbb{E}\big|\{H \in \widehat{X}_t : t H \in A\}\big| \\
&= t^{-1}\, t\, \Theta(A) = \Theta(A) .
\end{aligned}$$

Analogously, it can be shown that the random number $|\{H' \in t\widehat{X}_t : H' \in A\}|$ has a Poisson distribution with parameter $\Theta(A)$. Thus $t\widehat{X}_t$ is a Poisson hyperplane process with intensity measure Θ. □

As the PHT process evolves in time, each new hyperplane divides an infinite number of existing cells. In the following lemma, we consider a fixed polytope $P \in \mathcal{P}_d$ and its first division by a hyperplane. Then we determine the distributions of the birth time and lifetime of a single cell P of a PHT X_t that exists at time $t > 0$. In Section 2.3.2, we describe the evolution of the process $(X_t \wedge W, t > 0)$, which is the restriction to a window $W \in \mathcal{P}_d$. These results will be essential for a comparison of PHTs and STIT tessellations in forthcoming chapters and for the proof of several key theorems.

Note that by the translation invariance of Θ we have that $\Theta([\operatorname{int} P]) = \Theta([P])$ for $P \in \mathcal{P}_d$. Furthermore, for $P \in \mathcal{P}_d$, we define the probability measure

$$\Theta_{[P]}(\cdot) := \Theta([P])^{-1}\,\Theta(\cdot \cap [P]) \quad \text{on } \mathcal{H}.$$

Lemma 2.32. *Let Θ be a hyperplane measure of the form* (2.62), *satisfying* (2.63), *and let $(X_t, t > 0)$ be the PHT process driven by Θ.*

(a) *For $P \in \mathcal{P}_d$, define the random variable $(H, \beta) \in \widehat{X}^*$ by*

$$\beta := \min\{s > 0 : (H', s) \in \widehat{X}^*, H' \in [P]\}.$$

Then the law of (H, β) is $\Theta_{[P]} \otimes \mathsf{Expo}(\Theta([P]))$.

(b) *For fixed $t > 0$ and $P \in X_t$, define the birth time of P as*

$$\beta(P) := \max\{s < t : (H, s) \in \widehat{X}^*, H \cap P \in \mathcal{F}_{d-1}(P)\}.$$

Furthermore, define the random variable $(H(P,t), \beta^\dagger) \in \widehat{X}^*$ by

$$\beta^\dagger(P) := \min\{s > t : (H, s) \in \widehat{X}^*, H \in [P]\},$$

which is the time of the division of P. Then, given $P \in X_t$, the random variable $(H(P,t), \beta^\dagger(P) - t)$ has the conditional distribution $\Theta_{[P]} \otimes \mathsf{Expo}(\Theta([P]))$, and it is conditionally independent of $(X_s, s < t)$.

The random variable $\beta^\dagger(P) - \beta(P)$, which is the life time of $P \in X_t$, has the law $\mathsf{Expo}(\Theta([P]))$.

Proof. (a) For $t \geq 0$ and a nonnegative measurable $f : \mathcal{H} \to \mathbb{R}$, an application of the Mecke equation (19.4) and Definition 19.10 of a Poisson point process yield

$$\begin{aligned}
&\mathbb{E} \sum_{(H,\beta)\in\widehat{X}^*} \mathbb{1}\{\beta = \min\{s > 0 : (H,s) \in \widehat{X}^*, H \in [P]\}\}\,\mathbb{1}\{\beta > t\} f(H) \\
&= \int_{\mathbb{R}^+}\int_{\mathcal{H}} \mathbb{E}\,\mathbb{1}\{\beta = \min\{s > 0 : (H,s) \in \widehat{X}^* + \delta_{(H,\beta)}, H \in [P]\}\}\,\mathbb{1}\{\beta > t\} \\
&\quad \times f(H)\,\Theta(\mathrm{d}H)\,\lambda_+(\mathrm{d}\beta) \\
&= \int_{\mathbb{R}^+}\int_{\mathcal{H}} \mathbb{E}\,\mathbb{1}\{([P] \times (0,\beta)) \cap \widehat{X}^* = \emptyset\}\,\mathbb{1}\{H \in [P]\}\,\mathbb{1}\{\beta > t\} \\
&\quad \times f(H)\,\Theta(\mathrm{d}H)\,\lambda_+(\mathrm{d}\beta)
\end{aligned}$$

$$= \int_{\mathbb{R}^+} \int_{\mathcal{H}} e^{-\Theta([P])\beta} \mathbb{1}\{H \in [P]\} \mathbb{1}\{\beta > t\} f(H) \Theta(dH) \lambda_+(d\beta)$$
$$= e^{-\Theta([P])t} \int_{\mathcal{H}} f(H) \Theta_{[P]}(dH) .$$

(b) For $t' \geq 0$ and a nonnegative measurable $f : \mathcal{H} \to \mathbb{R}$ and $P \in \mathcal{P}_d$, the independence property in Definition 19.10 of Poisson point processes yields for the conditional expectation that

$$\mathbb{E}\Big(\sum_{(H,\beta)\in\widehat{X}^*} \mathbb{1}\{\beta = \min\{s > t : (H,s) \in \widehat{X}^*, H \in [P]\}\}$$
$$\times \mathbb{1}\{\beta - t > t'\} f(H) \mid P \in X_t \Big)$$
$$= \mathbb{E} \sum_{(H,\beta)\in\widehat{X}^*} \mathbb{1}\{\beta = \min\{s > t : (H,s) \in \widehat{X}^*, H \in [P]\}\} \mathbb{1}\{\beta - t > t'\} f(H) .$$

Then the calculations for $(H(P,t), \beta^\dagger - t)$ are analogous to those in the proof of (a). As shown above, under the condition $P \in X_t$, the remaining life time $\beta^\dagger - t$ of P after t is exponentially distributed with parameter $\Theta([P])$ and does not depend on t. Hence by the memoryless property of the exponential distribution the total life time $\beta^\dagger - \beta(P)$ of P has the same conditional distribution. The conditional independence, given $P \in X_t$, of $(H(P,t), \beta^\dagger - t)$ from $(X_s, s < t)$ follows from the independence property in Definition 19.10 of Poisson point processes. □

2.3.2 Rate kernel and generator

We now consider the restriction $X_t \wedge W$, introduced in Section 2.1.4, of the PHT $X_t, t > 0$, to a polytope $W \in \mathcal{P}_d$, and accordingly the process $(X_t \wedge W, t > 0)$. To describe the rate kernel and generator of the restricted PHT process, denote for fixed $W \in \mathcal{P}_d$,

$$\oslash^{\mathrm{PHT}}(H,T) := (T \setminus \{P \in T : H \in [\mathrm{int}P]\}) \cup \{P \cap H^+ : P \in T, H \in [\mathrm{int}P]\}$$
$$\cup \{P \cap H^- : P \in T, H \in [\mathrm{int}P]\}$$

with $T \in \mathcal{T}_W$, $t \geq 0$, and $H \in [W]$. Thus $\oslash^{\mathrm{PHT}}(H,T)$ is the resulting tessellation when a hyperplane H divides cells of the tessellation T. Let $\mathbb{P}^{\oslash\mathrm{PHT}}(T, \cdot)$ be the image measure of $\Theta_{[W]}$ under $\oslash^{\mathrm{PHT}}(\cdot, T) : [W] \to \mathcal{T}_W$. More explicitly, for a nonnegative measurable function $g : \mathcal{T}_W \to \mathbb{R}$,

$$\int_{\mathcal{T}_W} g(T') \mathbb{P}^{\oslash\mathrm{PHT}}(T, dT') = \int_{\mathcal{H}} g(\oslash^{\mathrm{PHT}}(H,T)) \Theta_{[W]}(dH) .$$

For the definition of the generator of a time-homogeneous pure jump Markov process, we refer to Section 14.2, page 28 middle, of [47].

Lemma 2.33. *Let $(X_t, t > 0)$ be the PHT process driven by Θ, and let $W \in \mathcal{P}_d$. Then the process $(X_t \wedge W, t > 0)$ of restricted PHTs is a pure jump Markov process with the rate kernel $\Theta([W])\, \mathbb{P}^{\oslash\text{PHT}}(T, \cdot)$ and the generator $\widehat{\mathbb{L}}^{\text{PHT}}$ given by*

$$\widehat{\mathbb{L}}^{\text{PHT}} g(T) = \int_{[W]} g(\oslash^{\text{PHT}}(H, T)) - g(T)\, \Theta(\mathrm{d}H)$$

for bounded and measurable $g : \mathcal{T}_W \to \mathbb{R}$ and $T \in \mathcal{T}_W$.

Proof. Since the underlying Poisson point process $\widehat{X}^*$ on $\mathcal{H} \times \mathbb{R}^+$ has intensity measure $\Theta \otimes \lambda_+$, the time intervals between two consecutive jumps of $(X_t \wedge W, t > 0)$ are i. i. d. exponentially distributed with parameter $\Theta([W])$. The jump transition kernel of this process is $\mathbb{P}^{\oslash\text{PHT}}(T, \cdot)$. Thus the rate kernel of $(X_t \wedge W, t > 0)$ is $\Theta([W])\, \mathbb{P}^{\oslash\text{PHT}}(T, \cdot)$; see page 281 of [89]. In (3.3) the notation $\uplus$ for the superposition of tessellations is introduced. Using the infinitesimal description of a pure jump process (see, for example, Corollary 15.29 of [21]) yields that for the PHT process in W,

$$\begin{aligned}
\widehat{\mathbb{L}}^{\text{PHT}} g(T) &= \lim_{s \downarrow 0} \frac{\mathbb{E} g(X_s \uplus T) - g(T)}{s} \\
&= \lim_{s \downarrow 0} \frac{(\Theta([W])s + o(s)) \int_{[W]} g(\oslash^{\text{PHT}}(H, T)) - g(T)\, \Theta_{[W]}(\mathrm{d}H)}{s} \\
&= \int_{[W]} g(\oslash^{\text{PHT}}(H, T)) - g(T)\, \Theta(\mathrm{d}H),
\end{aligned}$$

because $\Theta_{[W]}(\cdot) = \Theta([W])^{-1}\, \Theta(\cdot \cap [W])$. This completes the proof. □

2.3.3 The associated zonoid

Let $t > 0$, and let Θ be a hyperplane measure of the form (2.62), satisfying (2.63). The measure $t\Theta$ can be represented by a convex body $\Pi_{t\Theta} \in \mathcal{K}$, the so-called associated zonoid or Matheron zonoid. This body is defined by its support function

$$h(\Pi_{t\Theta}, u) := \frac{t}{2} \int_{S^{d-1}} |\langle u, v \rangle|\, \varphi(\mathrm{d}v), \quad u \in S^{d-1}, \tag{2.68}$$

and we call it the **associated zonoid** of the pair (t, Θ). In Section 5.2 of [82], where it is referred to as the Matheron zonoid, it is explained that the function on the right-hand side of (2.68) is indeed a support function and furthermore, that the measure $t\Theta$ is uniquely determined by the body $\Pi_{t\,\Theta}$. Note that t only acts as a scaling parameter. In fact, the scaling relation

$$\Pi_{t\,\Theta} = t\,\Pi_{\Theta}$$

holds for $t > 0$. Furthermore, by the point symmetry to the origin o, the support function of $\Pi_{t\,\Theta}$ admits the following geometric interpretation:

$$2\,h(\Pi_{t\,\Theta}, u) = V_1(\Pi_{t\,\Theta}|\mathrm{lin}(u))\,, \quad u \in S^{d-1}\,, \tag{2.69}$$

that is, $2\,h(\Pi_{t\,\Theta}, u)$ is the length of the orthogonal projection of $\Pi_{t\,\Theta}$ onto the linear subspace spanned by u.

The use of the term “zonoid” is justified by the fact that it is well known in convex geometry (see Theorem 3.5.3 of [171]) that convex bodies with a support function of the form (2.68) belong to the class of so-called zonoids. This is a particularly interesting class of convex sets consisting of all limits, with respect to the Hausdorff distance, of finite Minkowski sums of line segments.

In the class of PHTs, the distribution of a tessellation is uniquely determined by the intensity measure $t\,\Theta$ of the underlying Poisson hyperplane process, and hence this distribution is also determined by the associated zonoid $\Pi_{t\,\Theta}$ of the pair (t, Θ). Several characteristics of a PHT can be expressed efficiently by functionals of $\Pi_{t\,\Theta}$; see, for example, Theorem 2.38 and Corollary 2.39. Analogous connections hold in the class of STIT tessellations. Also in that class, the distribution of a tessellation is determined by the measure $t\,\Theta$ and thus by the associated zonoid $\Pi_{t\,\Theta}$.

We now determine the associated zonoids and explicit expressions of their intrinsic volumes for two particular cases of Θ, namely the motion-invariant measure and the Mondrian hyperplane measures. For the case where $\varphi = \varphi_{\mathrm{iso}}$ is the rotation invariant probability measure on S^{d-1}, we also write Θ_{iso} and call it the **motion-invariant hyperplane measure**. The associated zonoid is a ball with radius determined by

$$h(\Pi_{\Theta_{\mathrm{iso}}}, u) = h(\Pi_{\Theta_{\mathrm{iso}}}, e_1) = \frac{1}{2}\int\limits_{S^{d-1}} |v_1|\,\varphi_{\mathrm{iso}}(\mathrm{d}v)$$

for any $u \in S^{d-1}$ by rotation invariance, where $e_1 = (1, 0, \dots, 0)$, and v_1 stands for the first coordinate of v. Applying the standard slice integration formula for spheres (see, e. g., Corollary A.5 of [9]), we obtain

$$\begin{aligned}\frac{1}{2}\int\limits_{S^{d-1}} |v_1|\,\varphi_{\mathrm{iso}}(\mathrm{d}v) &= \frac{d-1}{2d}\frac{\kappa_{d-1}}{\kappa_d}\int\limits_{-1}^{1} |x|(1-x^2)^{\frac{d-3}{2}}\,\mathrm{d}x \\ &= \frac{d-1}{2d}\frac{\kappa_{d-1}}{\kappa_d}\frac{2}{d-1} = \frac{\kappa_{d-1}}{d\,\kappa_d}\end{aligned}$$

or, equivalently,

$$\Pi_{\Theta_{\text{iso}}} = \frac{\kappa_{d-1}}{d\,\kappa_d}\,B^d\,. \tag{2.70}$$

Another particular class is formed by the so-called **Mondrian hyperplane measures**. For the standard orthonormal basis $e_1,\dots,e_d$ in $\mathbb{R}^d$ corresponding to the coordinate axes and for each tuple $(p_1,\dots,p_d) \in (0,1)^d$ of weights with $\sum_{i=1}^d p_i = 1$, define the spherical directional distribution

$$\varphi := \sum_{i=1}^{d} \frac{p_i}{2}\,(\delta_{e_i} + \delta_{-e_i})$$

and by (2.62) a hyperplane measure Θ. For comments on this terminology, see the paragraph below equation (4.4). For such a Mondrian hyperplane measure, by (2.68),

$$h(\Pi_\Theta, u) = \frac{1}{2}\sum_{i=1}^{d} p_i\,|\langle u, e_i\rangle|\,, \quad u \in S^{d-1}\,.$$

By (3.36) of [171] and the one-to-one correspondence between the directional distribution and the associated zonoid it follows that

$$\Pi_\Theta = p_1\,K_1 + \dots + p_d\,K_d\,, \tag{2.71}$$

where $K_i := \{r\,e_i : r \in [-\frac{1}{2},\frac{1}{2}]\}$, $i \in \{1,\dots,d\}$, is the centered line segment in direction e_i having unit length. This means that Π_Θ is an orthogonal parallelepiped with sides parallel to the coordinate axes determined by $e_1,\dots,e_d$ and having side lengths $p_1,\dots,p_d$.

The next lemma provides well-known formulae for the intrinsic volumes of the d-dimensional unit ball and an orthogonal parallelepiped. They will repeatedly be used in the next sections.

Lemma 2.34. *Fix $k \in \{0,\dots,d\}$.*

(a) *We have that $V_k(B^d) = \binom{d}{k}\frac{\kappa_d}{\kappa_{d-k}}$.*

(b) *Let P be a d-dimensional orthogonal parallelepiped with side lengths $a_1,\dots,a_d \in (0,\infty)$. Then*

$$V_k(P) = \sum_{1\le i_1<i_2<\dots<i_k\le d} a_{i_1}\,a_{i_2}\cdots a_{i_k} \quad \textit{for } k > 0\,,$$

and $V_0(P) = 1$.

Proof. To show (a), let $r > 0$ and consider the r-parallel set $B^d + rB^d$ of B^d. On the one hand,

$$V_d(B^d + rB^d) = V_d((1+r)B^d) = (1+r)^d\,V_d(B^d) = \sum_{k=0}^{d}\binom{d}{k}\kappa_d\,r^k\,.$$

On the other hand, using the Steiner formula (1.3), it follows that

$$V_d(B^d + rB^d) = \sum_{k=0}^{d} \kappa_k V_{d-k}(B^d) r^k .$$

Comparing coefficients leads to $V_{d-k}(B^d) = \binom{d}{k}\frac{\kappa_d}{\kappa_k}$ or, equivalently, $V_k(B^d) = \binom{d}{k}\frac{\kappa_d}{\kappa_{d-k}}$. This completes the proof of (a).

Part (b) can be shown with the help of the Wills functional $g(K) := \sum_{j=0}^{d} V_j(K)$ for convex bodies $K \in \mathcal{K}$; see [65] or [171], page 310. This functional has the following representation as an integral:

$$g(K) = \int_{\mathbb{R}^d} e^{-\pi \operatorname{dist}(x,K)^2} \lambda_d(dx) , \tag{2.72}$$

where $\operatorname{dist}(x, K)$ denotes the Euclidean distance from $x \in \mathbb{R}^d$ to K. To see this, for a fixed convex body K, consider the image measure $\lambda^{\operatorname{dist}}(\cdot) := \lambda_d \circ \operatorname{dist}(\cdot, K)^{-1}$ on $[0, \infty)$ of the d-dimensional Lebesgue measure under the map $x \mapsto \operatorname{dist}(x, K)$. Then using Fubini's theorem, we have

$$\begin{aligned}\int_{\mathbb{R}^d} e^{-\pi \operatorname{dist}(x,K)^2} \lambda_d(dx) &= \int_{\mathbb{R}_+} e^{-\pi s^2} \lambda^{\operatorname{dist}}(ds)\\ &= \int_{\mathbb{R}_+} \int_s^\infty 2\pi r\, e^{-\pi r^2}\, dr\, \lambda^{\operatorname{dist}}(ds)\\ &= \int_0^\infty 2\pi r\, e^{-\pi r^2} \lambda^{\operatorname{dist}}([0, r])\, dr\\ &= \int_0^\infty 2\pi r\, e^{-\pi r^2} V_d(K + rB^d)\, dr .\end{aligned}$$

Applying the Steiner formula (1.3) to $V_d(K+rB^d)$ and calculating the integral with respect to r yield (2.72). The integral representation of the Wills functional can be used to show the following multiplicative property. Fix $k \in \{1, \ldots, d-1\}$ and let $L_1 := \mathbb{R}^k \times \{0\}^{d-k}$ and $L_2 := \{0\}^k \times \mathbb{R}^{d-k}$ be two orthogonal subspaces of $\mathbb{R}^d$. Then, for $C_1, C_2 \in \mathcal{K}$ with $C_1 \subset L_1$ and $C_2 \subset L_2$, the Minkowski sum $C_1 + C_2$ satisfies

$$g(C_1 + C_2) = g(C_1)\, g(C_2) . \tag{2.73}$$

Indeed, by the Pythagorean theorem we have that

$$\operatorname{dist}(x, C_1 + C_2)^2 = \operatorname{dist}(x|L_1, C_1)^2 + \operatorname{dist}(x|L_2, C_2)^2 ,$$

where $x|L_1 = (x_1, \ldots, x_k, 0, \ldots, 0)$ and $x|L_2 = (0, \ldots, 0, x_{k+1}, \ldots, x_d)$ are the orthogonal projections of x onto the subspaces L_1 and L_2, respectively. Since the values of the intrinsic volumes do not depend on the dimension of the ambient space,

$$\begin{aligned} g(C_1 + C_2) &= \int_{\mathbb{R}^d} e^{-\pi \operatorname{dist}(x, C_1+C_2)^2} \, \lambda_d(dx) \\ &= \int_{\mathbb{R}^{d-k}} \int_{\mathbb{R}^k} e^{-\pi(\operatorname{dist}(x|L_1, C_1)^2 + \operatorname{dist}(x|L_2, C_2)^2)} \, \lambda_k(d(x_1, \ldots, x_k)) \, \lambda_{d-k}(d(x_{k+1}, \ldots, x_d)) \\ &= g(C_1) \, g(C_2) . \end{aligned}$$

Now let P be a d-dimensional orthogonal parallelepiped with side lengths $a_1, \ldots, a_d \in (0, \infty)$. Since the intrinsic volumes and hence the Wills functional are motion invariant, we can assume without loss of generality that

$$P = a_1 K_1 + \cdots + a_d K_d$$

with segments K_i, analogously to (2.71). Thus, applying (2.73) with induction over the number of summands yields for $\alpha > 0$ that

$$g(\alpha P) = \prod_{j=1}^{d} g(\alpha a_j K_j) = \prod_{j=1}^{d} (1 + \alpha a_j) = \sum_{k=0}^{d} \mathbf{e}_k(a_1, \ldots, a_d) \, \alpha^k, \tag{2.74}$$

where $\mathbf{e}_k(a_1, \ldots, a_d) = \sum_{1 \le i_1 < i_2 < \cdots < i_k \le d} a_{i_1} a_{i_2} \cdots a_{i_k}$ is the kth elementary symmetric function of $a_1, \ldots, a_d$, and where we used that $g(\alpha a_j K_j) = V_0(\alpha a_j K_j) + V_1(\alpha a_j K_j) = 1 + \alpha a_j$ for $j \in \{1, \ldots, d\}$. On the other hand, the homogeneity of the intrinsic volumes implies that for all $K \in \mathcal{K}$ and $\alpha > 0$,

$$g(\alpha K) = \sum_{k=0}^{d} \alpha^k \, V_k(K) . \tag{2.75}$$

Comparing the coefficients of the powers of α in (2.74) and (2.75) completes the proof of part (b). □

2.3.4 Intersection processes and skeletons

For $t > 0$, consider the Poisson hyperplane process $\widehat{X}_t$ defined in (2.65), and let X_t be the induced tessellation of $\mathbb{R}^d$. We write $\widehat{X}_{t,\neq}^{d-k}$ for the set of all $(d-k)$-tuples of pairwise different elements of $\widehat{X}_t$. For $k \in \{0, \ldots, d-1\}$, the **intersection process of order** $d - k$ of $\widehat{X}_t$ is defined as the point process on $A(d, k)$, written as a counting measure,

$$\widehat{X}_t^{(d-k)} := \frac{1}{(d-k)!} \sum_{(H_1,\dots,H_{d-k})\in \widehat{X}_{t,\neq}^{d-k}} \mathbb{1}\{H_1 \cap \cdots \cap H_{d-k} \in A(d,k)\}\, \delta_{H_1\cap\cdots\cap H_{d-k}}\,. \tag{2.76}$$

This is the process of the k-dimensional affine subspaces generated by the intersection of $d-k$ pairwise different hyperplanes of $\widehat{X}_t$, which are in general position. In particular, the intersection process of order 1 is just the hyperplane process $\widehat{X}_t$ itself, whereas the intersection process of order d is the stationary point process of vertices of the tessellation X_t. More generally, for $k \in \{0,\dots,d-1\}$, the k-skeleton of the PHT X_t is the union of the affine subspaces in $\widehat{X}_t^{(d-k)}$, that is,

$$\operatorname{skel}_k X_t = \bigcup_{E\in \widehat{X}_t^{(d-k)}} E\,. \tag{2.77}$$

By the multivariate Mecke equation (19.4), applied to the Poisson hyperplane process $\widehat{X}_t$ with the intensity measure $t\,\Theta$, for all nonnegative measurable functions f : $\mathsf{N}(\mathcal{H}) \times A(d,k) \to \mathbb{R}$,

$$\begin{aligned} &\mathbb{E} \sum_{E\in \widehat{X}_t^{(d-k)}} f(\widehat{X}_t, E) \\ &\quad = \frac{t^{d-k}}{(d-k)!} \int_{\mathcal{H}^{d-k}} \mathbb{1}\left\{\bigcap_{i=1}^{d-k} H_i \in A(d,k)\right\} \\ &\qquad \times \mathbb{E} f\left(\widehat{X}_t + \sum_{i=1}^{d-k} \delta_{H_i}, \bigcap_{i=1}^{d-k} H_i\right) \Theta^{\otimes(d-k)}(\mathrm{d}(H_1,\dots,H_{d-k}))\,. \end{aligned} \tag{2.78}$$

Analogous integrals with respect to $\Theta^{\otimes(d-k)}$ appear also for STIT tessellations, for example, in applications of the Mecke-type Theorem 14.1 in Chapter 14.

Using expression (2.62) for Θ, the integral with respect to the product measure $\Theta^{\otimes(d-k)}$ can be transformed into an iterated integral with respect to the Lebesgue measure λ_{d-k} and the product measure $\varphi^{\otimes(d-k)}$ of the spherical directional distribution φ. A step of this transformation is provided by the following proposition, which is Lemma 4.2.7 of [82].

Lemma 2.35. *Let $k \in \{0,\dots,d-1\}$, and let f be a nonnegative measurable function on $\mathcal{H}^{d-k}$. If $u_1,\dots,u_{d-k} \in S^{d-1}$ are linearly independent vectors with $U := \operatorname{lin}\{u_1,\dots,u_{d-k}\}$, then*

$$\begin{aligned} &\int_{\mathbb{R}^{d-k}} f(H(u_1,r_1),\dots,H(u_{d-k},r_{d-k}))\,\lambda_{d-k}(\mathrm{d}(r_1,\dots,r_{d-k})) \\ &\quad = \nabla_{d-k}(u_1,\dots,u_{d-k}) \int_U f(u_1^{\perp}+x,\dots,u_{d-k}^{\perp}+x)\,\lambda_{d-k}(\mathrm{d}x)\,, \end{aligned}$$

where $\nabla_{d-k}(u_1,\dots,u_{d-k})$ denotes the $(d-k)$-volume of the parallelepiped spanned by $u_1,\dots,u_{d-k}$.

Noting that $u_1,\dots,u_{d-k} \in S^{d-1}$ are linearly independent vectors if and only if $\bigcap_{i=1}^{d-k} u_i^\perp$ is a k-dimensional linear subspace yields the following corollary for the intersections of hyperplanes.

Corollary 2.36. *Let $t > 0$, and let Θ be a hyperplane measure of the form (2.62), satisfying (2.63). Furthermore, let $k \in \{0,\dots,d-1\}$, and let $g : A(d,k) \to \mathbb{R}$ be a nonnegative measurable function. Then*

$$\int_{\mathcal{H}^{d-k}} \mathbb{1}\{H_1 \cap \cdots \cap H_{d-k} \in A(d,k)\}\, g(H_1 \cap \cdots \cap H_{d-k})\, \Theta^{\otimes(d-k)}(\mathrm{d}(H_1,\dots,H_{d-k}))$$
$$= \int_{(S^{d-1})^{d-k}} \int_{\operatorname{lin}\{u_1,\dots,u_{d-k}\}} \mathbb{1}\{u_1^\perp \cap \cdots \cap u_{d-k}^\perp \in G(d,k)\}\, \nabla_{d-k}(u_1,\dots,u_{d-k})$$
$$\times g((u_1^\perp \cap \cdots \cap u_{d-k}^\perp) + x)\, \lambda_{d-k}(\mathrm{d}x)\, \varphi^{\otimes(d-k)}(\mathrm{d}(u_1,\dots,u_{d-k}))\,.$$

The process $\widehat{X}_t^{(d-k)}$ is a stationary process of k-dimensional affine subspaces, but note that it is no longer a Poisson process. Its intensity measure can be disintegrated, which means that there are $\gamma_{k,t} > 0$, the intensity, and a probability measure χ_k on $G(d,k)$, the directional distribution, such that for all nonnegative measurable functions $g : A(d,k) \to \mathbb{R}$,

$$\mathbb{E} \sum_{E \in \widehat{X}_t^{(d-k)}} g(E) = \gamma_{k,t} \int_{G(d,k)} \int_{L^\perp} g(L+x)\, \lambda_{L^\perp}(\mathrm{d}x)\, \chi_k(\mathrm{d}L) \tag{2.79}$$

by Theorem 2.2.2 of [82] and the Campbell theorem 19.4. Note that by (2.67), t is a scaling factor, and therefore the directional distribution does not depend on t. The directional distribution χ_k of the intersection process of a Poisson hyperplane process is the same as the k-volume-weighted directional distribution χ_k of the k-skeleton assigned to the PHT generated by this hyperplane process.

Equation (2.79), together with Corollary 2.36, yields for all nonnegative measurable functions $g : A(d,k) \to \mathbb{R}$,

$$\gamma_{k,t} \int_{G(d,k)} g(L)\, \chi_k(\mathrm{d}L)$$
$$= \frac{t^{d-k}}{(d-k)!} \int_{(S^{d-1})^{d-k}} g(u_1^\perp \cap \cdots \cap u_{d-k}^\perp)$$
$$\times \nabla_{d-k}(u_1,\dots,u_{d-k})\, \varphi^{\otimes(d-k)}(\mathrm{d}(u_1,\dots,u_{d-k}))\,, \tag{2.80}$$

which is Theorem 4.2.8 of [82]. By Theorem 4.4.3 of [172] the intensity $\gamma_{k,t}$ is also the density $\eta_{k,t}$ of the intersection process, which is the mean total k-volume of the intersection process per unit volume. Thus

$$\eta_{k,t} - \gamma_{k,t} .$$

Note that $\gamma_{k,t}$ differs from $\gamma_{k,t}^{\text{PHT}}$, introduced in Section 11.2 as the intensity of the process of k-faces of the PHT X_t.

Furthermore, as already expressed in (2.77), the intersection process $\widehat{X}_t^{(d-k)}$ of order $d-k$ generates the k-skeleton of the PHT X_t. Hence the directional distribution in the typical point of $\text{skel}_k X_t$ and thus also the V_k-weighted directional distribution of the typical k-face of X_t equals χ_k given by (2.80). Because PHTs are face-to-face tessellations, their primitive k-skeletons coincide with their k-skeletons, and thus χ_k is the same as χ_k^{pr} defined in (2.50).

The density $\eta_{k,t}^{\text{PHT}}$, which is the mean total k-volume per unit volume of $\text{skel}_k X_t$ defined in (2.48), by (2.80) is

$$\begin{aligned}\eta_{k,t}^{\text{PHT}} &= \eta_{k,t} = \mathbb{E}\,\mathscr{H}^k([0,1]^d \cap \text{skel}_k X_t)\\ &= \frac{t^{d-k}}{(d-k)!} \int\limits_{(S^{d-1})^{d-k}} \nabla_{d-k}(u_1,\dots,u_{d-k})\,\varphi^{\otimes(d-k)}(\mathrm{d}(u_1,\dots,u_{d-k})) .\end{aligned} \tag{2.81}$$

Furthermore, Theorem 5.3.1 of [82] yields that this density equals the $(d-k)$th intrinsic volume of the zonoid of the pair (t,Θ), that is,

$$\eta_{k,t}^{\text{PHT}} = V_{d-k}(\Pi_{t\Theta}) . \tag{2.82}$$

2.3.5 Sections with linear subspaces

In contrast to intersection processes, which consist of the affine subspaces arising as intersections of elements of a Poisson hyperplane process, we consider now the section of a Poisson hyperplane process and the corresponding PHT with a fixed linear subspace. For a k-dimensional linear subspace $L \in G(d,k)$, $k \in \{1,\dots,d-1\}$, denote by

$$A(L,k-1) := \{A \in A(d,k-1) : A \subset L\}$$

the set of all $(k-1)$-dimensional affine subspaces of L. For $t > 0$ and a Poisson hyperplane process $\widehat{X}_t$, we symbolically write

$$\widehat{X}_t \cap L := \{H \cap L : H \in \widehat{X}_t , H \cap L \in A(L,k-1)\} .$$

The statements of the following theorem are taken from Section 5.6 of [82].

Theorem 2.37. *Let Θ be a hyperplane measure of the form* (2.62)*, satisfying* (2.63)*, and let $\widehat{X}_t$, for fixed $t > 0$, be the Poisson hyperplane process defined by* (2.65)*. Furthermore, let $L \in G(d,k)$, $k \in \{1,\dots,d-1\}$. Then $\widehat{X}_t \cap L$ is a stationary Poisson process of $(k-1)$-dimensional flats in L with the intensity measure $t\,\Theta_L$, where Θ_L is given by*

$$\int_{A(L,k-1)} f(A)\,\Theta_L(\mathrm{d}A) = \eta_{L,1}^{\mathrm{PHT}} \int_{S^{d-1}\cap L} \int_{\mathbb{R}} f(H(u,r)\cap L)\,\lambda_1(\mathrm{d}r)\,\varphi_L(\mathrm{d}u)$$

for all nonnegative measurable $f : A(L,k-1) \to \mathbb{R}$ *with*

$$\eta_{L,1}^{\mathrm{PHT}} = V_1(\Pi_\Theta|L)$$

and

$$\varphi_L(B) := \frac{1}{\eta_{L,1}^{\mathrm{PHT}}} \int_{S^{d-1}\setminus L^\perp} \mathbb{1}\left\{\frac{u|L}{\|u|L\|} \in B\right\} \|u|L\|\,\varphi(\mathrm{d}u)$$

for $B \in \mathcal{B}(S^{d-1}\cap L)$, *characterizes the spherical directional distribution. The density of* $\widehat{X}_t \cap L$, *is* $\eta_{L,t}^{\mathrm{PHT}} := t\,\eta_{L,1}^{\mathrm{PHT}}$. *Furthermore, the associated zonoid of the pair* (t,Θ_L) *is* $\Pi_{t\Theta}|L$, *the orthogonal projection of* $\Pi_{t\Theta}$ *to* L.

Now a relation between the typical k-face of a stationary PHT X_t and the typical cell of a section $X_t \cap L$ with a k-dimensional linear subspace L is given. This is the statement of Theorem 11.2.2 in [82]. Recall that χ_k defined in (2.79) is the directional distribution of the intersection process $\widehat{X}_t^{(d-k)}$ and thus also the directional distribution in the typical point of the k-skeleton $\mathrm{skel}_k X_t$.

Theorem 2.38. *Let* Θ *be a hyperplane measure of the form* (2.62), *satisfying* (2.63), *and let* $(X_t,\ t>0)$ *be the PHT process driven by* Θ. *Furthermore, fix* $k \in \{1,\dots,d-1\}$ *and* $t>0$.

Then the directional distribution $\mathbb{Q}_{k,t:h}^{\mathrm{PHT}}$ *of the typical k-face of the Poisson hyperplane tessellation* X_t *is given by*

$$\mathbb{Q}_{k,t:h}^{\mathrm{PHT}}(A) = \frac{V_{d-k}(\Pi_\Theta)}{\binom{d}{k} V_d(\Pi_\Theta)} \int_A V_k(\Pi_\Theta|L)\,\chi_k(\mathrm{d}L)$$

for $A \in \mathcal{B}(G(d,k))$.

Moreover, for $\mathbb{Q}_{k,t:h}^{\mathrm{PHT}}$*-almost all* $L \in G(d,k)$, *the conditional distribution* $\mathbb{Q}_{k,t|h=L}^{\mathrm{PHT}}$ *of the typical k-face of* X_t, *given the direction L, is equal to the distribution of the typical cell of the stationary PHT* $X_t \cap L$.

Choosing $k=1$ yields the following result for the distribution of the length of the typical edge in a stationary PHT. Note that in this case, $X_t \cap L$ is generated by a stationary Poisson point process on the line L with intensity $\eta_{L,t}^{\mathrm{PHT}} = t\,V_1(\Pi_\Theta|L)$.

Corollary 2.39. *With the assumptions of Theorem 2.38 and* $k=1$, *for* $\mathbb{Q}_{1,t:h}^{\mathrm{PHT}}$*-almost all* $L \in G(d,1)$, *the conditional distribution of the length of the typical 1-face of* X_t, *given its direction L, is the exponential distribution with parameter* $t\,V_1(\Pi_\Theta|L)$.

2.4 Notes

1. The k-primitives of a tessellation, $k \in \{0, \dots, d\}$, were introduced with different names in papers dealing with tessellations that are not face-to-face, for example, in [125, 138, 110, 193] and references therein.
   In [193] the term "primitives" was coined by Cowan, and the emphasis was on primitives, as opposed to components with internal structure, for general tessellations in $\mathbb{R}^3$ that are not face-to-face.
2. The standard definition of a face-to-face tessellation in $\mathbb{R}^d$ is given in (2.3). Up to now, it is an open question, whether the following definition is equivalent: A tessellation T in $\mathbb{R}^d$ is face-to-face if any $(d-1)$-dimensional intersection $P \cap P'$ of two cells $P, P' \in T$ is a $(d-1)$-face of both cells P, P'. For dimension $d = 2$, it is obvious, and for dimension $d = 3$, this is shown in Lemma 1.1 of [43].
3. In the proof of Lemma 2.16, several results and comments given in Section 1.1.5 of [137] are used. Note that the correction concerning Proposition 1.1.53 of [137], indicated on author's webpage, does not affect the validity of the proof of Lemma 2.16.
4. Equation (2.32) of Lemma 2.20 was shown for $d = 2$ in Theorem 4.1 of [124].
5. In Theorem 2.29 the exchange formula (2.61) is given for two sets of tessellation components and an arbitrary translation-covariant relation on these sets. So far, formulae of this type were known for sets of primitives. Mecke has shown a first result of this kind for $d = 2$ in Theorem 5.3 of [125]. Møller generalized this for tessellations in arbitrary dimension in Theorem 5.1 of [138]; see also Theorem 10.1.1 of [172] for face-to-face tessellations in $\mathbb{R}^d$. In all these papers the adjacency relation was regarded, which is a particular case of a translation-covariant relation; see Definition 16.1. Chiu replaced this adjacency relation by a neighborhood relation for face-to-face tessellations in $\mathbb{R}^d$; see Theorem 3 of [32].
6. Background for Section 2.3.3 on associated zonoids: The idea of associating a deterministic convex body with a random geometric system goes back to Matheron [121, 122], who introduced and studied the zonoid – called by him the Steiner compact – associated with a stationary Poisson hyperplane tessellation. It has also been successfully applied to Boolean models and to more general set-valued random processes; see [199, 200] and Section 4.6 of [172]. The common aspect is that the distribution of a number of random geometric objects is governed by an intensity parameter and an even probability measure on the sphere. This pair determines a deterministic zonoid, which can then be used to express expectations defined for the random system.
7. The result of Lemma 2.34 is well known and can be found in [172], equation (4.8), for part (a) and Theorem 4.2.1 of [96] for part (b). We included a proof for completeness and remark that our argument leading to the result of part (b) was inspired by [65, 105].

3 Iteration and superposition of tessellations

The iteration, also referred to as nesting in the literature, is an operation for tessellations that plays the central role in this book. In fact, the STIT tessellations we will study are those random tessellations whose distribution does not change under iteration after an appropriate scaling. Occasionally, we also consider the operation of superposition of tessellations. The random tessellations having a distribution that is stable under superposition are precisely the Poisson hyperplane tessellations.

3.1 Iteration

A tessellation $T \in \mathcal{T}$ of $\mathbb{R}^d$ is a countable subset of $\mathcal{P}_d$, the set of d-dimensional polytopes. To define the iteration of tessellations, a numbering of the cells is useful.

Lemma 3.1. *There exist measurable maps* $\mathsf{n}_i : \mathcal{T} \to \mathcal{P}_d$, $i \in \mathbb{N}_0$, *such that for all* $T \in \mathcal{T}$,

$$T = \{\mathsf{n}_i(T),\ i \in \mathbb{N}_0\}, \quad \text{and} \quad \mathsf{n}_i(T) \neq \mathsf{n}_j(T) \quad \text{if } i \neq j. \tag{3.1}$$

Proof. Denote $\mathsf{N}_{\mathcal{T}} := \{\delta(T, \cdot) : T \in \mathcal{T}\}$, where $\delta(T, \cdot) := \sum_{P \in T} \delta_P(\cdot)$; see (2.37). By Lemma 19.1 the map $i_s : \mathsf{N}_s(\mathcal{F}) \to \overline{\boldsymbol{\mathcal{F}}}_{\ell f}(\mathcal{F})$, which maps a simple counting measure on $\mathcal{F}$ onto its support, is a bimeasurable bijection, and hence $\mathsf{N}_{\mathcal{T}} \in \mathcal{N}$, because $\mathcal{T} \in \mathcal{B}(\overline{\boldsymbol{\mathcal{F}}}_{\ell f}(\mathcal{F}))$; see Lemma 18.2. Denote by $i_{s \llcorner \mathsf{N}_{\mathcal{T}}} : \mathsf{N}_{\mathcal{T}} \to \mathcal{T}$ the restriction of i_s to $\mathsf{N}_{\mathcal{T}}$. We can apply Lemma 19.2, which yields the proposition by taking $\mathsf{n}_i := \mathsf{m}_i \circ i_{s \llcorner \mathsf{N}_{\mathcal{T}}}^{-1} : \mathcal{T} \to \mathcal{P}_d$, $i \in \mathbb{N}_0$. ☐

We denote $\underline{\mathsf{n}} := (\mathsf{n}_i,\ i \in \mathbb{N}_0)$ and call it a **measurable numbering** of the cells. Recall that the symbol $\wedge$ for the restriction of a tessellation to a polytope was introduced in (2.20).

Definition 3.2. *For a tessellation* $T \in \mathcal{T}$, *a measurable numbering of cells* $\underline{\mathsf{n}} = (\mathsf{n}_i,\ i \in \mathbb{N}_0)$, *and a sequence of tessellations* $\underline{T} = (T^{(i)}, i \in \mathbb{N}_0) \in \mathcal{T}^{\mathbb{N}_0}$, ***the iteration of** T **with the numbering** $\underline{\mathsf{n}}$ **and** $\underline{T}$ is defined as the tessellation*

$$T \boxplus^{\underline{\mathsf{n}}} \underline{T} := \bigcup_{i \in \mathbb{N}_0} (T^{(i)} \wedge \mathsf{n}_i(T)). \tag{3.2}$$

By definition the iteration depends on the numbering $\underline{\mathsf{n}}$ of the cells of T, so that different numberings of cells of the tessellation T can yield different resulting tessellations after iteration.

Lemma 3.3. *For a fixed measurable numbering* $\underline{\mathsf{n}}$ *as in Lemma* 3.1, *the mapping* $(T, \underline{T}) \mapsto T \boxplus^{\underline{\mathsf{n}}} \underline{T}$ *is* $(\mathcal{B}(\mathcal{T}) \otimes \mathcal{B}(\mathcal{T})^{\otimes \mathbb{N}_0}, \mathcal{B}(\mathcal{T}))$*-measurable.*

Proof. Let $\underline{T} = (T^{(j)}, j \in \mathbb{N}_0)$. For simplicity of notation, we write $P_j := \mathsf{n}_j(T)$ and $P_i^{(j)} := \mathsf{n}_i(T^{(j)})$, $i, j \in \mathbb{N}_0$.

https://doi.org/10.1515/9783112222409-004

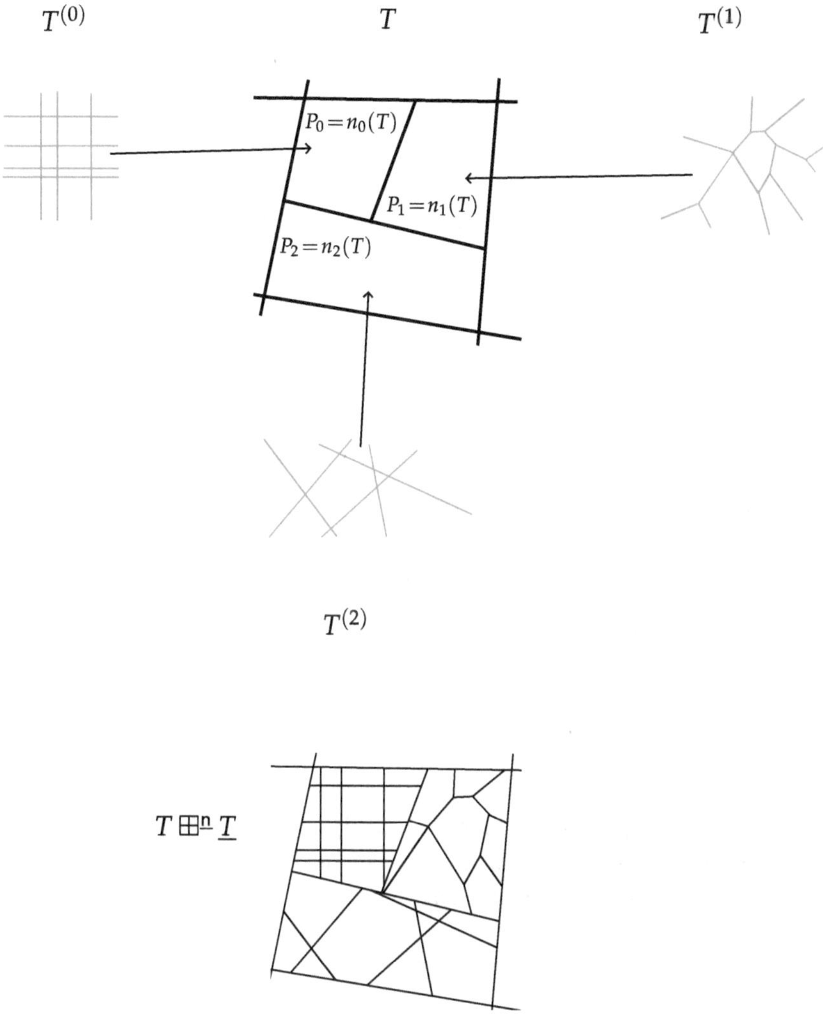

Figure 3.1: Iteration. Upper panel: T is the "frame tessellation", and its cells are numbered. Furthermore, $T^{(0)}, T^{(1)}, T^{(2)}$ are members of a sequence of tessellations, which are nested into the respective cells. Lower panel: Tessellation that results after iteration.

Referring to the generating system of $\mathcal{B}(\mathcal{T})$ given in Lemma 18.3, we consider for all $k \in \mathbb{N}_0$ and $A \in \mathcal{B}(\mathcal{P}_d)$,

$$\begin{aligned}&\{(T, \underline{T}) \in \mathcal{T} \times \mathcal{T}^{\mathbb{N}_0} : |(T \boxplus^{\mathrm{n}} \underline{T}) \cap A| \geq k\}\\ &= \bigcup_{\ell=1}^{k} \bigcup_{\substack{I \subset \mathbb{N}_0 \\ |I| = \ell}} \left\{(T, \underline{T}) \in \mathcal{T} \times \mathcal{T}^{\mathbb{N}_0} : \sum_{j \in I} |(T^{(j)} \wedge P_j) \cap A| = k\right\}.\end{aligned}$$

Now fix $\ell \in \{1,\dots,k\}$ and $I \subset \mathbb{N}_0$ with $|I| = \ell$ and denote by

$$[[k]] := \left\{(a_j, j \in I) \in \mathbb{N}^\ell : \sum_{j\in I} a_j = k\right\}$$

the set of all tuples of a_j that sum up to k. Then the measurability of the mapping $T \mapsto T \wedge P$, $T \in \mathcal{T}$, $P \in \mathcal{P}_d$, which is shown in Lemma 18.10, yields that

$$\begin{aligned}
&\left\{(T,\underline{T}) \in \mathcal{T} \times \mathcal{T}^{\mathbb{N}_0} : \sum_{j\in I} |(T^{(j)} \wedge P_j) \cap A| = k\right\} \\
&= \bigcup_{[[k]]} \bigcap_{j\in I} \{(T,\underline{T}) \in \mathcal{T} \times \mathcal{T}^{\mathbb{N}_0} : |(T^{(j)} \wedge P_j) \cap A| = a_j\} \\
&\in \mathcal{B}(\mathcal{T}) \otimes \mathcal{B}(\mathcal{T})^{\otimes \mathbb{N}_0} .
\end{aligned}$$

Hence, as a countable union of measurable sets,

$$\{(T,\underline{T}) \in \mathcal{T} \times \mathcal{T}^{\mathbb{N}_0} : |(T \boxplus^{\underline{n}} \underline{T}) \cap A| \geq k\} \in \mathcal{B}(\mathcal{T}) \otimes \mathcal{B}(\mathcal{T})^{\otimes \mathbb{N}_0} ,$$

which completes the proof. □

3.2 Superposition

We now turn to the superposition operation of tessellations. We will make it very briefly for completeness, as this operation will not be studied in this book. However, our aim is to emphasize the difference between the two operations. The **superposition of tessellations**, denoted $\uplus : \mathcal{T} \times \mathcal{T} \to \mathcal{T}$, is defined by

$$\partial(T_1 \uplus T_2) := \partial T_1 \cup \partial T_2 \tag{3.3}$$

for $T_1, T_2 \in \mathcal{T}$ or, equivalently,

$$T_1 \uplus T_2 := \{P_1 \cap P_2 : P_1 \in T_1, P_2 \in T_2, \operatorname{int}(P_1 \cap P_2) \neq \emptyset\} .$$

Intuitively, the images of the $(d-1)$-skeletons of two tessellations T_1 and T_2 are superposed; see Figure 3.2. Thus, in contrast to iteration, different cells of T_1 are not divided by cut-outs of a sequence of different tessellations, but all cells of T_1 are subdivided by one and the same tessellation T_2. Clearly, the same definition can also be used on $\mathcal{T}_W \times \mathcal{T}_W$ for $W \in \mathcal{P}_d$.

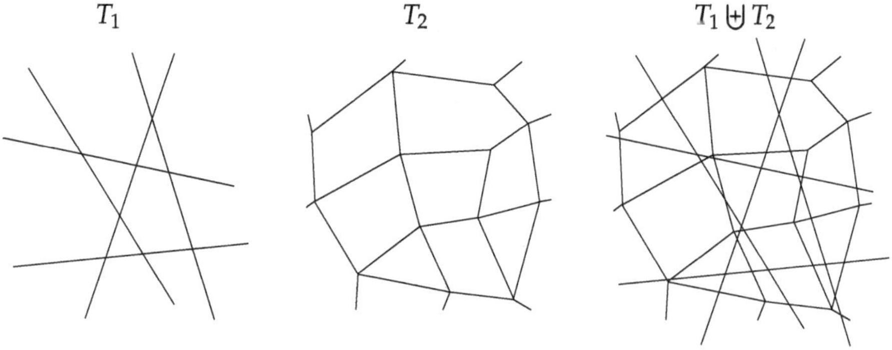

Figure 3.2: Superposition. Two tessellations T_1 and T_2 and the result $T_1 \uplus T_2$ of their superposition.

3.3 Iteration of random tessellations

Now we investigate the effect of iteration when this operation is applied to random tessellations. In particular, it is of interest how certain characteristics of the original tessellations determine characteristics of the resulting tessellation when a sequence of i. i. d. tessellations is nested into a random "frame tessellation".

Lemma 3.4. *Let X be a random tessellation of $\mathbb{R}^d$, let $\underline{\mathsf{n}} = (\mathsf{n}_i,\ i \in \mathbb{N}_0)$ be a measurable numbering of cells as in* (3.1), *and let $\underline{X} = (X^{(i)}, i \in \mathbb{N}_0)$ be a sequence of random tessellations of $\mathbb{R}^d$, all defined on the same probability space* $(\Omega, \mathbf{A}, \mathbb{P})$.

(a) *Then $X \boxplus^{\underline{\mathsf{n}}} \underline{X}$ is a random tessellation of $\mathbb{R}^d$.*

(b) *If $\underline{X} = (X^{(i)}, i \in \mathbb{N}_0)$ is a sequence of i. i. d. tessellations, also independent of X, then the distribution of $X \boxplus^{\underline{\mathsf{n}}} \underline{X}$ does not depend on the numbering $\underline{\mathsf{n}}$.*

(c) *If X is stationary and $\underline{X} = (X^{(i)}, i \in \mathbb{N}_0)$ is a sequence of i. i. d. random stationary tessellations, also independent of X, then $X \boxplus^{\underline{\mathsf{n}}} \underline{X}$ is a random stationary tessellation.*

(d) *If X and $\underline{X}$ satisfy the assumption of (c) and if all involved tessellations are isotropic, then $X \boxplus^{\underline{\mathsf{n}}} \underline{X}$ is a random stationary and isotropic tessellation.*

Proof. (a) This follows immediately from Lemma 3.3.

(b) Let $\underline{\mathsf{n}}$ and $\underline{\mathsf{n}}'$ be measurable numberings. Then for all $T \in \mathcal{T}$, there exists a bijection $\pi(\cdot, T) : \mathbb{N}_0 \to \mathbb{N}_0$ such that $\mathsf{n}'_i(T) = \mathsf{n}_{\pi(i,T)}(T)$ for all $i \in \mathbb{N}_0$. Under the assumption of (b), $(X^{(i)}, i \in \mathbb{N}_0) \overset{\mathrm{D}}{=} (X^{(\pi^{-1}(i,X))}, i \in \mathbb{N}_0)$, and the sequence on the right-hand side is also independent of X. Hence

$$
\begin{aligned}
(X \boxplus^{\underline{\mathsf{n}}'} \underline{X}) &= \bigcup_{i \in \mathbb{N}_0} (X^{(i)} \wedge \mathsf{n}'_i(X)) \\
&= \bigcup_{i \in \mathbb{N}_0} (X^{(i)} \wedge \mathsf{n}_{\pi(i,X)}(X))
\end{aligned}
$$

$$= \bigcup_{i \in \mathbb{N}_0} (X^{(\pi^{-1}(i,X))} \wedge \mathsf{n}_i(X))$$

$$\overset{D}{=} \bigcup_{i \in \mathbb{N}_0} (X^{(i)} \wedge \mathsf{n}_i(X))$$

$$= X \boxplus^{\underline{\mathsf{n}}} \underline{X} .$$

(c) Let $\underline{\mathsf{n}}$ be a measurable numbering. Then for all $T \in \mathcal{T}$ and $x \in \mathbb{R}^d$, there is a bijection $\pi(\cdot, T+x) : \mathbb{N}_0 \to \mathbb{N}_0$, depending on T and x, such that $\mathsf{n}_i(T)+x = \mathsf{n}_{\pi(i,T+x)}(T+x)$, $i \in \mathbb{N}_0$. Hence the assumptions of stationarity and independence, together with (b), yield

$$\begin{aligned}(X \boxplus^{\underline{\mathsf{n}}} \underline{X}) + x &= \bigcup_{i \in \mathbb{N}_0} (X^{(i)} \wedge \mathsf{n}_i(X)) + x \\ &= \bigcup_{i \in \mathbb{N}_0} ((X^{(i)} + x) \wedge (\mathsf{n}_i(X) + x)) \\ &\overset{D}{=} \bigcup_{i \in \mathbb{N}_0} (X^{(i)} \wedge \mathsf{n}_{\pi(i,X+x)}(X + x)) \\ &\overset{D}{=} \bigcup_{i \in \mathbb{N}_0} (X^{(i)} \wedge \mathsf{n}_i(X + x)) \\ &\overset{D}{=} \bigcup_{i \in \mathbb{N}_0} (X^{(i)} \wedge \mathsf{n}_i(X)) \\ &= X \boxplus^{\underline{\mathsf{n}}} \underline{X} .\end{aligned}$$

(d) This can be shown analogously to (c) replacing translations by rotations. □

When the assumptions of Lemma 3.4(b) are satisfied, the distribution of the iteration does not depend on the numbering $\underline{\mathsf{n}}$, and we use the notation

$$X \boxplus \underline{X} \quad \text{instead of } X \boxplus^{\underline{\mathsf{n}}} \underline{X} . \tag{3.4}$$

Furthermore, denoting $aP := \{ax : x \in P\}$, $P \in \mathcal{P}$, $a \in \mathbb{R}^+$, and correspondingly, for $T \in \mathcal{T}$ and $\underline{T} = (T^{(i)}, i \in \mathbb{N}_0) \in \mathcal{T}^{\mathbb{N}_0}$, denoting $aT := \{aP : P \in T\}$ and $a\underline{T} := (aT^{(i)}, i \in \mathbb{N}_0)$, we have that

$$(aX) \boxplus a\underline{X} \overset{D}{=} a\,(X \boxplus \underline{X}) . \tag{3.5}$$

Consider now a k-dimensional affine subspace $A \in A(d,k)$, $k \in \{1,\ldots,d-1\}$. The symbol $\boxplus_A$ is used for the iteration of tessellations within A. If the assumptions of Lemma 3.4(c) are satisfied, then

$$(X \boxplus \underline{X}) \cap A \overset{D}{=} (X \cap A) \boxplus_A (\underline{X} \cap A) , \tag{3.6}$$

since the assumed stationarity of the random tessellations guarantees that a. s. the sections with A are well-defined; see Lemma 2.30.

The operation $\boxplus$ cannot be associative in the usual sense, as it combines a single tessellation with a sequence of tessellations. However, we have the following kind of a modified associativity property.

Lemma 3.5. *Let X_1 be a random tessellation of $\mathbb{R}^d$, and let $\underline{X}_j = (X_j^{(i)}, i \in \mathbb{N}_0)$, $j \in \{2,3\}$, be independent sequences of i. i. d. random tessellations of $\mathbb{R}^d$, also independent of X_1. Furthermore, let $\underline{X_2 \boxplus \underline{X}_3} := (X_2^{(n)} \boxplus (X_3^{(n,\ell)}, \ell \in \mathbb{N}_0), n \in \mathbb{N}_0)$, where $(X_3^{(n,\ell)}, \ell \in \mathbb{N}_0)$, $n \in \mathbb{N}_0$, are i. i. d. sequences, all with the same distribution as $\underline{X}_3$. Then*

$$(X_1 \boxplus \underline{X}_2) \boxplus \underline{X}_3 \stackrel{\mathrm{D}}{=} X_1 \boxplus \underline{X_2 \boxplus \underline{X}_3}\,.$$

Proof. By Lemma 3.4(b) and the assumption of independence of the tessellations the distributions of $(X_1 \boxplus \underline{X}_2) \boxplus \underline{X}_3$ and $X_1 \boxplus \underline{X_2 \boxplus \underline{X}_3}$ do not depend on the chosen numbering of cells. Below, a cell $P_m \in (X_1 \boxplus \underline{X}_2)$ is rewritten as $P_\ell^{(i)}$, thus expressing that P_m arises as the intersection of a cell $P_i \in X_1$ and a cell $P_\ell \in X_2^{(i)}$. Correspondingly, the numbering of the sequence $\underline{X}_3$ by pairs of indices is adapted. We have

$$\begin{aligned}
(X_1 \boxplus \underline{X}_2) \boxplus \underline{X}_3 &\stackrel{\mathrm{D}}{=} \bigcup_{P_m \in (X_1 \boxplus \underline{X}_2)} X_3^{(m)} \wedge P_m \\
&\stackrel{\mathrm{D}}{=} \bigcup_{P_i \in X_1} \bigcup_{P_\ell^{(i)} \in X_2^{(i)} \wedge P_i} X_3^{(i,\ell)} \wedge P_\ell^{(i)} \\
&\stackrel{\mathrm{D}}{=} \bigcup_{P_i \in X_1} \bigcup_{P_\ell \in X_2^{(i)},\, P_i \cap P_\ell \in \mathcal{P}_d} X_3^{(i,\ell)} \wedge (P_i \cap P_\ell)
\end{aligned}$$

and

$$\begin{aligned}
X_1 \boxplus \underline{X_2 \boxplus \underline{X}_3} &\stackrel{\mathrm{D}}{=} \bigcup_{P_i \in X_1} (X_2^{(i)} \boxplus (X_3^{(i,\ell)}, \ell \in \mathbb{N}_0)) \wedge P_i \\
&\stackrel{\mathrm{D}}{=} \bigcup_{P_i \in X_1} \Big(\bigcup_{P_\ell \in X_2^{(i)}} X_3^{(i,\ell)} \wedge P_\ell \Big) \wedge P_i \\
&\stackrel{\mathrm{D}}{=} \bigcup_{P_i \in X_1} \bigcup_{P_\ell \in X_2^{(i)},\, P_i \cap P_\ell \in \mathcal{P}_d} X_3^{(i,\ell)} \wedge (P_i \cap P_\ell)\,.
\end{aligned}$$

Thus the proof is completed. □

In the stationary case, for the random tessellation arising as a result of an iteration, the density η_{d-1} and the $(d-1)$-volume weighted directional distribution ψ_o, defined in (2.48) and (2.51), respectively, are determined by the parameters of the ingredient tessellations.

Lemma 3.6. *Let X be a random stationary tessellation in $\mathbb{R}^d$ with density $\eta_{d-1} > 0$ and $(d-1)$-volume weighted directional distribution ψ_o. Furthermore, let $\underline{X} = (X^{(i)}, i \in \mathbb{N}_0)$ be a sequence of i. i. d. random stationary tessellations of $\mathbb{R}^d$, independent of X, where each $X^{(i)}$ has the density $\eta_{d-1}^{(1)} > 0$ and the $(d-1)$-volume weighted directional distribution $\psi_o^{(1)}$. Then $\eta_{d-1}^{X \boxplus \underline{X}}$ and $\psi_o^{X \boxplus \underline{X}}$ of $X \boxplus \underline{X}$ are given by*

$$\eta_{d-1}^{X \boxplus \underline{X}} \, \psi_o^{X \boxplus \underline{X}} = \eta_{d-1} \, \psi_o + \eta_{d-1}^{(1)} \, \psi_o^{(1)} , \tag{3.7}$$

and in particular,

$$\eta_{d-1}^{X \boxplus \underline{X}} = \eta_{d-1} + \eta_{d-1}^{(1)} . \tag{3.8}$$

Proof. For all $A \in \mathcal{B}(G(d,1))$, by (2.51) and definition (3.2) of iteration,

$$\begin{aligned}
&\eta_{d-1}^{X \boxplus \underline{X}} \, \psi_o^{X \boxplus \underline{X}}(A) \\
&= \mathbb{E} \sum_{F \in \mathcal{F}_{d-1}^{\mathrm{pr}}(X \boxplus \underline{X})} \mathbb{1}\{F^{\perp} \in A\} \, V_{d-1}(F \cap [0,1]^d) \\
&= \int_{\mathcal{T}} \int_{\mathcal{T}^{\mathbb{N}_0}} \Bigg(\sum_{F \in \mathcal{F}_{d-1}^{\mathrm{pr}}(T)} \mathbb{1}\{F^{\perp} \in A\} \, V_{d-1}(F \cap [0,1]^d) \\
&\quad + \sum_{P_i \in T} \sum_{F \in \mathcal{F}_{d-1}^{\mathrm{pr}}(T^{(i)})} \mathbb{1}\{F^{\perp} \in A\} \, V_{d-1}(F \cap [0,1]^d \cap P_i) \Bigg) \, \mathbb{P}_{\underline{X}}(\mathrm{d}\underline{T}) \, \mathbb{P}_X(\mathrm{d}T) \\
&= \eta_{d-1} \, \psi_o(A) + \mathbb{E}\Bigg(\sum_{P_i \in X} \lambda_d([0,1]^d \cap P_i) \Bigg) \eta_{d-1}^{(1)} \, \psi_o^{(1)}(A) \\
&= \eta_{d-1} \, \psi_o(A) + \eta_{d-1}^{(1)} \, \psi_o^{(1)}(A) ,
\end{aligned}$$

where for the last but one equation, definition (2.47) of the density is used. This shows (3.7). Choosing $A = G(d,1)$ yields (3.8). □

Note that Lemma 16.2 provides further relations for the iteration that are specific to the planar case.

If instead of an iteration, the superposition $X_1 \uplus X_2$ of two independent random stationary tessellations X_1 and X_2 is considered, then by definition (3.3) this is the union of the $(d-1)$-skeletons of these tessellations. Let $\eta_{d-1,i}$, $\psi_{o,i}$, $i \in \{1,2\}$, be the respective characteristics of these tessellations. Using (2.51), we easily see that the relations for the densities and the $(d-1)$-volume weighted directional distributions are analogous to (3.7) and (3.8), that is, $\eta_{d-1}^{X_1 \uplus X_2} \, \psi_o^{X_1 \uplus X_2} = \eta_{d-1,1} \, \psi_{o,1} + \eta_{d-1,2} \, \psi_{o,2}$, and in particular, $\eta_{d-1}^{X_1 \uplus X_2} = \eta_{d-1,1} + \eta_{d-1,2}$.

3.4 Stability under iteration and infinite divisibility

The stability of a distribution with respect to an operation applied to random variables and the corresponding infinite divisibility of a distribution are important concepts in probability theory. The reason for this is that such distributions typically appear as limiting distributions in distributional limit theorems naturally related to the respective operation. Probably, the best known examples are the general central limit theorem related to classical infinitely divisible distributions; see, for example, [60] or the Fisher–Tippet theorem related to max-stable distributions on the real line, as presented in [52].

Here we adapt the usual definitions of stability and infinite divisibility to the iteration of tessellations, which, according to Definition 3.2, is an operation applied to a tessellation and a *sequence* of tessellations. To arrive at a potential limit tessellation, an additional scaling of tessellations is necessary. For $T \in \mathcal{T}$ and a positive real number $a > 0$, it is defined by $a\,T := \{a\,P : P \in T\}$, where $a\,P := \{a\,x : x \in P\}$.

To describe a repeated iteration, let $m \geq 2$, let X be a random stationary tessellation, and let $\underline{X}_1, \ldots, \underline{X}_{m-1}$ be independent sequences of i. i. d. random tessellations with the same law as X, also independent of X. We introduce the notation

$$\boxplus^m X := (\boxplus^{m-1} X) \boxplus \underline{X}_{m-1} = (\ldots(X \boxplus \underline{X}_1) \boxplus \ldots) \boxplus \underline{X}_{m-1} \tag{3.9}$$

and call it the $(m-1)$-**fold iteration of** X. For simplicity of notation, we write $\boxplus^m X := X$ for $m = 1$.

We are now prepared to introduce the concept of stability under iteration. Although this is a property of distributions on the space of tessellations, for convenience, we formulate it for random tessellations.

Definition 3.7. *Let X be a random stationary tessellation of $\mathbb{R}^d$.*

(a) *Then X is a* ***tessellation stable under iteration****, called* ***STIT tessellation****, if and only if for all $m \in \mathbb{N}$,*

$$m\,(\boxplus^m X) \stackrel{\mathrm{D}}{=} X\,. \tag{3.10}$$

(b) *Furthermore, X is* ***infinitely divisible with respect to iteration*** *if and only if for all $m \in \mathbb{N}$, there exists a random stationary tessellation $X^{\langle m\rangle}$ such that*

$$\boxplus^m X^{\langle m\rangle} \stackrel{\mathrm{D}}{=} X\,. \tag{3.11}$$

If a random stationary tessellation X is STIT then it is also infinitely divisible. To see this, take $X^{\langle m\rangle} := m\,X$. Then (3.10) and (3.5) yield (3.11).

Even if random tessellations are of interest mainly for dimension $d \geq 2$, nevertheless, the following proposition for the one-dimensional case is still relevant when linear sections of higher-dimensional STIT tessellations are considered.

Lemma 3.8. *Let $\widehat{X}$ be a stationary Poisson point process on $\mathbb{R}^1$. Then the induced tessellation X of $\mathbb{R}^1$ is a STIT tessellation.*

Proof. Let $\widehat{X}$ be a stationary Poisson point process on $\mathbb{R}^1$, and let $\underline{\widehat{X}} = (\widehat{X}^{(i)}, i \in \mathbb{N}_0)$ be independent sequence of i. i. d. stationary Poisson point processes on $\mathbb{R}^1$, all with the same intensity $t > 0$. Denote the respective tessellations of $\mathbb{R}^1$ by X and $\underline{X} = (X'^{(i)}, i \in \mathbb{N}_0)$. It is sufficient to show that $\boxplus^m X$ has the same distribution as the tessellation induced by a Poisson point process with intensity $m\,t$, $m \in \mathbb{N}$. First note that $\boxplus^2 X$ is generated by the point process

$$\widehat{X} + \sum_{i \in \mathbb{N}_0} \widehat{X}^{(i)}(\cdot \cap P_i)\,,$$

where the intervals P_i, $i \in \mathbb{N}_0$, denote the cells of X. By the superposition theorem 19.13 for countably many independent Poisson point processes, $\sum_{i\in\mathbb{N}_0} \widehat{X}^{(i)}(\cdot \cap P_i)$ is a stationary Poisson point process on $\mathbb{R}^1$ with intensity t, and thus its distribution does not depend on $\widehat{X}$. Hence the distribution of $\boxplus^2 X$ is identical to that of a tessellation of $\mathbb{R}^1$ generated by a Poisson point process with intensity $2\,t$. Now induction over m completes the proof. □

The STIT property and the infinite divisibility of a random tessellation is inherited by sections with lower-dimensional flats as defined in (2.22). Recall that the symbol $\boxplus_A$ is used for the iteration of tessellations within a flat A.

Lemma 3.9. *Let X be a random stationary tessellation of $\mathbb{R}^d$, $k \in \{1, \ldots, d-1\}$, and let $A \in A(d, k)$ be a k-dimensional flat.*
(a) *If X is a STIT tessellation, then $X \cap A$ is a STIT tessellation of A with respect to $\boxplus_A$.*
(b) *If X is infinitely divisible with respect to $\boxplus$, then $X \cap A$ is a random tessellation of A, which is infinitely divisible with respect to $\boxplus_A$.*

Proof. The statements immediately follow by (3.6). □

STIT tessellations are the main topic of this book. In the following chapter, in Definition 4.2 a process is described, already called a STIT tessellation process in a window W, but it is not yet shown that it is indeed the restriction of a STIT tessellation in $\mathbb{R}^d$ to W. The proof is given in Theorem 6.4. In addition, a characterization of STIT tessellations is derived in Theorem 7.9.

If we consider the superposition $X_1 \uplus X_2 \uplus \cdots \uplus X_m$ of m independent random stationary tessellations instead of the iteration, then by definition (3.3) this is the union of the $(d-1)$-skeletons of these tessellations and thus the union of random closed sets. The superposition Theorem 19.13 for Poisson point processes yields that for i. i. d. stationary PHTs $X_1, \ldots, X_m$, we have that

$$m\,(X_1 \uplus X_2 \uplus \cdots \uplus X_m) \stackrel{D}{=} X_1\,,$$

which means that stationary PHTs are stable under superposition of tessellations.

3.5 Notes

1. The iteration of stationary and isotropic Poisson line tessellations in the plane is used in [180], pages 295–296, to construct an example that illustrates the following fact: On the intersection of the iterated tessellation with a fixed line, the distribution of the resulting point processes cannot be distinguished from the corresponding intersection with the skeleton of a Poisson line tessellation, provided that the intensities are equal.
   This idea was extended in conference presentations [135, 134] and then in paper [136]. There iteration and superposition of planar stationary tessellations are applied to construct a variety of tessellations for which the distribution of the typical cell coincides with that of a PHT. A particular focus is on the generated T-shaped vertices; see also Section 9.2 for more detailed comments.
   In [135], it is already conjectured that a repeated and rescaled iteration of appropriate tessellations converges to a limit tessellation with T-shaped vertices only.
   Furthermore, the iteration of general tessellations was already mentioned in the first edition (1987) of [33].
   In Section 7.13 of [5] and in the previous Russian edition (1989) of that book, the iteration was discussed. As already emphasized in Section 1, in personal communications in the 1980s, Ambartzumian posed the problem of investigating limits of repeated and rescaled iteration of random tessellations.
   In [91] and [194], iteration was used to construct planar ergodic tessellations with certain prescribed mean values and directional distributions, respectively. The purpose of these studies was to determine the joint range of the respective characteristics for planar random stationary tessellations.
   A systematic approach to the iteration of tessellations in $\mathbb{R}^d$, including a solid formal foundation, was provided in [110]. However, some proofs of key propositions are only contained in [109] and [107].
2. The proof of Lemma 3.3 is based on the proof of Theorem 3.1.1 in [107]. The propositions (c) and (d) of Lemma 3.4 are shown in Theorem 3.2.1 of [107] under the weaker assumption that $\underline{X} = (X^{(i)}, i \in \mathbb{N}_0)$ is an *exchangeable* sequence of random tessellations that are *jointly* stationary or *jointly* stationary and isotropic, respectively; see also Theorems 3.1 and 3.2 of [110]. This approach offers a wide range of operations on tessellation and includes the superposition of tessellations as a special case where $X^{(1)} = X^{(2)} = X^{(3)} = \cdots$. The iteration as considered in the present book appears in [110] under the name "nesting" as the particular case in which $\underline{X} = (X^{(i)}, i \in \mathbb{N}_0)$ is a sequence of i. i. d. random tessellations.
3. The results of Lemma 3.6 were shown for the two-dimensional case in Theorem 3 of [194]. In addition, that theorem contains further relations that are specific to the planar case; see Lemma 16.2.
   For general dimension, in [110], formulae are derived that relate the intensities and mean intrinsic volumes of typical k-primitives, $k \in \{1, \ldots, d-1\}$, of the original tessellations with those of the results of superposition and iteration, respectively.
   In [108] the iteration and superposition of planar Poisson–Voronoi tessellations and Poisson line tessellations are investigated, and some distributional results, based on simulations, are provided. A similar approach was used in [12] to model keratin filament networks.
   In [149] the iteration of random tessellation is used to model telecommunication systems, and simulation algorithms are presented. References to applications of STIT tessellations in this area are provided in Section 9.3.2.
4. Models in which the cells of a random stationary tessellation have an "interior structure" are considered in [74]. This includes the case that the cells are subdivided by other random tessellations, which means the iteration of tessellations. For a rather general class of functionals of this "internal structure", first and second moments are determined, and laws of large numbers and central limit theorems for growing observation windows are proved.
   In [75], first- and second-order formulae as well as central limit theorems for growing observation windows are shown with respect to certain functionals related to the iteration of tessellations, that

is, $X \boxplus \underline{X}$. It is assumed that the "initial tessellation" X is stationary and that $\underline{X} = (X^{(i)}, i \in \mathbb{N}_0)$ is a sequence of i. i. d. random stationary and isotropic tessellations in $\mathbb{R}^d$. The functionals are the total $(k-1)$ volumes, $k \in \{1, \ldots, d-1\}$, of the intersection of the skeleton $\mathrm{skel}_k(X)$ with the $(d-1)$-skeletons of the nested $X^{(i)}$. Central limit theorems are shown that focus on iteration using PHTs and Poisson–Voronoi tessellations and more general tessellations that satisfy several technical conditions.

5. In [144] the problem of whether repeated and rescaled iteration of i. i. d. stationary tessellations converges to a tessellation was addressed. It was shown that such a limit, if it exists, must have the STIT property, and further necessary conditions for the planar case were derived. At that time, however, it was not yet clear whether STIT tessellations exist. The existence was later shown in [145].
6. Although Theorem 7.9 provides a characterization of STIT tessellations, it is still an open problem whether there are random stationary tessellations that are infinitely divisible with respect to iteration and do not satisfy the STIT property.
7. The superposition of two stationary planar tessellations has already been considered in [170]. The superposition $\uplus$ of random tessellations is defined as the union of their $(d-1)$-skeletons, which are random closed sets. Thus, regarding infinite divisibility with respect to superposition, the well-established corresponding concept of infinite divisibility with respect to union of generic random closed sets comes into play; see [137] or [172]. However, there seem to be no special results that characterize all random tessellations that are infinitely divisible under superposition.

4 Description of the STIT tessellation process in a window

This chapter introduces the main topic of this book. We start with a description of STIT tessellation processes in polytopes $W \in \mathcal{P}_d$, which are called windows. Such a "local" definition is much easier to understand than a "global" one, which is given in Chapter 8. The local approach is justified by the property of spatial consistency, dealt with in Section 5.3. It states that the distribution of the random tessellation inside a window does not depend on the choice of a larger polytope in which the construction is performed. This property ensures the existence of STIT tessellation processes in $\mathbb{R}^d$. In this section, we already use the acronym STIT, although it is only in Theorem 6.4 that we show the tessellations we consider are indeed STable under ITeration in the sense of Definition 3.7.

4.1 Definition

The construction of a STIT tessellation process is based on a translation-invariant hyperplane measure Θ. Such a measure is essentially determined by the choice of an even probability measure φ on the unit sphere S^{d-1}, that is, φ is a probability measure on S^{d-1} satisfying $\varphi(A) = \varphi(-A)$ for all $A \in \mathcal{B}(S^{d-1})$.

As already mentioned in Section 2.3.1, throughout this book, we work with measures Θ on the measurable space $(\mathcal{H}, \mathcal{B}(\mathcal{H}))$ of hyperplanes in $\mathbb{R}^d$ that satisfy

$$\int_{\mathcal{H}} f(H)\,\Theta(\mathrm{d}H) = \int_{S^{d-1}} \int_{\mathbb{R}} f(H(u,r))\,\lambda_1(\mathrm{d}r)\,\varphi(\mathrm{d}u) \tag{4.1}$$

for all nonnegative measurable functions $f : \mathcal{H} \to \mathbb{R}$. This definition yields that Θ is translation invariant, that is, $\Theta(A + x) = \Theta(A)$ for all $A \in \mathcal{B}(\mathcal{H})$ and $x \in \mathbb{R}^d$, where $A + x := \{H + x : H \in A\}$. The probability measure φ appearing in (4.1) is called the **spherical directional distribution** associated with Θ.

A hyperplane measure Θ can also be characterized by its directional distribution φ_o on $G(d, 1)$, which is defined as

$$\varphi_o(A) := \varphi\Big(\bigcup_{L \in A} L \cap S^{d-1}\Big), \quad A \in \mathcal{B}(G(d,1)). \tag{4.2}$$

We remark that any nonzero translation-invariant hyperplane measure can be written as $a\,\Theta$, where Θ is given by (4.1), and $a > 0$; see (4.33) of [172]. Throughout the present book, we choose $a = 1$. As will become clear below, this is not an essential loss of generality, because for STIT tessellation processes, the factor a just causes a scaling of space or time, respectively; see Remark 4.4 and Corollary 6.3.

https://doi.org/10.1515/9783112222409-005

Convention 4.1. We assume that

$$\varphi \text{ is not concentrated on a great hypersphere of } S^{d-1} . \tag{4.3}$$

In [82] a spherical directional distribution that satisfies this convention is called nondegenerate. The convention guarantees that the construction of STIT tessellation processes described later yields a. s. random processes of tessellations; see Theorem 5.10.

Now we describe two particular choices of spherical directional distributions, which will play a role in this book:

(i) In the case that $\varphi = \varphi_{\mathrm{iso}}$ is the rotation-invariant probability measure on S^{d-1}, we write Θ_{iso} for the hyperplane measure with spherical directional distribution φ_{iso} and call it the **motion-invariant hyperplane measure**.

(ii) Another case is the class of **Mondrian hyperplane measures**. To introduce them, let $e_1, \dots, e_d$ be the standard orthonormal basis in $\mathbb{R}^d$. For each tuple $(p_1, \dots, p_d) \in (0,1)^d$ of weights with $\sum_{i=1}^d p_i = 1$, the spherical directional distribution

$$\varphi := \sum_{i=1}^{d} \frac{p_i}{2} (\delta_{e_i} + \delta_{-e_i}) \tag{4.4}$$

satisfies (4.3), and by (4.1) it defines a Mondrian hyperplane measure. This terminology is motivated by the fact that a STIT tessellation process driven by a Mondrian hyperplane measure is reminiscent of the famous paintings of the Dutch modernist painter Piet Mondrian. The term "Mondrian process" was coined in [168]. These Mondrian STIT tessellation processes are also used in the context of machine learning, for example, in [154, 155]; see also Section 9.3.

For a hyperplane measure Θ of the form (4.1), satisfying (4.3), and an open or closed set $B \subset \mathbb{R}^d$ with $0 < \Theta([B]) < \infty$, a probability distribution on the set $[B] = \{H \in \mathcal{H} : B \cap H \neq \emptyset\}$ of all hyperplanes hitting B is defined by

$$\Theta_{[B]}(\cdot) := \Theta([B])^{-1}\, \Theta(\cdot \cap [B]) . \tag{4.5}$$

In particular, for any compact set with nonempty interior and especially for any polytope $P \in \mathcal{P}_d$, we have $0 < \Theta([P]) < \infty$. Furthermore, by the translation invariance of Θ, for all $P \in \mathcal{P}_d$,

$$\Theta([P]) = \Theta([\operatorname{int} P]) \quad \text{and} \quad \Theta_{[P]}(\cdot) = \Theta_{[\operatorname{int} P]}(\cdot) .$$

Therefore we will mostly write $[P]$ instead of $[\operatorname{int} P]$, and even when the condition $H \in [\operatorname{int} P]$ is necessary for a hyperplane H to divide P, we abbreviate this by writing $H \in [P]$, neglecting thereby only sets of Θ-measure zero.

For a hyperplane measure Θ that satisfies Convention 4.1, a STIT tessellation process is determined in the following way, referring to the L and D rules mentioned in Section 1.1:

L-STIT The life time of a cell P depends on P only and is exponentially distributed with parameter $\Theta([P])$.

D-STIT The random hyperplane dividing a cell P depends on P only, and its law is given by $\Theta_{[P]}$.

This will now be formalized precisely. Let the window $W \in \mathcal{P}_d$ be fixed. In the following definition of the tessellation-valued process $(Y_{t,W},\ t \geq 0)$, the birth time of a hyperplane H is denoted by $\beta(H)$, and the birth time of a cell P is denoted by $\beta(P)$. Furthermore, a **division event for** P is written as (P, H, β). Its interpretation is that P is divided by the hyperplane $H \in [\operatorname{int} P]$ at time β, the birth time of H.

If T is a tessellation of $\mathbb{R}^d$ or of a window W (see Definitions 2.1 and 2.10, respectively), $P \in T$ is a cell, and $H \in [\operatorname{int} P]$ is a hyperplane that intersects the interior of P, then we denote by

$$\oslash(P, H, T) := (T \setminus \{P\}) \cup \{P \cap H^+, P \cap H^-\} \tag{4.6}$$

the tessellation that arises from T by subdividing the cell P into the two subcells $P \cap H^+$ and $P \cap H^-$ using the hyperplane H; see Figure 4.1. If the birth time $\beta(H)$ is assigned to the hyperplane H that divides P, then the triple $(P, H, \beta(H)) \in \mathcal{P}_d \times \mathcal{H} \times (0, \infty)$ is the corresponding division event. Moreover, $\beta(H)$ is also the birth time of the two daughter cells $P \cap H^+$ and $P \cap H^-$.

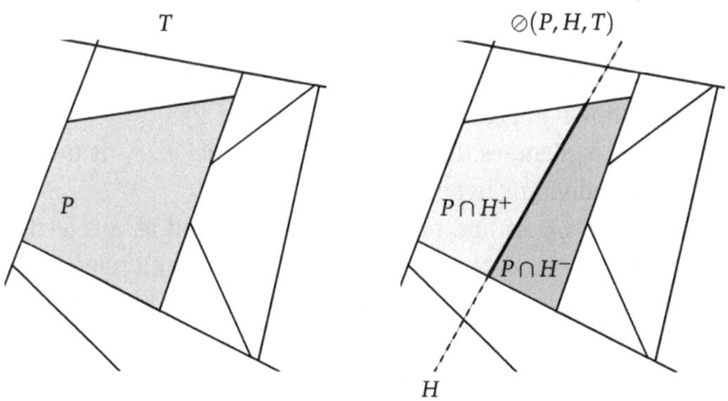

Figure 4.1: The operation $\oslash$. Left: A tessellation T with a cell P. Right: The cell P, and no other cell, is divided by a hyperplane H into two new cells. This yields the tessellation $\oslash(P, H, T)$.

There are several equivalent ways to define a STIT tessellation process in a window W; see Theorem 4.14. We start by presenting a constructive approach, which is useful for an immediate intuitive understanding and leads directly to a simulation algorithm.

Definition 4.2. *Let Θ be a hyperplane measure of the form* (4.1)*, satisfying* (4.3)*, and let $W \in \mathcal{P}_d$ be a d-dimensional polytope. Furthermore, let $\underline{\tau} = (\tau_i,\ i \in \mathbb{N})$ be a sequence of i. i. d. random variables, exponentially distributed with parameter 1, and let $\underline{\widetilde{H}} = (H_i,\ i \in \mathbb{N})$ be a sequence of i. i. d. hyperplanes, independent of $\underline{\tau}$, each one with law $\Theta_{[W]}$. The* ***STIT tessellation process*** *$(Y_{t,W},\ t \geq 0)$* ***in W, driven by the measure*** *Θ, is constructed as follows:*

(i) *$Y_{t,W} := \{W\}$ for $0 \leq t < \Theta([W])^{-1}\,\tau_1$, $P_1 := W$, and $\beta(P_1) := 0$.*

(ii) *If $t > 0$ and $Y_{t,W} = \{P_{i_1},\dots,P_{i_n}\}$, which means that $\beta(P_{i_k}) \leq t < \beta(P_{i_k}) + \Theta([P_{i_k}])^{-1}\,\tau_{i_k}$ for $k \in \{1,\dots,n\}$, define $j_n \in \{i_1,\dots,i_n\}$, σ_n, and H_n as follows:*

$$\sigma_n := \beta(P_{j_n}) + \Theta([P_{j_n}])^{-1}\,\tau_{j_n} := \min\{\beta(P_{i_k}) + \Theta([P_{i_k}])^{-1}\,\tau_{i_k} \;:\; k \in \{1,\dots,n\}\}$$

and

$$H_n := \widetilde{H}_\ell \quad \text{with } \ell := \min\{l > m : \widetilde{H}_m = H_{n-1},\ \widetilde{H}_l \in [\operatorname{int} P_{j_n}]\} \tag{4.7}$$

for $n \geq 2$, and $H_1 := \widetilde{H}_1$.
Then the random process $(Y_{t,W},\ t \geq 0)$ remains constant until time σ_n, and at that time, it jumps into the state

$$Y_{\sigma_n,W} := (\{P_{i_1},\dots,P_{i_n}\} \setminus \{P_{j_n}\}) \cup \{P_{2n}, P_{2n+1}\}$$

with $P_{2n} := P_{j_n} \cap H_n^+$, $P_{2n+1} := P_{j_n} \cap H_n^-$. The birth time of the dividing hyperplane and of the new cells is $\beta(H_n) = \beta(P_{2n}) = \beta(P_{2n+1}) := \sigma_n$.

This algorithm is visualized in Figure 4.2. Examples of states at different times of STIT tessellation processes are shown in Figures 4.3 and 4.4.

In part (ii) of this definition, σ_n denotes the time of the nth division, P_{j_n} is the cell divided at that time, and H_n is the dividing hyperplane.

In the context of a simulation algorithm, (4.7) can be interpreted as a rejection method: To divide a cell P_{j_n}, generate i. i. d. hyperplanes with law $\Theta_{[W]}$ and reject them until a hyperplane appears that intersects P_{j_n}. An elementary calculation shows that the distribution of the dividing hyperplane, under the condition that it intersects P_{j_n}, is $\Theta_{[P_{j_n}]}$. The use of a sequence $\underline{\widetilde{H}} = (\widetilde{H}_i,\ i \in \mathbb{N})$, together with the rule (4.7), ensures that the dividing hyperplanes of different cells are conditionally independent, given the respective cells.

If $\Theta = \Theta_{\text{iso}}$, then the parameter $\Theta_{\text{iso}}([P])$ of the exponential life time distribution of a cell P is $\Theta_{\text{iso}}([P]) = \overline{w}(P) = \frac{2\kappa_{d-1}}{d\,\kappa_d} V_1(P)$, where $\overline{w}$ denotes the mean width, and V_1 denotes the first intrinsic volume. For $d = 2$, this means that $\Theta_{\text{iso}}([P])$ is the perimeter of P divided by π.

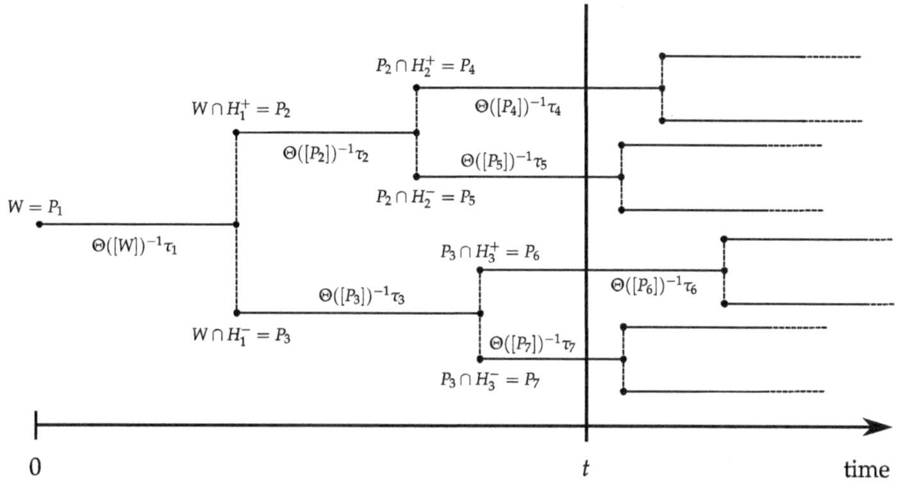

Figure 4.2: Example of the scheme of cell division inside a window W. The lengths of the horizontal segments depict the life times of the cells.

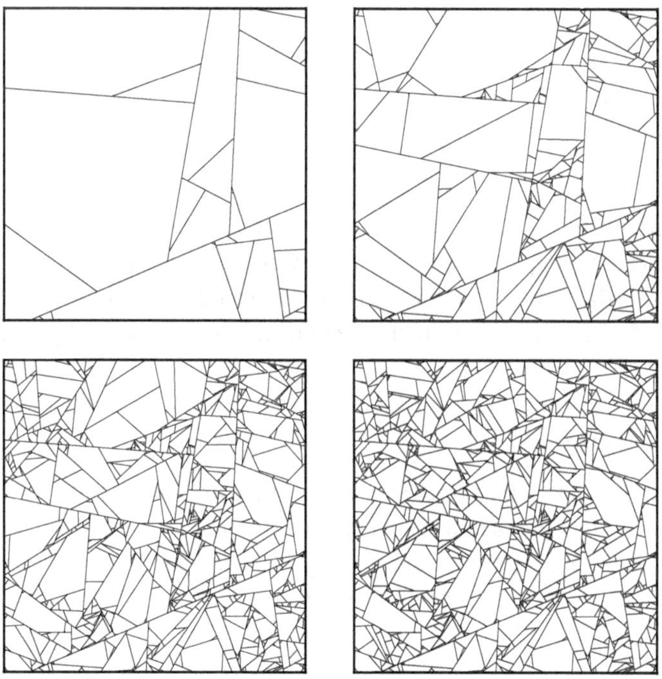

Figure 4.3: Simulation of a STIT tessellation process $(Y_{t,W},\ t \geq 0)$ driven by the motion-invariant hyperplane measure Θ_{iso} at times $t = 5; 10; 15; 20$.

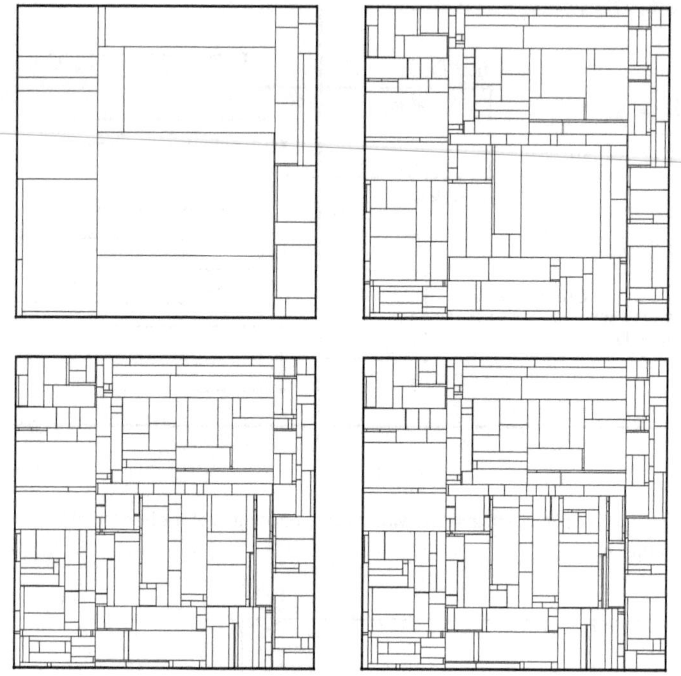

Figure 4.4: Simulation of a Mondrian STIT tessellation process $(Y_{t,W}, t \geq 0)$ with spherical directional distribution φ given in (4.4) with $p_1 = p_2 = 1/2$ at times $t = 5; 10; 15; 20$.

For the Mondrian hyperplane measure with the spherical directional distribution φ given in (4.4) and the case that $p_1 = \cdots = p_d = 1/d$, the value of $\Theta([P])$ is the average over the lengths of the d orthogonal projections of P onto the coordinate axes in $\mathbb{R}^d$. If, in particular, P is a cuboid with sides parallel to the axes and side lengths $b_1, \ldots, b_d$, then $\Theta([P]) = \frac{1}{d}\sum_{i=1}^{d} b_i$.

We add two remarks concerning the spherical directional distribution and the scaling of the hyperplane measure Θ.

Remark 4.3 (A caveat). *When simulating a random hyperplane with law $\Theta_{[W]}$, $W \in \mathcal{P}_d$, we have to take into account that the spherical directional distribution φ of* (4.1) *may not be used directly for the generation of the direction of the hyperplane. To see this, we recall from* (1.2) *that the support function of W is denoted by $h(W, \cdot)$. Then, for $A \in \mathcal{B}(S^{d-1})$, by* (4.1) *we have that*

$$\int_{\mathcal{H}} \mathbb{1}\{H \in [W],\, u(H) \in A\}\, \Theta(\mathrm{d}H)$$

$$= \int_{S^{d-1}} \int_{\mathbb{R}} \mathbb{1}\{H(u,r) \in [W],\, u \in A\}\, \lambda_1(\mathrm{d}r)\, \varphi(\mathrm{d}u)$$

$$= \int_{S^{d-1}} |h(W,u) - h(W,-u)| \, \mathbb{1}\{u \subset A\} \, \varphi(\mathrm{d}u)$$

$$= \int_{S^{d-1}} \mathbb{1}\{u \in A\} \, w(W,u) \, \varphi(\mathrm{d}u) \,,$$

where $w(W,\cdot) : S^{d-1} \to [0,\infty)$ stands for the width or breadth function of W. This shows that in the distribution $\Theta_{[W]}$ the spherical directional distribution φ has an additional density proportional to the width function $w(W,\cdot)$.

Remark 4.4. *As already mentioned, any nonzero translation-invariant hyperplane measure can be written as a Θ, where Θ is given by* (4.1), *and $a > 0$; see also (4.33) of* [172]. *Choosing the factor $a = 1$ for the hyperplane measure Θ given in* (4.1) *does not mean a substantial loss of generality. Reading Definition* 4.2, *it becomes evident that for an arbitrary $a > 0$, the STIT tessellation process driven by $a\,\Theta$ has the same distribution as the process $(Y_{at,W}, t \geq 0)$ when $(Y_{t,W}, t \geq 0)$ is driven by Θ; see also Corollary* 6.3.

Finally, the definition of a STIT tessellation process in a window is adapted to a STIT division process for an initial tessellation. For $W \in \mathcal{P}_d$, the set $\mathcal{T}_W$ of all tessellations of W is introduced in Definition 2.10.

Definition 4.5. *Let Θ be a hyperplane measure of the form* (4.1), *satisfying* (4.3), *and let $W \in \mathcal{P}_d$ be a d-dimensional polytope. Furthermore, let $T \in \mathcal{T}_W$ be a tessellation with cells $T = \{P_1, \ldots, P_m\}$, $m \in \mathbb{N}$, and let $(Y_{t,P_i}, t \geq 0)$ be independent STIT tessellation processes in P_i, $i \in \{1, \ldots, m\}$, all driven by the same hyperplane measure Θ. Then the process $(Y_{t,T}, t \geq 0)$ with*

$$Y_{t,T} := \bigcup_{P_i \in T} Y_{t,P_i} \,, \quad t \geq 0 \,, \tag{4.8}$$

is called the ***STIT division process driven by*** Θ ***with initial tessellation*** T.

4.2 Basic properties

We begin by introducing the set of division events and the concept of maximal polytopes that arise in the STIT process construction. Both will play an important role in the upcoming parts of this book. We then show that the random STIT tessellation process $(Y_{t,W}, t \geq 0)$, as defined in the previous section, is a pure jump Markov process on the space of tessellations of W, and we establish some related properties. At the same time, this leads to alternative equivalent descriptions of the STIT tessellation process.

Related to the construction of the STIT tessellation process $(Y_{t,W}, t \geq 0)$ in W, we define the **set of division events**

$$\mathcal{D}(Y_{t,W}, t \geq 0) := \{(P_{j_n}, H_n, \beta(H_n)) \in \mathcal{P}_d \times \mathcal{H} \times \mathbb{R}^+ : H_n \in [P_{j_n}], n \in \mathbb{N}\} \,, \tag{4.9}$$

where the indices correspond to those in (ii) of Definition 4.2. As already explained above, the interpretation of a division event is that at time $\beta(H_n) = \sigma_n$ the hyperplane H_n divides the cell P_{j_n}. Obviously, there is a one-to-one correspondence between the random process $(Y_{t,W}, t \geq 0)$ and the set $\mathcal{D}(Y_{t,W}, t \geq 0)$ of division events, which cause the jumps of $(Y_{t,W}, t \geq 0)$.

Each division event generates a $(d-1)$-polytope, arising as the intersection of the dividing hyperplane with the cell that gets divided. This is illustrated in Figure 4.5. Such a $(d-1)$-polytope will be referred to as a maximal $(d-1)$-polytope. It can additionally be marked by its birth time, which leads to the set of **birth time marked maximal $(d-1)$-polytopes in W**,

$$\mathcal{F}_{d-1,W}^{\max *} := \{(P \cap H, s) : (P, H, s) \in \mathcal{D}(Y_{t,W}, t \geq 0)\} . \tag{4.10}$$

Even though we speak of birth time marked maximal polytopes, this does not mean that $\mathcal{F}_{d-1,W}^{\max *}$ is a marked particle process when the birth times are not bounded. In this case the set of unmarked maximal polytopes is not locally finite.

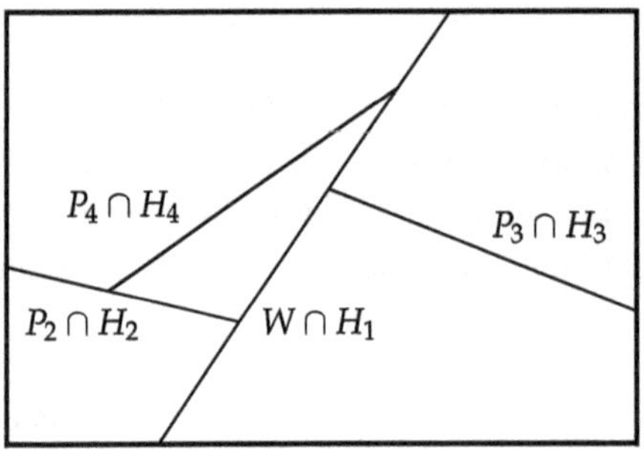

Figure 4.5: Maximal $(d-1)$-polytopes in a window. The window $W = P_1$ is divided by the hyperplane H_1, generating the maximal $(d-1)$-polytope $W \cap H_1$ and the new cells P_2 and P_3. Then P_2 is divided by H_2, and the maximal $(d-1)$-polytope $P_2 \cap H_2$ arises, together with the new cells P_4 and P_5, and so on.

For $k \in \{0, \ldots, d-2\}$, the set of all **birth time marked maximal k-polytopes in W** is defined as

$$\mathcal{F}_{k,W}^{\max *} := \bigcup_{(F,s) \in \mathcal{F}_{d-1,W}^{\max *}} (\mathcal{F}_k(F) \times \{s\}) . \tag{4.11}$$

Thus a maximal k-polytope is a k-dimensional face of a maximal $(d-1)$-polytope. Note that for $k \leq d-2$, a maximal k-polytope F can be a subset of the boundary ∂W of W. However, by the assumed translation invariance of the driving hyperplane measure Θ this a. s. cannot happen for the maximal polytopes of dimension $d-1$.

Furthermore, for $k \in \{0, \ldots, d-1\}$ and $t > 0$, we denote

$$\mathcal{F}_{k,W}^{\max *}(Y_{t,W}) := \{(F, s) \in \mathcal{F}_{k,W}^{\max *} : s \leq t\} \tag{4.12}$$

and

$$\mathcal{F}_{k,W}^{\max}(Y_{t,W}) := \{F \in \mathcal{P}_k : (F,s) \subset \mathcal{F}_{k,W}^{\max *}, s \le t\}. \tag{4.13}$$

For any fixed $t > 0$, the set $\mathcal{F}_{k,W}^{\max *}(Y_{t,W})$ can be considered as the support of a marked particle process, because in Theorem 4.7(d), it will be shown that the random number of cells of $Y_{t,W}$ is stochastically dominated by a geometrically distributed random variable, and hence the mean number of elements of $\mathcal{F}_{k,W}^{\max *}(Y_{t,W})$ is finite.

The maximal polytopes and their versions marked with birth times should be regarded as the basic building blocks of STIT tessellations, and they are the natural objects to be studied. Consequently, they will play an essential role throughout the book.

By construction it is obvious that for $k \in \{0, \ldots, d-1\}$ and $t > 0$,

$$\operatorname{skel}_k Y_{t,W} = \mathcal{F}_k(W) \cup \bigcup_{(F,s)\in\mathcal{F}_{k,W}^{\max *}(Y_{t,W})} F. \tag{4.14}$$

The following lemma shows why the term "maximal polytope" is justified to name the objects occurring in the division of cells.

Lemma 4.6. *Let $W \in \mathcal{P}_d$ be a d-dimensional polytope, let Θ be a hyperplane measure of the form* (4.1), *satisfying* (4.3), *and let $(Y_{t,W}, t \ge 0)$ be the STIT tessellation process in W, driven by the measure Θ. Then, for $k \in \{1, \ldots, d-1\}$ and $t > 0$,*

$$\operatorname{aff} F_1 \ne \operatorname{aff} F_2 \text{ a. s. } \quad \text{for } (F_1, s_1), (F_2, s_2) \in \mathcal{F}_{k,W}^{\max *}(Y_{t,W}),\ F_1 \ne F_2. \tag{4.15}$$

Furthermore, with the notation

$$F_{t,W}^{\max}(F) := \bigcup_{\substack{F'\in\mathcal{F}_k(Y_{t,W})\\ \operatorname{aff} F' = \operatorname{aff} F}} F' \quad \text{for } F \in \mathcal{F}_k(Y_{t,W}),$$

we have that

$$\mathcal{F}_{k,W}^{\max}(Y_{t,W}) = \{F_{t,W}^{\max}(F) : F \in \mathcal{F}_k(Y_{t,W}) \setminus \mathcal{F}_k(W)\} \text{ a. s.} \tag{4.16}$$

Proof. Property (4.15) follows by the translation invariance of the driving measure Θ and the independence of the dividing hyperplanes in the sequence $\widetilde{H} = (\widetilde{H}_i, i \in \mathbb{N})$. Definition (2.2) of skeletons, together with (4.14) and (4.15), yields (4.16). □

Note that the definition of maximal polytopes in (4.13) is consistent with (10.22), where a definition of maximal polytopes for general tessellations $T \in \mathcal{T}_W$ is given. This follows by (4.16) and the fact that the elements of $\mathcal{F}_{k,W}^{\max}(Y_{t,W})$ are connected. Furthermore, (4.15) indicates that for $\mathcal{F}_{k,W}^{\max}(Y_{t,W})$ even a stronger property holds.

In contrast to (4.13), in (8.7) and (8.8), we define maximal polytopes marked with their birth times for STIT tessellations in the space $\mathbb{R}^d$, that is, without any edge effect caused by a window W. Lemma 4.6 and the analogous Lemma 8.8 form the basis for Lemma 8.10, which connects both types of maximal polytopes by a limit for growing windows. Furthermore, in Lemma 11.4, it is shown that the birth time marked maximal polytopes $\mathcal{F}_{k,W}^{\max *}$ in W can be described, up to equivalence in distribution, as sections of the birth time marked maximal polytopes $\mathcal{F}_{k}^{\max *}$ in $\mathbb{R}^d$, restricted to W.

Now we derive some basic properties of the STIT tessellation process $(Y_{t,W}, t \geq 0)$ directly from the definition. They concern the strong Markov property, holding times of the states, and nonexplosion.

By Definition 4.2 the STIT tessellation process $(Y_{t,W}, t \geq 0)$ is constant between two jumps. The random time between two successive jumps is referred to as the holding time of the respective state of the process. By definition the life time of an individual cell P is exponentially distributed with parameter $\Theta([P])$. If we consider the time until the next jump, that is, the minimum of the remaining life times of the extant cells, then the sum of these parameters is essential. Thus, for $T \in \mathcal{T}_W$, denote

$$\theta(T) := \sum_{P \in T} \Theta([P]) . \tag{4.17}$$

Note that in the following theorem, we distinguish between the life time of a cell and the **remaining** life time of a cell after time s, under the condition that the cell exists at time s.

Theorem 4.7. *Let $W \in \mathcal{P}_d$, and let Θ be a hyperplane measure of the form (4.1), satisfying (4.3). Then the STIT tessellation process $(Y_{t,W}, t \geq 0)$ in W driven by the measure Θ has the following properties.*

(a) *Let $s > 0$. Under the condition that $Y_{s,W} = \{P_1, \dots, P_n\}$ consider the division events $(P_i, H_i, \beta(H_i)) \in \mathcal{D}(Y_{t,W}, t \geq 0)$, $i \in \{1, \dots n\}$. Then the joint conditional distribution of the dividing hyperplanes and of the remaining life times of the cells after time s, that is, of the random variables*

$$((H_i, \beta(H_i) - s), i \in \{1, \dots n\}) , \quad \textit{is} \quad \bigotimes_{i=1}^{n} (\Theta_{[P_i]} \otimes \mathsf{Expo}(\Theta([P_i]))) ,$$

and these random variables are conditionally independent of $(Y_{s',W}, 0 \leq s' < s)$, given $Y_{s,W} = \{P_1, \dots, P_n\}$.

(b) *The process $(Y_{t,W}, t \geq 0)$ is a time-homogeneous Markov process.*

(c) *For $s \geq 0$, the holding time of the state $Y_{s,W} = \{P_1, \dots, P_n\}$ is exponentially distributed with parameter $\theta(Y_{s,W})$. The probability that at the end of the holding time the cell P_i is the one which is divided is $\theta(Y_{s,W})^{-1}\, \Theta([P_i])$, $i \in \{1, \dots, n\}$.*

(d) *Let G be an integer-valued random variable, geometrically distributed with parameter $\mathrm{e}^{-t\Theta([W])}$. Then, for all $t > 0$ and $n \in \mathbb{N}_0$,*

$$\mathbb{P}(|Y_{t,W}| > n) \leq \mathbb{P}(G > n) , \tag{4.18}$$

that is, the random number of cells of $Y_{t,W}$ is stochastically dominated by a geometrically distributed random variable.

(e) *The process* $(Y_{t,W}, t \geq 0)$ *is a nonexploding pure jump process and has the strong Markov property for all stopping times with respect to the filtration adapted to* $(Y_{t,W}, t \geq 0)$.

Proof. In the proof, we use the same notation for the indices as introduced in Definition 4.2.

(a) For $s > 0$, let $Y_{s,W} = \{P_{i_1}, \ldots, P_{i_n}\}$ with birth times $\beta(P_{i_k})$, $k \in \{1, \ldots, n\}$. This means that $\beta(P_{i_k}) \leq s < \beta(P_{i_k}) + \Theta([P_{i_k}])^{-1}\tau_{i_k}$ for $k \in \{1, \ldots, n\}$. By Lemma 18.2 the space $\mathcal{T}_W$ of all tessellations of W is a Borel space. The birth times of the cells are real-valued, and hence by Theorem 8.5 of [89] the conditional distribution in equation (4.19) below, given $Y_{s,W}$ and $\beta(P_{i_k})$, $k \in \{1, \ldots, n\}$, exists. Due to the memoryless property of exponential distributions and the independence assumptions in the construction of STIT tessellation processes, the conditional joint distribution of the *remaining* life times after s of the extant cells is determined by

$$\begin{aligned}
&\mathbb{P}\left(\bigcap_{k=1}^{n}\{\beta(P_{i_k}) + \Theta([P_{i_k}])^{-1}\tau_{i_k} > s + s_{i_k}\} \,\Big|\, \bigcap_{k=1}^{n}\{\beta(P_{i_k}) \leq s < \beta(P_{i_k}) + \Theta([P_{i_k}])^{-1}\tau_{i_k}\}\right)\\
&\quad = \frac{\mathbb{P}(\bigcap_{k=1}^{n}\{\Theta([P_{i_k}])^{-1}\tau_{i_k} > s - \beta(P_{i_k}) + s_{i_k}\})}{\mathbb{P}(\bigcap_{k=1}^{n}\{\Theta([P_{i_k}])^{-1}\tau_{i_k} > s - \beta(P_{i_k})\})}\\
&\quad = \prod_{k=1}^{n}\mathbb{P}(\Theta([P_{i_k}])^{-1}\tau_{i_k} > s_{i_k})
\end{aligned} \tag{4.19}$$

for $s_{i_k} > 0$, $k \in \{1, \ldots, n\}$. For $Y_{s,W} = \{P_{i_1}, \ldots, P_{i_n}\}$, by Definition 4.2 the conditional independence of the dividing hyperplanes $H_1, \ldots, H_n$ from the life times is obvious.

(b) The conditional distribution calculated in the proof of (a) depends neither on s nor on the birth times $\beta(P_{i_k})$, $k \in \{1, \ldots, n\}$, but only on the P_{i_k}, $k \in \{1, \ldots, n\}$. This, together with the assumptions on the sequences $\underline{\tau} = (\tau_i, i \in \mathbb{N})$ and $\underline{\widetilde{H}} = (\widetilde{H}_i, i \in \mathbb{N})$ in Definition 4.2, yields the Markov property and the time homogeneity of the process.

(c) Choosing $s_{i_k} = s'$ for all $k \in \{1, \ldots, n\}$ in (4.19), we obtain for the nth jump time σ_n, introduced in Definition 4.2 (ii), that

$$\mathbb{P}(\sigma_n > s + s' | \sigma_n > s) = \exp\left(-\sum_{k=1}^{n}\Theta([P_{i_k}])s'\right).$$

Thus the holding time of the state $Y_{s,W} = \{P_{i_1}, \ldots, P_{i_n}\}$ is exponentially distributed with parameter $\theta(Y_{s,W}) = \sum_{k=1}^{n}\Theta([P_{i_k}])$, which is the distribution of the minimum of n independent random variables, exponentially distributed with the respective parameters $\Theta([P_{i_k}])$, $k \in \{1, \ldots, n\}$. The probability that this minimum is attained by the variable with index i_k is $\theta(Y_{s,W})^{-1}\Theta([P_{i_k}])$, and this is the probability of the event that at the end of the holding time of the process, the cell P_{i_k} is the one that is divided.

(d) As shown in the proof of (c), for $n \in \mathbb{N}$, the time between the $(n-1)$th and nth divisions is exponentially distributed with parameter $\theta(Y_{s,W}) = \sum_{k=1}^{n} \Theta([P_{i_k}])$, where P_{i_k}, $k \in \{1, \dots n\}$, are the cells after the $(n-1)$th division. Because

$$\sum_{k=1}^{n} \Theta([P_{i_k}]) \leq n\,\Theta([W]),$$

the process $(|Y_{t,W}|,\ t \geq 0)$ of the number of cells is stochastically dominated by a continuous-time linear birth process with birth rate $\Theta([W])$. The value of such a birth process at time $t > 0$ is geometrically distributed with parameter $e^{-t\,\Theta([W])}$, as proved on page 235 of [167]. This yields (4.18).

(e) The fact that $(Y_{t,W},\ t \geq 0)$ is a nonexploding pure jump process follows from (d). It can also directly be shown, because

$$\sum_{n \in \mathbb{N}} \left(\sum_{k=1}^{n} \Theta([P_{i_k}]) \right)^{-1} > \sum_{n \in \mathbb{N}} (n\Theta([W]))^{-1} = \infty,$$

and thus the sufficient condition of Proposition 13.5 of [89] for nonexplosion of pure jump Markov processes is satisfied. Finally, a pure jump Markov process has the strong Markov property, which is shown in Theorem 13.1 of [89]. □

By Theorem 4.7(e) the set $\mathcal{D}(Y_{t,W},\ t \geq 0)$ of division events defined in (4.9) has the locally finiteness property, and hence it can be understood as a point process with state space $\mathcal{P}_d \times \mathcal{H} \times \mathbb{R}^+$, the **point process of division events.**

There are several equivalent ways to describe the construction of a STIT tessellation process in a window. One of them, which is used in Section 13.4.1, is given by the following remark.

Remark 4.8. *By Theorem 4.7(c) and the independence assumptions in Definition 4.2 the STIT tessellation process in a window $W \in \mathcal{P}_d$, driven by Θ, can also be constructed as follows.*

(i) *The sequence $(T_m,\ m \in \mathbb{N})$, $T_m \in \mathcal{T}_W$, of states of the process can be generated without regarding the life times of the cells or the holding times of the Markov process: Put $T_1 := \{W\}$. Then, given a state $T_m = \{P_1, \dots, P_m\}$, choose a cell P_i at random with probability $(\sum_{j=1}^{m} \Theta([P_j]))^{-1}\,\Theta([P_i])$, $i \in \{1, \dots, m\}$, and divide the chosen cell by a random hyperplane H with law $\Theta_{[P_i]}$. This yields the next state $T_{m+1} := \oslash(P_i, H, \{P_1, \dots, P_n\})$, according to notation* (4.6).

(ii) *Let $(\tau_i,\ i \in \mathbb{N})$ be a sequence of i. i. d. random variables, exponentially distributed with parameter 1. For a given sequence $(T_m,\ m \in \mathbb{N})$, the holding time of the state T_m is $\theta(T_m)^{-1}\,\tau_m$. Thus the nth jump time of the process is $\sigma_n = \sum_{m=1}^{n} \theta(T_m)^{-1}\,\tau_m$, $n \in \mathbb{N}$. Put $\sigma_0 := 0$.*

(iii) *A version of the STIT tessellation process $(Y_{t,W},\ t \geq 0)$ in $W \in \mathcal{P}_d$ driven by Θ is given by*

$$Y_{t,W} := T_m \quad \text{for } \sigma_{m-1} \le t < \sigma_m .$$

The following lemma looks technical, but it expresses **the** key property of STIT tessellation processes in windows. It is formulated in Theorem 6.2 for the corresponding tessellation process in $\mathbb{R}^d$, and formulae (4.20) and (6.1) will be essential tools in many proofs. In particular, (6.1) is used in the proof of Theorem 6.4, which states that the tessellations we already call STIT are indeed stable under iteration. Intuitively, Lemma 4.9 means that after some stopping time, continuation of the STIT construction according to (ii) of Definition 4.2 is equivalent to the iteration of independent STIT tessellations into the cells extant at that stopping time.

Denote by $\mathfrak{I} := (\mathfrak{I}_t,\, t \ge 0)$ with $\mathfrak{I}_t := \sigma(\{Y_{s,W},\ 0 \le s \le t\})$ the filtration induced by the process $(Y_{t,W},\, t \ge 0)$. Recall that according to (4.8), for all $\sigma < t$, the random tessellation $Y_{t-\sigma,\, Y_{\sigma,W}}$ is the state of a STIT division process at time $t - \sigma$ with initial tessellation $Y_{\sigma,W}$.

Lemma 4.9. *Let $W \in \mathcal{P}_d$, and let Θ be a hyperplane measure of the form* (4.1), *satisfying* (4.3), *and let $(Y_{t,W},\, t \ge 0)$ be the STIT tessellation process in W, driven by the measure Θ. Then, for all $t > 0$ and stopping times $\sigma \le t$ with respect to $\mathfrak{I}$,*

$$Y_{t,W} \overset{\mathrm{D}}{=} Y_{t-\sigma,\, Y_{\sigma,W}} . \tag{4.20}$$

Proof. By Theorem 4.7(e) the process $(Y_{t,W},\, t \ge 0)$ has the strong Markov property, and hence (a) of that theorem can also applied to stopping times. Thus (4.20) follows by Definitions 4.5 and 4.2. □

Finally, we show that in the special case $d = 1$ the 0-skeleton of a STIT tessellation in W, which is a point process in the interval W, has the same distribution as a Poisson point process with appropriate intensity measure.

Lemma 4.10. *Let $d = 1$, let $W \in \mathcal{P}_1$ be a closed interval, and let Θ be given by* (4.1), *which implies $\varphi(\{-1\}) = \varphi(\{1\}) = 1/2$. Let $(Y_{t,W},\, t \ge 0)$ be the STIT tessellation process in W, driven by the measure Θ. Furthermore, let $\widehat{X}_{t,W}$ be a Poisson point process on W with intensity measure $t\,\lambda_1 \llcorner W$, and let $X_{t,W}$ be the induced tessellation of W. Then, for all $t > 0$,*

$$Y_{t,W} \overset{\mathrm{D}}{=} X_{t,W} .$$

Proof. As an auxiliary point process, consider the Poisson point process $\widehat{X}^*_W$ on $W \times \mathbb{R}^+$ with intensity measure $\lambda_1 \llcorner W \otimes \lambda_+$. Analogously to (2.64), this gives rise to the spatio-temporal process $(\widehat{X}^*_{t,W},\, t > 0)$ with

$$\widehat{X}^*_{t,W} := \{(H, s) \in \widehat{X}^*_W : s \le t\}, \quad t > 0 . \tag{4.21}$$

Let $t > 0$ be fixed. Then $\widehat{X}^*_{t,W}$ can be regarded as a marked point process with points H in W marked with their birth times $s \in (0, t]$. The projection of $\widehat{X}^*_{t,W}$ to W, that is, the deletion of the birth times, yields a Poisson point process

$$\widehat{X}_{t,W} := \{H : (H,s) \in \widehat{X}^*_{t,W}\} \tag{4.22}$$

with intensity measure $t\,\lambda_1 \llcorner W$, as in (2.65).

Now comparing Lemma 2.32 with Theorem 4.7 shows that for dimension $d = 1$, the PHT process and the STIT tessellation process in W are identically distributed. □

4.3 Rate kernel and generator

Let $W \in \mathcal{P}_d$, let Θ be a hyperplane measure of the form (4.1), satisfying (4.3), and let $(Y_{t,W},\ t \geq 0)$ be the STIT tessellation process in W, driven by the measure Θ as described in Definition 4.2. Clearly, this random process has the state space $\mathcal{T}_W$.

Now we determine the rate kernel and the generator of the pure jump Markov process $(Y_{t,W},\ t \geq 0)$. For $T \in \mathcal{T}_W$, recall from (4.17) the notation

$$\theta(T) := \sum_{P \in T} \Theta([P]) .$$

By Theorem 4.7(c) the function $\theta : \mathcal{T}_W \to \mathbb{R}$ is the **rate function** of $(Y_{t,W},\ t \geq 0)$ as defined on page 279 of [89]. Furthermore, the measure

$$\mu_T := \sum_{P \in T} \frac{\Theta([P])}{\theta(T)}\, \delta_P \otimes \Theta_{[P]} \tag{4.23}$$

is a convex combination of probability measures on $\mathcal{P}_d \times \mathcal{H}$, and μ_T has support $\bigcup_{P \in T}\{\{P\} \times [P]\}$. Because $\mathcal{T}_W$ is a Borel space (see Lemma 18.2), we have that $\{T\} \in \mathcal{B}(\mathcal{T}_W)$ for all $T \in \mathcal{T}_W$. Hence by Lemma 18.13 the map $\oslash(\cdot,\cdot,T) : \bigcup_{P \in T}\{\{P\} \times [P]\} \to \mathcal{T}_W$ defined in (4.6) is measurable in the first two variables for fixed $T \in \mathcal{T}_W$. For such $T \in \mathcal{T}_W$, define $\mathbb{P}^{\oslash}(T,\cdot)$ as the image measure of μ_T under the map $\oslash(\cdot,\cdot,T)$, that is,

$$\mathbb{P}^{\oslash}(T,\cdot) := \mu_T \circ \oslash(\cdot,\cdot,T)^{-1} . \tag{4.24}$$

More explicitly, for a nonnegative measurable function $g : \mathcal{T}_W \to \mathbb{R}$,

$$\begin{aligned} \int_{\mathcal{T}_W} g(T')\, \mathbb{P}^{\oslash}(T, \mathrm{d}T') &= \int_{\mathcal{P}_d \times \mathcal{H}} g(\oslash(P,H,T))\, \mu_T(\mathrm{d}(P,H)) \\ &= \sum_{P \in T} \frac{\Theta([P])}{\theta(T)} \int_{\mathcal{H}} g(\oslash(P,H,T))\, \Theta_{[P]}(\mathrm{d}H) . \end{aligned} \tag{4.25}$$

By Theorem 4.7(c), $\mathbb{P}^{\oslash}(\cdot,\cdot)$ is the **jump transition kernel** of $(Y_{t,W},\ t \geq 0)$ as defined on page 279 of [89], and hence the **rate kernel** is $\theta(\cdot)\, \mathbb{P}^{\oslash}(\cdot,\cdot)$.

In addition, the time-homogeneous pure jump Markov process $(Y_{t,W}, t \geq 0)$ can also be characterized by its generator. A general relation between the bounded rate kernel and the generator of a pure jump process with a measurable state space is given in Chapter 17 of [89]. However, note that the rate kernel for a STIT tessellation process is not bounded. Here we mainly refer to [47], where it is assumed that the state space is a Borel space and not necessarily a Polish space, as it is often required in many other references. By Lemma 18.2 the space $\mathcal{T}_W$ is indeed a Borel space.

Lemma 4.11. *Let $W \in \mathcal{P}_d$, let Θ be a hyperplane measure of the form* (4.1)*, satisfying* (4.3)*, and let $(Y_{t,W}, t \geq 0)$ be the STIT tessellation process in W, driven by the measure Θ. The generator $\widehat{\mathbb{L}}$ of $(Y_{t,W}, t \geq 0)$ is given by*

$$\begin{aligned}\widehat{\mathbb{L}}g(T) &= \sum_{P\in T} \int_{[P]} (g(\oslash(P,H,T)) - g(T))\,\Theta(\mathrm{d}H)\\ &= \int_{[W]} \sum_{P\in T} \mathbb{1}\{H \in [P]\}\,(g(\oslash(P,H,T)) - g(T))\,\Theta(\mathrm{d}H)\end{aligned} \tag{4.26}$$

for bounded and measurable $g : \mathcal{T}_W \to \mathbb{R}$ and $T \in \mathcal{T}_W$.

Proof. The statement can be shown referring to the definition of the generator of a time-homogeneous Markov process; see Section 14.2 of [47] and the infinitesimal description in Corollary 15.29 of [21] of a pure jump process. Recall that in (4.8) the STIT division process $(Y_{t,T}, t \geq 0)$ with initial tessellation $T \in \mathcal{T}_W$ was introduced. Now using Theorem 4.7 and (4.25) yields

$$\begin{aligned}\widehat{\mathbb{L}}g(T) &= \lim_{s\downarrow 0} \frac{\mathbb{E}g(Y_{s,T}) - g(T)}{s}\\ &= \lim_{s\downarrow 0} \frac{(\theta(T)\,s + o(s))\,(\int_{\mathcal{T}_W} g(T')\,\mathbb{P}^{\oslash}(T,\mathrm{d}T') - g(T))}{s}\\ &= \theta(T) \int_{\mathcal{T}_W} (g(T') - g(T))\,\mathbb{P}^{\oslash}(T,\mathrm{d}T')\\ &= \theta(T) \sum_{P\in T} \int_{[P]} \frac{\Theta([P])}{\theta(T)} (g(\oslash(P,H,T)) - g(T))\,\Theta_{[P]}(\mathrm{d}H)\\ &= \sum_{P\in T} \int_{[P]} g(\oslash(P,H,T)) - g(T)\,\Theta(\mathrm{d}H)\\ &= \sum_{P\in T} \int_{[W]} \mathbb{1}\{H \in [P]\}\,(g(\oslash(P,H,T)) - g(T))\,\Theta(\mathrm{d}H)\,.\end{aligned}$$

This completes the proof. □

The generator of the STIT tessellation process will be used in Part II of this monograph to associate natural classes of martingales to it. For that purpose, we also need to consider the **time-augmented** process $((Y_{t,W}, t),\ t \geq 0)$, where additionally a function depending on $t \geq 0$ can be taken into account. We denote by $\frac{\mathrm{d}^+}{\mathrm{d}t}$ the right-hand derivative of a function of t.

Lemma 4.12. *With the assumptions of Lemma 4.11 and* $\widehat{\mathbb{L}}$ *as described in* (4.26), *the generator* $\widehat{\mathbb{L}^*}$ *of the time-augmented process* $((Y_{t,W}, t),\ t \geq 0)$ *is given by*

$$\widehat{\mathbb{L}^*}(g(T)\,b(t)) = b(t)\,\widehat{\mathbb{L}}g(T) + g(T)\,\frac{\mathrm{d}^+}{\mathrm{d}t}b(t) \tag{4.27}$$

for bounded and measurable $g : \mathcal{T}_W \to \mathbb{R}$, $T \in \mathcal{T}_W$, *and right-hand differentiable* $b : [0,\infty) \to \mathbb{R}$.

Proof. With the same method as in the proof of Lemma 4.11, we obtain

$$\begin{aligned}
&\widehat{\mathbb{L}^*}(g(T)\,b(t)) \\
&= \lim_{s\downarrow 0} \frac{\mathbb{E}g(Y_{s,T})\,b(t+s) - g(T)\,b(t)}{s} \\
&= \lim_{s\downarrow 0} \frac{\mathbb{E}g(Y_{s,T})\,b(t+s) - g(T)\,b(t+s) + g(T)\,b(t+s) - g(T)\,b(t)}{s},
\end{aligned}$$

which yields (4.27). □

4.4 Summary of equivalent characterizations

While in Section 4.3 we considered the dynamics of STIT tessellation processes in incremental time intervals, in Lemma 4.13 below we describe the distribution of entire trajectories of these processes. Finally, we summarize equivalent definitions of the STIT tessellation process in Theorem 4.14.

Consider the measurable space $(\mathcal{T}_W, \mathcal{B}(\mathcal{T}_W))$ and on $\mathcal{T}_W^{[0,\infty)}$ the standard σ-algebra $\mathcal{B}^{\mathrm{fin}}(\mathcal{T}_W^{[0,\infty)})$ generated by the one-dimensional projections, which map a trajectory $(T_t,\ t \geq 0) \in \mathcal{T}_W^{[0,\infty)}$ onto T_s for fixed $s \geq 0$. This setting is sufficient for the statement that the finite-dimensional distributions determine the distribution of a stochastic process on $[0,\infty)$ with state space $(\mathcal{T}_W, \mathcal{B}(\mathcal{T}_W))$; see Proposition 4.2 of [89]. For $t > 0$, let $\mathcal{B}^{\mathrm{fin}}(\mathcal{T}_W^{[0,\infty)})_{[0,t]}$ be the sub-σ-algebra generated by the projections $(T_u,\ u \geq 0) \mapsto T_s$, $0 \leq s \leq t$. Note that if $(Y_{s,W},\ s \geq 0)$ is a STIT tessellation process in $W \in \mathcal{P}_d$, then for fixed $t > 0$ and $A \in \mathcal{B}^{\mathrm{fin}}(\mathcal{T}_W^{[0,\infty)})_{[0,t]}$, the event $\{(Y_{s,W},\ s \geq 0) \in A\}$ depends only on the trajectory $(Y_{s,W},\ 0 \leq s \leq t)$ until time t. Therefore, for $A \in \mathcal{B}^{\mathrm{fin}}(\mathcal{T}_W^{[0,\infty)})_{[0,t]}$ and $A_t := \{(m_s,\ 0 \leq s \leq t) \in \mathcal{T}_W^{[0,t]} : (m_s,\ s \geq 0) \in A\}$, we have $\mathbb{P}((Y_{s,W},\ s \geq 0) \in A) = \mathbb{P}((Y_{s,W},\ 0 \leq s \leq t) \in A_t)$. As the STIT tessellation process in W is a nonexploding pure jump process according to Theorem 4.7(e), the event A_t can be partitioned as follows:

$$A_t = \bigcup_{n=0}^{\infty} A_{t,n} \cup \widetilde{A}$$

with

$$A_{t,n} := \{(m_s,\ 0 \le s \le t) \in A_t \ : \ \text{exactly } n \text{ jumps in } [0,t]\}, \quad n \in \mathbb{N}_0\,,$$

where $\widetilde{A}$ is some null-event, that is, $\mathbb{P}((Y_{s,W},\ 0 \le s \le t) \in \widetilde{A}) = 0$. Furthermore, for fixed $n \ge 1$, there is the map $\mathcal{J} : A_{t,n} \to (\mathcal{T}_W \times [0,t])^n$ with

$$\mathcal{J}((m_s,\ 0 \le s \le t)) := ((m_{s_1}, s_1), \dots, (m_{s_n}, s_n))\,, \tag{4.28}$$

where $0 < s_1 < \cdots < s_n$ are the jump times, and $m_{s_1}, \dots, m_{s_n}$ are the states at the respective jump times. With a slight abuse of notation, we also write with the same symbol $\mathcal{J}(A_{t,n}) := \{\mathcal{J}(\mathbf{m}) : \mathbf{m} \in A_{t,n}\}$.

To simplify the following formulae, denote $T_0 := \{W\}$ and $T_j := \oslash(P_j, H_j, T_{j-1}), j \in \{1, \dots, n\}$, for $P_j \in T_{j-1}$ and $H_j \in [P_j]$.

Lemma 4.13. *Let $W \in \mathcal{P}_d$, and let Θ be a hyperplane measure of the form* (4.1), *satisfying* (4.3). *Then the distribution of the STIT tessellation process $(Y_{t,W},\ t \ge 0)$ in W, driven by the measure Θ, is given by $Y_{0,W} = \{W\}$ and for all $t > 0$ and $A \in \mathcal{B}^{\text{fin}}(\mathcal{T}_W^{[0,\infty)})_{[0,t]}$ by*

$$\begin{aligned}
&\mathbb{P}((Y_{s,W},\ s \ge 0) \in A\,) \\
&\quad = \mathrm{e}^{-t\,\Theta([W])}\ \mathbb{1}\{(Y_{s,W},\ s \ge 0) \in A \ \text{and}\ Y_{t,W} = \{W\}\} \\
&\qquad + \sum_{n=1}^{\infty} \int \cdots \int_{0<s_1<s_2<\cdots<s_n<t} \int_{[W]} \sum_{P_2 \in T_1} \int_{[P_2]} \cdots \sum_{P_n \in T_{n-1}} \int_{[P_n]} \\
&\qquad \mathrm{e}^{-(t-s_n)\,\theta(T_n)}\, \mathrm{e}^{-(s_n-s_{n-1})\,\theta(T_{n-1})} \dots \mathrm{e}^{-s_1\,\theta(\{W\})} \\
&\qquad \times \mathbb{1}\{((T_1, s_1), \dots, (T_n, s_n)) \in \mathcal{J}(A_{t,n})\}\, \Theta(\mathrm{d}H_n) \dots \Theta(\mathrm{d}H_1)\, \mathrm{d}s_n \dots \mathrm{d}s_1
\end{aligned}$$

with $s_0 = 0$ and $\mathcal{J}$ defined by (4.28).

Proof. Recall that σ_i denotes the ith jump time of the process $(Y_{s,W},\ s \ge 0)$ introduced in Definition 4.2. Put $\sigma_0 := 0$. Then

$$\begin{aligned}
\mathbb{P}((Y_{s,W},\ s \ge 0) \in A) &= \mathbb{P}((Y_{s,W},\ 0 \le s \le t) \in A_t) \\
&= \sum_{n=0}^{\infty} \mathbb{P}((Y_{s,W},\ 0 \le s \le t) \in A_{t,n}) \\
&= \mathbb{1}\{A_{t,0} \ne \emptyset\}\, \mathbb{P}(\sigma_1 > t) \\
&\quad + \sum_{n=1}^{\infty} \mathbb{P}(((Y_{\sigma_1,W}, \sigma_1), \dots, (Y_{\sigma_n,W}, \sigma_n)) \in J(A_{t,n}),\ \sigma_{n+1} > t)\,.
\end{aligned}$$

The joint distribution of $((Y_{\sigma_1,W}, \sigma_1), \dots, (Y_{\sigma_n,W}, \sigma_n))$ can be disintegrated iteratively according to Theorem 2.21. By the Markov property of the STIT tessellation process it is sufficient to consider the conditional distributions

$$\mathbb{P}_{(Y_{\sigma_i,W},\,\sigma_i)\,|\,(Y_{s_{i-1},W}=T_{i-1},\,\sigma_{i-1}=s_{i-1})}\,, \qquad i \in \{1,\dots,n\}\,.$$

To describe these conditional distributions, we apply Theorem 4.7 and use the rate kernel $\theta(T_{i-1})\,\mathbb{P}^{\oslash}(T_{i-1},\cdot)$ introduced in (4.25). This yields that for $B_i \in \mathcal{B}(\mathcal{T}_W)\otimes\mathcal{B}([0,t])$,

$$\begin{aligned}
&\int_{\mathcal{T}_W\times[0,\infty)} \mathbb{1}\{(T_i,s_i)\in B_i\}\,\mathbb{P}_{(Y_{\sigma_i,W},\,\sigma_i)\,|\,(Y_{s_{i-1},W}=T_{i-1},\,\sigma_{i-1}=s_{i-1})}(\mathrm{d}(T_i,s_i))\\
&\quad= \int_{s_{i-1}}^{t} \theta(T_{i-1})\,\mathrm{e}^{-(s_i-s_{i-1})\,\theta(T_{i-1})}\\
&\qquad\times \int_{\mathcal{H}} \sum_{P_i\in T_{i-1}} \frac{\Theta([P_i])}{\theta(T_{i-1})}\,\mathbb{1}\{(\oslash(P_i,H_i,T_{i-1}),s_i)\in B_i\}\,\Theta_{[P_i]}(\mathrm{d}H_i)\,\mathrm{d}s_i\\
&\quad= \int_{s_{i-1}}^{t} \mathrm{e}^{-(s_i-s_{i-1})\,\theta(T_{i-1})}\\
&\qquad\times \sum_{P_i\in T_{i-1}} \int_{\mathcal{H}} \mathbb{1}\{H_i\in[P_i]\}\,\mathbb{1}\{(\oslash(P_i,H_i,T_{i-1}),s_i)\in B_i\}\,\Theta(\mathrm{d}H_i)\,\mathrm{d}s_i\,.
\end{aligned}$$

Furthermore, the conditional probability that $\sigma_{n+1} > t$, given $(Y_{s_n,W} = T_n,\, \sigma_n = s_n)$, is $\exp(-(t-s_n)\,\theta(T_n))$. For fixed $n \ge 1$, we thus obtain

$$\begin{aligned}
&\mathbb{P}(((Y_{\sigma_1,W},\sigma_1),\dots,(Y_{\sigma_n,W},\sigma_n))\in\mathcal{J}(A_{t,n}),\ \sigma_{n+1}>t)\\
&\quad= \int\!\cdots\!\int_{0<s_1<s_2<\cdots<s_n<t} \int_{[W]} \sum_{P_2\in T_1}\int_{[P_2]} \cdots \sum_{P_n\in T_{n-1}}\int_{[P_n]}\\
&\qquad \mathrm{e}^{-(t-s_n)\,\theta(T_n)}\,\mathrm{e}^{-(s_n-s_{n-1})\,\theta(T_{n-1})}\dots\mathrm{e}^{-s_1\,\theta(\{W\})}\\
&\qquad\times \mathbb{1}\{((\oslash(W,H_1,\{W\}),s_1),\dots,(\oslash(P_n,H_n,T_{n-1}),s_n))\in\mathcal{J}(A_{t,n})\}\\
&\qquad \Theta(\mathrm{d}H_n)\dots\Theta(\mathrm{d}H_1)\,\mathrm{d}s_n\dots\mathrm{d}s_1\,.
\end{aligned}$$

Summarizing, this yields

$$\begin{aligned}
&\mathbb{P}((Y_{s,W},\,s\ge 0)\in A)\\
&\quad= \mathrm{e}^{-t\,\Theta([W])}\,\mathbb{1}\{(Y_{s,W},\,s\ge 0)\in A \text{ and } Y_{t,W}=\{W\}\}\\
&\qquad+ \sum_{n=1}^{\infty} \int\!\cdots\!\int_{0<s_1<s_2<\cdots<s_n<t} \int_{[W]} \sum_{P_2\in T_1}\int_{[P_2]} \cdots \sum_{P_n\in T_{n-1}}\int_{[P_n]}\\
&\qquad \mathrm{e}^{-(t-s_n)\,\theta(T_n)}\,\mathrm{e}^{-(s_n-s_{n-1})\,\theta(T_{n-1})}\dots\mathrm{e}^{-s_1\,\theta(\{W\})}\\
&\qquad\times \mathbb{1}\{((T_1,s_1),\dots,(T_n,s_n))\in\mathcal{J}(A_{t,n})\}\,\Theta(\mathrm{d}H_n)\dots\Theta(\mathrm{d}H_1)\,\mathrm{d}s_n\dots\mathrm{d}s_1
\end{aligned}$$

with $s_0 = 0$. The proof is thus complete. □

To summarize, we now have four equivalent definitions of STIT tessellation processes in a window. Regarding item (a) of the following theorem, note that in Remark 4.8 an alternative algorithm is given for the same construction of a STIT tessellation process introduced in Definition 4.2.

Theorem 4.14. *Let $W \in \mathcal{P}_d$, let Θ be a hyperplane measure of the form* (4.1), *satisfying* (4.3), *and let $(Y_{t,W}, t \geq 0)$ be a $\mathcal{T}_W$-valued stochastic process. Then the following statements are equivalent.*

(a) *$(Y_{t,W}, t \geq 0)$ is a STIT tessellation process in W, driven by Θ in the sense of Definition* 4.2.

(b) *$(Y_{t,W}, t \geq 0)$ is a time-homogeneous pure jump Markov process in $\mathcal{T}_W$ with $Y_{0,W} = \{W\}$ and rate kernel $\theta(\cdot)\,\mathbb{P}^{\oslash}(\cdot,\cdot)$; see* (4.17) *and* (4.24).

(c) *$(Y_{t,W}, t \geq 0)$ is a time-homogeneous pure jump Markov process in $\mathcal{T}_W$ with $Y_{0,W} = \{W\}$ and generator $\widehat{\mathbb{L}}$, given in Lemma* 4.11.

(d) *$(Y_{t,W}, t \geq 0)$ is a stochastic process with the distribution given in Lemma* 4.13.

Proof. By Theorem 4.7(c) and (4.24), (b) is a consequence of (a). By Theorems 13.3 and 13.4 of [89] the distribution of $(Y_{t,W}, t \geq 0)$ is uniquely determined by the rate kernel, and furthermore, it follows that (b) implies (a). The fact that (a) implies (c) is the content of Lemma 4.11. To show that (c) implies (a), first note that by Lemma 18.2 the measurable space $(\mathcal{T}_W, \mathcal{B}(\mathcal{T}_W))$ is a Borel space, which means that there exist a Borel set $B \subseteq [0,1]$ and a bimeasurable bijection $f : \mathcal{T}_W \to B$. Thus Proposition 15.38 of [21] can be applied, which holds for pure jump processes that have a Borel subset of $\mathbb{R}$ as their state space. Finally, Lemma 4.13 yields that (a) is equivalent to (d). This completes the argument. □

These four ways of defining the STIT tessellation process in a window W, together with the spatial consistency that we establish in Theorem 5.10, provide equivalent definitions of the STIT tessellation process in $\mathbb{R}^d$. A fifth description is given in Chapter 8 by a “global” construction.

4.5 Notes

1. The STIT tessellation model in a window, as described in Definition 4.2, was originally introduced in [145]. Essentially, Theorem 4.7 was shown ibidem.
2. In [168] a cell division model was applied in the context of machine learning (see Section 9.3.3), and the name “Mondrian process” was coined. It was developed independently, but it is exactly the STIT tessellation process driven by the Mondrian hyperplane measure (4.4) with $p_1 = \cdots = p_d$. That model was motivated by the search for a partition of a parallelepiped in arbitrary dimension, which corresponds to the partition of a line segment by a Poisson point process. In that paper, also the spatial consistency of the Mondrian model (see Definition 5.9(a), there referred to as “self-consistency”), is shown. No connection between the particular Mondrian tessellation and the general STIT tessellation model was indicated. In Section 7 of [168], it is mentioned that this Mondrian tessellation process has been extensively studied in combinatorics and computational geometry, where it is known as the “guillotine partition”.

3. Lemma 4.9 was shown for a fixed stopping time in Lemma 2 of [145].
4. The generator of the STIT tessellation process was first considered and applied in [174, 176, 177].
5. The description of the distribution of the STIT tessellation process in a window, given in Lemma 4.13, is a specification of propositions on page 1918 and Lemma 4.1 of [58], where a corresponding statement is shown for an essentially wider class of cell splitting processes.
6. The construction described in Remark 4.8 was provided in [116]. In Section 9 ibidem a further alternative construction of the STIT tessellation process in a window is described.

5 Fundamental properties of the STIT tessellation process in a window

This chapter begins with the study of the cells containing a fixed convex body $K \in \mathcal{K}$, the so-called K-cells. As a key result, it is shown that in any window $W \in \mathcal{P}_d$ with $K \subset W$, the process of the K-cells belonging to a PHT process and the one belonging to a STIT tessellation process in W are identically distributed if both tessellation processes are driven by the same hyperplane measure Θ.

This proposition also leads to the first step of the calculation of the capacity functional for the $(d-1)$-skeleton of the STIT tessellations in a window W, which in turn gives rise to the property of spatial consistency of these tessellations. This is the essential argument to show the existence of random stationary tessellations of the space $\mathbb{R}^d$ for which the restriction to any window has the same distribution as the STIT tessellation constructed in W.

Finally, the remarkable result is shown that in a class of cell division processes with rather general life time distributions and division rules, the STIT tessellations are the only ones that are spatially consistent.

The statements of this chapter prepare proofs of the fundamental features of STIT tessellations in $\mathbb{R}^d$, which are shown in Chapter 6. Moreover, the considerations in the present chapter facilitate a deeper understanding of the nature of the STIT tessellation process, and they can help to comprehend the idea of the global construction, which is described in Chapter 8.

5.1 Cells that contain a fixed convex body

We begin with an important relationship between STIT tessellation processes and PHT processes that play a role throughout the book. For a fixed convex body $K \in \mathcal{K}$, we consider the processes of the cells containing K, the so-called K-cells. If $K = \{o\}$, then these are the zero-cells of the respective tessellations. Note that in this context, it makes no difference whether a convex body K or any compact set C with $\operatorname{conv} C = K$ is considered, because the cells of the tessellations are convex. For simplicity, we restrict ourselves to $K \in \mathcal{K}$.

5.1.1 The PHT process

Recall from Section 2.3.1 the concept of a PHT process driven by a hyperplane measure Θ of the form (4.1), satisfying (4.3). Let $\widehat{X}^*$ be a Poisson point process on $\mathcal{H} \times \mathbb{R}^+$ with intensity measure $\Theta \otimes \lambda_+$, where λ_+ is the Lebesgue measure on $\mathbb{R}^+$. Define for $t > 0$,

$$\widehat{X}_t^* := \{(H, s) \in \widehat{X}^* : s \le t\}$$

https://doi.org/10.1515/9783112222409-006

and

$$\widehat{X}_t := \{H : (H,s) \in \widehat{X}_t^*\}.$$

For each $t > 0$, the Poisson hyperplane process $\widehat{X}_t$ induces the PHT X_t, and we thus obtain the process $(X_t, t > 0)$ of PHTs driven by Θ.

Fix a convex body $K \in \mathcal{K}$ and note that for $t > 0$,

$$\mathbb{P}(\partial X_t \cap K = \emptyset) = \mathrm{e}^{-t\,\Theta([K])}. \tag{5.1}$$

This follows from Definition 19.10(a) and the fact that the intensity measure of the Poisson hyperplane process $\widehat{X}_t$ is $t\,\Theta$.

Now we consider the cell $Z^{\mathrm{PHT}}_{\langle K\rangle,t} \in X_t$ for which $K \subseteq Z^{\mathrm{PHT}}_{\langle K\rangle,t}$ if such a cell exists, and we call it the K**-cell of** X_t; see Figure 5.1. Note that by the stationarity of the PHT X_t the K-cell is a. s. uniquely determined if it exists. Denote

$$\beta_K^{\mathrm{PHT}} := \min\{s > 0 : (H,s) \in \widehat{X}^*, H \in [K]\},$$

which is the random time of the first section of K by a hyperplane if the minimum is taken of a nonempty set; otherwise, $\beta_K^{\mathrm{PHT}} := \infty$. If β_K^{PHT} is finite, then denote by H_K^{PHT} the hyperplane of the a. s. unique pair $(H_K^{\mathrm{PHT}}, \beta_K^{\mathrm{PHT}}) \in \widehat{X}^*$. This yields a $\mathcal{P}_d$-valued stochastic process

$$(Z^{\mathrm{PHT}}_{\langle K\rangle,t}, 0 < t < \beta_K^{\mathrm{PHT}}) \quad \text{with } K \subseteq Z^{\mathrm{PHT}}_{\langle K\rangle,t} \in X_t,$$

the K**-cell process of** $(X_t, t > 0)$.

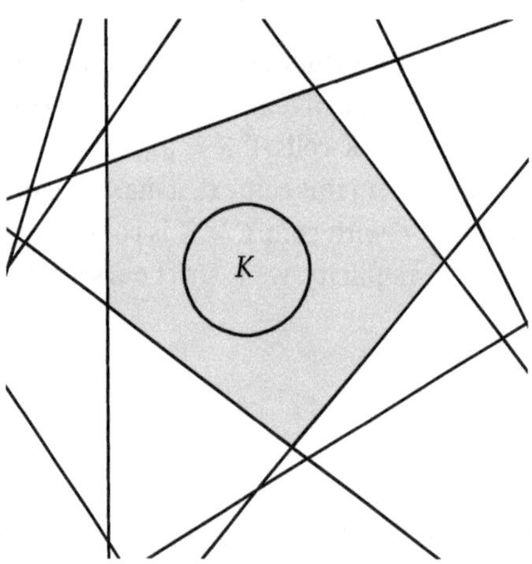

Figure 5.1: The K-cell $Z^{\mathrm{PHT}}_{\langle K\rangle,t}$ of a Poisson hyperplane tessellation X_t. According to the definition used in this volume, such a cell exists only under the condition that K is not hit by a hyperplane of $\widehat{X}_t^*$.

For a fixed $W \in \mathcal{P}_d$ with $K \subseteq W$, consider the process $(X_t \wedge W, t \geq 0)$ (see (2.20)) with $X_0 \wedge W := \{W\}$ of restricted Poisson hyperplane tessellations and for it, the K**-cell**

process of $(X_t,\ t \ge 0)$ **restricted to** W,

$$(Z^{\mathrm{PHT}}_{\langle K\rangle,t,W}\,,\ 0 \le t < \beta^{\mathrm{PHT}}_K) \tag{5.2}$$

with $Z^{\mathrm{PHT}}_{\langle K\rangle,0,W} := W$ and $Z^{\mathrm{PHT}}_{\langle K\rangle,t,W} := Z^{\mathrm{PHT}}_{\langle K\rangle,t} \cap W$ for $t > 0$.

The process $(Z^{\mathrm{PHT}}_{\langle K\rangle,t,W}\,,\ 0 \le t < \beta^{\mathrm{PHT}}_K)$ is a pure jump process because it alters its state only at a discrete set of jump times β_i, $i \in \mathbb{N}$, which are defined as follows:

$$\beta^{\mathrm{PHT}}_1 := \min\{s < \beta^{\mathrm{PHT}}_K : (H,s) \in \widehat{X}^*,\ H \in [W]\},$$

and if the minimum is taken of a nonempty set, then H^{PHT}_1 is the a. s. uniquely determined hyperplane such that $(H^{\mathrm{PHT}}_1, \beta^{\mathrm{PHT}}_1) \in \widehat{X}^*$. Then, for $i > 1$,

$$\beta^{\mathrm{PHT}}_i := \min\{s : \beta^{\mathrm{PHT}}_{i-1} < s < \beta^{\mathrm{PHT}}_K : (H,s) \in \widehat{X}^*,\ H \in [Z^{\mathrm{PHT}}_{\langle K\rangle,\beta_{i-1},W}]\},$$

and H^{PHT}_i is the a. s. uniquely determined hyperplane with $(H^{\mathrm{PHT}}_i, \beta^{\mathrm{PHT}}_i) \in \widehat{X}^*$. This definition is applied to all $i \in \mathbb{N}$ for which the minimum is taken of a nonempty set. Otherwise, β^{PHT}_i is not defined.

Let ν denote the random variable with values in $\mathbb{N}_0 \cup \{\infty\}$ equal to the maximal index i such that β^{PHT}_i is defined. If the first hyperplane H that intersects W also intersects K, then we put $\nu := 0$.

Lemma 5.1. *Let $K \in \mathcal{K}$ and $W \in \mathcal{P}_d$ with $K \subseteq W$.*

(a) *The process $(Z^{\mathrm{PHT}}_{\langle K\rangle,t}\,,\ 0 < t < \beta^{\mathrm{PHT}}_K)$ is a Markov process on $\mathcal{P}_d$. If $\Theta([K]) > 0$, then the law of β^{PHT}_K is $\mathrm{Expo}(\Theta([K]))$. Furthermore, $\beta^{\mathrm{PHT}}_K = \infty$ a. s. if and only if K is a singleton.*

(b) *The process $(Z^{\mathrm{PHT}}_{\langle K\rangle,t,W}\,,\ 0 \le t < \beta^{\mathrm{PHT}}_K)$ is a pure jump Markov process and thus has the strong Markov property. Furthermore, for $i < \nu$, given β_i and $Z^{\mathrm{PHT}}_{\langle K\rangle,\beta_i,W}$, the joint conditional distribution of $(H_{i+1}, \beta_{i+1} - \beta_i)$ is $\Theta_{[Z^{\mathrm{PHT}}_{\langle K\rangle,\beta_i,W}]} \otimes \mathrm{Expo}(\Theta([Z^{\mathrm{PHT}}_{\langle K\rangle,\beta_i,W}]))$, and thus the process is time-homogeneous.*

Proof. (a) The Markov property follows by the independence property in Definition 19.10(b). The law of β^{PHT}_K results from Definition 19.10(a). If K is a singleton, then (4.1) for the translation-invariant measure Θ implies that $\Theta([K]) = 0$. If $K \in \mathcal{K}$ but is not a singleton, then Convention 4.1 on Θ yields that $\Theta([K]) > 0$.

(b) It is shown in Theorem 13.1 of [89] that a pure jump Markov process has the strong Markov property. The joint conditional distribution of $(H_{i+1}, \beta_{i+1} - \beta_i)$ results from the strong Markov property and Lemma 2.32. □

This Lemma 5.1, together with Theorem 4.7, forms the base for Lemma 5.2, where it will be shown that the respective K-cell processes of PHTs and of STIT tessellations, both restricted to a window W, are identically distributed. In this context, the case $K = \{o\}$ is of particular interest. Adapting the notation of Section 2.2.3, we write

$$(Z^{\mathrm{PHT}}_{\langle o\rangle,t}\,,\ t > 0) \quad \text{with } o \in Z^{\mathrm{PHT}}_{\langle o\rangle,t} \text{ and } Z^{\mathrm{PHT}}_{\langle o\rangle,t} \in X_t \tag{5.3}$$

for the **zero-cell process of** $(X_t,\ t > 0)$. This process plays a central role in Chapter 8 for the global construction of STIT tessellation processes. Furthermore, a tail triviality property for this process is proven in Section 15.5.

5.1.2 The STIT tessellation process in a window

Now we deal with the K-cell process of a STIT tessellation process. Let Θ be a hyperplane measure of the form (4.1), satisfying (4.3), and let $(Y_{t,W},\ t \geq 0)$ be the STIT tessellation process in $W \in \mathcal{P}_d$, driven by the measure Θ, as described in Definition 4.2. Let $\mathcal{D}(Y_{t,W},\ t \geq 0)$ be the associated process of division events defined in (4.9). Recall that the division events are triples $(P, H, \beta) \in \mathcal{P}_d \times \mathcal{H} \times \mathbb{R}^+$, which express that a cell P of the STIT tessellation is divided by a hyperplane H at time β.

For a convex body $K \in \mathcal{K}$ with $K \subseteq W$, we consider the K-cell $Z^{\mathrm{STIT}}_{\langle K\rangle,t,W}$ of $Y_{t,W}$, that is, $K \subseteq Z^{\mathrm{STIT}}_{\langle K\rangle,t,W}$ if it exists. If $o \in W$ and $K = \{o\}$, then this is the zero-cell, denoted $Z^{\mathrm{STIT}}_{\langle o\rangle,t,W}$. Define

$$\beta^{\mathrm{STIT}}_{K,W} := \min\{\beta(H) : (P, H, \beta(H)) \in \mathcal{D}(Y_{t,W},\ t \geq 0),\ H \in [K]\},$$

which is the time at which K is intersected for the first time by a dividing hyperplane if the minimum is taken over a nonempty set; otherwise, $\beta^{\mathrm{STIT}}_{K,W} := \infty$. If $\beta^{\mathrm{STIT}}_{K,W}$ is finite, then $H^{\mathrm{STIT}}_{K,W}$ denotes the a. s. unique hyperplane with $\beta^{\mathrm{STIT}}_{K,W} = \beta(H^{\mathrm{STIT}}_{K,W})$ in the triple $(P, H^{\mathrm{STIT}}_{K,W}, \beta(H^{\mathrm{STIT}}_{K,W})) \in \mathcal{D}(Y_{t,W},\ t \geq 0)$. This gives rise to the process

$$(Z^{\mathrm{STIT}}_{\langle K\rangle,t,W},\ 0 \leq t < \beta^{\mathrm{STIT}}_{K,W}) \quad \text{with } K \subseteq Z^{\mathrm{STIT}}_{\langle K\rangle,t,W} \text{ and } Z^{\mathrm{STIT}}_{\langle K\rangle,t,W} \in Y_{t,W}\,. \tag{5.4}$$

By Definition 4.2 and Theorem 4.7 the process $(Z^{\mathrm{STIT}}_{\langle K\rangle,t,W},\ 0 \leq t < \beta^{\mathrm{STIT}}_{K,W})$ is a time-homogeneous pure jump Markov process. It terminates at time $\beta^{\mathrm{STIT}}_{K,W}$ when the set K is intersected by the hyperplane $H^{\mathrm{STIT}}_{K,W}$.

The result of the following lemma can be interpreted for the particular case $K = \{o\}$ in such a way that an observer sitting in the origin o and only observing the process of the interior of the zero-cell cannot distinguish between a PHT process and a STIT tessellation process.

Note that the distribution of $(H^{\mathrm{STIT}}_{K,W}, \beta(H^{\mathrm{STIT}}_{K,W}))$ does not depend on the window W for general $K \in \mathcal{K}$ if $K \subseteq W$.

Lemma 5.2. *Let $W \in \mathcal{P}_d$ and $K \in \mathcal{K}$ with $K \subseteq W$. Furthermore, let Θ be a hyperplane measure of the form* (4.1), *satisfying* (4.3), *let $(Y_{t,W},\ t \geq 0)$ be the STIT tessellation process in W, and let $(X_t,\ t > 0)$ the PHT process, both driven by Θ. Then*

$$(Z^{\mathrm{STIT}}_{\langle K\rangle,t,W},\ 0 \leq t < \beta^{\mathrm{STIT}}_{K,W}) \stackrel{\mathrm{D}}{=} (Z^{\mathrm{PHT}}_{\langle K\rangle,t,W},\ 0 \leq t < \beta^{\mathrm{PHT}}_{K})\,. \tag{5.5}$$

Moreover, if $\Theta([K]) > 0$, *then*

$$(H_{K,W}^{\text{STIT}}, \beta(H_{K,W}^{\text{STIT}})) \overset{\text{D}}{=} (H_K^{\text{PHT}}, \beta_K^{\text{PHT}})$$

with the law $\Theta_{[K]} \otimes \text{Expo}(\Theta([K]))$.

Proof. Both processes in (5.5) are pure jump Markov processes. The identity of their distributions follows by Theorem 4.7(a) and Lemma 5.1(b). In particular, at any division step, also the probabilities that the dividing hyperplanes intersect K are identical for STIT and PHT, and therefore $\beta_{K,W}^{\text{STIT}} \overset{\text{D}}{=} \beta_{K,W}^{\text{PHT}}$ with distribution $\text{Expo}(\Theta([K]))$. □

5.2 Capacity functional

So far, in particular in Theorem 4.14, we described the tessellation-valued continuous-time stochastic process $(Y_{t,W}, t \geq 0)$ for some fixed $W \in \mathcal{P}_d$. We now also fix a time $t > 0$ and study the distribution of the random tessellation $Y_{t,W}$ in W. By Lemma 18.9 the map $\text{skel}_{d-1} : \mathcal{T}_W \to \mathcal{F}$ is measurable, and therefore for a fixed time t, the $(d-1)$-skeleton $\partial Y_{t,W}$ is a random closed set. Thus the distribution of $\partial Y_{t,W}$ is completely characterized by its capacity functional; see (2.23) and Lemma 2.16. Moreover, the capacity functional will turn out to be a useful device to derive some of the fundamental properties of STIT tessellations. In particular, we use it in Section 5.3 to show the spatial consistency of STIT tessellations, which in turn implies the existence of STIT tessellations in the space $\mathbb{R}^d$. Furthermore, in Chapter 7 the limit of the capacity functional for a sequence of repeated and rescaled iterations of an arbitrary random stationary tessellation is investigated. This leads to a uniqueness result for STIT tessellations.

As a general assumption in this section, let Θ be a hyperplane measure of the form (4.1), satisfying (4.3). Furthermore, let $W \in \mathcal{P}_d$, and let $(Y_{t,W}, t \geq 0)$ be a STIT tessellation process in W, driven by Θ. To begin with, we consider the value of the capacity functional of $\partial Y_{t,W}$ for connected compact sets $C \in \mathcal{C}$.

Lemma 5.3. *If $C \in \mathcal{C}$ is connected and $W \in \mathcal{P}_d$ with $C \subseteq W$, then*

$$\mathbb{P}(\partial Y_{t,W} \cap C = \emptyset) = \mathrm{e}^{-t\,\Theta([C])}, \tag{5.6}$$

independently of the choice of $W \supseteq C$. For all $t > 0$ for which $C \subseteq t\,W$, we have

$$\mathbb{P}(t\,\partial Y_{t,W} \cap C = \emptyset) = \mathbb{P}(\partial Y_{1,W} \cap C = \emptyset)\,. \tag{5.7}$$

Proof. Note that $[C] = [\text{conv}\,C]$ for a connected set $C \in \mathcal{C}$. Thus equation (5.6) follows immediately by Lemma 5.2 and (5.1). To show the scaling property (5.7), observe that $t\,\partial Y_{t,W} \cap C = \emptyset$ if and only if $\partial Y_{t,W} \cap t^{-1}C = \emptyset$ and that equation (4.1) applied to $f(H) = \mathbb{1}\{H \in [t^{-1}C]\}$ yields $\Theta([t^{-1}C]) = t^{-1}\Theta([C])$. Hence, using (5.6), we obtain (5.7). □

To determine the distribution of $\partial Y_{t,W}$, it is sufficient to consider the capacity functional for sets $C \in \mathcal{R}$ of the convex ring; see Lemma 2.16. Here it is useful to represent $C \in \mathcal{R}$ as the union of its connected components, which are themselves elements of the convex ring $\mathcal{R}$. For each $C \in \mathcal{R}$, there exists $k \in \mathbb{N}$ such that $C = \bigcup_{i=1}^{k} C_i$, where all $C_i \in \mathcal{C}$ are closed and connected, and $C_i \cap C_j = \emptyset$ for $i \neq j$. For a partition $\{J_1, \dots, J_l\}$ of the index set $\{1, \dots, k\}$ into l nonempty sets, denote $D := \{D_1, \dots, D_l\}$ with $D_j := \bigcup_{i \in J_j} C_i, j \in \{1, \dots, l\}$. Denote by $\mathbf{D}(C)$ the set of all such partitions D of C with $l \in \{1, \dots, k\}$. Formally,

$$\mathbf{D}(C) := \Big\{ \{D_1, \dots, D_l\} : D_j = \bigcup_{i \in J_j} C_i, j \in \{1, \dots, l\}, \\ \{J_1, \dots, J_l\} \text{ a partition of } \{1, \dots, k\}, \, l \in \{1, \dots, k\} \Big\} . \tag{5.8}$$

For $\{D_1, D_2\} \in \mathbf{D}(C)$, denote the set of all hyperplanes that separate D_1 and D_2 by

$$[D_1|D_2] := \{ H \in \mathcal{H} : D_1 \subset H^+, D_2 \subset H^- \text{ or } D_2 \subset H^+, D_1 \subset H^- \} . \tag{5.9}$$

Recall the notation $Z^{\mathrm{PHT}}_{\langle K \rangle, t, W}$ from (5.2) and $Z^{\mathrm{STIT}}_{\langle K \rangle, t, W}$ from (5.4).

Lemma 5.4. *Let $C \in \mathcal{R}$, $C \subseteq W$. Then*

$$\begin{aligned} &\mathbb{P}(\partial Y_{t,W} \cap C = \emptyset) \\ &= e^{-t\Theta([\operatorname{conv} C])} + \sum_{\{D_1, D_2\} \in \mathbf{D}(C)} \Theta([D_1|D_2]) \\ &\quad \times \int_0^t e^{-s\Theta([\operatorname{conv} C])} \, \mathbb{P}(\partial Y_{t-s, Z_{\langle D_1 \rangle, s, W}} \cap D_1 = \emptyset) \, \mathbb{P}(\partial Y_{t-s, Z_{\langle D_2 \rangle, s, W}} \cap D_2 = \emptyset) \, ds . \end{aligned}$$

Proof. The propositions used in this proof are Lemmas 4.9 and 5.2. Let $C \in \mathcal{R}$, consisting of k pairwise disjoint connected components, be such that $C \subseteq W$. For $k = 1$, we have that $\mathbf{D}(C) = \{\{C\}\}$, and the formula reduces to (5.6) in Lemma 5.3. For $k > 1$, the event $\partial Y_{t,W} \cap C = \emptyset$ is decomposed into the event that the convex hull of C is not intersected and the event that the first intersection of $\operatorname{conv} C$ at time $\beta^{\mathrm{STIT}}_{\operatorname{conv} C, W}$ by the hyperplane $H^{\mathrm{STIT}}_{\operatorname{conv} C, W}$ separates D_1 and D_2, $\{D_1, D_2\} \in \mathbf{D}(C)$.

The idea of the proof is to use Lemma 5.2, with the notation introduced in Section 5.1.2, to the STIT tessellation process until time $\beta^{\mathrm{STIT}}_{\operatorname{conv} C, W}$. Then the Mecke equation (19.4) for the auxiliary Poisson point process $\widehat{X}^*$ of birth time marked hyperplanes is applied to the pair $(H^{\mathrm{STIT}}_{\operatorname{conv} C, W}, \beta^{\mathrm{STIT}}_{\operatorname{conv} C, W})$. Because $\beta^{\mathrm{STIT}}_{\operatorname{conv} C, W}$, as a jump time of the process $(Y_{t,W}, t \geq 0)$, is a stopping time, Lemma 4.9 can be used for the time interval $(\beta^{\mathrm{STIT}}_{\operatorname{conv} C, W}, t]$. To avoid overburden symbols, we use a simplified notation in the following equations, omitting the upper indices STIT, PHT, and sometimes also W, even when the dependence on the window has to be taken into account. Furthermore, we denote $\widehat{X}_{s-} := \{H : (H, r) \in \widehat{X}^*, r < s\}$. We have

$$
\begin{aligned}
&\mathbb{P}(\partial Y_{t,W} \cap C = \emptyset) \\
&= \mathbb{P}(\partial Y_{t,W} \cap \operatorname{conv} C = \emptyset) \\
&\quad + \sum_{\{D_1,D_2\}\in\mathbf{D}(C)} \mathbb{P}(\beta_{\operatorname{conv} C} < t,\ H_{\operatorname{conv} C} \in [D_1 \mid D_2], \\
&\quad \partial Y_{t-\beta_{\operatorname{conv} C},\, Z_{\langle D_1\rangle,\beta_{\operatorname{conv} C}}} \cap D_1 = \emptyset,\ \partial Y_{t-\beta_{\operatorname{conv} C},\, Z_{\langle D_2\rangle,\beta_{\operatorname{conv} C}}} \cap D_2 = \emptyset) \\
&= \mathbb{P}(\partial Y_t \cap \operatorname{conv} C = \emptyset) \\
&\quad + \sum_{\{D_1,D_2\}\in\mathbf{D}(C)} \mathbb{E} \sum_{(H,s)\in\widehat{X}^*} \mathbb{1}\{s < t,\ H \in [D_1 \mid D_2],\ \widehat{X}_{s-} \cap [\operatorname{conv} C] = \emptyset, \\
&\quad \partial Y_{t-s, Z_{\langle D_1\rangle,s}} \cap D_1 = \emptyset,\ \partial Y_{t-s, Z_{\langle D_2\rangle,s}} \cap D_2 = \emptyset\} \\
&= \mathrm{e}^{-t\,\Theta([\operatorname{conv} C])} \\
&\quad + \sum_{\{D_1,D_2\}\in\mathbf{D}(C)} \int_{\mathcal{H}} \mathbb{1}\{H \in [D_1|D_2]\} \\
&\quad \times \int_0^t \mathrm{e}^{-s\,\Theta([\operatorname{conv} C])}\, \mathbb{P}(\partial Y_{t-s, Z_{\langle D_1\rangle,s}} \cap D_1 = \emptyset)\, \mathbb{P}(\partial Y_{t-s, Z_{\langle D_2\rangle,s}} \cap D_2 = \emptyset)\, \mathrm{d}s\, \Theta(\mathrm{d}H) \\
&= \mathrm{e}^{-t\,\Theta([\operatorname{conv} C])} + \sum_{\{D_1,D_2\}\in\mathbf{D}(C)} \Theta([D_1|D_2]) \\
&\quad \times \int_0^t \mathrm{e}^{-s\,\Theta([\operatorname{conv} C])}\, \mathbb{P}(\partial Y_{t-s, Z_{\langle D_1\rangle,s}} \cap D_1 = \emptyset)\, \mathbb{P}(\partial Y_{t-s, Z_{\langle D_2\rangle,s}} \cap D_2 = \emptyset)\, \mathrm{d}s\,,
\end{aligned}
$$

which completes the proof. □

In the following lemma, we provide a more detailed formula for the capacity functional. This also implies a key result that the capacity functional of $\partial Y_{t,W}$ for $C \in \mathcal{R}$ does not depend on the choice of a window $W \supset C$. This will be used in the proofs of Theorems 5.10 and 7.1. In the following lemma, assumption (5.10) will be used. Below, in Remark 5.7, it is explained that this condition is not an essential restriction when the capacity functional is studied in the mentioned theorems.

Lemma 5.5. *Let $C \in \mathcal{R}$, $C \subseteq W$, and assume that*

$$\Theta([\operatorname{conv} C]) \neq \sum_{D_j \in D} \Theta([\operatorname{conv} D_j]) \quad \text{for all } D \in \mathbf{D}(C) \text{ with } |D| \geq 2\,. \tag{5.10}$$

Then there exist uniquely determined $d^{\mathrm{STIT}}(C,D) \in \mathbb{R}$, $D \in \mathbf{D}(C)$, depending also on Θ, but depending neither on $t > 0$ nor on W, such that

$$\mathbb{P}(\partial Y_{t,W} \cap C = \emptyset) = \sum_{D\in\mathbf{D}(C)} d^{\mathrm{STIT}}(C,D) \prod_{D_j \in D} \exp(-t\,\Theta([\operatorname{conv} D_j]))\,. \tag{5.11}$$

Furthermore, for all $D \in \mathbf{D}(C)$,

$$d^{\mathrm{STIT}}(a\,C, a\,D) = d^{\mathrm{STIT}}(C, D) \quad \textit{for all } a > 0 \textit{ such that } a\,C \subseteq W\,, \tag{5.12}$$

and

$$d^{\mathrm{STIT}}(C + x, D + x) = d^{\mathrm{STIT}}(C, D) \quad \textit{for all } x \in \mathbb{R}^d \textit{ for which } C + x \subseteq W\,. \tag{5.13}$$

Proof. We argue via induction over the number k of connected components of C. Again, for $k = 1$, the claim follows by (5.11) in Lemma 5.3. In this case, $d^{\mathrm{STIT}}(C, \{C\}) = 1$, and hence also (5.12) and (5.13) hold.

The induction hypothesis is that (5.11), (5.12), and (5.13) are satisfied for all $C \in \mathcal{R}$ with at most k connected components and for all $D \in \mathbf{D}(C)$. Now consider $C \in \mathcal{R}$ with $k+1$ connected components. If C consists of more than one connected component, then $\Theta([\operatorname{conv} C]) > 0$, because Θ satisfies Convention 4.1. Now the induction hypothesis can be applied, and we obtain by Lemma 5.4 that

$$\begin{aligned}
&\mathbb{P}(\partial Y_{t,W} \cap C = \emptyset)\\
&= \exp(-t\,\Theta([\operatorname{conv} C]))\\
&\quad + \sum_{\{D_1,D_2\}\in\mathbf{D}(C)} \Theta([D_1|D_2]) \sum_{B_1\in\mathbf{D}(D_1)} \sum_{B_2\in\mathbf{D}(D_2)} d^{\mathrm{STIT}}(D_1, B_1)\, d^{\mathrm{STIT}}(D_2, B_2)\\
&\quad \times \int_0^t \exp(-s\,\Theta([\operatorname{conv} C])) \prod_{S\in B_1\cup B_2} \exp(-(t-s)\,\Theta([\operatorname{conv} S]))\,\mathrm{d}s
\end{aligned} \tag{5.14}$$

$$\begin{aligned}
&= \exp(-t\,\Theta([\operatorname{conv} C]))\\
&\quad + \sum_{\{D_1,D_2\}\in\mathbf{D}(C)} \sum_{B_1\in\mathbf{D}(D_1)} \sum_{B_2\in\mathbf{D}(D_2)} d^{\mathrm{STIT}}(D_1, B_1)\, d^{\mathrm{STIT}}(D_2, B_2)\\
&\quad \times \frac{\Theta([D_1|D_2])}{-\Theta([\operatorname{conv} C]) + \sum_{S\in B_1\cup B_2} \Theta([\operatorname{conv} S])}\\
&\quad \times \left(\exp(-t\,\Theta([\operatorname{conv} C])) - \exp\left(-t \sum_{S\in B_1\cup B_2} \Theta([\operatorname{conv} S]) \right) \right),
\end{aligned} \tag{5.15}$$

where all the denominators are different from zero by assumption (5.10).

A rearrangement of the items proves (5.11), which shows that the coefficients d^{STIT} do depend neither on W nor on t and that they are uniquely determined. Notice that a set $B_1 \cup B_2$ can be generated by different $\{D_1, D_2\} \in \mathbf{D}(C)$, $B_1 \in \mathbf{D}(D_1)$, $B_2 \in \mathbf{D}(D_2)$.

By (4.1) we obtain that $\Theta([a\,D_1|a\,D_2]) = a\,\Theta([D_1|D_2])$ and $\Theta([(a\,A)]) = a\,\Theta([A])$ for all $a > 0$ and $A \in \mathcal{B}(\mathbb{R}^d)$. Thus (5.15) and the induction hypothesis yield (5.12). The translation invariance (5.13) follows from the induction hypothesis and the translation invariance of Θ. □

If $C \in \mathcal{C}$ consists of exactly two connected components, then the capacity functional can easily be calculated explicitly using (5.14), even if condition (5.10) is not satisfied.

Corollary 5.6. *If $C = C_1 \cup C_2 \subseteq W$ with exactly two connected components $C_1, C_2 \in \mathcal{C}$, then for $t > 0$,*

$$\mathbb{P}(\partial Y_{t,W} \cap C = \emptyset) = \begin{cases} e^{-t\,\Theta([\operatorname{conv} C])} + \Theta([C_1|C_2]) \frac{e^{-t\,\Theta([\operatorname{conv} C])} - e^{-t(\Theta([C_1])+\Theta([C_2]))}}{\Theta([C_1])+\Theta([C_2])-\Theta([\operatorname{conv} C])} \\ \qquad \text{if the denominator is different from zero,} \\ e^{-t\,\Theta([\operatorname{conv} C])}(1 + t\,\Theta([C_1|C_2])) \qquad \text{otherwise.} \end{cases}$$

Proof. If $C = C_1 \cup C_2 \subseteq W$ with exactly two connected components $C_1, C_2 \in \mathcal{C}$, then $\mathbf{D}(C) = \{\{C\}, \{C_1, C_2\}\}$. As shown at the beginning of the proof of Lemma 5.5, we have that $d^{\mathrm{STIT}}(C_i, \{C_i\}) = 1$ for $i \in \{1, 2\}$. Thus, if $\Theta([C_1]) + \Theta([C_2]) - \Theta([\operatorname{conv} C]) \neq 0$, then the result follows by (5.15). On the other hand, if $\Theta([C_1]) + \Theta([C_2]) - \Theta([\operatorname{conv} C]) = 0$, then the integrand in (5.14) equals $e^{-t\,\Theta([\operatorname{conv} C])}$, and this yields the result for this case. □

This result also indicates why condition (5.10) on C is essential for d^{STIT} to be independent of t. To provide an example where (5.10) is violated, choose in $\mathbb{R}^2$ the measure Θ as the Mondrian line measure with $p_1 = p_2 = 1/2$, defined by (4.4), and $C = [0,1]^2 \cup ([\frac{1}{2}, \frac{3}{2}] \times [\frac{3}{2}, \frac{5}{2}])$.

Remark 5.7. *Let Θ be a fixed hyperplane measure of the form* (4.1), *satisfying* (4.3). *If $C \in \mathcal{R}$ with $C \subset \operatorname{int} W$ does not satisfy* (5.10), *then we approximate it by its parallel sets $C + \varepsilon B^d$ with distance $\varepsilon > 0$. By* (4.1) *we have $\Theta([\operatorname{conv}(D_j + \varepsilon B^d)]) = \Theta([\operatorname{conv} D_j]) + 2\varepsilon$. Thus, for all $D \in \mathbf{D}(C)$ with $|D| \geq 2$, there is $\varepsilon_D > 0$ such that for all $\varepsilon < \varepsilon_D$, $\sum_{D_j \in D} \Theta([\operatorname{conv}(D_j + \varepsilon B^d)]) = 2\varepsilon\,|D| + \sum_{D_j \in D} \Theta([\operatorname{conv} D_j])$. Hence $C + \varepsilon B^d$ satisfies* (5.10) *for $\varepsilon > 0$ small enough. Because the capacity functional is continuous from above for sequences of compact sets (see Theorem 2.12 of* [172]*), it is sufficient for our purposes to consider $C \in \mathcal{R}$ that fulfil* (5.10).

Corollary 5.8. *If $C \in \mathcal{R}$ with $C \subseteq W$ satisfies* (5.10), *then*

$$\mathbb{P}(t\,\partial Y_{t,W} \cap C = \emptyset) = \mathbb{P}(\partial Y_{1,W} \cap C = \emptyset)$$

for all $t > 0$ such that $C \subseteq t\,W$.

Proof. Because $\mathbb{P}(t\,\partial Y_{t,W} \cap C = \emptyset) = \mathbb{P}(\partial Y_{t,W} \cap t^{-1}C = \emptyset)$, applying (5.11) with $t\,\Theta([\operatorname{conv} t^{-1} D_j]) = \Theta([\operatorname{conv} D_j])$ for all $D_j \in D$, $D \in \mathbf{D}(C)$, Lemma 5.5 yields the statement of the corollary. □

5.3 Spatial consistency

Consider the STIT tessellation process $\mathbf{Y}_W := (Y_{t,W}, t \geq 0)$ in a window $W \in \mathcal{P}_d$ driven by a hyperplane measure Θ, as defined in Definition 4.2. Although $Y_{t,W}$ is a random tessellation of W for each $t > 0$, it is not obvious that the tessellation of the window can be extended consistently to a tessellation of the whole space. The problem is whether there exists a tessellation Y_t of $\mathbb{R}^d$ such that its restriction $Y_t \wedge W$ has the same law as $Y_{t,W}$.

As explained in the introduction to Section 5.2, the $(d-1)$ skeleton $\partial Y_{t,W}$ is a random closed set. Thus the mentioned problem concerns the concept of spatial consistency of random closed sets, treated in its general form in Theorem 2.3.1 of [172].

Definition 5.9. (a) *Let $\{X_W, W \in \mathcal{P}_d\}$ be a family of random tessellations, where X_W is a tessellation of W. It is called* ***spatially consistent*** *if for all $W_1, W_2 \in \mathcal{P}_d$ with $W_1 \subset W_2$,*

$$X_{W_2} \wedge W_1 \stackrel{D}{=} X_{W_1}.$$

(b) *Let $\{\mathbf{X}_W := (X_{t,W}, t \geq 0), W \in \mathcal{P}_d\}$ be a family of tessellation-valued random processes, where $X_{t,W}$ is a tessellation of W. It is called* ***spatially consistent*** *if for all $W_1, W_2 \in \mathcal{P}_d$ with $W_1 \subset W_2$,*

$$(X_{t,W_2} \wedge W_1, t > 0) \stackrel{D}{=} (X_{t,W_1}, t > 0). \tag{5.16}$$

The property of spatial consistency of a random tessellation is illustrated in Figure 5.2.

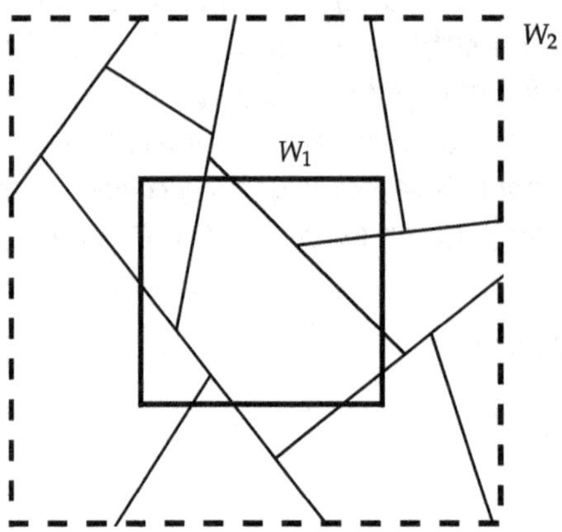

Figure 5.2: Spatial consistency of STIT tessellations. The distribution of the tessellation inside $W_1 \subset W_2$ does not depend on W_2, where the construction of the STIT tessellation process is performed.

5.3.1 Spatial consistency of STIT tessellations

Based on the propositions of Section 5.2 for the capacity functional of the skeleton of STIT tessellations, we can now derive their spatial consistency. In particular, the tessellation-valued random process $(Y_t, t > 0)$, whose existence is shown in part (d) of Theorem 5.10, will in Definition 6.1 be named **the STIT tessellation process driven by** Θ. In fact, it will be shown in Theorem 6.4 that for all $t > 0$, the random tessellation Y_t is stable with respect to iteration.

Theorem 5.10. *Let Θ be a hyperplane measure of the form* (4.1), *satisfying* (4.3), *and for $W \in \mathcal{P}_d$, let $(Y_{t,W}, t \geq 0)$ be the STIT tessellation process in W, driven by Θ; see Definition* 4.2.

(a) *For all $t > 0$, the family $\{Y_{t,W}, W \in \mathcal{P}_d\}$ is spatially consistent.*

(b) *For all $t > 0$, there exists a random stationary tessellation Y_t of $\mathbb{R}^d$ such that for all $W \in \mathcal{P}_d$,*

$$Y_t \wedge W \stackrel{D}{=} Y_{t,W} . \tag{5.17}$$

By (5.17) *the distribution of Y_t is uniquely defined.*

(c) *The family $\{(Y_{t,W}, t \geq 0), W \in \mathcal{P}_d\}$ of STIT tessellation processes is spatially consistent.*

(d) *There exists a Markov process $(Y_t, t > 0)$ of tessellations in $\mathbb{R}^d$ such that*

$$(Y_t \wedge W, t > 0) \stackrel{D}{=} (Y_{t,W}, t > 0) \tag{5.18}$$

for all $W \in \mathcal{P}_d$. The distribution of $(Y_t, t > 0)$ is uniquely determined by (5.18).

Proof. (a) Let $t > 0$, $W_1, W_2 \in \mathcal{P}_d$, and $W_1 \subset W_2$. Then, for all $C \in \mathcal{R}$ satisfying (5.10) and $C \subseteq W_1$, Lemma 5.5 yields that

$$\mathbb{P}(\partial(Y_{t,W_2} \wedge W_1) \cap C = \emptyset) = \mathbb{P}(\partial Y_{t,W_2} \cap C = \emptyset) = \mathbb{P}(\partial Y_{t,W_1} \cap C = \emptyset) .$$

Hence by Lemma 2.16 (c) it follows that $Y_{t,W_2} \wedge W_1 \stackrel{D}{=} Y_{t,W_1}$.

(b) Having (a), now Theorem 2.3.1 of [172] can be applied to the random closed sets $\mathrm{cl}(\partial Y_{t,W_i} \setminus \partial W_i)$ with $W_i := [-i, i]^d \in \mathcal{P}_d$, $i \in \mathbb{N}$. This yields, for fixed $t > 0$, the existence of a random closed set F with a uniquely determined distribution and $F \cap W_i \stackrel{D}{=} \mathrm{cl}(\partial Y_{t,W_i} \setminus \partial W_i)$. To prove that F is the $(d-1)$-skeleton of a random tessellation, consider the connected components of the complement F^c that are open polyhedral sets. For any fixed point $x \in \mathbb{R}^d$ and all $i \in \mathbb{N}$ such that $x \in \mathrm{int}\, W_i$, Lemma 5.3 yields that $\mathbb{P}(x \notin F) = \mathbb{P}(x \notin \partial Y_{t,W_i}) = 1$. Denote by $Z^F_{\langle x \rangle}$ the closure of the connected component of F^c that contains x. According to Lemma 5.2, we obtain $Z^F_{\langle x \rangle} \cap W_i \stackrel{D}{=} Z^{\mathrm{STIT}}_{\langle \{x\} \rangle, t, W_i} \stackrel{D}{=} Z^{\mathrm{PHT}}_{\langle \{x\} \rangle, t, W_i} = Z^{\mathrm{PHT}}_{\langle \{x\} \rangle, t} \cap W_i$ for all $i \in \mathbb{N}$. Because Θ satisfies Convention 4.1, we can apply Theorem 10.3.2 of [172], which shows that indeed $Z^{\mathrm{PHT}}_{\langle \{x\} \rangle, t}$ is a random polytope. Thus $\mathbb{P}(\exists i \in \mathbb{N} : Z^F_{\langle x \rangle, t} \subset \mathrm{int}\, W_i) = 1$. This shows that the closures of the connected components of F^c form a collection of random polytopes and their union equals $\mathbb{R}^d$. Property (b) of Definition 2.1 is guaranteed by using the connected components of F^c, and (c) of Definition 2.1 follows by Theorem 4.7. Summarizing, this shows that $F = \partial Y_t$ for some random tessellation Y_t. The stationarity of Y_t is a consequence of Lemma 5.5 and the translation invariance of the measure Θ.

(c) Consider the finite-dimensional distributions of the processes $(Y_{t,W}, t \geq 0)$, $W \in \mathcal{P}_d$. Let $n \in \mathbb{N}$, $0 < t_1 < \cdots < t_n$, and $W_1, W_2 \in \mathcal{P}_d$ with $W_1 \subset W_2$. By (a) we have that $Y_{t_1,W_2} \wedge W_1 \overset{D}{=} Y_{t_1,W_1}$. Then by Definition 4.5(b), Lemma 4.9, and the spatial consistency property shown in (a) we have

$$\begin{aligned} Y_{t_2,W_2} \wedge W_1 &\overset{D}{=} Y_{t_2-t_1,Y_{t_1,W_2}} \wedge W_1 \\ &\overset{D}{=} Y_{t_2-t_1,Y_{t_1,W_2}\wedge W_1} \\ &\overset{D}{=} Y_{t_2-t_1,Y_{t_1,W_1}} \\ &\overset{D}{=} Y_{t_2,W_1} \,. \end{aligned}$$

Hence, given Y_{t_1,W_1} and Y_{t_1,W_2}, respectively, the conditional distributions of $Y_{t_2,W_2} \wedge W_1$ and Y_{t_2,W_1} coincide. Thus the joint distributions of $(Y_{t_1,W_2} \wedge W_1, Y_{t_2,W_2} \wedge W_1)$ and of $(Y_{t_1,W_1}, Y_{t_2,W_1})$ are equal. Carrying this forward step by step to t_n, we obtain for the finite-dimensional distributions that

$$(Y_{t_1,W_2} \wedge W_1, \ldots, Y_{t_n,W_2} \wedge W_1) \overset{D}{=} (Y_{t_1,W_1}, \ldots, Y_{t_n,W_1}) \,.$$

This shows the consistency of the family of STIT tessellation processes $\{(Y_{t,W}, t \geq 0), W \in \mathcal{P}_d\}$, according to Definition 5.9(b).

(d) Based on the family $\{(Y_{t,W}, t \geq 0), W \in \mathcal{P}_d\}$ of STIT tessellation processes and their spatial consistency, we show the existence of a Markov process $(Y_t, t > 0)$ that satisfies (5.18). For this purpose, we construct an appropriate projective family of finite-dimensional distributions. Let $n \in \mathbb{N}$ and $0 < t_1 < \cdots < t_n$. We define $(Y_{t_1}, \ldots, Y_{t_n})$ as follows. Firstly, choose Y_{t_1} as a random tessellation that satisfies (5.17) for $t = t_1$. Then, for a family $\{(Y_{t,P}, t \geq 0), P \in Y_{t_1}\}$ of STIT tessellation processes, driven by Θ, which are conditionally independent, given Y_{t_1}, define

$$Y_{t_2} := \bigcup_{P \in Y_{t_1}} Y_{t_2-t_1,P} \,. \tag{5.19}$$

Thus by (5.19), the spatial consistency of STIT tessellations, shown in (a), Lemma 4.9, and Definition 4.5 (b) we obtain that

$$\begin{aligned} Y_{t_2} \wedge W &= \Big(\bigcup_{P \in Y_{t_1}} Y_{t_2-t_1,P} \Big) \wedge W \\ &\overset{D}{=} Y_{t_2-t_1, Y_{t_1}\wedge W} \\ &\overset{D}{=} Y_{t_2,W} \end{aligned}$$

for all $W \in \mathcal{P}_d$. This yields that the joint distributions of $(Y_{t_1} \wedge W, Y_{t_2} \wedge W)$ and $(Y_{t_1,W}, Y_{t_2,W})$ coincide for all $W \in \mathcal{P}_d$.

Construction (5.19) can then be applied consecutively for Y_{t_i}, $i \in \{3, \dots, n\}$, and hence

$$(Y_{t_1} \wedge W, \dots, Y_{t_n} \wedge W) \overset{\mathrm{D}}{=} (Y_{t_1,W}, \dots, Y_{t_n,W})$$

for all $W \in \mathcal{P}_d$. Now, because the families of the finite-dimensional distributions of the processes $(Y_{t,W}, t \geq 0)$, $W \in \mathcal{P}_d$, are projective and the distribution of each Y_{t_i} is uniquely determined by the distributions of the $Y_{t_i} \wedge W$, $W \in \mathcal{P}_d$, the family of the finite-dimensional distributions of $(Y_{t_1}, \dots, Y_{t_n})$, $n \in \mathbb{N}$, $0 < t_1 < \cdots < t_n$, is projective as well. Because $\mathcal{T}$ is a Borel space, as shown in Lemma 18.2, the Kolmogorov theorem (see Theorem 8.23 of [89]) yields that there exists a process $(Y_t, t > 0)$ satisfying (5.18). By Proposition 4.2 of [89] its distribution is uniquely determined.

The Markov property of the process $(Y_t, t > 0)$ follows from construction (5.19). □

In Theorem 6.4, it will be shown that for any $t > 0$, the tessellation Y_t has the STIT property (3.10), and this justifies calling $\mathbf{Y} = (Y_t, t > 0)$ a STIT tessellation process driven by the hyperplane measure Θ; see Definition 6.1.

There is also a "global construction" of the STIT tessellation process that is not based on windows. This is presented in Chapter 8.

5.3.2 Spatial consistency of cell division processes

In the present section, a broader class of tessellation-valued cell division processes within windows is considered, which are driven by rather general L and D rules as explained in Section 1.1. We derive necessary conditions for the spatial consistency of such processes. Our main result will be that STIT tessellation processes are the only members in this class of cell division processes that are spatially consistent.

To formulate a precise result, we define families of cell division processes inside windows $W \in \mathcal{P}_d$ analogously to Definition 4.2.

Definition 5.11. *Let $\{\Lambda_P, P \in \mathcal{P}_d\}$ be a family of probability measures on $\mathcal{H}$ with $\Lambda_P([P]) = 1$ for all $P \in \mathcal{P}_d$, and let $\{\mathsf{L}_P, P \in \mathcal{P}_d\}$ be a family of probability measures on $\mathbb{R}^+$. Furthermore, for $P \in \mathcal{P}_d$, let $\tau(P)$ be a random variable with law L_P. For $W \in \mathcal{P}_d$, the* ***cell division process*** *$(M_{t,W}, t \geq 0)$ in W, driven by these families of distributions, is defined by the following rules.*

(i) *$M_{t,W} := \{W\}$ for $0 \leq t < \tau(W)$, $P_1 := W$, and $\beta(P_1) := 0$.*

(ii) *For $t > 0$, let $M_{t,W} = \{P_{i_1}, \dots, P_{i_n}\}$ with respective birth times $\beta(P_{i_1}), \dots, \beta(P_{i_n})$ and life times $\tau(P_{i_1}), \dots, \tau(P_{i_n})$. Define $j_n \in \{i_1, \dots, i_n\}$ and σ_n such that $\sigma_n := \beta(P_{j_n}) + \tau(P_{j_n}) := \min\{\beta(P_{i_k}) + \tau(P_{i_k}) : k \in \{1, \dots, n\}\}$.*

The process remains constant until time σ_n, and at that time, it jumps into the state

$$M_{\sigma_n,W} := \oslash(P_{j_n}, H_n, M_{t,W}) = (\{P_{i_1}, \dots, P_{i_n}\} \setminus \{P_{j_n}\}) \cup \{P_{2n}, P_{2n+1}\}$$

with $P_{2n} := P_{j_n} \cap H_n^+$ and $P_{2n+1} := P_{j_n} \cap H_n^-$, where H_n is a random hyperplane with law $\Lambda_{P_{j_n}}$, and H_n is conditionally independent, given P_{j_n}, of the $n-1$ dividing hyperplanes used before, that is, of all hyperplanes H with $\beta(H) < \sigma_n$. Furthermore, H_n is independent of $\tau(P)$ for all cells P with $\beta(P) < \sigma_n$.
The joint birth time of the dividing hyperplane and of the new cells is defined as $\beta(H_n) = \beta(P_{2n}) = \beta(P_{2n+1}) := \sigma_n$. Given P_{2n} and P_{2n+1}, the assigned life times are conditionally independent random variables $\tau(P_{2n})$ and $\tau(P_{2n+1})$ withe laws $\mathsf{L}_{P_{2n}}$ and $\mathsf{L}_{P_{2n+1}}$, also independent of $\tau(P)$ for all cells P with $\beta(P) < \sigma_n$ and independent of all hyperplanes H with $\beta(H) < \sigma_n$.

Thus the joint distribution of the dividing hyperplane and the life time of a cell P is $\Lambda_P \otimes \mathsf{L}_P$. In other words, the L rule of the cell division process is given by $\{\mathsf{L}_P, P \in \mathcal{P}_d\}$, and the D rule by $\{\Lambda_P, P \in \mathcal{P}_d\}$. Note that the independence assumptions are essential in this definition.

Now we derive necessary conditions on $\{\Lambda_P, P \in \mathcal{P}_d\}$ and $\{\mathsf{L}_P, P \in \mathcal{P}_d\}$ for the spatial consistency of the generated cell division process. For $P_1, P_2 \in \mathcal{P}_d$ with $P_1 \subset P_2$, denote by

$$\mathcal{P}_d(P_1, P_2, \Lambda_{P_2}) := \{P \in \mathcal{P}_d : P_1 \subseteq P,\ P = P_2 \cap H^+ \text{ or } P = P_2 \cap H^-,\ H \in \operatorname{supp} \Lambda_{P_2}\}$$

the set of all polytopes containing P_1 that can arise after the first division of P_2 by a random hyperplane H with distribution Λ_{P_2}, given that $H \cap \operatorname{int} P_1 = \emptyset$. Note that the condition $H \in \operatorname{supp} \Lambda_{P_2}$ can indeed restrain the set $\mathcal{P}_d(P_1, P_2, \Lambda_{P_2})$, for example, if Λ_{P_2} is based on a Mondrian hyperplane measure. This plays a role in (5.23) below.

Lemma 5.12. *Let $\{\Lambda_P, P \in \mathcal{P}_d\}$ be a family of probability measures on $\mathcal{H}$ with $\Lambda_P([P]) = 1$ for all $P \in \mathcal{P}_d$, and let $\{\mathsf{L}_P, P \in \mathcal{P}_d\}$ be a family of probability measures on $\mathbb{R}^+$. Furthermore, assume that for all $P \in \mathcal{P}_d$,*

$$\operatorname{supp} \mathsf{L}_P = [0, \infty) \textit{ and } \mathsf{L}_P \textit{ has a continuous density } f_P \textit{ with respect to } \lambda_+ . \tag{5.20}$$

If the family $\{(M_{t,W}, t \geq 0),\ W \in \mathcal{P}_d\}$ of cell division processes driven by $\{\Lambda_P, P \in \mathcal{P}_d\}$ and $\{\mathsf{L}_P, P \in \mathcal{P}_d\}$ is spatially consistent in the sense of (5.16), then necessarily the following conditions must be satisfied.

Let $P_1, P_2 \in \mathcal{P}_d$ with $P_1 \subset P_2$. Then

$$\Lambda_{P_1} \ll \Lambda_{P_2}, \tag{5.21}$$

$$\Lambda_{P_1}(A) = \frac{\Lambda_{P_2}(A \cap [P_1])}{\Lambda_{P_2}([P_1])} \quad \textit{for all } A \in \mathcal{B}(\mathcal{H}) \textit{ if } \Lambda_{P_2}([P_1]) > 0 . \tag{5.22}$$

Furthermore, for the random variables $\tau(P)$ with law L_P, $P \in \mathcal{P}_d$,

$$\lim_{s \downarrow 0} \frac{\mathbb{P}(\tau(P_1) \leq t + s \mid \tau(P_1) > t)}{\mathbb{P}(\tau(P_3) \leq t - r + s \mid \tau(P_3) > t - r)} = \Lambda_{P_3}([P_1]) \tag{5.23}$$

for all $t > 0$ and $r \in [0, t)$, where $P_3 = P_2$ if $r = 0$, and $P_3 \in \mathcal{P}_d(P_1, P_2, \Lambda_{P_2})$ if $r > 0$.

Proof. Regarding condition (5.16) of spatial consistency, we consider the first division of a window $W_1 \in \mathcal{P}_d$ embedded in a larger window $W_2 \in \mathcal{P}_d$, that is, $W_1 \subset W_2$. In the process $(M_{t,W_1}, t \geq 0)$ the first division event is caused by a pair $(H_1, \tau(W_1))$ consisting of the dividing hyperplane H_1 at time $\tau(W_1)$. By definition this pair has the law $\Lambda_{W_1} \otimes \mathsf{L}_{W_1}$. Under spatial consistency, in the cell division process $(M_{t,W_2}, t \geq 0)$, this law of the first division of W_1 must remain.

Note that there can be several divisions of W_2 before the first division of W_1. Due to the continuity assumption (5.20), for all times $t > 0$ for which $\tau(W_1) > t$, any time $t - r$, $r \in [0, t)$, can be a division time of W_2. If W_2 is divided without dividing W_1, then a new polytope, say $P_3 \in \mathcal{P}_d(W_1, W_2, \Lambda_{W_2})$ with birth time r, appears with $W_1 \subset P_3 \subset W_2$. The probability that W_1 is divided by a random hyperplane that divides $P_3 \supset W_1$ is $\Lambda_{P_3}([W_1])$. Together with the independence assumptions in Definition 5.11, this yields the necessary property (5.23), which can be interpreted as the ratio of hazard rates with respect to the life times of W_1 and P_3.

Regarding the division rule Λ_{W_1} for the first division of W_1, condition (5.21) with $P_1 = W_1$ is a consequence of spatial consistency. Furthermore, for $A \in \mathcal{B}(\mathcal{H})$, the conditional probability $\Lambda_{P_2}(A \cap [W_1])/\Lambda_{P_2}([W_1])$ that a hyperplane that divides P_2 belongs to A, under the condition that it also divides W_1, must be equal to $\Lambda_{W_1}(A)$. This shows (5.22) for $P_1 = W_1$. □

In the following lemma, we show that properties (5.20)–(5.23) imply that there exists a hyperplane measure Λ such that the life time distribution of a cell P depends on $\Lambda([P])$ and that the law Λ_P of the dividing hyperplane is also determined by Λ. Thus the whole cell division process $(M_{t,W}, t \geq 0)$ in W is entirely controlled by the measure Λ.

Lemma 5.13. *Let $\{\Lambda_P, P \in \mathcal{P}_d\}$ and $\{\mathsf{L}_P, P \in \mathcal{P}_d\}$ be families of distributions satisfying conditions* (5.20)–(5.23). *Then there exists a measure Λ on $\mathcal{H}$ such that $\Lambda([P]) > 0$ for all $P \in \mathcal{P}_d$ and*

$$\Lambda_P(A) = \frac{\Lambda(A \cap [P])}{\Lambda([P])} \quad \textit{for all } P \in \mathcal{P}_d,\ A \in \mathcal{B}(\mathcal{H}). \tag{5.24}$$

The measure Λ satisfying (5.24) *is uniquely determined up to a constant factor. Furthermore, this factor can be chosen such that L_P is the exponential distribution with parameter $\Lambda([P])$, $P \in \mathcal{P}_d$.*

Proof. Let $(W_n \in \mathcal{P}_d, n \in \mathbb{N}_0)$ be a sequence of windows with $W_0 \subset W_1 \subset W_2 \subset \cdots$ and $\bigcup_{n\in\mathbb{N}} W_n = \mathbb{R}^d$. Then (5.21) yields that $\Lambda_{W_n}([W_0]) > 0$ for all $n \in \mathbb{N}_0$. Now define the function Λ on $\mathcal{B}(\mathcal{H})$ by

$$\Lambda(A) := \lim_{n\to\infty} \frac{\Lambda_{W_n}(A)}{\Lambda_{W_n}([W_0])}, \quad A \in \mathcal{B}(\mathcal{H}). \tag{5.25}$$

The existence of the limit is obvious if $\Lambda_{W_n}(A) = 0$ for all $n \in \mathbb{N}_0$. Otherwise, if there is $n_0 \in \mathbb{N}_0$ such that $\Lambda_{W_{n_0}}(A) > 0$, then for all $n \geq n_0$, consider the ratio of two subsequent values in (5.25). The assumption that $\Lambda_{W_{n+1}}$ is a probability measure on $[W_{n+1}]$ implies that $\Lambda_{W_{n+1}}(A) = \Lambda_{W_{n+1}}(A \cap [W_{n+1}])$. Then applying (5.22) with $P_1 := W_n$, $A := [W_0]$, and $P_2 := W_{n+1}$ and again using (5.22) for $P_1 := W_n$ and $P_2 := W_{n+1}$, we have

$$\begin{aligned}\frac{\Lambda_{W_{n+1}}(A)}{\Lambda_{W_{n+1}}([W_0])}\frac{\Lambda_{W_n}([W_0])}{\Lambda_{W_n}(A)} &= \frac{\Lambda_{W_{n+1}}(A \cap [W_{n+1}])}{\Lambda_{W_{n+1}}([W_0])}\frac{\Lambda_{W_n}([W_0])}{\Lambda_{W_n}(A)} \\ &= \frac{1}{\Lambda_{W_{n+1}}([W_n])}\frac{\Lambda_{W_{n+1}}(A \cap [W_{n+1}])}{\Lambda_{W_n}(A)} \\ &\geq \frac{1}{\Lambda_{W_{n+1}}([W_n])}\frac{\Lambda_{W_{n+1}}(A \cap [W_n])}{\Lambda_{W_n}(A)} \\ &= \frac{\Lambda_{W_{n+1}}([W_n])}{\Lambda_{W_{n+1}}([W_n])} = 1 .\end{aligned}$$

Hence the sequence on the right-hand side of (5.25) is monotone in n. Note that the limit in (5.25) depends on the choice of W_0, and therefore Λ is not necessarily uniquely determined.

Now we prove that the function Λ defined by (5.25) is a measure on $\mathcal{H}$. Obviously, Λ is nonnegative, and $\Lambda(\emptyset) = 0$. Let $A_1, A_2, \ldots \in \mathcal{B}(\mathcal{H})$ be a sequence of pairwise disjoint sets. Then by the monotone convergence theorem we have

$$\begin{aligned}\Lambda\left(\bigcup_{i=1}^{\infty} A_i\right) &= \lim_{n\to\infty}\frac{\Lambda_{W_n}(\bigcup_{i=1}^{\infty} A_i)}{\Lambda_{W_n}([W_0])} = \lim_{n\to\infty}\frac{\sum_{i=1}^{\infty}\Lambda_{W_n}(A_i)}{\Lambda_{W_n}([W_0])} = \sum_{i=1}^{\infty}\lim_{n\to\infty}\frac{\Lambda_{W_n}(A_i)}{\Lambda_{W_n}([W_0])} \\ &= \sum_{i=1}^{\infty}\Lambda(A_i) ,\end{aligned}$$

which is the required σ-additivity for Λ.

To show (5.24), let $P \in \mathcal{P}_d$ and $n_P \in \mathbb{N}_0$ be such that $P \subseteq W_{n_P}$. Then, by (5.21), $\Lambda_{W_{n_P}}([P]) > 0$. With the monotonicity shown above,

$$\Lambda([P]) = \lim_{n\to\infty}\frac{\Lambda_{W_n}([P])}{\Lambda_{W_n}([W_0])} \geq \frac{\Lambda_{W_{n_P}}([P])}{\Lambda_{W_{n_P}}([W_0])} > 0 .$$

Moreover, for all $n \geq n_P$, by (5.22) with $P_1 := \operatorname{conv}(P \cup W_0)$, $A := [W_0]$, and $P_2 := W_n$,

$$\frac{\Lambda_{W_n}([P])}{\Lambda_{W_n}([W_0])} \leq \frac{\Lambda_{W_n}([\operatorname{conv}(P \cup W_0)])}{\Lambda_{W_n}([W_0])} = \frac{1}{\Lambda_{\operatorname{conv}(P\cup W_0)}([W_0])} .$$

Hence

$$0 < \Lambda([P]) = \lim_{n\to\infty}\frac{\Lambda_{W_n}([P])}{\Lambda_{W_n}([W_0])} \leq \frac{1}{\Lambda_{\operatorname{conv}(P\cup W_0)}([W_0])} < \infty . \tag{5.26}$$

Furthermore, for all $A \in \mathcal{B}(\mathcal{H})$ and $n \geq n_P$, condition (5.22) yields

$$\Lambda_P(A) = \frac{\Lambda_{W_n}(A \cap [P])}{\Lambda_{W_n}([P])} = \frac{\Lambda_{W_n}(A \cap [P])}{\Lambda_{W_n}([W_0])} \frac{\Lambda_{W_n}([W_0])}{\Lambda_{W_n}([P])} .$$

Now by (5.25) and (5.26) we have

$$\begin{aligned}\Lambda_P(A) &= \lim_{n\to\infty} \frac{\Lambda_{W_n}(A \cap [P])}{\Lambda_{W_n}([W_0])} \frac{\Lambda_{W_n}([W_0])}{\Lambda_{W_n}([P])} = \lim_{n\to\infty} \frac{\Lambda_{W_n}(A \cap [P])}{\Lambda_{W_n}([W_0])} \lim_{n\to\infty} \frac{\Lambda_{W_n}([W_0])}{\Lambda_{W_n}([P])} \\ &= \frac{\Lambda(A \cap [P])}{\Lambda([P])} ,\end{aligned}$$

which shows (5.24).

To prove the uniqueness of Λ up to a constant factor, let Λ' be another measure on $\mathcal{H}$ satisfying (5.24). Then, for $A \in \mathcal{B}(\mathcal{H})$ and $P \in \mathcal{P}_d$,

$$\Lambda_P(A) = \frac{\Lambda(A \cap [P])}{\Lambda([P])} = \frac{\Lambda'(A \cap [P])}{\Lambda'([P])} . \tag{5.27}$$

Thus, for $P_1, P_2 \in \mathcal{P}_d$ with $P_1 \subset P_2$ and for $A := [P_1]$, we obtain

$$\frac{\Lambda([P_1])}{\Lambda'([P_1])} = \frac{\Lambda([P_2])}{\Lambda'([P_2])} ,$$

and hence there is a constant $c > 0$ such that $\Lambda'([P]) = c\,\Lambda([P])$ for all $P \in \mathcal{P}$. With (5.27), we obtain that

$$\Lambda'(A \cap [P]) = c\,\Lambda(A \cap [P])$$

for all $A \in \mathcal{B}(\mathcal{H})$ and $P \in \mathcal{P}_d$.

Then for all $A \in \mathcal{B}(\mathcal{H})$, a sequence $W_n \in \mathcal{P}_d$, $n \in \mathbb{N}_0$, with $W_0 \subset W_1 \subset W_2 \subset \cdots$, and $\bigcup_{n\in\mathbb{N}} W_n = \mathbb{R}^d$, we have

$$\Lambda'(A) = \lim_{n\to\infty} \Lambda'(A \cap [W_n]) = \lim_{n\to\infty} c\,\Lambda(A \cap [W_n]) = c\,\Lambda(A) ,$$

which shows that $\Lambda' = c\,\Lambda$.

We turn now toward the life time distributions L_P, $P \in \mathcal{P}_d$. Let $P_1, P_2 \in \mathcal{P}_d$ with $P_1 \subset P_2$, $t > 0, r > 0$, and $P_3 \in \mathcal{P}_d(P_1, P_2, \Lambda_{P_2})$. By (5.24) we have $\Lambda_{P_3}([P_1]) = \frac{\Lambda([P_1])}{\Lambda([P_3])}$.

Condition (5.23) yields that

$$\lim_{s\downarrow 0} \frac{\mathbb{P}(t < \tau(P_1) \le t+s)\, s\, \mathbb{P}(\tau(P_3) > t-r)}{s\, \mathbb{P}(\tau(P_1) > t)\, \mathbb{P}(t-r < \tau(P_3) \le t-r+s)} = \frac{\Lambda([P_1])}{\Lambda([P_3])} .$$

By assumption (5.20) there is a continuous density function f_P of L_P, $P \in \mathcal{P}_d$, and thus the limit as $s \downarrow 0$ yields

$$\frac{f_{P_1}(t)}{\mathbb{P}(\tau(P_1) > t)} \cdot \frac{\mathbb{P}(\tau(P_3) > t-r)}{f_{P_3}(t-r)} = \frac{\Lambda([P_1])}{\Lambda([P_3])} .$$

Denoting the hazard rate $h(P,t-r) := \frac{f_P(t-r)}{\mathbb{P}(\tau(P)>t-r)}$, $P \in \mathcal{P}_d$, we obtain that

$$\frac{h(P_1,t)}{h(P_3,t-r)} = \frac{\Lambda([P_1])}{\Lambda([P_3])},$$

and hence, for all $P \in \mathcal{P}_d$, the function $(t,r) \mapsto h(P,t-r)$ depends neither on r nor on t. Therefore there is $\alpha > 0$ such that $h(P,t) = \alpha\Lambda([P])$ for all $P \in \mathcal{P}_d$ and $t > 0$. Thus $f_P(t) = \alpha\Lambda([P])\,\mathbb{P}(\tau(P) > t)$ for all $t > 0$, and hence L_P is the exponential distribution with parameter $\alpha\Lambda([P])$. Because Λ is determined by the other conditions only up to a constant factor, we can choose $\alpha = 1$. □

To summarize: If a family $\{(M_{t,W}, t \geq 0),\ W \in \mathcal{P}_d\}$ of cell division processes as described in Definition 5.11, with the additional assumption (5.20), is spatially consistent, then there must be a hyperplane measure Λ that controls the processes $(M_{t,W}, t \geq 0)$, $W \in \mathcal{P}_d$, as described in Lemma 5.13.

In case of spatial consistency, there exists a process $(M_t, t > 0)$ such that $M_t \wedge W \stackrel{D}{=} M_{t,W}$ for all $W \in \mathcal{P}_d$, $t > 0$. To guarantee that M_t is a random stationary tessellation of $\mathbb{R}^d$, it is required that Λ is a translation-invariant hyperplane measure and that its spherical directional distribution satisfies Convention 4.1, that is, it is not concentrated on a great hypersphere of S^{d-1}.

Theorem 5.14. *Let $\{(M_{t,W}, t \geq 0),\ W \in \mathcal{P}_d\}$ be a family of cell division processes driven by $\{\Lambda_P, P \in \mathcal{P}_d\}$ and $\{\mathsf{L}_P, P \in \mathcal{P}_d\}$, satisfying* (5.20), *in the sense of Definition* 5.11. *Moreover, assume that this family is spatially consistent and that a hyperplane measure Λ, characterized in Lemma* 5.13, *is translation invariant and that its spherical directional distribution satisfies* (4.3). *Then there exist a hyperplane measure Θ of the form* (4.1) *and a constant $a > 0$ such that $\Lambda = a\,\Theta$ and*

$$(M_{t,W}, t > 0) \stackrel{D}{=} (Y_{at} \wedge W, t > 0), \tag{5.28}$$

where $(Y_t, t > 0)$ is the STIT tessellation process satisfying (5.18), *driven by Θ.*

Proof. If Λ is translation invariant, then there exist a hyperplane measure Θ of the form (4.1) and a constant $a > 0$ such that $\Lambda = a\,\Theta$; see (4.33) of [172]. A comparison of Definition 5.11 with Definition 4.2 indicates that it is sufficient to show that the life time distributions of the cells and the laws of the dividing hyperplanes for $(M_{t,W}, t > 0)$ and for $(Y_{at,W}, t > 0)$ are identical. Lemma 5.13 yields that $\Lambda_P = \Theta_{[P]}$ for all $P \in \mathcal{P}_d$ and that the life time is exponentially distributed with parameter $a\,\Theta([P])$. An application of Remark 4.4 or Corollary 6.3 completes the proof. □

To illustrate the proposition of Theorem 5.14, assume that the life time of a cell P depends on the volume $V_d(P)$ of P. It can be reasonable to consider for $d \geq 2$ a cell division process where the life time of P is exponentially distributed with parameter $V_d(P)$. If this process is to be spatially consistent, then by Lemma 5.13 for the corresponding measure Λ, we obtain that $\Lambda([P]) = V_d(P)$ for all $P \in \mathcal{P}_d$. Each compact convex set $K \in \mathcal{K}$ can be approximated from above by a sequence of d-dimensional polytopes, and hence continuity arguments yield that $\Lambda([K]) = V_d(K)$ for all $K \in \mathcal{K}$. Let S be a line segment, $S \in \mathcal{K}$. Then, for $P \in \mathcal{P}_d$ with $S \subset P$, equation (5.24) yields for $A = [S]$ that $\Lambda_P([S]) = \frac{V_d(S)}{V_d(P)} = 0$. Hence, for $P \in \mathcal{P}_d$,

$$\Lambda_P([P]) \leq \sum_{S \in \mathcal{F}_1(P)} \Lambda_P([S]) = 0\,,$$

which contradicts the condition $\Lambda_P([P]) = 1$. This shows that there is no cell division process in the sense of Definition 5.11 that is spatially consistent and where the life time of a cell P has the law $\mathrm{Expo}(V_d(P))$.

However, we point out that there can exist a cell division process, *without an initial tessellation at time* $t = 0$, with values in the space $\mathcal{T}$ of tessellations of $\mathbb{R}^d$, where the life times of the cells P are exponentially distributed with parameter $V_d(P)$ or other functionals of P. For some details and references, see Section 9.1.3. This indicates that spatial consistency is not a necessary condition for the existence of appropriate global cell division processes for given life time and division rules.

5.4 Notes

1. In [82] the concept of a K-cell is introduced differently. There, for a hyperplane process $\widehat{X}_t$ and a fixed convex body $K \in \mathcal{K}$, the K-cell is the cell of the hyperplane tessellation generated by $\widehat{X}_t \setminus \{H \in \widehat{X}_t : H \cap K \neq \emptyset\}$.
2. The capacity functional of the $(d-1)$-skeleton $\partial Y_{t,W}$ was determined in Section 4 of [145]. As a conclusion, the spatial consistency of STIT tessellations, part (b) of Theorem 5.10, was shown in Theorem 1 ibidem.
3. Section 5.3.2 is essentially the content of [140]. However, there conditions (5.20)–(5.23) were formulated as assumptions, and Lemma 5.12, which shows their necessity, was not yet proven.

6 Existence and fundamental properties of STIT tessellations in $\mathbb{R}^d$

In the previous Chapters 4 and 5, the STIT tessellation process in a fixed window $W \in \mathcal{P}_d$ was dealt with. The results of Theorem 5.10 on spatial consistency open up the vista to a stochastic process $(Y_t, t > 0)$ on the space of tessellations in $\mathbb{R}^d$. This perspective is pursued in the present chapter, studying essential properties of this process. In Theorem 6.4, it is ultimately shown that the tessellations, which were already referred to as STIT tessellations, indeed have the announced property of being STable under ITeration of tessellations in the sense of Definition 3.7(a). Subsequently, Section 6.2 deals with densities and directional distributions of the skeletons of STIT tessellations, and they are related to the corresponding features of PHTs. Furthermore, Section 6.3 is dedicated to sections of STIT tessellations by flats, which generate lower-dimensional STIT tessellations within these flats. Section 6.4 addresses linear transformations of STIT tessellations. As a result, for any STIT tessellation with a discrete directional distribution, a linear transformation is described, which transforms this tessellation into a tessellation that has the same distribution as the section of a higher-dimensional Mondrian STIT tessellation with a particular linear subspace. Finally, in Section 6.5, it is proven that an exponential transformation of the time axis transmutes the rescaled STIT tessellation process into a time-stationary stochastic process on the space of tessellations.

6.1 Key properties of the STIT tessellation process

In Definition 4.2 the tessellation-valued process $(Y_{t,W}, t \geq 0)$ within a window $W \in \mathcal{P}_d$, driven by a hyperplane measure Θ, was introduced and was already called a STIT tessellation process in W. Furthermore, according to Theorem 5.10, there exists a tessellation-valued process $(Y_t, t > 0)$ in the space $\mathbb{R}^d$ such that for all $t > 0$ and $W \in \mathcal{P}_d$, the restriction $Y_t \wedge W$ has the same distribution as the tessellation $Y_{t,W}$. This is the background for the following definition.

Definition 6.1. *Let Θ be a hyperplane measure of the form* (4.1), *satisfying* (4.3), *and let $\{(Y_{t,W}, t \geq 0), W \in \mathcal{P}_d\}$ be a family of STIT tessellation processes in W, driven by Θ. A tessellation-valued stochastic process $\mathbf{Y} = (Y_t, t > 0)$ is called a* ***STIT tessellation process driven by*** Θ *if it satisfies* (5.18), *that is,*

$$(Y_t \wedge W, t > 0) \overset{\mathrm{D}}{=} (Y_{t,W}, t > 0) \quad \text{for all } W \in \mathcal{P}_d .$$

We emphasize at this point that the STIT tessellation process $\mathbf{Y}$ is defined for $t > 0$ only, because there is no initial state at time $t = 0$ of the STIT tessellation process in $\mathbb{R}^d$. This is in contrast to the local STIT processes $(Y_{t,W}, t > 0)$, $W \in \mathcal{P}_d$, which can be extended to $t = 0$ by putting $Y_{0,W} := \{W\}$. A “global construction” of $(Y_t, t > 0)$ is

https://doi.org/10.1515/9783112222409-007

described in Chapter 8. Furthermore, Theorem 7.9 provides a uniqueness result stating that a stationary tessellation has the STIT property if and only if its distribution coincides with that of Y_t for suitably chosen $t > 0$ and Θ.

A STIT tessellation process driven by the motion-invariant hyperplane measure Θ_{iso} is a process of motion-invariant tessellations. This follows by Lemma 5.5, regarding that the coefficients $d^{\text{STIT}}(C, D)$ are also rotation invariant if $\Theta = \Theta_{\text{iso}}$. Furthermore, a STIT tessellation process driven by a Mondrian hyperplane measure Θ of the form (4.4) is called the **Mondrian STIT tessellation process.**

We now show some essential properties of the STIT tessellation process $(Y_t,\ t > 0)$ in $\mathbb{R}^d$, which are used as important tools in many proofs. The zero-cell of Y_t, $t > 0$, is denoted by $Z^{\text{STIT}}_{\langle o\rangle,t}$, and the typical cell by Z^{STIT}_t.

Theorem 6.2. *For a given hyperplane measure* Θ *of the form* (4.1), *satisfying* (4.3), *let* $\mathbf{Y} = (Y_t,\ t > 0)$ *be a STIT tessellation process driven by* Θ, *and let* $\underline{\mathbf{Y}} = (\mathbf{Y}^{(i)},\ i \in \mathbb{N}_0) = ((Y^{(i)}_t,\ t > 0),\ i \in \mathbb{N}_0)$ *be a sequence of independent copies of* $\mathbf{Y}$. *Then*

$$Y_{t+s} \overset{\text{D}}{=} Y_t \boxplus \underline{Y_s} \quad \textit{for all } t, s > 0, \textit{ where } \underline{Y_s} := \left(Y^{(i)}_s,\ i \in \mathbb{N}_0\right), \tag{6.1}$$

and

$$t\,Y_t \overset{\text{D}}{=} Y_1 \quad \textit{for all } t > 0\,. \tag{6.2}$$

Furthermore,

$$(Z^{\text{STIT}}_{\langle o\rangle,t}\,,\ t > 0) \overset{\text{D}}{=} (Z^{\text{PHT}}_{\langle o\rangle,t}\,,\ t > 0) \tag{6.3}$$

and

$$Z^{\text{STIT}}_t \overset{\text{D}}{=} Z^{\text{PHT}}_t \quad \textit{for all } t > 0\,, \tag{6.4}$$

where $(Z^{\text{PHT}}_{\langle o\rangle,t}\,,\ t > 0)$ *is the zero-cell process of a PHT process* $(X_t,\ t > 0)$ *driven by* Θ *(see* (2.66)*), and* Z^{PHT}_t *is the typical cell of the respective PHT* X_t.

Proof. With construction (5.19), which was used in the proof of Theorem 5.10, and the consistency property (5.18), we obtain that for $t, s > 0$, we have $Y_{t+s} \overset{\text{D}}{=} \bigcup_{P \in Y_t} Y_{s,P} \overset{\text{D}}{=} \bigcup_{P \in Y_t} (Y_s \wedge P) \overset{\text{D}}{=} Y_t \boxplus \underline{Y_s}$, where in the last step, Definition 3.2 is applied. This proves (6.1).

Relation (6.2) follows from Lemma 2.16, Corollary 5.8, and Theorem 5.10(b).

To prove (6.3), we use (5.5) for $K = \{o\}$ and the equality $Z^{\text{PHT}}_{\langle o\rangle,t,W} = Z^{\text{PHT}}_{\langle o\rangle,t} \cap W$ for all $W \in \mathcal{P}_d$. This yields

$$\begin{aligned}(Z^{\text{STIT}}_{\langle o\rangle,t} \cap W,\ t > 0) &= (Z_{\langle o\rangle}(Y_t) \cap W,\ t > 0)\\ &= (Z_{\langle o\rangle}(Y_t \wedge W) \cap W,\ t > 0)\\ &\overset{\text{D}}{=} (Z_{\langle o\rangle}(Y_{t,W}),\ t > 0)\end{aligned}$$

$$
\begin{aligned}
&= (Z^{\mathrm{STIT}}_{\langle o\rangle,t,W}, t > 0)\\
&\overset{\mathrm{D}}{=} (Z^{\mathrm{PHT}}_{\langle o\rangle,t,W}, t > 0)\\
&= (Z^{\mathrm{PHT}}_{\langle o\rangle,t} \cap W, t > 0)\,. \qquad (6.5)
\end{aligned}
$$

Now, for an indirect proof, assume that (6.3) does not hold. Then according to Lemma 1.1.67 and Proposition 1.1.69 of [137], there exist $n \in \mathbb{N}$, $0 < t_1 < \cdots < t_n$, $K_1, \ldots, K_n \in \mathcal{K}$, and $W \in \mathcal{P}_d$ such that $K_i \subseteq W$, $i \in \{1,\ldots,n\}$, with

$$
\begin{aligned}
\mathbb{P}(K_i \subseteq (Z^{\mathrm{STIT}}_{\langle o\rangle,t_i} \cap W), i \in \{1,\ldots,n\}) &= \mathbb{P}(K_i \subseteq Z^{\mathrm{STIT}}_{\langle o\rangle,t_i}, i \in \{1,\ldots,n\})\\
&\neq \mathbb{P}(K_i \subseteq Z^{\mathrm{PHT}}_{\langle o\rangle,t_i}, i \in \{1,\ldots,n\})\\
&= \mathbb{P}(K_i \subseteq (Z^{\mathrm{PHT}}_{\langle o\rangle,t_i} \cap W), i \in \{1,\ldots,n\})\,.
\end{aligned}
$$

However, this contradicts (6.5).

Finally, (6.4) follows directly by (6.3) and (2.40). □

As already mentioned in Section 4.1 and Remark 4.4, for a translation-invariant measure, we can choose Θ of the form (4.1) with the scaling factor 1, without losing substantial generality. This is shown in the following corollary.

Corollary 6.3. *Let $\mathbf{Y} = (Y_t, t > 0)$ be a STIT tessellation process, driven by a hyperplane measure Θ of the form* (4.1)*, satisfying* (4.3)*, and let $\mathbf{Y}^a = (Y^a_t, t > 0)$ be a STIT tessellation process, driven by the hyperplane measure $a\,\Theta$, $a > 0$. Then*

$$(Y_t, t > 0) \overset{\mathrm{D}}{=} (a\,Y^a_t, t > 0)\,.$$

Proof. If for a fixed window $W \in \mathcal{P}_d$, in Definition 4.2 the measure Θ is replaced by $a\,\Theta$, $a > 0$, then $(Y^a_{t,W}, t > 0) \overset{\mathrm{D}}{=} (Y_{at,W}, t > 0)$ because for all $P \in \mathcal{P}_d$, we have $(a\,\Theta)_{[P]} = \Theta_{[P]}$ (see (4.5)), and if $\mathrm{Expo}(\Theta([P]))$ is the law of an exponentially distributed random variable E, then $\mathrm{Expo}(a\,\Theta([P]))$ is the law of $a^{-1}E$. Furthermore, the spatial consistency yields that $(Y^a_t, t > 0) \overset{\mathrm{D}}{=} (Y_{at}, t > 0)$, and an application of (6.2) completes the proof. □

From (6.1) and (6.2) we can now deduce the key proposition, which justifies the acronym "STIT". Recall that this refers to "STability with respect to ITeration" of tessellations, which is the focus of this book.

Theorem 6.4. *Let Θ be a hyperplane measure of the form* (4.1)*, satisfying* (4.3)*, and let $(Y_t, t > 0)$ be a STIT tessellation process, driven by Θ. Then for any $t > 0$, the random tessellation Y_t is STIT in the sense of Definition* 3.7.

Property (6.1), together with the independence assumptions in Theorem 6.2, leads directly to the following result.

Corollary 6.5. *With the assumptions of Theorem* 6.4*, the process $\mathbf{Y} = (Y_t, t > 0)$ is a time-homogeneous Markov process on the space $(\mathcal{T}, \mathcal{B}(\mathcal{T}))$ of tessellations. Furthermore, the* *process* $(Z^{\mathrm{STIT}}_{\langle o\rangle,t},\ t > 0)$ *of zero-cells is a time-homogeneous Markov process on the space* $(\mathcal{P}_d, \mathcal{B}(\mathcal{P}_d))$ *of d-dimensional polytopes.*

Specifying the general notation introduced in (2.44), we write $\mathbb{Q}^{\mathrm{STIT}}_{d,t}$ and $\mathbb{Q}^{\mathrm{PHT}}_{d,t}$ for the distributions of the typical cell of a STIT tessellation Y_t with time parameter t and a PHT X_t with intensity t, respectively, assuming that both are driven by the same hyperplane measure Θ. Then equation (6.4) can be reformulated as

$$\mathbb{Q}^{\mathrm{STIT}}_{d,t} = \mathbb{Q}^{\mathrm{PHT}}_{d,t}\,.$$

Furthermore, by Corollary 2.23 this implies that the intensity measures of STIT tessellations and PHTs, considered as particle processes, coincide. Hence by the Campbell theorem 19.4 we immediately obtain an identity for functionals applied to the cells of the respective random tessellations.

Corollary 6.6. *Let* Θ *be a hyperplane measure of the form* (4.1)*, satisfying* (4.3)*, let* $(Y_t,\ t > 0)$ *be a STIT tessellation process, and let* $(X_t,\ t > 0)$ *be a PHT process, both driven by* Θ*. Then*

$$\mathbb{E}\sum_{P\in Y_t} f(P) = \mathbb{E}\sum_{P\in X_t} f(P)$$

for all nonnegative measurable functions $f : \mathcal{P}_d \to \mathbb{R}$ *and* $t > 0$.

Considering the random tessellations as particle processes, the intensity measures of Y_t *and of* X_t *coincide.*

In particular, this result enables the direct transfer of a number of propositions to STIT tessellations, which are known for Poisson hyperplane tessellations and are comprehensively presented in [82]. Note that we will find the result of Corollary 6.6 again in Corollary 11.3, where it is derived independently by other methods.

Despite the identically distributed typical cells, the STIT tessellations differ significantly from PHTs. Both the STIT tessellation processes $\mathbf{Y} = (Y_t,\ t > 0)$ and the PHT processes $\mathbf{X} = (X_t,\ t > 0)$ (see (2.66)) are cell division processes. A division event in $\mathbf{Y}$ leads to the division of a *single* cell, and this division is independent of the division of other cells. In contrast, a new hyperplane emerging in $\mathbf{X}$ divides *all the cells it hits*, and this causes a long-range dependence in the division of separate extant cells.

This difference is also reflected in the distinction between the operations iteration and superposition of tessellations, dealt with in Section 3 and illustrated in Figures 3.1 and 3.2. Let $t, s > 0$. Then by (6.1)

$$Y_{t+s} \stackrel{\mathrm{D}}{=} Y_t \boxplus \underline{Y_s}\,,$$

and for the PHT process, with the notation $\uplus$ for the superposition of tessellations introduced in (3.3),

$$X_{t+s} \stackrel{\mathrm{D}}{=} X_t \uplus X_s^{(1)},$$

where $\mathbf{X}^{(1)} = (X_t^{(1)}, t > 0)$ is an independent copy of $\mathbf{X}$. The last equation is a consequence of Theorem 19.13.

The above observations show that for dimensions $d \geq 2$, there is a clear difference between STIT tessellations and PHTs. In contrast, for the one-dimensional case, Lemma 4.10 and the spatial consistency (see Theorem 5.10) yield the following lemma.

Lemma 6.7. *Let $d = 1$, and let Θ be given by* (4.1), *which implies $\varphi(\{-1\}) = \varphi(\{1\}) = 1/2$, and let $(Y_t, t > 0)$ be a STIT tessellation process, driven by Θ. Furthermore, let $\widehat{X}_t$ be a stationary Poisson point process on $\mathbb{R}$ with intensity measure $t\,\lambda_1$, and let X_t be the induced tessellation on $\mathbb{R}$. Then, for all $t > 0$,*

$$Y_t \stackrel{\mathrm{D}}{=} X_t\,.$$

Note that it was also shown in Lemma 3.8 that a tessellation of $\mathbb{R}^1$, generated by a stationary Poisson point process, is a STIT tessellation.

Finally, we show that STIT tessellations satisfy the requirements of Convention 2.28.

Lemma 6.8. *For a given hyperplane measure Θ of the form* (4.1), *satisfying* (4.3), *let $\mathbf{Y} = (Y_t, t > 0)$ be a STIT tessellation process driven by Θ. Then, for all $t > 0$, the tessellation Y_t is stationary and satisfies Convention* 2.28.

Proof. First, the stationarity of Y_t for each $t > 0$ is a consequence of Lemma 5.5 with (5.13), the translation invariance of the hyperplane measure Θ, and Theorem 5.10. To show the finiteness of the densities and intensities, first note that for $k \in \{0,\ldots,d-1\}$, the spatial consistency of Y_t (see (5.17)) implies

$$[0,1]^d \cap \operatorname{skel}_k^{\mathrm{pr}} Y_t \subseteq \operatorname{skel}_k^{\mathrm{pr}} (Y_t \wedge [0,1]^d) \stackrel{\mathrm{D}}{=} \operatorname{skel}_k^{\mathrm{pr}} Y_{t,[0,1]^d}\,.$$

By (4.18) the number of cells of $Y_{t,[0,1]^d}$ is stochastically dominated by a geometrically distributed random variable with parameter $\mathrm{e}^{-t\,\Theta([0,1]^d)}$. We have to take into account that parts of the skeleton can also be located on the boundary of $[0,1]^d$ consisting of $2d$ faces of dimension $d-1$. Any k-primitive is the intersection of $d-k$ polytopes of dimension $d-1$, where these polytopes can be maximal polytopes of $Y_{t,[0,1]^d}$ according to (4.12) or faces of $[0,1]^d$. Hence, if G is a geometrically distributed random variable with parameter $\mathrm{e}^{-t\,\Theta([0,1]^d)}$, then for the number of k-primitives, we have

$$\mathbb{P}(|\mathcal{F}_k^{\mathrm{pr}}(Y_{t,[0,1]^d})| > n) \leq \mathbb{P}\left(\binom{G+1-2d}{d-k} > n\right).$$

Since

$$\binom{G+1-2d}{d-k} \leq (G+1-2d)^{d-k}$$

and all moments of the geometrically distributed random variable are finite, we obtain that $\gamma_k^{\mathrm{pr}} \leq \mathbb{E}|\mathcal{F}_k^{\mathrm{pr}}(Y_{t,[0,1]^d})| < \infty$. Furthermore, since $V_k(F) \leq \sup\{\mathcal{H}^k([0,1]^d \cap L) : L \in A(d,k)\}$, the densities $\eta_k^{\mathrm{pr}} = \mathbb{E}\mathcal{H}^k([0,1]^d \cap \mathrm{skel}_k^{\mathrm{pr}} Y_t)$ are finite as well.

The finiteness of all mean jth intrinsic volumes of the typical k-primitives $\overline{V}_{k,j}^{\mathrm{pr}}$, defined in (2.45), $j \in \{0,\dots,k\}$, $k \in \{0,\dots,d-1\}$, can be shown as follows. By (6.3) the zero-cells of STIT tessellations and PHTs coincide in law when their parameters are suitably matched. For PHT, Theorem 6.1.2 of [82] guarantees that the distribution of the radius of the smallest ball, centered at o and containing the zero-cell, has an exponential tail and hence finite moments of all orders. Monotonicity of intrinsic volumes then implies that all mean jth intrinsic volumes of the typical k-primitives $\overline{V}_{k,j}^{\mathrm{pr}}$ are finite for both models. □

6.2 Densities and directional distributions

By (2.38) and Lemma 2.27 the distribution of the typical cell of a random stationary tessellation determines the cell intensity γ_d, the surface density η_{d-1}, and the spherical directional distribution ψ in the typical point of the $(d-1)$-skeleton. Thus (6.4) yields that these characteristics of STIT tessellations Y_t are the same as those of PHTs X_t, $t > 0$, if the corresponding processes are driven by the same hyperplane measure Θ. For the definition of the PHT process, see Section 2.3.1.

Theorem 6.9. *Let Θ be a hyperplane measure of the form (4.1), satisfying (4.3), let $(Y_t, t > 0)$ be the STIT tessellation process, and let $(X_t, t > 0)$ be the PHT process, both driven by the same hyperplane measure Θ. Then, for all $t > 0$, for the densities η_{d-1}, and for the $(d-1)$-volume-weighted directional distributions $\chi_{d-1}^{\mathrm{STIT}}$ and $\chi_{d-1}^{\mathrm{PHT}}$ of the $(d-1)$-skeletons of the STIT tessellation Y_t and the PHT X_t, respectively, we have*

$$\eta_{d-1}(Y_t) = \eta_{d-1}(X_t) = t\,, \tag{6.6}$$

$$\chi_{d-1}^{\mathrm{STIT}} = \chi_{d-1}^{\mathrm{PHT}}\,. \tag{6.7}$$

By (2.52) this can also be reformulated as

$$\psi^{\mathrm{STIT}} = \psi^{\mathrm{PHT}} = \varphi\,. \tag{6.8}$$

Furthermore, for the densities η_{d-2} of the $(d-2)$-skeletons and the directional distributions $\chi_{d-2}^{\mathrm{STIT}}$ and $\chi_{d-2}^{\mathrm{PHT}}$ in the typical point of the $(d-2)$-skeletons (see (2.50) and (2.18)), we have

$$\eta_{d-2}(Y_t) = 2\eta_{d-2}(X_t)\,, \tag{6.9}$$

$$\chi_{d-2}^{\mathrm{STIT}} = \chi_{d-2}^{\mathrm{PHT}}\,. \tag{6.10}$$

Note that a relation between $\eta_k(Y_t)$ and $\eta_k(X_t)$ for all $k \in \{0, \dots, d-1\}$ is provided by equation (14.13), where that result is derived from the multivariate Mecke-type equation of Corollary 14.7 and results from Section 11.3.1.

Proof of Theorem 6.9. Equation (6.6) follows by (2.38) and Theorem 4.4.3 of [172], applied to the Poisson hyperplane process defined in (2.65).

Then (6.7) and the first equation in (6.8) are consequences of (2.58) and (2.52) since the typical cells of the STIT tessellation Y_t and the PHT X_t are identically distributed; see (6.4). To show the second equation in (6.8), we adapt the proof of Theorem 4.4.3 of [172] as follows. For $A \in \mathcal{B}(G(d,1))$, using (4.2), we have

$$\begin{aligned}
&\eta_{d-1}(X_t)\,\psi_o^{\mathrm{PHT}}(A) \\
&\quad = \mathbb{E} \sum_{H \in \widehat{X}_t} \int_{\mathbb{R}^d} \mathbb{1}\{x \in H \cap [0,1]^d\}\, \mathbb{1}\{H^\perp \in A\}\, \mathscr{H}^{d-1}(\mathrm{d}x) \\
&\quad = t \int_{\mathcal{H}} \mathscr{H}^{d-1}(H \cap [0,1]^d)\, \mathbb{1}\{H^\perp \in A\}\, \Theta(\mathrm{d}H) \\
&\quad = t \int_{S^{d-1}} \int_{\mathbb{R}} \mathscr{H}^{d-1}(H(u,r) \cap [0,1]^d)\, \mathbb{1}\{H(u,r)^\perp \in A\}\, \lambda_1(\mathrm{d}r)\, \varphi(\mathrm{d}u) \\
&\quad = t\, \varphi_o(A)\,.
\end{aligned}$$

Thus by (2.52) the second equation in (6.8) is shown.

By (2.18) we obtain that $\mathrm{skel}^{\mathrm{pr}}_{d-2}\, T = \mathrm{skel}_{d-2}\, T$ for all $T \in \mathcal{T}$, and hence $\eta^{\mathrm{pr}}_{d-2}(X) = \eta_{d-2}(X)$ for arbitrary random stationary tessellations X.

To prove (6.9) and (6.10), first recall that for a random stationary tessellation X and $A \in \mathcal{B}(G(d, d-2))$, by (2.50) we have

$$\eta_{d-2}(X)\,\chi_{d-2}(A) = \mathbb{E} \int_{\mathbb{R}^d} \mathbb{1}\{x \in [0,1]^d \cap \mathrm{skel}_{d-2}\, X\}\, \mathbb{1}\{\mathrm{aff}_o\, F(x) \in A\}\, \mathscr{H}^{d-2}(\mathrm{d}x)\,.$$

Consider the $(d-2)$-polytopes that belong to the $(d-2)$-skeleton. In a PHT, they are generated by the intersection of exactly two hyperplanes. Hence almost all $x \in [0,1]^d \cap \mathrm{skel}_{d-2}\, X_t$ belong to $(d-2)$-faces of exactly four cells; the exceptions are located on the relative boundaries of $(d-2)$-faces, and thus the set of all exceptions has $(d-2)$-volume zero. By Definition 4.2 of the construction of STIT tessellations almost all $x \in [0,1]^d \cap \mathrm{skel}_{d-2}\, Y_t$ belong to $(d-2)$-faces of exactly two cells, namely the two cells which are born by division of a common “mother” cell. The set of exceptions has again $(d-2)$-volume zero.

Similarly to the proof of Lemma 2.27, for a STIT tessellation Y_t with fixed $t > 0$ and the distribution $\mathbb{Q}^{\mathrm{STIT}}_{d,t}$ of the typical cell Z^{STIT}_t, we obtain

$$\begin{aligned}
&\eta_{d-2}(Y_t)\,\chi^{\mathrm{STIT}}_{d-2}(A) \\
&\quad = \mathbb{E} \int_{\mathbb{R}^d} \mathbb{1}\{x \in [0,1]^d \cap \mathrm{skel}_{d-2}\, Y_t\}\, \mathbb{1}\{\mathrm{aff}_o\, F(x) \in A\}\, \mathscr{H}^{d-2}(\mathrm{d}x)
\end{aligned}$$

$$= \frac{1}{2}\mathbb{E} \sum_{(c,P)\in\mathcal{Z}(Y_t)} \sum_{F\in\mathcal{F}_{d-2}(P)} V_{d-2}((F+c)\cap[0,1]^d)\,\mathbb{1}\{\mathrm{aff}_o F \in A\}$$
$$= \frac{1}{2}\gamma_d(Y_t) \int_{\mathbb{R}^d}\int_{\mathcal{P}_{d,c}} \sum_{F\in\mathcal{F}_{d-2}(P)} V_{d-2}((F+x)\cap[0,1]^d)\,\mathbb{1}\{\mathrm{aff}_o F \in A\}\,\mathbb{Q}_{d,t}^{\mathrm{STIT}}(\mathrm{d}P)\,\lambda_d(\mathrm{d}x)$$
$$= \frac{1}{2}\gamma_d(Y_t)\,\mathbb{E} \sum_{F\in\mathcal{F}_{d-2}(Z_t^{\mathrm{STIT}})} V_{d-2}(F)\,\mathbb{1}\{\mathrm{aff}_o F \in A\},$$

and, analogously for a Poisson hyperplane tessellation X_t,

$$\eta_{d-2}(X_t)\,\chi_{d-2}^{\mathrm{PHT}}(A) = \frac{1}{4}\gamma_d(X_t)\,\mathbb{E} \sum_{F\in\mathcal{F}_{d-2}(Z_t^{\mathrm{PHT}})} V_{d-2}(F)\,\mathbb{1}\{\mathrm{aff}_o F \in A\}.$$

Since $Z_t^{\mathrm{STIT}} \overset{\mathrm{D}}{=} Z_t^{\mathrm{PHT}}$ by (6.4), equation (2.38) yields that $\gamma_d(Y_t) = \gamma_d(X_t)$ and

$$\mathbb{E} \sum_{F\in\mathcal{F}_{d-2}(Z_t^{\mathrm{STIT}})} V_{d-2}(F)\,\mathbb{1}\{\mathrm{aff}_o F \in A\} = \mathbb{E} \sum_{F\in\mathcal{F}_{d-2}(Z_t^{\mathrm{PHT}})} V_{d-2}(F)\,\mathbb{1}\{\mathrm{aff}_o F \in A\},$$

which proves (6.9) and (6.10). □

Note that such simple links as shown in Theorem 6.9 do not hold for the k-skeletons if $k < d-2$. In these cases the structure of the skeletons of STIT tessellations is much more involved.

In Remark 17.4, relations between the intensities and distributions of the typical k-faces, $k \in \{0,\dots,d\}$, of STIT tessellations and PHT are provided, where the *multiplicities* of the k-faces in STIT tessellations are taken into account.

For PHTs, all densities and directional distributions can be expressed by integrals with respect to the directional distribution φ, as it is shown in Section 2.3.4. For $k = d-2$, this in particular yields

$$\eta_{d-2}(Y_t) = 2\eta_{d-2}(X_t) = t^2 \int_{S^{d-1}}\int_{S^{d-1}} |\sin\sphericalangle(u,u')|\,\varphi(\mathrm{d}u)\,\varphi(\mathrm{d}u') \quad \text{for all } t>0;$$

see also (16.18) for $d = 2$ and (17.6) for $d = 3$. Furthermore, for the directional distribution, we obtain

$$\chi_{d-2}^{\mathrm{STIT}}(A) = \chi_{d-2}^{\mathrm{PHT}}(A)$$
$$= \frac{\int_{S^{d-1}}\int_{S^{d-1}} \mathbb{1}\{\mathrm{lin}(u)^\perp \cap \mathrm{lin}(u')^\perp \in A\}\,|\sin\sphericalangle(u,u')|\,\varphi(\mathrm{d}u)\,\varphi(\mathrm{d}u')}{\int_{S^{d-1}}\int_{S^{d-1}} |\sin\sphericalangle(u,u')|\,\varphi(\mathrm{d}u)\,\varphi(\mathrm{d}u')}$$

for all $A \in \mathcal{B}(G(d,d-2))$.

6.3 Sections with affine subspaces

We now consider the tessellations that arise when STIT tessellations are intersected by affine subspaces of $\mathbb{R}^d$. Let $(Y_t, t > 0)$ be a STIT tessellation process, and let $L \in A(d,k)$, $k \in \{1,\dots,d-1\}$, be a k-dimensional affine subspace. Then for all $t > 0$, the section $Y_t \cap L$ (see (2.22)) is a. s. well-defined, and by Lemma 3.9(a) it is a STIT tessellation within L. Furthermore, the distribution of $Y_t \cap L$ is invariant under translations that leave L invariant. Since Y_t is stationary in space, the flat L can be translated, and therefore we can assume without loss of generality that $o \in L$ and thus $L \in G(d,k)$, which means that L is a linear subspace.

In this section, we use the notation $\mathcal{H}^{(L)} := A(L,k-1)$ for the set of $(k-1)$-dimensional affine subspaces of L. Recall that $\mathcal{H} := A(d,d-1)$ is the set of hyperplanes in $\mathbb{R}^d$. For a hyperplane measure Θ of the form (4.1), satisfying (4.3), and a linear subspace $L \in G(d,k)$, define the measure Θ_L on $\mathcal{H}^{(L)}$ by

$$\begin{aligned}\int\limits_{\mathcal{H}^{(L)}} f(H_L)\,\Theta_L(\mathrm{d}H_L) &:= \int\limits_{\mathcal{H}} f(H \cap L)\,\mathbb{1}\{H^{\perp} \not\subset L^{\perp}\}\,\Theta(\mathrm{d}H)\\ &= \int\limits_{S^{d-1}}\int\limits_{\mathbb{R}} f(H(u,r) \cap L)\,\mathbb{1}\{u \notin L^{\perp}\}\,\lambda_1(\mathrm{d}r)\,\varphi(\mathrm{d}u)\end{aligned} \tag{6.11}$$

for nonnegative measurable functions $f : \mathcal{H}^{(L)} \to \mathbb{R}$. In particular, for $P \subset L$, $P \in \mathcal{P}_k$, denote $[P]_L := \{H_L \in \mathcal{H}^{(L)} : H_L \cap P \neq \emptyset\}$. This yields that

$$\Theta_L([P]_L) = \Theta([P]).$$

Intuitively, Θ_L is the image measure of Θ under the mapping $H \mapsto H \cap L$ for $H \in \mathcal{H}$. Note that $H \cap L$ can be empty or equal to L and that the Θ-measure of sets of hyperplanes H which are parallel to L can be positive. Nevertheless, by assumption (4.3) on Θ it is guaranteed that Θ_L is a nonzero measure.

Note that this measure Θ_L is the same as that in Theorem 2.37, where it is described in more detail.

Lemma 6.10. *Let Θ be a hyperplane measure of the form* (4.1), *satisfying* (4.3), *and let $(Y_t, t > 0)$ be the STIT tessellation process driven by Θ. Then, for all $t > 0$, $L \in G(d,k)$, $k \in \{1,\dots,d-1\}$, and the measure Θ_L, we have*

$$Y_t \cap L \overset{\mathrm{D}}{=} Y_t(\Theta_L, L), \tag{6.12}$$

where $(Y_t(\Theta_L, L), t > 0)$ denotes the STIT tessellation process in L, driven by Θ_L.

Proof. First observe that by Lemma 3.9(a) the tessellation $Y_t \cap L$ is a STIT tessellation in L. For a window $W_L \subset L$, $W_L \in \mathcal{P}_k$, consider the pure jump Markov process $((Y_t \cap L) \wedge W_L, t > 0)$. The rate function and the jump transition kernel of this process can be determined as it was done in Section 4.3 for a STIT tessellation process in a window. The key argument is that for a cell $P \in (Y_t \cap L) \wedge W_L$, the life time is exponentially distributed with parameter $\Theta([P]) = \Theta_L([P]_L)$. This can be shown using the Markov property and (5.6), replacing there W by P. Hence the rate function is given by

$$\theta_{W_L}(T) = \sum_{P \in T \wedge W_L} \Theta_L([P]_L), \quad T \in \mathcal{T}_L .$$

Furthermore, the measure μ_T given in (4.23), which describes the random selection of the cell to be divided and the rule of the dividing affine subspace, now becomes

$$\mu_{W_L,T}(\cdot) = \sum_{P \in T \wedge W_L} \frac{\Theta_L([P]_L)}{\theta_{W_L}(T)} \delta_P \otimes \frac{\Theta_L(\cdot \cap [P]_L)}{\Theta_L([P]_L)}, \quad T \in \mathcal{T}_L .$$

Thus the jump transition kernel of $((Y_t \cap L) \wedge W_L, t > 0)$ coincides with that of the STIT tessellation process in W_L driven by the measure Θ_L. Analogously to Lemma 4.11, we can show that this also holds for the generator of the process. By Theorem 4.14 and the spatial consistency of STIT tessellations, Theorem 5.10, we obtain (6.12), and the proof is complete. □

We now analyze the densities and directional distributions of sectional STIT tessellations.

Lemma 6.11. *Let Θ be a hyperplane measure of the form* (4.1), *satisfying* (4.3), *let $(Y_t, t > 0)$ be the STIT tessellation process, and let $(X_t, t > 0)$ be the PHT process, both driven by the same hyperplane measure Θ. Then, for all $t > 0$ and any k-dimensional linear subspace $L \in G(d, k)$, $k \in \{1, \dots, d-1\}$, the densities η_{k-1} and the $(k-1)$-volume-weighted directional distributions χ_{k-1} of $Y_t \cap L$ and PHT $X_t \cap L$ are identical.*

Proof. First note that the zero-cell of $Y_t \cap L$ is the section of the zero-cell of Y_t with L. By (6.3) we have

$$Z_{\langle o \rangle}(Y_t \cap L) = Z^{\mathrm{STIT}}_{\langle o \rangle,t} \cap L \overset{\mathrm{D}}{=} Z^{\mathrm{PHT}}_{\langle o \rangle,t} \cap L = Z_{\langle o \rangle}(X_t \cap L) .$$

Hence (2.40) yields the identity of the distributions of the typical cells of $Y_t \cap L$ and $X_t \cap L$. Thus by Lemma 2.27 also the densities η_{k-1} and $(k-1)$-volume-weighted directional distributions χ_{k-1} of the respective random tessellations coincide. □

Note that Lemma 6.11 can also be shown by a combination of Theorem 2.37 with Lemma 6.10 and Theorem 6.9.

In the next step we apply results for PHTs to derive more concrete formulae for the densities and directional distributions of the intersections $Y_t \cap L$. For $u \in S^{d-1}$ and $L \in G(d, k)$, we denote by $\nabla_{d-k+1}(L^\perp, u)$ the $(d-k+1)$-dimensional volume of the parallelepiped spanned by an orthonormal basis of $L^\perp$ and the unit vector u. The set of all $(k-1)$-dimensional linear subspaces of L is denoted by $G_L(k, k-1)$.

Theorem 6.12. *For $d \geq 3$, let $(Y_t, t > 0)$ be a STIT tessellation process, driven by a hyperplane measure Θ of the form (4.1), satisfying (4.3). Furthermore, let $L \in G(d,k)$, $k \in \{2, \dots, d-1\}$, be a k-dimensional linear subspace of $\mathbb{R}^d$.*

Then for all $t > 0$, the $(k-1)$-skeleton of $Y_t \cap L$ has the $(k-1)$-volume-weighted directional distribution $\chi_{k-1}(Y_t \cap L, \cdot)$ and the density $\eta_{k-1}(Y_t \cap L)$, that is, the mean total $(k-1)$-volume of $\partial Y_t \cap L$ per unit k-volume in L, given by

$$\eta_{k-1}(Y_t \cap L)\chi_{k-1}(Y_t \cap L, A) = t \int_{S^{d-1}} \mathbb{1}\{(L \cap H(0,u)) \in A\} \nabla_{d-k+1}(L^{\perp}, u)\, \varphi(du) \tag{6.13}$$

for $A \in \mathcal{B}(G_L(k, k-1))$.

Moreover, any other STIT tessellation process within L with the parameters described in (6.13) has the same distribution as $(Y_t \cap L, t > 0)$.

Proof. By Lemma 6.11 the respective characteristics can be determined considering a stationary PHT X_t instead of Y_t. An application of Theorem 4.4.7 and Theorem 4.4.3 of [172], together with (2.80), which relates χ_{d-1} and φ, yields (6.13). By Theorem 7.9 any stationary STIT tessellation within L with the same characteristics η_{k-1} and φ has the same distribution as $Y_t \cap L$. □

For the isotropic case, we obtain an explicit formula for the intensities of tessellations arising from sections of STIT tessellations.

Lemma 6.13. *If $(Y_t, t > 0)$ is a process of isotropic STIT tessellations, driven by the motion invariant hyperplane measure Θ_{iso}, then for all $t > 0$ and all $L \in G(d,k)$, $d \geq 2$, $k \in \{1, \dots, d-1\}$, the random tessellation $Y_t \cap L$ is isotropic within L with the density*

$$\eta_{k-1}(Y_t \cap L) = t\, c_{k-1,d}\,,$$

where $c_{k-1,d}$ is defined in (1.6).

Proof. Denote by $(X_t, t > 0)$ a PHT process driven by Θ_{iso} and generated by a hyperplane process $(\widehat{X}_t, t > 0)$; see (2.65). Furthermore, choose $K \in \mathcal{K}$ with $K \subset L$ and $V_k(K) = 1$. Then, by (6.6) and the Campbell theorem 19.4,

$$\begin{aligned}\eta_{k-1}(Y_t \cap L) &= \eta_{k-1}(X_t \cap L) \\ &= \mathbb{E} \sum_{H \in \widehat{X}_t} V_{k-1}(H \cap K) \\ &= t \int_{\mathcal{H}} V_{k-1}(H \cap K)\, \Theta_{\text{iso}}(dH)\,.\end{aligned}$$

Thus an application of the Crofton formula (1.6) completes the proof. □

We consider now a general hyperplane measure Θ and focus on the particular case $k = 1$, that is, sections of a STIT tessellation with lines. The function $\eta_0(Y_t \cap \cdot) : G(d,1) \to [0,\infty)$ expresses the intensities of the section point processes $\partial Y_t \cap L$ as a function of L. For the skeletons of generic random stationary tessellations and, more generally, also for stationary surface processes, this function is referred to as the **rose of intersections**; see page 335 of [33].

Let Θ be a hyperplane measure of the form (4.1), satisfying (4.3), let $(Y_t,\ t > 0)$ be the STIT tessellation process, and let $(X_t,\ t > 0)$ be the PHT process, both driven by the same hyperplane measure Θ. Recall that $\Pi_{t\Theta}$ denotes the associated zonoid for the pair (t,Θ), defined by (2.68). For $u \in S^{d-1}$, let $\mathrm{lin}(u) \in G(d,1)$ denote the spanned one-dimensional linear subspace. Thus by Lemma 6.11, Theorem 6.12, and (2.68) the support function of $\Pi_{t\Theta}$ is related to the rose of intersections for the PHT X_t as well as for the STIT tessellation Y_t by

$$2\,h(\Pi_{t\Theta}, u) = \eta_0(X_t \cap \mathrm{lin}(u)) = \eta_0(Y_t \cap \mathrm{lin}(u)), \quad u \in S^{d-1}. \tag{6.14}$$

Applying Lemma 3.9(a) and Lemma 6.7, we obtain the following stronger proposition specifying the distribution of the section point process on one-dimensional linear subspaces.

Corollary 6.14. *Let $(Y_t,\ t > 0)$ be a STIT tessellation process, driven by a hyperplane measure Θ of the form (4.1), satisfying (4.3). The section point processes $\partial Y_t \cap \mathrm{lin}(u)$ on $\mathrm{lin}(u)$ is a stationary Poisson point process with intensity $2\,h(\Pi_{t\Theta}, u) = t\,V_1(\Pi_\Theta|\mathrm{lin}(u))$.*

6.4 Linear transformations of STIT tessellations

Now we will show that a linear transformation of a STIT tessellation in $\mathbb{R}^d$ yields again a STIT tessellation, and its driving measure is described.

In this section, all vectors are *column* vectors. For a matrix A, the transposed matrix is denoted A^T. Fix $n \in \mathbb{N}$ with $n \geq d$. For a $(d \times n)$-matrix A and a set $C \subseteq \mathbb{R}^d$, we write $\mathsf{A}^\mathrm{T} C := \{\mathsf{A}^\mathrm{T} x : x \in C\}$. Furthermore, $\mathsf{A}^\mathrm{T} T := \{\mathsf{A}^\mathrm{T} P : P \in T\}$ for a tessellation $T \in \mathcal{T}$. Now assume that A has rank d and denote by $v_1, \dots, v_d$ the rows of A. Furthermore, let $L_\mathsf{A} := \mathrm{lin}(v_1, \dots, v_d)$ be the linear subspace they span in $\mathbb{R}^n$. We denote by $\mathcal{H}^{(L_\mathsf{A})} := A(L_\mathsf{A}, d-1)$ the set of $(d-1)$-dimensional affine subspaces of $\mathbb{R}^n$ that are subsets of L_A. Recall that $\mathcal{H} := A(d, d-1)$ is the set of hyperplanes in $\mathbb{R}^d$.

Define the map $g_\mathsf{A} : \mathcal{H} \to \mathcal{H}^{(L_\mathsf{A})}$ by

$$g_\mathsf{A}(H) := \mathsf{A}^\mathrm{T} H, \quad H \in \mathcal{H}.$$

For a hyperplane measure Θ on $\mathcal{H}$, denote by $(Y_t(\Theta \circ g_\mathsf{A}^{-1}, L_\mathsf{A}),\ t > 0)$ the STIT tessellation process on L_A, driven by the image measure $\Theta \circ g_\mathsf{A}^{-1}$.

Lemma 6.15. *Let $(Y_t, t > 0)$ be the STIT tessellation process in $\mathbb{R}^d$, driven by a hyperplane measure Θ of the form (4.1), satisfying (4.3). Furthermore, for $n \geq d$, let A be a $(d{\times}n)$-matrix of rank d. Then for all $t > 0$,*

$$\mathsf{A}^{\mathrm{T}}\, Y_t \overset{\mathrm{D}}{=} Y_t(\Theta \circ g_{\mathsf{A}}^{-1}, L_{\mathsf{A}})\,. \tag{6.15}$$

This means that the linear transform $\mathsf{A}^{\mathrm{T}}\, Y_t$ of Y_t has the same distribution as the STIT tessellation process at time t in the linear subspace $L_{\mathsf{A}} \subseteq \mathbb{R}^n$, driven by the image measure $\Theta \circ g_{\mathsf{A}}^{-1}$.

Note that in the case $n = d$, we have that $L_{\mathsf{A}} = \mathbb{R}^d$, and thus

$$\mathsf{A}^{\mathrm{T}}\, Y_t \overset{\mathrm{D}}{=} Y_t(\Theta \circ g_{\mathsf{A}}^{-1}, \mathbb{R}^d)\,.$$

Proof of Lemma 6.15. Observe that $\mathsf{A}^{\mathrm{T}}\, \mathbb{R}^d = L_{\mathsf{A}}$ by the definition of L_{A}. Moreover, as A has rank d, the mapping $x \mapsto \mathsf{A}^{\mathrm{T}}\, x$, for $x \in \mathbb{R}^d$, is a bijection between $\mathbb{R}^d$ and L_{A}.

The linearity of the map implies that $g_{\mathsf{A}} : \mathcal{H} \to \mathcal{H}^{(L_{\mathsf{A}})}$ is also a bijection between the respective sets of affine subspaces.

Hence, because Y_t is a stationary tessellation, it follows that $\mathsf{A}^{\mathrm{T}}\, Y_t$ is a random tessellation of L_{A} with distribution that is invariant under translations in L_{A}.

By the mentioned bijection property, for each cell P_{A} of $\mathsf{A}^{\mathrm{T}}\, Y_t$, there is a unique polytope $P \in Y_t$ with $P_{\mathsf{A}} = \mathsf{A}^{\mathrm{T}}\, P$. This means that Y_t can be recovered uniquely from $\mathsf{A}^{\mathrm{T}}\, Y_t$, and hence also the distributions of the life time and the division rule of P_{A} are uniquely determined. Consider the division of P and the induced division of P_{A}. Because g_{A} is a linear transform, P is intersected by a hyperplane H if and only if its image $g_{\mathsf{A}}(H)$ intersects P_{A}. The life time of P is exponentially distributed with parameter $\Theta([P])$, and hence also the life time of P_{A} is exponentially distributed with parameter $\Theta([P]) = \Theta \circ g_{\mathsf{A}}^{-1}([\mathsf{A}^{\mathrm{T}}\, P]) = \Theta \circ g_{\mathsf{A}}^{-1}([P_{\mathsf{A}}])$. Let $\widetilde{\Theta}$ denote the division rule for P_{A} that is induced by the division rule $\Theta_{[P]}$ for P. Then, for $B \in \mathcal{B}(\mathcal{H}^{(L_{\mathsf{A}})}) \llcorner [P_{\mathsf{A}}]$,

$$\widetilde{\Theta}(B) = \Theta_{[P]}(g_{\mathsf{A}}^{-1}(B)) = \frac{\Theta \circ g_{\mathsf{A}}^{-1}(B)}{\Theta([P])} = \frac{\Theta \circ g_{\mathsf{A}}^{-1}(B)}{\Theta \circ g_{\mathsf{A}}^{-1}([P_{\mathsf{A}}])} = (\Theta \circ g_{\mathsf{A}}^{-1})_{[P_{\mathsf{A}}]}(B)\,.$$

Thus P_{A} has the same life time distribution and division rule as a cell in a STIT tessellation process on L_{A}, driven by $\Theta \circ g_{\mathsf{A}}^{-1}$. This shows (6.15), and the proof is completed. □

As an application of Lemma 6.15, we consider STIT tessellations with discrete directional distributions and their relation to Mondrian STIT tessellations. If $u_1, \dots, u_d$ are a base in $\mathbb{R}^d$ (not necessarily orthonormal) and Θ is the hyperplane measure of the form (4.1) with

$$\varphi := \frac{1}{2} \sum_{j=1}^{d} p_j \,(\delta_{u_j} + \delta_{-u_j})\,,$$

where $\sum_{j=1}^{d} p_j = 1$ and $p_j > 0$ for all $j \in \{1, \dots, d\}$, then for all $t > 0$, the cells of the STIT tessellation Y_t are "oblique" parallelepipeds. Thus by a linear transform of Y_t we can obtain a Mondrian STIT tessellation generated by the Mondrian hyperplane measure Θ_M of the form (4.1) with spherical directional distribution

$$\varphi_M := \frac{1}{2d} \sum_{i=1}^{d} (\delta_{e_i} + \delta_{-e_i});$$

see (4.4). The cells of that Mondrian STIT tessellation are orthogonal parallelepipeds.

More generally, if the spherical directional distribution φ of Θ is concentrated on $n > d$ directions, then a linear transform of the corresponding STIT tessellation has the same distribution as the section of a Mondrian STIT tessellation in $\mathbb{R}^n$ with a certain linear subspace. This is now formalized. Denote by $\mathcal{H}^{(n)} := A(n, n-1)$ the set of all hyperplanes in $\mathbb{R}^n$.

Theorem 6.16. *For $n \geq d$, let $\{u_1, \dots, u_n\} \subset S^{d-1}$ be such that $u_1, -u_1, \dots, u_n, -u_n$ are pairwise different and contain a base of $\mathbb{R}^d$. Furthermore, let $(Y_t, t > 0)$ be the STIT tessellation process in $\mathbb{R}^d$, driven by the hyperplane measure Θ of the form* (4.1) *with*

$$\varphi := \frac{1}{2} \sum_{j=1}^{n} p_j (\delta_{u_j} + \delta_{-u_j}),$$

where $\sum_{j=1}^{n} p_j = 1$ and $p_j > 0$ for all $j \in \{1, \dots, n\}$. Denote by A the $(d \times n)$-matrix with columns $p_1 u_1, \dots, p_n u_n$. Additionally, let $v_1, \dots, v_d$ be the rows of A, and let $L_{\mathsf{A}} := \mathrm{lin}(v_1, \dots, v_d)$ be the linear subspace they span in $\mathbb{R}^n$. Moreover, let $(Y_t(\Theta_M, \mathbb{R}^n), t > 0)$ be the Mondrian STIT tessellation process in $\mathbb{R}^n$ driven by the hyperplane measure Θ_M on $\mathcal{H}^{(n)}$ of the form (4.1) *with spherical directional distribution*

$$\varphi_M := \frac{1}{2n} \sum_{j=1}^{n} (\delta_{e_j} + \delta_{-e_j});$$

see (4.4). *Then for all $t > 0$,*

$$\mathsf{A}^{\mathrm{T}} Y_t \overset{\mathrm{D}}{=} Y_{n\,t}(\Theta_M, \mathbb{R}^n) \cap L_{\mathsf{A}}.$$

Note that in the case $d = n$, we have $L_{\mathsf{A}} = \mathrm{lin}(v_1, \dots, v_d) = \mathbb{R}^d$, and thus for all $t > 0$,

$$\mathsf{A}^{\mathrm{T}} Y_t \overset{\mathrm{D}}{=} Y_{d\,t}(\Theta_M, \mathbb{R}^d).$$

This means that in this case a certain linear transform in space and a time scaling can transmute the STIT tessellation to a Mondrian STIT tessellation.

Proof of Theorem 6.16. The proof is based on Lemmas 6.15 and 6.10. Consider the map $g_{\mathsf{A}} : \mathcal{H} \to \mathcal{H}^{(L_{\mathsf{A}})}$ given by

$$g_{\mathsf{A}}(H) := \mathsf{A}^{\mathrm{T}} H, \quad H \in \mathcal{H}.$$

The image measure is $\Theta \circ g_{\mathsf{A}}^{-1}$ on $\mathcal{H}^{(L_{\mathsf{A}})}$. Now by Lemma 6.15, for all $t > 0$,

$$\mathsf{A}^{\mathrm{T}} Y_t \overset{\mathrm{D}}{=} Y_t(\Theta \circ g_{\mathsf{A}}^{-1}, L_{\mathsf{A}}). \tag{6.16}$$

Furthermore, in $\mathbb{R}^n$, denote by $\Theta_{M,L_{\mathsf{A}}}$ the measure on $\mathcal{H}^{(L_{\mathsf{A}})}$ assigned to Θ_M and L_{A} according to (6.11). Then Lemma 6.10 yields that

$$Y_t(\Theta_M, \mathbb{R}^n) \cap L_{\mathsf{A}} \overset{\mathrm{D}}{=} Y_t(\Theta_{M,L_{\mathsf{A}}}, L_{\mathsf{A}}). \tag{6.17}$$

Regarding (6.16) and (6.17), it remains to compare the measure $\Theta \circ g_{\mathsf{A}}^{-1}$ with $\Theta_{M,L_{\mathsf{A}}}$. First note that the definition of the matrix A yields that $\mathsf{A}\, e_j = p_j\, u_j, j \in \{1, \dots, n\}$, and hence for all $x \in H(u_j, r), H(u_j, r) \in \mathcal{H}, r \in \mathbb{R}$, by the definition of A,

$$\langle \mathsf{A}^{\mathrm{T}} x, e_j \rangle = \langle x, \mathsf{A}\, e_j \rangle = \langle x, p_j\, u_j \rangle = p_j\, r.$$

Thus, because A has rank d,

$$\mathsf{A}^{\mathrm{T}} H(u_j, r) = H(e_j, p_j\, r) \cap L_{\mathsf{A}} \in \mathcal{H}^{(L_{\mathsf{A}})}.$$

It follows that for $B \in \mathcal{B}(\mathcal{H}^{(L_{\mathsf{A}})})$,

$$\begin{aligned}
\Theta \circ g_{\mathsf{A}}^{-1}(B) &= \Theta(\{H \in \mathcal{H} : \mathsf{A}^{\mathrm{T}} H \in B\}) \\
&= \sum_{j=1}^{n} p_j \int_{\mathbb{R}} \mathbb{1}\{\mathsf{A}^{\mathrm{T}} H(u_j, r) \in B\}\, \lambda_1(\mathrm{d}r) \\
&= \sum_{j=1}^{n} \int_{\mathbb{R}} \mathbb{1}\{H(e_j, p_j\, r) \cap L_{\mathsf{A}} \in B\}\, p_j\, \lambda_1(\mathrm{d}r) \\
&= \sum_{j=1}^{n} \int_{\mathbb{R}} \mathbb{1}\{H(e_j, r) \cap L_{\mathsf{A}} \in B\}\, \lambda_1(\mathrm{d}r) \\
&= n\, \Theta_M(\{H_n \in \mathcal{H}^{(n)} : (H_n \cap L_{\mathsf{A}}) \in B\}) \\
&= n\, \Theta_{M,L_{\mathsf{A}}}(B).
\end{aligned}$$

Hence

$$\Theta \circ g_{\mathsf{A}}^{-1} = n\, \Theta_{M,L_{\mathsf{A}}}. \tag{6.18}$$

By Corollary 6.3, $Y_t(n\, \Theta_{M,L_{\mathsf{A}}}, L_{\mathsf{A}}) \overset{\mathrm{D}}{=} Y_{nt}(\Theta_{M,L_{\mathsf{A}}}, L_{\mathsf{A}})$ for all $t > 0$. Thus, summarizing (6.16), (6.17), and (6.18), we complete the proof. □

6.5 Stationarity on the time axis

Regarding the STIT tessellation process $(Y_t,\ t > 0)$, we have by (6.2) that $t\,Y_t \overset{D}{=} Y_1$ for all $t > 0$. This means that the one-dimensional distributions of the process $(t\,Y_t,\ t > 0)$ do not depend on t. Nevertheless, this is not a time-stationary process. This can be seen by considering the transition probabilities of this process. For example, using (6.1), we have

$$\begin{aligned}&\mathbb{P}(2\,\partial Y_2 \cap B^d = \emptyset \mid \partial Y_1 \cap B^d = \emptyset)\\ &\quad = \mathbb{P}(2\,\partial(Y_1 \boxplus (Y_1^{(i)},\ i \in \mathbb{N}_0)) \cap B^d = \emptyset \mid \partial Y_1 \cap B^d = \emptyset)\\ &\quad = \mathbb{P}(2\,\partial Y_1^{(0)} \cap B^d = \emptyset),\end{aligned}$$

because $2\,\partial Y_1 \cap B^d = \emptyset$ if $\partial Y_1 \cap B^d = \emptyset$. Analogously, for the same time difference, we have

$$\begin{aligned}&\mathbb{P}(1000\,\partial Y_{1000} \cap B^d = \emptyset \mid 999\,\partial Y_{999} \cap B^d = \emptyset)\\ &\quad = \mathbb{P}(1000\,\partial(Y_{999} \boxplus (Y_1^{(i)},\ i \in \mathbb{N}_0)) \cap B^d = \emptyset \mid 999\,\partial Y_{999} \cap B^d = \emptyset)\\ &\quad = \mathbb{P}(1000\,\partial Y_1^{(0)} \cap B^d = \emptyset),\end{aligned}$$

and the difference of these two probabilities is obvious. On the other hand, a time-stationary process can be defined by an exponential time change. This resembles the well-known classical situation of the Wiener process and the Ornstein–Uhlenbeck process; see Example 7.16 in [98].

Theorem 6.17. *Let $(Y_t,\ t > 0)$ be a STIT tessellation process, driven by a hyperplane measure Θ of the form* (4.1), *satisfying* (4.3), *and let $a > 1$. Then*

$$(a^t\,Y_{a^t},\ t \in \mathbb{R})$$

is a time-stationary process.

Proof. As argued in the proof of Theorem 5.10, it is sufficient to consider the finite-dimensional distributions of the process $(a^t\,Y_{a^t},\ t \in \mathbb{R})$. This is due to Kolmogorov's theorem, Theorem 8.23 of [89] and the fact that the state space $\mathcal{T}$ of the process is a Borel space according to Lemma 18.2. Hence it is sufficient to show that for all $t > 0$, $n \in \mathbb{N}$, and $(s_1, \dots, s_n) \in \mathbb{R}^n$ with $s_1 < \cdots < s_n$, we have the equality in distribution

$$(a^{s_1+t}\,Y_{a^{s_1+t}}, \dots, a^{s_n+t}\,Y_{a^{s_n+t}}) \overset{D}{=} (a^{s_1}\,Y_{a^{s_1}}, \dots, a^{s_n}\,Y_{a^{s_n}})\,.$$

For simplicity, we show this only for $n = 2$, and the case of general $n \in \mathbb{N}$ is then straightforward.

First, with (6.2), we obtain that $a^t\,Y_{a^t a^{s_1}} \overset{D}{=} Y_{a^{s_1}}$ and $a^t\,Y_{a^t(a^{s_2}-a^{s_1})} \overset{D}{=} Y_{a^{s_2}-a^{s_1}}$. Then an application of (3.5) and property (6.1), together with Theorem 6.2, yields

$$\begin{aligned}
(a^{s_1+t} Y_{a^{s_1+t}}, a^{s_2+t} Y_{a^{s_2+t}}) &\overset{\mathrm{D}}{=} (a^{s_1} a^t Y_{a^{s_1} a^t}, a^{s_2} a^t (Y_{a^{s_1+t}} \boxplus \underline{Y_{a^{s_2+t} - a^{s_1+t}}})) \\
&\overset{\mathrm{D}}{=} (a^{s_1} a^t Y_{a^t a^{s_1}}, a^{s_2} (a^t Y_{a^t a^{s_1}} \boxplus a^t \underline{Y_{a^t (a^{s_2} - a^{s_1})}})) \\
&\overset{\mathrm{D}}{=} (a^{s_1} Y_{a^{s_1}}, a^{s_2} (Y_{a^{s_1}} \boxplus \underline{Y_{a^{s_2} - a^{s_1}}})) \\
&\overset{\mathrm{D}}{=} (a^{s_1} Y_{a^{s_1}}, a^{s_2} Y_{a^{s_2}}) .
\end{aligned}$$

This completes the argument. □

In Sections 15.4–15.6 the time-stationary process $(a^t Y_{a^t}, t \in \mathbb{R})$ is studied with respect to its mixing properties and relations to Bernoulli shifts and Bernoulli flows.

6.6 Notes

1. If a random stationary tessellation is STIT, then property (6.4) of the typical cell necessarily holds. At that time it was still an open problem whether STIT tessellations exist for dimensions $d > 1$.
2. The stationarity of STIT tessellations and the scaling property (6.2) were shown in Lemma 5 of [145]. Theorem 6.4, that is, the stability under iteration property, is Theorem 2 of [145].
3. Some of the results in Theorem 6.9 were shown in [146] for $d = 2$ and in [147] for $d = 2$ and $d = 3$.
4. In Lemma 5 of [144] the proposition of Corollary 6.14 was shown, for $d = 2$ and a random stationary STIT tessellation *if a STIT tessellation exists*.
5. Theorem 6.16 was first shown for the case $p_1 = \cdots = p_d = 1/d$ as Theorem 3.1 in [154] and then in [153], Proposition 17 in its more general form. In that paper a STIT tessellation with discrete directional distribution is called an "oblique Mondrian tessellation".
6. Theorem 6.17 is Theorem 1.1 of [114], where the time-stationarity of a STIT tessellation was dealt with.

7 Limit of rescaled iteration and uniqueness

The construction in windows, described in Definition 4.2, and the spatial consistency guarantee the existence of STIT tessellations in $\mathbb{R}^d$; see Theorem 6.4. In Section 7.2 a uniqueness theorem is shown. It says that for a fixed density of the $(d-1)$-skeleton and a given spherical directional distribution in the typical point of the $(d-1)$-skeleton, the distribution of a random stationary tessellation is uniquely determined if it is stable under iteration. Hence this distribution must coincide with that of the respective STIT tessellation.

This uniqueness result follows by the convergence theorem proved in the first section.

7.1 Weak limit and domain of attraction

We present now a solution of the problem posed by R. V. Ambartzumian in the 1980s: What happens in the limit when sequences of i. i. d. random stationary tessellations are consecutively iterated and rescaled? Even if the result after finitely many iteration steps is a random stationary tessellation, it is not obvious that a limit exists and, moreover, if some limiting object exists, whether this object is a random tessellation.

The result concerns the weak convergence of random tessellations. For this, it is an appropriate approach to describe a tessellation by its $(d-1)$-skeleton, which is an element of the space $\mathcal{F}$ of closed subsets of $\mathbb{R}^d$, and $\mathcal{F}$ is endowed with the Fell topology, generated by the set system in (18.1). Then using the corresponding Borel σ-algebra $\mathcal{B}(\mathcal{F})$ on $\mathcal{F}$, the distribution of a random tessellation is determined by the capacity functional of its $(d-1)$-skeleton; see Lemma 2.16. Furthermore, by Theorem 1.7.7 of [137] the weak convergence of a sequence of random closed sets is equivalent to the convergence of the respective capacity functionals. Applying this, we say that a sequence $(X_n,\, n \in \mathbb{N})$ of random tessellations **converges weakly** to a random tessellation X if

$$\lim_{n\to\infty} \mathbb{P}(\partial X_n \cap C = \emptyset) = \mathbb{P}(\partial X \cap C = \emptyset) \quad \text{for all } C \in \mathcal{C}\,.$$

In Section 5.2 the respective capacity functional for STIT tessellations was described. Now in Theorem 7.1 it will be shown that a sequence, constructed by repeated and rescaled iteration of an arbitrary random stationary tessellation, converges weakly to a STIT tessellation. This implies that the domain of attraction of STIT tessellations with respect to rescaled iteration is the set of all random stationary tessellations that satisfy Convention 2.28.

Recall the notation $\boxplus^n$ for the $(n-1)$-fold iteration of i. i. d. random tessellations, introduced in (3.9).

Theorem 7.1. *Let X be a random stationary tessellation in $\mathbb{R}^d$, satisfying Convention 2.28, with the density η_{d-1} of the $(d-1)$-skeleton and the spherical directional distribution ψ in*

https://doi.org/10.1515/9783112222409-008

the typical point of ∂X; see (2.52). Furthermore, let Y_t be the STIT tessellation according to Theorem 6.4 with Θ as in (4.1), with $\varphi = \psi$ and $t = \eta_{d-1}$. Then the sequence $(n(\boxplus^n X),\ n \in \mathbb{N})$ of tessellations converges weakly to Y_t as $n \to \infty$, formally,

$$\lim_{n\to\infty} \mathbb{P}(n\,\partial(\boxplus^n X) \cap C = \emptyset) = \mathbb{P}(\partial Y_t \cap C = \emptyset) \quad \textit{for all } C \in \mathcal{C}\,.$$

The proof of this theorem is prepared by some lemmas. Throughout this section, X denotes a random stationary tessellation in $\mathbb{R}^d$ satisfying Convention 2.28 with the directional distribution χ_{d-1} in the typical point of ∂X (see (2.50)) and the corresponding spherical directional distribution ψ of the orthogonal unit vector, defined in (2.52).

To begin with, we prove an integral geometric proposition. Recall that $\operatorname{aff}_o F \in G(d, d-1)$ is the $(d-1)$-dimensional linear subspace parallel to F, and $F^\perp \in G(d,1)$ describes the orthogonal direction to $F \in \mathcal{P}_{d-1}$. The orthogonal projection of C onto the respective subspace is denoted by $C|\operatorname{aff}_o F$ or $C|F^\perp$.

Lemma 7.2. *Let $d \geq 2$, and let $T \in \mathcal{T}$ be a tessellation of $\mathbb{R}^d$.*
(a) *For all pairs $F_1, F_2 \in \mathcal{F}^{\mathrm{pr}}_{d-1}(T)$ of $(d-1)$-primitives of T, $F_1 \neq F_2$, and $C \in \mathcal{C}$, we have*

$$\lim_{n\to\infty} n\lambda_d\left(\left(F_1 + \frac{1}{n}C\right) \cap \left(F_2 + \frac{1}{n}C\right)\right) = 0\,. \tag{7.1}$$

(b) *For all $F \in \mathcal{P}_{d-1}$ and $C \in \mathcal{K}$, we have*

$$\lim_{n\to\infty} n\lambda_d\left(F + \frac{1}{n}C\right) = \lambda_{d-1}(F)\,\lambda_1(C|F^\perp) = V_{d-1}(F)\,\lambda_1(C|F^\perp)\,. \tag{7.2}$$

Proof. (a) Let $F_1, F_2 \in \mathcal{F}^{\mathrm{pr}}_{d-1}(T)$, $F_1 \neq F_2$, and $C \in \mathcal{C}$. Fix some $r \in \mathbb{N}$ such that $C \subseteq rB^d$. By (2.9), $\operatorname{relint} F_1 \cap \operatorname{relint} F_2 = \emptyset$, which implies that $\dim(F_1 \cap F_2) \leq d-2$.

If $F_1 \cap F_2 = \emptyset$, then there is $n(r) \in \mathbb{N}$ such that $(F_1 + \frac{r}{n}B^d) \cap (F_2 + \frac{r}{n}B^d) = \emptyset$ for all $n \geq n(r)$, and (7.1) becomes evident. Otherwise, if $F_1 \cap F_2 \neq \emptyset$, then we denote by $H_i \in \mathcal{H}$ the hyperplane that contains F_i, $i \in \{1, 2\}$. If $H_1 = H_2$, then the fact that the relative interiors of F_1 and F_2 are disjoint, implies for the intersection of the parallel sets that the relative interiors of the primitives are negligible, that is, for all $n \in \mathbb{N}$,

$$\left(F_1 + \frac{r}{n}B^d\right) \cap \left(F_2 + \frac{r}{n}B^d\right) = \left(\bigcup_{F\in\mathcal{F}_{d-2}(F_1)} F + \frac{r}{n}B^d\right) \cap \left(\bigcup_{F\in\mathcal{F}_{d-2}(F_2)} F + \frac{r}{n}B^d\right).$$

To see this, assume that there is $x \in \operatorname{relint} F_1$ belonging to $(F_1 + \frac{r}{n}B^d) \cap (F_2 + \frac{r}{n}B^d)$ such that $\|x - x'\| > \frac{r}{n}$ for all $x' \in \bigcup_{F\in\mathcal{F}_{d-2}(F_1)} F + \frac{r}{n}B^d$. Then $(x + \frac{r}{n}B^d) \cap F_1 \subset \operatorname{relint} F_1$, hence $(x + \frac{r}{n}B^d) \cap F_2 = \emptyset$, and thus $x \notin F_2 + \frac{r}{n}B^d$, a contradiction. Now the Steiner formula (1.3) yields that for all r, $n \in \mathbb{N}$,

$$\lambda_d\left(\left(F_1 + \frac{r}{n}B^d\right) \cap \left(F_2 + \frac{r}{n}B^d\right)\right) \leq \kappa_2 \left(\frac{r}{n}\right)^2 \sum_{F\in\mathcal{F}_{d-2}(F_1)\cup\mathcal{F}_{d-2}(F_2)} V_{d-2}(F) + o\left(\frac{1}{n^2}\right),$$

where $\kappa_2 = \pi$ is the area of the two-dimensional unit ball, which is the area of the unit circle. This implies (7.1) for the case that $H_1 = H_2$.

Finally, consider the case that $F_1 \cap F_2 \neq \emptyset$ and $H_1 \neq H_2$. Then the assumption $F_1 \cap F_2 \neq \emptyset$ implies that H_1 and H_2 are not parallel. Let $u_1, u_2 \in S^{d-1}$ be their respective orthogonal directions. As we are looking for an upper bound for $\lambda_d((F_1 + \frac{r}{n}B^d) \cap (F_2 + \frac{r}{n}B^d))$, we can use the motion invariance of the Lebesgue measure to assume without loss of generality that $H_1 \cap H_2 = \mathbb{R}^{d-2} \times \{0\}^2$. This implies that $H_1 \cap H_2 = \operatorname{lin}(u_1, u_2)^\perp$. We obtain that for all $n \in \mathbb{N}$,

$$\begin{aligned}&\left(H_1 + \frac{r}{n}B^d\right) \cap \left(H_2 + \frac{r}{n}B^d\right)\\ &\quad = \operatorname{lin}(u_1, u_2)^\perp + \left(\left(H_1 + \frac{r}{n}B^d\right) \cap \left(H_2 + \frac{r}{n}B^d\right) \cap \operatorname{lin}(u_1, u_2)\right). \qquad (7.3)\end{aligned}$$

Elementary planar geometry yields that

$$\lambda_2\left(\left(H_1 + \frac{r}{n}B^d\right) \cap \left(H_2 + \frac{r}{n}B^d\right) \cap \operatorname{lin}(u_1, u_2)\right) = 4\left(\frac{r}{n}\right)^2 \nabla_2^{-1}(u_1, u_2),$$

where $\nabla_2(u_1, u_2)$ is the area of the parallelogram spanned by u_1, u_2. Next, we choose $s > 0$ such that $(F_1 + rB^d) \cup (F_2 + rB^d) \subset s\,[-1, 1]^d$. Then, using (7.3), we have

$$\begin{aligned}\lambda_d\left(\left(F_1 + \frac{r}{n}B^d\right) \cap \left(F_2 + \frac{r}{n}B^d\right)\right) &\le \lambda_d\left(\left(H_1 + \frac{r}{n}B^d\right) \cap \left(H_2 + \frac{r}{n}B^d\right) \cap s[-1,1]^d\right)\\ &\le 4\left(\frac{r}{n}\right)^2 \nabla_2^{-1}(u_1, u_2)\, 2^{d-2}\, s^{d-2}.\end{aligned}$$

Thus, by the assumption that $C \subseteq rB^d$,

$$\lim_{n\to\infty} n\lambda_d\left(\left(F_1 + \frac{1}{n}C\right) \cap \left(F_2 + \frac{1}{n}C\right)\right) \le \lim_{n\to\infty} 4n\left(\frac{r}{n}\right)^2 \nabla_2^{-1}(u_1, u_2)\, 2^{d-2}\, s^{d-2} = 0.$$

This completes the proof of part (a).

(b) To show (7.2), consider $F \in \mathcal{P}_{d-1}$ with orthogonal direction $F^\perp \in G(d, 1)$. Writing $F + \frac{1}{n}C = \bigcup_{x \in \frac{1}{n}C}(F + x)$, an application of Fubini's theorem yields

$$\lambda_{d-1}(F)\,\lambda_1\left(\frac{1}{n}C|F^\perp\right) \le \lambda_d\left(F + \frac{1}{n}C\right) \le \lambda_{d-1}\left(F + \left(\frac{1}{n}C|\operatorname{aff}_o F\right)\right)\lambda_1\left(\frac{1}{n}C|F^\perp\right),$$

where $\operatorname{aff}_o F$ is the $(d-1)$-dimensional subspace parallel to F. Thus

$$\lim_{n\to\infty} n\lambda_d\left(F + \frac{1}{n}C\right) = \lambda_{d-1}(F)\,\lambda_1(C|F^\perp) = V_{d-1}(F)\,\lambda_1(C|F^\perp),$$

because $\lambda_{d-1}(F) = V_{d-1}(F)$. This proves part (b). □

We turn now to random stationary tessellations. For a random stationary tessellation X, we use the marked point process $\mathcal{Z}(\mathcal{F}^{\mathrm{pr}}_{d-1}(X))$ of $(d-1)$-primitives as introduced in (2.41) for a fixed center function $c : \mathcal{F}_{d-1} \to \mathbb{R}^d$ and the respective family of Palm distributions $\mathbb{P}^{o,F}$, $F \in \mathcal{P}_{d-1,c}$, on $\mathcal{T}$. The mark distribution is the distribution $\mathbb{Q}^{\mathrm{pr}}_{d-1}$ of the typical $(d-1)$-primitive, defined in (2.44). For $C \in \mathcal{B}(\mathbb{R}^d)$, we denote $\langle C\rangle := \{F \in \mathcal{P}_{d-1} : F \cap C \neq \emptyset\}$. Furthermore, for the set $\mathcal{F}^{\mathrm{pr}}_{d-1}(X)$ of $(d-1)$-primitives, we define

$$\xi^{\mathrm{pr}}_X\langle C\rangle := \left|\mathcal{F}^{\mathrm{pr}}_{d-1}(X) \cap \langle C\rangle\right|, \tag{7.4}$$

which is the number of $(d-1)$-primitives of X that intersect C.

Lemma 7.3. *Let X be a random stationary tessellation of $\mathbb{R}^d$, satisfying Convention 2.28, with density $\eta_{d-1} \in \mathbb{R}^+$ and the spherical directional distribution ψ in the typical point of ∂X. Then for all $C \in \mathcal{K}$,*

$$\lim_{n\to\infty} n\,\mathbb{E}\xi^{\mathrm{pr}}_X\left\langle \frac{1}{n}C\right\rangle = \eta_{d-1} \int\limits_{S^{d-1}} \lambda_1(C|u)\,\psi(\mathrm{d}u), \tag{7.5}$$

which for $d = 1$ means

$$\lim_{n\to\infty} n\,\mathbb{E}\xi^{\mathrm{pr}}_X\left\langle \frac{1}{n}C\right\rangle = \eta_0\,\lambda_1(C),$$

and, furthermore,

$$\lim_{n\to\infty} n\,\mathbb{E}\left(\xi^{\mathrm{pr}}_X\left\langle \frac{1}{n}C\right\rangle \mathbb{1}\left\{\xi^{\mathrm{pr}}_X\left\langle \frac{1}{n}C\right\rangle \geq 2\right\}\right) = 0. \tag{7.6}$$

Proof. Let $C \in \mathcal{K}$ and fix some $r \in \mathbb{N}$ such that $C \subseteq rB^d$. First note that for all $n \in \mathbb{N}$ and $F \in \mathcal{P}_{d-1}$, an application of Steiner's formula (1.3) yields

$$\begin{aligned} n \int\limits_{\mathbb{R}^d} \mathbb{1}\left\{(F+x) \in \left\langle \frac{1}{n}C\right\rangle\right\} \lambda_d(\mathrm{d}x) &= n\,\lambda_d\left(-F + \frac{1}{n}C\right) \\ &\leq n\,\lambda_d\left(-F + \frac{r}{n}B^d\right) \\ &= \sum_{j=0}^{d-1} V_j(F)\, r^{d-j}\,\kappa_{d-j}\, n^{-d+j+1} \\ &\leq \sum_{j=0}^{d-1} V_j(F)\, r^{d-j}\,\kappa_{d-j}\,. \end{aligned}$$

By Convention 2.28 the integral

$$\int\limits_{\mathcal{P}_{d-1,c}} \sum_{j=0}^{d-1} V_j(F)\, r^{d-j}\, \kappa_{d-j}\, \mathbb{Q}_{d-1}^{\mathrm{pr}}(\mathrm{d}F) < \infty\,,$$

and hence the dominated convergence theorem can be applied. Furthermore, because $C \subseteq (C|\mathrm{aff}_o F) + (C|F^{\perp})$ for $C \in \mathcal{K}$, an application of Fubini's theorem yields

$$\begin{aligned} n\,\lambda_{d-1}(F)\,\lambda_1\Big(\frac{1}{n}C|F^{\perp}\Big) &\le n\,\lambda_d\Big(-F + \frac{1}{n}C\Big) \\ &\le n\,\lambda_d\Big(-F + \frac{1}{n}(C|\mathrm{aff}_o F) + \frac{1}{n}(C|F^{\perp})\Big) \\ &= n\,\lambda_{d-1}\Big(-F + \frac{1}{n}(C|\mathrm{aff}_o F)\Big)\,\lambda_1\Big(\frac{1}{n}(C|F^{\perp})\Big). \end{aligned}$$

Thus, using that $V_{d-1}(F) = \lambda_{d-1}(F)$, we obtain

$$\lim_{n\to\infty} n\,\lambda_d\Big(-F + \frac{1}{n}C\Big) = V_{d-1}(F)\,\lambda_1(C|F^{\perp})\,.$$

Now Campbell's theorem and equations (7.2), (2.53), and (2.55) yield that

$$\begin{aligned} \lim_{n\to\infty} n\,\mathbb{E}\xi_X^{\mathrm{pr}}\Big\langle\frac{1}{n}C\Big\rangle &= \lim_{n\to\infty} n\,\mathbb{E} \sum_{(x,F)\in\mathcal{Z}(\mathcal{F}_{d-1}^{\mathrm{pr}}(X))} \mathbb{1}\Big\{(F+x) \in \Big\langle\frac{1}{n}C\Big\rangle\Big\} \\ &= \lim_{n\to\infty} n\,\gamma_{d-1}^{\mathrm{pr}} \int\limits_{\mathcal{P}_{d-1,c}} \int\limits_{\mathbb{R}^d} \mathbb{1}\Big\{F + x \in \Big\langle\frac{1}{n}C\Big\rangle\Big\}\,\lambda_d(\mathrm{d}x)\,\mathbb{Q}_{d-1}^{\mathrm{pr}}(\mathrm{d}F) \\ &= \lim_{n\to\infty} n\,\gamma_{d-1}^{\mathrm{pr}} \int\limits_{\mathcal{P}_{d-1,c}} \lambda_d\Big(-F + \frac{1}{n}C\Big)\,\mathbb{Q}_{d-1}^{\mathrm{pr}}(\mathrm{d}F) \\ &= \gamma_{d-1}^{\mathrm{pr}} \int\limits_{\mathcal{P}_{d-1,c}} \lim_{n\to\infty} n\,\lambda_d\Big(-F + \frac{1}{n}C\Big)\,\mathbb{Q}_{d-1}^{\mathrm{pr}}(\mathrm{d}F) \\ &= \gamma_{d-1}^{\mathrm{pr}} \int\limits_{\mathcal{P}_{d-1,c}} V_{d-1}(F)\,\lambda_1(C|F^{\perp})\,\mathbb{Q}_{d-1}^{\mathrm{pr}}(\mathrm{d}F) \\ &= \eta_{d-1} \int\limits_{S^{d-1}} \lambda_1(C|u)\,\psi(\mathrm{d}u)\,, \end{aligned}$$

which is (7.5).

To show (7.6), observe that

$$\begin{aligned} &n \int\limits_{\mathbb{R}^d} \mathbb{1}\Big\{(F+x) \in \Big\langle\frac{1}{n}C\Big\rangle\Big\}\,\mathbb{1}\Big\{\xi_T^{\mathrm{pr}}\Big\langle\frac{1}{n}C - x\Big\rangle \ge 2\Big\}\,\lambda_d(\mathrm{d}x) \\ &\quad\le n \int\limits_{\mathbb{R}^d} \mathbb{1}\Big\{(F+x) \in \Big\langle\frac{1}{n}C\Big\rangle\Big\}\,\lambda_d(\mathrm{d}x)\,, \end{aligned}$$

and therefore, with the same arguments as in the proof of (7.5), the dominated convergence theorem can be applied. Thus the refined Campbell theorem in the form (2.43) yields

$$
\begin{aligned}
&\lim_{n\to\infty} n\,\mathbb{E}\Big(\xi_X^{\mathrm{pr}}\Big\langle \frac{1}{n}C\Big\rangle \mathbb{1}\Big\{\xi_X^{\mathrm{pr}}\Big\langle \frac{1}{n}C\Big\rangle \ge 2\Big\}\Big)\\
&= \lim_{n\to\infty} n\,\mathbb{E} \sum_{(x,F)\in\mathcal{Z}(\mathcal{F}_{d-1}^{\mathrm{pr}}(X))} \mathbb{1}\Big\{(F+x)\in\Big\langle \frac{1}{n}C\Big\rangle\Big\}\mathbb{1}\Big\{\xi_X^{\mathrm{pr}}\Big\langle \frac{1}{n}C\Big\rangle \ge 2\Big\}\\
&= \lim_{n\to\infty} n\,\gamma_{d-1}^{\mathrm{pr}} \int_{\mathcal{P}_{d-1,c}}\int_{\mathcal{T}}\int_{\mathbb{R}^d} \mathbb{1}\Big\{(F+x)\in\Big\langle \frac{1}{n}C\Big\rangle\Big\}\mathbb{1}\Big\{\xi_T^{\mathrm{pr}}\Big\langle \frac{1}{n}C-x\Big\rangle \ge 2\Big\}\\
&\quad \lambda_d(\mathrm{d}x)\,\mathbb{P}^{o,F}(\mathrm{d}T)\,\mathbb{Q}_{d-1}^{\mathrm{pr}}(\mathrm{d}F)\\
&= \gamma_{d-1}^{\mathrm{pr}} \int_{\mathcal{P}_{d-1,c}}\int_{\mathcal{T}} \lim_{n\to\infty} n \int_{\mathbb{R}^d} \mathbb{1}\Big\{(F+x)\in\Big\langle \frac{1}{n}C\Big\rangle\Big\}\mathbb{1}\Big\{\xi_T^{\mathrm{pr}}\Big\langle \frac{1}{n}C-x\Big\rangle \ge 2\Big\}\\
&\quad \lambda_d(\mathrm{d}x)\,\mathbb{P}^{o,F}(\mathrm{d}T)\,\mathbb{Q}_{d-1}^{\mathrm{pr}}(\mathrm{d}F)\\
&\le \gamma_{d-1}^{\mathrm{pr}} \int_{\mathcal{P}_{d-1,c}}\int_{\mathcal{T}} \lim_{n\to\infty} n \int_{\mathbb{R}^d} \mathbb{1}\Big\{(F+x)\in\Big\langle \frac{1}{n}C\Big\rangle\Big\}\\
&\quad \times \sum_{F'+y\neq F,\,(y,F')\in\mathcal{Z}(\mathcal{F}_{d-1}^{\mathrm{pr}}(T))} \mathbb{1}\Big\{(F'+y+x)\in\Big\langle \frac{1}{n}C\Big\rangle\Big\}\lambda_d(\mathrm{d}x)\,\mathbb{P}^{o,F}(\mathrm{d}T)\,\mathbb{Q}_{d-1}^{\mathrm{pr}}(\mathrm{d}F)\\
&= \gamma_{d-1}^{\mathrm{pr}} \int_{\mathcal{P}_{d-1,c}}\int_{\mathcal{T}} \lim_{n\to\infty} \sum_{F'+y\neq F,\,(y,F')\in\mathcal{Z}(\mathcal{F}_{d-1}^{\mathrm{pr}}(T))} n\lambda_d\Big(\Big(-F+\frac{1}{n}C\Big)\cap\Big(-F'-y+\frac{1}{n}C\Big)\Big)\\
&\quad \mathbb{P}^{o,F}(\mathrm{d}T)\,\mathbb{Q}_{d-1}^{\mathrm{pr}}(\mathrm{d}F)\\
&= \gamma_{d-1}^{\mathrm{pr}} \int_{\mathcal{P}_{d-1,c}}\int_{\mathcal{T}} \sum_{F'+y\neq F,\,(y,F')\in\mathcal{Z}(\mathcal{F}_{d-1}^{\mathrm{pr}}(T))} \lim_{n\to\infty} n\lambda_d\Big(\Big(-F+\frac{1}{n}C\Big)\cap\Big(-F'-y+\frac{1}{n}C\Big)\Big)\\
&\quad \mathbb{P}^{o,F}(\mathrm{d}T)\,\mathbb{Q}_{d-1}^{\mathrm{pr}}(\mathrm{d}F)\\
&= 0\,.
\end{aligned}
$$

For the sum in the last but third line, the number of items different from 0 is finite for all $T\in\mathcal{T}$. Thus the result follows by (7.1). □

The key proposition for the proof of Theorem 7.1 is a modified version of Dobrushin's lemma. This classical result says that for simple stationary point processes Φ on the real axis, $\mathbb{P}(\Phi(0,a]\ge 2) = o(a)$ as $a\to 0$, which means that in a short interval, with a probability close to 1, there is at most one point of the process; see Proposition 3.3.V of [45]. In Lemma 7.4, we prove an analogue for the stationary process of primitives of a random stationary tessellation X.

Lemma 7.4. *Let X be a random stationary tessellation of $\mathbb{R}^d$, satisfying Convention 2.28, with density $\eta_{d-1} \in \mathbb{R}^+$ and the spherical directional distribution ψ in the typical point of ∂X. Then for all $C \in \mathcal{K}$,*

$$\lim_{n\to\infty} n\,\mathbb{P}\Big(\xi_X^{\mathrm{pr}}\Big\langle \frac{1}{n}C\Big\rangle \geq 2\Big) = 0\,, \tag{7.7}$$

and

$$\lim_{n\to\infty} n\,\mathbb{P}\Big(\partial X \cap \frac{1}{n}C \neq \emptyset\Big) = \eta_{d-1} \int\limits_{S^{d-1}} \lambda_1(C|u)\,\psi(\mathrm{d}u)\,. \tag{7.8}$$

Proof. If $d = 1$, then $\mathcal{F}_{d-1}^{\mathrm{pr}}(X)$ is a stationary point process on the real line. In this case, (7.7) is the statement of the classical Dobrushin lemma (see Proposition 3.3.V of [45]), and (7.8) corresponds to Korolyuk's theorem (Proposition 3.3.IV of [45]). Now let $d \geq 2$. We have that

$$\lim_{n\to\infty} n\,\mathbb{P}\Big(\xi_X^{\mathrm{pr}}\Big\langle \frac{1}{n}C\Big\rangle \geq 2\Big) \leq \lim_{n\to\infty} n\,\mathbb{E}\Big(\xi_X^{\mathrm{pr}}\Big\langle \frac{1}{n}C\Big\rangle\, \mathbb{1}\Big\{\xi_X^{\mathrm{pr}}\Big\langle \frac{1}{n}C\Big\rangle \geq 2\Big\}\Big) = 0$$

by (7.6). Thus (7.7) is proved.

To show (7.8), we apply (7.7), (7.6), and finally (7.5), which yields

$$\begin{aligned}
\lim_{n\to\infty} n\,\mathbb{P}\Big(\partial X \cap \frac{1}{n}C \neq \emptyset\Big) &= \lim_{n\to\infty} n\,\Big(\mathbb{P}\Big(\xi_X^{\mathrm{pr}}\Big\langle \frac{1}{n}C\Big\rangle = 1\Big) + \mathbb{P}\Big(\xi_X^{\mathrm{pr}}\Big\langle \frac{1}{n}C\Big\rangle \geq 2\Big)\Big) \\
&= \lim_{n\to\infty} n\,\mathbb{P}\Big(\xi_X^{\mathrm{pr}}\Big\langle \frac{1}{n}C\Big\rangle = 1\Big) \\
&= \lim_{n\to\infty} n\,\Big(\mathbb{E}\xi_X^{\mathrm{pr}}\Big\langle \frac{1}{n}C\Big\rangle - \mathbb{E}\Big(\xi_X^{\mathrm{pr}}\Big\langle \frac{1}{n}C\Big\rangle\, \mathbb{1}\Big\{\xi_X^{\mathrm{pr}}\Big\langle \frac{1}{n}C\Big\rangle \geq 2\Big\}\Big)\Big) \\
&= \lim_{n\to\infty} n\,\mathbb{E}\xi_X^{\mathrm{pr}}\Big\langle \frac{1}{n}C\Big\rangle \\
&= \eta_{d-1} \int\limits_{S^{d-1}} \lambda_1(C|u)\,\psi(\mathrm{d}u)\,,
\end{aligned}$$

which is (7.8). □

In the next step, to investigate the capacity functionals for $(d-1)$-skeletons, we consider the probabilities $\mathbb{P}(n\partial X \cap C \neq \emptyset)$ and $\mathbb{P}(n\,\partial(\boxplus^n X) \cap C = \emptyset)$, $n \in \mathbb{N}$, for connected sets $C \in \mathcal{R}$. Then the results will be generalized inductively to sets $C \in \mathcal{R}$ with finitely many connected components.

Lemma 7.5. *Let X be a random stationary tessellation of $\mathbb{R}^d$, satisfying Convention 2.28, with density $\eta_{d-1} \in \mathbb{R}^+$ and the spherical directional distribution ψ in the typical point of ∂X. Define the hyperplane measure Θ by (4.1) with $\varphi = \psi$. If $C \in \mathcal{R}$ is connected, then*

$$\lim_{n\to\infty} n\,\mathbb{P}(n\partial X \cap C \neq \emptyset) = \eta_{d-1}\,\Theta([\operatorname{conv} C]), \tag{7.9}$$

and

$$\lim_{n\to\infty} \mathbb{P}(n\,\partial(\boxplus^n X) \cap C = \emptyset) = \lim_{n\to\infty} \mathbb{P}(n\partial X \cap C = \emptyset)^n = \mathrm{e}^{-\eta_{d-1}\Theta([\operatorname{conv} C])}. \tag{7.10}$$

Proof. To prove (7.9), note that $\lambda_1(C|u) = \lambda_1(\operatorname{conv} C|u)$ if $C \in \mathcal{R}$ is connected. Thus (7.8) and (4.1) yield

$$\lim_{n\to\infty} n\,\mathbb{P}(n\partial X \cap C \neq \emptyset) = \eta_{d-1} \int_{S^{d-1}} \lambda_1(\operatorname{conv} C|u)\,\psi(\mathrm{d}u) = \eta_{d-1}\,\Theta([\operatorname{conv} C]).$$

To show the first equation in (7.10), assume without loss of generality that $o \in C$. Then by (3.9)

$$\begin{aligned}
&\lim_{n\to\infty} \mathbb{P}(n\,\partial(\boxplus^n X) \cap C = \emptyset)\\
&= \lim_{n\to\infty} \mathbb{P}(n\partial X \cap C = \emptyset,\ n\partial X_1^{(0)} \cap C = \emptyset, \dots, n\partial X_{n-1}^{(0)} \cap C = \emptyset)\\
&= \lim_{n\to\infty} \mathbb{P}(n\partial X \cap C = \emptyset)^n .
\end{aligned}$$

For the second equation in (7.10), we can use the equality of the limits

$$\lim_{n\to\infty} n \ln \mathbb{P}(n\partial X \cap C = \emptyset) = -\lim_{n\to\infty} n\,\mathbb{P}(n\partial X \cap C \neq \emptyset),$$

which can be derived from the Taylor expansion of the logarithm and the fact that $\lim_{n\to\infty} \mathbb{P}(n\partial X \cap C \neq \emptyset) = 0$. Thus equation (7.9) implies that $\lim_{n\to\infty} \mathbb{P}(n\partial X \cap C = \emptyset)^n = \mathrm{e}^{-\eta_{d-1}\Theta([\operatorname{conv} C])}$. This completes the proof of Lemma 7.5. □

Now we prepare the proof for general sets $C \in \mathcal{R}$. Recall definition (5.8) of the set $\mathbf{D}(C)$ of all partitions of the family of connected components of C.

With a slight abuse of notation, we write $\operatorname{conv}(x_1, x_2)$ for the line segment joining two points $x_1, x_2 \in \mathbb{R}^d$. For $\{D_1, D_2\} \in \mathbf{D}(C)$, a $(d-1)$-dimensional polytope $F \in \mathcal{P}_{d-1}$ **separates** D_1 and D_2 if

$$\operatorname{conv}(x_1, x_2) \cap F \neq \emptyset \quad \text{for all } x_1 \in D_1,\ x_2 \in D_2 .$$

We denote by $D_1 \updownarrow D_2$ the set of all $(d-1)$-dimensional polytopes that separate D_1 and D_2, and define

$$\xi_X^{\mathrm{pr}}(D_1 \updownarrow D_2) := \left|\{(x, F) \in \mathcal{Z}(\mathcal{F}_{d-1}^{\mathrm{pr}}(X)) : F + x \in (D_1 \updownarrow D_2)\}\right|,$$

which is the number of $(d-1)$-primitives of the tessellation X that separate D_1 and D_2. Recall (5.9), where the notation $[D_1|D_2]$ was introduced for the set of hyperplanes that separate D_1 and D_2.

The following lemma relates the hyperplane measure Θ to the limit of the probability for a separation of D_1 and D_2 by a $(d-1)$-primitive of X. This result will be used in the inductive proof of Lemma 7.8

Lemma 7.6. *Let X be a random stationary tessellation of $\mathbb{R}^d$, satisfying Convention 2.28, with density $\eta_{d-1} \in \mathbb{R}^+$ and the spherical directional distribution ψ in the typical point of ∂X. Define the hyperplane measure Θ by (4.1) with $\psi = \varphi$. Furthermore, let $C \in \mathcal{R}$ and $\{D_1, D_2\} \in \mathbf{D}(C)$. Then*

$$\lim_{n\to\infty} n\,\mathbb{P}(\xi_{nX}^{\mathrm{pr}}\langle \operatorname{conv} C\rangle = 1,\ \xi_{nX}^{\mathrm{pr}}(D_1 \updownarrow D_2) = 1) = \eta_{d-1}\,\Theta([D_1|D_2])\,.$$

Proof. First observe that

$$\begin{aligned}
&\lim_{n\to\infty} n\,\mathbb{E} \sum_{(x,F)\in\mathcal{Z}(\mathcal{F}_{d-1}^{\mathrm{pr}}(X))} \mathbb{1}\{\xi_{nX}^{\mathrm{pr}}\langle \operatorname{conv} C\rangle \ge 2\}\,\mathbb{1}\{n(F+x) \in D_1 \updownarrow D_2\}\\
&\quad\le \lim_{n\to\infty} n\,\mathbb{E}\Big(\xi_X^{\mathrm{pr}}\Big\langle \frac{1}{n}\operatorname{conv} C\Big\rangle\,\mathbb{1}\Big\{\xi_X^{\mathrm{pr}}\Big\langle \frac{1}{n}\operatorname{conv} C\Big\rangle \ge 2\Big\}\Big) = 0
\end{aligned}$$

by (7.6). Hence

$$\begin{aligned}
&\lim_{n\to\infty} n\,\mathbb{P}(\xi_{nX}^{\mathrm{pr}}\langle \operatorname{conv} C\rangle = 1,\ \xi_{nX}^{\mathrm{pr}}(D_1 \updownarrow D_2) = 1)\\
&\quad= \lim_{n\to\infty} n\,\mathbb{E}\,\mathbb{1}\{\xi_{nX}^{\mathrm{pr}}\langle \operatorname{conv} C\rangle = 1,\ \xi_{nX}^{\mathrm{pr}}(D_1 \updownarrow D_2) = 1\}\\
&\quad= \lim_{n\to\infty} n\,\mathbb{E} \sum_{(x,F)\in\mathcal{Z}(\mathcal{F}_{d-1}^{\mathrm{pr}}(X))} \mathbb{1}\{\xi_{nX}^{\mathrm{pr}}\langle \operatorname{conv} C\rangle = 1\}\,\mathbb{1}\{n(F+x) \in D_1 \updownarrow D_2\}\\
&\quad= \lim_{n\to\infty} n\,\mathbb{E} \sum_{(x,F)\in\mathcal{Z}(\mathcal{F}_{d-1}^{\mathrm{pr}}(X))} \mathbb{1}\{n(F+x) \in D_1 \updownarrow D_2\}\,.
\end{aligned}$$

For $F \in \mathcal{P}_{d-1}$, we denote by $\operatorname{aff}_o F \in G(d, d-1)$ the $(d-1)$-dimensional linear subspace parallel to F and by $F^\perp \in G(d,1)$ the orthogonal complement of $\operatorname{aff}_o F$. Furthermore, $u(F) \in S^{d-1}$ denotes the intersection of $F^\perp$ with the upper unit hemisphere, and thus $H(u(F), z) \in \mathcal{H}$ is the hyperplane parallel to F with the signed distance $z \in \mathbb{R}$ to the origin. Below we apply the refined Campbell theorem in the form (2.43), rewriting $\mathbb{R}^d = \operatorname{aff}_o F + F^\perp$ for $F \in \mathcal{F}_{d-1}^{\mathrm{pr}}(X)$, and the substitution of z by z/n. Moreover, based on Convention 2.28, we have dominated convergence for the limit of integrals, which yields

$$\begin{aligned}
&\lim_{n\to\infty} n\,\mathbb{E} \sum_{(x,F)\in\mathcal{Z}(\mathcal{F}_{d-1}^{\mathrm{pr}}(X))} \mathbb{1}\{n(F+x) \in D_1 \updownarrow D_2\}\\
&\quad= \gamma_{d-1}^{\mathrm{pr}} \lim_{n\to\infty} n \int_{\mathcal{P}_{d-1,c}} \int_{z\in F^\perp} \int_{y\in\operatorname{aff}_o F} \mathbb{1}\{n(F+(y+z)) \in D_1 \updownarrow D_2\}\\
&\quad\lambda_{d-1}(\mathrm{d}y)\,\lambda_1(\mathrm{d}z)\,\mathbb{Q}_{d-1}^{\mathrm{pr}}(\mathrm{d}F)
\end{aligned}$$

$$= \gamma_{d-1}^{\mathrm{pr}} \int_{\mathcal{P}_{d-1,c}} \int_{z \in F^\perp} \int_{y \in \mathrm{aff}_o F} \lim_{n\to\infty} \mathbb{1}\left\{(F+y) + \frac{1}{n} z \in \frac{1}{n} D_1 \updownarrow \frac{1}{n} D_2\right\}$$

$$\lambda_{d-1}(\mathrm{d}y)\,\lambda_1(\mathrm{d}z)\,\mathbb{Q}_{d-1}^{\mathrm{pr}}(\mathrm{d}F)$$

$$= \gamma_{d-1}^{\mathrm{pr}} \int_{\mathcal{P}_{d-1,c}} \int_{\mathbb{R}} \lambda_{d-1}(F)\mathbb{1}\{H(u(F),z) \in [D_1|D_2]\}\,\lambda_1(\mathrm{d}z)\,\mathbb{Q}_{d-1}^{\mathrm{pr}}(\mathrm{d}F)\,.$$

Thus, by (2.55), (2.51), and (2.52),

$$\lim_{n\to\infty} n\,\mathbb{P}(\xi_{nX}^{\mathrm{pr}}\langle \mathrm{conv}\, C\rangle = 1,\ \xi_{nX}^{\mathrm{pr}}(D_1 \updownarrow D_2) = 1)$$

$$= \gamma_{d-1}^{\mathrm{pr}} \overline{V}_{d-1}^{\mathrm{pr}} \int_{S^{d-1}} \int_{\mathbb{R}} \mathbb{1}\{H(u,z) \in [D_1|D_2]\}\,\lambda_1(\mathrm{d}z)\,\psi(\mathrm{d}u)$$

$$= \eta_{d-1}\,\Theta([D_1|D_2])\,,$$

because $\lambda_{d-1}(F) = V_{d-1}(F)$, and it is assumed that $\psi = \varphi$. □

The following auxiliary lemma provides a relation that guarantees that the denominator of certain fractions, which appear in the proof of Lemma 7.8, is different from 0.

Lemma 7.7. *Let X be a random stationary tessellation of $\mathbb{R}^d$, satisfying Convention 2.28, with density $\eta_{d-1} \in \mathbb{R}^+$ and the spherical directional distribution ψ in the typical point of ∂X. Assume that for Θ given by (4.1), $\psi = \varphi$. If $C \in \mathcal{R}$ and condition (5.10) is satisfied, then*

$$\mathbb{P}(n\partial X \cap \mathrm{conv}\, C = \emptyset) - \prod_{S\in D} \mathbb{P}(n\partial X \cap \mathrm{conv}\, S = \emptyset) \neq 0$$

for all but a finite number of $n \in \mathbb{N}$ and all $D \in \mathbf{D}(C)$ with $|D| \geq 2$.

Proof. With $\mathbb{P}(n\partial X \cap \mathrm{conv}\, S = \emptyset) = 1 - \mathbb{P}(n\partial X \cap \mathrm{conv}\, S \neq \emptyset)$, we obtain

$$\prod_{S\in D} \mathbb{P}(n\partial X \cap \mathrm{conv}\, S = \emptyset) = 1 - \sum_{S\in D} \mathbb{P}(n\partial X \cap \mathrm{conv}\, S \neq \emptyset) + r(n)\,, \tag{7.11}$$

where the remainder term $r(n)$ is the sum over all items that contain at least two factors $\mathbb{P}(n\partial X \cap \mathrm{conv}\, S \neq \emptyset)\mathbb{P}(n\partial X \cap \mathrm{conv}\, S' \neq \emptyset)$, and hence $\lim_{n\to\infty} n\,r(n) = 0$ by (7.9). Moreover, (7.9) yields that

$$\lim_{n\to\infty} n\left(\mathbb{P}(n\partial X \cap \mathrm{conv}\, C = \emptyset) - \prod_{S\in D} \mathbb{P}(n\partial X \cap \mathrm{conv}\, S = \emptyset)\right)$$

$$= \lim_{n\to\infty} n\left(-\mathbb{P}(n\partial X \cap \mathrm{conv}\, C \neq \emptyset) + \sum_{S\in D} \mathbb{P}(n\partial X \cap \mathrm{conv}\, S \neq \emptyset)\right)$$

$$= -\eta_{d-1}\,\Theta([\mathrm{conv}\, C]) + \eta_{d-1} \sum_{S\in D} \Theta([\mathrm{conv}\, S]) \neq 0$$

by assumption (5.10). This completes the argument. □

After these preparations, now the essential ingredient for the proof of Theorem 7.1 can be shown.

Lemma 7.8. *Let X be a random stationary tessellation of $\mathbb{R}^d$, satisfying Convention 2.28, with density $\eta_{d-1} \in \mathbb{R}^+$ and the spherical directional distribution ψ in the typical point of ∂X. Define the hyperplane measure Θ by (4.1) with $\psi = \varphi$. Moreover, let $C \in \mathcal{R}$, satisfying the condition (5.10), with $\mathbf{D}(C)$ as defined in (5.8). Furthermore, for $D \in \mathbf{D}(C)$ let $d^{\mathrm{STIT}}(C,D)$ be defined as in Lemma 5.5 for the STIT tessellation process driven by Θ. Then there exists $q(C,X) \geq 0$ that does not depend on n, and for all but a finite number of $n \in \mathbb{N}$, there exist real numbers $d_n(C,D,X)$ and $q_n(C,X)$ such that*

$$\begin{aligned} &\mathbb{P}(n\partial(\boxplus^n X) \cap C = \emptyset) \\ &\quad = \sum_{D\in\mathbf{D}(C)} d_n(C,D,X) \prod_{D_j\in D} \mathbb{P}(n\partial X \cap \operatorname{conv} D_j = \emptyset)^n + q_n(C,X)\,. \end{aligned} \tag{7.12}$$

The coefficients $d_n(C,D,X)$ fulfil the following conditions:

$$d_\ell\left(C, D, \frac{n}{\ell}X\right) = d_n(C,D,X) \tag{7.13}$$

for all $\ell, n \in \mathbb{N}$, and

$$\lim_{n\to\infty} d_n(C,D,X) = d^{\mathrm{STIT}}(C,D)\,. \tag{7.14}$$

For the remainder term $q_n(C,X)$, we have

$$|q_n(C,X)| \leq q(C,X)\, n\, \mathbb{P}(\xi^{\mathrm{pr}}_{nX}\langle \operatorname{conv} C\rangle \geq 2) \tag{7.15}$$

for all but a finite number of $n \in \mathbb{N}$, where

$$q\left(C, \frac{n}{\ell}X\right) \leq q(C,X) \tag{7.16}$$

for all $n \in \mathbb{N}$, $\ell \in \{1, \dots n-1\}$, and finally,

$$\lim_{n\to\infty} q_n(C,X) = 0\,. \tag{7.17}$$

Proof. We prove the statement by induction over the number k of connected components of C. For $k = 1$, that is, if $C \in \mathcal{C}$ is connected, the set $\mathbf{D}(C)$ contains only one element, namely $\{C\}$. In this case, with analogous arguments as in the proof of (7.10), we obtain the form (7.12) with $d_n(C, \{C\}, X) = 1$ and $q_n(C,X) = 0$ for all $n \in \mathbb{N}$. Furthermore, we can choose $q(C,X) = 0$. Thus (7.10), (5.6), and (6.6) yield (7.14). Hence the statement of the lemma holds for $k = 1$.

As induction hypothesis, assume that (7.12), (7.13), (7.14), (7.15), (7.16), and (7.17) hold for all sets in $\mathcal{R}$ with at most k connected components. Now consider $C \in \mathcal{R}$ with $k+1$ connected components.

If $n\partial(\boxplus^n X) \cap C = \emptyset$ but $n\partial(\boxplus^n X) \cap \operatorname{conv} C \neq \emptyset$, then there is the smallest index $i \in \{1, \ldots, n-1\}$ in the representation (3.9) for which the skeletons of tessellations in the sequence $\underline{X}_i$ intersect $\operatorname{conv} C$. We put $i = 0$ if already $\partial X \cap \operatorname{conv} C \neq \emptyset$. This first division of $\operatorname{conv} C$ can induce a separation into two parts, say, D_1 and D_2, for some $\{D_1, D_2\} \in \mathbf{D}(C)$, but it can also cause another splitting of C. Accordingly, we can write the probability of the event $\{n\partial(\boxplus^n X) \cap C = \emptyset\}$ as a double sum of probabilities over $i \in \{0, \ldots, n-1\}$ and over all $\{D_1, D_2\} \in \mathbf{D}(C)$ plus a remainder term, denoted by $q_n'(C, X)$ below. Since the respective sets D_1 and D_2 have at most k connected components, the induction hypothesis can be applied. The asymptotic behavior of $q_n'(C, X)$ as $n \to \infty$ will be treated by the Dobrushin-type Lemma 7.4. This is formalized as follows:

$$
\begin{aligned}
&\mathbb{P}(n\partial(\boxplus^n X) \cap C = \emptyset) \\
&\quad = \mathbb{P}(n\partial X \cap \operatorname{conv} C = \emptyset)^n + \sum_{\{D_1, D_2\} \in \mathbf{D}(C)} \mathbb{P}(\xi_{nX}^{\mathrm{pr}}\langle \operatorname{conv} C\rangle = 1,\ \xi_{nX}^{\mathrm{pr}}(D_1 \updownarrow D_2) = 1) \\
&\qquad \times \Bigg(\sum_{i=0}^{n-2} \mathbb{P}(n\partial X \cap \operatorname{conv} C = \emptyset)^i\, \mathbb{P}(\partial(\boxplus^{n-i-1} nX) \cap D_1 = \emptyset) \\
&\qquad \times \mathbb{P}(\partial(\boxplus^{n-i-1} nX) \cap D_2 = \emptyset) + \mathbb{P}(n\partial X \cap \operatorname{conv} C = \emptyset)^{n-1} \Bigg) + q_n'(C, X),
\end{aligned}
$$

where

$$
\begin{aligned}
q_n'(C, X) &:= \sum_{D \in \mathbf{D}(C)} \mathbb{P}(\xi_{nX}^{\mathrm{pr}}\langle \operatorname{conv} C\rangle \geq 2,\ n\partial X \cap C = \emptyset,\ n\partial X \text{ induces partition } D) \\
&\qquad \times \Bigg(\sum_{i=0}^{n-2} \mathbb{P}(n\partial X \cap \operatorname{conv} C = \emptyset)^i \prod_{B \in D} \mathbb{P}(\partial(\boxplus^{n-i-1} nX) \cap B = \emptyset) \\
&\qquad + \mathbb{P}(n\partial X \cap \operatorname{conv} C = \emptyset)^{n-1} \Bigg) \\
&\leq |\mathbf{D}(C)|\, n\, \mathbb{P}(\xi_{nX}^{\mathrm{pr}}\langle \operatorname{conv} C\rangle \geq 2).
\end{aligned} \tag{7.18}
$$

Now the induction step will be shown for all properties (7.12)–(7.17).

Induction step for (7.12) **and** (7.13)

Observe that by (3.5) and the induction hypothesis, for $j \in \{1, 2\}$,

$$
\begin{aligned}
&\mathbb{P}(\partial(\boxplus^{n-i-1} nX) \cap D_j = \emptyset) \\
&\quad = \mathbb{P}\left((n-i-1)\partial\left(\boxplus^{n-i-1} \frac{n}{n-i-1} X \right) \cap D_j = \emptyset \right)
\end{aligned}
$$

$$= \sum_{B_j \in \mathbf{D}(D_j)} d_{n-i-1}\left(D_j, B_j, \frac{n}{n-i-1}X\right) \prod_{S_j \in B_j} \mathbb{P}(n\partial X \cap \operatorname{conv} S_j = \emptyset)^{n-i-1}$$
$$+ q_{n-i-1}\left(D_j, \frac{n}{n-i-1}X\right).$$

Applying the induction hypothesis (7.13), this yields that

$$\begin{aligned}
&\mathbb{P}(n\partial(\boxplus^n X) \cap C = \emptyset) \\
&\quad = \mathbb{P}(n\partial X \cap \operatorname{conv} C = \emptyset)^n \\
&\qquad + \sum_{\{D_1,D_2\} \in \mathbf{D}(C)} \mathbb{P}(\xi_{nX}^{\mathrm{pr}}\langle \operatorname{conv} C\rangle = 1,\ \xi_{nX}^{\mathrm{pr}}(D_1 \updownarrow D_2) = 1) \\
&\qquad \times \Bigg(\sum_{B_1 \in \mathbf{D}(D_1)} \sum_{B_2 \in \mathbf{D}(D_2)} d_n(D_1, B_1, X) d_n(D_2, B_2, X) \sum_{i=0}^{n-2} \mathbb{P}(n\partial X \cap \operatorname{conv} C = \emptyset)^i \\
&\qquad \times \prod_{S_1 \in B_1} \mathbb{P}(n\partial X \cap \operatorname{conv} S_1 = \emptyset)^{n-i-1} \prod_{S_2 \in B_2} \mathbb{P}(n\partial X \cap \operatorname{conv} S_2 = \emptyset)^{n-i-1} \\
&\qquad + \mathbb{P}(n\partial X \cap \operatorname{conv} C = \emptyset)^{n-1} \Bigg) + q_n(C, X)
\end{aligned} \tag{7.19}$$

with the remainder term $q_n(C,X)$ given by

$$\begin{aligned}
&q_n(C,X) := q_n'(C,X) \\
&\quad + \sum_{\{D_1,D_2\} \in \mathbf{D}(C)} \mathbb{P}(\xi_{nX}^{\mathrm{pr}}\langle \operatorname{conv} C\rangle = 1,\ \xi_{nX}^{\mathrm{pr}}(D_1 \updownarrow D_2) = 1) \sum_{i=0}^{n-2} \mathbb{P}(n\partial X \cap \operatorname{conv} C = \emptyset)^i \\
&\quad \times \Bigg(q_{n-i-1}\left(D_2, \frac{n}{n-i-1}X\right) \sum_{B_1 \in \mathbf{D}(D_1)} d_n(D_1, B_1, X) \prod_{S_1 \in B_1} \mathbb{P}(n\partial X \cap \operatorname{conv} S_1 = \emptyset)^{n-i-1} \\
&\quad + q_{n-i-1}\left(D_1, \frac{n}{n-i-1}X\right) \sum_{B_2 \in \mathbf{D}(D_2)} d_n(D_2, B_2, X) \prod_{S_2 \in B_2} \mathbb{P}(n\partial X \cap \operatorname{conv} S_2 = \emptyset)^{n-i-1} \\
&\quad + q_{n-i-1}\left(D_1, \frac{n}{n-i-1}X\right) q_{n-i-1}\left(D_2, \frac{n}{n-i-1}X\right) \Bigg).
\end{aligned} \tag{7.20}$$

By Lemma 7.7 for all but a finite number of $n \in \mathbb{N}$, we have

$$\prod_{S \in B_1 \cup B_2} \mathbb{P}(n\partial X \cap \operatorname{conv} S = \emptyset) \neq \mathbb{P}(n\partial X \cap \operatorname{conv} C = \emptyset),$$

and for these n, the summation over i in equation (7.19) yields

$$\begin{aligned}
&\mathbb{P}(n\partial(\boxplus^n X) \cap C = \emptyset) \\
&\quad = \mathbb{P}(n\partial X \cap \operatorname{conv} C = \emptyset)^n \\
&\qquad + \sum_{\{D_1,D_2\} \in \mathbf{D}(C)} \sum_{B_1 \in \mathbf{D}(D_1)} \sum_{B_2 \in \mathbf{D}(D_2)} \Big(d_n(D_1, B_1, X) d_n(D_2, B_2, X)
\end{aligned}$$

$$\times\, \mathbb{P}(\xi_{nX}^{\mathrm{pr}}\langle \operatorname{conv} C\rangle = 1,\ \xi_{nX}^{\mathrm{pr}}(D_1 \updownarrow D_2) = 1)$$
$$\times \frac{\mathbb{P}(n\partial X \cap \operatorname{conv} C = \emptyset)^n - \prod_{S\in B_1\cup B_2} \mathbb{P}(n\partial X \cap \operatorname{conv} S = \emptyset)^n}{\mathbb{P}(n\partial X \cap \operatorname{conv} C = \emptyset) - \prod_{S\in B_1\cup B_2} \mathbb{P}(n\partial X \cap \operatorname{conv} S = \emptyset)}\Bigg)$$
$$+\, q_n(C,X)\,.$$

This can be rewritten as

$$\begin{aligned}
&\mathbb{P}(n\partial(\boxplus^n X) \cap C = \emptyset)\\
&\quad= \mathbb{P}(n\partial X \cap \operatorname{conv} C = \emptyset)^n\\
&\qquad\times\Bigg(1 + \sum_{\{D_1,D_2\}\in\mathbf{D}(C)}\ \sum_{B_1\in\mathbf{D}(D_1)}\ \sum_{B_2\in\mathbf{D}(D_2)} d_n(D_1,B_1,X)\,d_n(D_2,B_2,X)\\
&\qquad\times \frac{\mathbb{P}(\xi_{nX}^{\mathrm{pr}}\langle \operatorname{conv} C\rangle = 1,\ \xi_{nX}^{\mathrm{pr}}(D_1 \updownarrow D_2) = 1)}{\mathbb{P}(n\partial X \cap \operatorname{conv} C = \emptyset) - \prod_{S\in B_1\cup B_2} \mathbb{P}(n\partial X \cap \operatorname{conv} S = \emptyset)}\Bigg)\\
&\qquad- \sum_{\{D_1,D_2\}\in\mathbf{D}(C)}\ \sum_{B_1\in\mathbf{D}(D_1)}\ \sum_{B_2\in\mathbf{D}(D_2)} d_n(D_1,B_1,X)\,d_n(D_2,B_2,X)\\
&\qquad\times \frac{\mathbb{P}(\xi_{nX}^{\mathrm{pr}}\langle \operatorname{conv} C\rangle = 1,\ \xi_{nX}^{\mathrm{pr}}(D_1 \updownarrow D_2) = 1)}{\mathbb{P}(n\partial X \cap \operatorname{conv} C = \emptyset) - \prod_{S\in B_1\cup B_2} \mathbb{P}(n\partial X \cap \operatorname{conv} S = \emptyset)}\\
&\qquad\times \prod_{S\in B_1\cup B_2} \mathbb{P}(n\partial X \cap \operatorname{conv} S = \emptyset)^n\\
&\qquad+ q_n(C,X)\,.
\end{aligned} \tag{7.21}$$

Note that for $D \in \mathbf{D}(C)$, there can be different partitions $\{D_1,D_2\} \in \mathbf{D}(C)$ and $B_1 \in \mathbf{D}(D_1)$ and $B_2 \in \mathbf{D}(D_2)$ such that $B_1 \cup B_2 = D$. However, the summands in (7.21) can be rearranged to obtain the desired form (7.12) with suitable coefficients $d_n(C,D,X)$, $D \in \mathbf{D}(C)$.

To prove the induction step for (7.13), it is sufficient to replace n by ℓ and X by $\frac{n}{\ell}X$, $\ell, n \in \mathbb{N}$, in the coefficients of $\mathbb{P}(n\partial X \cap \operatorname{conv} C = \emptyset)^n$ and of $\prod_{S\in B_1\cup B_2} \mathbb{P}(n\partial X \cap \operatorname{conv} S = \emptyset)^n$, respectively, in formula (7.21). Obviously, ℓ cancels out everywhere.

Induction step for (7.15)
Consider (7.20). The induction hypothesis for (7.15) and (7.16) yields that for all but a finite number of $n \in \mathbb{N}$, for $j \in \{1,2\}$ and $i \in \{0,\dots,n-2\}$,

$$\begin{aligned}
&\left|q_{n-i-1}\left(D_j, \frac{n}{n-i-1}X\right)\right|\\
&\quad\le q\left(D_j, \frac{n}{n-i-1}X\right)(n-i-1)\,\mathbb{P}(\xi^{\mathrm{pr}}_{(n-i-1)(\frac{n}{n-i-1}X)}\langle \operatorname{conv} D_j\rangle \ge 2)\\
&\quad\le q(D_j,X)\,n\,\mathbb{P}(\xi_{nX}^{\mathrm{pr}}\langle \operatorname{conv} D_j\rangle \ge 2)\,,
\end{aligned}$$

and with (7.13) and (7.12), also

$$\begin{aligned}
&\left|\sum_{B_j\in\mathbf{D}(D_j)} d_{n-i-1}\left(D_j, B_j, \frac{n}{n-i-1}X\right)\prod_{S_j\in B_j}\mathbb{P}(n\partial X\cap \operatorname{conv} S_j=\emptyset)^{n-i-1}\right|\\
&\quad=\left|\mathbb{P}\left((n-i-1)\partial\left(\boxplus^{n-i-1}\frac{n}{n-i-1}X\right)\cap D_j=\emptyset\right)-q_{n-i-1}\left(D_j,\frac{n}{n-i-1}X\right)\right|\\
&\quad\le 1+\left|q_{n-i-1}\left(D_j,\frac{n}{n-i-1}X\right)\right|.
\end{aligned}$$

Because $\mathbb{P}(\xi_{nX}^{\mathrm{pr}}\langle \operatorname{conv} D_j\rangle\ge 2)\le \mathbb{P}(\xi_{nX}^{\mathrm{pr}}\langle \operatorname{conv} C\rangle\ge 2)$, due to (7.18), it follows that

$$\begin{aligned}
&|q_n(C,X)|\\
&\quad\le q_n'(C,X)\\
&\qquad+\sum_{\{D_1,D_2\}\in\mathbf{D}(C)} n\,\mathbb{P}(\xi_{nX}^{\mathrm{pr}}\langle \operatorname{conv} C\rangle=1,\ \xi_{nX}^{\mathrm{pr}}(D_1\updownarrow D_2)=1)\\
&\qquad\times\big(q(D_2,X)n\,\mathbb{P}(\xi_{nX}^{\mathrm{pr}}\langle \operatorname{conv} D_2\rangle\ge 2)\,(1+q(D_1,X)n\,\mathbb{P}(\xi_{nX}^{\mathrm{pr}}\langle \operatorname{conv} D_1\rangle\ge 2))\\
&\qquad+q(D_1,X)n\,\mathbb{P}(\xi_{nX}^{\mathrm{pr}}\langle \operatorname{conv} D_1\rangle\ge 2)\,(1+q(D_2,X)n\,\mathbb{P}(\xi_{nX}^{\mathrm{pr}}\langle \operatorname{conv} D_2\rangle\ge 2))\\
&\qquad+q(D_1,X)\,q(D_2,X)n^2\,\mathbb{P}(\xi_{nX}^{\mathrm{pr}}\langle \operatorname{conv} D_1\rangle\ge 2)\mathbb{P}(\xi_{nX}^{\mathrm{pr}}\langle \operatorname{conv} D_2\rangle\ge 2)\big)\\
&\quad\le n\,\mathbb{P}(\xi_{nX}^{\mathrm{pr}}\langle \operatorname{conv} C\rangle\ge 2)\bigg(|\mathbf{D}(C)|+\sum_{\{D_1,D_2\}\in\mathbf{D}(C)} n\,\mathbb{P}(\xi_{nX}^{\mathrm{pr}}\langle \operatorname{conv} C\rangle\ge 1)\\
&\qquad\times\big(q(D_2,X)(1+q(D_1,X)n\,\mathbb{P}(\xi_{nX}^{\mathrm{pr}}\langle \operatorname{conv} C\rangle\ge 1))\\
&\qquad+q(D_1,X)(1+q(D_2,X)n\,\mathbb{P}(\xi_{nX}^{\mathrm{pr}}\langle \operatorname{conv} C\rangle\ge 1))\\
&\qquad+q(D_1,X)\,q(D_2,X)n\,\mathbb{P}(\xi_{nX}^{\mathrm{pr}}\langle \operatorname{conv} C\rangle\ge 1)\big)\bigg).
\end{aligned}$$

By (7.9) and (7.4) we have that $\lim_{n\to\infty} n\,\mathbb{P}(\xi_{nX}^{\mathrm{pr}}\langle \operatorname{conv} C\rangle\ge 1)$ exists and is finite. Choosing

$$\begin{aligned}
q(C,X) := |\mathbf{D}(C)| &+\sum_{\{D_1,D_2\}\in\mathbf{D}(C)}\sup_{m\in\mathbb{N}} m\,\mathbb{P}(\xi_{mX}^{\mathrm{pr}}\langle \operatorname{conv} C\rangle\ge 1)\\
&\times\Big(q(D_2,X)\Big(1+q(D_1,X)\sup_{m\in\mathbb{N}} m\,\mathbb{P}(\xi_{mX}^{\mathrm{pr}}\langle \operatorname{conv} C\rangle\ge 1)\Big)\\
&+q(D_1,X)\Big(1+q(D_2,X)\sup_{m\in\mathbb{N}} m\,\mathbb{P}(\xi_{mX}^{\mathrm{pr}}\langle \operatorname{conv} C\rangle\ge 1)\Big)\\
&+q(D_1,X)\,q(D_2,X)\sup_{m\in\mathbb{N}} m\,\mathbb{P}(\xi_{mX}^{\mathrm{pr}}\langle \operatorname{conv} C\rangle\ge 1)\Big),
\end{aligned}\tag{7.22}$$

the induction step for (7.15) is complete.

Induction step for (7.16)

For all but a finite number of $n\in\mathbb{N}$, $\ell\in\{1,\dots,n-1\}$, and $j\in\{1,2\}$, the induction hypothesis yields $q(D_j,\frac{n}{\ell}X)\le q(D_j,X)$. Furthermore, for all $m\in\mathbb{N}$, the stationarity of X yields

$$\mathbb{P}(\xi^{\mathrm{pr}}_{m\frac{n}{\ell}X}\langle \operatorname{conv} C\rangle \geq 1) = \mathbb{P}\left(\xi^{\mathrm{pr}}_{mX}\left\langle \frac{\ell}{n}\operatorname{conv} C\right\rangle \geq 1\right) \leq \mathbb{P}(\xi^{\mathrm{pr}}_{mX}\langle \operatorname{conv} C\rangle \geq 1),$$

which, together with (7.22), implies (7.16).

The induction step for (7.17)

This follows immediately from (7.15) and Lemma 7.4.

Induction step for (7.14)

By induction hypothesis, $\lim_{n\to\infty} d_n(D_j, B_j, X) = d^{\mathrm{STIT}}(D_j, B_j)$ for all $B_j \in \mathbf{D}(D_j), j \in \{1, 2\}$, $\{D_1, D_2\} \in \mathbf{D}(C)$. An application of (7.11), (7.9), and Lemma 7.6 yields

$$\begin{aligned}
&\lim_{n\to\infty} \frac{\mathbb{P}(\xi^{\mathrm{pr}}_{nX}\langle \operatorname{conv} C\rangle = 1,\ \xi^{\mathrm{pr}}_{nX}(D_1 \updownarrow D_2) = 1)}{\mathbb{P}(n\partial X \cap \operatorname{conv} C = \emptyset) - \prod_{S\in B_1\cup B_2}\mathbb{P}(n\partial X \cap \operatorname{conv} S = \emptyset)}\\
&= \lim_{n\to\infty} \frac{n\mathbb{P}(\xi^{\mathrm{pr}}_{nX}\langle \operatorname{conv} C\rangle = 1,\ \xi^{\mathrm{pr}}_{nX}(D_1 \updownarrow D_2) = 1)}{-n\mathbb{P}(n\partial X \cap \operatorname{conv} C \neq \emptyset) + \sum_{S\in B_1\cup B_2} n\mathbb{P}(n\partial X \cap \operatorname{conv} S \neq \emptyset) + n\,r(n)}\\
&= \frac{\Theta([D_1|D_2])}{-\Theta([\operatorname{conv} C]) + \sum_{S\in B_1\cup B_2}\Theta([\operatorname{conv} S])}\,.
\end{aligned}$$

Now we consider (7.21) and take the limit as $n \to \infty$ of the coefficients of $\mathbb{P}(n\partial X \cap \operatorname{conv} C = \emptyset)^n$ and of $\prod_{S\in B_1\cup B_2}\mathbb{P}(n\partial X \cap \operatorname{conv} S = \emptyset)^n$, respectively, using (7.10). This shows that in the limit the calculation of the $d(C, D, X)$ in the induction step is the same as the corresponding one of $d^{\mathrm{STIT}}(C, D)$ for the STIT tessellation in (5.15).

This concludes the proof of Lemma 7.8. □

Proof of Theorem 7.1. If X is a random stationary tessellation in $\mathbb{R}^d$ with the density η_{d-1} of the $(d-1)$-skeleton and the spherical directional distribution ψ in the typical point of ∂X, then $\eta_{d-1}^{n(\boxplus^n X)} = \eta_{d-1}$ by (3.8), and by (3.7) and (2.52) $\psi^{n(\boxplus^n X)} = \psi$ for all $n \in \mathbb{N}$.

Furthermore, if Y_t is the STIT tessellation according to Theorem 6.4 with Θ as in (4.1), then $\eta_{d-1}(Y_t) = t$ by (6.6), and $\psi^{\mathrm{STIT}} = \varphi$ by (6.8).

Hence a combination of Lemma 7.8 with (7.10) and with Lemma 5.5 completes the proof of the theorem for all $C \in \mathcal{R}$ that satisfy (5.10). Regarding Lemma 2.16 and Remark 5.7, this is sufficient for the proof of the convergence of the capacity functional for all $C \in \mathcal{C}$, and thus the proof of Theorem 7.1 is complete. □

7.2 Uniqueness of tessellations stable under iteration

As an important implication of Theorem 7.1, we obtain a uniqueness result for STIT tessellations. This means that for a given density η_{d-1} of the $(d-1)$-skeleton and a spherical directional distribution ψ in the typical point of this skeleton, the distribution of tessellations that are stable under iteration is uniquely determined. Hence the STIT tessellations that are characterized in Theorem 5.10 or by equivalent descriptions, for example, the global construction presented in Chapter 8, are the only tessellations that are stable under iteration.

Theorem 7.9. *Let* $(Y_t, t > 0)$ *be the STIT tessellation process described in Definition* 6.1, *driven by the hyperplane measure* Θ *of the form* (4.1) *with the spherical directional distribution* φ, *satisfying* (4.3). *Furthermore, let* X *be a random stationary tessellation of* $\mathbb{R}^d$, *satisfying Convention* 2.28, *and having the density* η_{d-1} *of the* $(d-1)$*-skeleton and the spherical directional distribution* ψ *in the typical point of* ∂X. *Then* X *has the STIT property* (3.10) *if and only if* $X \overset{D}{=} Y_t$ *with* $\eta_{d-1} = t$ *and* $\psi = \varphi$.

Proof. If $X \overset{D}{=} Y_t$ for some $t > 0$, then the STIT property (3.10) follows by Theorem 6.4. The equality $\eta_{d-1} = t$ follows from (6.6), and (6.8) yields the identity of the spherical directional distribution ψ with the probability measure φ of Θ in (4.1).

On the other hand, if X, with $\eta_{d-1} = t$, and the spherical directional distribution ψ in the typical point of ∂X, satisfies the STIT property (3.10), then for all $C \in \mathcal{C}$ satisfying (5.10), by Lemma 7.8

$$\mathbb{P}(\partial X \cap C = \emptyset) = \lim_{n\to\infty} \mathbb{P}(n\,\partial(\boxplus^n X) \cap C = \emptyset) = \mathbb{P}(\partial Y_t \cap C = \emptyset)\,.$$

Hence $X \overset{D}{=} Y_t$, which completes the proof. □

By Lemma 3.8 and Theorem 7.9 we obtain the following proposition for the one-dimensional case.

Lemma 7.10. *In* $\mathbb{R}^1$ *the tessellation generated by a point process* Φ *is STIT if and only if* Φ *is a stationary Poisson point process.*

7.3 Notes

1. The contents of Section 7.1 is basically that of Section 6 of [145]. Note that there Dobrushin's lemma was erroneously referred to as Korolyuk's theorem.
2. The uniqueness Theorem 7.9 is Corollary 2 of [145].
3. Lemma 7.10 was proven as Theorem 3 in [144].
   An alternative proof of the corresponding result, using the capacity functional of the skeleton, was given in Theorem 3.6 of [50] for splitting schemes on the sphere.
4. In Theorem 1 of [144], sequences of rescaled consecutive superpositions of i. i. d. stationary segment processes in $\mathbb{R}^2$ were considered. It was shown that theses sequences converge weakly to Poisson line processes.

8 Global construction of the STIT tessellation process

In Chapter 4, we have introduced the STIT tessellation process as a random process of tessellations $(Y_{t,W}, t \geq 0)$ within a window $W \in \mathcal{P}_d$. Subsequently, we used the spatial consistency property of this process to derive in Theorem 5.10 the existence of a STIT tessellation process $(Y_t, t > 0)$ in the full Euclidean space $\mathbb{R}^d$. In the present chapter, we describe a direct approach to a process of random tessellations of $\mathbb{R}^d$ using a global construction. We then show that this process has the same law as the STIT tessellation process $(Y_t, t > 0)$ obtained by spatial consistency. The principal obstacle of such an approach is the lack of an appropriate initial tessellation at time $t = 0$. This problem is resolved by first constructing a suitable random frame from a process of random polytopes. This process has the so-called big-bang phenomenon at time zero in the sense that the random polytopes converge to $\mathbb{R}^d$ as $t \to 0$. Within these polytopes, STIT division processes described in Chapter 4 are then launched in a further step to build the global tessellation process. The background to this construction is the results of Chapter 5, where we considered relationships between STIT tessellation and Poisson hyperplane tessellation processes.

8.1 Description of the construction

In this section, we fix a hyperplane measure Θ of the form (4.1), satisfying (4.3). Let $\widetilde{\Sigma}$ be a Poisson point process on the measurable space

$$(\mathbb{R}^d \times \mathbb{R}^- \times \mathsf{N}_s(\mathcal{H} \times \mathbb{R}^+), \mathcal{B}(\mathbb{R}^d) \otimes \mathcal{B}(\mathbb{R}^-) \otimes \mathcal{N}_s(\mathcal{H} \times \mathbb{R}^+))$$

with intensity measure $\lambda_d \otimes \lambda_- \otimes \mathbb{P}_{\widehat{X}^*}$, where $\mathbb{P}_{\widehat{X}^*}$ is the distribution of the Poisson process $\widehat{X}^*$ of birth time marked hyperplanes with intensity measure $\Theta \otimes \lambda_+$, as defined in Section 2.3.1. Furthermore, let $\widehat{X}_o^*$ be a Poisson process of birth time marked hyperplanes with distribution $\mathbb{P}_{\widehat{X}^*}$, independent of $\widetilde{\Sigma}$. Then define the point process

$$\Sigma := \widetilde{\Sigma} + \delta_{(o,0,\widehat{X}_o^*)} . \tag{8.1}$$

We interpret Σ as a collection of random *reference points* in $\mathbb{R}^d$, each one marked with a *priority* in $\mathbb{R}^-$ and with a *set of birth time marked hyperplanes*. The point process Σ determines a tessellation-valued process $(\widetilde{Y}_t, t > 0)$ as follows.

(i) In a first step the random process of zero-cells $(Z_{\langle o\rangle,t}, t > 0)$ is constructed, where $Z_{\langle o\rangle,t}$ is the zero-cell of $\widetilde{Y}_t$. To do this, we use the item $\delta_{(o,0,\widehat{X}_o^*)}$ of (8.1). The Poisson point process $\widehat{X}_o^*$ of birth time marked hyperplanes generates the process

$$(Z_{\langle o\rangle,t}, t > 0) := (Z^{\mathrm{PHT}}_{\langle o\rangle,t}, t > 0), \tag{8.2}$$

where the right-hand side was defined in (5.3).

https://doi.org/10.1515/9783112222409-009

This is a Markov process with values in $\mathcal{P}_d$. For all $t_0 > 0$, the process $(Z_{\langle o\rangle,t}, t \geq t_0)$ is a pure jump process. Using the set $\mathbb{Z}$ of integers as an index set, the jumps can be numbered as follows. Let $(\beta_i, i \in \mathbb{Z})$ denote the ordered sequence of jump times of $(Z_{\langle o\rangle,t}, t > 0)$ with $\beta_0 < 1 \leq \beta_1$. Note that this sequence has an accumulation point at time $t = 0$. We denote by $((H_i, \beta_i), i \in \mathbb{Z})$ the corresponding sequence of birth time marked hyperplanes from $\widehat{X}_o^*$.

(ii) In a second step, we derive a tessellation-valued Markov process $(S_t, t > 0)$ from the process of zero-cells. For $t > 0$, we define

$$S_t := \{Z_{\langle o\rangle,t}\} \cup \{\mathrm{cl}\,(Z_{\langle o\rangle,\beta_{i-1}} \setminus Z_{\langle o\rangle,\beta_i}) : \beta_i \leq t\}. \tag{8.3}$$

Thus, at time β_i, when the process of zero-cells jumps from $Z_{\langle o\rangle,\beta_{i-1}}$ to $Z_{\langle o\rangle,\beta_i}$, the process $(S_t, t > 0)$ jumps from the state $S_{\beta_{i-1}}$ into the state

$$\begin{aligned} S_{\beta_i} &:= (S_{\beta_{i-1}} \setminus \{Z_{\langle o\rangle,\beta_{i-1}}\}) \cup \{Z_{\langle o\rangle,\beta_i}, \mathrm{cl}\,(Z_{\langle o\rangle,\beta_{i-1}} \setminus Z_{\langle o\rangle,\beta_i})\} \\ &= \oslash(Z_{\langle o\rangle,\beta_{i-1}}, H_i, S_{\beta_{i-1}}). \end{aligned}$$

Two such jumps are illustrated in Figure 8.1. This means that for all $t > 0$, the zero-cell of S_t is $Z_{\langle o\rangle,t} = Z_{\langle o\rangle,t}^{\mathrm{PHT}}$ and the other cells of S_t are the polytopes segregated consecutively during the evolution of the zero-cell process. The assigned birth times are

$$\beta(\mathrm{cl}\,(Z_{\langle o\rangle,\beta_{i-1}} \setminus Z_{\langle o\rangle,\beta_i})) = \beta(Z_{\langle o\rangle,\beta_i}) := \beta_i.$$

Note that $(S_t, t > 0)$ is a process of tessellations of $\mathbb{R}^d$; see Lemmas 8.3 and 8.5.

(iii) Furthermore, for the construction of $(\widetilde{Y}_t, t > 0)$, now we also use the item $\widetilde{\Sigma}$ of (8.1). For a generic polytope $P \in \mathcal{P}_d$ with birth time $\beta > 0$, we construct an auxiliary

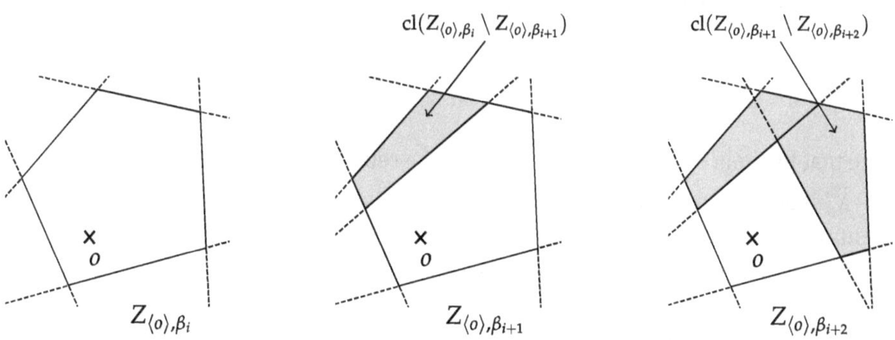

Figure 8.1: The zero-cell process. The jumps of the zero-cell process cause the jumps of the process $(S_t, t > 0)$, defined in step (ii) of the construction. In step (iii) it is described how the grey polytopes are subdivided independently after their separation from a larger zero-cell.

process $(\Delta_t(P),\ t \geq \beta)$ of tessellations of P that has the same distribution as a STIT tessellation process in P, up to a time shift by β. The initial state is

$$\Delta_\beta(P) := \{P\},$$

and the jumps into new states are defined by division events as follows. For all $P' \in \Delta_t(P)$ with $t > \beta$, there is an a. s. unique division event (P', H, s) determined by

$$\begin{aligned}
&(H, s) \in \pi, \\
&(x, r, \pi) \in \Sigma, \\
&x \in P', \\
&r = \max\{r' : (x', r', \pi') \in \Sigma,\ x' \in P'\}, \\
&s = \min\{s' > t : (H', s') \in \pi,\ H' \in [\operatorname{int} P']\}.
\end{aligned} \tag{8.4}$$

This means that a triple $(x, r, \pi) \in \Sigma$ is chosen, with its reference point $x \in P'$ and highest priority $r \in \mathbb{R}^-$ in the set of all triples that have their reference points in P'. Then from the pertaining realization π of the birth time marked Poisson hyperplane process a pair $(H, s) \in \pi$ is chosen that has the smallest birth time greater than t and satisfies $H \in [\operatorname{int} P']$.

At time s the cell P' is divided by H, and the process jumps into the new state $\Delta_s(P) := \oslash(P', H, \Delta_{s-}(P))$, where $\Delta_{s-}(P) := \lim_{s' \uparrow s} \Delta_{s'}(P)$ is the state of the process just before the jump. The birth times of the new cells are $\beta(P' \cap H^+) = \beta(P' \cap H^-) := s$. The set of all division events of $(\Delta_t(P),\ t \geq \beta)$ is denoted by $\mathcal{D}(\Delta_t(P),\ t \geq \beta)$.

(iv) Finally, using the cells of the auxiliary process $(S_t,\ t > 0)$, the construction of (iii) is applied to all of its cells. Regarding Figure 8.1, this means that a STIT tessellation process is constructed in each of the grey polytopes, launched at the birth times of them. Thus we define the process $(\widetilde{Y}_t,\ t > 0)$ by

$$\widetilde{Y}_t := \{Z_{\langle o \rangle, t}\} \cup \bigcup_{\beta_i \leq t} \Delta_t(\operatorname{cl}(Z_{\langle o \rangle, \beta_{i-1}} \setminus Z_{\langle o \rangle, \beta_i})). \tag{8.5}$$

Note that the item $\delta_{(o, 0, \widehat{X}_o^*)}$ in (8.1), which has the reference point o and priority 0, means that $\widehat{X}_o^*$ is always used to divide the zero cell – consistently with the rule (8.4) for division events and as described in step (i) of the construction.

The set of division events of the process $(\widetilde{Y}_t,\ t > 0)$ is

$$\begin{aligned}
\mathcal{D}(\Sigma) := &\left\{(Z_{\langle o \rangle, \beta_{i-1}}, H_i, \beta_i) : i \in \mathbb{Z}\right\} \\
&\cup \bigcup_{i \in \mathbb{Z}} \mathcal{D}(\Delta_t(\operatorname{cl}(Z_{\langle o \rangle, \beta_{i-1}} \setminus Z_{\langle o \rangle, \beta_i})),\ t \geq \beta_i);
\end{aligned} \tag{8.6}$$

compare with (4.9).

At a first glance, the complex structure of the process Σ in (8.1) may appear unnecessarily overburdened. However, the introduction of the auxiliary reference points $x \in \mathbb{R}^d$ and priorities $r \in \mathbb{R}^-$ allows for a useful formalization of the condition that for distinct cells, the subsequent division processes are independent.

Before we prove some properties of the global cell division process $(\widetilde{Y}_t, t > 0)$, we consider some essential features of the point process $\widetilde{\Sigma}$ in (8.1).

Lemma 8.1. *The Poisson point process $\widetilde{\Sigma}$ in* (8.1) *has the following properties.*

(a) *For any $r < 0$, the set $\{x : (x, r', \pi) \in \widetilde{\Sigma}, r < r' < 0\}$ of the reference points with priority greater than r forms a stationary Poisson point process on $\mathbb{R}^d$ with finite intensity $|r|$.*
(b) *The set of all reference points $\{x : (x, r, \pi) \in \widetilde{\Sigma}\}$ is a. s. countably infinite and dense in $\mathbb{R}^d$.*
(c) *Let $P \in \mathcal{P}_d$. Then the set $\{r \in \mathbb{R}^- : (x, r, \pi) \in \widetilde{\Sigma}, x \in P\}$ of all priorities of points in $\widetilde{\Sigma}$ with the reference point in P is a homogeneous Poisson point process on $(-\infty, 0)$ with intensity measure $\lambda_d(P)\,\lambda_-$.*
(d) *For $P_1, P_2 \in \mathcal{P}_d$ with $\operatorname{int} P_1 \cap \operatorname{int} P_2 = \emptyset$, the point processes $\{(x, r, \pi) \in \widetilde{\Sigma} : x \in P_1\}$ and $\{(x, r, \pi) \in \widetilde{\Sigma} : x \in P_2\}$ are independent.*

Proof. (a) This follows by the restriction Theorem 5.2 and the mapping Theorem 5.1 of [101].

(b) The representation

$$\{x : (x, r, \pi) \in \widetilde{\Sigma}\} = \bigcup_{n \in \mathbb{N}} \{x : (x, r, \pi) \in \widetilde{\Sigma}, -n < r < 0\}$$

yields that this set is a. s. countable. To show that it is a dense subset of $\mathbb{R}^d$, let $B(z, \varepsilon)$ be the ball centered at $z \in \mathbb{R}^d$ with radius $\varepsilon > 0$. Then

$$\begin{aligned} &\mathbb{P}(\widetilde{\Sigma}(B(z, \varepsilon) \times \mathbb{R}^- \times \mathsf{N}_s(\mathcal{H} \times \mathbb{R}^+)) > 0) \\ &= \lim_{n \to \infty} \mathbb{P}(\widetilde{\Sigma}(B(z, \varepsilon) \times (-n, 0) \times \mathsf{N}_s(\mathcal{H} \times \mathbb{R}^+)) > 0) \\ &= \lim_{n \to \infty} (1 - \exp(-n\lambda_d(B(z, \varepsilon)))) = 1 . \end{aligned}$$

(c) Here the argument is the same as in (a).

(d) This follows immediately from Definition 19.10 and part (c). □

Remark 8.2. *The Poisson point process $\widetilde{\Sigma}$ on $\mathbb{R}^d \times \mathbb{R}^- \times \mathsf{N}_s(\mathcal{H} \times \mathbb{R}^+)$ induces a point process $\widehat{\Sigma}$ on $\mathbb{R}^d \times \mathbb{R}^- \times \mathcal{H} \times \mathbb{R}^+$ defined by*

$$\widehat{\Sigma} := \{(x, r, H, s) : (x, r, \pi) \in \widetilde{\Sigma}, (H, s) \in \pi\} .$$

This point process $\widehat{\Sigma}$ is no longer a Poisson point process but a Cox process, that is, a Poisson point process with random intensity measure on $\mathbb{R}^d \times \mathbb{R}^- \times \mathcal{H} \times \mathbb{R}^+$. For the definition of *a Cox process or, synonymously, a doubly stochastic Poisson point process, see Section 3.2 of* [172] *or Section 13.2 of* [101].

The purpose of the subsequent lemmas is to show that $(\widetilde{Y}_t, t > 0)$ is indeed a tessellation-valued random process in $\mathbb{R}^d$. We begin by proving that as $t \to 0$, the zero-cells fill the space $\mathbb{R}^d$.

Lemma 8.3. *Let* $(Z_{\langle o\rangle,t}, t > 0)$ *be the process defined in* (8.2). *Then*

$$\bigcup_{t>0} Z_{\langle o\rangle,t} = \mathbb{R}^d \quad a.s.$$

Proof. By (5.6) we have that for $t > 0$ and $n \in \mathbb{N}$,

$$\mathbb{P}(nB^d \subseteq Z_{\langle o\rangle,t}) = e^{-t\,\Theta([nB^d])} = e^{-2tn}.$$

Hence by the continuity of probability measures

$$\begin{aligned}\mathbb{P}\Big(\mathbb{R}^d \subseteq \bigcup_{t>0} Z_{\langle o\rangle,t}\Big) &= \mathbb{P}\Big(\bigcup_{n\in\mathbb{N}} nB^d \subseteq \bigcup_{t>0} Z_{\langle o\rangle,t}\Big)\\ &= \mathbb{P}\Big(\bigcap_{n\in\mathbb{N}} \bigcup_{m\in\mathbb{N}} \{nB^d \subseteq Z_{\langle o\rangle,\frac{1}{m}}\}\Big)\\ &= \lim_{n\to\infty} \lim_{m\to\infty} \mathbb{P}(nB^d \subseteq Z_{\langle o\rangle,\frac{1}{m}}) = 1.\end{aligned}$$

This proves the claim. □

The previous lemma implies that the random tessellation $\widetilde{Y}_t$ satisfies $\bigcup_{P\in\widetilde{Y}_t} P = \mathbb{R}^d$ for all $t > 0$, and hence property (a) of Definition 2.1 holds. Obviously, also condition (b) of Definition 2.1 is satisfied, because by the definition of the cell division process the interiors of the generated cells are disjoint. With the following two lemmas, we prove the local finiteness of $\widetilde{Y}_t$, that is, $\widetilde{Y}_t$ also satisfies part (c) of Definition 2.1.

Lemma 8.4. *For* $i \in \mathbb{Z}$, *denote* $P_i := \mathrm{cl}(Z^{\mathrm{PHT}}_{\langle o\rangle,\beta_{i-1}} \setminus Z^{\mathrm{PHT}}_{\langle o\rangle,\beta_i})$. *Then*

$$(\Delta_t(P_i), t \geq \beta_i) \overset{\mathrm{D}}{=} (Y_{t-\beta_i,P_i}, t \geq \beta_i),$$

where $(Y_{s,P_i}, s \geq 0)$ *is the STIT tessellation process in* P_i, *driven by the measure* Θ, *as introduced in Definition* 4.2.

Proof. Referring to Definition 4.2, we have that $\Delta_{\beta_i}(P_i) = \{P_i\} = Y_{0,P_i}$. Let $P \in \Delta_t(P_i)$ with birth time $\beta(P)$. According to Lemma 2.32, the joint distribution of $(H, \beta(H) - \beta(P))$, that is, of the life time of P and of the dividing hyperplane, is $\Theta_{[P]} \otimes \mathsf{Expo}(\Theta([P]))$ for the process $(\Delta_t(P_i), t \geq \beta_i)$. By Theorem 4.7 it is identical with the corresponding distribution in the construction of $(Y_{t-\beta_i,P_i}, t \geq \beta_i)$. Furthermore, given $P \in \Delta_t(P_i)$, the random variable $(H, \beta(H) - t)$ is conditionally independent of $(\Delta_s(P_i), \beta_i \leq s < t)$ and all $P' \in \Delta_t(P_i)$,

$P' \neq P$. This follows from the independence property (b) in Definition 19.10 of Poisson point processes and the properties of the point process Σ in (8.1); see Lemma 8.1. The corresponding conditional independence property for the STIT tessellation process was also shown in Theorem 4.7. □

Lemma 8.5. *For all $t > 0$ and compact sets $C \subset \mathbb{R}^d$, the set $\{P \in \widetilde{Y}_t : P \cap C \neq \emptyset\}$ is a. s. finite.*

Proof. Fix $C \in \mathcal{C}$ and $t > 0$. Then, for the process $(S_t, t > 0)$ defined in (8.3), the set $\{P \in S_t : P \cap C \neq \emptyset\}$ is a. s. finite, because $C \in \mathcal{C}$ is intersected by an a. s. finite number of hyperplanes of $\widehat{X}_t$; see (2.65). Then, for all $P \in S_t$, Lemmas 8.4 and 4.9 yield that $\Delta_t(P)$ a. s. consists of finitely many cells. □

Now we show that $(\widetilde{Y}_t, t > 0)$ is a STIT tessellation process. We start with the restrictions to windows.

Lemma 8.6. *Let Θ be a hyperplane measure of the form* (4.1), *satisfying* (4.3), *let $W \in \mathcal{P}_d$ with $o \in \operatorname{int} W$, and let $(Y_{t,W}, t > 0)$ be the STIT tessellation process in W, driven by Θ. Then, for the process $(\widetilde{Y}_t, t > 0)$ defined in* (8.5), *we have*

$$(\widetilde{Y}_t \wedge W, t > 0) \stackrel{\mathrm{D}}{=} (Y_{t,W}, t > 0)\,.$$

Proof. Let $W \in \mathcal{P}_d$. Lemma 8.3 yields that a. s. there exists $t > 0$ such that $W \subset \widetilde{Z}_{\langle o\rangle,t}$, the zero-cell of $\widetilde{Y}_t$. For $(\widetilde{Y}_t, t > 0)$, the joint distribution of (H, β), the first hyperplane intersecting W and its birth time, is $\Theta_{[W]} \otimes \mathsf{Expo}(\Theta([W]))$, because it is determined by the law of the Poisson point process $\widehat{X}^*$, which in turn determines the process of zero-cells; see (8.2). Thus (H, β) is identically distributed as the respective pair in the construction of $(Y_{t,W}, t > 0)$; see Definition 4.2. Then, for the process of zero-cells in W, we apply Lemma 5.2, and for the division processes of the other cells in W, we use Lemma 8.4, which also implies the spatial consistency of the processes $(\Delta_t(P_i), t \geq \beta_i)$, $i \in \mathbb{Z}$. The conditional independence of the division processes in different cells is a consequence of Lemma 8.1(d). □

Hence Theorem 5.10(d) yields that the global construction indeed generates a STIT tessellation process.

Theorem 8.7. *Let Θ be a hyperplane measure of the form* (4.1), *satisfying* (4.3), *and let $(Y_t, t > 0)$ be the STIT tessellation process driven by Θ in the sense of Definition* 6.1. *If $(\widetilde{Y}_t, t > 0)$ is the tessellation-valued process as described in* (8.5), *then*

$$(\widetilde{Y}_t, t > 0) \stackrel{\mathrm{D}}{=} (Y_t, t > 0)\,.$$

This theorem enables to switch between the two descriptions of STIT tessellation processes in $\mathbb{R}^d$:

- the one of Chapter 4 based on local cell division processes in windows and the spatial consistency property,
- the direct global construction just developed, which is based on the point process Σ given by (8.1).

The global construction is essential for the proofs in Chapter 14.

8.2 Maximal polytopes

In view of the global construction explained in Section 8.1, the $(d-1)$-dimensional polytopes that divide a cell into two new cells can be regarded as the building blocks of STIT tessellations and thus form a natural class of objects to be investigated. We remark that a polytope F is maximal in the sense that a. s. the affine subspace $\operatorname{aff} F$ does not contain other $(d-1)$-dimensional parts of the skeleton of the STIT tessellation; see Lemma 8.8(b) for a formal statement. This property is the motivation to refer to these polytopes as $(d-1)$-dimensional maximal polytopes. These components will be studied thoroughly in the following parts of this book.

Regarding the cell division process, we define the set of **birth time marked maximal $(d-1)$-polytopes** as

$$\mathcal{F}_{d-1}^{\max *}(\Sigma) := \{(P \cap H, \beta) : (P, H, \beta) \in \mathcal{D}(\Sigma)\} \subset \mathcal{P}_{d-1} \times \mathbb{R}^+, \tag{8.7}$$

where Σ is defined by (8.1), and $\mathcal{D}(\Sigma)$ in (8.6). It is evident from the construction that there is a one-to-one correspondence between the collection of division events $\mathcal{D}(\Sigma)$ and the set $\mathcal{F}_{d-1}^{\max *}(\Sigma)$ of birth time marked maximal $(d-1)$-polytopes. Furthermore, it is a crucial observation that after its birth time, say $s > 0$, a maximal $(d-1)$-polytope $F = P \cap H$, which arises as the intersection of a hyperplane H with a cell P, can be divided further by other maximal $(d-1)$-polytopes generated in the ongoing STIT division processes in the two subcells $P \cap H^+$ and $P \cap H^-$. However, regardless of such further subdivisions, the pair (F, s) will be referred to as a birth time marked maximal $(d-1)$-polytope at all times $t \geq s$.

Lower-dimensional maximal polytopes are defined as the faces of maximal $(d-1)$-polytopes. Formally, for $k \in \{0, \dots, d-2\}$ the set of **birth time marked maximal k-polytopes** is defined as

$$\mathcal{F}_k^{\max *}(\Sigma) := \bigcup_{(F,\beta) \in \mathcal{F}_{d-1}^{\max *}(\Sigma)} (\mathcal{F}_k(F) \times \{\beta\}) \subset \mathcal{P}_k \times \mathbb{R}^+ . \tag{8.8}$$

Finally, for the STIT tessellation process $(Y_t, t > 0)$ generated by the point process Σ in (8.1) and for $t > 0$ and $k \in \{0, \dots, d-1\}$, we put

$$\mathcal{F}_k^{\max *}(Y_t) := \{(F, s) \in \mathcal{F}_k^{\max *}(\Sigma) : s \leq t\} \tag{8.9}$$

and

$$\mathcal{F}_k^{\max}(Y_t) := \{F : (F, s) \in \mathcal{F}_k^{\max *}(\Sigma),\ s \le t\}. \tag{8.10}$$

As anticipated above, the next lemma shows why the term “maximal polytope” is justified to name the objects occurring in the division of cells and that a STIT tessellation is completely defined by its maximal $(d-1)$-polytopes. It is an analogue of Lemma 4.6.

Lemma 8.8. *Let* Θ *be a hyperplane measure of the form* (4.1)*, satisfying* (4.3)*, and let* $(Y_t, t > 0)$ *be the STIT tessellation process driven by* Θ*. Let* $t > 0$ *and* $k \in \{0, \dots, d-1\}$*. Then*

(a) $\operatorname{skel}_k Y_t = \bigcup_{F \in \mathcal{F}_k^{\max}(Y_t)} F$ *a. s.*,

(b) $\operatorname{aff} F_1 \ne \operatorname{aff} F_2$ *a. s. for* $(F_1, s_1), (F_2, s_2) \in \mathcal{F}_k^{\max *}(\Sigma)$ *with* $F_1 \ne F_2$,

(c) *putting*

$$F_t^{\max}(F) := \bigcup_{F' \in \mathcal{F}_k(Y_t),\, \operatorname{aff} F' = \operatorname{aff} F} F' \quad \text{for } F \in \mathcal{F}_k(Y_t),$$

we have that $\mathcal{F}_k^{\max}(Y_t) = \{F_t^{\max}(F) : F \in \mathcal{F}_k(Y_t)\}$ *a. s.*

Proof. (a) From definition (2.2) of skeletons we have $\operatorname{skel}_k Y_t \supseteq \bigcup_{F \in \mathcal{F}_k^{\max}(Y_t)} F$. To show the equality of the sets, consider $(\operatorname{skel}_k Y_t) \setminus \bigcup_{F \in \mathcal{F}_k^{\max}(Y_t)} F$. By the global construction and the definition of $\mathcal{F}_k^{\max}(Y_t)$, which is based on the set $\mathcal{D}(\Sigma)$ of division events, for all $\varepsilon > 0$, we have that $(\operatorname{skel}_k Y_t) \setminus \bigcup_{F \in \mathcal{F}_k^{\max}(Y_t)} F = (\operatorname{skel}_k Y_\varepsilon) \setminus \bigcup_{F \in \mathcal{F}_k^{\max}(Y_\varepsilon)} F$. In other words, the difference set $(\operatorname{skel}_k Y_\varepsilon) \setminus \bigcup_{F \in \mathcal{F}_k^{\max}(Y_\varepsilon)} F$ remains constant for all $t > \varepsilon$, because the process evolves by division of polytopes.

Assume that the set $B_\varepsilon := (\operatorname{skel}_k Y_\varepsilon) \setminus \bigcup_{F \in \mathcal{F}_k^{\max}(Y_\varepsilon)} F$ is nonempty and hence of positive k-dimensional Hausdorff measure. Because B_ε is stationary, it follows that $\eta_{k,\varepsilon} := \mathbb{E}\mathcal{H}^k(B_\varepsilon \cap [0,1]^d) > 0$. The arguments above yield that $\eta_{k,\varepsilon_1} = \eta_{k,\varepsilon_2} > 0$ for all $\varepsilon_1, \varepsilon_2 > 0$, which contradicts the scaling property (6.2) of the STIT tessellation process. Hence $(\operatorname{skel}_k Y_t) \setminus \bigcup_{F \in \mathcal{F}_k^{\max}(Y_t)} F = \emptyset$ a. s.

(b) This property follows immediately by the translation invariance of the hyperplane measure Θ and the independence of the hyperplanes that appear in Σ and are used in the global construction.

(c) This follows by definition (2.2) of skeletons together with the results of parts (a) and (b). □

The statement in parts (a) and (b) of the previous lemma also implies that there is no unbounded $(d-1)$-dimensional part of a hyperplane in the $(d-1)$ skeleton of Y_t. Note that this does not contradict the fact that, for example, choosing a Mondrian hyperplane measure Θ and an arbitrarily large $a > 0$, in the window $W = [-a, a]^d$, there is, after the first division of W, at least one $(d-1)$-polytope in $\operatorname{skel}_{d-1} Y_t \cap W$, that is, a $(d-1)$-cuboid of side length $2a$. Such a polytope is always a cut-out of a $(d-1)$-dimensional polytope.

By Lemma 8.8 the $(d-1)$-skeleton $\partial Y_t = \mathrm{skel}_{d-1} Y_t$ of a STIT tessellation Y_t consists of maximal $(d-1)$-polytopes, independent of the construction method of Y_t.

Because of the global construction, each cell arises as a result of the division of another cell by a maximal $(d-1)$-polytope, and hence all cells have at least one $(d-1)$-face that is a maximal $(d-1)$-polytope. Formally, for all $t > 0$ and $P \in Y_t$,

$$\mathcal{F}_{d-1}(P) \cap \mathcal{F}_{d-1}^{\max}(Y_t) \neq \emptyset \quad \text{a. s.} \tag{8.11}$$

However, note that a maximal $(d-1)$-polytope that is a $(d-1)$-face of a cell P can be divided by other maximal polytopes of division processes outside of P.

Furthermore, the maximal $(d-1)$-polytopes of a STIT tessellation satisfy a **"mikado principle"**, which means that in a first step the maximal $(d-1)$-polytope can be removed whose relative interior is not intersected by other maximal polytopes. This deletion reunites two adjacent cells to their mother cell. This can be proceeded sequentially. Thus a realization of a STIT tessellation can completely be removed by deleting maximal polytopes subsequently in the way just described. Intuitively, this may be considered as a reverse global construction. Formally, for all $t > 0$, define

$${}^{0}\mathcal{F}_{d-1}^{\max}(Y_t) := \{F \in \mathcal{F}_{d-1}^{\max}(Y_t) : F' \cap \mathrm{relint}\, F = \emptyset \text{ for all } F' \in \mathcal{F}_{d-1}^{\max}(Y_t),\ F' \neq F\}$$

and for $n \in \mathbb{N}$,

$$\begin{aligned} {}^{n}\mathcal{F}_{d-1}^{\max}(Y_t) := \Bigg\{ & F \in \mathcal{F}_{d-1}^{\max}(Y_t) \setminus \bigcup_{i=0}^{n-1} {}^{i}\mathcal{F}_{d-1}^{\max}(Y_t) : \\ & F' \cap \mathrm{relint}\, F = \emptyset \text{ for all } F' \in \mathcal{F}_{d-1}^{\max}(Y_t) \setminus \bigcup_{i=0}^{n-1} {}^{i}\mathcal{F}_{d-1}^{\max}(Y_t),\ F' \neq F \Bigg\} . \end{aligned}$$

Lemma 8.9. *With the notation introduced above, for all $t > 0$, we have*

$$\mathcal{F}_{d-1}^{\max}(Y_t) = \bigcup_{i=0}^{\infty} {}^{i}\mathcal{F}_{d-1}^{\max}(Y_t) . \tag{8.12}$$

Proof. Let $F \in \mathcal{F}_{d-1}^{\max}(Y_t)$. Then there are r, s with $0 < r < s < t$, $P \in Y_r$, and a division event $(P, H, s) \in \mathcal{D}(\Sigma)$ such that $(P \cap H, s) = (F, s) \in \mathcal{F}_{d-1}^{\max *}(Y_t)$; see (8.7). Because $P \in \mathcal{P}_d$, by Theorem 4.7(e) there appear a. s. only finitely many maximal $(d-1)$-polytopes within P until time t. Hence there is a. s. a pair $(F_0, s_0) \in \mathcal{F}_{d-1}^{\max *}(Y_t)$ with $F_0 \subset P$, $s_0 \geq s$, and $s_0 = \max\{s' : (F', s') \in \mathcal{F}_{d-1}^{\max *}(Y_t),\ F' \subset P\}$, which means that F_0 is the last division before t of a cell inside P. Hence $F_0 \in {}^{0}\mathcal{F}_{d-1}^{\max}(Y_t)$. Then handling the remaining maximal $(d-1)$-polytopes within P in descending order of their birth times, for all those F', there exists $j \in \mathbb{N}$ such that $F' \in {}^{j}\mathcal{F}_{d-1}^{\max}(Y_t)$, and thus for F, there exists eventually $i \in \mathbb{N}$ such that $F \in {}^{i}\mathcal{F}_{d-1}^{\max}(Y_t)$, which proves (8.12). □

An example of a tessellation that does not satisfy the mikado principle is that constructed in [136] and illustrated in Figure 16.3. This tessellation is a particular case of a rather general model presented in [6]. There the growing boundaries of cells are generated by a sweeping line construction; see also Section 9.2.2.

Furthermore, we describe the type of the maximal $(d-2)$-polytopes. By Lemma 8.8(b,c), for all $F \in \mathcal{F}_{d-2}^{\max}(Y_t)$, there are a. s. exactly two maximal $(d-1)$-polytopes $F_1, F_2 \in \mathcal{F}_{d-1}^{\max}(Y_t)$ such that

$$F = F_1 \cap F_2 \quad \text{and} \quad F_1 \subset (\operatorname{aff} F_2)^+ \quad \text{or} \quad F_1 \subset (\operatorname{aff} F_2)^- . \tag{8.13}$$

This means, for example, that in a planar STIT tessellation, a. s. all vertices are T-vertices; see Section 16.2.1.

In the light of Lemma 8.8, it seems quite natural to introduce for STIT tessellations the sets of maximal k-polytopes, $k \in \{0, \dots, d-1\}$, as a further set of components in the sense of Definition 2.9. For a center function $c : \mathcal{P}_k \to \mathbb{R}^d$, we define

$$\mathcal{Z}(\mathcal{F}_k^{\max}(Y_t)) := \{(c(F), F - c(F)) : F \in \mathcal{F}_k^{\max}(Y_t)\} . \tag{8.14}$$

In accordance with (2.25), we denote by

$$\gamma_{k,t}^{\max} := \mathbb{E} \sum_{(x,F)\in\mathcal{Z}(\mathcal{F}_k^{\max}(Y_t))} \mathbb{1}\{x \in [0,1]^d\} \tag{8.15}$$

the intensity of maximal k-polytopes of Y_t. Furthermore, following (2.26), we write $\mathbb{Q}_{k,t}^{\max}$ for the distribution of the typical maximal k-polytope of Y_t. It is formally given by

$$\mathbb{Q}_{k,t}^{\max}(B) := \frac{1}{\gamma_{k,t}^{\max}} \mathbb{E} \sum_{(x,F)\in\mathcal{Z}(\mathcal{F}_k^{\max}(Y_t))} \mathbb{1}\{x \in [0,1]^d\} \, \mathbb{1}\{F \in B\} \tag{8.16}$$

for $B \in \mathcal{B}(\mathcal{P}_{k,c})$. By Lemma 8.8, (8.15), and (8.16) we obtain, analogously to (2.53), for the density of the k-skeleton of Y_t that

$$\eta_{k,t} = \mathbb{E}(\mathcal{H}^k([0,1]^d \cap \operatorname{skel}_k Y_t)) = \gamma_{k,t}^{\max} \int_{\mathcal{P}_{k,c}} V_k(F) \, \mathbb{Q}_{k,t}^{\max}(\mathrm{d}F) . \tag{8.17}$$

In (4.10) and (4.11) the sets $\mathcal{F}_{k,W}^{\max *}$ of birth time marked maximal k-polytopes *restricted to* $W \in \mathcal{P}_d$, $k \in \{0, \dots, d-1\}$, were defined. For some purposes in later chapters, it is now shown how a limit for growing windows yields a result for "global" maximal polytopes of $\mathcal{F}_k^{\max *}(\Sigma)$. The concepts of maximal polytopes arising from the construction in a window (see (4.10) and (4.11)) are related to those appearing in the global construction; see (8.7) and (8.8). Here the limit is considered for maximal polytopes F of $Y_{t,aW}$ that are completely contained in the interior of aW, that is, $F \subset \operatorname{int}(aW)$, or, equivalently, $F \cap \partial(aW) = \emptyset$ as $a \to \infty$.

Lemma 8.10. *Let Θ be a hyperplane measure of the form* (4.1), *satisfying* (4.3), *and let* $(Y_t, t > 0)$ *be the STIT tessellation process driven by Θ. Furthermore, fix $k \in \{0, \dots, d-1\}$ and a center function $c : \mathcal{P}_k \to \mathbb{R}^d$. Then, for all $W \in \mathcal{P}_d$ and all nonnegative measurable functions $g : \mathcal{P}_{k,c} \to \mathbb{R}$, we have*

$$\gamma_{k,t}^{\max} \int_{\mathcal{P}_{k,c}} g(F)\, \mathbb{Q}_{k,t}^{\max}(\mathrm{d}F) = \lim_{a\to\infty} \frac{1}{V_d(aW)} \mathbb{E} \sum_{F \in \mathcal{F}_{k,aW}^{\max}(Y_{t,aW})} \mathbb{1}\{F \cap \partial(aW) = \emptyset\}\, g(F - c(F)).$$

Proof. By the spatial consistency of STIT tessellations (see (5.17)) we have that $Y_t \wedge W \overset{D}{=} Y_{t,W}$. For the system of k-faces of cells, this implies that

$$\sum_{F \in \mathcal{F}_k(Y_t \wedge W)} \delta_F \overset{D}{=} \sum_{F \in \mathcal{F}_k(Y_{t,W})} \delta_F$$

and thus by Lemma 4.6 and Lemma 8.8(c) that

$$\sum_{F \in \mathcal{F}_k^{\max}(Y_t)} \delta_{F \cap W} \overset{D}{=} \sum_{F \in \mathcal{F}_{k,W}^{\max}(Y_{t,W})} \delta_F .$$

By the stationarity of the process of maximal polytopes we have $\mathbb{1}\{F \subset aW\} = \mathbb{1}\{F \subset aW\}\, \mathbb{1}\{F \cap \partial(aW) = \emptyset\}$ a. s. Furthermore, note that the translation covariance of the center function c implies that $F + x - c(F + x) = F - c(F)$ for all $x \in \mathbb{R}^d$. Hence the map $F \mapsto g(F - c(F))$ is translation invariant. Thus by Theorem 4.1.3 of [172] it follows that

$$\begin{aligned}
\gamma_{k,t}^{\max} \int_{\mathcal{P}_{k,c}} g(F)\, \mathbb{Q}_{k,t}^{\max}(\mathrm{d}F)
&= \lim_{a\to\infty} \frac{1}{V_d(aW)} \mathbb{E} \sum_{F \in \mathcal{F}_k^{\max}(Y_t)} \mathbb{1}\{F \subset aW\}\, g(F - c(F)) \\
&= \lim_{a\to\infty} \frac{1}{V_d(aW)} \mathbb{E} \sum_{F \in \mathcal{F}_k^{\max}(Y_t)} \mathbb{1}\{F \subset aW\}\, \mathbb{1}\{F \cap \partial(aW) = \emptyset\}\, g(F - c(F)) \\
&= \lim_{a\to\infty} \frac{1}{V_d(aW)} \mathbb{E} \sum_{F \in \mathcal{F}_{k,aW}^{\max}(Y_{t,aW})} \mathbb{1}\{F \cap \partial(aW) = \emptyset\}\, g(F - c(F)) .
\end{aligned}$$

This completes the proof. □

8.3 Notes

1. The global construction of a STIT tessellation process, described in Section 8.1, was presented in [129] and [130].

2. In Section 8.2 the set $\mathcal{F}_{d-1}^{\max}(Y_t)$ of maximal $(d-1)$-polytopes of a STIT tessellations Y_t is introduced as the set of all $(d-1)$ polytopes that emerge when a cell is divided by a hyperplane. By Lemma 8.8 this is consistent with definition (10.22) of maximal $(d-1)$-polytopes for general tessellations, there formulated for restrictions to windows. Already in [44], maximal segments in general planar tessellations were considered, and thus that paper formed the starting point for the definitions in higher dimensions; see also Note 5 in Chapter 16.

9 Modifications of STIT tessellations, related models, applications

In this chapter, we discuss models and literature related to the STIT tessellation model. Rather than providing a comprehensive survey, we focus on two classes of models that share key properties with STIT tessellations. In Section 9.1, we introduce various types of cell division processes, and in Section 9.2, we turn to planar tessellations involving only T-nodes. In the final section, we highlight potential fields of application where these models can be adapted. Notably, the Mondrian STIT tessellation model is already used in machine learning.

Our goal is to draw attention to alternative approaches that may be of interest both in the theory of random tessellations and in applications. There are many such approaches: some adopt constructions inspired by stochastic geometry, whereas others arise from diverse applications, including materials science, soft matter, geology, thin films, communication systems, and machine learning.

9.1 Cell division processes

A cell division process is defined by the rules that determine how the polytopes in a given tessellation are successively split.

Discrete time

A concise systematization of cell division processes in a window with discrete time is presented in [39], where these processes are classified by a "selection rule" and a "division rule". The selection rule specifies how one cell is randomly chosen from the finitely many existing cells, whereas the division rule describes the distribution governing the random partition of the chosen cell.

In a series of papers, in particular, [133, 37, 38, 39] and references therein, extensive ideas, results, and open problems were presented concerning discrete-time processes generated by consecutive division of cells according to predefined protocols. In a related paper [41] the focus is on the number of sides of cells in a planar tessellation – described by a planar graph – and their limit as the number of consecutive divisions goes to infinity.

In addition, we highlight several examples from the extensive literature on modeling fracture patterns and material fragmentation. In [67], several models for planar fracture patterns in a window are suggested, which include also random tessellations generated by cell division. Higher-dimensional versions of a one-dimensional fragmentation process – which is modeled by a Poisson point process on a line segment – are studied in [63]. In particular, it contains planar cell division processes with an area-weighted selection rule. In [28], cell division processes in a unit square of $\mathbb{R}^2$ or a unit

https://doi.org/10.1515/9783112222409-010

cube of $\mathbb{R}^3$, respectively, are studied, where the dividing lines or planes are orthogonal to the axes. Three different selection rules are taken into account. Asymptotic results concerning the size of maximal segments or polygons, respectively, are shown for these models. A planar cell division model where both the selection rule and the division rule for a cell depend on neighbors of the respective cell is defined and simulated in [77].

In contrast to the models discussed above, in which each step randomly divides only a single cell, there are other approaches that differ significantly. For example, in [85] a planar tessellation process is generated by dividing each of the extant cells simultaneously into four new cells by a pair of a horizontal and a vertical line. Also, for the so-called KD-trees, a deterministic protocol for the division of the cells is applied; see Section 21.2 of [161].

Continuous time

For cell division processes in continuous time, the analogue of the selection rule is the L-rule. It specifies the distribution of a cell's random lifetime. As described in Section 1.1, the resulting tessellation-valued process is thus determined by the L- and D-rules, which govern the lifetimes of the cells and their division rules, respectively. This approach does not necessarily require a finite number of cells and is therefore not limited to windows.

Now we turn our attention to continuous-time cell division processes in more detail.

9.1.1 In a window

Any state of a cell division process in a window $W \in \mathcal{P}_d$ has finitely many cells, and thus the process is a pure jump process with state space $\mathcal{T}_W$. Involving the full history into the description of a state, this augmented process is Markovian. Thus its distribution can be described by its initial state and rate kernel, which is the product of the rate function and jump transition kernel, as defined on page 279 of [89]. The rate kernel for the STIT model is given in Section 4.3 and that for the PHT process in Section 2.3.2. In a STIT tessellation process, at any time at most one cell is divided, whereas in a PHT process within a window, more cells can be divided simultaneously. In what follows, we will concentrate on tessellation-valued processes for which at any time, at most one cell is divided. This excludes, for example, the PHT processes.

For a general approach, fix a kernel μ from $\{(t, (T_s, s \in [0,t]), P) : t \geq 0, T_s \in \mathcal{T}_W, P \in T_t\}$ to $\mathcal{H}$ such that the support of the measure $\mu(t, (T_s, s \in [0,t]), P, \cdot)$ is a subset of $[\operatorname{int} P]$, that is, the set of all hyperplanes hitting the interior of P. The total masses of these measures assign weights to the cells $P \in T_t$, and these weights, divided by the sum over all weights, determine the probabilities for the selection of a cell to be divided. Then, with a suitable normalization, $\mu(t, (T_s, s \in [0,t]), P, \cdot)$ also yields the distribution of a random hyperplane that divides a cell P if it is divided. To formalize this, define the "cumulative" kernel $\hat{\mu}$ from $\{(t, (T_s, s \in [0,t])) : t \geq 0, T_s \in \mathcal{T}_W\}$ to $\mathcal{T}_W \times \mathcal{H}$ by

$$\hat{\mu}(t, (T_s, s \in [0,t]), \cdot) := \sum_{P \in T_t} \delta_P \otimes \mu(t, (T_s, s \in [0,t]), P, \cdot). \tag{9.1}$$

Now, to describe a pure jump process, define its rate function as

$$\hat{\mu}(t, (T_s, s \in [0,t]), T_t \times [W])$$

and the jump transition kernel by

$$\frac{\hat{\mu}(t, (T_s, s \in [0,t]), \cdot)}{\hat{\mu}(t, (T_s, s \in [0,t]), T_t \times [W])}.$$

It determines the random choice of a cell – in [39], for discrete time, called the selection rule – and also the distribution of the hyperplane dividing the chosen cell. As shown in Section 4.3, the STIT tessellation process belongs to this general class of cell division processes.

Additionally, as considered in [58], cells can also carry marks, called "colors". Then the life time and division rules for cells may also depend on their color. After each division of a mother cell, the two daughter cells are marked with colors by a rule, which can be incorporated in the division kernel.

A rather flexible special case can be described using a nonnegative measurable weight function $p : \mathcal{P}_d \times \mathcal{H} \to \mathbb{R}$, a hyperplane measure Θ of the form (4.1), satisfying (4.3), and a kernel of the form

$$\mu(t, (T_s, s \in [0,t]), P, \mathrm{d}H) := p(P,H)\, \Theta(\mathrm{d}H) \tag{9.2}$$

for $t \geq 0$, $(T_s, s \in [0,t]) \in \mathcal{T}_W^{[0,t]}$, $P \in T_t$, $H \in \mathcal{H}$. In this model, neither the life time nor the division rule of a cell $P \in T_t$ depends on the past, time, or other cells of T_t. In [179] the so-called shape driven processes are introduced. The main idea is that the division rule depends on the shape of the divided cell, and its size is only a scaling factor. This can be expressed in the form (9.2), choosing the function p such that for all $\alpha > 0$,

$$p(\alpha P, \alpha H) = \alpha\, p(P,H), \quad P \in \mathcal{P}_d,\, H \in [P]. \tag{9.3}$$

For this type of models, see also Section 9.1.3.

A simulation study for a variety of cell division models in a window of $\mathbb{R}^2$ is presented in [103] and [102]. For the life time distribution of cells, alternatively to the STIT rule, also exponentially distributed life times with the cell area as parameter, called L-AREA, are chosen. Moreover, apart from STIT division rule, also several other division methods are considered. One of them for the choice of a random direction of the dividing line, uses the same algorithm as for STIT, but then the translation of the line is not uniformly distributed in the corresponding interval but follows a truncated Gaussian distribution with its mean in the center of this interval and the standard variation SIGMA as an additional parameter. This division rule is called D-GAUSS and yields divisions of the cells preferably through the central part of the cells. As an alternative to this rule, the translation of the dividing line is chosen such that for the two emerging daughter cells, say P_1 and P_2, a function of the two isoperimetric quotients $\pi\, V_2(P_i)/V_1(P_i)^2$, $i \in \{1,2\}$, is maximized. In a further version of a division rule, which can be combined with the other ones, dividing lines are rejected if one of the four section angles with the boundary of the cell is smaller than a fixed threshold. These division rules aim to model the phenomenon in real crack structures that cells are divided mostly through their central region and that the cracks tend to meet the boundary of the divided cell almost orthogonally. All these models dealt with in [103] and [102] can also be described using a kernel of the form (9.2).

Some simulations with L-AREA and D-GAUSS in comparison to L-STIT and D-STIT are shown in Figures 9.1 and 9.2. Some statistics for them are presented in the respective Tables 9.1 and 9.2.

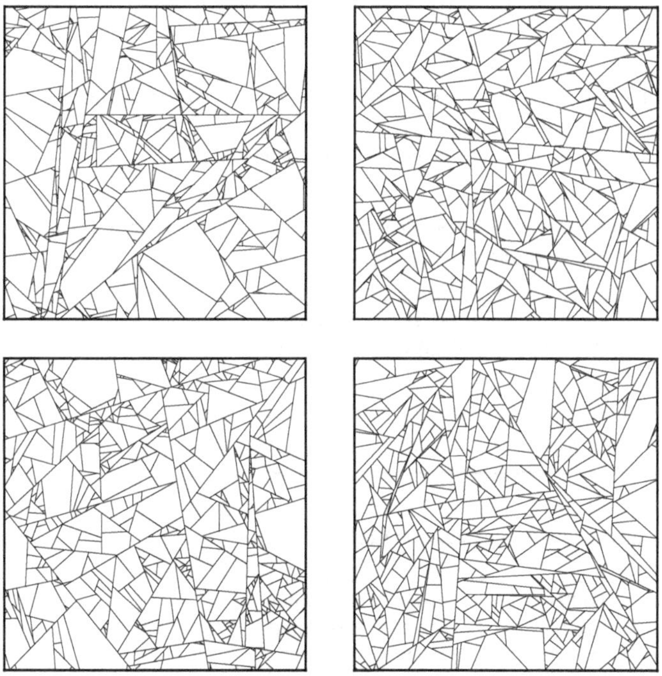

Figure 9.1: Comparison of different life time and division rules for isotropic directional distribution. The numbers of cells are approximately equal. Upper left: STIT tessellation. Upper right: L-AREA, D-STIT. Lower left: L-STIT, D-GAUSS with SIGMA = 0.10. Lower right: L-AREA, D-GAUSS with SIGMA = 0.10.

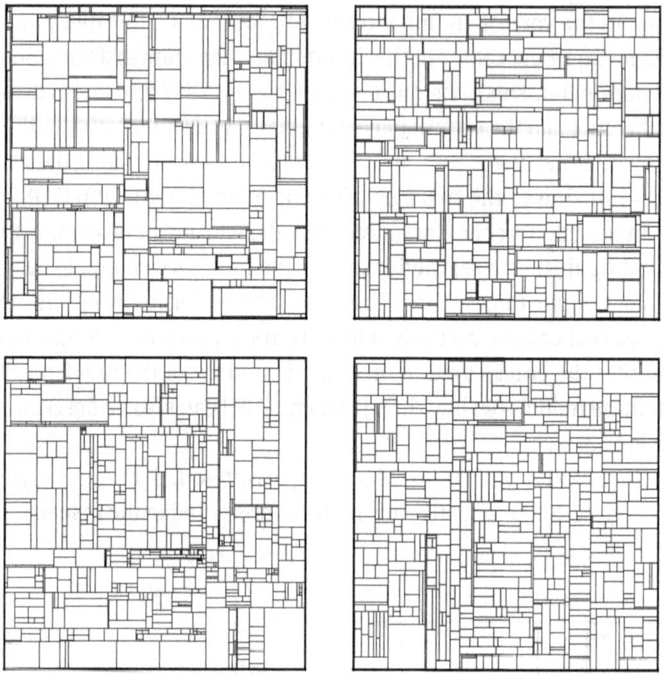

Figure 9.2: Comparison of different life time and division rules, for spherical directional distribution φ given in (4.4) with $p_1 = p_2 = 0.5$. Upper left: Mondrian STIT tessellation. Upper right: L-AREA, D-STIT. Lower left: L-STIT, D-GAUSS with SIGMA = 0.10. Lower right: L-AREA, D-GAUSS with SIGMA = 0.10.

Table 9.1: Statistics for the simulated tessellations shown in Figure 9.1. The exact value of the CV of area for an isotropic STIT tessellation in $\mathbb{R}^2$ is 1.98; see Section 9.5.1 of [33] or [141].

	STIT	L-AREA D-STIT	L-STIT D-GAUSS	L-AREA D-GAUSS
number of cells	538	529	541	559
CV of area	2.24	1.18	1.33	0.91
CV of perimeter	0.86	0.60	0.66	0.51

Table 9.2: Statistics for the simulated tessellations shown in Figure 9.2.

	STIT	L-AREA D-STIT	L-STIT D-GAUSS	L-AREA D-GAUSS
number of cells	513	511	591	522
CV of area	1.59	0.97	1.37	0.69
CV of perimeter	0.71	0.51	0.60	0.43

The abbreviation CV means the estimated coefficient of variation, which is the quotient of the estimated standard variation and the arithmetic mean of the cell areas or cell parameters, respectively. In [141] and [103], it was demonstrated that the CV of the cell areas is a simple but efficient statistical goodness-of-fit criterion for the adaption of a model to data of real crack structures.

Another particular cell division process is introduced and investigated in [27]. A square in $\mathbb{R}^2$ is split into rectangles, where both the life time distribution and the division rule depend on the shape of these rectangular cells.

Some particular cell division processes in windows can be related to special *fragmentation models* in the sense of [14]. There, for the cells only a scalar size parameter, for example, the volume, is taken into account, without regarding the geometry of the cells. Within a cuboidal window $W \in \mathcal{P}_d$ with its edges parallel to the axes of $\mathbb{R}^d$ and for a Mondrian hyperplane measure Θ consider the cell division process $(M_t,\ t \geq 0)$, where the life time of each cell P is exponentially distributed with parameter $V_d(P)$, and the division rule is the same as that for the STIT tessellation process driven by Θ. Define the map $B : \mathcal{T}_W \to (0, \infty)^{\mathbb{N}}$ by

$$B(T) := \left(V_d(P^1), \dots, V_d(P^m), 0, \dots\right), \quad V_d(P^1) \geq \dots \geq V_d(P^m)$$

for $T = \{P^1, \dots, P^m\}$.

Then, according to Definition 1.1 of [14], the process $(B(M_t),\ t \geq 0)$ is a self-similar fragmentation chain with index $\alpha = 1$ of self-similarity and the dislocation measure ν that is the law of $(\xi, 1-\xi, 0, 0, \dots)$, where ξ is a random variable, uniformly distributed on the interval $[\frac{1}{2}, 1]$. After each division, the volumes have to be ordered in a descending sequence. For more detail, see Section 5.3 of [120].

The term “fragmentation” is used quite broadly in the literature. In [62] and many references therein, a range of related concepts is surveyed. Notably, Figure 5 of that article presents various geometric models, including some cell division processes.

A further modification of the planar STIT model is introduced and applied in [139], where the directional distribution of dividing lines can be adjusted for each individual cell based on its size during the cell division process.

9.1.2 In $\mathbb{R}^d$, starting with an initial tessellation

A broad general model of “branching random tessellations” is introduced in [179] and [58]. These works establish a theoretical foundation for such processes, particularly showing how the strict independence assumptions in the STIT tessellation process can be relaxed.

Definition (9.1) of a cumulative kernel can be extended to tessellations $T_s, s \geq 0$, of the entire space $\mathbb{R}^d$. However, in that case, the resulting cell division process – if it exists – would no longer be a pure jump process. Moreover, unlike in the setting of a window, we face the challenge of how to initiate such a process. One way to address this is to start with an initial random tessellation X_0 and then apply the kernel to it. In Theorem 3.10 of [58], a sufficient condition on the kernel is provided for ensuring the existence of a corresponding cell division process. Additionally, that paper formulates several equivalent criteria for a tessellation-valued process to qualify as a cell division process driven by a cumulative kernel of the form (9.1). These criteria include an interpretation as a Gibbsian spatial system of interacting branching processes of cells. Consequently, [58] serves as a solid foundation for further research on general cell division processes.

9.1.3 In $\mathbb{R}^d$, global construction

Defining general cell division processes within a fixed window $W \in \mathcal{P}_d$ and simulating them do not pose particular difficulties. However, as shown in Section 5.3.2, except the STIT tessellation model, the distribution in any subregion of W depends on the choice of W. This is because all alternatives to the STIT construction lose the property of spatial consistency. Nonetheless, even if a cell division model is not spatially consistent, there can still exist a tessellation-valued process in the entire space $\mathbb{R}^d$ driven by specific L- and D-rules. We turn now our attention to this situation.

For certain cell division processes, a global construction is possible. In Section 8.1, this was demonstrated for the STIT tessellation process. By a *global construction* we refer to a description of a cell division process that is not confined to a window and does not rely on an initial tessellation. Thus at time $t = 0$ the process is not defined; however, for all $t > 0$, a random tessellation X_t is specified so that $(X_t, t > 0)$ forms a tessellation-valued process adhering to predefined lifetime and division rules.

The central idea behind the global constructions developed so far is to begin with an auxiliary zero-cell process $(Z_{\langle o \rangle,t}, t > 0)$. We may imagine that the tessellation-valued process, defined by the given rules, already exists. In particular, then there is a virtual sequence of zero-cells, which expands to full space $\mathbb{R}^d$ as time t approaches zero. This is formalized by an application of a "time-reversal" of the process of cell-dividing hyperplanes with earlier birth times. Once such a zero-cell process $(Z_{\langle o \rangle,t}, t > 0)$ has been established, the cells can be subdivided according to the lifetime and division rules within polytopes, following a procedure analogous to steps (ii), (iii), and (iv) in Section 8.1.

Apart from STIT tessellation processes, [179] provides a global construction for shape-driven processes with a kernel satisfying (9.2) and (9.3) and certain additional technical assumptions. One such requirement is the existence of a zero-cell distribution compatible with the kernel. Several properties and results for that model are established in [179].

Using a method related to the global construction of STIT tessellation processes – namely, employing the zero-cell process of a PHT process as the starting point – [120] describes another family of models. The key difference from the STIT model is that for the exponentially distributed lifetime of a cell, the rate parameter can be given by a fairly general functional of the cell. A sufficient condition on this functional is provided to guarantee the existence of such a cell division process. Moreover, it is shown that for the Mondrian hyperplane measure, the intrinsic volumes $V_j, j \in \{1, \dots, d\}$, of the cells can serve as lifetime parameters. Note that these models can also be described by a kernel of the form (9.2).

9.1.4 Non-Euclidean spaces

The STIT tessellation model has successfully been transferred from Euclidean to spherical spaces in [83], generalizing an earlier paper [50]. Expectations, variances, and covariances of spherical curvature measures induced by these so-called splitting tessellations are studied using tools from spherical integral geometry. Also, the spherical pair-correlation function of the $(d-1)$-dimensional Hausdorff measure is computed explicitly and compared to its analogue for Poisson great hypersphere tessellations. In addition, the typical cell distribution and the distribution of the typical spherical maximal face of any dimension $k \in \{1, \dots, d-1\}$ are expressed as mixtures of the related distributions of Poisson great hypersphere tessellations. A similar program has been initiated in [78] for splitting tessellations in hyperbolic space as well.

9.2 T-tessellations in $\mathbb{R}^2$

All vertices of planar STIT tessellations are T-shaped almost surely; see Section 16.2.1. The so-called T-tessellations are random tessellations in $\mathbb{R}^2$ that contain only T-shaped vertices, which means that each vertex has exactly three adjacent edges in the sense of Definition 16.1, and two of these edges are collinear. Moreover, it is required that the supporting lines of distinct maximal segments, the I-segments, are pairwise different. In [88] an upper bound is provided for the number of possible T-tessellations within a window $W \in \mathcal{P}_2$, based on the number of supporting lines that intersect W.

All T-tessellation models considered in this section are defined in $\mathbb{R}^2$. It remains an important open question whether and how these models can be generalized to higher dimensions.

9.2.1 Construction by split, merge, and flip

A general class of T-tessellations is defined and studied in [95]. The main idea for this model is that for its construction, three operations on tessellations are applied. The first one is cell *division*, called split, which is also used in the models considered in Section 9.1. The second operation is *merging* of pairs of cells that share a common 1-face. The third operation is the so-called *flip*. A flip at a T-vertex means that from the pair of adjacent collinear edges one is deleted, and instead the third adjacent edge is extended by a collinear one. These operations can only be applied if the result is again a T-tessellation. The flips can generate T-tessellations that do not satisfy the "mikado principle"; see Section 8.2 and Figure 16.3. In a window $W \in \mathcal{P}_2$ a T-tessellation is generated by finitely many admissible applications of divisions, merges, and flips. If in a construction step, there are several choices of admissible operations, then the choice can either be uniformly at random, or it can be chosen with respect to a weight function. This weight function may depend on the already generated tessellation, and it can prefer or penalize certain properties of the resulting T-tessellation. For example, we might restrict the variability of the cell areas or choose vertex angles so that they are preferentially close to $\pi/2$.

9.2.2 Sweeping line construction

The "point-based polygonal models for random graphs" introduced in [7, 8] and further developed in [6] include, in particular, a class of planar graphs forming the skeleton of a tessellation. Their construction relies on a sweeping line and produces isotropic planar T-tessellations. This class of tessellation models is also defined and investigated in [136] for arbitrary directional distributions.

The construction proceeds as follows. Consider a compact convex region $K \in \mathcal{K}$. A line with a fixed direction $u \in S^1$ sweeps over K, meaning that it is translated from one tangential position to the opposite one. Moreover, let $\widehat{X}$ be a stationary Poisson line process with intensity $\eta > 0$ and spherical directional distribution φ. Imagine the lines of $\widehat{X}$ drawn in light grey. At the initial position of the sweeping line, say $H(u, r_\ell)$, assume that $K \subset H(u, r_\ell)^+$. As the sweeping line moves, it colors black the portions of lines from $\widehat{X}$ that appear inside K. Essentially, once the sweeping line encounters an intersection point of two lines from $\widehat{X}$, one of these lines is *blocked*, whereas the other one *survives* and continues. The choice of which line survives is made by tossing a fair coin. Furthermore, at each position of the sweeping line and on each existing black line, new line segments may be initiated according to independent Poisson point processes. The intensities of these starting point processes and the directions of the new line segments depend on the direction of the supporting line and on φ.

This construction is tuned so that, as shown in [8, 136], spatial consistency in the sense of Definition 5.9 is satisfied. Moreover, the corresponding tessellation in $\mathbb{R}^2$ is stationary, and the distribution of the typical cell is the same as that for a Poisson line tessellation with suitably chosen intensity and directional distribution. It is straightforward to see (see Figure 16.3) that the mikado principle fails to hold with positive probability, implying that this construction cannot be used to generate a STIT tessellation.

9.2.3 Gilbert tessellation

The Gilbert model is named in honor of the author of the pioneering paper [59]. The construction starts with a point pattern in the plane in which each point is marked with a direction. This provides the origin for line segments that grow to both sides in the given direction with a finite constant speed. A segment is blocked when it meets another already existing segment, and a vertex emerges. If, for example, the point pattern is a stationary Poisson point process, then all vertices are of T-type. As a consequence of this construction, the Gilbert tessellation does not satisfy the "mikado principle" (see Section 8.2 and Figure 16.3), and hence it differs from the STIT tessellation.

In [106] the particular case was studied where the segments grow in the two directions parallel to the axes only. Furthermore, in that paper an approximating penetration model, also introduced by Gilbert, is investigated.

In [173] a law of large numbers, variance asymptotics, and a central limit theorem are shown for geometric functionals associated with a Gilbert tessellation based on a stationary Poisson point process. This is done by embedding the Gilbert model into the rich theory of geometric stabilization and stabilizing functionals.

In [25, 24] the authors study a modification they call "half-Gilbert tessellation". In that model, with horizontal and vertical directions, not all the growing segments are blocked when they meet an already existing line. Thus a random tessellation is generated that contains T- and X-vertices.

9.3 Fields of application and statistical aspects

Although STIT tessellation processes were initially motivated by purely theoretical considerations explained in Section 1.1, their geometric properties make them plausible candidates for modeling real crack patterns observed in nature, such as mud cracks or the craquelling of ceramic glazes. Moreover, certain road systems and communication networks sometimes exhibit structures that could be interpreted as tessellation skeletons. In any case, a thorough statistical verification and adaptation of the model are essential.

A promising application of STIT tessellations has recently been reported in the field of machine learning, specifically for subdividing datasets.

There is a substantial body of literature dealing with tessellation applications, particularly in various areas of materials science. However, here we focus on the publications that specifically involve STIT tessellations.

9.3.1 Materials science, crack structures, fragmentation

Several attempts have been made to apply STIT tessellations directly to observed crack patterns in materials.

A first study was presented in [141], featuring two-dimensional images of mud cracks in soil, ceramic craquelure, titanium-nitride coatings on cemented carbide, and muscle tissue. Preliminary statistical checks, using both the coefficient of variation CV of cell areas and the so-called linear contact distribution function, indicated that STIT tessellations cannot be applied directly for modeling. The primary deficiency lies in the high variability of cell sizes in STIT tessellations compared to the relatively "homogeneous" cell sizes observed in real crack structures. Hence modifications of the STIT model are required for real-world applications. Some of these possible alterations have been mentioned in Section 9.1.

In [139] the authors aim to adapt the STIT tessellation model to geological data, analyzing two-dimensional trace map data of fracture patterns. Aside from STIT tessellations, they also consider a model in which the directional distributions of the dividing lines depend on the size of the cells. Their statistical analysis indicates that the STIT model cannot be applied directly, and although the proposed modification improves the fit for one of the examined examples, it does not work as well for other fracture patterns.

A materials science sample – chromium/gold metallized polydimethylsiloxane – is investigated in [103] and [102]. Here the STIT model also does not sufficiently capture the observed crack patterns. Consequently, the lifetime distributions and division rules of the cell division process are modified; see Section 9.1. In [102] an optimization procedure is applied to 40 parameterized models, ultimately achieving a reasonably good fit.

It is important to note that in all of the above-mentioned papers, the improved results for modified STIT models are based on simulations rather than theoretical findings.

9.3.2 Street and communication systems

In [203] and [34], planar STIT tessellations are proposed as stochastic models for city street systems. These tessellations are used to simulate a particular aspect of wireless propagation in telecommunication systems, namely the path-loss exponent. A related simulation study on indoor wave propagation is presented in [148], where STIT and Poisson line tessellations are employed, primarily with two orthogonal directions.

A more advanced application of planar STIT tessellations for network modeling is provided in [150], where the underlying infrastructure is assumed to be represented by a planar STIT tessellation. To construct a multihierarchical network, the birth times of the *I-segments*, that is, the maximal 1-polytopes, arising from the STIT tessellation process are used. Those I-segments with birth times in the time interval $(0, t_1)$ form the highest hierarchical level. If the birth time lies in (t_1, t_2), then the I-segment belongs to an intermediate level, whereas those with larger birth times generate the lowest level of the network.

In [1] agricultural landscapes composed of fields and roads are modeled by T-tessellations. Model parameters are estimated to fit several real datasets, and goodness-of-fit is then evaluated. In [2] the statistical methodology behind this approach is further refined.

In Section 7 of [168], it is noted that the Mondrian STIT tessellation process has been extensively studied in combinatorics and computational geometry, where it is called the "guillotine partition". This model has a wide range of applications, including circuit design, approximation algorithms, and computer graphics. However, there is no explicit link to STIT tessellations in that paper.

9.3.3 Partition of data sets – machine learning

As discussed in the previous sections, employing the STIT tessellation model in areas such as materials science, geology, or communication networks typically requires modifications and adaptations. In contrast, this model can be applied directly and effectively in machine learning, especially for classification and regression tasks. Here the challenge is not purely geometric; the tessellation is instead used to partition datasets, which are often high-dimensional. In [57, 56, 153, 154, 155] a survey of the relevant literature shows this to be a highly active area of research.

To clarify the ideas further, we focus now on a specific scenario under investigation. Let $W \in \mathcal{P}_d$, and let $(U, V) \in W \times \mathbb{R}$ be a random variable. Consider a data set $\mathsf{D}_n := \{(U_1, V_1), \dots, (U_n, V_n)\}$, $n \in \mathbb{N}$, consisting of i. i. d. copies of (U, V) and a nonparametric regression model with deterministic function f and random variable ε, that is,

$$V = f(U) + \varepsilon \quad \text{with } \mathbb{E}\,\varepsilon = 0\,,\ \operatorname{var} \varepsilon < \infty\,.$$

An estimator for the unknown function $f : \mathbb{R}^d \to \mathbb{R}$ has to be constructed. The idea is to partition W and thus the data set D_n by a random tessellation $X \in \mathcal{T}_W$. For $x \in W$, the cell of X that contains x is denoted by $Z_{\langle x \rangle}$. Then an estimator is defined by

$$\widehat{f}(x, X) := \frac{1}{\sum_{i=1}^{n} \mathbb{1}\{U_i \in Z_{\langle x \rangle}\}} \sum_{i=1}^{n} \mathbb{1}\{U_i \in Z_{\langle x \rangle}\}\, V_i\,, \quad x \in W\,. \tag{9.4}$$

Furthermore, the estimator can be averaged over m partitions of D_n, that is, $\widehat{f}_m(x, X) := \frac{1}{m} \sum_{j=1}^{m} \widehat{f}(x, X_j)$ for m i. i. d. copies $X_1, \dots, X_m$ of X.

When the tessellation X is generated by a process of sequential cell division, the partition of D_n can be represented by a random graph, the so-called decision tree. The root of the tree is labeled by D_n. The leaves of the tree, corresponding to the cells $P \in X$, are labeled by $\{(U_i, V_i) : U_i \in P\}$. The function $\widehat{f}_m(\cdot, X)$ is then referred to as a random forest estimator. In this setting, the problem arises to find appropriate random tessellations X that minimize the mean quadratic error of $\widehat{f}_m(\cdot, X)$ with respect to f.

In [168] the authors employ a cell division model called the Mondrian process. It was developed independently, yet it turns out to coincide exactly with the STIT tessellation process driven by the Mondrian hyperplane measure (4.4) when $p_1 = \cdots = p_d$. Their motivation was to find partitions of parallelepipeds in any dimension, reminiscent of the way a Poisson point process partitions a line segment. In that paper, the property of spatial consistency – recall Definition 5.9 – termed "self consistency", is demonstrated, and its significance is highlighted. However, no connection was made between this particular Mondrian tessellation and the general STIT tessellation model.

A generalized version of the cell division model from [168] was examined in [54], where alternative directional distributions are also considered. A direct link between machine learning and STIT tessellations, including an explicit reference, was established in [57], where several STIT tessellation processes, called *random tessellation processes*, are explored in simulation studies as generalizations of the Mondrian STIT tessellation process.

A significant portion of this research systematically utilizes tools from stochastic geometry, in particular theoretical results pertaining to STIT tessellations. This line of work began with [154, 155, 153], where the authors prove a series of theorems regarding error bounds for kernel density and random forest regression estimators based on STIT tessellation processes. For instance, it is shown that estimator (9.4) attains minimax optimality for certain classes of functions f. Notably, this property leverages the fact that STIT tessellations have a Poisson typical cell. In addition to providing quality guarantees for these estimators, computational efficiency is crucial. A remarkable result by O'Reilly and Tran is presented in Section 6.4 of this book. In Theorem 6.16, it is shown how the partition of space induced by a STIT tessellation with discrete directional distribution can be transformed into a construction of the Mondrian STIT tessellation in a higher-dimensional space, which is then intersected with an appropriate linear subspace. Clearly, constructing a Mondrian STIT tessellation is less complex than building a general STIT tessellation. In this spirit, in [158], it is proposed to use randomly rotated Mondrian STIT tessellations to approximate a computationally much more complex isotropic STIT tessellation process.

Furthermore, in [11] the problem of dimension reduction is addressed, incorporating data adaptivity into the Mondrian STIT tessellation process through an iterative two-step procedure.

In summary, the following conventions and findings make STIT tessellations the preferred – indeed, in some cases the only – model of choice in various machine learning contexts:

- Typically, decision trees are binary trees. Hence, in consecutive data splitting, cell division processes as described in Sections 1.1 and 9.1.1 are well suited because at most one cell is divided into two new cells at each step.
- In the literature cited above, spatial consistency, also called "self-consistency", is highlighted as an advantageous property of the random tessellations used to partition the data. It is one of the properties that underlies computationally efficient updates to the estimator when new training data is received over time – called online learning – and ensures that the online estimator has the same distribution as

if you had all the data to begin with. Within the class of cell division processes discussed in Section 5.3.2, only STIT tessellation processes are spatially consistent.
- Apart from PHTs and STIT tessellations, there exist other random tessellations that also exhibit a Poisson typical cell; see Section 9.2. However, among the known models, STIT tessellations are unique in that they are generated by a cell division process.

Part II: **Geometry of STIT tessellations via martingale methods**

10 Associated martingales

In this second part of the monograph, we delve into moment formulae and asymptotic distributional properties of geometric functionals related to cells and maximal polytopes. As evidenced by the constructive description of STIT tessellations in Chapters 4 and 8, the fundamental constituents of these tessellations are the maximal polytopes. Consequently, they form a natural category of objects whose geometric and probabilistic properties should be investigated. The present chapter lays down the essential theoretical groundwork for this purpose. Our starting point is the Markovian description of STIT tessellation processes within a confined window $W \in \mathcal{P}_d$. Progressing further, we construct several classes of martingales arising as solutions to distinct associated martingale problems via Dynkin theory. These martingales turn out to be powerful and versatile tools to derive in the forthcoming chapters exact and asymptotic first- and second-order moment formulae for geometric functionals of maximal polytopes. They also constitute pivotal components in the proofs of the distributional limit theorems for these functionals if we replace W by $n\,W$ and consider the asymptotics as $n \to \infty$.

10.1 Extended generator

To prepare the construction of martingales associated with a STIT tessellation process in a window $W \in \mathcal{P}_d$, the generator $\widehat{\mathbb{L}}$ given in (4.26) needs to be extended. This extension is required to treat functions that are not bounded and therefore do not necessarily lie in the domain of $\widehat{\mathbb{L}}$. However, since the functions we consider can be represented as sums or sums of products of bounded functions, it can be shown that they belong to the domain of the so-called extended generator $\mathbb{L}$. Formally, this is done by localization arguments. The domain of the functions just mentioned is the set $\mathcal{T}_W$ of all tessellations of W, and by Lemma 18.2 it is a Borel space. In fact, this space is a measurable subset of the Polish space $\overline{\mathcal{F}}(\mathcal{F})$, but it is not a countable intersection of open sets, and consequently it is not clear whether it is itself a Polish space. Fortunately, for the purposes of martingale problems, the assumption of a Borel state space is sufficient. Thus, when discussing general Markov processes within the context of this book, we mainly rely on the monograph [47], where it is assumed that the state space is a Borel space and not necessarily a Polish space, as it is often required in the literature. In this section, we recall the relevant background material for general Markov processes, which in the forthcoming sections is specialized to geometric functionals related to a STIT tessellation process.

To start off, let $(\Omega, \mathbf{A}, \mathbb{P}, \mathfrak{F})$ be a filtered probability space, meaning that $\mathfrak{F} = (\mathfrak{F}_t, t \geq 0)$ is an increasing family of sub-σ-fields of $\mathbf{A}$: $\mathfrak{F}_s \subseteq \mathfrak{F}_t$ for $s \leq t$. In what follows, we will implicitly assume that the underlying probability space is rich enough to carry all the random objects we consider.

Definition 10.1. *A real-valued stochastic process $\mathbf{M} = (M_t, t \geq 0)$ is called an $\mathfrak{F}$-**martingale** if the following conditions are satisfied:*

https://doi.org/10.1515/9783112222409-012

(a) $\mathbf{M}$ *is adapted to* $\mathfrak{F}$, *that is,* M_t *is* $\mathfrak{F}_t$*-measurable for all* $t \geq 0$,
(b) $\mathbb{E}|M_t| < \infty$ *for all* $t \geq 0$,
(c) $\mathbb{E}[M_t \mid \mathfrak{F}_s] = M_s$ *a. s. for all* $0 \leq s \leq t$.

It is called a ***centered*** $\mathfrak{F}$***-martingale*** *if, in addition,*
(d) $\mathbb{E}[M_t] = 0$ *for all* $t \geq 0$.

In most cases, the martingales under consideration are centered. Accordingly, we will often omit the qualifier "centered" except when this property is explicitly invoked. Instances of noncentered martingales will be indicated explicitly.

The next lemma shows that with each Markov process, we can associate martingales arising as solutions of appropriate martingale problems. In particular, the martingale property of the random process considered in part (a) of the next result is often referred to as the **Dynkin martingale formula**. To present it and the forthcoming results, we introduce some further notation. By $C^1([0,\infty))$ we denote the set of all real-valued continuously differentiable functions on $[0,\infty)$. For a Borel space E, we denote by $B(E)$ the set of all bounded and measurable functions from E to $\mathbb{R}$. Finally, the tensor product of two functions $f, g \in B(E)$ is given by $f \otimes g : E \times E \to \mathbb{R}$, $(x,y) \mapsto f(x)g(y)$.

Lemma 10.2. *Let* E *be a Borel space, and let* $\mathbf{U} = (U_t, t \geq 0)$ *be a time-homogeneous Markov process taking values in* E *and having generator* $\mathcal{L}$ *with domain* $D(\mathcal{L}) \subseteq B(E)$. *Let* $\mathfrak{F} = (\mathfrak{F}_t, t \geq 0)$ *with* $\mathfrak{F}_t := \sigma(\{U_s, 0 \leq s \leq t\})$ *be the filtration induced by* $\mathbf{U}$.
(a) *Let* $g \in D(\mathcal{L})$. *Then the random process*

$$g(U_t) - g(U_0) - \int_0^t \mathcal{L}g(U_s)\,\mathrm{d}s\,, \quad t \geq 0\,,$$

is a centered martingale with respect to the filtration $\mathfrak{F}$.
(b) *Let* $g \in D(\mathcal{L})$ *and* $b \in C^1([0,\infty))$. *Then the random process* $(\overline{M}_t(g \otimes b), t \geq 0)$ *with*

$$\begin{aligned}\overline{M}_t(g \otimes b) &:= g(U_t)\,b(t) - g(U_0)\,b(0)\\ &\quad - \int_0^t \left(b(s)\,\mathcal{L}g(U_s) + g(U_s)\,\frac{\mathrm{d}}{\mathrm{d}s} b(s) \right) \mathrm{d}s\,,\end{aligned}$$

is a centered martingale with respect to the filtration $\mathfrak{F}$.

Proof. Statement (a) follows from (b) by taking $b \equiv 1$. Now we outline the proof of (b), referring to Section 14.2, page 28 of [47], and Proposition 14.13 of [47] for full details.

Let $(P_t, t \geq 0)$ be the transition semigroup of the Markov process $\mathbf{U}$ with generator $\mathcal{L}$:

$$P_t g(x) = \mathbb{E}\big[g(U_t) \mid U_0 = x\big]\,, \quad x \in E, \quad g \in B(E)\,.$$

Then the assumed time-homogeneity of $\mathbf{U}$ can be expressed by saying that

$$\mathbb{E}[g(U_{s+t}) \mid U_s = x] = \mathbb{E}[g(U_t) \mid U_0 = x] = P_t g(x) \tag{10.1}$$

for all $s, t \geq 0$, $x \in E$, and $g \in B(E)$. Recall that by $\mathfrak{F}_t$ we denote the σ-field generated by the process $(U_s, 0 \leq s \leq t)$ up to time t. Then for $0 \leq u \leq t$, we have that

$$\begin{aligned}
&\mathbb{E}\left[g(U_t)b(t) - g(U_0)b(0) - \int_0^t \left(b(s)\mathcal{L}g(U_s) + g(U_s)\frac{\mathrm{d}}{\mathrm{d}s}b(s) \right) \mathrm{d}s \mid \mathfrak{F}_u \right] \\
&\quad = g(U_u)b(u) - g(U_0)b(0) - \int_0^u \left(b(s)\mathcal{L}g(U_s) + g(U_s)\frac{\mathrm{d}}{\mathrm{d}s}b(s) \right) \mathrm{d}s \\
&\qquad + \mathbb{E}\left[g(U_t)b(t) - g(U_u)b(u) - \int_u^t \left(b(s)\mathcal{L}g(U_s) + g(U_s)\frac{\mathrm{d}}{\mathrm{d}s}b(s) \right) \mathrm{d}s \mid \mathfrak{F}_u \right]
\end{aligned}$$

a. s. by the standard properties of conditional expectations. It remains to show that the conditional expectation in the last line is a. s. equal to zero, that is,

$$\mathbb{E}\left[g(U_t)b(t) - g(U_u)b(u) - \int_u^t \left(b(s)\mathcal{L}g(U_s) + g(U_s)\frac{\mathrm{d}}{\mathrm{d}s}b(s) \right) \mathrm{d}s \mid \mathfrak{F}_u \right] = 0 \tag{10.2}$$

a. s. To see this, we rewrite this expression in terms of the transition semigroup $(P_t, t \geq 0)$:

$$\begin{aligned}
&\mathbb{E}\left[g(U_t)b(t) - g(U_u)b(u) - \int_u^t \left(b(s)\mathcal{L}g(U_s) + g(U_s)\frac{\mathrm{d}}{\mathrm{d}s}b(s) \right) \mathrm{d}s \mid \mathfrak{F}_u \right] \\
&\quad = b(t)\mathbb{E}[g(U_t) \mid \mathfrak{F}_u] - g(U_u)b(u) \\
&\qquad - \int_u^t \left(b(s)\mathbb{E}[\mathcal{L}g(U_s) \mid \mathfrak{F}_u] + \frac{\mathrm{d}}{\mathrm{d}s}b(s)\, \mathbb{E}[g(U_s) \mid \mathfrak{F}_u] \right) \mathrm{d}s \\
&\quad = b(t)P_{t-u}g(U_u) - g(U_u)b(u) - \int_u^t \left(b(s)P_{s-u}\mathcal{L}g(U_u) + P_{s-u}g(U_u)\frac{\mathrm{d}}{\mathrm{d}s}b(s) \right) \mathrm{d}s\,,
\end{aligned}$$

where we also used that $\mathbf{U}$ is a time-homogeneous Markov process; recall (10.1). We can now apply the Kolmogorov forward equation in Proposition 14.10 of [47], which in our situation implies that $P_{s-u}\mathcal{L}g(U_u) = \frac{\mathrm{d}}{\mathrm{d}s}P_{s-u}g(U_u)$. As a consequence, the last expression equals

$$b(t)P_{t-u}g(U_u) - g(U_u)b(u) - \int_u^t \left(b(s)\frac{\mathrm{d}}{\mathrm{d}s}P_{s-u}g(U_u) + P_{s-u}g(U_u)\,\frac{\mathrm{d}}{\mathrm{d}s}b(s) \right) \mathrm{d}s\,.$$

Integration by parts eventually leads to

$$\mathbb{E}\left[g(U_t)b(t) - g(U_u)b(u) - \int_u^t \left(b(s)\mathcal{L}g(U_s) + g(U_s)\frac{\mathrm{d}}{\mathrm{d}s}b(s)\right)\mathrm{d}s \mid \mathfrak{F}_u\right]$$

$$= b(t)P_{t-u}g(U_u) - g(U_u)b(u) - b(t)P_{t-u}g(U_u) + b(u)P_0\,g(U_u)$$

$$+ \int_u^t \left(P_{s-u}g(U_u)\,\frac{\mathrm{d}}{\mathrm{d}s}b(s) - P_{s-u}g(U_u)\,\frac{\mathrm{d}}{\mathrm{d}s}b(s)\right)\mathrm{d}s = 0$$

a. s., where we additionally used that $P_0\,g(x) = g(x)$ for all $x \in E$. This shows (10.2) and thus the martingale property of the random process $(\overline{M}_t(g \otimes b),\ t \geq 0)$. □

Recall that a family $(U_t,\ t \in J)$ of real-valued random variables, indexed by some set $J \neq \emptyset$, is **uniformly integrable** if

$$\lim_{c\to\infty} \sup_{t\in J} \mathbb{E}\big[|U_t|\,\mathbb{1}\{|U_t| > c\}\big] = 0\,.$$

A filtration $\mathfrak{F} = (\mathfrak{F}_t,\ t \geq 0)$ is a **right-continuous filtration** if

$$\mathfrak{F}_t = \bigcap_{s>t} \mathfrak{F}_s \quad \text{for all } t \geq 0\,.$$

Moreover, $\mathfrak{F}$ is **complete** if each $\mathfrak{F}_t$, $t \geq 0$, contains all events contained in some probability null event, that is, $\mathfrak{F}_t$ contains all $F \subset \Omega$ such that $F \subseteq F'$ for some $F' \in \mathbf{A}$ with $\mathbb{P}(F') = 0$. We call $\mathfrak{F}$ **augmented** if it is both right-continuous and complete. For convenience, to avoid technical difficulties, all filtrations we work with in this book are assumed to be augmented; see also Lemma 10.6.

We further recall that by an **($\mathfrak{F}$-)stopping time** we understand a random variable τ with values in $[0,\infty]$ satisfying $\{\tau \leq t\} \in \mathfrak{F}_t$ for all $t \geq 0$.

Definition 10.3. *A stochastic process $(M_t,\ t \geq 0)$ adapted to an augmented filtration $\mathfrak{F} = (\mathfrak{F}_t,\ t \geq 0)$ is called a* **local martingale** *if there exists an increasing sequence of $\mathfrak{F}$-stopping times $(\tau_n,\ n \in \mathbb{N})$ such that $\tau_n \uparrow \infty$ a. s., and for each $n \in \mathbb{N}$, the process $(M_t^{\tau_n},\ t \geq 0)$ given by $M_t^{\tau_n} := M_{\min\{t,\tau_n\}}$ is an $\mathfrak{F}$-martingale.*

Sequences $(\tau_n,\ n \in \mathbb{N})$ of stopping times satisfying the conditions of Definition 10.3 are referred to as **localizing sequences**.

Lemma 10.4. *Let $(M_t,\ t \geq 0)$ be a local martingale with respect to an augmented filtration $\mathfrak{F}$. If for all $a > 0$ and all sequences $(\varkappa_n,\ n \in \mathbb{N})$ of $\mathfrak{F}$-stopping times satisfying $\varkappa_n \leq a$ a. s., the family*

$$\{M_{\varkappa_n},\ n \in \mathbb{N}\}$$

is uniformly integrable, then $(M_t,\ t \geq 0)$ is an $\mathfrak{F}$-martingale.

Proof. Let $(M_t, t \geq 0)$ be a local martingale with respect to an augmented filtration $\mathfrak{F}$ and localizing sequence $(\tau_n, n \in \mathbb{N})$. Fix $0 \leq s < t$. Then

$$\mathbb{E}[M_{\min\{t,\tau_n\}} \mid \mathfrak{F}_s] = M_{\min\{s,\tau_n\}} \quad \text{a. s. for all } n \in \mathbb{N}\,, \tag{10.3}$$

since $(M_t, t \geq 0)$ is a local martingale. Putting $\varkappa_n := \min\{t, \tau_n\}$ and noting that $\varkappa_n \leq t$, the sequence of random variables $M_{\varkappa_n}$, $n \in \mathbb{N}$, is uniformly integrable by assumption. So, since $M_{\varkappa_n}$ converges to M_t a. s., $M_{\varkappa_n}$ also converges to M_t in the L^1-sense as $n \to \infty$. Since the conditional expectation is an L^1-contraction, we are allowed to take the limit as $n \to \infty$ on both sides of (10.3) to conclude that

$$\mathbb{E}[M_t \mid \mathfrak{F}_s] = M_s \quad \text{a. s.}$$

This proves the martingale property of $(M_t, t \geq 0)$. □

We also recall the concept of an **extended generator** from Definition 14.15 of [47].

Definition 10.5. *Let E be a Borel space, and let $\mathbf{U} = (U_t, t \geq 0)$ be a Markov process taking values in E. Let $\mathcal{D}(\mathbb{L})$ denote the set of all measurable functions $f : E \to \mathbb{R}$ for which there exists a measurable function $h : E \to \mathbb{R}$ such that for almost all $\omega \in \Omega$, the function $t \mapsto h(U_t(\omega))$ is measurable, $\int_0^t |h(U_s(\omega))|\,\mathrm{d}s < \infty$ a. s., and the process $(M_t, t \geq 0)$ defined as*

$$M_t := f(U_t) - f(U_0) - \int_0^t h(U_s)\,\mathrm{d}s\,, \quad t \geq 0\,,$$

*is a local martingale with respect to the augmented filtration induced by $\mathbf{U}$. In this case, we write $\mathbb{L}f := h$ and call $\mathbb{L}$ the **extended generator** of the process $\mathbf{U}$ with domain $\mathcal{D}(\mathbb{L})$.*

The domain of the extended generator $\mathbb{L}$ is in general larger than that of the classical generator, since it allows for unbounded functions and identifies them via martingale characterizations. In the applications below, our procedure will typically be as follows: starting from the generator $\widehat{\mathbb{L}}$ of the STIT tessellation process given in (4.26), we formally insert certain functionals of interest and use the Dynkin formula to construct a candidate process $(M_t(\,\cdot\,), t \geq 0)$. The essential step is then to verify that this process is indeed a martingale. Once this is established, the corresponding functional is thereby shown to belong to the domain of the extended generator $\mathbb{L}$. In other words, the martingale property provides a rigorous justification for the formal computations with $\widehat{\mathbb{L}}$.

Finally, our goal in this section is to show that the martingale property and Markov property of a stochastic process are preserved by augmentation of the induced filtration. For this, we first show that the conditional expectations of random variables measurable with respect to the original filtration do not change under augmentation. Let $(\Omega, \mathbf{A}, \mathbb{P})$ be a probability space, and let $\mathfrak{G} = (\mathfrak{G}_t, t \geq 0)$ be a filtration. Then the **augmented induced filtration** is defined by $\mathfrak{F} = (\mathfrak{F}_t, t \geq 0)$ with

$$\mathfrak{F}_t := \overline{\mathfrak{G}_t^+}, \quad t \geq 0,$$

where $\mathfrak{G}_t^+ := \bigcap_{s>t} \mathfrak{G}_s$, and $\overline{\mathfrak{G}_t^+}$ denotes the completion of $\mathfrak{G}_t^+$ by the sets $F \subset \Omega$ for which there is some $F' \in \Lambda$ with $F \subseteq F'$ and $\mathbb{P}(F') = 0$.

Lemma 10.6. *Let* $\mathbf{U} = (U_t,\ t \geq 0)$ *be a stochastic process with right-continuous sample paths taking values in a measurable space E, let* $\mathfrak{G} = (\mathfrak{G}_t,\ t \geq 0)$ *be its induced filtration, and denote by* $\mathfrak{F} = (\mathfrak{F}_t,\ t \geq 0)$ *the corresponding augmented induced filtration. Then, for every* $t \geq 0$ *and every integrable* $\mathfrak{G}_t$*-measurable random variable* $g : \Omega \to \mathbb{R}$*, we have*

$$\mathbb{E}[g \mid \mathfrak{F}_t] = \mathbb{E}[g \mid \mathfrak{G}_t] \quad a.s.$$

In particular, if $\mathbf{U}$ *is Markov with respect to* $\mathfrak{G}$*, then it is also Markov with respect to* $\mathfrak{F}$*, and if* $\mathbf{U}$ *is a martingale with respect to* $\mathfrak{G}$*, then it remains a martingale with respect to* $\mathfrak{F}$*.*

Proof. We embed our situation into the path-space framework of [191, Section 3.2]. Fix a finite time horizon $T > 0$ and consider the canonical process $X(t, \omega) := U_t(\omega)$ on the Skorokhod path space over $[0, T]$ with canonical filtration $(\mathcal{F}_t^T, 0 \leq t \leq T)$ as in [191]. In that notation, the filtration $(\mathfrak{G}_t, 0 \leq t \leq T)$ coincides with $(\mathcal{F}_t^T, 0 \leq t \leq T)$, and our right-continuous completed filtration $(\mathfrak{F}_t, 0 \leq t \leq T)$ coincides with the filtration $((\mathcal{F}_t^T)_{t+}^{\tau,\mu}, 0 \leq t \leq T)$ used in Proposition 3.5 of [191] for the special choice $\tau = 0$ and initial distribution μ given by the law of U_0.

Now let $t \in [0, T]$, and let g be a bounded and $\mathfrak{G}_t$-measurable random variable. In the notation of [191], this means that g is $\mathcal{F}_t^T$-measurable. Proposition 3.5, and in particular equality (3.96) on p. 197, implies that for all such g,

$$\mathbb{E}[g \mid (\mathcal{F}_t^T)_{t+}^{\tau,\mu}] = \mathbb{E}[g \mid \mathcal{F}_t^T] \quad \text{a. s. under } \mathbb{P}_{\tau,\mu}.$$

Translating back to our notation, $(\mathcal{F}_t^T)_{t+}^{\tau,\mu}$ is exactly $\mathfrak{F}_t$, whereas $\mathcal{F}_t^T$ is $\mathfrak{G}_t$, and $\mathbb{P}_{\tau,\mu}$ is just the law of $(U_s)_{0 \leq s \leq T}$ on the path space. Hence

$$\mathbb{E}[g \mid \mathfrak{F}_t] = \mathbb{E}[g \mid \mathfrak{G}_t] \quad \text{a. s.}$$

for every bounded $\mathfrak{G}_t$-measurable g. The extension to integrable g follows by a truncation argument.

Finally, if $\mathbf{U}$ is a martingale or a Markov process with respect to $\mathfrak{G}$, the defining conditional expectations involve only $\mathfrak{G}_s$-measurable random variables. Since the conditional expectations of such variables are unchanged when passing from $\mathfrak{G}_s$ to $\mathfrak{F}_s$, the martingale or Markov property holds with respect to $\mathfrak{F}$ as well. □

10.2 Martingales for cell functionals

In this section and the next one, our aim is to use the general results from the previous section to construct martingales naturally associated with a STIT tessellation process in a window. Specifically, in this section, we construct martingales that are linked to cell properties. In the following section, our focus shifts to martingales associated with functionals of maximal polytopes. In parallel, we address similar functionals for the process of restricted Poisson hyperplane tessellations and establish associated martingales for this process as well. The constructions outlined here, along with those detailed in Section 10.3, will enable us in Chapter 11 to unveil a range of remarkable connections between the STIT model and Poisson hyperplane tessellations.

Let $W \in \mathcal{P}_d$, and let Θ be a hyperplane measure of the form (4.1), satisfying (4.3). Furthermore, $(Y_{t,W}, t \geq 0)$ denotes the STIT tessellation process in W, driven by the measure Θ, and $\mathfrak{I} := (\mathfrak{I}_t, t \geq 0)$ is the augmented induced filtration. For $k \in \{0, 1, \dots, d\}$, we denote by $\mathcal{P}_{k,W} := \{P \in \mathcal{P}_k : P \subseteq W\}$ the set of all k-dimensional polytopes contained in W.

Theorem 10.7. *Let $W \in \mathcal{P}_d$, and let Θ be a hyperplane measure of the form (4.1), satisfying (4.3). Furthermore, let $\Psi : \mathcal{P}_{d,W} \to \mathbb{R}$ be bounded and measurable, and define $\Sigma_\Psi : \mathcal{T}_W \to \mathbb{R}$ by*

$$\Sigma_\Psi(T) := \sum_{P \in T} \Psi(P), \quad T \in \mathcal{T}_W. \tag{10.4}$$

(a) *Let $(Y_{t,W}, t \geq 0)$ be the STIT tessellation process in W, driven by the measure Θ. Then Σ_Ψ belongs to the domain $\mathcal{D}(\mathbb{L})$ of the extended generator $\mathbb{L}$,*

$$\mathbb{L}\Sigma_\Psi(T) = \int_{[W]} \sum_{P \in T} \mathbb{1}\{H \in [P]\} \left(\Psi(P \cap H^+) + \Psi(P \cap H^-) - \Psi(P)\right) \Theta(\mathrm{d}H)$$

for all $T \in \mathcal{T}_W$, and

$$M_t(\Psi) := \Sigma_\Psi(Y_{t,W}) - \Sigma_\Psi(Y_{0,W}) - \int_0^t \mathbb{L}\Sigma_\Psi(Y_{s,W})\,\mathrm{d}s, \quad t \geq 0, \tag{10.5}$$

is a centered $\mathfrak{I}$-martingale.

(b) *Let $(X_t \wedge W, t > 0)$ be the process of restricted Poisson hyperplane tessellations driven by Θ, and put $X_0 \wedge W := \{W\}$. Then Σ_Ψ belongs to the domain $\mathcal{D}(\mathbb{L}^{\mathrm{PHT}})$ of the extended generator $\mathbb{L}^{\mathrm{PHT}}$,*

$$\mathbb{L}^{\mathrm{PHT}}\Sigma_\Psi(T) = \mathbb{L}\Sigma_\Psi(T) \tag{10.6}$$

for all $T \in \mathcal{T}_W$, and

$$M_t^{\mathrm{PHT}}(\Psi) := \Sigma_\Psi(X_t \wedge W) - \Sigma_\Psi(X_0 \wedge W) - \int_0^t \mathbb{L}^{\mathrm{PHT}}\Sigma_\Psi(X_s \wedge W)\,\mathrm{d}s, \quad t \geq 0, \tag{10.7}$$

is a centered martingale with respect to the augmented filtration induced by this process.

Proof. (a) As introduced in Definition 4.2, let σ_n be the time of the nth jump of $(Y_{t,W},\ t \geq 0)$. Then the number of cells at that time is $|Y_{\sigma_n,W}| = n + 1$. Since Ψ is assumed to be bounded, there is a real number $b > 0$ such that $|\Psi(P)| \leq b$ for all $P \in \mathcal{P}_{d,W}$. Hence $|\Sigma_\Psi(Y_{0,W})| \leq b$, $|\Sigma_\Psi(Y_{\sigma_n,W})| \leq (n+1)b$, and

$$\sup_{0 \leq s \leq t} |\Sigma_\Psi(Y_{s,W})| \leq b\,|Y_{t,W}|\,.$$

Furthermore, for any $a > 0$, we have

$$\begin{aligned}
&\sup_{0 \leq t \leq a} \left| \int_0^t \mathbb{L}\Sigma_\Psi(Y_{s,W})\,\mathrm{d}s \right| \\
&\leq \sup_{0 \leq t \leq a} \int_0^t \sum_{P \in Y_{s,W}} \int_{[P]} \left|\Psi(P \cap H^-) + \Psi(P \cap H^+) - \Psi(P)\right| \Theta(\mathrm{d}H) \\
&\leq 3\,b\,a\,\Theta([W])\,|Y_{a,W}|\,.
\end{aligned}$$

By (4.18),

$$\mathbb{E}|Y_{a,W}|^k < \infty\,, \quad k \in \mathbb{N}\,, \tag{10.8}$$

and thus

$$\sup_{0 \leq t \leq a} \mathbb{E} \left| \int_0^t \mathbb{L}\Sigma_\Psi(Y_{s,W})\,\mathrm{d}s \right| < \infty\,. \tag{10.9}$$

According to Theorem 4.7, we have that $\sigma_n \uparrow \infty$ a. s. as $n \to \infty$. For $n \in \mathbb{N}$, let $(M_t^{\sigma_n}(\Psi),\ t \geq 0)$ denote the stopped process $M_t^{\sigma_n}(\Psi) := M_{\min\{\sigma_n,t\}}(\Psi)$. We have

$$M_t^{\sigma_n}(\Psi) = \Sigma_\Psi(Y_{\min\{\sigma_n,t\},W}) - \Sigma_\Psi(Y_{0,W}) - \int_0^{\min\{\sigma_n,t\}} \mathbb{L}\Sigma_\Psi(Y_{s,W})\,\mathrm{d}s\,.$$

Consider now the damped functional

$$\Sigma_\Psi^{\mathrm{damp}}(Y_{s,W}) := \frac{(\min\{|Y_{s,W}|, n+1\})^2}{|Y_{s,W}|^2}\,\Sigma_\Psi(Y_{s,W})\,. \tag{10.10}$$

This functional is bounded and therefore lies in the domain of the generator $\widehat{\mathbb{L}}$ defined in (4.26). By Lemma 10.2(a), applied with $g = \Sigma_\Psi^{\mathrm{damp}}$, the process

$$M_t(\Sigma_\Psi^{\mathrm{damp}}) := \Sigma_\Psi^{\mathrm{damp}}(Y_{t,W}) - \Sigma_\Psi^{\mathrm{damp}}(Y_{0,W}) - \int_0^t \widehat{\mathbb{L}}\Sigma_\Psi^{\mathrm{damp}}(Y_{s,W})\,\mathrm{d}s\,, \quad t \geq 0\,,$$

is a centered $\mathfrak{J}$-martingale.

For $s \leq \sigma_n$, we have $|Y_{s,W}| \leq n+1$, and hence the damping factor in (10.10) equals 1, so that

$$\Sigma_\Psi^{\mathrm{damp}}(Y_{s,W}) = \Sigma_\Psi(Y_{s,W})\,.$$

Moreover, although the two functionals coincide on $Y_{s,W}$ for $s \leq \sigma_n$, the generator values $\mathbb{L}\Sigma_\Psi(Y_{s,W})$ and $\widehat{\mathbb{L}}\Sigma_\Psi^{\mathrm{damp}}(Y_{s,W})$ need not coincide, since the generator at a state y involves the values of the functional at states obtained from y by one additional splitting step. In particular, a discrepancy can occur only when $|Y_{s,W}| = n+1$, which along a sample path happens only at jump times. Hence $\mathbb{L}\Sigma_\Psi(Y_{s,W})$ and $\widehat{\mathbb{L}}\Sigma_\Psi^{\mathrm{damp}}(Y_{s,W})$ can differ only at $s = \sigma_n$. Since the integrals are Lebesgue integrals, changing the integrand on such a set does not affect their values. Thus

$$M_t^{\sigma_n}(\Psi) = M_{\min\{\sigma_n,t\}}(\Sigma_\Psi^{\mathrm{damp}})$$

for all $t \geq 0$. Since $(M_t(\Sigma_\Psi^{\mathrm{damp}}),\ t \geq 0)$ is a martingale, each $(M_t^{\sigma_n}(\Psi),\ t \geq 0)$ is a martingale as well. Therefore $(M_t(\Psi),\ t \geq 0)$ is a local martingale with localizing sequence $(\sigma_n,\ n \geq 1)$.

To show that it is a martingale, it is sufficient to show that for all $a > 0$,

$$\mathbb{E} \sup_{0 \leq t \leq a} |M_t(\Psi)| < \infty\,, \tag{10.11}$$

as this implies that the condition of Lemma 10.4 is satisfied. To see this, note that

$$\begin{aligned}\sup_{0\leq t\leq a} |M_t(\Psi)| &\leq \sup_{0\leq t\leq a} |\Sigma_\Psi(Y_{t,W})| + |\Sigma_\Psi(Y_{0,W})| + \sup_{0\leq t\leq a}\left|\int_0^t \mathbb{L}\Sigma_\Psi(Y_{s,W})\,\mathrm{d}s\right| \\ &\leq b\,|Y_{a,W}| + b + 3\,b\,a\,\Theta([W])\,|Y_{a,W}|\,,\end{aligned}$$

where we used (10.9). In view of (10.8), we have $\mathbb{E}|Y_{a,W}| < \infty$, and thus

$$\mathbb{E} \sup_{0\leq t\leq a} |M_t(\Psi)| < \infty\,,$$

which shows (10.11) and hence the martingale property of $(M_t(\Psi),\ t \geq 0)$.

(b) Identity (10.6) holds because in both the STIT tessellation process and the restricted Poisson hyperplane tessellation process, the splitting rule is the same: whenever a hyperplane $H \in [P]$ divides a cell P, it is replaced by $P \cap H^+$ and $P \cap H^-$, and Σ_Ψ changes by $\Psi(P \cap H^+) + \Psi(P \cap H^-) - \Psi(P)$. Thus $\mathbb{L}^{\mathrm{PHT}}\Sigma_\Psi = \mathbb{L}\Sigma_\Psi$.

The remainder of the proof is identical to part (a), replacing $Y_{t,W}$ by $X_t \wedge W$. The only difference is that now

$$\mathbb{E}|X_a \wedge W| \leq e^{a\,\Theta([W])}$$

because the number of dividing hyperplanes up to time a is Poisson distributed with parameter $a\,\Theta([W])$ and each hyperplane can at most double the number of existing cells. This ensures the analogue of (10.11) and completes the proof. □

10.3 Martingales for maximal polytopes

10.3.1 Maximal (d – 1)-polytopes

We proceed with the construction of martingales associated with a STIT tessellation process. While the preceding section dealt with functionals related to cells, our focus shifts in this section to functionals of maximal ($d-1$)-polytopes, which might also carry additional markings given by their birth times. In this context, recall that the collection $\mathcal{F}^{\max *}_{d-1,W}$ represents the set of birth time marked maximal ($d-1$)-polytopes within a window $W \in \mathcal{P}_d$, as defined in equation (4.10).

For bounded and measurable functions $\phi : \mathcal{P}_{d-1,W} \to \mathbb{R}$ and $\phi_i : \mathcal{P}_{d-1,W} \to \mathbb{R}$, $i \in \{1,2\}$, let us define

$$\Sigma_{\phi,t,W} := \sum_{(F,s)\in\mathcal{F}^{\max *}_{d-1,W}} \mathbb{1}\{s \leq t\}\,\phi(F)\,, \quad t \geq 0\,, \tag{10.12}$$

$$\Sigma_{\phi_1,\phi_2,t,W} := \Sigma_{\phi_1,t,W}\,\Sigma_{\phi_2,t,W}\,, \quad t \geq 0\,, \tag{10.13}$$

and

$$\overline{\Sigma}_{\phi_1,\phi_2,t,W} := (\Sigma_{\phi_1,t,W} - b_1(t))(\Sigma_{\phi_2,t,W} - b_2(t))\,, \quad t \geq 0\,, \tag{10.14}$$

for $b_1, b_2 \in C^1([0,\infty))$. As introduced in (2.22), for a hyperplane $H \in [\operatorname{int} W]$, we denote by $Y_{s,W} \cap H := \{P \cap H : P \in Y_{s,W}\,,\, H \cap \operatorname{int} P \neq \emptyset\}$ the sectional tessellation $Y_{s,W}$ induces within $H \cap W$. In what follows, we will frequently use the notation

$$A_{\phi_1,\phi_2,s,W} := \int_{[W]} \sum_{F\in Y_{s,W}\cap H} \phi_1(F)\phi_2(F)\,\Theta(dH)\,, \quad s \geq 0\,. \tag{10.15}$$

If $\phi_1 = \phi_2 = \phi$, then we write $A_{\phi,s,W} := A_{\phi,\phi,s,W}$ for short.

Theorem 10.8. *Let $W \in \mathcal{P}_d$, let Θ be a hyperplane measure of the form (4.1), satisfying (4.3), and let $(Y_{t,W},\, t \geq 0)$ be the STIT tessellation process in W, driven by the measure Θ. For $\phi : \mathcal{P}_{d-1,W} \to \mathbb{R}$ and $\phi_i : \mathcal{P}_{d-1,W} \to \mathbb{R}$, $i \in \{1,2\}$, bounded and measurable, and $b_1, b_2 \in C^1([0,\infty))$ with $b_1(0) = b_2(0) = 0$, define $\Sigma_{\phi,s,W}$, $\Sigma_{\phi_i,s,W}$, $i \in \{1,2\}$, $\Sigma_{\phi_1,\phi_2,s,W}$, and $\overline{\Sigma}_{\phi_1,\phi_2,s,W}$*

as in (10.12), (10.13), *and* (10.14), *respectively. Then the following stochastic processes are centered* $\mathfrak{J}$*-martingales:*

(a) $$M_t(\phi) := \Sigma_{\phi,t,W} - \int_0^t \mathbb{L}\Sigma_{\phi,s,W}\,\mathrm{d}s, \quad t \geq 0, \tag{10.16}$$

with

$$\mathbb{L}\Sigma_{\phi,s,W} = \int_{[W]} \sum_{F\in Y_{s,W}\cap H} \phi(F)\,\Theta(\mathrm{d}H)\,; \tag{10.17}$$

(b) $$M_t(\phi_1,\phi_2) := \Sigma_{\phi_1,\phi_2,t,W} - \int_0^t \mathbb{L}\Sigma_{\phi_1,\phi_2,s,W}\,\mathrm{d}s, \quad t \geq 0, \tag{10.18}$$

with

$$\mathbb{L}\Sigma_{\phi_1,\phi_2,s,W} = A_{\phi_1,\phi_2,s,W} + \Sigma_{\phi_2,s,W}\,\mathbb{L}\Sigma_{\phi_1,s,W} + \Sigma_{\phi_1,s,W}\,\mathbb{L}\Sigma_{\phi_2,s,W}\,; \tag{10.19}$$

(c) $$\overline{M}_t(\overline{\Sigma}_{\phi_1,\phi_2}) := \overline{\Sigma}_{\phi_1,\phi_2,t,W} - \int_0^t \mathbb{L}\overline{\Sigma}_{\phi_1,\phi_2,s,W}\,\mathrm{d}s, \quad t \geq 0, \tag{10.20}$$

with

$$\begin{aligned}\mathbb{L}\overline{\Sigma}_{\phi_1,\phi_2,s,W} &= \mathbb{L}\Sigma_{\phi_1,\phi_2,s,W} - b_2(s)\,\mathbb{L}\Sigma_{\phi_1,s,W} - \Sigma_{\phi_1,s,W}\frac{\mathrm{d}}{\mathrm{d}s}b_2(s)\\ &\quad - b_1(s)\,\mathbb{L}\Sigma_{\phi_2,s,W} - \Sigma_{\phi_2,s,W}\frac{\mathrm{d}}{\mathrm{d}s}b_1(s) + \frac{\mathrm{d}}{\mathrm{d}s}(b_1(s)b_2(s))\,.\end{aligned} \tag{10.21}$$

Proof. (a) To identify the action of the extended generator on $\Sigma_{\phi,s,W}$, we compute $\widehat{\mathbb{L}\Sigma}_{\phi,s,W}$ formally using (4.26). This yields a candidate function h. According to Definition 10.5, since $\Sigma_{\phi,0,W} = 0$, it remains to check that

$$M_t(\phi) := \Sigma_{\phi,t,W} - \int_0^t h(Y_{s,W})\,\mathrm{d}s, \quad t \geq 0,$$

is a martingale. The martingale property is then verified by the damping argument, analogous to the proof of Theorem 10.7.

To present the details, we start by noting that, by Lemma 18.2, $\mathcal{T}_W$ is a Borel space, allowing us to build on the general results of Section 10.1. To apply Lemma 10.2(a), the functional $\Sigma_{\phi,t,W}$ has to be defined on $\mathcal{T}_W$. For this, put

$$\begin{aligned}\mathcal{F}^{\max}_{d-1,W}(T) := \{&F \subseteq \mathrm{cl}(\partial T \setminus \partial W) : \operatorname{aff} F \in A(d, d-1),\\ &\forall F' \subseteq \partial T,\ \operatorname{aff} F = \operatorname{aff} F' \Rightarrow F' \subseteq F \text{ or } F' \cap F = \emptyset\}\end{aligned} \tag{10.22}$$

for $T \in \mathcal{T}_W$. In other words, $\mathcal{F}^{\max}_{d-1,W}(T)$ is the set of maximal connected components of dimension $d-1$ of the $(d-1)$-skeleton of T. The map $\mathcal{F}^{\max}_{d-1,W} : \mathcal{T}_W \to \overline{\boldsymbol{\mathcal{F}}}_{\ell f}$ is measurable.

Note that a. s. $F \in \mathcal{F}^{\max}_{d-1,W}(Y_{t,W})$ for $t > 0$ if and only if there is some time $s \in (0, t]$ at which the maximal $(d-1)$-polytope F is born in the course of the STIT process within W, that is, if $(F, s) \in \mathcal{F}^{\max *}_{d-1,W}$. This is because the hyperplane measure Θ is assumed to be translation invariant, and hence in the STIT tessellation process, there are a. s. no two dividing hyperplanes of cells that span the same affine subspace.

Furthermore, we can extend the mapping ϕ from $\mathcal{P}_{d-1,W}$ to the convex ring $\mathcal{R}$ by putting $\phi(P) := 0$ for $P \in \mathcal{R} \setminus \mathcal{P}_{d-1,W}$, and analogously for ϕ_i, $i \in \{1, 2\}$. This extension remains bounded and measurable. This way, we obtain a functional $\Sigma_{\phi,W} : \mathcal{T}_W \to \mathbb{R}$ with

$$\Sigma_{\phi,W}(T) := \sum_{F \in \mathcal{F}^{\max}_{d-1,W}(T)} \phi(F) , \quad T \in \mathcal{T}_W .$$

In particular, $\Sigma_{\phi,W}(Y_{t,W}) = \Sigma_{\phi,t,W}$ a. s. with $\Sigma_{\phi,t,W}$ as defined in (10.12).

The martingale property of the stochastic process in (10.16) with the extended generator (10.17) can be shown as in the proof of Theorem 10.7(a), replacing ψ by ϕ, $\Sigma_\Psi(Y_{s,W})$ by $\Sigma_{\phi,s,W}$ and $\Sigma^{\text{damp}}_\Psi(Y_{s,W})$ by $\Sigma^{\text{damp}}_{\phi,s,W}$. To show the analogue of (10.11), assume that $|\phi(F)| \le b$ for some $b > 0$ and all $F \in \mathcal{P}_{d-1,W}$. Then $\Sigma_{\phi,0,W} = 0$, and

$$\sup_{0 \le t \le a} |\Sigma_{\phi,t,W}| \le b \left(|Y_{a,W}| - 1 \right) , \tag{10.23}$$

because in a STIT tessellation process in W, the number of maximal $(d-1)$-polytopes is by one smaller than the number of cells.

Furthermore, we define a functional $\mathbb{L}$ by applying the generator $\widehat{\mathbb{L}}$ in (4.26) to $\Sigma_{\phi,s,W}$ and then show that this yields the value of the extended generator:

$$\begin{aligned}
\mathbb{L}\Sigma_{\phi,s,W} &= \sum_{P \in Y_{s,W}} \int_{[P]} \left(\Sigma_{\phi,s,W} + \phi(P \cap H) - \Sigma_{\phi,s,W} \right) \Theta(\mathrm{d}H) \\
&= \sum_{P \in Y_{s,W}} \int_{[P]} \phi(P \cap H) \, \Theta(\mathrm{d}H) \\
&= \int_{[W]} \sum_{P \in Y_{s,W}} \mathbb{1}\{H \in [P]\} \, \phi(P \cap H) \, \Theta(\mathrm{d}H) \\
&= \int_{[W]} \sum_{F \in Y_{s,W} \cap H} \phi(F) \, \Theta(\mathrm{d}H) ,
\end{aligned}$$

which is (10.17). Moreover, for $a \ge 0$,

$$\begin{aligned}
\sup_{0 \le t \le a} \left| \int_0^t \mathbb{L}\Sigma_{\phi,s,W} \, \mathrm{d}s \right| &\le \sup_{0 \le t \le a} \int_0^t \int_{[W]} \sum_{F \in Y_{s,W} \cap H} |\phi(F)| \, \Theta(\mathrm{d}H) \, \mathrm{d}s \\
&\le a \, b \, \Theta([W]) \, |Y_{a,W}| .
\end{aligned} \tag{10.24}$$

By (10.8) we have that $\mathbb{E}|Y_{a,W}| < \infty$ for all $a \geq 0$, and hence

$$\mathbb{E} \sup_{0 \leq t \leq a} |M_t(\phi)| \leq b\,(\mathbb{E}|Y_{a,W}| - 1) + a\,b\,\Theta([W])\,\mathbb{E}|Y_{a,W}| < \infty .$$

This completes the proof of the martingale property of $(M_t(\phi),\ t \geq 0)$ in (10.16).

(b) Next, we apply the generator $\widehat{\mathbb{L}}$ formally to the functional $\Sigma_{\phi_1,s,W}\,\Sigma_{\phi_2,s,W}$, which provides a candidate expression for $\mathbb{L}\Sigma_{\phi_1,\phi_2,s,W}$. To justify that this expression indeed defines the action of the extended generator in the sense of Definition 10.5, we show that the corresponding Dynkin-type process is a martingale by the same damping and uniform integrability arguments as in part (a). We start by constructing the candidate function:

$$\begin{aligned}
&\mathbb{L}(\Sigma_{\phi_1,s,W}\,\Sigma_{\phi_2,s,W})\\
&= \sum_{P \in Y_{s,W}} \int_{[P]} \big((\Sigma_{\phi_1,s,W} + \phi_1(P \cap H))(\Sigma_{\phi_2,s,W} + \phi_2(P \cap H)) - \Sigma_{\phi_1,s,W}\,\Sigma_{\phi_2,s,W}\big)\,\Theta(\mathrm{d}H)\\
&= \sum_{P \in Y_{s,W}} \int_{[P]} \big(\phi_1(P \cap H)\phi_2(P \cap H) + \phi_1(P \cap H)\Sigma_{\phi_2,s,W} + \phi_2(P \cap H)\Sigma_{\phi_1,s,W}\big)\,\Theta(\mathrm{d}H)\\
&= \int_{[W]} \sum_{P \in Y_{s,W}} \mathbb{1}\{H \in [P]\}\big(\phi_1(P \cap H)\phi_2(P \cap H)\\
&\qquad + \phi_1(P \cap H)\Sigma_{\phi_2,s,W} + \phi_2(P \cap H)\Sigma_{\phi_1,s,W}\big)\,\Theta(\mathrm{d}H)\\
&= \int_{[W]} \sum_{F \in Y_{s,W} \cap H} \big(\phi_1(F)\phi_2(F) + \phi_1(F)\Sigma_{\phi_2,s,W} + \phi_2(F)\Sigma_{\phi_1,s,W}\big)\,\Theta(\mathrm{d}H)\\
&= A_{\phi_1,\phi_2,s,W} + \Sigma_{\phi_2,s,W}\,\mathbb{L}\Sigma_{\phi_1,s,W} + \Sigma_{\phi_1,s,W}\,\mathbb{L}\Sigma_{\phi_2,s,W}\,,
\end{aligned}$$

which is (10.19). Moreover,

$$\begin{aligned}
&\sup_{0 \leq t \leq a} \left| \int_0^t \mathbb{L}\Sigma_{\phi_1,\phi_2,s,W}\,\mathrm{d}s \right|\\
&\qquad \leq \sup_{0 \leq t \leq a} \int_0^t |A_{\phi_1,\phi_2,s,W} + \Sigma_{\phi_2,s,W}\,\mathbb{L}\Sigma_{\phi_1,s,W} + \Sigma_{\phi_1,s,W}\,\mathbb{L}\Sigma_{\phi_2,s,W}|\,\mathrm{d}s\\
&\qquad \leq a\,(b^2\,\Theta([W])\,|Y_{a,W}| + 2b\,(|Y_{a,W}| - 1)\,b\,|Y_{a,W}|\,\Theta([W]))\,,
\end{aligned}$$

using that $|\phi_i(F)| \leq b$ for some $b > 0$ for $i \in \{1,2\}$ and $F \in \mathcal{P}_{d-1,W}$.

The proof of the martingale property of the random process $(M_t(\phi_1,\phi_2),\ t \geq 0)$ defined at (10.18) follows the same line and uses the same damping of $\Sigma_{\phi_1,\phi_2,s,W}$ as in the proof of (10.16). The only difference is in the last part, where the analogue of (10.11) is shown. Namely, by assumption there is $b > 0$ such that $|\phi_i(F)| \leq b$ for $i \in \{1,2\}$ and all $F \in \mathcal{P}_{d-1,W}$. Then $\Sigma_{\phi_1,\phi_2,0,W} = 0$, and

$$\sup_{0\le t\le a} |\Sigma_{\phi_1,\phi_2,t,W}| \le b^2 \left(|Y_{a,W}| - 1\right)^2 .$$

Using the above bounds, we see that $\sup_{0\le t\le a} |M_t(\phi_1,\phi_2)|$ is bounded by a constant multiple of $1 + |Y_{a,W}|^2$. By (10.8), $\mathbb{E}|Y_{a,W}|^2 < \infty$ for all $a \ge 0$, and hence

$$\mathbb{E} \sup_{0\le t\le a} \left|M_t(\phi_1,\phi_2)\right| < \infty .$$

This completes the proof of the martingale property of the random process in (10.18).

(c) It remains to show that the random process $(\overline{M}_t(\overline{\Sigma}_{\phi_1,\phi_2}),\ t \ge 0)$ is a martingale. By straightforward computation,

$$\begin{aligned}\overline{M}_t(\overline{\Sigma}_{\phi_1,\phi_2}) = {} & \Sigma_{\phi_1,\phi_2,t,W} - \int_0^t \mathbb{L}\Sigma_{\phi_1,\phi_2,s,W}\,\mathrm{d}s \\ & - \Sigma_{\phi_1,t,W}\, b_2(t) + \int_0^t \left(b_2(s)\,\mathbb{L}\Sigma_{\phi_1,s,W} + \Sigma_{\phi_1,s,W} \frac{\mathrm{d}}{\mathrm{d}s} b_2(s)\right) \mathrm{d}s \\ & - \Sigma_{\phi_2,t,W}\, b_1(t) + \int_0^t \left(b_1(s)\,\mathbb{L}\Sigma_{\phi_2,s,W} + \Sigma_{\phi_2,s,W} \frac{\mathrm{d}}{\mathrm{d}s} b_1(s)\right) \mathrm{d}s \\ & + b_1(t) b_2(t) - \int_0^t \frac{\mathrm{d}}{\mathrm{d}s}(b_1(s) b_2(s))\,\mathrm{d}s .\end{aligned} \tag{10.25}$$

By (10.18) the term in the first line on the right-hand side of (10.25) is a martingale. Moreover, the term in the last line is zero, since $b_1(0) = b_2(0) = 0$. Since linear combinations of martingales are martingales as well, it is sufficient to show that the terms in the second and third lines of (10.25) are also martingales. This means that it is sufficient to show that for bounded and measurable $\phi : \mathcal{P}_{d-1,W} \to \mathbb{R}$ and $b \in C^1([0,\infty))$, the stochastic process

$$K_t(\Sigma_\phi, b) := \Sigma_{\phi,t,W}\, b(t) - \int_0^t \left(b(s)\,\mathbb{L}\Sigma_{\phi,s,W} + \Sigma_{\phi,s,W} \frac{\mathrm{d}}{\mathrm{d}s} b(s)\right) \mathrm{d}s , \quad t \ge 0 ,$$

is a centered $\mathfrak{I}$-martingale. To prove this, to Σ_ϕ the same damping as in (10.10) is applied. Then by Lemma 10.2(b) with $E = \mathcal{T}_W$, $U_t = Y_{t,W}$, $g = \Sigma_\phi^{\mathrm{damp}}$, and b we have that the stochastic process

$$K_t(\Sigma_\phi^{\mathrm{damp}}, b) := \Sigma_{\phi,t,W}^{\mathrm{damp}}\, b(t) - \int_0^t \left(b(s)\,\mathbb{L}\Sigma_{\phi,s,W}^{\mathrm{damp}} + \Sigma_{\phi,s,W}^{\mathrm{damp}} \frac{\mathrm{d}}{\mathrm{d}s} b(s)\right) \mathrm{d}s , \quad t \ge 0 ,$$

is a centered $\mathfrak{I}$-martingale. Because $K_t^{\sigma_n}(\Sigma_\phi, b) = K_{\min\{\sigma_n, t\}}(\Sigma_\phi^{\text{damp}}, b)$ for all $t \geq 0$ and $n \in \mathbb{N}$, we first conclude that $(K_t(\Sigma_\phi, b), t \geq 0)$ is a local martingale with respect to the sequence of jump times σ_n of $(Y_{t,W}, t \geq 0)$. To show that it is a martingale, it is sufficient to show that for all $a > 0$, the analogue to (10.11) holds. Since $b \in C^1([0,\infty))$, the function b and its derivative are bounded on the closed interval $[0,a]$. Moreover, as in the estimates for parts (a) and (b), $\sup_{0\leq t\leq a} |K_t(\Sigma_\phi, b)|$ can be bounded by a constant multiple of $1 + |Y_{a,W}|^2$. By (10.8) this yields

$$\mathbb{E} \sup_{0\leq t\leq a} \left|K_t(\Sigma_\phi, b)\right| < \infty\,,$$

and Lemma 10.4 implies that $(K_t(\Sigma_\phi, b), t \geq 0)$ is an $\mathfrak{I}$-martingale. This completes the proof of part (c). □

For later reference, we remark that for the derivative of the expectation, by calculations analogous to the proof of Lemma 4.11 we obtain

$$\begin{aligned}\frac{\mathrm{d}}{\mathrm{d}t}\mathbb{E}\Sigma_{\phi,t,W} &= \lim_{s\downarrow 0} \frac{\mathbb{E}\Sigma_{\phi,t+s,W} - \mathbb{E}\Sigma_{\phi,t,W}}{s} \\ &= \lim_{s\downarrow 0} \frac{\mathbb{E}[\Sigma_{\phi,t+s,W} - \Sigma_{\phi,t,W}]}{s} \\ &= \lim_{s\downarrow 0} \frac{\mathbb{E}[s \int_{[W]} \sum_{P\in Y_{t,W}} \mathbb{1}\{H \in [P]\}\, \phi(P\cap H)\, \Theta(\mathrm{d}H) + o(s)]}{s} \\ &= \mathbb{E}\, \mathbb{L}\Sigma_{\phi,t,W}\,. \end{aligned} \tag{10.26}$$

10.3.2 Birth time marked maximal (*d* – 1)-polytopes

In upcoming chapters, we will come across scenarios where, in contrast to the previous section, a more general framework is required. Specifically, we will face situations where we need to address geometric properties of maximal $(d-1)$-polytopes that also take into account their respective birth times during the STIT construction.

Recall definition (4.10):

$$\mathcal{F}_{d-1,W}^{\max *}(Y_{t,W}) := \left\{(F,s) \in \mathcal{F}_{d-1,W}^{\max *} : s \leq t\right\}.$$

Note that $(\mathcal{F}_{d-1,W}^{\max *}(Y_{t,W}), t \geq 0)$ is a pure jump Markov process, but it is neither time homogeneous nor time augmented as the process considered in Lemma 10.2(b). To derive the associated generator and corresponding martingales, we first introduce an appropriate Borel space. For this purpose, we use the notation $\mathcal{F}_{d-1,W}^{\max}(T)$ for the collection of all maximal $(d-1)$-polytopes of a tessellation $T \in \mathcal{T}_W$; see (10.22). Moreover, let $\mathcal{T}_W^*$ denote the space of all tessellations in W for which each maximal $(d-1)$-polytope is equipped with a measurable mark taking values in $(0,\infty)$. For $T^* \in \mathcal{T}^*_W$, we write $\mathcal{F}^{\max *}_{d-1,W}(T^*)$ for the corresponding set of marked maximal $(d-1)$-polytopes.

Observe that for all $T^* \in \mathcal{T}^*_W$,

$$\mathcal{F}^{\max *}_{d-1,W}(T^*) \in \overline{\mathcal{F}}_{ef}(\mathcal{P}_{d-1,W} \times (0,\infty)) \subset \overline{\mathcal{F}}_{ef}(\mathcal{F} \times (0,\infty)).$$

Moreover, the space $\overline{\mathcal{F}}_{ef}(\mathcal{P}_{d-1,W} \times (0,\infty))$ is a measurable subset of the Polish space $\overline{\mathcal{F}}(\mathcal{F} \times (0,\infty))$; see the proof of Lemma 10.1.2 and Lemma 3.1.4 in [172].

For $\mathcal{S}^* \in \overline{\mathcal{F}}_{ef}(\mathcal{P}_{d-1,W} \times (0,\infty))$, we define the projection onto the first coordinate by

$$\mathcal{S} := \{F \in \mathcal{P}_{d-1,W} : (F,s) \in \mathcal{S}^*\}.$$

We further work with the space

$$\overline{\mathcal{F}}_{ef}(\mathcal{F})^{\mathcal{T},W,*} := \left\{\mathcal{S}^* \in \overline{\mathcal{F}}_{ef}(\mathcal{P}_{d-1,W} \times (0,\infty)) : \partial W \cup \bigcup_{F \in \mathcal{S}, F \subset W} F \in \partial \mathcal{T}_W\right\},$$

which, roughly speaking, consists of finite collections of marked $(d-1)$-polytopes in W whose unions form the skeletons of tessellations in $W \in \mathcal{P}_d$.

To verify that $\overline{\mathcal{F}}_{ef}(\mathcal{F})^{\mathcal{T},W,*}$ is a Borel space, we first note that the projection map $\mathcal{S}^* \mapsto \mathcal{S}$ is measurable. By Theorem 12.2.7 of [172] the indicator function given by $\mathbb{1}\{F \in \mathcal{P}_{d-1,W}\}$, $F \in \mathcal{F}$, is measurable, and by Theorem 3.6.2 of [172] the operation of taking the union of finitely many sets is measurable. Consequently, the mapping

$$\mathcal{S} \mapsto \partial W \cup \bigcup_{F \in \mathcal{S}, F \subset W} F$$

from $\overline{\mathcal{F}}_{ef}(\mathcal{F})$ to $\mathcal{F}$ is measurable. Since $\partial \mathcal{T}_W$ is a measurable subset of $\mathcal{F}$ by Lemma 18.4, it follows that $\overline{\mathcal{F}}_{ef}(\mathcal{F})^{\mathcal{T},W,*}$ is a measurable subset of the Polish space $\overline{\mathcal{F}}(\mathcal{F} \times (0,\infty))$. We therefore conclude that $\overline{\mathcal{F}}_{ef}(\mathcal{F})^{\mathcal{T},W,*}$ is indeed a Borel space.

Let $\mathsf{T}^* : \overline{\mathcal{F}}_{ef}(\mathcal{F})^{\mathcal{T},W,*} \to \mathcal{T}^*_W$ be the map satisfying $T^* = \mathsf{T}^*(\mathcal{F}^{\max *}_{d-1,W}(T^*))$ for $T^* \in \mathcal{T}^*_W$. Accordingly, with a slight abuse of notation, in this section, we write $Y^*_{t,W} := \mathsf{T}^*(\mathcal{F}^{\max *}_{d-1,W}(Y_{t,W}))$ for a STIT tessellation at time t whose all $(d-1)$-dimensional maximal polytopes are marked with their birth times. Let $\mathfrak{I}^* = (\mathfrak{I}^*_t, t \geq 0)$ be the augmented filtration induced by $(Y^*_{t,W}, t \geq 0)$.

First, we determine the rate kernel and the generator of this process. The rate function of the pure jump Markov process $(Y^*_{t,W}, t \geq 0)$ is the same as that of $(Y_{t,W}, t \geq 0)$, namely $\theta : \mathcal{T}_W \to \mathbb{R}$ with

$$\theta(T) := \sum_{P \in T} \Theta([P]);$$

see (4.17) and Theorem 4.7(c). To determine the time-dependent jump transition kernel, first note that also the cell that is divided at a jump time is selected according to the same mechanism as for $(Y_{t,W}, t \geq 0)$. We can therefore use the measure introduced in (4.23), which is given by

$$\mu_T = \sum_{P \in T} \frac{\Theta([P])}{\theta(T)} \delta_P \otimes \Theta_{[P]} .$$

The measure μ_T has support $\bigcup_{P \in T}\{\{P\} \times [P]\}$.

Now, for all $s > 0$, define the map $\oslash_s(\cdot, \cdot, T^*) : \bigcup_{P \in T}\{\{P\} \times [P]\} \to \mathcal{T}_W^*$ by

$$\oslash_s(P, H, T^*) = \mathsf{T}^*\big(\mathcal{F}_{d-1,W}^{\max *}(T^*) \cup \{(P \cap H, s)\}\big) .$$

Furthermore, for $T^* \in \mathcal{T}_W^*$, define $\mathbb{P}^{\oslash_s}(T^*, \cdot)$ as the image measure of μ_T under the map $\oslash_s(\cdot, \cdot, T^*)$, that is,

$$\mathbb{P}^{\oslash_s}(T^*, \cdot) := \mu_T \circ \oslash_s(\cdot, \cdot, T^*)^{-1} . \tag{10.27}$$

More explicitly, for a nonnegative measurable function $g : \mathcal{T}_W^* \to \mathbb{R}$,

$$\begin{aligned}
\int_{\mathcal{T}_W^*} g(T'^*)\, \mathbb{P}^{\oslash_s}(T^*, \mathrm{d}T'^*) &= \int_{\mathcal{P}_d \times \mathcal{H}} g(\oslash_s(P, H, T^*))\, \mu_T(\mathrm{d}(P, H)) \\
&= \sum_{P \in T} \frac{\Theta([P])}{\theta(T)} \int_{\mathcal{H}} g(\oslash_s(P, H, T^*))\, \Theta_{[P]}(\mathrm{d}H) .
\end{aligned}$$

Thus $\mathbb{P}^{\oslash_s}(\cdot, \cdot)$ is the jump transition kernel of $(Y_{t,W}^*, t \geq 0)$. Consequently, the rate kernel is $\theta(\cdot)\, \mathbb{P}^{\oslash_s}(\cdot, \cdot)$.

In analogy to the proof of Lemma 4.11, we derive a family $(\widehat{\mathbb{L}}_s^*, s > 0)$ of time-dependent generators for the Markov process $(Y_{t,W}^*, t \geq 0)$ as follows. For $s > 0$ and bounded and measurable $g : \mathcal{T}_W^* \to \mathbb{R}$,

$$\begin{aligned}
\widehat{\mathbb{L}}_s^* g(T^*) &= \lim_{\varepsilon \downarrow 0} \frac{\mathbb{E}[g(Y_{s+\varepsilon,W}^*) \mid Y_{s,W}^* = T^*] - g(T^*)}{\varepsilon} \\
&= \lim_{\varepsilon \downarrow 0} \frac{(\theta(T)\, \varepsilon + o(\varepsilon))\, (\int_{\mathcal{T}_W^*} g(T'^*)\, \mathbb{P}^{\oslash_s}(T^*, \mathrm{d}T'^*) - g(T^*))}{\varepsilon} \\
&= \theta(T) \int_{\mathcal{T}_W^*} (g(T'^*) - g(T^*))\, \mathbb{P}^{\oslash_s}(T^*, \mathrm{d}T'^*) \\
&= \theta(T) \sum_{P \in T} \int_{[P]} \frac{\Theta([P])}{\theta(T)} (g(\oslash_s(P, H, T^*)) - g(T^*))\, \Theta_{[P]}(\mathrm{d}H) \\
&= \sum_{P \in T} \int_{[P]} g(\oslash_s(P, H, T^*)) - g(T^*)\, \Theta(\mathrm{d}H) \\
&= \sum_{P \in T} \int_{[W]} \mathbb{1}\{H \in [P]\}(g(\oslash_s(P, H, T^*)) - g(T^*))\, \Theta(\mathrm{d}H) ,
\end{aligned}$$

where T is the unmarked tessellation underlying T^*. In particular, for all nonnegative bounded and measurable functions $f : \overline{\mathcal{F}}_{ef}(\mathcal{F})^{\mathcal{T},W,*} \to \mathbb{R}$,

$$\mathbb{L}_s^* f(\mathcal{F}_{d-1}^{\max *}(T^*)) := \int_{[W]} \sum_{F \in T \cap H} (f(\mathcal{F}_{d-1}^{\max *}(T^*) \cup \{(F, s)\}) - f(\mathcal{F}_{d-1}^{\max *}(T^*))) \,\Theta(dH) . \tag{10.28}$$

After these preparations, we can now derive the following birth time marked analogue of Theorem 10.8(a).

Theorem 10.9. *Let $W \in \mathcal{P}_d$, let Θ be a hyperplane measure of the form* (4.1), *satisfying* (4.3), *and let $(Y_{t,W}, t \geq 0)$ be the STIT tessellation process in W, driven by the measure Θ. For a measurable function $\phi^* : \mathcal{P}_{d-1,W} \times \mathbb{R}^+ \to \mathbb{R}$ that is bounded on $\mathcal{P}_{d-1,W} \times (0, a]$ and for all $a > 0$ and $t \geq 0$, define*

$$\Sigma_{\phi^*,t,W} := \sum_{(F,s) \in \mathcal{F}_{d-1,W}^{\max *}(Y_{t,W})} \phi^*(F, s) .$$

Then the stochastic process

$$M_t^*(\phi^*) := \Sigma_{\phi^*,t,W} - \int_0^t \int_{[W]} \sum_{F \in Y_{s,W} \cap H} \phi^*(F, s) \,\Theta(dH) \,ds , \quad t \geq 0 ,$$

is a centered $\mathfrak{I}^*$-martingale.

Proof. Fix $t \geq 0$. We work on the Borel state space $\overline{\mathcal{F}}_{ef}(\mathcal{F})^{\mathcal{T},W,*}$ and consider the pure jump Markov process $\mathbf{Y}^* = (Y_{s,W}^*, s \geq 0)$ introduced above. For each $s > 0$, define the measurable functional

$$f_s : \mathcal{T}_W^* \to \mathbb{R}, \quad f_s(T^*) := \sum_{(F,u) \in \mathcal{F}_{d-1,W}^{\max *}(T^*)} \phi^*(F, u) \,\mathbb{1}\{u \leq s\} .$$

Note that $f_t(Y_{t,W}^*) = \Sigma_{\phi^*,t,W}$ and $f_0(Y_{0,W}^*) = 0$.

Next, we define a candidate time-dependent extended generator. Using the jump rate $\theta(T) = \sum_{P \in T} \Theta([P])$ and the time-s jump kernel $\mathbb{P}^{\oslash_s}(T^*, \cdot)$ defined in (10.27), the time-dependent generator acting on f_s is

$$\begin{aligned} \mathbb{L}_s^* f_s(T^*) &= \sum_{P \in T} \int_{[P]} (f_s(\oslash_s(P, H, T^*)) - f_s(T^*)) \,\Theta(dH) \\ &= \int_{[W]} \sum_{F \in T \cap H} \phi^*(F, s) \,\Theta(dH) , \end{aligned}$$

because adding the marked facet (F, s) increases f_s exactly by $\phi^*(F, s)$; compare with the computation preceding (10.28). In particular, for the process value $T^* = Y^*_{s,W}$, we have

$$\mathbb{L}^*_s f_s(Y^*_{s,W}) = \int_{[W]} \sum_{F \in Y_{s,W} \cap H} \phi^*(F, s)\, \Theta(\mathrm{d}H)\,. \tag{10.29}$$

Let $(\sigma_n)_{n\in\mathbb{N}}$ be the jump times of $(Y_{s,W})$ such that $\sigma_n \uparrow \infty$ a. s. and $|Y_{\sigma_n,W}| = n + 1$. For each $n \in \mathbb{N}$, define the bounded "damped" version

$$f^{\text{damp}}_{s,n}(T^*) := \frac{(\min\{|T|,\, n+1\})^2}{|T|^2} f_s(T^*)\,,$$

where $|T|$ denotes the number of cells of the underlying tessellation of T^*. The damping factor is understood to be 1 when $|T| = 0$. Since for fixed $s > 0$, the function ϕ^* is bounded on $\mathcal{P}_{d-1,W} \times (0, s]$, say by $b_s < \infty$, and since the number of maximal $(d-1)$-faces of a tessellation T is at most $|T|-1$, we have $|f_s(T^*)| \le b_s(|T|-1)$. An inspection of the function $k \mapsto (\min\{k, n+1\})^2 k^{-2}(k-1)$ on $\mathbb{N}$ shows that

$$\sup_{k\in\mathbb{N}} \frac{(\min\{k, n+1\})^2}{k^2}(k-1) < \infty\,,$$

so $f^{\text{damp}}_{s,n}$ is indeed bounded for each fixed $s > 0$ and $n \in \mathbb{N}$. Hence it belongs to the domain of the classical times-s generator, and the corresponding Dynkin process

$$M^{*,\text{damp}}_{t,n} := f^{\text{damp}}_{t,n}(Y^*_{t,W}) - f^{\text{damp}}_{0,n}(Y^*_{0,W}) - \int_0^t \widehat{\mathbb{L}}^*_s f^{\text{damp}}_{s,n}(Y^*_{s,W})\,\mathrm{d}s\,, \quad t \ge 0\,,$$

is a centered $\mathfrak{I}^*$-martingale for each $n \in \mathbb{N}$. Moreover, for $t \le \sigma_n$, we have $f^{\text{damp}}_{t,n}(Y^*_{t,W}) = f_t(Y^*_{t,W}) = \Sigma_{\phi^*,t,W}$, and similarly $\widehat{\mathbb{L}}^*_s f^{\text{damp}}_{s,n}(Y^*_{s,W}) = \mathbb{L}^*_s f_s(Y^*_{s,W})$ for $s \le \sigma_n$. Hence

$$M^{*,\text{damp}}_{t,n} = \Sigma_{\phi^*,\min\{t,\sigma_n\},W} - \int_0^{\min\{t,\sigma_n\}} \mathbb{L}^*_s f_s(Y^*_{s,W})\,\mathrm{d}s =: M^*_{\min\{t,\sigma_n\}}(\phi)\,.$$

Therefore $(M^*_t(\phi),\, t \ge 0)$ is a local $\mathfrak{I}^*$-martingale with localizing sequence $(\sigma_n,\, n \in \mathbb{N})$.

To prove that this is actually a martingale, fix $a > 0$ and let $b_a := \sup\{|\phi^*(F, u)| : F \in \mathcal{P}_{d-1,W},\, 0 < u \le a\}$. We have $b_a < \infty$ by assumption. Then, as in (10.23),

$$\sup_{0\le t\le a} |\Sigma_{\phi^*,t,W}| \le b_a\,(|Y_{a,W}| - 1)\,,$$

since the number of maximal $(d-1)$-polytopes is one less than the number of cells. Moreover, using (10.29),

$$\sup_{0\le t\le a}\left|\int_0^t \mathbb{L}_s^* f_s(Y_{s,W}^*)\,\mathrm{d}s\right| \le \int_0^a \int_{[W]} \sum_{F\in Y_{s,W}\cap H} |\phi^*(F,s)|\,\Theta(\mathrm{d}H)\,\mathrm{d}s$$

$$< a\,b_a\,\Theta([W])\,|Y_{a,W}|\,.$$

Taking the expectations and using $\mathbb{E}|Y_{a,W}| < \infty$ from (10.8) yield

$$\mathbb{E}\sup_{0\le t\le a}|M_t^*(\phi)| < \infty\,.$$

By Lemma 10.4 the local martingale $M_t^*(\phi)$ is an $\mathfrak{I}^*$-martingale. Finally, combining (10.29) with the definition gives exactly

$$M_t^*(\phi) = \Sigma_{\phi^*,t,W} - \int_0^t \int_{[W]} \sum_{F\in Y_{s,W}\cap H} \phi^*(F,s)\,\Theta(\mathrm{d}H)\,\mathrm{d}s\,, \quad t\ge 0\,,$$

which is centered since $M_0^*(\phi) = 0$. This completes the proof. □

10.3.3 Maximal k-polytopes with $k \in \{0,\dots,d-2\}$

The previous two sections were dedicated to maximal polytopes of dimension $d-1$. We now shift our focus to lower-dimensional maximal polytopes of dimension $k \in \{0,\dots, d-2\}$. Let us recall that the collection $\mathcal{F}_{k,W}^{\max *}$ represents the set of birth time marked maximal polytopes of dimension k within a window $W \in \mathcal{P}_d$, as defined in (4.11). Essentially, a functional of the form $\sum_{F\in\mathcal{F}_k(F')}\phi(F)$ with $F' \in \mathcal{F}_{d-1,W}^{\max}$ and $\phi : \mathcal{P}_{k,W} \to \mathbb{R}$ could be treated similarly to the one presented in (10.16). However, it is important to note that this functional might not be bounded in all cases. Addressing this technical challenge is the objective of the following lemma. We remark that even though we deal with maximal polytopes of dimension $k \le d-2$, we will retain the same notation as in (10.16) and (10.17). The dimension k will always be clear from the domain of the function ϕ.

Lemma 10.10. *Let $W \in \mathcal{P}_d$, let Θ be a hyperplane measure of the form* (4.1), *satisfying* (4.3), *and let $(Y_{t,W},\, t\ge 0)$ be the STIT tessellation process in W, driven by the measure Θ. For $k \in \{0,\dots,d-2\}$ and a bounded and measurable function $\phi : \mathcal{P}_{k,W} \to \mathbb{R}$, define*

$$\Sigma_{\phi,t,W} := \sum_{(F',s)\in\mathcal{F}_{d-1,W}^{\max *}} \mathbb{1}\{s\le t\} \sum_{F\in\mathcal{F}_k(F')} \phi(F)\,\mathbb{1}\{F\not\subseteq \partial W\}\,, \quad t\ge 0\,.$$

Then the following stochastic process is a centered $\mathfrak{I}$-martingale:

$$M_t(\phi) := \Sigma_{\phi,t,W} - \int_0^t \mathbb{L}\Sigma_{\phi,s,W}\,\mathrm{d}s\,, \quad t\ge 0\,.$$

In particular, $\Sigma_{\phi,\cdot,W}$ belongs to the domain $\mathcal{D}(\mathbb{L})$ of the extended generator with

$$\mathbb{L}\Sigma_{\phi,s,W} = \int_{[W]} \sum_{P\in Y_{s,W}} \mathbb{1}\{H \in [P]\} \sum_{F\in\mathcal{F}_k(P\cap H)} \phi(F)\, \mathbb{1}\{F \not\subseteq \partial W\}\, \Theta(\mathrm{d}H)$$
$$= \int_{[W]} \sum_{F'\in Y_{s,W}\cap H} \sum_{F\in\mathcal{F}_k(F')} \phi(F)\, \mathbb{1}\{F \not\subseteq \partial W\}\, \Theta(\mathrm{d}H), \quad s \ge 0.$$

Proof. The expression for $\mathbb{L}\Sigma_{\phi,s,W}$ follows by the same computation as in (10.17), with ϕ replaced by $\sum_{F\in\mathcal{F}_k(F')} \phi(F)\, \mathbb{1}\{F \not\subseteq \partial W\}$. To conclude that $\Sigma_{\phi,s,W}$ belongs to $\mathcal{D}(\mathbb{L})$, it remains to show that $M_t(\phi)$ is an $\mathfrak{I}$-martingale, which we do next.

Let $m := |\mathcal{F}_{d-1}(W)|$ be the number of $(d-1)$-faces of W. If there are n cells after $n-1$ divisions, then the number of $(d-1)$-faces of each cell can be at most $m+n-1$. A rough upper bound for the number of k-faces of a maximal $(d-1)$-polytope F that divides such a cell is $\binom{m+n-1}{d-k-1}$, because a k-dimensional face of F appears when F intersects exactly $d-k-1$ faces of dimension $d-1$ of the cell. Assume that $|\phi(F)| \le b$ for some $b > 0$ and all $F \in \mathcal{P}_{k,W}$. From here the idea of the proof is the same in the proof of (10.16). The only difference is that (10.23) has to be replaced by

$$\sup_{0\le t\le a} |\Sigma_{\phi,t,W}| \le b\,(|Y_{a,W}|-1)\binom{m+|Y_{a,W}|-1}{d-k-1}, \tag{10.30}$$

and instead of (10.24), we now have

$$\sup_{0\le t\le a}\left|\int_0^t \mathbb{L}\Sigma_{\phi,s,W}\,\mathrm{d}s\right| \le \sup_{0\le t\le a} \int_0^t \int_{[W]} \sum_{F'\in Y_{s,W}\cap H} \sum_{F\in\mathcal{F}_k(F')} |\phi(F)|\,\Theta(\mathrm{d}H)\,\mathrm{d}s$$
$$\le a\,b\,\Theta([W])\,|Y_{a,W}|\binom{m+|Y_{a,W}|-1}{d-k-1}$$
$$\le a\,b\,\Theta([W])\,|Y_{a,W}|\,(m+|Y_{a,W}|-1)^{d-k-1}.$$

Moreover, by (10.8) we have that $\mathbb{E}|Y_{a,W}|^{\ell} < \infty$ for all $a \ge 0$ and $\ell \in \mathbb{N}$, which completes the proof. □

10.4 Martingales for intrinsic volumes of maximal $(d-1)$-polytopes

Our objective now is to use the outcomes derived in the preceding section for specific geometric functionals associated with maximal $(d-1)$-polytopes, with a particular focus on the intrinsic volumes. Formally, let $W \in \mathcal{P}_d$, and let Θ be a hyperplane measure of the form (4.1), satisfying (4.3). We denote by $(Y_{t,W},\ t \ge 0)$ the STIT tessellation process in W, driven by the measure Θ. Then, for $t > 0$, $j \in \{0,\dots,d-1\}$, and a bounded and measurable function $f : G(d,1) \to \mathbb{R}$, we put

$$\Sigma^f_{V_j,t,W} := \sum_{(F,s)\in\mathcal{F}^{\max *}_{d-1,W}} \mathbb{1}\{s \le t\} f(F^\perp)\, V_j(F)\,, \tag{10.31}$$

$$\overline{\Sigma}^f_{V_j,t,W} := \Sigma^f_{V_j,t,W} - b(t)\,, \quad b(t) := \mathbb{E}\Sigma^f_{V_j,t,W}\,, \tag{10.32}$$

and

$$\Sigma_{V_j,t,W} := \Sigma^f_{V_j,t,W}\,, \quad \overline{\Sigma}_{V_j,t,W} := \overline{\Sigma}^f_{V_j,t,W} \quad \text{if } f \equiv 1\,. \tag{10.33}$$

Regarding the function $b(t) := \mathbb{E}\Sigma^f_{V_j,t,W}$, by (10.26) and (10.17) we have that

$$\frac{\mathrm{d}}{\mathrm{d}t}\mathbb{E}\Sigma^f_{V_j,t,W} = \mathbb{E}\,\mathbb{L}\Sigma^f_{V_j,t,W} = \mathbb{E}\int_{[W]} \sum_{F\in Y_{t,W}\cap H} f(H^\perp)\, V_j(F)\, \Theta(\mathrm{d}H)\,. \tag{10.34}$$

For any $H \in [W]$, the sectional tessellation $Y_{t,W} \cap H$ has the same distribution as $Y_t \cap H$ restricted to $W \cap H$. By Lemma 3.9(a), $Y_t \cap H$ itself is a STIT tessellation within the hyperplane H and therefore satisfies (4.18). Hence $\frac{\mathrm{d}}{\mathrm{d}t}\mathbb{E}\Sigma^f_{V_j,t,W}$ takes a finite value for each $t > 0$ but is not necessarily bounded in t.

In cell division processes, there is a connection between the intrinsic volumes of maximal polytopes and those of cells. It is worth highlighting that this relationship is independent of both the life time distribution and any specific division rule. For simplicity, we will outline this connection here solely within the context of STIT tessellations, avoiding additional notation.

Lemma 10.11. *Let $W \in \mathcal{P}_d$, $t \ge 0$, and $j \in \{0, \dots, d-1\}$. Then*

$$\Sigma_{V_j,t,W} = \Big(\sum_{P\in Y_{t,W}} V_j(P)\Big) - V_j(W)\,. \tag{10.35}$$

Proof. We proceed by induction with respect to the number n of divisions of the window W. Clearly, $\sum_{P\in Y_{0,W}} V_j(P) = V_j(W)$. After the nth division event at time σ_n, $n \in \mathbb{N}$, when a cell $P' \in Y_{\sigma_{n-1},W}$ is divided by a hyperplane H, the additivity of V_j implies that

$$\begin{aligned}\sum_{P\in Y_{\sigma_n,W}} V_j(P) &= \Big(\sum_{P\in Y_{\sigma_{n-1},W}} V_j(P)\Big) - V_j(P') + V_j(P'\cap H^+) + V_j(P'\cap H^-)\\ &= \Big(\sum_{P\in Y_{\sigma_{n-1},W}} V_j(P)\Big) + V_j(P'\cap H)\,.\end{aligned}$$

Induction with respect to n shows that

$$\sum_{P\in Y_{t,W}} V_j(P) = \sum_{(F,s)\in\mathcal{F}^{\max *}_{d-1,W}} \mathbb{1}\{s \le t\}\, V_j(F) + V_j(W) = \Sigma_{V_j,t,W} + V_j(W)\,,$$

which completes the proof. ☐

In this section, we show that the intrinsic volume functionals introduced above, as well as certain products of them, belong to the domain of the extended generator. Equivalently, we construct the associated martingales, as summarized in the forthcoming Corollary 10.12. For this purpose, we compute the values of the extended generator acting on these functionals, which will also be crucial for the second-order analysis in Chapter 12.

By (10.17) we have that

$$\mathbb{L}\Sigma^f_{V_j,t,W} = \int_{[W]} \sum_{F\in Y_{t,W}\cap H} f(H^\perp)\, V_j(F)\, \Theta(\mathrm{d}H)\,. \tag{10.36}$$

Now put

$$A^f_{j,t,W} := \int_{[W]} \sum_{F\in Y_{t,W}\cap H} f(H^\perp)^2\, V_j(F)^2\, \Theta(\mathrm{d}H)\,.$$

Then, by (10.19),

$$\mathbb{L}(\Sigma^f_{V_j,t,W})^2 = A^f_{j,t,W} + 2\,\Sigma^f_{V_j,t,W}\,\mathbb{L}\Sigma^f_{V_j,t,W}\,.$$

In the particular case $j = d-1$, by the additivity of V_{d-1}, (4.1) and Fubini's theorem yield that

$$\mathbb{L}\Sigma^f_{V_{d-1},t,W} = \int_{[W]} V_{d-1}(W\cap H) f(H^\perp)\,\Theta(\mathrm{d}H) = V_d(W) \int_{S^{d-1}} f(\mathrm{lin}\, u)\,\varphi(\mathrm{d}u)\,. \tag{10.37}$$

In particular, unlike the case $j \le d-2$, here the generator value does not depend on the tessellation state. Hence by (10.34)

$$\mathbb{L}\Sigma^f_{V_{d-1},t,W} = \mathbb{E}\mathbb{L}\Sigma^f_{V_{d-1},t,W} = \frac{\mathrm{d}}{\mathrm{d}t}\mathbb{E}\Sigma^f_{V_{d-1},t,W} = \mathbb{L}\mathbb{E}\Sigma^f_{V_{d-1},t,W}\,. \tag{10.38}$$

Thus

$$\mathbb{L}(\Sigma^f_{V_{d-1},t,W})^2 = A^f_{d-1,t,W} + 2\,\Sigma^f_{V_{d-1},t,W}\,\frac{\mathrm{d}}{\mathrm{d}t}\mathbb{E}\Sigma^f_{V_{d-1},t,W}\,. \tag{10.39}$$

We note in passing that (10.38), in conjunction with (10.37), implies that

$$\mathbb{E}\Sigma^f_{V_{d-1},t,W} = t\,V_d(W) \int_{S^{d-1}} f(\mathrm{lin}\, u)\,\varphi(\mathrm{d}u)\,, \tag{10.40}$$

which for the special choice $f \equiv 1$ yields the relation $\mathbb{E}\Sigma_{V_{d-1},t,W} = t\,V_d(W)$ we already encountered for the particular case $V_d(W) = 1$ in Theorem 6.9; see especially (6.6). As a specification of (10.15), we use the notation

$$A_{i,j,t,W} := \int_{[W]} \sum_{F \in Y_{t,W} \cap H} V_i(F)\, V_j(F)\, \Theta(\mathrm{d}H) \tag{10.41}$$

for $i,j \in \{0,\dots,d-1\}$. In particular, $A_{j,j,t,W} = A^f_{j,t,W}$ if $f \equiv 1$. Then by (10.19) it follows that

$$\mathbb{L}(\Sigma_{V_i,t,W}\, \Sigma_{V_j,t,W}) = A_{i,j,t,W} + \Sigma_{V_j,t,W}\, \mathbb{L}\Sigma_{V_i,t,W} + \Sigma_{V_i,t,W}\, \mathbb{L}\Sigma_{V_j,t,W}\,.$$

Let us now discuss the case $\Theta = \Theta_{\text{iso}}$ of the motion-invariant hyperplane measure. For $j \in \{0,\dots,d-1\}$, the rotation invariance of Θ_{iso} and the particular choice $f \equiv 1$ imply that the functionals do not depend on the direction of the maximal polytope. We can thus use Crofton's formula (1.6), which says that

$$\int_{[P]} V_j(P \cap H)\, \Theta_{\text{iso}}(\mathrm{d}H) = c_{j,d}\, V_{j+1}(P)$$

with $c_{j,d}$ given by (1.6). Then, by (10.35) and (10.36),

$$\mathbb{L}\Sigma_{V_j,t,W} = c_{j,d} \sum_{P \in Y_{t,W}} V_{j+1}(P) = c_{j,d}\left(\Sigma_{V_{j+1},t,W} + V_{j+1}(W)\right). \tag{10.42}$$

We are now in a position to apply Theorem 10.8. This shows that the functionals and products considered above are indeed contained in the domain of the extended generator and yields the martingales listed in the following corollary. They will play an essential role in Chapter 12.

Corollary 10.12. *Let $W \in \mathcal{P}_d$, let Θ be a hyperplane measure of the form* (4.1), *satisfying* (4.3), *and let $(Y_{t,W},\ t \geq 0)$ be the STIT tessellation process in W, driven by the measure Θ. Then for $i,j \in \{0,\dots,d-1\}$ and bounded and measurable $f : G(d,1) \to \mathbb{R}$, the following stochastic processes are centered $\mathfrak{I}$-martingales:*

(a) $\overline{\Sigma}^f_{V_{d-1},t,W},\quad t \geq 0.$ (10.43)

(b) $(\overline{\Sigma}^f_{V_{d-1},t,W})^2 - \int_0^t A^f_{d-1,s,W}\, \mathrm{d}s, \quad t \geq 0.$ (10.44)

(c) $\Sigma^f_{V_j,t,W} - \int_0^t \int_{[W]} \sum_{F \in (Y_{s,W} \cap H)} f(H^\perp)\, V_j(F)\, \Theta(\mathrm{d}H)\mathrm{d}s, \quad t \geq 0.$

(d) *Moreover, in the motion-invariant case, that is, for $\Theta = \Theta_{\text{iso}}$, the random process*

$$\begin{aligned}&\overline{\Sigma}_{V_i,t,W}\overline{\Sigma}_{V_j,t,W} - \int_0^t A_{i,j,s,W}\, \mathrm{d}s \\ &\quad - \int_0^t \left(c_{i,d}\, \overline{\Sigma}_{V_{i+1},s,W}\overline{\Sigma}_{V_j,s,W} + c_{j,d}\, \overline{\Sigma}_{V_i,s,W}\overline{\Sigma}_{V_{j+1},s,W}\right) \mathrm{d}s\,, \quad t \geq 0\,,\end{aligned} \tag{10.45}$$

with the coefficients $c_{i,d}$ given in (1.6), *is a centered $\mathfrak{I}$-martingale as well.*

Proof. As a preparation, we show that the mapping $t \mapsto \mathbb{E}\Sigma^f_{V_{d-1},t,W}$ for a general hyperplane measure Θ, satisfying (4.3), and the mappings $t \mapsto \mathbb{E}\Sigma_{V_j,t,W}$ for $j \in \{0,\dots,d-2\}$ for the particular choice $\Theta = \Theta_{\text{iso}}$ belong to $C^1([0,\infty))$. For $\mathbb{E}\Sigma^f_{V_{d-1},t,W}$, this is clear from the explicit formula (10.40). To show that the mapping $t \mapsto \mathbb{E}\Sigma_{V_j,t,W}$ is in $C^1([0,\infty))$ for $j \in \{0,\dots,d-2\}$, by (10.26) it is sufficient to prove that $t \mapsto \mathbb{E}\mathbb{L}\Sigma_{V_j,t,W}$ is continuous. Since $\Theta = \Theta_{\text{iso}}$, we can use (10.42) to deduce that for $j = d-2$,

$$\mathbb{E}\mathbb{L}\Sigma_{V_{d-2},t,W} = c_{d-2,d}\left(\mathbb{E}\Sigma_{V_{d-1},t,W} + V_{d-1}(W)\right),$$

which is continuous in t as $t \mapsto \mathbb{E}\Sigma_{V_{d-1},t,W}$ is in $C^1([0,\infty))$. Next, for $j = d-3$, we use again (10.42) to see that

$$\mathbb{E}\mathbb{L}\Sigma_{V_{d-3},t,W} = c_{d-3,d}\left(\mathbb{E}\Sigma_{V_{d-2},t,W} + V_{d-2}(W)\right),$$

which is again continuous, since $t \mapsto \mathbb{E}\Sigma_{V_{d-2},t,W}$ is in $C^1([0,\infty))$, as we have just seen. Repeating this argument inductively shows that $t \mapsto \mathbb{E}\mathbb{L}\Sigma_{V_j,t,W}$ is continuous and hence that $t \mapsto \mathbb{E}\Sigma_{V_j,t,W}$ is in $C^1([0,\infty))$ for $j \in \{0,\dots,d-2\}$ if $\Theta = \Theta_{\text{iso}}$. In fact, we will see in Theorem 12.2 that $\mathbb{E}\Sigma_{V_j,t,W}$ is a polynomial in t.

(a) For $s \geq 0$, by (10.38),

$$\mathbb{L}\overline{\Sigma}^f_{V_{d-1},s,W} = \mathbb{L}\Sigma^f_{V_{d-1},s,W} - \mathbb{L}\mathbb{E}\Sigma^f_{V_{d-1},s,W} = 0,$$

and thus by (10.16)

$$\overline{\Sigma}^f_{V_{d-1},t,W} - \int_0^t \mathbb{L}\overline{\Sigma}^f_{V_{d-1},s,W}\,ds = \overline{\Sigma}^f_{V_{d-1},t,W}, \quad t \geq 0,$$

is an $\mathfrak{I}$-martingale.

(b) By (10.20) and (10.21) a martingale for the functional $(\overline{\Sigma}^f_{V_{d-1},t,W})^2$ is

$$\begin{aligned}
&(\overline{\Sigma}^f_{V_{d-1},t,W})^2 - \int_0^t \mathbb{L}(\overline{\Sigma}^f_{V_{d-1},s,W})^2\,ds \\
&= (\overline{\Sigma}^f_{V_{d-1},t,W})^2 - \int_0^t \left(\mathbb{L}(\Sigma^f_{V_{d-1},s,W})^2 - 2\Sigma^f_{V_{d-1},s,W}\frac{d}{ds}\mathbb{E}\Sigma^f_{V_{d-1},s,W}\right) ds \\
&\quad + \int_0^t \left(2\,(\mathbb{E}\Sigma^f_{V_{d-1},s,W})\,\mathbb{L}\Sigma^f_{V_{d-1},s,W} - \frac{d}{ds}(\mathbb{E}\Sigma^f_{V_{d-1},s,W})^2\right) ds \\
&= (\overline{\Sigma}^f_{V_{d-1},t,W})^2 - \int_0^t A^f_{d-1,s,W}\,ds, \quad t \geq 0,
\end{aligned}$$

where the last equality follows from (10.39), (10.26), and (10.38).

(c) This follows from the martingale property of (10.16) and (10.36).

(d) By identities (10.20) and (10.21) and then by (10.19) a martingale for the functional $\overline{\Sigma}_{V_i,t,W}\Sigma_{V_j,t,W}$ is given by

$$\begin{aligned}
&\overline{\Sigma}_{V_i,t,W}\overline{\Sigma}_{V_j,t,W} - \int_0^t \mathbb{L}(\overline{\Sigma}_{V_i,s,W}\overline{\Sigma}_{V_j,s,W})\,\mathrm{d}s \\
&= \overline{\Sigma}_{V_i,t,W}\overline{\Sigma}_{V_j,t,W} - \int_0^t \Big(\mathbb{L}(\Sigma_{V_i,s,W}\Sigma_{V_j,s,W}) - (\mathbb{E}\Sigma_{V_j,s,W})\,\mathbb{L}\Sigma_{V_i,s,W} - \Sigma_{V_i,s,W}\frac{\mathrm{d}}{\mathrm{d}s}\mathbb{E}\Sigma_{V_j,s,W} \\
&\quad - (\mathbb{E}\Sigma_{V_i,s,W})\,\mathbb{L}\Sigma_{V_j,s,W} - \Sigma_{V_j,s,W}\frac{\mathrm{d}}{\mathrm{d}s}\mathbb{E}\Sigma_{V_i,s,W} + \frac{\mathrm{d}}{\mathrm{d}s}(\mathbb{E}\Sigma_{V_i,s,W}\,\mathbb{E}\Sigma_{V_j,s,W})\Big)\,\mathrm{d}s \\
&= \overline{\Sigma}_{V_i,t,W}\overline{\Sigma}_{V_j,t,W} - \int_0^t \Big(A_{i,j,s,W} + \Sigma_{V_j,s,W}\,\mathbb{L}\Sigma_{V_i,s,W} + \Sigma_{V_i,s,W}\,\mathbb{L}\Sigma_{V_j,s,W} \\
&\quad - (\mathbb{E}\Sigma_{V_j,s,W})\,\mathbb{L}\Sigma_{V_i,s,W} - \Sigma_{V_i,s,W}\frac{\mathrm{d}}{\mathrm{d}s}\mathbb{E}\Sigma_{V_j,s,W} \\
&\quad - (\mathbb{E}\Sigma_{V_i,s,W})\,\mathbb{L}\Sigma_{V_j,s,W} - \Sigma_{V_j,s,W}\frac{\mathrm{d}}{\mathrm{d}s}\mathbb{E}\Sigma_{V_i,s,W} + \frac{\mathrm{d}}{\mathrm{d}s}(\mathbb{E}\Sigma_{V_i,s,W}\,\mathbb{E}\Sigma_{V_j,s,W})\Big)\,\mathrm{d}s \\
&= \overline{\Sigma}_{V_i,t,W}\overline{\Sigma}_{V_j,t,W} - \int_0^t A_{i,j,s,W}\,\mathrm{d}s \\
&\quad - \int_0^t ((\mathbb{L}\Sigma_{V_i,s,W} - \mathbb{E}\mathbb{L}\Sigma_{V_i,s,W})\overline{\Sigma}_{V_j,s,W} + \overline{\Sigma}_{V_i,s,W}(\mathbb{L}\Sigma_{V_j,s,W} - \mathbb{E}\mathbb{L}\Sigma_{V_j,s,W}))\,\mathrm{d}s\,,
\end{aligned}$$

$t \geq 0$, where the last equality follows from (10.26). Finally, if we take $\Theta = \Theta_{\mathrm{iso}}$, then the martingale property of (10.45) follows by (10.42). □

10.5 Martingales for Hausdorff measures on skeletons

In the preceding section, we examined sums involving intrinsic volumes of maximal $(d-1)$-polytopes. In this section, our focus shifts to the consideration of the k-dimensional volume or Hausdorff measure derived from the k-skeleton of $Y_{t,W}$, where $k \in \{0,\dots, d-1\}$. By the latter we mean the random measure on $(\mathbb{R}^d, \mathcal{B}(\mathbb{R}^d))$ defined by

$$\begin{aligned}
\mathcal{H}^k_{t,W}(\cdot) &:= \sum_{(F,s)\in\mathcal{F}^{\max*}_{k,W}} \mathbb{1}\{s \leq t\}\,\mathbb{1}\{F \not\subseteq \partial W\}\,\mathcal{H}^k(\cdot \cap F) \\
&= \sum_{(F,s)\in\mathcal{F}^{\max*}_{k,W}} \mathbb{1}\{s \leq t\}\,\mathbb{1}\{F \not\subseteq \partial W\}\,\lambda_F(\cdot)\,,
\end{aligned} \tag{10.46}$$

where $(Y_{t,W},\ t \ge 0)$ is a STIT tessellation process in $W \in \mathcal{P}_d$ driven by some hyperplane measure Θ, and where we used (4.14). For $k = d-1$ and $B \in \mathcal{B}(\mathbb{R}^d)$, $\mathscr{H}^k_{t,W}(B)$ is in fact a special case of the functional $\Sigma_{\phi,t,W}$ from Section 10.3.1, whereas for $k \in \{0,\dots,d-2\}$ and $B \in \mathcal{B}(\mathbb{R}^d)$, the quantity $\mathscr{H}^k_{t,W}(B)$ is in fact a special case of the functional $\Sigma_{\phi,t,W}$ from Section 10.3.3 with $\phi(F) = \mathscr{H}^k(B \cap F)$. By Lemma 10.10 and (10.37), for $k \in \{0,\dots,d-1\}$ and $B \in \mathcal{B}(\mathbb{R}^d)$, the extended generator for this functional is

$$\mathbb{L}\mathscr{H}^k_{s,W}(B) = \int_{[W]} \sum_{F' \in Y_{s,W}\cap H} \sum_{F \in \mathcal{F}_k(F')} \mathbb{1}\{F \not\subseteq \partial W\}\, \mathscr{H}^k(B \cap F)\, \Theta(\mathrm{d}H) \tag{10.47}$$

for $s \ge 0$. To deal with products of two such random measures, we introduce the notation

$$\begin{aligned} A^{\mathscr{H}}_{i,j,s,W}(B_1,B_2) := \int_{[W]} \sum_{F' \in Y_{s,W}\cap H} &\Bigg(\sum_{F \in \mathcal{F}_i(F')} \mathbb{1}\{F \not\subseteq \partial W\}\, \mathscr{H}^i(B_1 \cap F)\Bigg) \\ &\times \Bigg(\sum_{F \in \mathcal{F}_j(F')} \mathbb{1}\{F \not\subseteq \partial W\}\, \mathscr{H}^j(B_2 \cap F)\Bigg) \Theta(\mathrm{d}H), \end{aligned} \tag{10.48}$$

where $B_1, B_2 \in \mathcal{B}(\mathbb{R}^d)$, $i,j \in \{0,\dots,d-1\}$, and $s \ge 0$.

Corollary 10.13. *Let $W \in \mathcal{P}_d$, let Θ be a hyperplane measure of the form* (4.1)*, satisfying* (4.3)*, and let $(Y_{t,W},\ t \ge 0)$ be the STIT tessellation process in W, driven by the measure Θ. Then, for fixed $i,j \in \{0,\dots,d-1\}$ and $B_1, B_2 \in \mathcal{B}(\mathbb{R}^d)$, the functional $\mathscr{H}^i_{s,W}(B_1)\mathscr{H}^j_{s,W}(B_2)$ belongs to the domain $\mathcal{D}(\mathbb{L})$ of the extended generator $\mathbb{L}$ with*

$$\begin{aligned} &\mathbb{L}(\mathscr{H}^i_{s,W}(B_1)\, \mathscr{H}^j_{s,W}(B_2)) \\ &\quad = A^{\mathscr{H}}_{i,j,s,W}(B_1,B_2) + \mathscr{H}^i_{s,W}(B_1)\, \mathbb{L}\mathscr{H}^j_{s,W}(B_2) + \mathscr{H}^j_{s,W}(B_2)\, \mathbb{L}\mathscr{H}^i_{s,W}(B_1), \end{aligned}$$

and the stochastic process

$$\mathscr{H}^i_{t,W}(B_1)\, \mathscr{H}^j_{t,W}(B_2) - \int_0^t \mathbb{L}(\mathscr{H}^i_{s,W}(B_1)\, \mathscr{H}^j_{s,W}(B_2))\, \mathrm{d}s, \quad t \ge 0, \tag{10.49}$$

is a centered $\mathfrak{I}$-martingale.

Proof. We formally apply (10.19) to the unbounded functionals

$$\phi_1(F') := \sum_{F \in \mathcal{F}_i(F')} \mathbb{1}\{F \not\subseteq \partial W\}\, \mathscr{H}^i(B_1 \cap F), \quad F' \in \mathcal{P}_{d-1,W},$$

and ϕ_2, which is defined the same way, but with i replaced by j and B_1 replaced by B_2. This yields the form of $\mathbb{L}$. It remains to show that this is indeed an extended generator and that $\mathscr{H}^i_{s,W}(B_1)\, \mathscr{H}^j_{s,W}(B_2)$ belongs to the domain $\mathcal{D}(\mathbb{L})$.

For $i \in \{0,\dots,d-1\}, F \in \mathcal{P}_{i,W}$, and $B \in \mathcal{B}(\mathbb{R}^d)$, we have that $\mathcal{H}^i(B \cap F) \leq V_i(F)$, which yields the boundedness on $\mathcal{P}_{i,W}$. This, together with (10.30), leads to

$$\sup_{0\leq s\leq a} \mathcal{H}^i_{s,W}(B_1)\mathcal{H}^j_{s,W}(B_2) \leq b_i\,(|Y_{a,W}|-1)\binom{m+|Y_{a,W}|-1}{d-i-1}\, b_j\,(|Y_{a,W}|-1)\binom{m+|Y_{a,W}|-1}{d-j-1},$$

where $b_i := \sup_{F\in\mathcal{P}_{i,W}} V_i(F)$, b_j is defined analogously, and $m := |\mathcal{F}_{d-1}(W)|$ is the number of $(d-1)$-faces of W. With this upper bound, we can now use the same method with the damped process as in the proof of Theorem 10.8 to show the martingale property of the random process (10.49). □

10.6 Notes

1. The theory of martingales associated with STIT tessellation processes was first developed and systematically applied in the series of publications [174, 175, 176, 177, 178]. Some technical details have later been clarified in [83], where so-called random splitting tessellations on the d-dimensional unit sphere were introduced and investigated by martingale methods. Our exposition largely follows this line of work, refining the arguments to align with the objectives of this book.
2. The definition of local martingales varies across the literature. Some authors define them in terms of bounded stopping times, whereas others via localization sequences. We follow [47].
3. There are various criteria under which a local martingale is a martingale. For example, bounded local martingales are martingales. However, the martingales we deal with are unbounded, and hence weaker criteria have to be used. One such criterion states that a local martingale $(X_t,\ t\geq 0)$ is a true martingale if and only if it belongs to class DL; see Proposition IV.1.7 in [166] or Problem 5.19 of [90]. This means that for each $t \geq 0$, the family of random variables $\{X_\tau\}$, where τ is a stopping time such that $\tau \leq t$ a. s., is uniformly integrable. We only used the implication that the uniform integrability of a stopped process family implies the martingale property; see Lemma 10.4.
4. The proof of Lemma 10.6 is inspired by that of Theorem 2.11 in [191], which considers a significantly more general setting. More broadly, our approach to birth time marked $(d-1)$-polytopes in Section 10.3.2 is based on the framework for handling time-inhomogeneous Markov processes in [191].
5. The martingales constructed in this chapter can be viewed as concrete instances of the Doob–Meyer decomposition for pure jump processes. In general, given a pure jump Markov process and a functional F of its state, the process

$$F(Y_t) - F(Y_0) - \int_0^t \mathbb{L}F(Y_s)\,\mathrm{d}s$$

is a local martingale, where $\mathbb{L}$ denotes the (extended) generator. In this sense the compensator of the jump process associated with F is precisely the integral of $\mathbb{L}F$. Our proofs essentially verify that the functionals of interest, such as sums of intrinsic volumes, Hausdorff measures of skeletons, and their products, belong to the domain of the extended generator, so that the corresponding martingale-compensator decomposition is valid. This connection highlights that the martingale approach adopted here is closely related to the general theory of compensators for point processes and semimartingales; see Section I.2.c in [86]. This will be further employed in Chapter 13.

6. Let $\Sigma_{V_{d-1},t,W}$ denote the total surface content of the STIT tessellation at time t, driven by a hyperplane measure Θ, inside a window $W \in \mathcal{P}_d$. We have shown in Corollary 10.12(a) that the random process $\Sigma := (\overline{\Sigma}_{V_{d-1},t,W},\ t \geq 0)$ is a martingale with respect to the filtration $\mathfrak{I}$. For any $\vartheta \in \mathbb{R}$, we may thus consider the Doléans-Dade exponential

$$\mathcal{E}_t(\vartheta) := \exp\big(\vartheta\, \overline{\Sigma}_{V_{d-1},t,W} - \Psi_t(\vartheta)\big), \quad t \geq 0, \tag{10.50}$$

where

$$\Psi_t(\vartheta) := \int_0^t \int_{[W]} \sum_{P \in Y_{s,W}} \big(e^{\vartheta\, V_{d-1}(P\cap H)} - 1 - \vartheta\, V_{d-1}(P \cap H)\big)\, \Theta(dH)\, ds$$

is the predictable cumulant compensator associated with Σ; see Section 8.8 of [98]. Since all jump sizes of Σ are bounded by $B(W) := \sup_{H\in[W]} V_{d-1}(W \cap H) < \infty$ and Θ is locally finite, the integral defining $\Psi_t(\vartheta)$ is finite for all $\vartheta \in \mathbb{R}$. Thus $\mathcal{E}_t(\vartheta)$ is well defined, and the random process $(\mathcal{E}_t(\vartheta), t \geq 0)$ is a mean-one $\mathfrak{I}$-martingale. An application of this martingale property will be discussed in the notes of Chapter 12.

11 Relations between STIT and Poisson hyperplane tessellations

In this chapter, we use some of the martingales constructed in Chapter 10 to establish fundamental relationships between STIT and Poisson hyperplane tessellations. The strong connection between these two models has already been encountered in Section 5.1, where it was demonstrated that the zero-cells, K-cells, and typical cells of both tessellation models – when confined to a window W – are identically distributed. In Section 6.1, analogous statements were obtained for the respective global tessellations. In the current chapter, we derive further results in this direction by employing martingale methods, which encompass the previously mentioned results as particular cases. These new tools enable us to address not only cells but also maximal polytopes and faces of dimension $k \in \{1, \dots, d-1\}$. Special attention is dedicated to the case $k = 1$, for which explicit distributional results are derived.

11.1 An identity for cell functionals

Our first result is a statement on the identity of the intensity measures of the cells of STIT tessellations and Poisson hyperplane tessellations in a window $W \in \mathcal{P}_d$. To state it, let us recall that for a bounded and measurable function $\Psi : \mathcal{P}_{d,W} \to \mathbb{R}$, we write

$$\Sigma_\Psi(T) = \sum_{P \in T} \Psi(P), \quad T \in \mathcal{T}_W;$$

see (10.4).

Theorem 11.1. *Let $W \in \mathcal{P}_d$, and let Θ be a hyperplane measure of the form* (4.1), *satisfying* (4.3). *Let $(Y_{t,W}, t \geq 0)$ be the STIT tessellation process in W, driven by the measure Θ, and $(X_t \wedge W, t \geq 0)$ be the process of restricted Poisson hyperplane tessellations driven by Θ with $X_0 \wedge W := \{W\}$. Then*

$$\mathbb{E}\Sigma_\Psi(Y_{t,W}) = \mathbb{E}\Sigma_\Psi(X_t \wedge W) \quad \textit{for all } t \geq 0. \tag{11.1}$$

Proof. The expectations of the martingales in (10.5) and (10.7) are both zero, and $\Sigma_\Psi(Y_{0,W}) = \Sigma_\Psi(X_0 \wedge W)$. Thus, by (10.6), the proof is complete. □

Remark 11.2. *Theorem* 11.1 *can alternatively be concluded from Corollary* 6.6 *by choosing the function f there as $f(P) := \mathbb{1}\{P \cap W \in \mathcal{P}_d\}\, \Psi(P \cap W)$, $P \in \mathcal{P}_d$.*

Starting with Theorem 11.1, we can get back the result of Theorem 6.2, which says that the typical cells of STIT tessellations and Poisson hyperplane tessellations are identically distributed if their driving characteristics are appropriately chosen. To demonstrate this, recall that according to the general definitions in (2.42) and (2.43), we denote

https://doi.org/10.1515/9783112222409-013

by $\gamma_{d,t}^{\mathrm{STIT}}$ and $\mathbb{Q}_{d,t}^{\mathrm{STIT}}$ the cell intensity and distribution of the typical cell of the STIT tessellation Y_t in $\mathbb{R}^d$, respectively. Correspondingly, $\gamma_{d,t}^{\mathrm{PHT}}$ and $\mathbb{Q}_{d,t}^{\mathrm{PHT}}$ denote the cell intensity and distribution of the typical cell of the Poisson hyperplane tessellation X_t, implicitly assuming that both Y_t and X_t are driven by the same hyperplane measure and both cells are defined using the same center function $c : \mathcal{P}_d \to \mathbb{R}^d$.

Corollary 11.3. *Let Θ be a hyperplane measure of the form* (4.1)*, satisfying* (4.3)*, and let $(Y_t, t > 0)$ be the STIT tessellation process driven by Θ. Furthermore, let $(X_t, t > 0)$ be the process of Poisson hyperplane tessellations driven by Θ. Then*

$$\gamma_{d,t}^{\mathrm{STIT}} = \gamma_{d,t}^{\mathrm{PHT}} \quad \text{and} \quad \mathbb{Q}_{d,t}^{\mathrm{STIT}} = \mathbb{Q}_{d,t}^{\mathrm{PHT}}$$

for all $t > 0$.

Proof. By (2.29) and by (11.1), for $A \in \mathcal{B}(\mathcal{P}_{d,c})$, we have

$$\begin{aligned}\gamma_{d,t}^{\mathrm{STIT}} \mathbb{Q}_{d,t}^{\mathrm{STIT}}(A) &= \lim_{a\to\infty} \frac{1}{V_d(aW)} \mathbb{E} \sum_{P\in Y_t} \mathbb{1}\{P \subseteq aW\}\, \mathbb{1}\{P - c(P) \in A\} \\ &= \lim_{a\to\infty} \frac{1}{V_d(aW)} \mathbb{E} \sum_{P\in X_t} \mathbb{1}\{P \subseteq aW\}\, \mathbb{1}\{P - c(P) \in A\} \\ &= \gamma_{d,t}^{\mathrm{PHT}} \mathbb{Q}_{d,t}^{\mathrm{PHT}}(A)\,.\end{aligned}$$

Choosing $A = \mathcal{P}_{d,c}$ yields $\gamma_{d,t}^{\mathrm{STIT}} = \gamma_{d,t}^{\mathrm{PHT}}$ and completes the proof. □

11.2 Relationships for maximal polytopes

We now turn from cells to maximal polytopes and their faces. Recall from (4.10) and (4.11) that for all $k \in \{0, \dots, d-1\}$,

$$\mathcal{F}_{k,W}^{\max *} = \bigcup_{(F,s)\in\mathcal{F}_{d-1,W}^{\max *}} (\mathcal{F}_k(F) \times \{s\})$$

is the set of birth time marked maximal polytopes of dimension k of a STIT tessellation process in a window $W \in \mathcal{P}_d$. Note that by definition, for $k < d-1$, a maximal k-polytope F can in principle be located on the boundary of W, that is, $F \subseteq \partial W$ can hold for some $(F, s) \in \mathcal{F}_{k,W}^{\max *}$. For comparison with k-faces of Poisson hyperplane tessellations, it turns out that we need to consider only those maximal polytopes F for which $F \not\subseteq \partial W$. For a bounded and measurable function $\phi : \mathcal{P}_{k,W} \to \mathbb{R}$, let us define the functional

$$\Sigma_{\phi,t,W} := \sum_{(F,s)\in\mathcal{F}_{k,W}^{\max *}} \mathbb{1}\{s \le t\}\, \mathbb{1}\{F \not\subseteq \partial W\}\, \phi(F)\,.$$

Our next result shows that the distribution of $\Sigma_{\phi,t,W}$ can alternatively be defined using the global STIT tessellation process $(Y_t, t > 0)$ as defined in Chapter 8, which is then restricted to W. Thus, as long as distributional properties of $\Sigma_{\phi,t,W}$ are studied, we are allowed to freely switch between the "local" definition of $\Sigma_{\phi,t,W}$ and its version based on the global tessellation process.

Lemma 11.4. *Let $W \in \mathcal{P}_d$, let Θ be a hyperplane measure of the form* (4.1), *satisfying* (4.3), *and let $(Y_{t,W},\ t \geq 0)$ be the STIT tessellation process in W driven by Θ. Further, let $(Y_t,\ t > 0)$ be the STIT tessellation process in $\mathbb{R}^d$ driven by Θ. Fix $k \in \{0, \dots, d-1\}$ and let $\phi : \mathcal{P}_{k,W} \to \mathbb{R}$ be a bounded and measurable function. Then, for $t > 0$,*

$$\begin{aligned}\Sigma_{\phi,t,W} &= \sum_{(F,s)\in\mathcal{F}_{k,W}^{\max*}} \mathbb{1}\{s \leq t\}\, \mathbb{1}\{F \not\subseteq \partial W\}\, \phi(F)\\ &\overset{D}{=} \sum_{(F,s)\in\mathcal{F}_{k}^{\max*}} \mathbb{1}\{s \leq t\}\, \mathbb{1}\{F \cap W \in \mathcal{P}_k\}\, \phi(F \cap W)\,,\end{aligned}$$

where the set $\mathcal{F}_k^{\max*}$ of birth time marked maximal k-polytopes is defined in (8.8).

Proof. By Theorem 5.10(b) we have that $Y_{t,W} \overset{D}{=} Y_t \wedge W$. For the k-faces of the cells, this yields

$$\{F \in \mathcal{F}_k(Y_{t,W}) : F \not\subseteq \partial W\} \overset{D}{=} \{F \cap W : F \in \mathcal{F}_k(Y_t),\ F \cap W \in \mathcal{P}_k\}\,.$$

Thus, using Lemma 4.6 and Lemma 8.8(c), we have that

$$\{F \in \mathcal{F}_{k,W}^{\max}(Y_{t,W}) : F \not\subseteq \partial W\} \overset{D}{=} \{F \cap W : F \in \mathcal{F}_k^{\max}(Y_t),\ F \cap W \in \mathcal{P}_k\}\,,$$

which completes the proof. □

The following theorem can be regarded as the lower-dimensional analogue of Theorem 11.1 and provides a relation for intensity measures of maximal k-polytopes of a STIT tessellation and k-faces of a Poisson hyperplane tessellation, driven by the same hyperplane measure.

Theorem 11.5. *Let $W \in \mathcal{P}_d$, let Θ be a hyperplane measure of the form* (4.1), *satisfying* (4.3), *and let $(Y_{t,W},\ t \geq 0)$ be the STIT tessellation process in W driven by Θ. Furthermore, let $(X_t,\ t > 0)$ be the process of Poisson hyperplane tessellations driven by Θ. Then, for $k \in \{0, \dots, d-1\}$, $t > 0$, and a bounded and measurable function $\phi : \mathcal{P}_{k,W} \to \mathbb{R}$, we have*

$$\mathbb{E}\Sigma_{\phi,t,W} = (d-k)\, 2^{d-k-1} \int_0^t \frac{1}{s}\, \mathbb{E} \sum_{F\in\mathcal{F}_k(X_s\wedge W)} \mathbb{1}\{F \not\subseteq \partial W\}\, \phi(F)\, \mathrm{d}s\,.$$

Remark 11.6. *Note that the integral in the last theorem is finite for all $t > 0$, since the number of k-faces in $X_s \wedge W$ grows at most linearly in the number of hyperplanes intersecting W, which in turn is almost surely finite.*

Proof of Theorem 11.5. Using first Lemma 10.10 and then applying Theorem 11.1, we see that

$$\begin{aligned}
\mathbb{E}\Sigma_{\phi,t,W} &= \mathbb{E}\int_0^t \mathbb{L}\Sigma_{\phi,s,W}\,\mathrm{d}s \\
&= \int_0^t \int_{[W]} \mathbb{E}\sum_{P\in Y_{s,W}} \mathbb{1}\{H\in[P]\} \sum_{F\in\mathcal{F}_k(P\cap H)} \mathbb{1}\{F\not\subseteq\partial W\}\,\phi(F)\,\Theta(\mathrm{d}H)\,\mathrm{d}s \\
&= \int_0^t \int_{[W]} \mathbb{E}\sum_{P\in Y_s\wedge W} \mathbb{1}\{H\in[P]\} \sum_{F\in\mathcal{F}_k(P\cap H)} \mathbb{1}\{F\not\subseteq\partial W\}\,\phi(F)\,\Theta(\mathrm{d}H)\,\mathrm{d}s \\
&= \int_0^t \int_{[W]} \mathbb{E}\sum_{P\in X_s\wedge W} \mathbb{1}\{H\in[P]\} \sum_{F\in\mathcal{F}_k(P\cap H)} \mathbb{1}\{F\not\subseteq\partial W\}\,\phi(F)\,\Theta(\mathrm{d}H)\,\mathrm{d}s \\
&= \int_0^t \frac{1}{s}\int_{[W]} \mathbb{E}\sum_{P\in X_s\wedge W} \mathbb{1}\{H\in[P]\} \sum_{F\in\mathcal{F}_k(P\cap H)} \mathbb{1}\{F\not\subseteq\partial W\}\,\phi(F)\,(s\,\Theta)(\mathrm{d}H)\,\mathrm{d}s\,.
\end{aligned}$$

To proceed, define

$$g(H,\widehat{X}_s) := \sum_{P\in X_s\wedge W} \mathbb{1}\{H\in[P]\} \sum_{F\in\mathcal{F}_k(P\cap H)} \mathbb{1}\{F\not\subseteq\partial W\}\,\phi(F)\,,$$

and apply the Mecke equation (19.12) to the Poisson hyperplane process with intensity measure $s\,\Theta$. This yields

$$\int_{[W]} \mathbb{E}g(H,\widehat{X}_s)\,(s\,\Theta)(\mathrm{d}H) = \mathbb{E}\sum_{G\in\widehat{X}_s} g(G,\widehat{X}_s)\,.$$

It follows that

$$\begin{aligned}
&\int_0^t \frac{1}{s}\int_{[W]} \mathbb{E}\sum_{P\in X_s\wedge W} \mathbb{1}\{H\in[P]\} \sum_{F\in\mathcal{F}_k(P\cap H)} \mathbb{1}\{F\not\subseteq\partial W\}\,\phi(F)\,(s\,\Theta)(\mathrm{d}H)\,\mathrm{d}s \\
&\quad = \int_0^t \frac{1}{s}\,\mathbb{E}\sum_{G\in\widehat{X}_s}\sum_{P\in X_s\wedge W} \mathbb{1}\{G\in[P]\} \sum_{F\in\mathcal{F}_k(P\cap G)} \mathbb{1}\{F\not\subseteq\partial W\}\,\phi(F)\,\mathrm{d}s\,.
\end{aligned}$$

Each k-face F is counted once for every pair (P,G) such that P is a cell adjacent to F and G is a hyperplane supporting one of the $(d-1)$-faces of P that contains F. In a Poisson hyperplane tessellation,

- F is adjacent to exactly 2^{d-k} cells,
- within each such cell, it is contained in exactly $d-k$ of its $(d-1)$-faces,
- and each $(d-1)$-face is shared by exactly two adjacent cells.

Thus each F is counted exactly $2^{d-k} \cdot (d-k) \cdot \frac{1}{2} = (d-k)\,2^{d-k-1}$ times. Therefore

$$\begin{aligned}\mathbb{E} \sum_{G\in\widehat{X}_s} \sum_{P\in X_s\wedge W} \mathbb{1}\{G \in [P]\} &\sum_{F\in\mathcal{F}_k(P\cap G)} \mathbb{1}\{F \not\subseteq \partial W\}\,\phi(F)\\ &= (d-k)\,2^{d-k-1}\,\mathbb{E} \sum_{F\in\mathcal{F}_k(X_s\wedge W)} \mathbb{1}\{F \not\subseteq \partial W\}\,\phi(F)\,.\end{aligned}$$

Plugging this back yields

$$\begin{aligned}\mathbb{E}\Sigma_{\phi,t,W} &= \int_0^t \frac{1}{s}\,\mathbb{E} \sum_{G\in\widehat{X}_s} \sum_{P\in X_s\wedge W} \mathbb{1}\{G \in [P]\} \sum_{F\in\mathcal{F}_k(P\cap G)} \mathbb{1}\{F \not\subseteq \partial W\}\,\phi(F)\,\mathrm{d}s\\ &= (d-k)\,2^{d-k-1} \int_0^t \frac{1}{s}\,\mathbb{E} \sum_{F\in\mathcal{F}_k(X_s\wedge W)} \mathbb{1}\{F \not\subseteq \partial W\}\,\phi(F)\,\mathrm{d}s\,,\end{aligned}$$

and the proof is complete. □

Choosing $\phi(F) = V_0(F) = 1$ in Theorem 11.5 yields the following result on the mean number of maximal k-polytopes. We report here the formula only in the isotropic case, for which the most explicit result is available. For general driving measures, the formula involves the associated zonoid we introduced in Section 2.3.3; see Corollary 11.13.

Corollary 11.7. *With the assumptions of Theorem 11.5, in the isotropic model, that is, if $\Theta = \Theta_{\mathrm{iso}}$, the mean number of maximal k-polytopes of $Y_{t,W}$ that are not contained in the boundary of W is*

$$\mathbb{E}\Sigma_{1,t,W} = 2^{d-k-1} \sum_{j=d-k}^{d} \binom{j-1}{d-k-1} \left(\frac{\kappa_{d-1}}{d\,\kappa_d}\right)^j \kappa_j\, V_j(W)\, t^j\,.$$

Proof. We obtain the result from Theorem 11.5 by choosing $\phi(F) = V_0(F) = 1$. More precisely, Theorem 10.3.4 of [172] (see also Lemma 2.4.6 in [82]) yields that

$$\mathbb{E} \sum_{F\in\mathcal{F}_k(X_s\wedge W)} \mathbb{1}\{F \not\subseteq \partial W\} = \sum_{j=d-k}^{d} \binom{j}{d-k} \left(\frac{\kappa_{d-1}}{d\,\kappa_d}\right)^j \kappa_j\, s^j\, V_j(W)\,.$$

Inserting this into Theorem 11.5 gives

$$\begin{aligned}\mathbb{E}\Sigma_{1,t,W} &= (d-k)\,2^{d-k-1} \sum_{j=d-k}^{d} \binom{j}{d-k} \left(\frac{\kappa_{d-1}}{d\,\kappa_d}\right)^j \kappa_j\, V_j(W) \int_0^t s^{j-1}\,\mathrm{d}s\\ &= (d-k)\,2^{d-k-1} \sum_{j=d-k}^{d} \binom{j}{d-k} \left(\frac{\kappa_{d-1}}{d\,\kappa_d}\right)^j \kappa_j\, V_j(W)\, \frac{t^j}{j}\\ &= 2^{d-k-1} \sum_{j=d-k}^{d} \binom{j-1}{d-k-1} \left(\frac{\kappa_{d-1}}{d\,\kappa_d}\right)^j \kappa_j\, V_j(W)\, t^j\,,\end{aligned}$$

using that the intensity measure of the Poisson hyperplane process generating X_s equals $s\,\Theta_{\text{iso}}$; see (2.65). □

Analogously to Section 11.1, it is possible to derive from Theorem 11.5 a relation for the typical maximal polytopes of STIT tessellations Y_t in $\mathbb{R}^d$. We recall that the set of birth time marked maximal k-polytopes, $k \in \{0,\dots,d-1\}$, of STIT tessellations – without restriction to a window – was defined in (8.7) and (8.8). For all $t > 0$, by stationarity the intensity $\gamma_{k,t}^{\max}$ and the distribution $\mathbb{Q}_{k,t}^{\max}$ of the typical maximal k-polytope of Y_t are given by (8.15) and (8.16), respectively. Maximal polytopes will extensively be studied in Chapter 14. Here we rely on a description of typical maximal polytopes by the limit relation provided in Lemma 8.10, which is similar to the one we already used in the proof of Corollary 11.3. Correspondingly, $\gamma_{k,t}^{\text{PHT}}$ and $\mathbb{Q}_{k,t}^{\text{PHT}}$ denote the k-face intensity and distribution of the typical k-face of the Poisson hyperplane tessellation X_t, driven by the same hyperplane measure. As in the case of cells, we assume in this section that the typical maximal k-polytope of Y_t and the typical k-face of X_t are defined using a common center function $c : \mathcal{P}_k \to \mathbb{R}^d$.

Theorem 11.8. *Let Θ be a hyperplane measure of the form* (4.1), *satisfying* (4.3), *and let $(Y_t,\, t > 0)$ be the STIT tessellation process driven by Θ. Furthermore, let $(X_t,\, t > 0)$ be the process of Poisson hyperplane tessellations driven by Θ. Then, for $k \in \{1,\dots,d-1\}$ and $t > 0$, we have that*

$$\gamma_{k,t}^{\max} = \frac{(d-k)\,2^{d-k-1}}{d}\gamma_{k,t}^{\text{PHT}} \tag{11.2}$$

and

$$\mathbb{Q}_{k,t}^{\max}(A) = \int_0^t \frac{ds^{d-1}}{t^d}\,\mathbb{Q}_{k,s}^{\text{PHT}}(A)\,\mathrm{d}s\,, \quad A \in \mathcal{B}(\mathcal{P}_{k,c})\,. \tag{11.3}$$

Proof. For $W \in \mathcal{P}_d$, $k \in \{1,\dots,d-1\}$, the fixed center function $c : \mathcal{P}_k \to \mathbb{R}^d$, $A \in \mathcal{B}(\mathcal{P}_{k,c})$, and $a > 0$, by (2.42) we have that

$$\mathbb{E}\sum_{F\in\mathcal{F}_k(X_s\wedge aW)} \mathbb{1}\{F\cap\partial(aW)=\emptyset\}\,\mathbb{1}\{F-c(F)\in A\} \le \mathbb{E}\big|\{F\in\mathcal{F}_k(X_s\wedge aW)\}\big|\,.$$

Moreover, there exists a constant $C = C(d,k,\Theta,W) \in (0,\infty)$, only depending on the parameters in brackets, such that for all $s > 0$ and $a \ge 1$,

$$\mathbb{E}\,|\mathcal{F}_k(X_s\wedge aW)| \le C\,s^d\,a^d\,\Theta([W])\,.$$

Indeed, in an arrangement of n hyperplanes in $\mathbb{R}^d$ in general position, the number of k-faces is bounded above by a polynomial in n of degree d, and in our case, $n = |\widehat{X}_s \cap [aW]|$ is a Poisson random variable with mean proportional to $s\,\Theta([aW]) = s\,a\,\Theta([W])$. Thus

$$\frac{1}{s}\frac{1}{V_d(aW)}\mathbb{E}\sum_{F\in\mathcal{F}_k(X_s\wedge aW)}\mathbb{1}\{F\cap\partial(aW)=\emptyset\}\,\mathbb{1}\{F-c(F)\in A\}$$
$$\leq\frac{1}{s}\frac{1}{V_d(aW)}\,C\,s^d\,a^d\,\Theta([W])=C's^{d-1}$$

with another constant $C'=C'(d,k,\Theta,W)\in(0,\infty)$. From Lemma 8.10 and Theorem 11.5, replacing $\mathbb{1}\{F\not\subseteq\partial W\}\,\phi(F)$ with $\mathbb{1}\{F\cap\partial(aW)=\emptyset\}\,\mathbb{1}\{F-c(F)\in A\}$ and using the dominated convergence theorem, it follows that

$$\begin{aligned}
&\gamma_{k,t}^{\max}\mathbb{Q}_{k,t}^{\max}(A)\\
&=\lim_{a\to\infty}\frac{1}{V_d(aW)}\mathbb{E}\sum_{F\in\mathcal{F}_{k,aW}^{\max}(Y_{t,aW})}\mathbb{1}\{F\cap\partial(aW)=\emptyset\}\,\mathbb{1}\{F-c(F)\in A\}\\
&=\lim_{a\to\infty}\frac{(d-k)2^{d-k-1}}{V_d(aW)}\int_0^t\frac{1}{s}\mathbb{E}\sum_{F\in\mathcal{F}_k(X_s\wedge aW)}\mathbb{1}\{F\cap\partial(aW)=\emptyset\}\,\mathbb{1}\{F-c(F)\in A\}\,\mathrm{d}s\\
&=(d-k)2^{d-k-1}\int_0^t\frac{1}{s}\lim_{a\to\infty}\frac{1}{V_d(aW)}\\
&\quad\times\mathbb{E}\sum_{F\in\mathcal{F}_k(X_s\wedge aW)}\mathbb{1}\{F\cap\partial(aW)=\emptyset\}\,\mathbb{1}\{F-c(F)\in A\}\,\mathrm{d}s\\
&=(d-k)2^{d-k-1}\int_0^t\frac{1}{s}\gamma_{k,s}^{\mathrm{PHT}}\,\mathbb{Q}_{k,s}^{\mathrm{PHT}}(A)\,\mathrm{d}s\,.
\end{aligned}$$

Choosing $A=\mathcal{P}_{k,c}$, we obtain

$$\gamma_{k,t}^{\max}=(d-k)2^{d-k-1}\int_0^t\frac{1}{s}\left(\frac{s}{t}\right)^d\gamma_{k,t}^{\mathrm{PHT}}\,\mathrm{d}s=\frac{(d-k)2^{d-k-1}}{d}\gamma_{k,t}^{\mathrm{PHT}},$$

and thus

$$\begin{aligned}
\gamma_{k,t}^{\max}\mathbb{Q}_{k,t}^{\max}(A)&=(d-k)2^{d-k-1}\int_0^t\frac{1}{s}\left(\frac{s}{t}\right)^d\gamma_{k,t}^{\mathrm{PHT}}\,\mathbb{Q}_{k,s}^{\mathrm{PHT}}(A)\,\mathrm{d}s\\
&=\gamma_{k,t}^{\max}\int_0^t\frac{ds^{d-1}}{t^d}\,\mathbb{Q}_{k,s}^{\mathrm{PHT}}(A)\,\mathrm{d}s\,.
\end{aligned}$$

This completes the argument. □

Remark 11.9. *Remark* 17.4 *provides explicit formulae relating the intensities of k-faces and distributions of the typical k-face in Poisson hyperplane tessellations to the corresponding characteristics of the multiplicity-weighted k-faces in STIT tessellations.*

Relation (11.3) expresses the fact that the distribution of the typical maximal k-polytope of a STIT tessellation Y_t is a mixture of distributions of typical k-dimensional faces of Poisson hyperplane tessellations X_s, $0 < s < t$. The mixture is controlled by the beta distribution on the time interval $(0, t)$ with parameters d and 1, whose density is given by $s \mapsto d\, s^{d-1}\, t^{-d}\, \mathbb{1}\{0 < s < t\}$.

We also report here the following result about the directional distribution of the typical maximal k-polytope of a STIT tessellation Y_t. By the latter we understand the probability measure $\mathbb{Q}^{\max}_{k,t:h}$ on the Grassmannian $G(d, k)$ of k-dimensional linear subspaces of $\mathbb{R}^d$ given by

$$\mathbb{Q}^{\max}_{k,t:h}(A) = \int_{\mathcal{P}_{k,c}} \mathbb{1}\{\operatorname{aff}_o F \in A\}\, \mathbb{Q}^{\max}_{k,t}(\mathrm{d}F)\,, \quad A \in \mathcal{B}(G(d,k))\,. \tag{11.4}$$

We emphasize that $\mathbb{Q}^{\max}_{k,t:h}$ is independent of the choice of the center function behind $\mathbb{Q}^{\max}_{k,t}$. Moreover, $\mathbb{Q}^{\max}_{k,t:h}$ is also independent of the time parameter t because the typical maximal k-polytope of Y_t has the same distribution as that of Y_1, say, up to a rescaling by $1/t$. Similarly, we denote by $\mathbb{Q}^{\mathrm{PHT}}_{k,t:h}$ the directional distribution of the typical k-face of a Poisson hyperplane tessellation with intensity measure $t\,\Theta$. Clearly, $\mathbb{Q}^{\mathrm{PHT}}_{k,t:h}$ is independent of t.

Corollary 11.10. *Let Θ be a hyperplane measure of the form* (4.1), *satisfying* (4.3). *Let $(Y_t, t > 0)$ be the STIT tessellation process driven by Θ. Furthermore, let $(X_t, t > 0)$ be the process of Poisson hyperplane tessellations driven by Θ. Then $\mathbb{Q}^{\max}_{k,t:h} = \mathbb{Q}^{\mathrm{PHT}}_{k,t:h}$ for all $k \in \{1, \dots, d-1\}$ and $t > 0$.*

Proof. From (11.3) and the fact that $\mathbb{Q}^{\mathrm{PHT}}_{k,s:h} = \mathbb{Q}^{\mathrm{PHT}}_{k,t:h}$ for all $0 < s < t$ it follows that

$$\begin{aligned}
\mathbb{Q}^{\max}_{k,t:h}(A) &= \int_{\mathcal{P}_{k,c}} \mathbb{1}\{\operatorname{aff}_o F \in A\}\, \mathbb{Q}^{\max}_{k,t}(\mathrm{d}F) \\
&= \int_0^t \frac{d s^{d-1}}{t^d} \int_{\mathcal{P}_{k,c}} \mathbb{1}\{\operatorname{aff}_o F \in A\}\, \mathbb{Q}^{\mathrm{PHT}}_{k,s}(\mathrm{d}F)\, \mathrm{d}s \\
&= \mathbb{Q}^{\mathrm{PHT}}_{k,t:h}(A) \int_0^t \frac{d s^{d-1}}{t^d}\, \mathrm{d}s \\
&= \mathbb{Q}^{\mathrm{PHT}}_{k,t:h}(A)
\end{aligned}$$

for all $A \in \mathcal{B}(G(d,k))$. □

Remark 11.11. *The distribution $\mathbb{Q}^{\mathrm{PHT}}_{k,t:h}$ or, equivalently $\mathbb{Q}^{\max}_{k,t:h}$, according to Theorem 1 in* [81], *can be expressed in terms of the so-called associated zonoid we introduced in Section* 2.3.3. *For the special case $k = 1$, we refer to* (11.12).

11.3 Maximal polytopes and associated zonoids

In the previous sections, we have established fundamental connections between STIT tessellations and Poisson hyperplane tessellations. We will now demonstrate how these can be used to derive formulae for typical maximal polytopes of STIT tessellations. In particular, the mean intrinsic volumes of the typical k-dimensional maximal polytope of a STIT tessellation can be expressed in terms of an associated zonoid. This will eventually allow us to study extremal problems via classical isoperimetric inequalities.

11.3.1 Mean intrinsic volumes

Let Θ be a hyperplane measure and $t > 0$, and recall from Section 2.3.3 the definition of the zonoid $\Pi_{t\Theta}$ associated with the pair (t, Θ). By means of this auxiliary convex body we can express the mean intrinsic volumes of typical maximal polytopes via the intrinsic volumes of $\Pi_{t\Theta}$.

Corollary 11.12. *Let Θ be a hyperplane measure of the form* (4.1), *satisfying* (4.3), *and let $(Y_t, t > 0)$ be the STIT tessellation process driven by Θ. For $0 \le j \le k \le d-1$, the mean jth intrinsic volume of the typical k-dimensional maximal polytope of Y_t is*

$$\int_{\mathcal{P}_{k,c}} V_j(F)\, \mathbb{Q}_{k,t}^{\max}(\mathrm{d}F) = \frac{d}{t^j(d-j)} \frac{\binom{d-j}{d-k}}{\binom{d}{k}} \frac{V_{d-j}(\Pi_\Theta)}{V_d(\Pi_\Theta)} .$$

In particular, if $\Theta = \Theta_{\mathrm{iso}}$, then

$$\int_{\mathcal{P}_{k,c}} V_j(F)\, \mathbb{Q}_{k,t}^{\max}(\mathrm{d}F) = \frac{d^{j+1}}{t^j(d-j)} \binom{k}{j} \left(\frac{\kappa_d}{\kappa_{d-1}}\right)^j \frac{1}{\kappa_j},$$

and if Θ is the Mondrian hyperplane measure with weights $p_1 = \cdots = p_d = 1/d$, then

$$\int_{\mathcal{P}_{k,c}} V_j(F)\, \mathbb{Q}_{k,t}^{\max}(\mathrm{d}F) = \frac{d^{j+1}}{t^j(d-j)} \binom{k}{j}.$$

Proof. By Theorem 3.2.1 and Theorem 5.5.3 of [82], Theorem 11.8, and the homogeneity of the intrinsic volumes we have that

$$\begin{aligned}\int_{\mathcal{P}_{k,c}} V_j(F)\, \mathbb{Q}_{k,t}^{\max}(\mathrm{d}F) &= \int_0^t \frac{ds^{d-1}}{t^d} \frac{\binom{d-j}{d-k} V_{d-j}(s\Pi_\Theta)}{\binom{d}{k} V_d(s\Pi_\Theta)} \,\mathrm{d}s \\ &= \frac{\binom{d-j}{d-k}}{\binom{d}{k}} \frac{V_{d-j}(\Pi_\Theta)}{V_d(\Pi_\Theta)} \int_0^t \frac{ds^{d-1}}{t^d} \frac{s^{d-j}}{s^d} \,\mathrm{d}s\end{aligned}$$

$$= \frac{d\,t^{-j}}{d-j}\frac{\binom{d-j}{d-k}}{\binom{d}{k}}\frac{V_{d-j}(\Pi_\Theta)}{V_d(\Pi_\Theta)}\,. \tag{11.5}$$

In the motion-invariant case, $\Pi_\Theta = \Pi_{\Theta_{\text{iso}}}$ is a ball of radius $\frac{\kappa_{d-1}}{d\kappa_d}$; see (2.70). Thus, for $k \in \{0,\dots,d\}$, according to Lemma 2.34(a), we have

$$V_k(\Pi_{\Theta_{\text{iso}}}) = \left(\frac{\kappa_{d-1}}{d\,\kappa_d}\right)^k V_k(B^d) = \binom{d}{k}\left(\frac{\kappa_{d-1}}{d\,\kappa_d}\right)^k \frac{\kappa_d}{\kappa_{d-k}}\,. \tag{11.6}$$

The result thus follows from (11.5) by taking into account that $\frac{\binom{d-j}{d-k}\binom{d}{j}}{\binom{d}{k}} = \binom{k}{j}$.

If Θ is the Mondrian hyperplane measure with $p_1 = \dots = p_d = 1/d$, then

$$V_k(\Pi_\Theta) = \binom{d}{k}\frac{1}{d^k} \tag{11.7}$$

for $k \in \{0,\dots,d\}$ by (2.71) and Lemma 2.34(b). The result then follows from (11.5) in the same way. □

11.3.2 Intensity

Our next goal is to generalize Corollary 11.7 to the case of a general directional distribution, but with the restriction to a spherical observation window.

Corollary 11.13. *Let Θ be a hyperplane measure of the form* (4.1), *satisfying* (4.3), *and let $(Y_t,\ t > 0)$ be the STIT tessellation process driven by Θ. Then, for $k \in \{0,\dots,d-1\}$, the mean number $\mathbb{E}\nu_k(t,\Theta,RB^d)$ of maximal k-polytopes of Y_t having nonempty intersection with the interior of a d-dimensional ball RB^d of radius $R > 0$ is*

$$\mathbb{E}\nu_k(t,\Theta,RB^d) = 2^{d-k-1}\sum_{j=d-k}^{d}\binom{j-1}{d-k-1}\kappa_j R^j\,V_j(\Pi_\Theta)\,t^j\,.$$

Proof. Fix a polytope $W \in \mathcal{P}_d$ satisfying $RB^d \subset W$ and consider the STIT tessellation process $(Y_{t,W},\ t \geq 0)$ within W. By the spatial consistency of this process it follows that $\mathbb{E}\nu_k(t,\Theta,RB^d) = \mathbb{E}\Sigma_{\phi,t,W}$ with the function $\phi : \mathcal{P}_{k,W} \to \mathbb{R}$ given by $\phi(F) := \mathbb{1}\{F \cap \operatorname{int} RB^d \neq \emptyset\}$. We can now argue as in the proof of Corollary 11.7, using identity (8.5) of [82] in conjunction with Theorem 11.5. This yields

$$\begin{aligned}\mathbb{E}\nu_k(t,\Theta,RB^d) &= \mathbb{E}\Sigma_{\phi,t,W}\\ &= (d-k)\,2^{d-k-1}\sum_{j=d-k}^{d}\binom{j}{d-k}\kappa_j R^j\int_0^t \frac{1}{s}V_j(\Pi_{s\Theta})\,ds\end{aligned}$$

$$= (d-k)\,2^{d-k-1} \sum_{j=d-k}^{d} \binom{j}{d-k} \kappa_j R^j \int_0^t s^{j-1} V_j(\Pi_\Theta)\, ds$$

$$= (d-k)\,2^{d-k-1} \sum_{j=d-k}^{d} \binom{j}{d-k} \kappa_j R^j \, V_j(\Pi_\Theta)\, \frac{t^j}{j}$$

$$= 2^{d-k-1} \sum_{j=d-k}^{d} \binom{j-1}{d-k-1} \kappa_j R^j \, V_j(\Pi_\Theta)\, t^j ,$$

where we used the fact that the intensity measure of the Poisson hyperplane process that generates X_s is equal to $s\,\Theta$ (see Equation (2.65)) together with the scaling relation $V_j(\Pi_{s\,\Theta}) = s^j V_j(\Pi_\Theta)$. □

We may represent, for $k \in \{0, \dots, d-1\}$ and $t > 0$, the **intensity of maximal k-polytopes** as

$$\gamma_{k,t}^{\max}(\Theta) = \lim_{R\to\infty} \frac{\mathbb{E}\nu_k(t, \Theta, RB^d)}{V_d(RB^d)} .$$

Thus $\gamma_{k,t}^{\max}(\Theta)$ is the mean number of maximal k-polytopes per unit volume, which can be approximated by large-window averages. Note that in this section, we write $\gamma_{k,t}^{\max}(\Theta)$ instead of $\gamma_{k,t}^{\max}$ to emphasize the dependence on the driving hyperplane measure. According to Corollary 11.13, it follows that

$$\gamma_{k,t}^{\max}(\Theta) = \lim_{R\to\infty} \frac{2^{d-k-1}}{\kappa_d R^d} \sum_{j=d-k}^{d} \binom{j-1}{d-k-1} \kappa_j R^j \, V_j(\Pi_\Theta)\, t^j$$

$$= 2^{d-k-1} \binom{d-1}{d-k-1} V_d(\Pi_\Theta)\, t^d . \tag{11.8}$$

11.3.3 Selected extremal problems

In this section, we study exemplarily selected extremal problems related to maximal polytopes and their intensities in a STIT tessellation process $(Y_t, t > 0)$ in $\mathbb{R}^d$, driven by a hyperplane measure Θ of the form (4.1), satisfying (4.3). We start with the following inequality for the mean number $\mathbb{E}\nu_k(t, \Theta, RB^d)$ of maximal k-polytopes having nonempty intersection with a ball of radius R; recall Corollary 11.13.

Theorem 11.14. *Let Θ be a hyperplane measure of the form* (4.1), *satisfying* (4.3), *and let $(Y_t, t > 0)$ be the STIT tessellation process driven by Θ. Then, for $k \in \{0, \dots, d-1\}$,*

$$\mathbb{E}\nu_k(t, \Theta, RB^d) \le \mathbb{E}\nu_k(t, \Theta_{\text{iso}}, RB^d)$$

for any fixed $t > 0$ and $R > 0$, with equality if and only if $\Theta = \Theta_{\text{iso}}$.

Proof. Let us recall that the isoperimetric inequality for the intrinsic volumes of convex bodies says that for all $K \in \mathcal{K}$ and $0 < i \le j \le d$,

$$\left(\frac{\kappa_{d-i}}{\binom{d}{i}} V_i(K)\right)^j \ge \kappa_d^{j-i}\left(\frac{\kappa_{d-j}}{\binom{d}{j}} V_j(K)\right)^i$$

with equality if and only if K is a ball, provided that $V_i(K) > 0$; see equation (14.31) in [172]. We apply this inequality with $i = 1$ to the convex body $K = \Pi_{t\,\Theta}$:

$$\left(\frac{\kappa_{d-1}}{d} V_1(\Pi_{t\,\Theta})\right)^j \ge \kappa_d^{j-1}\frac{\kappa_{d-j}}{\binom{d}{j}} V_j(\Pi_{t\,\Theta}).$$

Using $V_1(\Pi_{t\,\Theta}) = tV_1(\Pi_\Theta)$ and $V_j(\Pi_{t\,\Theta}) = t^j V_j(\Pi_\Theta)$, the inequality is equivalent to

$$V_j(\Pi_\Theta) \le \binom{d}{j}\frac{\kappa_d}{\kappa_{d-j}}\left(\frac{\kappa_{d-1}}{d\,\kappa_d} V_1(\Pi_\Theta)\right)^j = V_j(\Pi_{\Theta_{\text{iso}}})\, V_1(\Pi_\Theta)^j, \tag{11.9}$$

where we used (11.6) in the last step. Applying this inequality to each of the summands appearing in the expression for $\mathbb{E}\nu_k(t,\Theta,RB^d)$ in Corollary 11.13, we find that

$$\begin{aligned}\mathbb{E}\nu_k(t,\Theta,RB^d) &= 2^{d-k-1}\sum_{j=d-k}^{d}\binom{j-1}{d-k-1}\kappa_j R^j\, V_j(\Pi_\Theta)\, t^j \\ &\le 2^{d-k-1}\sum_{j=d-k}^{d}\binom{j-1}{d-k-1}\kappa_j R^j\, V_j(\Pi_{\Theta_{\text{iso}}})\, t^j = \mathbb{E}\nu_k(t,\Theta_{\text{iso}},RB^d).\end{aligned}$$

This proves the claim including the equality cases. □

Applying inequality (11.9) for $j = d$ to representation (11.8) for the maximal k-polytope intensity $\gamma_{k,t}^{\max}(\Theta)$, we immediately arrive at the following result.

Corollary 11.15. *Let Θ be a hyperplane measure of the form* (4.1), *satisfying* (4.3), *and let $(Y_t, t > 0)$ be the STIT tessellation process driven by Θ. Then, for $k \in \{0,\dots,d-1\}$,*

$$\gamma_{k,t}^{\max}(\Theta) \le \gamma_{k,t}^{\max}(\Theta_{\text{iso}})$$

for any fixed $t > 0$, with equality if and only if $\Theta = \Theta_{\text{iso}}$.

Next, we deal with the expected jth intrinsic volume of the typical maximal k-polytope of a STIT tessellation Y_t, $t > 0$, in $\mathbb{R}^d$, $k \in \{1,\dots,d-1\}$, $j \in \{0,\dots,k\}$, which is driven by a hyperplane measure Θ. To emphasize the role of the hyperplane measure, we introduce the shorthand notation

$$\overline{V}_{k,j,t}^{\max}(\Theta) := \int_{\mathcal{P}_{k,c}} V_j(F)\, \mathbb{Q}_{k,t}^{\max}(\mathrm{d}F).$$

We claim that the expected jth intrinsic volume of the typical maximal k-polytope attains its minimal value precisely in the isotropic case.

Corollary 11.16. *Let Θ be a hyperplane measure of the form (4.1), satisfying (4.3), and let $(Y_t, t > 0)$ be the STIT tessellation process driven by Θ. Then, for $k \in \{0, \ldots, d-1\}$ and $j \in \{0, \ldots, k\}$, we have*

$$\overline{V}^{\max}_{k,j,t}(\Theta) \geq \overline{V}^{\max}_{k,j,t}(\Theta_{\text{iso}})$$

for any fixed $t > 0$, with equality if and only if $\Theta = \Theta_{\text{iso}}$.

Proof. By Corollary 11.12 we have

$$\overline{V}^{\max}_{k,j,t}(\Theta) = \frac{d}{t^j(d-j)} \frac{\binom{d-j}{d-k}}{\binom{d}{k}} \frac{V_{d-j}(\Pi_\Theta)}{V_d(\Pi_\Theta)}.$$

Now we can use the fact that for any convex body $K \subset \mathbb{R}^d$, we have the inequality

$$\frac{V_{d-j}(K)}{V_d(K)} \geq \frac{\binom{d}{j}}{\kappa_j} \left(\frac{d\kappa_d}{\kappa_{d-1}} \frac{1}{V_1(K)} \right)^j,$$

with equality if and only if K is a ball; see Lemma 18.1.1 in [82]. Applying this to $K = \Pi_\Theta$ and invoking the fact that $V_1(\Pi_\Theta) = 1$, we conclude that

$$\begin{aligned}\overline{V}^{\max}_{k,j,t}(\Theta) &\geq \frac{d}{t^j(d-j)} \frac{\binom{d-j}{d-k}}{\binom{d}{k}} \frac{\binom{d}{j}}{\kappa_j} \left(\frac{d\kappa_d}{\kappa_{d-1}} \right)^j \\ &= \frac{d}{t^j(d-j)} \binom{k}{j} \frac{1}{\kappa_j} \left(\frac{d\kappa_d}{\kappa_{d-1}} \right)^j = \overline{V}^{\max}_{k,j,t}(\Theta_{\text{iso}}),\end{aligned}$$

where the last equality is again a consequence of Corollary 11.12. This completes the proof. □

The next extremum problem we treat is of a rather different nature and is concerned with an inclusion probability for the typical maximal k-polytope. The goal is to demonstrate how the relation between STIT tessellations and Poisson hyperplane tessellations can be used to translate known results for Poisson hyperplane tessellations to the STIT model. To formulate it, we need to introduce some further concepts and notation. We start by recalling that the surface area measure $S_{d-1}(K, A)$ of a convex body $K \in \mathcal{K}$ evaluated at $A \in \mathcal{B}(S^{d-1})$ is the surface content of the set of all boundary points of K that have an outer unit normal in A. Given an even measure φ on S^{d-1} that is not concentrated on a great subsphere, Minkowski's existence theorem, Theorem 14.3.1 in [172], ensures that there exists a uniquely determined convex body $K \in \mathcal{K}$ such that $S_{d-1}(K, \cdot) = \varphi(\cdot)$. To apply this to STIT tessellations, we fix $t > 0$ and a hyperplane measure Θ of the form (4.1), satisfying (4.3), whose spherical directional distribution is denoted by φ. Then the **Blaschke body** $B_{t\Theta}$ associated with the pair (t,Θ) is defined as the convex body whose surface area measure is given by $t\varphi$: $S_{d-1}(B_{t\Theta},\cdot) = t\varphi(\cdot)$. The Blaschke body $B_{t\Theta}$ is related to the zonoid $\Pi_{t\Theta}$ by the concept of the projection body. We recall that the projection body ΞK of $K \in \mathcal{K}$ is the convex body with support function $h(\Xi K, u) = V_{d-1}(K|u^\perp)$, $u \in S^{d-1}$. Using the projection body, we have that $\Pi_{t\Theta} = \Xi B_{t\Theta}$. Furthermore, for $L \in G(d,k)$, we denote by $B_{t\Theta}^{\cap L}$ the unique convex body in L whose surface area measure with L as ambient space equals $t\pi_L\varphi$, where

$$\pi_L\varphi(A) := \int\limits_{S^{d-1}\setminus L^\perp} \mathbb{1}\left\{\frac{u|L}{\|u|L\|} \in A\right\} \|u|L\|\,\varphi(du), \quad A \in \mathcal{B}(S^{d-1} \cap L),$$

is the spherical projection of φ onto L. Finally, we recall from (11.4) that $\mathbb{Q}_{k,t:h}^{\max}$ stands for the directional distribution of the typical maximal k-polytope of Y_t.

Corollary 11.17. *Let Θ be a hyperplane measure of the form* (4.1), *satisfying* (4.3), *and let $(Y_t,\ t>0)$ be the STIT tessellation process in $\mathbb{R}^d$ driven by Θ. Fix $k \in \{0,\dots,d-1\}$, and for each $L \in G(d,k)$, let K_L be a convex body in L satisfying $V_k(K_L)=1$. Then, for $\mathbb{Q}_{k,t:h}^{\max}$-almost all $L \in G(d,k)$, the probability $p_{d,k,t,\Theta,L}$ that the typical maximal k-polytope of Y_t contains a translate of K_L satisfies*

$$p_{d,k,t,\Theta,L} \le \int\limits_0^t \frac{ds^{d-1}}{t^d} \exp\left(-2k\, V_k(B_{s\Theta}^{\cap L})^{1-\frac{1}{k}}\right) ds,$$

with equality if and only if K_L is homothetic to $B_{t\Theta}^{\cap L}$.

Proof. Denote by $\mathbb{Q}_{k,t|h=L}^{\max}$ the conditional distribution of the typical maximal k-polytope of Y_t, given that its direction is equal to $L \in G(d,k)$. Correspondingly, $\mathbb{Q}_{k,s|h=L}^{\mathrm{PHT}}$ stands for the conditional distribution of the typical k-face of a Poisson hyperplane tessellation X_s, $0<s<t$, with given direction L. Then disintegration of (11.3) with respect to $\mathbb{Q}_{k,t|h}^{\max}$ shows that

$$\mathbb{Q}_{k,t|h=L}^{\max}(A) = \int\limits_0^t \frac{ds^{d-1}}{t^d}\, \mathbb{Q}_{k,s|h=L}^{\mathrm{PHT}}(A)\, ds, \quad A \in \mathcal{B}(\mathcal{P}_L),$$

where $\mathcal{P}_L$ is the space of polytopes in L, and hence

$$p_{d,k,t,\Theta,L} = \int\limits_0^t \frac{ds^{d-1}}{t^d} p_{d,k,s,\Theta,L}^{\mathrm{PHT}}\, ds \tag{11.10}$$

for $\mathbb{Q}_{k,t:h}^{\max}$-almost all $L \in G(d,k)$, where $p_{d,k,s,\Theta,L}^{\mathrm{PHT}}$ is the probability that the typical k-face of a Poisson hyperplane tessellation X_s whose underlying Poisson process has intensity measure $s\,\Theta$ contains a translate of K_L. We can now apply Theorem 3 of [81] to see that

$$p^{\mathrm{PHT}}_{d,k,s,\Theta,L} \le \exp(-2k\, V_k(B^{\cap L}_{s\Theta})^{1-\frac{1}{k}}),$$

where equality holds if and only if K_L is homothetic to $B^{\cap L}_{s\Theta}$. The result thus follows from (11.10). □

In the isotropic case $\Theta = \Theta_{\mathrm{iso}}$, $B^{\cap L}_{s\Theta_{\mathrm{iso}}}$ is a k-dimensional ball in $L \in G(d,k)$ of radius $s^{1/(k-1)}$, which implies $V_k(B^{\cap L}_{s\Theta_{\mathrm{iso}}}) = \kappa_k s^{k/(k-1)}$ and

$$V_k(B^{\cap L}_{s\Theta})^{1-\frac{1}{k}} = \kappa_k^{1-\frac{1}{k}} s,$$

and hence

$$p_{d,k,t,\Theta_{\mathrm{iso}},L} \le \int_0^t \frac{ds^{d-1}}{t^d} \exp(-2k\kappa_k^{1-\frac{1}{k}} s)\, \mathrm{d}s,$$

independent of the choice of $L \in G(d,k)$. For example, if $d = 3$ and $k = 2$, then we can express $p_{3,2,t,\Theta,L}$ as

$$p_{3,2,t,\Theta_{\mathrm{iso}},L} = \frac{3}{t^3}\int_0^t s^2 e^{-as}\, \mathrm{d}s = \frac{3}{t^3}\left[\frac{2}{a^3} - e^{-at}\left(\frac{t^2}{a} + \frac{2t}{a^2} + \frac{2}{a^3}\right)\right], \quad a := 4\sqrt{\pi},$$

whereas the corresponding probability for Poisson hyperplane tessellations is just $p^{\mathrm{PHT}}_{3,2,t,\Theta_{\mathrm{iso}},L} = e^{-at}$.

11.4 Distributional results for the typical I-segment

After having determined the expected intrinsic volumes of the typical k-dimensional maximal polytope and the intensity of maximal k-polytopes of a STIT tessellation for general $k \in \{1,\dots,d-1\}$, we turn now to the special case $k = 1$, that is, we consider the typical maximal 1-polytope. We will refer to the latter as the **typical I-segment**. In this particular situation a detailed description of its distribution $\mathbb{Q}^{\max}_{1,t}$ with respect to a fixed center function c on the space $\mathcal{P}_{1,c}$ of centered line segments in $\mathbb{R}^d$ is available. It is uniquely determined by the distribution of the direction and conditional distribution of the length of the I-segment with a given direction. In what follows, all typical line segments are understood with respect to the same center function c.

Theorem 11.18. *Let Θ be a hyperplane measure of the form* (4.1), *satisfying* (4.3), *and let φ be the associated spherical directional distribution. Let $(Y_t,\ t > 0)$ be the STIT tessellation process driven by Θ. Then, for all nonnegative measurable functions $f : G(d,1) \to \mathbb{R}$, $r \ge 0$, and $t > 0$, we have that*

$$\int_{\mathcal{P}_{1,c}} f(\mathrm{aff}_o F)\,\mathbb{1}\{V_1(F) > r\}\,\mathbb{Q}_{1,t}^{\max}(\mathrm{d}F) = \frac{V_{d-1}(\Pi_\Theta)}{t^d\,V_d(\Pi_\Theta)}$$

$$\times \int_{G(d,1)} \int_0^t s^{d-1} \exp(-r\,s\,V_1(\Pi_\Theta|L))\,\mathrm{d}s\, f(L)\, V_1(\Pi_\Theta|L)\,\chi_1(\mathrm{d}L) \tag{11.11}$$

with χ_1 *given by* (2.80).

Proof. Let $(X_t,\ t > 0)$ be a process of Poisson hyperplane tessellations driven by Θ. According to Corollary 11.10, for any $t > 0$, the directional distribution $\mathbb{Q}_{1,t:h}^{\max}$ on $G(d,1)$ of the typical I-segment of a STIT tessellation Y_t coincides with the directional distribution $\mathbb{Q}_{1,t:h}^{\mathrm{PHT}}$ of the typical edge of a Poisson hyperplane tessellation X_t. Thus

$$\int_{\mathcal{P}_{1,c}} f(\mathrm{aff}_o F)\,\mathbb{1}\{V_1(F) > r\}\,\mathbb{Q}_{1,t}^{\max}(\mathrm{d}F)$$

$$= \int_{G(d,1)} \int_0^\infty f(L)\,\mathbb{1}\{u > r\}\,\mathbb{Q}_{1,t:\ell|h=L}^{\max}(\mathrm{d}u)\,\mathbb{Q}_{1,t:h}^{\max}(\mathrm{d}L)$$

$$= \int_{G(d,1)} \int_0^\infty f(L)\,\mathbb{1}\{u > r\}\,\mathbb{Q}_{1,t:\ell|h=L}^{\max}(\mathrm{d}u)\,\mathbb{Q}_{1,t:h}^{\mathrm{PHT}}(\mathrm{d}L),$$

where $\mathbb{Q}_{1,t:\ell|h=L}^{\max}$ is the conditional length distribution of the typical I-segment of Y_t, given its direction equals $L \in G(d,1)$. According to Theorem 2.38, we have that

$$\mathbb{Q}_{1,s:h}^{\mathrm{PHT}}(A) = \frac{V_{d-1}(\Pi_\Theta)}{d\,V_d(\Pi_\Theta)} \int_A V_1(\Pi_\Theta|L)\,\chi_1(\mathrm{d}L) \tag{11.12}$$

for all $A \in \mathcal{B}(G(d,1))$ and $s > 0$. Moreover, disintegrating (11.3) with respect to $\mathbb{Q}_{1,t:h}^{\mathrm{PHT}} = \mathbb{Q}_{1,t:h}^{\max}$, we get

$$\mathbb{Q}_{1,t:\ell|h=L}^{\max}(B) = \int_0^t \frac{ds^{d-1}}{t^d}\,\mathbb{Q}_{1,s:\ell|h=L}^{\mathrm{PHT}}(B)\,\mathrm{d}s$$

for $B \in \mathcal{B}((0,\infty))$ and $\mathbb{Q}_{1,t:h}^{\max}$-almost all $L \in G(d,1)$, implying that

$$\int_{\mathcal{P}_{1,c}} f(\mathrm{aff}_o F)\,\mathbb{1}\{V_1(F) > r\}\,\mathbb{Q}_{1,t}^{\max}(\mathrm{d}F)$$

$$= \frac{V_{d-1}(\Pi_\Theta)}{t^d V_d(\Pi_\Theta)} \int_0^t s^{d-1} \int_{G(d,1)} f(L)\,\mathbb{Q}_{1,s:\ell|h=L}^{\mathrm{PHT}}((r,\infty))\,V_1(\Pi_\Theta|L)\,\chi_1(\mathrm{d}L)\,\mathrm{d}s\,.$$

We can now use once again Theorem 2.38 to conclude that for given $L \in G(d,1)$, $\mathbb{Q}^{\mathrm{PHT}}_{1,s:\ell|h=L}$ is the exponential distribution with parameter $sV_1(\Pi_\Theta|L)$. Indeed, this follows from the fact that the intersection of X_s with L is a homogeneous Poisson point process on L with intensity $s\,\Theta([e(L)])$, where $e(L)$ is a unit length segment on L, together with the observation that

$$\Theta([e(L)]) = \int_{S^{d-1}} |\langle u(L), u\rangle|\, \varphi(\mathrm{d}u) = V_1(\Pi_\Theta|L)\,, \tag{11.13}$$

where $u(L) \in S^{d-1}$ satisfies $\mathrm{lin}(u) = L$; see (2.69). As a consequence,

$$\begin{aligned}&\int_{\mathcal{P}_{1,c}} f(\mathrm{aff}_o(F))\, \mathbb{1}\{V_1(F) > r\}\, \mathbb{Q}^{\max}_{1,t}(\mathrm{d}F)\\ &\quad= \frac{V_{d-1}(\Pi_\Theta)}{t^d V_d(\Pi_\Theta)} \int_0^t s^{d-1} \int_{G(d,1)} f(L)\, \exp(-r\,s\,V_1(\Pi_\Theta|L))\, V_1(\Pi_\Theta|L)\, \chi_1(\mathrm{d}L)\, \mathrm{d}s\,,\end{aligned}$$

which proves the claim. □

We note that the inner integral in (11.11) with respect to s can be expressed by means of the lower incomplete gamma function $\Gamma(s,x) := \int_0^x a^{s-1}\mathrm{e}^{-a}\,\mathrm{d}a$, $x, s > 0$. Indeed,

$$\int_0^t s^{d-1} \exp(-r\,s\,V_1(\Pi_\Theta|L))\, \mathrm{d}s = \left(\frac{1}{r\,V_1(\Pi_\Theta|L)}\right)^d \Gamma(d, r\,V_1(\Pi_\Theta|L)\,t)\,.$$

Theorem 11.18 shows that for STIT tessellations, the marginal distribution of the length of the typical I-segment is a mixture of exponential distributions. Hence the density and moments of this length distribution can be calculated as the mixture of the respective quantities for exponential distributions. We report here the results for the isotropic case and for the Mondrian distribution.

Corollary 11.19. *In the isotropic case, for $t > 0$, the distribution of the length of the typical I-segment of Y_t has the Lebesgue density*

$$\begin{aligned}q^{\max}_{1:\ell}(r) &= \frac{2\kappa_{d-1}}{t^d \kappa_d} \int_0^t s^d \exp\left(-2\,\frac{\kappa_{d-1}}{d\kappa_d}\,s\,r\right) \mathrm{d}s\\ &= \left(\frac{\kappa_d}{2\,t\,\kappa_{d-1}}\right)^d \left(\frac{d}{r}\right)^{d+1} \Gamma\left(d+1, \frac{2\,\kappa_{d-1}\,t}{d\,\kappa_d}\,r\right), \quad r > 0\,.\end{aligned}$$

Proof. Since $\Pi_{\Theta_{\mathrm{iso}}}$ is a ball of radius $\frac{\kappa_{d-1}}{d\kappa_d}$, $V_1(\Pi_{\Theta_{\mathrm{iso}}}|L) = 2\frac{\kappa_{d-1}}{d\,\kappa_d}$ for all $L \in G(d,1)$, and $V_d(\Pi_{\Theta_{\mathrm{iso}}})^{-1} V_{d-1}(\Pi_{\Theta_{\mathrm{iso}}})\, V_1(\Pi_{\Theta_{\mathrm{iso}}}|L) = d$ follows from (11.6). Thus, choosing $f \equiv 1$ in (11.11) and taking the derivative with respect to r, we can complete the proof. □

In particular, for $d = 2$ and $d = 3$, we obtain

$$q_{1:\ell}^{\max}(r) = \frac{1}{t^2 r^3}\left(\pi^2 - (\pi^2 + 2\pi tr + 2t^2r^2)\,\mathrm{e}^{-\frac{2}{\pi}tr}\right) \qquad \text{for } d = 2\,,$$
$$q_{1:\ell}^{\max}(r) = \frac{3}{t^3 r^4}\left(48 - (48 + 24tr + 6t^2r^2 + t^3r^3)\,\mathrm{e}^{-\frac{1}{2}tr}\right) \quad \text{for } d = 3\,;$$

compare with (16.54) and Corollary 17.12, respectively, as well as with Figure 11.1. Integration of these densities yields that the mean segment lengths are π/t in the planar case $d = 2$ and $3/t$ for $d = 3$. For general space dimensions $d \ge 2$, we obtain

$$\begin{aligned}
\int_0^\infty r q_{1:\ell}^{\max}(r)\,\mathrm{d}r &= \frac{2\,\kappa_{d-1}}{t^d\,\kappa_d}\int_0^\infty r\int_0^t s^d\,\exp\!\left(-2\,\frac{\kappa_{d-1}}{d\kappa_d}\,s\,r\right)\mathrm{d}s\,\mathrm{d}r\\
&= \frac{2\,\kappa_{d-1}}{t^d\,\kappa_d}\int_0^t s^d\int_0^\infty r\,\exp\!\left(-2\,\frac{\kappa_{d-1}}{d\kappa_d}\,s\,r\right)\mathrm{d}r\,\mathrm{d}s\\
&= \frac{2\,\kappa_{d-1}}{t^d\,\kappa_d}\left(\frac{d\kappa_d}{2\,\kappa_{d-1}}\right)^2\int_0^t s^{d-2}\int_0^\infty u\,\mathrm{e}^{-u}\,\mathrm{d}u\,\mathrm{d}s\\
&= \frac{d^2\kappa_d}{2\,(d-1)\kappa_{d-1}\,t}
\end{aligned}$$

for the expected length of the typical I-segment, which is consistent with the special case $k = j = 1$ of Corollary 11.12. In a similar way, for the variance of the length of the typical I-segment, we have

$$\int_0^\infty r^2 q_{1:\ell}^{\max}(r)\,\mathrm{d}r - \left(\int_0^\infty r\,q_{1:\ell}^{\max}(r)\,\mathrm{d}r\right)^2 = \begin{cases} +\infty\,, & d = 2\,,\\ \frac{d^3(d^2-2d+2)\kappa_d^2}{4(d-2)(d-1)^2\kappa_{d-1}^2\,t^2}\,, & d \ge 3\,. \end{cases}$$

For example, in the spatial case $d = 3$, we get the value $24/t^2$. In general, from the explicit length density formula we easily see that for the length of the typical I-segment, only the

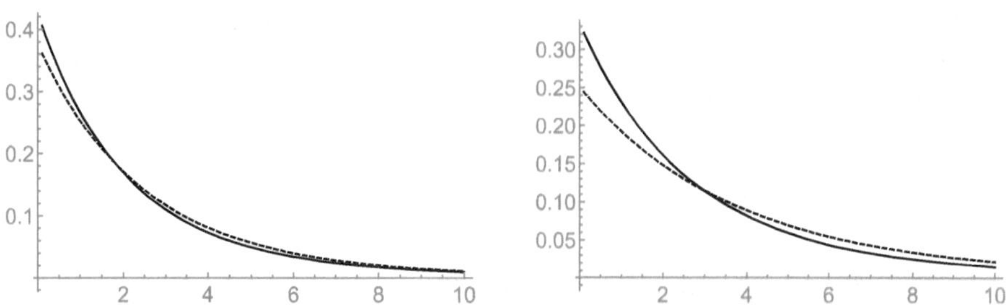

Figure 11.1: The densities $q_{1:\ell}^{\max}(r)$ for $d = 2$ (solid) and $d = 3$ (dashed). Left: Isotropic case. Right: Mondrian distribution. We have chosen $t = 1$.

moments of order 1 up to $d-1$ are finite. This phenomenon persists for STIT tessellations Y_t in $\mathbb{R}^d$ driven by a general hyperplane measure Θ of the form (4.1), satisfying (4.3). Indeed, for $m \geq 0$ and $t > 0$, we have that

$$\begin{aligned}
&\int_0^\infty V_1(F)^m \, \mathbb{Q}_{1,t}^{\max}(\mathrm{d}F) \\
&= m \int_0^\infty r^{m-1} \int_{\mathcal{P}_{1,c}} \mathbb{1}\{V_1(F) > r\} \, \mathbb{Q}_{1,t}^{\max}(\mathrm{d}F) \, \mathrm{d}r \\
&= \frac{m\, V_{d-1}(\Pi_\Theta)}{t^d\, V_d(\Pi_\Theta)} \int_{G(d,1)} V_1(\Pi_\Theta|L) \int_0^t s^{d-1} \int_0^\infty r^{m-1} \mathrm{e}^{-rsV_1(\Pi_\Theta|L)} \, \mathrm{d}r \, \mathrm{d}s \, \chi_1(\mathrm{d}L) \\
&= \frac{m!\, V_{d-1}(\Pi_\Theta)}{t^d\, V_d(\Pi_\Theta)} \int_{G(d,1)} V_1(\Pi_\Theta|L)^{1-m} \int_0^t s^{d-1-m} \, \mathrm{d}s \, \chi_1(\mathrm{d}L),
\end{aligned}$$

where we used Theorem 11.18 and Fubini's theorem. Now we see that the inner integral with respect to s is finite if and only if $m < d$, that is,

$$\int_0^\infty V_1(F)^m \, \mathbb{Q}_{1,t}^{\max}(\mathrm{d}F) = \begin{cases} \frac{m!\, t^{-m}\, V_{d-1}(\Pi_\Theta)}{(d-m) V_d(\Pi_\Theta)} \int_{G(d,1)} V_1(\Pi_\Theta|L)^{1-m} \, \chi_1(\mathrm{d}L), & m < d, \\ +\infty, & m \geq d. \end{cases} \tag{11.14}$$

We turn now to the Mondrian distribution.

Corollary 11.20. *If Θ is the Mondrian hyperplane measure given in* (4.4) *with $p_1 = \cdots = p_d = 1/d$, then the distribution of the length of the typical maximal I-segment of Y_t, $t > 0$, has the Lebesgue density*

$$q_{1:\ell}^{\max}(r) = \frac{1}{t^d} \int_0^t s^d \exp\left(-\frac{r}{d} s\right) \mathrm{d}s = \frac{1}{t^d} \left(\frac{d}{r}\right)^{d+1} \Gamma\left(d+1, \frac{r}{d} t\right), \quad r > 0.$$

Proof. In this case, χ_1 is the discrete uniform distribution on $\{\mathrm{lin}\, e_1, \ldots, \mathrm{lin}\, e_d\}$, and for each of these directions, we have that $V_1(\Pi_\Theta|L) = 1/d$. Furthermore, $V_{d-1}(\Pi_\Theta)/V_d(\Pi_\Theta) = d^2$ by (11.7). The result follows now from (11.11) by choosing $f \equiv 1$ and then taking the derivative with respect to r. □

For $d = 2$ and $d = 3$, this yields the densities

$$q_{1:\ell}^{\max}(r) = \frac{2}{t^2 r^3} (8 - (r^2 t^2 + 4rt + 8)\, \mathrm{e}^{-rt/2}) \quad \text{for } d = 2,$$

$$q_{1:\ell}^{\max}(r) = \frac{3}{t^3 r^4} (162 - (r^3 t^3 + 9r^2 t^2 + 54rt + 162)\, \mathrm{e}^{-rt/3}) \quad \text{for } d = 3;$$

see Figure 11.1. For the expected length of the typical I-segment, in dimension $d \geq 2$, we obtain

$$\begin{aligned}\int_0^\infty r\, q_{1:\ell}^{\max}(r)\,\mathrm{d}r &= \frac{1}{t^d}\int_0^\infty r \int_0^t s^d \exp\Big(-\frac{r}{d}s\Big)\,\mathrm{d}s\,\mathrm{d}r \\ &= \frac{1}{t^d}\int_0^t s^d \int_0^\infty r\exp\Big(-\frac{r}{d}s\Big)\,\mathrm{d}r\,\mathrm{d}s \\ &= \frac{d^2}{t^d}\int_0^t s^{d-2}\int_0^\infty u\,\mathrm{e}^{-u}\,\mathrm{d}u\,\mathrm{d}s = \frac{d^2}{(d-1)\,t}\end{aligned}$$

in accordance with Corollary 11.12. For example, for $d=2$ and $d=3$, we get the mean values $4/t$ and $9/(2t)$, respectively. Moreover, for the variance, we similarly arrive at

$$\int_0^\infty r^2 q_{1:\ell}^{\max}(r)\,\mathrm{d}r - \Bigg(\int_0^\infty r\,q_{1:\ell}^{\max}(r)\,\mathrm{d}r\Bigg)^2 = \begin{cases} +\infty\,, & d=2\,,\\ \frac{d^3(d^2-2d+2)}{(d-2)(d-1)^2t^2}\,, & d\geq 3\,,\end{cases}$$

with the specific value $135/(4t^2)$ for $d=3$.

Remark 11.21. *Let us denote the length density in Corollary* 11.19 *or Corollary* 11.20 *by* $q_{1:\ell}^{\max}(r;t)$ *to emphasize its dependence on the time parameter* t*. These densities satisfy the scaling identity*

$$q_{1:\ell}^{\max}(r;t) = t\,q_{1:\ell}^{\max}(r\,t;1)\,,$$

which expresses the self-similarity of STIT tessellations under the time-space scaling $t\,Y_t \overset{d}{=} Y_1$*. Consequently, all moments of the I-segment length distribution scale as* t^{-m} *for the* m*th moment.*

11.5 Relationships for birth time marked maximal $(d-1)$-polytopes

In Section 11.2, we have seen that the distribution of the typical maximal $(d-1)$-polytope of a STIT tessellation in $\mathbb{R}^d$ can be represented as a mixture of distributions of typical $(d-1)$-dimensional faces of Poisson hyperplane tessellations. In the same spirit, we will now establish a similar representation for the birth time marked version. We start with the following birth time marked analogue to Theorem 11.5 for $k=d-1$.

Theorem 11.22. *Let* $W \in \mathcal{P}_d$, Θ *be a hyperplane measure of the form* (4.1), *satisfying* (4.3), *and let* $(Y_{t,W},\ t\geq 0)$ *be the STIT tessellation process in* W *driven by* Θ*. Furthermore, let*

$(X_t, t > 0)$ be the process of Poisson hyperplane tessellations driven by Θ. For $t > 0$, let $\phi^* : \mathcal{P}_{d-1,W} \times [0,t] \to \mathbb{R}$ be a bounded and measurable function, and define

$$\Sigma_{\phi^*,t,W} := \sum_{(F,s)\in\mathcal{F}^{\max*}_{d-1,W}} \mathbb{1}\{s \le t\}\,\phi^*(F,s)$$

Then

$$\mathbb{E}\Sigma_{\phi^*,t,W} = \int_0^t \frac{1}{s}\,\mathbb{E}\sum_{F\in\mathcal{F}_{d-1}(X_s\wedge W)} \phi^*(F,s)\,\mathrm{d}s\,.$$

Remark 11.23. *The existence of the integral in Theorem 11.22 is guaranteed, since ϕ^* is bounded and the expected number of facets of $X_s \wedge W$ is almost surely finite for bounded W.*

Proof of Theorem 11.22. We can argue as in the proof of Theorem 11.5. Indeed, we first use Theorem 10.9 and then Theorem 11.1 to see that

$$\begin{aligned}
\mathbb{E}\Sigma_{\phi^*,t,W} &= \mathbb{E}\int_0^t \int_{[W]} \sum_{F\in Y_{s,W}\cap H} \phi^*(F,s)\,\Theta(\mathrm{d}H)\,\mathrm{d}s \\
&= \int_0^t \int_{[W]} \mathbb{E}\sum_{P\in Y_{s,W}} \mathbb{1}\{H\in[P]\} \sum_{F\in\mathcal{F}_{d-1}(P\cap H)} \phi^*(F,s)\,\Theta(\mathrm{d}H)\,\mathrm{d}s \\
&= \int_0^t \int_{[W]} \mathbb{E}\sum_{P\in Y_s\wedge W} \mathbb{1}\{H\in[P]\} \sum_{F\in\mathcal{F}_{d-1}(P\cap H)} \phi^*(F,s)\,\Theta(\mathrm{d}H)\,\mathrm{d}s \\
&= \int_0^t \int_{[W]} \mathbb{E}\sum_{P\in X_s\wedge W} \mathbb{1}\{H\in[P]\} \sum_{F\in\mathcal{F}_{d-1}(P\cap H)} \phi^*(F,s)\,\Theta(\mathrm{d}H)\,\mathrm{d}s \\
&= \int_0^t \frac{1}{s}\int_{[W]} \mathbb{E}\sum_{P\in X_s\wedge W} \mathbb{1}\{H\in[P]\} \sum_{F\in\mathcal{F}_{d-1}(P\cap H)} \phi^*(F,s)\,(s\,\Theta)(\mathrm{d}H)\,\mathrm{d}s \\
&= \int_0^t \frac{1}{s}\,\mathbb{E}\sum_{F\in\mathcal{F}_{d-1}(X_s\wedge W)} \phi^*(F,s)\,\mathrm{d}s\,,
\end{aligned}$$

where the last line follows by the Mecke equation in Theorem 19.11 applied to the Poisson hyperplane process with intensity measure $s\,\Theta$. This completes the argument. □

We can use the previous result to deduce the announced relation for the typical birth time marked maximal $(d-1)$-polytope. For $t > 0$, consider the point process $\mathcal{F}^{\max*}_{d-1}(Y_t)$ of birth time marked maximal $(d-1)$-polytopes of the STIT tessellation process

$(Y_s, s > 0)$, driven by a hyperplane measure Θ, which are born until time t. The typical element of this process with respect to some fixed spatial center function $c : \mathcal{P}_{d-1} \to \mathbb{R}^d$ is called the typical birth time marked maximal $(d-1)$-polytope of Y_t, and its distribution is denoted by $\mathbb{Q}^{\max *}_{d-1,t}$. Formally, for $A^* \in \mathcal{B}(\mathcal{P}_{d-1,c} \times \mathbb{R}^+)$,

$$\mathbb{Q}^{\max *}_{d-1,t}(A^*) = \frac{1}{\gamma^{\max}_{d-1,t}} \mathbb{E} \sum_{(F,s) \in \mathcal{F}^{\max *}_{d-1}} \mathbb{1}\{s \le t\} \mathbb{1}\{c(F) \in B\} \mathbb{1}\{(F - c(F), s) \in A^*\},$$

where $B \subset \mathbb{R}^d$ is an arbitrary Borel set of unit volume. We also recall that $\mathbb{Q}^{\mathrm{PHT}}_{d-1,s}$ stands for the distribution of the typical $(d-1)$-dimensional face of a Poisson hyperplane tessellation in $\mathbb{R}^d$ with intensity measure $s\,\Theta$ of the underlying hyperplane process, with respect to the same center function c. With this notation, we obtain the birth time marked extension of Theorem 11.8 for $k = d-1$.

Corollary 11.24. *Let Θ be a hyperplane measure of the form (4.1), satisfying (4.3). Let $(Y_t, t > 0)$ be the STIT tessellation process driven by Θ. Furthermore, let $(X_t, t > 0)$ be the process of Poisson hyperplane tessellations driven by Θ. The distribution $\mathbb{Q}^{\max *}_{d-1,t}$ of the typical birth time marked $(d-1)$-dimensional maximal polytope of Y_t satisfies*

$$\mathbb{Q}^{\max *}_{d-1,t}(A^*) = \int_0^t \frac{ds^{d-1}}{t^d} (\mathbb{Q}^{\mathrm{PHT}}_{d-1,s} \otimes \delta_s)(A^*)\, \mathrm{d}s$$

for $A^* \in \mathcal{B}(\mathcal{P}_{d-1,c} \times \mathbb{R}^+)$.

Proof. The argument is the same as that in the proof of Theorem 11.8. In particular, for $A \in \mathcal{B}(\mathcal{P}_{d-1,c})$ and $B \in \mathcal{B}(\mathbb{R}^+)$, we have at

$$\begin{aligned}
&\gamma^{\max}_{d-1,t}\, \mathbb{Q}^{\max *}_{d-1,t}(A \times B) \\
&\quad = \lim_{a\to\infty} \frac{1}{V_d(aW)} \mathbb{E} \sum_{(F,s) \in \mathcal{F}^{\max *}_{d-1,aW}} \mathbb{1}\{s \le t\} \mathbb{1}\{F \cap \partial(aW) = \emptyset\} \\
&\qquad \times \mathbb{1}\{F - c(F) \in A,\ s \in B\} \\
&\quad = \lim_{a\to\infty} \frac{1}{V_d(aW)} \int_0^t \frac{1}{s} \mathbb{E} \sum_{F \in \mathcal{F}_{d-1}(X_s \wedge aW)} \mathbb{1}\{F \cap \partial(aW) = \emptyset\} \\
&\qquad \times \mathbb{1}\{F - c(F) \in A,\ s \in B\}\, \mathrm{d}s \\
&\quad = \int_0^t \frac{1}{s} \mathbb{1}\{s \in B\} \lim_{a\to\infty} \frac{1}{V_d(aW)} \\
&\qquad \times \mathbb{E} \sum_{F \in \mathcal{F}_{d-1}(X_s \wedge aW)} \mathbb{1}\{F \cap \partial(aW) = \emptyset\}\, \mathbb{1}\{F - c(F) \in A\}\, \mathrm{d}s
\end{aligned}$$

$$= \int_0^t \frac{1}{s}\, \mathbb{1}\{s \in B\}\, \gamma_{d-1,s}^{\mathrm{PHT}}\, \mathbb{Q}_{d-1,s}^{\mathrm{PHT}}(A)\, \mathrm{d}s$$

$$= \int_0^t \frac{1}{s} \gamma_{d-1,s}^{\mathrm{PHT}} (\mathbb{Q}_{d-1,s}^{\mathrm{PHT}} \otimes \delta_s)(A \times B)\, \mathrm{d}s\,.$$

Since the σ-field $\mathcal{B}(\mathcal{P}_{d-1,c} \times \mathbb{R}^+)$ is generated by sets of the form $A \times B$ with $A \in \mathcal{B}(\mathcal{P}_{d-1,c})$ and $B \in \mathcal{B}(\mathbb{R}^+)$, the proof is complete. □

The following result describes the density of the birth time distribution of the typical maximal I-segment. The result will later also be a special case of Corollary 14.13(a), which is derived by different methods.

Corollary 11.25. *The birth time β of the typical maximal $(d-1)$-polytope of a STIT tessellation Y_t with $t > 0$ has the density*

$$q_{1:\beta}^{\max}(s) = \frac{ds^{d-1}}{t^d}\,, \qquad s \in (0,t)\,.$$

In particular, the expected birth time and its variance are

$$\mathbb{E}\beta = \frac{d}{d+1}\, t \quad \text{and} \quad \operatorname{var}(\beta) = \frac{d}{(d+2)(d+1)^2}\, t^2\,.$$

Proof. Take $A^* = \mathcal{P}_{d-1,c} \times [0,u]$ with $u > 0$ in Corollary 11.24. This gives

$$\mathbb{P}(\beta \le u) = \mathbb{Q}_{d-1,t}^{\max *}(\mathcal{P}_{d-1,c} \times [0,u]) = \int_0^u \frac{ds^{d-1}}{t^d}\, \mathrm{d}s\,,$$

and so β has the density $\frac{ds^{d-1}}{t^d}$ on $(0,t)$, from which the formulae for $\mathbb{E}\beta$ and $\operatorname{var}(\beta)$ immediately follow. □

Finally, in the isotropic case, we determine the joint distribution of length and birth time of the typical maximal I-segment. Also, several conditional distributions are determined.

Corollary 11.26. *Let Y_t be a stationary and isotropic STIT tessellation in $\mathbb{R}^d$ with $t > 0$. Denote by ℓ the length of the typical I-segment and by β its birth time. Then (ℓ, β) has the joint density*

$$q_{1:\ell,\beta}^{\max}(r,s) = \frac{2\kappa_{d-1}}{\kappa_d}\, \frac{s^d}{t^d}\, \exp\!\left(-\frac{2\kappa_{d-1}}{d\kappa_d}\, rs\right), \quad r > 0\,, \quad 0 < s < t\,.$$

Moreover, for the conditional density of ℓ given $\beta = s \in (0,t)$, we have

$$q_{1:\ell|\beta=s}^{\max}(r) = \frac{2s\kappa_{d-1}}{d\kappa_d}\, \exp\!\left(-\frac{2s\kappa_{d-1}}{d\kappa_d}\, r\right), \quad r > 0\,.$$

Finally, the conditional density of β given $\ell = r > 0$ is the truncated gamma density on $(0,t)$ with shape $d+1$ and rate $\frac{2\kappa_{d-1}}{d\,\kappa_d}\,r$, namely

$$q^{\max}_{1:\beta|\ell=r}(s) = \frac{s^d}{\Gamma(d+1, \frac{2\kappa_{d-1}}{d\,\kappa_d}\,rt)}\left(\frac{2\kappa_{d-1}}{d\,\kappa_d}\,r\right)^{d+1}\exp\left(-\frac{2r\kappa_{d-1}}{d\kappa_d}\,s\right), \quad 0 < s < t\,,$$

where $\Gamma(\cdot\,,\cdot)$ denotes the lower incomplete gamma function.

Proof. By Corollary 11.25 the birth time β of the typical I-segment has the density

$$q^{\max}_{1:\beta}(s) = \frac{ds^{d-1}}{t^d}\,, \quad 0 < s < t\,. \tag{11.15}$$

Using this in conjunction with Corollary 11.24, we see that, conditionally on $\beta = s$, the distribution of the typical I-segment coincides with that of the typical edge in a Poisson hyperplane tessellation driven by $s\,\Theta_{\mathrm{iso}}$. From the discussion before (11.13) in the proof of Theorem 11.18 it follows that its length has the exponential distribution with parameter $s\,V_1(\Pi_{\Theta_{\mathrm{iso}}}|L) = \frac{2\kappa_{d-1}}{d\kappa_d}\,s$, where $L \in G(d,1)$ is an arbitrary line. Thus

$$q^{\max}_{1:\ell|\beta=s}(r) = \frac{2s\kappa_{d-1}}{d\kappa_d}\exp\left(-\frac{2s\kappa_{d-1}}{d\kappa_d}\,r\right), \quad r > 0\,. \tag{11.16}$$

Since in the isotropic case the conditional law of ℓ given $\beta = s$ is independent of the direction, we obtain the joint density of (ℓ,β) by multiplication of (11.15) and (11.16):

$$\begin{aligned} q^{\max}_{1:\ell,\beta}(r,s) &= q^{\max}_{1:\ell|\beta=s}(r)\,q^{\max}_{1:\beta}(s) \\ &= \left(\frac{2s\kappa_{d-1}}{d\kappa_d}\exp\left(-\frac{2s\kappa_{d-1}}{d\kappa_d}\,r\right)\right)\left(\frac{ds^{d-1}}{t^d}\right) \\ &= \frac{2\kappa_{d-1}}{\kappa_d}\frac{s^d}{t^d}\exp\left(-\frac{2\kappa_{d-1}}{d\kappa_d}\,rs\right) \end{aligned}$$

for $r > 0$ and $0 < s < t$, as claimed. Integrating this over $s \in (0,t)$ yields the length density

$$q^{\max}_{1:\ell}(r) = \frac{2\kappa_{d-1}}{\kappa_d t^d}\int_0^t s^d\exp\left(-\frac{2s\kappa_{d-1}}{d\kappa_d}\,r\right)ds\,.$$

Thus

$$q^{\max}_{1:\beta|\ell=r}(s) = \frac{q^{\max}_{1:\ell,\beta}(r,s)}{q^{\max}_{1:\ell}(r)}\,, \quad 0 < s < t\,,$$

which is the stated truncated gamma density. □

Specializing to $d = 2$, we have

$$\begin{aligned} q^{\max}_{1:\ell,\beta}(r,s) &= \frac{4}{\pi}\frac{s^2}{t^2}\exp\Big(-\frac{2}{\pi}rs\Big), && r>0,\quad 0<s<t,\\ q^{\max}_{1:\ell\,|\,\beta=s}(r) &= \frac{2s}{\pi}\exp\Big(-\frac{2s}{\pi}r\Big), && r>0,\\ q^{\max}_{1:\beta\,|\,\ell=r}(s) &= \frac{(\frac{2r}{\pi})^3\,s^2}{\Gamma(3,\frac{2r}{\pi}t)}\exp\Big(-\frac{2r}{\pi}s\Big), && 0<s<t,\\ q^{\max}_{1:\beta}(s) &= \frac{2s}{t^2}, && 0<s<t, \end{aligned}$$

where $\Gamma(3, \frac{2r}{\pi}t) = 2 - 2e^{-\frac{2rt}{\pi}}(1 + \frac{2rt}{\pi} + \frac{2r^2t^2}{\pi^2})$.

For $d = 3$, we obtain

$$\begin{aligned} q^{\max}_{1:\ell,\beta}(r,s) &= \frac{3}{2}\frac{s^3}{t^3}\exp\Big(-\frac{1}{2}rs\Big), && r>0,\quad 0<s<t,\\ q^{\max}_{1:\ell\,|\,\beta=s}(r) &= \frac{s}{2}\exp\Big(-\frac{s}{2}r\Big), && r>0,\\ q^{\max}_{1:\beta\,|\,\ell=r}(s) &= \frac{(\frac{r}{2})^4\,s^3}{\Gamma(4,\frac{r}{2}t)}\exp\Big(-\frac{r}{2}s\Big), && 0<s<t,\\ q^{\max}_{1:\beta}(s) &= \frac{3s^2}{t^3}, && 0<s<t, \end{aligned}$$

with $\Gamma(4, \frac{r}{2}t) = 6 - 6e^{-\frac{rt}{2}}(1 + \frac{rt}{2} + \frac{1}{2}(\frac{rt}{2})^2 + \frac{1}{6}(\frac{rt}{2})^3)$.

We remark that the topic of birth time marked maximal polytopes will be picked up again in Section 14.5. In particular, in Lemma 14.16, we will prove by different methods an extension of Corollary 11.24 for typical maximal polytopes of arbitrary dimension $k \in \{1, \dots, d-1\}$ and even allow for an additional weighting by their intrinsic volumes.

11.6 Notes

1. The presentation in this chapter mainly follows the paper [177]. However, the argument leading to a result analogous to Theorem 11.1 relied on the uniqueness of the solution to an operator integral equation, a property common to both STIT tessellations and Poisson hyperplane tessellations. Some technicals details left out in [177] were later added in [83], where a similar relationship was derived for the cell intensity measure of a so-called splitting tessellation on the d-dimensional unit sphere. The latter can be regarded as the spherical analogue of a STIT tessellation in $\mathbb{R}^d$.
2. In the planar case $d = 2$ and in the spatial case $d = 3$, the mean values for the length of the typical I-segment are in accordance with the values obtained in [128, 146] and [187, 189], respectively. The method there was based on the stochastic stability of STIT tessellation under the operation of rescaled iteration of tessellations. Combined with general intersection formulae for random stationary tessellations, this results in a difference equation that can ultimately be solved explicitly. How-

ever, this method appears to be restricted to low-dimensional cases. This approach is described also in Section 16.2 for the planar and Section 17.2.3 for the spatial case.

3. The length density $q_{1:\ell}^{\max}(r)$ of the typical I-segment in a planar stationary and isotropic STIT tessellation has been calculated for the first time in [128] by an entirely different method based on Palm theory and the stochastic stability of STIT tessellation under the operation of rescaled iteration. We refer the reader to Section 16.3, where this is explained. Corresponding results for the spatial case $d = 3$ have been found by similar arguments in [189] and are described in detail in Section 17.3.2. However, this approach appears to be limited to I-segments and does not extend to higher-dimensional maximal polytopes, as in Theorem 11.8. The reason is that the argument would essentially require that the full distribution of the volume of the typical cell in a stationary Poisson hyperplane tessellation is available. The mathematical analysis of this distribution remains a longstanding open problem in stochastic geometry and is currently only approachable through simulation studies.
4. The associated zonoid and the closely related Blaschke body have been widely used to study geometric extreme value problems, particularly for Poisson hyperplane tessellations. Using methods from Fourier analysis, Davidson [68] obtained the result of Corollary 11.15 for $d = 2$ and $k = 1$ with the STIT tessellation replaced by a Poisson hyperplane tessellation. The general case goes back to the work [190], which also uses the isoperimetric inequality for convex bodies. We emphasize that the results presented in Section 11.3.3, particularly Corollary 11.17, are selective and should be viewed as representative examples of extremum results obtainable via the associated zonoid or the Blaschke body, in conjunction with the link between STIT tessellations and Poisson hyperplane tessellations, see [82].
5. The problem of exceedances and order statistics for the inradii of cells in a stationary and isotropic STIT tessellation in the plane was treated in [31]. Using the Chen–Stein method for Poisson approximation it was shown that, for an appropriate choice of the growth of the observation window and the threshold for the exceedances, the number of exceedances is asymptotically Poisson distributed. More precisely, for fixed $t > 0$ and $\tau > 0$, define $v_\varrho(\tau) := \frac{1}{2t}(\log\varrho - \log\tau)$ and $W_\varrho := \frac{\sqrt{\pi\varrho}}{t}[-\frac{1}{2}, \frac{1}{2}]^2$, $\varrho > 0$. Then consider the random variable

$$N_\varrho := \sum_{P\in Y_t} \mathbb{1}\{c(P) \in W_\varrho, R(P) > v_\varrho(\tau)\}, \quad \varrho > 0,$$

where $c(P)$ is the a. s. uniquely defined center of the largest disc inside of P, and $R(P)$ denotes its radius. Theorem 1.1 of [31] states that, as $\varrho \to \infty$, the random variable N_ϱ converges in distribution and in the total-variation distance to a Poisson random variable with mean τ. If M_ϱ denotes the largest inradius of a cell of Y_t with center in W_ϱ, then the choice $\tau = \mathrm{e}^{-u}$ directly implies that the distribution of M_ϱ converges to a Gumbel distribution after suitable rescaling:

$$\lim_{\varrho\to\infty} \mathbb{P}(2tM_\varrho - \log\varrho \le u) = \mathrm{e}^{-\mathrm{e}^{-u}}, \quad u \in \mathbb{R}.$$

These results match those for stationary and isotropic Poisson line tessellations in [30]. In the light of Theorem 11.1, this phenomenon might not come as a surprise. However, transferring results on the distribution of exceedances from Poisson hyperplane tessellations to STITs is far from trivial, as spatial mixing properties must also be considered. Since STIT tessellations have stronger spatial mixing properties than Poisson hyperplane tessellations as we show in Section 15.3, it is natural to expect that the respective results for Poisson hyperplane tessellations also hold for the STIT model. We expect that also the general results of [159] for Poisson hyperplane tessellations in arbitrary dimensions and for various other size functionals can be transferred to STIT tessellations.

12 Second-order analysis

The purpose of this chapter is a detailed second-order analysis of the total intrinsic volumes of the maximal $(d-1)$-polytopes, with a special focus on the total surface content. More precisely, we provide explicit asymptotic and nonasymptotic expressions for their variances and covariances. Our main device is the martingale property of various random processes associated with a STIT tessellation process established in Chapter 10, which is combined with integral-geometric tools. For first-order moments, which we briefly consider at the beginning of this chapter for completeness, the classical Crofton formula is sufficient, whereas for second-order moments, we need an integral-geometric transformation formula of Blaschke–Petkantschin type. Since these geometric tools require the motion-invariant hyperplane measure, many of the considerations in this chapter are restricted to stationary and isotropic STIT tessellations. Besides second-order moments, also the correlation structure of the random surface measure induced by a stationary and isotropic STIT tessellation is investigated in detail. On this level, we develop a link between the skeleton of a STIT tessellation and a particular Boolean model of $(d-1)$-polytopes. Finally, for stationary and isotropic STIT tessellations in the plane, we investigate the second-order structure of the vertex point process as well.

12.1 Geometric functionals in general convex windows

The construction of the STIT tessellation process $(Y_{t,W}, t \geq 0)$ in a window has been defined for polytopes $W \in \mathcal{P}_d$. By the spatial consistency property described in Chapter 6, or alternatively the global construction of Chapter 8, there is a stochastic process $(Y_t, t > 0)$ of tessellations in $\mathbb{R}^d$ such that $(Y_{t,W}, t > 0) \overset{D}{=} (Y_t \wedge W, t > 0)$ for all $W \in \mathcal{P}_d$. In the present chapter, we also consider the restrictions $(Y_t \wedge W, t > 0)$ of the STIT process $(Y_t, t > 0)$ to general d-dimensional convex domains $W \in \mathcal{K}_d$. By slight abuse of notation, let us write

$$\mathcal{F}_{d-1,W}^{\max*} := \{(F \cap W, s) : \exists (F, s) \in \mathcal{F}_{d-1}^{\max*}, F \cap \operatorname{int} W \neq \emptyset\}$$

for the collection of birth time marked traces of all $(d-1)$-dimensional maximal polytopes having nonempty intersection with the interior of W. By spatial consistency of STIT tessellations this definition coincides with the earlier one in (4.10). We emphasize that although the sets F for members $(F, s) \in \mathcal{F}_{d-1,W}^{\max*}$ are still convex subsets of $\operatorname{aff} F$, they are not necessarily $(d-1)$-dimensional polytopes. The latter property holds if W satisfies $W \in \mathcal{P}_d$. However, we emphasize that the intrinsic volumes of such F are still well-defined in this setup. Putting

$$\mathcal{F}_{d-1,W}^{\max*}(Y_t) := \{(F, s) \in \mathcal{F}_{d-1,W}^{\max*} : s \leq t\}$$

https://doi.org/10.1515/9783112222409-014

for $t > 0$ and $W \in \mathcal{K}_d$, this allows us to extend the definitions of the geometric functionals $\Sigma^f_{V_j,t,W}$, $j \in \{0,\dots,d-1\}$, originally introduced for $W \in \mathcal{P}_d$ in (10.31), to the general case $W \in \mathcal{K}_d$ as follows:

$$\Sigma^f_{V_j,t,W} := \sum_{(F,s)\in\mathcal{F}^{\max*}_{d-1,W}(Y_t)} V_j(F)f(F^\perp)\,, \tag{12.1}$$

where $f : G(d,1) \to \mathbb{R}$ is bounded and measurable. It is convenient to put $\Sigma^f_{V_j,0,W} := 0$. The following result will allow us to transfer the results for these functionals obtained for polytopal windows $W \in \mathcal{P}_d$ to $W \in \mathcal{K}_d$ by an approximation argument.

Lemma 12.1. *Let Θ be a hyperplane measure of the form* (4.1)*, satisfying* (4.3)*, let $(Y_t,\, t > 0)$ be the STIT tessellation process driven by Θ, and let $f : G(d,1) \to \mathbb{R}$ be a bounded and measurable function, $j \in \{0,\dots,d-1\}$, and $t > 0$. Furthermore, fix $W \in \mathcal{K}_d$, and let $(W_n,\, n \in \mathbb{N})$ be a sequence of polytopes $W_n \in \mathcal{P}_d$ satisfying $W_n \subseteq W \subseteq W_n + \frac{1}{n}B^d$. Then*

$$\lim_{n\to\infty} \sum_{(F,s)\in\mathcal{F}^{\max*}_{d-1}(Y_t)} V_j(F\cap W_n)f(F^\perp) = \sum_{(F,s)\in\mathcal{F}^{\max*}_{d-1}(Y_t)} V_j(F\cap W)f(F^\perp) \tag{12.2}$$

a. s. and in L^p for any $p \geq 1$.

Proof. Given $W \in \mathcal{K}_d$, the existence of a sequence $(W_n,\, n \in \mathbb{N})$ of polytopes $W_n \in \mathcal{P}_d$ satisfying $W_n \subseteq W \subseteq W_n + \frac{1}{n}B^d$ is ensured by Theorem 1.8.16 of [171]. Moreover, the intrinsic volumes V_j are continuous functionals on $\mathcal{K}$ with respect to the Hausdorff metric. Furthermore, the stationarity of the STIT tessellation Y_t yields that $\{F\cap W : \exists (F,s) \in \mathcal{F}^{\max*}_{d-1},\, F\cap \operatorname{int} W \neq \emptyset\} = \{F\cap W : \exists (F,s) \in \mathcal{F}^{\max*}_{d-1},\, F\cap W \neq \emptyset\}$ a. s. This implies the a. s. convergence in (12.2).

To show the L^p-convergence for $p \geq 1$, we observe that

$$\begin{aligned}\mathbb{E}&\Big|\sum_{(F,s)\in\mathcal{F}^{\max*}_{d-1}(Y_t)} V_j(F\cap W_n)f(F^\perp)\Big|^p\\ &\leq \Big(\sup_{L\in G(d,1)} |f(L)|^p\Big)\Big(\sup_{H\in[W_n]} V_{d-1}(W_n\cap H)^p\Big)\,\mathbb{E}|\mathcal{F}^{\max*}_{d-1,W_n}(Y_t)|^p < \infty\,,\end{aligned}$$

where we bounded $\mathbb{E}|\mathcal{F}^{\max*}_{d-1,W_n}(Y_t)|^p$ using (4.18), noting that the number of cells of Y_{t,W_n} is exactly one more than the number of maximal $(d-1)$-polytopes in $\mathcal{F}^{\max*}_{d-1,W_n}(Y_t)$. Using $W_n \subseteq W$ for $n \in \mathbb{N}$, it follows that

$$\sup_{n\in\mathbb{N}} \mathbb{E}|\Sigma^f_{V_j,t,W_n}|^p < \infty\,. \tag{12.3}$$

As this holds for all $p \geq 1$, it follows that the sequence of random variables $|\Sigma^f_{V_j,t,W_n}|^p$, $n \in \mathbb{N}$, is uniformly integrable for all $p \geq 1$. Since a. s. convergence implies convergence in probability, Theorem 5.12 of [89] now yields the desired L^p-convergence in (12.2). □

As anticipated above, the previous result allows us to formulate results concerning first- and second-order moments, such as $\mathbb{E}\Sigma_{V_j,t,W}$, $\operatorname{var}\Sigma_{V_j,t,W}$ and $\operatorname{cov}(\Sigma_{V_i,t,W},\Sigma_{V_j,t,W})$ for $i,j \in \{0,\dots,d-1\}$, not only for polytopes $W \in \mathcal{P}_d$, but also for general convex windows $W \in \mathcal{K}_d$, and especially for balls.

12.2 Mean total intrinsic volumes

Let $(Y_t,\ t>0)$ be a STIT tessellation process driven by some hyperplane measure Θ. We start by recalling that, as in (12.1) with $f \equiv 1$, we write $\Sigma_{V_j,t,W}$, $j \in \{0,\dots,d-1\}$, for the sum over the intrinsic volumes of order j of the intersections of all maximal $(d-1)$-polytopes of Y_t, $t>0$, with $W \in \mathcal{K}_d$. In particular, $\Sigma_{V_0,t,W}$ is the number of maximal $(d-1)$-polytopes having nonempty intersection with the interior of W, whereas $\Sigma_{V_{d-1},t,W}$ is the total surface content Y_t leaves within W. It will turn out to be convenient to formally extend this definition to the case $j=d$ by putting $\Sigma_{V_d,t,W} := 0$ for all $t>0$.

Our first result provides an expression for $\mathbb{E}\Sigma_{V_j,t,W}$ if $\Theta = \Theta_{\mathrm{iso}}$ is the motion-invariant hyperplane measure or, equivalently, if the STIT tessellation is stationary and isotropic.

Theorem 12.2. *Let $W \in \mathcal{K}_d$, and let $(Y_t,\ t>0)$ be the STIT tessellation process driven by the motion-invariant hyperplane measure Θ_{iso}. Then, for $t>0$ and $j \in \{0,\dots,d-1\}$,*

$$\mathbb{E}\,\Sigma_{V_j,t,W} = \sum_{\ell=1}^{d-j} \frac{t^\ell}{\ell!}\left(\prod_{i=0}^{\ell-1} c_{j+i,d}\right) V_{j+\ell}(W) \tag{12.4}$$

with the constants $c_{j,d}$ introduced in (1.6).

Proof. By the martingale property of the centered stochastic process in Corollary 10.12(c), from (10.42) we first obtain

$$\begin{aligned}\mathbb{E}\,\Sigma_{V_j,t,W} &= \mathbb{E}\int_0^t \mathbb{L}\Sigma_{V_j,s,W}\,\mathrm{ds}\\ &= c_{j,d}\int_0^t \mathbb{E}\,\Sigma_{V_{j+1},s,W} + V_{j+1}(W)\,\mathrm{ds}\\ &= t\,c_{j,d}\,V_{j+1}(W) + c_{j,d}\int_0^t \mathbb{E}\,\Sigma_{V_{j+1},s,W}\,\mathrm{ds}\,.\end{aligned}$$

We may now recursively apply this argument to $\mathbb{E}\,\Sigma_{V_{j+1},s,W}$ and to the subsequently appearing terms until we arrive at the expression $\mathbb{E}\Sigma_{V_d,t,W}$ under the integral. However, $\Sigma_{V_d,t,W} \equiv 0$, implying that

$$\mathbb{E}\,\Sigma_{V_j,t,W} = t\,c_{j,d}\,V_{j+1}(W) + c_{j,d}\int_0^t s_1\,c_{j+1,d}\,V_{j+2}(W)\,\mathrm{d}s_1 + \cdots$$
$$+\, c_{j,d}\cdots c_{d-2,d}\int_0^t\int_0^{s_1}\cdots\int_0^{s_{d-j-1}} s_{d-j}\,c_{d-1,d}\,V_d(W)\,\mathrm{d}s_{d-j}\ldots\mathrm{d}s_1\,.$$

This yields the statement. □

We remark that taking $j = d-1$ in the previous theorem, we see that

$$\mathbb{E}\,\Sigma_{V_{d-1},t,W} = t\,c_{d-1,d}\,V_d(W) = t\,V_d(W)\,,$$

which coincides with the result (6.6) in the isotropic case.

Given (12.4), we may also consider the asymptotics of $\mathbb{E}\,\Sigma_{V_j,t,aW}$ for general $j \in \{0,\ldots,d-1\}$ as $a \to \infty$. Clearly, the term with $\ell = d-j$ in (12.4) is dominating the sum by the homogeneity property of intrinsic volumes. Thus, using the notation $\sim$ for asymptotic equivalence as introduced in Chapter 1, in the isotropic case, we have that

$$\begin{aligned}\mathbb{E}\,\Sigma_{V_j,t,aW} &\sim \frac{t^{d-j}}{(d-j)!}\left(\prod_{i=0}^{d-j-1} c_{j+i,d}\right)V_d(aW)\\ &= \frac{a^d t^{d-j}}{(d-j)!}\left(\frac{\Gamma(\frac{d}{2})}{\Gamma(\frac{d+1}{2})}\right)^{d-j}\frac{\Gamma(\frac{d+1}{2})}{\Gamma(\frac{j+1}{2})}\,V_d(W)\end{aligned} \tag{12.5}$$

as $a \to \infty$, where to obtain the last equality, we have used definition (1.6) of the constants $c_{j+i,d}$ to evaluate the telescopic product.

12.3 A general variance formula

We start in this section the analysis of second-order moments related to the maximal $(d-1)$-polytopes of STIT tessellations. Our first result is a general variance formula for the total surface content, which does not require the driving hyperplane measure Θ to be isotropic. More generally, we allow that the contribution of each $(d-1)$-dimensional polytope is additionally weighted by a function only depending on its direction; recall (12.1). To state the result, for a pair of points $(x,y) \in \mathbb{R}^d \times \mathbb{R}^d$, we denote by $\Theta([xy]) := \Theta([\mathrm{conv}\{x,y\}])$ the Θ-measure of the set of hyperplanes that intersect the line segment connecting x with y.

Theorem 12.3. *Let $W \in \mathcal{K}_d$, and let Θ be a hyperplane measure of the form* (4.1), *satisfying* (4.3). *Furthermore, let $(Y_t,\ t > 0)$ be the STIT tessellation process driven by the measure Θ. For $\Sigma^f_{V_{d-1},t,W}$ with bounded and measurable $f : G(d,1) \to \mathbb{R}$, we have*

$$\operatorname{var}\Sigma^f_{V_{d-1},t,W} = \int_{[W]} f^2(H^\perp) \int_{W\cap H}\int_{W\cap H} \frac{1-\exp(-t\,\Theta([xy]))}{\Theta([xy])} \lambda_{d-1}(\mathrm{d}x)\,\lambda_{d-1}(\mathrm{d}y)\,\Theta(\mathrm{d}H)\,,$$

which is finite for all $t > 0$.

Proof. First of all, the finiteness of the variance of $\Sigma^f_{V_{d-1},t,W}$ for $t > 0$ is a direct consequence of (12.3). To derive the integral expression for $\operatorname{var}\Sigma^f_{V_{d-1},t,W}$, we first assume that $W \in \mathcal{P}_d$ and then use the martingale property of the centered random process (10.44) derived in Corollary 10.12. After taking expectations, this yields

$$\begin{aligned}\operatorname{var}\Sigma^f_{V_{d-1},t,W} &= \mathbb{E}\int_0^t A^f_{d-1,s,W}\,\mathrm{d}s\\ &= \mathbb{E}\int_0^t\int_{[W]} \sum_{F\in Y_{s,W}\cap H} \left(V_{d-1}(F)f(H^\perp)\right)^2 \Theta(\mathrm{d}H)\,\mathrm{d}s\,. \end{aligned}\tag{12.6}$$

In a next step, we rewrite $V_{d-1}(F)^2$ as the double integral $\int_{W\cap H}\int_{W\cap H} \mathbb{1}\{(x,y)\in F^2\}\lambda_{d-1}(\mathrm{d}x)\,\lambda_{d-1}(\mathrm{d}y)$ and use Fubini's theorem to see that the last expression is the same as

$$\begin{aligned}&\mathbb{E}\int_0^t\int_{[W]} f^2(H^\perp) \sum_{F\in Y_{s,W}\cap H} \int_{W\cap H}\int_{W\cap H} \mathbb{1}\{(x,y)\in F^2\}\,\lambda_{d-1}(\mathrm{d}x)\,\lambda_{d-1}(\mathrm{d}y)\,\Theta(\mathrm{d}H)\,\mathrm{d}s\\ &= \int_0^t\int_{[W]} f^2(H^\perp) \int_{W\cap H}\int_{W\cap H} \mathbb{E}\sum_{F\in Y_{s,W}\cap H} \mathbb{1}\{(x,y)\in F^2\}\,\lambda_{d-1}(\mathrm{d}x)\,\lambda_{d-1}(\mathrm{d}y)\,\Theta(\mathrm{d}H)\,\mathrm{d}s\,.\end{aligned}$$

To proceed, we notice that two points $x,y \in W\cap H$ belong to the same cell F of the sectional tessellation $Y_{s,W}\cap H$ if and only if the line segment connecting x with y does not intersect the tessellation skeleton $\partial Y_{s,W}$. As a consequence,

$$\mathbb{E}\sum_{F\in Y_{s,W}\cap H} \mathbb{1}\{(x,y)\in F^2\} = \mathbb{P}(\operatorname{conv}\{x,y\}\cap\partial Y_{s,W} = \emptyset) = \exp(-s\,\Theta([x,y]))\,,$$

where for the last equality, we applied Lemma 5.3 with $C = \operatorname{conv}\{x,y\}$. Thus carrying out eventually the integration with respect to s yields

$$\operatorname{var}\Sigma^f_{V_{d-1},t,W} = \int_0^t\int_{[W]} f^2(H^\perp) \int_{W\cap H}\int_{W\cap H} \mathbb{P}(\operatorname{conv}\{x,y\}\cap\partial Y_{s,W} = \emptyset)\, \lambda_{d-1}(\mathrm{d}x)\,\lambda_{d-1}(\mathrm{d}y)\,\Theta(\mathrm{d}H)\,\mathrm{d}s$$

$$= \int_{[W]} f^2(H^\perp) \int_{W\cap H} \int_{W\cap H} \int_0^t \exp(-s\,\Theta([xy]))\,ds$$
$$\lambda_{d-1}(dx)\,\lambda_{d-1}(dy)\,\Theta(dH)$$
$$= \int_{[W]} f^2(H^\perp) \int_{W\cap H} \int_{W\cap H} \frac{1-\exp(-t\,\Theta([xy]))}{\Theta([xy])}$$
$$\lambda_{d-1}(dx)\,\lambda_{d-1}(dy)\,\Theta(dH)\,.$$

Having proved the statement for polytopes $W \in \mathcal{P}_d$, we can now apply Lemma 12.1 to extend the result to general $W \in \mathcal{K}_d$. This completes the proof. □

We note that although $\operatorname{var}\Sigma^f_{V_{d-1},t,W} < \infty$ for all $t > 0$, the variance $\operatorname{var}\Sigma^f_{V_{d-1},t,W}$ as a function of t is not necessarily uniformly bounded. For example, if we take $\Theta = \Theta_{\mathrm{iso}}$ and $f \equiv 1$, then, as we will see later, $\operatorname{var}\Sigma_{V_{d-1},t,W}$ is uniformly bounded if and only if $d \geq 3$; see Remark 12.14 and also compare with Remark 13.21 for the Mondrian case.

12.4 Variances and covariances in the isotropic case

12.4.1 A general formula

In this section, we deal with the variance of $\Sigma_{V_j,t,W}$, the sum of the jth intrinsic volumes of maximal $(d-1)$-polytopes of $Y_{t,W}$, and, more generally, with the covariances of the functionals $\Sigma_{V_i,t,W}$ and $\Sigma_{V_j,t,W}$, $i,j \in \{0,\dots,d-1\}$, for STIT tessellations that are stationary and isotropic. To present the result, we recall the definition of $A_{i,j,s,W}$ from (10.41) and for measurable functions $g : [0,t] \to \mathbb{R}$ and $n \in \mathbb{N}$, introduce the notation

$$I^n(g,t) := \int_0^t \int_0^{s_1} \cdots \int_0^{s_{n-1}} g(s_n)\,ds_n \dots ds_2\,ds_1 = \frac{1}{(n-1)!} \int_0^t (t-s)^{n-1} g(s)\,ds \tag{12.7}$$

for the n-fold iterated integral of g whenever it is well defined. Here (12.7) is known as Cauchy's formula for repeated integration. In what follows, we will also use the convention that $\binom{0}{0} = 1$ and that an empty product $\prod_{\ell=0}^{-1}(\cdots)$, formally ranging from $\ell = 0$ to $\ell = -1$, is interpreted as 1.

Theorem 12.4. *Let $W \in \mathcal{K}_d$, and let Θ_{iso} be the motion-invariant hyperplane measure. Furthermore, let $(Y_t,\ t > 0)$ be the STIT tessellation process driven by the measure Θ_{iso}. Then, for $i,j \in \{0,\dots,d-1\}$ and $t > 0$,*

$$\operatorname{cov}(\Sigma_{V_i,t,W}, \Sigma_{V_j,t,W})$$
$$= \sum_{m=0}^{d-i-1} \sum_{n=0}^{d-j-1} \binom{m+n}{m} \left(\prod_{\ell=0}^{m-1} c_{i+\ell,d}\right)\left(\prod_{\ell=0}^{n-1} c_{j+\ell,d}\right) I^{m+n+1}(\mathbb{E}A_{i+m,j+n,\cdot,W},\, t)\,.$$

In particular,

$$\operatorname{var}\Sigma_{V_i,t,W} = \sum_{m,n=0}^{d-i-1}\binom{m+n}{m}\left(\prod_{\ell=0}^{m-1}c_{i+\ell,d}\right)\left(\prod_{\ell=0}^{n-1}c_{i+\ell,d}\right)I^{m+n+1}(\mathbb{E}A_{i+m,i+n,\cdot,W},t)$$

with the constants $c_{k,d}$ introduced in (1.6).

Proof. Assume that $W \in \mathcal{P}_d$. The covariance formula follows by taking the expectations in (10.45). By construction the process there is a centered martingale, so its expectation vanishes. Hence

$$\begin{aligned}\operatorname{cov}(\Sigma_{V_i,t,W},\Sigma_{V_j,t,W}) &= \int_0^t \mathbb{E}A_{i,j,s,W}\,\mathrm{d}s\\ &\quad + \int_0^t c_{i,d}\operatorname{cov}(\Sigma_{V_{i+1},s,W},\Sigma_{V_j,s,W}) + c_{j,d}\operatorname{cov}(\Sigma_{V_i,s,W},\Sigma_{V_{j+1},s,W})\,\mathrm{d}s\,. \end{aligned} \tag{12.8}$$

Now we apply the same argument repeatedly to the covariances under the second integral. This leads to a recursion, which terminates once we arrive at a term involving $\Sigma_{V_d,s,W} \equiv 0$. The variance formula arises by specializing to the case $i = j$. Finally, the general case with $W \in \mathcal{K}_d$ follows from Lemma 12.1. □

Taking $i = d-1$ in the variance formula of the previous theorem, we obtain

$$\operatorname{var}\Sigma_{V_{d-1},t,W} = I^1(\mathbb{E}A_{d-1,d-1,\cdot,W},t) = \mathbb{E}\int_0^t\int_{[W]}\sum_{F\in Y_{s,W}\cap H} V_{d-1}(F)^2\,\Theta_{\mathrm{iso}}(\mathrm{d}H)\,\mathrm{d}s\,,$$

an expression we already encountered in greater generality at (12.6) in the proof of Theorem 12.3.

Moreover, taking $d = 2$ and $i = j = 0$ or $i = 1, j = 0$, we arrive at

$$\begin{aligned}\operatorname{var}\Sigma_{V_0,t,W} &= I^1(\mathbb{E}A_{0,0,\cdot,W},t) + 2\,\frac{2}{\pi}I^2(\mathbb{E}A_{1,0,\cdot,W},t) + 2\left(\frac{2}{\pi}\right)^2 I^3(\mathbb{E}A_{1,1,\cdot,W},t)\\ &=: T_1 + T_2 + T_3\end{aligned} \tag{12.9}$$

for the variance of the number of I-segments or maximal 1-polytopes in W and

$$\operatorname{cov}(\Sigma_{V_1,t,W},\Sigma_{V_0,t,W}) = I^1(\mathbb{E}A_{1,0,\cdot,W},t) + \frac{2}{\pi}I^2(\mathbb{E}A_{1,1,\cdot,W},t) =: U_1 + U_2 \tag{12.10}$$

for the covariance of the total edge length and the number of I-segments of a stationary and isotropic STIT tessellation in W with time parameter t in the plane. To evaluate the first term T_1 in (12.9), we use the definition of $A_{0,0,\cdot,W}$. Since $V_0(F) = 1$ a. s. for each segment $F \in Y_{s,W} \cap H$, $H \in [W]$, we obtain

$$
\begin{aligned}
T_1 &= \mathbb{E}\int_0^t \int_{[W]} \sum_{F\in Y_{s,W}\cap H} V_0(F)^2\,\Theta_{\mathrm{iso}}(\mathrm{d}H)\,\mathrm{d}s \\
&= \mathbb{E}\int_0^t \int_{[W]} \big|\{F \in Y_{s,W}\cap H\}\big|\,\Theta_{\mathrm{iso}}(\mathrm{d}H)\,\mathrm{d}s \\
&= \int_0^t \int_{[W]} (\gamma_0(Y_s\cap H)\,V_1(W\cap H)+1)\,\Theta_{\mathrm{iso}}(\mathrm{d}H)\,\mathrm{d}s\,,
\end{aligned}
$$

where $\gamma_0(Y_s \cap H)$ denotes the intensity of the sectional Poisson point processes $Y_s \cap H$ induced by Y_s on the line H. By the intersection property of STIT tessellations in Lemma 6.11, $\gamma_0(Y_s \cap H) = \frac{2s}{\pi}$, and we arrive at

$$
\begin{aligned}
T_1 &= \int_0^t \int_{[W]} \left(\frac{2s}{\pi}V_1(W\cap H)+1\right)\Theta_{\mathrm{iso}}(\mathrm{d}H)\,\mathrm{d}s \\
&= \int_0^t \left(\frac{2s}{\pi}V_2(W)+\frac{2}{\pi}V_1(W)\right)\mathrm{d}s \\
&= \frac{t^2}{\pi}V_2(W)+\frac{2t}{\pi}V_1(W)\,,
\end{aligned}
\tag{12.11}
$$

where we used Crofton's formula (1.6) multiple times. Similarly, for the term T_2 in (12.9), Crofton's formula yields the expression

$$
\begin{aligned}
T_2 &= \frac{4}{\pi}\int_0^t (t-s)\,\mathbb{E}\int_{[W]} \sum_{F\in Y_{s,W}\cap H} V_1(F)\,\Theta_{\mathrm{iso}}(\mathrm{d}H)\,\mathrm{d}s \\
&= \frac{4}{\pi}\int_0^t (t-s)\int_{[W]} V_1(W\cap H)\,\Theta_{\mathrm{iso}}(\mathrm{d}H)\,\mathrm{d}s \\
&= \frac{4}{\pi}\int_0^t (t-s)V_2(W)\,\mathrm{d}s \\
&= \frac{2t^2}{\pi}V_2(W)\,,
\end{aligned}
\tag{12.12}
$$

and for U_1 in (12.10), we have

$$
U_1 = \int_0^t \int_{[W]} \mathbb{E}\sum_{F\in Y_{s,W}\cap H} V_1(F)\,\Theta_{\mathrm{iso}}(\mathrm{d}H)\,\mathrm{d}s
$$

$$= \int_0^t \int_{[W]} V_1(W \cap H)\, \Theta_{\mathrm{iso}}(\mathrm{d}H)\, \mathrm{d}s$$
$$= tV_2(W)\,.$$

However, to bring the terms T_3 and U_2 in (12.9) and (12.10) into an accessible form, we need an integral-geometric transformation formula of Blaschke–Petkantschin type, which is interesting in its own right and which will be developed in the next section. Using this device, we will come back to the terms T_3 and U_2 in Section 12.4.4.

12.4.2 An integral-geometric transformation formula

In this section, we develop the announced integral-geometric transformation formula, which will be concluded from the classical affine Blaschke–Petkantschin formula, for which we refer to Theorem 7.2.7 of [172]. Since our formula will be used also in the study of the pair correlation function of the random surface measure in Section 12.7, we present the result directly in full generality.

Proposition 12.5. *Let $B_1, B_2 \in \mathcal{B}(\mathbb{R}^d)$ be Borel sets, and let $h : B_1 \times B_2 \to \mathbb{R}$ be a nonnegative measurable function. Then*

$$\int_{\mathcal{H}} \int_{B_1 \cap H} \int_{B_2 \cap H} h(x,y)\, \lambda_{d-1}(\mathrm{d}x)\, \lambda_{d-1}(\mathrm{d}y)\, \Theta_{\mathrm{iso}}(\mathrm{d}H)$$
$$= \frac{(d-1)\kappa_{d-1}}{d\kappa_d} \int_{B_1} \int_{B_2} \frac{h(x,y)}{\|x-y\|}\, \lambda_d(\mathrm{d}x)\, \lambda_d(\mathrm{d}y)\,. \tag{12.13}$$

Proof. We use the notation and results of Sections 7.1 and 13.2 in [172]. Denote by ν_1 the rotation-invariant probability measure on the Grassmannian $G(d,1)$ of all one-dimensional linear subspaces in $\mathbb{R}^d$. Let the measure μ_1 on $A(d,1)$, the set of all lines in $\mathbb{R}^d$, be defined by

$$\int_{A(d,1)} g(L)\, \mu_1(\mathrm{d}L) = \int_{G(d,1)} \int_{L^\perp} g(L+x)\, \lambda_{d-1}(\mathrm{d}x)\, \nu_1(\mathrm{d}L) \tag{12.14}$$

for nonnegative measurable functions $g : A(d,1) \to \mathbb{R}$. Furthermore, for a fixed hyperplane $H \in \mathcal{H}$, let $x_1 \in \mathbb{R}^d$ be such that $H - x_1 \in G(d, d-1)$, and write $G(H - x_1, 1) := \{L \in G(d,1) : L \subset H - x_1\}$ for the set of all lines contained in the hyperplane $H - x_1$. Let $\nu_1^{H-x_1}$ be the probability measure on $G(H - x_1, 1)$ that is invariant under the rotations of $\mathbb{R}^d$ mapping $H - x_1$ into itself. Then, in analogy to the measure μ_1 on $A(d,1)$, define the measure μ_1^H on $A(H,1) := \{L \in A(d,1) : L \subset H\}$ by

$$\int_{A(H,1)} g(L)\,\mu_1^H(\mathrm{d}L) = \int_{G(H-x_1,1)} \int_{L^\perp \cap H} g(L+x_1+x_2)\,\lambda_{d-2}(\mathrm{d}x_2)\,\nu_1^{H-x_1}(\mathrm{d}L)$$

for nonnegative measurable functions $g : A(H,1) \to \mathbb{R}$. Then by the remark at the end of Section 7.1 of [172] we have the relation

$$\int_{A(d,1)} g(L)\,\mu_1(\mathrm{d}L) = \int_{\mathcal{H}} \int_{A(H,1)} g(L)\,\mu_1^H(\mathrm{d}L)\,\Theta_{\mathrm{iso}}(\mathrm{d}H) \tag{12.15}$$

for nonnegative measurable functions $g : A(d,1) \to \mathbb{R}$.

The affine Blaschke–Petkantschin formula (see Theorem 7.2.7 of [172] with $q = 1$) says that

$$\begin{aligned}&\int_{\mathbb{R}^d}\int_{\mathbb{R}^d} g(x,y)\,\lambda_d(\mathrm{d}x)\,\lambda_d(\mathrm{d}y)\\ &= \frac{d\kappa_d}{2} \int_{A(d,1)} \int_L \int_L g(x,y)\|x-y\|^{d-1}\,\lambda_1(\mathrm{d}x)\,\lambda_1(\mathrm{d}y)\,\mu_1(\mathrm{d}L)\end{aligned} \tag{12.16}$$

for nonnegative measurable functions $g : (\mathbb{R}^d)^2 \to \mathbb{R}$. For $k > -d$ and

$$g(x,y) := \mathbb{1}\{x \in B_1\}\mathbb{1}\{y \in B_2\}\|x-y\|^k h(x,y),$$

with h as in the statement of the proposition, we obtain

$$\begin{aligned}&\int_{B_1}\int_{B_2} \|x-y\|^k h(x,y)\,\lambda_d(\mathrm{d}x)\,\lambda_d(\mathrm{d}y)\\ &= \frac{d\kappa_d}{2} \int_{A(d,1)} \int_{B_1\cap L} \int_{B_2\cap L} \|x-y\|^{d-1+k} h(x,y)\,\lambda_1(\mathrm{d}x)\,\lambda_1(\mathrm{d}y)\,\mu_1(\mathrm{d}L),\end{aligned} \tag{12.17}$$

where the condition $k > -d$ ensures the integrability of $\|x-y\|^k$ on the left-hand side. For the particular choice $k = -1$, this yields

$$\begin{aligned}&\int_{B_1}\int_{B_2} \frac{h(x,y)}{\|x-y\|}\,\lambda_d(\mathrm{d}x)\,\lambda_d(\mathrm{d}y)\\ &= \frac{d\kappa_d}{2} \int_{A(d,1)} \int_{B_1\cap L} \int_{B_2\cap L} \|x-y\|^{d-2} h(x,y)\,\lambda_1(\mathrm{d}x)\,\lambda_1(\mathrm{d}y)\,\mu_1(\mathrm{d}L).\end{aligned} \tag{12.18}$$

On the other hand, for a fixed hyperplane $H \in \mathcal{H}$, (12.16) can also be applied if we replace d by $d-1$ and μ_1 by μ_1^H. This yields an analogue to (12.17) by substituting B_1 by $B_1 \cap H$ and B_2 by $B_2 \cap H$ and choosing $k = 0$:

$$\int_{B_1\cap H}\int_{B_2\cap H} h(x,y)\,\lambda_{d-1}(\mathrm{d}x)\,\lambda_{d-1}(\mathrm{d}y)$$
$$= \frac{(d-1)\kappa_{d-1}}{2}\int_{A(H,1)}\int_{B_1\cap H\cap L}\int_{B_2\cap H\cap L} \|x-y\|^{d-2}h(x,y)\,\lambda_1(\mathrm{d}x)\,\lambda_1(\mathrm{d}y)\,\mu_1^H(\mathrm{d}L)\,.$$

Integrating both sides with respect to Θ_{iso}, by (12.15) we obtain

$$\int_{\mathcal{H}}\int_{B_1\cap H}\int_{B_2\cap H} h(x,y)\,\lambda_{d-1}(\mathrm{d}x)\,\lambda_{d-1}(\mathrm{d}y)\,\Theta_{\text{iso}}(\mathrm{d}H)$$
$$= \frac{(d-1)\kappa_{d-1}}{2}\int_{\mathcal{H}}\int_{A(H,1)}\int_{B_1\cap H\cap L}\int_{B_2\cap H\cap L} \|x-y\|^{d-2}$$
$$\times\, h(x,y)\,\lambda_1(\mathrm{d}x)\,\lambda_1(\mathrm{d}y)\,\mu_1^H(\mathrm{d}L)\,\Theta_{\text{iso}}(\mathrm{d}H)$$
$$= \frac{(d-1)\kappa_{d-1}}{2}\int_{A(d,1)}\int_{B_1\cap L}\int_{B_2\cap L} \|x-y\|^{d-2}h(x,y)\,\lambda_1(\mathrm{d}x)\,\lambda_1(\mathrm{d}y)\,\mu_1(\mathrm{d}L)\,. \qquad (12.19)$$

Comparing (12.18) and (12.19), we conclude that

$$\int_{\mathcal{H}}\int_{B_1\cap H}\int_{B_2\cap H} h(x,y)\,\lambda_{d-1}(\mathrm{d}x)\,\lambda_{d-1}(\mathrm{d}y)\,\Theta_{\text{iso}}(\mathrm{d}H) = \frac{(d-1)\kappa_{d-1}}{d\kappa_d}\int_{B_1}\int_{B_2}\frac{h(x,y)}{\|x-y\|}\,\lambda_d(\mathrm{d}x)\,\lambda_d(\mathrm{d}y)\,,$$

and the proof is complete. □

12.4.3 Variance of the total surface content

In the present section, using the integral-geometric transformation formula derived in Proposition 12.5, we deduce a simple expression for the variance of the total surface content of a stationary and isotropic STIT tessellation in a window. Here and in the further variance and covariance formulae we develop below, we will always present two formulae for each of the (co)variances under investigation, both reflecting in a different way the influence of the geometry of the underlying observation window $W \in \mathcal{K}_d$. For the second of these representations, we need the following well-known notion from convex geometry.

Definition 12.6. *For the rotation-invariant probability measure φ_{iso} on S^{d-1} and for compact sets $C \in \mathcal{C}$ with* $\operatorname{int} C \neq \emptyset$, *let $\bar{c}_C : [0,\infty) \to \mathbb{R}$ with*

$$\bar{c}_C(r) := \int_{S^{d-1}} V_d(C\cap(C+ru))\,\varphi_{\text{iso}}(\mathrm{d}u)\,,\qquad r\geq 0\,,$$

be the ***isotropized set-covariogram function*** *of C.*

Remark 12.7. *Explicit formulae for the isotropized set-covariogram function $\bar{\mathfrak{c}}_C$ are available only in rare cases. For example, if $C = RB^d$ is the d-dimensional centered ball of radius $R > 0$, then*

$$\bar{\mathfrak{c}}_{RB^d}(r) = 2\kappa_{d-1}R^d \int_{r/(2R)}^{1} (1-s^2)^{\frac{d-1}{2}}\, ds\,, \quad 0 \le r \le 2R\,,$$

as we can easily see by a slice-integration argument. In particular, for $d = 2$ and $d = 3$, we get, for $0 \le r \le 2R$,

$$\bar{\mathfrak{c}}_{RB^2}(r) = 2R^2 \arccos\frac{r}{2R} - Rr\sqrt{1-\frac{r^2}{4R^2}}\,,$$

$$\bar{\mathfrak{c}}_{RB^3}(r) = \frac{4\pi R^3}{3}\left(1-\frac{3r}{4R}+\frac{r^3}{16R^3}\right);$$

see, for example, page 66 of [13].

After these preparations, we now draw the desired consequences. The first one is an explicit formula for the variance of the total surface content within a window $W \in \mathcal{K}_d$ of a stationary and isotropic STIT tessellation in $\mathbb{R}^d$ with time parameter $t > 0$.

Theorem 12.8. *Let $W \in \mathcal{K}_d$, and let Θ_{iso} be the motion-invariant hyperplane measure. Furthermore, let $(Y_t, t > 0)$ be the STIT tessellation process driven by the measure Θ_{iso}. Then, for all $t > 0$,*

$$\operatorname{var}\Sigma_{V_{d-1},t,W} = \frac{d-1}{2}\int_W\int_W \frac{1-\exp(-\frac{2\kappa_{d-1}}{d\kappa_d}t\|x-y\|)}{\|x-y\|^2}\,\lambda_d(dx)\,\lambda_d(dy) \tag{12.20}$$

$$= \frac{d(d-1)\kappa_d}{2}\int_0^\infty \bar{\mathfrak{c}}_W(r)\, r^{d-3}\left(1-\exp\left(-\frac{2\kappa_{d-1}}{d\kappa_d}tr\right)\right) dr\,. \tag{12.21}$$

Proof. According to Crofton's formula (1.6), we have

$$\begin{aligned}
\Theta_{\text{iso}}([xy]) &= \int_{\mathcal{H}} \mathbb{1}\{H \cap \operatorname{conv}\{x,y\} \neq \emptyset\}\,\Theta_{\text{iso}}(dH)\\
&= \int_{\mathcal{H}} V_0(H \cap \operatorname{conv}\{x,y\})\,\Theta_{\text{iso}}(dH)\\
&= \frac{\Gamma(\frac{d}{2})}{\Gamma(\frac{1}{2})\Gamma(\frac{d+1}{2})} V_1(\operatorname{conv}\{x,y\})\\
&= \frac{2\kappa_{d-1}}{d\kappa_d}\|x-y\|\,,
\end{aligned} \tag{12.22}$$

since $\mathbb{1}\{H \cap \operatorname{conv}\{x,y\} \neq \emptyset\} = V_0(H \cap \operatorname{conv}\{x,y\}) \in \{0,1\}$ for all $H \in \mathcal{H}$. Using Theorem 12.3 with $f \equiv 1$ and applying (12.13) with $B_1 = B_2 = W$ and with

$$h(x,y) := \frac{1-\exp(-t\,\Theta_{\mathrm{iso}}([xy]))}{\Theta_{\mathrm{iso}}([xy])} = \frac{1-\exp(-\frac{2\kappa_{d-1}}{d\kappa_d}t\|x-y\|)}{\frac{2\kappa_{d-1}}{d\kappa_d}\|x-y\|},$$

we get

$$\begin{aligned}
&\operatorname{var}\Sigma_{V_{d-1},t,W}\\
&= \int_{[W]}\int_{W\cap H}\int_{W\cap H} \frac{1-\exp(-\frac{2\kappa_{d-1}}{d\kappa_d}t\|x-y\|)}{\frac{2\kappa_{d-1}}{d\kappa_d}\|x-y\|}\,\lambda_{d-1}(\mathrm{d}x)\,\lambda_{d-1}(\mathrm{d}y)\,\Theta_{\mathrm{iso}}(\mathrm{d}H)\\
&= \frac{d-1}{2}\int_W\int_W \frac{1-\exp(-\frac{2\kappa_{d-1}}{d\kappa_d}t\|x-y\|)}{\|x-y\|^2}\,\lambda_d(\mathrm{d}x)\,\lambda_d(\mathrm{d}y)\,.
\end{aligned}$$

This proves the first formula. The second one follows from the identity

$$\int_W\int_W g(\|x-y\|)\,\lambda_d(\mathrm{d}x)\,\lambda_d(\mathrm{d}y) = d\kappa_d\int_0^\infty \bar{c}_W(r)\,g(r)\,r^{d-1}\,\mathrm{d}r \tag{12.23}$$

for nonnegative measurable functions $g : [0,\infty) \to \mathbb{R}$. Indeed, putting $z = x - y$, we have

$$\begin{aligned}
&\int_W\int_W g(\|x-y\|)\,\lambda_d(\mathrm{d}x)\,\lambda_d(\mathrm{d}y)\\
&\quad= \int_{\mathbb{R}^d} g(\|z\|)\,\lambda_d(\{(x,y)\in W\times W : x-y=z\})\,\lambda_d(\mathrm{d}z)\\
&\quad= \int_{\mathbb{R}^d} g(\|z\|)\,\lambda_d(W\cap(W+z))\,\lambda_d(\mathrm{d}z)\,.
\end{aligned}$$

Introducing spherical coordinates and using Fubini's theorem and the definition of the isotropoized set-covariogram function, we obtain

$$\begin{aligned}
&\int_{\mathbb{R}^d} g(\|z\|)\,\lambda_d(W\cap(W+z))\,\lambda_d(\mathrm{d}z)\\
&\quad= d\kappa_d\int_0^\infty\int_{S^{d-1}} g(r)\,\lambda_d(W\cap(W+ru))\,\varphi_{\mathrm{iso}}(\mathrm{d}u)\,\mathrm{d}r\\
&\quad= d\kappa_d\int_0^\infty g(r)r^{d-1}\int_{S^{d-1}} \lambda_d(W\cap(W+ru))r^{d-1}\,\varphi_{\mathrm{iso}}(\mathrm{d}u)\,\mathrm{d}r
\end{aligned}$$

$$= d\kappa_d \int_0^\infty \bar{c}_W(r)\, g(r)\, r^{d-1}\, \mathrm{d}r\,,$$

which is (12.23). Applying this transformation formula yields

$$\operatorname{var} \Sigma_{V_{d-1},t,W} = \frac{d(d-1)\kappa_d}{2} \int_0^\infty \bar{c}_W(r)\, \frac{1-\exp(-\frac{2\kappa_{d-1}}{d\kappa_d}tr)}{r^2}\, r^{d-1}\, \mathrm{d}r$$

$$= \frac{d(d-1)\kappa_d}{2} \int_0^\infty \bar{c}_W(r)\, r^{d-3}\Big(1-\exp\Big(-\frac{2\kappa_{d-1}}{d\kappa_d}tr\Big)\Big)\, \mathrm{d}r\,.$$

This completes the proof. □

Remark 12.9. *A fully explicit expression of the considered variance can be given in dimension $d = 3$ for the restriction of a stationary and isotropic STIT tessellation to a centered ball RB^3 of radius $R > 0$. In this case, using the result of Remark 12.7, we arrive at*

$$\operatorname{var} \Sigma_{V_2,t,RB^3} = \frac{4\pi^2}{3t^4}\big(t^2R^2(12 - 8tR + 3t^2R^2) + 24(1 + tR)\mathrm{e}^{-tR} - 24\big)$$

for the variance of the total surface content of a stationary and isotropic STIT tessellation in $\mathbb{R}^3$ within RB^3. Similar formulae are available for all odd space dimensions $d \geq 3$, since in this case the isotropized covariance function $\bar{c}_{RB^d}(r)$ is a polynomial in r of degree d.

12.4.4 Further variances and covariances in dimension $d = 2$

The second consequence we want to draw from the Blaschke–Petkantschin-type formula are explicit formulae for the variance of the total number and the covariance of the total number and the total length of I-segments of a planar stationary and isotropic STIT tessellation with time parameter $t \geq 0$ within a window $W \in \mathcal{K}_2$. This continues and completes the discussion concerning the terms T_3 and U_2 in (12.9) and (12.10) started at the end of Section 12.4.1.

Theorem 12.10. *Let $W \in \mathcal{K}_2$, and let Θ_{iso} be the motion-invariant line measure in $\mathbb{R}^2$. Furthermore, let $(Y_t,\, t > 0)$ be the planar STIT tessellation process driven by the measure Θ_{iso}.*
(a) *For $t > 0$, we have*

$$\operatorname{var} \Sigma_{V_0,t,W}$$
$$= \frac{2t}{\pi} V_1(W) + \frac{3t^2}{\pi} V_2(W)$$
$$+ \frac{4}{\pi^3} \int_W \int_W \int_0^t (t-s)^2\, \frac{\exp(-\frac{2s}{\pi}\|x-y\|)}{\|x-y\|}\, \mathrm{d}s\, \lambda_2(\mathrm{d}x)\, \lambda_2(\mathrm{d}y)$$

$$= \frac{2t}{\pi} V_1(W) + \frac{3t^2}{\pi} V_2(W) + \frac{8}{\pi^2} \int_0^\infty \bar{c}_W(r) \int_0^t (t-s)^2 \exp\Big(-\frac{2rs}{\pi}\Big)\, ds\, dr\,.$$

(b) *For $t > 0$, we have*

$$\begin{aligned}
&\operatorname{cov}(\Sigma_{V_1,t,W}, \Sigma_{V_0,t,W}) \\
&= tV_2(W) + \frac{2}{\pi^2} \int_W \int_W \int_0^t (t-s) \frac{\exp(-\frac{2s}{\pi}\|x-y\|)}{\|x-y\|}\, ds\, \lambda_2(dx)\, \lambda_2(dy) \\
&= tV_2(W) + \frac{4}{\pi} \int_0^\infty \bar{c}_W(r) \int_0^t (t-s) \exp\Big(-\frac{2rs}{\pi}\Big)\, ds\, dr\,.
\end{aligned}$$

Proof. We start with the formula for $\operatorname{var} \Sigma_{V_0,t,W}$. Recalling (12.9) and putting this together with (12.11) and (12.12), we see that

$$\operatorname{var} \Sigma_{V_0,t,W} = T_1 + T_2 + T_3 = \frac{2t}{\pi} V_1(W) + \frac{3t^2}{\pi} V_2(W) + \frac{8}{\pi^2} I^3(\mathbb{E}A_{1,1,\cdot,W}, t)\,.$$

To determine the last term, we argue as in the proof of Theorem 12.3:

$$\begin{aligned}
I^3(\mathbb{E}A_{1,1,\cdot,W}, t) &= \frac{1}{2} \int_0^t (t-s)^2 \mathbb{E} \int_{[W]} \sum_{F \in Y_{s,W} \cap H} V_1(F)^2\, \Theta_{\mathrm{iso}}(dH)\, ds \\
&= \frac{1}{2} \int_0^t (t-s)^2 \int_{[W]} \int_{W \cap H} \int_{W \cap H} \mathbb{E} \sum_{F \in Y_{s,W} \cap H} \mathbb{1}\{(x,y) \in F^2\} \\
&\qquad \lambda_1(dx)\, \lambda_1(dy)\, \Theta_{\mathrm{iso}}(dH)\, ds \\
&= \frac{1}{2} \int_{[W]} \int_{W \cap H} \int_{W \cap H} \int_0^t (t-s)^2 \exp\Big(-\frac{2s}{\pi}\|x-y\|\Big) \\
&\qquad ds\, \lambda_1(dx)\, \lambda_1(dy)\, \Theta_{\mathrm{iso}}(dH)\,.
\end{aligned}$$

We can now apply the Blaschke–Petkantschin-type transformation formula of Proposition 12.5 to see that this is the same as

$$\frac{1}{2\pi} \int_W \int_W \int_0^t (t-s)^2 \frac{\exp(-\frac{2s}{\pi}\|x-y\|)}{\|x-y\|}\, ds\, \lambda_2(dx)\, \lambda_2(dy)\,.$$

As a result,

$$\operatorname{var}\Sigma_{V_0,t,W} = \frac{2t}{\pi}V_1(W) + \frac{3t^2}{\pi}V_2(W) + \frac{4}{\pi^3}\int_W\int_W\int_0^t (t-s)^2\,\frac{\exp(-\frac{2s}{\pi}\|x-y\|)}{\|x-y\|}\,\mathrm{d}s\,\lambda_2(\mathrm{d}x)\,\lambda_2(\mathrm{d}y)\,,$$

which proves the first formula for $\operatorname{var}\Sigma_{V_0,t,W}$. The second formula follows by using the transformation formula (12.23):

$$\operatorname{var}\Sigma_{V_0,t,W} = \frac{2t}{\pi}V_1(W) + \frac{3t^2}{\pi}V_2(W) + \frac{8}{\pi^2}\int_0^\infty \bar{c}_W(r)\int_0^t (t-s)^2\exp\Big(-\frac{2rs}{\pi}\Big)\,\mathrm{d}s\,\mathrm{d}r\,.$$

This completes the proof of part (a).

Similarly, for $\operatorname{cov}(\Sigma_{V_1,t,W},\Sigma_{V_0,t,W})$, recalling (12.10), we have that

$$\operatorname{cov}(\Sigma_{V_1,t,W},\Sigma_{V_0,t,W}) = U_1 + U_2 = tV_2(W) + \frac{2}{\pi}I^2(\mathbb{E}A_{1,1,\cdot,W},t)\,.$$

Again, as in the proof of Theorem 12.3, we can rewrite $I^2(\mathbb{E}A_{1,1,\cdot,W},t)$ as follows:

$$\begin{aligned} &I^2(\mathbb{E}A_{1,1,\cdot,W},t)\\ &= \int_0^t (t-s)\int_{[W]}\mathbb{E}\sum_{F\in Y_{s,W}\cap H} V_1(F)^2\,\Theta_{\mathrm{iso}}(\mathrm{d}H)\,\mathrm{d}s\\ &= \int_{[W]}\int_{W\cap H}\int_{W\cap H}\int_0^t (t-s)\exp\Big(-\frac{2s}{\pi}\|x-y\|\Big)\,\mathrm{d}s\,\lambda_1(\mathrm{d}x)\,\lambda_1(\mathrm{d}y)\,\Theta_{\mathrm{iso}}(\mathrm{d}H)\\ &= \frac{1}{\pi}\int_W\int_W\int_0^t (t-s)\,\frac{\exp(-\frac{2s}{\pi}\|x-y\|)}{\|x-y\|}\,\mathrm{d}s\,\lambda_2(\mathrm{d}x)\,\lambda_2(\mathrm{d}y)\,, \end{aligned}$$

where we applied Proposition 12.5 once again. As a result, we arrive at

$$\begin{aligned} &\operatorname{cov}(\Sigma_{V_1,t,W},\Sigma_{V_0,t,W})\\ &\quad = tV_2(W) + \frac{2}{\pi^2}\int_W\int_W\int_0^t (t-s)\,\frac{\exp(-\frac{2s}{\pi}\|x-y\|)}{\|x-y\|}\,\mathrm{d}s\,\lambda_2(\mathrm{d}x)\,\lambda_2(\mathrm{d}y)\,, \end{aligned}$$

which is the first formula for $\operatorname{cov}(\Sigma_{V_1,t,W},\Sigma_{V_0,t,W})$ in part (b). The second one follows after an application of (12.23). □

12.5 Asymptotic variances and covariances in the isotropic model

In this section, we consider the variances and covariances of the intrinsic volume functionals $\Sigma_{V_j,t,aW}$ for $j \in \{0,\dots,d-1\}$ in a family of growing windows aW with $W \in \mathcal{K}$ and as $a \to \infty$. We emphasize at this point once again that although in this book, we only deal with STIT tessellation processes $(Y_{t,W},\ t \geq 0)$ in polytopal windows $W \in \mathcal{P}_d$, the expectation and variance of the geometric functionals $\Sigma_{V_j,t,W}$ are well defined for general $W \in \mathcal{K}_d$ by Lemma 12.1.

12.5.1 Total surface content

We start our analysis with the total surface content. Let $a > 0$ and $W \in \mathcal{K}_d$. Then the scaling property (6.2) of STIT tessellations implies that

$$\operatorname{var} \Sigma_{V_{d-1},t,aW} = a^{2(d-1)} \operatorname{var} \Sigma_{V_{d-1},at,W} \,. \tag{12.24}$$

The asymptotically relevant geometric properties of W will be captured by the following two notions, for which we refer to Chapter 9 of [123] and to [172], page 363, respectively.

Definition 12.11. *Let $W \in \mathcal{K}_d$. Then, for $\alpha < d$,*

$$E_\alpha(W) := \int_W \int_W \|x-y\|^{-\alpha} \, \lambda_d(\mathrm{d}x)\, \lambda_d(\mathrm{d}y) \tag{12.25}$$

*denotes the α-**energy** of W. Moreover,*

$$I_s(W) := \frac{d\kappa_d}{2} \int_{A(d,1)} V_1(W \cap L)^s \, \mu_1(\mathrm{d}L) \tag{12.26}$$

*is the **chord power integral** of W of order $s \geq 1$. Here μ_1 stands for the motion-invariant measure on $A(d,1)$ as in (12.14).*

The relationship between energies and chord power integrals is described in the following lemma.

Lemma 12.12. *Let $W \in \mathcal{K}_d$ and $\alpha < d$. Then $E_\alpha(W) = \frac{2}{(d+1-\alpha)(d-\alpha)} I_{d-\alpha+1}(W)$.*

Proof. We use (12.17) with $B_1 = B_2 = W$, $h \equiv 1$, and $k = -\alpha$ to see that

$$E_\alpha(W) = \int_W \int_W \|x-y\|^{-\alpha} \, \lambda_d(\mathrm{d}x)\, \lambda_d(\mathrm{d}y)$$

$$= \frac{d\kappa_d}{2} \int\limits_{A(d,1)} \int\limits_{W\cap L} \int\limits_{W\cap L} \|x-y\|^{d-1-\alpha}\, \lambda_1(\mathrm{d}x)\, \lambda_1(\mathrm{d}y)\, \mu_1(\mathrm{d}L)$$

$$= \frac{d\kappa_d}{(d+1-\alpha)(d-\alpha)} \int\limits_{A(d,1)} V_1(W\cap L)^{d+1-\alpha}\, \mu_1(\mathrm{d}L)$$

$$= \frac{2}{(d+1-\alpha)(d-\alpha)} I_{d+1-\alpha}(W)\,.$$

This completes the proof. □

We are now prepared to present our result on the asymptotic variance growth for the total surface content. An important distinction arises between the cases $d = 2$ and $d \geq 3$, which will frequently appear in this chapter and the next one.

Theorem 12.13. *Consider the STIT tessellation process $(Y_t,\, t > 0)$ driven by the motion-invariant hyperplane measure Θ_{iso}, and let $W \in \mathcal{K}_d$. Then, for $d = 2$,*

$$\operatorname{var} \Sigma_{V_1,t,aW} \sim \pi\, V_2(W)\, a^2 \log a \quad \text{as } a \to \infty\,, \tag{12.27}$$

whereas for $d \geq 3$,

$$\operatorname{var} \Sigma_{V_{d-1},t,aW} \sim \frac{d-1}{2} E_2(W)\, a^{2(d-1)} = \frac{1}{d-2} I_{d-1}(W)\, a^{2(d-1)} \quad \text{as } a \to \infty \tag{12.28}$$

for any fixed $t > 0$.

Proof. We start from the variance representation (12.21) together with the scaling property (12.24), which for $d = 2$ gives

$$\operatorname{var} \Sigma_{V_1,t,aW} = \pi a^2 \int\limits_{S^1} \int\limits_0^\infty \mathfrak{c}_W(ru)\, r^{-1} \left(1 - \exp\left(-\frac{2}{\pi} tar\right)\right) \mathrm{d}r\, \varphi_{\text{iso}}(\mathrm{d}u)\,. \tag{12.29}$$

By Proposition 4-3-1 b) of [122] the covariogram $\mathfrak{c}_W(ru) = V_2(W \cap (W + ru))$ is differentiable with respect to r and satisfies

$$\frac{\mathrm{d}}{\mathrm{d}r} \mathfrak{c}_W(ru) = -V_1(W \cap (W + ru) \mid u^\perp)\,, \quad u \in S^1\,.$$

Hence for r close to 0, we have the first-order expansion

$$\mathfrak{c}_W(ru) = V_2(W) - r\, V_1(W \mid u^\perp) + \rho_u(r)\,, \tag{12.30}$$

where $\rho_u(r)/r \to 0$ as $r \downarrow 0$, uniformly in $u \in S^1$. Moreover, $\mathfrak{c}_W(ru) = 0$ for $r > w(u)$, where $w(u) \leq \operatorname{diam} W$ denotes the width of W in direction u.

Substituting (12.30) into (12.29), we obtain the decomposition

$$
\begin{aligned}
\operatorname{var}\Sigma_{V_1,t,aW} &= \pi a^2 V_2(W)\int_{S^1}\int_0^{w(u)} r^{-1}\Big(1-\exp\Big(-\frac{2}{\pi}tar\Big)\Big)\,\mathrm{d}r\,\varphi_{\mathrm{iso}}(\mathrm{d}u) \\
&\quad - \pi a^2\int_{S^1}\int_0^{w(u)} V_1(W\mid u^\perp)\Big(1-\exp\Big(-\frac{2}{\pi}tar\Big)\Big)\,\mathrm{d}r\,\varphi_{\mathrm{iso}}(\mathrm{d}u) \\
&\quad + \pi a^2\int_{S^1}\int_0^{w(u)} \frac{\rho_u(r)}{r}\Big(1-\exp\Big(-\frac{2}{\pi}tar\Big)\Big)\,\mathrm{d}r\,\varphi_{\mathrm{iso}}(\mathrm{d}u) \\
&=: S_1(a)+S_2(a)+S_3(a)\,.
\end{aligned}
$$

We analyze the three terms separately. For any fixed $c > 0$ and $w > 0$, we have the asymptotic relation

$$
\begin{aligned}
\int_0^w r^{-1}(1-\exp(-car))\,\mathrm{d}r &= \int_0^{aw}\frac{1-\mathrm{e}^{-cs}}{s}\,\mathrm{d}s \\
&= \int_0^1\frac{1-\mathrm{e}^{-cs}}{s}\,\mathrm{d}s + \log(aw) - \int_1^{aw}\frac{\mathrm{e}^{-cs}}{s}\,\mathrm{d}s \\
&= \log a + O(1)
\end{aligned} \tag{12.31}
$$

as $a \to \infty$, where the $O(1)$-term is bounded uniformly in $w \in (0, w_0]$ for some $w_0 > 0$. Using (12.31) with $c = 2t/\pi$, $w = w(u)$, and $w_0 = \operatorname{diam} W$, we obtain

$$
S_1(a) = \pi a^2 V_2(W)(\log a + O(1))\,.
$$

Next, since $0 \le 1-\exp(-\frac{2}{\pi}tar) \le 1$ for all $r > 0$, we get

$$
|S_2(a)| \le \pi a^2\int_{S^1} V_1(W\mid u^\perp)\,w(u)\,\varphi_{\mathrm{iso}}(\mathrm{d}u) = O(a^2)\,,
$$

which is of lower order compared to $a^2\log a$. Finally, because of $\rho_u(r)/r \to 0$ uniformly in $u \in S^1$, the third term satisfies $S_3(a) = O(a^2)$ as well. Combining these estimates, we conclude that

$$
\operatorname{var}\Sigma_{V_1,t,aW} \sim \pi\,V_2(W)\,a^2\log a\,, \quad a\to\infty\,,
$$

which proves (12.27).

To prove (12.28) for $d \ge 3$, we use (12.20) and the scaling property (12.24) to see that

$$
\begin{aligned}
\operatorname{var}\Sigma_{V_{d-1},t,aW} &= a^{2(d-1)}\operatorname{var}\Sigma_{V_{d-1},at,W} \\
&= a^{2(d-1)}\frac{d-1}{2}\int_W\int_W\frac{1-\exp(-\frac{2\kappa_{d-1}}{d\kappa_d}at\|x-y\|)}{\|x-y\|^2}\,\lambda_d(\mathrm{d}x)\,\lambda_d(\mathrm{d}y)
\end{aligned}
$$

$$\sim a^{2(d-1)}\frac{d-1}{2}\int\limits_W\int\limits_W \frac{1}{\|x-y\|^2}\,\lambda_d(\mathrm{d}x)\,\lambda_d(\mathrm{d}y)$$
$$= a^{2(d-1)}\frac{d-1}{2}E_2(W) \quad \text{as } a\to\infty .$$

The second formula follows from Lemma 12.12 applied with $\alpha = 2$. □

Remark 12.14. *We emphasize that the proof for the asymptotic variance formula for $d \geq 3$ does not carry over to the planar case $d = 2$. The reason for this is that for $d = 2$, the energy integral $\int_W \int_W \|x-y\|^{-2}\,\lambda_2(\mathrm{d}x)\,\lambda_2(\mathrm{d}y)$ diverges for any $W \in \mathcal{K}_2$. In particular, this implies in the isotropic case that $\sup_{t\geq 0} \operatorname{var}\Sigma_{V_{d-1},t,W} = \infty$ for $d = 2$, whereas $\sup_{t\geq 0} \operatorname{var}\Sigma_{V_{d-1},t,W} < \infty$ for $d \geq 3$.*

The asymptotic behavior of the variance of the total surface content is related to the volume of the growing window as follows. For $d = 2$, we have that

$$\frac{\operatorname{var}\Sigma_{V_1,t,aW}}{V_2(aW)} \sim \pi \log a\,,$$

which diverges to infinity at a logarithmic speed. On the other hand, for $d \geq 3$, we have

$$\frac{\operatorname{var}\Sigma_{V_{d-1},t,aW}}{V_d(aW)} \sim \frac{d-1}{2}\frac{E_2(W)}{V_d(W)}\,a^{d-2}\,,$$

which diverges to infinity at a polynomial speed. Thus, since $\operatorname{var}\Sigma_{V_1,t,aW}$ grows faster than the area or volume of aW, this suggests the presence of long-range spatial dependencies in stationary and isotropic STIT tessellations Y_t. The growth behavior of $\log a$ reveals that in the planar case, these dependencies are considerably weaker than in higher dimensions, where the growth follows an order of a^{d-2}. This is also reflected in the limit theorems in Chapter 13, which highlight differences across dimensions.

12.5.2 Refined results for the total edge length in dimension $d = 2$

In this section, we refine the asymptotic variance relation (12.27) for the total edge length in dimension $d = 2$ by providing a second-order term of the asymptotic expansion. The result will be used in Chapter 13 to prove functional limit theorems for the total edge length process. In this section, we focus on the isotropic case.

Theorem 12.15. *Consider the STIT tessellation process $(Y_t,\, t > 0)$ in the plane $\mathbb{R}^2$ driven by the motion-invariant line measure Θ_{iso}, and let $W \in \mathcal{K}_2$. Then, for any fixed $t > 0$,*

$$\operatorname{var}\Sigma_{V_1,t,aW} = \pi V_2(W)a^2\log a + S_t(W)a^2 + O(a) \quad \text{as } a\to\infty\,,$$

where $S_t(W)$ is given by

$$S_t(W) := \pi V_2(W)\Big(\log t + \log\frac{2}{\pi} + \gamma - 1\Big) + \pi \int_{[W]} V_1(W \cap L)\log V_1(W \cap L)\,\Theta_{\mathrm{iso}}(\mathrm{d}L),$$

and $\gamma \approx 0{,}577\ldots$ *is the Euler–Mascheroni constant.*

The proof of Theorem 12.15 relies on the following analytical result, in which again γ stands for the Euler–Mascheroni constant.

Lemma 12.16. *For* $0 < u < v < \infty$, $t > 0$, *and* $a > 0$, *define*

$$I(u,v,t,a) := \frac{\pi}{2}\int_u^v\int_u^v \frac{1 - \mathrm{e}^{-\frac{2}{\pi}at|x-y|}}{|x-y|}\,\mathrm{d}x\,\mathrm{d}y.$$

Then, for fixed u, v, and t,

$$I(u,v,t,a) = \pi(v-u)\Big(\log a + \log t + \log\frac{2}{\pi} + \gamma - 1 + \log(v-u)\Big) + O(a^{-1})$$

as $a \to \infty$.

Proof. We apply the shift $-y$ to x, then substitute $\frac{2}{\pi}atx$ by x', and finally replace x' by x again to see that

$$\begin{aligned} I(u,v,t,a) &= \frac{\pi}{2}\int_u^v\int_{u-y}^{v-y} \frac{1 - \mathrm{e}^{-\frac{2}{\pi}at|x|}}{|x|}\,\mathrm{d}x\,\mathrm{d}y \\ &= \frac{\pi}{2}\int_u^v\int_0^{\frac{2}{\pi}at(v-y)} \frac{1-\mathrm{e}^{-x}}{x}\,\mathrm{d}x\,\mathrm{d}y + \frac{\pi}{2}\int_u^v\int_0^{\frac{2}{\pi}at(y-u)} \frac{1-\mathrm{e}^{-x}}{x}\,\mathrm{d}x\,\mathrm{d}y \\ &= \frac{\pi}{2at}\int_0^{\frac{2}{\pi}at(v-u)}\int_0^y \frac{1-\mathrm{e}^{-x}}{x}\,\mathrm{d}x\,\mathrm{d}y \\ &= \frac{\pi^2}{2at}\int_0^{\frac{2}{\pi}at(v-u)} \mathrm{Ein}(y)\,\mathrm{d}y, \end{aligned}$$

where $\mathrm{Ein}(y)$ is the exponential integral $\mathrm{Ein}(y) := \int_0^y \frac{1-\mathrm{e}^{-x}}{x}\,\mathrm{d}x, y > 0$. This special function is known to satisfy the estimate

$$\log y + \gamma + \frac{\mathrm{e}^{-y}}{2}\log\Big(1 + \frac{2}{y}\Big) \le \mathrm{Ein}(y) \le \log y + \gamma + \mathrm{e}^{-y}\log\Big(1 + \frac{1}{y}\Big)$$

for $y > 0$. This follows from inequality (6.8.1) of [152] after noting that the function Ein is connected to the function $E_1(y) := \int_y^\infty \frac{e^{-x}}{x}\,dx$ by $\mathrm{Ein}(y) = E_1(y) + \log y + \gamma$. Putting $f(z) := \log(1 + \frac{1}{z})$, $z > 0$, and then defining

$$S(u, v, t, a) := \pi(v - u)\Big(\log t + \log a + \log\frac{2}{\pi} + \gamma - 1 + \log(v - u)\Big),$$

$$T_1(u, v, t, a) := \frac{\pi^2}{4ta}\int_0^{\frac{2}{\pi}t(v-u)a} f\Big(\frac{x}{2}\Big)\,e^{-x}\,dx\,,$$

$$T_2(u, v, t, a) := \frac{\pi^2}{4ta}\int_0^{\frac{2}{\pi}t(v-u)a} f(x)\,e^{-x}\,dx\,,$$

we arrive at the inequality

$$S(u, v, t, a) + T_1(u, v, t, a) \le I(u, v, t, a) \le S(u, v, t, a) + T_2(u, v, t, a)\,.$$

The observation that

$$\int_0^\infty f\Big(\frac{x}{2}\Big)\,e^{-x}\,dx < \infty \quad \text{and} \quad \int_0^\infty f(x)\,e^{-x}\,dx < \infty$$

completes the proof. □

Proof of Theorem 12.15. Using Theorem 12.3 with $d = 2$ together with the scaling relation (12.24), we see that

$$\operatorname{var}\Sigma_{V_1,t,aW} = \frac{\pi a^2}{2}\int_{[W]}\int_{W\cap L}\int_{W\cap L}\frac{1 - e^{-\frac{2}{\pi}ta\|x-y\|}}{\|x-y\|}\,\lambda_1(dx)\,\lambda_1(dy)\,\Theta_{\mathrm{iso}}(dL)\,.$$

Since the value of $I(u, v, t, a)$ in Lemma 12.16 with respect to u and v depends only on the length of the interval (u, v), we can apply the result to the inner double integral, where the length of the respective interval is $V_1(W \cap L)$. This gives

$$\begin{aligned}\operatorname{var}\Sigma_{V_1,t,aW} = \pi a^2 \int_{[W]} V_1(W\cap L)\Big(&\log a + \log t + \log\frac{2}{\pi}\\ &+ \gamma - 1 + \log V_1(W\cap L)\Big)\,\Theta_{\mathrm{iso}}(dL) + O(a)\,.\end{aligned}$$

The result follows by an application of Crofton's formula (1.6). □

Remark 12.17. *For a disc rB^2 of radius $r > 0$, we can compute the integral term in $S_t(rB^2)$ as follows:*

$$\begin{aligned}\int_{[rB^2]} V_1(rB^2 \cap L) \log V_1(rB^2 \cap L)\, \Theta_{\mathrm{iso}}(\mathrm{d}L) &= \int_{-r}^{r} 2\sqrt{r^2 - u^2} \log(2\sqrt{r^2 - u^2})\, \mathrm{d}u \\ &= 4r^2 \int_0^{\pi/2} \cos^2(t) \log(2r\cos(t))\, \mathrm{d}t \\ &= 4r^2 \log(2r) \int_0^{\pi/2} \cos^2(t)\, \mathrm{d}t + 4r^2 \int_0^{\pi/2} \cos^2(t) \log(\cos(t))\, \mathrm{d}t\,, \end{aligned} \tag{12.32}$$

where we applied the substitution $u = r \sin t$. The first integral is just $\int_0^{\pi/2} \cos^2(t)\, \mathrm{d}t = \pi/4$. For the second integral, we use identity (4.387.9) of [61] *with $n = 1$. This gives $\int_0^{\pi/2} \cos^2(t) \log(\cos(t))\, \mathrm{d}t = \frac{\pi}{4}(\frac{1}{2} - \log 2)$. Plugging this back into* (12.32)*, we see that the integral in $S_t(rB^2)$ equals*

$$\begin{aligned}\int_{[rB^2]} V_1(rB^2 \cap L) \log V_1(rB^2 \cap L)\, \Theta_{\mathrm{iso}}(\mathrm{d}L) &= \pi r^2 \log(2r) + 4r^2 \cdot \frac{\pi}{4}\left(\frac{1}{2} - \log 2\right) \\ &= \pi r^2 \left(\frac{1}{2} + \log r\right). \end{aligned} \tag{12.33}$$

Thus

$$S_t(rB^2) = \pi^2 r^2 \left(\log t + \log \frac{2}{\pi} + \gamma - \frac{1}{2} + \log r\right).$$

The integral term $\int_{[W]} V_1(W \cap L) \log V_1(W \cap L)\, \Theta_{\mathrm{iso}}(\mathrm{d}L)$ in $S_t(W)$, $W \in \mathcal{K}_2$, admits an entropic interpretation. For a nonnegative measurable function $f : A(2,1) \to \mathbb{R}$ on the space $A(2,1)$ of lines in $\mathbb{R}^2$ satisfying $\int_{A(2,1)} f(L)\, \Theta_{\mathrm{iso}}(\mathrm{d}L) = 1$, we define the **entropy** of f as

$$\mathrm{Ent}(f) := -\int_{A(2,1)} f(L) \log f(L)\, \Theta_{\mathrm{iso}}(\mathrm{d}L)\,. \tag{12.34}$$

Choosing for f the function $f_W(L) := V_2(W)^{-1} V_1(W \cap L)$, $L \in A(2,1)$, we arrive at

$$\begin{aligned}\mathrm{Ent}(f_W) &= -\frac{1}{V_2(W)} \int\limits_{A(2,1)} V_1(W\cap L)\log\frac{V_1(W\cap L)}{V_2(W)}\,\Theta_{\mathrm{iso}}(\mathrm{d}L)\\ &= -\frac{1}{V_2(W)} \int\limits_{A(2,1)} V_1(W\cap L)\log V_1(W\cap L)\,\Theta_{\mathrm{iso}}(\mathrm{d}L)\\ &\quad + \frac{\log V_2(W)}{V_2(W)} \int\limits_{A(2,1)} V_1(W\cap L)\,\Theta_{\mathrm{iso}}(\mathrm{d}L)\\ &= -\frac{1}{V_2(W)} \int\limits_{A(2,1)} V_1(W\cap L)\log V_1(W\cap L)\,\Theta_{\mathrm{iso}}(\mathrm{d}L) + \log V_2(W),\end{aligned}$$

where we used Crofton's formula (1.6) in the last step. Thus

$$\int\limits_{[W]} V_1(W\cap L)\log V_1(W\cap L)\,\Theta_{\mathrm{iso}}(\mathrm{d}L) = V_2(W)(\log V_2(W) - \mathrm{Ent}(f_W)),$$

and the term $S_t(W)$ in Theorem 12.15 can be represented as

$$S_t(W) = \pi V_2(W)\Big(\log\Big(\frac{2t}{\pi}V_2(W)\Big) + \gamma - 1 - \mathrm{Ent}(f_W)\Big). \tag{12.35}$$

12.5.3 Further results in dimension $d = 2$

After having investigated the asymptotic variance of the total surface content in a sequence of growing windows aW in general space dimensions $d \geq 2$, we turn now to the variance of the total number of I-segments and the covariance between the total edge length and the number of I-segments in dimension $d = 2$.

Theorem 12.18. *Consider the STIT tessellation process $(Y_t,\, t > 0)$ in the plane $\mathbb{R}^2$ driven by the motion-invariant line measure Θ_{iso}, and let $W \in \mathcal{K}_2$. Then, for any fixed $t > 0$,*

$$\mathrm{var}\,\Sigma_{V_0,t,aW} \sim \frac{4}{\pi}V_2(W)t^2a^2\log a \quad \textit{as } a \to \infty, \tag{12.36}$$

and

$$\mathrm{cov}(\Sigma_{V_1,t,aW}, \Sigma_{V_0,t,aW}) \sim 2V_2(W)ta^2\log a \quad \textit{as } a \to \infty.$$

Proof. From Theorem 12.10(a) we have

$$\begin{aligned}\mathrm{var}\,\Sigma_{V_0,t,aW} &= \mathrm{var}\,\Sigma_{V_0,at,W}\\ &= \frac{2at}{\pi}V_1(W) + \frac{3a^2t^2}{\pi}V_2(W) + \frac{8}{\pi^2}\int\limits_0^\infty \bar{c}_W(r)\int\limits_0^{at}(at-s)^2\exp\Big(-\frac{2rs}{\pi}\Big)\,\mathrm{d}s\,\mathrm{d}r\end{aligned}$$

$$= \frac{2at}{\pi} V_1(W) + \frac{3a^2t^2}{\pi} V_2(W) + \frac{8a^3}{\pi^2} \int_0^\infty \bar{c}_W(r) \int_0^t (t-s)^2 \exp\Big(-\frac{2ars}{\pi}\Big)\, \mathrm{d}s\, \mathrm{d}r\,.$$

To deal with the integral term, we can use the same argument as in the proof of Theorem 12.13. Namely, writing $w(u) = \sup\{r \geq 0 : V_2(W \cap (W + ru)) > 0\}$ for $u \in S^1$, we have that

$$\int_0^\infty \bar{c}_W(r) \int_0^t (t-s)^2 \exp\Big(-\frac{2ars}{\pi}\Big)\, \mathrm{d}s\, \mathrm{d}r$$
$$\sim V_2(W) \int_{S^1} \int_0^{w(u)} \int_0^t (t-s)^2 \exp\Big(-\frac{2ars}{\pi}\Big)\, \mathrm{d}s\, \mathrm{d}r\, \varphi_{\mathrm{iso}}(\mathrm{d}u)$$

as $a \to \infty$. Letting $a > 1/w(u)$ and substituting ar by r, we get

$$\int_0^{w(u)} \int_0^t (t-s)^2 \exp\Big(-\frac{2ars}{\pi}\Big)\, \mathrm{d}s\, \mathrm{d}r$$
$$= \frac{1}{a} \int_0^1 \int_0^t (t-s)^2 \exp\Big(-\frac{2rs}{\pi}\Big)\, \mathrm{d}s\, \mathrm{d}r + \frac{1}{a} \int_1^{aw(u)} \int_0^t (t-s)^2 \exp\Big(-\frac{2rs}{\pi}\Big)\, \mathrm{d}s\, \mathrm{d}r\,.$$

The first of these integral terms is bounded in a, whereas for the second one, we have

$$\frac{1}{a} \int_1^{aw(u)} \Big[\frac{\pi t^2}{2r} - \frac{\pi^2 t}{2r^2} + \frac{\pi^3}{4r^3}\big(1 - \mathrm{e}^{-\frac{2rt}{\pi}}\big)\Big] \mathrm{d}r = \frac{1}{a}\Big(\frac{\pi t^2}{2} \log a + O(1)\Big),$$

where $O(1)$ stands for a function that stays bounded in a, uniformly for all $u \in S^1$. Plugging this back into the expression for $\operatorname{var} \Sigma_{V_0,t,aW}$, we arrive at

$$\operatorname{var} \Sigma_{V_0,t,aW} \sim \frac{2at}{\pi} V_1(W) + \frac{3a^2t^2}{\pi} V_2(W) + \frac{8a^3}{\pi^2} \cdot V_2(W) \cdot \frac{1}{a} \frac{\pi t^2}{2} \log a$$
$$\sim \frac{4}{\pi} V_2(W) t^2 a^2 \log a$$

as $a \to \infty$, proving the first part of the theorem.

Similarly, we can use Theorem 12.10(b) to see that

$$\operatorname{cov}(\Sigma_{V_1,t,aW}, \Sigma_{V_0,t,aW})$$
$$= a \operatorname{cov}(\Sigma_{V_1,at,W}, \Sigma_{V_0,at,W})$$
$$= atV_2(W) + \frac{4a}{\pi} \int_0^\infty \bar{c}_W(r) \int_0^{at} (at - s) \exp\Big(-\frac{2rs}{\pi}\Big)\, \mathrm{d}s\, \mathrm{d}r$$

$$= atV_2(W) + \frac{4a^3}{\pi}\int_0^{\infty}\bar{c}_W(r)\int_0^t (t-s)\exp\Big(-\frac{2ars}{\pi}\Big)\,\mathrm{d}s\,\mathrm{d}r$$

$$\sim atV_2(W) + \frac{4a^2}{\pi}V_2(W)\int_{S^1}\Bigg[\int_0^1\int_0^t (t-s)\mathrm{e}^{-\frac{2rs}{\pi}}\,\mathrm{d}s\,\mathrm{d}r$$

$$+\int_1^{aw(u)}\Big[\frac{\pi t}{2r}-\frac{\pi^2}{4r^2}\big(1-\mathrm{e}^{-\frac{2rt}{\pi}}\big)\Big]\mathrm{d}r\Bigg]\,\varphi_{\mathrm{iso}}(\mathrm{d}u)$$

$$= atV_2(W) + \frac{4a^2}{\pi}V_2(W)\int_{S^1}\Big[O(1)+\Big(\frac{\pi t}{2}\log a + O(1)\Big)\Big]\,\varphi_{\mathrm{iso}}(\mathrm{d}u),$$

where again each $O(1)$-term stands for a function that stays bounded in a, uniformly for all $u \in S^1$. As a result,

$$\mathrm{cov}(\Sigma_{V_1,t,aW},\Sigma_{V_0,t,aW}) \sim \frac{4a^2}{\pi}V_2(W)\cdot\frac{\pi t}{2}\log a = 2V_2(W)ta^2\log a$$

as $a \to \infty$, proving the second part of the theorem. □

Remark 12.19. *From the previous theorem we conclude that*

$$\varrho(\Sigma_{V_1,t,aW},\Sigma_{V_0,t,aW}) := \frac{\mathrm{cov}(\Sigma_{V_1,t,aW},\Sigma_{V_0,t,aW})}{\sqrt{\mathrm{var}(\Sigma_{V_1,t,aW})\,\mathrm{var}(\Sigma_{V_0,t,aW})}} \longrightarrow 1, \quad a \to \infty.$$

This means that the two random variables $\Sigma_{V_1,t,aW}$ and $\Sigma_{V_0,t,aW}$ become asymptotically perfectly correlated, that is, asymptotically, one is a linear function of the other. This is well reflected in the joint limit theorem, Theorem 13.15, in which the limiting random variables are linear functions of each other.

12.5.4 Dimension $d \geq 3$

With a method similar to that we used in Section 12.5.3 in the planar case $d = 2$, we can in general space dimensions $d \geq 3$ derive the following relations for the asymptotic covariances $\mathrm{cov}(\Sigma_{V_i,t,aW},\Sigma_{V_j,t,aW})$ as $a \to \infty$. We emphasize that for $d \geq 3$, the growth rate is different from the one we encountered in Theorem 12.18 for the planar case $d = 2$.

Theorem 12.20. *Consider the STIT tessellation process $(Y_t,\, t > 0)$ in $\mathbb{R}^d$ with $d \geq 3$ driven by the motion-invariant hyperplane measure Θ_{iso}. Let $W \in \mathcal{K}_d$ and $i,j \in \{0,\dots,d-1\}$. Then, for any fixed $t > 0$,*

$$\operatorname{cov}(\Sigma_{V_i,t,aW},\Sigma_{V_j,t,aW}) \sim \frac{d-1}{2}\frac{t^{2(d-1)-(i+j)}}{(2(d-1)-(i+j))!}\binom{2(d-1)-(i+j)}{d-i-1} \times\left(\prod_{\ell=0}^{d-i-1} c_{i+\ell,d}\right)\left(\prod_{\ell=0}^{d-j-1} c_{j+\ell,d}\right)E_2(W)a^{2(d-1)}$$

as $a\to\infty$ and with the constants $c_{i,j}$ given by (1.6). *In particular,*

$$\operatorname{var}\Sigma_{V_i,t,aW} \sim \frac{d-1}{2}\frac{t^{2(d-i-1)}}{(2(d-i-1))!}\binom{2(d-i-1)}{d-i-1}\left(\prod_{\ell=0}^{d-i-1} c_{i+\ell,d}\right)^2 E_2(W)a^{2(d-1)}\,.$$

Proof. We start by observing that the scaling property (6.2) of STIT tessellations implies that

$$\operatorname{cov}(\Sigma_{V_i,t,aW},\Sigma_{V_j,t,aW}) = a^{i+j}\operatorname{cov}(\Sigma_{V_i,at,W},\Sigma_{V_j,at,W})$$

and that the latter covariance is given by Theorem 12.4 as

$$\operatorname{cov}(\Sigma_{V_i,at,W},\Sigma_{V_j,at,W}) = \sum_{m=0}^{d-i-1}\sum_{n=0}^{d-j-1}\binom{m+n}{m}\left(\prod_{\ell=0}^{m-1} c_{i+\ell,d}\right)\left(\prod_{\ell=0}^{n-1} c_{j+\ell,d}\right)I^{m+n+1}(\mathbb{E}A_{i+m,j+n,\cdot,W},at)\,.$$

Now fixing $m\in\{0,\dots,d-i-1\}$ and $n\in\{0,\dots,d-j-1\}$ and applying definition (10.41) of $A_{i+m,j+n,\cdot,W}$, we find that for $a>1/t$,

$$\begin{aligned}
I^{n+m+1}&(\mathbb{E}A_{i+m,j+n,\cdot,W},at)\\
&= \frac{1}{(n+m)!}\int_0^{at}(at-s)^{n+m}\mathbb{E}A_{i+m,j+n,s,W}\,\mathrm{d}s\\
&= \frac{a^{n+m+1}}{(n+m)!}\int_0^{t}(t-s)^{n+m}\mathbb{E}A_{i+m,j+n,as,W}\,\mathrm{d}s\\
&= \frac{a^{n+m+1}}{(n+m)!}\int_0^{1/a}(t-s)^{n+m}\mathbb{E}A_{i+m,j+n,as,W}\,\mathrm{d}s\\
&\quad+\frac{a^{n+m+1}}{(n+m)!}\int_{1/a}^{t}(t-s)^{n+m}\mathbb{E}A_{i+m,j+n,as,W}\,\mathrm{d}s =: S_1+S_2\,.
\end{aligned}\tag{12.37}$$

In what follows, we deal with the terms S_1 and S_2 separately, starting with S_2. Writing $Z_{as,H}$ for the typical cell of the sectional tessellation $Y_{as}\cap H$, we can use Lemma 2.24 and the Cauchy–Schwarz inequality to see that

$$\begin{aligned}
S_2 &= \frac{a^{n+m+1}}{(n+m)!}\int_{1/a}^{t}(t-s)^{n+m}\int_{[W]}\mathbb{E}\sum_{F\in Y_{as,W}\cap H}V_{i+m}(F)V_{j+n}(F)\,\Theta_{\mathrm{iso}}(\mathrm{d}H)\,\mathrm{d}s\\
&\le \frac{a^{n+m+1}}{(n+m)!}\int_{1/a}^{t}(t-s)^{n+m}\int_{[W]}\mathbb{E}\sum_{\substack{F\in Y_{as}\cap H\\ F\cap W\neq\emptyset}}V_{i+m}(F)V_{j+n}(F)\,\Theta_{\mathrm{iso}}(\mathrm{d}H)\,\mathrm{d}s\\
&= \frac{a^{n+m+1}}{(n+m)!}\int_{1/a}^{t}(t-s)^{n+m}\\
&\quad\times\int_{[W]}\frac{\mathbb{E}[V_{i+m}(Z_{as,H})V_{j+n}(Z_{as,H})V_{d-1}(Z_{as,H}+(-(W\cap H)))]}{\mathbb{E}V_{d-1}(Z_{as,H})}\,\Theta_{\mathrm{iso}}(\mathrm{d}H)\,\mathrm{d}s\\
&\le \frac{a^{n+m+1}}{(n+m)!}\int_{1/a}^{t}(t-s)^{n+m}\\
&\quad\times\int_{[W]}\frac{\sqrt{\mathbb{E}[V_{i+m}(Z_{as,H})^2V_{j+n}(Z_{as,H})^2]\mathbb{E}[V_{d-1}(Z_{as,H}+(-(W\cap H)))^2]}}{\mathbb{E}V_{d-1}(Z_{as,H})}\\
&\quad\Theta_{\mathrm{iso}}(\mathrm{d}H)\,\mathrm{d}s\,.
\end{aligned}$$

By the scaling property (6.2) of STIT tessellations and the homogeneity of intrinsic volumes we have that

$$\begin{aligned}
\sqrt{\mathbb{E}[V_{i+m}(Z_{as,H})^2V_{j+n}(Z_{as,H})^2]} &= \left(\frac{1}{as}\right)^{i+m+j+n}\sqrt{\mathbb{E}[V_{i+m}(Z_{1,H})^2V_{j+n}(Z_{1,H})^2]}\\
&=: \left(\frac{1}{as}\right)^{i+m+j+n} c_1(i,j,n,m)
\end{aligned}$$

and similarly

$$\mathbb{E}V_{d-1}(Z_{as,H}) = \left(\frac{1}{as}\right)^{d-1}\mathbb{E}V_{d-1}(Z_{1,H}) =: \left(\frac{1}{as}\right)^{d-1} c_2\,.$$

We emphasize that $\mathbb{E}[V_{i+m}(Z_{1,H})^2V_{j+n}(Z_{1,H})^2]$ and $\mathbb{E}V_{d-1}(Z_{1,H})$ do not depend on the choice of H, since the random tessellation Y_1 is stationary and isotropic. Moreover, we have that $c_2 > 0$. Thus

$$\begin{aligned}
S_2 \le{}& \frac{a^{2-i-j-d}}{(n+m)!}\int_{1/a}^{t}\frac{(t-s)^{n+m}}{s^{i+m+j+n+d-1}}\int_{[W]}\frac{c_1(i,j,n,m)}{c_2}\\
&\times\sqrt{\mathbb{E}[V_{d-1}(Z_{as,H}+(-(W\cap H)))^2]}\,\Theta_{\mathrm{iso}}(\mathrm{d}H)\,\mathrm{d}s
\end{aligned}$$

$$= \frac{a^{2-i-j-2d}}{(n+m)!} \frac{c_1(i,j,n,m)}{c_2} \int_{1/a}^{t} \frac{(t-s)^{n+m}}{s^{i+m+j+n+2d-1}}$$

$$\times \int_{[asW]} \sqrt{\mathbb{E}[V_{d-1}(Z_{1,H} + (-(asW \cap H)))^2]}\, \Theta_{\mathrm{iso}}(\mathrm{d}H)\, \mathrm{d}s\,,$$

where we used once again the scaling property of STIT tessellations.

Let us write B_W for the smallest ball containing W and $R_W > 0$ for its radius. Then

$$\begin{aligned} V_{d-1}(Z_{1,H} + (-(asW \cap H)))^2 &\le V_{d-1}(Z_{1,H} + (asB_W \cap H))^2 \\ &= \left(\sum_{\ell=0}^{d-1} (asR_W)^{d-\ell-1} \kappa_{d-\ell-1} V_\ell(Z_{1,H}) \right)^2 \\ &\le 2^{d-1} \sum_{\ell=0}^{d-1} (asR_W)^{2(d-\ell-1)} \kappa_{d-\ell-1}^2 V_\ell(Z_{1,H})^2\,, \end{aligned}$$

where we applied the Steiner formula (1.3) within H and the elementary inequality $(b_0 + \cdots + b_{d-1})^2 \le 2^{d-1}(b_0^2 + \cdots + b_{d-1}^2)$ for $b_0, \ldots, b_{d-1} \ge 0$ in the last step. Since $\mathbb{E}[V_\ell(Z_{1,H})^2] < \infty$ for $\ell \in \{0, \ldots, d-1\}$, we see that, as $a \to \infty$, the expectation of the last expression is of order $O((as)^{2(d-1)})$ as long as $s > 1/a$. Moreover, $\Theta_{\mathrm{iso}}([asW]) = O(V_1(asW)) = O(as)$. Thus

$$\begin{aligned} S_2 &= O\left(a^{2-i-j-2d} \int_{1/a}^{t} \frac{(t-s)^{n+m}}{s^{i+m+j+n+2d-1}} \cdot \sqrt{(as)^{2(d-1)}} \cdot as\, \mathrm{d}s \right) \\ &= O(a^{2-i-j-d}\, a^{i+m+j+n+d-2}) = O(a^{n+m}) \end{aligned}$$

as $a \to \infty$.

Next, recall from (12.37) the definition of S_1 and observe that

$$\begin{aligned} S_1 &= \frac{a^{n+m+1}}{(n+m)!} \int_0^{1/a} (t-s)^{n+m} \int_{[W]} \mathbb{E} \sum_{F \in Y_{as,W} \cap H} V_{i+m}(F) V_{j+n}(F)\, \Theta_{\mathrm{iso}}(\mathrm{d}H)\, \mathrm{d}s \\ &\le \frac{a^{n+m+1}}{(n+m)!} c_3(W) \int_0^{1/a} (t-s)^{n+m} \int_{[W]} \mathbb{E} \sum_{F \in Y_{as} \cap H} \mathbb{1}\{F \cap (W \cap H) \ne \emptyset\}\, \Theta_{\mathrm{iso}}(\mathrm{d}H)\, \mathrm{d}s\,, \end{aligned}$$

where $c_3(W) := \max_{H \in [W]} V_{i+m}(W \cap H) V_{j+n}(W \cap H)$. We now apply Lemma 2.24 once again to see that

$$\mathbb{E} \sum_{F \in Y_{as} \cap H} \mathbb{1}\{F \cap (W \cap H) \ne \emptyset\} = \frac{\mathbb{E} V_{d-1}(Z_{as,H} + (-(W \cap H)))}{\mathbb{E} V_{d-1}(Z_{as,H})}\,.$$

As above, we have that

$$\mathbb{E}V_{d-1}(Z_{as,H}+(-(W\cap H)))=O((as)^{d-1})\quad\text{and}\quad\mathbb{E}V_{d-1}(Z_{as,H})=c_2(as)^{-(d-1)},$$

which yields

$$\int\limits_{[W]}\mathbb{E}\sum_{F\in Y_{as}\cap H}\mathbb{1}\{F\cap(W\cap H)\neq\emptyset\}\,\Theta_{\mathrm{iso}}(\mathrm{d}H)=O((as)^{-(d-1)}\cdot(as)^{d-1})=O(1)$$

as $a\to\infty$, provided that $s<1/a$. As a consequence,

$$S_1=O\left(a^{n+m+1}\int\limits_0^{1/a}(t-s)^{n+m}\,\mathrm{d}s\right)=O\left(a^{n+m+1}\cdot\frac{1}{a}\right)=O(a^{n+m})$$

as $a\to\infty$.

The bounds for S_1 and S_2 show that all terms in the covariance representation with $(m,n)\neq(d-i-1,d-j-1)$ are of order

$$O(a^{i+j}\cdot a^{(d-i-1)+(d-j-1)-1})=O(a^{2d-3}),\tag{12.38}$$

which is negligible in comparison to the term corresponding to the pair $(m,n)=(d-i-1,d-j-1)$, which is of order $O(a^{2(d-1)})$.

However, for this term, which is given by

$$S_3:=\binom{2(d-1)-(i+j)}{d-i-1}\left(\prod_{\ell=0}^{d-i-1}c_{i+\ell,d}\right)\left(\prod_{\ell=0}^{d-j-1}c_{j+\ell,d}\right)\times I^{2(d-1)-(i+j)+1}(\mathbb{E}A_{d-1,d-1,\cdot,W},at),$$

we can be much more precise. Indeed, using the variance representation (12.20), we see that

$$\begin{aligned}
&I^{2(d-1)-(i+j)+1}(\mathbb{E}A_{d-1,d-1,\cdot,W},at)\\
&=I^{2(d-1)-(i+j)}(\operatorname{var}\Sigma_{V_{d-1},\cdot,W},at)\\
&=\frac{1}{(2(d-1)-(i+j)-1)!}\int\limits_0^{at}(at-s)^{2(d-1)-(i+j)-1}\\
&\quad\times\frac{d-1}{2}\int\limits_W\int\limits_W\frac{1-e^{-\frac{2\kappa_{d-1}}{d\kappa_d}s\|x-y\|}}{\|x-y\|^2}\,\lambda_d(\mathrm{d}x)\,\lambda_d(\mathrm{d}y)\,\mathrm{d}s\\
&=\frac{d-1}{2}\,\frac{a^{2(d-1)-(i+j)}}{(2(d-1)-(i+j)-1)!}\int\limits_W\int\limits_W\\
&\quad\times\int\limits_0^t(t-s)^{2(d-1)-(i+j)-1}\frac{1-e^{-\frac{2\kappa_{d-1}}{d\kappa_d}as\|x-y\|}}{\|x-y\|^2}\,\mathrm{d}s\,\lambda_d(\mathrm{d}x)\,\lambda_d(\mathrm{d}y)
\end{aligned}$$

$$= \frac{d-1}{2}\frac{a^{2(d-1)-(i+j)}}{(2(d-1)-(i+j)-1)!}\int_W\int_W \frac{1}{2(d-1)-(i+j)}\frac{t^{2(d-1)-(i+j)}}{\|x-y\|^2} + O\left(\frac{1}{a}\right)\lambda_d(\mathrm{d}x)\,\lambda_d(\mathrm{d}y),$$

where in the last step, we expanded the exponential function into a series and carried out the integration for the first term, all further summands only contributing to the error term $O(1/a)$. Plugging this back into the expression for $\mathrm{cov}(\Sigma_{V_i,t,aW}, \Sigma_{V_j,t,aW})$ and recalling the definition of the 2-energy $E_2(W)$ of W from (12.25), we finally arrive at

$$\begin{aligned}
\mathrm{cov}(\Sigma_{V_i,t,aW}, \Sigma_{V_j,t,aW}) &\sim a^{i+j} S_3 \\
&\sim a^{i+j}\frac{d-1}{2}\frac{(at)^{2(d-1)-(i+j)}}{(2(d-1)-(i+j))!}\binom{2(d-1)-(i+j)}{d-i-1} \\
&\quad\times\left(\prod_{\ell=0}^{d-i-1} c_{i+\ell,d}\right)\left(\prod_{\ell=0}^{d-j-1} c_{j+\ell,d}\right)E_2(W) \\
&= \frac{d-1}{2}\frac{t^{2(d-1)-(i+j)}}{(2(d-1)-(i+j))!}\binom{2(d-1)-(i+j)}{d-i-1} \\
&\quad\times\left(\prod_{\ell=0}^{d-i-1} c_{i+\ell,d}\right)\left(\prod_{\ell=0}^{d-j-1} c_{j+\ell,d}\right)E_2(W)a^{2(d-1)}.
\end{aligned}$$

This completes the proof of the covariance asymptotics. The variance asymptotics follows from this by choosing $i = j$. □

Remark 12.21. *Using Lemma 12.12, the result of Theorem 12.20 can alternatively be expressed in terms of the chord power integral $I_{d-1}(W)$ of W.*

Remark 12.22. *From Theorem 12.20 we conclude that for any fixed $t > 0$ and $i,j \in \{0,\dots,d-1\}$,*

$$\varrho(\Sigma_{V_i,t,aW}, \Sigma_{V_j,t,aW}) = \frac{\mathrm{cov}(\Sigma_{V_i,t,aW}, \Sigma_{V_j,t,aW})}{\sqrt{\mathrm{var}(\Sigma_{V_i,t,aW})\,\mathrm{var}(\Sigma_{V_j,t,aW})}} \longrightarrow 1, \quad a\to\infty.$$

This means that all pairs of intrinsic volume functionals $\Sigma_{V_i,t,aW}$ and $\Sigma_{V_j,t,aW}$ become asymptotically perfectly correlated as the observation window grows. As in the planar case, this is in line with the limit theorem in Theorem 13.25 we derive in the next chapter.

12.6 Extremal variances and covariances

It is a natural question to ask for which convex bodies $W \in \mathcal{K}_d$ the variance $\mathrm{var}\,\Sigma_{V_{d-1},t,W}$ of the total surface content of a stationary and isotropic STIT tessellation Y_t in $\mathbb{R}^d$ attains its maximal value if the volume of W and the time parameter $t > 0$ are kept fixed. In space dimensions $d \geq 3$, we may consider a similar question for the asymptotic covariance

$$c_{i,j,t,W} := \lim_{a\to\infty} a^{-2(d-1)} \operatorname{cov}(\Sigma_{V_i,t,aW}, \Sigma_{V_j,t,aW}) \tag{12.39}$$

between $\Sigma_{V_i,t,aW}$ and $\Sigma_{V_j,t,aW}$ for $i,j \in \{0,\dots,d-1\}$ and ask for the maximal value $c_{i,j,t,W}$ can take as a function of W. We remark that the limit in (12.39) indeed exists as a consequence of Theorem 12.20. Also note that the analogous question in dimension $d = 2$ is not interesting, since in this case the asymptotic variances and covariances only depend on the area but not on the shape of W; see Theorem 12.18. We summarize the answers to two extremum problems formulated in the following result.

Theorem 12.23. *Fix $t > 0$ and $W \in \mathcal{K}_d$ and consider a stationary and isotropic STIT tessellation Y_t in $\mathbb{R}^d$ with time parameter t. Further, let $B^d(W)$ be a d-dimensional ball satisfying $V_d(B^d(W)) = V_d(W)$.*
(a) *For general space dimensions $d \geq 2$, we have that*

$$\operatorname{var} \Sigma_{V_{d-1},t,W} \leq \operatorname{var} \Sigma_{V_{d-1},t,B^d(W)}\,.$$

(b) *For space dimensions $d \geq 3$, we have that*

$$c_{i,j,t,W} \leq c_{i,j,t,B^d(W)}\,.$$

Equality in (a) and (b) holds if and only if $W = \mathrm{Isom}(B^d(W))$ for some isometry $\mathrm{Isom} : \mathbb{R}^d \to \mathbb{R}^d$, up to a set of λ_d-measure zero.

To prove Theorem 12.23, we use the sharp **Riesz rearrangement inequality**. We rephrase here a special case of this inequality in a slightly more general setting, to which the statement and proof of Theorem 2 of [23] can be extended; see also Theorem 2 of [22] and the discussion around Lemma 6.1 of [15]. To formulate it, let $f : \mathbb{R}^d \to \mathbb{R}$ be a nonnegative Lebesgue-integrable function, and let f^* be the **symmetric decreasing rearrangement** of f with respect to the origin, that is,

$$f^*(x) := \int_0^\infty \mathbb{1}_{\{f>s\}^*}(x)\,\mathrm{d}s\,, \quad x \in \mathbb{R}^d\,,$$

where $\{f > s\}^*$ denotes the ball centered at the origin and has the same volume as the level set $\{f > s\}$. In what follows, we say that the function f is nonzero if $\lambda_d(\{f > 0\}) > 0$.

Proposition 12.24. *Let $f, g : \mathbb{R}^d \to [0,\infty)$ be Lebesgue integrable, let $K : [0,\infty) \to [0,\infty]$ be decreasing, and define*

$$I_K(f,g) := \int_{\mathbb{R}^d}\int_{\mathbb{R}^d} f(x)g(y)K(\|x-y\|)\,\lambda_d(\mathrm{d}x)\,\lambda_d(\mathrm{d}y)\,.$$

Then $I_K(f,g) \le I_K(f^*,g^*)$. Moreover, if K is strictly decreasing, f and g are nonzero, and $I_K(f^*,g^*) < \infty$, then $I_K(f,g) = I_K(f^*,g^*)$ if and only if there is an isometry $\mathrm{Isom} : \mathbb{R}^d \to \mathbb{R}^d$ *such that $f = f^* \circ \mathrm{Isom}$ and $g = g^* \circ \mathrm{Isom}$ λ_d-almost everywhere.*

Proof of Theorem 12.23. To prove part (a), we recall from (12.20) that

$$\operatorname{var}\Sigma_{V_{d-1},t,W} = \frac{(d-1)}{2}\int\limits_W\int\limits_W \frac{1-\exp(-\frac{2\kappa_{d-1}}{d\kappa_d}t\|x-y\|)}{\|x-y\|^2}\,\lambda_d(\mathrm{d}x)\,\lambda_d(\mathrm{d}y)\,.$$

Our goal is to apply the sharp Riesz rearrangement inequality with $f = g = \mathbb{1}_W$ and

$$K(\|x-y\|) := \frac{(d-1)}{2}\frac{1-\exp(-\frac{2\kappa_{d-1}}{d\kappa_d}t\|x-y\|)}{\|x-y\|^2}\,.$$

Since $W \in \mathcal{K}_d$, the functions f and g are integrable. Moreover,

$$K(r) := \frac{(d-1)}{2}\frac{1-\exp(-\frac{2\kappa_{d-1}}{d\kappa_d}tr)}{r^2}\,, \qquad r > 0\,,$$

is strictly decreasing in r. The proof of part (a) of the theorem is now completed by an application of Proposition 12.24 and by noticing that $f^* = g^* = \mathbb{1}_{B^d(W)}$:

$$\operatorname{var}\Sigma_{V_{d-1},t,W} = I_K(\mathbb{1}_W,\mathbb{1}_W) \le I_K(\mathbb{1}_W^*,\mathbb{1}_W^*) = I_K(\mathbb{1}_{B^d(W)},\mathbb{1}_{B^d(W)}) = \operatorname{var}\Sigma_{V_{d-1},t,B^d(W)}\,.$$

To see that (b) holds, we recall from Theorem 12.20 that

$$c_{i,j,t,W} = c(i,j,d)t^{2(d-1)-(i+j)}E_2(W)$$

for some constant $c(i,j,d)$ only depending on i, j and the space dimension d. To the 2-energy $E_2(W)$ of W, we can again apply Proposition 12.24 with $f = g = \mathbb{1}_W$ and $K(\|x-y\|) := \|x-y\|^{-2}$ to conclude the result as in part (a).

Finally, the equality cases in (a) and (b) are also covered by Proposition 12.24, since in all cases the kernel K is even strictly decreasing and both f and g are nonzero as $V_d(W) > 0$ by assumption. □

A similar question can also be asked for the geometric prefactor $S_t(W)$, $W \in \mathcal{K}_2$, in the second-order term appearing in the refined variance asymptotics of Theorem 12.15.

Theorem 12.25. *Fix $t > 0$ and $W \in \mathcal{K}_2$ and consider a stationary and isotropic STIT tessellation Y_t in the plane $\mathbb{R}^2$ with time parameter t. Further, let $B^2(W)$ be a disc satisfying $V_2(B^2(W)) = V_2(W)$. Then*

$$S_t(W) \le S_t(B^2(W))\,.$$

Proof. We use the entropic representation

$$S_t(W) = \pi V_2(W)\Big(\log\Big(\frac{2t}{\pi}V_2(W)\Big) + \gamma - 1 - \mathrm{Ent}(f_W)\Big)$$

of $S_t(W)$ established in (12.35) with the entropy defined at (12.34) and the function f_W given by $f_W(L) = V_2(W)^{-1}V_1(W \cap L)$, $L \in A(2,1)$. Next, we employ the entropy inequality

$$\mathrm{Ent}(f_W) \geq \mathrm{Ent}(f_{B^2(W)})$$

from Theorem 3.6 of [201] or equation (7.4) of [205]. Since $V_2(B^2(W)) = V_2(W)$ by assumption, it follows that $S_t(W) \leq S_t(B^2(W))$, as claimed. □

12.7 Second moment measure and pair correlation function of the surface measure

12.7.1 A general framework

We start by introducing some notation and by setting up a general framework, which will get specialized in the subsequent sections to different situations. For that purpose, let Φ and Ψ be two random stationary measures on $\mathbb{R}^d$ with positive and finite intensities $\gamma_\Phi := \mathbb{E}\Phi(B)$ and $\gamma_\Psi := \mathbb{E}\Psi(B)$, respectively, where $B \in \mathcal{B}(\mathbb{R}^d)$ is an arbitrary Borel set with $\lambda_d(B) = 1$. The **cross-covariance measure** $\mathrm{Cross}_{\Phi,\Psi}$ of Φ and Ψ is defined as

$$\mathrm{Cross}_{\Phi,\Psi}(B_1 \times B_2) := \mathrm{cov}(\Phi(B_1), \Psi(B_2)), \quad B_1, B_2 \in \mathcal{B}(\mathbb{R}^d).$$

By stationarity of Φ and Ψ we can define the **reduced cross-covariance measure** $\widehat{\mathrm{Cross}}_{\Phi,\Psi}$ on $\mathbb{R}^d$ by the relation

$$\mathrm{Cross}_{\Phi,\Psi}(B_1 \times B_2) = \int_{B_1}\int_{B_2 - x} \widehat{\mathrm{Cross}}_{\Phi,\Psi}(\mathrm{d}y)\,\lambda_d(\mathrm{d}x).$$

The reduced cross-covariance measure is related to the **reduced mixed moment measure** $\mu^{(2)}_{\Phi,\Psi}$ of Φ and Ψ by

$$\widehat{\mu}^{(2)}_{\Phi,\Psi}(B) := \widehat{\mathrm{Cross}}_{\Phi,\Psi}(B) + \gamma_\Phi\,\gamma_\Psi\,\lambda_d(B), \quad B \in \mathcal{B}(\mathbb{R}^d). \tag{12.40}$$

If Φ and Ψ are also isotropic, then the reduced mixed moment measure is rotation invariant, and we may equivalently describe it through the radial **reduced mixed moment function**

$$K_{\Phi,\Psi}(r) := \gamma_\Phi^{-1}\,\gamma_\Psi^{-1}\,\widehat{\mu}^{(2)}_{\Phi,\Psi}(rB^d), \quad r \geq 0. \tag{12.41}$$

Finally, we define the **cross-correlation function** by

$$g_{\Phi,\Psi}(r) := \frac{1}{d\kappa_d r^{d-1}} \frac{\mathrm{d}K_{\Phi,\Psi}(r)}{\mathrm{d}r}, \qquad r > 0,$$

of Φ and Ψ, provided that the derivative exists. The value of the function $g_{\Phi,\Psi}(r)$ describes how the reduced mixed moment measure is distributed over spherical shells of radius r. Equivalently, $g_{\Phi,\Psi}(r)$ gives the relative likelihood, compared to a stationary Poisson point process, of finding mass of Φ at distance r from a typical point of Ψ. In the special case that $\Psi = \Phi$, we write

$$\begin{aligned}
&\mathrm{Cov}_\Phi(B_1 \times B_2) && \text{instead of } \mathrm{Cross}_{\Phi,\Phi}(B_1 \times B_2),\\
&\widehat{\mathrm{Cov}}_\Phi(B) && \text{instead of } \widehat{\mathrm{Cross}}_{\Phi,\Phi}(B),\\
&\widehat{\mu}^{(2)}_\Phi(B) && \text{instead of } \widehat{\mu}^{(2)}_{\Phi,\Phi}(B),\\
&K_\Phi(r) && \text{instead of } K_{\Phi,\Phi}(r),\\
&g_\Phi(r) && \text{instead of } g_{\Phi,\Phi}(r)
\end{aligned}$$

for $B_1, B_2, B \in \mathcal{B}(\mathbb{R}^d)$ and $r > 0$ and call

$$\begin{aligned}
&\mathrm{Cov}_\Phi && \textbf{the covariance measure of } \Phi,\\
&\widehat{\mathrm{Cov}}_\Phi && \textbf{the reduced covariance measure of } \Phi,\\
&\widehat{\mu}^{(2)}_\Phi && \textbf{the reduced second moment measure of } \Phi,\\
&K_\Phi && \textbf{the } K\textbf{-function of } \Phi,\\
&g_\Phi && \textbf{the pair correlation function of } \Phi.
\end{aligned}$$

We remark that the K-function and the pair correlation function are widely used tools in spatial statistics, stochastic geometry, and statistical physics for studying the second-order structure of stationary and isotropic random sets, point processes, or random measures. They quantify second-order information by measuring the expected mass of the random measure Φ contained in a ball of radius r centered at a typical point of its support.

It is possible to give a representation of the reduced mixed moment measure $\widehat{\mu}^{(2)}_{\Phi,\Psi}$ and the reduced mixed moment function $K_{\Phi,\Psi}$ in the spirit of Palm distributions, which is sometimes easier to work with directly. Namely, for $B \in \mathcal{B}(\mathbb{R}^d)$, we have that

$$\widehat{\mu}^{(2)}_{\Phi,\Psi}(B) = \mathbb{E} \int_A \Psi(B + x)\, \Phi(\mathrm{d}x) = \mathbb{E} \int_A \Phi(B + x)\, \Psi(\mathrm{d}x),$$

where $A \in \mathcal{B}(\mathbb{R}^d)$ is an arbitrary set with Lebesgue measure one, for example, $A = [0,1]^d$; see equation (2.22) of [13]. The measure $\widehat{\mu}_{\Phi,\Psi}$ describes Ψ as seen from a typical point of

Φ or, equivalently, Φ as seen from a typical point of Ψ. This Palm-type representation is valid without an isotropy assumption. If isotropy holds, then it naturally reduces to the radial formulation in (12.41). Using such a representation, we can write

$$K_{\Phi,\Psi}(r) = \frac{1}{\gamma_\Phi \gamma_\Psi} \mathbb{E} \int_A \Psi(rB^d + x)\, \Phi(\mathrm{d}x) = \frac{1}{\gamma_\Phi \gamma_\Psi} \mathbb{E} \int_A \Phi(rB^d + x)\, \Psi(\mathrm{d}x)\,. \tag{12.42}$$

12.7.2 Second moment measure of STIT tessellations

In this section, we determine the covariance measure and the pair correlation function of the random surface measure

$$\mathscr{H}_t^{d-1}(\,\cdot\,) := \mathscr{H}^{d-1}(\,\cdot\, \cap \partial Y_t)$$

induced by the STIT tessellation Y_t in $\mathbb{R}^d$, driven by a hyperplane measure Θ, and governed by the time parameter $t > 0$. We note that $\mathscr{H}_t^{d-1}(\,\cdot\,)$ is the stationary counterpart of the random measure $\mathscr{H}_{t,W}^{d-1}(\,\cdot\,)$ from (10.46), but without restriction to a window W. By applying the general definitions from Section 12.7.1 to the random measure $\Phi = \mathscr{H}_t^{d-1}$ we can introduce

the covariance measure	$\mathrm{Cov}_{d-1,t} := \mathrm{Cov}_{\mathscr{H}_t^{d-1}}$,
the reduced covariance measure	$\widehat{\mathrm{Cov}}_{d-1,t} := \widehat{\mathrm{Cov}}_{\mathscr{H}_t^{d-1}}$,
the reduced second moment measure	$\widehat{\mu}^{(2)}_{d-1,t} := \widehat{\mu}^{(2)}_{\mathscr{H}_t^{d-1}}$,
the K-function	$K_{d-1,t} := K_{\mathscr{H}_t^{d-1}}$,
the pair correlation function	$g_{d-1,t} := g_{\mathscr{H}_t^{d-1}}$

of $\mathscr{H}_t^{d-1}(\,\cdot\,)$. We remark that $K_{d-1,t}$ and $g_{d-1,t}$ are considered only if $\Theta = \Theta_{\mathrm{iso}}$, in which case the random measure $\mathscr{H}_t^{d-1}(\,\cdot\,)$ is stationary and isotropic; see Figure 12.1.

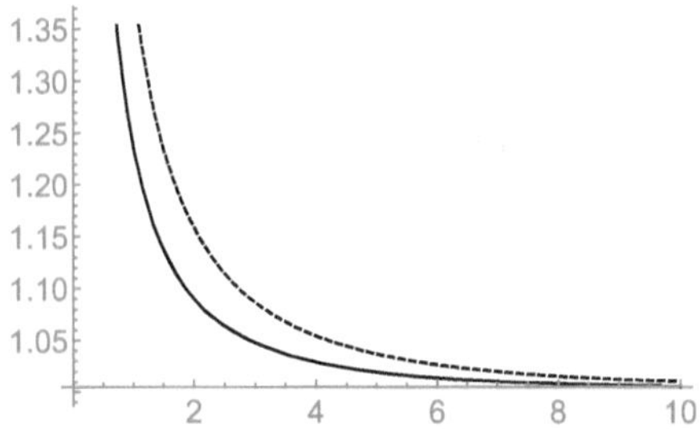

Figure 12.1: Pair correlation function $g_{d-1,t}$ for $d = 2$ (solid) and $d = 3$ (dashed) with $t = 1$.

Theorem 12.26. *Let Θ be a hyperplane measure of the form (4.1), satisfying (4.3), and let $(Y_t, t > 0)$ be the STIT tessellation process driven by the measure Θ. Then for any fixed $t > 0$, the covariance measure of $\mathcal{H}^{d-1}(\cdot \cap \partial Y_t)$ is given by*

$$\mathrm{Cov}_{d-1,t}(B_1 \times B_2) = \int_{\mathcal{H}} \int_{B_1\cap H} \int_{B_2\cap H} \frac{1-\mathrm{e}^{-t\,\Theta([xy])}}{\Theta([xy])} \lambda_{d-1}(\mathrm{d}x)\,\lambda_{d-1}(\mathrm{d}y)\,\Theta(\mathrm{d}H)$$

for bounded $B_1, B_2 \in \mathcal{B}(\mathbb{R}^d)$. If $\Theta = \Theta_{\mathrm{iso}}$, then the pair correlation function $g_{d-1,t}$ is given by

$$g_{d-1,t}(r) = 1 + \frac{d-1}{2t^2r^2}\left(1 - \mathrm{e}^{-\frac{2\kappa_{d-1}}{d\kappa_d}tr}\right), \quad r > 0\,.$$

Proof. For given bounded sets $B_1, B_2 \in \mathcal{B}(\mathbb{R}^d)$, let $W \in \mathcal{P}_d$ be such that $B_1, B_2 \subset W$. By the spatial consistency of STIT tessellations (see Theorem 5.10) it follows that

$$\mathrm{Cov}_{d-1,t}(B_1 \times B_2) = \mathrm{cov}(\mathcal{H}^{d-1}_{t,W}(B_1), \mathcal{H}^{d-1}_{t,W}(B_2))\,,$$

where $\mathcal{H}^{d-1}_{t,W}(\cdot)$ is defined in (10.46).

For $A^{\mathcal{H}}_{d-1,d-1,s,W}(B_1, B_2)$ defined in (10.48), analogously to the proof of Theorem 12.3, we have that

$$\begin{aligned}
&\mathbb{E}\int_0^t A^{\mathcal{H}}_{d-1,d-1,s,W}(B_1, B_2)\,\mathrm{d}s \\
&= \mathbb{E}\int_0^t \int_{[W]} \sum_{F\in Y_{s,W}\cap H} \mathcal{H}^{d-1}(B_1\cap F)\,\mathcal{H}^{d-1}(B_2\cap F)\,\Theta(\mathrm{d}H)\,\mathrm{d}s \\
&= \mathbb{E}\int_0^t \int_{[W]} \sum_{F\in Y_{s,W}\cap H} \int_{B_1\cap H}\int_{B_2\cap H} \mathbb{1}\{(x,y)\in F^2\}\,\lambda_{d-1}(\mathrm{d}x)\,\lambda_{d-1}(\mathrm{d}y)\,\Theta(\mathrm{d}H)\,\mathrm{d}s \\
&= \int_0^t \int_{[W]} \int_{B_1\cap H}\int_{B_2\cap H} \mathbb{P}(\mathrm{conv}\{x,y\}\cap\partial Y_{s,W} = \emptyset)\,\lambda_{d-1}(\mathrm{d}x)\,\lambda_{d-1}(\mathrm{d}y)\,\Theta(\mathrm{d}H)\,\mathrm{d}s \\
&= \int_0^t \int_{[W]} \int_{B_1\cap H}\int_{B_2\cap H} \mathrm{e}^{-s\,\Theta([xy])}\,\lambda_{d-1}(\mathrm{d}x)\,\lambda_{d-1}(\mathrm{d}y)\,\Theta(\mathrm{d}H)\,\mathrm{d}s \\
&= \int_{[W]} \int_{B_1\cap H}\int_{B_2\cap H} \frac{1-\mathrm{e}^{-t\,\Theta([xy])}}{\Theta([xy])}\,\lambda_{d-1}(\mathrm{d}x)\,\lambda_{d-1}(\mathrm{d}y)\,\Theta(\mathrm{d}H)\,.
\end{aligned}$$

Furthermore, by (10.47), (2.48), and (6.6) we have

$$\mathbb{E}\int_0^t \mathscr{H}_{s,W}^{d-1}(B_1)\,\mathbb{L}\mathscr{H}_{s,W}^{d-1}(B_2)\,\mathrm{d}s = \int_{[W]} \mathscr{H}^{d-1}(B_2\cap H)\,\Theta(\mathrm{d}H)\int_0^t \mathbb{E}\mathscr{H}_{s,W}^{d-1}(B_1)\,\mathrm{d}s$$
$$= \frac{t^2}{2}\lambda_d(B_1)\int_{[W]} \mathscr{H}^{d-1}(B_2\cap H)\,\Theta(\mathrm{d}H)$$
$$= \frac{t^2}{2}\lambda_d(B_1)\lambda_d(B_2)\,. \tag{12.43}$$

Now the martingale property of the centered random process in (10.49) can be used to determine $\mu_{d-1,t}^{(2)}(B_1\times B_2)$. Namely, taking expectations yields that

$$\mu_{d-1,t}^{(2)}(B_1\times B_2) = \mathbb{E}\int_0^t A_{d-1,d-1,s,W}^{\mathscr{H}}(B_1,B_2)\,\mathrm{d}s$$
$$+\,\mathbb{E}\int_0^t \mathscr{H}_{s,W}^{d-1}(B_1)\,\mathbb{L}\mathscr{H}_{s,W}^{d-1}(B_2) + \mathscr{H}_{s,W}^{d-1}(B_2)\,\mathbb{L}\mathscr{H}_{s,W}^{d-1}(B_1)\,\mathrm{d}s$$
$$= \int_{[W]}\int_{B_1\cap H}\int_{B_2\cap H} \frac{1-\mathrm{e}^{-t\,\Theta([xy])}}{\Theta([xy])}\,\lambda_{d-1}(\mathrm{d}x)\,\lambda_{d-1}(\mathrm{d}y)\,\Theta(\mathrm{d}H)$$
$$+\,t^2\lambda_d(B_1)\lambda_d(B_2)\,.$$

In view of (12.40), this completes the proof of the first part of the theorem.

For the second part, we take $\Theta = \Theta_{\mathrm{iso}}$ and use (12.22), which says that $\Theta_{\mathrm{iso}}([xy]) = \frac{2\kappa_{d-1}}{d\kappa_d}\|x-y\|$. Then the Blaschke–Petkantschin-type integral-geometric transformation formula from Proposition 12.5 can be applied to see that

$$\mathrm{Cov}_{d-1,t}(B_1\times B_2)$$
$$= \frac{d-1}{2}\int_{B_1}\int_{B_2} \frac{1-\mathrm{e}^{-\frac{2\kappa_{d-1}}{d\kappa_d}t\|x-y\|}}{\|x-y\|^2}\,\lambda_d(\mathrm{d}x)\,\lambda_d(\mathrm{d}y)$$
$$= t^2\int_{\mathbb{R}^d}\int_{\mathbb{R}^d} \mathbb{1}\{x\in B_1, x+h\in B_2\}\frac{d-1}{2}\frac{1-\mathrm{e}^{-\frac{2\kappa_{d-1}}{d\kappa_d}t\|h\|}}{t^2\|h\|^2}\,\lambda_d(\mathrm{d}h)\,\lambda_d(\mathrm{d}x)\,.$$

Hence by definition (12.41) of the reduced second-moment function we have

$$K_{d-1,t}(r) = d\kappa_d\int_0^r\left(1+\frac{(d-1)(1-\mathrm{e}^{-\frac{2\kappa_{d-1}}{d\kappa_d}ts})}{2t^2s^2}\right)s^{d-1}\,\mathrm{d}s\,,\quad r\ge 0\,,$$

which yields

$$\begin{aligned} g_{d-1,t}(r) &= \frac{d\kappa_d}{d\kappa_d r^{d-1}}\left(1 + \frac{(d-1)(1-e^{-\frac{2\kappa_{d-1}}{d\kappa_d}tr})}{2t^2r^2}\right)r^{d-1} \\ &= 1 + \frac{d-1}{2t^2r^2}(1 - e^{-\frac{2\kappa_{d-1}}{d\kappa_d}tr}), \quad r \geq 0. \end{aligned}$$

This concludes the proof of the second part of the theorem. □

Remark 12.27. *As a particular consequence of Theorem* 12.26, *for* Θ_{iso}, *the variance formulae* (12.20) *and* (12.21) *for the total surface content can be recovered using that*

$$\operatorname{var}\Sigma_{V_{d-1},t,W} = t^2 d\kappa_d \int_0^\infty \overline{c}_W(r)\,(g_{d-1,t}(r) - 1)\,\mathrm{d}r$$

from equation (7.25) in [33].

Remark 12.28. *The pair correlation function* $g_{d-1,t}$ *satisfies the scaling property*

$$g_{d-1,t}(r) = g_{d-1,1}(tr), \quad t > 0, \quad r > 0. \tag{12.44}$$

This is consistent with the scaling property of STIT tessellations. Formally, note that by (6.2)

$$t^{2(d-1)}Cov_{d-1,t}\left(\frac{1}{t}B_1 \times \frac{1}{t}B_2\right) = Cov_{d-1,1}(B_1 \times B_2), \quad B_1, B_2 \in \mathcal{B}(\mathbb{R}^d), \quad t > 0.$$

Thus the definition of the reduced covariance measure yields

$$\begin{aligned} &\int_{\mathbb{R}^d} t^d\, \widehat{Cov}_{d-1,t}\left(\frac{1}{t}(B_2 - x)\right)\mathbb{1}\{x \in B_1\}\,\lambda_d(\mathrm{d}x) \\ &\quad = \int_{\mathbb{R}^d} \widehat{Cov}_{d-1,1}(B_2 - x)\mathbb{1}\{x \in B_1\}\,\lambda_d(\mathrm{d}x), \end{aligned}$$

and hence relation (12.44) *follows by the definition of the pair correlation function.*

The result of Theorem 12.26 can be compared with the pair correlation function $g^{\text{PHT}}_{d-1,t}(r)$ of the surface measure of a stationary and isotropic Poisson hyperplane tessellation X_t driven by $t\,\Theta_{\text{iso}}$; see (2.66). In this case, we have that

$$g^{\text{PHT}}_{d-1,t}(r) = 1 + \frac{(d-1)\kappa_{d-1}}{d\,\kappa_d\,tr}, \quad r > 0; \tag{12.45}$$

compare with (12.68) in the notes at the end of this chapter and (10.22) in [82]. Using the Mecke equation for Poisson point processes, this can also be seen as follows by means of the corresponding reduced second-moment function $K^{\text{PHT}}_{d-1,t}$. Under the Palm distribution with respect to the random surface measure, the typical point is located at the origin on an additionally inserted random hyperplane H_0 whose unit normal vector is uniformly distributed on the sphere S^{d-1} and which is independent of the remaining Poisson hyperplane tessellation X_t. Thus

$$\begin{aligned} tK^{\mathrm{PHT}}_{d-1,t}(r) &= \mathscr{H}^{d-1}(H_0 \cap rB^d) + \mathbb{E}\mathscr{H}^{d-1}(\partial X_t \cap rB^d) \\ &= \kappa_{d-1} r^{d-1} + t\kappa_d r^d , \end{aligned}$$

which leads to (12.45) after differentiation with respect to r and division by $d\kappa_d r^{d-1}$.

Remark 12.29. *In the sparse regime as $t \downarrow 0$, the pair correlation function $g_{d-1,t}$ of a stationary and isotropic STIT tessellation in $\mathbb{R}^d$ can be connected to* (12.45). *Indeed, developing the exponential function in a Taylor series, we obtain*

$$\begin{aligned} g_{d-1,t}(r) &= 1 + \frac{d-1}{2t^2r^2}\left(1 - e^{-\frac{2\kappa_{d-1}}{d\kappa_d} tr}\right) \\ &= 1 + \frac{(d-1)\kappa_{d-1}}{d\kappa_d\, rt} + \frac{d-1}{4}\left(\frac{2\kappa_{d-1}}{d\kappa_d}\right)^2 + O(tr) \end{aligned}$$

as $t \downarrow 0$. We thus conclude that the first-order term coincides with the pair correlation function of a stationary and isotropic Poisson hyperplane tessellation.

12.7.3 Comparison with a Boolean model of (d – 1)-polytopes

Recall that $\mathcal{P}_{d-1,c}$ is the space of $(d-1)$-dimensional polytopes in $\mathbb{R}^d$ such that their center with respect to a fixed center function c is at the origin. Let $\mathbb{Q}$ be a probability measure on $\mathcal{P}_{d-1,c}$, and let Φ_γ be a stationary Poisson point process in $\mathbb{R}^d$ with intensity $\gamma > 0$. By Ψ_γ we denote an independent $\mathbb{Q}$-marking of Φ_γ that consists of pairs (x, F) with $x \in \Phi_\gamma$ and F that are independent copies of a random $(d-1)$-polytope with distribution $\mathbb{Q}$. We note that by the marking theorem for Poisson point processes, Proposition 5.5 of [101], Ψ_γ is a Poisson process on the product space $\mathbb{R}^d \times \mathcal{P}_{d-1,c}$ with intensity measure $\gamma\lambda_d \otimes \mathbb{Q}$. The random union set

$$\bigcup_{(x,F)\in\Psi_\gamma} (F + x)$$

is referred to as the **Boolean model** with intensity γ and grain distribution $\mathbb{Q}$. We introduce the random stationary surface measure

$$\mathscr{H}^{d-1}_{\mathrm{BM}}(\cdot) := \sum_{(x,F)\in\Psi_\gamma} h^{F+x}_{d-1}(\cdot), \quad h^{F+x}_{d-1}(\cdot) := \mathscr{H}^{d-1}(\cdot \cap (F + x)) .$$

Moreover, we let $\mathrm{Cov}^{\mathrm{BM}}$ be the **covariance measure** of $\mathscr{H}^{d-1}_{\mathrm{BM}}(\cdot)$ as defined in Section 12.7.1 with $\Phi = \mathscr{H}^{d-1}_{\mathrm{BM}}$. Similarly, for a STIT tessellation Y_t with $t > 0$, driven by some hyperplane measure Θ of the form (4.1), satisfying (4.3), the random surface measure of Y_t, in view of Lemma 8.8, can be represented as

$$\mathscr{H}_t^{d-1}(\cdot) = \sum_{F\in\mathcal{F}_{d-1}^{\max}(Y_t)} h_{d-1}^F(\cdot), \quad h_{d-1}^F(\cdot) := \mathscr{H}^{d-1}(\cdot\cap F),$$

and we denote its covariance measure by $\mathrm{Cov}_{d-1,t}$, which again is defined in Section 12.7.1 by taking $\Phi = \mathscr{H}_t^{d-1}$ there; see also the discussion before Theorem 12.26. The next result provides formulae for $\mathrm{Cov}^{\mathrm{BM}}$ and $\mathrm{Cov}_{d-1,t}$ in terms of γ and $\mathbb{Q}$, or t and $\mathbb{Q}_{d-1,t}^{\max}$, the distribution of the typical maximal $(d-1)$-polytope of Y_t, respectively.

Theorem 12.30. (a) *Consider a Boolean model with intensity $\gamma > 0$ and grain distribution $\mathbb{Q}$ on $\mathcal{P}_{d-1,c}$ as described above. Then*

$$Cov^{BM}(B_1\times B_2) = \gamma\int_{\mathbb{R}^d}\int_{\mathcal{P}_{d-1,c}} h_{d-1}^{F+x}(B_1)\, h_{d-1}^{F+x}(B_2)\,\mathbb{Q}(\mathrm{d}F)\,\lambda_d(\mathrm{d}x)$$

for $B_1, B_2 \in \mathcal{B}(\mathbb{R}^d)$.

(b) *Let Θ be a hyperplane measure of the form* (4.1), *satisfying* (4.3), *and let* $(Y_t,\ t>0)$ *be the STIT tessellation process driven by the measure Θ. Then, for any fixed $t > 0$, we have that*

$$Cov_{d-1,t}(B_1\times B_2) = \gamma_{d-1,t}^{\max}\int_{\mathbb{R}^d}\int_{\mathcal{P}_{d-1,c}} h_{d-1}^{F+x}(B_1)\, h_{d-1}^{F+x}(B_2)\,\mathbb{Q}_{d-1,t}^{\max}(\mathrm{d}F)\,\lambda_d(\mathrm{d}x)$$

for $B_1, B_2 \in \mathcal{B}(\mathbb{R}^d)$, where $\gamma_{d-1,t}^{\max}$ is the intensity of maximal $(d-1)$-polytopes in Y_t, and $\mathbb{Q}_{d-1,t}^{\max}$ denotes the distribution of the typical maximal $(d-1)$-polytope of Y_t.

Proof. Fix bounded Borel sets $B_1, B_2 \in \mathcal{B}(\mathbb{R}^d)$. Then

$$\begin{aligned}\mathrm{Cov}^{\mathrm{BM}}(B_1\times B_2) &= \mathbb{E}\Big[\sum_{(x,F),(y,G)\in\Psi_\gamma} h_{d-1}^{F+x}(B_1)h_{d-1}^{G+y}(B_2)\Big]\\ &\quad-\Big(\mathbb{E}\sum_{(x,F)\in\Psi_\gamma} h_{d-1}^{F+x}(B_1)\Big)\Big(\mathbb{E}\sum_{(x,F)\in\Psi_\gamma} h_{d-1}^{F+x}(B_2)\Big).\end{aligned} \tag{12.46}$$

The first of these expectations can be split up as follows:

$$\mathbb{E}\Big[\sum_{((x,F),(y,G))\in\Psi_{\gamma,\neq}^2} h_{d-1}^{F+x}(B_1)h_{d-1}^{G+y}(B_2)\Big] + \mathbb{E}\Big[\sum_{(x,F)\in\Psi_\gamma} h_{d-1}^{F+x}(B_1)h_{d-1}^{F+x}(B_2)\Big], \tag{12.47}$$

where $\Psi_{\gamma,\neq}^2$ denotes the collection of pairs of distinct particles of Ψ_γ. To the first term, we can apply the multivariate Mecke equation for Poisson point processes of Theorem 19.11. Recalling that the intensity measure of Ψ_γ is γ times the product measure of λ_d and $\mathbb{Q}$, this gives

$$\mathbb{E}\Big[\sum_{((x,F),(y,G))\in\Psi^2_{\gamma,\neq}} h^{F+x}_{d-1}(B_1)h^{G+y}_{d-1}(B_2)\Big]$$

$$= \gamma^2 \int_{\mathbb{R}^d}\int_{\mathcal{P}_{d-1,c}}\int_{\mathbb{R}^d}\int_{\mathcal{P}_{d-1,c}} h^{F+x}_{d-1}(B_1)h^{G+y}_{d-1}(B_2)\,\mathbb{Q}(dG)\,\lambda_d(dy)\,\mathbb{Q}(dF)\,\lambda_d(dx)$$

$$= \Big(\gamma\int_{\mathbb{R}^d}\int_{\mathcal{P}_{d-1,c}} h^{F+x}_{d-1}(B_1)\,\mathbb{Q}(dF)\,\lambda_d(dx)\Big)\Big(\gamma\int_{\mathbb{R}^d}\int_{\mathcal{P}_{d-1,c}} h^{G+y}_{d-1}(B_2)\,\mathbb{Q}(dG)\,\lambda_d(dy)\Big).$$

Similarly,

$$\Big(\mathbb{E}\sum_{(x,F)\in\Psi_\gamma} h^{F+x}_{d-1}(B_1)\Big)\Big(\mathbb{E}\sum_{(x,F)\in\Psi_\gamma} h^{F+x}_{d-1}(B_2)\Big)$$

$$= \Big(\gamma\int_{\mathbb{R}^d}\int_{\mathcal{P}_{d-1,c}} h^{F+x}_{d-1}(B_1)\,\mathbb{Q}(dF)\,\lambda_d(dx)\Big)\Big(\gamma\int_{\mathbb{R}^d}\int_{\mathcal{P}_{d-1,c}} h^{G+y}_{d-1}(B_2)\,\mathbb{Q}(dG)\,\lambda_d(dy)\Big).$$

As a consequence, the last term in (12.46) cancels with the first term in (12.47), and we arrive at

$$\mathrm{Cov}^{\mathrm{BM}}(B_1\times B_2) = \mathbb{E}\Big[\sum_{(x,F)\in\Psi_\gamma} h^{F+x}_{d-1}(B_1)h^{F+x}_{d-1}(B_2)\Big].$$

Applying now the Mecke equation (19.5) for Poisson processes, we find that

$$\mathrm{Cov}^{\mathrm{BM}}(B_1\times B_2) = \gamma\int_{\mathbb{R}^d}\int_{\mathcal{P}_{d-1,c}} h^{F+x}_{d-1}(B_1)h^{F+x}_{d-1}(B_2)\,\mathbb{Q}(dF)\,\lambda_d(dx), \tag{12.48}$$

which completes the proof of part (a).

To prove part (b), we fix bounded Borel sets $B_1, B_2 \in \mathcal{B}(\mathbb{R}^d)$ and $W \in \mathcal{P}_d$ satisfying $B_1, B_2 \subset W$. We can now apply the martingale property of the centered stochastic process (10.49) with $i = j = d-1$ in Corollary 10.13 to see that

$$\begin{aligned}
&\mathrm{Cov}_{d-1,t}(B_1\times B_2)\\
&\quad= \mathbb{E}[\mathcal{H}^{d-1}_{t,W}(B_1)\mathcal{H}^{d-1}_{t,W}(B_2)] - \mathbb{E}\mathcal{H}^{d-1}_{t,W}(B_1)\,\mathbb{E}\mathcal{H}^{d-1}_{t,W}(B_2)\\
&\quad= \mathbb{E}\int_0^t A_{d-1,d-1,s,W}(B_1,B_2) + \mathcal{H}^{d-1}_{s,W}(B_1)\,\mathbb{L}\mathcal{H}^{d-1}_{s,W}(B_2) + \mathcal{H}^{d-1}_{s,W}(B_2)\,\mathbb{L}\mathcal{H}^{d-1}_{s,W}(B_1)\,ds\\
&\qquad - \mathbb{E}\mathcal{H}^{d-1}_{t,W}(B_1)\,\mathbb{E}\mathcal{H}^{d-1}_{t,W}(B_2).
\end{aligned}$$

Using now twice (12.43), it follows that

$$\mathbb{E}\int_0^t \mathscr{H}_{s,W}^{d-1}(B_1)\,\mathbb{L}\mathscr{H}_{s,W}^{d-1}(B_2) + \mathscr{H}_{s,W}^{d-1}(B_2)\,\mathbb{L}\mathscr{H}_{s,W}^{d-1}(B_1)\,\mathrm{d}s = t^2\lambda_d(B_1)\lambda_d(B_2)\,,$$

a term which cancels with $\mathbb{E}\mathscr{H}_{t,W}^{d-1}(B_1)\,\mathbb{E}\mathscr{H}_{t,W}^{d-1}(B_2)$. Thus

$$\begin{aligned}\mathrm{Cov}_{d-1,t}(B_1 \times B_2) &= \mathbb{E}\int_0^t A_{d-1,d-1,s,W}(B_1, B_2)\,\mathrm{d}s \\ &= \int_0^t \int_{[W]} \mathbb{E} \sum_{F\in Y_{s,W}\cap H} h_{d-1}^F(B_1)h_{d-1}^F(B_2)\,\Theta(\mathrm{d}H)\,\mathrm{d}s\,, \end{aligned} \tag{12.49}$$

a formula we already encountered implicitly in the proof of Theorem 12.26. In the next step, we apply Theorem 11.1 to the expectation, which implies that

$$\begin{aligned}\mathrm{Cov}_{d-1,t}(B_1 \times B_2) &= \mathbb{E}\int_0^t A_{d-1,d-1,s,W}(B_1, B_2)\,\mathrm{d}s \\ &= \int_0^t \int_{[W]} \mathbb{E} \sum_{F\in(X_s\wedge W)\cap H} h_{d-1}^F(B_1)h_{d-1}^F(B_2)\,\Theta(\mathrm{d}H)\,\mathrm{d}s\,, \end{aligned}$$

where X_s is a Poisson hyperplane tessellation induced by a Poisson process on the space of hyperplanes in $\mathbb{R}^d$ with intensity measure $s\,\Theta$. Moreover,

$$\begin{aligned}\int_{[W]} \mathbb{E} \sum_{F\in(X_1\wedge W)\cap H} h_{d-1}^F(B_1)h_{d-1}^F(B_2)\,\Theta(\mathrm{d}H) &= \mathbb{E}\sum_{P\in X_1\wedge W}\int_{[P]} h_{d-1}^{P\cap H}(B_1)h_{d-1}^{P\cap H}(B_2)\,\Theta(\mathrm{d}H) \\ &= \mathbb{E}\sum_{F\in\mathcal{F}_{d-1}(X_1\wedge W)} h_{d-1}^F(B_1)h_{d-1}^F(B_2)\,. \end{aligned}$$

Replacing Θ by $s\,\Theta$, $0 < s \le t$, we arrive at

$$\int_{[W]} \mathbb{E} \sum_{F\in(X_s\wedge W)\cap H} h_{d-1}^F(B_1)h_{d-1}^F(B_2)\,\Theta(\mathrm{d}H) = \frac{1}{s}\,\mathbb{E}\sum_{F\in\mathcal{F}_{d-1}(X_s\wedge W)} h_{d-1}^F(B_1)h_{d-1}^F(B_2)\,,$$

and hence

$$\mathrm{Cov}_{d-1,t}(B_1 \times B_2) = \int_0^t \frac{1}{s}\,\mathbb{E}\sum_{F\in\mathcal{F}_{d-1}(X_s\wedge W)} h_{d-1}^F(B_1)h_{d-1}^F(B_2)\,\mathrm{d}s\,.$$

The next step consists in applying Theorem 11.5 with $k = d-1$ to conclude that

$$\mathrm{Cov}_{d-1,t}(B_1 \times B_2) = \mathbb{E}\sum_{(F,s)\in\mathcal{F}_{d-1,W}^{\max *}} \mathbb{1}\{s \le t\}h_{d-1}^F(B_1)h_{d-1}^F(B_2)$$

$$= \mathbb{E} \sum_{F \in \mathcal{F}_{d-1}^{\max}(Y_t)} h_{d-1}^F(B_1) h_{d-1}^F(B_2)\,,$$

where $\mathcal{F}_{d-1}^{\max}(Y_t)$ denotes the stationary particle process in $\mathbb{R}^d$ consisting of all maximal $(d-1)$-polytopes of Y_t. As such, $\mathcal{F}_{d-1}^{\max}(Y_t)$ has intensity $\gamma_{d-1,t}^{\max}$ and grain distribution $\mathbb{Q}_{d-1,t}^{\max}$. Thus Campbell's theorem (Theorem 19.4) yields the identity

$$\mathrm{Cov}_{d-1,t}(B_1 \times B_2) = \gamma_{d-1,t}^{\max} \int_{\mathbb{R}^d} \int_{\mathcal{P}_{d-1,c}} h_{d-1}^{F+x}(B_1) h_{d-1}^{F+x}(B_2)\, \mathbb{Q}_{d-1,t}^{\max}(\mathrm{d}F)\, \lambda_d(\mathrm{d}x)\,,$$

which completes the proof of part (b). □

Remark 12.31. *Part (a) of Theorem* 12.30 *remains valid for all Boolean models for which* h_{d-1}^F *is well defined. This is the case if the grain distribution of the Boolean model is concentrated on the class of* $\mathcal{H}^{d-1}$*-rectifiable and* $\mathcal{H}^{d-1}$*-measurable closed sets in* $\mathbb{R}^d$*; see* [204].

Remark 12.32. *Representation* (12.48) *of the covariance measure for the Boolean model was a consequence of the Mecke equation for Poisson processes of Theorem* 19.11, *whereas representation* (12.49) *of the covariance measure for the STIT model was derived using martingale tools. In Chapter* 14, *we derive a Mecke-type equation for maximal polytopes of STIT tessellations. As an application, we will provide in Section* 14.2 *an alternative approach to Theorem* 12.30*(b) by means of this device. This yields deeper insights into the analogy between the Boolean and STIT models on the level of second-order quantities.*

From the previous theorem we can draw the following interesting conclusion, which says that the covariance measure of the random surface measure of a STIT tessellation coincides with the corresponding covariance measure of a particular Boolean model of $(d-1)$-polytopes. As a special case, the pair correlation functions of the surface measures of both models must coincide. In other words, this means that these second-order quantities are not sensitive enough to discriminate between the two random set models.

Corollary 12.33. *Let* Θ *be a hyperplane measure of the form* (4.1), *satisfying* (4.3), *and let* $(Y_t, t > 0)$ *be the STIT tessellation process driven by the measure* Θ. *For given* $t > 0$, *consider a Boolean model of* $(d-1)$*-polytopes in* $\mathbb{R}^d$ *with intensity* $\gamma = \gamma_{d-1,t}^{\max}$ *and grain distribution* $\mathbb{Q} = \mathbb{Q}_{d-1,t}^{\max}$. *Then* $\mathrm{Cov}^{BM} = \mathrm{Cov}_{d-1,t}$. *In particular, if* $\Theta = \Theta_{\mathrm{iso}}$, *then the pair correlation functions of the corresponding surface measures coincide.*

12.8 Pair correlation function of the vertex point process in dimension $d = 2$

In this section, we consider a stationary and isotropic STIT tessellation Y_t with time parameter $t > 0$ in the plane $\mathbb{R}^2$. By $\mathsf{V}_t := \mathsf{V}(Y_t)$ we denote its vertex point process. This is

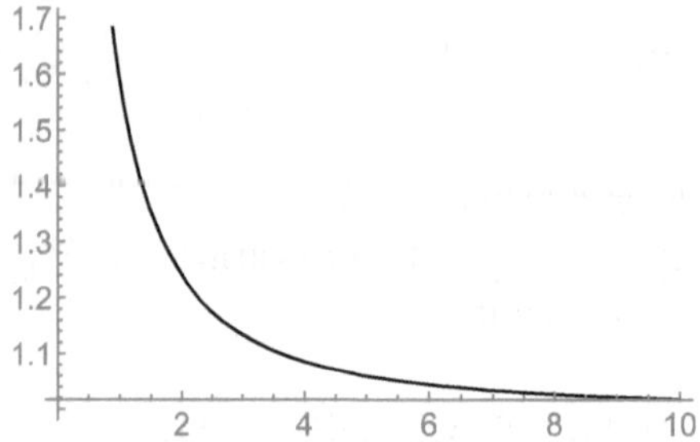

Figure 12.2: Pair correlation function $g_{0,t}$ for $t = 1$.

a stationary and isotropic point process in $\mathbb{R}^2$ with intensity $\gamma_0 = \frac{2t^2}{\pi}$. Our focus lies on the second-order structure of the point process V_t. More precisely, the main result of this section is an explicit formula for the pair correlation function of V_t; see also Figure 12.2. More generally, as in Section 12.7.1 with $\Phi = \mathsf{V}_t$ there, we introduce

the covariance measure	$\mathrm{Cov}_{0,t} := \mathrm{Cov}_{\mathsf{V}_t}$,
the reduced covariance measure	$\widehat{\mathrm{Cov}}_{0,t} := \widehat{\mathrm{Cov}}_{\mathsf{V}_t}$,
the reduced second moment measure	$\widehat{\mu}^{(2)}_{0,t} := \widehat{\mu}^{(2)}_{\mathsf{V}_t}$,
the K-function	$K_{0,t} := K_{\mathsf{V}_t}$,
the pair correlation function	$g_{0,t} := g_{\mathsf{V}_t}$

of V_t.

Theorem 12.34. *Let $(Y_t,\ t > 0)$ be the STIT tessellation process in $\mathbb{R}^2$ driven by the motion-invariant measure Θ_{iso}. Then for any fixed $t > 0$, we have that*

$$g_{0,t}(r) = 1 + \frac{2}{t^2r^2} - \frac{\pi}{t^3r^3} + \frac{\pi^2}{4t^4r^4} - \left(\frac{1}{2t^2r^2} - \frac{\pi}{2t^3r^3} + \frac{\pi^2}{4t^4r^4}\right)\exp\left(-\frac{2}{\pi}rt\right)$$

for $r > 0$.

Proof. Let $B_1, B_2 \in \mathcal{B}(\mathbb{R}^2)$ be bounded Borel sets, and let $W \in \mathcal{P}_2$ be such that $B_1, B_2 \subset W$. Further, let $(Y_{t,W},\ t \geq 0)$ with $Y_{t,W} \overset{\mathrm{D}}{=} Y_t \wedge W$ be the STIT tessellation process localized to W. For $0 \leq s \leq t$ and $i \in \{1,2\}$, define

$$\zeta^{B_i}(e) := \sum_{x \in \mathcal{F}_0(e)} \mathbb{1}\{x \in B_i\}, \quad (e,u) \in \mathcal{F}^{\max *}_{1,W}, \quad u \leq s, \tag{12.50}$$

$$\Sigma_{\Theta^{B_i}_{\mathrm{iso}},s,W} := \sum_{(e,u) \in \mathcal{F}^{\max *}_{1,W}} \mathbb{1}\{u \leq s\} \int_{[e]} \mathbb{1}\{e \cap H \cap B_i \neq \emptyset\}\, \Theta_{\mathrm{iso}}(\mathrm{d}H). \tag{12.51}$$

Then recalling notation (10.15), we have

$$A_{\zeta^{B_i},s,W} = \int_{[W]} \sum_{e \in Y_{s,W} \cap H} \sum_{x \in \mathcal{F}_0(e)} \mathbb{1}\{x \in B_i\}\, \Theta_{\mathrm{iso}}(\mathrm{d}H).$$

For Θ_{iso}-almost every H, each vertex of the sectional tessellation $Y_{s,W} \cap H$ lying in B_i is shared by exactly two adjacent intervals of $Y_{s,W} \cap H$. Hence the inner double sum counts twice the number of such vertices. Equivalently,

$$2 \sum_{(e,u) \in \mathcal{F}_{1,W}^{\max*}} \mathbb{1}\{e \cap H \cap B_i \neq \emptyset\}$$

counts the same quantity, since each such vertex arises as the unique intersection point $e \cap H$ of a maximal segment e of $Y_{s,W}$ with H. Integration with respect to Θ_{iso} yields

$$\begin{aligned}
A_{\zeta^{B_i},s,W} &= \int_{[W]} \sum_{e \in Y_{s,W} \cap H} \sum_{x \in \mathcal{F}_0(e)} \mathbb{1}\{x \in B_i\} \, \Theta_{\text{iso}}(\mathrm{d}H) \\
&= 2 \int_{[W]} \sum_{(e,u) \in \mathcal{F}_{1,W}^{\max*}} \mathbb{1}\{e \cap H \cap B_i \neq \emptyset\} \, \Theta_{\text{iso}}(\mathrm{d}H) \\
&= 2 \sum_{(e,u) \in \mathcal{F}_{1,W}^{\max*}} \int_{[W]} \mathbb{1}\{e \cap H \cap B_i \neq \emptyset\} \, \Theta_{\text{iso}}(\mathrm{d}H) \\
&= 2 \sum_{(e,u) \in \mathcal{F}_{1,W}^{\max*}} \int_{[e]} \mathbb{1}\{e \cap H \cap B_i \neq \emptyset\} \, \Theta_{\text{iso}}(\mathrm{d}H) \\
&= 2 \Sigma_{\Theta_{\text{iso}}^{B_i},s,W} \,.
\end{aligned} \tag{12.52}$$

Noting that $\mathsf{V}_t(B_i) = \Sigma_{\zeta^{B_i},t,W}$ for $i \in \{1,2\}$, according to our general sum definition (10.12), we can now apply Theorem 10.8(c) with $b_1(t) = \mathbb{E}\Sigma_{\zeta^{B_1},t,W}$ and $b_2(t) = \mathbb{E}\Sigma_{\zeta^{B_2},t,W}$. After taking expectations, this leads to

$$\begin{aligned}
\operatorname{cov}(\mathsf{V}_t(B_1), \mathsf{V}_t(B_2)) &= \mathbb{E} \int_0^t A_{\zeta^{B_1},\zeta^{B_2},s,W} \, \mathrm{d}s \\
&\quad + 2 \int_0^t \operatorname{cov}(\Sigma_{\Theta_{\text{iso}}^{B_1},s,W}, \Sigma_{\zeta^{B_2},s,W}) + \operatorname{cov}(\Sigma_{\zeta^{B_1},s,W}, \Sigma_{\Theta_{\text{iso}}^{B_2},s,W}) \, \mathrm{d}s \,,
\end{aligned}$$

where we also used (12.52). The same argument can now be applied to the covariances $\operatorname{cov}(\Sigma_{\Theta_{\text{iso}}^{B_1},s,W}, \Sigma_{\zeta^{B_2},s,W})$ and $\operatorname{cov}(\Sigma_{\zeta^{B_1},s,W}, \Sigma_{\Theta_{\text{iso}}^{B_2},s,W})$ under the integral sign:

$$\begin{aligned}
\operatorname{cov}(\Sigma_{\Theta_{\text{iso}}^{B_1},s,W}, \Sigma_{\zeta^{B_2},s,W}) &= \mathbb{E} \int_0^s A_{\Theta_{\text{iso}}^{B_1},\zeta^{B_2},u,W} \, \mathrm{d}u \\
&\quad + \mathbb{E} \int_0^s \overline{A}_{\Theta_{\text{iso}}^{B_1},u,W} \overline{\Sigma}_{\zeta^{B_2},u,W} + \overline{A}_{\zeta^{B_1},u,W} \overline{\Sigma}_{\Theta_{\text{iso}}^{B_2},u,W} \, \mathrm{d}u \\
&= \mathbb{E} \int_0^s A_{\Theta_{\text{iso}}^{B_1},\zeta^{B_2},u,W} \, \mathrm{d}u + 2 \int_0^s \operatorname{cov}(\Sigma_{\Theta_{\text{iso}}^{B_1},u,W}, \Sigma_{\Theta_{\text{iso}}^{B_2},u,W}) \, \mathrm{d}u \,,
\end{aligned}$$

where we used (12.52) again together with the fact that $\overline{A}_{\Theta^{B_1}_{\text{iso}},\cdot,W} \equiv 0$. In the previous computation, we used the shorthand notation $\overline{A}_{\Theta^{B_1}_{\text{iso}},u,W}$ for the centered random variable $A_{\Theta^{B_1}_{\text{iso}},u,W} - \mathbb{E}A_{\Theta^{B_1}_{\text{iso}},u,W}$. In the same way, we see that

$$\operatorname{cov}(\Sigma_{\zeta^{B_1},s,W}, \Sigma_{\Theta^{B_2}_{\text{iso}},s,W}) = \mathbb{E}\int_0^s A_{\zeta^{B_1},\Theta^{B_2}_{\text{iso}},u,W}\,du + 2\int_0^s \operatorname{cov}(\Sigma_{\Theta^{B_1}_{\text{iso}},u,W}, \Sigma_{\Theta^{B_2}_{\text{iso}},u,W})\,du\,.$$

Finally, we can apply Theorem 10.8(b) to conclude that

$$\operatorname{cov}(\Sigma_{\Theta^{B_1}_{\text{iso}},u,W}, \Sigma_{\Theta^{B_2}_{\text{iso}},u,W}) = \mathbb{E}\int_0^u A_{\Theta^{B_1}_{\text{iso}},\Theta^{B_2}_{\text{iso}},v,W}\,dv\,,$$

where again we used the fact that $\overline{A}_{\Theta^{B_1}_{\text{iso}},\cdot,W} \equiv 0$ and $\overline{A}_{\Theta^{B_2}_{\text{iso}},\cdot,W} \equiv 0$. Putting all these identities together, we arrive at

$$\begin{aligned}\operatorname{cov}(\mathsf{V}_t(B_1), \mathsf{V}_t(B_2)) &= \mathbb{E}\int_0^t A_{\zeta^{B_1},\zeta^{B_2},s,W}\,ds\\ &\quad + 2\,\mathbb{E}\int_0^t\int_0^s A_{\zeta^{B_1},\Theta^{B_2}_{\text{iso}},u,W} + A_{\Theta^{B_1}_{\text{iso}},\zeta^{B_2},u,W}\,du\,ds\\ &\quad + 8\,\mathbb{E}\int_0^t\int_0^s\int_0^u A_{\Theta^{B_1}_{\text{iso}},\Theta^{B_2}_{\text{iso}},v,W}\,dv\,du\,ds\,.\end{aligned} \tag{12.53}$$

To evaluate the terms of the form $\mathbb{E}A_{\phi,s,W}$, where $\phi : \mathcal{P}_1 \to \mathbb{R}$ is bounded and measurable and has support in $\{e \in \mathcal{P}_1 : e \subset W\}$, we use the fact that the intersection of a STIT tessellation Y_s with a line H induces a homogeneous Poisson point process on $W \cap H$ with intensity $\frac{2s}{\pi}$; see Lemma 6.11. Together with the multivariate Mecke equation for Poisson processes of Theorem 19.11, this gives

$$\begin{aligned}\mathbb{E}A_{\phi,s,W} &= \mathbb{E}\int_{[W]} \sum_{e\in Y_{s,W}\cap H} \phi(e)\,\Theta_{\text{iso}}(dH)\\ &= \frac{1}{2}\int_{[W]} \mathbb{E}\sum_{\substack{x,y\in\mathcal{F}_0(Y_{s,W}\cap H)\\ x\neq y}} \phi(\operatorname{conv}\{x,y\})\\ &\quad \times \mathbb{1}\{\operatorname{conv}\{x,y\}\cap\mathcal{F}_0(Y_{s,W}\cap H) = \{x,y\}\}\,\Theta_{\text{iso}}(dH)\\ &= \frac{2s^2}{\pi^2}\int_{[W]}\int_{W\cap H}\int_{W\cap H} \phi(\operatorname{conv}\{x,y\})\\ &\quad \times \mathbb{P}(\operatorname{conv}\{x,y\}\cap\partial Y_{s,W} = \emptyset)\,\lambda_2(dx)\,\lambda_2(dy)\,\Theta_{\text{iso}}(dH)\end{aligned}$$

$$= \frac{2s^2}{\pi^3} \int_W \int_W \phi(\mathrm{conv}\{x,y\}) \frac{e^{-\frac{2s}{\pi}\|x-y\|}}{\|x-y\|} \lambda_2(dx)\,\lambda_2(dy)\,, \tag{12.54}$$

where we used Lemma 5.3 with $C = \mathrm{conv}\{x,y\}$ and employed the Blaschke–Petkantschin-type formula of Proposition 12.5. Moreover, we have used the fact that $(\lambda_2 \otimes \lambda_2)(\{(x,y) \in \partial Y_{s,W} \times \partial Y_{s,W}\}) = 0$ a. s. Applying this repeatedly to the terms in (12.53) and using definition (12.7) of the iterated integral, we find that

$$\begin{aligned}
&\mathrm{cov}(\mathsf{V}_t(B_1), \mathsf{V}_t(B_2)) \\
&= \frac{2}{\pi^3} \int_W \int_W \zeta^{B_1}(\mathrm{conv}\{x,y\}) \zeta^{B_2}(\mathrm{conv}\{x,y\}) \frac{I^1(s^2 e^{-\frac{2s}{\pi}\|x-y\|}, t)}{\|x-y\|} \lambda_2(dx)\,\lambda_2(dy) \\
&+ \frac{4}{\pi^3} \int_W \int_W (\zeta^{B_1}(\mathrm{conv}\{x,y\}) \Theta_{\mathrm{iso}}([\mathrm{conv}\{x,y\} \cap B_2]) \\
&+ \Theta_{\mathrm{iso}}([\mathrm{conv}\{x,y\} \cap B_1]) \zeta^{B_2}(\mathrm{conv}\{x,y\})) \frac{I^2(s^2 e^{-\frac{2s}{\pi}\|x-y\|}, t)}{\|x-y\|} \lambda_2(dx)\,\lambda_2(dy) \\
&+ \frac{16}{\pi^3} \int_W \int_W \Theta_{\mathrm{iso}}([\mathrm{conv}\{x,y\} \cap B_1]) \Theta_{\mathrm{iso}}([\mathrm{conv}\{x,y\} \cap B_2]) \\
&\times \frac{I^3(s^2 e^{-\frac{2s}{\pi}\|x-y\|}, t)}{\|x-y\|} \lambda_2(dx)\,\lambda_2(dy)\,.
\end{aligned}$$

Note that in each of the double integrals the integration domain W can be replaced by $\mathbb{R}^2$ in view of the spatial consistency of STIT tessellations and since we assumed that $B_1, B_2 \subset W$. This, together with the substitution $y = x + z$ and the translated measure

$$\zeta^{B-x}(\mathrm{conv}\{0,z\}) := \zeta^B(\mathrm{conv}\{x, x+z\})\,, \tag{12.55}$$

where we write $\zeta^B(\mathrm{conv}\{u,v\}) := \mathbb{1}\{u \in B\} + \mathbb{1}\{v \in B\}$ also for an arbitrary line segment with vertices $u, v \in \mathbb{R}^d$, yields

$$\begin{aligned}
&\mathrm{cov}(\mathsf{V}_t(B_1), \mathsf{V}_t(B_2)) \\
&= \frac{2}{\pi^3} \int_{\mathbb{R}^2} \int_{\mathbb{R}^2} \zeta^{B_1-x}(\mathrm{conv}\{0,z\}) \zeta^{B_2-x}(\mathrm{conv}\{0,z\}) \frac{I^1(s^2 e^{-\frac{2s}{\pi}\|z\|}, t)}{\|z\|} \lambda_2(dx)\,\lambda_2(dz) \\
&+ \frac{4}{\pi^3} \int_{\mathbb{R}^2} \int_{\mathbb{R}^2} (\zeta^{B_1-x}(\mathrm{conv}\{0,z\}) \Theta_{\mathrm{iso}}([\mathrm{conv}\{0,z\} \cap (B_1 - x)]) \\
&+ \Theta_{\mathrm{iso}}([\mathrm{conv}\{0,z\} \cap (B_2 - x)]) \zeta^{B_2-x}(\mathrm{conv}\{0,z\})) \frac{I^2(s^2 e^{-\frac{2s}{\pi}\|z\|}, t)}{\|z\|} \lambda_2(dx)\,\lambda_2(dz) \\
&+ \frac{16}{\pi^3} \int_{\mathbb{R}^2} \int_{\mathbb{R}^2} \Theta_{\mathrm{iso}}([\mathrm{conv}\{0,z\} \cap (B_1 - x)]) \Theta_{\mathrm{iso}}([\mathrm{conv}\{0,z\} \cap (B_2 - x)]) \\
&\times \frac{I^3(s^2 e^{-\frac{2s}{\pi}\|z\|}, t)}{\|z\|} \lambda_2(dx)\,\lambda_2(dz)
\end{aligned}$$

$$
\begin{aligned}
&= \frac{2}{\pi^3} \int_{\mathbb{R}^2}\int_{\mathbb{R}^2} \zeta^{B_1-x}(\operatorname{conv}\{0,z\})\zeta^{B_2-x}(\operatorname{conv}\{0,z\}) \frac{I^1(s^2 e^{-\frac{2s}{\pi}\|z\|}, t)}{\|z\|}\,\lambda_2(dx)\,\lambda_2(dz) \\
&+ \frac{8}{\pi^4} \int_{\mathbb{R}^2}\int_{\mathbb{R}^2} (\zeta^{B_1-x}(\operatorname{conv}\{0,z\})\lambda_{\operatorname{conv}\{0,z\}}(B_1 - x) \\
&+ \lambda_{\operatorname{conv}\{0,z\}}(B_2 - x)\zeta^{B_2-x}(\operatorname{conv}\{0,z\})) \frac{I^2(s^2 e^{-\frac{2s}{\pi}\|z\|}, t)}{\|z\|}\,\lambda_2(dx)\,\lambda_2(dz) \\
&+ \frac{64}{\pi^5} \int_{\mathbb{R}^2}\int_{\mathbb{R}^2} \lambda_{\operatorname{conv}\{0,z\}}(B_1 - x)\lambda_{\operatorname{conv}\{0,z\}}(B_2 - x) \frac{I^3(s^2 e^{-\frac{2s}{\pi}\|z\|}, t)}{\|z\|}\,\lambda_2(dx)\,\lambda_2(dz),
\end{aligned}
$$

where $\lambda_{\operatorname{conv}\{0,z\}}$ stands for the Lebesgue measure on the line segment $\operatorname{conv}\{0,z\}$. To reduce the covariance measure, denote further by $\Delta_{\operatorname{conv}}\{0,z\} := \delta_0 + \delta_z$ the sum of the point measures concentrated at the two endpoints 0 and z of $\operatorname{conv}\{0,z\}$. Then for bounded $B \in \mathcal{B}(\mathbb{R}^2)$,

$$
\begin{aligned}
\widehat{\operatorname{Cov}}_{0,t}(B) = &\frac{2}{\pi^3} \int_{\mathbb{R}^2} \frac{I^1(s^2 e^{-\frac{2s}{\pi}\|z\|}, t)}{\|z\|} \int_{\operatorname{conv}\{0,z\}} \int_{\operatorname{conv}\{0,z\}} \mathbb{1}\{v - w \in B\} \\
&\Delta_{\operatorname{conv}\{0,z\}}(dv)\,\Delta_{\operatorname{conv}\{0,z\}}(dw)\,\lambda_2(dz) \\
&+ \frac{16}{\pi^4} \int_{\mathbb{R}^2} \frac{I^2(s^2 e^{-\frac{2s}{\pi}\|z\|}, t)}{\|z\|} \int_{\operatorname{conv}\{0,z\}} \int_{\operatorname{conv}\{0,z\}} \mathbb{1}\{v - w \in B\} \\
&\Delta_{\operatorname{conv}\{0,z\}}(dv)\,\lambda_{\operatorname{conv}\{0,z\}}(dw)\,\lambda_2(dz) \\
&+ \frac{64}{\pi^5} \int_{\mathbb{R}^2} \frac{I^3(s^2 e^{-\frac{2s}{\pi}\|z\|}, t)}{\|z\|} \int_{\operatorname{conv}\{0,z\}} \int_{\operatorname{conv}\{0,z\}} \mathbb{1}\{v - w \in B\} \\
&\lambda_{\operatorname{conv}\{0,z\}}(dv)\,\lambda_{\operatorname{conv}\{0,z\}}(dw)\,\lambda_2(dz).
\end{aligned}
$$

Noting that

$$
\begin{aligned}
&\int_{\operatorname{conv}\{0,z\}} \int_{\operatorname{conv}\{0,z\}} \mathbb{1}\{v - w \in B\}\,\Delta_{\operatorname{conv}\{0,z\}}(dv)\,\Delta_{\operatorname{conv}\{0,z\}}(dw) \\
&= \mathbb{1}\{z \in B\} + \mathbb{1}\{-z \in B\} + 2 \cdot \mathbb{1}\{0 \in B\}
\end{aligned}
$$

and

$$
\begin{aligned}
&\int_{\operatorname{conv}\{0,z\}} \int_{\operatorname{conv}\{0,z\}} \mathbb{1}\{v - w \in B\}\,\Delta_{\operatorname{conv}\{0,z\}}(dv)\,\lambda_{\operatorname{conv}\{0,z\}}(dw) \\
&= 2\lambda_{\operatorname{conv}\{0,z\}}(B) \qquad (12.56)
\end{aligned}
$$

with $\lambda_{\mathrm{conv}\{-z,z\}}$ being the Lebesgue measure on the line segment $\mathrm{conv}\{-z, z\}$, we see that

$$\begin{aligned}\widehat{\mathrm{Cov}}_{0,t}(B) &= \frac{2}{\pi^3}\int_{\mathbb{R}^2}\big(\mathbb{1}\{z \in B\} + \mathbb{1}\{-z \in B\} + 2\mathbb{1}\{0 \in B\}\big)\,\frac{I^1(s^2 e^{-\frac{2s}{\pi}\|z\|}, t)}{\|z\|}\,\lambda_2(\mathrm{d}z)\\ &\quad + \frac{32}{\pi^4}\int_{\mathbb{R}^2}\lambda_{\mathrm{conv}\{0,z\}}(B)\,\frac{I^2(s^2 e^{-\frac{2s}{\pi}\|z\|}, t)}{\|z\|}\,\lambda_2(\mathrm{d}z)\\ &\quad + \frac{64}{\pi^5}\int_{\mathbb{R}^2}\frac{I^3(s^2 e^{-\frac{2s}{\pi}\|z\|}, t)}{\|z\|}\int_{\mathrm{conv}\{0,z\}}\int_{\mathrm{conv}\{0,z\}}\mathbb{1}\{v - w \in B\}\\ &\quad \lambda_{\mathrm{conv}\{0,z\}}(\mathrm{d}v)\,\lambda_{\mathrm{conv}\{0,z\}}(\mathrm{d}w)\,\lambda_2(\mathrm{d}z)\,.\end{aligned}$$

Dividing by $\gamma_0^2 = \frac{4t^4}{\pi^2}$, subtracting $\lambda_2(B)$, and choosing for B a centered disc of radius $r \geq 0$, we conclude the following formula for the reduced second moment function:

$$\begin{aligned}K_{0,t}(r) &= \pi r^2 + \frac{1}{\pi t^4}\int_{\mathbb{R}^2}\big(\mathbb{1}\{\|z\| \leq r\} + 1\big)\,\frac{I^1(s^2 e^{-\frac{2s}{\pi}\|z\|}, t)}{\|z\|}\,\lambda_2(\mathrm{d}z)\\ &\quad + \frac{8}{\pi^2 t^4}\int_{\mathbb{R}^2}\min\{r, \|z\|\}\,\frac{I^2(s^2 e^{-\frac{2s}{\pi}\|z\|}, t)}{\|z\|}\,\lambda_2(\mathrm{d}z)\\ &\quad + \frac{32}{\pi^3 t^4}\int_{\mathbb{R}^2}\frac{I^3(s^2 e^{-\frac{2s}{\pi}\|z\|}, t)}{\|z\|}\int_0^{\min\{\|z\|,r\}}(\|z\| - s)\,\mathrm{d}s\,\lambda_2(\mathrm{d}z)\\ &= \pi r^2 + \frac{2}{t^4}\int_0^\infty\big(\mathbb{1}\{\varrho \leq r\} + 1\big)\,I^1(s^2 e^{-\frac{2s}{\pi}\varrho}, t)\,\mathrm{d}\varrho\\ &\quad + \frac{16}{\pi t^4}\int_0^\infty\min\{r, \varrho\}\,I^2(s^2 e^{-\frac{2s}{\pi}\varrho}, t)\,\mathrm{d}\varrho\\ &\quad + \frac{64}{\pi^2 t^4}\int_0^\infty I^3(s^2 e^{-\frac{2s}{\pi}\varrho}, t)\Big(\varrho\min\{r, \varrho\} - \frac{1}{2}\min\{r, \varrho\}^2\Big)\,\mathrm{d}\varrho\,,\end{aligned}$$

where we have passed to polar coordinates in the last step. As a result, we find that

$$\begin{aligned}K_{0,t}(r) &= \pi r^2 + \int_0^r \frac{4}{t^4}I^1(s^2 e^{-\frac{2s}{\pi}\varrho}, t) + \frac{16}{\pi t^4}\varrho I^2(s^2 e^{-\frac{2s}{\pi}\varrho}, t) + \frac{32}{\pi^2 t^4}\varrho^2 I^3(s^2 e^{-\frac{2s}{\pi}\varrho}, t)\,\mathrm{d}\varrho\\ &\quad + \int_r^\infty \frac{2}{t^4}I^1(s^2 e^{-\frac{2s}{\pi}\varrho}, t) + \frac{16}{\pi t^4}r I^2(s^2 e^{-\frac{2s}{\pi}\varrho}, t) + \frac{64}{\pi^2 t^4}\Big(\varrho r - \frac{r^2}{2}\Big)I^3(s^2 e^{-\frac{2s}{\pi}\varrho}, t)\,\mathrm{d}\varrho\,.\end{aligned}$$

We can now differentiate with respect to r by using the Leibniz integral rule:

$$\begin{aligned}\frac{\mathrm{d}K_{0,t}(r)}{\mathrm{d}r} &= 2\pi r + \frac{4}{t^4}I^1(s^2\mathrm{e}^{-\frac{2s}{\pi}r},t) + \frac{16}{\pi t^4}rI^2(s^2\mathrm{e}^{-\frac{2s}{\pi}r},t) + \frac{32}{\pi^2 t^4}r^2 I^3(s^2\mathrm{e}^{-\frac{2s}{\pi}r},t)\\
&\quad - \frac{2}{t^4}I^1(s^2\mathrm{e}^{-\frac{2s}{\pi}r},t) - \frac{16}{\pi t^4}rI^2(s^2\mathrm{e}^{-\frac{2s}{\pi}r},t) + \frac{16}{\pi t^4}\int_r^\infty I^2(s^2\mathrm{e}^{-\frac{2s}{\pi}\varrho},t)\,\mathrm{d}\varrho\\
&\quad - \frac{32}{\pi^2 t^4}r^2 I^3(s^2\mathrm{e}^{-\frac{2s}{\pi}r},t) + \frac{64}{\pi^2 t^4}\int_r^\infty (\varrho - r)I^3(s^2\mathrm{e}^{-\frac{2s}{\pi}\varrho},t)\mathrm{d}\varrho\\
&= 2\pi r + \frac{2}{t^4}I^1(s^2\mathrm{e}^{-\frac{2s}{\pi}r},t) + \frac{16}{\pi t^4}\int_r^\infty I^2(s^2\mathrm{e}^{-\frac{2s}{\pi}\varrho},t)\,\mathrm{d}\varrho\\
&\quad + \frac{64}{\pi^2 t^4}\int_r^\infty (\varrho - r)I^3(s^2\mathrm{e}^{-\frac{2s}{\pi}\varrho},t)\,\mathrm{d}\varrho\,.\end{aligned}$$

Dividing this expression by $2\pi r$ and using definition (12.7) of the iterated integrals lead to

$$\begin{aligned}g_{0,t}(r) &= \frac{1}{2\pi r}\frac{\mathrm{d}K_{0,t}(r)}{\mathrm{d}r}\\
&= 1 + \frac{1}{\pi t^4}\int_0^t s^2\mathrm{e}^{-\frac{2s}{\pi}r}\,\mathrm{d}s + \frac{8}{\pi^2 r t^4}\int_0^t\int_r^\infty (t-s)s^2\mathrm{e}^{-\frac{2s}{\pi}\varrho}\,\mathrm{d}\varrho\,\mathrm{d}s\\
&\quad + \frac{32}{\pi^3 r t^4}\int_0^t\int_r^\infty (\varrho - r)(t-s)^2 s^2\mathrm{e}^{-\frac{2s}{\pi}\varrho}\,\mathrm{d}\varrho\,\mathrm{d}s\\
&= 1 + \frac{1}{\pi t^4}\cdot\frac{\pi}{4r^3}(\pi^2 - (\pi^2 + 2\pi r t + 2r^2t^2)\mathrm{e}^{-\frac{2rt}{\pi}})\\
&\quad + \frac{8}{\pi^2 r t^4}\cdot\frac{\pi^3}{8r^3}(rt - \pi + (\pi + rt)\mathrm{e}^{-\frac{2rt}{\pi}})\\
&\quad + \frac{32}{\pi^3 r t^4}\cdot\frac{\pi^3}{32r^3}(2t^2r^2 - 2\pi r t + \pi^2 - \pi^2\mathrm{e}^{-\frac{2rt}{\pi}})\\
&= 1 + \frac{2}{t^2r^2} - \frac{\pi}{t^3r^3} + \frac{\pi^2}{4t^4r^4} - \left(\frac{1}{2t^2r^2} - \frac{\pi}{2t^3r^3} + \frac{\pi^2}{4t^4r^4}\right)\mathrm{e}^{-\frac{2rt}{\pi}}\,.\end{aligned}$$

This completes the proof. □

The result of the previous theorem should be compared to the pair correlation function $g_{0,t}^{\mathrm{PHT}}$ of the vertex point process $\mathsf{V}_t^{\mathrm{PHT}}$ of a stationary and isotropic Poisson line tessellation X_t in $\mathbb{R}^2$ induced by a Poisson point process on the space of lines in $\mathbb{R}^2$ with intensity measure $t\,\Theta_{\mathrm{iso}}$; see (2.66). In this case, we have that

$$g_{0,t}^{\mathrm{PHT}}(r) = 1 + \frac{4}{\pi t r}, \quad r > 0\,; \tag{12.57}$$

compare with (12.68) in the notes at the end of this chapter and with (10.24) in [82]. Using the Mecke equation (19.5) for Poisson point processes, this can also be seen as follows by means of the corresponding reduced second-moment function $K_{0,t}^{\mathrm{PHT}}$. Under the Palm distribution with respect to the vertex point process, the typical point is located at the origin, generated by an additionally inserted pair (H_1, H_2) of random lines, independent of the remaining Poisson line tessellation. Denoting by $\gamma_0^{\mathrm{PHT}} = t^2/\pi$ the intensity of $\mathsf{V}_t^{\mathrm{PHT}}$, we thus find that

$$\begin{aligned}\gamma_0^{\mathrm{PHT}} K_{0,t}^{\mathrm{PHT}}(r) &= \left(1 + \frac{8}{\pi} rt\right) + \mathbb{E}\mathcal{H}^0(\mathcal{F}_0(X_t \wedge rB^2)) \\ &= 1 + \frac{8}{\pi} rt + \gamma_0^{\mathrm{PHT}} \pi r^2,\end{aligned}$$

where the term $1 + \frac{8}{\pi} rt$ counts the origin and the expected number of intersection points on $H_1 \cap rB^2$ and $H_2 \cap rB^2$ induced by X_t. This leads to (12.57) after differentiation with respect to r and division by $2\pi r$.

Instead of a process of lines, we consider now a Boolean model of line segments in $\mathbb{R}^2$. We denote its intensity by γ^{BM} and assume that its segments have a uniformly distributed orientation and a common length distribution L on $[0, \infty)$. The mean segment length is denoted by $\overline{s} := \int_0^\infty s\, L(\mathrm{d}s)$. To avoid trivial situations, we assume that $\gamma^{\mathrm{BM}} \in (0, \infty)$ and $\overline{s} \in (0, \infty)$. It has been shown in [182] that the K-function $K_{1,L}^{\mathrm{BM}}$ of the stationary and isotropic random length measure induced by such a Boolean segment process satisfies

$$K_{1,L}^{\mathrm{BM}}(r) = \pi r^2 + \frac{1}{\gamma^{\mathrm{BM}} \overline{s}^2} \left(\int_0^r s^2\, L(\mathrm{d}s) + \int_r^\infty (2rs - r^2)\, L(\mathrm{d}s) \right), \quad r > 0. \tag{12.58}$$

The corresponding pair correlation function is thus given by

$$g_{1,L}^{\mathrm{BM}}(r) = 1 + \frac{1}{\pi \gamma^{\mathrm{BM}}\, \overline{s}^2 r} \int_r^\infty (s - r)\, L(\mathrm{d}s), \quad r > 0; \tag{12.59}$$

see also Example 8.4(b) of [33] and Example 3.21 of [13]. The Boolean segment process induces a stationary and isotropic point process in $\mathbb{R}^2$ arising from the intersection of its segments. By equation (7) in [160] this intersection point process has intensity

$$\gamma_\cap^{\mathrm{BM}} = \frac{(\gamma^{\mathrm{BM}})^2\, \overline{s}^2}{\pi},$$

and its K-function $K_{0,L}^{\mathrm{BM}}$ is given by

$$K_{0,L}^{\mathrm{BM}}(r) = \pi r^2 + \frac{4}{\gamma^{\mathrm{BM}}\, \overline{s}^2} \left(\int_0^r s^2\, L(\mathrm{d}s) + \int_r^\infty (2rs - r^2)\, L(\mathrm{d}s) \right), \quad r > 0. \tag{12.60}$$

Indeed, by the alternative representation (12.42) of the K-function we need to multiply the prefactor of the integral term in (12.58) by

$$2 \cdot \frac{2}{\pi}\gamma^{\mathrm{BM}}\,\overline{s} \cdot \frac{\gamma^{\mathrm{BM}}\,\overline{s}}{\gamma_{\cap}^{\mathrm{BM}}} = 4 \qquad (12.61)$$

to perform the transition from $K_{1,L}^{\mathrm{BM}}$ to $K_{0,L}^{\mathrm{BM}}$. Here

- the factor 2 represents the fact that the typical point of the intersection process is located at the intersection of precisely two line segments;
- the factor $\frac{2}{\pi}\gamma^{\mathrm{BM}}\,\overline{s}$ is the intensity of the homogeneous point process arising from the intersection of the Boolean segment process with fixed line or line segment; see equation (8.38) in [33];
- the fraction $\gamma^{\mathrm{BM}}\,\overline{s}/\gamma_{\cap}^{\mathrm{BM}}$ takes into account the exchange of the intensities in the definition of the K-function when replacing the random length measure, whose length density is $\gamma^{\mathrm{BM}}\,\overline{s}$, with the intersection point process having intensity $\gamma_{\cap}^{\mathrm{BM}}$.

It follows from (12.60) by differentiation that the pair correlation function of the intersection point process is

$$g_{0,L}^{\mathrm{BM}}(r) = 1 + \frac{4}{\pi\gamma^{\mathrm{BM}}\,\overline{s}^2 r}\int_r^{\infty}(s-r)\,L(\mathrm{d}s)\,, \quad r > 0\,. \qquad (12.62)$$

Let us assume now that $\gamma^{\mathrm{BM}} = \gamma_{1,t}^{\max} = \frac{t^2}{\pi}$ for some $t > 0$ and that the length distribution L has the density

$$q_{1:\ell}^{\max}(r) = \frac{1}{t^2 r^3}\left(\pi^2 - (\pi^2 + 2\pi t r + 2t^2 r^2)\mathrm{e}^{-\frac{2}{\pi}tr}\right), \quad r > 0\,, \qquad (12.63)$$

with respect to the Lebesgue measure on $(0,\infty)$. It follows from Corollaries 11.19 and 12.33 that the random length measure induced by such a Boolean segment process has the same pair correlation function $g_{1,L}^{\mathrm{BM}}$ as the random length measure on the edge skeleton of a stationary and isotropic STIT tessellation in the plane with time parameter t:

$$g_{1,L}^{\mathrm{BM}}(r) = g_{1,t}(r) = 1 + \frac{1}{2t^2 r^2}\left(1 - \mathrm{e}^{-\frac{2}{\pi}rt}\right), \quad r > 0\,. \qquad (12.64)$$

In this context, it is natural to ask whether the pair correlation function $g_{0,L}^{\mathrm{BM}}$ of the intersection point process coincides with the pair correlation function $g_{0,t}$ of the vertex point process of the planar stationary and isotropic STIT tessellation with time parameter t. Substituting (12.63) into the expression for $g_{0,L}^{\mathrm{BM}}(r)$, we obtain

$$g_{0,L}^{\mathrm{BM}}(r) = 1 + \frac{2}{r^2 t^2}\left(1 - \mathrm{e}^{-\frac{2}{\pi}rt}\right), \quad r > 0\,.$$

Thus a comparison with Theorem 12.34 reveals that

$$g_{0,L}^{\mathrm{BM}}(r) \neq g_{0,t}(r) \quad \text{for } r > 0\,, \tag{12.65}$$

which implies that – unlike the pair correlation function of the random length measure – the pair correlation function of the vertex or intersection point process is sufficiently sensitive to distinguish between a stationary and isotropic planar STIT tessellation and a stationary and isotropic Boolean segment process with intensity γ and length distribution L, as described above.

Remark 12.35. *As in Remark 12.29, we can consider the expansion of the pair correlation function $g_{0,t}$ in Theorem 12.34 as $t \downarrow 0$. Developing the exponential function in a Taylor series, we obtain*

$$g_{0,t}(r) = 1 + \frac{7}{3\pi rt} - \frac{11}{6\pi^2} + O(rt)$$

as $t \downarrow 0$. In contrast, the pair correlation function $g_{0,t}^{\mathrm{PHT}}$ of a stationary and isotropic Poisson line tessellation behaves like $1 + \frac{4}{\pi rt}$; see (12.57). Thus in the sparse regime the vertex process of the STIT tessellation exhibits weaker clustering than the vertex process of the Poisson line tessellation.

12.9 Cross-correlation function for the length measure and the vertex point process in dimension $d = 2$

As in the previous section, we consider a stationary and isotropic STIT tessellation Y_t with time parameter $t > 0$ in the plane $\mathbb{R}^2$. By V_t we denote the stationary and isotropic point process of vertices of Y_t, and by $\mathscr{H}_t^1$ we denote the stationary and isotropic random length measure on the skeleton ∂Y_t induced by Y_t. We recall that V_t has the intensity $\gamma_0 = \frac{2t^2}{\pi}$, whereas $\mathscr{H}_t^1$ has the density t. In this section, we are interested in the correlation structure between the point process V_t and the random measure $\mathscr{H}_t^1$. Our main result is an explicit expression for the cross-correlation function $g_{0,1,t}$ between V_t and $\mathscr{H}_t^1$; see also Figure 12.3. As in Section 12.7.1 with $\Phi = \mathsf{V}_t$ and $\Psi = \mathscr{H}_t^1$ there, we introduce

the cross covariance measure	$\mathrm{Cross}_{0,1,t} := \mathrm{Cross}_{\mathsf{V}_t,\mathscr{H}_t^1}$,
the reduced cross-covariance measure	$\widehat{\mathrm{Cross}}_{0,1,t} := \widehat{\mathrm{Cross}}_{\mathsf{V}_t,\mathscr{H}_t^1}$,
the reduced mixed moment measure	$\widehat{\mu}^{(2)}_{0,1,t} := \widehat{\mu}^{(2)}_{\mathsf{V}_t,\mathscr{H}_t^1}$,
the mixed K-function	$K_{0,1,t} := K_{\mathsf{V}_t,\mathscr{H}_t^1}$,
the mixed pair correlation function	$g_{0,1,t} := g_{\mathsf{V}_t,\mathscr{H}_t^1}$

of two random measures V_t and $\mathscr{H}_t^1$.

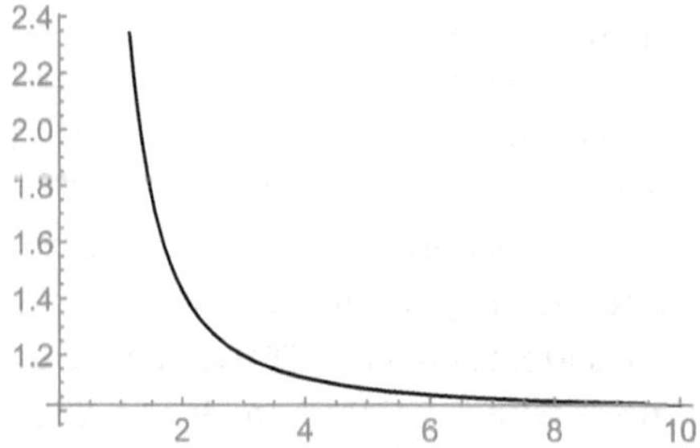

Figure 12.3: Cross-correlation function $g_{0,1,t}$ for $t = 1$.

Theorem 12.36. *Let $(Y_t,\ t > 0)$ be the STIT tessellation process in $\mathbb{R}^2$ driven by the motion-invariant measure Θ_{iso}. Then for any fixed $t > 0$, we have that*

$$g_{0,1,t}(r) = 1 + \frac{2}{t^2r^2} - \frac{\pi}{4t^3r^3} - \left(\frac{1}{2t^2r^2} - \frac{\pi}{4t^3r^3}\right)e^{-\frac{2}{\pi}rt}$$

for $r > 0$.

Proof. Let $B_1, B_2 \in \mathcal{B}(\mathbb{R}^2)$ be bounded Borel sets, and let $W \in \mathcal{P}_2$ be such that $B_1, B_2 \subset W$. Further, let $(Y_{t,W},\ t \geq 0)$ with $Y_{t,W} \overset{\mathrm{D}}{=} Y_t \wedge W$ be the STIT tessellation process localized to W. For $0 \leq s \leq t$ and $i \in \{1,2\}$, recall from (12.50) and (12.51) the definitions of $\zeta^{B_i}(e)$ and $\Sigma_{\Theta_{\mathrm{iso}}^{B_i},s,W}$, respectively. Further, for $B \in \mathcal{B}(\mathbb{R}^2)$, define

$$\begin{aligned} \mathscr{H}^B(e) &:= \mathscr{H}^1(e \cap B), \quad e \in \mathcal{P}_1, \\ \mathscr{H}^1_{t,W}(B) &:= \sum_{e \in \mathcal{F}_1^{\max}(Y_{t,W})} \mathscr{H}^1(e \cap B), \end{aligned}$$

and observe that $\mathscr{H}_t^1(B_2) = \Sigma_{\mathscr{H}^{B_2},t,W}$. Now apply Theorem 10.8(c) with $b_1(t) = \mathbb{E}\mathrm{V}_t(B_1) = \mathbb{E}\Sigma_{\zeta^{B_1},t,W}$ and $b_2(t) = \mathbb{E}\mathscr{H}^1_{t,W}(B_2)$ to see that

$$\operatorname{cov}(\mathrm{V}_t(B_1), \mathscr{H}_t^1(B_2)) = \mathbb{E}\int_0^t A_{\zeta^{B_1},\mathscr{H}^{B_2},s,W}\,\mathrm{d}s + 2\int_0^t \operatorname{cov}(\Sigma_{\Theta_{\mathrm{iso}}^{B_1},s,W}, \Sigma_{\mathscr{H}^{B_2},s,W})\,\mathrm{d}s,$$

where we additionally used (12.52). Applying Theorem 10.8(b), we can express $\operatorname{cov}(\Sigma_{\Theta_{\mathrm{iso}}^{B_1},s,W}, \Sigma_{\mathscr{H}^{B_2},s,W})$ as

$$\operatorname{cov}(\Sigma_{\Theta_{\mathrm{iso}}^{B_1},s,W}, \Sigma_{\mathscr{H}^{B_2},s,W}) = \mathbb{E}\int_0^s A_{\Theta_{\mathrm{iso}}^{B_1},\mathscr{H}^{B_2},u,W}\,\mathrm{d}u,$$

and it follows that

$$\operatorname{cov}(\mathrm{V}_t(B_1), \mathscr{H}_t^1(B_2)) = \mathbb{E}\int_0^t A_{\zeta^{B_1},\mathscr{H}^{B_2},s,W}\,\mathrm{d}s + 2\mathbb{E}\int_0^t\int_0^s A_{\Theta_{\mathrm{iso}}^{B_1},\mathscr{H}^{B_2},u,W}\,\mathrm{d}u\,\mathrm{d}s.$$

To evaluate the integrands further, we use twice (12.54) together with (12.55) to see that

$$\begin{aligned}
&\operatorname{cov}(\mathsf{V}_t(B_1), \mathscr{H}_t^1(B_2)) \\
&= \frac{2}{\pi^3} \int_W \int_W \zeta^{B_1}(\operatorname{conv}\{x,y\}) \mathscr{H}^{B_2}(\operatorname{conv}\{x,y\}) \frac{I^1(s^2 e^{-\frac{2s}{\pi}\|x-y\|}, t)}{\|x-y\|} \lambda_2(dx)\,\lambda_2(dy) \\
&\quad + \frac{4}{\pi^3} \int_W \int_W \Theta_{\mathrm{iso}}^{B_1}([\operatorname{conv}\{x,y\}]) \mathscr{H}^{B_2}(\operatorname{conv}\{x,y\}) \frac{I^2(s^2 e^{-\frac{2s}{\pi}\|x-y\|}, t)}{\|x-y\|} \lambda_2(dx)\,\lambda_2(dy) \\
&= \frac{2}{\pi^3} \int_{\mathbb{R}^2} \int_{\mathbb{R}^2} \zeta^{B_1}(\operatorname{conv}\{x,y\}) \lambda_{\operatorname{conv}\{x,y\}}(B_2) \frac{I^1(s^2 e^{-\frac{2s}{\pi}\|x-y\|}, t)}{\|x-y\|} \lambda_2(dx)\,\lambda_2(dy) \\
&\quad + \frac{8}{\pi^4} \int_{\mathbb{R}^2} \int_{\mathbb{R}^2} \lambda_{\operatorname{conv}\{x,y\}}(B_1) \lambda_{\operatorname{conv}\{x,y\}}(B_2) \frac{I^2(s^2 e^{-\frac{2s}{\pi}\|x-y\|}, t)}{\|x-y\|} \lambda_2(dx)\,\lambda_2(dy),
\end{aligned}$$

where, as in the proof of Theorem 12.34, $\lambda_{\operatorname{conv}\{x,y\}}$ is the Lebesgue measure on the line segment $\operatorname{conv}\{x,y\}$. Next, we apply the substitution $y = x + z$:

$$\begin{aligned}
&\operatorname{cov}(\mathsf{V}_t(B_1), \mathscr{H}_t^1(B_2)) \\
&= \frac{2}{\pi^3} \int_{\mathbb{R}^2} \int_{\mathbb{R}^2} \zeta^{B_1-x}(\operatorname{conv}\{0,z\}) \lambda_{\operatorname{conv}\{0,z\}}(B_2 - x) \frac{I^1(s^2 e^{-\frac{2s}{\pi}\|z\|}, t)}{\|z\|} \lambda_2(dx)\,\lambda_2(dz) \\
&\quad + \frac{8}{\pi^4} \int_{\mathbb{R}^2} \int_{\mathbb{R}^2} \lambda_{\operatorname{conv}\{0,z\}}(B_1 - x) \lambda_{\operatorname{conv}\{0,z\}}(B_2 - x) \frac{I^2(s^2 e^{-\frac{2s}{\pi}\|z\|}, t)}{\|z\|} \lambda_2(dx)\,\lambda_2(dz).
\end{aligned}$$

From this expression we can read off the reduced cross-covariance measure $\widehat{\mathrm{Cross}}_{0,1,t}$. Namely, for bounded $B \in \mathcal{B}(\mathbb{R}^2)$, we have that

$$\begin{aligned}
\widehat{\mathrm{Cross}}_{0,1,t}(B) = \frac{2}{\pi^3} \int_{\mathbb{R}^2} \frac{I^1(s^2 e^{-\frac{2s}{\pi}\|z\|}, t)}{\|z\|} & \int_{\operatorname{conv}\{0,z\}} \int_{\operatorname{conv}\{0,z\}} \mathbb{1}\{v - w \in B\} \\
& \Delta_{\operatorname{conv}\{0,z\}}(dv)\, \lambda_{\operatorname{conv}\{0,z\}}(dw)\, \lambda_2(dz) \\
& + \frac{8}{\pi^4} \int_{\mathbb{R}^2} \frac{I^2(s^2 e^{-\frac{2s}{\pi}\|z\|}, t)}{\|z\|} \int_{\operatorname{conv}\{0,z\}} \int_{\operatorname{conv}\{0,z\}} \mathbb{1}\{v - w \in B\} \\
& \lambda_{\operatorname{conv}\{0,z\}}(dv)\, \lambda_{\operatorname{conv}\{0,z\}}(dw)\, \lambda_2(dz),
\end{aligned}$$

where again $\Delta_{\operatorname{conv}\{0,z\}} = \delta_0 + \delta_z$ is the sum of the point measures concentrated at the two vertices of $\operatorname{conv}\{0, z\}$. Invoking (12.56), we arrive at

$$\widehat{\text{Cross}}_{0,1,t}(B) = \frac{4}{\pi^3} \int_{\mathbb{R}^2} \lambda_{\text{conv}\{0,z\}}(B) \frac{I^1(s^2 e^{-\frac{2s}{\pi}\|z\|}, t)}{\|z\|} \lambda_2(\mathrm{d}z)$$
$$+ \frac{8}{\pi^4} \int_{\mathbb{R}^2} \frac{I^2(s^2 e^{-\frac{2s}{\pi}\|z\|}, t)}{\|z\|} \int_{\text{conv}\{0,z\}} \int_{\text{conv}\{0,z\}} \mathbb{1}\{v - w \in B\}$$
$$\lambda_{\text{conv}\{0,z\}}(\mathrm{d}v)\, \lambda_{\text{conv}\{0,z\}}(\mathrm{d}w)\, \lambda_2(\mathrm{d}z).$$

Dividing this expression by $t\gamma_0 = \frac{2t^3}{\pi}$ and plugging in for B a centered disc of radius $r \geq 0$, we conclude that

$$K_{0,1,t}(r) = \pi r^2 + \frac{2}{t^3\pi^2} \int_{\mathbb{R}^2} \min\{r, \|z\|\} \frac{I^1(s^2 e^{-\frac{2s}{\pi}\|z\|}, t)}{\|z\|} \lambda_2(\mathrm{d}z)$$
$$+ \frac{8}{t^3\pi^3} \int_{\mathbb{R}^2} \frac{I^2(s^2 e^{-\frac{2s}{\pi}\|z\|}, t)}{\|z\|} \int_0^{\min\{r,\|z\|\}} (\|z\| - s)\, \mathrm{d}s\, \lambda_2(\mathrm{d}z)$$
$$= \pi r^2 + \frac{4}{t^3\pi} \int_0^\infty \min\{r, \varrho\}\, I^1(s^2 e^{-\frac{2s}{\pi}\varrho}, t)\, \mathrm{d}\varrho$$
$$+ \frac{16}{t^3\pi^2} \int_0^\infty I^2(s^2 e^{-\frac{2s}{\pi}\varrho}, t) \Big(\varrho \min\{r, \varrho\} - \frac{1}{2} \min\{r, \varrho\}^2\Big)\, \mathrm{d}\varrho,$$

where we passed to polar coordinates in the last step. Thus

$$K_{0,1,t}(r) = \pi r^2 + \frac{4}{t^3\pi} \int_0^r \varrho I^1(s^2 e^{-\frac{2s}{\pi}\varrho}, t) + \frac{8}{t^3\pi^2} \varrho^2 I^2(s^2 e^{-\frac{2s}{\pi}\varrho}, t)\, \mathrm{d}\varrho$$
$$+ \frac{4}{t^3\pi} \int_r^\infty r I^1(s^2 e^{-\frac{2s}{\pi}\varrho}, t) + \frac{16}{t^3\pi^2} \Big(\varrho r - \frac{r^2}{2}\Big) I^2(s^2 e^{-\frac{2s}{\pi}\varrho}, t)\, \mathrm{d}\varrho.$$

Using the Leibniz integral rule, the derivative of $K_{0,1,t}(r)$ equals

$$\frac{\mathrm{d}K_{0,1,t}(r)}{\mathrm{d}r} = 2\pi r + \frac{4}{t^3\pi} r I^1(s^2 e^{-\frac{2s}{\pi}r}, t) + \frac{8}{t^3\pi^2} r^2 I^2(s^2 e^{-\frac{2s}{\pi}r}, t)$$
$$- \frac{4}{t^3\pi} r I^1(s^2 e^{-\frac{2s}{\pi}r}, t) + \frac{4}{t^3\pi} \int_r^\infty I^1(s^2 e^{-\frac{2s}{\pi}\varrho}, t)\, \mathrm{d}\varrho$$
$$- \frac{8}{t^3\pi^2} r^2 I^2(s^2 e^{-\frac{2s}{\pi}r}, t) + \frac{16}{t^3\pi^2} \int_r^\infty (\varrho - r) I^2(s^2 e^{-\frac{2s}{\pi}\varrho}, t)\, \mathrm{d}\varrho$$
$$= 2\pi r + \frac{4}{t^3\pi} \int_r^\infty I^1(s^2 e^{-\frac{2s}{\pi}\varrho}, t)\, \mathrm{d}\varrho + \frac{16}{t^3\pi^2} \int_r^\infty (\varrho - r) I^2(s^2 e^{-\frac{2s}{\pi}\varrho}, t)\, \mathrm{d}\varrho.$$

Consequently, using definition (12.7) of the iterated integrals, we have

$$\begin{aligned} g_{0,1,t}(r) &= \frac{1}{2\pi r}\frac{\mathrm{d}K_{0,1,t}(r)}{\mathrm{d}r} \\ &= 1 + \frac{2}{t^3\pi^2 r}\int_0^t\int_r^\infty s^2 \mathrm{e}^{-\frac{2s}{\pi}\varrho}\,\mathrm{d}\varrho\,\mathrm{d}s + \frac{8}{t^3\pi^3 r}\int_0^t\int_r^\infty (\varrho - r)(t-s)s^2\mathrm{e}^{-\frac{2s}{\pi}\varrho}\,\mathrm{d}\varrho\,\mathrm{d}s \\ &= 1 + \frac{2}{t^3\pi^2 r}\cdot\frac{\pi^2}{8r^2}(\pi - (\pi + 2rt)\mathrm{e}^{-\frac{2rt}{\pi}}) + \frac{16}{t^3\pi^2}\cdot\frac{\pi^3}{16r^2}(2rt - \pi + \pi\mathrm{e}^{-\frac{2rt}{\pi}}) \\ &= 1 + \frac{2}{t^2r^2} - \frac{\pi}{4t^3r^3} - \left(\frac{1}{2t^2r^2} - \frac{\pi}{4t^3r^3}\right)\mathrm{e}^{-\frac{2rt}{\pi}}, \end{aligned}$$

which completes the argument. □

As in the previous sections, the result of Theorem 12.36 can be compared to the cross-correlation function $g_{0,1,t}^{\mathrm{PHT}}$ of the length measure and the vertex point process of a stationary and isotropic Poisson line tessellation X_t in $\mathbb{R}^2$ with length density t; see (2.66). In this case, we have that

$$g_{0,1,t}^{\mathrm{PHT}}(r) = 1 + \frac{2}{\pi t r}, \qquad r > 0\,; \tag{12.66}$$

see (12.68) in the notes at the end of this chapter. Again, using the multivariate Mecke equation for Poisson point processes, this can also be seen as follows by means of the corresponding reduced mixed moment function $K_{0,1,t}^{\mathrm{PHT}}$. Under the Palm distribution with respect to the vertex point process, the typical point is located at the origin on an additionally inserted pair (H_1, H_2) of random lines and which is independent of the remaining Poisson line tessellation X_t. Recalling that the random length measure has the intensity t, we thus find that

$$\begin{aligned} tK_{0,1,t}^{\mathrm{PHT}}(r) &= \mathcal{H}^1(H_1 \cap rB^2) + \mathcal{H}^1(H_2 \cap rB^2) + \mathbb{E}\mathcal{H}^1(\partial X_t \cap rB^2) \\ &= 4r + t\pi r^2\,. \end{aligned}$$

This leads to (12.66) after differentiation with respect to r and division by $2\pi r$.

Similarly as at the end of Section 12.8, one can also compare the cross-correlation function $g_{0,1,t}$ of a stationary and isotropic STIT tessellation in the plane with the cross-correlation function related to a Boolean model of line segments in $\mathbb{R}^2$. For that purpose, fix $\gamma^{\mathrm{BM}} \in (0,\infty)$ and let L be a probability measure on $[0,\infty)$ satisfying $\overline{s} = \int_0^\infty s\,L(\mathrm{d}s) \in (0,\infty)$. We then consider a stationary and isotropic Boolean model of line segments in $\mathbb{R}^2$ with intensity γ^{BM} and where the common length distribution of the segments is given by L. The reduced mixed moment function $K_{0,1,L}$ between the stationary and isotropic random length measure and the stationary and isotropic intersection point process induced by such a Boolean segment process satisfies

$$K_{0,1,L}^{\mathrm{BM}}(r) = \pi r^2 + \frac{2}{\gamma^{\mathrm{BM}}\,\overline{s}^2}\left(\int_0^r s^2\,L(\mathrm{d}s) + \int_r^\infty (2rs - r^2)\,L(\mathrm{d}s)\right), \quad r > 0\,.$$

Indeed, by the alternative representation (12.42) of the reduced mixed moment function we need to multiply the prefactor of the integral term in (12.58) by

$$\frac{2}{\pi}\gamma^{\mathrm{BM}}\,\overline{s}\cdot\frac{\gamma^{\mathrm{BM}}\,\overline{s}}{\gamma_\cap^{\mathrm{BM}}} = 2$$

to perform the transition from $K_{1,L}^{\mathrm{BM}}$ to $K_{0,1,L}^{\mathrm{BM}}$. Here, as in (12.61),

- $\frac{2}{\pi}\gamma^{\mathrm{BM}}\,\overline{s}$ represents the intensity of the homogeneous point process arising from the intersection of the Boolean segment process with a fixed line or line segment;
- $\gamma^{\mathrm{BM}}\,\overline{s}/\gamma_\cap^{\mathrm{BM}}$ takes into account the exchange of the intensities when replacing the random length measure with length density $\gamma^{\mathrm{BM}}\,\overline{s}$ by the intersection point process having intensity $\gamma_\cap^{\mathrm{BM}}$.

Consequently, the corresponding cross-correlation function is given by

$$g_{0,1,L}^{\mathrm{BM}}(r) = 1 + \frac{2}{\pi\gamma^{\mathrm{BM}}\,\overline{s}^2 r}\int_r^\infty (s - r)\,L(\mathrm{d}s)\,, \quad r > 0\,.$$

We note that by comparing with (12.58) and (12.60) or with (12.59) and (12.62) that the only difference between the K-functions $K_{1,L}^{\mathrm{BM}}$, $K_{0,L}^{\mathrm{BM}}$, and $K_{0,1,L}^{\mathrm{BM}}$ and the cross or pair correlation functions $g_{1,L}^{\mathrm{BM}}$, $g_{0,L}^{\mathrm{BM}}$, and $g_{0,1,L}^{\mathrm{BM}}$ are the prefactors 1, 2, and 4 in front of the integral expressions, respectively, a phenomenon which can also be observed for the corresponding functions related to a stationary and isotropic Poisson line process; see (12.45), (12.57), and (12.66). Let us assume now that $\gamma^{\mathrm{BM}} = \gamma_{1,t}^{\max} = \frac{t^2}{\pi}$ for some $t > 0$ and that the length distribution L of the segments has the density on $(0, \infty)$ given by (12.63). In that case, we obtain

$$g_{0,1,L}^{\mathrm{BM}}(r) = 1 + \frac{1}{r^2 t^2}\left(1 - \mathrm{e}^{-\frac{2}{\pi}rt}\right), \quad r > 0\,,$$

which shows that $g_{0,1,L}^{\mathrm{BM}}(r) \neq g_{0,1,t}(r)$ according to Theorem 12.36. This is in line with (12.65) and demonstrates that also the cross-correlation function is a device that is sensitive enough to distinguish between a stationary and isotropic planar STIT tessellation and a stationary and isotropic Boolean segment process with intensity γ^{BM} and length distribution L with density (12.63).

Remark 12.37. *As in Remarks* 12.29 *and* 12.35, *we consider the small time asymptotics of the cross-correlation function $g_{0,1,t}(r)$ in Theorem* 12.36. *We have*

$$g_{0,1,t}(r) = 1 + \frac{1}{t^2 r^2} + \frac{3}{2\pi t r} - \frac{4}{3\pi^2} + O(rt)$$

as $t \downarrow 0$.

12.10 Notes

1. The results in this chapter were originally established in [176] for the total surface content. The paper [174] contains the explicit variance formulae in the planar case $d = 2$, and in [175] the covariance relations in higher-dimensional STIT tessellations with $d \geq 3$ have been derived. We remark that the paper [174] more generally deals with STIT tessellations in the plane driven by line measures in the rather broad class of diffuse and locally finite line measures, which are not necessarily translation invariant. In this book, we restrict our focus to the stationary case and often even to the isotropic case, for simplicity.
2. The paper [175] contains also a detailed analysis of the second-order structure of so-called face measures. For a STIT tessellation Y_t, $t > 0$, in $\mathbb{R}^d$ with $d \geq 3$, these are defined as

$$F_{j;t} := \sum_{F \in \mathcal{F}_{d-1}^{\max}(Y_t)} v_j^F \quad \text{with } v_j^F(\cdot) = \sum_{G \in \mathcal{F}_j(F)} V_j(\cdot \cap G) .$$

Note that $F_{j;t}(W)$ is the total j-volume in $W \in \mathcal{K}_d$ of the j-skeleton induced by Y_t. It is important to note that, in general, the total mass of a face measure is not necessarily related to the cumulative intrinsic volume functional $\Sigma_{V_j,t,W}$ of the same order. For instance, $\Sigma_{V_0,t,W}$ counts the number of $(d-1)$-dimensional maximal polytopes in Y_t within W, whereas the total number of vertices, $F_{0,t}(W)$, is not deterministically related to it. Theorem 2 of [175] yields an explicit formula for the covariance measure between $F_{i;t}$ and $F_{j;t}$ for arbitrary $i, j \in \{0, 1, \ldots, d-1\}$:

$$\begin{aligned} \mathrm{Cov}_{F_{d-1-i;t}, F_{d-1-j;t}} &= \sum_{m=0}^{i} \sum_{n=0}^{j} \binom{i+j-m-n}{i-m} \left(\prod_{k=m+1}^{i} c_{d-k,d} \right) \left(\prod_{\ell=n+1}^{j} c_{d-\ell,d} \right) \\ &\quad \times \int_{\mathbb{R}^d} \int_{\mathcal{P}_{d-1}} \left[v_{d-1-m}^{F+x} \otimes v_{d-1-n}^{F+x} \right] I^{i+j-m-n} \left(\gamma_{d-1,\cdot}^{\max} \mathbb{Q}_{d-1,\cdot}^{\max}, t \right)(\mathrm{d}F) \lambda_d(\mathrm{d}x) . \end{aligned} \tag{12.67}$$

where $I^r(\gamma_{d-1,\cdot}^{\max} \mathbb{Q}_{d-1,\cdot}^{\max}, t)(\mathrm{d}F) = \frac{1}{(r-1)!} \int_0^t (t-s)^{r-1} \gamma_{d-1,s}^{\max} \mathbb{Q}_{d-1,s}^{\max}(\mathrm{d}F) \mathrm{d}s$ with $r = i + j - m - n$. The result of Theorem 12.30(b) can be regarded as a special case of this formula with $i = j = 0$. In principle, using arguments similar to the ones in the proof of Theorem 12.20, (12.67) can be used to deduce the leading-order term of the asymptotic covariance between $F_{i;t}(aW)$ and $F_{j;t}(aW)$ as $a \to \infty$.
3. The result of Lemma 12.12 is classical in integral geometry and for $d = 2$ and $d = 3$ goes at least back to Blaschke; see § 8 and § 32 in [18]. For general space dimensions, the formula can be found on page 374 of [172]. Moreover, for $\alpha = 2 - d$, it recovers the classical representation of the Newton potential of the Lebesgue measure restricted to a convex body $W \in \mathcal{K}_d$.
4. The Blaschke–Petkantschin formulae form one of the cornerstones of integral geometry. Their origin goes back to the classical work of Blaschke and Petkantschin in the 1930s, who investigated transformation rules for multiple integrals under changes of geometric variables. The central idea is to decompose the Lebesgue measure on a product space, such as $(\mathbb{R}^d)^k$, into a product of measures on lower-dimensional geometric entities, for example, on linear or affine subspaces, or spheres or simplices spanned by the given points. Such decompositions allow us to integrate over directions, offsets, and local configurations rather than over absolute positions, which is essential in stochastic and convex geometry. Modern treatments and extensions can be found in [172, Chapter 7], where the formulae are derived systematically using invariant measures on Grassmannians.
In the present context, we have used a special case of these ideas in the proof of Proposition 12.5. An alternative way to derive this formula is as follows. Parameterize hyperplanes by $H(u, s) = \{z \in \mathbb{R}^d : \langle z, u \rangle = s\}$ with $u \in S^{d-1}$, $s \in \mathbb{R}$, and write $\Theta_{\text{iso}}(\mathrm{d}H) = \varphi_{\text{iso}}(\mathrm{d}u)\, \mathrm{d}s$. For $B \in \mathcal{B}(\mathbb{R}^d)$, we may write

$$\lambda_{d-1}\big(B \cap H(u, s)\big) = \lim_{\varepsilon \downarrow 0} \frac{1}{2\varepsilon} \int_B \mathbb{1}\left\{ \left| \langle x, u \rangle - s \right| \leq \varepsilon \right\} \lambda_d(\mathrm{d}x) .$$

Apply this twice and integrate over s to get

$$\int_{\mathbb{R}} \mathbb{1}\{|\langle x,u\rangle - s| \le \varepsilon\}\, \mathbb{1}\{|\langle y,u\rangle - s| \le \varepsilon\}\, \mathrm{d}s = \left(2\varepsilon - |\langle x-y,u\rangle|\right)_+ .$$

Hence, with $z := x - y$,

$$\begin{aligned}&\int_{\mathcal{H}} \int_{B_1\cap H} \int_{B_2\cap H} h(x,y)\, \lambda_{d-1}(\mathrm{d}x)\, \lambda_{d-1}(\mathrm{d}y)\, \Theta_{\mathrm{iso}}(\mathrm{d}H)\\ &\quad = \lim_{\varepsilon\downarrow 0} \frac{1}{4\varepsilon^2} \int_{S^{d-1}} \int_{B_1} \int_{B_2} h(x,y) \left(2\varepsilon - |\langle z,u\rangle|\right)_+ \lambda_d(\mathrm{d}x)\, \lambda_d(\mathrm{d}y)\, \varphi_{\mathrm{iso}}(\mathrm{d}u) .\end{aligned}$$

For fixed $z \ne 0$, the set $\{u \in S^{d-1} : |\langle z,u\rangle| \le t\}$ is a spherical belt of half-thickness $t/\|z\|$. Its content equals

$$\mathcal{H}^{d-1}\left(\{|\langle z,u\rangle| \le t\}\right) = (d-1)\kappa_{d-1} \int_{-t/\|z\|}^{t/\|z\|} \left(1 - s^2\right)^{\frac{d-3}{2}} \mathrm{d}s = (d-1)\kappa_{d-1} \frac{2t}{\|z\|} + o(t)$$

as $t \downarrow 0$. Hence, dividing by $d\kappa_d$, we get

$$\varphi_{\mathrm{iso}}\left(\{u : |\langle z,u\rangle| \le t\}\right) = \frac{(d-1)\kappa_{d-1}}{d\kappa_d} \cdot \frac{2t}{\|z\|} + o(t) .$$

Therefore

$$\frac{1}{2\varepsilon} \int_{S^{d-1}} \left(1 - \frac{|\langle z,u\rangle|}{2\varepsilon}\right)_+ \varphi_{\mathrm{iso}}(\mathrm{d}u) = \frac{1}{(2\varepsilon)^2} \int_0^{2\varepsilon} \varphi_{\mathrm{iso}}\left(\{|\langle z,u\rangle| \le t\}\right) \mathrm{d}t ,$$

which tends to $\frac{(d-1)\kappa_{d-1}}{d\kappa_d} \cdot \frac{1}{\|z\|}$ as $\varepsilon \downarrow 0$. Plugging this back and using that $h \ge 0$ yield

$$\int_{\mathcal{H}} \int_{B_1\cap H} \int_{B_2\cap H} h(x,y)\, \lambda_{d-1}(\mathrm{d}x)\, \lambda_{d-1}(\mathrm{d}y)\, \Theta_{\mathrm{iso}}(\mathrm{d}H) = \frac{(d-1)\kappa_{d-1}}{d\kappa_d} \int_{B_1} \int_{B_2} \frac{h(x,y)}{\|x-y\|} \lambda_d(\mathrm{d}x)\, \lambda_d(\mathrm{d}y) ,$$

which is (12.13).

5. The refined variance asymptotics in Theorem 12.15 originate from the PhD thesis [49]. They further include explicit formulae for the second-order term $S_t(W)$ when W is an ellipse or a rectangle. For an ellipse of the form

$$E_{a,b} = \left\{(x,y) \in \mathbb{R}^2 : \frac{x^2}{a^2} + \frac{y^2}{b^2} \le 1\right\}$$

with $a,b > 0$, we obtain

$$\int_{[E_{a,b}]} V_1(E_{a,b} \cap L) \log V_1(E_{a,b} \cap L)\, \Theta_{\mathrm{iso}}(\mathrm{d}L) = \pi ab\left(\log \min\{a,b\} + \frac{1}{2} - \tilde{c}_{\max\{a,b\},\min\{a,b\}}\right),$$

where the constant $\tilde{c}_{u,v}$, defined for $0 < v < u$, admits the representation

$$\tilde{c}_{u,v} = \frac{1}{2\pi} \sum_{j=1}^{\infty} (-1)^{j+1} \frac{\kappa_{2j}}{j\,\kappa_{2j-1}} \left(\frac{v^2}{u^2} - 1\right)^j .$$

It can in fact be shown that $\tilde{c}_{u,v} = \log(\frac{u+v}{2u})$, which implies

$$\int\limits_{[E_{a,b}]} V_1(E_{a,b} \cap L) \log V_1(E_{a,b} \cap L)\, \Theta_{\text{iso}}(\mathrm{d}L) = \pi ab\left(\log\left(\frac{2ab}{a+b}\right) + \frac{1}{2}\right).$$

In particular, choosing $a = b = r > 0$ recovers (12.33).

6. Entropy and entropy inequalities for distributions of random lines were studied and established in [201, 205]. We remark that the definition of $\mathrm{Ent}(f)$ in [201] is slightly different from that in (12.35). However, the difference only comes from an alternative normalization of the invariant line measure Θ_{iso}.
7. It was shown in the PhD thesis [49] that the total surface content in dimension $d \geq 2$ of a stationary STIT tessellation Y_t, $t > 0$, driven by a nondegenerate hyperplane measure Θ, has finite moments of all orders. Much stronger exponential moment bounds and concentration inequalities for the total surface content can be concluded using the martingale property of the Doléans-Dade exponential defined in (10.50). Indeed, recall that each jump size is bounded by $B(W) := \sup_{H \in [W]} V_{d-1}(W \cap H) < \infty$. Using the elementary inequality

$$e^{\vartheta x} - 1 - \vartheta x \leq \frac{e^{\vartheta B(W)} - 1 - \vartheta B(W)}{B(W)} x, \quad x \in \left[0, B(W)\right], \quad \vartheta \geq 0,$$

it follows that

$$\Psi_t(\vartheta) \leq \Psi_t^{\uparrow}(\vartheta) := \frac{e^{\vartheta B(W)} - 1 - \vartheta B(W)}{B(W)} t\, V_d(W).$$

Hence $\mathbb{E}[\exp(\vartheta \overline{\Sigma}_{V_{d-1},t,W})] \leq \exp(\Psi_t^{\uparrow}(\vartheta))$ for all $\vartheta \geq 0$. In particular, since $\Sigma_{V_{d-1},t,W} \geq 0$, the total surface content has finite exponential moments of all orders. By Markov's inequality this implies that for all $x > 0$ and $\vartheta \geq 0$,

$$\mathbb{P}(\overline{\Sigma}_{V_{d-1},t,W} \geq x) \leq \exp\left(-\vartheta x + \frac{e^{\vartheta B(W)} - 1 - \vartheta B(W)}{B(W)} t\, V_d(W)\right).$$

Optimizing over ϑ leads to the Bernstein–type concentration inequality

$$\mathbb{P}(\overline{\Sigma}_{V_{d-1},t,W} \geq x) \leq \exp\left(-\frac{x^2}{2(tB(W)V_d(W) + B(W)x/3)}\right), \quad x > 0,$$

and the same bound holds for the lower tail.
8. The pair correlation function for the random length measure of a stationary and isotropic Poisson line processes in the plane can be found in (8.27) of [33]. A significantly more general result, covering a broad class of random measures in arbitrary dimensions and Poisson hyperplane processes in spaces of constant curvature -1, 0, or $+1$, is presented in Theorem 3 of [79]. More precisely, the authors considered, among other quantities, for $i, j \in \{0, \ldots, d-1\}$, the cross-correlation function g_{ij} between the i- and the j-dimensional Hausdorff measure on the i- or j-skeleton of an isometry-invariant Poisson hyperplane tessellation of intensity $t > 0$ in a d-dimensional space of constant curvature $K \in \{-1, 0, +1\}$, respectively. Using the modified sine function

$$\mathrm{sn}_K(r) := \begin{cases} \sin(r), & K = +1, \\ r, & K = 0, \\ \sinh(r), & K = -1, \end{cases}$$

it was shown that

$$g_{ij}(r) = 1 + \sum_{n=1}^{\min\{d-i,d-j,d-1\}} n! \binom{d-i}{n} \binom{d-j}{n} \frac{d-n}{d-n+1} \frac{\kappa_{d-n}}{\kappa_{d-n+1}}$$
$$\times \left(\frac{d\kappa_d}{(d+1)\kappa_{d+1}} \right)^{n-1} \frac{1}{(t \operatorname{sn}_K(r))^n} \tag{12.68}$$

for $r > 0$ if $K \in \{-1, 0\}$ and $0 < r < \pi$ in the case $K = +1$.

9. A special case of Corollary 12.33 for $d = 2$ in the stationary and isotropic setting was first established in [195]. In its general form, it is taken from [175].
10. The concept of the cross-correlation function $g_{\Phi,\Psi}$ for two stationary (weighted) random measures Φ and Ψ originates from [183]. Among other results, the authors of that paper proved that $g_{\Phi,\Psi}(r) = g_{\Psi,\Phi}(r)$ for all $r > 0$. We use a similar symmetry relation in our alternative representation (12.42) for the reduced mixed moment function $K_{\Phi,\Psi}$.
11. The idea of using the sharp Riesz rearrangement inequality to deduce inequalities for variances and covariances for geometric functionals of random sets has been inspired by the paper [15], where this method has been used for geometric functionals of Poisson k-flat processes in constant curvature spaces. This approach can also be used to derive an inequality for the covariance between the total edge length and the total I-segment count, with an explicit expression given in Theorem 12.10(b). However, the same argument seems to fail for the variance of the I-segment count, since the variance in this case also depends on the first intrinsic volume $V_1(W)$ of W, which decreases if W is replaced by B_W. An alternative way of proving part (b) of Theorem 12.23 is to rewrite, up to constants, the 2-energy $E_2(W)$ as the chord-power integral $I_{d-1}(W)$ by means of Lemma 12.12 and then to use Theorem 8.6.6 of [172] to conclude that $I_{d-1}(W) \leq I_{d-1}(B_W)$.
12. An alternative approach to higher moments of the total surface content has been developed in the PhD thesis [49], following a suggestion of Lothar Heinrich (personal communication) in the planar case. This method relies on a detailed analysis of the capacity functional

$$T_{Y_{t,W}}(\cdot) := \mathbb{P}(\partial Y_{t,W} \cap \cdot = \emptyset)$$

and applies to stationary, though not necessarily isotropic, STIT tessellations $Y_{t,W}$ in $\mathbb{R}^d$ with time parameter $t \geq 0$ inside a convex window $W \in \mathcal{K}_d$. Writing $B^d_\varepsilon(x)$ for the closed ball of radius $\varepsilon > 0$ centered at $x \in \mathbb{R}^d$, we obtain

$$\mathbb{E}\Sigma^2_{V_{d-1},t,W} = \lim_{\varepsilon \downarrow 0} \frac{1}{4\varepsilon^2} \int_{W+B^d_\varepsilon(o)} \int_{W+B^d_\varepsilon(o)} \Big[1 - 2\big(1 - T_{Y_{t,W}}\big(B^d_\varepsilon(x) \cap W\big)\big)$$
$$+ 1 - T_{Y_{t,W}}\big(\big(B^d_\varepsilon(x) \cup B^d_\varepsilon(y)\big) \cap W\big)\Big] \lambda_d(\mathrm{d}y)\, \lambda_d(\mathrm{d}x).$$

In the isotropic setting, this representation was then used to provide an alternative approach to Theorem 12.8.
13. In [118] the authors apply a general covariance inequality from [69, 202] in terms of β-mixing coefficients to deduce upper variance bounds for a rather general class of geometric functionals that can be associated with a stationary, not necessarily isotropic STIT tessellation. However, in all known cases the resulting upper bounds are rather far from the actual order of the variances. On the other hand, it also delivers bounds in cases where currently no other methods are available. An example is the number of k-dimensional faces, $k \in \{0, 1, \dots, d\}$, having their center in a convex window.

13 Limit theorems

In this chapter, we develop a comprehensive limit theory for a class of geometric functionals associated with maximal $(d-1)$-polytopes of STIT tessellation processes within a sequence of growing convex windows, where we often focus on the stationary and isotropic case, for which usually the most explicit results are available. It turns out that a central role in this theory is played by the surface content, which we consider in two closely related yet distinct scenarios:

(a) The *total surface content*, which is induced by the collection of all $(d-1)$-dimensional maximal polytopes.

(b) The *residual surface content*, which is induced by $(d-1)$-dimensional maximal polytopes whose birth time exceeds a fixed threshold.

The limit theorems for these two versions of the surface content exhibit qualitatively different behaviors. Although the residual surface content admits Gaussian fluctuations, the fluctuations of the total surface content are Gaussian only in dimension $d = 2$, whereas non-Gaussian fluctuations emerge in all higher space dimensions $d \geq 3$. Here by fluctuations we mean the centered and scaled deviations of a random variable or process from its deterministic first-order behavior. Gaussian fluctuations refer to convergence to a normal law, whereas non-Gaussian fluctuations do not. This phenomenon is mainly caused by the $(d-1)$-dimensional maximal polytopes born at the very first stages in the global STIT construction. Moreover, we do not only consider the surface content as a scalar-valued random variable but also as a continuous-time stochastic process and provide various functional limit theorems. In addition to the total surface content, we establish functional limit theorems for the vector of intrinsic volumes of $(d-1)$-dimensional maximal polytopes, which we again regard as a continuous-time stochastic process. The leading-order fluctuations for this broad class of geometric functionals are entirely determined by those of the total surface content through a simple mechanism driven by Crofton's formula. A functional central limit theorem for martingales is in the background of many results developed in this chapter.

13.1 Preliminaries

13.1.1 A functional central limit theorem for martingales

In this section, we reformulate a functional central limit theorem for continuous-time martingales. To begin with, we briefly compile some concepts and notation used in the formulation of this result.

Let $(M_t, t \geq 0)$ be a not necessarily centered martingale with respect to an augmented filtration $\mathfrak{F} := (\mathfrak{F}_t, t \geq 0)$. It is **square-integrable** if

https://doi.org/10.1515/9783112222409-015

$$\mathbb{E}[M_t^2] < \infty \quad \text{for all } t \geq 0\,.$$

A process $\mathbf{M} = (M_t, t \geq 0)$ is a **locally square-integrable local martingale** if it is a local martingale and there exists a localizing sequence $(\tau_n, n \in \mathbb{N})$ with $\tau_n \uparrow \infty$ a. s. such that for each $n \in \mathbb{N}$, the stopped process $(M_t^{\tau_n}, t \geq 0)$ is a square-integrable $\mathfrak{F}$-martingale; see Definition 10.3.

In this chapter, we will frequently use Doob's L^2-maximal inequality from Theorem 7.31 of [98], which says that

$$\mathbb{E}\Big(\sup_{0\leq s\leq t}|M_s|\Big)^2 \leq 4\mathbb{E}[M_t^2] \tag{13.1}$$

for a square-integrable martingale $(M_t,\ t \geq 0)$.

For two real-valued stochastic processes $\mathbf{M}_1 = (M_{1,t},\ t \geq 0)$ and $\mathbf{M}_2 = (M_{2,t},\ t \geq 0)$, denote their product by $\mathbf{M}_1\,\mathbf{M}_2 := (M_{1,t}\,M_{2,t},\ t \geq 0)$. According to Theorem 8.12 of [98], if $\mathbf{M}_1 = (M_{1,t},\ t \geq 0)$ and $\mathbf{M}_2 = (M_{2,t},\ t \geq 0)$ are locally square-integrable local martingales, then there exists – up to stochastic equivalence – a unique predictable stochastic process $\langle \mathbf{M}_1, \mathbf{M}_2\rangle = (\langle M_{1,\cdot}, M_{2,\cdot}\rangle_t,\ t \geq 0)$, normalized by the condition $\langle M_{1,\cdot}, M_{2,\cdot}\rangle_0 = 0$, such that

$$\mathbf{M}_1\,\mathbf{M}_2 - \langle \mathbf{M}_1, \mathbf{M}_2\rangle \quad \text{is a local martingale}\,.$$

The random process $\langle \mathbf{M}_1, \mathbf{M}_2\rangle$ is called the **predictable quadratic covariation** of $\mathbf{M}_1$ and $\mathbf{M}_2$. It is predictable and of locally finite variation. Here we recall that a càdlàg process has finite variation if a. s. on each compact interval, the total variation of its path is finite. If $\mathbf{M}_1 = \mathbf{M}_2 = \mathbf{M}$, then we write $\langle \mathbf{M}\rangle := \langle \mathbf{M}, \mathbf{M}\rangle$ for the **predictable quadratic variation** of $\mathbf{M}$. If $\mathbf{M}$ is continuous, then $\langle \mathbf{M}\rangle$ is continuous and increasing with $\langle \mathbf{M}\rangle_0 = 0$. The predictable quadratic covariation is bilinear in the sense that

$$\begin{aligned}\langle a_1\mathbf{M}_1 + a_2\mathbf{M}_2, b_1\mathbf{N}_1 + b_2\mathbf{N}_2\rangle &= a_1b_1\langle \mathbf{M}_1, \mathbf{N}_1\rangle + a_1b_2\langle \mathbf{M}_1, \mathbf{N}_2\rangle \\ &\quad + a_2b_1\langle \mathbf{M}_2, \mathbf{N}_1\rangle + a_2b_2\langle \mathbf{M}_2, \mathbf{N}_2\rangle\end{aligned} \tag{13.2}$$

for processes $\mathbf{M}_1, \mathbf{M}_2, \mathbf{N}_1, \mathbf{N}_2$ and scalars $a_1, a_2, b_1, b_2 \in \mathbb{R}$. If $\mathbf{A}$ is a continuous finite-variation process, then

$$\langle \mathbf{M}_i, \mathbf{A}\rangle \equiv 0 \tag{13.3}$$

for $i \in \{1,2\}$; see page 233 of [98]. In particular, if $\mathbf{A}$ has constant (random) paths, then $\langle \mathbf{A}\rangle \equiv 0$ and $\langle \mathbf{M}_i, \mathbf{A}\rangle \equiv 0$. We also need the following converse of the previously mentioned property.

Lemma 13.1. *Let* $\mathbf{M} = (M_t,\ t \geq 0)$ *be a locally square-integrable local* $\mathfrak{F}$*-martingale. If its predictable quadratic variation is identically zero, that is,* $\langle \mathbf{M}\rangle_t = 0$ *for all* $t \geq 0$*, then* $\mathbf{M}$ *is a. s. constant in* t.

Proof. First suppose that $\mathbf{M}$ is a square-integrable martingale. Set $N_t := M_t - M_0$, $t \geq 0$. Then $\mathbf{N} := (N_t, t \geq 0)$ is a square-integrable martingale with $\langle \mathbf{N} \rangle = \langle \mathbf{M} \rangle \equiv 0$ and $N_0 = 0$. By the definition of the predictable quadratic variation, the process $N_t^2 - \langle N \rangle_t = N_t^2$, $t \geq 0$, is a martingale. Since $N_0^2 = 0$ a. s., we have

$$\mathbb{E}[N_t^2] = \mathbb{E}[N_0^2] = 0 \quad \text{for all } t \geq 0 .$$

Moreover, as $N_t^2 \geq 0$, we conclude that $N_t^2 = 0$ a. s. for all $t \geq 0$, and hence $N_t = 0$ a. s. for all t. Thus $M_t = M_0$ a. s. for all $t \geq 0$.

Now let $\mathbf{M}$ be merely a locally square-integrable local martingale with $\langle \mathbf{M} \rangle \equiv 0$. Choose a localizing sequence $(\tau_n,\ n \in \mathbb{N})$ with $\tau_n \uparrow \infty$ such that the stopped process $\mathbf{M}^{\tau_n}$ is a square-integrable martingale for each n. By the first part of the proof we obtain

$$M_{t \wedge \tau_n} = M_0 \quad \text{a. s. for all } t \geq 0 \text{ and } n \in \mathbb{N} .$$

Since $\tau_n \uparrow \infty$ a. s., this implies $M_t = M_0$ a. s. for all $t \geq 0$. □

We will also use the following time-change rule for predictable quadratic covariations.

Lemma 13.2. *If $\alpha : [0, \infty) \to [0, \infty)$ is deterministic and increasing with $\alpha(0) = 0$, then for locally square-integrable martingales $\mathbf{M}_1$ and $\mathbf{M}_2$, we have*

$$\langle \mathbf{M}_1 \circ \alpha,\ \mathbf{M}_2 \circ \alpha \rangle_t = \langle \mathbf{M}_1, \mathbf{M}_2 \rangle_{\alpha(t)} \quad \textit{for all } t \geq 0 .$$

Proof. Define the time-changed processes $\widetilde{\mathbf{M}}_i := \mathbf{M}_i \circ \alpha$, $i \in \{1, 2\}$, and the predictable quadratic covariation $A_t := \langle \mathbf{M}_1, \mathbf{M}_2 \rangle_{\alpha(t)}$, $t \geq 0$. Then for all $t \geq 0$,

$$\widetilde{M}_{1,t} \widetilde{M}_{2,t} - A_t = M_{1,\alpha(t)} M_{2,\alpha(t)} - \langle \mathbf{M}_1, \mathbf{M}_2 \rangle_{\alpha(t)} = \big(\mathbf{M}_1 \mathbf{M}_2 - \langle \mathbf{M}_1, \mathbf{M}_2 \rangle\big) \circ \alpha(t) .$$

Since $\mathbf{M}_1 \mathbf{M}_2 - \langle \mathbf{M}_1, \mathbf{M}_2 \rangle$ is a local martingale, its composition with the deterministic increasing time change α is again a local martingale with respect to the filtration $(\mathfrak{F}_{\alpha(t)})_{t \geq 0}$. Thus $\widetilde{\mathbf{M}}_1 \widetilde{\mathbf{M}}_2 - \mathbf{A}$ is a local martingale. By the uniqueness of the predictable quadratic covariation it follows that $\mathbf{A}$ coincides, up to stochastic equivalence, with $\langle \widetilde{\mathbf{M}}_1, \widetilde{\mathbf{M}}_2 \rangle$, that is,

$$\langle \mathbf{M}_1 \circ \alpha,\ \mathbf{M}_2 \circ \alpha \rangle_t = \langle \mathbf{M}_1, \mathbf{M}_2 \rangle_{\alpha(t)} \quad \text{for all } t \geq 0 .$$

This completes the argument. □

In particular, for a locally square-integrable martingale $\mathbf{M}$, Lemma 13.2 gives

$$\langle \mathbf{M} \circ \alpha \rangle_t = \langle \mathbf{M} \rangle_{\alpha(t)} \cdot$$

If, in addition, $\langle \mathbf{M} \rangle$ is absolutely continuous with derivative $a(\cdot)$ and α is absolutely continuous with derivative $\frac{\mathrm{d}}{\mathrm{d}s}\alpha$, then $\langle \mathbf{M} \rangle_t = \int_0^t a(u)\,\mathrm{d}u$, and substituting $u = \alpha(s)$ gives

$$\langle \mathbf{M} \circ \alpha \rangle_t = \int_0^{\alpha(t)} a(u)\, \mathrm{d}u = \int_0^t a(\alpha(s)) \frac{\mathrm{d}}{\mathrm{d}s} \alpha(s)\, \mathrm{d}s \tag{13.4}$$

by the chain rule.

The stochastic processes we consider belong to the **Skorokhod space** $\mathsf{D}^m[0,\infty)$ of functions $g : [0,\infty) \to \mathbb{R}^m$ for some $m \in \mathbb{N}$, which are right-continuous and have left-sided limits. They are referred to as càdlàg functions. To derive functional limit theorems for random processes with values in $\mathsf{D}^m[0,\infty)$, it is necessary to supply that space with a suitable topology. To do so, we write Incr for the set of all continuous and strictly increasing bijections $\iota : [0,\infty) \to [0,\infty)$ with $\iota(0) = 0$ and $\iota(t) \to \infty$ as $t \to \infty$. Then we say that a sequence $(g_n, n \in \mathbb{N})$ of functions in $\mathsf{D}^m[0,\infty)$ converges to $g \in \mathsf{D}^m[0,\infty)$ if there is a sequence $(\iota_n, n \in \mathbb{N}) \in \mathrm{Incr}^{\mathbb{N}}$ such that

$$\lim_{n\to\infty} \sup_{t\le N} |\iota_n(t) - t| = 0 \quad \text{for all } N \in \mathbb{N}$$

and

$$\lim_{n\to\infty} \sup_{t\le N} \|g_n(\iota_n(t)) - g(t)\| = 0 \quad \text{for all } N \in \mathbb{N}\,;$$

see the definition, which is part of Theorem VI.1.14 of [86]. The same result ensures that the topology induced by this convergence, the so-called **Skorokhod topology**, turns $\mathsf{D}^m[0,\infty)$ into a Polish space. Informally speaking, two càdlàg functions are "close" to each other in the space $\mathsf{D}^m[0,\infty)$ if there is a reparameterization of the time-axis that is uniformly close to the identity function and when applied to one of the functions, it brings it close to the other one.

We will write $g(s-) := \lim_{s'\uparrow s} g(s')$ with the convention that $g(0-) = g(0)$ and denote

$$\mathcal{J}(g,t) := \sup\{\|g(s) - g(s-)\|,\ 0 \le s \le t\}\,,$$

the size of the largest jump of a function $g \in \mathsf{D}^m[0,\infty)$ on the interval $[0,t]$, $t > 0$. To simplify the notation, we omit the index m in the case $m = 1$. Moreover, we indicate by $\mathsf{C}[0,\infty) \subset \mathsf{D}[0,\infty)$ the space of continuous functions on $[0,\infty)$ supplied with the usual topology of local uniform convergence; see VI.§ 1a of [86].

Next, we recall that the **standard Wiener process** $(\mathcal{W}_t, t \ge 0)$ is the continuous centered Gaussian process with covariance function $\mathrm{cov}(\mathcal{W}_s, \mathcal{W}_t) = \min(s,t)$ for $s,t \ge 0$. In particular, for each fixed $t > 0$, the random variable $\mathcal{W}_t$ is normally distributed with variance t, which we indicate by writing $\mathcal{W}_t \sim \mathsf{Norm}(0,t)$. Let $g \in L^2[0,t]$ be a square-integrable function on the interval $[0,t]$, $t > 0$. If g is simple and takes the form $g(s) = a_i$ for $s \in [t_{i-1}, t_i)$ with $a_i \in \mathbb{R}$, $i \in \{1,\dots,n\}$, $0 = t_0 < t_1 < \cdots < t_n = t$, and some $n \in \mathbb{N}$, then we define the Wiener integral

$$\int_0^t g(s)\,\mathrm{d}\mathcal{W}_s := \sum_{i=1}^n a_i\,(\mathcal{W}_{t_i} - \mathcal{W}_{t_{i-1}})\,.$$

Then $\mathbb{E}\int_0^t g(s)\,\mathrm{d}\mathcal{W}_s = 0$. Moreover, we have isometry property

$$\mathrm{cov}\left(\int_0^t g_1(s)\,\mathrm{d}\mathcal{W}_s, \int_0^t g_2(s)\,\mathrm{d}\mathcal{W}_s\right) = \int_0^t g_1(s)g_2(s)\,\mathrm{d}s$$

for two simple functions g_1 and g_2, which shows that the mapping $g \mapsto \int_0^t g(s)\,\mathrm{d}\mathcal{W}_s$ is an isometric linear embedding from $L^2[0,t]$ to $L^2(\Omega, \mathbf{A}, \mathbb{P})$ with their respective L^2-norms, where $(\Omega, \mathbf{A}, \mathbb{P})$ is the probability space on which all the random elements we consider are defined. In particular, if $g \in L^2[0,t]$ and $(g_n,\ n \in \mathbb{N})$ is a sequence of simple functions converging to g in $L^2[0,t]$, then the sequence of random variables $\int_0^t g_n(s)\,\mathrm{d}\mathcal{W}_s$ is a Cauchy sequence in $L^2(\Omega, \mathbf{A}, \mathbb{P})$. Thus the $L^2(\Omega, \mathbf{A}, \mathbb{P})$-limit

$$\int_0^t g(s)\,\mathrm{d}\mathcal{W}_s := \lim_{n\to\infty} \int_0^t g_n(s)\,\mathrm{d}\mathcal{W}_s$$

is well defined and called the **Wiener integral** of g. If, moreover, g is locally square-integrable, then we may define the stochastic process $(\int_0^t g(s)\,\mathrm{d}\mathcal{W}_s,\ t \geq 0)$, which is a continuous centered Gaussian process with covariance

$$\mathrm{cov}\left(\int_0^t g(u)\,\mathrm{d}\mathcal{W}_u, \int_0^s g(u)\,\mathrm{d}\mathcal{W}_u\right) = \int_0^{\min(s,t)} g^2(u)\,\mathrm{d}u\,, \quad s,t \geq 0\,; \tag{13.5}$$

see Theorem 4.11 of [98]. In particular, for fixed $t > 0$, the random variable $\int_0^t g(s)\,\mathrm{d}\mathcal{W}_s$ has the normal distribution $\mathrm{Norm}(0, \int_0^t g(s)^2\,\mathrm{d}s)$.

After these preparations, we can now rephrase a version of the martingale functional central limit theorem, combining results of [76, 198].

Theorem 13.3. *Let $(\mathbf{M}_n,\ n \in \mathbb{N})$ be a sequence of locally square-integrable local martingales in $\mathsf{D}[0,\infty)$. Assume that $M_n(0) = 0$ for all $n \in \mathbb{N}$ and that for all $t > 0$,*

$$\lim_{n\to\infty} \mathbb{E}\mathcal{J}(\langle \mathbf{M}_n \rangle, t) = 0 \tag{13.6}$$

and

$$\lim_{n\to\infty} \mathbb{E}(\mathcal{J}(\mathbf{M}_n, t)^2) = 0\,. \tag{13.7}$$

If there is a locally square-integrable function $g : [0,\infty) \to \mathbb{R}$ such that

$$\langle \mathbf{M}_n \rangle_t \overset{\text{law}}{\Longrightarrow} \int_0^t g(s)^2 \,\mathrm{d}s \quad \textit{as } n \to \infty \tag{13.8}$$

for all t > 0, then

$$\mathbf{M}_n \overset{\text{law}}{\Longrightarrow} \left(\int_0^t g(s)\,\mathrm{d}\mathcal{W}_s,\ t \geq 0 \right) \quad \textit{in } \mathsf{D}[0,\infty) \quad \textit{as } n \to \infty,$$

where $(\mathcal{W}_t, t \geq 0)$ *is a standard Wiener process.*

13.1.2 The functional S(*f*, *W*)

Let Θ be a hyperplane measure of the form (4.1), satisfying (4.3), and denote by φ its spherical directional distribution. By $(Y_t, t > 0)$ we denote the STIT tessellation process driven by the measure Θ. The cell intensity of Y_1 is indicated by $\gamma_{d,1}$, and $\mathbb{Q}_{d,1}$ stands for the distribution of the typical cell of Y_1, which is defined with respect to some fixed center function. For a convex body $W \in \mathcal{K}_d$ and a bounded and measurable function $f : G(d,1) \to \mathbb{R}$, let us introduce the functional

$$\mathsf{S}(f,W) := \gamma_{d,1}\, V_d(W) \int_{\mathcal{P}_{d,c}} \int_{[P]} f^2(H^\perp) V_{d-1}^2(P \cap H)\, \Theta(\mathrm{d}H)\, \mathbb{Q}_{d,1}(\mathrm{d}P). \tag{13.9}$$

It will play an essential role in the limit theory for STIT tessellation processes we develop in this chapter. By Remark 2.18, together with the translation invariance of the hyperplane measure Θ and of V_{d-1}, the definition of $\mathsf{S}(f,W)$ does not depend on the choice of the center function.

We start with a first reformulation of $\mathsf{S}(f,W)$, which is easier to handle and relates the intersection of the typical cell of Y_1 and a hyperplane with the typical cell of the sectional tessellation in that hyperplane. Formally, by Lemma 3.9(a), for $H \in \mathcal{H}$, the sectional tessellation $Y_1 \cap H$ is a stationary STIT tessellation within H, where stationarity then refers to translations of $\mathbb{R}^d$ that fix H. Denote by γ_{d-1}^H the cell intensity of $Y_1 \cap H$ and by $\mathbb{Q}_{d-1}^H$ the distribution of its $(d-1)$-dimensional typical cell with respect to some fixed center function $c : \mathcal{P}_{d-1} \to H$.

Lemma 13.4. *Let* $W \in \mathcal{K}_d$, *and let* Θ *be a hyperplane measure of the form* (4.1), *satisfying* (4.3). *Then for any bounded and measurable function* $f : G(d,1) \to \mathbb{R}$,

$$\begin{aligned}\mathsf{S}(f,W) &= \int_{[W]} f^2(H^\perp) V_{d-1}(W \cap H) \frac{\int_{\mathcal{P}_{d-1,c}} V_{d-1}^2(F)\, \mathbb{Q}_{d-1}^H(\mathrm{d}F)}{\int_{\mathcal{P}_{d-1,c}} V_{d-1}(F)\, \mathbb{Q}_{d-1}^H(\mathrm{d}F)}\, \Theta(\mathrm{d}H) \\ &= V_d(W) \int_{\mathcal{S}^{d-1}} f^2(\operatorname{lin} u)\, \gamma_{d-1}^{u^\perp} \int_{\mathcal{P}_{d-1,c}} V_{d-1}^2(F)\, \mathbb{Q}_{d-1}^{u^\perp}(\mathrm{d}F)\, \varphi(\mathrm{d}u).\end{aligned} \tag{13.10}$$

Proof. We start by noting that the second equality follows by translation invariance of Θ, stationarity of Y_1, and by (2.38). Next, starting with (13.10), for fixed $u \in S^{d-1}$, we apply the Campbell-type formula of Lemma 2.20 to the sectional tessellation $Y_1 \cap u^\perp$ to see that

$$\gamma_{d-1}^{u^\perp} \int_{\mathcal{P}_{d-1,c}} V_{d-1}^2(F)\,\mathbb{Q}_{d-1}^{u^\perp}(\mathrm{d}F) = \mathbb{E} \sum_{F \in Y_1 \cap u^\perp} V_{d-1}^2(F)\,\mathbb{1}\{c(F) \in B_{u^\perp}\}$$
$$= \mathbb{E} \sum_{P \in Y_1} V_{d-1}^2(P \cap u^\perp)\,\mathbb{1}\{c(P \cap u^\perp) \in B_{u^\perp}\},$$

where $B_{u^\perp} \in \mathcal{B}(u^\perp)$ is an arbitrary Borel set of $(d-1)$-volume one in $u^\perp$. We use now once more Lemma 2.20, but this time for the random tessellation Y_1, to conclude that the last expression is the same as

$$\gamma_{d,1} \int_{\mathcal{P}_{d,c}} \int_{\mathbb{R}^d} V_{d-1}^2((P+x) \cap u^\perp)\mathbb{1}\{c((P+x) \cap u^\perp) \in B_{u^\perp}\}\,\lambda_d(\mathrm{d}x)\,\mathbb{Q}_{d,1}(\mathrm{d}P).$$

In the next step, we decompose $\mathbb{R}^d$ as the direct sum $u^\perp \oplus \operatorname{lin}(u)$ and write $x = z + ru \in \mathbb{R}^d$ with $z \in u^\perp$ and $r \in \mathbb{R}$. Then

$$\gamma_{d-1}^{u^\perp} \int_{\mathcal{P}_{d-1,c}} V_{d-1}^2(F)\,\mathbb{Q}_{d-1}^{u^\perp}(\mathrm{d}F)$$
$$= \gamma_{d,1} \int_{\mathcal{P}_{d,c}} \int_{\mathbb{R}} \int_{u^\perp} V_{d-1}^2((P+(z+ru)) \cap u^\perp)\mathbb{1}\{c((P+(z+ru)) \cap u^\perp) \in B_{u^\perp}\}$$
$$\lambda_{d-1}(\mathrm{d}z)\,\lambda_1(\mathrm{d}r)\,\mathbb{Q}_{d,1}(\mathrm{d}P)$$
$$= \gamma_{d,1} \int_{\mathcal{P}_{d,c}} \int_{\mathbb{R}} V_{d-1}^2((P+ru) \cap u^\perp) \int_{u^\perp} \mathbb{1}\{c((P+(z+ru)) \cap u^\perp) \in B_{u^\perp}\}$$
$$\lambda_{d-1}(\mathrm{d}z)\,\lambda_1(\mathrm{d}r)\,\mathbb{Q}_{d,1}(\mathrm{d}P)$$
$$= \gamma_{d,1} \int_{\mathcal{P}_{d,c}} \int_{\mathbb{R}} V_{d-1}^2((P+ru) \cap u^\perp)\,\lambda_1(\mathrm{d}r)\,\mathbb{Q}_{d,1}(\mathrm{d}P),$$

since the value of $V_{d-1}^2((P+(z+ru)) \cap u^\perp)$ is independent of z, and $B_{u^\perp}$ was assumed to have λ_{d-1}-measure one. Plugging this into (13.10) and applying the substitution $s = -r$, we conclude that

$$V_d(W) \int_{S^{d-1}} f^2(\operatorname{lin} u)\,\gamma_{d-1}^{u^\perp} \int_{\mathcal{P}_{d-1,c}} V_{d-1}^2(F)\,\mathbb{Q}_{d-1}^{u^\perp}(\mathrm{d}F)\,\varphi(\mathrm{d}u)$$
$$= \gamma_{d,1}\,V_d(W) \int_{S^{d-1}} f^2(\operatorname{lin} u) \int_{\mathcal{P}_{d,c}} \int_{\mathbb{R}} V_{d-1}^2((P+ru) \cap u^\perp)\,\lambda_1(\mathrm{d}r)\,\mathbb{Q}_{d,1}(\mathrm{d}P)\,\varphi(\mathrm{d}u)$$

$$= \gamma_{d,1} V_d(W) \int_{\mathcal{P}_{d,c}} \int_{S^{d-1}} \int_{\mathbb{R}} f^2(\operatorname{lin} u)\, V_{d-1}^2(P \cap (u^\perp + su))\, \lambda_1(\mathrm{d}s)\, \varphi(\mathrm{d}u)\, \mathbb{Q}_{d,1}(\mathrm{d}P)$$

$$= \gamma_{d,1} V_d(W) \int_{\mathcal{P}_{d,c}} \int_{[P]} f^2(H^\perp) V_{d-1}^2(P \cap H)\, \Theta(\mathrm{d}H)\, \mathbb{Q}_{d,1}(\mathrm{d}P),$$

which is the original definition (13.9) of $\mathsf{S}(f, W)$. This proves the first equality. □

The following lemma provides yet another formula for $\mathsf{S}(f, W)$, which allows us to evaluate $\mathsf{S}(f, W)$ for the motion invariant and the Mondrian hyperplane measure. To present it, we denote by Π_Θ the zonoid associated with the pair $(1, \Theta)$ as introduced in (2.68). For a convex body $W \in \mathcal{K}_d$ containing the origin in its interior, the **polar body** of W is defined as $W^\circ := \{x \in \mathbb{R}^d : \langle x, y\rangle \le 1 \text{ for all } y \in W\}$. We also recall that $e_1, \dots, e_d$ stands for the standard orthonormal basis in $\mathbb{R}^d$.

Lemma 13.5. *Let $W \in \mathcal{K}_d$, and let Θ be a hyperplane measure of the form* (4.1), *satisfying* (4.3). *Then for any bounded and measurable function $f : G(d, 1) \to \mathbb{R}$,*

$$\mathsf{S}(f, W) = V_d(W) \frac{(d-1)!}{2^{d-1}} \int_{S^{d-1}} f^2(\operatorname{lin} u) V_{d-1}((\Pi_\Theta | u^\perp)^\circ)\, \varphi(\mathrm{d}u), \tag{13.11}$$

where $\Pi_\Theta | u^\perp$ stands for the orthogonal projection of Π_Θ onto the hyperplane $u^\perp$, and $(\Pi | u^\perp)^\circ$ is the polar body relative to $u^\perp$. If $\Theta = \Theta_{\mathrm{iso}}$, then this reduces to

$$\mathsf{S}(f, W) = V_d(W)(d-1)!\, \pi^{d-1} \frac{\Gamma(\frac{d+1}{2})^{d-2}}{\Gamma(\frac{d}{2})^{d-1}} \int_{S^{d-1}} f^2(\operatorname{lin} u)\, \varphi_{\mathrm{iso}}(\mathrm{d}u).$$

Finally, if Θ is the Mondrian hyperplane measure with weights $p_1 = \dots = p_d = 1/d$, then

$$\mathsf{S}(f, W) = V_d(W)\, d^{d-2}\, 2^{d-1} \sum_{i=1}^{d} f^2(\operatorname{lin} e_i).$$

Proof. We evaluate the integrals in Lemma 13.4 using results on Poisson hyperplane tessellations.

Let X_1 be a Poisson hyperplane tessellation in $\mathbb{R}^d$ induced by Poisson point process on $\mathcal{H}$ with intensity measure Θ; see (2.66). Then for all $u \in S^{d-1}$, by Corollary 11.3 the typical cell of $Y_1 \cap u^\perp$ and that of $X_1 \cap u^\perp$ are identically distributed. Next, the intersection of the Poisson hyperplane tessellation X_1 with the hyperplane $u^\perp$ is again a Poisson hyperplane tessellation within $u^\perp$ having cell intensity $\gamma_{d-1}^{u^\perp}$. By Lemma 5.6.1 of [82] the zonoid associated with the sectional Poisson hyperplane tessellation $X_1 \cap u^\perp$ is $\Pi_\Theta | u^\perp$. Thus by Theorem 5.5.3 of [82] and by (2.38)

$$\int_{\mathcal{P}_{d-1,c}} V_{d-1}(F)\, \mathbb{Q}_{d-1}^{u^\perp}(\mathrm{d}F) = \frac{1}{\gamma_{d-1}^{u^\perp}} = \frac{1}{V_{d-1}(\Pi_\Theta | u^\perp)}.$$

Furthermore, by Theorems 6.1.4 and 6.2.6 of [82] we have that

$$\int_{\mathcal{P}_{d-1,c}} V_{d-1}^2(F)\,\mathbb{Q}_{d-1}^{u^\perp}(\mathrm{d}F) = \frac{1}{\gamma_{d-1}^{u^\perp}}\,\frac{(d-1)!}{2^{d-1}}\,V_{d-1}\big((\Pi_\Theta|u^\perp)^\circ\big),$$

which yields (13.11) in view of Lemma 13.4.

In the isotropic case, by (2.70) the associated zonoid is $\Pi_{\Theta_{\mathrm{iso}}} = \frac{\kappa_{d-1}}{d\,\kappa_d}B^d$, the centered ball with radius $\frac{\kappa_{d-1}}{d\,\kappa_d}$. Hence $(\Pi_\Theta|u^\perp)^\circ$ is the centered ball with radius $\frac{d\,\kappa_d}{\kappa_{d-1}}$ within $u^\perp$, and thus

$$V_{d-1}\big((\Pi_\Theta|u^\perp)^\circ\big) = \frac{(d\kappa_d)^{d-1}}{\kappa_{d-1}^{d-1}}\,\kappa_{d-1} = d^{d-1}\pi^{d-1}\frac{\Gamma(\frac{d+1}{2})^{d-2}}{\Gamma(\frac{d+2}{2})^{d-1}} = 2^{d-1}\pi^{d-1}\frac{\Gamma(\frac{d+1}{2})^{d-2}}{\Gamma(\frac{d}{2})^{d-1}}\,.$$

This yields the formula for $\mathsf{S}(f, W)$ in the isotropic case.

Finally, if Θ is the Mondrian hyperplane measure with weights $p_1 = \cdots = p_d = 1/d$, then $\Pi_\Theta = \frac{1}{d}[-\frac{1}{2}, \frac{1}{2}]^d$ by (2.71). Moreover,

$$\begin{aligned}\int_{S^{d-1}} f^2(\mathrm{lin}\,u)V_{d-1}\big((\Pi_\Theta|u^\perp)^\circ\big)\,\varphi(\mathrm{d}u) &= \frac{1}{d}\sum_{i=1}^{d} f^2(\mathrm{lin}\,e_i)V_{d-1}\big((\Pi_\Theta|e_i^\perp)^\circ\big)\\ &= \frac{1}{d}V_{d-1}\big((\Pi_\Theta|\mathbb{R}^{d-1})^\circ\big)\sum_{i=1}^{d} f^2(\mathrm{lin}\,e_i)\\ &= \frac{1}{d}V_{d-1}\left(\left(\frac{1}{d}\left[-\frac{1}{2},\frac{1}{2}\right]^{d-1}\right)^\circ\right)\sum_{i=1}^{d} f^2(\mathrm{lin}\,e_i)\end{aligned}$$

with the identification $\mathbb{R}^{d-1} = e_1^\perp$, since the value of $V_{d-1}((\Pi_\Theta|e_i^\perp)^\circ)$ does not depend on the index i. Writing $B_1^{d-1} = \{x = (x_1, \ldots, x_{d-1}) \in \mathbb{R}^{d-1} : |x_1| + \cdots + |x_{d-1}| \le 1\}$ for the unit ball with respect to the 1-norm on $\mathbb{R}^{d-1}$ and using the definition of the polar body, we see that

$$\begin{aligned}V_{d-1}\left(\left(\frac{1}{d}\left[-\frac{1}{2},\frac{1}{2}\right]^{d-1}\right)^\circ\right) &= V_{d-1}(2dB_1^{d-1})\\ &= (2d)^{d-1}V_{d-1}(B_1^{d-1}) = (2d)^{d-1}\frac{2^{d-1}}{(d-1)!}\,.\end{aligned}$$

Plugging this back into the definition of $\mathsf{S}(f, W)$, we arrive at

$$\begin{aligned}\mathsf{S}(f, W) &= V_d(W)\,\frac{(d-1)!}{2^{d-1}}\int_{S^{d-1}} f^2(\mathrm{lin}\,u)V_{d-1}\big((\Pi_\Theta|u^\perp)^\circ\big)\,\varphi(\mathrm{d}u)\\ &= V_d(W)\,\frac{(d-1)!}{2^{d-1}}\,\frac{1}{d}\,(2d)^{d-1}\,\frac{2^{d-1}}{(d-1)!}\sum_{i=1}^{d} f^2(\mathrm{lin}\,e_i)\end{aligned}$$

$$= V_d(W)\, d^{d-2}\, 2^{d-1} \sum_{i=1}^{d} f^2(\mathrm{lin}\, e_i)\,,$$

which completes the proof. □

Remark 13.6. *As a byproduct, the results of this section allow us to compute the expected dth-order chord-power integral of the typical cell Z_d^{PHT} of a stationary and isotropic Poisson hyperplane tessellation in $\mathbb{R}^d$ with intensity one. Indeed, recalling Definition* 12.11 *and then applying Lemma* 12.12 *and Proposition* 12.5, *we have*

$$\mathbb{E}I_d(Z_d^{\mathrm{PHT}}) = \frac{2}{d(d-1)}\,\mathbb{E}E_1(Z_d^{\mathrm{PHT}}) = \frac{d^2\kappa_d}{2\kappa_{d-1}}\,\mathbb{E}\int_{[Z_d^{\mathrm{PHT}}]} V_{d-1}(Z_d^{\mathrm{PHT}} \cap H)^2\, \Theta_{\mathrm{iso}}(dH)\,.$$

Since the integral term in the last line is the same as $\frac{1}{\gamma_{d,1}}\mathrm{S}(1, [0,1]^d)$, *Lemma* 13.5, *together with the fact that* $\gamma_{d,1} = V_d(\Pi_{\Theta_{\mathrm{iso}}}) = \kappa_d(\frac{\kappa_{d-1}}{d\kappa_d})^d$, *which follows from* (2.70), *yields*

$$\begin{aligned}\mathbb{E}I_d(Z_d^{\mathrm{PHT}}) &= \frac{d^2\kappa_d}{2\kappa_{d-1}}\frac{1}{\kappa_d}\left(\frac{d\kappa_d}{\kappa_{d-1}}\right)^d (d-1)!\,\pi^{d-1}\frac{\Gamma(\frac{d+1}{2})^{d-2}}{\Gamma(\frac{d}{2})^{d-1}}\\ &= d\,2^{d-1}\pi^{d-\frac{1}{2}}\,d!\left(\frac{\Gamma(\frac{d+1}{2})}{\Gamma(\frac{d}{2})}\right)^{2d-1}.\end{aligned}$$

To the best of our knowledge, explicit formulae for the expected value of other chord-power integrals of Z_d^{PHT} *are only known for orders* 1 *and* 3*:*

$$\mathbb{E}I_1(Z_d^{\mathrm{PHT}}) = \frac{d\kappa_d}{2}\,\mathbb{E}V_d(Z_d^{\mathrm{PHT}}) = d\,2^{d-1}\pi^{\frac{d}{2}}\left(\frac{\Gamma(\frac{d+1}{2})}{\Gamma(\frac{d}{2})}\right)^d,$$

which follows from Theorem 5.5.3 in [82], *and*

$$\begin{aligned}\mathbb{E}I_3(Z_d^{\mathrm{PHT}}) &= \frac{d(d+1)}{2}\,\mathbb{E}V_d^2(Z_d^{\mathrm{PHT}}) = \frac{d(d+1)}{2}\,\mathbb{E}V_d(Z_d^{\mathrm{PHT}})\,\mathbb{E}V_d(Z_o^{\mathrm{PHT}})\\ &= d\,2^{d-1}\pi^d(d+1)!\left(\frac{\Gamma(\frac{d+1}{2})}{\Gamma(\frac{d}{2})}\right)^{2d}.\end{aligned}$$

In the last equation, Z_o^{PHT} *is the zero cell of the Poisson hyperplane tessellation, and the equality is a consequence of Theorems 6.1.6 and 6.2.6 in* [82]. *Note that the same result holds for the typical cell of a stationary and isotropic STIT tessellation in* $\mathbb{R}^d$ *with time parameter one by Corollary* 11.3.

13.2 Gaussian fluctuations for the residual surface content

Consider a STIT tessellation Y_t in $\mathbb{R}^d$ with driving measure Θ of the form (4.1), satisfying (4.3), and let $W \in \mathcal{P}_d$. We observe now Y_t for fixed t within the sequence of growing windows nW, $n \in \mathbb{N}$, and consider the total surface content $\Sigma_{V_{d-1},t,nW}$. The question concerning the fluctuations of $\Sigma_{V_{d-1},t,nW}$ as $n \to \infty$ can be approached in two closely related, but significantly different regimes. Namely, we can consider

(a) either the *total surface content* induced by the collection of all $(d-1)$-dimensional maximal polytopes in nW

(b) or the *residual surface content* induced by those $(d-1)$-dimensional maximal polytopes in nW whose birth time is above a fixed positive threshold $s_0 > 0$.

In the present section, we focus on the residual surface content and argue that the analysis of this case cannot be extended to the total surface content, which formally corresponds to the case where $s_0 = 0$. The latter functional will be analyzed in detail in the subsequent sections separately for $d = 2$ and $d \geq 3$, leading to results of qualitatively different nature.

In this chapter, we will use the following convention: Whenever we base distributional limit theorems for maximal $(d-1)$-polytopes in growing windows nW with $n \to \infty$ on a global STIT tessellation process $(Y_t, t > 0)$, for bounded and measurable $\phi : \mathcal{P}_{d-1} \to \mathbb{R}$, we understand the functional $\Sigma_{\phi,t,nW}$ based on the corresponding local process $(Y_{t,nW}, t \geq 0)$ (see (10.12)), as the restriction $Y_t \wedge nW$ has the same distribution as $Y_{t,nW}$ and a. s. no maximal $(d-1)$-polytope can be contained in the boundary $\partial(nW)$ of nW. Throughout this section, we write $\overline{\Sigma}_{\phi,t,nW}$ for the centered random variable $\Sigma_{\phi,t,nW} - \mathbb{E}\Sigma_{\phi,t,nW}$ and recall (10.12).

Theorem 13.7. *Let $W \in \mathcal{P}_d$, and let Θ be a hyperplane measure of the form* (4.1), *satisfying* (4.3). *Furthermore, let $(Y_t, t > 0)$ be the STIT tessellation process driven by the measure Θ. Then for all $s_0 > 0$ and $t > 0$, the random variable*

$$\frac{1}{n^{d/2}}(\overline{\Sigma}_{V_{d-1},s_0+t,nW} - \overline{\Sigma}_{V_{d-1},s_0,nW})$$

converges in law to the normal distribution $\mathrm{Norm}(0, \mathrm{S}(1,W)\int_{s_0}^{s_0+t} s^{-(d-1)}\,\mathrm{d}s)$ *as $n \to \infty$, where $\mathrm{S}(1,W)$ is explicitly given by* (13.9) *or alternatively by* (13.10) *or* (13.11) *with $f \equiv 1$.*

We emphasize that Theorem 13.7 cannot be extended to $s_0 \downarrow 0$, as would be of interest and potentially would lead to a Gaussian limit theorem for the centered and normalized total surface content. The problem is that the variance integral $\int_{s_0}^{s_0+t} s^{-(d-1)}\,\mathrm{d}s$ diverges at $s_0 = 0$ for all $d \geq 2$. The reason for this phenomenon is the fact that the leading-order fluctuations of the total surface content are of larger order than the volume order n^d appearing in Theorem 13.7. They arise in nW at the very beginning of the STIT construction, more precisely, within the time interval $(0, \frac{1}{\log n}]$ if $d = 2$ or $(0, \frac{\log n}{n}]$ if $d \geq 3$, and are thus asymptotically not visible after any fixed time horizon $s_0 > 0$.

We analyze and discuss this phenomenon in depth in the forthcoming sections of this chapter.

Theorem 13.7 is a direct consequence of a much stronger result, namely the following *functional* central limit theorem for the weighted residual surface content *process*.

Theorem 13.8. *Let $W \in \mathcal{P}_d$, and let Θ be a hyperplane measure of the form* (4.1), *satisfying* (4.3). *Let $(Y_t,\ t > 0)$ be the STIT tessellation process driven by the measure Θ. Furthermore, let $f : G(d,1) \to \mathbb{R}$ be a bounded and measurable function. For fixed $s_0 > 0$ and $n \in \mathbb{N}$, define the residual surface content process*

$$\Delta^f_{s_0,t,nW} := \frac{1}{n^{d/2}}(\overline{\Sigma}^f_{V_{d-1},s_0+t,nW} - \overline{\Sigma}^f_{V_{d-1},s_0,nW}), \quad t \geq 0 .$$

Then

$$(\Delta^f_{s_0,t,nW},\ t \geq 0) \overset{\text{law}}{\Longrightarrow} \left(\int_0^t g(s)\, \mathrm{d}\mathcal{W}_s,\ t \geq 0\right) \quad \text{on the space } \mathsf{D}[0,\infty)\,, \tag{13.12}$$

where $g(s) = \sqrt{\mathsf{S}(f,W)}\,(s+s_0)^{-\frac{d-1}{2}}$, $(\mathcal{W}_s,\ s \geq 0)$ is a standard Wiener process, and $\mathsf{S}(f,W)$ is given by (13.9) *or alternatively by* (13.10) *or* (13.11).

Remark 13.9. *The limiting process in Theorem* 13.8 *can equivalently be characterized as a time-changed Wiener process. Namely, putting $\upsilon(t) := \mathsf{S}(f,W)\int_{s_0}^{s_0+t} s^{-(d-1)}\,\mathrm{d}s$, we can rephrase* (13.12) *by saying that*

$$(\Delta^f_{s_0,t,nW},\ t \geq 0) \overset{\text{law}}{\Longrightarrow} (\mathcal{W}_{\upsilon(t)},\ t \geq 0) \quad \text{on the space } \mathsf{D}[0,\infty)\,. \tag{13.13}$$

This follows from a comparison of the covariance functions of the two centred and continuous Gaussian processes defined by (13.12) *and* (13.13), *respectively. Indeed, by* (13.5) *the covariance function of the centered Gaussian process defined by* (13.12) *is given by*

$$\operatorname{cov}\left(\int_0^t g(s)\,\mathrm{d}\mathcal{W}_s, \int_0^u g(s)\,\mathrm{d}\mathcal{W}_s\right) = \mathsf{S}(f,W)\int_0^t \frac{\mathrm{d}s}{(s+s_0)^{d-1}} = \mathsf{S}(f,W)\int_{s_0}^{s_0+t} \frac{\mathrm{d}s}{s^{d-1}}$$

for $0 < t < u$. On the other hand, the covariance function of the centered Gaussian process in (13.13) *is*

$$\operatorname{cov}(\mathcal{W}_{\upsilon(t)}, \mathcal{W}_{\upsilon(u)}) = \upsilon(t) = \mathsf{S}(f,W)\int_{s_0}^{s_0+t} \frac{\mathrm{d}s}{s^{d-1}}\,,$$

since $\upsilon(t)$ is a strictly increasing function in t.

Proof of Theorem 13.8. To prove the theorem, our goal is to apply the functional central limit theorem for martingales, Theorem 13.3, to the sequence of stochastic processes

$(\Delta^f_{s_0,t,nW}, t \geq 0)$, $n \in \mathbb{N}$. Recall that $\mathfrak{I} = (\mathfrak{I}_t, t > 0)$ denotes the augmented filtration induced by the STIT tessellation process $(Y_t, t > 0)$. Then note that for each $n \in \mathbb{N}$, the stochastic process $(\overline{\Sigma}^f_{V_{d-1},s_0+t,nW}, t \geq 0)$ is an $\mathfrak{I}$-martingale by (10.43). Moreover, since the random constant processes $(\overline{\Sigma}^f_{V_{d-1},s_0,nW}, t \geq 0)$, $n \in \mathbb{N}$, trivially are $\mathfrak{I}$-martingales as well, we conclude that for each $n \in \mathbb{N}$,

$$\Delta^f_{s_0,t,nW} = n^{-d/2}(\overline{\Sigma}^f_{V_{d-1},s_0+t,nW} - \overline{\Sigma}^f_{V_{d-1},s_0,nW}), \quad t \geq 0,$$

is an $\mathfrak{I}$-martingale satisfying $\Delta^f_{s_0,0,nW} = 0$ and having càdlàg paths. Next,

$$\begin{aligned}\mathbb{E}(\Delta^f_{s_0,t,nW})^2 &\leq n^{-d}(\mathbb{E}(\overline{\Sigma}^f_{V_{d-1},\, s_0+t,\, nW})^2 + \mathbb{E}(\overline{\Sigma}^f_{V_{d-1},\, s_0,\, nW})^2) \\ &= n^{-d}(\operatorname{var}\overline{\Sigma}^f_{V_{d-1},\, s_0+t,\, nW} + \operatorname{var}\overline{\Sigma}^f_{V_{d-1},\, s_0,\, nW}),\end{aligned}$$

which by Theorem 12.3 is finite for all $t \geq 0$. Hence $(\Delta^f_{s_0,t,nW}, t \geq 0)$ is a square-integrable martingale for each $n \in \mathbb{N}$.

The rest of the proof consists of checking conditions (13.6)–(13.8) of Theorem 13.3.

Step 1: Predictable quadratic variation and condition (13.6)

By (10.44) the predictable quadratic variation of $(\overline{\Sigma}^f_{V_{d-1},\, s_0+t,\, nW}, t \geq 0)$ satisfies

$$\langle \overline{\Sigma}^f_{V_{d-1},\, s_0+\cdot,\, nW}\rangle_t = \int_0^t A^f_{d-1,\, s_0+s,\, nW}\, ds, \quad t \geq 0.$$

Since the second term in $\Delta^f_{s_0,\cdot,nW}$ has constant paths, its predictable quadratic variation and the predictable quadratic covariation with any locally square-integrable martingale vanish according to (13.3). Therefore by (13.2)

$$\langle \Delta^f_{s_0,\cdot,nW}\rangle_t = \frac{1}{n^d}\int_0^t A^f_{d-1,\, s_0+s,\, nW}\, ds = \frac{1}{n^d}\int_{s_0}^{s_0+t} A^f_{d-1,\, s,\, nW}\, ds, \quad t \geq 0. \tag{13.14}$$

In particular, the mapping $t \mapsto \langle \Delta^f_{s_0,\cdot,nW}\rangle_t$ is continuous and nondecreasing for each fixed n and hence has no jumps. Consequently,

$$\mathcal{J}(\langle \Delta^f_{s_0,\cdot,nW}\rangle, t) = 0, \quad t \geq 0,$$

and condition (13.6) of Theorem 13.3 is satisfied.

Step 2: Proof of (13.7)

Each jump of $(\Delta^f_{s_0,t,nW}, t \geq 0)$ occurs when a single cell splits. The $(d-1)$-volume of the newly created maximal $(d-1)$-polytope is bounded by the $(d-1)$-volume of a central section of the cell's circumball, hence by $\kappa_{d-1}R(P)^{d-1}$ with $R(P)$ the circumball radius of the divided cell P. In the STIT construction, cells only get subdivided, so every cell at time $s_0 + t$ is contained in some ancestor cell at time s_0. Therefore, for fixed $n \in \mathbb{N}$,

$$\mathcal{J}(\Delta^f_{s_0,\cdot,nW}, t) \le \frac{1}{n^{d/2}}\Big(\sup_{L\in G(d,1)} |f(L)|\Big)\,\kappa_{d-1} \sup_{P\in Y_{s_0,nW}} R(P)^{d-1}.$$

By Theorems 6.1.2 and 6.2.13 in [82] the circumball radius of the typical cell of a Poisson hyperplane tessellation X_{s_0} with intensity s_0 and induced by the hyperplane measure Θ has moments of all orders. Since the typical cells of X_{s_0} and Y_{s_0} are identically distributed,

$$\mathbb{E}[R(Z^{\mathrm{STIT}}_{s_0})^m] < \infty \tag{13.15}$$

for all $m \in \mathbb{N}$. Applying now Hölder's inequality leads to

$$\begin{aligned}
&\lim_{n\to\infty} \mathbb{E}\mathcal{J}(\Delta^f_{s_0,\cdot,nW}, t)^2 \\
&\le \lim_{n\to\infty} \big(\mathbb{E}\mathcal{J}(\Delta^f_{s_0,\cdot,nW}, t)^3\big)^{2/3} \\
&\le \lim_{n\to\infty} \Bigg(\mathbb{E}\Big(\frac{1}{n^{d/2}}\Big(\sup_{L\in G(d,1)} |f(L)|\Big)\,\kappa_{d-1} \sup_{P\in Y_{s_0,nW}} R(P)^{d-1}\Big)^3\Bigg)^{2/3} \\
&= \Big(\sup_{L\in G(d,1)} |f(L)|\Big)^2 \kappa_{d-1}^2 \lim_{n\to\infty} n^{-d}\Big(\mathbb{E}\Big(\sup_{P\in Y_{s_0,nW}} R(P)^{d-1}\Big)^3\Big)^{2/3}.
\end{aligned}$$

It remains to argue that the limit in the last line is zero. Using the tail-integral identity $\mathbb{E}X^3 = 3\int_0^\infty r^2\,\mathbb{P}(X \ge r)\,\mathrm{d}r$ for a random variable $X \ge 0$ and then the union bound together with the Campbell-type formula from Lemma 2.20, we obtain

$$\begin{aligned}
&\lim_{n\to\infty} n^{-d}\Big(\mathbb{E}\Big(\sup_{P\in Y_{s_0,nW}} R(P)^{d-1}\Big)^3\Big)^{2/3} \\
&= \lim_{n\to\infty} n^{-d/3}\Bigg(3\int_0^\infty r^2 n^{-d}\,\mathbb{P}\Big(\sup_{P\in Y_{s_0,nW}} R(P)^{d-1} \ge r\Big)\mathrm{d}r\Bigg)^{2/3} \\
&\le \lim_{n\to\infty} n^{-d/3}\Bigg(3\int_0^\infty r^2 n^{-d}\,\mathbb{E}\Big(\sum_{P\in Y_{s_0}} \mathbb{1}\{R(P)^{d-1} \ge r\}\,\mathbb{1}\{P \cap nW \ne \emptyset\}\Big)\,\mathrm{d}r\Bigg)^{2/3} \\
&= \lim_{n\to\infty} n^{-d/3}\Bigg(3\gamma_{d,s_0}\int_0^\infty r^2 \int_{\mathcal{P}_{d,c}} \mathbb{1}\{R(P)^{d-1} \ge r\}\,V_d\Big(W + \frac{1}{n}P\Big)\,\mathbb{Q}^{\mathrm{STIT}}_{d,s_0}(\mathrm{d}P)\,\mathrm{d}r\Bigg)^{2/3} \\
&\le \lim_{n\to\infty} n^{-d/3}\Bigg(3\gamma_{d,s_0}\int_0^\infty r^2 \int_{\mathcal{P}_{d,c}} \mathbb{1}\{R(P)^{d-1} \ge r\}V_d(W + 2R(P)B^d)\,\mathbb{Q}^{\mathrm{STIT}}_{d,s_0}(\mathrm{d}P)\,\mathrm{d}r\Bigg)^{2/3},
\end{aligned}$$

where γ_{d,s_0} is the cell intensity, and $\mathbb{Q}^{\text{STIT}}_{d,s_0}$ is the distribution of the typical cell of Y_{s_0}. The Steiner formula (1.3) implies that

$$V_d(W + 2R(P)B^d) = \sum_{i=0}^{d} V_i(W)\kappa_{d-i}(2R(P))^{d-i},$$

and hence

$$\begin{aligned}
&\int_0^\infty r^2 \int_{\mathcal{P}_{d,c}} \mathbb{1}\{R(P)^{d-1} \geq r\} V_d(W + 2R(P)B^d)\, \mathbb{Q}^{\text{STIT}}_{d,s_0}(\mathrm{d}P)\,\mathrm{d}r \\
&= \sum_{i=0}^{d} V_i(W)\kappa_{d-i} \int_{\mathcal{P}_{d,c}} (2R(P))^{d-i} \int_0^\infty r^2\, \mathbb{1}\{R(P)^{d-1} \geq r\}\,\mathrm{d}r\, \mathbb{Q}^{\text{STIT}}_{d,s_0}(\mathrm{d}P) \\
&= \frac{1}{3}\sum_{i=0}^{d} V_i(W) 2^{d-i}\kappa_{d-i} \int_{\mathcal{P}_{d,c}} R(P)^{(d-i)+3(d-1)}\, \mathbb{Q}^{\text{STIT}}_{d,s_0}(\mathrm{d}P) \\
&= \frac{1}{3}\sum_{i=0}^{d} V_i(W) 2^{d-i}\kappa_{d-i}\, \mathbb{E}[R(Z^{\text{STIT}}_{s_0})^{(d-i)+3(d-1)}],
\end{aligned}$$

which is finite according to (13.15). Consequently,

$$\begin{aligned}
&\lim_{n\to\infty} n^{-d}\Big(\mathbb{E}\Big(\sup_{P\in Y_{s_0,nW}} R(P)^{d-1}\Big)^3\Big)^{2/3} \\
&\leq \Big(\gamma_{d,s_0}\sum_{i=0}^{d} V_i(W)2^{d-i}\kappa_{d-i}\,\mathbb{E}[R(Z^{\text{STIT}}_{s_0})^{(d-i)+3(d-1)}]\Big) \lim_{n\to\infty} n^{-d/3} = 0,
\end{aligned}$$

which is condition (13.7).

Step 3: Proof of (13.8)
Finally, to prove (13.8), we consider the limit of $\langle \Delta^f_{s_0,\cdot,nW}\rangle_t$, given in (13.14), as $n \to \infty$. Using the scaling property (6.2) and the homogeneity of V_{d-1}, we find that

$$\begin{aligned}
\frac{1}{n^d}A^f_{d-1,s,nW} &= \frac{1}{n^d}\int_{[nW]} f^2(H^\perp) \sum_{F\in Y_{s,nW}\cap H} V^2_{d-1}(F)\,\Theta(\mathrm{d}H) \\
&\overset{\mathrm{D}}{=} \frac{1}{n^d}\int_{[nW]} f^2(H^\perp) \sum_{P\in Y_{1,snW}} V^2_{d-1}((s^{-1}P)\cap H)\,\Theta(\mathrm{d}H) \\
&= \frac{1}{n^d}\frac{1}{s}\int_{[snW]} f^2(H^\perp) \sum_{P\in Y_{1,snW}} V^2_{d-1}((s^{-1}P)\cap s^{-1}H)\,\Theta(\mathrm{d}H) \\
&= \frac{1}{n^d}\frac{1}{s^{2(d-1)+1}}\int_{[snW]} f^2(H^\perp) \sum_{P\in Y_{1,snW}} V^2_{d-1}(P\cap H)\,\Theta(\mathrm{d}H)
\end{aligned}$$

$$= \frac{1}{n^d} \frac{1}{s^{2(d-1)+1}} \sum_{P \in Y_{1,snW}} \int_{[snW]} f^2(H^\perp)\, V_{d-1}^2(P \cap H)\, \Theta(\mathrm{d}H)$$

$$= \frac{1}{(sn)^d} \frac{1}{s^{d-1}} \sum_{P \in Y_{1,snW}} \int_{[P]} f^2(H^\perp)\, V_{d-1}^2(P \cap H)\, \Theta(\mathrm{d}H)\,.$$

To determine the limit of $\frac{1}{n^d} A^f_{d-1,s,nW}$ as $n \to \infty$, we introduce

$$A^{f,\mathrm{lo}}_{d-1,s,nW} := \frac{1}{s^{d-1}} \sum_{P \in Y_1} \mathbb{1}\{P \subseteq snW\} \int_{[P]} f^2(H^\perp)\, V_{d-1}^2(P \cap H)\, \Theta(\mathrm{d}H)\,,$$

$$A^{f,\mathrm{up}}_{d-1,s,nW} := \frac{1}{s^{d-1}} \sum_{P \in Y_1} \mathbb{1}\{P \cap snW \neq \varnothing\} \int_{[P]} f^2(H^\perp)\, V_{d-1}^2(P \cap H)\, \Theta(\mathrm{d}H)\,,$$

so that a. s.

$$\frac{1}{(sn)^d} A^{f,\mathrm{lo}}_{d-1,s,nW} \le \frac{1}{n^d} A^f_{d-1,s,nW} \le \frac{1}{(sn)^d} A^{f,\mathrm{up}}_{d-1,s,nW}\,. \tag{13.16}$$

To apply the ergodic theorem in Corollary 15.28(c) to the upper bound in the last expression, we need to verify that

$$T := \int_{\mathcal{P}_{d,c}} \left| \int_{[P]} f^2(H^\perp)\, V_{d-1}^2(P \cap H)\, \Theta(\mathrm{d}H) \right| V_d(P + B^d)\, \mathbb{Q}_{d,1}(\mathrm{d}P) < \infty\,.$$

To do so, we write $R = R(Z_1^{\mathrm{STIT}})$ for the radius of the circumball of the typical cell of Y_1, which has the same distribution as the corresponding quantity associated with the typical cell of a Poisson hyperplane tessellation X_1 driven by Θ. Then since the typical cell is contained in a translate of a ball RB^d of radius R,

$$T \le \Big(\sup_{L \in G(d,1)} f^2(L) \Big)\, \mathbb{E}[\kappa_{d-1}^2 R^{2(d-1)}\, \Theta([RB^d])\, V_d((R+1)B^d)]$$

$$= \Big(\sup_{L \in G(d,1)} f^2(L) \Big)\, \kappa_d \kappa_{d-1}^2 \Theta([B^d])\, \mathbb{E}[R^{2(d-1)+1}(R+1)^d]\,.$$

Since by (13.15) the random variable R has finite moments of all orders, it follows that $T < \infty$, and we can apply Corollary 15.28(c) to conclude that the expression in the upper bound of (13.16) satisfies

$$\frac{1}{(sn)^d} A^{f,\mathrm{up}}_{d-1,s,nW} \overset{\mathbb{P}}{\Longrightarrow} \frac{\gamma_{d,1}}{s^{d-1}} V_d(W) \int_{\mathcal{P}_{d,c}} \int_{[P]} f^2(H^\perp)\, V_{d-1}^2(P \cap H)\, \Theta(\mathrm{d}H)\, \mathbb{Q}_{d,1}(\mathrm{d}P) \tag{13.17}$$

as $n \to \infty$. To the lower bound in (13.16), we apply Corollary 15.28(b), which implies that, as $n \to \infty$,

$$\frac{1}{(sn)^d} A^{f,\mathrm{lo}}_{d-1,s,nW} \overset{\mathbb{P}}{\Longrightarrow} \frac{\gamma_{d,1}}{s^{d-1}} V_d(W) \int_{\mathcal{P}_{d,c}} \int_{[P]} f^2(H^\perp)\, V^2_{d-1}(P \cap H)\, \Theta(\mathrm{d}H)\, \mathbb{Q}_{d,1}(\mathrm{d}P)\,. \tag{13.18}$$

Combining (13.16) with (13.17) and (13.18), we conclude that also $n^{-d} A^f_{d-1,s,nW}$ converges in probability as $n \to \infty$ to the common limit of $(sn)^{-d} A^{f,\mathrm{lo}}_{d-1,s,nW}$ and $(sn)^{-d} A^{f,\mathrm{up}}_{d-1,s,nW}$, that is,

$$\frac{1}{n^d} A^f_{d-1,s,nW} \overset{\mathbb{P}}{\Longrightarrow} \frac{\gamma_{d,1}}{s^{d-1}} V_d(W) \int_{\mathcal{P}_{d,c}} \int_{[P]} f^2(H^\perp)\, V^2_{d-1}(P \cap H)\, \Theta(\mathrm{d}H)\, \mathbb{Q}_{d,1}(\mathrm{d}P)$$

as $n \to \infty$. We remark that this convergence takes place even uniformly in s. In fact, this is a consequence of the scaling property of STIT tessellations (see (6.2)), implying that, in distribution, all instances of the left-hand side for different values of s are just rescalings of the same object with time parameter 1, say. Thus, in terms of the considered convergence in probability to a deterministic limit, we actually deal with a single asymptotic statement. Thus for $t \geq 0$, as $n \to \infty$, we have

$$\begin{aligned}
\langle \Delta^f_{s_0,\cdot,nW} \rangle_t &= \int_{s_0}^{s_0+t} \frac{1}{n^d} A^f_{d-1,s,nW}\, \mathrm{d}s \\
&\overset{\mathbb{P}}{\Longrightarrow} \int_{s_0}^{s_0+t} \frac{1}{s^{d-1}}\, \mathrm{d}s\, V_d(W)\, \gamma_{d,1} \int_{\mathcal{P}_{d,c}} \int_{[P]} f^2(H^\perp)\, V^2_{d-1}(P \cap H)\, \Theta(\mathrm{d}H)\, \mathbb{Q}_{d,1}(\mathrm{d}P) \\
&= \mathrm{S}(f, W) \int_{s_0}^{s_0+t} \frac{1}{s^{d-1}}\, \mathrm{d}s \\
&= \int_0^t g(s)^2\, \mathrm{d}s
\end{aligned} \tag{13.19}$$

with the function g as in the statement of the theorem.

After having established (13.6)–(13.8), we can finally apply Theorem 13.3 to conclude the functional convergence in law in (13.12). This completes the proof. □

13.3 Gaussian fluctuations in dimension $d = 2$

Starting with this section, we will focus on the *total* edge length in the planar case and on the *total* surface content for higher-dimensional STIT tessellations in Section 13.4. The difference to the residual edge length or surface content studied in Section 13.2 above is that now we – at least asymptotically – take into account all edges or $(d-1)$-dimensional maximal polytopes, not only those born after a fixed threshold $s_0 > 0$.

13.3.1 Total edge length

In this section, our focus lies on the centered functional $\overline{\Sigma}^f_{V_1,t,nW}$, defined by (10.32) with $t \geq 0$, $n \in \mathbb{N}$, and $W \in \mathcal{P}_2$, of a STIT tessellation Y_t in the plane. Note that for $f \equiv 1$, this is indeed the total length of the 1-skeleton of Y_t, whereas for a general function f, each I-segment is additionally weighted by a factor only depending on its direction. To present the next result, for $n \in \mathbb{N}$, define the nonlinear time flow

$$\tau(t,n) := \exp((\log n - \log\log n)(t-1)) = \left(\frac{\log n}{n}\right)^{1-t}, \quad t \geq 0, \tag{13.20}$$

and the random process

$$L^{f,\tau}_{t,nW} := \frac{1}{n\sqrt{\log n}} \overline{\Sigma}^f_{V_1,\tau(t,n),nW}, \quad t \geq 0. \tag{13.21}$$

Let us interpret the nonlinear time flow τ geometrically. For fixed n, the map $t \mapsto \tau(t,n)$ is strictly increasing with $\tau(1,n) = 1$ and inverse

$$t = 1 + \frac{\log \tau(t,n)}{\log n - \log\log n}.$$

Compared to the linear clock, the nonlinear time flow has the following effect:

- Events that occur very early under the original time t, where few but long I-segments are born, are placed later under the nonlinear clock $\tau(t,n)$, namely closer to 1 from below;
- On the other hand, events that occur late under the original time t, where many short I-segments appear, are placed earlier under the nonlinear clock $\tau(t,n)$, again closer to 1 but this time from above.

Equivalently, equal increments in time t correspond to multiplicative changes in the nonlinear time flow, since

$$\frac{\mathrm{d}}{\mathrm{d}t}\tau(t,n) = \tau(t,n)\,(\log n - \log\log n).$$

So roughly speaking, under the nonlinear time flow, long I-segments tend to be born later, and short I-segments earlier in comparison with the linear clock.

Theorem 13.10. *Let $W \in \mathcal{P}_2$, and let Θ be a line measure of the form* (4.1), *satisfying* (4.3). *Further, let $(Y_t,\ t > 0)$ be the planar STIT tessellation process driven by Θ, and let $f : G(2,1) \to \mathbb{R}$ be a bounded and measurable function. Then*

$$(L^{f,\tau}_{t,nW},\ t \geq 0) \overset{\text{law}}{\Longrightarrow} \left(\sqrt{\mathsf{S}(f,W)}\,\mathcal{W}_t,\ t \geq 0\right) \quad \textit{as } n \to \infty \quad \textit{in} \quad \mathsf{D}[0,\infty),$$

where $(\mathcal{W}_t,\ t \geq 0)$ is a standard Wiener process, and $\mathsf{S}(f,W)$ is given by (13.9) *or alternatively by* (13.10) *or* (13.11).

Proof. The strategy of the proof consists in defining the auxiliary random processes $(M_{n,t},\ t \geq 0)$, $n \in \mathbb{N}$, by

$$\begin{aligned} M_{n,t} &:= \frac{1}{n\sqrt{\log n - \log\log n}} \left(\overline{\Sigma}^f_{V_1,\tau(t,n),nW} - \overline{\Sigma}^f_{V_1,\tau(0,n),nW}\right) \\ &= \sqrt{\frac{\log n}{\log n - \log\log n}} \left(L^{f,\tau}_{t,nW} - L^{f,\tau}_{0,nW}\right), \quad t \geq 0, \end{aligned}$$

and then applying Theorem 13.3 to conclude a functional central limit theorem for $(M_{n,t},\ t \geq 0)$. The final step is to show that the difference processes

$$\left(\sqrt{\frac{\log n - \log\log n}{\log n}}\, M_{n,t} - L^{f,\tau}_{t,nW},\ t \geq 0\right)$$

converge in law to the zero-process in $\mathsf{D}[0,\infty)$ as $n \to \infty$.

Step 1: Functional limit theorem for $(M_{n,t},\ t \geq 0)$
First, the local square-integrability of the random processes $(M_{n,t},\ t \geq 0)$, $n \in \mathbb{N}$, follows from Theorem 12.3. Next, for any $n \in \mathbb{N}$, the predictable quadratic variation process of $(M_{n,t},\ t \geq 0)$ can be determined as follows. Define $\ell_n := \log n - \log\log n$, $c_n := 1/(n\sqrt{\ell_n})$, and $X_u := \overline{\Sigma}^f_{V_1,u,nW}$ for $u \geq 0$. Then

$$\langle M_{n,\cdot}\rangle_t = c_n^2\,\langle X_{\tau(\cdot,n)}\rangle_t, \quad t \geq 0;$$

see the discussion around (13.4). Applying the time change defined by (13.20), we have

$$\langle X_{\tau(\cdot,n)}\rangle_t = \langle X_\cdot\rangle_{\tau(t,n)} = \int_0^{\tau(t,n)} A^f_{1,s,nW}\,\mathrm{d}s = \int_0^t \tau'(s,n)\,A^f_{1,\tau(s,n),nW}\,\mathrm{d}s,$$

where the derivative $\tau'(s,n)$ refers to s, and where we applied $\langle X_\cdot\rangle_{\tau(s,n)} = A^f_{1,\tau(s,n),nW}$ from the analogue of (13.14) with $s_0 = 0$ in the last step. It follows that

$$\langle M_{n,\cdot}\rangle_t = c_n^2 \int_0^t \tau'(s,n)\, A^f_{1,\tau(s,n),nW}\, \mathrm{d}s$$
$$= \frac{1}{n^2 \ell_n} \int_0^t \tau(s,n)\, \ell_n\, A^f_{1,\tau(s,n),nW}\, \mathrm{d}s$$
$$= \int_0^t \frac{\tau(s,n)}{n^2} A^f_{1,\tau(s,n),nW}\, \mathrm{d}s\,,$$

where we also used that $\tau'(s,n) = \tau(s,n)\,\ell_n$. Although $(A^f_{1,t,nW}, t \geq 0)$ is a pure jump process, the predictable quadratic variation process $(\langle M_{n,\cdot}\rangle_t, t \geq 0)$ has no jumps, and hence condition (13.6) is satisfied.

A jump of $(M_{n,t}, t \geq 0)$ is caused by the division of precisely one tessellation cell. The length of the maximal segment arising by this division cannot be larger than the length $2nR(W)$ of the central slice of the circumcircle of the window nW by a line. Hence

$$\lim_{n\to\infty} \mathbb{E}\mathcal{J}(M_{n,\cdot}, t)^2 \leq \lim_{n\to\infty} \frac{4}{n^2(\log n - \log\log n)} \Big(\sup_{L\in G(2,1)} |f(L)|\Big)^2 n^2 R(W)^2 = 0\,,$$

which shows that condition (13.7) is satisfied as well. Repeating the argument leading to (13.19), we see that

$$\frac{\tau(s,n)}{n^2} A^f_{1,\tau(s,n),nW} \overset{\mathbb{P}}{\Longrightarrow} \mathsf{S}(f,W) \tag{13.22}$$

as $n \to \infty$, uniformly in s. As in the proof of Theorem 13.8, the uniformity in s comes from the scaling property of STIT tessellations (see (6.2)), which implies that, in distribution, all instances of the left-hand side for different values of s are just rescalings of the same object with time parameter 1, say. Thus, in terms of the considered convergence in probability to a deterministic limit, we are in fact just dealing with a single asymptotic statement. Consequently, by (13.22) it follows that

$$\langle M_{n,\cdot}\rangle_t \overset{\mathbb{P}}{\Longrightarrow} \int_0^t \mathsf{S}(f,W)\,\mathrm{d}s = t\,\mathsf{S}(f,W) \quad \text{as } n \to \infty\,.$$

We are now prepared to apply the martingale functional central limit theorem, Theorem 13.3, yielding the convergence in law as $n \to \infty$ of the sequence of random processes $(M_{n,t}, t \geq 0)$ to the random process

$$\int_0^t \sqrt{\mathsf{S}(f,W)}\,\mathrm{d}\mathcal{W}_s = \sqrt{\mathsf{S}(f,W)}\,\mathcal{W}_t\,, \quad t \geq 0\,,$$

in $\mathsf{D}[0,\infty)$. Moreover, since $\log n - \log\log n$ and $\log n$ are asymptotically equivalent, we also have that the sequence of random processes

$$\left(\sqrt{\frac{\log n - \log\log n}{\log n}}\, M_{n,t},\ t \geq 0\right), \qquad n \in \mathbb{N},$$

converges in law to the same limit process in $\mathsf{D}[0,\infty)$.

Step 2: Passing to the process $(L^{f,\tau}_{t,nW},\ t \geq 0)$

Finally, to pass to a limit theorem for the sequence of random processes $(L^{f,\tau}_{t,nW},\ t \geq 0)$, $n \in \mathbb{N}$, we consider the difference

$$\begin{aligned}
L^{f,\tau}_{t,nW} &- \sqrt{\frac{\log n - \log\log n}{\log n}}\, M_{n,t} \\
&= \frac{1}{n\sqrt{\log n}} \overline{\Sigma}^f_{V_1,\tau(t,n),nW} \\
&\quad - \sqrt{\frac{\log n - \log\log n}{\log n}} \cdot \frac{1}{n\sqrt{\log n - \log\log n}} \big(\overline{\Sigma}^f_{V_1,\tau(t,n),nW} - \overline{\Sigma}^f_{V_1,\tau(0,n),nW}\big) \\
&= \frac{1}{n\sqrt{\log n}} \overline{\Sigma}^f_{V_1,\tau(t,n),nW} - \frac{1}{n\sqrt{\log n}} \overline{\Sigma}^f_{V_1,\tau(t,n),nW} + \frac{1}{n\sqrt{\log n}} \overline{\Sigma}^f_{V_1,\tau(0,n),nW} \\
&= \frac{1}{n\sqrt{\log n}} \overline{\Sigma}^f_{V_1,\tau(0,n),nW},
\end{aligned}$$

which is independent of the time parameter t. It follows that

$$\mathbb{E} \sup_{t\geq 0} \left| L^{f,\tau}_{t,nW} - \sqrt{\frac{\log n - \log\log n}{\log n}}\, M_{n,t} \right|^2 = \frac{1}{n^2 \log n} \operatorname{var} \Sigma^f_{V_1,\frac{\log n}{n},nW}.$$

Using the scaling property analogous to (12.24) for the variance of the total edge length and then Theorem 12.3, we have

$$\begin{aligned}
&\operatorname{var} \Sigma^f_{V_1,\frac{\log n}{n},nW} \\
&\quad = \frac{n^2}{\log^2 n} \operatorname{var} \Sigma^f_{V_1,1,\log n\, W} \\
&\quad = \frac{n^2}{\log^2 n} \int\limits_{[(\log n)W]} f^2(H^\perp) \int\limits_{((\log n)W)\cap H} \int\limits_{((\log n)W)\cap H} \frac{1 - \exp(-\Theta([xy]))}{\Theta([xy])} \\
&\qquad \lambda_1(\mathrm{d}x)\,\lambda_1(\mathrm{d}y)\,\Theta(\mathrm{d}H).
\end{aligned}$$

Next, we apply the change of variables $x = (\log n)u$ and $y = (\log n)v$ along each line H. Then $\Theta([xy]) = \log n\, \Theta([uv])$, and the last integral becomes

$$= \frac{n^2}{\log^2 n} \int_{[W]} f^2(H^\perp) \int_{W\cap H} \int_{W\cap H} \frac{1-\exp(-\Theta([uv])\log n)}{\log n\, \Theta([uv])}$$
$$\times \log^3 n\, \lambda_1(du)\, \lambda_1(dv)\, \Theta(dH)$$
$$= n^2 \int_{[W]} f^2(H^\perp) \int_{W\cap H} \int_{W\cap H} \frac{1-\exp(-\Theta([uv])\log n)}{\Theta([uv])} \lambda_1(du)\, \lambda_1(dv)\, \Theta(dH)\,.$$

As a consequence,

$$\mathbb{E} \sup_{t\geq 0} \left| L_{t,nW}^{f,\tau} - \sqrt{\frac{\log n - \log\log n}{\log n}} M_{n,t} \right|^2$$
$$\leq \Big(\sup_{L\in G(d,1)} f^2(L) \Big) \int_{[W]} \int_{W\cap H} \int_{W\cap H} \frac{1-\exp(-\Theta([xy])\log n)}{\Theta([xy])\log n} \lambda_1(dx)\, \lambda_1(dy)\, \Theta(dH)\,.$$

Since the function $z \mapsto \frac{1-\exp(-z)}{z}$ is bounded by 1 for $z > 0$, we can apply the dominated convergence theorem. Because of

$$\lim_{n\to\infty} \frac{1-\exp(-\Theta([xy])\log n)}{\Theta([xy])\log n} = 0\,, \tag{13.23}$$

we have that

$$\lim_{n\to\infty} \mathbb{E} \sup_{t\geq 0} \left| L_{t,nW}^{f,\tau} - \sqrt{\frac{\log n - \log\log n}{\log n}} M_{n,t} \right|^2 = 0\,. \tag{13.24}$$

This yields that

$$\sup_{t\geq 0} \left| L_{t,nW}^{f,\tau} - \sqrt{\frac{\log n - \log\log n}{\log n}} M_{n,t} \right|^2 \xRightarrow{\mathbb{P}} 0\,,$$

and hence the difference vanishes in $\mathsf{D}[0,\infty)$. As a result, the sequences of random processes

$$\left(\sqrt{\frac{\log n - \log\log n}{\log n}} M_{n,t},\ t \geq 0 \right), \quad n \in \mathbb{N}\,,$$

and

$$(L_{t,nW}^{f,\tau},\ t \geq 0)\,, \quad n \in \mathbb{N}\,,$$

converge in law in $\mathsf{D}[0,\infty)$ to the same limit process $(\sqrt{\mathsf{S}(f,W)}\,\mathcal{W}_t,\ t \geq 0)$. This completes the argument. □

The previous functional limit theorem yields the following central limit theorem for the total edge length of a stationary STIT tessellation in the plane as a special case.

Corollary 13.11. *Let $W \in \mathcal{P}_2$, and let Θ be a line measure of the form* (4.1), *satisfying* (4.3). *Further, let $(Y_t,\ t > 0)$ be the planar STIT tessellation process driven by Θ. Then, for any fixed $t > 0$,*

$$\frac{1}{n\sqrt{\log n}}\,\overline{\Sigma}_{V_1,t,nW} \overset{\text{law}}{\Longrightarrow} L \quad \text{as } n \to \infty\,,$$

where L is a centered Gaussian random variable with variance $\mathsf{S}(1,W)$ given by (13.9) *or alternatively by* (13.10) *or* (13.11). *In particular, if $\Theta = \Theta_{\text{iso}}$, then $\mathsf{S}(1,W) = \pi V_2(W)$.*

Proof. Taking $f \equiv 1$ and $t = 1$ in Theorem 13.10 and noting that $\tau(1,n) = 1$ for $n \in \mathbb{N}$, we conclude that

$$\frac{1}{n\sqrt{\log n}}\,\overline{\Sigma}_{V_1,1,nW} \overset{\text{law}}{\Longrightarrow} \sqrt{\mathsf{S}(1,W)}\,\mathcal{W}_1 \quad \text{as } n \to \infty\,, \tag{13.25}$$

where $\mathcal{W}_1 \sim \mathsf{Norm}(0,1)$ is the value of a standard Wiener process at time 1. Putting $L := \sqrt{\mathsf{S}(1,W)}\,\mathcal{W}_1 \sim \mathsf{Norm}(0,\mathsf{S}(1,W))$ gives the result for $t = 1$.

For general $t > 0$, we write

$$\frac{1}{n\sqrt{\log n}}\,\overline{\Sigma}_{V_1,t,nW} \overset{\text{D}}{=} \frac{1}{nt\sqrt{\log n}}\,\overline{\Sigma}_{V_1,1,ntW} = \frac{\sqrt{\log(nt)}}{\sqrt{\log n}}\,\frac{1}{nt\sqrt{\log(nt)}}\,\overline{\Sigma}_{V_1,1,ntW}\,.$$

Since $\sqrt{\log(nt)}/\sqrt{\log n} \to 1$ as $n \to \infty$, the result follows from (13.25) applied with n replaced by nt. Finally, if $\Theta = \Theta_{\text{iso}}$, then we conclude that $\mathsf{S}(1,W) = \pi V_2(W)$ by Lemma 13.5. □

Since the convergence in (13.23) is of order $1/\log n$, the error term (13.24) decays logarithmically as well. In this light, we may conjecture that the speed of convergence in the central limit theorem for the total edge length is rather slow. Our next result confirms this intuition in the isotropic case by providing a logarithmically decaying lower bound on the so-called d_2-distance between the rescaled total edge length and the Gaussian random variable L. Denoting by $\mathsf{C}^2_{\le 1}$ the space of twice continuously differentiable functions $h : \mathbb{R} \to \mathbb{R}$ with $\sup_{x\in\mathbb{R}} |h''(x)| \le 1$, for two real-valued random variables X and Y, the d_2-distance is defined as

$$d_2(X,Y) := \sup_{h \in \mathsf{C}^2_{\le 1}} \big|\mathbb{E}h(X) - \mathbb{E}h(Y)\big|\,.$$

Note that convergence in d_2-distance implies convergence in distribution. Namely, if $d_2(X_n,X) \to 0$ as $n \to \infty$ for random variables $X, X_1, X_2, \ldots$, then $X_n \overset{\text{law}}{\Longrightarrow} X$.

Corollary 13.12. *Let $W \in \mathcal{P}_2$, and let $(Y_t,\ t > 0)$ be the planar STIT tessellation process driven by the motion-invariant measure Θ_{iso}. Then, for any fixed $t > 0$,*

$$d_2\Big(\frac{1}{n\sqrt{\log n}}\,\overline{\Sigma}_{V_1,t,nW}, L\Big) \ge \frac{c}{\log n}\,, \quad n \ge n_0\,,$$

where L is a centered Gaussian random variable with variance $\pi V_2(W)$, and $c \in (0,\infty)$ and $n_0 \in \mathbb{N}$ are constants only depending on W and t.

Proof. The function $h_0(x) = x^2/2$, $x \in \mathbb{R}$, belongs to the class $C^2_{\le 1}$. Thus

$$d_2\Big(\frac{1}{n\sqrt{\log n}}\overline{\Sigma}_{V_1,t,nW}, L\Big) \ge \Big|\mathbb{E}h_0\Big(\frac{1}{n\sqrt{\log n}}\overline{\Sigma}_{V_1,t,nW}\Big) - \mathbb{E}h_0(L)\Big|$$
$$= \frac{1}{2}\Big|\frac{1}{n^2\log n}\operatorname{var}\Sigma_{V_1,t,nW} - \operatorname{var} L\Big|$$
$$= \frac{1}{2}\Big|\frac{1}{n^2\log n}\operatorname{var}\Sigma_{V_1,t,nW} - \pi V_2(W)\Big|.$$

We can now apply Theorem 12.15 to see that

$$\Big|\frac{1}{n^2\log n}\operatorname{var}\Sigma_{V_1,t,nW} - \pi V_2(W)\Big| = \frac{|S_t(W)n^2 + O(n)|}{n^2\log n} \ge \frac{|S_t(W)|}{2\log n}$$

for all $n \ge n_0$, where $n_0 \in (0,\infty)$ is a suitable constant only depending on W and on t. As a result,

$$d_2\Big(\frac{1}{n\sqrt{\log n}}\overline{\Sigma}_{V_1,t,nW}, L\Big) \ge \frac{|S_t(W)|}{4\log n}, \qquad n \ge n_0,$$

which completes the proof. □

Now we present a version of Theorem 13.10 with $f \equiv 1$ on a linear time scale, but which puts – for technical reasons – the initial time after the n-dependent and asymptotically vanishing threshold $1/\log n$. We remark that this is in contrast to the residual edge length process, for which the initial time stays fixed and is not allowed to converge to zero with n. In fact, we shift the time argument and still include all I-segments with birth time in $(0, t+\frac{1}{\log n})$. We show that this leads to a constant limit process, the interpretation of which will be explained below.

Corollary 13.13. *Let $W \in \mathcal{P}_2$, and let Θ be a line measure of the form* (4.1), *satisfying* (4.3). *Further, let $(Y_t, t > 0)$ be the planar STIT tessellation process driven by Θ. For $n \in \mathbb{N}$, define the random process*

$$L_{t,nW} := \frac{1}{n\sqrt{\log n}}\overline{\Sigma}_{V_1,t+\frac{1}{\log n},nW}, \qquad t \ge 0.$$

Then

$$(L_{t,nW}, t \ge 0) \overset{\text{law}}{\Longrightarrow} (L, t \ge 0) \quad \textit{as } n \to \infty \quad \textit{in} \quad \mathsf{D}[0,\infty),$$

where L is a centered Gaussian random variable with variance $\mathsf{S}(1, W)$ given by (13.9) *or alternatively by* (13.10) *or* (13.11).

Proof. For $n \in \mathbb{N}$, we define the n-dependent time change

$$\sigma(t,n) := 1 - \frac{\log(t + \frac{1}{\log n})}{\log(\frac{\log n}{n})}, \quad t \geq 0 .$$

Then recalling (13.20), we have

$$t + \frac{1}{\log n} = \tau(\sigma(t,n),n) \quad \text{and} \quad \sigma(t) := \lim_{n\to\infty} \sigma(t,n) = 1$$

locally uniformly in t. Note that for each $n \in \mathbb{N}$, the function $\sigma(\,\cdot\,,n) : [0,\infty) \to [0,\infty)$ is continuous and strictly increasing and that the function $\sigma(\,\cdot\,)$ is continuous as well. Thus, choosing $f \equiv 1$ and recalling definition (13.21) of $L^{f,\tau}_{t,nW}$, from the functional convergence established in Theorem 13.10 in combination with Lemma 2.3 of [99], which ensures that functional convergence is preserved under appropriate time changes, it follows that

$$\begin{aligned}(L^{1,\tau}_{\sigma(t,n),nW},\, t \geq 0) &= \left(\frac{1}{n\sqrt{\log n}}\,\overline{\Sigma}_{V_1,\tau(\sigma(t,n),n),nW},\, t \geq 0\right)\\ &\overset{\text{law}}{\Longrightarrow} \sqrt{\mathsf{S}(1,W)}\,(\mathcal{W}_{\sigma(t)},\, t \geq 0) = \sqrt{\mathsf{S}(1,W)}\,(\mathcal{W}_1,\, t \geq 0)\end{aligned}$$

as $n \to \infty$ in $\mathsf{D}[0,\infty)$, where $(\mathcal{W}_t,\, t \geq 0)$ is a standard Wiener process. In other words, this means that

$$\begin{aligned}(L_{t,nW},\, t \geq 0) &= \left(\frac{1}{n\sqrt{\log n}}\,\overline{\Sigma}_{V_1,t+\frac{1}{\log n},nW},\, t \geq 0\right)\\ &= \left(\frac{1}{n\sqrt{\log n}}\,\overline{\Sigma}_{V_1,\tau(\sigma(t,n),n),nW},\, t \geq 0\right)\\ &\overset{\text{law}}{\Longrightarrow} \sqrt{\mathsf{S}(1,W)}\,(\mathcal{W}_1,\, t \geq 0)\\ &= (L,\, t \geq 0)\end{aligned}$$

as $n \to \infty$ in $\mathsf{D}[0,\infty)$. This proves the claim. □

The previous result shows that the variance of the total edge length in nW is of order $n^2 \log n$ and that it is almost entirely generated in the very early stages of the global STIT tessellation process, specifically within the short time interval $(0, 1/\log n]$. We refer to this as the **big-bang phase** of the STIT construction. After this phase, that is, for times larger than $1/\log n$, the additional variance grows only on the smaller order n^2. Hence late fluctuations are negligible compared to those accumulated during the big-bang phase.

This dichotomy is reflected in Corollary 13.13, where the limit process is constant in time: the big-bang fluctuations dominate from the outset and then remain “frozen” throughout the subsequent evolution. In contrast, Theorem 13.10 uses a nonlinear time flow that stretches the big-bang phase and compresses the later stages. Under this scaling, the early fluctuations unfold dynamically and give rise to a nondegenerate Wiener process in the limit.

A closely related phenomenon occurs for branching random walks. Here the population evolves on a Galton–Watson tree: each individual produces a random number of children, and along each edge an i. i. d. displacement increment is added. For a vertex v in generation $n \in \mathbb{N}$, S_v is the sum of the n increments along the path from the root to v. Then we can consider the classical additive martingale

$$W_n(\theta) := \frac{1}{m(\theta)^n} \sum_{|v|=n} e^{-\theta S_v}, \quad n \geq 0, \tag{13.26}$$

where $m(\theta) := \mathbb{E}[\sum_{|v|=1} e^{-\theta S_v}]$, and θ is an admissible parameter with $m(\theta) < \infty$. It is known from Theorem 1 of [17] that $(W_n(\theta), n \geq 0)$ converges a. s. and in L^1 under mild conditions. The variance of $W_n(\theta)$ is essentially determined by the first few branching generations. Intuitively, once the population has grown large, the effect of additional displacements cancels out on the martingale scale, and no new leading-order randomness is created. Thus the macroscopic randomness of the branching random walks is already "frozen" after the initial generations.

This is directly analogous to the big-bang phase of STIT tessellations: the creation of the first long I-segments within the time interval $(0, 1/\log n]$ already produces the leading-order variance of order $n^2 \log n$. Later insertions of short I-segments contribute only on the smaller order n^2, just as later generations in the branching random walk contribute only refinements of an already random limit.

Remark 13.14. *Keeping the linear time scale and looking only at the increment after the n-dependent initial time* $1/\log n$, *we may define the random process*

$$\widetilde{L}_{t,nW} := \frac{1}{n\sqrt{\log n}} (\overline{\Sigma}_{V_1, t+\frac{1}{\log n}, nW} - \overline{\Sigma}_{V_1, \frac{1}{\log n}, nW}), \quad t \geq 0 .$$

Then

$$(\widetilde{L}_{t,nW}, t \geq 0) \overset{\text{law}}{\Longrightarrow} (0, t \geq 0) \quad \textit{in } \mathsf{D}[0,\infty), \quad n \to \infty .$$

To see this, write $\tau(s,n) = (\frac{\log n}{n})^{1-s}$ *and* $L^{1,\tau}_{s,nW} = \frac{1}{n\sqrt{\log n}} \overline{\Sigma}_{V_1, \tau(s,n), nW}$. *With* $\sigma(t,n) := 1 - \frac{\log(t+\frac{1}{\log n})}{\log(\frac{\log n}{n})}$, *we have* $\tau(\sigma(t,n),n) = t + \frac{1}{\log n}$ *and* $\tau(\sigma(0,n),n) = \frac{1}{\log n}$. *Hence*

$$\widetilde{L}_{t,nW} = L^{1,\tau}_{\sigma(t,n),nW} - L^{1,\tau}_{\sigma(0,n),nW} .$$

By Theorem 13.10 *and the random time-change lemma, Lemma 2.3 in* [99], *already used in the previous proof,* $L^{1,\tau}_{\sigma(t,n),nW} \overset{\text{law}}{\Longrightarrow} \sqrt{\mathsf{S}(1,W)}\,\mathcal{W}_1$ *uniformly in t on compact intervals, while* $L^{1,\tau}_{\sigma(0,n),nW} \overset{\text{law}}{\Longrightarrow} \sqrt{\mathsf{S}(1,W)}\,\mathcal{W}_1$. *Thus the difference converges to* 0.

13.3.2 Total edge length and number of I-segments

Consider a STIT tessellation process $(Y_t,\ t > 0)$ in $\mathbb{R}^2$ driven by the motion-invariant line measure Θ_{iso} and $W \in \mathcal{P}_2$. From the previous section let us recall the definition of the total edge length process: for $n \in \mathbb{N}$,

$$L_{t,nW} := \frac{1}{n\sqrt{\log n}}\,\overline{\Sigma}_{V_1,t+\frac{1}{\log n},nW}\,, \qquad t \ge 0\,,$$

within the scaled window nW, where we again put the initial time after the asymptotically vanishing threshold $1/\log n$. Moreover, we introduce the random process

$$C_{t,nW} := \frac{1}{n\sqrt{\log n}}\,\overline{\Sigma}_{V_0,t+\frac{1}{\log n},nW}\,, \qquad t \ge 0\,,$$

which counts the number of I-segments in nW or, equivalently, the number of cells of $Y_{t,nW}$ minus 1. The next theorem describes the *joint* asymptotic fluctuations of both processes in the Skorokhod space $\mathsf{D}^2[0,\infty)$.

Theorem 13.15. *Let $W \in \mathcal{P}_2$, and let $(Y_t,\ t > 0)$ be the planar STIT tessellation process driven by the motion-invariant measure Θ_{iso}. Then*

$$((L_{t,nW}, C_{t,nW}),\ t \ge 0) \overset{\text{law}}{\Longrightarrow} \left(\left(L, \frac{2}{\pi}tL\right),\ t \ge 0\right) \quad \text{as } n \to \infty \quad \text{in } \mathsf{D}^2[0,\infty)\,,$$

where L is a centered Gaussian random variable with variance $\pi V_2(W)$.

Proof. For $n \in \mathbb{N}$, define the auxiliary random process

$$M_{n,t} := \overline{\Sigma}_{V_0,t,nW} - \int_0^t \overline{\mathbb{L}\Sigma_{V_0,s,nW}}\,\mathrm{d}s\,, \quad t \ge 0\,,$$

where $\overline{\mathbb{L}\Sigma_{V_0,s,nW}} := \mathbb{L}\Sigma_{V_0,s,nW} - \mathbb{E}\mathbb{L}\Sigma_{V_0,s,nW}$, and by (10.36)

$$\mathbb{L}\Sigma_{V_0,s,nW} = \int_{[nW]} \sum_{F \in Y_{s,nW} \cap H} V_0(F)\,\Theta_{\mathrm{iso}}(\mathrm{d}H)\,.$$

By Corollary 10.12(c) the random process

$$\left(\Sigma_{V_0,t,nW} - \int_0^t \mathbb{L}\Sigma_{V_0,s,nW}\,\mathrm{d}s,\ t \ge 0\right)$$

is a centered martingale, and hence $(M_{n,t},\ t \ge 0)$ is a centered martingale as well. Thus

$$\mathbb{E}M_{n,t}^2 = \operatorname{var}\Sigma_{V_0,t,nW} - 2\int_0^t \mathbb{E}[\overline{\Sigma}_{V_0,t,nW}\,\overline{\mathbb{L}\Sigma_{V_0,s,nW}}]\,ds + \int_0^t\int_0^t \mathbb{E}[\overline{\mathbb{L}\Sigma_{V_0,s,nW}}\,\overline{\mathbb{L}\Sigma_{V_0,u,nW}}]\,du\,ds\,.$$

By (10.42) we have

$$\mathbb{L}\Sigma_{V_0,s,nW} = \frac{2}{\pi}(\Sigma_{V_1,s,nW} + V_1(nW)) \quad \text{and hence} \quad \overline{\mathbb{L}\Sigma_{V_0,s,nW}} = \frac{2}{\pi}\overline{\Sigma}_{V_1,s,nW}\,.$$

Using this together with the tower property of conditional expectations, we obtain

$$\begin{aligned}
\mathbb{E}M_{n,t}^2 &= \operatorname{var}\Sigma_{V_0,t,nW} - \frac{4}{\pi}\int_0^t \mathbb{E}[\overline{\Sigma}_{V_0,t,nW}\,\overline{\Sigma}_{V_1,s,nW}]\,ds \\
&\quad + \frac{4}{\pi^2}\int_0^t\int_0^t \mathbb{E}[\overline{\Sigma}_{V_1,s,nW}\,\overline{\Sigma}_{V_1,u,nW}]\,du\,ds \\
&= \operatorname{var}\Sigma_{V_0,t,nW} - \frac{4}{\pi}\int_0^t \mathbb{E}\big[\mathbb{E}[\overline{\Sigma}_{V_0,t,nW}\,\overline{\Sigma}_{V_1,s,nW}|\mathfrak{I}_s]\big]\,ds \\
&\quad + \frac{8}{\pi^2}\int_0^t\int_0^s \mathbb{E}[\overline{\Sigma}_{V_1,s,nW}\,\overline{\Sigma}_{V_1,u,nW}]\,du\,ds \\
&= \operatorname{var}\Sigma_{V_0,t,nW} - \frac{4}{\pi}\int_0^t \mathbb{E}\big[\overline{\Sigma}_{V_1,s,nW}\,\mathbb{E}[\overline{\Sigma}_{V_0,t,nW}|\mathfrak{I}_s]\big]\,ds \\
&\quad + \frac{8}{\pi^2}\int_0^t\int_0^s \mathbb{E}[\overline{\Sigma}_{V_1,s,nW}\,\overline{\Sigma}_{V_1,u,nW}]\,du\,ds\,,
\end{aligned}$$

where we also used symmetry of the double integral.

Since $(\overline{\Sigma}_{V_1,t,nW},\ t \geq 0)$ is a martingale by Corollary 10.12(a), we have a. s.

$$\mathbb{E}[\overline{\Sigma}_{V_1,t,nW} \mid \mathfrak{I}_s] = \overline{\Sigma}_{V_1,s,nW}\,, \qquad s < t\,.$$

Moreover, from the martingale property of $(M_{n,t},\ t \geq 0)$ and $\overline{\mathbb{L}\Sigma_{V_0,s,nW}} = \frac{2}{\pi}\overline{\Sigma}_{V_1,s,nW}$ it follows that for $s < t$,

$$\begin{aligned}
\mathbb{E}[\overline{\Sigma}_{V_0,t,nW} \mid \mathfrak{I}_s] - \overline{\Sigma}_{V_0,s,nW} &= \mathbb{E}\left[\int_0^t \overline{\mathbb{L}\Sigma_{V_0,u,nW}}\,du - \int_0^s \overline{\mathbb{L}\Sigma_{V_0,u,nW}}\,du \,\Big|\, \mathfrak{I}_s\right] \\
&= \int_s^t \mathbb{E}[\overline{\mathbb{L}\Sigma_{V_0,u,nW}} \mid \mathfrak{I}_s]\,du
\end{aligned}$$

$$= \frac{2}{\pi}\int_s^t \mathbb{E}[\overline{\Sigma}_{V_1,u,nW} \mid \mathfrak{I}_s]\,du$$
$$= \frac{2}{\pi}\int_s^t \overline{\Sigma}_{V_1,s,nW}\,du$$
$$= \frac{2}{\pi}(t-s)\,\overline{\Sigma}_{V_1,s,nW} \tag{13.27}$$

a. s. Thus the expression for $\mathbb{E}M_{n,t}^2$ simplifies to

$$\begin{aligned}\mathbb{E}M_{n,t}^2 &= \mathfrak{var}\,\Sigma_{V_0,t,nW} - \frac{4}{\pi}\int_0^t \mathbb{E}[\overline{\Sigma}_{V_1,s,nW}\,\overline{\Sigma}_{V_0,s,nW}]\,ds - \frac{8}{\pi^2}\int_0^t (t-s)\mathbb{E}\overline{\Sigma}^2_{V_1,s,nW}\,ds \\ &\quad + \frac{8}{\pi^2}\int_0^t\int_0^s \mathbb{E}[\overline{\Sigma}_{V_1,s,nW}\,\overline{\Sigma}_{V_1,u,nW}]\,du\,ds \\ &= \mathfrak{var}\,\Sigma_{V_0,t,nW} - \frac{4}{\pi}\int_0^t \mathfrak{cov}(\overline{\Sigma}_{V_1,s,nW}, \overline{\Sigma}_{V_0,s,nW})\,ds\,.\end{aligned}$$

The cancellation of the last two terms in the expression for $\mathbb{E}M_{n,t}^2$ follows from the martingale property of $(\overline{\Sigma}_{V_1,t,nW}, t \geq 0)$ and Fubini's theorem. To see this, we recall that for a square-integrable martingale $(X_t, t \geq 0)$ and for $u \leq s$, we have $\mathbb{E}[X_s X_u] = \mathbb{E}[\mathbb{E}[X_s \mid \mathfrak{I}_u]\,X_u] = \mathbb{E}[X_u^2]$. Applying this to the double integral term yields

$$\int_0^t\int_0^s \mathbb{E}[\overline{\Sigma}_{V_1,s,nW}\,\overline{\Sigma}_{V_1,u,nW}]\,du\,ds = \int_0^t\int_0^s \mathbb{E}[\overline{\Sigma}^2_{V_1,u,nW}]\,du\,ds = \int_0^t (t-u)\,\mathbb{E}[\overline{\Sigma}^2_{V_1,u,nW}]\,du\,,$$

where the last equation is obtained by changing the integration order on the triangular domain $0 \leq u \leq s \leq t$ and integration with respect to s.

Now Theorem 12.10 implies that

$$\begin{aligned}\mathbb{E}M_{n,t}^2 &= \frac{2t}{\pi}V_1(nW) + \frac{3t^2}{\pi}V_2(nW) + \frac{8}{\pi^2}\int_0^\infty \mathfrak{c}_{nW}(r)\int_0^t (t-s)^2 e^{-\frac{2}{\pi}rs}\,ds\,dr \\ &\quad - \frac{4}{\pi}\int_0^t\left(sV_2(nW) + \frac{4}{\pi}\int_0^\infty \mathfrak{c}_{nW}(r)\int_0^s (s-u)e^{-\frac{2}{\pi}rs}\,du\,dr\right)ds \\ &= \frac{2t}{\pi}V_1(nW) + \frac{t^2}{\pi}V_2(nW) + \frac{8}{\pi^2}\int_0^\infty \mathfrak{c}_{nW}(r)\int_0^t (t-s)^2 e^{-\frac{2}{\pi}rs}\,ds\,dr \\ &\quad - \frac{16}{\pi^2}\int_0^\infty \mathfrak{c}_{nW}(r)\int_0^t\int_0^s (s-u)e^{-\frac{2}{\pi}rs}\,du\,ds\,dr\,.\end{aligned}$$

Since

$$2\int_0^t\int_0^s (s-u)e^{-\frac{2}{\pi}rs}\,du\,ds = \int_0^t (t-s)^2 e^{-\frac{2}{\pi}rs}\,ds\,,$$

the last two terms cancel, and we obtain

$$\mathbb{E}M_{n,t}^2 = \frac{2t}{\pi}V_1(nW) + \frac{t^2}{\pi}V_2(nW) = \frac{2nt}{\pi}V_1(W) + \frac{n^2t^2}{\pi}V_2(W)\,. \tag{13.28}$$

Next, we define for $n \in \mathbb{N}$ the auxiliary random process

$$N_{n,t} := \frac{1}{n\sqrt{\log n}}M_{n,t+\frac{1}{\log n}}\,, \quad t \geq 0\,.$$

Doob's L^2-maximal inequality (13.1) and (13.28) yield

$$\begin{aligned}\mathbb{E}\Big(\sup_{0\le s\le t}|N_{n,s}|\Big)^2 &\le 4\,\mathbb{E}N_{n,t}^2\\ &= \frac{1}{n^2\log n}\,\mathbb{E}M^2_{n,t+\frac{1}{\log n}}\\ &= \frac{1}{\pi n^2\log n}\left(2V_1(W)n\Big(t+\frac{1}{\log n}\Big) + n^2V_2(W)\Big(t+\frac{1}{\log n}\Big)^2\right).\end{aligned}$$

Taking the limit as $n \to \infty$, we conclude that

$$\lim_{n\to\infty}\mathbb{E}\Big(\sup_{0\le s\le t}N_{n,s}\Big)^2 = 0$$

for all $t > 0$, showing that the sequence of random processes $(N_{n,t},\ t \geq 0)$, $n \in \mathbb{N}$, converges in law in $\mathsf{D}[0,\infty)$ to the zero process as $n \to \infty$.

Next, note that

$$\begin{aligned}N_{n,t} &= \frac{1}{n\sqrt{\log n}}\overline{\Sigma}_{V_0,t+\frac{1}{\log n},nW} - \frac{1}{n\sqrt{\log n}}\int_0^{t+\frac{1}{\log n}}\overline{\mathbb{L}\Sigma_{V_0,s,nW}}\,ds\\ &= \frac{1}{n\sqrt{\log n}}\overline{\Sigma}_{V_0,t+\frac{1}{\log n},nW} - \frac{2}{\pi}\int_0^{t+\frac{1}{\log n}}\frac{1}{n\sqrt{\log n}}\overline{\Sigma}_{V_1,s,nW}\,ds\end{aligned}$$

$$= \frac{1}{n\sqrt{\log n}}\overline{\Sigma}_{V_0,t+\frac{1}{\log n},nW} - \frac{2}{\pi}\int_{-\frac{1}{\log n}}^{t} \frac{1}{n\sqrt{\log n}}\overline{\Sigma}_{V_1,s+\frac{1}{\log n},nW}\,\mathrm{d}s$$

$$= C_{t,nW} - \frac{2}{\pi}\int_{-\frac{1}{\log n}}^{t} L_{s,nW}\,\mathrm{d}s\,.$$

According to Corollary 13.13, the sequence of random processes $(L_{s,nW},\ s \geq 0)$ converges in law to the random constant process $(L,\ s \geq 0)$ in $\mathsf{D}[0,\infty)$. Moreover, by (11.10) in [53] it follows that in this situation, integration is a continuous functional on the Skorokhod space $\mathsf{D}[0,\infty)$. As a consequence, the sequence of random processes $((L_{t,nW}, C_{t,nW}),\ t \geq 0)$ jointly converges, as $n \to \infty$, in law in $\mathsf{D}^2[0,\infty)$ to the bivariate random constant process

$$\left(L, \frac{2}{\pi}\int_0^t L\,\mathrm{d}s,\ t \geq 0\right) = \left(\left(L, \frac{2}{\pi}tL\right),\ t \geq 0\right).$$

This completes the proof. □

An immediate consequence of the previous functional limit theorem is the following central limit theorem for the number of I-segments in a sequence of growing windows.

Corollary 13.16. *Let $W \in \mathcal{P}_2$, and let $(Y_t,\ t \geq 0)$ be the planar STIT tessellation process driven by the motion-invariant measure Θ_{iso}. Then, for all $t > 0$,*

$$\frac{1}{n\sqrt{\log n}}\overline{\Sigma}_{V_0,t,nW} \overset{\text{law}}{\Longrightarrow} \frac{2}{\pi}tL \quad \text{as } n \to \infty\,,$$

where L is a centered Gaussian random variable with variance $\pi V_2(W)$.

Proof. Let $\varepsilon := 1/\log n$. By the scaling property (6.2) of STIT tessellations and the fact that the intrinsic volume V_0 is homogeneous of degree zero we have

$$\Sigma_{V_0,t,nW} \overset{\mathrm{D}}{=} \Sigma_{V_0,t+\varepsilon,\frac{t}{t+\varepsilon}nW}\,.$$

Thus, writing $m_n := \frac{t}{t+\varepsilon}n$, we have

$$\frac{1}{n\sqrt{\log n}}\overline{\Sigma}_{V_0,t,nW} \overset{\mathrm{D}}{=} \left(\frac{t+\varepsilon}{t}\sqrt{\frac{\log m_n}{\log n}}\right)\frac{1}{m_n\sqrt{\log m_n}}\overline{\Sigma}_{V_0,t+\varepsilon,m_nW}\,.$$

The term in brackets converges to one as $n \to \infty$, whereas according to Theorem 13.15, the remaining expression converges in law to $\frac{2}{\pi}tL$. This yields the result. □

As explained at the end of the previous section, the leading-order fluctuations of the total edge length $L_{t,nW}$ are generated entirely during the big-bang phase $(0, 1/\log n]$. In contrast, the fluctuations of the number $C_{t,nW}$ of I-segments in nW persist for all t and grow linearly in t, with magnitude proportional to the big-bang fluctuation of $L_{t,nW}$. This is reflected in the asymptotic variances: By Theorems 12.13 and 12.18 the asymptotic variance of $C_{t,nW}$ exhibits a nonvanishing quadratic dependence on t, whereas the asymptotic variance of $L_{t,nW}$ does not depend on t.

Moreover, the fluctuations of $C_{t,nW}$ are fully determined by those of $L_{t,nW}$. This mechanism will reappear in higher dimensions in the next section. Summarizing, we obtain the following intuitive picture in the large-n regime: If L denotes the limiting fluctuation of $L_{t,nW}$ created during the big-bang phase, then for $t \geq 1/\log n$,

$$L_{t,nW} \approx L, \quad C_{t,nW} \approx \frac{2}{\pi} t L,$$

in the sense of convergence in distribution after the $n\sqrt{\log n}$-normalization. Thus, although $L_{t,nW}$ "freezes" at its big-bang fluctuation, $C_{t,nW}$ continues to evolve linearly in t with slope determined by the same random seed L.

A closely related phenomenon appears already for supercritical branching processes (and hence also for branching random walks when we only track the population size). Let N_t be the number of particles alive at time t in a supercritical branching random walk with Malthusian parameter $\mu > 0$. By the Kesten–Stigum theorem [93], $e^{-\mu t} N_t \to \xi$ a. s., where ξ is the limit of the additive martingale (13.26). For the cumulative number of particles over a time horizon of length u, we find

$$e^{-\mu t} \int_t^{t+u} N_s \, ds \overset{\mathbb{P}}{\Longrightarrow} \frac{1 - e^{-\mu u}}{\mu} \xi$$

as $t \to \infty$. Since the right-hand side is of order $u\xi$ for small u, time integration reveals a linear growth proportional to ξ. In spirit, this mirrors the STIT case, where the total edge length $L_{n,t}$ freezes at its big-bang fluctuation L, whereas the number of I-segments evolves approximately linearly in t as $\frac{2}{\pi} tL$.

Remark 13.17. *If, instead, we define the incremental versions*

$$L^{\text{inc}}_{t,nW} := \frac{1}{n\sqrt{\log n}} \left(\overline{\Sigma}_{V_1, t+\frac{1}{\log n}, nW} - \overline{\Sigma}_{V_1, \frac{1}{\log n}, nW}\right),$$

$$C^{\text{inc}}_{t,nW} := \frac{1}{n\sqrt{\log n}} \left(\overline{\Sigma}_{V_0, t+\frac{1}{\log n}, nW} - \overline{\Sigma}_{V_0, \frac{1}{\log n}, nW}\right),$$

then $(L^{\text{inc}}_{t,nW}, t \geq 0) \overset{\text{law}}{\Longrightarrow} (0, t \geq 0)$ *and* $(C^{\text{inc}}_{t,nW}, t \geq 0) \overset{\text{law}}{\Longrightarrow} (0, t \geq 0)$ *in* $\mathsf{D}[0,\infty)$. *Heuristically, the* $n^2 \log n$*-scale fluctuations are entirely created during the initial big-bang phase* $(0, 1/\log n]$, *and what happens afterward contributes only order* n^2, *which is negligible in* L^2 *at the* $n\sqrt{\log n}$ *scale.*

13.4 Non-Gaussian fluctuations in dimension $d \geq 3$

13.4.1 Total surface content: the Mondrian case

The goal of this section and the next one is to study the fluctuations of the total surface content within a sequence of expanding observation windows nW, induced by a stationary STIT tessellation Y_t with time parameter $t > 0$ in $\mathbb{R}^d$ for $d \geq 3$. We will see that – in sharp contrast to the planar case $d = 2$ studied in Section 13.3 – these fluctuations exhibit non-Gaussian behavior. Fundamentally, this effect can be attributed to the fact that for $d \geq 3$, the variance of $\overline{\Sigma}_{V_{d-1},t,nW}$ is of order $O(n^{2(d-1)})$ (see (12.28)), whereas the variance of the residual surface content $\overline{\Sigma}_{V_{d-1},t,nW} - \overline{\Sigma}_{V_{d-1},s_0,nW}$ for some initial time $s_0 > 0$ is only of order $O(n^d)$, as established in Theorem 13.8.

Thus, for $d \geq 3$, the largest maximal $(d-1)$-polytopes generated in the initial phase of the STIT tessellation process contribute predominantly to the overall variance and dominate the fluctuations. We do not establish this result in full generality for all non-degenerate hyperplane measures Θ and all observation windows W but instead restrict our analysis to two specific cases, the Mondrian case in this section and the isotropic case in the next one. In both cases, we choose a cubic observation window. Here we consider Θ as the Mondrian hyperplane measure with weights

$$p_1 = \cdots = p_d = \frac{1}{d}$$

(recall (4.4)) and take the d-dimensional unit cube as the observation window

$$W := \left[-\frac{1}{2}, \frac{1}{2}\right]^d .$$

In this setup the argument can be presented most clearly, as explicit computations can be carried out in all steps of the proof. Throughout the remainder of this section, the symbols Θ and W are understood accordingly. The main result can now be stated as follows.

Theorem 13.18. *Fix $d \geq 3$ and consider the Mondrian STIT tessellation process $(Y_t,\ t > 0)$ with equal weights $p_1 = \cdots = p_d = 1/d$. Then for $W = [-\frac{1}{2}, \frac{1}{2}]^d$ and $t = 1$, we have that*

$$a^{-(d-1)} \overline{\Sigma}_{V_{d-1},1,aW} \overset{\text{law}}{\Longrightarrow} S \quad \text{as } a \to \infty ,$$

where S is a non-Gaussian centered random variable with variance

$$\operatorname{var} S = d\, 2^{d-1} \int_0^1 \cdots \int_0^1 \frac{\prod_{i=1}^{d-1}(1 - z_i)}{\sum_{i=1}^{d-1} z_i} \, \mathrm{d}z_1 \ldots \mathrm{d}z_{d-1} < \infty .$$

Remark 13.19. *For small space dimensions d, the variance of S can be determined explicitly. For example,*

$$12\int_0^1\int_0^1 \frac{(1-z_1)(1-z_2)}{z_1+z_2}\,dz_1\,dz_2 = 4(4\log 2 - 1)$$

if $d = 3$, *or*

$$32\int_0^1\int_0^1\int_0^1 \frac{(1-z_1)(1-z_2)(1-z_3)}{z_1+z_2+z_3}\,dz_1\,dz_2\,dz_3 = \frac{4}{5}(81\log 3 - 112\log 2 - 5)$$

if $d = 4$.

Remark 13.20. *There are alternative representations for the variance integral in Theorem* 13.18. *Namely, we can use that*

$$\frac{1}{\sum_{i=1}^{d-1} z_i} = \int_0^\infty e^{-t\sum_{i=1}^{d-1} z_i}\,dt$$

to factorize the $(d-1)$*-fold integral. Together with* $\int_0^1 (1-z)\,e^{-tz}\,dz = \frac{1}{t^2}(t-1+e^{-t})$, *we obtain*

$$\operatorname{var} S = d\,2^{d-1}\int_0^\infty \left(\frac{t-1+e^{-t}}{t^2}\right)^{d-1} dt\,.$$

Alternatively, using the substitution $u = e^{-t}$, *we get the representation*

$$\operatorname{var} S = d\,2^{d-1}\int_0^1 \left(\frac{u-1-\log u}{(\log u)^2}\right)^{d-1} \frac{1}{u}\,du\,.$$

Proof of Theorem 13.18. The proof is based on the martingale convergence theorem for square-integrable martingales. To be precise, according to Corollary 10.12, the random process $\Sigma := (\overline{\Sigma}_{V_{d-1},a,W},\ a \ge 0)$ is a martingale with respect to the augmented filtration $\mathfrak{I} = (\mathfrak{I}_a,\ a \ge 0)$ induced by the STIT tessellation process $(Y_{a,W},\ a \ge 0)$ within W. The first step is to show that Σ is square-integrable for $d \ge 3$.

Step 1: Square integrability of the surface content martingale Σ

The goal is to prove that

$$\sup_{a\ge 0} \mathbb{E}\overline{\Sigma}^2_{V_{d-1},a,W} < \infty\,. \tag{13.29}$$

To see this, recall that for any $a \ge 0$,

$$\begin{aligned}\mathbb{E}\overline{\Sigma}^2_{V_{d-1},a,W} &= \operatorname{var}\Sigma_{V_{d-1},a,W}\\ &= \int\limits_{[W]}\int\limits_{W\cap H}\int\limits_{W\cap H}\frac{1-\exp(-a\,\Theta([xy]))}{\Theta([xy])}\,\lambda_{d-1}(\mathrm{d}x)\,\lambda_{d-1}(\mathrm{d}y)\,\Theta(\mathrm{d}H) \end{aligned}\tag{13.30}$$

according to Theorem 12.3. Since $W = [-\frac{1}{2}, \frac{1}{2}]^d$ and Θ is concentrated on the set of those $H \in [W]$ for which $H = e_i^\perp + ue_i$ with $i \in \{1, \dots, d\}$ and some $u \in [-\frac{1}{2}, \frac{1}{2}]$ by the definition of Θ each intersection $W \cap H$ is a copy of a $(d-1)$-dimensional unit cube in $\mathbb{R}^d$. As a consequence, the value of the inner double integral does not depend on the choice of such H. Moreover,

$$\Theta([W]) = \frac{1}{d}\sum_{i=1}^{d}\int\limits_{\mathbb{R}}\mathbb{1}\{(e_i^\perp + ue_i)\cap W \neq \emptyset\}\,du = \frac{1}{d}\sum_{i=1}^{d}\int\limits_{-\frac{1}{2}}^{\frac{1}{2}}\mathrm{d}z = 1\,,$$

and we obtain

$$\begin{aligned}\mathbb{E}\overline{\Sigma}^2_{V_{d-1},a,W} &= \int\limits_{[-\frac{1}{2},\frac{1}{2}]^{d-1}\times\{0\}}\ \int\limits_{[-\frac{1}{2},\frac{1}{2}]^{d-1}\times\{0\}}\frac{1-\exp(-a\,\Theta([xy]))}{\Theta([xy])}\,\lambda_{d-1}(\mathrm{d}x)\,\lambda_{d-1}(\mathrm{d}y)\\ &\leq \int\limits_{[-\frac{1}{2},\frac{1}{2}]^{d-1}\times\{0\}}\ \int\limits_{[-\frac{1}{2},\frac{1}{2}]^{d-1}\times\{0\}}\frac{1}{\Theta([xy])}\,\lambda_{d-1}(\mathrm{d}x)\,\lambda_{d-1}(\mathrm{d}y)\,. \end{aligned}\tag{13.31}$$

It remains to show that the integral in the last line is finite. To see this, we start by noting that for $x = (x_1, \dots, x_{d-1}, 0)$ and $y = (y_1, \dots, y_{d-1}, 0)$ in $[-\frac{1}{2}, \frac{1}{2}]^{d-1} \times \{0\}$,

$$\Theta([xy]) = \frac{1}{d}\sum_{i=1}^{d-1}|x_i - y_i|\,.$$

In other words, this means that (13.31) is the same as d times the moment of order -1 of the 1-distance $\|X - Y\|_1 := \sum_{i=1}^{d-1}|X_i - Y_i|$ of two uniformly distributed random points $X = (X_1, \dots, X_{d-1})$ and $Y = (Y_1, \dots, Y_{d-1})$ in a $(d-1)$-dimensional unit cube, say $[-\frac{1}{2}, \frac{1}{2}]^{d-1}$:

$$\int\limits_{[-\frac{1}{2},\frac{1}{2}]^{d-1}\times\{0\}}\ \int\limits_{[-\frac{1}{2},\frac{1}{2}]^{d-1}\times\{0\}}\frac{1}{\Theta([xy])}\,\lambda_{d-1}(\mathrm{d}x)\,\lambda_{d-1}(\mathrm{d}y) = \mathbb{E}\frac{d}{\|X - Y\|_1}\,.$$

Since the coordinates $X_1, \dots, X_{d-1}$ of X and $Y_1, \dots, Y_{d-1}$ of Y are independent and uniformly distributed on $[0, 1]$, for each $i \in \{1, \dots, d-1\}$, the random variable $|X_i - Y_i|$ has the density $z_i \mapsto 2(1 - z_i)$ on $[0, 1]$. Thus

$$\mathbb{E}\frac{d}{\|X - Y\|_1} = d\,2^{d-1}\int\limits_0^1\cdots\int\limits_0^1\frac{\prod_{i=1}^{d-1}(1 - z_i)}{\sum_{i=1}^{d-1}z_i}\,\mathrm{d}z_1\dots\mathrm{d}z_{d-1}\,.\tag{13.32}$$

To show that the $(d-1)$-fold integral is finite, we apply the arithmetic-geometric mean inequality, which says that

$$\frac{1}{d-1}\sum_{i=1}^{d-1} z_i \geq \left(\prod_{i=1}^{d-1} z_i\right)^{1/(d-1)}.$$

Hence

$$\begin{aligned}
\mathbb{E}\frac{d}{\|X-Y\|_1} &\leq \frac{d\,2^{d-1}}{d-1}\int_0^1 \cdots \int_0^1 \frac{\prod_{i=1}^{d-1}(1-z_i)}{(\prod_{i=1}^{d-1} z_i)^{1/(d-1)}}\,\mathrm{d}z_1 \dots \mathrm{d}z_{d-1} \\
&= \frac{d\,2^{d-1}}{d-1}\int_0^1 \cdots \int_0^1 \prod_{i=1}^{d-1} \frac{1-z_i}{z_i^{1/(d-1)}}\,\mathrm{d}z_1 \dots \mathrm{d}z_{d-1} \\
&\leq 2^d \int_0^1 \cdots \int_0^1 \prod_{i=1}^{d-1} \frac{1}{z_i^{1/(d-1)}}\,\mathrm{d}z_1 \dots \mathrm{d}z_{d-1} \\
&= 2^d \left(\int_0^1 \frac{1}{z^{1/(d-1)}}\,\mathrm{d}z\right)^{d-1} \\
&= 2^d\left(1+\frac{1}{d-2}\right)^{d-1},
\end{aligned}$$

where we used that $\frac{d}{d-1} \leq 2$, $1-z_i \leq 1$ for each $i \in \{1,\dots,d-1\}$, and in the last line, our assumption $d \geq 3$. This proves (13.29). At the same time the above computations also yield

$$\lim_{a\to\infty} \operatorname{var} \Sigma_{V_{d-1},a,W} = \mathbb{E}\frac{d}{\|X-Y\|_1},$$

and since the latter is given by (13.32), we conclude that

$$\lim_{a\to\infty} \operatorname{var} \Sigma_{V_{d-1},a,W} = d\,2^{d-1}\int_0^1 \cdots \int_0^1 \frac{\prod_{i=1}^{d-1}(1-z_i)}{\sum_{i=1}^{d-1} z_i}\,\mathrm{d}z_1 \dots \mathrm{d}z_{d-1}. \tag{13.33}$$

Step 2: Martingale convergence and construction of the limiting random variable *S*
The next step starts with the observation that the scaling property (6.2) of STIT tessellations together with the homogeneity of V_{d-1} implies that

$$a^{-(d-1)}\overline{\Sigma}_{V_{d-1},1,aW} \overset{\mathrm{D}}{=} \overline{\Sigma}_{V_{d-1},a,W}. \tag{13.34}$$

By (13.29) the random process $\Sigma = (\overline{\Sigma}_{V_{d-1},a,W},\ a \geq 0)$ is a square-integrable martingale. Consequently, by the martingale convergence theorem (see Corollary 9.23 of [89]) there exists a centered square-integrable random variable $S := S(W)$ such that

$$S = \lim_{a\to\infty} \overline{\Sigma}_{V_{d-1},a,W} \tag{13.35}$$

a. s. and in L^2. The latter and (13.33) imply that

$$\operatorname{var} S = \lim_{a\to\infty} \mathbb{E}\overline{\Sigma}^2_{V_{d-1},a,W} = d\,2^{d-1} \int_0^1 \cdots \int_0^1 \frac{\prod_{i=1}^{d-1}(1-z_i)}{\sum_{i=1}^{d-1} z_i}\, dz_1 \ldots dz_{d-1}\,.$$

In particular, $a^{-(d-1)}\overline{\Sigma}_{V_{d-1},1,aW}$ converges in distribution as $a \to \infty$ to the random variable S.

Step 3: Non-Gaussian tails of the limiting random variable S

It remains to show that S cannot be a Gaussian random variable. We do this by analyzing the tail decay of S and argue that the tails of S are much heavier in comparison to those of a centered Gaussian distribution. To do so, we use the description of the STIT tessellation process as explained in Remark 4.8, fix $k \in \mathbb{N}$, and introduce the following three events:

- B_k is the event that exactly k hyperplanes are born during the STIT construction in W in the time interval $[0,1]$, all of which are orthogonal to the coordinate axis E_1 spanned by e_1;
- $B_k^{(1)}$ is the event that the first $k+1$ divisions of W in the STIT construction are orthogonal to the coordinate axis E_1 spanned by e_1;
- $B_k^{(2)}$ is the event that exactly k hyperplanes are born during the STIT construction in W in the time interval $[0,1]$.

Then

$$\mathbb{P}(B_k) \geq \mathbb{P}(B_k^{(2)} \mid B_k^{(1)})\mathbb{P}(B_k^{(1)})\,. \tag{13.36}$$

To determine the probability of the event $B_k^{(1)}$, we argue inductively. Of course, the probability that the first division of W arises by a hyperplane orthogonal to E_1 is just $1/d$, since we have assumed the weights of the Mondrian hyperplane measure to be all equal to $1/d$. Suppose now that during the STIT construction within W, precisely $m < k+1$ cell divisions have already been realized, all of which are orthogonal to E_1. This corresponds to an event C_m, say. We determine the probability that given C_m, the next division will be orthogonal to E_1 as well. Under C_m, the cube W is divided into $m+1$ orthogonal parallelepipeds $W_1, \ldots, W_{m+1} \subset W$. For $j \in \{1, \ldots, m+1\}$, the side lengths of W_j are $\ell_j, 1, \ldots, 1$ for some $\ell_j \in (0,1)$. By the definition of the Mondrian hyperplane measure Θ we thus have that

$$\begin{aligned}\Theta([W_j]) &= \frac{1}{d}\sum_{i=1}^{d}\int_{\mathbb{R}} \mathbb{1}\{(e_i^\perp + ze_i)\cap W_j \neq \emptyset\}\,dz \\ &= \frac{1}{d}\bigg(\int_{\mathbb{R}} \mathbb{1}\{(e_1^\perp + ze_1)\cap W_j \neq \emptyset\}\,dz + \sum_{i=2}^{d}\int_{\mathbb{R}} \mathbb{1}\{(e_i^\perp + ze_i)\cap W_j \neq \emptyset\}\,dz\bigg) \\ &= \frac{1}{d}(\ell_j + d - 1)\,. \end{aligned} \tag{13.37}$$

Given C_m, the probability that the $(m+1)$th cell division in the STIT construction divides cell W_j is

$$\begin{aligned}\frac{\Theta([W_j])}{\Theta([W_1]) + \cdots + \Theta([W_{m+1}])} &= \frac{\ell_j + d - 1}{(\ell_1 + d - 1) + \cdots + (\ell_{m+1} + d - 1)} \\ &= \frac{\ell_j + d - 1}{1 + (m+1)(d-1)}\,,\end{aligned}$$

since $\ell_1 + \cdots + \ell_{m+1} = 1$. Moreover, given that W_j gets divided, the probability that the hyperplane that divides W_j is orthogonal to E_1 is

$$\frac{\frac{1}{d}\ell_j}{\Theta([W_j])} = \frac{\ell_j}{\ell_j + d - 1}\,.$$

As a consequence, the conditional probability of C_{m+1} given C_m is the same as

$$\begin{aligned}\mathbb{P}(C_{m+1} \mid C_m) &= \sum_{j=1}^{m+1} \frac{\ell_j + d - 1}{1 + (m+1)(d-1)}\,\frac{\ell_j}{\ell_j + d - 1} \\ &= \sum_{j=1}^{m+1} \frac{\ell_j}{1 + (m+1)(d-1)} \\ &= \frac{1}{1 + (m+1)(d-1)}\,,\end{aligned}$$

where we used once again that $\ell_1 + \cdots + \ell_{m+1} = 1$. Since the sequence of events C_m is decreasing, we arrive at

$$\begin{aligned}\mathbb{P}(B_k^{(1)}) &= \mathbb{P}(C_1 \cap \cdots \cap C_{k+1}) \\ &= \mathbb{P}(C_1)\mathbb{P}(C_2 \mid C_1)\cdots\mathbb{P}(C_{k+1}|C_k) = \prod_{i=1}^{k+1}\frac{1}{1 + i(d-1)}\,.\end{aligned}$$

To deal with the conditional probability of the event $B_k^{(2)}$, given $B_k^{(1)}$, recall from (4.17) the definition of the θ-value of a tessellation $T \in \mathcal{T}_W$. Suppose, as above, that T_m is a tessellation of W into $m+1$ orthogonal parallelepipeds $W_1, \ldots, W_{m+1}$ and that, for some $j \in \{1, \ldots, m+1\}$, $H \in [W_j]$ is a hyperplane orthogonal to E_1. The tessellation T_{m+1} arises from T_m by split of the cell W_j using the hyperplane H, that is, $T_{m+1} = \oslash(W_j, H, T_m)$. We

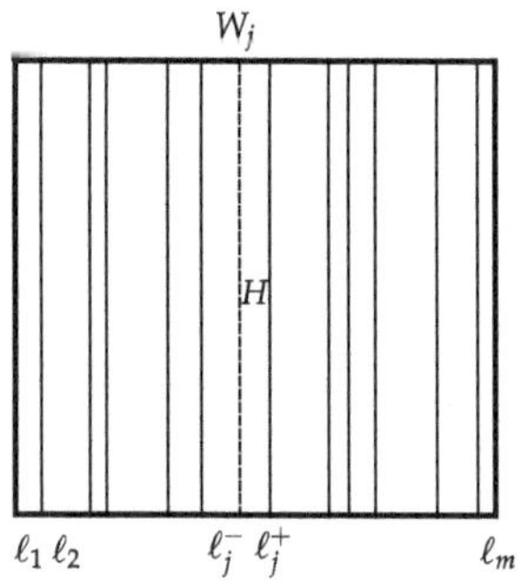

Figure 13.1: Illustration of the argument leading to $\theta(T_{m+1})$.

denote by W_j^+ and W_j^- the two subcells into which W_j gets divided by H and by ℓ_j, ℓ_j^+, and ℓ_j^- their side lengths on E_1, respectively. Then using (13.37), we have that

$$\begin{aligned}\theta(T_{m+1}) &= \sum_{P \in T_{m+1}} \Theta([P]) \\ &= \theta(T_m) - \Theta([W_j]) + \Theta([W_j^+]) + \Theta([W_j^-]) \\ &= \theta(T_m) - \frac{1}{d}(\ell_j + d - 1) + \frac{1}{d}(\ell_j^+ + d - 1) + \frac{1}{d}(\ell_j^- + d - 1) \\ &= \theta(T_m) + 1 - \frac{1}{d},\end{aligned}$$

since $\ell_j^+ + \ell_j^- = \ell_j$; see Figure 13.1. Also, if T_0 is the tessellation consisting of the single cell W, then $\theta(T_0) = 1$. Conditional on $B_k^{(1)}$, the number of divisions up to time t evolves like a linear birth process, also called Yule process, with rate $1 - 1/d$. In particular, the total number of divisions in $[0, 1]$ has the same distribution as the state of this birth process at time 1. Since the latter random variable follows a geometric distribution with parameter $e^{-(1-\frac{1}{d})}$ (see page 235 of [167]), we have that

$$\mathbb{P}(B_k^{(2)} \mid B_k^{(1)}) = e^{-(1-\frac{1}{d})}(1 - e^{-(1-\frac{1}{d})})^k$$

Plugging all this back into (13.36), we see that

$$\mathbb{P}(B_k) \geq e^{-(1-\frac{1}{d})}(1 - e^{-(1-\frac{1}{d})})^k \prod_{i=1}^{k+1} \frac{1}{1 + i(d-1)}.$$

Finally, we observe that $(k+1)! \leq (k+1)^{k+1}$ implies that

$$\prod_{i=1}^{k+1} \frac{1}{1 + i(d-1)} \geq \prod_{i=1}^{k+1} \frac{1}{i + i(d-1)} = \frac{1}{d^{k+1}} \frac{1}{(k+1)!} \geq \left(\frac{1}{d(k+1)}\right)^{k+1},$$

which leads to the lower bound

$$\mathbb{P}(B_k) \geq e^{-(1-\frac{1}{d})}(1 - e^{-(1-\frac{1}{d})})^k \left(\frac{1}{d(k+1)}\right)^{k+1} =: A_k.$$

Taking logarithms, we obtain

$$\log A_k = -\left(1 - \frac{1}{d}\right) + k\log(1 - e^{-(1-\frac{1}{d})}) - (k+1)\log d - (k+1)\log(k+1)\,.$$

All terms except the last are $O(k)$, while $-(k+1)\log(k+1) = -J(k\log k)$, as $k \to \infty$, where we recall that J stands for a function that is bounded from above and below by a constant multiple of the argument. Thus we arrive at

$$\mathbb{P}(B_k) \geq e^{-J(k\log k)}\,. \tag{13.38}$$

Conditionally on the event B_k, let $H_1, \ldots, H_k \in [W]$ be the hyperplanes orthogonal to E_1 that are born within the time interval $[0,1]$. These hyperplanes partition the cube W into $k+1$ orthogonal parallelepipeds $W_1, \ldots, W_{k+1}$ with side lengths $\ell_1, \ldots, \ell_{k+1}$ along the E_1-axis. On B_k the centered surface content can therefore be written as

$$S \stackrel{D}{=} (k-1) + \sum_{i=1}^{k+1} S_i\,, \tag{13.39}$$

where, conditional on the tessellation of W generated by $H_1, \ldots, H_k$, the random variables $S_1, \ldots, S_{k+1}$ are independent with $S_i \stackrel{D}{=} S(W_i)$. The shift $k-1$ corresponds to

$$V_{d-1}(H_1 \cap W) + \cdots + V_{d-1}(H_k \cap W) - V_d(W)\,,$$

the first k summands being the total $(d-1)$-volume of the k inserted hyperplanes and the second the centering term $-\mathbb{E}\Sigma_{V_{d-1},1,W}$.

Formally, for each $i \in \{1, \ldots, k+1\}$, we have

$$S_i = \lim_{a\to\infty} \overline{\Sigma}_{V_{d-1},a,W_i}\,,$$

as in (13.35). Given B_k, the conditional variance of S_i is

$$\begin{aligned}\operatorname{var}(S_i \mid B_k) &= \lim_{a\to\infty} \operatorname{var} \Sigma_{V_{d-1},a,W_i}\\ &= \int\limits_{[W_i]} \int\limits_{W_i\cap H} \int\limits_{W_i\cap H} \frac{1}{\Theta([xy])}\,\lambda_{d-1}(dx)\,\lambda_{d-1}(dy)\,\Theta(dH)\,.\end{aligned}$$

By decomposing according to the orientation of H we obtain

$$\begin{aligned}\operatorname{var}(S_i \mid B_k) &= \frac{\ell_i}{d} \int\limits_{[-\frac12,\frac12]^{d-1}\times\{0\}} \int\limits_{[-\frac12,\frac12]^{d-1}\times\{0\}} \frac{1}{\Theta([xy])}\,\lambda_{d-1}(dx)\,\lambda_{d-1}(dy)\\ &+ \frac{d-1}{d} \int\limits_{[0,\ell_i]\times[-\frac12,\frac12]^{d-2}\times\{0\}} \int\limits_{[0,\ell_i]\times[-\frac12,\frac12]^{d-2}\times\{0\}} \frac{1}{\Theta([xy])}\,\lambda_{d-1}(dx)\,\lambda_{d-1}(dy)\,.\end{aligned} \tag{13.40}$$

The first term equals $\frac{\ell_i}{d}$ var$S$, since

$$\begin{aligned}
&\frac{\ell_i}{d} \int\limits_{[-\frac{1}{2},\frac{1}{2}]^{d-1}\times\{0\}} \int\limits_{[-\frac{1}{2},\frac{1}{2}]^{d-1}\times\{0\}} \frac{1}{\Theta([xy])} \lambda_{d-1}(\mathrm{d}x)\,\lambda_{d-1}(\mathrm{d}y) \\
&= \frac{\ell_i}{d} d\, 2^{d-1} \int_0^1 \cdots \int_0^1 \frac{\prod_{i=1}^{d-1}(1-z_i)}{\sum_{i=1}^{d-1} z_i}\, \mathrm{d}z_1 \ldots \mathrm{d}z_{d-1} \\
&= \frac{\ell_i}{d} \operatorname{var} S\,.
\end{aligned}$$

For the second, we use that X_1 and Y_1 are uniformly distributed on $[0, \ell_i]$ and X_j and Y_j are uniformly distributed on $[0, 1]$ for $j \in \{2, \ldots, d\}$. Noting that in this case the density of $|X_1 - Y_1|$ is given by $z_1 \mapsto \frac{2}{\ell_i^2}(\ell_i - z_1)$, this gives

$$\begin{aligned}
&\frac{d-1}{d} \int\limits_{[0,\ell_i]\times[-\frac{1}{2},\frac{1}{2}]^{d-2}\times\{0\}} \int\limits_{[0,\ell_i]\times[-\frac{1}{2},\frac{1}{2}]^{d-2}\times\{0\}} \frac{1}{\Theta([xy])} \lambda_{d-1}(\mathrm{d}x)\,\lambda_{d-1}(\mathrm{d}y) \\
&= \frac{d-1}{d} \ell_i^2\, d \int_0^{\ell_i} \int_0^1 \cdots \int_0^1 \frac{2(\ell_i - z_1)}{\ell_i^2} \frac{\prod_{i=2}^{d-1} 2(1-z_i)}{\sum_{i=1}^{d-1} z_i}\, \mathrm{d}z_{d-1} \ldots \mathrm{d}z_2\, \mathrm{d}z_1 \\
&= \ell_i^2 (d-1) 2^{d-1} \int_0^1 \cdots \int_0^1 \frac{\prod_{i=1}^{d-1}(1-z_i)}{z_1\ell_i + z_2 + \cdots + z_{d-1}}\, \mathrm{d}z_1 \ldots \mathrm{d}z_{d-1}\,,
\end{aligned}$$

where in the last step, we applied the substitution $z_1 = u\ell_i$ and then replaced u by the symbol z_1 again. Since $\ell_i \in (0, 1)$, the denominator satisfies $z_1\ell_i + z_2 + \cdots + z_{d-1} \geq \ell_i(z_1 + \cdots + z_{d-1})$, and thus

$$\begin{aligned}
&\frac{d-1}{d} \int\limits_{[0,\ell_i]\times[-\frac{1}{2},\frac{1}{2}]^{d-2}\times\{0\}} \int\limits_{[0,\ell_i]\times[-\frac{1}{2},\frac{1}{2}]^{d-2}\times\{0\}} \frac{1}{\Theta([xy])} \lambda_{d-1}(\mathrm{d}x)\,\lambda_{d-1}(\mathrm{d}y) \\
&\leq \ell_i (d-1) 2^{d-1} \int_0^1 \cdots \int_0^1 \frac{\prod_{i=1}^{d-1}(1-z_i)}{\sum_{i=1}^{d-1} z_i}\, \mathrm{d}z_1 \ldots \mathrm{d}z_{d-1} \\
&= \frac{d-1}{d} \ell_i \operatorname{var} S\,.
\end{aligned}$$

Putting these estimates back into (13.40), we obtain

$$\operatorname{var}(S_i \mid B_k) \leq \ell_i \left(\frac{1}{d} + \frac{d-1}{d} \right) \operatorname{var} S = \ell_i \operatorname{var} S\,.$$

Because $\ell_1 + \cdots + \ell_{k+1} = 1$, it follows that

$$\operatorname{var}\left(\sum_{i=1}^{k+1} S_i \mid B_k \right) \leq \operatorname{var} S < \infty\,.$$

By Chebyshev's inequality,

$$\mathbb{P}\left(\sum_{i=1}^{k+1} S_i \geq -2\sqrt{\operatorname{var} S} \mid B_k\right) \geq 1 - \frac{\operatorname{var}(\sum_{i=1}^{k+1} S_i \mid B_k)}{4 \operatorname{var} S} \geq \frac{3}{4}, \tag{13.41}$$

uniformly in k.

Now let $k = \lceil N + 1 + 2\sqrt{\operatorname{var} S}\rceil$. Combining (13.38), (13.39), and (13.41), we arrive at

$$\begin{aligned}\mathbb{P}(S \geq N) &\geq \mathbb{P}\left(\sum_{i=1}^{k+1} S_i \geq -2\sqrt{\operatorname{var} S} \mid B_k\right)\mathbb{P}(B_k)\\ &\geq \frac{3}{4}\,\mathrm{e}^{-J(N \log N)}\\ &= \mathrm{e}^{-J(N \log N)}\end{aligned}$$

as $N \to \infty$, where the constant factor $\frac{3}{4}$ is absorbed into the $J(\cdot)$ notation. Since Gaussian random variables $G \sim \mathrm{Norm}(0, \sigma^2)$ with $\sigma^2 > 0$ satisfy

$$\mathbb{P}(G \geq N) = \mathrm{e}^{-J(N^2)}, \quad N \to \infty,$$

we conclude that S has significantly heavier tails than a Gaussian distribution. In particular, S cannot be Gaussian. This completes the proof. □

Remark 13.21. *We emphasize that the proof of the previous theorem cannot be extended to the case $d = 2$. In particular, in the planar case the integral on the right-hand side of* (13.33) *diverges, implying that the martingale convergence theorem is not applicable. In fact, in the planar case, the intersections $W \cap H$ in the variance formula* (13.30) *are just copies of the interval* $[0,1]$. *However, for X and Y independent and uniformly distributed on* $[0,1]$,

$$\mathbb{E}\frac{1}{\|X - Y\|_1} = \int_0^1\int_0^1 \frac{1}{|x - y|}\,\mathrm{d}x\,\mathrm{d}y = \infty,$$

implying that in the Mondrian case, $\operatorname{var} \Sigma_{V_1,t,W} < \infty$ for each $t \geq 0$, but $\sup_{t\geq 0} \operatorname{var} \Sigma_{V_1,t,W} = \infty$. This parallels the same phenomenon in the isotropic case, which was treated in Chapter 12; see especially Remark 12.14.

Theorem 13.18 deals with a STIT tessellation with time parameter $t = 1$. The next result extends this limit theorem to any $t > 0$.

Corollary 13.22. *Fix $d \geq 3$ and consider the Mondrian STIT tessellation process $(Y_t, t > 0)$ with equal weights $p_1 = \cdots = p_d = 1/d$. Then for all $t > 0$ and for $W = [-\frac{1}{2}, \frac{1}{2}]^d$,*

$$a^{-(d-1)}\overline{\Sigma}_{V_{d-1},t,aW} \overset{\text{law}}{\Longrightarrow} S \quad \text{as } a \to \infty,$$

where S is the same non-Gaussian centered random variable as in Theorem 13.18, *with variance given by* (13.33).

Proof. The scaling property (6.2) of STIT tessellations, together with the homogeneity of V_{d-1}, implies that

$$a^{-(d-1)}\overline{\Sigma}_{V_{d-1},t,aW} \stackrel{\mathrm{D}}{=} \overline{\Sigma}_{V_{d-1},ta,W}\,.$$

Since $ta \to \infty$ as $a \to \infty$ for any fixed t, the limit law is the same as in Theorem 13.18. In particular, the distribution of the limit does not depend on t and equals that of S. □

13.4.2 Total surface content: the isotropic case

The purpose of the present section is to extend the result of Theorem 13.18 to the isotropic case by means of a perturbation argument. Roughly speaking, the idea is to show that with sufficiently high probability we can construct a configuration that conceptually mimics that of the Mondrian case, which in turn was sufficient to establish that the limiting random variable has non-Gaussian tail decay. To ease comparison with the proof of Theorem 13.18, we will use the same notation. In particular, we choose again the d-dimensional cube as the observation window.

Theorem 13.23. *Fix $d \geq 3$ and consider the STIT tessellation process $(Y_t,\, t > 0)$ driven by the motion-invariant hyperplane measure Θ_{iso}. Then, for $W = [-\frac{1}{2}, \frac{1}{2}]^d$,*

$$a^{-(d-1)}\overline{\Sigma}_{V_{d-1},1,aW} \stackrel{\text{law}}{\Longrightarrow} S \quad \text{as } a \to \infty\,,$$

where S is a non-Gaussian centered random variable with variance

$$\operatorname{var} S = \frac{d-1}{2} \int\limits_W \int\limits_W \|x-y\|^{-2}\, \lambda_d(\mathrm{d}x)\, \lambda_d(\mathrm{d}y) < \infty\,.$$

Proof. As in the proof of Theorem 13.18, we start by proving that

$$\sup_{a \geq 0} \mathbb{E}\overline{\Sigma}^2_{V_{d-1},a,W} < \infty\,. \tag{13.42}$$

To see this, we use Theorem 12.8, which yields

$$\begin{aligned}
\mathbb{E}\overline{\Sigma}^2_{V_{d-1},a,W} &= \operatorname{var} \Sigma_{V_{d-1},a,W} \\
&= \frac{d-1}{2} \int\limits_W \int\limits_W \frac{1-\exp(-\frac{2\kappa_{d-1}}{d\kappa_d} a\|x-y\|)}{\|x-y\|^2}\, \lambda_d(\mathrm{d}x)\, \lambda_d(\mathrm{d}y)
\end{aligned}$$

$$\leq \frac{d-1}{2} \int_W \int_W \|x-y\|^{-2}\,\lambda_d(\mathrm{d}x)\,\lambda_d(\mathrm{d}y)\,.$$

According to Lemma 12.12, since $d \geq 3$, we have that

$$\int_W \int_W \|x-y\|^{-2}\,\lambda_d(\mathrm{d}x)\,\lambda_d(\mathrm{d}y) = \frac{1}{(d-1)(d-2)} I_{d-1}(W) < \infty\,,$$

where we recall that $I_{d-1}(W)$ is the chord power integral of W of order $d-1$. This proves (13.42).

We can now apply the martingale convergence theorem (see Corollary 9.23 of [89]) together with the scaling relation (13.34) to conclude that $a^{-(d-1)}\overline{\Sigma}_{V_{d-1},1,aW} \overset{\mathrm{D}}{=} \overline{\Sigma}_{V_{d-1},a,W}$ converges a. s. and in L^2 as $a \to \infty$ to a centered random variable $S = S(W)$ with variance

$$\operatorname{var} S = \lim_{a\to\infty} \operatorname{var} \Sigma_{V_{d-1},a,W} = \frac{d-1}{2} \int_W \int_W \|x-y\|^{-2}\,\lambda_d(\mathrm{d}x)\,\lambda_d(\mathrm{d}y)\,.$$

To show that S is non-Gaussian, we analyze the tail decay of S again and prove that this is much heavier than for any centered Gaussian random variable. We fix $k \in \mathbb{N}$, denote $E_1 = \operatorname{lin} e_1$ with $e_1 = (1, 0, \ldots, 0)$, partition $E_1 \cap W$ into $2k$ intervals $I_1, \ldots, I_{2k}$ of length $\frac{1}{2k}$, and define the orthogonal parallelepipeds $P_j := I_j \times [-\frac{1}{2}, \frac{1}{2}]^{d-1}$ for $j \in \{1, \ldots, 2k\}$. Then we introduce the following event B_k (see the left panel of Figure 13.2):

- exactly k hyperplanes $H_1, \ldots, H_k$ are born during time $[0, 1]$;
- each intersection $H_i \cap W$ is contained in P_{2i} for $i \in \{1, \ldots, k\}$ and does not intersect the two facets orthogonal to E_1 that bound P_{2i}.

Note that the second item implies the following two crucial geometric properties:

1. Any two intersections $W \cap H_{i_1}$ and $W \cap H_{i_2}$ with $i_1, i_2 \in \{1, \ldots, k\}$ and $i_1 \neq i_2$ are disjoint;
2. The intersection volumes satisfy $V_{d-1}(W \cap H_i) \geq 1$ for $i \in \{1, \ldots, k\}$.

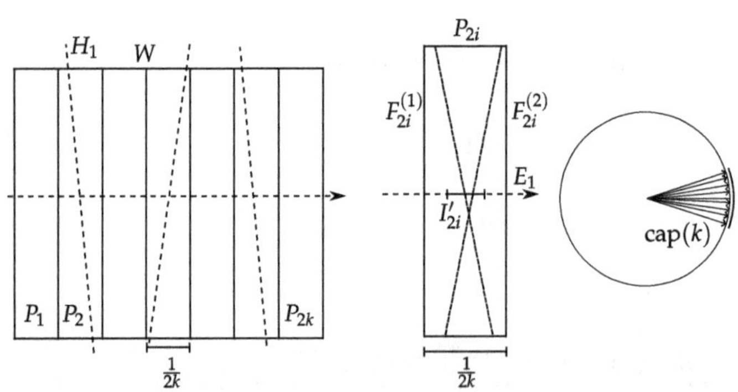

Figure 13.2: Left: Illustration of the event B_k. Right: Illustration of the argument leading to (13.43).

To estimate from below the probability of the event B_k, let $F_{2i}^{(1)}$ and $F_{2i}^{(2)}$ be the facets orthogonal to E_1 that bound the parallelepiped P_{2i}, $i \in \{1, \dots, k\}$. We start by dealing with the quantity

$$\theta_k := \Theta_{\text{iso}}\big(\{H \in [P_{2i}] : H \cap F_{2i}^{(1)} = \emptyset \text{ and } H \cap F_{2i}^{(2)} = \emptyset\}\big) .$$

Let $I'_{2i} \subset I_{2i}$ be the interval of length $\frac{1}{2}\lambda_1(I_{2i}) = \frac{1}{4k}$ having the same center as I_{2i}. Let $\mathsf{cap}(k)$ be the spherical cap centered at e_1 with angular radius $\frac{\pi}{2} - \arctan(4k\sqrt{d-1})$. The choice of the constant $4k\sqrt{d-1}$ is minimal in the following sense: write the unit normal of H as $u = (u_1, u_\perp)$ with $u_1 = \langle u, e_1\rangle$ and remaining coordinates gathered in $u_\perp := u|e_1^\perp$. If the normal vector makes an angle α with e_1, then $\tan\alpha = \|u_\perp\|/|u_1|$. For $x, x' \in H$, we have $u_1(x_1 - x_1') + \langle u_\perp, x_\perp - x_\perp'\rangle = 0$, and hence

$$|x_1 - x_1'| \leq \|x_\perp - x_\perp'\| \tan\alpha ,$$

where we write $x = (x_1, x_\perp)$ and $x' = (x_1', x_\perp')$. Since the facets orthogonal to E_1 have half-diagonal $\frac{1}{2}\sqrt{d-1}$, any hyperplane intersecting I'_{2i} can drift in the e_1–direction by at most $\tan\alpha \cdot \frac{1}{2}\sqrt{d-1}$ across the facet. On the other hand, the distance from $\partial I'_{2i}$ to ∂I_{2i} equals $\frac{1}{8k}$ on each side. Thus, to guarantee that a hyperplane intersecting I'_{2i} cannot reach either of the two facets orthogonal to E_1, we require

$$\tan\alpha \frac{1}{2}\sqrt{d-1} \leq \frac{1}{8k}, \quad \text{that is,} \ \tan\alpha \leq \frac{1}{4k\sqrt{d-1}},$$

which is exactly enforced by $\alpha \leq \arctan(\frac{1}{4k\sqrt{d-1}}) = \frac{\pi}{2} - \arctan(4k\sqrt{d-1})$.

Then any hyperplane $H \in [P_{2i}]$ satisfying $H \cap I'_{2i} \neq \emptyset$ and $H^\perp \cap \mathsf{cap}(k) \neq \emptyset$ does not intersect the facets $F_{2i}^{(1)}$ and $F_{2i}^{(2)}$; see the right panel of Figure 13.2. Thus

$$\begin{aligned}&\{H \in [P_{2i}] : H \cap F_{2i}^{(1)} = \emptyset \text{ and } H \cap F_{2i}^{(2)} = \emptyset\}\\ &\quad\supseteq \{H \in [P_{2i}] : H \cap I'_{2i} \neq \emptyset,\ H^\perp \cap \mathsf{cap}(k) \neq \emptyset\} .\end{aligned}$$

Using Lemma 2.3 of [10] to bound from below the spherical Lebesgue measure of $\mathsf{cap}(k)$, we obtain

$$\begin{aligned}\theta_k &\geq \lambda_1(I'_{2i})\, \varphi_{\text{iso}}(\mathsf{cap}(k))\\ &\geq \frac{1}{4k}\frac{1}{(d-1)\kappa_{d-1}}\frac{1}{2^d}\left(\frac{\pi}{2} - \arctan(4k\sqrt{d-1})\right)^{d-1}\\ &\geq \frac{1}{4k}\frac{1}{(d-1)\kappa_{d-1}}\frac{1}{2^d}\left(\frac{1}{8k\sqrt{d-1}}\right)^{d-1}\\ &= J\left(\frac{1}{k^d}\right),\end{aligned} \tag{13.43}$$

writing again $J(\cdot)$ for a function bounded from above and below by multiples of the argument, and where we used that $\frac{\pi}{2} - \arctan x = \arctan(1/x) \geq \frac{1}{2x}$ for $x > 0$.

To provide an estimate from below for $\mathbb{P}(B_k)$, we introduce

- the event $B_k^{(1)}$ that the first k hyperplanes $H_1, \dots, H_k$ appearing in the STIT construction satisfy $H_i \in [P_{2i}]$ and are such that H_i does not intersect the two facets orthogonal to E_1 that bound P_{2i} for $i \in \{1, \dots, k\}$;
- the event $B_k^{(2)}$ that exactly k hyperplanes are born during time $[0, 1]$.

For the probability of $B_k^{(1)}$, note that at each step i the next hyperplane is chosen, conditional on the already selected hyperplanes $H_1, \dots, H_{i-1}$, with law proportional to Θ_{iso}. Hence

$$\mathbb{P}(H_i \in [P_{2i}], \text{ and } H_i \text{ avoids the two bounding facets} \mid H_1, \dots, H_{i-1}) \geq \frac{\theta_k}{\Theta_{\text{iso}}([W])},$$

since the normalizing denominator at step i is at most $\Theta_{\text{iso}}([W])$. Therefore, by iterating this conditional bound,

$$\begin{aligned}\mathbb{P}(B_k^{(1)}) &= \prod_{i=1}^{k} \mathbb{P}(H_i \in [P_{2i}] \text{ and avoids the facets} \mid H_1, \dots, H_{i-1}) \\ &\geq \left(\frac{\theta_k}{\Theta_{\text{iso}}([W])} \right)^k \geq J\left(\frac{1}{k^{dk}} \right),\end{aligned}$$

where we used (13.43) in the last step.

Next, in the isotropic case and conditionally on $B_k^{(1)}$, each cell division by a hyperplane H_i increases the θ-value of the tessellation by at least $\Theta_{\text{iso}}([W \cap H_i]) \geq \Theta_{\text{iso}}([\frac{1}{2}B^{d-1}])$, where B^{d-1} denotes the $(d-1)$-dimensional unit ball in $E_1^{\perp}$. Therefore

$$\mathbb{P}(B_k^{(2)} \mid B_k^{(1)}) \geq \exp(-\Theta_{\text{iso}}([W]))\,(1 - \exp(-c))^k$$

with the constant $c := \Theta_{\text{iso}}([\frac{1}{2}B^{d-1}]) > 0$ independent of k.

Putting these estimates together, we conclude that

$$\mathbb{P}(B_k) \geq \mathbb{P}(B_k^{(2)} \mid B_k^{(1)})\, \mathbb{P}(B_k^{(1)}) \geq J\left((1 - \mathrm{e}^{-c})^k \frac{1}{k^{dk}} \right) = \mathrm{e}^{-J(k \log k)} \tag{13.44}$$

as $k \to \infty$.

Conditionally on B_k and hence on the random hyperplanes $H_1, \dots, H_k$, we let $W_1, \dots, W_{k+1}$ be the $k+1$ cells into which W is partitioned. For each $i \in \{1, \dots, k+1\}$, let

$$S_i := S(W_i) := \lim_{a \to \infty} \overline{\Sigma}_{V_{d-1}, a, W_i}$$

be the a. s. and L^2-limit given by the martingale convergence theorem. Conditionally on B_k, the random variables $S_1, \dots, S_{k+1}$ are independent. Moreover,

$$\begin{aligned}\operatorname{var}(S_i \mid B_k) &= \frac{d-1}{2} \int_{W_i}\int_{W_i} \|x-y\|^{-2}\, \lambda_d(\mathrm{d}x)\, \lambda_d(\mathrm{d}y) \\ &= \frac{d-1}{2} E_2(W_i)\,, \quad i \in \{1,\dots,k+1\}\,,\end{aligned}$$

where $E_2(\cdot)$ denotes the 2-energy; see (12.25).

By the construction of B_k we have

$$V_{d-1}(W \cap H_1) + \cdots + V_{d-1}(W \cap H_k) \geq k\,,$$

so that

$$\mathbb{P}(S \geq z \mid B_k) \geq \mathbb{P}\big(k - V_d(W) + (S_1 + \cdots + S_{k+1}) \geq z \mid B_k\big)\,, \quad z \in \mathbb{R}\,,$$

where $-V_d(W) = -\mathbb{E}\Sigma_{V_{d-1},1,W}$ is the centering term for the total surface content at time 1. Furthermore,

$$\begin{aligned}\operatorname{var}\Big(\sum_{i=1}^{k+1} S_i \mid B_k\Big) &= \frac{d-1}{2} \sum_{i=1}^{k+1} E_2(W_i) \\ &\leq \frac{d-1}{2} E_2\Big(\bigcup_{i=1}^{k+1} W_i\Big) \\ &= \frac{d-1}{2} E_2(W) \\ &= \operatorname{var} S\,,\end{aligned}$$

since E_2 is superadditive, that is, for $K, L \in \mathcal{K}_d$,

$$E_2(K \cup L) = E_2(K) + E_2(L) + 2\int_K\int_L \|x-y\|^{-2}\, \lambda_d(\mathrm{d}x)\, \lambda_d(\mathrm{d}y) \geq E_2(K) + E_2(L)\,.$$

By Chebyshev's inequality we thus obtain

$$\mathbb{P}(S_1 + \cdots + S_{k+1} \geq -2\sqrt{\operatorname{var} S} \mid B_k) \geq 1 - \frac{\operatorname{var}(\sum_{i=1}^{k+1} S_i \mid B_k)}{4 \operatorname{var} S} \geq \frac{3}{4}\,.$$

Choosing $k := \lceil N + V_d(W) + 2\sqrt{\operatorname{var} S}\rceil$, we conclude that

$$\begin{aligned}\mathbb{P}(S \geq N) &\geq \mathbb{P}(S \geq N \mid B_{\lceil N+V_d(W)+2\sqrt{\operatorname{var} S}\rceil})\, \mathbb{P}(B_{\lceil N+V_d(W)+2\sqrt{\operatorname{var} S}\rceil}) \\ &\geq \mathbb{P}(S_1 + \cdots + S_{\lceil N+V_d(W)+2\sqrt{\operatorname{var} S}\rceil+1} \geq -2\sqrt{\operatorname{var} S} \mid B_{\lceil N+V_d(W)+2\sqrt{\operatorname{var} S}\rceil}) \\ &\quad \times \mathbb{P}(B_{\lceil N+V_d(W)+2\sqrt{\operatorname{var} S}\rceil}) \\ &\geq \frac{3}{4} \mathbb{P}(B_{\lceil N+V_d(W)+2\sqrt{\operatorname{var} S}\rceil}) \\ &= \mathrm{e}^{-J(N \log N)}\end{aligned}$$

as $N \to \infty$ by (13.44). Since this tail bound decays more slowly than for any centered Gaussian random variable, the proof is complete. □

Theorem 13.23 shows that the normalized total surface content $a^{-(d-1)}\Sigma_{V_{d-1},1,aW}$ of a stationary and isotropic STIT tessellation at time $t = 1$ in $\mathbb{R}^d$ with $d \geq 3$ converges in distribution to a non-Gaussian limit random variable S. In the next step, we show that the total surface content, considered as a stochastic process in time, converges to the constant process $t \mapsto S$ if the time origin is shifted to $\frac{\log n}{n}$. This is the analogue of Corollary 13.13 and demonstrates that the leading fluctuations of the total surface content in nW for $d \geq 3$ are generated during the initial phase $[0, \frac{\log n}{n}]$, which we refer to as the **big-bang phase** of the STIT construction. We emphasize that this setup differs from that in Section 13.2, where only the surface content of those $(d-1)$-dimensional maximal polytopes is considered that are born after a fixed time threshold $s_0 > 0$, independent of n.

Corollary 13.24. *Fix $d \geq 3$ and let $(Y_t, t > 0)$ be the STIT tessellation process driven by the motion-invariant hyperplane measure Θ_{iso}. For $n \in \mathbb{N}$ and $W = [-\frac{1}{2}, \frac{1}{2}]^d$, define*

$$S_{t,nW} := n^{-(d-1)}\Sigma_{V_{d-1},t+\frac{\log n}{n},nW}, \qquad t \geq 0 .$$

Then

$$(S_{t,nW}, t \geq 0) \overset{\text{law}}{\Longrightarrow} (S, t \geq 0) \quad \text{as } n \to \infty \quad \text{in } \mathsf{D}[0,\infty),$$

where S is the non-Gaussian random variable from Theorem 13.23.

Proof. The proof is divided into two steps.

Step 1: Tightness

We start by proving that the sequence of random processes $(S_{t,nW}, t \geq 0)$, $n \in \mathbb{N}$, is tight in $\mathsf{D}[0,\infty)$. For that purpose, we use Theorem 4.13 of [86], which says that the required tightness property follows from

(a) the tightness in $\mathbb{R}$ of the sequence of random variables $S_{0,nW}$, $n \in \mathbb{N}$,
(b) the tightness in $\mathsf{D}[0,\infty)$ of the sequence of predictable quadratic variation processes $(\langle S_{\cdot,nW}\rangle_t, t \geq 0)$, $n \in \mathbb{N}$, and
(c) the concentration of the limit laws of the sequence of predictable quadratic variation processes $(\langle S_{\cdot,nW}\rangle_t, t \geq 0)$, $n \in \mathbb{N}$, on the space $\mathsf{C}[0,\infty)$ of continuous functions on $[0,\infty)$.

We remark that (b) and (c) together are commonly referred to as C-tightness of $(\langle S_{\cdot,nW}\rangle_t, t \geq 0)$, $n \in \mathbb{N}$, in the literature; see, for example, Definition 3.25 of [86].

To show (a), let $M > 0$, use Chebyshev's inequality, and then Theorem 12.13 to see that

$$\mathbb{P}(|S_{0,nW}| \geq M) \leq \frac{\operatorname{var} S_{0,nW}}{M^2} \leq \frac{\operatorname{var} \Sigma_{V_{d-1},\frac{\log n}{n},nW}}{M^2 n^{2(d-1)}} \leq \frac{C}{M^2}$$

for some constant $C > 0$ and all $n \in \mathbb{N}$. Since the last expression can be made arbitrarily small by choosing M large enough, the tightness property in (a) follows.

To verify (b), we start by observing that

$$\langle S_{\cdot,nW} \rangle_t = \frac{1}{n^{2(d-1)}} \langle \overline{\Sigma}_{V_{d-1},\cdot+\frac{\log n}{n},nW} \rangle_t = \frac{1}{n^{2(d-1)}} \int_{\frac{\log n}{n}}^{t+\frac{\log n}{n}} A_{d-1,s,nW}\, \mathrm{d}s\,,$$

as already established in (13.14). To show that

$$\lim_{n\to\infty} \langle S_{\cdot,nW} \rangle_t = 0 \tag{13.45}$$

for all $t \geq 0$ in probability, it is sufficient to prove that $\mathbb{E}\langle S_{\cdot,nW} \rangle_t \to 0$ as $n \to \infty$. Using the same arguments as in the proof of Theorem 12.3, we find that

$$\begin{aligned}
\mathbb{E}\langle S_{\cdot,nW} \rangle_t &= \frac{1}{n^{2(d-1)}} \int_{\frac{\log n}{n}}^{t+\frac{\log n}{n}} \mathbb{E} A_{d-1,s,nW}\, \mathrm{d}s \\
&= \frac{1}{n^{2(d-1)}} \int_{\frac{\log n}{n}}^{t+\frac{\log n}{n}} \mathbb{E} \int_{[nW]} \sum_{F \in Y_{s,nW} \cap H} V_{d-1}(F)^2\, \Theta_{\mathrm{iso}}(\mathrm{d}H)\, \mathrm{d}s \\
&= \frac{1}{n^{2(d-1)}} \int_{\frac{\log n}{n}}^{t+\frac{\log n}{n}} \int_{[nW]} \int_{nW\cap H} \int_{nW\cap H} \mathrm{e}^{-s\Theta_{\mathrm{iso}}([xy])}\, \lambda_{d-1}(\mathrm{d}x)\, \lambda_{d-1}(\mathrm{d}y)\, \Theta_{\mathrm{iso}}(\mathrm{d}H)\, \mathrm{d}s \\
&= n \int_{[W]} \int_{W\cap H} \int_{W\cap H} \int_{\frac{\log n}{n}}^{t+\frac{\log n}{n}} \mathrm{e}^{-sn\Theta_{\mathrm{iso}}([xy])}\, \mathrm{d}s\, \lambda_{d-1}(\mathrm{d}x)\, \lambda_{d-1}(\mathrm{d}y)\, \Theta_{\mathrm{iso}}(\mathrm{d}H) \\
&= \int_{[W]} \int_{W\cap H} \int_{W\cap H} \frac{1}{n^{\Theta_{\mathrm{iso}}([xy])}} \frac{1-\mathrm{e}^{-nt\Theta_{\mathrm{iso}}([xy])}}{\Theta_{\mathrm{iso}}([xy])}\, \lambda_{d-1}(\mathrm{d}x)\, \lambda_{d-1}(\mathrm{d}y)\, \Theta_{\mathrm{iso}}(\mathrm{d}H)\,.
\end{aligned}$$

In the isotropic case, $\Theta_{\mathrm{iso}}([xy]) = \frac{2\kappa_{d-1}}{d\kappa_d}\|x-y\| =: c_d\|x-y\|$, and we obtain with the help of Proposition 12.5 that

$$\mathbb{E}\langle S_{\cdot,nW} \rangle_t \leq \frac{d-1}{2} \int_W \int_W \frac{1}{n^{c_d\|x-y\|}} \frac{1}{\|x-y\|^2}\, \lambda_d(\mathrm{d}x)\, \lambda_d(\mathrm{d}y)\,.$$

Now let RB^d be a d-dimensional ball centered at o with radius $R > 0$ such that $W \subseteq RB^d$. Then

$$\int_W \int_W \frac{1}{n^{c_d \|x-y\|}} \frac{1}{\|x-y\|^2} \lambda_d(\mathrm{d}x)\, \lambda_d(\mathrm{d}y) \le \int_{RB^d} \int_{RB^d} \frac{1}{n^{c_d \|x-y\|}} \frac{1}{\|x-y\|^2} \lambda_d(\mathrm{d}x)\, \lambda_d(\mathrm{d}y),$$

and the affine Blaschke–Petkantschin formula (12.16) yields

$$\begin{aligned} &\int_{RB^d} \int_{RB^d} \frac{1}{n^{c_d \|x-y\|}} \frac{1}{\|x-y\|^2} \lambda_d(\mathrm{d}x)\, \lambda_d(\mathrm{d}y) \\ &\quad = \frac{d\kappa_d}{2} \int_{A(d,1)} \int_{RB^d \cap L} \int_{RB^d \cap L} \frac{\|x-y\|^{d-3}}{n^{c_d \|x-y\|}} \lambda_1(\mathrm{d}x)\, \lambda_1(\mathrm{d}y)\, \mu_1(\mathrm{d}L). \end{aligned}$$

Since each intersection $RB^d \cap L$ is an interval on $L \in A(d,1)$ of length at most $2R$, it follows that

$$\begin{aligned} &\int_{A(d,1)} \int_{RB^d \cap L} \int_{RB^d \cap L} \frac{\|x-y\|^{d-3}}{n^{c_d \|x-y\|}} \lambda_1(\mathrm{d}x)\, \lambda_1(\mathrm{d}y)\, \mu_1(\mathrm{d}L) \\ &\quad \le \mu_1(\{L \in A(d,1) : RB^d \cap L \ne \emptyset\}) \int_0^{2R} \int_0^{2R} \frac{|x-y|^{d-3}}{n^{c_d |x-y|}} \,\mathrm{d}x\, \mathrm{d}y \\ &\quad = (2R)^{d-1} \mu_1(\{L \in A(d,1) : RB^d \cap L \ne \emptyset\}) \int_0^1 \int_0^1 \frac{|v-w|^{d-3}}{n^{b_d |v-w|}} \,\mathrm{d}v\, \mathrm{d}w \end{aligned}$$

with $b_d := 2c_d R$, where we applied the substitution $x = 2Rv$ and $y = 2Rw$ in the last step.

To further evaluate the integral in the last expression, we use the fact that

$$\int_0^1 \int_0^1 f(|v-w|)\, \mathrm{d}v\, \mathrm{d}w = 2 \int_0^1 (1-u) f(u)\, \mathrm{d}u$$

for all measurable functions $f : \mathbb{R} \to \mathbb{R}$ for which $(1-u)|f(u)|$ is integrable over $[0,1]$. Applying this transformation with $f(u) = u^{d-3} n^{-b_d u}$, we have that

$$\int_0^1 \int_0^1 \frac{|v-w|^{d-3}}{n^{b_d |v-w|}} \,\mathrm{d}v\, \mathrm{d}w = 2 \int_0^1 (1-u) \frac{u^{d-3}}{n^{b_d u}} \,\mathrm{d}u = 2 \int_0^1 (1-u) u^{d-3} \mathrm{e}^{-b_d \log n\, u} \,\mathrm{d}u.$$

We can now split the integral into two terms:

$$\int_0^1 (1-u) u^{d-3} \mathrm{e}^{-b_d \log n\, u} \,\mathrm{d}u = \int_0^1 u^{d-3} \mathrm{e}^{-b_d \log n\, u} \,\mathrm{d}u - \int_0^1 u^{d-2} \mathrm{e}^{-b_d \log n\, u} \,\mathrm{d}u.$$

Using the lower incomplete gamma function $\Gamma(s,x) = \int_0^x a^{s-1}e^{-a}\,da$, $x, s > 0$, we can thus write

$$\int_0^1 (1-u)u^{d-3}e^{-b_d \log n\, u}\,du = \frac{\Gamma(d-2, b_d \log n)}{(b_d \log n)^{d-2}} - \frac{\Gamma(d-1, b_d \log n)}{(b_d \log n)^{d-1}}.$$

Summarizing, we have derived the inequality

$$\begin{aligned}\mathbb{E}\langle S_{\cdot,nW}\rangle_t &\leq \frac{C_1}{\log^{d-1} n}(\Gamma(d-2, C_2 \log n)\log n - \Gamma(d-1, C_2 \log n)) \\ &\leq \frac{C_3}{\log^{d-2} n},\end{aligned}$$

in which $C_1, C_2, C_3 \in (0,\infty)$ are suitable constants only depending on d and W. Taking the limit as $n \to \infty$ and noting that the expression on the right-hand side tends to zero because of our assumption that $d \geq 3$, we conclude that (13.45) holds in expectation and hence in probability. We can now use Lemma 4.3 of [111] to conclude that the sequence $((\langle S_{\cdot,nW}\rangle_t,\ t \geq 0), n \in \mathbb{N}$, is relatively compact in $\mathrm{D}[0,\infty)$ and hence tight by Prohorov's theorem, Theorem 23.2 of [89]. This proves (b) and at the same time (c), since the zero-function is an element of $\mathrm{C}[0,\infty)$.

Step 2: Identification of the limit

From the tightness property just established it follows that every sequence $(n_k,\ k \in \mathbb{N})$ with $n_k \to \infty$ admits a subsequence, still denoted $(n_k,\ k \in \mathbb{N})$, such that $(S_{t,n_k W},\ t \geq 0)$ converges in distribution in $\mathrm{D}[0,\infty)$ to some càdlàg process $\mathbf{X} = (X_t,\ t \geq 0)$. By Corollaries VI.6.29 and IX.1.19 of [86] and by (13.45), $\mathbf{X}$ is a local martingale whose predictable quadratic variation is identically zero. Hence by Lemma 13.1 $\mathbf{X}$ is a. s. constant in t, say $X_t \equiv X$ for all $t \geq 0$ with some random variable X.

To identify the law of X, note that

$$S_{0,nW} = n^{-(d-1)}\overline{\Sigma}_{V_{d-1},\frac{\log n}{n},nW} \overset{\mathrm{D}}{=} \overline{\Sigma}_{V_{d-1},\log n,W} \overset{\text{law}}{\Longrightarrow} S$$

as $n \to \infty$. Thus X must have the same distribution as S, which by Theorem 13.23 is non-Gaussian and centered. Consequently, any subsequential limit is the random constant process $(S, t \geq 0)$. Since this limit does not depend on the subsequence, we conclude that $(S_{t,nW},\ t \geq 0) \overset{\text{law}}{\Longrightarrow} (S,\ t \geq 0)$ in $\mathrm{D}[0,\infty)$ as $n \to \infty$. □

13.4.3 Intrinsic volumes of maximal polytopes

After having studied the fluctuations of the total surface content in the previous section, we turn now to the intrinsic volumes of maximal polytopes for stationary and isotropic

STIT tessellations. More precisely, let again $W = [-\frac{1}{2}, \frac{1}{2}]^d$, and let $(Y_t, t > 0)$ be the STIT tessellation process driven by the motion-invariant hyperplane measure Θ_{iso}. For $j \in \{0, \dots, d-1\}$, we recall that

$$\Sigma_{V_j,t,W} := \sum_{(F,s)\in\mathcal{F}^{\max *}_{d-1,W}} \mathbb{1}\{s \le t\}\, V_j(F)$$

is the sum over the intrinsic volumes of order j of all maximal $(d-1)$-polytopes in W born until time $t \ge 0$. Moreover, for $n \in \mathbb{N}$, we define the random processes

$$S_{V_j,t,nW} := n^{-(d-1)}\overline{\Sigma}_{V_j,t+\frac{\log n}{n},nW}, \quad t \ge 0\,.$$

In particular, $S_{V_{d-1},t,nW} = S_{t,nW}$ is the total surface content process as considered in Corollary 13.24. We emphasize that the time shift $t + \frac{\log n}{n}$ is of crucial importance as it places the time origin right after the big-bang phase in which the dominating fluctuations of $\overline{\Sigma}_{V_{d-1},t,nW}$ arise. In contrast to this behavior of the surface process, for $j \in \{0, \dots, d-2\}$, the fluctuations of the functionals $\overline{\Sigma}_{V_j,t,nW}$ arise polynomially in time and are asymptotically completely controlled by those of $\overline{\Sigma}_{V_{d-1},t,nW}$. This is described by the main result of this section through a joint functional limit theorem for the d-dimensional vector of random processes $(S_{V_j,t,nW}, t \ge 0)$. It can be considered as the higher-dimensional analogue of Theorem 13.15.

Theorem 13.25. *Fix $d \ge 3$ and consider the STIT tessellation process $(Y_t, t > 0)$ driven by the motion-invariant hyperplane measure Θ_{iso}. Then, for $W = [-\frac{1}{2}, \frac{1}{2}]^d$,*

$$((S_{V_{d-1},t,nW}, S_{V_{d-2},t,nW}, \dots, S_{V_0,t,nW}), t \ge 0) \overset{\text{law}}{\Longrightarrow} ((S,\, q_2 tS,\, \dots,\, q_d t^{d-1} S), t \ge 0)$$

as $n \to \infty$ in $\mathrm{D}^d[0, \infty)$, where $q_j := \frac{1}{(j-1)!} c_{d-2,d} \cdots c_{d-j,d}$ for $j \in \{2, \dots, d\}$. Here $c_{k,d}$ is the constant given by (1.6), and S is the non-Gaussian random variable from Theorem 13.23.

Proof. Let $n \in \mathbb{N}$ and for each $j \in \{0, \dots, d-1\}$, define the auxiliary random process $\mathbf{M}_j = (M_{j,t}, t \ge 0)$ by

$$M_{j,t} := \overline{\Sigma}_{V_j,t,nW} - \int_0^t \overline{\mathbb{L}\Sigma_{V_j,s,nW}}\, \mathrm{d}s\,, \quad t \ge 0\,,$$

where by (10.36)

$$\mathbb{L}\Sigma_{V_j,s,nW} = \int_{[W]} \sum_{F\in(Y_{s,nW}\cap H)} V_j(F)\, \Theta_{\text{iso}}(\mathrm{d}H)$$

and $\overline{\mathbb{L}\Sigma_{V_j,s,nW}} = \mathbb{L}\Sigma_{V_j,s,nW} - \mathbb{E}\mathbb{L}\Sigma_{V_j,s,nW}$. By (10.42) we have

$$\mathbb{L}\Sigma_{V_j,s,nW} = c_{j,d}\left(\Sigma_{V_{j+1},s,nW} + V_{j+1}(nW)\right),$$

and hence $\overline{\mathbb{L}\Sigma_{V_j,s,nW}} = c_{j,d}\overline{\Sigma}_{V_{j+1},s,nW}$. Thus by Corollary 10.12(c) the random process

$$M_{j,t} = \overline{\Sigma}_{V_j,t,nW} - c_{j,d}\int_0^t \overline{\Sigma}_{V_{j+1},s,nW}\,\mathrm{d}s\,, \quad t \geq 0\,, \tag{13.46}$$

is a centered martingale adapted to the augmented filtration $(\mathfrak{I}_t,\ t > 0)$ induced by the STIT process $(Y_t,\ t > 0)$.

We will now show that, in the large-n asymptotics, the second moments of the random processes $\mathbf{M}_j$ with $j \in \{0,\dots,d-2\}$ are negligible when compared with those of $\mathbf{M}_{d-1}$. To see this, we first square both sides of (13.46) and take expectations:

$$\begin{aligned}
\mathbb{E}M_{j,t}^2 &= \operatorname{var}\Sigma_{V_j,t,nW} - 2c_{j,d}\int_0^t \mathbb{E}[\overline{\Sigma}_{V_j,t,nW}\,\overline{\Sigma}_{V_{j+1},s,nW}]\,\mathrm{d}s \\
&\quad + c_{j,d}^2\int_0^t\int_0^t \mathbb{E}[\overline{\Sigma}_{V_{j+1},s,nW}\,\overline{\Sigma}_{V_{j+1},u,nW}]\,\mathrm{d}u\,\mathrm{d}s \\
&= \operatorname{var}\Sigma_{V_j,t,nW} - 2c_{j,d}\int_0^t \mathbb{E}[\overline{\Sigma}_{V_{j+1},s,nW}\,\mathbb{E}[\overline{\Sigma}_{V_j,t,nW}\mid\mathfrak{I}_s]]\,\mathrm{d}s \\
&\quad + 2c_{j,d}^2\int_0^t\int_s^t \mathbb{E}[\overline{\Sigma}_{V_{j+1},s,nW}\,\overline{\Sigma}_{V_{j+1},u,nW}]\,\mathrm{d}u\,\mathrm{d}s\,.
\end{aligned} \tag{13.47}$$

From the martingale property of $\mathbf{M}_j$ it follows that the conditional expectation of the process satisfies

$$\mathbb{E}[\overline{\Sigma}_{V_j,t,nW}\mid\mathfrak{I}_s] = \overline{\Sigma}_{V_j,s,nW} + c_{j,d}\int_s^t \mathbb{E}[\overline{\Sigma}_{V_{j+1},u,nW}\mid\mathfrak{I}_s]\,\mathrm{d}u$$

almost surely; compare with (13.27). Plugging this into (13.47), we see that

$$\begin{aligned}
\mathbb{E}M_{j,t}^2 &= \operatorname{var}\Sigma_{V_j,t,nW} - 2c_{j,d}\int_0^t \mathbb{E}[\overline{\Sigma}_{V_j,s,nW}\,\overline{\Sigma}_{V_{j+1},s,nW}]\,\mathrm{d}s \\
&\quad - 2c_{j,d}^2\int_0^t\int_s^t \mathbb{E}[\mathbb{E}[\overline{\Sigma}_{V_{j+1},u,nW}\mid\mathfrak{I}_s]\,\overline{\Sigma}_{V_{j+1},s,nW}]\,\mathrm{d}u\,\mathrm{d}s \\
&\quad + 2c_{j,d}^2\int_0^t\int_s^t \mathbb{E}[\overline{\Sigma}_{V_{j+1},s,nW}\,\overline{\Sigma}_{V_{j+1},u,nW}]\,\mathrm{d}u\,\mathrm{d}s
\end{aligned}$$

$$= \operatorname{var} \Sigma_{V_j,t,nW} - 2c_{j,d} \int_0^t \mathbb{E}[\overline{\Sigma}_{V_j,s,nW}\, \overline{\Sigma}_{V_{j+1},s,nW}]\, \mathrm{d}s$$

$$= \operatorname{var} \Sigma_{V_j,t,nW} - 2c_{j,d} \int_0^t \operatorname{cov}(\Sigma_{V_j,s,nW}, \Sigma_{V_{j+1},s,nW})\, \mathrm{d}s$$

$$= \int_0^t \mathbb{E}A_{j,s,nW}\, \mathrm{d}s\,, \tag{13.48}$$

where the last identity follows from (12.8). However, in the proof of Theorem 12.20, we have seen that $\mathbb{E}A_{j,s,nW} = o(n^{2(d-1)})$ as $n \to \infty$ for $j \in \{0, \dots, d-2\}$. Indeed, $\mathbb{E}A_{j,s,nW}$ is the sum of the terms S_1 and S_2 there, which were shown to be of order $O(n^{2d-3})$; see especially (12.38). Consequently,

$$\int_0^t \mathbb{E}A_{j,s,nW}\, \mathrm{d}s = O(n^{2d-3}) = o(n^{2(d-1)})\,,$$

and from (13.48) we conclude that $\operatorname{var} M_{j,t} = o(n^{2(d-1)})$ for $j \in \{0, \dots, d-2\}$. In contrast, for $j = d-1$, we know from Theorem 12.3 that $\operatorname{var} M_{d-1,t}$ is of order $O(n^{2(d-1)})$. This confirms that the contributions of $M_{j,t}$ with $j \le d-2$ are negligible compared to those of $M_{d-1,t}$.

We are now in the position to apply Doob's L^2-maximal inequality (13.1), which yields

$$\mathbb{E}\Big(\sup_{0\le s\le t} |M_{j,s}|^2\Big) \le 4 \operatorname{var} M_{j,t} = O(n^{2d-3}) = o(n^{2(d-1)}) \tag{13.49}$$

as $n \to \infty$ for all $j \in \{0, \dots, d-2\}$.

Next, we recall that the normalization of the random processes $(S_{V_j,t,nW}, t \ge 0)$ involves the factor $n^{-(d-1)}$. From (13.46) and the variance estimate (13.49) it follows that after normalization, the martingale contribution vanishes in L^2 as $n \to \infty$. Thus we may replace $S_{V_j,t,nW}$ by $c_{j,d} \int_0^t S_{V_{j+1},s,nW}\, \mathrm{d}s$ for each $j \in \{0, \dots, d-2\}$ with an error term tending to zero as $n \to \infty$. Applying this recursion step by step, starting from $j = d-2$ and going downward, the limiting behavior of the whole vector is reduced to that of $(S_{V_{d-1},t,nW}, t \ge 0)$. The latter is provided by Corollary 13.24, so the proof is complete. □

As explained in the previous sections of this chapter, the leading-order fluctuations of the total surface content $S_{V_{d-1},t,nW}$ in nW only arise during the big-bang phase. In contrast, the fluctuations of the other intrinsic volume processes $S_{V_j,t,nW}$ with $j \in \{0, \dots, d-2\}$ in nW arise after that phase and grow polynomially in time at a rate proportional to t^{d-1-j} times the initial big-bang fluctuation of $S_{V_{d-1},t,nW}$. This is well reflected by the fact that the asymptotic variance of $\Sigma_{V_j,t,nW}$ exhibits a nonvanishing polynomial dependence on t, whereas the asymptotic variance of the total surface content does not depend on t.

Moreover, we see that the fluctuations of all the processes $S_{V_j,t,nW}$ with $j \in \{0,\dots,d-2\}$ are completely determined by those of $S_{V_{d-1},t,nW}$, a phenomenon we already encountered at the end of Section 13.3 in the planar case.

Summarizing, we have the following simple and intuitive picture for the total surface content $S_{V_{d-1},t,nW}$ and the intrinsic volume processes $S_{V_j,t,nW}$ with $j \in \{0,\dots,d-2\}$ in the large-n asymptotics, which parallels in a sense the two-dimensional situation discussed in Section 13.3: $S_{V_{d-1},t,nW}$ roughly behaves like S, whereas $S_{V_j,t,nW}$ behaves like $q_{d-j}\, t^{d-1-j} S$ for $t \geq \frac{\log n}{n}$ and $j \in \{0,\dots,d-2\}$, where $q_j = \frac{1}{(j-1)!} c_{d-2,d} \cdots c_{d-j,d}$.

13.4.4 The isotropic limit as a random field

For stationary and isotropic STIT tessellations in $\mathbb{R}^d$ with $d \geq 3$, we have established in Theorem 13.23 the existence of a non-Gaussian random variable S, which has turned out to underlie a large spectrum of limit theorems. So far, we have focussed on the case where S arises by expansion of the unit cube $W = [-\frac{1}{2}, \frac{1}{2}]^d$. However, in view of the consistency property of STIT tessellations, it is natural to consider the limit

$$S(K) := \lim_{a\to\infty} a^{-(d-1)} \overline{\Sigma}_{V_{d-1},1,aK} \tag{13.50}$$

of the normalized total surface content within aK as $a \to \infty$ for arbitrary convex bodies $K \in \mathcal{K}_d$. This gives rise to a **random field** $(S(K),\ K \in \mathcal{K}_d)$, indexed by $\mathcal{K}_d$, whose properties are in the focus of this section. In what follows, we use the notation $\|X\|^p_{L^p(\mathbb{P})} := \mathbb{E}|X|^p$ for random variables X and $p \geq 1$.

Theorem 13.26. *Let $d \geq 3$, and let $(Y_t,\ t > 0)$ be a stationary and isotropic STIT tessellation process driven by the motion-invariant hyperplane measure Θ_{iso}. For each convex body $K \in \mathcal{K}_d$, define $S(K)$ by (13.50). Then we have the following:*

(a) *For every $K \in \mathcal{K}_d$, the limit in (13.50) exists a. s. and in L^2. Moreover, $\mathbb{E}S(K) = 0$.*

(b) *For all $K, L \in \mathcal{K}_d$,*

$$\operatorname{var} S(K) = \frac{d-1}{2} \int_K \int_K \frac{1}{\|x-y\|^2} \lambda_d(dx)\, \lambda_d(dy)\,, \tag{13.51}$$

$$\operatorname{cov}(S(K), S(L)) = \frac{d-1}{2} \int_K \int_L \frac{1}{\|x-y\|^2} \lambda_d(dx)\, \lambda_d(dy)\,. \tag{13.52}$$

In particular, $\operatorname{var} S(K) < \infty$.

(c) *For every translation vector $u \in \mathbb{R}^d$, every rotation $\varrho \in SO(d)$, and every $b > 0$,*

$$S(\varrho K + u) \overset{D}{=} S(K) \quad \text{and} \quad S(bK) \overset{D}{=} b^{d-1} S(K)\,, \quad K \in \mathcal{K}_d\,.$$

(d) *If $K_n, K \in \mathcal{K}_d$ satisfy $K_n \to K$ as $n \to \infty$ in the Hausdorff metric, then*

$$\lim_{n\to\infty} \|S(K_n) - S(K)\|^2_{L^2(\mathbb{P})} = 0 .$$

Consequently, $K \mapsto S(K)$ defines a centered set-indexed random field on $\mathcal{K}_d$ with covariance kernel (13.52).

Proof. (a) Fix $K \in \mathcal{K}_d$. As in the proof of Theorem 13.23, carried out there for the case that W is the unit cube, the variance formula in the isotropic case from Theorem 12.8 gives, for every $a > 0$,

$$\begin{aligned}\mathbb{E}(\overline{\Sigma}_{V_{d-1},a,K})^2 &= \frac{d-1}{2} \int_K \int_K \frac{1 - \exp(-\frac{2\kappa_{d-1}}{d\kappa_d} a\|x-y\|)}{\|x-y\|^2} \lambda_d(\mathrm{d}x)\,\lambda_d(\mathrm{d}y) \\ &\le \frac{d-1}{2} \int_K \int_K \frac{1}{\|x-y\|^2} \lambda_d(\mathrm{d}x)\,\lambda_d(\mathrm{d}y) < \infty ,\end{aligned}$$

because $d \ge 3$ makes the kernel $\|x-y\|^{-2}$ locally integrable. Hence

$$\sup_{a>0} \mathbb{E}(\overline{\Sigma}_{V_{d-1},a,K})^2 < \infty .$$

By the martingale convergence theorem, Corollary 9.23 of [89], together with the scaling relation (13.34), the process $a \mapsto a^{-(d-1)}\,\overline{\Sigma}_{V_{d-1},1,aK}$ converges a. s. and in L^2 as $a \to \infty$ to a centered limit $S(K)$. This proves the existence and also that $\mathbb{E}S(K) = 0$.

(b) By the dominated convergence theorem, using the bound established in the proof of part (a) and the second-order formula above, we have

$$\operatorname{var} S(K) = \lim_{a\to\infty} \mathbb{E}(a^{-(d-1)}\overline{\Sigma}_{V_{d-1},1,aK})^2 = \frac{d-1}{2} \int_K \int_K \frac{1}{\|x-y\|^2} \lambda_d(\mathrm{d}x)\,\lambda_d(\mathrm{d}y) ,$$

which is (13.51). For the covariance, let $K, L \in \mathcal{K}_d$ and consider the centered limits $S(K)$ and $S(L)$ obtained as above. Using Theorem 12.26 and then arguing as in the proof of Theorem 12.8, we find

$$\operatorname{cov}(\Sigma_{V_{d-1},a,K}, \Sigma_{V_{d-1},a,L}) = \frac{d-1}{2} \int_K \int_L \frac{1 - \exp(-\frac{2\kappa_{d-1}}{d\kappa_d} a\|x-y\|)}{\|x-y\|^2} \lambda_d(\mathrm{d}x)\,\lambda_d(\mathrm{d}y) ,$$

which, together with the scaling relation (13.34), implies

$$\operatorname{cov}(S(K), S(L)) = \frac{d-1}{2} \int_K \int_L \frac{1}{\|x-y\|^2} \lambda_d(\mathrm{d}x)\,\lambda_d(\mathrm{d}y)$$

by the dominated convergence theorem, since $d \ge 3$. This is (13.52).

(c) Let $u \in \mathbb{R}^d$, $\varrho \in SO(d)$, and $b > 0$. The stationarity and isotropy of the STIT tessellation driven by Θ_{iso} imply

$$\overline{\Sigma}_{V_{d-1},1,\,a(\varrho K+u)} \overset{\mathrm{D}}{=} \overline{\Sigma}_{V_{d-1},1,\,aK} \quad \text{for all } a > 0\,,$$

and hence $S(\varrho K + u) \overset{\mathrm{D}}{=} S(K)$. The scaling relation (13.34) gives, for $b > 0$,

$$a^{-(d-1)}\,\overline{\Sigma}_{V_{d-1},1,\,a(bK)} = (ab)^{-(d-1)}\,\overline{\Sigma}_{V_{d-1},1,\,abK}\; b^{d-1} \overset{\text{law}}{\Longrightarrow} b^{d-1}\,S(K)$$

as $a \to \infty$, which proves $S(bK) \overset{\mathrm{D}}{=} b^{d-1}S(K)$.

(d) Fix $R > 0$ and suppose without loss of generality that $K_n, K \subset RB^d$ with $K_n \to K$ in the Hausdorff metric. By (13.52) and bilinearity,

$$\begin{aligned}\|S(K_n) - S(K)\|^2_{L^2(\mathbb{P})} \\ = \frac{d-1}{2}\int_{\mathbb{R}^d}\int_{\mathbb{R}^d} \frac{(\mathbb{1}_{K_n}(x) - \mathbb{1}_K(x))(\mathbb{1}_{K_n}(y) - \mathbb{1}_K(y))}{\|x-y\|^2}\,\lambda_d(\mathrm{d}x)\,\lambda_d(\mathrm{d}y)\,.\end{aligned} \tag{13.53}$$

Since $K_n \to K$ in Hausdorff metric, $\mathbb{1}_{K_n} \to \mathbb{1}_K$ pointwise almost everywhere, and all sets are contained in RB^d. The kernel $\|x-y\|^{-2}$ is integrable on $(RB^d) \times (RB^d)$ for $d \geq 3$, so by dominated convergence theorem the right-hand side in (13.53) tends to 0. This proves the claimed L^2-continuity. □

In part (d) of the previous theorem, we have seen the first L^2-continuity property of the random field $(S(K), K \in \mathcal{K}_d)$. We now provide more refined regularity properties. We denote by δ^{Hausd} the Hausdorff distance and write $\|f\|_{L^p(\lambda_d)} := (\int_{\mathbb{R}^d} |f|^p\,\lambda_d(\mathrm{d}x))^{1/p}$ for the L^p-norm of a function $f : \mathbb{R}^d \to \mathbb{R}$ with respect to the d-dimensional Lebesgue measure. Moreover, we recall that $A \triangle B = (A \setminus B) \cup (B \setminus A)$, $A, B \in \mathcal{B}(\mathbb{R}^d)$, stands for the symmetric difference.

Theorem 13.27. *Let $d \geq 3$ and $\Theta = \Theta_{\mathrm{iso}}$. For $K \in \mathcal{K}_d$, let $S(K)$ be the centered limit from Theorem 13.26. Then we have the following regularity properties:*

(a) *There exists a constant $C_d \in (0,\infty)$ only depending on d such that for all $K, L \in \mathcal{K}_d$,*

$$\|S(K) - S(L)\|_{L^2(\mathbb{P})} \leq C_d\,\lambda_d(K \triangle L)^{\frac{d+2}{2d}}\,. \tag{13.54}$$

(b) *Fix $R > 0$ and consider $\mathcal{K}_R := \{K \in \mathcal{K}_d : K \subset RB^d\}$. Then there exists a constant $C_{d,R} \in (0,\infty)$ only depending on d and R such that for all $K, L \in \mathcal{K}_R$,*

$$\|S(K) - S(L)\|_{L^2(\mathbb{P})} \leq C_{d,R}\,\delta^{\mathrm{Hausd}}(K,L)^{\frac{d+2}{2d}}\,. \tag{13.55}$$

Proof. (a) Let $f := \mathbb{1}_{K \triangle L}$ and observe that for all $x, y \in \mathbb{R}^d$,

$$\begin{aligned}&(\mathbb{1}_K(x) - \mathbb{1}_L(x))(\mathbb{1}_K(y) - \mathbb{1}_L(y)) \\ &\quad = f(x)f(y) - 2(\mathbb{1}_{K\setminus L}(x)\mathbb{1}_{L\setminus K}(y) + \mathbb{1}_{L\setminus K}(x)\mathbb{1}_{K\setminus L}(y))\,.\end{aligned}$$

By (13.53) it follows that

$$\mathbb{E}(S(K)-S(L))^2 = \frac{d-1}{2}\int_{\mathbb{R}^d}\int_{\mathbb{R}^d}\frac{f(x)f(y)}{\|x-y\|^2}\,\lambda_d(dx)\,\lambda_d(dy) - (d-1)\int_{K\setminus L}\int_{L\setminus K}\frac{1}{\|x-y\|^2}\,\lambda_d(dx)\,\lambda_d(dy) \le \frac{d-1}{2}\int_{\mathbb{R}^d}\int_{\mathbb{R}^d}\frac{f(x)f(y)}{\|x-y\|^2}\,\lambda_d(dx)\,\lambda_d(dy)\,.$$

By the Hardy–Littlewood–Sobolev inequality in [104], applied with $\alpha = 2$, $d \ge 3$, and $q = p = 2d/(d+2) \in (1,2)$ there we have

$$\int_{\mathbb{R}^d}\int_{\mathbb{R}^d}\frac{f(x)f(y)}{\|x-y\|^2}\,\lambda_d(dx)\,\lambda_d(dy) \le \widetilde{C}_d\,\|f\|^2_{L^p(\lambda_d)} = \widetilde{C}_d\,\lambda_d(K\triangle L)^{\frac{2}{p}} = \widetilde{C}_d\,\lambda_d(K\triangle L)^{\frac{d+2}{d}}\,,$$

which yields (13.54) with $C_d := \sqrt{\frac{d-1}{2}\widetilde{C}_d}$; see [104] for the exact value of $\widetilde{C}_d$.

(b) If $K, L \subset RB^d$ and $\delta^{\mathrm{Hausd}}(K,L) \le \varepsilon$, then $K \triangle L \subset [(\partial K) + \varepsilon B^d] \cup [(\partial L) + \varepsilon B^d]$, and from Steiner's formula (1.3) it follows that for $\varepsilon < 1$,

$$\lambda_d(K\triangle L) \le \lambda_d((\partial K) + \varepsilon B^d) + \lambda_d((\partial L) + \varepsilon B^d) \le 2c_d \max\{\mathcal{H}^{d-1}(\partial K), \mathcal{H}^{d-1}(\partial L)\}\varepsilon$$

for some constant $c_d \in (0,\infty)$ only depending on d. Since K and L are convex and satisfy $K, L \subseteq RB^d$, it follows that $\max\{\mathcal{H}^{d-1}(\partial K), \mathcal{H}^{d-1}(\partial L)\} \le \mathcal{H}^{d-1}(\partial(RB^d)) = d\kappa_d R^{d-1}$. On the other hand, if $\varepsilon \ge 1$, then we have the trivial bound $\lambda_d(K\triangle L) \le \lambda_d(2RB^d) = \kappa_d(2R)^d$. Combining this with (a) gives

$$\|S(K)-S(L)\|_{L^2(\mathbb{P})} \le C_{d,R}\,\varepsilon^{\frac{d+2}{2d}}$$

for all $\varepsilon > 0$ with a suitable constant $C_{d,R} \in (0,\infty)$. This proves (13.55). □

Remark 13.28. *The covariance kernel of the random field $(S(K),\ K \in \mathcal{K}_d)$ is given by*

$$\mathrm{cov}(S(K), S(L)) = \frac{d-1}{2}\int_K\int_L \frac{1}{\|x-y\|^2}\,\lambda_d(dx)\,\lambda_d(dy)\,,$$

that is, by the 2-energy between the indicator functions of K and L. This is closely analogous to the covariance structure of the Gaussian free field in $d \ge 3$, whose covariance is given by the $(d-2)$-energy; see [181]. By Theorem 13.26(c) the field $(S(K),\ K \in \mathcal{K}_d)$ satisfies $S(bK) \overset{D}{=} b^{d-1}S(K)$ for $b > 0$, so S is self-similar of order $d-1$ under dilations of the index set. This contrasts with the Gaussian free field in $d \ge 3$, which is self-similar of order $1-d/2$ under the scaling $f(x) \mapsto f(x/b)$. The main difference is that $(S(K),\ K \in \mathcal{K}_d)$ cannot be Gaussian in view of Theorem 13.23.

13.5 Notes

1. The first limit theorems for the total and residual surface content of STIT tessellations were developed in [178]. The results in Sections 13.2 and 13.4.1 are built on the approach of [178], with additional clarifications and technical refinements. The Gaussian limit theory for the total edge length and the number of I-segments for planar STIT tessellations presented in Section 13.3 has been established in [174], whereas the limit theorems for higher-dimensional STIT tessellations of Section 13.4.3 are essentially taken from [175].
2. There is more than one natural topology on the Skorokhod space $\mathsf{D}^m[0,\infty)$. Skorokhod introduced four distinct topologies on $\mathsf{D}^m[0,\infty)$: the J_1-, J_2-, M_1-, and M_2-topologies. In this book, we focus exclusively on the J_1-topology, the strongest among the Skorokhod topologies and commonly referred to as *the* Skorokhod topology in the literature. We follow this convention and direct the reader to the monographs [53, 197] for further background material and to the survey paper [92] for a very accessible and well-illustrated introduction to the different Skorokhod topologies. We also note that there exist natural non-Skorokhod topologies on $\mathsf{D}^m[0,\infty)$, such as the S-topology introduced in [87].
3. It is rather common to introduce the Skorokhod topology on the space $\mathsf{D}[0,T]$ of càdlàg functions $g : [0,T] \to \mathbb{R}$ only for some finite time horizon $T > 0$. In this context, it is natural to say that a sequence of functions $f_n \in \mathsf{D}[0,\infty)$, $n \in \mathbb{N}$, converges to some $f \in \mathsf{D}[0,\infty)$ if for every $T > 0$, the restriction of f_n to $[0,T]$ converges as $n \to \infty$ to the restriction of the limiting function f. However, this condition is too restrictive; instead, we must consider only those intervals $[0,T]$ for which T is a continuity point of f. In the same context, we remark that it is tempting to use the fact that $\mathsf{D}^m[0,\infty) = (\mathsf{D}[0,\infty))^m$ for $m \in \mathbb{N}$ and to introduce the Skorokhod topology on $\mathsf{D}^m[0,\infty)$ as the product topology of the Skorokhod topologies on each factor. However, we can show that the resulting topology is weaker than the Skorokhod topology directly constructed on $\mathsf{D}^m[0,\infty)$. This difference arises because the product topology allows for different time parameterizations in each component, whereas the stronger topology on $\mathsf{D}^m[0,\infty)$ enforces a common time change across all components. Again, we refer the reader to [53, 92, 197] for more detail.
4. The functional central limit theorem for continuous-time martingales, presented as Theorem 13.3, combines Theorem 5.1 of [76] and Theorem 2.1 of [198]. Additional results in this direction and a broader discussion of limit theorems for stochastic processes can be found in the comprehensive monographs [53, 86].

Part III: **Implications of the global construction**

14 A Mecke-type equation and applications to maximal polytopes

The key result of this chapter, Theorem 14.1, provides a formula for STIT tessellations that is similar in its structure to the Mecke equation in Theorem 19.11 for Poisson point processes. In its general form, Theorem 14.1 is rather intricate, but it allows for several special cases, which are presented subsequently. The proof uses the global construction of STIT tessellations, which in turn is based on the Poisson point process $\widetilde{\Sigma}$; see Section 8.1.

In Section 14.3 the maximal k-polytopes, $k \in \{0, \dots, d-2\}$, which arise as the intersections of $d-k$ maximal $(d-1)$-polytopes in the STIT tessellation process, will be studied in detail. Using the Mecke-type equation for STIT processes, propositions on maximal polytopes of STIT tessellations and their "internal structure" will be derived in the rest of this chapter.

14.1 A Mecke-type equation for the global construction

The propositions in this chapter refer to the global construction of a STIT tessellation process $\mathbf{Y} = (Y_t,\ t > 0)$ as described in Chapter 8. Thus the point process Σ with law $\mathbb{P}_\Sigma$ introduced in (8.1) plays the central role. Furthermore, the set $\mathcal{F}_{d-1}^{\max *}(\Sigma)$ of birth time marked maximal $(d-1)$-polytopes (see (8.7)) and the set $\mathcal{D}(\Sigma)$ of division events defined in (8.6) are derived from Σ. To simplify the notation, in this chapter, $\mathbb{P}^{\max *}$ denotes the distribution of $\mathcal{F}_{d-1}^{\max *}(\Sigma)$, and $\mathbb{P}_{Y_t}$ is the distribution of Y_t. The distribution of the process $\mathbf{Y}$ is $\mathbb{P}_\mathbf{Y}$. We have $\int g(y_t)\, \mathbb{P}_{Y_t}(dy_t) = \int g(y_t)\, \mathbb{P}_\mathbf{Y}(d\mathbf{y})$ for all nonnegative measurable $g : \mathcal{T} \to \mathbb{R}$ and $t > 0$.

For $(F, s) \in \mathcal{F}_{d-1}^{\max *}(\Sigma)$, let $P(F)$ denote the uniquely determined cell satisfying $(P(F), H, s) \in \mathcal{D}(\Sigma)$ and $F = P(F) \cap H$, that is, the cell divided at time s by the hyperplane H. Furthermore, for $P \in \mathcal{P}_d$, we define

$$\mathcal{F}_{d-1}^{\max *}(\Sigma) \wedge P := \{(P \cap F, s) : (F, s) \in \mathcal{F}_{d-1}^{\max *}(\Sigma),\ F \cap \operatorname{int} P \neq \emptyset\}$$

as the restriction of $\mathcal{F}_{d-1}^{\max *}(\Sigma)$ to P. For a realization of $\mathcal{F}_{d-1}^{\max *}(\Sigma)$, we use the symbol m^*. Furthermore, for a realization m^* and $s > 0$, the uniquely determined realization of the STIT tessellation at time s is written as $Y_s(m^*)$. For $t > 0$, the set of "time-shifted" maximal polytopes is denoted $m^*_{(+t)} := \{(F, s+t) : (F, s) \in m^*\}$. Correspondingly, $Y_s(\sigma)$ denotes the STIT tessellation at time s generated by a realization σ of Σ.

The following statement contains a rather general function g, which may, for any division event at some time $t > 0$, depend on the past, that is, $(Y_s,\ s \le t)$, on the division event itself, that is, the cell P divided by some hyperplane H, and furthermore, on the future development of the process in the two daughter cells of P.

https://doi.org/10.1515/9783112222409-017

Theorem 14.1. *Let Θ be a hyperplane measure of the form (4.1), satisfying (4.3), and let $(Y_t,\ t>0)$ be the STIT tessellation process generated by the point process Σ of (8.1). Then, for all nonnegative measurable functions $g : \{\mathcal{T}^{(0,t]} : t>0\} \times \mathsf{N}_s(\mathcal{P}_{d-1}\times\mathbb{R}^+)\times\mathcal{P}_d\times\mathcal{P}_{d-1}\times\mathbb{R}^+ \to \mathbb{R}$, we have the following* ***Mecke-type equation****:*

$$\mathbb{E}\sum_{(F,t)\in\mathcal{F}_{d-1}^{\max*}(\Sigma)} g((Y_s,\ s\le t),\ \mathcal{F}_{d-1}^{\max*}(\Sigma)\wedge P(F),\ P(F),\ F,\ t)$$
$$= \int_0^\infty \int_{\mathcal{T}^{\mathbb{R}^+}} \sum_{P\in y_t} \int_{[P]} \Bigg(\int_{\mathsf{N}_s(\mathcal{P}_{d-1}\times\mathbb{R}^+)} \int_{\mathsf{N}_s(\mathcal{P}_{d-1}\times\mathbb{R}^+)}$$
$$g((y_s,\ s\le t),\{(P\cap H,t)\}\cup(m^{*(1)}_{(+t)}\wedge(P\cap H^+))\cup(m^{*(2)}_{(+t)}\wedge(P\cap H^-)),P,P\cap H,t)$$
$$\mathbb{P}^{\max*}(\mathrm{d}m^{*(1)})\,\mathbb{P}^{\max*}(\mathrm{d}m^{*(2)})\Bigg)\,\Theta(\mathrm{d}H)\,\mathbb{P}_Y(\mathrm{d}y)\,\mathrm{d}t\,. \tag{14.1}$$

Proof. With the notation of (8.1), for a given $(x,r,\psi)\in\Sigma$ and $(H,t)\in\psi$, according to the global construction, there can a. s. be either no or exactly one cell P that is divided by H at time t. For better readability of the following formulae, we introduce abbreviations for $P\in\mathcal{P}_d$, not containing the origin, $H\in\mathcal{H}$ and $t\in\mathbb{R}^+$. Assume that P is a cell with birth time $\beta(P)$, as defined in (iii) of the description of the global construction in Section 8.1. Then

$$\begin{aligned}
&\mathbb{1}(P,\Sigma,(x,r,\psi),(H,t))\\
&\quad := \mathbb{1}\{(x,r,\psi)\in\Sigma\}\,\mathbb{1}\{(H,t)\in\psi\}\,\mathbb{1}\{x\in P,\ o\notin P\}\\
&\qquad \times\mathbb{1}\{r=\max\{r'\in\mathbb{R}^- : (x',r',\psi')\in\Sigma,\ x'\in P\}\}\\
&\qquad \times\mathbb{1}\{t=\min\{s>\beta(P) : (H',s)\in\psi,\ H'\in[\operatorname{int}P]\}\}\,.
\end{aligned}\tag{14.2}$$

Furthermore, for the division of the zero-cell, the Poisson process $\widehat{X}_o^*$ of birth time marked hyperplanes was introduced in Section 8.1; see (8.1). With this notation,

$$\begin{aligned}
\mathbb{1}(P,o,\widehat{X}_o^*,(H,t)) &:= \mathbb{1}\{(H,t)\in\widehat{X}_o^*\}\,\mathbb{1}\{o\in P\}\,\mathbb{1}\{H\in[\operatorname{int}P]\}\\
&\quad\times\mathbb{1}\{t=\min\{s>\beta(P) : (H',s)\in\widehat{X}_o^*\}\}\,.
\end{aligned}\tag{14.3}$$

The terms are 1 if P is divided by H at time t and 0 otherwise.

For the mapping $\Sigma\mapsto\mathcal{F}_{d-1}^{\max*}(\Sigma)$, the transformation formula for image measures yields

$$\begin{aligned}
A &:= \int_{\mathsf{N}_s(\mathcal{P}_{d-1}\times\mathbb{R}^+)} \sum_{(F,t)\in m^*} g((Y_s(m^*),\ s\le t),m^*\wedge P(F),P(F),F,t)\,\mathbb{P}^{\max*}(\mathrm{d}m^*)\\
&= \int_{\mathsf{N}_s(\mathbb{R}^d\times\mathbb{R}^-\times\mathsf{N}_s(\mathcal{H}\times\mathbb{R}^+))} \sum_{(F,t)\in m^*(\sigma)} g((Y_s(\sigma),\ s\le t),m^*(\sigma)\wedge P(F),P(F),F,t)\,\mathbb{P}_\Sigma(\mathrm{d}\sigma)\,.
\end{aligned}$$

A maximal polytope of dimension $d-1$ appears when a cell is divided by a hyperplane H. In the formulae below, the term $P \in \mathbf{Y}(\sigma)$ means that there exists a time $t > 0$ such that $P \in Y_t(\sigma)$, that is, P is a cell extant at time t of the STIT tessellation process generated by a realization σ of Σ.

Recalling that in (8.1) the point process Σ is defined as a sum, $\Sigma := \widetilde{\Sigma} + \delta_{(0,0,\widehat{X}_o^*)}$, we obtain

$$\begin{aligned}A = &\int\limits_{\mathsf{N}_s(\mathbb{R}^d\times\mathbb{R}^-\times\mathsf{N}_s(\mathcal{H}\times\mathbb{R}^+))}\ \int\limits_{\mathsf{N}_s(\mathcal{H}\times\mathbb{R}^+)} \sum_{P\in\mathcal{P}_d} \mathbb{1}\{P \in \mathbf{Y}(\widetilde{\sigma} + \delta_{(0,0,\pi)})\}\\ &\times\Big(\sum_{(x,r,\psi)\in\widetilde{\sigma}}\ \sum_{(H,t)\in\psi} \mathbb{1}(P,\widetilde{\sigma},(x,r,\psi),(H,t)) + \sum_{(H,t)\in\pi} \mathbb{1}(P,o,\pi,(H,t))\Big)\\ &\times g((Y_s(\widetilde{\sigma}+\delta_{(0,0,\pi)}),\ s\le t), m^*(\widetilde{\sigma}+\delta_{(0,0,\pi)}) \wedge P, P, P\cap H, t)\,\mathbb{P}_{\widehat{X}^*}(\mathrm{d}\pi)\,\mathbb{P}_{\widetilde{\Sigma}}(\mathrm{d}\widetilde{\sigma})\,.\end{aligned}$$

Applying the Mecke equation (19.5) to the Poisson point process $\widetilde{\Sigma}$ yields

$$\begin{aligned}A = &\int\limits_{\mathsf{N}_s(\mathbb{R}^d\times\mathbb{R}^-\times\mathsf{N}_s(\mathcal{H}\times\mathbb{R}^+))}\ \int\limits_{\mathsf{N}_s(\mathcal{H}\times\mathbb{R}^+)}\ \int\limits_{\mathbb{R}^-}\ \int\limits_{\mathbb{R}^d}\ \int\limits_{\mathsf{N}_s(\mathcal{H}\times\mathbb{R}^+)} \sum_{P\in\mathcal{P}_d} \mathbb{1}\{P \in \mathbf{Y}(\widetilde{\sigma} + \delta_{(x,r,\psi)} + \delta_{(0,0,\pi)})\}\\ &\times\Big(\sum_{(H,t)\in\psi} \mathbb{1}(P,\widetilde{\sigma}+\delta_{(x,r,\psi)},(x,r,\psi),(H,t)) + \sum_{(H,t)\in\pi} \mathbb{1}(P,o,\pi,(H,t))\Big)\\ &\times g((Y_s(\widetilde{\sigma}+\delta_{(x,r,\psi)}+\delta_{(0,0,\pi)}),\ s\le t), m^*(\widetilde{\sigma}+\delta_{(x,r,\psi)}+\delta_{(0,0,\pi)}) \wedge P, P, P\cap H, t)\\ &\quad\mathbb{P}_{\widehat{X}^*}(\mathrm{d}\psi)\,\lambda_d(\mathrm{d}x)\,\lambda_-(\mathrm{d}r)\,\mathbb{P}_{\widehat{X}^*}(\mathrm{d}\pi)\,\mathbb{P}_{\widetilde{\Sigma}}(\mathrm{d}\widetilde{\sigma})\,.\end{aligned}$$

Next, we apply the Mecke equation (19.5) again, this time with respect to ψ and π. This yields

$$\begin{aligned}A = &\int\limits_{\mathsf{N}_s(\mathbb{R}^d\times\mathbb{R}^-\times\mathsf{N}_s(\mathcal{H}\times\mathbb{R}^+))}\ \int\limits_{\mathsf{N}_s(\mathcal{H}\times\mathbb{R}^+)}\ \int\limits_{\mathbb{R}^-}\ \int\limits_{\mathbb{R}^d}\ \int\limits_{\mathsf{N}_s(\mathcal{H}\times\mathbb{R}^+)} (A_1 + A_2)\\ &\mathbb{P}_{\widehat{X}^*}(\mathrm{d}\psi)\,\lambda_d(\mathrm{d}x)\,\lambda_-(\mathrm{d}r)\,\mathbb{P}_{\widehat{X}^*}(\mathrm{d}\pi)\,\mathbb{P}_{\widetilde{\Sigma}}(\mathrm{d}\widetilde{\sigma})\end{aligned} \tag{14.4}$$

with

$$\begin{aligned}A_1 := &\int_0^\infty \int_{\mathcal{H}} \sum_{P\in\mathcal{P}_d} \mathbb{1}\{P \in \mathbf{Y}(\widetilde{\sigma} + \delta_{(x,r,\psi+\delta_{(H,t)})} + \delta_{(0,0,\pi)})\}\\ &\times \mathbb{1}(P,\widetilde{\sigma}+\delta_{(x,r,\psi+\delta_{(H,t)})},(x,r,\psi+\delta_{(H,t)}),(H,t))\\ &\times g((Y_s(\widetilde{\sigma}+\delta_{(x,r,\psi+\delta_{(H,t)})}+\delta_{(0,0,\pi)}),\ s\le t),\\ &m^*(\widetilde{\sigma}+\delta_{(x,r,\psi+\delta_{(H,t)})}+\delta_{(0,0,\pi)}) \wedge P, P, P\cap H, t)\,\Theta(\mathrm{d}H)\,\mathrm{d}t\end{aligned}$$

and

$$A_2 := \int_0^\infty \int_{\mathcal{H}} \sum_{P \in \mathcal{P}_d} \mathbb{1}\{P \in \mathbf{Y}(\widetilde{\sigma} + \delta_{(x,r,\psi)} + \delta_{(o,0,\pi+\delta_{(H,t)})})\} \, \mathbb{1}(P, o, \pi + \delta_{(H,t)}, (H,t))$$
$$\times g((Y_s(\widetilde{\sigma} + \delta_{(x,r,\psi)} + \delta_{(o,0,\pi+\delta_{(H,t)})}), s \le t),$$
$$m^*(\widetilde{\sigma} + \delta_{(x,r,\psi)} + \delta_{(o,0,\pi+\delta_{(H,t)})}) \wedge P, P, P \cap H, t) \, \Theta(\mathrm{d}H)\, \mathrm{d}t\,.$$

Note that the (x,r)-value of $(x,r,\psi + \delta_{(H,t)})$ is the same as that of (x,r,ψ). We use the abbreviation

$$\mathbb{1}(x \in P,\ r = \max,\ \widetilde{\sigma})$$
$$:= \mathbb{1}\{x \in P,\ o \notin P\} \, \mathbb{1}\{r \ge \max\{r' \in \mathbb{R}^- : (x', r', \psi') \in \widetilde{\sigma}, x' \in P\}\}\,.$$

Furthermore, we have that

(i) $P \in \mathbf{Y}(\widetilde{\sigma} + \delta_{(x,r,\psi+\delta_{(H,t)})} + \delta_{(o,0,\pi)})$ and $\mathbb{1}(P, \widetilde{\sigma} + \delta_{(x,r,\psi+\delta_{(H,t)})}, (x,r,\psi + \delta_{(H,t)}), (H,t)) = 1$

if and only if

(ii) $P \in Y_t(\widetilde{\sigma} + \delta_{(x,r,\psi)} + \delta_{(o,0,\pi)})$, $o \notin P$ and $\mathbb{1}(x \in P,\ r = \max,\ \widetilde{\sigma}) = 1$, and P is divided by H at time t using the *additional* pair (H,t).

If $t > 0$, $P \in Y_t(\sigma)$, and $H \in [P]$, then let $m^*(\sigma, \mathsf{D}(P,H,t))$ denote the realization of the maximal polytopes of the STIT tessellation process that until time t coincides with $\mathbf{Y}(\sigma)$, at time t the cell P is divided by an extra hyperplane H, and after time t the global construction is continued based on σ. Note that the division of P by H has an impact on the construction after time t. Analogously, define $(Y_s(\sigma, \mathsf{D}(P,H,t)),\ s \le t)$. With this notation, it follows that

$$A_1 = \int_0^\infty \int_{\mathcal{H}} \sum_{P \in \mathcal{P}_d} \mathbb{1}\{P \in Y_t(\widetilde{\sigma} + \delta_{(x,r,\psi)} + \delta_{(o,0,\pi)})\} \, \mathbb{1}\{o \notin P\} \, \mathbb{1}\{H \in [P]\}$$
$$\times \mathbb{1}(x \in P,\ r = \max,\ \widetilde{\sigma}) \, g((Y_s(\widetilde{\sigma} + \delta_{(x,r,\psi)} + \delta_{(o,0,\pi)}, \mathsf{D}(P,H,t)),\ s \le t),$$
$$m^*(\widetilde{\sigma} + \delta_{(x,r,\psi)} + \delta_{(o,0,\pi)}, \mathsf{D}(P,H,t)) \wedge P, P, P \cap H, t) \, \Theta(\mathrm{d}H)\, \mathrm{d}t$$

and, correspondingly,

$$A_2 = \int_0^\infty \int_{\mathcal{H}} \sum_{P \in \mathcal{P}_d} \mathbb{1}\{P \in Y_t(\widetilde{\sigma} + \delta_{(x,r,\psi)} + \delta_{(o,0,\pi)})\} \, \mathbb{1}\{o \in P\} \, \mathbb{1}\{H \in [P]\}$$
$$\times g((Y_s(\widetilde{\sigma} + \delta_{(x,r,\psi)} + \delta_{(o,0,\pi)}, \mathsf{D}(P,H,t)),\ s \le t),$$
$$m^*((\widetilde{\sigma} + \delta_{(x,r,\psi)} + \delta_{(o,0,\pi)}, \mathsf{D}(P,H,t))) \wedge P, P, P \cap H, t) \, \Theta(\mathrm{d}H)\, \mathrm{d}t\,.$$

Plugging this into (14.4) and applying then the Mecke equation (19.5) to the Poisson point process $\widetilde{\Sigma}$, but not to $\widehat{X}^*$, we conclude that

$$A = \int\limits_{\mathsf{N}_s(\mathbb{R}^d\times\mathbb{R}^-\times\mathsf{N}_s(\mathcal{H}\times\mathbb{R}^+))} \int\limits_{\mathsf{N}_s(\mathcal{H}\times\mathbb{R}^+)} \Bigg(\int\limits_0^\infty \int\limits_{\mathcal{H}} \sum_{P\in\mathcal{P}_d} \mathbb{1}\{P \in Y_t(\tilde{\sigma} + \delta_{(o,0,\pi)})\}\mathbb{1}\{H \in [P]\}$$

$$\times \Bigg(\mathbb{1}\{o \in P\} + \sum_{(x,r,\psi)\in\tilde{\sigma}} \mathbb{1}(x \in P,\, r = \max,\, \tilde{\sigma}) \Bigg)$$

$$\times\, g((Y_s(\tilde{\sigma} + \delta_{(o,0,\pi)}), \mathsf{D}(P,H,t)),\, s \le t),$$

$$m^*(\tilde{\sigma} + \delta_{(o,0,\pi)}, \mathsf{D}(P,H,t)) \wedge P, P, P \cap H, t)\, \Theta(\mathrm{d}H)\, \mathrm{d}t \Bigg) \mathbb{P}_{\widehat{X}^*}(\mathrm{d}\pi)\, \mathbb{P}_{\widetilde{\Sigma}}(\mathrm{d}\tilde{\sigma}) .$$

Note that for all $P \in Y_t(\tilde{\sigma} + \delta_{(o,0,\pi)})$, we have $\sum_{(x,r,\psi)\in\tilde{\sigma}} \mathbb{1}(x \in P,\, r = \max,\, \tilde{\sigma}) = \mathbb{1}\{o \notin P\}$ a. s. by the definition of $\widetilde{\Sigma}$. Hence

$$\mathbb{1}\{o \in P\} + \sum_{(x,r,\psi)\in\tilde{\sigma}} \mathbb{1}(x \in P,\, r = \max,\, \tilde{\sigma}) = 1 .$$

Now we use once more the transformation theorem for image measures. Notice that the cell P is divided at time t by the hyperplane H into two daughter cells and that within these two daughter cells, two independent STIT tessellation processes are realized. The independence of these processes follows from Theorem 8.7 and (6.1). This yields

$$A = \int\limits_0^\infty \int\limits_{\mathcal{T}^{\mathbb{R}^+}} \int\limits_{\mathsf{N}_s(\mathcal{P}_{d-1}\times\mathbb{R}^+)} \int\limits_{\mathsf{N}_s(\mathcal{P}_{d-1}\times\mathbb{R}^+)} \int\limits_{\mathcal{H}} \sum_{P\in y_t} g((y_s, , s \le t),$$

$$\{(P \cap H, t)\} \cup (m^{*(1)}_{(+t)} \wedge (P \cap H^+)) \cup (m^{*(2)}_{(+t)} \wedge (P \cap H^-)), P, P \cap H, t)$$

$$\times\, \mathbb{1}\{H \in [P]\}\, \Theta(\mathrm{d}H)\, \mathbb{P}^{\max *}(\mathrm{d}m^{*(1)})\, \mathbb{P}^{\max *}(\mathrm{d}m^{*(2)})\, \mathbb{P}_{\mathbf{Y}}(\mathrm{d}\mathbf{y})\, \mathrm{d}t$$

$$= \int\limits_0^\infty \int\limits_{\mathcal{T}^{\mathbb{R}^+}} \sum_{P\in y_t} \int\limits_{[P]} \int\limits_{\mathsf{N}_s(\mathcal{P}_{d-1}\times\mathbb{R}^+)} \int\limits_{\mathsf{N}_s(\mathcal{P}_{d-1}\times\mathbb{R}^+)} g((y_s, , s \le t),$$

$$\{(P \cap H, t)\} \cup (m^{*(1)}_{(+t)} \wedge (P \cap H^+)) \cup (m^{*(2)}_{(+t)} \wedge (P \cap H^-)), P, P \cap H, t)$$

$$\mathbb{P}^{\max *}(\mathrm{d}m^{*(1)})\, \mathbb{P}^{\max *}(\mathrm{d}m^{*(2)})\, \Theta(\mathrm{d}H)\, \mathbb{P}_{\mathbf{Y}}(\mathrm{d}\mathbf{y})\, \mathrm{d}t ,$$

which completes the proof. □

Theorem 14.1 shares some similarities with the Mecke equation in Theorem 19.11 for Poisson point processes. Obviously, $\mathcal{F}^{\max *}_{d-1}(\Sigma)$ is not a Poisson point process, but formally, the left-hand side of (14.1) has the same structure as the left-hand side of the Mecke equation for Poisson point processes. Moreover, on the right-hand side of (14.1), we see that an additional hyperplane H is introduced at time t using the intensity measure $\Theta \otimes \lambda_+$, which is similar to the right-hand side of the Mecke equation. The main differences are that in (14.1), for *each* cell $P \in y_t$, a hyperplane is added and furthermore, after the division of a cell P by a hyperplane H, realizations, denoted $m^{*(1)}$ and $m^{*(2)}$, of independent copies of $\mathcal{F}^{\max *}_{d-1}(\Sigma)$ are needed to continue the process in time.

Now we derive a much simpler but useful particular case of Theorem 14.1.

Corollary 14.2. *Let Θ be a hyperplane measure of the form* (4.1), *satisfying* (4.3), *and let* $(Y_t, t > 0)$ *be the STIT tessellation process generated by the point process Σ of* (8.1). *Then, for all nonnegative measurable functions $f : \mathcal{P}_{d-1} \times \mathbb{R}^+ \to \mathbb{R}$, we have that*

$$\mathbb{E} \sum_{(F,t)\in\mathcal{F}_{d-1}^{\max *}(\Sigma)} f(F,t) = \int_0^\infty \int_{\mathcal{T}} \int_{\mathcal{H}} \sum_{P\in y_t} f(P\cap H, t)\, \mathbb{1}\{H \in [P]\}\, \Theta(\mathrm{d}H)\, \mathbb{P}_{Y_t}(\mathrm{d}y_t)\, \mathrm{d}t\,.$$

Proof. Choose the function g in (14.1) as

$$g((Y_s,\, s \le t),\, \mathcal{F}_{d-1}^{\max *}(\Sigma) \wedge P(F),\, P(F),\, F,\, t) := f(F,t)\,.$$

This yields

$$\mathbb{E} \sum_{(F,t)\in\mathcal{F}_{d-1}^{\max *}(\Sigma)} f(F,t) = \int_0^\infty \int_{\mathcal{T}} \sum_{P\in y_t} \int_{[P]} f(P\cap H, t)\, \Theta(\mathrm{d}H)\, \mathbb{P}_{Y_t}(\mathrm{d}y_t)\, \mathrm{d}t\,,$$

which completes the proof. □

The following theorem and corollary can be interpreted as multivariate versions of the proposition above, analogously to the multivariate Mecke equation (19.4). To make the formulae not too convoluted, in the theorem, only the bivariate case is proven. We use the notation $\oslash(P,H,T)$ defined in (4.6) for the division of a cell P in a tessellation T by a hyperplane H and the notation $\boxplus$ introduced by (3.4) for the iteration of tessellations.

Theorem 14.3. *Let Θ be a hyperplane measure of the form* (4.1), *satisfying* (4.3), *and let* $(Y_t, t > 0)$ *be the STIT tessellation process generated by the point process Σ of* (8.1). *Then, for fixed $t > 0$ and all nonnegative measurable functions $g : \mathcal{P}_{d-1}^2 \times (\mathbb{R}^+)^2 \times \mathcal{T} \to \mathbb{R}$, we have the following bivariate Mecke-type equation:*

$$\begin{aligned}
&\mathbb{E} \sum_{(F_2,s_2)\in\mathcal{F}_{d-1}^{\max *}(\Sigma)} \sum_{(F_1,s_1)\in\mathcal{F}_{d-1}^{\max *}(\Sigma)} \mathbb{1}\{s_1 < s_2 \le t\}\, g(F_1, F_2, s_1, s_2, Y_t) \\
&= \int_0^t \int_0^{s_2} \int_{\mathcal{T}^{\mathbb{N}_0}} \int_{\mathcal{T}^{\mathbb{N}_0}} \int_{\mathcal{T}} \sum_{P_1\in y_{s_1}} \sum_{P_2\in\oslash(P_1,H_1,y_{s_1})\boxplus \underline{y_{s_2-s_1}}} \int_{[P_2]} \int_{[P_1]} \\
&\quad g(P_1\cap H_1,\, P_2\cap H_2,\, s_1,\, s_2,\, \oslash(P_2, H_2, \oslash(P_1, H_1, y_{s_1}) \boxplus \underline{y_{s_2-s_1}}) \boxplus \underline{y_{t-s_2}}) \\
&\quad \Theta(\mathrm{d}H_1)\, \Theta(\mathrm{d}H_2)\, \mathbb{P}_{Y_{s_1}}(\mathrm{d}y_{s_1})\, \mathbb{P}_{Y_{s_2-s_1}}^{\otimes\mathbb{N}_0}(\mathrm{d}\underline{y_{s_2-s_1}})\, \mathbb{P}_{Y_{t-s_2}}^{\otimes\mathbb{N}_0}(\mathrm{d}\underline{y_{t-s_2}})\, \mathrm{d}s_1\, \mathrm{d}s_2\,.
\end{aligned}$$

Proof. This proof is based on the same ideas as the proof of Theorem 14.1. The principal difference is that now, as a consequence of the application of the multivariate Mecke equation for Poisson point processes, the cell P_2 belongs to a tessellation generated by Σ *and an additional hyperplane* H_1 that divides P_1 at time s_1.

Recall the notation $\mathbb{1}(P, \Sigma, (x, r, \psi), (H, t))$ and $\mathbb{1}(P, o, \widehat{X}_o^*, (H, t))$ introduced in (14.2) and (14.3), respectively. These terms are 1 if P is divided by H at time t and 0 otherwise.

For the mapping $\Sigma \mapsto \mathcal{F}_{d-1}^{\max *}(\Sigma)$, the transformation formula for image measures yields

$$A := \int\limits_{\mathsf{N}_s(\mathcal{P}_{d-1}\times\mathbb{R}^+)} \sum_{(F_2,s_2)\in m^*} \sum_{(F_1,s_1)\in m^*} \mathbb{1}\{s_1 < s_2 \le t\}\, g(F_1, F_2, s_1, s_2, Y_t(m^*))\, \mathbb{P}^{\max *}(dm^*)$$

$$= \int\limits_{\mathsf{N}_s(\mathbb{R}^d\times\mathbb{R}^-\times\mathsf{N}_s(\mathcal{H}\times\mathbb{R}^+))} \sum_{(F_2,s_2)\in m^*(\sigma)} \sum_{(F_1,s_1)\in m^*(\sigma)} \mathbb{1}\{s_1 < s_2 \le t\}$$
$$\times\, g(F_1, F_2, s_1, s_2, Y_t(\sigma))\, \mathbb{P}_\Sigma(d\sigma)\,.$$

A maximal polytope of dimension $d - 1$ appears when a cell is divided by a hyperplane H. In the formulae below the term $P \in \mathbf{Y}(\sigma)$ means that there exists a time $t > 0$ such that $P \in Y_t(\sigma)$, that is, P is a cell extant at time t of the STIT tessellation process generated by a realization σ of Σ.

Recalling that in (8.1) the point process Σ is defined as the sum $\Sigma := \widetilde{\Sigma} + \delta_{(o,0,\widehat{X}_o^*)}$, we obtain that A is equal to

$$\int\limits_{\mathsf{N}_s(\mathbb{R}^d\times\mathbb{R}^-\times\mathsf{N}_s(\mathcal{H}\times\mathbb{R}^+))} \int\limits_{\mathsf{N}_s(\mathcal{H}\times\mathbb{R}^+)}$$
$$\sum_{P_1\in\mathcal{P}_d} \mathbb{1}\{P_1 \in \mathbf{Y}(\tilde{\sigma} + \delta_{(o,0,\pi)})\} \sum_{P_2\in\mathcal{P}_d} \mathbb{1}\{P_2 \in \mathbf{Y}(\tilde{\sigma} + \delta_{(o,0,\pi)})\}$$
$$\times \Bigg(\sum_{(x_1,r_1,\psi_1)\in\tilde{\sigma}} \sum_{(H_1,s_1)\in\psi_1} \mathbb{1}(P_1, \tilde{\sigma}, (x_1, r_1, \psi_1), (H_1, s_1)) + \sum_{(H_1,s_1)\in\pi} \mathbb{1}(P_1, o, \pi, (H_1, s_1)) \Bigg)$$
$$\times \Bigg(\sum_{(x_2,r_2,\psi_2)\in\tilde{\sigma}} \sum_{(H_2,s_2)\in\psi_2} \mathbb{1}(P_2, \tilde{\sigma}, (x_2, r_2, \psi_2), (H_2, s_2)) + \sum_{(H_2,s_2)\in\pi} \mathbb{1}(P_2, o, \pi, (H_2, s_2)) \Bigg)$$
$$\times\, \mathbb{1}\{s_1 < s_2 \le t\}\, g(P_1 \cap H_1, P_2 \cap H_2, s_1, s_2, Y_t(\tilde{\sigma} + \delta_{(o,0,\pi)}))\, \mathbb{P}_{\widehat{X}^*}(d\pi)\, \mathbb{P}_{\widetilde{\Sigma}}(d\tilde{\sigma})\,.$$

Multiplying the product of the two sums yields four summands and thus has the form $A = B_1 + B_2 + B_3 + B_4$. Here we consider only one term in all detail; the other three can be treated analogously. Denote

$$B_1 := \int\limits_{\mathsf{N}_s(\mathbb{R}^d\times\mathbb{R}^-\times\mathsf{N}_s(\mathcal{H}\times\mathbb{R}^+))} \int\limits_{\mathsf{N}_s(\mathcal{H}\times\mathbb{R}^+)}$$
$$\sum_{P_1\in\mathcal{P}_d} \mathbb{1}\{P_1 \in \mathbf{Y}(\tilde{\sigma} + \delta_{(o,0,\pi)})\} \sum_{P_2\in\mathcal{P}_d} \mathbb{1}\{P_2 \in \mathbf{Y}(\tilde{\sigma} + \delta_{(o,0,\pi)})\}$$
$$\times \sum_{(x_1,r_1,\psi_1)\in\tilde{\sigma}} \sum_{(H_1,s_1)\in\psi_1} \mathbb{1}(P_1, \tilde{\sigma}, (x_1, r_1, \psi_1), (H_1, s_1))$$
$$\times \sum_{(x_2,r_2,\psi_2)\in\tilde{\sigma}} \sum_{(H_2,s_2)\in\psi_2} \mathbb{1}(P_2, \tilde{\sigma}, (x_2, r_2, \psi_2), (H_2, s_2))\, \mathbb{1}\{s_1 < s_2 \le t\}$$
$$\times\, g(P_1 \cap H_1, P_2 \cap H_2, s_1, s_2, Y_t(\tilde{\sigma} + \delta_{(o,0,\pi)}))\, \mathbb{P}_{\widehat{X}^*}(d\pi)\, \mathbb{P}_{\widetilde{\Sigma}}(d\tilde{\sigma})\,.$$

We partition B_1 into two summands $B_1^{=} := \mathbb{1}\{(x_1,r_1,\psi_1) = (x_2,r_2,\psi_2)\}\, B_1$ and $B_1^{\neq} := \mathbb{1}\{(x_1,r_1,\psi_1) \neq (x_2,r_2,\psi_2)\}\, B_1$, accepting a slight abuse of notation. Note that $(x_1,r_1,\psi_1) = (x_2,r_2,\psi_2)$ is a. s. equivalent to $x_1 = x_2$ by the definition of $\widetilde{\Sigma}$. First consider

$$\begin{aligned} B_1^{=} = &\int_{\mathsf{N}_s(\mathbb{R}^d\times\mathbb{R}^-\times\mathsf{N}_s(\mathcal{H}\times\mathbb{R}^+))} \int_{\mathsf{N}_s(\mathcal{H}\times\mathbb{R}^+)} \\ &\sum_{P_1\in\mathcal{P}_d} \mathbb{1}\{P_1 \in \mathbf{Y}(\widetilde{\sigma} + \delta_{(o,0,\pi)})\} \sum_{P_2\in\mathcal{P}_d} \mathbb{1}\{P_2 \in \mathbf{Y}(\widetilde{\sigma} + \delta_{(o,0,\pi)})\} \\ &\times \sum_{(x_1,r_1,\psi_1)\in\widetilde{\sigma}} \sum_{(H_1,s_1)\in\psi_1} \mathbb{1}(P_1, \widetilde{\sigma}, (x_1,r_1,\psi_1), (H_1,s_1)) \\ &\times \sum_{(H_2,s_2)\in\psi_1} \mathbb{1}(P_2, \widetilde{\sigma}, (x_1,r_1,\psi_1), (H_2,s_2))\, \mathbb{1}\{s_1 < s_2 \le t\} \\ &\times g(P_1\cap H_1,\, P_2\cap H_2,\, s_1,\, s_2,\, Y_t(\widetilde{\sigma} + \delta_{(o,0,\pi)}))\, \mathbb{P}_{\widehat{X}^*}(\mathrm{d}\pi)\, \mathbb{P}_{\widetilde{\Sigma}}(\mathrm{d}\widetilde{\sigma})\,. \end{aligned}$$

Note that in this case, $P_2 \subset P_1$ by the global construction based on $\widetilde{\Sigma}$. An application of the Mecke equation (19.5) to the Poisson point process $\widetilde{\Sigma}$ and thereafter of the multivariate Mecke equation (19.4) to the Poisson point process ψ_1 yields that

$$\begin{aligned} B_1^{=} = &\int_{\mathsf{N}_s(\mathbb{R}^d\times\mathbb{R}^-\times\mathsf{N}_s(\mathcal{H}\times\mathbb{R}^+))} \int_{\mathsf{N}_s(\mathcal{H}\times\mathbb{R}^+)} \int_{\mathbb{R}^-} \int_{\mathbb{R}^d} \int_{\mathsf{N}_s(\mathcal{H}\times\mathbb{R}^+)} \int_{\mathbb{R}^+} \int_{\mathbb{R}^+} \int_{[P_2]} \int_{[P_1]} \\ &\sum_{P_1\in\mathcal{P}_d} \mathbb{1}\{P_1 \in \mathbf{Y}(\widetilde{\sigma} + \delta_{(x_1,r_1,\psi_1+\delta_{(H_1,s_1)}+\delta_{(H_2,s_2)})} + \delta_{(o,0,\pi)})\} \\ &\times \sum_{P_2\in\mathcal{P}_d} \mathbb{1}\{P_2 \in \mathbf{Y}(\widetilde{\sigma} + \delta_{(x_1,r_1,\psi_1+\delta_{(H_1,s_1)}+\delta_{(H_2,s_2)})} + \delta_{(o,0,\pi)})\} \\ &\times \mathbb{1}(P_1, \widetilde{\sigma} + \delta_{(x_1,r_1,\psi_1+\delta_{(H_1,s_1)}+\delta_{(H_2,s_2)})}, (x_1, r_1, \psi_1 + \delta_{(H_1,s_1)} + \delta_{(H_2,s_2)}), (H_1, s_1)) \\ &\times \mathbb{1}(P_2, \widetilde{\sigma} + \delta_{(x_1,r_1,\psi_1+\delta_{(H_1,s_1)}+\delta_{(H_2,s_2)})}, (x_1, r_1, \psi_1 + \delta_{(H_1,s_1)} + \delta_{(H_2,s_2)}), (H_2, s_2)) \\ &\times \mathbb{1}\{s_1 < s_2 \le t\} \\ &\times g(P_1\cap H_1,\, P_2\cap H_2,\, s_1,\, s_2,\, Y_t(\widetilde{\sigma} + \delta_{(x_1,r_1,\psi_1+\delta_{(H_1,s_1)}+\delta_{(H_2,s_2)})} + \delta_{(o,0,\pi)})) \\ &\Theta(\mathrm{d}H_1)\, \Theta(\mathrm{d}H_2)\, \mathrm{d}s_1\, \mathrm{d}s_2\, \mathbb{P}_{\widehat{X}^*}(\mathrm{d}\psi_1)\, \lambda_d(\mathrm{d}x_1)\, \lambda_-(\mathrm{d}r_1)\, \mathbb{P}_{\widehat{X}^*}(\mathrm{d}\pi)\, \mathbb{P}_{\widetilde{\Sigma}}(\mathrm{d}\widetilde{\sigma})\,. \end{aligned}$$

We use the abbreviation

$$\begin{aligned} &\mathbb{1}(x \in P,\, r = \max,\, \widetilde{\sigma}) \\ &\quad := \mathbb{1}\{x \in P,\, o \notin P\}\, \mathbb{1}\{r \ge \max\{r' \in \mathbb{R}^- : (x', r', \psi') \in \widetilde{\sigma}, x' \in P\}\}\,. \end{aligned}$$

Furthermore, for $s_1 < s_2$, we have a. s. that

(i) $P_1 \in \mathbf{Y}(\widetilde{\sigma} + \delta_{(x_1,r_1,\psi_1+\delta_{(H_1,s_1)}+\delta_{(H_2,s_2)})} + \delta_{(o,0,\pi)})$ and
$\mathbb{1}(P_1, \widetilde{\sigma} + \delta_{(x_1,r_1,\psi_1+\delta_{(H_1,s_1)}+\delta_{(H_2,s_2)})}, (x_1, r_1, \psi_1 + \delta_{(H_1,s_1)} + \delta_{(H_2,s_2)}), (H_1, s_1)) = 1$

if and only if

(ii) $P_1 \in Y_{s_1}(\tilde{\sigma} + \delta_{(x_1,r_1,\psi_1)} + \delta_{(0,0,\pi)})$ and $\mathbb{1}(x_1 \in P_1,\, r_1 = \max,\, \tilde{\sigma}) = 1$, and P_1 is divided by H_1 at time s_1 using the *additional* pair (H_1, s_1).

Note that the division of P_1 by H_1 has an impact on the construction after time s_1. If $P_1 \in Y_{s_1}(\tilde{\sigma})$, then the notation

$$\tilde{\sigma} \,\&\, \oslash(P_1, H_1, Y_{s_1})$$

means that $\tilde{\sigma}$ is supplemented such that P_1 is divided by H_1 at time s_1. After this division, the tessellation valued process is continued with $\tilde{\sigma}$. Thus, for $s_2 > s_1$, we have to consider in $B_1^=$ the cells P_2 for which $P_2 \in Y_{s_2}((\tilde{\sigma} + \delta_{(x_1,r_1,\psi_1)} + \delta_{(0,0,\pi)}) \,\&\, \oslash(P_1, H_1, Y_{s_1}))$ and $\mathbb{1}(x_1 \in P_2,\, r_1 = \max,\, \tilde{\sigma}) = 1$ and, furthermore, P_2 is divided by H_2 at time s_2 using the *additional* pair (H_2, s_2). It follows that

$$\begin{aligned}
B_1^= = & \int\limits_{\mathsf{N}_s(\mathbb{R}^d\times\mathbb{R}^-\times\mathsf{N}_s(\mathcal{H}\times\mathbb{R}^+))} \;\int\limits_{\mathsf{N}_s(\mathcal{H}\times\mathbb{R}^+)} \;\int\limits_{\mathbb{R}^-} \int\limits_{\mathbb{R}^d} \;\int\limits_{\mathsf{N}_s(\mathcal{H}\times\mathbb{R}^+)} \;\int\limits_{\mathbb{R}^+} \int\limits_{\mathbb{R}^+} \int\limits_{\mathcal{H}} \int\limits_{\mathcal{H}} \\
& \sum_{P_1 \in \mathcal{P}_d} \mathbb{1}\{P_1 \in Y_{s_1}(\tilde{\sigma} + \delta_{(x_1,r_1,\psi_1)} + \delta_{(0,0,\pi)})\} \\
& \times \mathbb{1}\{H_1 \in [P_1]\}\, \mathbb{1}(x_1 \in P_1,\, r_1 = \max,\, \tilde{\sigma}) \\
& \times \sum_{P_2 \in \mathcal{P}_d} \mathbb{1}\{P_2 \in Y_{s_2}((\tilde{\sigma} + \delta_{(x_1,r_1,\psi_1)} + \delta_{(0,0,\pi)}) \,\&\, \oslash(P_1, H_1, Y_{s_1}))\} \\
& \times \mathbb{1}\{H_2 \in [P_2]\}\, \mathbb{1}(x_1 \in P_2,\, r_1 = \max,\, \tilde{\sigma})\, \mathbb{1}\{s_1 < s_2 \le t\} \\
& \times g(P_1 \cap H_1,\, P_2 \cap H_2,\, s_1,\, s_2, \\
& Y_t((\tilde{\sigma} + \delta_{(x_1,r_1,\psi_1)} + \delta_{(0,0,\pi)}) \,\&\, \oslash(P_1, H_1, Y_{s_1}) \,\&\, \oslash(P_2, H_2, Y_{s_2}(\oslash(P_1, H_1, Y_{s_1}))))) \\
& \Theta(\mathrm{d}H_2)\, \Theta(\mathrm{d}H_1)\, \mathrm{d}s_1\, \mathrm{d}s_2\, \mathbb{P}_{\tilde{X}^*}(\mathrm{d}\psi_1)\, \lambda_d(\mathrm{d}x_1)\, \lambda_-(\mathrm{d}r_1)\, \mathbb{P}_{\tilde{X}^*}(\mathrm{d}\pi)\, \mathbb{P}_{\tilde{\Sigma}}(\mathrm{d}\tilde{\sigma}),
\end{aligned}$$

where we used the acronym $Y_{s_2}(\oslash(P_1, H_1, Y_{s_1}))$ for the state of the STIT tessellation process at time s_2 where the construction is amended by the additional division event $\oslash(P_1, H_1, Y_{s_1})$. Now by the Mecke equation (19.5) for the Poisson point process $\tilde{\Sigma}$, we conclude that

$$\begin{aligned}
B_1^= = & \int\limits_{\mathsf{N}_s(\mathbb{R}^d\times\mathbb{R}^-\times\mathsf{N}_s(\mathcal{H}\times\mathbb{R}^+))} \;\int\limits_{\mathsf{N}_s(\mathcal{H}\times\mathbb{R}^+)} \;\int\limits_{\mathbb{R}^+} \int\limits_{\mathbb{R}^+} \int\limits_{\mathcal{H}} \int\limits_{\mathcal{H}} \\
& \sum_{P_1 \in \mathcal{P}_d} \mathbb{1}\{P_1 \in Y_{s_1}(\tilde{\sigma} + \delta_{(0,0,\pi)})\} \sum_{(x_1,r_1,\psi_1) \in \tilde{\sigma}} \mathbb{1}(x_1 \in P_1,\, r_1 = \max,\, \tilde{\sigma}) \\
& \times \sum_{P_2 \in \mathcal{P}_d} \mathbb{1}\{P_2 \in Y_{s_2}((\tilde{\sigma} + \delta_{(0,0,\pi)}) \,\&\, \oslash(P_1, H_1, Y_{s_1}))\} \\
& \times \mathbb{1}(x_1 \in P_2,\, r_1 = \max,\, \tilde{\sigma})\, \mathbb{1}\{s_1 < s_2 \le t\}\, \mathbb{1}\{H_1 \in [P_1]\}\, \mathbb{1}\{H_2 \in [P_2]\}
\end{aligned}$$

$$\times g(P_1 \cap H_1, P_2 \cap H_2, s_1, s_2, \\ Y_t((\widetilde{\sigma} + \delta_{(o,0,\pi)}) \,\&\, \oslash (P_1, H_1, Y_{s_1}) \,\&\, \oslash (P_2, H_2, Y_{s_2}(\oslash(P_1, H_1, Y_{s_1}))))) \\ \Theta(dH_2)\, \Theta(dH_1)\, ds_1\, ds_2\, \mathbb{P}_{\widehat{X}^*}(d\pi)\, \mathbb{P}_{\widetilde{\Sigma}}(d\widetilde{\sigma}). \tag{14.5}$$

To deal with $B_1^{\neq}$, we apply the multivariate Mecke equation (19.4) to the Poisson point process $\widetilde{\Sigma}$, which yields

$$B_1^{\neq} = \int\limits_{N_s(\mathbb{R}^d \times \mathbb{R}^- \times N_s(\mathcal{H}\times\mathbb{R}^+))} \int\limits_{N_s(\mathcal{H}\times\mathbb{R}^+)} \int\limits_{\mathbb{R}^-}\int\limits_{\mathbb{R}^-}\int\limits_{\mathbb{R}^d}\int\limits_{\mathbb{R}^d} \int\limits_{N_s(\mathcal{H}\times\mathbb{R}^+)} \int\limits_{N_s(\mathcal{H}\times\mathbb{R}^+)} \\ \sum_{P_1 \in \mathcal{P}_d} \mathbb{1}\{P_1 \in \mathbf{Y}(\widetilde{\sigma} + \delta_{(x_1,r_1,\psi_1)} + \delta_{(x_2,r_2,\psi_2)} + \delta_{(o,0,\pi)})\} \\ \times \sum_{P_2 \in \mathcal{P}_d} \mathbb{1}\{P_2 \in \mathbf{Y}(\widetilde{\sigma} + \delta_{(x_1,r_1,\psi_1)} + \delta_{(x_2,r_2,\psi_2)} + \delta_{(o,0,\pi)})\} \\ \times \sum_{(H_1,s_1)\in\psi_1} \mathbb{1}(P_1, \widetilde{\sigma} + \delta_{(x_1,r_1,\psi_1)} + \delta_{(x_2,r_2,\psi_2)}, (x_1,r_1,\psi_1), (H_1,s_1)) \\ \times \sum_{(H_2,s_2)\in\psi_2} \mathbb{1}(P_2, \widetilde{\sigma} + \delta_{(x_1,r_1,\psi_1)} + \delta_{(x_2,r_2,\psi_2)}, (x_2,r_2,\psi_2), (H_2,s_2))\, \mathbb{1}\{s_1 < s_2 \le t\} \\ \times g(P_1 \cap H_1, P_2 \cap H_2, s_1, s_2, Y_t(\widetilde{\sigma} + \delta_{(x_1,r_1,\psi_1)} + \delta_{(x_2,r_2,\psi_2)} + \delta_{(o,0,\pi)})) \\ \mathbb{P}_{\widehat{X}^*}(d\psi_1)\, \mathbb{P}_{\widehat{X}^*}(d\psi_2)\, \lambda_d(dx_1)\, \lambda_d(dx_2)\, \lambda_-(dr_1)\, \lambda_-(dr_2)\, \mathbb{P}_{\widehat{X}^*}(d\pi)\, \mathbb{P}_{\widetilde{\Sigma}}(d\widetilde{\sigma}).$$

Next, we apply the Mecke equation (19.5), this time with respect to ψ_1 and ψ_2 separately. We obtain

$$B_1^{\neq} = \int\limits_{N_s(\mathbb{R}^d \times \mathbb{R}^- \times N_s(\mathcal{H}\times\mathbb{R}^+))} \int\limits_{N_s(\mathcal{H}\times\mathbb{R}^+)} \int\limits_{\mathbb{R}^-}\int\limits_{\mathbb{R}^-}\int\limits_{\mathbb{R}^d}\int\limits_{\mathbb{R}^d} \int\limits_{N_s(\mathcal{H}\times\mathbb{R}^+)} \int\limits_{N_s(\mathcal{H}\times\mathbb{R}^+)} \int\limits_{\mathbb{R}^+}\int\limits_{\mathbb{R}^+}\int\limits_{[P_2]}\int\limits_{[P_1]} \\ \sum_{P_1 \in \mathcal{P}_d} \mathbb{1}\{P_1 \in \mathbf{Y}(\widetilde{\sigma} + \delta_{(x_1,r_1,\psi_1+\delta_{(H_1,s_1)})} + \delta_{(x_2,r_2,\psi_2+\delta_{(H_2,s_2)})} + \delta_{(o,0,\pi)})\} \\ \times \sum_{P_2 \in \mathcal{P}_d} \mathbb{1}\{P_2 \in \mathbf{Y}(\widetilde{\sigma} + \delta_{(x_1,r_1,\psi_1+\delta_{(H_1,s_1)})} + \delta_{(x_2,r_2,\psi_2+\delta_{(H_2,s_2)})} + \delta_{(o,0,\pi)})\} \\ \times \mathbb{1}(P_1, \widetilde{\sigma} + \delta_{(x_1,r_1,\psi_1+\delta_{(H_1,s_1)})} + \delta_{(x_2,r_2,\psi_2+\delta_{(H_2,s_2)})}, (x_1,r_1,\psi_1+\delta_{(H_1,s_1)}), (H_1,s_1)) \\ \times \mathbb{1}(P_2, \widetilde{\sigma} + \delta_{(x_1,r_1,\psi_1+\delta_{(H_1,s_1)})} + \delta_{(x_2,r_2,\psi_2+\delta_{(H_2,s_2)})}, (x_2,r_2,\psi_2+\delta_{(H_2,s_2)}), (H_2,s_2)) \\ \times \mathbb{1}\{s_1 < s_2 \le t\} g(P_1 \cap H_1, P_2 \cap H_2, s_1, s_2, \\ Y_t(\widetilde{\sigma} + \delta_{(x_1,r_1,\psi_1+\delta_{(H_1,s_1)})} + \delta_{(x_2,r_2,\psi_2+\delta_{(H_2,s_2)})} + \delta_{(o,0,\pi)})) \\ \Theta(dH_1)\, \Theta(dH_2)\, ds_1\, ds_2\, \mathbb{P}_{\widehat{X}^*}(d\psi_1)\, \mathbb{P}_{\widehat{X}^*}(d\psi_2) \lambda_d(dx_1)\, \lambda_d(dx_2) \\ \lambda_-(dr_1)\, \lambda_-(dr_2)\, \mathbb{P}_{\widehat{X}^*}(d\pi)\, \mathbb{P}_{\widetilde{\Sigma}}(d\widetilde{\sigma}).$$

Note that the (x,r)-value of $(x,r,\psi+\delta_{(H,t)})$ is the same as that of (x,r,ψ). We use the abbreviation

$$\mathbb{1}(x \in P, r = \max, \widetilde{\sigma}) \\ := \mathbb{1}\{x \in P, o \notin P\}\, \mathbb{1}\{r \ge \max\{r' \in \mathbb{R}^- : (x',r',\psi') \in \widetilde{\sigma}, x' \in P\}\}.$$

Furthermore, for $s_2 > s_1$, we have a. s. that

(i) $P_1 \in \mathbf{Y}(\tilde{\sigma} + \delta_{(x_1,r_1,\psi_1+\delta_{(H_1,s_1)})} + \delta_{(x_2,r_2,\psi_2+\delta_{(H_2,s_2)})} + \delta_{(o,0,\pi)})$ and $\mathbb{1}(P_1, \tilde{\sigma} + \delta_{(x_1,r_1,\psi_1+\delta_{(H_1,s_1)})} + \delta_{(x_2,r_2,\psi_2+\delta_{(H_2,s_2)})}, (x_1, r_1, \psi_1 + \delta_{(H_1,s_1)}), (H_1, s_1)) = 1$

if and only if

(ii) $P_1 \in Y_{s_1}(\tilde{\sigma} + \delta_{(x_1,r_1,\psi_1)} + \delta_{(x_2,r_2,\psi_2)} + \delta_{(o,0,\pi)})$ and $\mathbb{1}(x_1 \in P_1, r_1 = \max, \tilde{\sigma}) = 1$, and P_1 is divided by H_1 at time s_1 using the *additional* pair (H_1, s_1).

Note that the division of P_1 by H_1 has an impact on the construction after time s_1. Again, we use the notation $\tilde{\sigma}\,\&\oslash(P_1, H_1, Y_{s_1})$ for $P_1 \in Y_{s_1}(\tilde{\sigma})$, which means that $\tilde{\sigma}$ is supplemented such that P_1 is divided by H_1 at time s_1. After this division the tessellation-valued process is continued with $\tilde{\sigma}$.

Thus, for $s_2 > s_1$, we have to consider in $B_1^{\neq}$ the cells P_2 for which $P_2 \in Y_{s_2}((\tilde{\sigma} + \delta_{(x_1,r_1,\psi_1)} + \delta_{(x_2,r_2,\psi_2)} + \delta_{(o,0,\pi)})\,\&\oslash(P_1, H_1, Y_{s_1}))$ and $\mathbb{1}(x_2 \in P_2, r_2 = \max, \tilde{\sigma} + \delta_{(x_1,r_1,\psi_1)}) = 1$, and P_2 is divided by H_2 at time s_2 using the *additional* pair (H_2, s_2). With this notation, it follows that

$$
\begin{aligned}
B_1^{\neq} = & \int\limits_{\mathsf{N}_s(\mathbb{R}^d\times\mathbb{R}^-\times\mathsf{N}_s(\mathcal{H}\times\mathbb{R}^+))} \int\limits_{\mathsf{N}_s(\mathcal{H}\times\mathbb{R}^+)} \int\limits_{\mathbb{R}^-}\int\limits_{\mathbb{R}^-}\int\limits_{\mathbb{R}^d}\int\limits_{\mathbb{R}^d} \int\limits_{\mathsf{N}_s(\mathcal{H}\times\mathbb{R}^+)} \int\limits_{\mathsf{N}_s(\mathcal{H}\times\mathbb{R}^+)} \int\limits_{\mathbb{R}^+}\int\limits_{\mathbb{R}^+}\int\limits_{\mathcal{H}}\int\limits_{\mathcal{H}} \\
& \sum_{P_1\in\mathcal{P}_d} \mathbb{1}\{P_1 \in Y_{s_1}(\tilde{\sigma} + \delta_{(x_1,r_1,\psi_1)} + \delta_{(o,0,\pi)})\} \\
& \times \mathbb{1}\{H_1 \in [P_1]\}\, \mathbb{1}(x_1 \in P_1, r_1 = \max, \tilde{\sigma}) \\
& \times \sum_{P_2\in\mathcal{P}_d} \mathbb{1}\{P_2 \in Y_{s_2}((\tilde{\sigma} + \delta_{(x_1,r_1,\psi_1)} + \delta_{(x_2,r_2,\psi_2)} + \delta_{(o,0,\pi)})\,\&\oslash(P_1, H_1, Y_{s_1}))\} \\
& \times \mathbb{1}\{H_2 \in [P_2]\}\, \mathbb{1}(x_2 \in P_2, r_2 = \max, \tilde{\sigma} + \delta_{(x_1,r_1,\psi_1)})\, \mathbb{1}\{s_1 < s_2 \le t\} \\
& \times g(P_1 \cap H_1, P_2 \cap H_2, s_1, s_2, Y_t((\tilde{\sigma} + \delta_{(x_1,r_1,\psi_1)} + \delta_{(x_2,r_2,\psi_2)} + \delta_{(o,0,\pi)}) \\
& \&\oslash(P_1, H_1, Y_{s_1}))\,\&\oslash(P_2, H_2, Y_{s_2}(\oslash(P_1, H_1, Y_{s_1})))) \\
& \Theta(\mathrm{d}H_2)\, \Theta(\mathrm{d}H_1)\, \mathrm{d}s_1\, \mathrm{d}s_2\, \mathbb{P}_{\widehat{X}^*}(\mathrm{d}\psi_1)\, \mathbb{P}_{\widehat{X}^*}(\mathrm{d}\psi_2) \\
& \lambda_d(\mathrm{d}x_1)\, \lambda_d(\mathrm{d}x_2)\, \lambda_-(\mathrm{d}r_1)\, \lambda_-(\mathrm{d}r_2)\, \mathbb{P}_{\widehat{X}^*}(\mathrm{d}\pi)\, \mathbb{P}_{\tilde{\Sigma}}(\mathrm{d}\tilde{\sigma})\,.
\end{aligned}
$$

Now by the multivariate Mecke equation (19.4) for the Poisson point process $\tilde{\Sigma}$ we conclude that

$$
\begin{aligned}
B_1^{\neq} = & \int\limits_{\mathsf{N}_s(\mathbb{R}^d\times\mathbb{R}^-\times\mathsf{N}_s(\mathcal{H}\times\mathbb{R}^+))} \int\limits_{\mathsf{N}_s(\mathcal{H}\times\mathbb{R}^+)} \int\limits_{\mathbb{R}^+}\int\limits_{\mathbb{R}^+}\int\limits_{\mathcal{H}}\int\limits_{\mathcal{H}} \\
& \sum_{P_1\in\mathcal{P}_d} \mathbb{1}\{P_1 \in Y_{s_1}(\tilde{\sigma} + \delta_{(o,0,\pi)})\} \sum_{(x_1,r_1,\psi_1)\in\tilde{\sigma}} \mathbb{1}(x_1 \in P_1, r_1 = \max, \tilde{\sigma}) \\
& \times \sum_{P_2\in\mathcal{P}_d} \mathbb{1}\{P_2 \in Y_{s_2}((\tilde{\sigma} + \delta_{(o,0,\pi)})\,\&\oslash(P_1, H_1, Y_{s_1}))\} \\
& \times \sum_{(x_2,r_2,\psi_2)\in\tilde{\sigma}} \mathbb{1}(x_2 \in P_2, r_2 = \max, \tilde{\sigma})\, \mathbb{1}\{x_1 \neq x_2\}
\end{aligned}
$$

$$
\begin{aligned}
&\times \mathbb{1}\{s_1 < s_2 \le t\}\, \mathbb{1}\{H_1 \in [P_1]\}\, \mathbb{1}\{H_2 \in [P_2]\} \\
&\times g(P_1 \cap H_1, P_2 \cap H_2, s_1, s_2, \\
&Y_t((\widetilde{\sigma} + \delta_{(o,0,\pi)}) \,\&\, \oslash (P_1, H_1, Y_{s_1})) \,\&\, \oslash (P_2, H_2, Y_{s_2}(\oslash(P_1, H_1, Y_{s_1})))) \\
&\Theta(\mathrm{d}H_2)\, \Theta(\mathrm{d}H_1)\, \mathrm{d}s_1\, \mathrm{d}s_2\, \mathbb{P}_{\widehat{X}^*}(\mathrm{d}\pi)\, \mathbb{P}_{\widetilde{\Sigma}}(\mathrm{d}\widetilde{\sigma})\,.
\end{aligned}
$$

Regarding the expression for $B_1^{=}$ given in (14.5) and the last one for $B_1^{\neq}$, observe that by the definition of $\widetilde{\Sigma}$, for all pairs P_1, P_2 with $P_1 \in Y_{s_1}(\widetilde{\sigma}+\delta_{(o,0,\pi)})$ and $P_2 \in Y_{s_2}(\widetilde{\sigma}+\delta_{(o,0,\pi)})$, we have

$$
\begin{aligned}
&\sum_{(x_1,r_1,\psi_1)\in\widetilde{\sigma}} \mathbb{1}(x_1 \in P_1, r_1 = \max, \widetilde{\sigma})\, \mathbb{1}(x_1 \in P_2, r_1 = \max, \widetilde{\sigma}) \\
&\quad + \sum_{(x_1,r_1,\psi_1)\in\widetilde{\sigma}} \mathbb{1}(x_1 \in P_1, r_1 = \max, \widetilde{\sigma}) \sum_{(x_2,r_2,\psi_2)\in\widetilde{\sigma}} \mathbb{1}(x_2 \in P_2, r_2 = \max, \widetilde{\sigma})\, \mathbb{1}\{x_1 \neq x_2\} \\
&= \mathbb{1}\{o \notin P_1, o \notin P_2\} \quad \text{a.s.}
\end{aligned}
\tag{14.6}
$$

With the same methods, the other summands B_2, B_3, B_4 are treated. Considering B_2 and B_3, the Mecke equation (19.5) is applied first to $\widetilde{\Sigma}$ and then separately to ψ and to π, respectively. Calculating B_4, the multivariate Mecke equation (19.4) is applied to π only, and the procedure is somehow analogous to that for $B_1^{=}$. The respective sums for B_2, B_3, and B_4, analogous to (14.6), are $\mathbb{1}\{o \notin P_1, o \in P_2\}$, $\mathbb{1}\{o \in P_1, o \notin P_2\}$, and $\mathbb{1}\{o \in P_1, o \in P_2\}$.

Thus the sum $A = B_1 + B_2 + B_3 + B_4$ includes all cells $P_1 \in Y_{s_1}(\widetilde{\sigma} + \delta_{(o,0,\pi)})$ and the items $P_2 \in Y_{s_2}((\widetilde{\sigma} + \delta_{(o,0,\pi)}) \,\&\, \oslash (P_1, H_1, Y_{s_1}))$. Those can be expressed by $P_2 \in \oslash(P_1, H_1, Y_{s_1}) \boxplus \underline{Y_{s_2-s_1}}$. Furthermore, using the operation $\boxplus$ of iteration, the tessellation $Y_t((\widetilde{\sigma} + \delta_{(o,0,\pi)} \,\&\, \oslash (P_1, H_1, Y_{s_1}))\,\&\, \oslash (P_2, H_2, Y_{s_2}(\oslash(P_1, H_1, Y_{s_1}))))$ can be written as $\oslash(P_2, H_2, \oslash(P_1, H_1, Y_{s_1}) \boxplus \underline{Y_{s_2-s_1}}) \boxplus \underline{Y_{t-s_2}}$. Finally, the transformation formula for the image measure with respect to the mapping $\Sigma \mapsto \mathbf{Y}(\Sigma)$ completes the proof. □

Induction over the number of birth time marked maximal $(d-1)$-polytopes yields the following **multivariate Mecke-type equation.**

Corollary 14.4. *Let Θ be a hyperplane measure of the form* (4.1), *satisfying* (4.3), *and let $(Y_t, t > 0)$ be the STIT tessellation process generated by the point process Σ of* (8.1). *Furthermore, for fixed $j \in \mathbb{N}$, denote by* (F, s) *the set of all tuples $((F_1, s_1), \ldots, (F_j, s_j)) \in \mathcal{F}_{d-1}^{\max *}(\Sigma)^j$ with $0 < s_1 < \cdots < s_j$. Then, for fixed $t > 0$ and all nonnegative measurable functions $f : (\mathcal{P}_{d-1} \times \mathbb{R}^+)^j \times \mathcal{T} \to \mathbb{R}$, we have that*

$$
\begin{aligned}
&\mathbb{E} \sum_{(\mathsf{F},\mathsf{s})} f((F_1, s_1), \ldots, (F_j, s_j), Y_t)\, \mathbb{1}\{s_j \le t\} \\
&\quad = \int\limits_{0<s_1<s_2<\cdots<s_j\le t} \!\!\cdots\! \int \int\limits_{\mathcal{T}} \int\limits_{(\mathcal{T}^{\mathbb{N}_0})} \cdots \int\limits_{(\mathcal{T}^{\mathbb{N}_0})} \int\limits_{\mathcal{H}^j} \\
&\qquad \sum_{P_1 \in \mathcal{Y}_{s_1}} \; \sum_{P_2 \in \oslash(P_1, H_1, \mathcal{Y}_{s_1}) \boxplus \underline{\mathcal{Y}_{s_2-s_1}}} \cdots \sum_{P_j \in \oslash(P_{j-1}, H_{j-1}, \oslash(\ldots) \boxplus \underline{\mathcal{Y}_{s_{j-1}-s_{j-2}}}) \boxplus \underline{\mathcal{Y}_{s_j-s_{j-1}}}}
\end{aligned}
$$

$$
\begin{aligned}
& f((P_1 \cap H_1, s_1), \ldots, (P_j \cap H_j, s_j), \\
& \oslash (P_j, H_j, \oslash(P_{j-1}, H_{j-1}, \oslash(\ldots) \boxplus \underline{y_{s_j - s_{j-1}}}) \boxplus \underline{y_{t-s_j}})) \\
& \times \mathbb{1}\{H_1 \in [P_1]\} \cdots \mathbb{1}\{H_j \in [P_j]\}\, \Theta^{\otimes j}(\mathrm{d}(H_1, \ldots, H_j)) \\
& \mathbb{P}_{\underline{Y_{t-s_j}}}(\mathrm{d}\underline{y_{t-s_j}})\, \mathbb{P}_{\underline{Y_{s_j-s_{j-1}}}}(\mathrm{d}\underline{y_{s_j-s_{j-1}}}) \ldots \mathbb{P}_{\underline{Y_{s_2-s_1}}}(\mathrm{d}\underline{y_{s_2-s_1}})\, \mathbb{P}_{Y_{s_1}}(\mathrm{d}y_{s_1})\, \mathrm{d}s_1 \ldots \mathrm{d}s_j .
\end{aligned}
$$

14.2 Covariance measure of the $(d-1)$-skeleton revisited

As an application of Theorem 14.3, we present in this section an alternative proof of Theorem 12.30(b) for the covariance measure of the Hausdorff measure on the $(d-1)$-skeleton of a STIT tessellation, as already announced in Remark 12.32. We recall that Theorem 12.30(b) was derived by means of martingale methods, whereas the corresponding result of Theorem 12.30(a) for the Boolean model of $(d-1)$-polytopes was based on the Mecke equation for Poisson point processes. The present alternative approach is essentially based on the Mecke-type equation for maximal polytopes of STIT tessellations and thus provides deeper insight into the analogy between the Boolean and STIT models on the level of second-order quantities.

To start with, fix $B_1, B_2 \in \mathcal{B}(\mathbb{R}^d)$ and $t > 0$. We use the notation of Theorem 12.30 and recall that $h^F_{d-1}(B) := \mathcal{H}^{d-1}(F \cap B)$, $B \in \mathcal{B}(\mathbb{R}^d)$. Taking into account the birth times of maximal $(d-1)$-polytopes, analogously to (12.46) and (12.47), we have that

$$
\begin{aligned}
& \mathrm{Cov}_{d-1,t}(B_1 \times B_2) \\
& = \mathbb{E} \sum_{(F,s) \in \mathcal{F}^{\max *}_{d-1}(\Sigma)} \mathbb{1}\{s \le t\}\, h^F_{d-1}(B_1)\, h^F_{d-1}(B_2) \\
& \quad + \mathbb{E} \sum_{((F_1,s_1),(F_2,s_2)) \in \mathcal{F}^{\max *}_{d-1}(\Sigma)^2} \mathbb{1}\{s_1 \neq s_2,\, s_1 \le t,\, s_2 \le t\}\, h^{F_1}_{d-1}(B_1)\, h^{F_2}_{d-1}(B_2) \\
& \quad - \Bigg(\mathbb{E} \sum_{(F,s) \in \mathcal{F}^{\max *}_{d-1}(\Sigma)} \mathbb{1}\{s \le t\}\, h^F_{d-1}(B_1) \Bigg) \Bigg(\mathbb{E} \sum_{(F,s) \in \mathcal{F}^{\max *}_{d-1}(\Sigma)} \mathbb{1}\{s \le t\}\, h^F_{d-1}(B_2) \Bigg) .
\end{aligned}
$$

We will show now that the second and third summands in the last expression are in fact equal, and hence only the first one remains. To do so, we first apply Theorem 14.3 to the function g given by $g(F_1, F_2, s_1, s_2, Y_t) := h^{F_1}_{d-1}(B_1)\, h^{F_2}_{d-1}(B_2)$. This yields

$$
\begin{aligned}
& \mathbb{E} \sum_{((F_1,s_1),(F_2,s_2)) \in \mathcal{F}^{\max *}_{d-1}(\Sigma)^2} \mathbb{1}\{s_1 \neq s_2,\, s_1 \le t,\, s_2 \le t\}\, h^{F_1}_{d-1}(B_1)\, h^{F_2}_{d-1}(B_2) \\
& \quad = 2\,\mathbb{E} \sum_{((F_1,s_1),(F_2,s_2)) \in \mathcal{F}^{\max *}_{d-1}(\Sigma)^2} \mathbb{1}\{s_1 < s_2 \le t\}\, h^{F_1}_{d-1}(B_1)\, h^{F_2}_{d-1}(B_2) \\
& \quad = 2 \iint_{0<s_1<s_2<t} \int_{\mathcal{T}} \int_{\mathcal{T}^{\mathbb{N}_0}} \sum_{P_1 \in y_{s_1}} \int_{\mathcal{H}} h^{P_1 \cap H_1}_{d-1}(B_1)\, \mathbb{1}\{H_1 \in [P_1]\}
\end{aligned}
$$

$$\times \sum_{P_2 \in \oslash(P_1, H_1, y_{s_1}) \boxplus \underline{y_{s_2-s_1}}} \int_{\mathcal{H}} h_{d-1}^{P_2 \cap H_2}(B_2)\, \mathbb{1}\{H_2 \in [P_2]\}\, \Theta(\mathrm{d}H_2)\, \Theta(\mathrm{d}H_1)$$

$$\mathbb{P}^{\otimes \mathbb{N}_0}_{Y_{s_2-s_1}}(\mathrm{d}\underline{y_{s_2-s_1}})\, \mathbb{P}_{Y_{s_1}}(\mathrm{d}y_{s_1})\, \mathrm{d}s_1\, \mathrm{d}s_2\,,$$

where we remark that the factor 2 stems from the fact that we order the birth times s_1 and s_2 of the maximal polytopes F_1 and F_2 according to $s_1 < s_2 < t$. The next step consists in observing that by Fubini's theorem

$$\int_{\mathcal{H}} h_{d-1}^{P \cap H}(B)\, \mathbb{1}\{H \in [P]\}\, \Theta(\mathrm{d}H) = \lambda_d(P \cap B) \tag{14.7}$$

for $B \in \mathcal{B}(\mathbb{R}^d)$ and $P \in \mathcal{P}_d$. Since $\oslash(P_1, H_1, y_{s_1}) \boxplus \underline{y_{s_2-s_1}}$ is a tessellation of $\mathbb{R}^d$, it follows that

$$\begin{aligned}
&\sum_{P_2 \in \oslash(P_1, H_1, y_{s_1}) \boxplus \underline{y_{s_2-s_1}}} \int_{\mathcal{H}} h_{d-1}^{P_2 \cap H_2}(B_2)\, \mathbb{1}\{H_2 \in [P_2]\}\, \Theta(\mathrm{d}H_2) \\
&\quad = \sum_{P_2 \in \oslash(P_1, H_1, y_{s_1}) \boxplus \underline{y_{s_2-s_1}}} \lambda_d(P_2 \cap B_2) = \lambda_d(B_2)\,.
\end{aligned}$$

Similarly, using (14.7) once again and the fact that y_{s_1} is also a tessellation of $\mathbb{R}^d$, we have that

$$\sum_{P_1 \in y_{s_1}} \int_{\mathcal{H}} h_{d-1}^{P_1 \cap H_1}(B_1)\, \mathbb{1}\{H_1 \in [P_1]\}\, \Theta(\mathrm{d}H_1) = \lambda_d(B_1)\,.$$

We thus arrive at

$$\begin{aligned}
&\mathbb{E} \sum_{((F_1,s_1),(F_2,s_2)) \in \mathcal{F}_{d-1}^{\max *}(\Sigma)^2} \mathbb{1}\{s_1 \neq s_2,\ s_1 \le t,\ s_2 \le t\}\, h_{d-1}^{F_1}(B_1)\, h_{d-1}^{F_2}(B_2) \\
&\quad = 2\lambda_d(B_1)\, \lambda_d(B_2) \iint_{0<s_1<s_2<t} \mathrm{d}s_1\, \mathrm{d}s_2 \\
&\quad = t^2 \lambda_d(B_1)\, \lambda_d(B_2)\,.
\end{aligned}$$

On the other hand, Corollary 14.2 applied with $f(F,s) := \mathbb{1}\{s \le t\}\, h_{d-1}^{F}(B)$ yields that for $B \in \mathcal{B}(\mathbb{R}^d)$,

$$\begin{aligned}
&\mathbb{E} \sum_{(F,s) \in \mathcal{F}_{d-1}^{\max *}(\Sigma)} \mathbb{1}\{s \le t\}\, h_{d-1}^{F}(B) \\
&\quad = \int_0^t \int_{\mathcal{T}} \int_{\mathcal{H}} \sum_{P \in y_s} h_{d-1}^{P \cap H}(B)\, \mathbb{1}\{H \in [P]\}\, \Theta(\mathrm{d}H)\, \mathbb{P}_{Y_s}(\mathrm{d}y_s)\, \mathrm{d}s
\end{aligned}$$

$$= \int_0^t \int_{\mathcal{T}} \sum_{P \in y_s} \lambda_d(P \cap B)\, \mathbb{P}_{Y_s}(\mathrm{d}y_s)\, \mathrm{d}s$$
$$= t\,\lambda_d(B)\,,$$

again using (14.7) and that y_s is a tessellation of $\mathbb{R}^d$.

Summarizing these calculations and using the notation

$$\mathcal{Z}(\mathcal{F}_{d-1}^{\max}(Y_t)) := \{(c(F), F - c(F)) : F \in \mathcal{F}_{d-1}^{\max}(Y_t)\}$$

for some fixed center function c, introduced in (8.14), we obtain

$$\begin{aligned}
\mathrm{Cov}_{d-1,t}(B_1 \times B_2) &= \mathbb{E} \sum_{(F,s) \in \mathcal{F}_{d-1}^{\max *}(\Sigma)} \mathbb{1}\{s \le t\}\, h_{d-1}^F(B_1)\, h_{d-1}^F(B_2) \\
&= \mathbb{E} \sum_{(x,F) \in \mathcal{Z}(\mathcal{F}_{d-1}^{\max}(Y_t))} h_{d-1}^{F+x}(B_1)\, h_{d-1}^{F+x}(B_2) \\
&= \gamma_{d-1,t}^{\max} \int_{\mathbb{R}^d} \int_{\mathcal{P}_{d-1,c}} h_{d-1}^{F+x}(B_1)\, h_{d-1}^{F+x}(B_2)\, \mathbb{Q}_{d-1,t}^{\max}(\mathrm{d}F)\, \lambda_d(\mathrm{d}x)\,,
\end{aligned}$$

where we used for the last equation the Campbell theorem 19.4 and the fact that the point process $\mathcal{Z}(\mathcal{F}_{d-1}^{\max}(Y_t))$ has intensity measure $\gamma_{d-1,t}^{\max}\,\lambda_d \otimes \mathbb{Q}_{d-1,t}^{\max}$; see Theorem 19.7. This is the result of Theorem 12.30(b).

14.3 Maximal polytopes and their internal structure

Maximal polytopes were already studied in Section 8.2 and in Part II. Here we add some results, which can be shown by applications of the multivariate Mecke-type equation for STIT tessellations developed in Section 14.1.

For $k \in \{0, \ldots, d-2\}$, the maximal k-polytopes were defined in (8.8) as the k-faces of maximal $(d-1)$-polytopes of a STIT tessellation. Thus by construction these maximal k-polytopes arise as intersections of $d-k$ maximal $(d-1)$-polytopes.

Note that the intersections of $d-k$ maximal $(d-1)$-polytopes in a STIT tessellation process are closely related to the so-called intersection processes of order $d-k$ of Poisson hyperplane processes, which are described in Section 2.3.4 and dealt with thoroughly in [82].

However, in dimension $d \ge 3$, not all intersections of $d-k$ maximal $(d-1)$-polytopes of a STIT tessellation are maximal k-polytopes, even if the intersection has dimension k. To see this, consider, for example, three maximal $(d-1)$-polytopes F_1, F_2, F_3 such that $F_1 \cap F_2 \cap F_3 \neq \emptyset$, with F_2 and F_3 located in different half-spaces generated by the hyperplane aff F_1; see the right panel of Figure 2.1. Hence the sets of polytopes that generate a maximal k-polytope must satisfy additional conditions. This will be elaborated in detail in the proof of Lemma 14.6. To prepare this, we introduce some supplementary notation.

We refer to the global construction of the STIT tessellation process based on the point process Σ given by (8.1) and the corresponding set $\mathcal{F}_{d-1}^{\max*}(\Sigma)$ of birth time marked maximal $(d-1)$-polytopes defined in (8.7). Furthermore, recall that $\mathcal{D}(\Sigma)$ is introduced in (8.6) as the set of the division events, which are the triples of the form $(P, H, s) \in \mathcal{P}_d \times \mathcal{H} \times \mathbb{R}^+$, indicating that a cell P is divided by a hyperplane H at time s and thus generating the birth time marked maximal $(d-1)$-polytope $(P \cap H, s) \in \mathcal{F}_{d-1}^{\max*}(\Sigma)$.

Regarding the $(d-k)$-tuples

$$(\mathbf{F}, \mathbf{s}) := ((F_1, s_1), \dots, (F_{d-k}, s_{d-k})) \in \mathcal{F}_{d-1}^{\max*}(\Sigma)^{d-k} \tag{14.8}$$

of birth time marked maximal $(d-1)$-polytopes that yield a maximal k-polytope, we write

$$(\mathbf{F}, \mathbf{s})_k \quad \text{if and only if} \quad s_1 < \dots < s_{d-k} \quad \text{and} \quad \overline{F} := \bigcap_{i=1}^{d-k} F_i \in \mathcal{F}_k^{\max}(\Sigma)\,. \tag{14.9}$$

For all $(\mathbf{F}, \mathbf{s})_k$, we call $(\overline{F}, \mathbf{s})$ a maximal k-polytope of the STIT tessellation process, marked with its birth time tuple $\mathbf{s} := (s_1, \dots, s_{d-k})$.

For a realization m^* of $\mathcal{F}_{d-1}^{\max*}(\Sigma)$ and $t > 0$, the notation $(\mathbf{F}, \mathbf{s})_k \in (m_t^*)^{d-k}$ means that $(\mathbf{F}, \mathbf{s})_k \in (m^*)^{d-k}$ and $s_{d-k} \le t$.

Thus by definition (8.16), for a fixed center function c,

$$\begin{aligned}
\gamma_{k,t}^{\max} & \int_{\mathcal{P}_{k,c}} g(F)\, \mathbb{Q}_{k,t}^{\max}(\mathrm{d}F) \\
&= \frac{1}{\lambda_d(B)} \int_{\mathcal{T}} \sum_{(x,F)\in \mathcal{Z}(\mathcal{F}_k^{\max}(y_t))} \mathbb{1}\{x \in B\}\, g(F)\, \mathbb{P}_{Y_t}(\mathrm{d}y_t) \\
&= \frac{1}{\lambda_d(B)} \int_{\mathsf{N}_s(\mathcal{P}_{d-1}\times\mathbb{R}^+)} \sum_{(\mathbf{F},\mathbf{s})_k\in(m_t^*)^{d-k}} \mathbb{1}\{c(\overline{F}) \in B\}\, g(\overline{F} - c(\overline{F}))\, \mathbb{P}^{\max*}(\mathrm{d}m^*)
\end{aligned} \tag{14.10}$$

for all nonnegative measurable functions $g : \mathcal{P}_{k,c} \to \mathbb{R}$ and sets $B \in \mathcal{B}(\mathbb{R}^d)$ with $0 < \lambda_d(B) < \infty$.

Lemma 14.5. *For each $(\mathbf{F}, \mathbf{s})_k$ defined in (14.9), there is a uniquely determined corresponding tuple*

$$(\mathbf{P}, \mathbf{H}, \mathbf{s})_k := ((P_1, H_1, s_1), \dots, (P_{d-k}, H_{d-k}, s_{d-k}))\,,$$

of division events $(P_i, H_i, s_i) \in \mathcal{D}(\Sigma)$ with $F_i = P_i \cap H_i$, $i \in \{1, \dots, d-k\}$, which generates this birth time marked maximal $(d-1)$-polytope. Furthermore,

(a) $P_1 \supset P_2 \supset \dots \supset P_{d-k}$, *and*

(b) $P_j \cap H_i \in \mathcal{F}_{d-1}(P_j)$, *for $i \in \{1, \dots, j-1\}$, $j \in \{2, \dots, d-k\}$.*

Proof. Let $(\mathbf{F},\mathbf{s})_k = ((F_1,s_1),\dots,(F_{d-k},s_{d-k}))$. Then by the definition of $\mathcal{F}^{\max *}_{d-1}(\Sigma)$ in (8.7) and of $\mathcal{F}^{\max}_k(\Sigma)$ in (8.8) the assumption $\overline{F} \in \mathcal{F}^{\max}_k(\Sigma)$ implies that there are a unique cell P_{d-k} and a unique hyperplane H_{d-k} such that $\overline{F} \in \mathcal{F}_k(F_{d-k}) = \mathcal{F}_k(P_{d-k} \cap H_{d-k})$. Including the birth time, it follows that $(P_{d-k}, H_{d-k}, s_{d-k}) \in \mathcal{D}(\Sigma)$. Furthermore, $F_i \cap P_{d-k} \in \mathcal{F}_{d-1}(P_{d-k})$ for $i \in \{1,\dots,d-k-1\}$, because it is required that $\overline{F} = \bigcap_{i=1}^{d-k} F_i \in \mathcal{F}^{\max}_k(\Sigma)$. By the way, note that this does not necessarily mean that $F_i \in \mathcal{F}_{d-1}(P_{d-k})$. In the next step, consider (F_{d-k-1}, s_{d-k-1}). In turn, there are a unique cell P_{d-k-1} and a unique hyperplane H_{d-k-1} such that $F_{d-k-1} = P_{d-k-1} \cap H_{d-k-1}$. Including the birth time, it follows that $(P_{d-k-1}, H_{d-k-1}, s_{d-k-1}) \in \mathcal{D}(\Sigma)$. Since $F_{d-k-1} \cap P_{d-k} \in \mathcal{F}_{d-1}(P_{d-k})$, we have that also $P_{d-k} \cap H_{d-k-1} \in \mathcal{F}_{d-1}(P_{d-k})$. Moreover, $P_{d-k} \subseteq P_{d-k-1}$, and even $P_{d-k} \subseteq P_{d-k-1} \cap H^+_{d-k-1}$ or $P_{d-k} \subseteq P_{d-k-1} \cap H^-_{d-k-1}$. However, note that in general "$\subseteq$" cannot be replaced by "$=$". Also, with the same arguments as above, $F_i \cap P_{d-k-1} \in \mathcal{F}_{d-1}(P_{d-k-1})$ for $i \in \{1,\dots,d-k-2\}$. Thus by recursion statements (a) and (b) of the lemma can be deduced. □

For a tuple $(\mathbf{P},\mathbf{H},\mathbf{s})_k$, we write $\overline{H} := \bigcap_{i=1}^{d-k} H_i$. By condition (14.9) we have that $\overline{H} \in A(d,k)$. When integrating with respect to hyperplane tuples, we use the abbreviation

$$\Theta^{\otimes(d-k)}(\mathrm{d}\mathbf{H}) := \Theta^{\otimes(d-k)}(\mathrm{d}(H_1,\dots,H_{d-k}))\,.$$

It deserves special attention that for the maximal k-polytopes of the STIT tessellation process, an *ordered* system of $d-k$ birth time marked hyperplanes is used. This is in contrast to the intersection process of order $d-k$ of a Poisson hyperplane process, dealt with in Lemma 14.16, where the $(d-k)!$ permutations of $H_1,\dots,H_{d-k}$ must be taken into account.

In the following proposition, for a fixed time $t > 0$ and a fixed dimension $k \in \{0,\dots,d-1\}$, we consider the set of all birth time marked maximal k-polytopes $(\overline{F},\mathbf{s})$ and the trace of the $(d-1)$-skeleton of the STIT tessellation on them. Formally, for all realizations m^* of $\mathcal{F}^{\max *}_{d-1}(\Sigma)$ and all $(\mathbf{F},\mathbf{s})_k \in (m^*_t)^{d-k}$,

$$m_t \sqcap \overline{F} := \operatorname{cl}\Bigg(\bigcup_{(F,s)\in m^*_t \setminus \{(F_1,s_1),\dots,(F_{d-k},s_{d-k})\}} F \cap \operatorname{relint} \overline{F}\Bigg), \tag{14.11}$$

which is the topological closure of the intersection of the relative interior of $\overline{F}$ with the other maximal $(d-1)$-polytopes of $m_t := \{F \in \mathcal{P}_{d-1} : (F,s) \in m^*_t\}$. Thus $m_t \sqcap \overline{F}$ describes the tessellation structure induced by m^*_t in the relative interior of the maximal k-polytope $\overline{F}$. The closure is taken to have that $m_t \sqcap \overline{F} \in \mathcal{F} \cup \{\emptyset\}$, where $\mathcal{F}$ is the space of nonempty closed subsets of $\mathbb{R}^d$. Note that for $k = 0$, the maximal k-polytope $\overline{F}$ is a singleton, and hence $m_t \sqcap \overline{F} = \emptyset$.

Lemma 14.6. *Let Θ be a hyperplane measure of the form* (4.1), *satisfying* (4.3), *and let $(Y_t, t > 0)$ be the STIT tessellation process generated by the point process Σ of* (8.1). *Then, for $t > 0$, $k \in \{0,\dots,d-1\}$, and all nonnegative measurable functions $g : \mathcal{P}_k \times (0,t)^{d-k} \times (\mathcal{F} \cup \{\emptyset\}) \to \mathbb{R}$, we have that*

$$\int_{\mathsf{N}_s(\mathcal{P}_{d-1}\times\mathbb{R}^+)} \sum_{(\mathbf{F},\mathbf{s})_k\in(m_t^*)^{d-k}} g(\overline{F},\mathbf{s},m_t\sqcap\overline{F})\,\mathbb{P}^{\max*}(\mathrm{d}m^*)$$
$$= 2^{d-k-1}\int\limits_{0<s_1<s_2<\cdots<s_{d-k}<t}\!\!\cdots\!\int \int_{\mathcal{T}}\int_{\mathcal{H}^{d-k}}\int_{\mathcal{T}}\cdots\int_{\mathcal{T}}\sum_{P\in y_{s_{d-k}}}$$
$$g\left(P\cap\overline{H},\mathbf{s},P\cap\overline{H}\cap\left(\bigcup_{i=1}^{d-k-1}\partial y_{t-s_i}^{(i)}\cup\partial y_{t-s_{d-k}}^{+}\cup\partial y_{t-s_{d-k}}^{-}\right)\right)$$
$$\mathbb{P}_{Y_{t-s_1}}(\mathrm{d}y_{t-s_1}^{(1)})\ldots\mathbb{P}_{Y_{t-s_{d-k-1}}}(\mathrm{d}y_{t-s_{d-k-1}}^{(d-k-1)})\,\mathbb{P}_{Y_{t-s_{d-k}}}(\mathrm{d}y_{t-s_{d-k}}^{+})\,\mathbb{P}_{Y_{t-s_{d-k}}}(\mathrm{d}y_{t-s_{d-k}}^{-})$$
$$\mathbb{1}\{P\cap\overline{H}\neq\emptyset\}\,\mathbb{1}\{\overline{H}\in A(d,k)\}\,\Theta^{\otimes(d-k)}(\mathrm{d}\mathbf{H})\,\mathbb{P}_{Y_{s_{d-k}}}(\mathrm{d}y_{s_{d-k}})\,\mathrm{d}s_1\ldots\mathrm{d}s_{d-k}\,.$$

Proof. With the notation and conventions introduced at the beginning of this section,

$$\int_{\mathsf{N}_s(\mathcal{P}_{d-1}\times\mathbb{R}^+)} \sum_{(\mathbf{F},\mathbf{s})_k\in(m_t^*)^{d-k}} g(\overline{F},\mathbf{s},m_t\sqcap\overline{F})\,\mathbb{P}^{\max*}(\mathrm{d}m^*)$$
$$= \int_{\mathsf{N}_s(\mathcal{P}_{d-1}\times\mathbb{R}^+)} \sum_{(\mathbf{F},\mathbf{s})\in\mathcal{F}_{d-1}^{\max*}(m_t^*)^{d-k}} \mathbb{1}\{0<s_1<\cdots<s_{d-k}\le t\}\,\mathbb{1}\{\overline{F}\in\mathcal{F}_k^{\max}(m_t^*)\}$$
$$\times\, g(\overline{F},\mathbf{s},m_t\sqcap\overline{F})\,\mathbb{P}^{\max*}(\mathrm{d}m^*)\,.$$

For the mapping $m_t^*\mapsto Y(m_t^*)=Y_t$, the transformation theorem for image measures can be applied to the integral with respect to $\mathbb{P}^{\max*}$, and then Corollary 14.4 can be used, choosing there $j=d-k$ and

$$f((F_1,s_1),\ldots,(F_{d-k},s_{d-k}),Y_t)$$
$$= \mathbb{1}\left\{\overline{F}=\bigcap_{i=1}^{d-k}F_i\in\mathcal{F}_k^{\max}(Y_t)\right\}\,\mathbb{1}\{0<s_1<s_2<\cdots<s_{d-k}<t\}\,g(\overline{F},\mathbf{s},m_t(Y_t)\sqcap\overline{F})\,.$$

As described in Lemma 14.5, the condition $\overline{F}=\bigcap_{i=1}^{d-k}F_i\in\mathcal{F}_k^{\max}(\Sigma)$ implies several restrictions to $(\mathbf{F},\mathbf{s})_k$ and $(\mathbf{P},\mathbf{H},\mathbf{s})_k$, respectively, that have to be taken into account.

To avoid overwhelming formalization, we prove the statement only for $k=d-1$ and $k=d-2$. This will show the method for further recursion until $k=0$.

For $k=d-1$, by Corollary 14.4 we obtain

$$\int_{\mathsf{N}_s(\mathcal{P}_{d-1}\times\mathbb{R}^+)} \sum_{(\mathbf{F},\mathbf{s})_{d-1}\in m_t^*} g(\overline{F},\mathbf{s},m_t\sqcap\overline{F})\,\mathbb{P}^{\max*}(\mathrm{d}m^*)$$
$$= \int_0^t\int_{\mathcal{T}}\int_{\mathcal{T}^{\mathbb{N}_0}}\int_{\mathcal{H}}\sum_{P_1\in y_{s_1}} g(P_1\cap H_1,\,s_1,\,m_t(\oslash(P_1,H_1,y_{s_1})\boxplus\underline{y_{t-s_1}})\sqcap(P_1\cap H_1))$$
$$\times\,\mathbb{1}\{H_1\in[P_1]\}\,\Theta(\mathrm{d}H_1)\,\underline{\mathbb{P}_{Y_{t-s_1}}}(\underline{\mathrm{d}y_{t-s_1}})\,\mathbb{P}_{\underline{Y_{s_1}}}(\mathrm{d}y_{s_1})\,\mathrm{d}s_1\,.$$

Now we consider $m_t(\oslash(P_1, H_1, y_{s_1}) \boxplus \underline{y_{t-s_1}}) \sqcap (P_1 \cap H_1)$. The hyperplane H_1 divides the cell P_1, resulting in two new cells $P_1 \cap H_1^+$ and $P_1 \cap H_1^-$. For the internal structure on $P_1 \cap H_1$ at time t, only $Y_{t-s_1}^+ \wedge (P_1 \cap H_1^+)$ and $Y_{t-s_1}^- \wedge (P_1 \cap H_1^-)$ must be taken into account, where $Y_{t-s_1}^+$ and $Y_{t-s_1}^-$ are independent copies of Y_{t-s_1}. Thus

$$m_t(\oslash(P_1, H_1, y_{s_1}) \boxplus \underline{y_{t-s_1}}) \sqcap (P_1 \cap H_1) = P_1 \cap H_1 \cap (\partial y_{t-s_1}^+ \cup \partial y_{t-s_1}^-),$$

which yields the statement of the lemma for $k = d - 1$.

In the next step, choose $k = d - 2$. By Corollary 14.4

$$\int\limits_{\mathsf{N}_s(\mathcal{P}_{d-1} \times \mathbb{R}^+)} \sum_{(\mathbf{F}, \mathbf{s})_{d-2} \in (m_t^*)^2} g(\overline{F}, \mathbf{s}, m_t \sqcap \overline{F})\, \mathbb{P}^{\max *}(dm^*)$$

$$\begin{aligned} = \int\limits_{0<s_1<s_2\le t} \int\limits_{\mathcal{T}} \int\limits_{\mathcal{T}^{\mathbb{N}_0}} \int\limits_{\mathcal{T}^{\mathbb{N}_0}} \int\limits_{\mathcal{H}^2} \sum_{P_1 \in y_{s_1}} \sum_{P_2 \in \oslash(P_1, H_1, y_{s_1}) \boxplus \underline{y_{s_2-s_1}}} & \mathbb{1}\{P_2 \cap H_1 \cap H_2 \in \mathcal{F}_{d-2}^{\max}(\oslash(P_2, H_2, \oslash(P_1, H_1, y_{s_1}) \boxplus \underline{y_{s_2-s_1}}))\} \\ & \times g(P_2 \cap H_1 \cap H_2, (s_1, s_2), \\ & m_t(\oslash(P_2, H_2, \oslash(P_1, H_1, y_{s_1}) \boxplus \underline{y_{s_2-s_1}}) \boxplus \underline{y_{t-s_j}}) \sqcap (P_2 \cap H_1 \cap H_2)) \\ & \times \mathbb{1}\{H_1 \in [P_1]\}\, \mathbb{1}\{H_2 \in [P_2]\}\, \Theta^{\otimes 2}(d(H_1, H_2)) \\ & \mathbb{P}_{\underline{Y_{t-s_2}}}(d\underline{y_{t-s_2}})\, \mathbb{P}_{\underline{Y_{s_2-s_1}}}(d\underline{y_{s_2-s_1}})\, \mathbb{P}_{Y_{s_1}}(dy_{s_1})\, ds_1\, ds_2, \end{aligned}$$

where we used that the condition

$$P_1 \cap H_1 \cap P_2 \cap H_2 \in \mathcal{F}_{d-2}^{\max}(\oslash(P_2, H_2, \oslash(P_1, H_1, y_{s_1}) \boxplus \underline{y_{s_2-s_1}}))$$

implies $P_2 \subset P_1$; see Lemma 14.5.

Furthermore, also due to $P_2 \subset P_1$, after the division of P_1 by H_1, generating $P_1 \cap H_1^+$ and $P_1 \cap H_1^-$, the sum $\sum_{P_2 \in \oslash(P_1, H_1, y_{s_1}) \boxplus \underline{y_{s_2-s_1}}}$ can be partitioned into $\sum_{P_2 \in y_{s_2-s_1}^+ \wedge (P_1 \cap H_1^+)}$ and $\sum_{P_2 \in y_{s_2-s_1}^- \wedge (P_1 \cap H_1^-)}$, where $y_{s_2-s_1}^+$ and $y_{s_2-s_1}^-$ are realizations of i. i. d. copies $Y_{s_2-s_1}^+$ and $Y_{s_2-s_1}^-$ of $Y_{s_2-s_1}$.

Consider $\sum_{P_2 \in y_{s_2-s_1}^+ \wedge (P_1 \cap H_1^+)}$. By Lemma 14.5(b) we have that $P_2 \cap H_1 \in \mathcal{F}_{d-1}(P_2)$. Hence there is a cell $P \in y_{s_2-s_1}^+$ such that $P_2 = P \cap H_1^+$, and thus $P_2 \cap H_1 \cap H_2 = P \cap H_1 \cap H_2$. The internal structure appearing on $P \cap H_1 \cap H_2$ at time t is

$$\begin{aligned} & m_t(\oslash(P_2, H_2, \oslash(P_1, H_1, y_{s_1}) \boxplus \underline{y_{s_2-s_1}}) \boxplus \underline{y_{t-s_j}}) \sqcap (P_2 \cap H_1 \cap H_2) \\ & = P \cap H_1 \cap H_2 \cap (\partial y_{t-s_1}^{(1)} \cup \partial y_{t-s_2}^+ \cup \partial y_{t-s_2}^-), \end{aligned}$$

where $y_{t-s_1}^{(1)}$ stands for the realization of the STIT tessellation that divides $P_1 \cap H_1^-$ in the time interval (s_1, t), and $y_{t-s_2}^+, y_{t-s_2}^-$ stand for the realizations that divide $P_2 \cap H_2^+$ and $P_2 \cap H_2^-$, respectively, in the time interval (s_2, t).

The complementary sum $\sum_{P_2 \in y^-_{s_2-s_1} \wedge (P_1 \cap H_1^-)}$ yields just the same terms as above, only substituting $P_1 \cap H_1^+$ by $P_1 \cap H_1^-$. Putting this together, the factor $2 = 2^{d-k-1}$ arises. Thus the lemma is shown for the case $k = d - 2$.

With the same method, the lemma can be shown for general $k \in \{0, \dots, d-1\}$. Lemma 14.5 always plays a crucial role. Considering all combinations of $P_i \cap H_i^+$ and $P_i \cap H_i^-$ for $i \in \{1, \dots, d-k-1\}$, we obtain the factor 2^{d-k-1}. This completes the proof. □

If the function g in Lemma 14.6 depends on the birth time marked maximal k-polytope only and not on its internal structure, then the restriction to a fixed time $t > 0$ can be omitted, and the result can be modified as follows.

Corollary 14.7. *Let Θ be a hyperplane measure of the form* (4.1), *satisfying* (4.3), *and let* $(Y_t, t > 0)$ *be the STIT tessellation process generated by the point process Σ of* (8.1). *Then, for $k \in \{0, \dots, d-1\}$ and all nonnegative measurable functions $g : \mathcal{P}_k \times (\mathbb{R}^+)^{d-k} \to \mathbb{R}$,*

$$\int_{\mathsf{N}_s(\mathcal{P}_{d-1}\times\mathbb{R}^+)} \sum_{(\mathbf{F},\mathbf{s})_k \in (m^*)^{d-k}} g(\overline{F}, \mathbf{s})\, \mathbb{P}^{\max *}(dm^*)$$
$$= 2^{d-k-1} \int\cdots\int_{0<s_1<s_2<\cdots<s_{d-k}} \int_{\mathcal{T}} \int_{\mathcal{H}^{d-k}} \sum_{P \in y_{s_{d-k}}} g(P \cap \overline{H}, \mathbf{s})\, \mathbb{1}\{P \cap \overline{H} \neq \emptyset\}$$
$$\times\, \mathbb{1}\{\overline{H} \in A(d,k)\}\, \Theta^{\otimes(d-k)}(d\mathbf{H})\, \mathbb{P}_{Y_{s_{d-k}}}(dy_{s_{d-k}})\, ds_1 \dots ds_{d-k}\,.$$

As an application of this corollary, we obtain a formula for the densities $\eta_k(Y_t)$ of the k-skeletons $\operatorname{skel}_k Y_t$, $k \in \{0, \dots, d-1\}$.

Choosing $g(\overline{F}, \mathbf{s}) := \mathcal{H}^k(\overline{F} \cap [0,1]^d)\, \mathbb{1}\{s_{d-k} \le t\}$, by Corollary 2.36 we obtain an expression for $\Theta^{\otimes(d-k)}(d\mathbf{H})$ in terms of the spherical directional distribution $\varphi^{\otimes(d-k)}(d(u_1, \dots, u_{d-k}))$. Furthermore, (8.17), (11.8), and Corollary 11.12 yield a relation to the associated zonoid Π_Θ. Thus

$$\begin{aligned}
\eta_k(Y_t) &= \frac{2^{d-k-1}\, t^{d-k}}{(d-k)!} \int_{\mathcal{H}^{d-k}} \mathcal{H}^k(\overline{H} \cap [0,1]^d)\, \mathbb{1}\{\overline{H} \in A(d,k)\}\, \Theta^{\otimes(d-k)}(d\mathbf{H}) \\
&= \frac{2^{d-k-1}\, t^{d-k}}{(d-k)!} \int_{(S^{d-1})^{d-k}} \int_{\operatorname{lin}\{u_1,\dots,u_{d-k}\}} \mathbb{1}\{u_1^\perp \cap \dots \cap u_{d-k}^\perp \in G(d,k)\} \\
&\quad \times \nabla_{d-k}(u_1, \dots, u_{d-k})\, \varphi^{\otimes(d-k)}(d(u_1, \dots, u_{d-k})) \\
&= 2^{d-k-1}\, t^{d-k}\, V_{d-k}(\Pi_\Theta)\,.
\end{aligned} \tag{14.12}$$

Note that $\nabla_{d-k}(u_1, \dots, u_{d-k}) = 0$ if $u_1^\perp \cap \dots \cap u_{d-k}^\perp \notin G(d,k)$.

Comparing (14.12) with the result given in (2.82) for PHT, we obtain that for $k \in \{0, \dots, d-1\}$,

$$\eta_k(Y_t) = 2^{d-k-1}\, \eta_{k,t}^{\mathrm{PHT}}\,, \tag{14.13}$$

thus extending the results of Theorem 6.9, equations (6.6) and (6.9). Furthermore, $\eta_0(Y_t) = \gamma_{0,t}^{\max}$, which is also given in (14.14).

To point to the difference between maximal 0-polytopes and 0-primitives (see Figure 2.1), we apply Corollary 14.7 and complement it by a formula for the intensity of 0-primitives.

Lemma 14.8. *Let Θ be a hyperplane measure of the form* (4.1), *satisfying* (4.3), *and let* $(Y_t, t > 0)$ *be the STIT tessellation process in* $\mathbb{R}^d$ *driven by* Θ. *Then for* $t > 0$, $\gamma_{0,t}^{\max}$ *defined in* (8.15), *and* $\gamma_{0,t}^{\mathrm{pr}}$ *introduced in* (2.42),

$$\gamma_{0,t}^{\max} = t^d \frac{2^{d-1}}{d!} \int\limits_{S^{d-1}} \cdots \int\limits_{S^{d-1}} \nabla_d(u_1, \dots, u_d)\, \varphi(du_1) \dots \varphi(du_d) \tag{14.14}$$

and

$$\gamma_{0,t}^{\mathrm{pr}} = t^d \int\limits_{S^{d-1}} \cdots \int\limits_{S^{d-1}} \nabla_d(u_1, \dots, u_d)\, \varphi(du_1) \dots \varphi(du_d). \tag{14.15}$$

Proof. Because $\gamma_{0,t}^{\max} = \eta_0(Y_t)$, (14.14) follows from the second line of (14.12) for $k = 0$.

To show (14.15), first note that each 0-primitive of Y_t is the intersection of d maximal $(d-1)$-polytopes of Y_t, and each zero-dimensional intersection of maximal $(d-1)$-polytopes is a 0-primitive.

In analogy to (14.8), we denote by $(\mathbf{F}, \mathbf{s})^{\mathrm{pr}} := ((F_1, s_1), \dots, (F_{d-1}, s_d)) \in \mathcal{F}_{d-1}^{\max *}(\Sigma)^d$ the d-tuples of birth time marked maximal $(d-1)$-polytopes with $0 < s_1 < \dots < s_d \le t$. However, in contrast to (14.9), we do not require that $\overline{F} = \bigcap_{i=1}^d F_i \in \mathcal{F}_0^{\max}(\Sigma)$ but only that $\overline{F} = \bigcap_{i=1}^d F_i \in \mathcal{P}_0$, which means that $\overline{F}$ is a singleton in $\mathbb{R}^d$, and its only element is not necessarily a vertex of a maximal $(d-1)$-polytope.

For those $(\mathbf{F}, \mathbf{s})^{\mathrm{pr}}$, the uniquely determined corresponding tuple of division events that generate these birth time marked maximal $(d-1)$-polytopes is denoted by $(\mathbf{P}, \mathbf{H}, \mathbf{s})^{\mathrm{pr}} := ((P_1, H_1, s_1), \dots, (P_d, H_d, s_d))$ with $(P_i, H_i, s_i) \in \mathcal{D}(\Sigma)$ and $F_i = P_i \cap H_i$, $i \in \{1, \dots, d\}$. For such a tuple, we write $\overline{H} := \bigcap_{i=1}^d H_i$, and we have $\overline{H} \in A(d, 0)$, which means that $\overline{H}$ is a singleton.

Thus the hyperplane H_1 divides the cell P_1. As a result of this division, $F_1 = P_1 \cap H_1$, and there appear two new cells $P_1 \cap H_1^+$ and $P_1 \cap H_1^-$. Hence the maximal polytope F_2 can be located either in $P_1 \cap H_1^+$ or in $P_1 \cap H_1^-$. In both cases, for F_3, there are three possible locations, namely if $F_2 \subset P_1 \cap H_1^+$, then F_3 is either in $P_1 \cap H_1^+ \cap H_2^+$, or in $P_1 \cap H_1^+ \cap H_2^-$, or in $P_1 \cap H_1^-$. Analogously, there are three cases if $F_2 \subset P_1 \cap H_1^-$. This can now be continued inductively. Thus there are $d!$ combinations of the possible choices for the division steps. Now Corollary 14.7 and (14.12) can be adapted, replacing the factor 2^{d-1} by $d!$. This yields

$$\gamma_{0,t}^{\mathrm{pr}} = \int\limits_{\mathsf{N}_s(\mathcal{P}_{d-1} \times \mathbb{R}^+)} \sum_{((F_1,s_1),\dots,(F_d,s_d)) \in (m^*)^d} \mathbb{1}\left\{\bigcap_{i=1}^d F_i \in [0,1]^d\right\} \mathbb{1}\{0 < s_1 < \dots < s_d \le t\}\, \mathbb{P}^{\max *}(dm^*)$$

$$= d!\,\frac{t^d}{d!} \int\limits_{\mathcal{H}^d} \mathbb{1}\{\overline{H} \in [0,1]^d\}\, \mathbb{1}\{\overline{H} \in A(d,0)\}\, \Theta^{\otimes d}(\mathrm{d}\mathbf{H})$$

$$= t^d \int\limits_{S^{d-1}} \cdots \int\limits_{S^{d-1}} \nabla_d(u_1,\ldots,u_d)\, \varphi(\mathrm{d}u_1)\ldots\varphi(\mathrm{d}u_d)\,.$$

This completes the argument. □

For $d = 2$, we obtain $\gamma_{0,t}^{\max} = \gamma_{0,t}^{\mathrm{pr}}$, which is consistent with (16.18) and (16.25). For $d = 3$, Lemma 14.8 results in $\gamma_{0,t}^{\max} = \frac{2}{3}\gamma_{0,t}^{\mathrm{pr}}$. This means that in a spatial STIT tessellation in the mean, one third of the 0-primitives are not maximal 0-polytopes; see also Section 17.1 and Lemma 17.5.

14.4 Typical weighted maximal polytopes

Now the focus is on the intrinsic volumes of maximal polytopes of STIT tessellations. Let Θ be a hyperplane measure of the form (4.1) satisfying (4.3), and let $(Y_t,\, t > 0)$ be the STIT tessellation process generated by the point process Σ of (8.1). For $k \in \{0,\ldots,d-1\}$ and $t > 0$, consider the set $\mathcal{F}_k^{\max *}(Y_t)$ of birth time marked maximal k-polytopes of the STIT tessellation Y_t; see (8.9). By $\mathcal{F}_k^{\max}(Y_t)$ we denote the same set of polytopes, but without the birth time marking. If a maximal k-polytope is described as in (14.9) by $(\mathbf{F},\mathbf{s})_k$, then $(\overline{F}, s_{d-k}) \in \mathcal{F}_k^{\max *}(Y_t)$ if $s_{d-k} \le t$. Recall that the uniquely determined corresponding tuple of division events that generate $(\mathbf{F},\mathbf{s})_k$ is denoted by $(\mathbf{P},\mathbf{H},\mathbf{s})_k := ((P_1,H_1,s_1),\ldots,(P_{d-k},H_{d-k},s_{d-k}))$ with $(P_i,H_i,s_i) \in \mathcal{D}(\Sigma)$ and $F_i = P_i \cap H_i$, $i \in \{1,\ldots,d-k\}$; see Lemma 14.5. For such a tuple, we write $\overline{H} := \bigcap_{i=1}^{d-k} H_i \in A(d,k)$.

The intensity $\gamma_{k,t}^{\max}$ of maximal k-polytopes and the distribution $\mathbb{Q}_{k,t}^{\max}$ on $\mathcal{P}_{k,c}$ of the centred typical maximal k-polytope are given by (8.15) and (8.16), respectively. Now we introduce the **mean jth intrinsic volume of the typical maximal k-polytope** for $j \in \{0,\ldots,k\}$ by

$$\overline{V}_{k,j,t}^{\max} := \int\limits_{\mathcal{P}_{k,c}} V_j(F)\, \mathbb{Q}_{k,t}^{\max}(\mathrm{d}F)\,. \tag{14.16}$$

Note that the value of $\overline{V}_{k,j,t}^{\max}$ does not depend on the choice of the center function c on $\mathcal{P}_k$, because the intrinsic volumes are motion-invariant functionals.

In Corollary 11.12 the value of $\overline{V}_{k,j,t}^{\max}$ is determined in terms of intrinsic volumes of the associated zonoid Π_Θ.

The **mean sum of V_j-values of the maximal k-polytopes of Y_t with center in a unit volume** is defined as

$$\rho_{k,j,t} := \gamma_{k,t}^{\max}\, \overline{V}_{k,j,t}^{\max}\,. \tag{14.17}$$

For $k \in \{0, \dots, d-1\}$ and $j \in \{0, \dots, k\}$, the scaling property (6.2) of STIT tessellations yields that $\gamma_{k,t}^{\max} = t^d \gamma_{k,1}^{\max}$, and the homogeneity of the intrinsic volumes implies $\overline{V}_{k,j,t}^{\max} = t^{-j}\, \overline{V}_{k,j,1}^{\max}$. Thus

$$\rho_{k,j,t} = t^{d-j} \rho_{k,j,1} \,. \tag{14.18}$$

For a fixed center function $c : \mathcal{P}_k \to \mathbb{R}^d$, by Corollary 14.7 and an integration over the time coordinates $s_1, \dots, s_{d-k-1}$,

$$\rho_{k,j,t} = \frac{2^{d-k-1}}{\lambda_d(B)\,(d-k-1)!} \int_0^t \int_{\mathcal{H}^{d-k}} \int_{\mathcal{T}} \sum_{P \in y_s} \mathbb{1}\{c(P \cap \overline{H}) \in B\}\, V_j(P \cap \overline{H})$$
$$\mathbb{P}_{Y_s}(\mathrm{d}y_s)\, \mathbb{1}\{\overline{H} \in A(d,k)\}\, \Theta^{\otimes(d-k)}(\mathrm{d}\mathbf{H})\, s^{d-k-1}\, \mathrm{d}s$$

for $B \in \mathcal{B}(\mathbb{R}^d)$ with $0 < \lambda_d(B) < \infty$, where we substituted s_{d-k} by s.

It will be useful to have a more concise representation for $\rho_{k,j,1}$, which in combination with (14.18) yields a formula for all $t > 0$. To prepare this and further results, we consider typical V_j-weighted maximal k-polytopes, marked with the $(d-k)$-tuple of the birth times of the adjacent maximal $(d-1)$-polytopes and additionally, with their internal structure. Recall the notation $(\mathbf{F}, \mathbf{s})_k$ introduced in (14.9) and $m_t \sqcap \overline{F}$ defined in (14.11). For a fixed time $t > 0$ and a center function $c : \mathcal{P}_k \to \mathbb{R}^d$, consider the mapping

$$m^* \mapsto \{(c(\overline{F}), \overline{F} - c(\overline{F}), \mathbf{s}, (m_t \sqcap \overline{F}) - c(\overline{F})) : (\mathbf{F}, \mathbf{s})_k \in (m_t^*)^{d-k}\} \,. \tag{14.19}$$

By $\mathbb{P}_{c,t}^{\max *}$ we denote the image measure of $\mathbb{P}^{\max *}$ under this mapping. Note that unlike Lemma 14.6, where a proposition for functions depending on $\overline{F}, \mathbf{s}, m_t \sqcap \overline{F}$ was shown, based on the multivariate Mecke equation (19.5) for Poisson point processes, now we will use the refined Campbell theorem 19.8 for marked point processes. By that theorem the probability measure $\mathbb{P}_{c,t}^{\max *}$ can be disintegrated, that is, there exists a probability measure $\mathbb{Q}_{(\mathsf{F},\boldsymbol{\beta},\Gamma),t}$ on $\mathcal{P}_{k,c} \times (0,t)^{d-k} \times \mathcal{F}$ such that for all nonnegative measurable functions $g : \mathbb{R}^d \times \mathcal{P}_{k,c} \times (0,t)^{d-k} \times \mathcal{F} \to \mathbb{R}$,

$$\int_{\mathsf{N}_s(\mathbb{R}^d \times \mathcal{P}_{k,c} \times (0,t)^{d-k} \times \mathcal{F})} \sum_{(x,F,\mathbf{s},C) \in \mathrm{m}} g(x, F, \mathbf{s}, C)\, \mathbb{P}_{c,t}^{\max *}(\mathrm{dm})$$
$$= \gamma_{k,t}^{\max} \int_{\mathbb{R}^d} \int_{\mathcal{P}_{k,c} \times (0,t)^{d-k} \times \mathcal{F}} g(x, F, \mathbf{s}, C)\, \mathbb{Q}_{(\mathsf{F},\boldsymbol{\beta},\Gamma),t}(\mathrm{d}(F, \mathbf{s}, C))\, \lambda_d(\mathrm{d}x) \,. \tag{14.20}$$

Note that $\mathbb{Q}_{k,t}^{\max} = \mathbb{Q}_{\mathsf{F},t}$, where $\mathbb{Q}_{k,t}^{\max}$ is defined in (8.16) as the distribution of the typical maximal k-polytope, and $\mathbb{Q}_{\mathsf{F},t}$ is a marginal distribution of $\mathbb{Q}_{(\mathsf{F},\boldsymbol{\beta},\Gamma),t}$. Hence

$$\gamma_{k,t}^{\max} \int_{\mathcal{P}_{k,c}} g(F)\, \mathbb{Q}_{k,t}^{\max}(\mathrm{d}F)$$

$$
\begin{aligned}
&= \frac{1}{\lambda_d(B)} \int_{\mathcal{T}} \sum_{(x,F)\in \mathcal{Z}(\mathcal{F}_k^{\max}(y_t))} \mathbb{1}\{x \in B\}\, g(F)\, \mathbb{P}_{Y_t}(\mathrm{d}y_t) \\
&= \frac{1}{\lambda_d(B)} \int_{\mathsf{N}_s(\mathcal{P}_{d-1}\times\mathbb{R}^+)} \sum_{(\mathbf{F},\mathbf{s})_k\in(m_t^*)^{d-k}} \mathbb{1}\{c(\overline{F}) \in B\}\, g(\overline{F} - c(\overline{F}))\, \mathbb{P}^{\max*}(\mathrm{d}m^*) \\
&= \frac{1}{\lambda_d(B)} \int_{\mathsf{N}_s(\mathbb{R}^d\times\mathcal{P}_{k,c}\times(0,t)^{d-k}\times\mathcal{F})} \sum_{(x,F,\mathbf{s},C)\in\mathrm{m}} \mathbb{1}\{x \in B\}\, g(F)\, \mathbb{P}_{c,t}^{\max*}(\mathrm{dm})
\end{aligned}
$$

for all nonnegative measurable functions $g : \mathcal{P}_{k,c} \to \mathbb{R}$ and sets $B \in \mathcal{B}(\mathbb{R}^d)$ with $0 < \lambda_d(B) < \infty$.

We call a center function $c : \mathcal{K} \to \mathbb{R}^d$ **affine covariant** if $c(aK) = a\,c(K)$ for all $K \in \mathcal{K}$ and $a \in \mathbb{R}^+$. Obviously, the usual center functions like the circumcentre have this property.

Lemma 14.9. *Let Θ be a hyperplane measure of the form* (4.1), *satisfying* (4.3), *and let $(Y_t, t > 0)$ be the STIT tessellation process generated by the point process Σ of* (8.1). *Furthermore, let $c : \mathcal{K} \to \mathbb{R}^d$ be an affine-covariant center function. Then for all $B \in \mathcal{B}(\mathbb{R}^d)$ with $0 < \lambda_d(B) < \infty$, $k \in \{0,\dots,d-1\}$, and $j \in \{0,\dots,k\}$,*

$$
\begin{aligned}
\rho_{k,j,1} = \frac{2^{d-k-1}}{\lambda_d(B)\,(d-k-1)!\,(d-j)} \int_{\mathcal{H}^{d-k}} \int_{\mathcal{T}} \sum_{P\in y_1} \mathbb{1}\{c(P\cap\overline{H}) \in B\} \\
\times V_j(P\cap\overline{H})\, \mathbb{1}\{\overline{H} \in A(d,k)\}\, \mathbb{P}_{Y_1}(\mathrm{d}y_1)\, \Theta^{\otimes(d-k)}(\mathrm{d}\mathbf{H}).
\end{aligned}
$$

Proof. The key is the use of the scaling property (6.2), which says that $\mathbb{P}_{sY_s} = \mathbb{P}_{Y_1}$. First note that by (14.20) for $g(x,F,\mathbf{s},C) = \frac{1}{\lambda_d(s_{d-k}^{-1}B)} \mathbb{1}\{x \in s_{d-k}^{-1}B\}\, V_j(F)$ and a fixed $B \in \mathcal{B}(\mathbb{R}^d)$ with $0 < \lambda_d(B) < \infty$, with (14.19), the translation invariance of V_j, and Fubini's theorem,

$$
\begin{aligned}
&\int_{\mathsf{N}_s(\mathcal{P}_{d-1}\times\mathbb{R}^+)} \sum_{(\mathbf{F},\mathbf{s})_k\in(m_t^*)^{d-k}} \frac{1}{\lambda_d(s_{d-k}^{-1}B)} \mathbb{1}\{c(\overline{F}) \in s_{d-k}^{-1}B\}\, V_j(\overline{F})\, \mathbb{P}^{\max*}(\mathrm{d}m^*) \\
&= \int_{\mathsf{N}_s(\mathbb{R}^d\times\mathcal{P}_{k,c}\times(0,t)^{d-k}\times\mathcal{F})} \sum_{(x,F,\mathbf{s},C)\in\mathrm{m}} \frac{1}{\lambda_d(s_{d-k}^{-1}B)} \mathbb{1}\{x \in s_{d-k}^{-1}B\}\, V_j(F)\, \mathbb{P}_{c,t}^{\max*}(\mathrm{dm}) \\
&= \gamma_{k,t}^{\max} \int_{\mathbb{R}^d} \int_{\mathcal{P}_{k,c}\times(0,t)^{d-k}\times\mathcal{F}} \frac{1}{\lambda_d(s_{d-k}^{-1}B)} \mathbb{1}\{x \in s_{d-k}^{-1}B\}\, V_j(F)\, \mathbb{Q}_{(\mathrm{F},\boldsymbol{\beta},\Gamma),t}(\mathrm{d}(F,\mathbf{s},C))\, \lambda_d(\mathrm{d}x) \\
&= \gamma_{k,t}^{\max} \int_{\mathbb{R}^d} \int_{\mathcal{P}_{k,c}\times(0,t)^{d-k}\times\mathcal{F}} \frac{1}{\lambda_d(B)} \mathbb{1}\{x \in B\}\, V_j(F)\, \mathbb{Q}_{(\mathrm{F},\boldsymbol{\beta},\Gamma),t}(\mathrm{d}(F,\mathbf{s},C))\, \lambda_d(\mathrm{d}x) \\
&= \int_{\mathsf{N}_s(\mathbb{R}^d\times\mathcal{P}_{k,c}\times(0,t)^{d-k}\times\mathcal{F})} \sum_{(x,F,\mathbf{s},C)\in\mathrm{m}} \frac{1}{\lambda_d(B)} \mathbb{1}\{x \in B\}\, V_j(F)\, \mathbb{P}_{c,t}^{\max*}(\mathrm{dm}) \\
&= \frac{1}{\lambda_d(B)} \int_{\mathsf{N}_s(\mathcal{P}_{d-1}\times\mathbb{R}^+)} \sum_{(\mathbf{F},\mathbf{s})_k\in(m_t^*)^{d-k}} \mathbb{1}\{c(\overline{F}) \in B\}\, V_j(\overline{F})\, \mathbb{P}^{\max*}(\mathrm{d}m^*).
\end{aligned}
\tag{14.21}
$$

For simplicity of notation, we further substitute s_{d-k} by s. Thus by (14.10), (14.21), Corollary 14.7, an integration with respect to the time coordinates $s_1, \dots, s_{d-k-1}$, the affine covariance of c, and the homogeneity of the intrinsic volume V_j we have

$$\begin{aligned}
\rho_{k,j,1} &= \gamma_{k,1}^{\max} \int_{\mathcal{P}_{k,c}} V_j(F)\, \mathbb{Q}_{k,1}^{\max}(\mathrm{d}F) \\
&= \frac{1}{\lambda_d(B)} \int_{\mathsf{N}_s(\mathcal{P}_{d-1}\times\mathbb{R}^+)} \sum_{(\mathbf{F},\mathbf{s})_k \in (m_t^*)^{d-k}} \mathbb{1}\{c(\overline{F}) \in B\}\, V_j(\overline{F})\, \mathbb{P}^{\max *}(\mathrm{d}m^*) \\
&= \int_{\mathsf{N}_s(\mathcal{P}_{d-1}\times\mathbb{R}^+)} \sum_{(\mathbf{F},\mathbf{s})_k \in (m_t^*)^{d-k}} \frac{1}{\lambda_d(\frac{1}{s}B)} \mathbb{1}\left\{c(\overline{F}) \in \frac{1}{s}B\right\} V_j(\overline{F})\, \mathbb{P}^{\max *}(\mathrm{d}m^*) \\
&= \frac{2^{d-k-1}}{(d-k-1)!} \int_0^1 \int_{\mathcal{H}^{d-k}} \int_{\mathcal{T}} \sum_{P \in y_s} \frac{1}{\lambda_d(\frac{1}{s}B)} \mathbb{1}\left\{c(P \cap \overline{H}) \in \frac{1}{s}B\right\} V_j(P \cap \overline{H}) \\
&\qquad \mathbb{P}_{Y_s}(\mathrm{d}y_s)\, \mathbb{1}\{\overline{H} \in A(d,k)\}\, \Theta^{\otimes(d-k)}(\mathrm{d}\mathbf{H})\, s^{d-k-1}\, \mathrm{d}s \\
&= \frac{2^{d-k-1}}{(d-k-1)!} \int_0^1 \int_{\mathcal{H}^{d-k}} \int_{\mathcal{T}} \sum_{P \in y_1} \frac{1}{\lambda_d(\frac{1}{s}B)} \mathbb{1}\left\{c\left(\frac{1}{s}P \cap \overline{H}\right) \in \frac{1}{s}B\right\} V_j\left(\frac{1}{s}P \cap \overline{H}\right) \\
&\qquad \mathbb{P}_{Y_1}(\mathrm{d}y_1)\, \mathbb{1}\{\overline{H} \in A(d,k)\}\, \Theta^{\otimes(d-k)}(\mathrm{d}\mathbf{H})\, s^{d-k-1}\, \mathrm{d}s \\
&= \int_0^1 \frac{s^d\, 2^{d-k-1}}{\lambda_d(B)\,(d-k-1)!} \int_{\mathcal{H}^{d-k}} \int_{\mathcal{T}} \sum_{P \in y_1} \mathbb{1}\{c(P \cap s\overline{H}) \in B\}\, s^{-j}\, V_j(P \cap s\overline{H}) \\
&\qquad \mathbb{P}_{Y_1}(\mathrm{d}y_1)\, \mathbb{1}\{\overline{H} \in A(d,k)\}\, \Theta^{\otimes(d-k)}(\mathrm{d}\mathbf{H})\, s^{d-k-1}\, \mathrm{d}s\,.
\end{aligned}$$

Now the substitution $H_i' := s H_i$, $i \in \{1, \dots, d-k\}$, yields $\Theta^{\otimes(d-k)}(\mathrm{d}\mathbf{H}) = s^{-d+k}\, \Theta^{\otimes(d-k)}(\mathrm{d}\mathbf{H}')$ and $\overline{H'} = s\overline{H}$. Then

$$\begin{aligned}
\rho_{k,j,1} = \frac{2^{d-k-1}}{(d-k-1)!} \int_0^1 \int_{\mathcal{H}^{d-k}} \frac{1}{\lambda_d(B)} \int_{\mathcal{T}} \sum_{P \in y_1} \mathbb{1}\{c(P \cap \overline{H'}) \in B\}\, V_j(P \cap \overline{H'}) \\
\mathbb{P}_{Y_1}(\mathrm{d}y_1)\, \mathbb{1}\{\overline{H'} \in A(d,k)\}\, \Theta^{\otimes(d-k)}(\mathrm{d}\mathbf{H}')\, s^{d-j-1}\, \mathrm{d}s\,,
\end{aligned}$$

and the integration with respect to s completes the proof. □

For the particular case that $\Theta = \Theta_{\text{iso}}$, the motion-invariant hyperplane measure, and $k = d-1$, we can derive an explicit expression for $\rho_{k,j,1}$ from the results of Section 12.2.

Lemma 14.10. *Let $(Y_t, t > 0)$ be the STIT tessellation process driven by be the motion-invariant hyperplane measure Θ_{iso}. Then for $j \in \{0, \dots, d-1\}$,*

$$\rho_{d-1,j,1} = \frac{1}{(d-j)!} \left(\frac{\Gamma(\frac{d}{2})}{\Gamma(\frac{d+1}{2})} \right)^{d-j} \frac{\Gamma(\frac{d+1}{2})}{\Gamma(\frac{j+1}{2})}\,.$$

Proof. First recall that $\Sigma_{V_j,t,W}$ is defined in (10.33) as the sum over the V_j-values of the maximal $(d-1)$-polytopes of the STIT tessellation $Y_{t,W}$ in $W \in \mathcal{P}_d$. For this functional and growing windows, we can apply Theorem 4.1.3(b) of [172] and Lemma 8.10, which yield that

$$\rho_{d-1,j,1} = \lim_{a\to\infty} \frac{\mathbb{E}\,\Sigma_{V_j,1,aW}}{a^d V_d(W)} .$$

On the other hand, by (12.5) for Θ_{iso}

$$\lim_{a\to\infty} \frac{\mathbb{E}\,\Sigma_{V_j,1,aW}}{a^d V_d(W)} = \frac{1}{(d-j)!}\left(\frac{\Gamma(\frac{d}{2})}{\Gamma(\frac{d+1}{2})}\right)^{d-j} \frac{\Gamma(\frac{d+1}{2})}{\Gamma(\frac{j+1}{2})} ,$$

which completes the proof. □

Now we turn to **typical weighted maximal k-polytopes of Y_t with intrinsic volumes $V_j, j \in \{0,\dots,k\}$, as weights and marked with the $(d-k)$-tuples of their birth times and their internal structure.** Using the notation $\overline{V}^{\max}_{k,j,t}$ introduced in (14.16), by (14.20) their distribution $\mathbb{Q}^{(j)}_{(\mathrm{F},\boldsymbol{\beta},\Gamma),t}$ is defined as

$$\begin{aligned}
&(\overline{V}^{\max}_{k,j,t})^{-1} \int\limits_{\mathsf{N}_s(\mathbb{R}^d\times\mathcal{P}_{k,c}\times(0,t)^{d-k}\times\mathcal{F})} \sum_{(x,F,\mathbf{s},C)\in \mathrm{m}} V_j(F) g(x,F,\mathbf{s},C)\, \mathbb{P}^{\max *}_{c,t}(\mathrm{dm}) \\
&= \gamma^{\max}_{k,t} \int\limits_{\mathbb{R}^d} \int\limits_{\mathcal{P}_{k,c}\times(0,t)^{d-k}\times\mathcal{F}} g(x,F,\mathbf{s},C)\, \mathbb{Q}^{(j)}_{(\mathrm{F},\boldsymbol{\beta},\Gamma),t}(\mathrm{d}(F,\mathbf{s},C))\, \lambda_d(\mathrm{d}x)
\end{aligned} \tag{14.22}$$

for nonnegative measurable functions $g : \mathbb{R}^d \times \mathcal{P}_{k,c} \times (0,t)^{d-k} \times \mathcal{F} \to \mathbb{R}$. Obviously, $\mathbb{Q}^{(0)}_{(\mathrm{F},\boldsymbol{\beta},\Gamma),t} = \mathbb{Q}_{(\mathrm{F},\boldsymbol{\beta},\Gamma),t}$, which was introduced in (14.20).

An application of the transformation formula for integrals, with the mapping (14.19), to the left-hand side of (14.22), and the definition (14.17) of $\rho_{k,j,t}$ yield the following lemma, which will be used in some proofs in Sections 14.5 and 14.6.

Lemma 14.11. *For $d \ge 2$, let Θ be a hyperplane measure of the form* (4.1), *satisfying* (4.3), *and let $(Y_t, t>0)$ be the STIT tessellation process generated by the point process Σ of* (8.1). *Then for $k \in \{0,\dots,d-1\}$, $j \in \{0,\dots,k\}$, $t>0$, and all nonnegative measurable functions $g : \mathcal{P}_k \times (0,t)^{d-k} \times \mathcal{F} \to \mathbb{R}$,*

$$\begin{aligned}
\rho_{k,j,t} &\int\limits_{\mathcal{P}_{k,c}\times(0,t)^{d-k}\times\mathcal{F}} g(F,\mathbf{s},C)\, \mathbb{Q}^{(j)}_{(\mathrm{F},\boldsymbol{\beta},\Gamma),t}(\mathrm{d}(F,\mathbf{s},C)) \\
&= \int\limits_{\mathsf{N}_s(\mathcal{P}_{d-1}\times\mathbb{R}^+)} \sum_{(\mathbf{F},\mathbf{s})_k\in(m_t^*)^{d-k}} V_j(\overline{F})\, \mathbb{1}\{c(\overline{F}) \in [0,1]^d\} \\
&\qquad \times g(\overline{F}-c(\overline{F}), \mathbf{s}, (m_t \sqcap \overline{F}) - c(\overline{F}))\, \mathbb{P}^{\max *}(\mathrm{d}m^*) ,
\end{aligned}$$

where $\mathbb{P}^{\max *}$ denotes the distribution of $\mathcal{F}^{\max *}_{d-1}(\Sigma)$.

14.5 Joint distribution of maximal polytopes and their birth times

In Section 11.5, relations between the birth time marked maximal $(d-1)$-polytopes of STIT tessellations and birth time marked $(d-1)$-faces of PHTs were established. Now we study the joint distribution of typical weighted maximal k-polytopes and their birth time tuple, that is, $\mathbb{Q}^{(j)}_{(\mathrm{F},\boldsymbol{\beta}),t}$, which is a marginal distribution of $\mathbb{Q}^{(j)}_{(\mathrm{F},\boldsymbol{\beta},\Gamma),t}$ introduced in (14.22). To begin with, we determine the marginal distribution $\mathbb{Q}^{(j)}_{\boldsymbol{\beta},t}$ of the joint birth time distribution for the typical V_j-weighted maximal k-polytope of Y_t.

Theorem 14.12. *Let $d \geq 2$, let Θ be a hyperplane measure of the form* (4.1)*, satisfying* (4.3)*, and let $(Y_t, t > 0)$ be the STIT tessellation process driven by Θ. Furthermore, fix $k \in \{0,\dots,d-1\}$, $j \in \{0,\dots,k\}$, $t > 0$, and an affine-covariant center function c. Then the distribution $\mathbb{Q}^{(j)}_{\boldsymbol{\beta},t}$ of the birth times $\boldsymbol{\beta} = (\beta_1,\dots,\beta_{d-k})$ of the typical V_j-weighted maximal k-polytope has the density with respect to the Lebesgue measure λ_{d-k} on $\mathbb{R}^{d-k}$ given by*

$$(s_1,\dots,s_{d-k}) \mapsto (d-j)(d-k-1)!\,\frac{s_{d-k}^{k-j}}{t^{d-j}}\,\mathbb{1}\{0 < s_1 < \cdots < s_{d-k} < t\}\,.$$

Proof. For a nonnegative measurable function $g : (0,t)^{d-k} \to \mathbb{R}$, by Lemma 14.11 and Corollary 14.7 we have that

$$\begin{aligned}
&\int\limits_{(0,t)^{d-k}} g(\mathbf{s})\,\mathbb{Q}^{(j)}_{\boldsymbol{\beta},t}(\mathrm{d}\mathbf{s}) \\
&= \int\limits_{\mathcal{P}_{k,c}\times(0,t)^{d-k}\times\mathcal{F}} g(\mathbf{s})\,\mathbb{Q}^{(j)}_{(\mathrm{F},\boldsymbol{\beta},\Gamma),t}(\mathrm{d}(F,\mathbf{s},C)) \\
&= 2^{d-k-1}(\rho_{k,j,t})^{-1} \\
&\quad \int\limits_0^t \int\limits_{\mathcal{T}} \int\cdots\int\limits_{0<s_1<s_2<\cdots<s_{d-k-1}<s_{d-k}} \int\limits_{\mathcal{H}^{d-k}} \sum_{P\in y_{s_{d-k}}} \mathbb{1}\{c(P\cap\overline{H}) \in [0,1]^d\}\, V_j(P\cap\overline{H})\, g(\mathbf{s}) \\
&\quad \times \mathbb{1}\{\overline{H} \in A(d,k)\}\,\Theta^{\otimes(d-k)}(\mathrm{d}\mathbf{H})\,\mathrm{d}s_1\dots\mathrm{d}s_{d-k-1}\,\mathbb{P}_{Y_{s_{d-k}}}(\mathrm{d}y_{s_{d-k}})\,\mathrm{d}s_{d-k}\,.
\end{aligned}$$

Now we apply the scaling property (6.2), which implies that $\mathbb{P}_{s_{d-k}Y_{s_{d-k}}} = \mathbb{P}_{Y_1}$, and transform the integrals accordingly, where we use that the center function c is affine covariant. This yields, similarly to the calculations in the proof of Lemma 14.9, that

$$\begin{aligned}
&\int\limits_{(0,t)^{d-k}} g(\mathbf{s})\,\mathbb{Q}^{(j)}_{\boldsymbol{\beta},t}(\mathrm{d}\mathbf{s}) \\
&= 2^{d-k-1}(\rho_{k,j,t})^{-1} \int\cdots\int\limits_{0<s_1<s_2<\cdots<s_{d-k}<t} \int\limits_{\mathcal{T}} \int\limits_{\mathcal{H}^{d-k}} \sum_{P\in y_1} \mathbb{1}\Big\{c\Big(\frac{1}{s_{d-k}}P\cap\overline{H}\Big) \in [0,1]^d\Big\} \\
&\quad \times V_j\Big(\frac{1}{s_{d-k}}P\cap\overline{H}\Big)\, g(\mathbf{s})\,\mathbb{1}\{\overline{H} \in A(d,k)\}\,\Theta^{\otimes(d-k)}(\mathrm{d}\mathbf{H})\,\mathbb{P}_{Y_1}(\mathrm{d}y_1)\,\mathrm{d}s_1\dots\mathrm{d}s_{d-k}
\end{aligned}$$

$$= 2^{d-k-1}(\rho_{k,j,t})^{-1} \int_{\mathcal{T}} \int_{\mathcal{H}^{d-k}} \sum_{P\in y_1} \mathbb{1}\{c(P\cap\overline{H}) \in [0,1]^d\}\, V_j(P\cap\overline{H})\, \mathbb{1}\{\overline{H} \in A(d,k)\}$$

$$\Theta^{\otimes(d-k)}(\mathrm{d}\mathbf{H})\, \mathbb{P}_{Y_1}(\mathrm{d}y_1) \int\cdots\int_{0<s_1<s_2<\cdots<s_{d-k}<t} g(\mathbf{s})\, s_{d-k}^{k-j}\, \mathrm{d}s_1 \ldots \mathrm{d}s_{d-k}\,.$$

An application of Lemma 14.9, together with the scaling relation (14.18), completes the proof. □

It is remarkable that the joint probability density function of the birth times depends only on the very last birth time. This is expressed more precisely in the following proposition. Note that the result for the particular case $k = d-1$ and $j = 0$ is shown in Corollary 11.25 by an alternative method.

Corollary 14.13. *Let $d \geq 2$, let Θ be a hyperplane measure of the form (4.1), satisfying (4.3), and let $(Y_t,\, t>0)$ be the STIT tessellation process driven by Θ. Furthermore, fix $k \in \{0,\ldots,d-1\}$, $j \in \{0,\ldots,k\}$, $t > 0$, and an affine-covariant center function c.*

(a) *The marginal distribution $\mathbb{Q}^{(j)}_{\beta_{d-k},t}$ of the last birth time of the typical V_j-weighted maximal k-polytope has the density with respect to the Lebesgue measure on $\mathbb{R}$ given by*

$$s_{d-k} \mapsto (d-j)\,\frac{s_{d-k}^{d-j-1}}{t^{d-j}}\,\mathbb{1}\{0 < s_{d-k} < t\}\,.$$

(b) *For all $s_{d-k} < t$, the conditional distribution $\mathbb{Q}^{(j)}_{(\beta_1,\ldots,\beta_{d-k-1}),t|\beta_{d-k}=s_{d-k}}$ of the birth times $(\beta_1,\ldots,\beta_{d-k-1})$ of the typical V_j-weighted maximal k-polytope, given $\beta_{d-k} = s_{d-k}$, has the density with respect to the Lebesgue measure on $\mathbb{R}^{d-k-1}$ given by*

$$(s_1,\ldots,s_{d-k-1}) \mapsto (d-k-1)!\, s_{d-k}^{-(d-k-1)}\, \mathbb{1}\{0 < s_1 < \cdots < s_{d-k}\}\,.$$

In particular, this conditional distribution does not depend on j, and it is the uniform distribution on the $(d-k-1)$-simplex $\{(s_1,\ldots,s_{d-k-1}) \in \mathbb{R}^{d-k-1} : 0 < s_1 < \cdots < s_{d-k-1} < s_{d-k}\}$.

Furthermore, the distribution $\mathbb{Q}^{(j)}_{(\mathrm{F},\beta_{d-k}),t}$ and the conditional distribution $\mathbb{Q}^{(j)}_{\mathrm{F},t|\beta_{d-k}=s_{d-k}}$ can be determined.

Corollary 14.14. *Let $d \geq 2$, let Θ be a hyperplane measure of the form (4.1), satisfying (4.3), and let $(Y_t,\, t>0)$ be the STIT tessellation process driven by Θ. Furthermore, fix $k \in \{0,\ldots,d-1\}$, $j \in \{0,\ldots,k\}$, $t > 0$, and an affine-covariant center function c. Then for all nonnegative measurable functions $g : \mathcal{P}_{k,c} \times (0,t) \to \mathbb{R}$,*

$$\int_{\mathcal{P}_{k,c}\times(0,t)} g(F,s)\, \mathbb{Q}^{(j)}_{(\mathrm{F},\beta_{d-k}),t}(\mathrm{d}(F,s))$$

$$= \frac{2^{d-k-1}\, t^{d-j}}{\rho_{k,j,t}\,(d-k-1)!\,(d-j)} \int_{(0,t)} \int_{\mathcal{T}} \int_{\mathcal{H}^{d-k}} \sum_{P \in y_s} V_j(P \cap \overline{H})\, \mathbb{1}\{c(P \cap \overline{H}) \in [0,1]^d\}$$
$$\times\, g\big((P \cap \overline{H}) - c(P \cap \overline{H}), s\big)\, \mathbb{1}\{\overline{H} \in A(d,k)\} \Theta^{\otimes(d-k)}(\mathrm{d}\mathbf{H})\, \mathbb{P}_{Y_s}(\mathrm{d}y_s)\, s^{-(k-j)}\, \mathbb{Q}^{(j)}_{\beta_{d-k},t}(\mathrm{d}s)\,.$$

Consequently, for $\mathbb{Q}^{(j)}_{\beta_{d-k},t}$-almost all $s \in (0,t)$, the conditional distribution $\mathbb{Q}^{(j)}_{\mathsf{F},t|\beta_{d-k}=s}$ is given by

$$\int_{\mathcal{P}_{k,c}} \tilde{g}(F)\, \mathbb{Q}^{(j)}_{\mathsf{F},t|\beta_{d-k}=s}(\mathrm{d}F)$$
$$= \frac{2^{d-k-1}\, t^{d-j}\, s^{-(k-j)}}{\rho_{k,j,t}\,(d-k-1)!\,(d-j)} \int_{\mathcal{T}} \int_{\mathcal{H}^{d-k}} \sum_{P \in y_s} V_j(P \cap \overline{H})\, \mathbb{1}\{c(P \cap \overline{H}) \in [0,1]^d\}$$
$$\times\, \tilde{g}\big((P \cap \overline{H}) - c(P \cap \overline{H})\big)\, \mathbb{1}\{\overline{H} \in A(d,k)\}\, \Theta^{\otimes(d-k)}(\mathrm{d}\mathbf{H})\, \mathbb{P}_{Y_s}(\mathrm{d}y_s)$$

for all nonnegative measurable $\tilde{g} : \mathcal{P}_{k,c} \to \mathbb{R}$.

Proof. By Lemma 14.11, Corollary 14.7, and Corollary 14.13(a),

$$\int_{\mathcal{P}_{k,c}\times(0,t)} g(F,s)\, \mathbb{Q}^{(j)}_{(\mathsf{F},\beta_{d-k}),t}(\mathrm{d}(F,s))$$
$$= \frac{2^{d-k-1}}{\rho_{k,j,t}\,(d-k-1)!}$$
$$\int_0^t \int_{\mathcal{T}} \int_{\mathcal{H}^{d-k}} \sum_{P \in y_s} V_j(P \cap \overline{H})\, \mathbb{1}\{c(P \cap \overline{H}) \in [0,1]^d\}\, g\big((P \cap \overline{H}) - c(P \cap \overline{H}), s\big)$$
$$\times\, \mathbb{1}\{\overline{H} \in A(d,k)\}\, \Theta^{\otimes(d-k)}(\mathrm{d}\mathbf{H})\, \mathbb{P}_{Y_s}(\mathrm{d}y_s)\, s^{d-k-1}\, \mathrm{d}s$$
$$= \frac{2^{d-k-1}\, t^{d-j}}{\rho_{k,j,t}\,(d-k-1)!\,(d-j)}$$
$$\int_{(0,t)} \int_{\mathcal{T}} \int_{\mathcal{H}^{d-k}} \sum_{P \in y_s} V_j(P \cap \overline{H})\, \mathbb{1}\{c(P \cap \overline{H}) \in [0,1]^d\}\, g\big((P \cap \overline{H}) - c(P \cap \overline{H}), s\big)$$
$$\times\, \mathbb{1}\{\overline{H} \in A(d,k)\}\, \Theta^{\otimes(d-k)}(\mathrm{d}\mathbf{H})\, \mathbb{P}_{Y_s}(\mathrm{d}y_s)\, s^{-(k-j)}\, \mathbb{Q}^{(j)}_{\beta_{d-k},t}(\mathrm{d}s)\,,$$

and the statement for $\mathbb{Q}^{(j)}_{(\mathsf{F},\beta_{d-k}),t}$ follows. Applying the disintegration formula (2.34) to the joint distribution $\mathbb{Q}^{(j)}_{(\mathsf{F},\beta_{d-k}),t}$ yields the result for the conditional distribution $\mathbb{Q}^{(j)}_{\mathsf{F},t|\beta_{d-k}=s}$ for $\mathbb{Q}^{(j)}_{\beta_{d-k},t}$-almost all s. □

An application of the results obtained so far yields the following conditional independence property for STIT tessellation processes. To formulate it, let $\mathbb{Q}^{(j)}_{\mathsf{F},t|\beta_{d-k}=s_{d-k}}$ and $\mathbb{Q}^{(j)}_{(\beta_1,\dots,\beta_{d-k-1}),t|\beta_{d-k}=s_{d-k}}$ denote the conditional distributions, as indicated by their indexes, pertaining to $\mathbb{Q}^{(j)}_{(\mathsf{F},\boldsymbol{\beta}),t}$.

Theorem 14.15. *Let $d \geq 2$, let Θ be a hyperplane measure of the form (4.1), satisfying (4.3), and let $(Y_t, t > 0)$ be the STIT tessellation process driven by Θ. Moreover, let $k \in \{0, \dots, d-1\}$, $j \in \{0, \dots, k\}$, $t > 0$, and let c be an affine-covariant center function. Then for all nonnegative measurable functions $g : \mathcal{P}_{k,c} \times (0,t)^{d-k} \to \mathbb{R}$,*

$$\int\limits_{\mathcal{P}_{k,c}\times(0,t)^{d-k}} g(F,\mathbf{s})\,\mathbb{Q}^{(j)}_{(F,\beta_1,\dots,\beta_{d-k}),t}(\mathrm{d}(F,\mathbf{s}))$$
$$= \int\limits_{(0,t)} \int\limits_{(0,t)^{d-k-1}} \int\limits_{\mathcal{P}_{k,c}} g(F,\mathbf{s})\,\mathbb{Q}^{(j)}_{F,t|\beta_{d-k}=s_{d-k}}(\mathrm{d}F)$$
$$\mathbb{Q}^{(j)}_{(\beta_1,\dots,\beta_{d-k-1}),t|\beta_{d-k}=s_{d-k}}(\mathrm{d}(s_1,\dots,s_{d-k-1}))\,\mathbb{Q}^{(j)}_{\beta_{d-k},t}(\mathrm{d}s_{d-k}),$$

which means the conditional independence of the typical V_j-weighted maximal k-polytope $\overline{F}$ and $(\beta_1, \dots, \beta_{d-k-1})$, given the last birth time $\beta_{d-k} = s_{d-k}$.

Proof. It is sufficient to consider functions g of the form

$$g(F,\mathbf{s}) = g_1(F)\,g_2(s_1,\dots,s_{d-k-1})\,g_3(s_{d-k}),$$

where $g_1 : \mathcal{P}_{k,c} \to \mathbb{R}$, $g_2 : (0,t)^{d-k-1} \to \mathbb{R}$, and $g_3 : (0,t) \to \mathbb{R}$ are nonnegative and measurable. Then the proposition for general g follows by a standard integration-theoretic procedure. As in the proof of Theorem 14.12, we have

$$\int\limits_{\mathcal{P}_{k,c}\times(0,t)^{d-k}} g(F,\mathbf{s})\,\mathbb{Q}^{(j)}_{(F,\beta_1,\dots,\beta_{d-k}),t}(\mathrm{d}(F,\mathbf{s}))$$
$$= 2^{d-k-1}(\rho_{k,j,t})^{-1} \int_0^t \int\limits_{0<s_1<s_2<\dots<s_{d-k-1}<s_{d-k}} \cdots \int \int\limits_{\mathcal{T}} \int\limits_{\mathcal{H}^{d-k}} \sum_{P \in y_{s_{d-k}}} g_1((P\cap\overline{H}) - c(P\cap\overline{H}))$$
$$\times\, \mathbb{1}\{c(P\cap\overline{H}) \in [0,1]^d\}\, V_j(P\cap\overline{H})\,\mathbb{1}\{\overline{H} \in A(d,k)\}\,\Theta^{\otimes(d-k)}(\mathrm{d}\mathbf{H})\,\mathbb{P}_{Y_{s_{d-k}}}(\mathrm{d}y_{s_{d-k}})$$
$$g_2(s_1,\dots,s_{d-k-1})\,\mathrm{d}s_1\dots\mathrm{d}s_{d-k-1}\,g_3(s_{d-k})\,\mathrm{d}s_{d-k}.$$

Now applying Corollaries 14.13 and 14.14, we obtain that this is equal to

$$\int\limits_{(0,t)} \int\limits_{(0,t)^{d-k-1}} \int\limits_{\mathcal{P}_{k,c}} \frac{(d-k-1)!(d-j)\,s_{d-k}^{k-j}}{t^{d-j}}\, g_1(F)\,\mathbb{Q}^{(j)}_{F,t|\beta_{d-k}=s_{d-k}}(\mathrm{d}F)$$
$$\times\, g_2(s_1,\dots,s_{d-k-1})\,\frac{1}{(d-k-1)!}\,s_{d-k}^{d-k-1}\,\mathbb{Q}^{(j)}_{(\beta_1,\dots,\beta_{d-k-1}),t|\beta_{d-k}=s_{d-k}}(\mathrm{d}(s_1,\dots,s_{d-k-1}))$$
$$\times\, g_3(s_{d-k})\,\frac{t^{d-j}}{(d-j)}\,s_{d-k}^{-(d-j-1)}\,\mathbb{Q}^{(j)}_{\beta_{d-k},t}(\mathrm{d}s_{d-k})$$

$$= \int_{(0,t)} \int_{(0,t)^{d-k-1}} \int_{\mathcal{P}_{k,c}} g(F,\mathbf{s})\, \mathbb{Q}^{(j)}_{\mathsf{F},t|\beta_{d-k}=s_{d-k}}(\mathrm{d}F)$$

$$\mathbb{Q}^{(j)}_{(\beta_1,\dots,\beta_{d-k-1}),t|\beta_{d-k}=s_{d-k}}(\mathrm{d}(s_1,\dots,s_{d-k-1}))\, \mathbb{Q}^{(j)}_{\beta_{d-k},t}(\mathrm{d}s_{d-k})\,.$$

Thus the proof is complete. □

With the probability density functions given in Corollary 14.13, the equation in Theorem 14.15 can also be written as

$$\int_{\mathcal{P}_{k,c}\times(0,t)^{d-k}} g(F,\mathbf{s})\, \mathbb{Q}^{(j)}_{(\mathsf{F},\beta_1,\dots,\beta_{d-k}),t}(\mathrm{d}(F,\mathbf{s}))$$

$$= (d-k-1)!\frac{(d-j)}{t^{d-j}} \int\cdots\int_{0<s_1<s_2<\dots<s_{d-k}<t} \int_{\mathcal{P}_{k,c}} g(F,\mathbf{s})\, \mathbb{Q}^{(j)}_{\mathsf{F},t|\beta_{d-k}=s_{d-k}}(\mathrm{d}F)$$

$$\times\, s_{d-k}^{k-j}\, \mathrm{d}s_1 \dots \mathrm{d}s_{d-k}\,.$$

A relation to Poisson hyperplane tessellations

In Chapter 11, several relationships between maximal polytopes of STIT tessellations and faces of PHTs were shown. In Theorem 11.8, this is done for typical maximal k-polytopes and k-faces, respectively, $k \in \{1,\dots,d-1\}$. Corollary 11.24 provides a corresponding result for birth time marked maximal $(d-1)$-polytopes and $(d-1)$-faces, respectively. Note that these propositions, which are formulated for integrals over indicator functions, can be directly generalized to equations for integrals of nonnegative measurable functions. Thus these statements can be used to show formulae also for V_j-weighted maximal polytopes of STIT tessellations.

Now, as a generalization of the above-mentioned results, we prove a relation between typical birth time marked maximal k-polytopes of STIT tessellations and k-faces of PHTs, $k \in \{1,\dots,d-1\}$. Furthermore, we explicitly include the weighting with the intrinsic volumes $V_j, j \in \{0,\dots,k\}$.

Recall that $\gamma^{\mathrm{PHT}}_{k,t}$ denotes the intensity of the k-faces of the PHT X_t, and $\mathbb{Q}^{\mathrm{PHT}}_{k,t}$ is the distribution of the typical k-face, respectively. Furthermore, the mean jth intrinsic volume $\overline{V}^{\max}_{k,j,t}$ of the typical maximal k-polytope of a STIT tessellation for $j \in \{0,\dots,k\}$ is defined in (14.16), and $\mathbb{Q}^{(j)}_{(\mathsf{F},\beta_{d-k}),t}$ is a marginal distribution of $\mathbb{Q}^{(j)}_{(\mathsf{F},\boldsymbol{\beta},\Gamma),t}$ introduced in (14.22).

Lemma 14.16. *Let Θ be a hyperplane measure of the form* (4.1), *satisfying* (4.3). *Let $(Y_t, t > 0)$ be the STIT tessellation process driven by Θ. Furthermore, let $(X_t,\ t > 0)$ be the process of Poisson hyperplane tessellations driven by Θ. Then for $k \in \{1,\dots,d-1\}, j \in \{0,\dots,k\}$, an affine-covariant center function c on $\mathcal{P}_k$, a nonnegative measurable function $g : \mathcal{P}_{k,c} \times (0,t) \to \mathbb{R}$, and all $t > 0$,*

$$\overline{V}^{\max}_{k,j,t} \int_{\mathcal{P}_{k,c}\times(0,t)} g(F,s)\, \mathbb{Q}^{(j)}_{(\mathsf{F},\beta_{d-k}),t}(\mathrm{d}(F,s)) = \frac{d}{t^d}\int_0^t s^{d-1} \int_{\mathcal{P}_{k,c}} V_j(F)\, g(F,s)\, \mathbb{Q}^{\mathrm{PHT}}_{k,s}(\mathrm{d}F)\, \mathrm{d}s\,.$$

Proof. To begin with, we use the first equation in the proof of Corollary 14.14. Then by Corollary 6.6 we can replace $\sum_{P\in Y_s}$ by $\sum_{P\in X_s}$. This yields

$$\begin{aligned}\int\limits_{\mathcal{P}_{k,c}\times(0,t)} & g(F,s)\,\mathbb{Q}^{(j)}_{(F,\beta_{d-k}),t}(\mathrm{d}(F,s))\\ &= \frac{2^{d-k-1}}{\rho_{k,j,t}\,(d-k-1)!}\\ &\quad\times\int_0^t\int\limits_{\mathcal{H}^{d-k}} \mathbb{E}\Big(\sum_{P\in X_s} V_j(P\cap\overline{H})\,\mathbb{1}\{c(P\cap\overline{H})\in[0,1]^d\}\,g((P\cap\overline{H})-c(P\cap\overline{H}),s)\Big)\\ &\quad\times\mathbb{1}\{\overline{H}\in A(d,k)\}\,\Theta^{\otimes(d-k)}(\mathrm{d}\mathbf{H})\,s^{d-k-1}\,\mathrm{d}s\,. \end{aligned}\tag{14.23}$$

Now to transform the inner integral with respect to $\Theta^{\otimes(d-k)}$, we apply the version (2.78) of the multivariate Mecke equation (19.5) to the Poisson hyperplane process $\widehat{X}_s$ with intensity measure $s\,\Theta$. For the function f in (2.78), put

$$f(\widehat{X}_s,E) := \sum_{F\subset E,\,F\in\mathcal{F}_k(X_s)} V_j(F)\,\mathbb{1}\{c(F)\in[0,1]^d\}\,g(F-c(F),s)$$

for the k-flats $E\in\widehat{X}_s^{(d-k)}$ of the intersection process of order $d-k$. By (2.77) we have that $\operatorname{skel}_k X_s=\bigcup_{E\in\widehat{X}_s^{(d-k)}}E$, and hence the sum $\sum_{E\in\widehat{X}_s^{(d-k)}}\sum_{F\subset E,\,F\in\mathcal{F}_k(X_s)}$ can be replaced by the sum $\sum_{F\in\mathcal{F}_k(X_s)}$. We obtain

$$\begin{aligned}\int\limits_{\mathcal{H}^{d-k}} & \mathbb{E}\Big(\sum_{P\in X_s} V_j(P\cap\overline{H})\,\mathbb{1}\{c(P\cap\overline{H})\in[0,1]^d\}\,g((P\cap\overline{H})-c(P\cap\overline{H}),s)\Big)\\ &\quad\times\mathbb{1}\{\overline{H}\in A(d,k)\}\,\Theta^{\otimes(d-k)}(\mathrm{d}\mathbf{H})\\ &= \frac{(d-k)!}{s^{d-k}}\,\mathbb{E}\Big(\sum_{F\in\mathcal{F}_k(X_s)} V_j(F)\,\mathbb{1}\{c(F)\in[0,1]^d\}\,g(F-c(F),s)\Big)\\ &= \frac{(d-k)!}{s^{d-k}}\,\gamma^{\mathrm{PHT}}_{k,s}\int\limits_{\mathcal{P}_{k,c}} V_j(F)\,g(F,s)\,\mathbb{Q}^{\mathrm{PHT}}_{k,s}(\mathrm{d}F)\,, \end{aligned}\tag{14.24}$$

where the last equation follows from the definition of the distribution of the typical k-face of the PHT X_s; see (2.26).

Furthermore, the scaling properties (2.67) and (6.2) of PHTs and of STIT tessellations, respectively, imply that $s^d\,\gamma^{\mathrm{PHT}}_{k,1}=\gamma^{\mathrm{PHT}}_{k,s}$ and $t^d\,\gamma^{\max}_{k,1}=\gamma^{\max}_{k,t}$. By (11.2) we have $\gamma^{\max}_{k,t}=\frac{(d-k)2^{d-k-1}}{d}\gamma^{\mathrm{PHT}}_{k,t}$. Inserting this into (14.23) and using definition (14.17) of $\rho_{k,j,t}$ yield

$$\int\limits_{\mathcal{P}_{k,c}\times(0,t)} g(F,s)\,\mathbb{Q}^{(j)}_{(F,\beta_{d-k}),t}(\mathrm{d}(F,s))$$

$$= \frac{2^{d-k-1}(d-k)}{\rho_{k,j,t}} \int_0^t \frac{1}{s} \gamma_{k,s}^{\mathrm{PHT}} \int_{\mathcal{P}_{k,c}} V_j(F)\, g(F,s)\, \mathbb{Q}_{k,s}^{\mathrm{PHT}}(\mathrm{d}F)\, \mathrm{d}s$$

$$= \frac{2^{d-k-1}(d-k)}{\rho_{k,j,t}} \int_0^t s^{d-1} \gamma_{k,1}^{\mathrm{PHT}} \int_{\mathcal{P}_{k,c}} V_j(F)\, g(F,s)\, \mathbb{Q}_{k,s}^{\mathrm{PHT}}(\mathrm{d}F)\, \mathrm{d}s$$

$$= \frac{d}{\rho_{k,j,t}} \gamma_{k,1}^{\max} \int_0^t s^{d-1} \int_{\mathcal{P}_{k,c}} V_j(F)\, g(F,s)\, \mathbb{Q}_{k,s}^{\mathrm{PHT}}(\mathrm{d}F)\, \mathrm{d}s$$

$$= (\rho_{k,j,t})^{-1} \gamma_{k,t}^{\max} \frac{d}{t^d} \int_0^t s^{d-1} \int_{\mathcal{P}_{k,c}} V_j(F)\, g(F,s)\, \mathbb{Q}_{k,s}^{\mathrm{PHT}}(\mathrm{d}F)\, \mathrm{d}s$$

$$= \left(\overline{V}_{k,j,t}^{\max}\right)^{-1} \frac{d}{t^d} \int_0^t s^{d-1} \int_{\mathcal{P}_{k,c}} V_j(F)\, g(F,s)\, \mathbb{Q}_{k,s}^{\mathrm{PHT}}(\mathrm{d}F)\, \mathrm{d}s\,,$$

which completes the proof. □

In particular, for $k = 1$ and $j = 0$, we have $\overline{V}_{1,0,t}^{\max} = 1$. Then for $a > 0$, $A \in \mathcal{B}(G(d,1))$, $r \in (0,t)$, and

$$g(F,s) := \mathbb{1}\{V_1(F) \le a\}\, \mathbb{1}\{\mathrm{aff}_o(F) \in A\}\, \mathbb{1}\{s \le r\}\,,$$

we obtain the following relation for the joint distribution of length, direction, and birth time of the typical maximal 1-polytope of Y_t:

$$\int_{\mathcal{P}_{1,c}\times(0,t)} \mathbb{1}\{V_1(F) \le a\}\, \mathbb{1}\{\mathrm{aff}_o(F) \in A\}\, \mathbb{1}\{s \le r\}\, \mathbb{Q}^{(0)}_{(\mathrm{F},\beta_{d-1}),t}(\mathrm{d}(F,s))$$
$$= \frac{d}{t^d} \int_0^r s^{d-1} \int_{\mathcal{P}_{1,c}} \mathbb{1}\{V_1(F) \le a\}\, \mathbb{1}\{\mathrm{aff}_o(F) \in A\}\, \mathbb{Q}_{1,s}^{\mathrm{PHT}}(\mathrm{d}F)\, \mathrm{d}s\,.$$

The result of Lemma 14.16 can be extended to a relation that includes the whole tuple $(\beta_1, \ldots, \beta_{d-k})$ of birth times of the typical V_j-weighted maximal k-polytope of Y_t.

Corollary 14.17. *With the assumptions of Lemma* 14.16, *for a nonnegative measurable function* $g : \mathcal{P}_{k,c} \times (0,t)^{d-k} \to \mathbb{R}$,

$$\int_{\mathcal{P}_{k,c}\times(0,t)^{d-k}} g(F,\mathbf{s})\, \mathbb{Q}^{(j)}_{(\mathrm{F},\beta_1,\ldots,\beta_{d-k}),t}(\mathrm{d}(F,\mathbf{s}))$$
$$= \frac{(d-k-1)!\, d}{t^d\, \overline{V}_{k,j,t}^{\max}} \int\cdots\int_{0<s_1<s_2<\cdots<s_{d-k}<t} \int_{\mathcal{P}_{k,c}} V_j(F)\, g(F,\mathbf{s})\, \mathbb{Q}_{k,s_{d-k}}^{\mathrm{PHT}}(\mathrm{d}F)$$
$$\times\, s_{d-k}^k\, \mathrm{d}s_1 \ldots \mathrm{d}s_{d-k-1}\, \mathrm{d}s_{d-k}\,.$$

Proof. By Theorem 14.15 and Corollary 14.13(b)

$$\int\limits_{\mathcal{P}_{k,c}\times(0,t)^{d-k}} g(F,\mathbf{s})\,\mathbb{Q}^{(j)}_{(F,\beta_1,\ldots,\beta_{d-k}),t}(\mathrm{d}(F,\mathbf{s}))$$

$$= \int\limits_{(0,t)}\int\limits_{\mathcal{P}_{k,c}}\int\limits_{(0,t)^{d-k-1}} g(F,\mathbf{s})\,\mathbb{Q}^{(j)}_{(\beta_1,\ldots,\beta_{d-k-1}),t|\beta_{d-k}=s_{d-k}}(\mathrm{d}(s_1,\ldots,s_{d-k-1}))$$

$$\mathbb{Q}^{(j)}_{F,t|\beta_{d-k}=s_{d-k}}(\mathrm{d}F)\,\mathbb{Q}^{(j)}_{\beta_{d-k},t}(\mathrm{d}s_{d-k})$$

$$= (d-k-1)! \int\limits_{\mathcal{P}_{k,c}\times(0,t)} \int\limits_{0<s_1<s_2<\cdots<s_{d-k-1}<s_{d-k}}\!\!\!\cdots\!\int g(F,\mathbf{s})\,\mathrm{d}s_1\ldots\mathrm{d}s_{d-k-1}$$

$$\times\, s_{d-k}^{-(d-k-1)}\,\mathbb{Q}^{(j)}_{(F,\beta_{d-k}),t}(\mathrm{d}(F,s_{d-k}))$$

$$= \frac{(d-k-1)!\,d}{t^d\,\overline{V}^{\max}_{k,j,t}} \int\limits_0^t \int\limits_{\mathcal{P}_{k,c}} \int\limits_{0<s_1<s_2<\cdots<s_{d-k-1}<s_{d-k}}\!\!\!\cdots\!\int g(F,\mathbf{s})\,s_{d-k}^k\,V_j(F)$$

$$\mathrm{d}s_1\ldots\mathrm{d}s_{d-k-1}\,\mathbb{Q}^{\mathrm{PHT}}_{k,s_{d-k}}(\mathrm{d}F)\,\mathrm{d}s_{d-k}\,,$$

where the last equation follows by Lemma 14.16. □

14.6 Number of internal vertices of I-segments

Now we turn to an application of the results, considering the maximal 1-polytopes, also referred to as I-segments, of a STIT tessellation Y_t in $\mathbb{R}^d$ for $d \geq 2$. Such segments may have vertices in their relative interior, which can arise already at the birth of the segment if $d \geq 3$, and also thereafter subject to further subdivision of adjacent cells. In the planar case $d = 2$, a maximal segment is always born without internal vertices. The following theorem provides the distributions of the numbers of internal vertices of the typical I-segment and of the typical length weighted I-segment. Formally, for $t > 0$ and $j \in \{0,1\}$, for the typical V_j-weighted maximal 1-polytope, we define the probabilities

$$\mathsf{p}_{1,j}(n) := \int \mathbb{1}\{|C| = n\}\,\mathbb{Q}^{(j)}_{(F,\boldsymbol{\beta},\Gamma),t}(\mathrm{d}(F,\mathbf{s},C))\,, \quad n \in \mathbb{N}_0\,,$$

with the distribution $\mathbb{Q}^{(j)}_{(F,\boldsymbol{\beta},\Gamma),t}$ defined in (14.22) and here used for $k = 1$. The variable C stands for the “internal structure” defined in (14.11), which here is a finite point set.

Theorem 14.18. *For $d \geq 2$, let Θ be a hyperplane measure of the form* (4.1), *satisfying* (4.3), *and let $(Y_t,\, t > 0)$ be the STIT tessellation process driven by Θ. Then for all $n \in \mathbb{N}_0$, the probabilities $\mathsf{p}_{1,0}(n)$ and $\mathsf{p}_{1,1}(n)$ depend neither on Θ nor on t and they are given by*

$$\mathsf{p}_{1,0}(n) = 2\int\limits_0^1 \frac{s^2(2-2s)^n}{(2-s)^{n+1}}\,\mathrm{d}s \quad \textit{and} \quad \mathsf{p}_{1,1}(n) = (n+1)\int\limits_0^1 \frac{s^2(2-2s)^n}{(2-s)^{n+2}}\,\mathrm{d}s$$

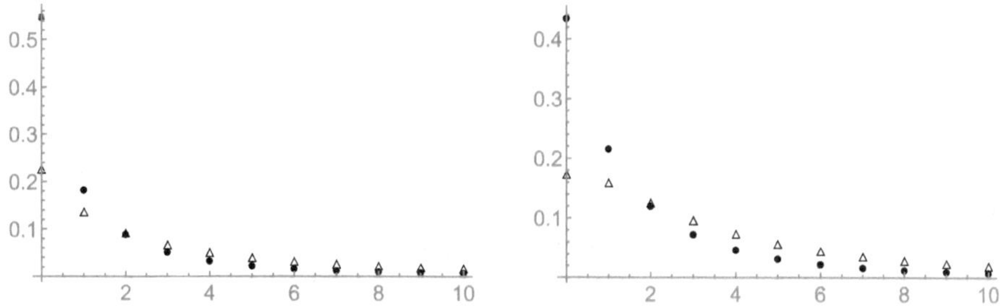

Figure 14.1: The probabilities $\mathrm{p}_{1,0}(n)$ (dots) and $\mathrm{p}_{1,1}(n)$ (triangles), $n \in \{0,\dots,10\}$. Left: $d = 2$. Right: $d = 3$.

for $d = 2$, *and for* $d \geq 3$, *by*

$$\mathrm{p}_{1,0}(n) = d(d-2)! \int_0^1 \int_0^{s_{d-1}} \dots \int_0^{s_2} \frac{s_{d-1}^2 (d - 2s_{d-1} - s_{d-2} - \dots - s_1)^n}{(d - s_{d-1} - s_{d-2} - \dots - s_1)^{n+1}} \, \mathrm{d}s_1 \dots \mathrm{d}s_{d-1}$$

and

$$\begin{aligned} &\mathrm{p}_{1,1}(n) \\ &\quad = (n+1)(d-1)! \int_0^1 \int_0^{s_{d-1}} \dots \int_0^{s_2} \frac{s_{d-1}^2 (d - 2s_{d-1} - s_{d-2} - \dots - s_1)^n}{(d - s_{d-1} - s_{d-2} - \dots - s_1)^{n+2}} \, \mathrm{d}s_1 \dots \mathrm{d}s_{d-1}. \end{aligned}$$

Proof. By Lemmas 14.11 and 14.6, for $k = 1$ and $j \in \{0, 1\}$,

$$\begin{aligned} &\int_{\mathcal{P}_{1,c} \times (0,t)^{d-k} \times \mathcal{F}} \mathbb{1}\{|C| = n\} \, \mathbb{Q}^{(j)}_{(\mathrm{F},\boldsymbol{\beta},\Gamma),t}(\mathrm{d}(F, \mathbf{s}, C)) \\ &= (\rho_{1,j,t})^{-1} \int_{\mathsf{N}_s(\mathcal{P}_{d-1} \times \mathbb{R}^+)} \sum_{(\mathbf{F},\mathbf{s})_1 \in (m_t^*)^{d-1}} V_j(\overline{F}) \\ &\quad \times \mathbb{1}\{c(\overline{F}) \in [0,1]^d\} \, \mathbb{1}\{|m_t \cap \overline{F}| = n\} \, \mathbb{P}^{\max *}(\mathrm{d}m^*) \\ &= 2^{d-2} (\rho_{1,j,t})^{-1} \int \dots \int_{0 < s_1 < s_2 < \dots < s_{d-1} < t} \int_{\mathcal{T}} \int_{\mathcal{H}^{d-1}} \int_{\mathcal{T}^d} \sum_{P \in y_{s_{d-1}}} V_j(P \cap \overline{H}) \, \mathbb{1}\{c(P \cap \overline{H}) \in [0,1]^d\} \\ &\quad \times \mathbb{1}\left\{ \left| P \cap \overline{H} \cap \left(\bigcup_{i=1}^{d-2} \partial y^{(i)}_{t-s_i} \cup \partial y^+_{t-s_{d-1}} \cup \partial y^-_{t-s_{d-1}} \right) \right| = n \right\} \\ &\quad \mathbb{P}_{Y_{t-s_1}}(\mathrm{d}y^{(1)}_{t-s_1}) \dots \mathbb{P}_{Y_{t-s_{d-2}}}(\mathrm{d}y^{(d-2)}_{t-s_{d-2}}) \, \mathbb{P}_{Y_{t-s_{d-1}}}(\mathrm{d}y^+_{t-s_{d-1}}) \, \mathbb{P}_{Y_{t-s_{d-1}}}(\mathrm{d}y^-_{t-s_{d-1}}) \\ &\quad \mathbb{1}\{\overline{H} \in A(d,1)\} \, \Theta^{\otimes(d-1)}(\mathrm{d}\mathbf{H}) \, \mathbb{P}_{Y_{s_{d-1}}}(\mathrm{d}y_{s_{d-1}}) \, \mathrm{d}s_1 \dots \mathrm{d}s_{d-1}. \end{aligned} \tag{14.25}$$

By Corollary 6.14 the term

$$\overline{H} \cap \left(\bigcup_{i=1}^{d-2} \partial Y^{(i)}_{t-s_i} \cup \partial Y^{+}_{t-s_{d-1}} \cup \partial Y^{-}_{t-s_{d-1}} \right)$$

in this expression describes the superposition of d independent Poisson point processes on the line $\overline{H}$ with a law invariant under translations on this line. Due to the stationarity of STIT tessellations, the intensities of these point processes depend only on the direction of this line. Denote by $\overline{H}_o \in G(d,1)$ the line parallel to $\overline{H}$ and containing the origin o. The intensity of the point process $\overline{H} \cap \partial Y_{t-s}$ is $(t-s)\, V_1(\Pi_\Theta | \overline{H}_o)$, $0 < s < t$; see Corollary 6.14. We denote $\gamma(\overline{H}_o) := V_1(\Pi_\Theta | \overline{H}_o)$. Hence by the superposition theorem 19.13 the intensity of the Poisson point process $\overline{H} \cap (\bigcup_{i=1}^{d-2} \partial Y^{(i)}_{t-s_i} \cup \partial Y^{+}_{t-s_{d-1}} \cup \partial Y^{-}_{t-s_{d-1}})$ equals $\gamma(\overline{H}_o)\, a(\mathbf{s})$ with

$$a(\mathbf{s}) := \sum_{i=1}^{d-2} (t - s_i) + 2(t - s_{d-1}) = d\,t - 2s_{d-1} - \sum_{i=1}^{d-2} s_i\,.$$

Thus, for any polytope $P \in \mathcal{P}_d$, the number of points of

$$P \cap \overline{H} \cap \left(\bigcup_{i=1}^{d-2} \partial Y^{(i)}_{t-s_i} \cup \partial Y^{+}_{t-s_{d-1}} \cup \partial Y^{-}_{t-s_{d-1}} \right)$$

follows a Poisson distribution with parameter $V_1(P \cap \overline{H})\, \gamma(\overline{H}_o)\, a(\mathbf{s})$. Hence (14.25) can be written as

$$\begin{aligned}
&\int_{\mathcal{P}_{k,c} \times (0,t)^{d-k} \times \mathcal{F}} \mathbb{1}\{|C| = n\}\, \mathbb{Q}^{(j)}_{(\mathrm{F},\boldsymbol{\beta},\Gamma),t}(\mathrm{d}(\overline{F}, \mathbf{s}, C)) \\
&= 2^{d-2} (\rho_{1,j,t})^{-1} \int \cdots \int_{0<s_1<s_2<\cdots<s_{d-1}<t} \int_{\mathcal{H}^{d-1}} \int_{\mathcal{T}} \sum_{P \in y_{s_{d-1}}} V_j(P \cap \overline{H})\, \mathbb{1}\{c(P \cap \overline{H}) \in [0,1]^d\} \\
&\quad \times \frac{(V_1(P \cap \overline{H})\, \gamma(\overline{H}_o)\, a(\mathbf{s}))^n}{n!} \exp(-V_1(P \cap \overline{H})\, \gamma(\overline{H}_o)\, a(\mathbf{s}))\, \mathbb{P}_{Y_{s_{d-1}}}(\mathrm{d}y_{s_{d-1}}) \\
&\quad \mathbb{1}\{\overline{H} \in A(d,1)\}\, \Theta^{\otimes(d-1)}(\mathrm{d}\mathbf{H})\, \mathrm{d}s_1 \ldots \mathrm{d}s_{d-1}\,.
\end{aligned} \tag{14.26}$$

Now regarding the two inner integrals, we can proceed analogously to the proof of Lemma 14.16 and pass to the PHT $X_{s_{d-1}}$. We replace $\sum_{P \in Y_{s_{d-1}}}$ by $\sum_{P \in X_{s_{d-1}}}$ and apply (14.24). This yields

$$\begin{aligned}
I_j := \frac{2^{d-2}}{\rho_{1,j,t}} &\int_{\mathcal{H}^{d-1}} \mathbb{E}\Bigg(\sum_{P \in X_{s_{d-1}}} V_j(P \cap \overline{H})\, \mathbb{1}\{c(P \cap \overline{H}) \in [0,1]^d\} \\
&\times \frac{(V_1(P \cap \overline{H})\, \gamma(\overline{H}_o)\, a(\mathbf{s}))^n}{n!} \exp(-V_1(P \cap \overline{H})\, \gamma(\overline{H}_o)\, a(\mathbf{s})) \Bigg) \\
&\times \mathbb{1}\{\overline{H} \in A(d,1)\}\, \Theta^{\otimes(d-1)}(\mathrm{d}\mathbf{H})
\end{aligned}$$

$$= \frac{2^{d-2}(d-1)!}{\rho_{1,j,t}\, s_{d-1}^{d-1}} \gamma_{1,s_{d-1}}^{\mathrm{PHT}} \int_{\mathcal{P}_{1,c}} V_j(F) \frac{(V_1(F)\,\gamma(\mathrm{aff}_o F)\, a(\mathbf{s}))^n}{n!} \exp(-V_1(F)\,\gamma(\mathrm{aff}_o F)\, a(\mathbf{s})) \mathbb{Q}_{1,s_{d-1}}^{\mathrm{PHT}}(\mathrm{d}F).$$

Furthermore, the scaling properties (2.67) and (6.2) of PHTs and STIT tessellations, respectively, imply that $s^d \gamma_{k,1}^{\mathrm{PHT}} = \gamma_{k,s}^{\mathrm{PHT}}$ and $t^d \gamma_{k,1}^{\max} = \gamma_{k,t}^{\max}$.

Using the scaling property (14.18) of $\rho_{k,j,t}$, we obtain

$$\frac{2^{d-2}\,(d-1)!}{\rho_{1,j,t}\, s_{d-1}^{d-1}} \gamma_{1,s_{d-1}}^{\mathrm{PHT}} = \frac{2^{d-2}\,(d-1)!}{t^{d-j}\,\rho_{1,j,1}} s_{d-1}\, \gamma_{1,1}^{\mathrm{PHT}}.$$

To determine the integral I_j, we apply the disintegration of $\mathbb{Q}_{1,s_{d-1}}^{\mathrm{PHT}}$ according to (2.36) and obtain

$$\begin{aligned} I_j &= \frac{2^{d-2}\,(d-1)!}{t^{d-j}\,\rho_{1,j,1}} s_{d-1}\, \gamma_{1,1}^{\mathrm{PHT}} \int_{G(d,1)} \int_{\mathcal{P}_{1,c}} V_j(F) \\ &\quad \times \frac{(V_1(F)\,\gamma(L)\, a(\mathbf{s}))^n}{n!} \exp(-V_1(F)\,\gamma(L)\, a(\mathbf{s}))\, \mathbb{Q}_{1,s_{d-1}|h=L}^{\mathrm{PHT}}(\mathrm{d}F)\, \mathbb{Q}_{1,s_{d-1}:h}^{\mathrm{PHT}}(\mathrm{d}L). \end{aligned}$$

By Corollary 2.39, for a given direction $L \in G(d,1)$, the conditional distribution of the length of the typical 1-face of the PHT $X_{s_{d-1}}$ is the exponential distribution with parameter $s_{d-1}\, V_1(\Pi_\Theta|L) = s_{d-1}\,\gamma(L)$. Thus the inner integral can be calculated explicitly.

For $j = 0$, definition (14.17) yields that $\rho_{1,0,1} = \gamma_{1,1}^{\max}$, and thus

$$I_0 = \frac{2^{d-2}\,(d-1)!}{t^d\, \gamma_{1,1}^{\max}} \gamma_{1,1}^{\mathrm{PHT}} \frac{s_{d-1}^2\, a(\mathbf{s})^n}{(s_{d-1} + a(\mathbf{s}))^{n+1}} = \frac{d\,(d-2)!}{t^d} \frac{s_{d-1}^2\, a(\mathbf{s})^n}{(s_{d-1} + a(\mathbf{s}))^{n+1}},$$

where we used that $\gamma_{1,t}^{\max} = \frac{(d-1)2^{d-2}}{d} \gamma_{1,t}^{\mathrm{PHT}}$; see (11.2).

For $j = 1$,

$$I_1 = \frac{2^{d-2}\,(d-1)!}{t^{d-1}\,\rho_{1,1,1}} \gamma_{1,1}^{\mathrm{PHT}} \frac{s_{d-1}^2\, a(\mathbf{s})^n\,(n+1)}{(s_{d-1} + a(\mathbf{s}))^{n+2}} \int_{G(d,1)} \gamma(L)^{-1}\, \mathbb{Q}_{1,s_{d-1}:h}^{\mathrm{PHT}}(\mathrm{d}L).$$

By Lemma 14.9 for $k = j = 1$ and $s = 1$, using (14.24) and then the disintegration of $\mathbb{Q}_{1,1}^{\mathrm{PHT}}$ according to (2.36), we obtain

$$\begin{aligned} \rho_{1,1,1} &= \frac{2^{d-2}}{(d-2)!\,(d-1)} \\ &\quad \times \int_{\mathcal{H}^{d-1}} \mathbb{E} \sum_{P \in Y_1} \mathbb{1}\{c(P \cap \overline{H}) \in [0,1]^d\}\, V_1(P \cap \overline{H})\, \mathbb{1}\{\overline{H} \in A(d,1)\}\, \Theta^{\otimes(d-1)}(\mathrm{d}\mathbf{H}) \\ &= 2^{d-2}\, \gamma_{1,1}^{\mathrm{PHT}} \int_{\mathcal{P}_{1,c}} V_1(F)\, \mathbb{Q}_{1,1}^{\mathrm{PHT}}(\mathrm{d}F) \end{aligned}$$

$$= 2^{d-2}\gamma_{1,1}^{\mathrm{PHT}} \int_{G(d,1)} \int_{\mathcal{P}_{1,c}} V_1(F)\,\mathbb{Q}_{1,1|h=L}^{\mathrm{PHT}}(\mathrm{d}F)\,\mathbb{Q}_{1,1:h}^{\mathrm{PHT}}(\mathrm{d}L)$$

$$= 2^{d-2}\gamma_{1,1}^{\mathrm{PHT}} \int_{G(d,1)} \gamma(L)^{-1}\,\mathbb{Q}_{1,1:h}^{\mathrm{PHT}}(\mathrm{d}L)\,,$$

where the last equation follows from the fact that by Corollary 2.39, for a given direction $L \in G(d,1)$, the conditional distribution of the length of the typical 1-face of the PHT X_1 is the exponential distribution with parameter $\gamma(L)$.

This yields

$$I_1 = \frac{(d-1)!}{t^{d-1}} \frac{s_{d-1}^2\, a(\mathbf{s})^n\,(n+1)}{(s_{d-1}+a(\mathbf{s}))^{n+2}}\,.$$

To complete the proof, it remains to insert this into the right-hand side of (14.26) and to substitute s_i by $\frac{1}{t}s_i$, $i \in \{1,\dots,d-1\}$. □

As a particular case, for $d = 3$, we obtain the following values:

$$\begin{aligned}
\mathrm{p}_{1,0}(0) &= \frac{189}{8}\ln 3 - 26\ln 2 - \frac{15}{2} &&\approx 0.43289\,,\\
\mathrm{p}_{1,0}(1) &= \frac{1593}{16}\ln 3 - 107\ln 2 - 35 &&\approx 0.21384\,,\\
\mathrm{p}_{1,0}(2) &= \frac{5319}{16}\ln 3 - 350\ln 2 - \frac{245}{2} &&\approx 0.11841\,,\\
\mathrm{p}_{1,0}(3) &= \frac{31617}{32}\ln 3 - 1025\ln 2 - \frac{4499}{12} &&\approx 0.07075\,.
\end{aligned}$$

These results correspond to (17.40) and (17.41). Exemplary values for the length-weighted typical maximal segment are

$$\begin{aligned}
\mathrm{p}_{1,1}(0) &= 5 + 18\ln 2 - \frac{63}{4}\ln 3 &&\approx 0.173506\,,\\
\mathrm{p}_{1,1}(1) &= 28 + 90\ln 2 - \frac{657}{8}\ln 3 &&\approx 0.159712\,.
\end{aligned}$$

The fact that $\mathrm{p}_{1,0}(n)$ and $\mathrm{p}_{1,1}(n)$ do not depend on the time t is consistent with the scaling property (6.2) of a STIT tessellation, because the number of internal vertices on a maximal 1-polytope does not change when the tessellation is rescaled. Theorem 14.18 can also be used to compute the moments of the respective distributions. In Corollary 17.14 the moments are determined for $d = 3$ and the typical maximal 1-polytope. Here we derive the results for the expectations in arbitrary dimension.

Corollary 14.19. *With the assumptions of Theorem* 14.18 *and dimension* $d \geq 2$, *let* $\mathrm{v}_1^{(j)}$, $j \in \{0,1\}$, *be random variables with distributions given by* $\mathrm{p}_{1,j}$. *Then*

$$\mathbb{E}\nu_1^{(0)} = \frac{d^2 - d + 2}{2(d-1)} \quad \textit{and} \quad \mathbb{E}\nu_1^{(1)} = \begin{cases} +\infty\,, & d = 2\,, \\ \frac{d^2-2d+4}{d-2}\,, & d \geq 3\,. \end{cases}$$

Proof. We only prove the formula for $\mathbb{E}\nu_1^{(0)}$; for $\mathbb{E}\nu_1^{(1)}$, it follows similarly. Plugging in the values for $\mathrm{p}_{1,0}(n)$, $n \geq 0$, from Theorem 14.18 and using Fubini's theorem, we obtain

$$\begin{aligned}
\mathbb{E}\nu_1^{(0)} &= \sum_{n=0}^{\infty} n\,\mathrm{p}_{1,0}(n) \\
&= d(d-2)! \int_0^1 \int_0^{s_{d-1}} \cdots \int_0^{s_2} \frac{s_{d-1}^2}{d - s_{d-1} - s_{d-2} - \cdots - s_1} \\
&\quad \times \sum_{n=0}^{\infty} n \left(\frac{d - 2s_{d-1} - s_{d-2} - \cdots - s_1}{d - s_{d-1} - s_{d-2} - \cdots - s_1} \right)^n \mathrm{d}s_1 \ldots \mathrm{d}s_{d-1} \\
&= d(d-2)! \int_0^1 \int_0^{s_{d-1}} \cdots \int_0^{s_2} \frac{s_{d-1}^2}{d - s_{d-1} - s_{d-2} - \cdots - s_1} \\
&\quad \times \frac{(d - s_{d-1} - s_{d-2} - \cdots - s_1)(d - 2s_{d-1} - s_{d-2} - \cdots - s_1)}{s_{d-1}^2} \,\mathrm{d}s_1 \ldots \mathrm{d}s_{d-1} \\
&= d(d-2)! \int_0^1 \int_0^{s_{d-1}} \cdots \int_0^{s_2} (d - 2s_{d-1} - s_{d-2} - \cdots - s_1)\,\mathrm{d}s_1 \ldots \mathrm{d}s_{d-1}\,,
\end{aligned}$$

where we used that $\sum_{n=0}^{\infty} n\,a^n = \frac{a}{(a-1)^2}$ for $|a| < 1$. We rewrite the iterated integral in the last line as

$$\begin{aligned}
& d \int_0^1 \int_0^{s_{d-1}} \cdots \int_0^{s_2} \mathrm{d}s_1 \ldots \mathrm{d}s_{d-1} - 2 \int_0^1 s_{d-1} \int_0^{s_{d-1}} \cdots \int_0^{s_2} \mathrm{d}s_1 \ldots \mathrm{d}s_{d-1} \\
& \quad - \sum_{i=2}^{d-2} \int_0^1 \int_0^{s_{d-1}} \cdots \int_0^{s_{i+1}} s_i \int_0^{s_i} \cdots \int_0^{s_2} \mathrm{d}s_1 \ldots \mathrm{d}s_{i-1}\,\mathrm{d}s_i\,\mathrm{d}s_{i+1} \ldots \mathrm{d}s_{d-1} \\
& \quad - \int_0^1 \int_0^{s_{d-1}} \cdots \int_0^{s_2} s_1\,\mathrm{d}s_1 \ldots \mathrm{d}s_{d-1}\,.
\end{aligned}$$

Using the iterated integral $I^n(g,t)$ introduced in (12.7), rewrite this as

$$\begin{aligned}
& d\,I^{d-1}(1,1) - 2 \int_0^1 s_{d-1} I^{d-2}(1, s_{d-1})\,\mathrm{d}s_{d-1} \\
& \quad - \sum_{i=2}^{d-2} \int_0^1 \int_0^{s_{d-1}} \cdots \int_0^{s_{i+1}} s_i I^{i-1}(1, s_i)\,\mathrm{d}s_i \ldots \mathrm{d}s_{d-1} - I^{d-1}(s_{d-1}, 1)
\end{aligned}$$

$$= \frac{d}{(d-2)!}\int_0^1 (1-s)^{d-2}\,ds - 2\int_0^1 \frac{s_{d-1}}{(d-3)!}\int_0^{s_{d-1}} (s_{d-1}-s)^{d-3}\,ds\,ds_{d-1}$$
$$- \sum_{i=2}^{d-2}\int_0^1\int_0^{s_{d-1}}\cdots\int_0^{s_{i+1}} \frac{s_i}{(i-2)!}\int_0^{s_i}(s_i-s)^{i-2}\,ds\,ds_i\ldots ds_{d-1}$$
$$- \frac{1}{(d-2)!}\int_0^1 (1-s)^{d-2}s\,ds$$
$$= \frac{d}{(d-1)!} - \frac{2}{(d-2)!}\int_0^1 s_{d-1}^{d-1}\,ds_{d-1} - \sum_{i=2}^{d-2}\frac{1}{(i-1)!}\int_0^1\int_0^{s_{d-1}}\cdots\int_0^{s_{i+1}} s_i^i\,ds_i\ldots ds_{d-1}$$
$$- \frac{1}{d!}.$$

Since

$$\int_0^1\int_0^{s_{d-1}}\cdots\int_0^{s_{i+1}} s_i^i\,ds_i\ldots ds_{d-1} = I^{d-i}(s^i,1)$$
$$= \frac{1}{(d-i-1)!}\int_0^1 (1-s)^{d-i-1}s^i\,ds$$
$$= \frac{1}{(d-i-1)!}\,\frac{(d-i-1)!i!}{d!} = \frac{i!}{d!},$$

we arrive at

$$\mathbb{E}v_1^{(0)} = d(d-2)!\left(\frac{d}{(d-1)!} - \frac{2(d-1)}{d!} - \frac{1}{d!}\sum_{i=1}^{d-2} i\right)$$
$$= d(d-2)!\left(\frac{d}{(d-1)!} - \frac{2(d-1)}{d!} - \frac{1}{d!}\frac{(d-2)(d-1)}{2}\right)$$
$$= \frac{d^2}{d-1} - 2 - \frac{d-2}{2} = \frac{d^2-d+2}{2(d-1)},$$

as claimed. □

Note that $\mathbb{E}v_1^{(0)} = 2$ for $d = 2$ and $d = 3$, and this value strictly increases in d for $d > 3$. In contrast, for the mean number of internal vertices on the typical length-weighted maximal segment, we have that $\mathbb{E}v_1^{(1)} = 7$ for $d = 3$, $\mathbb{E}v_1^{(1)} = 6$ for $d = 4$, $\mathbb{E}v_1^{(1)} = 6\frac{1}{3}$ for $d = 5$, and $\mathbb{E}v_1^{(1)} = 7$ for $d = 6$, and this value strictly increases for $d > 4$.

14.7 Notes

1. Theorem 14.1 combines Theorem 3.1 of [143] with Theorem 3.1 of [58]. In Theorem 3.1 of [58] the function g used in (14.1) is allowed to depend only on the evolution *in the past* of a given time t, expressed by $(Y_s, s \leq t)$ for $(F, t) \in \mathcal{F}_{d-1}^{\max *}(\Sigma)$, and g may not depend on the development *after* the time t, put by the term $\mathcal{F}_{d-1}^{\max *}(\Sigma) \wedge P(F, t)$. In contrast, in Theorem 3.1 of [143] the setting is converse: There the function g may depend on the future but not on the past.
2. Most of the material in the present chapter was provided in [143]. Theorem 14.12 extends and unifies results for the special case $k = d - 1$ and $j = 0$ in [177] and $d = 3$, $k = 1$, and $j \in \{0, 1\}$ in [189]. Lemma 14.6 and Corollary 14.7 are Proposition 3.2 and Corollary 3.3 of [143], respectively.
3. The relations to PHTs given in Lemma 14.16 and Corollary 14.17 were shown in Corollary 2, Theorem 3, and Corollary 3 of [177].
4. In the planar case $d = 2$ the probabilities $\mathrm{p}_{1,0}(n)$ are known from [131, 185], whereas for $d = 3$, the formula for $\mathrm{p}_{1,0}(n)$ is established in [189] by different methods.

15 Mixing properties

In this chapter, we study different aspects of weak dependence and ergodicity of STIT tessellations. Firstly, we compile concepts of spatial mixing properties for random stationary tessellations. Then we show that STIT tessellations possess some of these properties. Furthermore, for a rescaled time-stationary STIT tessellation process, we deal with its mixing behavior in time and also with its representation as a factor of a Bernoulli shift or flow, respectively.

In Section 15.1, we summarize basic definitions and concepts. Some of them refer to the group of translations acting on $\mathbb{R}^d$, and the stronger ones refer to the stochastic dependence between the behavior of a tessellation inside a window W_1 and outside a larger window $W_2 \supset W_1$, and the asymptotics for growing W_2.

In Section 15.2, we show that STIT tessellations are β-mixing in space and discuss some implications. The key concept in the proofs of the main result is *encapsulation.*

The ergodic properties of STIT and Poisson hyperplane tessellations are compared in Section 15.3.

Section 15.4 provides the result that for any fixed window $W \in \mathcal{P}_d$, the rescaled time-stationary STIT tessellation process $(a^t Y_{a^t} \wedge W,\ t \in \mathbb{R})$ is β-mixing in time. The process of zero-cells of a STIT tessellation process plays a central role in the proof. Furthermore, in Section 15.5, we show a version of tail triviality of the zero-cell process, in this case, looking *backward* in time and thus focusing on the behavior of large zero-cells at small times $t > 0$.

The Ornstein–Weiss theory concerning Bernoulli shifts, Bernoulli flows, and factor maps is applied to the rescaled time-stationary STIT tessellation process in Section 15.6. The results highlight further aspects of weak dependencies in space and time of the STIT tessellation process.

15.1 General concepts

In this section, some classical definitions of mixing conditions are adapted to random tessellations in the Euclidean space $\mathbb{R}^d$. We follow the concepts presented in [19], [46], and especially in [69].

Definition 15.1. *Let* $(\Omega, \mathbf{A}, \mathbb{P})$ *be a probability space, and let* Γ *be a group of automorphisms on* Ω*. The quadruple* $(\Omega, \mathbf{A}, \mathbb{P}, \Gamma)$ *is called a* ***dynamical system*** *if all* $\alpha \in \Gamma$ *are measure preserving, that is,* $\mathbb{P} \circ \alpha^{-1} = \mathbb{P}$ *for all* $\alpha \in \Gamma$*.*

In the first part of this chapter, we consider the space $\Omega = \mathcal{T}$ of tessellations with the group of translations acting on $\mathbb{R}^d$. Then in Section 15.6, for tessellation-valued processes with discrete time, we choose $\Omega = \mathcal{T}^{\mathbb{Z}}$ with the group of shifts by units of time or, for continuous time, $\Omega = \mathcal{T}^{\mathbb{R}}$ with the group of shifts acting on the real axis.

The central issue in this chapter is the concept of β-mixing.

https://doi.org/10.1515/9783112222409-018

Definition 15.2. *Let $(\Omega, \mathbf{A}, \mathbb{P})$ be a probability space, and let* $\mathbf{B}$ *and* $\mathbf{C}$ *be sub-σ-algebras of* $\mathbf{A}$. *The **β-mixing coefficient of** $(\mathbf{B}, \mathbf{C})$ is defined as*

$$\beta(\mathbf{B}, \mathbf{C}) := \frac{1}{2} \sup_{(\overline{C}, \overline{B})} \sum_{i=1}^{I} \sum_{j=1}^{J} \left| \mathbb{P}(C_i \cap B_j) - \mathbb{P}(C_i)\, \mathbb{P}(B_j) \right| , \tag{15.1}$$

where the supremum is taken over all pairs of finite partitions $\overline{C}$ of Ω with $\overline{C} = \{C_i, i \in \{1, \ldots I\}\}$ such that $C_i \in \mathbf{C}$, and $\overline{B}$ of Ω with $\overline{B} = \{B_j, j \in \{1, \ldots J\}\}$, $B_j \in \mathbf{B}$, $I, J \in \mathbb{N}$.

The β-mixing coefficient is also referred to as the **absolute regularity coefficient** or **weak Bernoulli coefficient.**

The following lemma provides, under additional assumptions, an alternative representation of $\beta(\mathbf{B}, \mathbf{C})$ and a useful upper bound.

Lemma 15.3. *With the notation of Definition 15.2, assume that the σ-algebra $\mathbf{A}$ has a countable generating system and that a regular conditional distribution $\mathbb{P}(\cdot\,|\mathbf{B})$ on $\mathbf{C}$ exists.*

(a) *Then $\beta(\mathbf{B}, \mathbf{C}) = \mathbb{E} \sup_{C \in \mathbf{C}} |\mathbb{P}(C|\mathbf{B}) - \mathbb{P}(C)|$.*

(b) *Let $D \in \mathbf{B}$ with $\mathbb{P}(D) > 0$ and $\varepsilon > 0$. If $|\mathbb{P}(C|B) - \mathbb{P}(C)| < \varepsilon$ for all $C \in \mathbf{C}$ and $B \in \mathbf{B}$ with $B \subseteq D$, $\mathbb{P}(B) > 0$, then*

$$\beta(\mathbf{B}, \mathbf{C}) < \varepsilon\, \mathbb{P}(D) + \mathbb{P}(D^c) .$$

Proof. (a) An application of Proposition 3.22, equation (5), of [19] yields the statement.

(b) We use ideas of a proof of [19], page 114. Let $D \in \mathbf{B}$ with $\mathbb{P}(D) > 0$ and $\varepsilon > 0$. First, we show that the following two statements are equivalent:

(i) $\forall C \in \mathbf{C}\ \forall B \in \mathbf{B} : B \subseteq D, \mathbb{P}(B) > 0 \Rightarrow |\mathbb{P}(C|B) - \mathbb{P}(C)| < \varepsilon$.

(ii) $\exists D' \in \mathbf{B}, \mathbb{P}(D') = 0, \forall C \in \mathbf{C}\ \forall \omega \in D \setminus D' : |\mathbb{P}(C|\mathbf{B})(\omega) - \mathbb{P}(C)| < \varepsilon$.

To see that (ii) implies (i), let $B \subseteq D$ with $\mathbb{P}(B) > 0$ and $C \in \mathbf{C}$. Then by (ii)

$$\begin{aligned} \left|\mathbb{P}(C|B) - \mathbb{P}(C)\right| &= \frac{1}{\mathbb{P}(B)} \left|\mathbb{P}(C \cap B) - \mathbb{P}(B)\mathbb{P}(C)\right| \\ &= \frac{1}{\mathbb{P}(B)} \left| \int_B \left(\mathbb{P}(C|\mathbf{B})(\omega) - \mathbb{P}(C)\right) \mathbb{P}(d\omega) \right| \\ &\le \frac{1}{\mathbb{P}(B)} \int_B \left|\mathbb{P}(C|\mathbf{B})(\omega) - \mathbb{P}(C)\right| \mathbb{P}(d\omega) \\ &< \varepsilon . \end{aligned}$$

Vice versa, we prove that (ii) follows from (i): If there is $D'' \subseteq D$, $D'' \in \mathbf{B}$, with $\mathbb{P}(D'') > 0$ such that $|\mathbb{P}(C|\mathbf{B})(\omega) - \mathbb{P}(C)| \ge \varepsilon$ for all $\omega \in D''$, then there is $B \in \mathbf{B}$ with $B \subseteq D$ and $\mathbb{P}(B) > 0$ such that

(1) $\mathbb{P}(C|\mathbf{B})(\omega) - \mathbb{P}(C) \geq \varepsilon$ for all $\omega \in B$,
or
(2) $\mathbb{P}(C) - \mathbb{P}(C|\mathbf{B})(\omega) \geq \varepsilon$ for all $\omega \in B$.

In case (1),

$$\varepsilon \geq \frac{1}{\mathbb{P}(B)} \int_B \mathbb{P}(C|\mathbf{B})(\omega) - \mathbb{P}(C)\, \mathbb{P}(d\omega)$$
$$= \frac{1}{\mathbb{P}(B)} (\mathbb{P}(C \cap B) - \mathbb{P}(B)\mathbb{P}(C)) = \mathbb{P}(C|B) - \mathbb{P}(C).$$

Case (2) can be dealt with analogously. This yields the existence of $B \in \mathbf{B}$ with $B \subseteq D$ and $\mathbb{P}(B) > 0$ such that $|\mathbb{P}(C|B) - \mathbb{P}(C)| \geq \varepsilon$. Thus the equivalence of (i) and (ii) is shown.

Now consider the random variable $V := \sup_{C \in \mathbf{C}} |\mathbb{P}(C|\mathbf{B}) - \mathbb{P}(C)|$. By (a), because $0 \leq V \leq 1$ a. s., the assumption on D in (b) and its equivalent expression (ii) yield

$$\beta(\mathbf{B}, \mathbf{C}) = \mathbb{E}V < \varepsilon\, \mathbb{P}(D) + \mathbb{P}(D^c),$$

which completes the proof. □

15.1.1 σ-algebras on the space of tessellations

Now we adapt the usual concepts of ergodicity and mixing to the measurable space $(\mathcal{T}, \mathcal{B}(\mathcal{T}))$ of tessellations in $\mathbb{R}^d$, $d \geq 1$, and the group of translations acting on it. A measurable set $A \in \mathcal{B}(\mathcal{T})$ is called (translation-)**invariant** if

$$A = A + x \quad \text{for all } x \in \mathbb{R}^d,$$

where the translation on the right-hand side is defined in (2.19). Denote by $\mathcal{I}(\mathcal{T})$ the set of all invariant subsets of $\mathcal{T}$. It is easily checked that $\mathcal{I}(\mathcal{T}) \subseteq \mathcal{B}(\mathcal{T})$ is a sub-σ-algebra, the **invariant σ-algebra**.

Furthermore, the definition of the tail σ-algebra is based on the system of measurable sets defined by their behavior *outside* a window. This concept is used for random measures and point processes in $\mathbb{R}^d$; see Section 12.3 of [46]. For $W \in \mathcal{P}_d$, define the σ-algebra

$$\mathcal{B}_{W^c}(\mathcal{T}) := \sigma(\{\{T \in \mathcal{T} : \partial T \cap C = \emptyset\} : C \subset W^c,\, C \in \mathcal{C}\}). \tag{15.2}$$

Obviously, $\mathcal{B}_{W^c}(\mathcal{T}) \subset \mathcal{B}(\mathcal{T})$. Moreover, if $W_1 \subseteq W_2$, then $\mathcal{B}_{W_2^c}(\mathcal{T}) \subseteq \mathcal{B}_{W_1^c}(\mathcal{T})$.

As a counterpart, define the σ-algebra of sets of tessellations determined by restraints *inside* a window $W \in \mathcal{P}_d$,

$$\mathcal{B}_W(\mathcal{T}) := \sigma(\{\{T \in \mathcal{T} : \partial T \cap C = \emptyset\} : C \subseteq W,\, C \in \mathcal{C}\}). \tag{15.3}$$

By definition $\mathcal{B}_W(\mathcal{T}) \subset \mathcal{B}(\mathcal{T})$, and $\mathcal{B}_{W_1}(\mathcal{T}) \subseteq \mathcal{B}_{W_2}(\mathcal{T})$ if $W_1 \subseteq W_2$.

Now let $(W_n\,,\, n \in \mathbb{N})$ be an increasing sequence of windows $W_n \in \mathcal{P}_d$ such that

$$\mathbb{R}^d = \bigcup_{n\in\mathbb{N}} W_n \quad \text{and} \quad W_n \subset \operatorname{int} W_{n+1} \quad \text{for all } n \in \mathbb{N}\,. \tag{15.4}$$

Then by Corollary 18.8

$$\sigma\Big(\bigcup_{n\in\mathbb{N}} \mathcal{B}_{W_n}(\mathcal{T})\Big) = \mathcal{B}(\mathcal{T})\,.$$

The following approximation result will be used in proofs of the main results. The symmetric difference of two sets A and B is denoted by $A \triangle B := (A \setminus B) \cup (B \setminus A)$.

Lemma 15.4. *Let $\mathbb{P}$ be a probability measure on $(\mathcal{T}, \mathcal{B}(\mathcal{T}))$, and let $(W_n\,,\, n \in \mathbb{N})$ be an increasing sequence of windows satisfying* (15.4). *Then every element of $\mathcal{B}(\mathcal{T})$ can be approximated by some set in $\bigcup_{n\in\mathbb{N}} \mathcal{B}_{W_n}(\mathcal{T})$ in the sense that*

$$\mathcal{B}(\mathcal{T}) = \big\{A \in \mathcal{B}(\mathcal{T}) : \forall \varepsilon > 0\, \exists n \in \mathbb{N}\, \exists A_n \in \mathcal{B}_{W_n}(\mathcal{T}) : \mathbb{P}(A \triangle A_n) < \varepsilon\big\}\,. \tag{15.5}$$

Proof. Denote $\mathsf{A} := \{A \in \mathcal{B}(\mathcal{T}) : \forall \varepsilon > 0\, \exists n \in \mathbb{N}\, \exists A_n \in \mathcal{B}_{W_n}(\mathcal{T}) : \mathbb{P}(A \triangle A_n) < \varepsilon\}$. Obviously,

$$\big\{\{T \in \mathcal{T} : \partial T \cap C = \emptyset\} : C \in \mathcal{C}\big\} \subseteq \mathsf{A}\,,$$

and thus, by Corollary 18.8, $\sigma(\mathsf{A}) = \mathcal{B}(\mathcal{T})$. Now it is sufficient to show that A is a $\cap$-stable Dynkin-system and hence a σ-algebra. If $A, B \in \mathsf{A}$ and $\varepsilon > 0$, then there are $m, n \in \mathbb{N}$, $A_m \in \mathcal{B}_{W_m}(\mathcal{T})$, and $B_n \in \mathcal{B}_{W_n}(\mathcal{T})$ such that $\mathbb{P}(A \triangle A_m) < \varepsilon/2$ and $\mathbb{P}(B \triangle B_n) < \varepsilon/2$. Assume, without loss of generality, that $m \le n$. Then $A_m \cap B_n \in \mathcal{B}_{W_n}(\mathcal{T})$ and $\mathbb{P}((A \cap B) \triangle (A_m \cap B_n)) < \varepsilon$, which means that $A \cap B \in \mathsf{A}$. As $\mathcal{T} \in \mathcal{B}_{W_n}(\mathcal{T})$ for all $n \in \mathbb{N}$, also $\mathcal{T} \in \mathsf{A}$. If $A, B \in \mathsf{A}$, $A \subseteq B$, and $\varepsilon > 0$, then there are $m, n \in \mathbb{N}$, $A_m \in \mathcal{B}_{W_m}(\mathcal{T})$, and $B_n \in \mathcal{B}_{W_n}(\mathcal{T})$ such that $\mathbb{P}(A \triangle A_m) < \varepsilon/2$ and $\mathbb{P}(B \triangle B_n) < \varepsilon/2$. For $k := \max\{m, n\}$, we have $B_n \setminus A_m \in \mathcal{B}_{W_k}(\mathcal{T})$ and $\mathbb{P}((B \setminus A) \triangle (B_n \setminus A_m)) < \varepsilon$. Finally, let $(A_i\,,\, i \in \mathbb{N})$ be a sequence of pairwise disjoint sets $A_i \in \mathsf{A}$, and let $\varepsilon > 0$. Then there is $\ell \in \mathbb{N}$ such that $\mathbb{P}(\bigcup_{i=\ell+1}^{\infty} A_i) < \varepsilon/2$, and for all $i \in \{1, \ldots, \ell\}$ there are $n_i \in \mathbb{N}$ and $B_i \in \mathcal{B}_{W_{n_i}}(\mathcal{T})$ with $\mathbb{P}(A_i \triangle B_i) < \varepsilon/(2\ell)$. For $k := \max\{n_1, \ldots, n_\ell\}$, we have $\bigcup_{i=1}^{\ell} B_i \in \mathcal{B}_{W_k}(\mathcal{T})$ and $\mathbb{P}((\bigcup_{i=1}^{\infty} A_i) \triangle (\bigcup_{i=1}^{\ell} B_{n_i})) < \varepsilon$. This completes the proof. □

Let $(W_n\,,\, n \in \mathbb{N})$ be an increasing sequence of windows satisfying (15.4). Then for all windows $W \in \mathcal{P}_d$, there exists m such that $W \subseteq W_m$ and thus $\mathcal{B}_{W_m^c}(\mathcal{T}) \subseteq \mathcal{B}_{W^c}(\mathcal{T})$. Hence $\bigcap_{n=1}^{\infty} \mathcal{B}_{W_n^c}(\mathcal{T}) \subseteq \mathcal{B}_{W^c}(\mathcal{T})$. This yields

$$\bigcap_{n=1}^{\infty} \mathcal{B}_{W_n^c}(\mathcal{T}) \subseteq \bigcap_{W\in\mathcal{P}_d} \mathcal{B}_{W^c}(\mathcal{T})\,,$$

and with the other obvious inclusion, we have $\bigcap_{n=1}^{\infty} \mathcal{B}_{W_n^c}(\mathcal{T}) = \bigcap_{W \in \mathcal{P}_d} \mathcal{B}_{W^c}(\mathcal{T})$. In analogy to Definition 12.3.IV of [46], for point processes or random measures, we define the tail σ-algebra on the space of tessellations.

Definition 15.5. *The* ***tail σ-algebra on*** $\mathcal{T}$ *is defined as*

$$\mathcal{B}_\infty(\mathcal{T}) := \bigcap_{W \in \mathcal{P}_d} \mathcal{B}_{W^c}(\mathcal{T}) = \bigcap_{n=1}^{\infty} \mathcal{B}_{W_n^c}(\mathcal{T}),$$

where $(W_n, n \in \mathbb{N})$ *is an increasing sequence of windows satisfying* (15.4).

Clearly, the tail σ-algebra does not depend on the choice of the increasing sequence of windows.

Now consider the dynamical system of the measurable space $(\mathcal{T}, \mathcal{B}(\mathcal{T}))$ with the group of translations acting on $\mathcal{T}$ and a probability measure $\mathbb{P}$ invariant under translations. Some measures of spatial dependence, which we deal with in this volume, are described in the following definition. The relations between them are provided by Corollary 15.10 and Lemma 15.13.

Definition 15.6. *The probability space* $(\mathcal{T}, \mathcal{B}(\mathcal{T}), \mathbb{P})$ *endowed with the group of translations acting on* $\mathcal{T}$ *and* $\mathbb{P}$ *that is invariant with respect to translations is called*

(a) ***ergodic*** *if* $\mathbb{P}(A) \in \{0,1\}$ *for all* $A \in \mathcal{I}(\mathcal{T})$, *that is, the invariant σ-algebra is trivial with respect to* $\mathbb{P}$,

(b) ***mixing*** *if*

$$\lim_{\|x\| \to \infty} \mathbb{P}(A \cap (B + x)) = \mathbb{P}(A)\,\mathbb{P}(B) \tag{15.6}$$

for all $A, B \in \mathcal{B}(\mathcal{T})$,

(c) ***tail $\mathbb{P}$-trivial*** *if* $\mathbb{P}(A) \in \{0,1\}$ *for all* $A \in \mathcal{B}_\infty(\mathcal{T})$.

If X is a random stationary tessellation with law $\mathbb{P}_X$ and $(\mathcal{T}, \mathcal{B}(\mathcal{T}), \mathbb{P}_X)$ with the group of translations is ergodic, mixing, or tail trivial, then we also call X an ergodic, mixing, or tail trivial tessellation, respectively.

As a special case of Theorem 9.3.2 of [172], we obtain the following criterion for the mixing property.

Lemma 15.7. *A random stationary tessellation* X *with law* $\mathbb{P}_X$ *is mixing if and only if*

$$\lim_{\|x\| \to \infty} \mathbb{P}(\partial X \cap C_1 = \emptyset, \partial X \cap (C_2 + x) = \emptyset) = \mathbb{P}(\partial X \cap C_1 = \emptyset)\,\mathbb{P}(\partial X \cap C_2 = \emptyset)$$

for all $C_1, C_2 \in \mathcal{C}$.

Furthermore, a sufficient condition for the $\mathbb{P}$-triviality of the tail σ-algebra can be shown.

Lemma 15.8. *Let $\mathbb{P}$ be a translation invariant probability measure on $(\mathcal{T}, \mathcal{B}(\mathcal{T}))$. Assume that for every window $W_1 \in \mathcal{P}_d$, all $A \in \mathcal{B}_{W_1}(\mathcal{T})$, and all $\varepsilon > 0$, there exists a window $W \in \mathcal{P}_d$, depending on (W_1, A, ε), such that $W_1 \subset \operatorname{int} W$ and*

$$\left|\mathbb{P}(A \cap B) - \mathbb{P}(A)\mathbb{P}(B)\right| < \varepsilon \quad \text{for all } B \in \mathcal{B}_{W^c}(\mathcal{T}) . \tag{15.7}$$

Then the tail σ-algebra $\mathcal{B}_\infty(\mathcal{T})$ is $\mathbb{P}$-trivial.

Proof. Let $A \in \mathcal{B}_\infty(\mathcal{T})$, and let $(W_n, n \in \mathbb{N})$ be an increasing sequence of windows satisfying (15.4). Let $\varepsilon > 0$ be fixed. Since $A \in \mathcal{B}(\mathcal{T})$, by (15.5) there are $k \in \mathbb{N}$ and $A_k \in \mathcal{B}_{W_k}(\mathcal{T})$ such that

$$\mathbb{P}(A \triangle A_k) < \varepsilon . \tag{15.8}$$

Hypothesis (15.7) implies that there exists a window W, depending on (W_k, A_k, ε), such that $W_k \subset \operatorname{int} W$ and $|\mathbb{P}(A_k \cap B) - \mathbb{P}(A_k)\mathbb{P}(B)| < \varepsilon$ for all $B \in \mathcal{B}_{W^c}(\mathcal{T})$. Now choose $n \geq k$ such that $W \subseteq W_n$ and hence $\mathcal{B}_{W_n^c}(\mathcal{T}) \subseteq \mathcal{B}_{W^c}(\mathcal{T})$. Thus

$$\left|\mathbb{P}(A_k \cap B) - \mathbb{P}(A_k)\mathbb{P}(B)\right| < \varepsilon \quad \text{for all } B \in \mathcal{B}_{W_n^c}(\mathcal{T}) .$$

Because $A \in \mathcal{B}_\infty(\mathcal{T}) \subset \mathcal{B}_{W_n^c}(\mathcal{T})$, we obtain that $|\mathbb{P}(A_k \cap A) - \mathbb{P}(A_k)\mathbb{P}(A)| < \varepsilon$, and by (15.8) we deduce that $|\mathbb{P}(A \cap A) - \mathbb{P}(A)\mathbb{P}(A)| < 4\,\varepsilon$ for all $\varepsilon > 0$. Thus $\mathbb{P}(A) = 1$ or $\mathbb{P}(A) = 0$. □

For the invariant σ-algebra $\mathcal{I}(\mathcal{T})$ with respect to translations acting on $\mathcal{T}$ and the tail σ-algebra $\mathcal{B}_\infty(\mathcal{T})$, the same inclusion relation can be shown, which is known for stochastic processes on the time axis. To do so, for a probability measure $\mathbb{P}$ on $(\mathcal{T}, \mathcal{B}(\mathcal{T}))$ and $A, B \in \mathcal{B}(\mathcal{T})$, we write

$$A = B \bmod \mathbb{P} \quad \text{if } \mathbb{P}(A \triangle B) = 0 .$$

Also, for a pair $\mathcal{B}', \mathcal{B}''$ of sub-σ-algebras of $\mathcal{B}(\mathcal{T})$, we write $\mathcal{B}' \subseteq \mathcal{B}'' \bmod \mathbb{P}$ if for all $A \in \mathcal{B}'$, there exists $B \in \mathcal{B}''$ such that $A = B \bmod \mathbb{P}$.

Lemma 15.9. *If $\mathbb{P}$ is a translation-invariant probability measure on $(\mathcal{T}, \mathcal{B}(\mathcal{T}))$, then*

$$\mathcal{I}(\mathcal{T}) \subseteq \mathcal{B}_\infty(\mathcal{T}) \bmod \mathbb{P} . \tag{15.9}$$

Proof. Denote $\mathbf{1} := (1, \dots, 1) \in \mathbb{R}^d$, so $a\,\mathbf{1} = (a, \dots, a)$ for $a \in \mathbb{R}$. Fix the sequence $(W_n = [-n, n]^d, n \in \mathbb{N})$. Note that $W_n - m\,\mathbf{1} \subset W_n^c$ if $m > 2n$.

Let $B \in \mathcal{I}(\mathcal{T})$. Then by Lemma 15.4, for all $n \in \mathbb{N}$, there exist $k = k(n) > n$ and $B_k \in \mathcal{B}_{W_k}(\mathcal{T})$ such that $\mathbb{P}(B \triangle B_k) < 2^{-n}$. Since $B_k \in \sigma(\{\{T \in \mathcal{T} : \partial T \cap C = \emptyset\} : C \subseteq W_k, C \in \mathcal{C}\})$, we have that for $N > 2k$,

$$\begin{aligned} B_k - N\,\mathbf{1} &\in \sigma(\{\{T \in \mathcal{T} : \partial T \cap C = \emptyset\} : C \subseteq W_k - N\,\mathbf{1},\ C \in \mathcal{C}\}) \\ &\subseteq \sigma(\{\{T \in \mathcal{T} : \partial T \cap C = \emptyset\} : C \subset W_k^c,\ C \in \mathcal{C}\}) = \mathcal{B}_{W_k^c}(\mathcal{T}) . \end{aligned}$$

Because $\mathbb{P}(B \triangle (B - N\mathbf{1})) = 0$ and $\mathbb{P}((B - N\mathbf{1}) \triangle (B_k - N\mathbf{1})) = \mathbb{P}(B \triangle B_k) < 2^{-n}$, we obtain that for all $N > 2k$,

$$\mathbb{P}(B \triangle (B_k - N\mathbf{1})) \le \mathbb{P}(B \triangle (B - N\mathbf{1})) + 2^{-n} = 2^{-n}.$$

Now choose $N = k^2$, $k > 2$, and use $k^2 > 2k$ to get

$$\mathbb{P}(B \triangle (B_k - k^2\mathbf{1})) \le 2^{-n} \quad \text{and} \quad B_k - k^2\mathbf{1} \in \mathcal{B}_{W_k^c}(\mathcal{T}).$$

For $m \ge 1$, define

$$A_m := \bigcup_{n>m} (B_{k(n)} - k(n)^2\mathbf{1}).$$

This sequence of sets satisfies $\mathbb{P}(B \triangle A_m) \le \sum_{n>m} 2^{-n} = 2^{-m}$, $A_m \in \mathcal{B}_{W^c_{k(m)}}(\mathcal{T})$, and A_m decreases with m. We conclude that $A := \bigcap_{m>1} A_m$ satisfies $\mathbb{P}(B \triangle A) = 0$ and $A \in \mathcal{B}_\infty(\mathcal{T})$. This shows (15.9). □

Thus we obtain the following proposition.

Corollary 15.10. *If the tail σ-algebra $\mathcal{B}_\infty(\mathcal{T})$ is $\mathbb{P}$-trivial, then $(\mathcal{T}, \mathcal{B}(\mathcal{T}), \mathbb{P})$ is ergodic with respect to translations acting on $\mathcal{T}$.*

15.1.2 β-mixing coefficient for random tessellations

The intuitive idea for defining mixing properties in space is to require that the dependence of parts of a point process or a random measure or a random set in different regions diminishes with a growing distance between these regions. This distance can be expressed by the norm of a translation vector in $\mathbb{R}^d$; see the definition of mixing in (15.6). Regarding stronger mixing properties for random tessellations, we apply an approach introduced in [69]. Two windows, one contained in the other one, are used, and the lager window is growing. For this setting, the decrease of the dependence of the behavior inside the small window and the behavior outside the growing large window is investigated. There are several concepts of mixing properties, which are dealt with in detail in [19]. Here we focus on the concept of the β-mixing coefficient introduced in Definition 15.2.

Definition 15.11. *Let X be a random stationary tessellation with distribution $\mathbb{P}_X$ on $(\mathcal{T}, \mathcal{B}(\mathcal{T}))$. For all $a, b \in \mathbb{R}^+$ with $a < b$, let $W_1 = [-a, a]^d \subset W_2 = [-b, b]^d$ be a pair of windows.*

(a) *The **β-mixing coefficient** of X is defined as*

$$\beta(a, b) := \beta(\mathcal{B}_{W_1}(\mathcal{T}), \mathcal{B}_{W_2^c}(\mathcal{T})), \tag{15.10}$$

where $\mathcal{B}_{W_1}(\mathcal{T})$ *and* $\mathcal{B}_{W_2^c}(\mathcal{T})$ *are defined in* (15.2) *and* (15.3)*, respectively, and the* β*-mixing coefficient for pairs of* σ*-algebras is given in* (15.1).

(b) *The random stationary tessellation X is* β***-mixing*** *if*

$$\lim_{b\to\infty} \beta(a,b) = 0 \quad \text{for all } a > 0\,. \tag{15.11}$$

The following proposition provides a criterion, which is useful for a proof of the convergence of the β-mixing coefficient.

Theorem 15.12. *With the notation of Definition* 15.11*, for all* $a > 0$,

$$\lim_{b\to\infty} \beta(a,b) = 0$$

if and only if for all $\varepsilon > 0$*, there exist* $b > a$ *and* $D \in \mathcal{B}_{W_1}(\mathcal{T})$ *such that*

(i) $\mathbb{P}_X(D) > 1 - \varepsilon$

and

(ii) *for all* $B \in \mathcal{B}_{W_2^c}(\mathcal{T})$ *and* $A \in \mathcal{B}_{W_1}(\mathcal{T})$ *with* $A \subseteq D$ *and* $\mathbb{P}_X(A) > 0$,

$$\left|\mathbb{P}_X(B \mid A) - \mathbb{P}_X(B)\right| < \varepsilon\,. \tag{15.12}$$

Proof. The proof of Theorem 3.32(II) of [19] can be adapted straightforwardly. To begin with, we show that (i) and (ii) imply $\lim_{b\to\infty} \beta(a,b) = 0$. According to Lemma 18.2, the space $\mathcal{T}$ of tessellations is a Borel space. Hence the σ-algebra $\mathcal{B}(\mathcal{T})$ has a countable generating system, and a regular conditional distribution $\mathbb{P}_X(\cdot\,|\mathcal{B}_{W_1}(\mathcal{T}))$ on $\mathcal{B}_{W_2^c}(\mathcal{T})$ exists; see Theorem 8.5 of [89]. Thus Lemma 15.3 can be applied, and by assumption (15.12) with $\mathbb{P}_X(D) > 1 - \varepsilon$ we have that $\beta(a,b) < 2\,\varepsilon$. The assumption that for all $\varepsilon > 0$, there exists an appropriate b implies $\lim_{b\to\infty} \beta(a,b) = 0$.

On the other hand, if $\lim_{b\to\infty} \beta(a,b) = 0$, then for all $\varepsilon > 0$, there is $b > a$ such that $\beta(a,b) < \varepsilon^2$. For those a, b, define the event

$$D := \left\{T \in \mathcal{T} : \sup_{B \in \mathcal{B}_{W_2^c}(\mathcal{T})} \left|\mathbb{P}_X(B|\mathcal{B}_{W_1}(\mathcal{T}))(T) - \mathbb{P}_X(B)\right| \le \varepsilon\right\}.$$

Note that $D \in \mathcal{B}_{W_1}(\mathcal{T})$ by the measurability of the conditional probability $\mathbb{P}_X(B|\mathcal{B}_{W_1}(\mathcal{T}))$. Using Lemma 15.3(a) and the Markov inequality, we obtain $\mathbb{P}_X(D^c) < \varepsilon$, and hence $\mathbb{P}_X(D) > 1 - \varepsilon$. Then, proceeding as in the proof of Lemma 15.3(b), that there (ii) implies (i), property (15.12) is shown. □

The relations between the properties ergodicity, tail triviality, mixing of Definition 15.6, and β-mixing defined in Definition 15.11(b) are summarized in the following lemma. They correspond to the analogous relations for processes on the time axis.

Lemma 15.13. *Let X be a random stationary tessellation with distribution* $\mathbb{P}_X$.

(a) *If X is β-mixing, then the tail σ-algebra $\mathcal{B}_\infty(\mathcal{T})$ is $\mathbb{P}_X$-trivial.*
(b) *If the tail σ-algebra $\mathcal{B}_\infty(\mathcal{T})$ is $\mathbb{P}_X$-trivial, then X is mixing.*
(c) *If X is mixing, then the σ-algebra $\mathcal{I}(\mathcal{T})$ of invariant events is $\mathbb{P}_X$-trivial, which means that X is ergodic.*

Proof. (a) Let $B \in \mathcal{B}_\infty(\mathcal{T}) \subseteq \mathcal{B}(\mathcal{T})$ and $B_m \in \mathcal{B}_{W_m}(\mathcal{T})$ for some $m \in \mathbb{N}$. By the assumption, for all $\delta > 0$, there is $n(\delta) \geq m$ such that $\beta(m, n) < \delta$ for all $n \geq n(\delta)$, and hence

$$\left|\mathbb{P}_X(B_m \cap B) - \mathbb{P}_X(B_m)\mathbb{P}_X(B)\right| = 0\,.$$

By (15.5), for all $\varepsilon > 0$, there are $m \in \mathbb{N}$ and $B_m \in \mathcal{B}_{W_m}(\mathcal{T})$ such that $\mathbb{P}_X(B \triangle B_m) < \varepsilon$, and hence

$$\mathbb{P}_X(B) - \varepsilon < \mathbb{P}_X(B_m \cap B) = \mathbb{P}_X(B_m)\mathbb{P}_X(B) \leq \mathbb{P}_X(B)^2 + \varepsilon\mathbb{P}_X(B)\,,$$

and thus necessarily $\mathbb{P}_X(B) = 0$ or $\mathbb{P}_X(B) = 1$.

(b) The proof Proposition 12.3.V of [46], there given for random measures, can straightforwardly be adapted to the σ-algebras for random tessellations considered here.

(c) For $A \in \mathcal{I}(\mathcal{T})$, choose $A = B$ in (15.6), which yields $\mathbb{P}_X(A) = \mathbb{P}_X(A)^2$ and thus the statement. □

15.2 STIT tessellations are β-mixing in space

In this section, we show that the STIT tessellations are β-mixing in space and derive upper bounds for the β-mixing coefficient. This is prepared by introducing the so-called encapsulation concept and some results on it. Based on this, the main propositions are shown in Section 15.2.2. In the last part a spatial version of an ergodic theorem is provided and applied to maximal polytopes of STIT tessellations.

15.2.1 Encapsulation for STIT tessellations

To find upper bounds for the β-mixing coefficient $\beta(a, b)$ defined in (15.10), we consider the event that the smaller window $W_1 = [-a, a]^d$ is separated from the boundary of the larger window $W_2 = [-b, b]^d$ by faces of the zero-cell of the STIT tessellation. In particular, this means that the separation appears before the interior of W_1 is divided by a hyperplane in the construction. This event is referred to as *encapsulation*. It is illustrated in Figure 15.1. The key idea is that, under the condition that an encapsulation occurs, the continuation of the STIT tessellation process inside W_1 is conditionally independent of the process outside W_2.

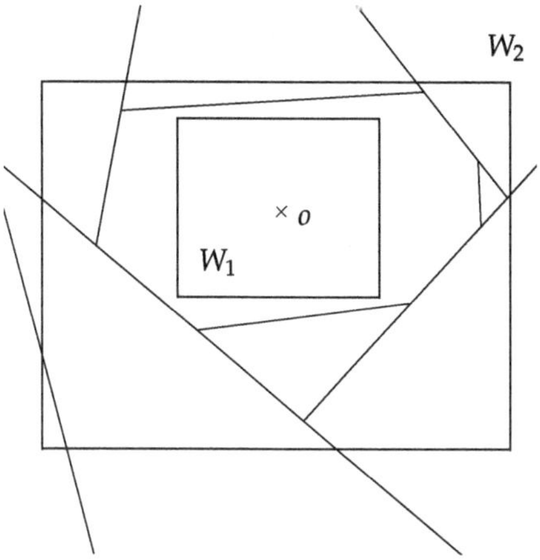

Figure 15.1: Encapsulation: The boundaries of the windows W_1 and W_2 are separated by the skeleton of a tessellation, and W_1 is not hit by this skeleton.

Let $(Y_t, t > 0)$ be a STIT tessellation process driven by the hyperplane measure Θ of the form (4.1), satisfying (4.3). Denoting by $Z^{\mathrm{STIT}}_{\langle o\rangle,t}$ the zero-cell of Y_t, the **encapsulation time of** $W_1 \subset W_2$, W_1, $W_2 \in \mathcal{P}_d$, $o \in \operatorname{int} W_1$, is defined as

$$S(W_1, W_2) := \inf\{t > 0 : W_1 \subseteq Z^{\mathrm{STIT}}_{\langle o\rangle,t} \subset W_2\} \tag{15.13}$$

with the convention that $\inf \emptyset = \infty$.

After encapsulation, the set of cells intersecting W_1 is disjoint from the set of cells intersecting W_2^c. Thus, conditioned on $Y_{S(W_1,W_2)}$, by (6.1) the processes inside W_1 and outside W_2 are conditionally independent. In this section, we write, with a slight abuse of notation, $Y_s \wedge W_1 = W_1$ instead of $Y_s \wedge W_1 = \{W_1\}$ to avoid an accumulation of brackets. Note that $\{Y_s \wedge W_1 = W_1\}$ describes the event that until time $s > 0$, there is no division of W_1 by a hyperplane or, equivalently, $W_1 \subseteq Z^{\mathrm{STIT}}_{\langle o\rangle,s}$.

We start with some technical lemmas, which at the end yield lower bounds for encapsulation probabilities.

Lemma 15.14. *Let $W_1, W_2 \in \mathcal{P}_d$ with $W_1 \subset W_2$ and $o \in \operatorname{int} W_1$, $0 < s < t$, and let $B \in \mathcal{B}_{W_2^c}(\mathcal{T})$ and $A \in \mathcal{B}_{W_1}(\mathcal{T})$ with $\mathbb{P}(Y_t \in A) > 0$. Then*

$$\begin{aligned}&\mathbb{P}(Y_t \in B, S(W_1, W_2) \le s, Y_s \wedge W_1 = W_1 \mid Y_t \in A)\\ &\quad = \mathbb{P}(Y_t \in B, S(W_1, W_2) \le s, Y_s \wedge W_1 = W_1)\, \frac{\mathbb{P}(Y_{t-s} \in A)}{\mathbb{P}(Y_t \in A)}.\end{aligned} \tag{15.14}$$

Proof. Let $W_1 \subset W_2$ with $o \in \operatorname{int} W_1$, and let $B \in \mathcal{B}_{W_2^c}(\mathcal{T})$ and $A \in \mathcal{B}_{W_1}(\mathcal{T})$ with $\mathbb{P}(Y_t \in A) > 0$. We apply (6.1) in the form $Y_t \overset{\mathrm{D}}{=} Y_s \boxplus \underline{Y_{t-s}}$ with $\underline{Y_{t-s}} := (Y^{(i)}_{t-s}, i \in \mathbb{N}_0)$ and assume that $Y^{(0)}_{t-s}$ is nested into the zero-cell $Z^{\mathrm{STIT}}_{\langle o\rangle,s}$ of Y_s. Below we use the independence of the random tessellations in $\underline{Y_{t-s}}$ and their independence from Y_s. If the event $(S(W_1, W_2) \le s, Y_s \wedge W_1 = W_1)$ occurs, then also the event $(W_1 \subseteq Z^{\mathrm{STIT}}_{\langle o\rangle,s} \subset W_2)$ does occur because

$Z^{\mathrm{STIT}}_{\langle o\rangle,s} \subseteq Z^{\mathrm{STIT}}_{\langle o\rangle,S(W_1,W_2)}$. If $W_1 \subseteq Z^{\mathrm{STIT}}_{\langle o\rangle,s} \subset W_2$, then the event $Y_s \boxplus \underline{Y_{t-s}} \in A$ depends on $Y^{(0)}_{t-s}$ only, and the event $Y_s \boxplus \underline{Y_{t-s}} \in B$ depends only on Y_s and $Y^{(i)}_{t-s}$, $i > 0$. Hence

$$\begin{aligned}
&\mathbb{P}(Y_t \in B,\, S(W_1,W_2) \le s, Y_s \wedge W_1 = W_1 \mid Y_t \in A) \\
&= \frac{1}{\mathbb{P}(Y_t \in A)} \mathbb{P}(Y_t \in B,\, S(W_1,W_2) \le s,\, Y_s \wedge W_1 = W_1,\, Y_t \in A) \\
&= \frac{1}{\mathbb{P}(Y_t \in A)} \mathbb{P}(Y_s \boxplus \underline{Y_{t-s}} \in B,\, S(W_1,W_2) \le s,\, Y_s \wedge W_1 = W_1,\, Y_s \boxplus \underline{Y_{t-s}} \in A) \\
&= \frac{1}{\mathbb{P}(Y_t \in A)} \mathbb{P}(Y_s \boxplus \underline{Y_{t-s}} \in B,\, S(W_1,W_2) \le s,\, Y_s \wedge W_1 = W_1)\, \mathbb{P}(Y^{(0)}_{t-s} \in A) \\
&= \frac{\mathbb{P}(Y_{t-s} \in A)}{\mathbb{P}(Y_t \in A)} \mathbb{P}(Y_t \in B,\, S(W_1,W_2) \le s,\, Y_s \wedge W_1 = W_1)\,,
\end{aligned}$$

and thus (15.14) is shown. □

For fixed $0 < s < t$ and $W_1 \in \mathcal{P}_d$, the event that the process $(Y_t \wedge W_1,\, t > 0)$ has no jump in the time interval $(t-s, t]$ is denoted

$$I := \{Y_t \wedge W_1 = Y_{t-s} \wedge W_1\}\,. \tag{15.15}$$

This event will be used frequently in the following calculations.

Lemma 15.15. *For* $0 < s < t$ *and* $A \in \mathcal{B}_{W_1}(\mathcal{T})$ *with* $\mathbb{P}(Y_t \in A) > 0$,

$$\mathbb{P}(I \mid Y_t \in A) \le \frac{\mathbb{P}(Y_{t-s} \in A)}{\mathbb{P}(Y_t \in A)} \le \frac{1}{\mathbb{P}(I \mid Y_{t-s} \in A)}\,.$$

Proof. If $A \in \mathcal{B}_{W_1}(\mathcal{T})$ with $\mathbb{P}(Y_t \in A) > 0$, then also $\mathbb{P}(Y_{t-s} \in A) > 0$, because $\mathbb{P}(Y_{t-s} \in A) \ge \mathbb{P}(\{Y_{t-s} \in A\} \cap I) = \mathbb{P}(\{Y_t \in A\} \cap I) = \mathbb{P}(I \mid Y_t \in A)\,\mathbb{P}(Y_t \in A) > 0$. Hence, also the conditional probability $\mathbb{P}(I \mid Y_{t-s} \in A)$ is well-defined. Furthermore, dividing the terms of $\mathbb{P}(Y_{t-s} \in A) \ge \mathbb{P}(\{Y_{t-s} \in A\} \cap I) = \mathbb{P}(\{Y_t \in A\} \cap I)$ by $\mathbb{P}(Y_t \in A)$ yields the first inequality. Analogously,

$$\mathbb{P}(Y_t \in A) \ge \mathbb{P}(\{Y_{t-s} \in A\} \cap I)\,,$$

which yields $\mathbb{P}(Y_t \in A)/\mathbb{P}(Y_{t-s} \in A) \ge \mathbb{P}(I \mid Y_{t-s} \in A)$, and taking the reciprocal on both sides completes the proof. □

Recall the notation $\theta(T) := \sum_{P \in T} \Theta([P])$, introduced in (4.17) for $T \in \mathcal{T}_W$.

Lemma 15.16. *For all* $W_1 \in \mathcal{P}_d$, $t > 0$, *and* $\varepsilon > 0$, *there exists* $\ell > 0$ *such that* $\mathbb{P}(\theta(Y_t \wedge W_1) < \ell) > 1 - \varepsilon$.

Proof. By Theorem 4.7 the process $(Y_t \wedge W_1\,,\, t > 0)$ has no explosion, and therefore $\theta(Y_t \wedge W_1)$ is a. s. finite for all $t > 0$. Hence

$$\mathbb{P}\left(\bigcup_{n=1}^{\infty}\{\theta(Y_t \wedge W_1) < n\}\right) = \lim_{n\to\infty} \mathbb{P}(\theta(Y_t \wedge W_1) < n) = 1\,,$$

which proves the statement. □

Now we will combine equation (15.14) with the bounds for $\mathbb{P}(Y_{t-s} \in A)/\mathbb{P}(Y_t \in A)$ provided by Lemma 15.15.

Lemma 15.17. *Let $W_1, W_2 \in \mathcal{P}_d$ with $W_1 \subset W_2$ and $o \in \operatorname{int} W_1$, $0 < s < t$, and let $B \in \mathcal{B}_{W_2^c}(\mathcal{T})$ and $A \in \mathcal{B}_{W_1}(\mathcal{T})$ with $\mathbb{P}(Y_t \in A) > 0$. Then, with I defined in* (15.15),

$$\begin{aligned}
&\left|\mathbb{P}(Y_t \in B, S(W_1, W_2) \le s, Y_s \wedge W_1 = W_1 \mid Y_t \in A) - \mathbb{P}(Y_t \in B)\right| \\
&\quad \le \max\{(\mathbb{P}(I \mid Y_{t-s} \in A)^{-1} - 1)\,\mathbb{P}(Y_t \in B)\,; \\
&\qquad (1 - \mathbb{P}(I \mid Y_t \in A))\,\mathbb{P}(Y_t \in B) \\
&\qquad + \mathbb{P}(I \mid Y_t \in A)\,(1 - \mathbb{P}(S(W_1, W_2) \le s,\ Y_s \wedge W_1 = W_1))\}\,.
\end{aligned}$$

Proof. First note that by the additivity and monotonicity of $\mathbb{P}$

$$\begin{aligned}
&\mathbb{P}(Y_t \in B) - (1 - \mathbb{P}(S(W_1, W_2) \le s,\ Y_s \wedge W_1 = W_1)) \\
&\quad \le \mathbb{P}(Y_t \in B,\ S(W_1, W_2) \le s, Y_s \wedge W_1 = W_1) \\
&\quad \le \mathbb{P}(Y_t \in B)\,.
\end{aligned}$$

Multiplication by $\frac{\mathbb{P}(Y_{t-s}\in A)}{\mathbb{P}(Y_t \in A)}$ and (15.14) yield

$$\begin{aligned}
&\frac{\mathbb{P}(Y_{t-s} \in A)}{\mathbb{P}(Y_t \in A)}\,(\mathbb{P}(Y_t \in B) - (1 - \mathbb{P}(S(W_1, W_2) \le s,\ Y_s \wedge W_1 = W_1))) \\
&\quad \le \mathbb{P}(Y_t \in B,\ S(W_1, W_2) \le s, Y_s \wedge W_1 = W_1 \mid Y_t \in A) \\
&\quad \le \frac{\mathbb{P}(Y_{t-s} \in A)}{\mathbb{P}(Y_t \in A)}\,\mathbb{P}(Y_t \in B)\,.
\end{aligned}$$

Thus by Lemma 15.15

$$\begin{aligned}
&\mathbb{P}(I \mid Y_t \in A)\,(\mathbb{P}(Y_t \in B) - (1 - \mathbb{P}(S(W_1, W_2) \le s,\ Y_s \wedge W_1 = W_1))) \\
&\quad \le \mathbb{P}(Y_t \in B,\ S(W_1, W_2) \le s, Y_s \wedge W_1 = W_1 \mid Y_t \in A) \\
&\quad \le \mathbb{P}(I \mid Y_{t-s} \in A)^{-1}\,\mathbb{P}(Y_t \in B)\,.
\end{aligned}$$

Subtracting $\mathbb{P}(Y_t \in B)$, we obtain

$$\begin{aligned}
&(\mathbb{P}(I \mid Y_t \in A) - 1)\,\mathbb{P}(Y_t \in B) \\
&\qquad - \mathbb{P}(I \mid Y_t \in A)\,(1 - \mathbb{P}(S(W_1, W_2) \le s,\ Y_s \wedge W_1 = W_1)) \\
&\quad \le \mathbb{P}(Y_t \in B,\ S(W_1, W_2) \le s, Y_s \wedge W_1 = W_1 \mid Y_t \in A) - \mathbb{P}(Y_t \in B) \\
&\quad \le (\mathbb{P}(I \mid Y_{t-s} \in A)^{-1} - 1)\,\mathbb{P}(Y_t \in B)\,.
\end{aligned}$$

Because the term in the first line is less than or equal to zero, the result for the absolute value follows. □

The next lemma prepares the application of Lemma 15.3 to the proof that STIT tessellations are β-mixing.

Lemma 15.18. *Let $W_1, W_2 \in \mathcal{P}_d$ with $W_1 \subset W_2$ and $o \in \operatorname{int} W_1$. For $D \in \mathcal{B}_{W_1}(\mathcal{T})$ and fixed $t > 0$, denote $\mathcal{U} := \{A \in \mathcal{B}_{W_1}(\mathcal{T}) : A \subseteq D,\ \mathbb{P}(Y_t \in A) > 0\}$. Then, for all $s \in (0,t)$ and I defined in* (15.15),

$$
\begin{aligned}
&\sup_{(A,B)\in\mathcal{U}\times\mathcal{B}_{W_2^c}(\mathcal{T})} \left|\mathbb{P}(Y_t \in B \mid Y_t \in A) - \mathbb{P}(Y_t \in B)\right| \\
&\quad \le \sup_{A\in\mathcal{U}} (1 - \mathbb{P}(S(W_1,W_2) \le s, Y_s \wedge W_1 = W_1 \mid Y_t \in A)) \\
&\quad + \sup_{(A,B)\in\mathcal{U}\times\mathcal{B}_{W_2^c}(\mathcal{T})} \max\{(\mathbb{P}(I \mid Y_{t-s} \in A)^{-1} - 1)\,\mathbb{P}(Y_t \in B)\,; \\
&\quad (1 - \mathbb{P}(I \mid Y_t \in A))\,\mathbb{P}(Y_t \in B) \\
&\quad + \mathbb{P}(I \mid Y_t \in A)\,(1 - \mathbb{P}(S(W_1,W_2) \le s, Y_s \wedge W_1 = W_1))\}\,.
\end{aligned}
$$

Proof. By the triangle inequality

$$
\begin{aligned}
&\sup_{(A,B)\in\mathcal{U}\times\mathcal{B}_{W_2^c}(\mathcal{T})} \left|\mathbb{P}(Y_t \in B \mid Y_t \in A) - \mathbb{P}(Y_t \in B)\right| \\
&\le \sup_{(A,B)\in\mathcal{U}\times\mathcal{B}_{W_2^c}(\mathcal{T})} \big|\mathbb{P}(Y_t \in B \mid Y_t \in A) \\
&\quad - \mathbb{P}(Y_t \in B, S(W_1,W_2) \le s, Y_s \wedge W_1 = W_1 \mid Y_t \in A)\big| \\
&\quad + \sup_{(A,B)\in\mathcal{U}\times\mathcal{B}_{W_2^c}(\mathcal{T})} \left|\mathbb{P}(Y_t \in B, S(W_1,W_2) \le s, Y_s \wedge W_1 = W_1 \mid Y_t \in A) - \mathbb{P}(Y_t \in B)\right|.
\end{aligned}
$$

Now we apply the inequality $\mathbb{P}(B' \mid A') - \mathbb{P}(B' \cap C \mid A') \le 1 - \mathbb{P}(C \mid A')$ to $B' = \{Y_t \in B\}$, $A' = \{Y_t \in A\}$, and $C = \{S(W_1,W_2) \le s, Y_s \wedge W_1 = W_1\}$, and then Lemma 15.17 to conclude the proof. □

Now we derive lower bounds for the conditional probabilities of the event I, which appeared in Lemma 15.18.

Lemma 15.19. *Let $W_1 \in \mathcal{P}_d$ with $o \in \operatorname{int} W_1$, and $0 < s < t$. Then for all $\ell > 0$ and $D := \{T \in \mathcal{T} : \theta(T \wedge W_1) < \ell\}$ and for all $A \in \mathcal{B}_{W_1}(\mathcal{T})$ with $A \subseteq D$, $\mathbb{P}(Y_t \in A) > 0$, we have*

$$
\mathbb{P}(I \mid Y_{t-s} \in A) > \mathrm{e}^{-s\ell} \quad \text{and} \quad \mathbb{P}(I \mid Y_t \in A) > \mathrm{e}^{-s\ell}
$$

with I defined in (15.15).

Proof. By Theorem 4.7(c) the holding time of the state $Y_{t-s} \wedge W_1$ until the next jump of the process $(Y_t \wedge W_1\,,\, t > 0)$ is exponentially distributed with parameter $\theta(Y_{t-s} \wedge W_1)$. By the definition of D we have that $\theta(T \wedge W_1) < \ell$ for $T \in A \subseteq D$, and hence $\mathbb{P}(I \mid Y_{t-s} \in A) > e^{-s\ell}$.

To show the second inequality, note that $\theta(Y_t \wedge W_1)$ is monotone in t. This follows from the fact that for $P, P_1, P_2 \in \mathcal{P}_d$ with $P = P_1 \cup P_2$, we have $[P] = [P_1] \cup [P_2]$, and hence $\Theta([P]) \leq \Theta([P_1]) + \Theta([P_2])$. Thus, if $Y_t \in A$, then $\theta(Y_{t-s} \wedge W_1) \leq \theta(Y_t \wedge W_1) < \ell$. This implies that $\mathbb{P}(I \mid Y_t \in A) > e^{-s\ell}$. □

For $W_1 := [-a, a]^d$ and $W_2 := [-b, b]^d$ with $0 < a < b$, denote by

$$\mathcal{P}(a, b) := \{P \in \mathcal{P}_d : W_1 \subseteq P \subseteq W_2\}$$

the set of all polytopes P whose boundaries ∂P separate the boundaries ∂W_1 and ∂W_2.

Lemma 15.20. *Let Θ be a hyperplane measure of the form* (4.1), *satisfying* (4.3). *Then for all $a > 0$, $s > 0$, and $\varepsilon \in (0, 1)$, there exists $b > a$ such that*

$$\mathbb{P}(Z^{\mathrm{STIT}}_{\langle o \rangle, s} \in \mathcal{P}(a, b)) > (1 - \varepsilon)\, e^{-s\,\Theta([W_1])} .$$

Proof. For fixed $a > 0$ and $s > 0$, by (5.6) and the spatial consistency property (5.17)

$$\begin{aligned} e^{-s\,\Theta([W_1])} &= \mathbb{P}(W_1 \subseteq Z^{\mathrm{STIT}}_{\langle o \rangle, s}) \\ &= \mathbb{P}\Big(\bigcup_{n \in \mathbb{N}} \{W_1 \subseteq Z^{\mathrm{STIT}}_{\langle o \rangle, s} \subseteq [-n, n]^d\}\Big) \\ &= \lim_{n \to \infty} \mathbb{P}(W_1 \subseteq Z^{\mathrm{STIT}}_{\langle o \rangle, s} \subseteq [-n, n]^d) . \end{aligned}$$

This yields the statements. □

Example 15.21. For $d = 2$ and the motion-invariant hyperplane measure Θ_{iso},

$$\mathbb{P}(Z^{\mathrm{STIT}}_{\langle o \rangle, s} \in \mathcal{P}(a, b)) \geq \exp\Big(-\frac{4as}{\pi}\Big)\Big(1 - \exp\Big(-\frac{2\,s}{\pi}\,(\sqrt{2}\,\sqrt{a^2 + b^2} - b - a)\Big)\Big)^4 .$$

This can be shown as follows. For $W_1 := [-a, a]^2$ and $W_2 := [-b, b]^2$ with $0 < a < b$, denote by F_i', F_i, $i \in \{1, \dots, 4\}$, the pairs of parallel sides of W_1 and W_2 of lengths $2a$ and $2b$, respectively, and distance $b - a$. Furthermore, recall from (5.9) that $[F_i' | F_i]$ describes the set of all hyperplanes that separate F_i' and F_i. By Crofton's formula (1.6) and Sylvester's "Seilliniensatz" of integral geometry (see Section I.3.3 of [169]),

$$\begin{aligned} \Theta_{\mathrm{iso}}([F_i' | F_i]) &= \frac{1}{\pi}\,(2\sqrt{(a + b)^2 + (b - a)^2} - 2(a + b)) \\ &= \frac{2}{\pi}\,(\sqrt{2}\,\sqrt{a^2 + b^2} - b - a) . \end{aligned}$$

Moreover, $\Theta_{\mathrm{iso}}([W_1]) = 4a/\pi$.

Now we use that $Z^{\mathrm{STIT}}_{\langle o\rangle,s} \overset{\mathrm{D}}{=} Z^{\mathrm{PHT}}_{\langle o\rangle,s}$ by (6.3), and thus we can consider a Poisson line process with intensity measure $s\,\Theta_{\mathrm{iso}}$. If there is no line in $[W_1]$ and at least one line in each of the $[F_i'|F_i]$, $i \in \{1,\dots,4\}$, then $Z^{\mathrm{PHT}}_{\langle o\rangle,s} \in \mathcal{P}(a,b)$. Because the five sets of lines $[W_1]$ and $[F_i'|F_i]$, $i \in \{1,\dots,4\}$, are pairwise disjoint, the independence property of Poisson point processes (see Definition 19.10(b)) yields the result.

Example 15.22. Let $d \geq 2$, and let Θ be the Mondrian hyperplane measure with the spherical directional distribution φ, given in (4.4), with $p_1 = \cdots = p_d = 1/d$. Then by analogous considerations as in Example 15.21 we obtain the exact value of $\mathbb{P}(Z^{\mathrm{STIT}}_{\langle o\rangle,s} \in \mathcal{P}(a,b))$:

$$\begin{aligned}\mathbb{P}(Z^{\mathrm{STIT}}_{\langle o\rangle,s} \in \mathcal{P}(a,b)) &= \left(\exp\left(-\frac{2as}{d}\right)\right)^d \left(1-\exp\left(-\frac{2s(b-a)}{d}\right)\right)^d \\ &= \left(\exp\left(-\frac{2sa}{d}\right) - \exp\left(-\frac{2sb}{d}\right)\right)^d .\end{aligned}$$

15.2.2 The β-mixing coefficient in space

After the preliminaries in the previous section, we are now prepared to provide an upper bound for the β-mixing coefficient, where the random variable $\theta(Y_t \wedge W_1)$ defined in (4.17) is essential. Thereafter, we refer to Theorem 15.12 and show that an appropriate event $D \in \mathcal{B}_{W_1}(\mathcal{T})$ exists, which implies the convergence of $\beta(a,b)$ to zero for fixed a and $b \to \infty$. Finally, a simple form of an upper bound for $\beta(a,b)$ is derived. For STIT tessellations, the mixing coefficient $\beta(a,b)$ of Y_t does also depend on $t>0$, but as time will be fixed, we do not indicate t explicitly in the notation.

Theorem 15.23. *Let $(Y_t, t>0)$ be the STIT tessellation process driven by the hyperplane measure Θ of the form (4.1), satisfying (4.3). Furthermore, for $0<a<b$, let $W_1 = [-a,a]^d$ and $W_2 = [-b,b]^d$, and let $\beta(a,b)$ be defined as in (15.10). Then for all $0<s<t$ and $\ell>0$, we have that for the β-mixing coefficient of Y_t,*

$$\begin{aligned}\beta(a,b) \leq \mathbb{P}(\theta(Y_t\wedge W_1)\geq \ell) + \mathbb{P}(\theta(Y_t \wedge W_1) < \ell)\,\big(1-\mathbb{P}(Z^{\mathrm{STIT}}_{\langle o\rangle,s} \in \mathcal{P}(a,b))\,\mathrm{e}^{-s\ell} \\ + \max\{\mathrm{e}^{s\ell}-1;\ 2-\mathrm{e}^{-s\ell} - \mathbb{P}(Z^{\mathrm{STIT}}_{\langle o\rangle,s} \in \mathcal{P}(a,b))\}\big)\,.\end{aligned}$$

Proof. Let $0<s<t$, $\ell>0$, and $D := \{\theta(Y_t\wedge W_1) < \ell\}$. Then by Lemmas 15.18, 15.14, and 15.19 and by the inequality

$$\mathbb{P}(S(W_1,W_2)\leq s,\ Y_s \wedge W_1 = W_1) \geq \mathbb{P}(Z^{\mathrm{STIT}}_{\langle o\rangle,s} \in \mathcal{P}(a,b))$$

we obtain that

$$\sup_{(A,B)\in\mathcal{U}\times\mathcal{B}_{W_2^c}(\mathcal{T})} \left|\mathbb{P}(Y_t \in B \mid Y_t \in A) - \mathbb{P}(Y_t\in B)\right|$$

$$\le 1 - \mathbb{P}(Z^{\mathrm{STIT}}_{\langle o\rangle,s} \in \mathcal{P}(a,b))\,\mathrm{e}^{-s\ell} + \max\{\mathrm{e}^{s\ell} - 1;\ 1 - \mathrm{e}^{-s\ell} + 1 - \mathbb{P}(Z^{\mathrm{STIT}}_{\langle o\rangle,s} \in \mathcal{P}(a,b))\}\,. \tag{15.16}$$

Lemma 15.3(b) yields

$$\beta(a,b) \le \mathbb{P}_{Y_t}(D^c) + \mathbb{P}_{Y_t}(D) \sup_{(A,B)\in\mathcal{U}\times\mathcal{B}_{W_2^c}(\mathcal{T})} \left|\mathbb{P}(Y_t \in B \mid Y_t \in A) - \mathbb{P}(Y_t \in B)\right|,$$

which, together with (15.16), completes the proof. □

To simplify the expression, the maximum of two nonnegative items in the upper bound given in Theorem 15.23 can be replaced by the sum of these items. Thus

$$\begin{aligned}\beta(a,b) &\\ \le \mathbb{P}(&\theta(Y_t \wedge W_1) \ge \ell)\\ &+ \mathbb{P}(\theta(Y_t \wedge W_1) < \ell)\,(2 + \mathrm{e}^{s\ell} - \mathrm{e}^{-s\ell} - (1 + \mathrm{e}^{-s\ell})\,\mathbb{P}(Z^{\mathrm{STIT}}_{\langle o\rangle,s} \in \mathcal{P}(a,b)))\,.\end{aligned} \tag{15.17}$$

For a given hyperplane measure Θ, this upper bound can then be minimized by appropriately choosing $\ell > 0$ and $s \in (0,t)$. Note that for given Θ and $a > 0$, the set $\mathcal{P}(a,b)$ can be empty if b is not large enough; see also Lemma 15.20.

The following result states that for STIT tessellations, condition (15.12) of Theorem 15.12 is satisfied.

Theorem 15.24. *Let $(Y_t, t > 0)$ be a STIT tessellation process driven by a hyperplane measure Θ of the form (4.1), satisfying (4.3). For $0 < a < b$, denote $W_1 = [-a,a]^d$ and $W_2 = [-b,b]^d$. Then for all $a > 0$, $t > 0$, and $\varepsilon > 0$, there exist $b > a$ and $D \in \mathcal{B}_{W_1}(\mathcal{T})$ such that*

(i) $\mathbb{P}(Y_t \in D) > 1 - \varepsilon$

and

(ii) *for all $B \in \mathcal{B}_{W_2^c}(\mathcal{T})$ and $A \in \mathcal{B}_{W_1}(\mathcal{T})$ with $A \subseteq D$ and $\mathbb{P}(Y_t \in A) > 0$,*

$$\left|\mathbb{P}(Y_t \in B \mid Y_t \in A) - \mathbb{P}(Y_t \in B)\right| < \varepsilon\,.$$

Proof. Let $a, t, \varepsilon > 0$ and $W_1 = [-a,a]^d$. By Lemma 15.16 there exists $\ell > 0$ such that $\mathbb{P}(Y_t \in D) > 1 - \varepsilon$ for $D := \{\theta(Y_t \wedge W_1) < \ell\}$. For such ℓ, choose $s \in (0,t)$ satisfying the following three conditions:

$$\begin{aligned}\mathrm{e}^{s\ell} - 1 &< \frac{\varepsilon}{2},\\ \mathrm{e}^{-s\ell} &> \left(1 - \frac{\varepsilon}{3}\right)^{1/3},\\ \mathrm{e}^{-s\,\Theta([W_1])} &> \left(1 - \frac{\varepsilon}{3}\right)^{1/3}.\end{aligned}$$

The existence of such s is obvious. Furthermore, Lemma 15.20 yields that $b > a$ can be chosen such that $\mathbb{P}(Z^{\mathrm{STIT}}_{\langle o\rangle,s} \in \mathcal{P}(a,b)) > (1-\frac{\varepsilon}{3})^{1/3}\,\mathrm{e}^{-s\Theta([W_1])}$. Inserting this into (15.16) completes the proof. □

Now we are prepared to show the main result of this section. In the following theorem, (a) follows immediately from (b). Nevertheless, we present (a) as a direct corollary of the statements above. The result in (b) means that the decay of the β-mixing coefficient is at least polynomial in the parameter b.

Theorem 15.25. *Let $(Y_t,\, t > 0)$ be a STIT tessellation process driven by the hyperplane measure Θ of the form* (4.1), *satisfying* (4.3). *For $0 < a < b$, denote $W_1 = [-a,a]^d$ and $W_2 = [-b,b]^d$. Then*

(a) *for all $a > 0$ and $t > 0$,*

$$\lim_{b\to\infty} \beta(a,b) = 0\,,$$

which means that STIT tessellations are β-mixing in space in the sense of Definition 15.11(b).

(b) *for all $a > 0$, $t > 0$, and $q \in (0,1)$, there exist constants $\rho(\Theta,d,t,a,q) < \infty$ and $b(\Theta,d,t,a,q) < \infty$ such that for $b > b(\Theta,d,t,a,q)$,*

$$\beta(a,b) \le \rho(\Theta,d,t,a,q)\, b^{-q}\,.$$

Proof. (a) This follows immediately by Theorems 15.24 and 15.12.

(b) Let $a > 0$, $t > 0$, and $q \in (0,1)$. Furthermore, fix some υ and z with $0 < \upsilon < z < 1$ and $z - \upsilon \ge q$. Moreover, choose $n \in \mathbb{N}$ such that $\upsilon n > q$. Then by taking $\ell := b^{\upsilon}$ and $s := b^{-z}$ in (15.17) we get that

$$\begin{aligned}\beta(a,b) \le\ & \mathbb{P}(\theta(Y_t \wedge W_1) \ge b^{\upsilon}) + 2 + \exp(b^{\upsilon-z}) - \exp(-b^{\upsilon-z}) \\ & - (1+\exp(-b^{\upsilon-z}))\,\mathbb{P}(Z^{\mathrm{STIT}}_{\langle o\rangle,b^{-z}} \in \mathcal{P}(a,b))\,.\end{aligned} \tag{15.18}$$

By Theorem 4.7(d) the random number $|Y_t \wedge W_1|$ of cells is dominated by a geometrically distributed random variable. Because $\theta(Y_t \wedge W_1) \le |Y_t \wedge W_1|\,\Theta([W_1])$, for all $n \in \mathbb{N}$, the moment $\mathbb{E}\,\theta(Y_t \wedge W_1)^n$ exists and is finite, and the Markov inequality yields

$$\mathbb{P}(\theta(Y_t \wedge W_1) \ge b^{\upsilon}) \le b^{-\upsilon n}\,\mathbb{E}\,\theta(Y_t \wedge W_1)^n \le b^{-q}\,\mathbb{E}\,\theta(Y_t \wedge W_1)^n\,. \tag{15.19}$$

To derive a lower bound for $\mathbb{P}(Z^{\mathrm{STIT}}_{\langle o\rangle,b^{-z}} \in \mathcal{P}(a,b))$, we use ideas and facts from the proof of Theorem 10.3.2 of [172]. Condition (4.3) on Θ guarantees that $o \in \operatorname{int}\operatorname{conv}\operatorname{supp}\varphi$. By Steinitz' theorem (see Theorem 1.3.10 of [171]) there exist a number $k \le 2d$ and distinct points $u_1,\dots,u_k \in \operatorname{supp}\varphi$ such that $o \in \operatorname{int}\operatorname{conv}\{u_1,\dots,u_k\}$. Hence the zero-cell induced by $\{H(u_1,2a),\dots,H(u_k,2a)\}$ is bounded and contains W_1. Furthermore, there are $\delta > 0$ and $b_0 > a$ such that for the δ-neighborhoods U_i of u_i, $i \in \{1,\dots,k\}$, we have that

(i) $U_1, \dots, U_k$ are pairwise disjoint, and
(ii) $o \in \operatorname{int} \operatorname{conv} \{v_1, \dots, v_k\}$ for all $(v_1, \dots, v_k) \in U_1 \times \dots \times U_k$, and the zero-cell induced by $\{H(v_1, 2a), \dots, H(v_k, 2a)\}$ is contained in $[-b_0, b_0]^d$, which means that the set of all those zero-cells is uniformly bounded.

For $c > 1$ and $i \in \{1, \dots, k\}$, define the sets of hyperplanes

$$A_i := \{H(v_i, p) \in \mathcal{H} : v_i \in U_i, p \in (2a, 2ac)\}.$$

Since $u_i \in \operatorname{supp} \varphi$, it follows that $\varphi(U_i) > 0$. Denote $m := \min\{\varphi(U_1), \dots, \varphi(U_k)\}$. Then $\Theta(A_i) \geq 2ma(c-1)$ for all $i \in \{1, \dots, k\}$. Furthermore, if $(H_1, \dots, H_k) \in A_1 \times \dots \times A_k$, then the zero-cell induced by $\{H_1, \dots, H_k\}$ is an element of $\mathcal{P}(a, c b_0)$.

Now we use that $Z^{\mathrm{STIT}}_{\langle o \rangle, b^{-z}} \overset{\mathrm{D}}{=} Z^{\mathrm{PHT}}_{\langle o \rangle, b^{-z}}$ by (6.3), and thus we can consider a Poisson line process with intensity measure $b^{-z}\,\Theta$. If there is no line in $[W_1]$ and at least one line in each of the A_i, $i \in \{1, \dots, k\}$, then $Z^{\mathrm{PHT}}_{\langle o \rangle, b^{-z}} \in \mathcal{P}(a, c b_0)$. Because the sets of lines $[W_1]$ and A_i, $i \in \{1, \dots, k\}$, are pairwise disjoint, the independence property of Poisson point processes (see Definition 19.10(b)) yields that

$$\begin{aligned} \mathbb{P}(Z^{\mathrm{STIT}}_{\langle o \rangle, b^{-z}} \in \mathcal{P}(a, c b_0)) &= \mathbb{P}(Z^{\mathrm{PHT}}_{\langle o \rangle, b^{-z}} \in \mathcal{P}(a, c b_0)) \\ &\geq \mathrm{e}^{-\Theta([W_1])\, b^{-z}} \left(1 - \mathrm{e}^{-2ma(c-1)\, b^{-z}}\right)^{2d}, \end{aligned}$$

where we used that $k \leq 2d$.

Summarizing, we obtain that for $b = c b_0$,

$$\begin{aligned} \beta(a, b) \leq{}& b^{-vn}\, \mathbb{E}\,\theta(Y_t \wedge W_1)^n + 2 + \mathrm{e}^{b^{v-z}} - \mathrm{e}^{-b^{v-z}} \\ &- \left(1 + \mathrm{e}^{-b^{v-z}}\right) \mathrm{e}^{-\Theta([W_1])\, b^{-z}} \left(1 - \mathrm{e}^{-2ma \frac{b-b_0}{b_0} b^{-z}}\right)^{2d}. \end{aligned} \tag{15.20}$$

The Taylor expansion for the exponential terms yields

$$\mathrm{e}^{b^{v-z}} - \mathrm{e}^{-b^{v-z}} = 2 b^{v-z} + o(b^{v-z}) \leq 2 b^{-q} + o(b^{-q}) \quad \text{as } b \to \infty. \tag{15.21}$$

Furthermore, with the remainder terms of the Taylor expansion of other exponential functions, there are $\delta_i \in (0,1)$, $i \in \{1, 2\}$, such that

$$\begin{aligned} &2 - \left(1 + \mathrm{e}^{-b^{v-z}}\right) \mathrm{e}^{-\Theta([W_1])\, b^{-z}} \left(1 - \mathrm{e}^{-2ma \frac{b-b_0}{b_0} b^{-z}}\right)^{2d} \\ &= 2 - \left(2 - b^{v-z}\, \mathrm{e}^{-\delta_1 b^{v-z}}\right)\left(1 - \Theta([W_1])\, b^{-z}\, \mathrm{e}^{-\delta_2 \Theta([W_1])\, b^{-z}}\right)\left(1 - \mathrm{e}^{-2ma \frac{b-b_0}{b_0} b^{-z}}\right)^{2d} \\ &\leq 2 - (2 - b^{v-z})(1 - \Theta([W_1])\, b^{-z}) \left(1 - \mathrm{e}^{-2ma \frac{b-b_0}{b_0} b^{-z}}\right)^{2d} \\ &= 2\left(1 - \left(1 - \mathrm{e}^{-2ma \frac{b-b_0}{b_0} b^{-z}}\right)^{2d}\right) + 2\,\Theta([W_1])\, b^{-z} \left(1 - \mathrm{e}^{-2ma \frac{b-b_0}{b_0} b^{-z}}\right)^{2d} \\ &\quad + b^{v-z} \left(1 - \mathrm{e}^{-2ma \frac{b-b_0}{b_0} b^{-z}}\right)^{2d} - b^{v-z}\, \Theta([W_1])\, b^{-z} \left(1 - \mathrm{e}^{-2ma \frac{b-b_0}{b_0} b^{-z}}\right)^{2d} \end{aligned}$$

$$\le 2(1-(1-e^{-2ma\frac{b-b_0}{b_0}b^{-z}})^{2d}) + 2\Theta([W_1])b^{-z} + b^{\upsilon-z}$$
$$- b^{\upsilon-z}\Theta([W_1])b^{-z}(1-e^{-2ma\frac{b-b_0}{b_0}b^{-z}})^{2d}.$$

Now for all $r > 0$, there exists a constant k, depending on m, a, b_0 such that

$$e^{-2ma\frac{b-b_0}{b_0}b^{-z}} = e^{-\frac{2ma}{b_0}b^{1-z}}e^{2mab^{-z}} \le e^{-\frac{2ma}{b_0}b^{1-z}}e^{2mab_0^{-z}} \le kb^{-r}.$$

This yields that as $b \to \infty$,

$$\begin{aligned}
&2(1-(1-e^{-2ma\frac{b-b_0}{b_0}b^{-z}})^{2d}) + 2\Theta([W_1])b^{-z} + b^{\upsilon-z} \\
&\quad - b^{\upsilon-z}\Theta([W_1])b^{-z}(1-e^{-2ma\frac{b-b_0}{b_0}b^{-z}})^{2d} \\
&\quad \le 4dkb^{-r} + o(b^{-r}) + 2\Theta([W_1])b^{-(z-\upsilon)} \\
&\qquad - \Theta([W_1])b^{-(2z-\upsilon)} + 2d\,\Theta([W_1])b^{-(2z-\upsilon)}kb^{-r} + o(b^{-r}).
\end{aligned} \tag{15.22}$$

Choosing $r > q$ and summarizing (15.20), (15.21), and (15.22), we complete the proof. □

Note that the upper bounds for $\beta(a, b)$ given above are based on rough upper bounds for some terms in the formulae. Therefore also stronger results are possible. However, we would expect that also with respect to mixing properties, the STIT tessellations are somehow between PHTs and Poisson-Voronoi tessellations; see also Section 15.3. Note that since by Theorem 15.25 and Lemma 15.13 the tail σ-algebra for a STIT tessellation is trivial, the STIT tessellation has a short-range correlation in the sense of Exercise 12.3.4 in [46].

15.2.3 An ergodic theorem

In this section, we start with ergodic theorems for random stationary and ergodic tessellations. Using the fact that STIT tessellations are ergodic by (15.11) and Lemma 15.13, a particular ergodic theorem follows for the maximal polytopes of these tessellations. This proposition is used in the proof of Theorem 13.8.

For a random tessellation X in $\mathbb{R}^d$ and a family $(U_C,\ C \in \mathcal{R} \cup \{\emptyset\})$ of measurable maps $U_C : \mathcal{T} \to \mathbb{R}$, consider the real-valued stochastic process $(U_C(X),\ C \in \mathcal{R} \cup \{\emptyset\})$ with parameter space $\mathcal{R} \cup \{\emptyset\}$, the convex ring augmented with the empty set. It is called **additive** if for all $C_1, C_2 \in \mathcal{R}$,

$$U_{C_1 \cup C_2}(X) + U_{C_1 \cap C_2}(X) = U_{C_1}(X) + U_{C_2}(X)$$

and $U_\emptyset = 0$. It is called **translation covariant** if for all $C \in \mathcal{R}$ and $x \in \mathbb{R}^d$,

$$U_{C+x}(X + x) = U_C(X),$$

where the translation of tessellations is defined in the usual way by (2.19). Furthermore, it is called **integrably bounded** if there exists a real-valued integrable random variable V such that

$$|U_K(X)| \leq V \quad \text{a. s.} \quad \text{for all } K \in \mathcal{K} \text{ with } K \subseteq [0,1]^d.$$

Denote by $\partial^+[0,1]^d := [0,1]^d \setminus [0,1)^d$ the "upper-right"' boundary of the unit cube. Recall the notation $\mathcal{I}(\mathcal{T})$ for the invariant σ-algebra, introduced in Section 15.1.1.

Theorem 15.26. *Let X be a random stationary and ergodic tessellation, and let $(U_C(X),\ C \in \mathcal{R} \cup \{\emptyset\})$ be a real-valued, integrably bounded, additive, and translation-covariant stochastic process with parameter space $\mathcal{R} \cup \{\emptyset\}$. Then for $K \in \mathcal{K}$ with $o \in \operatorname{int} K$,*

$$\lim_{r\to\infty} \frac{U_{rK}(X)}{V_d(rK)} = \mathbb{E}\big(U_{[0,1]^d}(X) - U_{\partial^+[0,1]^d}(X)\big) \quad \textit{a. s.}$$

Proof. By Theorem 9.3.1 of [172]

$$\lim_{r\to\infty} \frac{U_{rK}(X)}{V_d(rK)} = \mathbb{E}\big(U_{[0,1]^d}(X) - U_{\partial^+[0,1]^d}(X) | \mathcal{I}(\mathcal{T})\big) \quad \text{a. s.}$$

If X is ergodic, then by Definition 15.6(b) the invariant σ-algebra $\mathcal{I}(\mathcal{T})$ is $\mathbb{P}_X$-trivial, and hence

$$\mathbb{E}\big(U_{[0,1]^d}(X) - U_{\partial^+[0,1]^d}(X) | \mathcal{I}(\mathcal{T})\big) = \mathbb{E}\big(U_{[0,1]^d}(X) - U_{\partial^+[0,1]^d}(X)\big) \quad \text{a. s.},$$

which completes the proof. □

The theorem is now applied to random stationary tessellations and their A-components. Note that in the following theorem the functional ϕ does not have to be additive, but for (15.24) and (15.25), only convergence in probability is shown.

Theorem 15.27. *Let X be a random stationary and ergodic tessellation, $k \in \{0, \dots, d\}$. For fixed $\mathsf{A} : \mathcal{T} \to \overline{\mathcal{F}}_{\ell f}(\mathcal{P}_k)$, consider the set of* A*-components of X, as introduced in Definition 2.9. For a fixed center function c, let $\mathbb{Q}_{\mathsf{A}}$ be the distribution of the typical* A*-component defined in (2.26). Furthermore, let $K \in \mathcal{K}$ with $o \in \operatorname{int} K$, and let $\phi : \mathcal{P}_k \to \mathbb{R}$ be a translation-invariant, measurable, and $\mathbb{Q}_{\mathsf{A}}$-integrable functional. Then*

(a) *for the point process $\mathcal{Z}(\mathsf{A}(X))$ of centered* A*-components of X (see (2.24)),*

$$\lim_{r\to\infty} \frac{\sum_{(x,\mathsf{a})\in\mathcal{Z}(\mathsf{A}(X))} \mathbb{1}\{x \in rK\}\, \phi(\mathsf{a})}{V_d(rK)} = \gamma_{\mathsf{A}} \int_{\mathcal{P}_{k,c}} \phi(\mathsf{a})\, \mathbb{Q}_{\mathsf{A}}(d\mathsf{a}) \quad \textit{a. s.}, \tag{15.23}$$

(b)

$$\frac{\sum_{\mathsf{a}\in\mathsf{A}(X)} \mathbb{1}\{\mathsf{a} \subseteq rK\}\, \phi(\mathsf{a})}{V_d(rK)} \overset{\mathbb{P}}{\Longrightarrow} \gamma_{\mathsf{A}} \int_{\mathcal{P}_{k,c}} \phi(\mathsf{a})\, \mathbb{Q}_{\mathsf{A}}(d\mathsf{a}) \quad \textit{as } r \to \infty, \tag{15.24}$$

(c) *if additionally* $\int_{\mathcal{P}_{k,c}} |\phi(\mathsf{a})|\, V_d(\mathsf{a} + B^d)\, \mathbb{Q}_{\mathsf{A}}(d\mathsf{a}) < \infty$, *then*

$$\frac{\sum_{\mathsf{a}\in\mathsf{A}(X)} \mathbb{1}\{\mathsf{a} \cap rK \neq \emptyset\}\, \phi(\mathsf{a})}{V_d(rK)} \xrightarrow{\mathbb{P}} \gamma_{\mathsf{A}} \int_{\mathcal{P}_{k,c}} \phi(\mathsf{a})\, \mathbb{Q}_{\mathsf{A}}(d\mathsf{a}) \quad \text{as } r \to \infty. \tag{15.25}$$

Proof. (a) Recall that A is translation covariant by definition. Put

$$U_C(X) := \sum_{(x,\mathsf{a})\in\mathcal{Z}(\mathsf{A}(X))} \mathbb{1}\{x \in C\}\, \phi(\mathsf{a}) \quad \text{for } C \in \mathcal{R} \cup \{\emptyset\}.$$

Then $(U_C(X), C \in \mathcal{R} \cup \{\emptyset\})$ is a real-valued, additive, and translation-covariant stochastic process with parameter space $\mathcal{R} \cup \{\emptyset\}$. It is also integrably bounded because for all $K \subseteq [0,1]^d$, $K \in \mathcal{R}$,

$$|U_K(X)| \le \sum_{(x,\mathsf{a})\in\mathcal{Z}(\mathsf{A}(X))} \mathbb{1}\{x \in K\}\, |\phi(\mathsf{a})| \le \sum_{(x,\mathsf{a})\in\mathcal{Z}(\mathsf{A}(X))} \mathbb{1}\{x \in [0,1]^d\}\, |\phi(\mathsf{a})|,$$

and by (2.26) and the $\mathbb{Q}_{\mathsf{A}}$-integrability of ϕ,

$$\mathbb{E} \sum_{(x,\mathsf{a})\in\mathcal{Z}(\mathsf{A}(X))} \mathbb{1}\{x \in [0,1]^d\}\, |\phi(\mathsf{a})| = \gamma_{\mathsf{A}} \int_{\mathcal{P}_{k,c}} |\phi(\mathsf{a})|\, \mathbb{Q}_{\mathsf{A}}(d\mathsf{a}) < \infty.$$

Here by the stationarity of X we have that $U_{\partial^+[0,1]^d}(X) = 0$ a. s., and thus Theorem 15.26 yields the statement.

(b) Because the functional ϕ is assumed to be translation invariant, the value of $\int_{\mathcal{P}_{k,c}} \phi(\mathsf{a})\, \mathbb{Q}_{\mathsf{A}}(d\mathsf{a})$ does not depend on the choice of the center function c, as mentioned in Remark 2.18. Choose a center function $c : \mathcal{P}_k \to \mathbb{R}^d$ such that $c(\mathsf{a}) \in \operatorname{relint} \mathsf{a}$ for all $\mathsf{a} \in \mathsf{A}$, for example, the Steiner point. Then

$$\begin{aligned}
&\frac{\sum_{\mathsf{a}\in\mathsf{A}(X)} \mathbb{1}\{\mathsf{a} \subseteq rK\}\, \phi(\mathsf{a})}{V_d(rK)} \\
&= \frac{\sum_{(x,\mathsf{a})\in\mathcal{Z}(\mathsf{A}(X))} \mathbb{1}\{\mathsf{a} + x \subseteq rK\}\, \phi(\mathsf{a})}{V_d(rK)} \\
&= \frac{\sum_{(x,\mathsf{a})\in\mathcal{Z}(\mathsf{A}(X))} \mathbb{1}\{x \in rK\}\, \phi(\mathsf{a})}{V_d(rK)} - \frac{\sum_{(x,\mathsf{a})\in\mathcal{Z}(\mathsf{A}(X))} \mathbb{1}\{x \in rK,\, \mathsf{a} + x \nsubseteq rK\}\, \phi(\mathsf{a})}{V_d(rK)}.
\end{aligned}$$

Note that by the $\mathbb{Q}_{\mathsf{A}}$-integrability of ϕ the expectation of $\sum_{(x,\mathsf{a})\in\mathcal{Z}(\mathsf{A}(X))} \mathbb{1}\{x \in rK\}\, \phi(\mathsf{a})/V_d(rK)$ is finite, and by (2.28) it does not depend on r. Hence all the terms that occur in the above equations and also their limits as $r \to \infty$ are a. s. finite. By (15.23)

$$\frac{\sum_{(x,\mathsf{a})\in\mathcal{Z}(\mathsf{A}(X))} \mathbb{1}\{x \in rK\}\, \phi(\mathsf{a})}{V_d(rK)} \xrightarrow{\mathbb{P}} \gamma_{\mathsf{A}} \int_{\mathcal{P}_{k,c}} \phi(\mathsf{a})\, \mathbb{Q}_{\mathsf{A}}(d\mathsf{a}) \quad \text{as } r \to \infty,$$

because the a. s. convergence implies the convergence in probability. By (2.28) and (2.29)

$$\frac{1}{V_d(rK)}\mathbb{E}\sum_{(x,\mathrm{a})\in\mathcal{Z}(\mathrm{A}(X))}\mathbb{1}\{x\in rK,\ \mathrm{a}+x\not\subseteq rK\}\,\phi(\mathrm{a})$$
$$=\frac{1}{V_d(rK)}\Bigg(\mathbb{E}\sum_{(x,\mathrm{a})\in\mathcal{Z}(\mathrm{A}(X))}\mathbb{1}\{x\in rK\}\,\phi(\mathrm{a})$$
$$-\,\mathbb{E}\sum_{(x,\mathrm{a})\in\mathcal{Z}(\mathrm{A}(X))}\mathbb{1}\{x\in rK,\ \mathrm{a}+x\subseteq rK\}\,\phi(\mathrm{a})\Bigg)\to 0\quad\text{as } r\to\infty,$$

which is the convergence in mean, and this implies the convergence in probability. Because the convergence in probability is inherited to the difference, statement (15.24) is shown.

(c) As in the proof of (b), choose a center function $c:\mathcal{P}_k\to\mathbb{R}^d$ such that $c(\mathrm{a})\in$ relint a for all $\mathrm{a}\in\mathrm{A}$. Then

$$\frac{\sum_{\mathrm{a}\in\mathrm{A}(X)}\mathbb{1}\{\mathrm{a}\cap rK\neq\emptyset\}\,\phi(\mathrm{a})}{V_d(rK)}$$
$$=\frac{\sum_{(x,\mathrm{a})\in\mathcal{Z}(\mathrm{A}(X))}\mathbb{1}\{x\in rK\}\,\phi(\mathrm{a})}{V_d(rK)}$$
$$+\frac{\sum_{(x,\mathrm{a})\in\mathcal{Z}(\mathrm{A}(X))}\mathbb{1}\{x\notin rK,\ (\mathrm{a}+x)\cap rK\neq\emptyset\}\,\phi(\mathrm{a})}{V_d(rK)}.$$

By (2.30) the $\mathbb{Q}_{\mathrm{A}}$-integrability of ϕ and the additional assumption that

$$\int_{\mathcal{P}_{k,c}}|\phi(\mathrm{a})|\,V_d(\mathrm{a}+B^d)\,\mathbb{Q}_{\mathrm{A}}(d\mathrm{a})<\infty,$$

the limit as $r\to\infty$ of the expectation of $\sum_{\mathrm{a}\in\mathrm{A}(X)}\mathbb{1}\{\mathrm{a}\cap rK\neq\emptyset\}\,\phi(\mathrm{a})/V_d(rK)$ is finite, and so are the limits of the expectations of all other terms in the equations above. By (2.28) and (2.30)

$$\frac{1}{V_d(rK)}\mathbb{E}\sum_{(x,\mathrm{a})\in\mathcal{Z}(\mathrm{A}(X))}\mathbb{1}\{x\notin rK,\ (\mathrm{a}+x)\cap rK\neq\emptyset\}\,\phi(\mathrm{a})$$
$$=\frac{1}{V_d(rK)}\Bigg(\mathbb{E}\sum_{(x,\mathrm{a})\in\mathcal{Z}(\mathrm{A}(X))}\mathbb{1}\{(\mathrm{a}+x)\cap rK\neq\emptyset\}\,\phi(\mathrm{a})$$
$$-\,\mathbb{E}\sum_{(x,\mathrm{a})\in\mathcal{Z}(\mathrm{A}(X))}\mathbb{1}\{x\in rK\}\,\phi(\mathrm{a})\Bigg)\to 0\quad\text{as } r\to\infty,$$

and we can proceed as in the proof of (b). □

The results of Theorem 15.27 are now applied to STIT tessellations and their maximal k-polytopes. Evidently, it can also be written down for primitives or for other

A-components. Recall the notation $\mathcal{F}_k^{\max}(Y_t) := \{F : (F,s) \in \mathcal{F}_k^{\max *}(\Sigma),\ s \le t\}$ introduced in (8.10) for the process of maximal k-polytopes of Y_t and $\gamma_{k,t}^{\max}$ defined in (8.15) for the corresponding intensity.

Corollary 15.28. *Let $(Y_t, t > 0)$ be a STIT tessellation process driven by the hyperplane measure Θ of the form (4.1), satisfying (4.3), and let $W \in \mathcal{P}_d$ with $o \in \operatorname{int} W$. Furthermore, let $k \in \{0, \dots, d\}$, $t > 0$, a center function $c : \mathcal{P}_k \to \mathbb{R}^d$, the point process $\mathcal{Z}(\mathcal{F}_k^{\max}(Y_t))$ of centered maximal k-polytopes of Y_t, and a translation-invariant, measurable, and $\mathbb{Q}_k^{\max}$-integrable functional $\phi : \mathcal{P}_k \to \mathbb{R}$ be given. Then*

(a) $$\lim_{a\to\infty} \frac{\sum_{(x,F)\in\mathcal{Z}(\mathcal{F}_k^{\max}(Y_t))} \mathbb{1}\{x \in aW\}\,\phi(F)}{V_d(aW)} = \gamma_{k,t}^{\max} \int_{\mathcal{P}_{k,c}} \phi(F)\,\mathbb{Q}_k^{\max}(\mathrm{d}F) \quad \textit{a. s.},$$

(b) $$\frac{\sum_{F\in\mathcal{F}_k^{\max}(Y_t)} \mathbb{1}\{F \subseteq aW\}\,\phi(F)}{V_d(aW)} \overset{\mathbb{P}}{\Longrightarrow} \gamma_{k,t}^{\max} \int_{\mathcal{P}_{k,c}} \phi(F)\,\mathbb{Q}_k^{\max}(\mathrm{d}F) \quad \textit{as } a \to \infty,$$

(c) *if additionally* $\int_{\mathcal{P}_{k,o}} |\phi(F)|\,V_d(F + B^d)\,\mathbb{Q}_k^{\max}(\mathrm{d}F) < \infty$, *then*

$$\frac{\sum_{F\in\mathcal{F}_k^{\max}(Y_t)} \mathbb{1}\{F \cap aW \ne \emptyset\}\,\phi(F)}{V_d(aW)} \overset{\mathbb{P}}{\Longrightarrow} \gamma_{k,t}^{\max} \int_{\mathcal{P}_{k,c}} \phi(F)\,\mathbb{Q}_k^{\max}(\mathrm{d}F) \quad \textit{as } a \to \infty.$$

Proof. By Theorem 15.25, Corollary 15.10, and Lemma 15.13 the STIT tessellations are ergodic, and hence Theorem 15.27 can be applied. □

Note that for the evaluation of the sums on the left-hand sides in (a) and (c), information about Y_t outside aW is used.

15.3 Comparison of STIT and Poisson hyperplane tessellations

Mixing and ergodic properties of PHTs are dealt with in detail in Chapter 7 of [82]. Intuitively, because the skeletons of PHTs are built by hyperplanes, which are unbounded, and the skeletons of STIT tessellations consisting of bounded maximal polytopes (see Lemma 8.8(a,b)), we would expect that PHTs have somehow weaker mixing properties than STIT tessellations. Formally, this difference can be presented considering mixing and triviality of the tail σ-algebra as given in Definition 15.6. Also recall Definition 15.5 of the tail σ-algebra. For PHT, the event that the unit ball B_d is hit by a hyperplane belongs to the tail σ-algebra. This is due to the observation that at an arbitrary large distance from the origin, we can still check whether a hyperplane hits B_d or does not. This implies that the tail σ-algebra is not trivial with respect to the distribution of PHT. The corresponding result is a subject of the following lemma.

Let Θ be a hyperplane measure of the form (4.1), and let φ be its spherical directional distribution. Then φ is called **subspace-free** if the φ-value of each great subsphere of S^{d-1} is zero.

Lemma 15.29. *Let Θ be a hyperplane measure of the form* (4.1), *satisfying* (4.3). *The PHT X_t in $\mathbb{R}^d$, generated by a Poisson hyperplane process $\widehat{X}_t$ (see* (2.65)*) with intensity measure $t\,\Theta$, has the following properties:*
(a) *It is mixing if and only if the spherical directional distribution φ is subspace-free;*
(b) *The tail σ-algebra $\mathcal{B}_\infty(\mathcal{T})$ is not trivial with respect to the distribution $\mathbb{P}_{X_t}$ of X_t.*

Proof. (a) This is Theorem 7.1.4 of [82].

(b) Recall definition (2.4) of the set $\mathcal{F}_{d-1}^{\text{pr}}(T)$ of $(d-1)$-primitives of a tessellation T in $\mathbb{R}^d$ and the measurability of the mapping $T \mapsto \mathcal{F}_{d-1}^{\text{pr}}(T)$, which is shown in Lemma 18.5(b). Fix the sequence $(W_n = [-n,n]^d\,,\ n \in \mathbb{N})$ satisfying (15.4), and consider the set of tessellations

$$A := \{T \in \mathcal{T} : \forall n \in \mathbb{N}\ \exists F \in \mathcal{F}_{d-1}^{\text{pr}}(T) : F \subset W_n^c\,,\ \operatorname{aff} F \cap B^d \neq \emptyset\}\,.$$

By construction, $A \in \mathcal{B}_{W^c}(\mathcal{T})$ for all $W \in \mathcal{P}_d$, and hence $A \in \mathcal{B}_\infty(\mathcal{T})$. Note that $T \in A$ does not necessarily imply that $\partial T \cap B^d \neq \emptyset$. On the other hand, the set $\mathcal{T}^{\text{PHT}}$ of all realizations of PHT in $\mathbb{R}^d$ can be characterized by

$$\mathcal{T}^{\text{PHT}} = \Big\{T \in \mathcal{T} : \partial T = \bigcup_{F \in \mathcal{F}_{d-1}^{\text{pr}}(T)} \operatorname{aff} F\Big\}\,.$$

By (5.1) we obtain that

$$\mathbb{P}_{X_t}(A) = \mathbb{P}_{X_t}(A \cap \mathcal{T}^{\text{PHT}}) = \mathbb{P}(\partial X_t \cap B^d \neq \emptyset) = 1 - \mathrm{e}^{-t\,\Theta([B^d])} \notin \{0,1\}\,,$$

which completes the proof. □

In contrast, in Theorem 15.25, it was shown that the STIT tessellations are β-mixing, which by Lemma 15.13 implies the triviality of the tail-σ-algebra $\mathcal{B}_\infty(\mathcal{T})$ with respect to their distributions.

Moreover, it is known that Poisson–Voronoi tessellations have rather strong mixing properties. In Theorem 2.2 of [69], it was shown that the $\beta(a,b)$-coefficient, $0 < a < b$, defined in (15.10), has an exponential decay rate for fixed a as $b \to \infty$. A detailed formula for an upper bound is provided in (16.84).

Note that in [64] an exponential decay of $\beta(a,b)$ was shown also for the so-called β- and Gaussian–Delaunay tessellations.

Summarizing, the results shown so far indicate that with respect to mixing properties, the STIT tessellations are "in between" PHTs and Poisson–Voronoi tessellations.

15.4 The β-mixing coefficient in time

It was shown in Theorem 6.17 that the process

$$(a^t\, Y_{a^t}\,,\ t \in \mathbb{R})$$

is time-stationary if $(Y_t,\ t > 0)$ is a STIT tessellation process driven by some hyperplane measure Θ of the form (4.1), satisfying (4.3), and $a > 1$. Now we show that for all $W \in \mathcal{P}_d$, the time-stationary process $(a^t\, Y_{a^t} \wedge W,\ t \in \mathbb{R})$ with the restriction to W is β-mixing with respect to the time coordinate. We recall the usual definition of the β mixing coefficient in time and provide a criterion for a process to be β-mixing. Recall definition (15.1) of the β-mixing coefficient for a pair of σ-algebras.

Definition 15.30. *Let $(X_t,\ t \in \mathbb{R})$ be a random time-stationary process with values in $\mathcal{T}$ or, for a fixed $W \in \mathcal{P}_d$, in $\mathcal{T}_W$, respectively. For $t \in \mathbb{R}$, define the σ-algebras $\mathbf{A}_{-\infty}^t := \sigma(X_s,\ s \le t)$ and $\mathbf{A}_t^\infty := \sigma(X_s,\ s \ge t)$.*

(a) *For all $t > 0$, the β-**mixing coefficient in time** of $(X_t,\ t \in \mathbb{R})$ is defined as*

$$\beta(t) := \beta(\mathbf{A}_{-\infty}^0, \mathbf{A}_t^\infty)\,.$$

(b) *The process $(X_t,\ t \in \mathbb{R})$ is β-**mixing in time** or absolutely regular in time if*

$$\lim_{t\to\infty} \beta(t) = 0\,. \tag{15.26}$$

Analogously to Theorem 15.12, we can show a criterion for the convergence of the β-mixing coefficient.

Theorem 15.31. *With the notation of Definition 15.30, we have that*

$$\lim_{t\to\infty} \beta(t) = 0$$

if and only if for all $\varepsilon > 0$, there exist $t > 0$ and $D \in \mathbf{A}_{-\infty}^0$ such that

$$\begin{aligned}&\text{(i) } \mathbb{P}(D) > 1-\varepsilon\\ &\quad \textit{and}\\ &\text{(ii) } \textit{for all } B \in \mathbf{A}_t^\infty \textit{ and } A \in \mathbf{A}_{-\infty}^0 \textit{ with } A \subseteq D \textit{ and } \mathbb{P}(A) > 0,\ s\\ &\quad |\mathbb{P}(B \mid A) - \mathbb{P}(B)| < \varepsilon\,.\end{aligned} \tag{15.27}$$

Proof. By Lemma 15.3 and assumption (15.27) with $\mathbb{P}(D) > 1-\varepsilon$, we have that $\beta(t) < 2\varepsilon$. The assumption that for all $\varepsilon > 0$, there exists an appropriate $t > 0$ implies that $\lim_{t\to\infty} \beta(t) = 0$.

To show the other implication, first note that by definition (15.1) of the β-mixing coefficient for a pair of σ-algebras this coefficient is monotone in the sense that $\beta(\mathbf{B}, \mathbf{C}') \le \beta(\mathbf{B}, \mathbf{C})$ if $\mathbf{C}' \subseteq \mathbf{C}$. Hence $\beta(t)$ is decreasing in $t \in \mathbb{R}$. Thus for the proof, it is sufficient to consider the discrete-time process $(X_k,\ k \in \mathbb{Z})$. Furthermore, according to Lemma 18.2, the space $\mathcal{T}$ or $\mathcal{T}_W$, respectively, of tessellations is a Borel space, and thus $\mathcal{T}^{\mathbb{Z}}$ or $\mathcal{T}_W^{\mathbb{Z}}$, respectively, is a Borel space too. Hence the σ-algebra $\mathcal{B}(\mathcal{T}^{\mathbb{Z}})$ or $\mathcal{B}(\mathcal{T}_W^{\mathbb{Z}})$, respectively, has a countable generating system, and a regular conditional distribution $\mathbb{P}(\cdot\,|\mathbf{A}_{-\infty}^0)$ on $\mathbf{A}_k^\infty$, $k \in \mathbb{N}_0$, exists; see Theorem 8.5 of [89].

Now if $\lim_{k\to\infty}\beta(k)=0$, then for all $\varepsilon>0$, there is $k\in\mathbb{N}_0$ such that $\beta(k)\le\varepsilon^2$. For such k, define the event

$$D:=\left\{\omega\in\Omega:\ \sup_{B\in\mathbf{A}_k^\infty}\left|\mathbb{P}(B|\mathbf{A}_{-\infty}^0)(\omega)-\mathbb{P}(B)\right|\le\varepsilon\right\}.$$

We have $D\in\mathbf{A}_{-\infty}^0$ by the measurability property of the conditional probability $\mathbb{P}(B|\mathbf{A}_{-\infty}^0)$. By Lemma 15.3(a) and the Markov inequality, $\mathbb{P}(D^c)<\varepsilon$, and hence $\mathbb{P}(D)>1-\varepsilon$. Then, proceeding as in the proof of Lemma 15.3(b), where it is shown that there (ii) implies (i), property (15.27) is shown. □

This is now applied to STIT tessellation processes in windows $W\in\mathcal{P}_d$. Intuitively, this is based on the observation that "again and again" $(a^t\,Y_{a^t})\wedge W=\{W\}$. If the origin $o\in\operatorname{int}W$, then this means that the sequence of times at which the zero-cells of $(a^t\,Y_{a^t},\ t\in\mathbb{R})$ contain W goes to infinity. Related aspects are dealt with in Sections 15.5.2 and 15.6.2.

Lemma 15.32. *Let $(Y_t,\ t>0)$ be a STIT tessellation process driven by the hyperplane measure Θ of the form* (4.1), *satisfying* (4.3), *and let $a>1$ and $W\in\mathcal{P}_d$. Then for the process $((a^t\,Y_{a^t})\wedge W,\ t\in\mathbb{R})$, conditions (i) and (ii) of* (15.27) *are satisfied.*

Proof. Because the STIT tessellations are space-stationary, we can choose $W\in\mathcal{P}_d$ with $o\in\operatorname{int}W$. Let $r>0$ be such that $W\subset r\,B^d$. For $\varepsilon\in(0,1)$ such that $\varepsilon\,B^d\subset W$, define

$$D:=\left\{\omega\in\Omega:\partial Y_1(\omega)\cap\frac{\varepsilon}{4}\,B^d=\emptyset\right\}\in\mathbf{A}_{-\infty}^0.$$

By (5.6) and the spatial consistency property (5.17), $\mathbb{P}(D)=\mathrm{e}^{-\varepsilon/2}>1-\varepsilon/2$. Choose $t_0>0$ such that $\varepsilon\,a^{t_0}/4>r$. Then for all $t\ge t_0$ and $\omega\in D$, we have that $a^t\,Y_1(\omega)\wedge W=\{W\}$.

For $t\ge t_0$ and $B\in\mathbf{A}_t^\infty$, denote

$$B':=\{(a^s\,Y_{a^s}(\omega)\wedge W,\ s\ge t):\ \omega\in B\}.$$

Because $\mathbf{A}_t^\infty$ is generated by $(a^s\,Y_{a^s}(\cdot)\wedge W,\ s\ge t)$, we have

$$B=\{\omega\in\Omega:\ (a^s\,Y_{a^s}(\omega)\wedge W,\ s\ge t)\in B'\}.$$

Now for all $A\subseteq D$ with $\mathbb{P}(A)>0$, applying (6.1), (3.5), and the definition of the iteration $\boxplus$ given in (3.4), together with the independence property of the nested tessellations, we have

$$\begin{aligned}\mathbb{P}(B\,|\,A)&=\mathbb{P}((a^s\,Y_{a^s}\wedge W,\ s\ge t)\in B'\,|\,A)\\&=\mathbb{P}((a^s\,(Y_1\boxplus\underline{Y_{a^s-1}})\wedge W,\ s\ge t)\in B'\,|\,A)\\&=\mathbb{P}(((a^s\,Y_1\wedge W)\boxplus(a^s\,\underline{Y_{a^s-1}}\wedge W),\ s\ge t)\in B'\,|\,A)\\&=\mathbb{P}((a^s\,Y_{a^s-1}\wedge W,\ s\ge t)\in B').\end{aligned}$$

This conditional probability does not depend on the choice of A, and thus for all $A \subseteq D$ with $\mathbb{P}(A) > 0$, we obtain $\mathbb{P}(B \mid A) = \mathbb{P}(B \mid D)$. Hence

$$\begin{aligned} |\mathbb{P}(B \mid A) - \mathbb{P}(B)| &= |\mathbb{P}(B \mid D) - \mathbb{P}(B \cap D) - \mathbb{P}(B \cap D^c)| \\ &\le |\mathbb{P}(B \mid D) - \mathbb{P}(B \mid D)\mathbb{P}(D)| + \mathbb{P}(B \cap D^c) \\ &= \mathbb{P}(B \mid D)(1 - \mathbb{P}(D)) + \mathbb{P}(B \cap D^c) \\ &< \varepsilon , \end{aligned}$$

which is condition (15.27). □

Summarizing, we have shown the following result.

Theorem 15.33. *Let* $(Y_t, t > 0)$ *be a STIT tessellation process, driven by a hyperplane measure* Θ *of the form* (4.1), *satisfying* (4.3). *Furthermore, let* $a > 1$ *and* $W \in \mathcal{P}_d$. *Then the process* $(a^t Y_{a^t} \wedge W, t \in \mathbb{R})$ *is* β*-mixing in time in the sense of Definition* 15.30*(b).*

15.5 The zero-cell process

The process of zero-cells of a STIT-tessellation process plays a central role in the global construction described in Chapter 8.

15.5.1 Triviality of the past tail σ-field

Now we show that the tail-σ-algebra pertaining to the *time-reversed sequence* of zero-cells is trivial. Throughout this section, we always refer to a probability measure $\mathbb{P}$ that induces the distribution $\mathbb{P}_{\mathbf{Y}}$ of the STIT tessellation process $\mathbf{Y} = (Y_t, t > 0)$, driven by a hyperplane measure Θ of the form (4.1), satisfying (4.3).

Recall that $(Z^{\mathrm{STIT}}_{\langle o\rangle,t}, t > 0)$ denotes the zero-cell process of a STIT tessellation process $(Y_t, t > 0)$. Furthermore, $(Z^{\mathrm{STIT}}_{\langle o\rangle,t}, t > 0) \overset{\mathrm{D}}{=} (Z^{\mathrm{PHT}}_{\langle o\rangle,t}, t > 0)$ by (6.3) of Theorem 6.2, which means that the zero-cell process of a PHT process has the same distribution as that of a STIT tessellation process if both processes are driven by the same hyperplane measure Θ. Let $\hat{X}^*$ denote the corresponding Poisson point process on $\mathcal{H} \times \mathbb{R}^+$ of birth time marked hyperplanes, as introduced in Section 2.3.1. For simplicity of the notation,

$$(Z_{\langle o\rangle,t}, t > 0) := (Z^{\mathrm{STIT}}_{\langle o\rangle,t}, t > 0) .$$

This zero-cell process is a Markov process with values in $\mathcal{P}_d$. For all $t_0 > 0$, the process $(Z_{\langle o\rangle,t}, t \ge t_0)$ is a pure jump process. Using the set $\mathbb{Z}$ of integers as the index set, the jumps can be numbered. As in Section 8.1(i), let $(\beta_i, i \in \mathbb{Z})$ denote the ordered sequence of jump times of $(Z_{\langle o\rangle,t}, t > 0)$ with $\beta_0 < 1 \le \beta_1$ and by $((H_i, \beta_i), i \in \mathbb{Z})$ the corresponding sequence of birth time marked hyperplanes from $\widehat{X}^*$ that cause the jumps. Denote by $(Z_{\langle o\rangle,i}\,,\ i \in \mathbb{Z})$ the sequence of zero cells with $Z_{\langle o\rangle,i} := Z_{\langle o\rangle,\beta_i}$.

To introduce an appropriate tail-σ-algebra, consider the "time-reversed" process

$$(\hat{Z}_n,\ n \in \mathbb{N}_0) \quad \text{with } \hat{Z}_n := Z_{\langle o\rangle,(-n)}$$

and denote by $\mathcal{B}_n := \sigma(\hat{Z}_n)$, $n \in \mathbb{N}_0$, the σ-algebra generated by $\hat{Z}_n$. Then the corresponding tail-σ-algebra is

$$\mathcal{B}_\infty := \bigcap_{m\in\mathbb{N}_0} \bigvee_{k\geq m} \mathcal{B}_k\,,$$

where $\bigvee$ stands for the σ-algebra generated by the union of the respective σ-algebras. Note that the events of the tail-σ-algebra do not depend on the behavior of the zero-cell process *after* any time $t > 0$.

The tail σ-algebra $\mathcal{B}_\infty$ is also referred to as "the past tail σ-filed" of the sequence $(Z_{\langle o\rangle,i}\,,\ i \in \mathbb{Z})$; see Definition 2.11 of [19].

Now the $\mathbb{P}$-triviality of $\mathcal{B}_\infty$ is shown. With the terminology of Definition 2.13 in [19], this means that the sequence $(Z_{\langle o\rangle,i}\,,\ i \in \mathbb{Z})$ is "regular".

Theorem 15.34. *The tail-σ-algebra $\mathcal{B}_\infty$ of the process $(\hat{Z}_n,\ n \in \mathbb{N}_0)$ is $\mathbb{P}$-trivial, that is, $\mathbb{P}(B) \in \{0,1\}$ for all $B \in \mathcal{B}_\infty$.*

Proof. Let $B \in \mathcal{B}_\infty$ and fix $\varepsilon > 0$. First, note that

$$\mathcal{B}_\infty \subseteq \bigvee_{m\in\mathbb{N}_0} \mathcal{B}_m = \bigvee_{m\in\mathbb{N}_0} \bigvee_{0\leq k\leq m} \mathcal{B}_k\,.$$

Thus, because the sequence of σ-algebras $\bigvee_{0\leq k\leq m} \mathcal{B}_k$ is growing in $m \in \mathbb{N}$, by Lemma 4.16 of [89] there are $m \in \mathbb{N}$ and $B_m \in \bigvee_{k\leq m} \mathcal{B}_k$ such that $\mathbb{P}(B \,\triangle\, B_m) < \varepsilon$. In the following, we consider m and B_m as fixed.

The idea of the proof is to construct times $s < t$, two polytopes $W_1 \subset W_2$ containing the origin o, and events E, E_m, A, A_n, $n > m$, such that

(i) the event $E \cap E_m$ implies that $W_1 \subseteq \hat{Z}_m \subseteq W_2$ and that the jump time $\beta_{-m} \geq t$, and
(ii) the event $A \cap A_n$ implies that $W_2 \subseteq \hat{Z}_n$ and that the jump time $\beta_{-n} \leq s$.

Then B_m and all events $C \in \bigvee_{k\geq n} \mathcal{B}_k$ are conditionally independent, given the "separation" described by (i) and (ii). As also $B \in \bigvee_{k\geq n} \mathcal{B}_k$ and the probabilities of the constructed events are appropriately chosen, it will be shown that B is "approximately independent of itself", and hence its probability is either 0 or 1.

This is now formalized. Choose a pair $(t, r) \in (0,1) \times (0,\infty)$ such that

$$r > -\frac{(m+1)^2}{2} \ln\left(1 - (1-\varepsilon)^{\frac{1}{2(m+1)}}\right) \quad \text{and} \quad t < -\frac{1}{2r} \ln(1-\varepsilon)\,.$$

For this pair, define the families of intervals

$$a_i := \left((i-1)\frac{r}{m+1}, i\frac{r}{m+1}\right) \quad \text{and} \quad \triangle_i := \left(t+(i-1)\frac{1-t}{m+1}, t+i\frac{1-t}{m+1}\right),$$

$i \in \{1,\dots,m+1\}$. These intervals partition $(0,r)$ and the time interval $(t,1)$, respectively. Now consider the event

$$E_m := \{\hat{X}^* \cap (\{H(u,x) \in \mathcal{H} : x \in (-r,r)\} \times (0,t)) = \emptyset\}$$
$$\cap \bigcap_{i=1}^{m+1} \{\hat{X}^* \cap (\{H(u,x) \in \mathcal{H} : x \in a_i \cup (-a_i)\} \times \triangle_i) \neq \emptyset\},$$

which means that the ball $r B^d$ is not intersected by a hyperplane of $\hat{X}^*$ until time t, and then in each of the $m+1$ time intervals $\triangle_i := (t+(i-1)\frac{1-t}{m+1}, t+i\frac{1-t}{m+1})$, there is at least one hyperplane of $\hat{X}^*$ with a distance to the origin in the interval $a_i := ((i-1)\frac{r}{m+1}, i\frac{r}{m+1})$, $i \in \{1,\dots,m+1\}$. This guarantees that there are at least $m+1$ jumps of the zero-cell process in the time interval $(t,1)$, and thus $\beta_{-m} \geq t$. The definition of a Poisson point process yields that

$$\mathbb{P}(E_m) = \exp(-tr)\left(1-\exp\left(-\frac{2(1-t)\,r}{m^2}\right)\right)^m > 1-\varepsilon,$$

where the lower bound on the right-hand side is guaranteed by the choice of the pair (t,r) described above.

In the next step, we choose a window $W_1 \in \mathcal{P}_d$ with $r B^d \subset W_1$ and $o \in \operatorname{int} W_1$. Recall the definition of the encapsulation time $S(W_1, W_2)$ given in (15.13). By Lemma 15.20 there exists $W_2 \in \mathcal{P}_d$ with $W_1 \subset W_2$ such that for the event $E := \{S(W_1, W_2) \leq t\}$, we have that

$$\mathbb{P}(E) = \mathbb{P}(S(W_1, W_2) \leq t) > 1-\varepsilon.$$

Furthermore, choose $s \in (0,t)$ such that for the event $A := \{[W_2] \cap \hat{X}_s = \emptyset\}$, which means that at time s the window W_2 is contained in the zero-cell, we have

$$\mathbb{P}(A) = \exp(-s\,\Theta([W_2])) > 1-\varepsilon.$$

Moreover, having fixed s, for $n \in \mathbb{N}_0$, let A_n denote the event that in the time interval $(s,1)$, at most n hyperplanes of $\hat{X}^*$ intersect W_2. Then choose $n > m$ such that

$$\mathbb{P}(A_n) = \sum_{i=0}^{n} \frac{((1-s)\,\Theta([W_2]))^i}{i!} \exp(-(1-s)\,\Theta([W_2])) > 1-\varepsilon.$$

Note that $A \cap A_n \subseteq \{W_2 \subseteq \hat{Z}_n\}$.

By the definition of Poisson point processes the restricted point processes on disjoint sets, namely $\hat{X}^* \cap ([W_2] \times (t,1))$ and $\hat{X}^* \cap ([W_2]^c \times (0,s))$, are independent. Hence, for

$B_m \in \bigvee_{k\le m} \mathcal{B}_k$, $C \in \bigvee_{k\ge n} \mathcal{B}_k$, and $D := E \cap E_m \cap A \cap A_n$, we have that the events B_m and C are conditionally independent under the condition D, and thus

$$\begin{aligned}|\mathbb{P}(B_m \cap C) - \mathbb{P}(B_m)\,\mathbb{P}(C)| &= |\mathbb{P}(B_m \cap C \cap D) + \mathbb{P}(B_m \cap C \cap D^c) - \mathbb{P}(B_m)\,\mathbb{P}(C)| \\ &\le \left|\frac{\mathbb{P}(B_m \cap D)\,\mathbb{P}(C \cap D)}{\mathbb{P}(D)} - \mathbb{P}(B_m)\,\mathbb{P}(C)\right| + \mathbb{P}(B_m \cap C \cap D^c)\,.\end{aligned}$$

With the construction and the assumptions on the events above, we have that $\mathbb{P}(D) > 1-4\,\varepsilon$, which yields $(\mathbb{P}(B_m)-4\,\varepsilon)(\mathbb{P}(C)-4\,\varepsilon) \le \mathbb{P}(B_m \cap D)\mathbb{P}(C \cap D) \le \mathbb{P}(B_m)\mathbb{P}(C)$. Furthermore, $(1-4\,\varepsilon)\,\mathbb{P}(B_m)\,\mathbb{P}(C) \le \mathbb{P}(B_m)\,\mathbb{P}(C)\,\mathbb{P}(D) \le \mathbb{P}(B_m)\,\mathbb{P}(C)$, and also $\mathbb{P}(B_m \cap C \cap D^c) \le \mathbb{P}(D^c) < 4\,\varepsilon$. Hence, for $\varepsilon < 1/4$,

$$\begin{aligned}&\frac{1}{\mathbb{P}(D)}|\mathbb{P}(B_m \cap D)\,\mathbb{P}(C \cap D) - \mathbb{P}(B_m)\mathbb{P}(C)\,\mathbb{P}(D)| + \mathbb{P}(B_m \cap C \cap D^c) \\ &\quad < \frac{1}{1-4\,\varepsilon}\max\{4\,\varepsilon\,\mathbb{P}(B_m)\,\mathbb{P}(C)\,,\ 4\,\varepsilon(\mathbb{P}(B_m) + \mathbb{P}(C)) - 16\,\varepsilon^2\} + 4\varepsilon \\ &\quad \le \frac{\max\{4\,\varepsilon\,,\ 8\,\varepsilon - 16\,\varepsilon^2\} + 4\,\varepsilon}{1-4\,\varepsilon} \\ &\quad \le \frac{12\,\varepsilon}{1-4\,\varepsilon}\,.\end{aligned}$$

Because $B \in \mathcal{B}_\infty \subseteq \bigvee_{k\ge n} \mathcal{B}_k$, we can replace C by B, and we have

$$|\mathbb{P}(B_m \cap B) - \mathbb{P}(B_m)\,\mathbb{P}(B)| < \frac{12\,\varepsilon}{1-4\,\varepsilon}\,.$$

Then $\mathbb{P}(B \triangle B_m) < \varepsilon$ yields that $|\mathbb{P}(B \cap B) - \mathbb{P}(B)\,\mathbb{P}(B)| < \frac{12\,\varepsilon}{1-4\,\varepsilon} + 2\,\varepsilon$ for all $\varepsilon > 0$, and hence $\mathbb{P}(B) \in \{0, 1\}$. This completes the proof. □

Example 15.35. Let $g : \mathcal{P}_d \to [0,\infty)$ be a nonnegative measurable function on the set of d-dimensional polytopes. Then

$$\mathbb{P}\left(\text{there exists } m \in \mathbb{Z} \text{ such that } \sum_{i\le m} g(Z_{\langle o\rangle,\beta_i})^{-1} < \infty\right) \in \{0, 1\}\,.$$

In [120], this was used to investigate the existence of cell division processes with life times of the cells P that are exponentially distributed with parameter $g(P)$ instead of $\Theta([P])$. This is also mentioned in Section 9.1.3.

15.5.2 Regenerative property

Let $(Y_t,\ t > 0)$ be a STIT tessellation process driven by a hyperplane measure Θ of the form (4.1), satisfying (4.3), and fix $a > 1$. Then by Theorem 6.17 the rescaled process $(a^t\, Y_{a^t}\,,\ t \in \mathbb{R})$ is time-stationary. In this section and in the rest of the present chapter, the *discrete-time* tessellation-valued process

$$(a^j Y_{a^j}, j \in \mathbb{Z}) \tag{15.28}$$

plays a central role. For a random tessellation X, $k \in \mathbb{N}$, and sequences $\underline{X_i}$, $i \in \{1, \dots k\}$ of i. i. d. random tessellations, we use the notation

$$X \boxplus_{i=1}^k \underline{X_i} := ((\dots((X \boxplus \underline{X_1}) \boxplus \underline{X_2}) \boxplus \dots) \boxplus \underline{X_k}). \tag{15.29}$$

By (6.1), for $m \in \mathbb{Z}$ and $k \in \mathbb{N}$, we obtain that

$$Y_{a^{m+k}} \overset{\mathrm{D}}{=} Y_{a^m} \boxplus_{i=1}^k \underline{Y_{a^{m+i}-a^{m+i-1}}}.$$

Furthermore, (6.2) yields

$$a^{m+k} Y_{a^{m+k}} \overset{\mathrm{D}}{=} (a^k Y_1) \boxplus_{i=1}^k \frac{a^{k+1-i}}{a-1} \underline{Y_1^i}, \tag{15.30}$$

where $\underline{Y_1^i} := (Y^{(\ell),i}, \ell \in \mathbb{N}_0)$, $i \in \{1, \dots k\}$, are independent sequences of i. i. d. copies of the STIT tessellation Y_1, also independent of Y_1.

The zero-cell process pertaining to (15.28) is

$$(a^j Z^{\mathrm{STIT}}_{\langle o \rangle, a^j}, j \in \mathbb{Z}).$$

This is a discrete-time Markov process; see Corollary 6.5. Now, consider the 0–1-valued discrete-time stochastic process defined by

$$\mathsf{B}_j^K := \mathbb{1}\{K \subset a^j Z^{\mathrm{STIT}}_{\langle o \rangle, a^j}\}, \quad j \in \mathbb{Z},$$

for fixed $K \in \mathcal{K}$ with $o \in \operatorname{int} K$. This process does not inherit the Markov property, but it has the following regenerative property.

Lemma 15.36. *The process* $(\mathsf{B}_j^K, j \in \mathbb{Z})$ *satisfies the conditional independence property at the 1-values, which means that for all* $k, n \in \mathbb{N}$, $0 < j_1 < \dots < j_n$, *and all* $(b_1, \dots, b_k, b_{-j_1}, \dots b_{-j_n}) \in \{0,1\}^{k+n}$,

$$\begin{aligned} &\mathbb{P}(\mathsf{B}_1^K = b_1, \dots, \mathsf{B}_k^K = b_k \mid \mathsf{B}_0^K = 1, \mathsf{B}_{-j_1}^K = b_{-j_1}, \dots, \mathsf{B}_{-j_n}^K = b_{-j_n}) \\ &\quad = \mathbb{P}(\mathsf{B}_1^K = b_1, \dots, \mathsf{B}_k^K = b_k \mid \mathsf{B}_0^K = 1). \end{aligned} \tag{15.31}$$

The transition probabilities at the 1-values are

$$\mathbb{P}(\mathsf{B}_n^K = 1 \mid \mathsf{B}_0^K = 1) = \exp(-(1 - a^{-n})\, \Theta([K]))$$

for all $n \in \mathbb{N}$.

Proof. We apply (15.30) and assume that always the tessellation with the upper index (0), that is, $Y^{(0),i}$, is nested into the cell containing the origin. Thus under the condition $\mathsf{B}_0^K = 1$, which means that $K \subset Z_{\langle o\rangle,1}^{\mathrm{STIT}}$, the probability of $\{\mathsf{B}_1^K = b_1, \dots, \mathsf{B}_k^K = b_k\}$ only depends on the sequence $Y^{(0),1}, \dots, Y^{(0),k}$ of STIT tessellations independent of $(Y_t,\ t \in (0,1])$. Hence the considered conditional probability does not depend on the values $\mathsf{B}_{-j_1}^K = b_{-j_1}, \dots, \mathsf{B}_{-j_n}^K = b_{-j_n}$. This shows (15.31).

Furthermore, by (6.1), under the condition $\mathsf{B}_0^K = 1$, the probability for $\mathsf{B}_n^K = 1$ is the probability that K is a subset of the zero-cell of $a^n Y_{a^n-1}$. By (5.6) and the spatial consistency of STIT tessellations we have that because $\Theta([a^{-n}K]) = a^{-n}\Theta([K])$, this probability equals $\exp(-(1-a^{-n})\,\Theta([K]))$. This completes the proof. □

Choosing $b_{-j_1} = \dots = b_{-j_n} = 1$ in (15.31), we obtain the equation

$$\begin{aligned} &\mathbb{P}(\mathsf{B}_1^K = b_1, \dots, \mathsf{B}_k^K = b_k \mid \mathsf{B}_0^K = 1, \mathsf{B}_{-j_1}^K = 1, \dots, \mathsf{B}_{-j_n}^K = 1) \\ &\quad = \mathbb{P}(\mathsf{B}_1^K = b_1, \dots, \mathsf{B}_k^K = b_k \mid \mathsf{B}_0^K = 1)\,, \end{aligned}$$

which shows a **regenerative property at the** 1**-values** of the process $(\mathsf{B}_j^K,\ j \in \mathbb{Z})$ as considered in Section 3.7 of [167].

15.6 Bernoulli shifts

Now we turn to an alternative conception of characterizing weak dependencies in time-stationary discrete-time stochastic processes. A key reference for this approach is [156].

15.6.1 Bernoulli shifts and their factors

Let $(\Omega, \mathbf{A}, \mathbb{P})$ be a probability space, and let $\alpha : \Omega \to \Omega$ a measurable measure-preserving transformation, that is, $\mathbb{P}(\alpha^{-1}(B)) = \mathbb{P}(B)$ for all $B \in \mathbf{A}$. Thus, with the abelian group $\{\alpha^j : j \in \mathbb{Z}\}$ of transformations, we obtain a dynamical system in the sense of Definition 15.1. For short, we denote it by $(\Omega, \mathbf{A}, \mathbb{P}, \alpha)$.

Recall that a dynamical system $(\Omega, \mathbf{A}, \mathbb{P}, \alpha)$ is

- **mixing** if $\lim_{n\to\infty} \mathbb{P}((\alpha^{-n}A) \cap B) = \mathbb{P}(A)\,\mathbb{P}(B)$ for all $A, B \in \mathbf{A}$,
- **ergodic** if $\mathbb{P}(B \triangle \alpha^{-1}B) = 0$ implies $\mathbb{P}(B)\,\mathbb{P}(B^c) = 0$, $B \in \mathbf{A}$, which is equivalent to the $\mathbb{P}$-triviality of the σ-algebra of α-invariant sets.

A mixing dynamical system is ergodic.

Note that in Section 15.1.1, these definitions were adapted to the group of translations acting on $\mathbb{R}^d$ and $\mathcal{T}$, respectively.

Definition 15.37. *Let $(\Omega, \mathbf{A}, \mathbb{P}, \alpha)$ and $(\Omega', \mathbf{A}', \mathbb{P}', \alpha')$ be dynamical systems with measurable measure-preserving transformations α and α'.*

(a) *A **factor map** from $(\Omega, \mathbf{A}, \mathbb{P}, \alpha)$ to $(\Omega', \mathbf{A}', \mathbb{P}', \alpha')$ is a measurable map* $\mathrm{b} : \Omega \to \Omega'$ *such that* $\mathrm{b} \circ \alpha = \alpha' \circ \mathrm{b}$ *$\mathbb{P}$-a. s. and* $\mathbb{P}(\mathrm{b}^{-1}(B')) = \mathbb{P}'(B')$ *for all* $B' \in \mathbf{A}'$.
If a factor map b *is a bimeasurable bijection, which means that also its inverse* b^{-1} *is measurable, then it is called an **isomorphism**.*

(b) *The dynamical system $(\Omega', \mathbf{A}', \mathbb{P}', \alpha')$ is called a **factor** of $(\Omega, \mathbf{A}, \mathbb{P}, \alpha)$ if a factor map from $(\Omega, \mathbf{A}, \mathbb{P}, \alpha)$ to $(\Omega', \mathbf{A}', \mathbb{P}', \alpha')$ exists.*

(c) *The dynamical systems $(\Omega, \mathbf{A}, \mathbb{P}, \alpha)$ and $(\Omega', \mathbf{A}', \mathbb{P}', \alpha')$ are **isomorphic** if there exists an isomorphism between them.*

A particular class of dynamical systems are the **shift systems.** Let $(\mathcal{S}, \mathbf{S})$ be a measurable space. Define the shift transformation $\alpha : \mathcal{S}^{\mathbb{Z}} \to \mathcal{S}^{\mathbb{Z}}$ by

$$(\alpha(S_i, i \in \mathbb{Z}))_j := S_{j+1}, \quad j \in \mathbb{Z}.$$

If a probability measure $\mathbb{P}_{\mathbf{S}}$ on $(\mathcal{S}^{\mathbb{Z}}, \mathbf{S}^{\otimes \mathbb{Z}})$ is preserved by α, then the dynamical system $(\mathcal{S}^{\mathbb{Z}}, \mathbf{S}^{\otimes \mathbb{Z}}, \mathbb{P}_{\mathbf{S}}, \alpha)$ is a two-sided shift system.

For a probability measure $\mathbb{P}$ on $(\mathcal{S}, \mathbf{S})$, consider the special shift system

$$(\mathcal{S}^{\mathbb{Z}}, \mathbf{S}^{\otimes \mathbb{Z}}, \mathbb{P}^{\otimes \mathbb{Z}}, \alpha),$$

where the product measure $\mathbb{P}^{\otimes \mathbb{Z}}$ is defined in the usual way via cylinder sets. If $(\mathcal{S}, \mathbf{S}, \mathbb{P})$ is a Lebesgue probability space (see Definition 18.14), then according to [156], Part I, Section 9, the shift system is called a **generalized two-sided Bernoulli shift.**

In this section, we consider the generalized two-sided Bernoulli shifts

$$((\mathcal{T}_W^{\mathbb{N}_0})^{\mathbb{Z}}, \mathcal{B}((\mathcal{T}_W^{\mathbb{N}_0})^{\mathbb{Z}}), (\mathbb{P}_{Y_1 \wedge W}^{\otimes \mathbb{N}_0})^{\otimes \mathbb{Z}}, \alpha) \tag{15.32}$$

and

$$((\mathcal{T}^{\mathbb{N}_0})^{\mathbb{Z}}, \mathcal{B}((\mathcal{T}^{\mathbb{N}_0})^{\mathbb{Z}}), (\mathbb{P}_{Y_1}^{\otimes \mathbb{N}_0})^{\otimes \mathbb{Z}}, \alpha),$$

respectively, for a STIT tessellation Y_1 and a window $W \in \mathcal{P}_d$.

Indeed, by Lemma 18.15 the completed probability spaces with $\Omega = \mathcal{T}$ or $\Omega = \mathcal{T}_W$ for $W \in \mathcal{P}_d$ and a probability measure $\mathbb{P}$ are Lebesgue probability spaces. Furthermore, the completed product space of a countable family of Lebesgue probability spaces is a Lebesgue probability space; see Theorem 2.4.3 of [84]. Thus

$$(\mathcal{T}_W^{\mathbb{N}_0}, \mathcal{B}(\mathcal{T}_W^{\mathbb{N}_0}), \mathbb{P}_{Y_1 \wedge W}^{\otimes \mathbb{N}_0})$$

and

$$(\mathcal{T}^{\mathbb{N}_0}, \mathcal{B}(\mathcal{T}^{\mathbb{N}_0}), \mathbb{P}_{Y_1}^{\otimes \mathbb{N}_0})$$

are Lebesgue probability spaces if they are completed with respect to the distributions $\mathbb{P}_{Y_1 \wedge W}^{\otimes \mathbb{N}_0}$ or $\mathbb{P}_{Y_1}^{\otimes \mathbb{N}_0}$, respectively, for sequences of i. i. d. STIT tessellations with law $\mathbb{P}_{Y_1}$. In the following, we will tacitly assume that all probability spaces are completed.

15.6.2 The STIT tessellation process in a window

Let $(Y_t, t > 0)$ be a STIT tessellation process, driven by a hyperplane measure Θ of the form (4.1), satisfying (4.3). Furthermore, let $a > 1$. By Theorem 6.17 the process $(a^t Y_{a^t}, t \in \mathbb{R})$ is time-stationary. In this section, we consider the tessellation-valued discrete-time stochastic process

$$\mathbf{U} := (U_m, m \in \mathbb{Z}) := (a^m Y_{a^m}, m \in \mathbb{Z}) . \tag{15.33}$$

The main result relates the shift system associated with the process $\mathbf{U} \wedge W$ for some fixed $W \in \mathcal{P}_d$ to the Bernoulli shift (15.32).

Consider $\underline{\mathbf{Y}} := (\underline{Y}^m, m \in \mathbb{Z}) \in (\mathcal{T}^{\mathbb{N}_0})^{\mathbb{Z}}$ with $\underline{Y}^m = (Y^{(n),m}, n \in \mathbb{N}_0)$, where $Y^{(n),m}$ are i. i. d. with $Y^{(n),m} \stackrel{D}{=} Y_1$, $m \in \mathbb{Z}$, $n \in \mathbb{N}_0$, and assume the independence of $\mathbf{U}$. It will be convenient to arrange the elements in an array,

$$\begin{array}{cccccccc}
\dots & Y^{(0),-3} & Y^{(0),-2} & Y^{(0),-1} & Y^{(0),0} & Y^{(0),1} & Y^{(0),2} & \dots \\
\dots & Y^{(1),-3} & Y^{(1),-2} & Y^{(1),-1} & Y^{(1),0} & Y^{(1),1} & Y^{(1),2} & \dots \\
\dots & Y^{(2),-3} & Y^{(2),-2} & Y^{(2),-1} & Y^{(2),0} & Y^{(2),1} & Y^{(2),2} & \dots \\
 & \vdots & \vdots & \vdots & \vdots & \vdots & \vdots &
\end{array} \tag{15.34}$$

Adapting (15.30) to this notation, we have that for all $m \in \mathbb{Z}$,

$$(U_{m+k}, k \in \mathbb{N}_0) \stackrel{D}{=} \left(a^k U_m \boxplus_{i=1}^k \frac{a^{k+1-i}}{a-1} \underline{Y}^{m+i}, k \in \mathbb{N}_0 \right). \tag{15.35}$$

Note that $\underline{Y}^{m+i}$ is the $(m+i)$th column in the array (15.34). In what follows, we assume that the tessellation with the upper index (0), that is, $Y^{(0),m}$ or $T^{(0),m}$, $m \in \mathbb{Z}$, is always nested into the cell that contains the origin.

For a given window $W \in \mathcal{P}_d$, a key property of $\mathbf{U} \wedge W$ is that the scaling by a^k, $k \in \mathbb{Z}$, causes that only “finitely many states of the past” appear and “the history of the process diminishes from the window”. This means that there are infinitely many times $\ell_i \in \mathbb{Z}$, such that $U_{\ell_i} \wedge W = \{W\}$, and after ℓ_i, by (15.35) the distribution of the future of the process can be described using i. i. d. copies of Y_1. This fact was also used in Section 15.4, where it was shown that the rescaled STIT tessellation process in a window W is β-mixing in time.

Denote the shift of a sequence

$$\alpha_{\mathcal{T}_W} : \mathcal{T}_W^{\mathbb{Z}} \to \mathcal{T}_W^{\mathbb{Z}} \quad \text{by } (\alpha_{\mathcal{T}_W}(T^k, k \in \mathbb{Z}))^m := T^{m+1}$$

for all $m \in \mathbb{Z}$. Correspondingly, for the shift of the columns in the array (15.34),

$$\alpha : (\mathcal{T}_W^{\mathbb{N}_0})^{\mathbb{Z}} \to (\mathcal{T}_W^{\mathbb{N}_0})^{\mathbb{Z}} \quad \text{by } (\alpha(\underline{T}^k, k \in \mathbb{Z}))^m := \underline{T}^{m+1},$$

which is the shift transformation in (15.32).

Theorem 15.38. *Let $(Y_t, t > 0)$ be a STIT tessellation process, driven by a hyperplane measure Θ of the form* (4.1), *satisfying* (4.3). *Furthermore, let $a > 1$, let $W \in \mathcal{P}_d$ with $o \in \operatorname{int} W$, and let $\mathbf{U}$ be the discrete-time tessellation-valued process defined in* (15.33). *Then the shift system*

$$(\mathcal{T}_W^{\mathbb{Z}}, \mathcal{B}(\mathcal{T}_W^{\mathbb{Z}}), \mathbb{P}_{\mathbf{U}\wedge W}, \alpha_{\mathcal{T}_W})$$

is a factor of the generalized two-sided Bernoulli shift

$$((\mathcal{T}_W^{\mathbb{N}_0})^{\mathbb{Z}}, \mathcal{B}((\mathcal{T}_W^{\mathbb{N}_0})^{\mathbb{Z}}), (\mathbb{P}_{Y_1\wedge W}^{\otimes\mathbb{N}_0})^{\otimes\mathbb{Z}}, \alpha),$$

that is, there exists a factor map $\mathrm{b} : (\mathcal{T}_W^{\mathbb{N}_0})^{\mathbb{Z}} \to \mathcal{T}_W^{\mathbb{Z}}$ satisfying

$$\alpha_{\mathcal{T}_W} \circ \mathrm{b} = \mathrm{b} \circ \alpha \quad (\mathbb{P}_{Y_1\wedge W}^{\otimes\mathbb{N}_0})^{\otimes\mathbb{Z}}\text{-a. s.} \tag{15.36}$$

and

$$(\mathbb{P}_{Y_1\wedge W}^{\otimes\mathbb{N}_0})^{\otimes\mathbb{Z}} \circ \mathrm{b}^{-1} = \mathbb{P}_{\mathbf{U}\wedge W}. \tag{15.37}$$

To prepare the proof, we show some preliminary results. For short, we write $(\ell_i \downarrow, i \in \mathbb{N})$ for a strictly decreasing sequence of $\ell_i \in \mathbb{Z}^-$. Recall that $T \wedge W = \{W\}$ if and only if $\partial T \cap \operatorname{int} W = \emptyset$ for a tessellation $T \in \mathcal{T}$.

We start with some results for probabilities that the zero-cell of the STIT tessellation Y_1 dilated by a growing factor contains a fixed window W.

Lemma 15.39. *Let $a > 1$, let $W \in \mathcal{P}_d$ with $o \in \operatorname{int} W$, and let $(Y^{(0),j}, j \in \mathbb{Z}^-)$ be a sequence of independent copies of the STIT tessellation Y_1. Then*

$$\mathbb{P}(\partial Y^{(0),j} \cap bW = \emptyset) = \mathbb{P}(\partial Y_1 \cap W = \emptyset)^b \quad \textit{for all } j \in \mathbb{Z}^- \textit{ and } b > 0, \tag{15.38}$$

and

$$\mathbb{P}\left(\forall j \in \mathbb{Z}^- : \frac{a^{-j+1}}{a-1}\partial Y^{(0),j} \cap W = \emptyset\right) = \mathbb{P}(\partial Y_1 \cap W = \emptyset) > 0. \tag{15.39}$$

Furthermore, there exists a. s. a sequence $(\ell_i \downarrow, i \in \mathbb{N})$, $\ell_i \in \mathbb{Z}^-$, *such that for all* $i \in \mathbb{N}$ *and* $j \in \mathbb{Z}^-$, *we have that*

$$\frac{a^{-j+1}}{a-1} \partial Y^{(0),\ell_i+j} \cap W = \emptyset ,$$

that is,

$$\mathbb{P}\left(\exists (\ell_i \downarrow, i \in \mathbb{N})\, \forall i \in \mathbb{N}\, \forall j \in \mathbb{Z}^- : \frac{a^{-j+1}}{a-1} \partial Y^{(0),\ell_i+j} \cap W = \emptyset \right) = 1 . \tag{15.40}$$

Proof. The measurability of the events considered in this lemma follows from Lemma 18.7 and (18.1). For all $j \in \mathbb{Z}^-$ and $b > 0$, by (5.6), the spatial consistency (5.18) of STIT tessellations, and (4.1) we have

$$\mathbb{P}(\partial Y^{(0),j} \cap bW = \emptyset) = \mathrm{e}^{-\Theta([bW])} = \mathrm{e}^{-b\,\Theta([W])} = \mathbb{P}(\partial Y_1 \cap W = \emptyset)^b ,$$

which shows (15.38).

Furthermore,

$$\begin{aligned}
&\mathbb{P}\left(\forall j \in \mathbb{Z}^- : \frac{a^{-j+1}}{a-1} \partial Y^{(0),j} \cap W = \emptyset \right) \\
&= \lim_{m \to -\infty} \mathbb{P}\left(\forall j \in \{m, \dots, 0\} : \frac{a^{-j+1}}{a-1} \partial Y^{(0),j} \cap W = \emptyset \right) \\
&= \lim_{m \to -\infty} \prod_{j=m}^{0} \exp\left(-\frac{a-1}{a^{-j+1}} \Theta([W]) \right) \\
&= \lim_{m \to -\infty} \exp\left(-\frac{a-1}{a} \Theta([W]) \sum_{j=m}^{0} a^j \right) \\
&= \exp(-\Theta([W])) \\
&= \mathbb{P}(\partial Y_1 \cap W = \emptyset) ,
\end{aligned}$$

which proves (15.39).

To show (15.40), define the event

$$A := \left\{ (Y^{(0),j}, j \in \mathbb{Z}^-) \in \mathcal{T}^{\mathbb{Z}^-} : \frac{a^{-j+1}}{a-1} \partial Y^{(0),j} \cap W = \emptyset \text{ for all } j \in \mathbb{Z}^- \right\} .$$

By (15.39) we have $\mathbb{P}_{Y_1}^{\otimes \mathbb{Z}^-}(A) > 0$.

Consider the Bernoulli shift $(\mathcal{T}^{\mathbb{Z}}, \mathbb{P}_{Y_1}^{\otimes \mathbb{Z}}, \alpha_{\mathcal{T}})$ with $\alpha_{\mathcal{T}} : \mathcal{T}^{\mathbb{Z}} \to \mathcal{T}^{\mathbb{Z}}$ and $(\alpha_{\mathcal{T}}(Y^{(0),k}, k \in \mathbb{Z}))^m := Y^{(0),m+1}$ for all $m \in \mathbb{Z}$. Since Bernoulli shifts are ergodic, the Birkhoff ergodic theorem (see Theorem 25.6 of [89]) yields

$$\lim_{N \to \infty} \frac{1}{N} \sum_{i=0}^{N-1} \mathbb{1}\{\alpha_{\mathcal{T}}^{-i}(Y^{(0),j}, j \in \mathbb{Z}^-) \in A\} = \mathbb{P}_{Y_1}^{\otimes \mathbb{Z}^-}(A) > 0 \quad \mathbb{P}_{Y_1}^{\otimes \mathbb{Z}^-}\text{-a. s.}$$

Hence $\mathbb{P}_{Y_1}^{\otimes\mathbb{Z}^-}$-a. s. there exists a strictly decreasing sequence $(\ell_i\downarrow, i\in\mathbb{N})$ of negative integers such that $\alpha^{\ell_i}(Y^{(0),j}, j\in\mathbb{Z}^-)\in A$ for all $i\in\mathbb{N}$. This is (15.40). □

Now we prepare the definition of a factor map $\mathrm{b}=(\mathrm{b}_k, k\in\mathbb{Z})$. For fixed $W\in\mathcal{P}_d$, let $\underline{\mathbf{T}}=(\underline{T}^m, m\in\mathbb{Z})\in(\mathcal{T}_W^{\mathbb{N}_0})^{\mathbb{Z}}$ with $\underline{T}^m=(T^{(n),m}, n\in\mathbb{N}_0)$, represented in a array corresponding to (15.34).

By (15.40), for $(\mathbb{P}_{Y_1\wedge W}^{\otimes\mathbb{N}_0})^{\otimes\mathbb{Z}}$-almost all $\underline{\mathbf{T}}\in(\mathcal{T}_W^{\mathbb{N}_0})^{\mathbb{Z}}$, there exists a sequence $(\ell_i\downarrow, i\in\mathbb{N})$ of negative integers, depending on $\underline{\mathbf{T}}$, such that

$$\frac{a^{-j+1}}{a-1}T^{(0),\ell_i+j}\wedge W=\{W\}\quad\text{for all } i\in\mathbb{N}\text{ and } j\in\mathbb{Z}^-.$$

Denote by $T^{(0)}:=(T^{(0),m}, m\in\mathbb{Z})$ the first row of $\underline{\mathbf{T}}$ in the array corresponding to (15.34), and define the index set

$$J(\underline{\mathbf{T}}):=\left\{m\in\mathbb{Z}:\frac{a^{-j+1}}{a-1}T^{(0),m+j}\wedge W=\{W\}\text{ for all } j\in\mathbb{Z}^-\right\}. \tag{15.41}$$

The elements of $J(\underline{\mathbf{T}})$ will be used to partition the array $\underline{\mathbf{T}}$ into blocks of columns and this partition is the starting point for the definition of a factor map. For all $k\in\mathbb{Z}$, define the nonnegative integer

$$d(k):=d(k,\underline{\mathbf{T}}):=k-\max\{m\in J(\underline{\mathbf{T}}): m\le k\}, \tag{15.42}$$

which is a. s. well defined, because (15.40) guarantees that $(\mathbb{P}_{Y_1\wedge W}^{\otimes\mathbb{N}_0})^{\otimes\mathbb{Z}}$-a. s. for all $k\in\mathbb{Z}$, there is $m\in J(\underline{\mathbf{T}})$ such that $m\le k$. Thus the mappings

$$\underline{\mathbf{T}}\quad\mapsto\quad(\mathbb{1}\{k\in J(\underline{\mathbf{T}})\}, k\in\mathbb{Z})\quad\mapsto\quad(d(k), k\in\mathbb{Z}) \tag{15.43}$$

from $(\mathcal{T}_W^{\mathbb{N}_0})^{\mathbb{Z}}$ to $\{0,1\}^{\mathbb{Z}}$ and then further to $\mathbb{N}_0^{\mathbb{Z}}$ are well-defined.

The maps b_k, $k\in\mathbb{Z}$, will be defined now using the number $d(k)$ and iteration of tessellations. For a fixed array $\underline{\mathbf{T}}=(\underline{T}^m, m\in\mathbb{Z})\in(\mathcal{T}_W^{\mathbb{N}_0})^{\mathbb{Z}}$, we tacitly assume that for all the tessellations, a numbering of their cells is fixed and thus all the considered iterations are defined by (3.2). However, for simplicity of notation, we do not indicate these numberings here, as they do not play a role when the maps b_k are applied to i. i. d. STIT tessellations; see (3.4). Adapting the notation of (15.29), we define for all $k\in\mathbb{Z}$,

$$\mathrm{b}_k(\underline{\mathbf{T}}):=\{W\}\boxplus_{i=1}^{d(k)}\frac{a^{d(k)+1-i}}{a-1}\underline{T}^{k-d(k)+i}, \tag{15.44}$$

putting $\mathrm{b}_k(\underline{\mathbf{T}}):=\{W\}$ for $k\in J(\underline{\mathbf{T}})$. Note that the $\underline{T}^{k-d(k)+i}$ are *columns* of $\underline{\mathbf{T}}$ in the array corresponding to (15.34). The measurability of the above-mentioned maps follows by Lemmas 18.10 and 3.3 from the measurability of the events considered in Lemma 15.39.

In the next step, we prove a kind of a recursion formula for the maps b_k, $k\in\mathbb{Z}$.

Lemma 15.40. *Let* $(\mathsf{b}_k,\ k \in \mathbb{Z})$ *be the sequence of maps defined by* (15.44). *Then, for* $(\mathbb{P}_{Y_1\wedge W}^{\otimes\mathbb{N}_0})^{\otimes\mathbb{Z}}$*-almost all* $\underline{\mathbf{T}} \in (\mathcal{T}_W^{\mathbb{N}_0})^{\mathbb{Z}}$,

$$\mathsf{b}_k(\underline{\mathbf{T}}) = \left(a^{k-k'}\mathsf{b}_{k'}(\underline{\mathbf{T}}) \boxplus_{i=1}^{k-k'} \frac{a^{k-k'+1-i}}{a-1}\,\underline{T}^{k'+i}\right) \wedge W \tag{15.45}$$

for all $k', k \in \mathbb{Z}$ *such that* $k' < k$. *Furthermore, for all* $k \in \mathbb{Z}$,

$$\mathsf{b}_k(\underline{\mathbf{T}}) = \{W\} \quad \text{if and only if} \quad k \in J(\underline{\mathbf{T}})\,. \tag{15.46}$$

Proof. By the definitions of $\mathsf{b}_{k'}$ and of iteration

$$\begin{aligned}
&\left(a^{k-k'}\mathsf{b}_{k'}(\underline{\mathbf{T}}) \boxplus_{i=1}^{k-k'} \frac{a^{k-k'+1-i}}{a-1}\,\underline{T}^{k'+i}\right) \wedge W \\
&= \left(a^{k-k'}\left(\{W\} \boxplus_{i=1}^{d(k')} \frac{a^{d(k')+1-i}}{a-1}\,\underline{T}^{k'-d(k')+i}\right) \boxplus_{\ell=1}^{k-k'} \frac{a^{k-k'+1-\ell}}{a-1}\,\underline{T}^{k'+\ell}\right) \wedge W \\
&= \left(\{W\} \boxplus_{i=1}^{d(k')} \frac{a^{k-k'+d(k')+1-i}}{a-1}\,\underline{T}^{k'-d(k')+i}\right) \boxplus_{\ell=1}^{k-k'} \frac{a^{k-k'+1-\ell}}{a-1}\,\underline{T}^{k'+\ell} \\
&= \{W\} \boxplus_{i=1}^{k-k'+d(k')} \frac{a^{k-k'+d(k')+1-i}}{a-1}\,\underline{T}^{k'-d(k')+i},
\end{aligned}$$

where in the last step the index ℓ is shifted to $i := d(k') + \ell$.

If $\max\{m \in J(\underline{\mathbf{T}}) : m \le k'\} = \max\{m \in J(\underline{\mathbf{T}}) : m \le k\}$ and hence $k - d(k) = k' - d(k')$, then

$$\begin{aligned}
\{W\} \boxplus_{i=1}^{k-k'+d(k')} \frac{a^{k-k'+d(k')+1-i}}{a-1}\,\underline{T}^{k'-d(k')+i} &= \{W\} \boxplus_{i=1}^{d(k)} \frac{a^{d(k)+1-i}}{a-1}\,\underline{T}^{k-d(k)+i} \\
&= \mathsf{b}_k(\underline{\mathbf{T}})\,.
\end{aligned}$$

Otherwise, if $m_k := \max\{m \in J(\underline{\mathbf{T}}) : m \le k\} > \max\{m \in J(\underline{\mathbf{T}}) : m \le k'\}$, then substitute $\ell := k' - d(k') - m_k + i$ and observe that by the definition of $J(\underline{\mathbf{T}})$

$$\begin{aligned}
\{W\} \boxplus_{\ell=k'-d(k')-m_k+1}^{0} \frac{a^{k-m_k-\ell+1}}{a-1}\,\underline{T}^{m_k+\ell} &= \{W\} \boxplus_{\ell=k'-d(k')-m_k+1}^{0} a^{k-m_k}\frac{a^{-\ell+1}}{a-1}\,\underline{T}^{m_k+\ell} \\
&= \{W\}\,.
\end{aligned}$$

Using this result, we obtain that

$$\begin{aligned}
&\{W\} \boxplus_{i=1}^{k-k'+d(k')} \frac{a^{k-k'+d(k')+1-i}}{a-1}\,\underline{T}^{k'-d(k')+i} \\
&= \{W\} \boxplus_{\ell=k'-d(k')-m_k+1}^{k-m_k} \frac{a^{k-m_k-\ell+1}}{a-1}\,\underline{T}^{m_k+\ell} \\
&= \left(\{W\} \boxplus_{\ell=k'-d(k')-m_k+1}^{0} \frac{a^{k-m_k-\ell+1}}{a-1}\,\underline{T}^{m_k+\ell}\right) \boxplus_{\ell=1}^{k-m_k} \frac{a^{k-m_k-\ell+1}}{a-1}\,\underline{T}^{m_k+\ell}
\end{aligned}$$

$$= \{W\} \boxplus_{\ell=1}^{k-m_k} \frac{a^{k-m_k-\ell+1}}{a-1} \underline{T}^{m_k+\ell}$$
$$= \{W\} \boxplus_{\ell=1}^{d(k)} \frac{a^{d(k)+1-\ell}}{a-1} \underline{T}^{k-d(k)+\ell}$$
$$= \mathsf{b}_k(\underline{\mathbf{T}}) .$$

Thus (15.45) is shown.

To prove (15.46), first note that by definition $\mathsf{b}_k(\underline{\mathbf{T}}) = \{W\}$ if $k \in J(\underline{\mathbf{T}})$. On the other hand, assume that $k \in \mathbb{Z} \setminus J(\underline{\mathbf{T}})$. Then there exists $\ell \in \mathbb{Z}^-$ such that $\frac{a^{-\ell+1}}{a-1} T^{(0),k+\ell} \wedge W \neq \{W\}$. Hence also

$$\left(a^{-\ell+1} \mathsf{b}_{k+\ell-1}(\underline{\mathbf{T}}) \boxplus \frac{a^{-\ell+1}}{a-1} \underline{T}^{k+\ell} \right) \wedge W \neq \{W\} .$$

Then by (15.45) with $k' = k + \ell - 1$

$$\mathsf{b}_k(\underline{\mathbf{T}}) = \left(\left(a^{-\ell+1} \mathsf{b}_{k+\ell-1}(\underline{\mathbf{T}}) \boxplus \frac{a^{-\ell+1}}{a-1} \underline{T}^{k+\ell} \right) \boxplus_{i=1}^{-\ell} \frac{a^{-\ell+1-i}}{a-1} \underline{T}^{k+\ell+i} \right) \wedge W \neq \{W\} ,$$

which completes the proof. □

By (15.39) and (15.46) we obtain that for all $k \in \mathbb{Z}$,

$$\mathbb{P}(\mathsf{b}_k(\underline{\mathbf{Y}}) = \{W\}) = \mathbb{P}(\partial Y_1 \cap W = \emptyset) = \mathbb{P}(U_k \wedge W = \{W\}) . \tag{15.47}$$

Now we are prepared for the proof of the main result of this section.

Proof of Theorem 15.38. We show that $\mathsf{b} := (\mathsf{b}_k, k \in \mathbb{Z}) : (\mathcal{T}_W^{\mathbb{N}_0})^{\mathbb{Z}} \to \mathcal{T}_W^{\mathbb{Z}}$, defined in (15.44), satisfies (15.36) and (15.37) and hence is a factor map.

By definition (15.43), for all $k \in \mathbb{Z}$ and $(\mathbb{P}_{Y_1 \wedge W}^{\otimes \mathbb{N}_0})^{\otimes \mathbb{Z}}$-almost all $\underline{\mathbf{T}}$, we have that

$$d(k+1, \underline{\mathbf{T}}) = d(k, \alpha(\underline{\mathbf{T}})) ,$$

and hence

$$\begin{aligned}
\alpha_{\mathcal{T}_W}(\mathsf{b}(\underline{\mathbf{T}}))_k &= \mathsf{b}_{k+1}(\underline{\mathbf{T}}) \\
&= \{W\} \boxplus_{i=1}^{d(k+1,\underline{\mathbf{T}})} \frac{a^{d(k+1,\underline{\mathbf{T}})+1-i}}{a-1} \underline{T}^{k-d(k+1,\underline{\mathbf{T}})+i} \\
&= \{W\} \boxplus_{i=1}^{d(k,\alpha(\underline{\mathbf{T}}))} \frac{a^{d(k,\alpha(\underline{\mathbf{T}}))+1-i}}{a-1} \underline{T}^{k-d(k,\alpha(\underline{\mathbf{T}}))+i} \\
&= \mathsf{b}_k(\alpha(\underline{\mathbf{T}})) ,
\end{aligned} \tag{15.48}$$

which proves the commuting property (15.36).

Now we show (15.37). We write $\underline{\mathbf{Y}} \wedge W := (Y^{(n),m} \wedge W,\, n \in \mathbb{N}_0,\, m \in \mathbb{Z})$. By the time-stationarity of $(U_m,\, m \in \mathbb{Z})$ and the shift invariance of the Bernoulli shift, it is sufficient to show that for all $\ell \in \mathbb{N}_0$ and $B_n \in \mathcal{B}(\mathcal{T}_W)$, $n \in \{0,\dots,\ell\}$,

$$\mathbb{P}(\mathrm{b}_n(\underline{\mathbf{Y}} \wedge W) \in B_n \,, n \in \{0,\dots,\ell\}) = \mathbb{P}(U_n \wedge W \in B_n \,, n \in \{0,\dots,\ell\}) \,.$$

Regarding (15.35) and (15.45), it is even sufficient to prove that for all $B \in \mathcal{B}(\mathcal{T}_W)$,

$$\mathbb{P}(\mathrm{b}_0(\underline{\mathbf{Y}} \wedge W) \in B) = \mathbb{P}(U_0 \wedge W \in B) \,. \tag{15.49}$$

By Theorem 15.33 the process $\mathbf{U} \wedge W$ is β-mixing in time and hence is ergodic. The Birkhoff ergodic theorem (see Theorem 25.6 of [89]) yields

$$\lim_{N\to\infty} \frac{1}{N} \sum_{i=0}^{N-1} \mathbb{1}\{\partial U_{-i} \cap W = \emptyset\} = \mathbb{P}(\partial U_0 \cap W = \emptyset) > 0 \quad \mathbb{P}\text{-a. s.}$$

Hence, using the symbol $\uparrow$ for increasing sequences in $\mathbb{N}$,

$$\mathbb{P}(\exists (n_i \uparrow,\, i \in \mathbb{N})\, \forall i \in \mathbb{N} : \partial U_{-n_i} \cap W = \emptyset) = 1$$

and in particular,

$$\mathbb{P}(\exists k \in \mathbb{Z}^- : \partial U_k \cap W = \emptyset) = \mathbb{P}(\exists k \in \mathbb{Z}^- : U_k \wedge W = \{W\}) = 1 \,.$$

Partitioning the event as

$$\begin{aligned} &\{\exists k \in \mathbb{Z}^- : U_k \wedge W = \{W\}\} \\ &\quad = \bigcup_{\ell \in \mathbb{Z}^-} \{U_\ell \wedge W = \{W\}, U_j \wedge W \neq \{W\}, j \in \{\ell+1,\dots,0\}\} \,, \end{aligned}$$

we can write

$$\begin{aligned} &\mathbb{P}(U_0 \wedge W \in B) \\ &\quad = \sum_{\ell=-\infty}^{0} \mathbb{P}(U_0 \wedge W \in B \,, U_j \wedge W \neq \{W\}, j \in \{\ell+1,\dots,0\} \,|\, U_\ell \wedge W = \{W\}) \\ &\quad \times \mathbb{P}(U_\ell \wedge W = \{W\}) \,. \end{aligned} \tag{15.50}$$

If $U_\ell \wedge W = \{W\}$ for some $\ell \in \mathbb{Z}^-$, then by (15.35)

$$(U_{l+1} \wedge W, \dots, U_0 \wedge W) \overset{\mathrm{D}}{=} \left(\{W\} \boxplus \frac{a}{a-1} \underline{Y}^{\ell+1}, \dots, \{W\} \boxplus_{i=1}^{-\ell} \frac{a^{-\ell+1-i}}{a-1} \underline{Y}^{\ell+i} \right).$$

On the other hand, if $\mathrm{b}_\ell(\underline{\mathbf{Y}} \wedge W) = \{W\}$ for some $\ell \in \mathbb{Z}^-$, then $\ell \in J(\underline{\mathbf{Y}} \wedge W)$ by (15.46). Thus definition (15.44) yields that

$$\{\mathrm{b}_{\ell+1}(\underline{\mathbf{Y}} \wedge W) \neq \{W\}, \ldots, \mathrm{b}_0(\underline{\mathbf{Y}} \wedge W) \neq \{W\}\}$$
$$= \left\{\{W\} \boxplus \frac{a}{a-1} Y^{\ell+1} \neq \{W\}, \ldots, \{W\} \boxplus_{i=1}^{-\ell} \frac{a^{-\ell+1-i}}{a-1} Y^{\ell+i} \neq \{W\}\right\}.$$

Now with (15.50) and (15.47), we arrive at

$$\mathbb{P}(U_0 \wedge W \in B) = \sum_{\ell=-\infty}^{0} \mathbb{P}(\mathrm{b}_0(\underline{\mathbf{Y}} \wedge W) \in B, \mathrm{b}_j(\underline{\mathbf{Y}} \wedge W) \neq \{W\},$$
$$j \in \{\ell+1, \ldots, 0\} \mid \mathrm{b}_\ell(\underline{\mathbf{Y}} \wedge W) = \{W\})\, \mathbb{P}(\mathrm{b}_\ell(\underline{\mathbf{Y}} \wedge W) = \{W\})$$
$$= \mathbb{P}(\mathrm{b}_0(\underline{\mathbf{Y}} \wedge W) \in B).$$

Thus (15.49) is shown, and the proof is complete. □

Remark 15.41. *The construction in the proof of Theorem* 15.38 *implies that the factor map* b *is finitary with null anticipation, which means that for all* $W \in \mathcal{P}_d$ *and* $k \in \mathbb{Z}$, *the coordinate* $\mathrm{b}_k(\underline{\mathbf{Y}})$ *depends a. s. only on a finite number of columns of* $\underline{\mathbf{Y}}$ *with indices in* $\{n(k, \underline{\mathbf{Y}}), \ldots, k\}$, $n(k, \underline{\mathbf{Y}}) \le k$.

Corollary 15.42. *With the assumptions of Theorem* 15.38, *the shift system*

$$(\mathcal{T}_W^{\mathbb{Z}}, \mathcal{B}(\mathcal{T}_W^{\mathbb{Z}}), \mathbb{P}_{\mathbf{U} \wedge W}, \alpha_{\mathcal{T}_W})$$

is isomorphic to a generalized Bernoulli shift.

Proof. By Lemma 18.15 the completed probability space $(\mathcal{T}_W, \mathcal{B}(\mathcal{T}_W)_{\mathbb{P}}, \mathbb{P})$ is a Lebesgue probability space. With the probability measure $\mathbb{P}_{Y_1 \wedge W}$, it has the only atom $\{W\}$ with probability $\mathbb{P}_{Y_1 \wedge W}(\{W\}) = \mathrm{e}^{-\Theta([W])} < 1$. Due to the translation invariance of the measure Θ, defined in (4.1), and the construction of STIT tessellations in W (see Definition 4.2), there are no other atoms with respect to $\mathbb{P}_{Y_1 \wedge W}$. Thus it has infinite entropy; see page 48 of [156]. Hence, by Theorem 6 in Section 9 of [156], which states that a factor of a generalized Bernoulli shift of infinite entropy is isomorphic to a generalized Bernoulli shift, we obtain that $(\mathcal{T}_W^{\mathbb{Z}}, \mathcal{B}(\mathcal{T}_W^{\mathbb{Z}}), \mathbb{P}_{\mathbf{U} \wedge W}, \alpha_{\mathcal{T}_W})$ is isomorphic to a generalized Bernoulli shift. □

15.6.3 The STIT tessellation process in $\mathbb{R}^d$

The result of Theorem 15.38 can be extended to the STIT tessellation process in $\mathbb{R}^d$. The argument will be the a. s. spatial consistency property (15.57) of a sequence of tessellations and thus a stronger proposition than in Theorem 5.10, where spatial consistency in distribution is dealt with. In the present section, construction (15.44) of factor maps for windows plays the key role.

For a sequence of tessellations, denote the shift by

$$\alpha_{\mathcal{T}} : \mathcal{T}^{\mathbb{Z}} \to \mathcal{T}^{\mathbb{Z}} \quad \text{with } (\alpha_{\mathcal{T}}(T^k, k \in \mathbb{Z}))^m := T^{m+1}$$

for all $m \in \mathbb{Z}$. Furthermore, let $\underline{\mathbf{T}} = (\underline{T}^m, m \in \mathbb{Z}) \in (\mathcal{T}^{\mathbb{N}_0})^{\mathbb{Z}}$, analogously to array (15.34). Correspondingly, for the shift of the columns in array (15.34),

$$\alpha : (\mathcal{T}^{\mathbb{N}_0})^{\mathbb{Z}} \to (\mathcal{T}^{\mathbb{N}_0})^{\mathbb{Z}} \quad \text{with } (\alpha(\underline{T}^k, k \in \mathbb{Z}))^m := \underline{T}^{m+1}.$$

Theorem 15.43. *Let $(Y_t, t > 0)$ be a STIT tessellation process, driven by a hyperplane measure Θ of the form* (4.1), *satisfying* (4.3). *Furthermore, let $a > 1$, and let $\mathbf{U}$ be the scaled discrete-time process defined in* (15.33). *Then the shift system*

$$(\mathcal{T}^{\mathbb{Z}}, \mathcal{B}(\mathcal{T}^{\mathbb{Z}}), \mathbb{P}_{\mathbf{U}}, \alpha_{\mathcal{T}})$$

is a factor of the generalized Bernoulli shift

$$((\mathcal{T}^{\mathbb{N}_0})^{\mathbb{Z}}, \mathcal{B}((\mathcal{T}^{\mathbb{N}_0})^{\mathbb{Z}}), (\mathbb{P}_{Y_1}^{\otimes \mathbb{N}_0})^{\otimes \mathbb{Z}}, \alpha),$$

that is, there exists a factor map $\mathsf{b} : (\mathcal{T}^{\mathbb{N}_0})^{\mathbb{Z}} \to \mathcal{T}^{\mathbb{Z}}$ satisfying

$$\alpha_{\mathcal{T}} \circ \mathsf{b} = \mathsf{b} \circ \alpha \quad (\mathbb{P}_{Y_1}^{\otimes \mathbb{N}_0})^{\otimes \mathbb{Z}}\text{-a. s.} \tag{15.51}$$

and

$$(\mathbb{P}_{Y_1}^{\otimes \mathbb{N}_0})^{\otimes \mathbb{Z}} \circ \mathsf{b}^{-1} = \mathbb{P}_{\mathbf{U}}. \tag{15.52}$$

Proof. For an array $\underline{\mathbf{T}} = (\underline{T}^m, m \in \mathbb{Z}) \in (\mathcal{T}^{\mathbb{N}_0})^{\mathbb{Z}}$, corresponding to (15.34), denote by $T^{(0)} := (T^{(0),m}, m \in \mathbb{Z})$ the first row of $\underline{\mathbf{T}}$. Throughout the whole proof, a measurable numbering $\mathsf{n} := (\mathsf{n}_i, i \in \mathbb{N}_0)$, satisfying (3.1), of tessellation cells is fixed with the convention that the origin $o \in \mathsf{n}_0(T)$ for all $T \in \mathcal{T}$. Furthermore, consider a sequence $(W_n, n \in \mathbb{N})$ of windows $W_n \in \mathcal{P}_d$ such that $o \in \operatorname{int} W_1$, $W_n \subset W_{n+1}$, and $\bigcup_{n \in \mathbb{N}} W_n = \mathbb{R}^d$. Then for all $n \in \mathbb{N}$, define the index set

$$J(\underline{\mathbf{T}}, W_n) := \left\{ m \in \mathbb{Z} : \frac{a^{-j+1}}{a-1} T^{(0),m+j} \wedge W_n = \{W_n\} \text{ for all } j \in \mathbb{Z}^- \right\}, \tag{15.53}$$

analogously to (15.41). By definition it is evident that $J(\underline{\mathbf{T}}, W_{n+1}) \subseteq J(\underline{\mathbf{T}}, W_n)$, $n \in \mathbb{N}$.

According to Lemma 15.39, for $(\mathbb{P}_{Y_1}^{\otimes \mathbb{N}_0})^{\otimes \mathbb{Z}}$-almost all $\underline{\mathbf{T}}$, we have that for all $n \in \mathbb{N}$, the set $J(\underline{\mathbf{T}}, W_n) \cap \mathbb{Z}^-$ is infinite and hence contains a strictly decreasing sequence in $\mathbb{Z}^-$. Now fix $\underline{\mathbf{T}}$ with this property and denote

$$d(k,n) := k - \max\{m \in J(\underline{\mathbf{T}}, W_n) : m \le k\}, \quad k \in \mathbb{Z};$$

compare with (15.42). Then, analogously to (15.44), for $n \in \mathbb{N}$ and $k \in \mathbb{Z}$, define

$$\mathsf{b}_k(\underline{\mathbf{T}}, W_n) := \{W_n\} \boxplus_{i=1}^{d(k,n)} \frac{a^{d(k,n)+1-i}}{a-1} \underline{T}^{k-d(k,n)+i}. \tag{15.54}$$

In the next step, we show that for all $k \in \mathbb{Z}$ and $n \in \mathbb{N}$,

$$\mathsf{b}_k(\underline{\mathbf{T}}, W_{n+1}) \wedge W_n = \mathsf{b}_k(\underline{\mathbf{T}}, W_n). \tag{15.55}$$

To see this, fix $k \in \mathbb{Z}$ and $n \in \mathbb{N}$. If $d(k,n+1) = d(k,n)$, then (15.55) is evident by definition (15.54). Now consider the case $d(k,n+1) > d(k,n)$. Denote

$$m_i := \max\{m \in J(\underline{\mathbf{T}}, W_i) : m \le k\}, \quad i \in \{n, n+1\}.$$

Using (15.54) and substituting the index i by $j := -d(m_n, n+1) + i = m_{n+1} - m_n + i$, we have

$$\begin{aligned} \mathsf{b}_{m_n}(\underline{\mathbf{T}}, W_{n+1}) &= \{W_{n+1}\} \boxplus_{i=1}^{d(m_n,n+1)} \frac{a^{d(m_n,n+1)+1-i}}{a-1} \underline{T}^{m_n - d(m_n,n+1)+i} \\ &= \{W_{n+1}\} \boxplus_{j=-d(m_n,n+1)+1}^{0} \frac{a^{-j+1}}{a-1} \underline{T}^{m_n+j}. \end{aligned}$$

Regarding the definition of $J(\underline{\mathbf{T}}, W_n)$ in (15.53), we obtain that $\mathsf{b}_{m_n}(\underline{\mathbf{T}}, W_{n+1}) \wedge W_n = \{W_n\}$. Now if $k > m_n$, then by Lemma 15.40

$$\mathsf{b}_k(\underline{\mathbf{T}}, W_{n+1}) = \left(a^{k-m_n} \mathsf{b}_{m_n}(\underline{\mathbf{T}}) \boxplus_{i=1}^{k-m_n} \frac{a^{k-m_n+1-i}}{a-1} \underline{T}^{m_n+i} \right) \wedge W_{n+1}.$$

Because $a > 1$, we obtain that $a^{k-m_n} \mathsf{b}_{m_n}(\underline{\mathbf{T}}) \wedge W_n = \{W_n\}$, and hence, again by the definition of $J(\underline{\mathbf{T}}, W_n)$, $\mathsf{b}_k(\underline{\mathbf{T}}, W_{n+1}) \wedge W_n = \{W_n\}$. Thus (15.55) is shown for all $k \in \mathbb{Z}$.

We can conclude that for all $k \in \mathbb{Z}$, there exists a tessellation $\mathsf{b}_k(\underline{\mathbf{T}}) \in \mathcal{T}$ such that $\mathsf{b}_k(\underline{\mathbf{T}}) \wedge W_n = \mathsf{b}_k(\underline{\mathbf{T}}, W_n)$. Formally, the $(d-1)$-skeleton of $\mathsf{b}_k(\underline{\mathbf{T}})$ can be described as the union of the $(d-1)$-skeletons of $\mathsf{b}_k(\underline{\mathbf{T}}, W_n)$ in the interiors of W_n, $n \in \mathbb{N}$, that is,

$$\partial \mathsf{b}_k(\underline{\mathbf{T}}) := \bigcup_{n \in \mathbb{N}} (\partial \mathsf{b}_k(\underline{\mathbf{T}}, W_n) \cap \operatorname{int} W_n). \tag{15.56}$$

By (15.54) and (15.56) a map $\mathsf{b} := (\mathsf{b}_k, k \in \mathbb{Z}) : (\mathcal{T}^{\mathbb{N}_0})^{\mathbb{Z}} \to \mathcal{T}^{\mathbb{Z}}$ is defined, and it is shown that for all $k \in \mathbb{Z}$ and $n \in \mathbb{N}$,

$$\mathsf{b}_k(\underline{\mathbf{Y}}) \wedge W_n = \mathsf{b}_k(\underline{\mathbf{Y}}, W_n) \quad \text{a. s.} \tag{15.57}$$

It remains to verify that b is indeed a factor map. To prove (15.51), we apply (15.56) and property (15.48) for restrictions to windows. This yields that a. s.

$$\begin{aligned} \partial(\alpha_{\mathcal{T}}(\mathsf{b}(\underline{\mathbf{Y}})))^m &= \partial \mathsf{b}_{m+1}(\underline{\mathbf{Y}}) \\ &= \bigcup_{n \in \mathbb{N}} (\partial \mathsf{b}_{m+1}(\underline{\mathbf{Y}}, W_n) \cap \operatorname{int} W_n) \end{aligned}$$

$$= \bigcup_{n\in\mathbb{N}} (\partial \mathsf{b}_m(\alpha\,\underline{\mathbf{Y}}, W_n) \cap \operatorname{int} W_n)$$
$$= \partial \mathsf{b}_m(\alpha\,\underline{\mathbf{Y}}) \,.$$

Finally, (15.52) is shown. By (15.57) and (15.37)

$$(\mathsf{b}_k(\underline{\mathbf{Y}}) \wedge W_n \,,\, k \in \mathbb{Z}) \stackrel{\mathrm{D}}{=} (U_k \wedge W_n \,,\, k \in \mathbb{Z})$$

for all $n \in \mathbb{N}$, where **U** is defined in (15.33). Using Theorem 2.3.1 of [172] analogously as in the proof of Theorem 5.10(d), the spatial consistency property yields that

$$(\mathsf{b}_k(\underline{\mathbf{Y}}) \,,\, k \in \mathbb{Z}) \stackrel{\mathrm{D}}{=} (U_k \,,\, k \in \mathbb{Z}) \,,$$

which completes the proof. □

15.7 Notes

1. The book [19] provides an extensive and detailed presentation of strong mixing conditions. The connections with ergodic theory and central limit theorems are also dealt with. Furthermore, this book contains a wide list of references and historical remarks.
2. For random measures, and thus also for point processes, the concepts of ergodicity, mixing, and tail triviality are described in Chapter 12 of [46], also with references to original publications.
3. The approach for β-mixing of random tessellations, using two windows $W_1 \subset W_2$ and the corresponding σ-algebras $\mathcal{B}_{W_1}(\mathcal{T})$ and $\mathcal{B}_{W_2^c}(\mathcal{T})$, given in Definition 15.11 and then used in Section 15.2, is based on the seminal papers [69, 71].
4. In [100], it was firstly shown that the STIT tessellation Y_1 is mixing in space with the decay rate

   $$\left|\mathbb{P}\big(\partial Y_1 \cap C_1 = \emptyset,\ \partial Y_1 \cap (C_2 + x) = \emptyset\big) - \mathbb{P}(\partial Y_1 \cap C_1 = \emptyset)\,\mathbb{P}(\partial Y_1 \cap C_2 = \emptyset)\right| = O\big(\|x\|^{-1}\big)$$

   for all $C_1, C_2 \in \mathcal{C}$. An alternative proof of this proposition was given in [51].
   In [115] the concept of encapsulation was introduced, and the tail triviality of STIT tessellations was proved. In fact, this was even the proof that STIT tessellations are α-mixing in space; for the definition of α-mixing, see Definition 3.3 of [19].
   The stronger β-mixing property in space of STIT tessellations, presented in Theorem 15.25, was proved in [117] with an additional restriction on the hyperplane measure Θ.
5. Part (b) of Lemma 15.29 is Lemma 9 in [115].
6. In [26, 29], sufficient weak dependency conditions are formulated, which allow us to derive results for order statistics for certain functionals of cells of tessellations. All these results refer somehow to an underlying point process and make use of the mixing properties of this point process.
   In this spirit, also for stationary and isotropic STIT tessellation in the plane, the exceedances and order statistics for the inradii of cells are investigated in [31] using mixing properties of these tessellations; see also Note 5 in Chapter 11.
7. The concept of strong mixing forms a fundamental basis for proving laws of large numbers and central limit theorems (CLTs) in general; see [19]. So far, however, no CLTs for STIT tessellations have been established using this method; the results presented in Chapter 13 rely on martingale methods.

In Theorem 3.1 and conditions (3.4), (3.5), and (3.6) of [69], sufficient criteria are formulated that yield CLTs for functionals of random stationary tessellations. These criteria require the finiteness of specific moments of the considered functionals, together with a polynomial decay of sufficiently small order in $(b-a)$ for the β-mixing coefficient $\beta(a,b)$, as introduced in (15.10). This approach was applied in [71, 75] to several models in stochastic geometry. In [64] the so-called β-, β'-, and Gaussian–Delaunay tessellations are investigated. For the β- and Gaussian cases, CLTs are established via conditions (6.10), (6.11), and (6.12) of that paper, in analogy to conditions (3.4), (3.5), and (3.6) of [69]. As far as these conditions are concerned, a polynomial decay of $\beta(a,b)$ in $(b-a)$ can be sufficient if the degree of the polynomial is suitably small. However, the upper bound for the decay of $\beta(a,b)$ in Theorem 15.25 is a polynomial with a degree restricted to $(-1,0)$, which does not satisfy the above requirements.

It remains an open question whether the upper bound from Theorem 15.23 can be refined to achieve a sufficiently small polynomial degree. Because the approach in the proof of Theorem 15.23 critically depends on the encapsulation probability $\mathbb{P}(Z^{\mathrm{STIT}}_{\langle o\rangle,s} \in \mathcal{P}(a,b))$, it may be necessary to consider alternative methods. Finally, note that Theorems 13.7 and 13.8 prove CLTs for functionals of the maximal polytopes that are born only after a fixed threshold time $s_0 > 0$. Hence, it might be reasonable to adapt the β-mixing coefficient – and its upper bound – to accommodate such a condition.

8. Based on results for the spatial β-mixing coefficient and using results of [69], referring back to [202], in [118] an upper bound for the asymptotic variance of additive functionals of STIT tessellations was derived. However, in the motion-invariant case, this bound turned out to be considerably weaker than those obtained before in Corollary 2 of [174] and Theorem 6.1 of [176] and presented in Chapter 12.
9. In Theorem 1.2 of [114], it was shown that the rescaled stationary STIT tessellation process is mixing in time, which is a weaker result than Theorem 15.33, where β-mixing is proved.
10. Theorem 15.34 is Lemma 1 of [120]. The regenerative property of the rescaled zero-cell process of STIT tessellations, presented in Lemma 15.36, was dealt with in Lemma 2 of [119]. In Proposition 3 of that paper, it was also shown that this process is a Bernoulli flow.
11. For Theorem 15.38 and Corollary 15.42, we refer to [114]. Theorem 15.43 was proved in [113]. There it is also shown that the filtration associated with the nonanticipating factor map is standard in Vershik's sense; see [192]. These results concern the rescaled STIT tessellation process with discrete time.

    Furthermore, in [114, 113] the transition from discrete time to continuous time is addressed. This leads to the concept of a **Bernoulli flow**. A flow is a continuous-time dynamical system with a group of measure-preserving transformations α^t, $t \in \mathbb{R}$, such that $\alpha^t \circ \alpha^s = \alpha^{t+s}$ for all $s,t \in \mathbb{R}$. A Bernoulli flow is defined in Section 12 of [156] as a flow $(\Omega, \mathbf{A}, \mathbb{P}, \{\alpha^t, t \in \mathbb{R}\})$ for which the shift system $(\Omega, \mathbf{A}, \mathbb{P}, \alpha^1)$ is a generalized Bernoulli shift. To apply Theorem 4 in Section 12 of [156] on the isomorphy of all Bernoulli flows of infinite entropy, it is essential to construct appropriate *Lebesgue* probability spaces.

    Thus by Corollary 1.5 of [114] for a STIT tessellation process $(Y_t, t > 0)$ and for all $W \in \mathcal{P}_d$, the rescaled continuous-time process $((a^t Y_{a^t}) \wedge W, t \in \mathbb{R})$ is isomorphic to a Bernoulli flow of infinite entropy and is isomorphic to any other Bernoulli flow of infinite entropy defined on a Lebesgue probability space. Also, by Theorem 1.3 of [113] the rescaled continuous-time process $(a^t Y_{a^t}, t \in \mathbb{R})$ is isomorphic to a Bernoulli flow.
12. Apparently there are only a few results in the literature relating the concepts of strong mixing and Bernoulli shifts or flows.

    For time-stationary processes with a finite state space, β-mixing, or synonymously absolute regularity, is equivalent to the weak Bernoulli condition in the sense of [55], and this corresponds to condition (15.12); see also page 114 of [19].

In Theorems 2.10 and 2.11 of [112] a relation is shown between Rosenblatt's strong mixing condition, which is also called α-mixing, and the isomorphy to Bernoulli shifts. Note that β-mixing implies α-mixing; see Proposition 3.11 in [19].

For further comments and references, see Section 13.20 of [19]. Thus it will be of interest whether from the fact that STIT tessellations are β-mixing in space it can be concluded that the group of translations acting on $\mathbb{R}^d$ is a Bernoulli $\mathbb{R}^d$-action on STIT tessellations. In the final section of [117] a conjecture is formulated on the Bernoulli property of STIT with respect to translations, with a reference to the results in [157].

Part IV: **STIT tessellations in 2D and 3D**

16 Planar STIT tessellations

In this part, we focus on planar and spatial STIT tessellations. Of course, the results of the previous parts can also be applied to the dimensions $d = 2$ and $d = 3$. Proofs of these results for $d = 2$ and $d = 3$ are only given here if they are based on a new idea or if another interesting method is used. In any case, reference is made to the corresponding results in the previous parts. Additional results arise from a deeper and more detailed analysis of the combinatorial and topological structure of STIT tessellations, which is possible in two and three dimensions.

This part of the book is in a way self-contained. Almost all definitions and notation are explained again. In addition, illustrative descriptions are given where possible. Therefore the readers who are not primarily interested in tessellations in general dimensions can use this part to get an overview of the investigation methods and the results for planar and spatial STIT tessellations.

16.1 Notation and iteration

This section deals with general random stationary tessellations in $\mathbb{R}^2$. Let T be a planar tessellation according to Definition 2.1. Following (2.1), the sets of k-faces of the tessellation cells are $\mathcal{F}_k(T)$, $k \in \{0, 1, 2\}$, where $\mathcal{F}_2(T)$ is the set of cells. Furthermore, various effects where a tessellation is not face-to-face will be included in the considerations. A definition of face-to-face tessellations in $\mathbb{R}^d$ is given in (2.3). For planar tessellations, the difference between face-to-face and not face-to-face can be explained as follows. Let $\mathsf{z}_1, \mathsf{z}_2 \in \mathcal{F}_2(T)$ be two cells of a tessellation T with $\dim(\mathsf{z}_1 \cap \mathsf{z}_2) = 1$. If T is face-to-face, then this intersection $\mathsf{z}_1 \cap \mathsf{z}_2$ is a 1-face of both cells. This is not necessarily the case with a tessellation that is not face-to-face. Planar STIT tessellations are not face-to-face, which can be seen in Figures 1.3 and 4.3. For tessellations that are not face-to-face, further types of segments can be considered in addition to the 1-faces and 1-primitives. A refined classification of these components forms the basis for further specific results. To pursue this, a more descriptive notation system for the components of a tessellation is necessary. Such a system will now be developed, initially adapted to planar tessellations that are not face-to-face. In Chapter 17, we will extend it to tessellations in $\mathbb{R}^3$. In any case, references to related notation in the previous parts will be given.

We start with the **primitives** of a tessellation T. The set of k-primitives $\mathcal{F}_k^{\mathrm{pr}}(T)$ is defined in Section 2.1.2, and some properties were given there. To illustrate this notation, we can say that a k-primitive is a k-dimensional polytope that has no structure of the tessellation in its relative interior. The primitives of a planar tessellation are referred to as **vertices** ($k = 0$), **edges** ($k = 1$), and **cells** ($k = 2$). To avoid overburden notation, in this chapter, we use the abbreviations

$$\mathsf{V}(T) := \mathcal{F}_0^{\mathrm{pr}}(T), \quad \mathsf{E}(T) := \mathcal{F}_1^{\mathrm{pr}}(T), \quad \mathsf{Z}(T) := \mathcal{F}_2^{\mathrm{pr}}(T) = T$$

https://doi.org/10.1515/9783112222409-020

for the sets of vertices, edges, and cells of a tessellation T, respectively.

For tessellations that are not face-to-face, the k-dimensional faces can differ from the k-primitives. For example, a cell can have more vertices of a tessellation on its boundary than it has 0-faces, as shown in Figure 16.1.

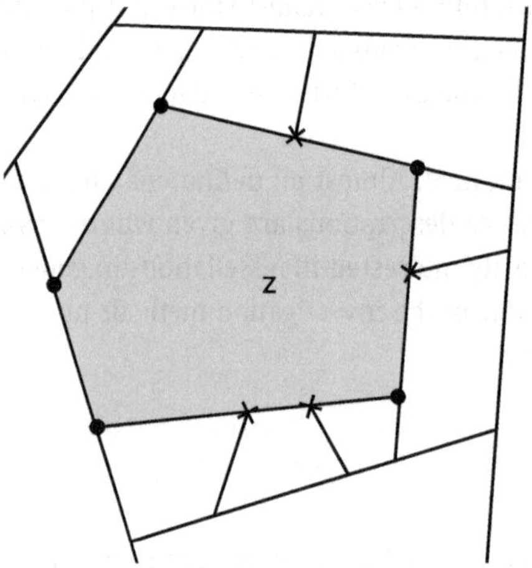

Figure 16.1: A cell z in a tessellation that is not face-to-face. This cell has five 0-faces and nine vertices on its boundary and therefore also five 1-faces and nine edges.

Further effects are that an edge is not necessarily a 1-face of a cell. Moreover, an element of $\mathcal{F}_1(T)$ can be a 1-face of one or two cells. These multiplicities will play a role later in our investigations. For example, in (16.28), we will consider 1-faces marked with their multiplicity and call them J-segments.

Henceforth, as a generic term, we will use the sets $\mathsf{A}(T)$ of **A-components** of T. An A-component is a k-dimensional polytope, $k \in \{0, 1, 2\}$. A definition for $\mathsf{A}(T)$ is given in (2.9). For example, $\mathsf{A}(T)$ can be $\mathsf{V}(T)$, $\mathsf{E}(T)$, or $\mathsf{Z}(T)$, that is, a set of primitives or $\mathcal{F}_0(T)$, $\mathcal{F}_1(T)$, the sets of 0- or 1-faces of the planar tessellation T. Furthermore, $\mathsf{A}(T)$ can also be a subset of one of these sets. If the connection between a set A of tessellation components and the tessellation T is clear, then we write for short A instead of $\mathsf{A}(T)$. Subsets of A are denoted by $\mathsf{A}\langle\cdot\rangle$, where the term in the angle brackets is a suitably chosen symbol introduced in an ad hoc manner. For example, we use the symbol $\mathsf{E}\langle B\rangle$ for the subclass of all edges with direction in $B \subseteq G(2,1)$ or write $\mathsf{V}\langle\text{new}\rangle$ for the set of "newly generated" vertices in an iterated tessellation as in the proof of Lemma 16.2.

For a random stationary tessellation X, the **intensity of A-components** γ_{A} is defined in (2.25). Let $c : \mathcal{P}_k \to \mathbb{R}^2$ be a center function for k-dimensional components, $k \in \{0, 1, 2\}$. Such a function is measurable and translation-covariant, which means that $c(P + x) = c(P) + x$ for all $P \in \mathcal{P}_k$ and $x \in \mathbb{R}^2$. Then

$$\gamma_{\mathsf{A}} = \mathbb{E} \sum_{\mathsf{a} \in \mathsf{A}(X)} \mathbb{1}\{c(\mathsf{a}) \in [0,1]^2\}, \tag{16.1}$$

which is also called the mean number of A-components per unit area. For all A-components considered for STIT tessellations in the following, their intensities always satisfy $\gamma_{\mathsf{A}} < \infty$. This is in accordance with Convention 2.28. With the notation from Section 2.2, this leads for the primitives to

$$\gamma_{\mathsf{V}} = \gamma_0^{\mathrm{pr}}, \quad \gamma_{\mathsf{E}} = \gamma_1^{\mathrm{pr}}, \quad \gamma_{\mathsf{Z}} = \gamma_2^{\mathrm{pr}}$$

for the **mean numbers of vertices, edges, and cells per unit area**, respectively.

The **skeletons** of k-faces and k-primitives, respectively, were defined in Section 2.1 for tessellations in $\mathbb{R}^d$, and some properties were given in (2.16)–(2.18). For a planar tessellation T, the 1-skeleton

$$\mathrm{skel}_1\, T = \mathrm{skel}_1^{\mathrm{pr}}\, T = \partial T = \bigcup_{\mathsf{e}\in\mathsf{E}(T)} \mathsf{e}$$

is of main interest. By (2.47) and (2.49), for a random stationary planar tessellation X, the 1-skeleton $\mathrm{skel}_1 X = \mathrm{skel}_1^{\mathrm{pr}} X = \partial X$ has the **length density**

$$\eta_1 = \mathbb{E} \sum_{\mathsf{e}\in\mathsf{E}(X)} V_1(\mathsf{e} \cap [0,1]^2),$$

also called the **mean total length of tessellation edges per unit area.** For the 0-skeleton of a planar tessellation, we have that $\eta_0 = \gamma_{\mathsf{V}}$.

Typical components, their distributions, and their mean values were introduced in Sections 2.2.2 and 2.2.3 for random stationary tessellations in $\mathbb{R}^d$. Recall definition (2.26) for $d = 2$: For any set A of k-dimensional components, $k \in \{0,1,2\}$, of a random stationary tessellation X and a center function $c : \mathcal{P}_k \to \mathbb{R}^2$, the probability measure $\mathbb{Q}_{\mathsf{A}}$ on $\mathcal{P}_{k,c}$ given by

$$\mathbb{Q}_{\mathsf{A}}(A) = \frac{1}{\gamma_{\mathsf{A}}} \mathbb{E} \sum_{\mathsf{a}\in\mathsf{A}(X)} \mathbb{1}\{c(\mathsf{a}) \in [0,1]^2\}\, \mathbb{1}\{\mathsf{a} - c(\mathsf{a}) \in A\}, \quad A \in \mathcal{B}(\mathcal{P}_{k,c}), \tag{16.2}$$

is the **distribution of the typical A-component.** While the intensity γ_{A} does not depend on the choice of the center function c, the distribution $\mathbb{Q}_{\mathsf{A}}$ depends on c. However, since in this chapter, we only deal with translation-invariant functionals of typical components, this dependency disappears, and a reference to a specific center function is not relevant in the following; see also Remark 2.18.

For the tessellation cells,

$$\mathbb{Q}_{\mathsf{Z}}(A) = \frac{1}{\gamma_{\mathsf{Z}}} \mathbb{E} \sum_{\mathsf{z}\in\mathsf{Z}(X)} \mathbb{1}\{c(\mathsf{z}) \in [0,1]^2\}\, \mathbb{1}\{\mathsf{z} - c(\mathsf{z}) \in A\}, \quad A \in \mathcal{B}(\mathcal{P}_{2,c}),$$

denotes the **distribution of the typical cell.** Using the notation from (2.44) in Section 2.2, we have $\mathbb{Q}_{\mathsf{Z}} = \mathbb{Q}_2^{\mathrm{pr}}$.

Following (2.45), we define the **mean area of the typical cell** as

$$a_Z := \int_{\mathcal{P}_{2,c}} V_2(P)\, \mathbb{Q}_Z(\mathrm{d}P) = \frac{1}{\gamma_Z}\, \mathbb{E} \sum_{z\in Z(X)} V_2(z)\mathbb{1}\{c(z) \in [0,1]^2\},$$

the **mean perimeter of the typical cell** as

$$\begin{aligned}\ell_Z := \int_{\mathcal{P}_{2,c}} \sum_{S\in\mathcal{F}_1(P)} V_1(S)\, \mathbb{Q}_Z(\mathrm{d}P) &= \frac{1}{\gamma_Z}\, \mathbb{E} \sum_{z\in Z(X)} \sum_{S\in\mathcal{F}_1(z)} V_1(S)\mathbb{1}\{c(z) \in [0,1]^2\} \\ &= \frac{2}{\gamma_Z}\, \mathbb{E} \sum_{z\in Z(X)} V_1(z)\mathbb{1}\{c(z) \in [0,1]^2\},\end{aligned}$$

and the **mean number of k-faces of the typical cell**, $k \in \{0,1\}$, as

$$\nu_k(Z) := \int_{\mathcal{P}_{2,c}} \sum_{F\in\mathcal{F}_k(P)} 1\, \mathbb{Q}_Z(\mathrm{d}P) = \frac{1}{\gamma_Z}\, \mathbb{E} \sum_{z\in Z(X)} \sum_{F\in\mathcal{F}_k(z)} \mathbb{1}\{c(z) \in [0,1]^2\},$$

where we recall that $\mathcal{F}_k(P)$ is the set of all k-faces of P.

Similarly, for the tessellation edges E,

$$\mathbb{Q}_\mathsf{E}(A) = \frac{1}{\gamma_\mathsf{E}}\, \mathbb{E} \sum_{e\in \mathsf{E}(X)} \mathbb{1}\{c(e) \in [0,1]^2\}\, \mathbb{1}\{e - c(e) \in A\}, \quad A \in \mathcal{B}(\mathcal{P}_{1,c}),$$

is the **distribution of the typical edge**, $\mathbb{Q}_\mathsf{E} = \mathbb{Q}_1^{\mathrm{pr}}$, and

$$\ell_\mathsf{E} := \int_{\mathcal{P}_{1,c}} V_1(S)\, \mathbb{Q}_\mathsf{E}(\mathrm{d}S) = \frac{1}{\gamma_\mathsf{E}}\, \mathbb{E} \sum_{e\in \mathsf{E}(X)} V_1(e)\mathbb{1}\{c(e) \in [0,1]^2\} \tag{16.3}$$

is the **mean length of the typical edge.**

The adjacencies between different components in the tessellation structure are an important aspect for further investigations. Therefore we now introduce the concept of adjacent components.

Definition 16.1. *Let* $\mathsf{A}(X)$ *and* $\mathsf{B}(X)$ *be two sets of components of a random stationary tessellation* X*. A component* $\mathsf{a} \in \mathsf{A}(X)$ *is said to be* ***adjacent*** *to a component* $\mathsf{b} \in \mathsf{B}(X)$ *if* $\mathsf{a} \subseteq \mathsf{b}$ *or* $\mathsf{b} \subseteq \mathsf{a}$*. We define the* ***mean number of*** B***-components adjacent to the typical*** A***-component*** *by*

$$\mu_{\mathsf{A},\mathsf{B}} := \frac{1}{\gamma_\mathsf{A}}\, \mathbb{E} \sum_{\mathsf{a}\in \mathsf{A}(X)} \mathbb{1}\{c(\mathsf{a}) \in [0,1]^2\} \sum_{\mathsf{b}\in \mathsf{B}(X)} \mathbb{1}\{\mathsf{a} \subseteq \mathsf{b} \text{ or } \mathsf{b} \subseteq \mathsf{a}\}. \tag{16.4}$$

For example, $\mu_{\mathsf{Z},\mathsf{V}}$ is the mean number of vertices of the typical cell, and $\mu_{\mathsf{V},\mathsf{Z}}$ is the mean number of cells adjacent to the typical vertex. We have to distinguish attentively between $\mu_{Z,V}$ and $\mu_{V,Z}$ and also between $\mu_{Z,V}$ and $\nu_0(Z)$, which is the mean number of 0-faces of the typical cell. In a tessellation that is not face-to-face, a cell can have vertices on its boundary that are not 0-faces of this cell, as shown in Figure 16.1. Hence $\mu_{Z,V} \geq \nu_0(Z)$, and only for a face-to-face tessellation, $\mu_{Z,V} = \nu_0(Z)$.

In Section 2.2.4, relations on $A(T) \times B(T)$ for a tessellation T were introduced. The adjacency between two components defined above can be seen as an example of such a relation:

$$\mathsf{a} \sim_{adj} \mathsf{b} := \mathsf{a} \subseteq \mathsf{b} \quad \text{or} \quad \mathsf{b} \subseteq \mathsf{a}\,.$$

It satisfies the required covariance property (2.59) concerning translations

$$\mathsf{a} \sim_{adj} \mathsf{b} \implies (\mathsf{a} + x) \sim_{adj} (\mathsf{b} + x) \tag{16.5}$$

for all $\mathsf{a} \in A(T)$, $\mathsf{b} \in B(T)$, $x \in \mathbb{R}^2$. Using (2.31) and (2.32) together with (2.33), we can also write the mean adjacency number $\mu_{A,B}$ as

$$\mu_{A,B} = \int_{\mathcal{T}} \sum_{\mathsf{b} \in B(T)} \mathbb{1}\{\mathsf{a}^o \sim_{adj} \mathsf{b}\}\, \mathbb{P}_A^o(\mathrm{d}T)\,, \tag{16.6}$$

where a^o is the A-component of T with $c(\mathsf{a}^o) = o$. Note that $\mathbb{P}_A^o$ is concentrated on $\mathcal{T}^A := \{T \in \mathcal{T} : \exists \mathsf{a} \in A(T) : c(\mathsf{a}) = o\}$. Relation (16.6) explains the naming of $\mu_{A,B}$ as the mean number of B-components adjacent to the typical A-component. The **exchange formula** given in Theorem 2.29 applied with $f(\mathsf{a}, X) = h(\mathsf{b}, X) = 1$ yields

$$\mathbb{E} \sum_{\mathsf{a} \in A(X)} \mathbb{1}\{c(\mathsf{a}) \in [0,1]^d\} \sum_{\mathsf{b} \in B(X)} \mathbb{1}\{\mathsf{a} \sim_{adj} \mathsf{b}\} = \mathbb{E} \sum_{\mathsf{b} \in B(X)} \mathbb{1}\{c(\mathsf{b}) \in [0,1]^d\} \sum_{\mathsf{a} \in A(X)} \mathbb{1}\{\mathsf{a} \sim_{adj} \mathsf{b}\}\,,$$

which in view of (16.4), can be rewritten as

$$\gamma_A\, \mu_{A,B} = \gamma_B\, \mu_{B,A}\,. \tag{16.7}$$

For example, applying (16.7) to the adjacencies between vertices and cells, the mean number of vertices of the typical cell and the mean number of cells adjacent to the typical vertex are related by $\gamma_V\, \mu_{V,Z} = \gamma_Z\, \mu_{Z,V}$.

Further investigations in this chapter deal with different segment types and their directions. For this purpose, the directional distribution of a segment is defined as a measure on $\mathcal{H}_o = G(2,1)$ as follows. For any line segment S, there exists a unique line $\operatorname{aff}_o S \in \mathcal{H}_o$ through the origin that is parallel to S. We identify the direction of S with that line $\operatorname{aff}_o S$ and not with the line orthogonal to S. Hence a segment S has a direction in $A \in \mathcal{B}(\mathcal{H}_o)$ if $\operatorname{aff}_o S \in A$. For $\mathscr{H}^1$-almost all $x \in \mathbb{R}^2$ on the 1-skeleton ∂X of a random tessellation X, there exists a unique tessellation edge $\mathsf{e}(x) \in E(X)$ with $x \in \mathsf{e}(x)$ and direction $\operatorname{aff}_o \mathsf{e}(x)$. Using this, analogously to (2.50), for a random stationary planar tessellation X, the probability measure χ on $\mathcal{H}_o$ given by

$$\chi(A) := \frac{1}{\eta_1} \mathbb{E} \int_{\mathbb{R}^2} \mathbb{1}\{x \in [0,1]^2 \cap \partial X\} \, \mathbb{1}\{\mathrm{aff}_o \, \mathrm{e}(x) \in A\} \, \mathscr{H}^1(\mathrm{d}x)$$
$$= \frac{1}{\eta_1} \mathbb{E} \sum_{\mathrm{e} \in \mathrm{E}(X)} \mathbb{1}\{\mathrm{aff}_o \, \mathrm{e} \in A\} \, V_1(\mathrm{e} \cap [0,1]^2), \quad A \in \mathcal{B}(\mathcal{H}_o), \tag{16.8}$$

is the **directional distribution in the typical point of the 1-skeleton** ∂X, also referred to as the length weighted directional distribution of the 1-skeleton ∂X. The term $\eta_1 \chi(A)$ equals the mean total length of edges with direction in $A \in \mathcal{B}(\mathcal{H}_o)$ per unit area. The measure χ coincides with the measure χ_1^{pr} in (2.50) for $d = 2$. For arbitrary dimension d, we have used $\chi_{d-1} := \chi_{d-1}^{\mathrm{pr}}$. In this chapter, for $d = 2$, we omit the lower index to enhance readability.

We will study distributions of different functionals of edges and later also of other segment types. For that purpose, the following style of notation is used. The probability measure $\mathbb{Q}_{\mathrm{E}:h}$ on $\mathcal{H}_o$ denotes the **directional distribution of the typical edge**, which for a center function $c : \mathcal{P}_1 \to \mathbb{R}^2$ and $A \in \mathcal{B}(\mathcal{H}_o)$ is

$$\mathbb{Q}_{\mathrm{E}:h}(A) := \int_{\mathcal{P}_{1,c}} \mathbb{1}\{\mathrm{aff}_o \, S \in A\} \, \mathbb{Q}_{\mathrm{E}}(\mathrm{d}S)$$
$$= \frac{1}{\gamma_{\mathrm{E}}} \mathbb{E} \sum_{\mathrm{e} \in \mathrm{E}(X)} \mathbb{1}\{\mathrm{aff}_o \, \mathrm{e} \in A\} \, \mathbb{1}\{c(\mathrm{e}) \in [0,1]^2\}.$$

With definition (2.46) for $\mathbb{Q}_{1:h}^{\mathrm{pr}}$, we have that $\mathbb{Q}_{\mathrm{E}:h} = \mathbb{Q}_{1:h}^{\mathrm{pr}}$. Furthermore, $\mathbb{Q}_{\mathrm{E}:h,\ell}$ is the **joint distribution of direction and length of the typical edge**, defined by

$$\mathbb{Q}_{\mathrm{E}:h,\ell}(A, B) := \int_{\mathcal{P}_{1,c}} \mathbb{1}\{\mathrm{aff}_o \, S \in A\} \, \mathbb{1}\{V_1(S) \in B\} \, \mathbb{Q}_{\mathrm{E}}(\mathrm{d}S)$$

for $A \in \mathcal{B}(\mathcal{H}_o)$, $B \in \mathcal{B}((0, \infty))$. In this and the forthcoming chapter, the symbols h and ℓ will always stand for direction and length, respectively. Furthermore, $\mathbb{Q}_{\mathrm{E}:\ell|h=L}$ denotes the **conditional length distribution of the typical edge, given its direction** $L \in \mathcal{H}_o = G(2,1)$. The existence of that conditional distribution follows for $\mathbb{Q}_{\mathrm{E}:h}$-almost all $L \in \mathcal{H}_o$ with (2.35) in Corollary 2.22.

Even for a random stationary planar tessellation, there exists a rich and diverse set of mean values and distributions. We start our considerations with a suitable selection of three mean values and two directional distributions. From these five characteristics a variety of other quantities can be calculated. Using the notation introduced above for a random stationary planar tessellation X, this quintuple consists of

γ_{V} – the mean number of vertices of X per unit area,
γ_{Z} – the mean number of cell-centroids of X per unit area,
η_1 – the mean total length of edges of X per unit area,

χ – the distribution of the direction in the typical point of the 1-skeleton ∂X,
$\mathbb{Q}_{E:h}$ – the directional distribution of the typical edge of X.

In the following lemma the behavior of this quintuple $(\gamma_V, \gamma_Z, \eta_1; \chi, \mathbb{Q}_{E:h})$ under the operation of **iteration** of tessellations is investigated. The operation iteration (also called nesting) is introduced in Sections 3.1 and 3.3. For a short description, we could say that the cells of a frame tessellation X are subdivided by cut-outs of tessellations $X^{(i)}$, $i \in \mathbb{N}_0$, which are indexed by the cells of the frame. In other words, this means that the tessellations $X^{(i)}$ are *nested* in the cells of the frame tessellation X. The result is the iterated tessellation $X \boxplus \underline{X}$ with $\underline{X} = (X^{(i)}, i \in \mathbb{N}_0)$; see (3.4) and Figure 3.1. We will now show how the five characteristics of the iterated tessellation are determined by the respective characteristics of the ingredients, assuming that the tessellations $X^{(i)}$, $i \in \mathbb{N}_0$, are independent and identically distributed.

Lemma 16.2. *Let $(\gamma_V, \gamma_Z, \eta_1; \chi, \mathbb{Q}_{E:h})$ be the characteristics of a stationary planar tessellation X, and let $\underline{X} = (X^{(i)}, i \in \mathbb{N}_0)$ be a sequence of stationary and i. i. d. planar tessellations, independent of X, with characteristics $(\gamma_V^{(1)}, \gamma_Z^{(1)}, \eta_1^{(1)}; \chi^{(1)}, \mathbb{Q}_{E:h}^{(1)})$. Then the iterated tessellation $X \boxplus \underline{X}$ is stationary with characteristics $(\gamma_V^{X\boxplus\underline{X}}, \gamma_Z^{X\boxplus\underline{X}}, \eta_1^{X\boxplus\underline{X}}; \chi^{X\boxplus\underline{X}}, \mathbb{Q}_{E:h}^{X\boxplus\underline{X}})$ given by*

$$\gamma_V^{X\boxplus\underline{X}} = \gamma_V + \gamma_V^{(1)} + 2\eta_1\eta_1^{(1)} \int_{\mathcal{H}_o}\int_{\mathcal{H}_o} |\sin\sphericalangle(L, L')|\, \chi(dL)\, \chi^{(1)}(dL'), \tag{16.9}$$

$$\gamma_Z^{X\boxplus\underline{X}} = \gamma_Z + \gamma_Z^{(1)} + \eta_1\eta_1^{(1)} \int_{\mathcal{H}_o}\int_{\mathcal{H}_o} |\sin\sphericalangle(L, L')|\, \chi(dL)\, \chi^{(1)}(dL'), \tag{16.10}$$

$$\eta_1^{X\boxplus\underline{X}} = \eta_1 + \eta_1^{(1)}, \tag{16.11}$$

$$\eta_1^{X\boxplus\underline{X}}\chi^{X\boxplus\underline{X}} = \eta_1\chi + \eta_1^{(1)}\chi^{(1)}, \tag{16.12}$$

and

$$\begin{aligned}
&(\gamma_V^{X\boxplus\underline{X}} + \gamma_Z^{X\boxplus\underline{X}})\, \mathbb{Q}_{E:h}^{X\boxplus\underline{X}}(A) \\
&\quad = (\gamma_V + \gamma_Z)\, \mathbb{Q}_{E:h}(A) + (\gamma_V^{(1)} + \gamma_Z^{(1)})\, \mathbb{Q}_{E:h}^{(1)}(A) \\
&\qquad + 2\eta_1\eta_1^{(1)} \int_{\mathcal{H}_o}\int_{A} |\sin\sphericalangle(L, L')|\, \chi(dL)\, \chi^{(1)}(dL') \\
&\qquad + \eta_1\eta_1^{(1)} \int_{A}\int_{\mathcal{H}_o} |\sin\sphericalangle(L, L')|\, \chi(dL)\, \chi^{(1)}(dL'), \quad A \in \mathcal{B}(\mathcal{H}_o).
\end{aligned} \tag{16.13}$$

Proof. For planar tessellations, Lemma 3.6 yields relationships (16.11) and (16.12).

The key argument to prove (16.9) is the following. The expression

$$\eta_1\eta_1^{(1)} \int_{\mathcal{H}_o}\int_{\mathcal{H}_o} |\sin\sphericalangle(L, L')|\, \chi(dL)\, \chi^{(1)}(dL') \tag{16.14}$$

is equal to the intensity of the point process of intersection points generated by two superposed independent stationary segment processes with densities $\eta_1, \eta_1^{(1)}$ and directional distributions $\chi, \chi^{(1)}$. This follows from formula (8.52) in [33] for the more general case of fiber processes.

The set of vertices of the iterated tessellation $X \boxplus \underline{X}$ is the union of vertices belonging to X, vertices belonging to the cut-outs of $\underline{X}$ in the interior of the cells of X, plus "newly generated" vertices $\mathsf{V}\langle\text{new}\rangle$. These "new" vertices arise on the 1-skeleton of the frame tessellation X by nesting the tessellations $X^{(i)}$ in the cells of X, as it can be seen in Figure 3.1.

By the method used in the proof of Lemma 3.6, for the intensity of vertices in the iterated tessellation, we obtain

$$\begin{aligned}
\gamma_{\mathsf{V}}^{X\boxplus\underline{X}} &= \mathbb{E} \sum_{\mathsf{v}\in\mathsf{V}(X\boxplus\underline{X})} \mathbb{1}\{\mathsf{v} \in [0,1]^2\} \\
&= \int_{\mathcal{T}} \int_{\mathcal{T}^{\mathbb{N}_0}} \Bigg(\sum_{\mathsf{v}\in\mathsf{V}(T)} \mathbb{1}\{\mathsf{v} \in [0,1]^2\} + \sum_{\mathsf{z}_i\in\mathsf{Z}(T)} \sum_{\mathsf{v}\in\mathsf{V}(T^{(i)})} \mathbb{1}\{(\mathsf{v} \cap \mathsf{z}_i) \in [0,1]^2\} \\
&\quad + \sum_{\mathsf{z}_i\subset\mathsf{Z}(T)} \sum_{\mathsf{e}\in\mathsf{E}(T^{(i)})} \mathbb{1}\{(\mathsf{e} \cap \partial\mathsf{z}_i) \in [0,1]^2\} \Bigg) \mathbb{P}_{\underline{X}}(\mathrm{d}\underline{T})\, \mathbb{P}_X(\mathrm{d}T) \\
&= \gamma_{\mathsf{V}} + \mathbb{E}\Bigg(\sum_{\mathsf{z}_i\in\mathsf{Z}(X)} \lambda_2([0,1]^2 \cap \mathsf{z}_i) \Bigg) \gamma_{\mathsf{V}}^{(1)} + \gamma_{\mathsf{V}\langle\text{new}\rangle} \\
&= \gamma_{\mathsf{V}} + \gamma_{\mathsf{V}}^{(1)} + \gamma_{\mathsf{V}\langle\text{new}\rangle} \,.
\end{aligned}$$

Each "new" vertex is created by the intersection of an edge of the frame tessellation X and an edge of a nested tessellation $X^{(i)}$. Because each edge of X has two adjacent cells, where tessellations are nested in, for the intensity of new vertices in the iterated tessellation, by (16.14) we obtain

$$\gamma_{\mathsf{V}\langle\text{new}\rangle} = 2\eta_1\eta_1^{(1)} \int_{\mathcal{H}_o} \int_{\mathcal{H}_o} |\sin \sphericalangle(L, L')|\, \chi(\mathrm{d}L)\, \chi^{(1)}(\mathrm{d}L') \,.$$

This yields (16.9).

To show (16.10), the following mean value relations for general planar tessellations are used, which are given in Section 9.3 of [33]:

$$\gamma_{\mathsf{Z}} = \gamma_{\mathsf{E}} - \gamma_{\mathsf{V}} \quad \text{and} \quad \gamma_{\mathsf{E}} = \frac{1}{2} \gamma_{\mathsf{V}} \mu_{\mathsf{V},\mathsf{E}} \,.$$

The second relationship follows from the adjacency relation (16.7) and the fact that each edge is adjacent to exactly two vertices, its 0-faces. Hence we have $\mu_{\mathsf{E},\mathsf{V}} = 2$. Furthermore, all new vertices in $X \boxplus \underline{X}$ have a. s. three emanating edges. This yields, analogously to the proof of (16.9),

$$\gamma_{\mathsf{V}}^{X \boxplus \underline{X}} \, \mu_{\mathsf{V,E}}^{X \boxplus \underline{X}} = \mathbb{E} \sum_{\mathsf{v} \in \mathsf{V}(X \boxplus \underline{X})} \mathbb{1}\{\mathsf{v} \in [0,1]^2\} \sum_{\mathsf{e} \in \mathsf{E}(X \boxplus \underline{X})} \mathbb{1}\{\mathsf{v} \sim_{adj} \mathsf{e}\}$$
$$= \gamma_{\mathsf{V}} \, \mu_{\mathsf{VE}} + \gamma_{\mathsf{V}}^{(1)} \, \mu_{\mathsf{VE}}^{(1)} + 3\gamma_{\mathsf{V}\langle\mathrm{new}\rangle} \,.$$

Applying (16.9), we obtain

$$\gamma_{\mathsf{Z}}^{X \boxplus \underline{X}} = \frac{1}{2} \gamma_{\mathsf{V}}^{X \boxplus \underline{X}} \, \mu_{\mathsf{V,E}}^{X \boxplus \underline{X}} - \gamma_{\mathsf{V}}^{X \boxplus \underline{X}}$$
$$= \frac{1}{2} \big(\gamma_{\mathsf{V}} \, \mu_{\mathsf{VE}} + \gamma_{\mathsf{V}}^{(1)} \, \mu_{\mathsf{VE}}^{(1)} + 3\gamma_{\mathsf{V}\langle\mathrm{new}\rangle}\big) - \big(\gamma_{\mathsf{V}} + \gamma_{\mathsf{V}}^{(1)} + \gamma_{\mathsf{V}\langle\mathrm{new}\rangle}\big)$$
$$= \gamma_{\mathsf{Z}} + \gamma_{\mathsf{Z}}^{(1)} + \frac{1}{2} \gamma_{\mathsf{V}\langle\mathrm{new}\rangle} \,,$$

which is equal to (16.10).

Finally, we prove (16.13). For $A \in \mathcal{B}(\mathcal{H}_o)$, we denote by $\mathsf{E}\langle A \rangle$ the subclass of all edges with direction in A. The intensity of edges with direction in A is $\gamma_{\mathsf{E}\langle A \rangle} = \gamma_{\mathsf{E}} \, \mathbb{Q}_{\mathsf{E}:h}(A) = (\gamma_{\mathsf{V}} + \gamma_{\mathsf{Z}}) \, \mathbb{Q}_{\mathsf{E}:h}(A)$. The adjacency relation (16.7), together with $\mu_{\mathsf{E}\langle A \rangle,\mathsf{V}} = 2$, yields

$$\gamma_{\mathsf{E}\langle A \rangle} = \frac{1}{2} \gamma_{\mathsf{V}} \, \mu_{\mathsf{V},\mathsf{E}\langle A \rangle} \,.$$

For the new vertices of the iterated tessellation, it is known that two of the three emanating edges are located on the 1-skeleton of X and the third one is part of the 1-skeleton of an $X^{(i)}$, which means that

$$\frac{1}{2} \gamma_{\mathsf{V}\langle\mathrm{new}\rangle} \, \mu_{\mathsf{V}\langle\mathrm{new}\rangle,\mathsf{E}\langle A \rangle} = 2\eta_1 \eta_1^{(1)} \int_{\mathcal{H}_o} \int_A |\sin \sphericalangle(L, L')| \, \chi(\mathrm{d}L) \, \chi^{(1)}(\mathrm{d}L')$$
$$+ \eta_1 \eta_1^{(1)} \int_A \int_{\mathcal{H}_o} |\sin \sphericalangle(L, L')| \, \chi(\mathrm{d}L) \, \chi^{(1)}(\mathrm{d}L') \,.$$

An analogous partition of $(\gamma_{\mathsf{V}}^{X \boxplus \underline{X}} + \gamma_{\mathsf{Z}}^{X \boxplus \underline{X}}) \, \mathbb{Q}_{\mathsf{E}:h}^{X \boxplus \underline{X}}(A) = \frac{1}{2} \gamma_{\mathsf{V}}^{X \boxplus \underline{X}} \, \mu_{\mathsf{V},\mathsf{E}\langle A \rangle}^{X \boxplus \underline{X}}$ as in the proof of (16.10) yields (16.13). □

Finally, note that using relations (16.11) and (16.12), we can see that in an iterated tessellation the mean total edge length and the directional distribution in the typical point of the 1-skeleton are the same as for the superposition of two independent tessellations with densities $\eta_1, \eta_1^{(1)}$ and directional distributions $\chi, \chi^{(1)}$, respectively. The superposition of tessellations is defined in (3.3).

16.2 Geometric characteristics, intensities, and mean values

Following the previous introduction to general tessellations, this chapter now deals with STIT tessellations in the plane. Illustrations of STIT tessellations with different directional distributions in the typical point of the 1-skeleton are given in Figure 1.3. In this section, our study starts with the investigation of some feasible and meaningful quantities, like intensities, densities, and mean values for STIT tessellations. Relations between those values are of interest, because they illustrate links between different components of the tessellation. First, the primitives are considered, and then different types of segments are classified and analyzed.

16.2.1 Primitives

Let $\mathbf{Y} = (Y_t, t > 0)$ be the planar STIT tessellation process driven by a translation-invariant line measure Θ given in (4.1), satisfying (4.3) in Convention 4.1. In the planar case, this convention means that the corresponding directional distribution φ is not concentrated on one direction. This assumption guarantees that the cells of the structure constructed with measure Θ are bounded and that the structure is a. s. a random tessellation. The tessellations Y_t are always stationary, which means that their distribution is invariant under translations of the plane. In this section, the mean values and other quantities of Y_t for fixed $t > 0$ are investigated. To establish the results, we mainly use that the tessellations are stable under iteration and that the typical cell of a STIT tessellation Y_t has the same distribution as the typical cell of a Poisson hyperplane tessellation X_t, introduced in Section 2.3.1, if both tessellations X_t and Y_t are based on the same measure Θ. These are among the key properties of a STIT tessellation described in Section 6.1. Some of the results for planar STIT tessellations we present follow immediately from the more general considerations for STIT tessellations in $\mathbb{R}^d$ developed in previous parts of this book. Proofs will be given here again if interesting different methods are used.

One component of the measure Θ is the spherical directional distribution φ given in (4.1), which for a planar STIT tessellation is an even probability measure on the unit sphere S^1. According to (4.2), the measure φ can be replaced by φ_o, which is the corresponding measure on $\mathcal{H}_o = G(2,1)$, the set of all lines through the origin. Recall that the definition of the directional distribution φ_0 given in (4.2) uses the orthogonal linear subspace as the direction. By (6.6) and (6.8) in Theorem 6.9 we obtain that a planar STIT tessellation Y_t based on Θ has the **length density**

$$\eta_1 = t\,, \tag{16.15}$$

and the **directional distribution in the typical point of the 1-skeleton** ∂Y_t, given in (16.8), for $A \in \mathcal{B}(\mathcal{H}_o)$, is

$$\chi(A) = \varphi_o(A^\perp)\,, \tag{16.16}$$

where $A^\perp = \{H^\perp : H \in A\}$. Recall that χ coincides with the measure $\chi_1^{\mathrm{pr}} = \chi_1$ in (2.50) for $d = 2$.

In this chapter, we will refer to the STIT tessellation Y_t for $t > 0$ as the STIT tessellation Y_t with directional distribution χ in the typical point of ∂Y_t. Furthermore, the value

$$t\chi(A) = t\,\varphi_o(A^{\perp})$$

can be interpreted as the mean total edge length per unit area of all edges of Y_t having direction in $A \in \mathcal{B}(\mathcal{H}_o)$. By Theorem 7.9 we obtain that for any probability measure χ on $(\mathcal{H}_o, \mathcal{B}(\mathcal{H}_o))$ that is not concentrated on a single direction and for any $t > 0$, there exists a stationary planar STIT tessellation Y_t with measure $t\chi$.

The function $s_{t\chi} : \mathcal{H}_o \to (0, \infty)$, with respect to $t > 0$ and the measure χ, defined as

$$s_{t\chi}(L) := t \int_{\mathcal{H}_o} |\sin \sphericalangle(L, L')|\,\chi(\mathrm{d}L')\,, \quad L \in \mathcal{H}_o\,, \tag{16.17}$$

is called the **rose of intersections**. This naming can be explained as follows. Applying Corollary 6.14 and Theorem 6.12 for $d = 2$, we obtain that for a line $L \in \mathcal{H} = A(2,1)$, the section $\partial Y_t \cap L$ of the 1-skeleton of a stationary planar STIT tessellation Y_t with L generates a stationary Poisson point process on L. This point process has intensity $s_{t\chi}(L_o) = \eta_0(Y_t \cap L_o)$, where $L_o \in \mathcal{H}_o = G(2,1)$ is the line parallel to L. Notice that $|\sin \sphericalangle(L, L')| = \nabla_2(u(L), u(L'))$, where $u(L), u(L') \in S^1$ are unit vectors in the upper half-sphere with directions $L, L' \in \mathcal{H}_o$, respectively.

The constant

$$\zeta_{t\chi} := t^2 \int_{\mathcal{H}_o}\int_{\mathcal{H}_o} |\sin \sphericalangle(L, L')|\,\chi(\mathrm{d}L')\,\chi(\mathrm{d}L) = t \int_{\mathcal{H}_o} s_{t\chi}(L)\,\chi(\mathrm{d}L) \tag{16.18}$$

was already used in the proof of Lemma 16.2 in (16.14) with $\eta_1 = \eta_1^{(1)} = t$. We will see later on, that this constant reflects the impact of the directional distribution χ on the mean values for typical components in a STIT tessellation. Furthermore, to highlight the dependence on t, we will more frequently use

$$\begin{aligned} s_{t\chi}(L) &= t\, s_{\chi}(L)\,, \quad L \in \mathcal{H}_o\,,\\ \zeta_{t\chi} &= t^2\, \zeta_{\chi}\,. \end{aligned}$$

In this part, too, relations between STIT tessellations and Poisson hyperplane tessellations play an important role. The Poisson hyperplane process $\widehat{X}_t$ with intensity measure $t\,\Theta$ is introduced in Section 2.3. For all $t > 0$, this process induces a Poisson hyperplane tessellation X_t. Section 2.3.4 deals with the intersection process $\widehat{X}_t^{(d-k)}$ of order $d-k$, $k \in \{0, \dots, d-1\}$, for $\widehat{X}_t$. This is a process of k-dimensional affine subspaces generated by the intersection of $d-k$ pairwise different hyperplanes of $\widehat{X}_t$. In particular, the intersection process of order 1 is just the hyperplane process $\widehat{X}_t$ itself, whereas the intersection process of order d is the stationary point process of vertices of the tessellation X_t.

In (2.79) and (2.80), it is shown that the probability measure χ_k on $G(d,k)$, which is the distribution of the direction in the typical point of $\widehat{X}_t^{(d-k)}$, is also the directional distribution in the typical point of $\mathrm{skel}_k X_t$. Furthermore, the mean total k-volume per unit volume of $\widehat{X}_t^{(d-k)}$ is equal to the mean total k-volume per unit volume of $\mathrm{skel}_k X_t$.

The **associated zonoid** $\Pi_{t\Theta}$ of the pair(t, Θ) is defined in (2.68). Equation (2.82) provides a relation between the densities of the intersection processes and the intrinsic volumes of the associated zonoid of $\widehat{X}_t$ with intensity measure $t\,\Theta$. This implies the following for $d = 2$.

Let Y_t be a planar STIT tessellation, and let X_t be a Poisson line tessellation, both based on the measure $t\,\Theta$. The characteristics of Y_t are χ, η_1, and γ_V, and those of X_t are χ^{PHT}, η_1^{PHT}, and γ_V^{PHT}. Theorem 6.9 yields

$$\chi = \chi_1^{\mathrm{PHT}}, \quad \eta_1 = \eta_1^{\mathrm{PHT}} = t, \quad \text{and} \quad \gamma_V = 2\,\gamma_V^{\mathrm{PHT}}. \tag{16.19}$$

Using (6.14) and (2.69), the rose of intersections $s_{t\chi}(\cdot)$ is related to the associated zonoid as follows:

$$s_{t\chi}(L) = 2\,h(\Pi_{t\Theta}, u(L)) = V_1(\Pi_{t\Theta}|L), \quad L \in \mathcal{H}_0, \tag{16.20}$$

where $h(\Pi_{t\Theta}, \cdot)$ is the support function of $\Pi_{t\Theta}$, and $u(L) \in S^1$ stands for a unit vector in the upper half-sphere with direction L. Hence $s_{t\chi}(L)$ equals the length of the orthogonal projection of $\Pi_{t\Theta}$ onto the linear subspace L spanned by u.

The constant $\zeta_{t\chi}$ is also related to the associated zonoid. Applying (2.80), $\frac{1}{2}\,\zeta_{t\chi}$ is equal to the intensity of the intersection process of order 2. Hence we obtain by (2.82) that

$$\zeta_{t\chi} = 2\,V_2(\Pi_{t\Theta}). \tag{16.21}$$

In the isotropic case, where the directional distribution χ in the typical point of the 1-skeleton is invariant under rotations, the associated zonoid $\Pi_{t\Theta} = t\,\Pi_{\Theta_{\mathrm{iso}}}$ is according to (2.70) a circle with radius $\frac{1}{\pi}\,t$. Thus, for a stationary and isotropic planar STIT tessellation Y_t, we have that

$$s_{t\chi}(L) \equiv \frac{2}{\pi}\,t \quad \text{and} \quad \zeta_{t\chi} = \frac{2}{\pi}\,t^2. \tag{16.22}$$

Another particular case are STIT tessellations driven by the Mondrian hyperplane measure given in (4.4). In this case the directional distribution χ in the typical point of the 1-skeleton is

$$\chi = p_1\,\delta_{L(e_1)} + p_2\,\delta_{L(e_2)}, \tag{16.23}$$

where the vectors e_1, e_2 form the standard orthonormal basis in $\mathbb{R}^2$, $L(e_1), L(e_2) \in \mathcal{H}_0$ are lines parallel to e_1, e_2, the measure δ_L is the Dirac measure on $\mathcal{B}(\mathcal{H}_0)$ with mass 1 at L, and $p_1, p_2 \in (0, 1)$ with $p_1 + p_2 = 1$. A simulation of a planar Mondrian STIT tessellation is shown in Figure 1.3. With (2.71) the associated zonoid Π_Θ is in this case a rectangle with edges parallel to e_1 and e_2 and with edge lengths p_1, p_2. In the Mondrian case, for Y_t, we obtain

$$\begin{aligned} s_{t\chi}(L) &= t\left(p_1\left|\sin(\sphericalangle(L,e_1))\right| + p_2\left|\sin(\sphericalangle(L,e_2))\right|\right), \quad L \in \mathcal{H}_o, \\ \zeta_{t\chi} &= 2t^2 p_1 p_2 . \end{aligned} \tag{16.24}$$

Two obvious important features of a random planar STIT tessellation result directly from its conception as a cell division process:

1. Applying (8.13), for $d = 2$, we obtain that the vertices of a random planar STIT tessellation are a. s. T-shaped vertices. These **T-vertices** have three emanating edges, and two of them are collinear, as we can see in Figure 1.3.
2. The cells of a random planar STIT tessellation are not necessarily in a face-to-face position, because the intersection of two neighboring cells with a common edge is not necessarily a 1-face of both of the cells.

The quintuple $(\gamma_V, \gamma_Z, \eta_1; \chi, \mathbb{Q}_{E:h})$ analyzed in the previous section for a planar tessellation is now specified for a STIT tessellation Y_t in $\mathbb{R}^2$. We drop t in these symbols to simplify the notation as long as we consider Y_t for a fixed $t > 0$. On the other hand, we use $s_{t\chi}(L) = t\, s_\chi(L)$ and $\zeta_{t\chi} = t^2 \zeta_\chi$ to show the dependence on t.

We have already determined two of the members of this quintuple, the length density η_1 in (16.15) and the measure χ in (16.16) for a STIT tessellation. Using the stability property (3.10) of Y_t, we obtain results for the remaining three characteristics.

Lemma 16.3. *For a planar STIT tessellation Y_t with directional distribution χ in the typical point of ∂Y_t, given in* (16.8)*, and with the constant ζ_χ (see* (16.18)*), the intensity of vertices γ_V and the intensity of cells γ_Z are*

$$\gamma_V = t^2 \zeta_\chi \quad \text{and} \quad \gamma_Z = \frac{1}{2} t^2 \zeta_\chi . \tag{16.25}$$

The directional distribution $\mathbb{Q}_{E:h}$ of the typical edge is given by

$$\begin{aligned} \mathbb{Q}_{E:h}(A) &= \frac{1}{\zeta_\chi} \int_{\mathcal{H}_o} \int_A \left|\sin \sphericalangle(L, L')\right| \chi(\mathrm{d}L)\, \chi(\mathrm{d}L') \\ &= \frac{1}{\zeta_\chi} \int_A s_\chi(L)\, \chi(\mathrm{d}L), \quad A \in \mathcal{B}(\mathcal{H}_o), \end{aligned} \tag{16.26}$$

where s_χ is the rose of intersections defined in (16.17).

Proof. Let Y_t be a STIT tessellation with vertex and cell intensities γ_V and γ_Z and length density $\eta_1 = t$. Then the tessellation $2Y_t \overset{D}{=} Y_{\frac{1}{2}t}$ (see (6.2)) has the corresponding intensities $\frac{1}{4}\gamma_V$ and $\frac{1}{4}\gamma_Z$ and density $\frac{1}{2}t$. The tessellation Y_t is stable under iteration. From (6.1) and (6.2) in Theorem 6.2 it follows that $2Y_t \boxplus 2\underline{Y}_t \overset{D}{=} Y_t$. Thus by (16.8) relations (16.25) are a consequence of (16.9) and (16.10). The stability property, together with (16.13) and (16.25), leads to the result for $\mathbb{Q}_{E:h}$. □

In a planar STIT tessellation, all vertices are likewise 0-primitives and maximal polytopes of dimension 0. Recall that in Section 8.2 the maximal $(d-1)$-polytopes are introduced as the cell-dividing polytopes for general dimension d. Furthermore, a maximal k-polytope, $k < d-1$, is a k-face of a maximal $(d-1)$-polytope. Using this, the result for γ_V in (16.25) is in line with that for $\gamma_{0,t}^{\mathrm{pr}} = \gamma_{0,t}^{\max}$ in Lemma 14.8 for $d = 2$.

From (16.26) we can conclude that the directional distribution $\mathbb{Q}_{E:h}$ of the typical edge in a STIT tessellation does not depend on t. This can also be explained by the scaling property (6.2) of STIT tessellations and by the fact that the directional distribution is invariant under scaling of the tessellation.

Now for a Poisson line tessellation X_t, we consider the directional distribution $\mathbb{Q}_{E:h}^{\mathrm{PHT}}$ of the typical edge. A Poisson line tessellation is face-to-face, and hence edges and 1-faces coincide. Thus, for X_t based on $t\,\Theta$, using (11.12) with $d = 2$, we obtain

$$\mathbb{Q}_{E:h}^{\mathrm{PHT}}(A) = \frac{V_1(\Pi_\Theta)}{2\,V_2(\Pi_\Theta)} \int_A V_1(\Pi_\Theta|L)\,\chi_1^{\mathrm{PHT}}(\mathrm{d}L)\,.$$

The intrinsic volumes of the associated zonoid are

$$\begin{aligned} V_1(\Pi_{t\Theta}) &= \eta_1^{\mathrm{PHT}} = t \quad \text{(see (2.82) and (16.19))}\,,\\ V_2(\Pi_{t\Theta}) &= \frac{1}{2}\,\zeta_{t\chi_1^{\mathrm{PHT}}} \quad \text{(see (16.21))}\,. \end{aligned} \tag{16.27}$$

Together with (16.19) and (16.20), this gives the following result.

Corollary 16.4. *For a STIT tessellation Y_t and a stationary Poisson line tessellation X_t, both with the same directional distribution χ in the typical point of their 1-skeleton, the distributions of the direction of typical edge coincide,*

$$\mathbb{Q}_{E:h} = \mathbb{Q}_{E:h}^{\mathrm{PHT}}\,.$$

Further results for other intensities and mean values now follow immediately from the mean value relations for general random stationary tessellations in the plane given in Section 9.3.4 in [33].

Corollary 16.5. *Let Y_t be a STIT tessellation with directional distribution χ in the typical point of ∂Y_t, given in (16.8), and with constant ζ_χ; see (16.18). Then*

$$\begin{aligned} \gamma_E &= \frac{3}{2}\,t^2\,\zeta_\chi && \textit{(intensity of edges)},\\ \ell_E &= \frac{2}{3\,t\,\zeta_\chi} && \textit{(mean length of the typical edge)},\\ \ell_Z &= \frac{4}{t\,\zeta_\chi} && \textit{(mean perimeter of the typical cell)}, \end{aligned}$$

$$a_Z = \frac{2}{t^2 \zeta_\chi} \qquad \textit{(mean area of the typical cell)},$$

$$\mu_{V,E} = \mu_{V,Z} = 3 \qquad \textit{(mean number of edges or cells adjacent to the typical vertex)},$$

$$\mu_{Z,V} = \mu_{Z,E} = 6 \qquad \textit{(mean number of vertices or edges adjacent to the typical cell)},$$

$$\nu_0(Z) = \nu_1(Z) = 4 \qquad \textit{(mean number of 0-faces or 1-faces of the typical cell)}.$$

16.2.2 Classification of segments

We will now look at three different types of segments in a planar STIT tessellation. An illustration of them is given in Figure 16.2. The first type are the **edges**, which are the 1-primitives. They were introduced in Section 16.1. Recall that an edge is a segment on the 1-skeleton of the tessellation bounded by two vertices and has no vertex in its relative interior. The set of all edges is denoted by

$$\mathrm{E}(Y_t) = \mathcal{F}_1^{\mathrm{pr}}(Y_t)\,.$$

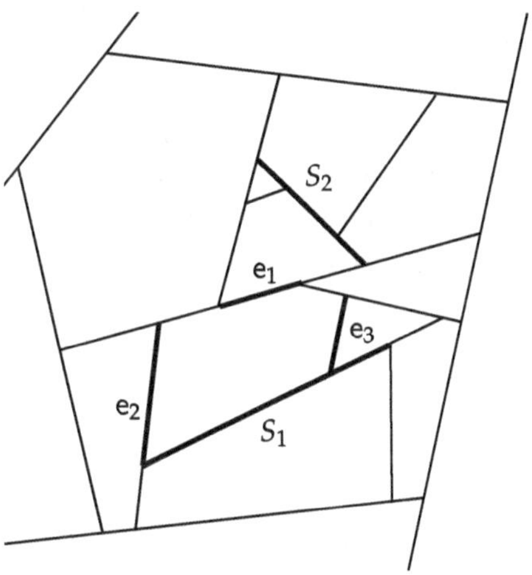

Figure 16.2: Different types of segments in a planar tessellation: e_1 is an edge, e_2 is an edge and a J-segment with multiplicity 1, e_3 is an edge, a J-segment with multiplicity 2, and an I-segment, S_1 is a J-segment with multiplicity 1, and S_2 is an I-segment.

In contrast, there exist the **I-segments** that are the maximal polytopes of dimension one. In Section 8.2, maximal polytopes of a STIT tessellation in $\mathbb{R}^d$ are defined. Based on the construction of a STIT tessellation, they can be described as follows. A maximal polytope is a $(d-1)$-dimensional polytope that divides a cell into two new cells. Thus, for $d = 2$, each division of a cell generates a maximal 1-polytope, which is an I-segment. After the birth of such an I-segment, further vertices can arise in its relative interior through subsequent divisions of the adjacent cells. According to Lemma 8.8, an I-segment is the maximal union of collinear and connected edges, which a. s. cannot be enlarged by another adjacent and collinear edge of the tessellation. The set of all I-segments is denoted by

$$\mathrm{I}(Y_t) := \mathcal{F}_1^{\max}(Y_t)\,.$$

The definition of the third type of segments needs special attention. Recall that any 1-face of a planar STIT tessellation can be a 1-face of one or two cells. According to that, a **J-segment** is defined as a 1-face S marked with its **multiplicity** $m(S) \in \{1,2\}$. The multiplicity $m(S)$ is equal to the number of cells $z \in Z$ that have S as a 1-face. This leads to the following definition of the set of all J-segments:

$$\mathrm{J}(Y_t) := \{(S, m(S)) \,:\, S \in \mathcal{F}_1(Y_t),\ m(S) = |\{z \in Z(Y_t) \,:\, S \in \mathcal{F}_1(z)\}|\}\,. \tag{16.28}$$

In accordance with Remark 19.3, we can consider $\mathrm{J}(Y_t)$ as a marked point process on $\mathcal{P}_1$ with mark space $M = \{1,2\}$.

We can state that the J- and I-segments may have tessellation vertices in their relative interior. Any point of the 1-skeleton that is not a tessellation vertex belongs to exactly one edge, to exactly one I-segment, and to one J-segment with multiplicity 2 or to two different J-segments, each with multiplicity 1. The relative interiors of two different edges are disjoint as well as the relative interiors of two different I-segments. In contrast to this, two J-segments $(S,1)$, $(S',1)$ with $S \neq S'$ can have common interior points. In a face-to-face tessellation, all J-segments have multiplicity 2.

We have already used γ_{E} to denote the intensity of edges and $\mathbb{Q}_{\mathrm{E}}$ to denote the distribution of the typical edge. Similarly, by (16.1) we can define the intensity γ_{I} of I-segments and by (16.2) the distribution $\mathbb{Q}_{\mathrm{I}}$ of the typical I-segment of a STIT tessellation Y_t.

Regarding the multiplicities, the intensity γ_{J} of the J-segments of Y_t is defined as

$$\gamma_{\mathrm{J}} := \mathbb{E} \sum_{(S,m(S))\in \mathrm{J}(Y_t)} \mathbb{1}\{c(S) \in [0,1]^2\}\, m(S) = \mathbb{E} \sum_{S\in\mathcal{F}_1(Y_t)} \mathbb{1}\{c(S) \in [0,1]^2\}\, m(S)\,.$$

The typical J-segment has the distribution $\mathbb{Q}_{\mathrm{J}}$ given by

$$\mathbb{Q}_{\mathrm{J}}(A) := \frac{1}{\gamma_{\mathrm{J}}} \mathbb{E} \sum_{(S,m(S))\in \mathrm{J}(Y_t)} \mathbb{1}\{c(S) \in [0,1]^2\}\, \mathbb{1}\{S - c(S) \in A\}\, m(S)\,, \tag{16.29}$$

$A \in \mathcal{B}(\mathcal{P}_{1,c})$. In other words, the typical J-segment is the typical multiplicity weighted 1-face. Hence, in a tessellation that is not face-to-face, the distribution of the typical J-segment can be different from the distribution of the typical 1-face. Analogously to (16.3), the mean length of the typical I-segment is denoted by ℓ_{I}, and

$$\ell_{\mathrm{J}} := \int_{\mathcal{P}_{1,c}} V_1(S)\, \mathbb{Q}_{\mathrm{J}}(dS) = \frac{1}{\gamma_{\mathrm{J}}} \mathbb{E} \sum_{(S,m(S))\in \mathrm{J}(Y_t)} V_1(S)\, \mathbb{1}\{c(S) \in [0,1]^2\}\, m(S)$$

is the mean length of the typical J-segment.

Using this segment classification, further mean values for a planar STIT tessellation can be given. Corollary 16.5 yields for the edges that

$$\gamma_{\mathrm{E}} = \frac{3}{2}\, t^2\, \zeta_\chi \quad \text{and} \quad \ell_{\mathrm{E}} = \frac{2}{3\, t\, \zeta_\chi}\,.$$

The following applies to J- and I-segments.

Lemma 16.6. *Let Y_t be a STIT tessellation with directional distribution χ in the typical point of ∂Y_t, given in* (16.8), *and with constant ζ_χ; see* (16.18). *The intensities of J-segments and I-segments are*

$$\gamma_{\mathrm{J}} = 2\, t^2\, \zeta_\chi\,, \quad \gamma_{\mathrm{I}} = \frac{1}{2}\, t^2\, \zeta_\chi\,,$$

and the mean lengths of the typical J-segment and the typical I-segment are

$$\ell_{\mathrm{J}} = \frac{1}{t\, \zeta_\chi}\,, \quad \ell_{\mathrm{I}} = \frac{2}{t\, \zeta_\chi}\,.$$

Proof. Each vertex of a planar STIT tessellation is a. s. the endpoint (0-face) of exactly one I-segment, because all vertices are T-type vertices. An I-segment has two endpoints, and therefore $\gamma_{\mathrm{I}} = \frac{1}{2}\gamma_{\mathrm{V}}$. Hence the result for γ_{I} follows by (16.25). The equality $\eta_1 = t$, given in (16.15), and the relation $\gamma_{\mathrm{I}}\, \ell_{\mathrm{I}} = \eta_1$, according to (8.17) for $k = 1$, provide the result for ℓ_{I}.

Section 6.1 describes the key properties of STIT tessellations. One of these is that the typical cell of the STIT tessellation Y_t has the same distribution as the typical cell of a Poisson line tessellation X_t if both tessellations have the same directional distribution χ. This is part of Theorem 6.2. Therefore $\gamma_{\mathrm{Z}} = \gamma_{\mathrm{Z}}^{\mathrm{PHT}} = \frac{1}{2}\, t^2 \zeta_\chi$ by Lemma 16.3. The relation $\gamma_{\mathrm{V}} - \gamma_{\mathrm{E}} + \gamma_{\mathrm{Z}} = 0$, which is valid for general planar stationary tessellations (see (9.8) in [33] or Theorem 10.1.6 in [172]), together with $\gamma_{\mathrm{V}}^{\mathrm{PHT}} = \frac{1}{2}\, \gamma_{\mathrm{V}} = \frac{1}{2}\, t^2 \zeta_\chi$, given in (16.19), yields

$$\gamma_{\mathrm{E}}^{\mathrm{PHT}} = t^2\, \zeta_\chi$$

for the intensity of edges in X_t. A stationary Poisson line tessellation is face-to-face, and thus all 1-faces are edges. The exchange formula (2.61) and the relations between Y_t and X_t enable us to conclude for the intensity of J-segments in Y_t:

$$\begin{aligned}
\gamma_{\mathrm{J}} &= \mathbb{E} \sum_{(S,m(S))\in \mathrm{J}(Y_t)} \mathbb{1}\{c(S) \in [0,1]^d\}\, m(S) \\
&= \mathbb{E} \sum_{S\in \mathcal{F}_1(Y_t)} \mathbb{1}\{c(S) \in [0,1]^2\} \sum_{z\in Z(Y_t)} \mathbb{1}\{S \in \mathcal{F}_1(z)\} \\
&= \mathbb{E} \sum_{z\in Z(Y_t)} \mathbb{1}\{c(z) \in [0,1]^2\} \sum_{S\in \mathcal{F}_1(Y_t)} \mathbb{1}\{S \in \mathcal{F}_1(z)\}
\end{aligned}$$

$$= \mathbb{E} \sum_{z \in Z(X_t)} \mathbb{1}\{c(z) \in [0,1]^2\} \sum_{e \in E(X_t)} \mathbb{1}\{e \in \mathcal{F}_1(z)\}$$

$$= \mathbb{E} \sum_{e \in E(X_t)} \mathbb{1}\{c(e) \in [0,1]^2\} \sum_{z \in Z(X_t)} \mathbb{1}\{e \in \mathcal{F}_1(z)\}$$

$$= 2\gamma_E^{PHT} = 2t^2\zeta_\chi , \tag{16.30}$$

because an edge in a Poisson line tessellation is always a 1-face of two neighboring cells. Finally, with $\gamma_J \ell_J = 2t$, we obtain the result for ℓ_J. Here the factor 2 arises as the collection of all J-segments covers twice the 1-skeleton of Y_t. □

The result for the intensity of I-segments in Y_t coincides with (11.2) in Theorem 11.8 for $d = 2$, that is, $\gamma_I = \frac{1}{2}\gamma_{1,t}^{PHT} = \frac{1}{2}\gamma_E^{PHT}$ with the notation used above.

Recall that J- and I-segments can have vertices in their relative interiors, whereby the relative interior of a segment is the segment without its two 0-faces. The random number of internal vertices of the typical J-segment of a planar STIT tessellation will be studied in Section 16.6. The distribution of the number of vertices in the relative interior of the maximal 1-polytope is given in Theorem 14.18 for arbitrary dimension d as an application of the Mecke-type equation in Theorem 14.1. In Section 16.4.2, internal vertices of the typical I-segment for $d = 2$ are considered again together with questions inherent in the planar case.

Here we are initially interested in the corresponding mean values, which can be expressed by special adjacencies. Similarly to (16.4), let

$$\mu_{\text{relint I,V}} := \frac{1}{\gamma_I} \mathbb{E} \sum_{S \in I(Y_t)} \mathbb{1}\{c(S) \in [0,1]^2\} \sum_{v \in V(Y_t)} \mathbb{1}\{v \in \text{relint}\, S\}$$

be the mean number of vertices in the relative interior of the typical I-segment, and let

$$\mu_{\text{relint J,V}} := \frac{1}{\gamma_J} \mathbb{E} \sum_{(S,m(S)) \in J(Y_t)} \mathbb{1}\{c(S) \in [0,1]^2\}\, m(S) \sum_{v \in V(Y_t)} \mathbb{1}\{v \in \text{relint}\, S\}$$

be the mean number of vertices in the relative interior of the typical J-segment. In both definitions the adjacency relation

$$v \sim_{adj} \text{relint}\, S \quad \text{if and only if} \quad v \in \text{relint}\, S$$

between a vertex v and a segment S of the tessellation is used. Note that only J-segments with multiplicity 1 can have internal vertices.

Lemma 16.7. *In a planar STIT tessellation Y_t with arbitrary directional distribution, the mean numbers of vertices in the relative interior of the typical I-segment and in the relative interior of the typical J-segment are given by*

$$\mu_{\text{relint I,V}} = 2 \quad \text{and} \quad \mu_{\text{relint J,V}} = \frac{1}{2} .$$

Proof. The result for the typial I-segment follows directly from equation (16.7):

$$\gamma_{\mathrm{I}}\,\mu_{\mathrm{relint\,I,V}} = \gamma_{\mathrm{V}}\,\mu_{\mathrm{V,relint\,I}} = \gamma_{\mathrm{V}}\,,$$

because each vertex of a planar STIT tessellation is a. s. a T-vertex. Consequently, the vertex is a 0-face of one I-segment, and likewise it is located in the relative interior of another I-segment. Hence $\mu_{\mathrm{V,relint\,I}} = 1$.

To show the result for the J-segments, the exchange formula in Theorem 2.29 with the components $\mathsf{A}(Y_t) = \mathcal{F}_1(Y_t)$, $\mathsf{B}(Y_t) = \mathsf{V}(Y_t)$, the functions $f(S, Y_t) = \sum_{\mathrm{z}\in \mathsf{Z}(Y_t)} \mathbb{1}\{S \in \mathcal{F}_1(\mathrm{z})\}$ and $h(\mathrm{v}, Y_t) = 1$, and with the adjacency relation, defined in (16.5), yields

$$\begin{aligned}
&\gamma_{\mathrm{J}}\,\mu_{\mathrm{relint\,J,V}}\\
&\quad= \mathbb{E} \sum_{(S,m(S))\in \mathsf{J}(Y_t)} \mathbb{1}\{c(S) \in [0,1]^2\}\, m(S) \sum_{\mathrm{v}\in \mathsf{V}(Y_t)} \mathbb{1}\{\mathrm{v} \sim_{adj} \mathrm{relint}\, S\}\\
&\quad= \mathbb{E} \sum_{S\in \mathcal{F}_1(Y_t)} \mathbb{1}\{c(S) \in [0,1]^2\} \sum_{\mathrm{z}\in \mathsf{Z}(Y_t)} \mathbb{1}\{S \in \mathcal{F}_1(\mathrm{z})\} \sum_{\mathrm{v}\in \mathsf{V}(Y_t)} \mathbb{1}\{\mathrm{v} \sim_{adj} \mathrm{relint}\, S\}\\
&\quad= \mathbb{E} \sum_{\mathrm{v}\in \mathsf{V}(Y_t)} \mathbb{1}\{\mathrm{v} \in [0,1]^2\} \sum_{S\in \mathcal{F}_1(Y_t)} \sum_{\mathrm{z}\in \mathsf{Z}(Y_t)} \mathbb{1}\{S \in \mathcal{F}_1(\mathrm{z})\}\, \mathbb{1}\{\mathrm{v} \sim_{adj} \mathrm{relint}\, S\}\\
&\quad= \mathbb{E} \sum_{\mathrm{v}\in \mathsf{V}(Y_t)} \mathbb{1}\{\mathrm{v} \in [0,1]^2\} \sum_{(S,m(S))\in \mathsf{J}(Y_t)} m(S)\, \mathbb{1}\{\mathrm{v} \sim_{adj} \mathrm{relint}\, S\}\\
&\quad= \gamma_{\mathrm{V}}\,\mu_{\mathrm{V,relint\,J}}\\
&\quad= \gamma_{\mathrm{V}}\,.
\end{aligned}$$

In the last step, we used the fact that each vertex is a. s. T-shaped vertex and is therefore located in the relative interior of a J-segment with multiplicity 1. Consequently, it follows that $\mu_{\mathrm{V,relint\,J}} = 1$. The relations $\gamma_{\mathrm{I}} = \frac{1}{2}\gamma_{\mathrm{V}}$ and $\gamma_{\mathrm{J}} = 2\gamma_{\mathrm{V}}$ from Lemma 16.6 and (16.25), respectively, complete the proof. □

The result for the typical I-segments agrees for $d = 2$ with a result given in Corollary 14.19 with $\mu_{\mathrm{relint\,I,V}} = \mathbb{E}\, \nu_1^{(0)}$.

At the end of this section, we will mention a property that makes it possible to recognize when a tessellation containing only T-shaped vertices cannot be an STIT tessellation. Such T-tessellations are considered in Section 9.2. Furthermore, in Section 8.2, it is shown that the maximal $(d-1)$-polytopes of a STIT tessellation in $\mathbb{R}^d$ satisfy a so-called "mikado principle". Put simply, this means that a realization of a STIT tessellation can completely be removed by deleting maximal polytopes subsequently in the following way. In a first step the maximal $(d-1)$-polytope can be removed whose relative interior is not intersected by other maximal polytopes. This deletion reunites two adjacent cells to their mother cell. This can be proceeded sequentially. The mikado principle results for $d = 2$ in the following property.

Corollary 16.8. *Any cell of a planar STIT tessellation Y_t has a. s. at least one 1-face that is also an I-segment.*

This result is in line with (8.11) for $d = 2$.

Figure 16.3 shows a tessellation cell z, where all 0-faces are T-shaped vertices. Furthermore, no 1-face of this cell z is equal to a maximal 1-polytope. This situation contradicts the mikado principle. Thus a STIT tessellation cannot contain such configuration of I-segments.

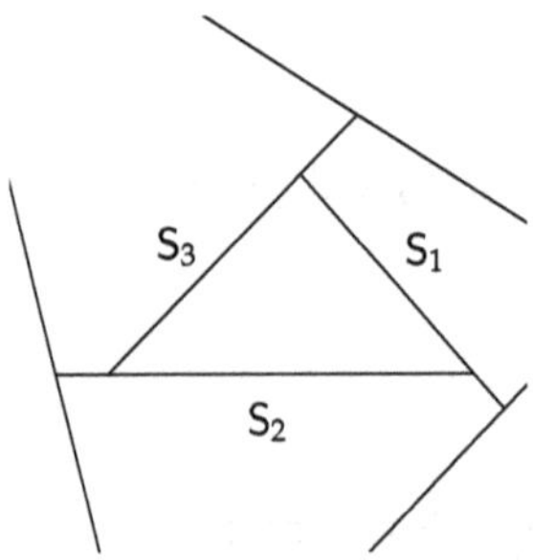

Figure 16.3: This configuration of segments S_1, S_2, S_3, which forms a triangular cell with T-shaped vertices, cannot occur in a planar STIT tessellation, because it does not comply with the mikado principle. For all S_1, S_2, S_3 their relative interior is intersected by another of the three segments.

16.3 Distributions of direction and length of different segment types

A central topic in random tessellation theory is the study of distributions of special components of the tessellation. These distributions contain information to characterize the tessellation structure and to specify properties of the tessellation components. In Section 6.1, it was shown that the typical cell of a STIT tessellation Y_t has the same distribution as the typical cell of a Poisson hyperplane tessellation X_t if both tessellations X_t and Y_t are based on the same measure $t\,\Theta$. The difference between the tessellations Y_t and X_t arises from the mutual arrangement of the cells.

In this chapter, distributions of length and direction for different types of segments in planar STIT tessellations will be considered. Furthermore, for the STIT tessellation process, several distributions concerning the birth time of I-segments are studied. Recall that the I-segments are the maximal polytopes in the planar case. In Section 14.5 the joint distribution of maximal polytopes and their birth times is given for general dimension d. To show the result for planar STIT tessellations in Section 16.4.1, a method specific to the planar case is used.

We proceed in a similar way when analyzing the vertices in the relative interior of an I-segment. The expectation of this random number for the typical I-segment is calculated in Lemma 16.7. We are now interested in the distribution of the number of vertices in the relative interior of the typical I-segment. This distribution is given in Theorem 14.18 for general dimensions. The result is proven once again for dimension $d = 2$ in Section 16.4.2 with considerations that apply to planar STIT tessellations.

The notational style for various distributions of the typical edge, given in Section 16.1, can also be adapted to the different segment types. For $\mathsf{A} \in \{\mathsf{E}, \mathsf{J}, \mathsf{I}\}$, let the measure $\mathbb{Q}_{\mathsf{A}:h,\ell}$ on $\mathcal{H}_o \times (0,\infty)$ be the **joint distribution of direction and length for the typical segment of type** A, that is,

$$\mathbb{Q}_{\mathsf{A}:h,\ell}(A,B) := \int\limits_{\mathcal{P}_{1,c}} \mathbb{1}\{\mathrm{aff}_o\, S \in A\}\, \mathbb{1}\{V_1(S) \in B\}\, \mathbb{Q}_{\mathsf{A}}(\mathrm{d}S)$$

for $A \in \mathcal{B}(\mathcal{H}_o)$ and $B \in \mathcal{B}((0,\infty))$. The directional distribution for the typical segment of type A is denoted by $\mathbb{Q}_{\mathsf{A}:h}$.

Remark 16.9. *Note that J-segments are not components in the strict sense of Definition 2.9. However, if a J-segment $(S, m(S))$ is considered as a 1-face $S \in \mathcal{F}_1(Y_t)$ with associated multiplicity $m(S)$ and these multiplicities are included in the formulae, then all three segment types can be treated in a similar manner. For brevity, we then write this as* $\mathsf{A} \in \{\mathsf{E}, \mathsf{J}, \mathsf{I}\}$.

For the typical edge of a planar STIT tessellation Y_t, the directional distribution $\mathbb{Q}_{\mathsf{E}:h}$ is already given in (16.26). In Corollary 16.4, it was shown that this distribution is the same as the directional distribution of the typical edge in a Poisson line tessellation X_t if X_t and Y_t have the same directional distribution χ. Later it will be proven that also the typical J- and I-segments of a STIT tessellation have the same directional distribution as the typical edge.

In addition, conditional length distributions are also considered below. For $\mathbb{Q}_{\mathsf{A}:h}$-almost all $L \in \mathcal{H}_o$, let $\mathbb{Q}_{\mathsf{A}:\ell|h=L}$ be the conditional length distribution of the typical A-segment, given its direction L. The existence of such a conditional length distribution follows from (2.35) in Corollary 2.22.

Most of the results will be given by the corresponding probability density functions, pdf for short, which is denoted by q with appropriate subscript. For example, $q_{\mathsf{A}:h,\ell}$ is the pdf of the joint distribution $\mathbb{Q}_{\mathsf{A}:h,\ell}$ of direction and length for the typical segment of type A with respect to the product measure $\chi \otimes \lambda_+$.

16.3.1 I-segments

We begin our studies with the I-segments, the maximal polytopes in a planar STIT tessellation Y_t. Relation (11.3) in Theorem 11.8 states that for $d = 2$, the following applies to the distribution of the typical I-segment in Y_t:

$$\mathbb{Q}_{\mathsf{I},t}(A) = \mathbb{Q}_{1,t}^{\max}(A) = \int\limits_0^t \frac{2s}{t^2}\, \mathbb{Q}_{1,s}^{\mathrm{PHT}}(A)\, \mathrm{d}s\,, \quad A \in \mathcal{B}(\mathcal{P}_{1,c})\,,$$

where $\mathbb{Q}_{1,s}^{\mathrm{PHT}}$ denotes the distribution of the typical edge in the Poisson line tessellation X_s. The main results in this section are consistent with those in Section 11.4, but for the proofs, special properties of planar STIT tessellations are used. This also leads to further results in the planar case. We start the investigations with the joint distribution of direction and length of the typical I-segment.

Theorem 16.10. *Let Y_t be a planar STIT tessellation with directional distribution χ in the typical point of ∂Y_t, given in (16.8), with rose of intersections $s_{t\chi}$ (see (16.17)), and with the constant $\zeta_{t\chi}$, given in (16.18). The joint distribution $\mathbb{Q}_{\mathrm{I}:h,\ell}$ of direction and length of the typical I-segment of Y_t has pdf $q_{\mathrm{I}:h,\ell}$ on $\mathcal{H}_o \times (0,\infty)$ given by*

$$q_{\mathrm{I}:h,\ell}(L,r) = \frac{2t}{\zeta_{t\chi} s_{t\chi}(L)} \int_0^{s_{t\chi}(L)} a^2 \mathrm{e}^{-ar}\,\mathrm{d}a\,, \quad (L,r) \in \mathcal{H}_o \times (0,\infty)\,, \tag{16.31}$$

with respect to the product measure $\chi \otimes \lambda_+$.

Remark 16.11. (1) *The result for the pdf $q_{\mathrm{I}:h,\ell}$ agrees with Theorem 11.18 for $d = 2$ and the particular choice $f(\mathrm{aff}_o F) = \mathbb{1}\{\mathrm{aff}_o F \in A\}$ for $A \in \mathcal{B}(\mathcal{H}_o)$. Moreover, the relations $V_1(\Pi_\Theta) = 1$ and $V_2(\Pi_\Theta) = \frac{1}{2}\zeta_\chi$ follow from* (16.27) *together with* (16.19), *and $V_1(\Pi_\Theta|L) = s_\chi(L)$ for $L \in \mathcal{H}_o = G(2,1)$ is given in* (16.20).

(2) *By integration with respect to r on the right-hand side of equation* (16.31) *we obtain that the directional distribution $\mathbb{Q}_{\mathrm{I}:h}$ of the typical I-segment has the pdf*

$$q_{\mathrm{I}:h}(L) = \frac{t}{\zeta_{t\chi}} s_{t\chi}(L) = \frac{1}{\zeta_\chi} s_\chi(L)\,, \quad L \in \mathcal{H}_o\,, \tag{16.32}$$

on $\mathcal{H}_o$ with respect to χ. Recall that $s_{t\chi} = t\, s_\chi$ and $\zeta_{t\chi} = t^2 \zeta_\chi$. Using (16.26), *for $A \in \mathcal{B}(\mathcal{H}_o)$, this results in*

$$\begin{aligned}\mathbb{Q}_{\mathrm{I}:h}(A) &= \frac{t}{\zeta_{t\chi}} \int_A s_{t\chi}(L)\,\chi(\mathrm{d}L) = \frac{1}{\zeta_\chi} \int_A s_\chi(L)\,\chi(\mathrm{d}L) \\ &= \mathbb{Q}_{\mathrm{E}:h}(A)\,.\end{aligned} \tag{16.33}$$

Together with Corollary 16.4, *this is consistent with $\mathbb{Q}^{\max}_{k,t:h} = \mathbb{Q}^{\mathrm{PHT}}_{k,t:h}$ in Corollary* 11.10 *for $d = 2$ and $k = 1$. This yields $\mathbb{Q}_{\mathrm{I}:h} = \mathbb{Q}^{\mathrm{PHT}}_{1,t:h}$, where $\mathbb{Q}^{\mathrm{PHT}}_{1,t:h}$ is the directional distribution of the typical edge in the stationary Poisson line tessellation X_t.*

To prepare the proof of Theorem 16.10, we will first introduce the so-called **remaining I-segment**. To illustrate, this is the segment starting from the typical point of the 1-skeleton ∂Y_t and stopping at the upper point (or at the right end point in the horizontal case), which is a 0-face of the I-segment containing the typical point; see also Figure 16.4.

To give a formal definition, we use that for $\mathscr{H}^1$-almost all $x \in \partial Y_t$, there exists a unique I-segment $I(x) \in \mathrm{I}(Y_t)$ with $x \in I(x)$. Only the vertices belong to two different I-segments. Hence the remaining I-segment $I_R(x)$ corresponding to x is defined for $x \in \partial Y_t \setminus \mathrm{V}(Y_t)$ as follows:

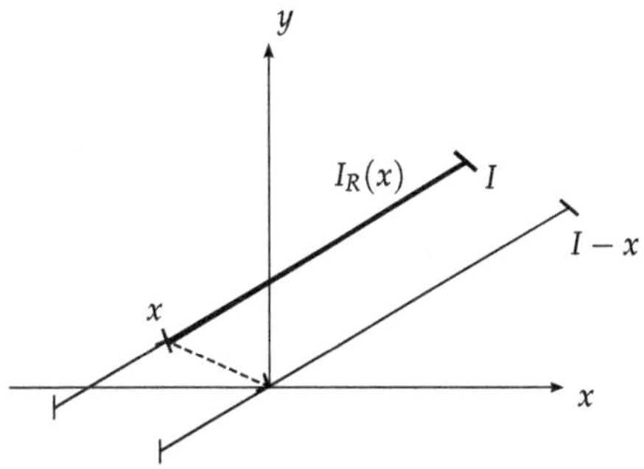

Figure 16.4: The remaining I-segment $I_R(x)$ related to the typical point x of the 1-skeleton with $x \in I$.

$$I_R(x) := \begin{cases} ((I(x)-x) \cap \{(x_1,x_2) \in \mathbb{R}^2 : x_2 \geq 0\}) + x\,, & I(x) \text{ not horizontal}\,, \\ ((I(x)-x) \cap \{(x_1,0) \in \mathbb{R}^2 : x_1 \geq 0\}) + x\,, & I(x) \text{ horizontal}\,. \end{cases} \tag{16.34}$$

The distribution $\mathbb{Q}_{I_R}$ of the remaining I-segment related to the typical point of the 1-skeleton ∂Y_t is given by

$$\mathbb{Q}_{I_R}(A) = \frac{1}{t}\,\mathbb{E} \int\limits_{\partial Y_t} \mathbb{1}\{x \in [0,1]^2\}\, \mathbb{1}\{I_R(x) \in A\}\, \mathscr{H}^1(\mathrm{d}x) \tag{16.35}$$

for $A \subseteq \mathcal{P}_1$.

In the following, we use the term *remaining I-segment* as an abbreviation for remaining I-segment related to the typical point of the 1-skeleton ∂Y_t.

By (16.8) it immediately follows that the directional distribution $\mathbb{Q}_{I_R:h}$ of the remaining I-segment coincides with the directional distribution χ of the tessellation Y_t.

Lemma 16.12. *For a planar STIT tessellation Y_t with directional distribution χ in the typical point of ∂Y_t, given in* (16.8)*, and with rose of intersections $s_{t\chi}$ (see* (16.17)*), the joint distribution $\mathbb{Q}_{I_R:h,\ell}$ of direction and length of the remaining I-segment has the pdf $q_{I_R:h,\ell}$ on $\mathcal{H}_o \times (0,\infty)$ of the form*

$$q_{I_R:h,\ell}(L,r) = \frac{1}{s_{t\chi}(L)} \int\limits_0^{s_{t\chi}(L)} a\,\mathrm{e}^{-ar}\,\mathrm{d}a\,, \quad (L,r) \in \mathcal{H}_o \times (0,\infty)\,, \tag{16.36}$$

with respect to the product measure $\chi \otimes \lambda_+$.

Proof. For a segment S of Y_t with Euclidean length $V_1(S)$ and direction $\operatorname{aff}_o S \in \mathcal{H}_o$, the so-called $t\chi$-**length** $\mathcal{L}_{t\chi}(S)$ is

$$\mathcal{L}_{t\chi}(S) := s_{t\chi}(\operatorname{aff}_o S)\, V_1(S)\,. \tag{16.37}$$

Recall that for a line $L \in \mathcal{H} = A(2,1)$, the section $\partial Y_t \cap L$ of the 1-skeleton of Y_t with L generates a stationary Poisson point process on L. Furthermore, this point process has intensity $s_{t\chi}(L_o)$, where $L_o \in \mathcal{H}_o = G(2,1)$ is the line parallel to L. Using this, we can say that for a STIT tessellation Y_t, the section point process $\partial Y_t \cap L$ is a stationary Poisson point process on L with $t\chi$-intensity 1, that is, the mean number of section points in a segment on L with $t\chi$-length 1 is equal to 1. More generally, for $a > 0$, the section point process $\partial(\frac{1}{a}Y_t) \cap L$ has $t\chi$-intensity a. Hence the $t\chi$-distance between the section points is exponentially distributed with parameter a. Furthermore, in the proof, we use that the $t\chi$-distance from a randomly chosen point on L to the next section point is also exponentially distributed with parameter a.

For $A \in \mathcal{B}(\mathcal{H}_o)$ and $r > 0$, we define by

$$G_{\mathrm{I}_R}(A,r) := \int_{\mathcal{P}_1} \mathbb{1}\{\mathrm{aff}_o\, S \in A\}\, \mathbb{1}\{\mathcal{L}_{t\chi}(S) > r\}\, \mathbb{Q}_{\mathrm{I}_R}(\mathrm{d}S) \tag{16.38}$$

the probability that the remaining I-segment of Y_t has a direction in A and its $t\chi$-length is greater than r. With the scaling property $t\,Y_t \overset{\mathrm{D}}{=} Y_1$ for all $t > 0$, which is given in (6.2), and $\mathbb{1}\{\mathcal{L}_{t\chi}(S) > r\} = \mathbb{1}\{\mathcal{L}_{\chi}(S) > \frac{r}{t}\}$ according to the definition of the $t\chi$-length in (16.37), it follows that $G_{\mathrm{I}_R}(A,r)$ is independent of t.

For $a, b > 0$, let us consider now the iterated tessellation $Y_{at} \boxplus \underline{Y_{bt}}$ with Y_{at} as the frame tessellation, whose cells are subdivided by tessellations with the distribution of Y_{bt}. By Theorem 6.2, $Y_{at} \boxplus \underline{Y_{bt}}$ has the same distribution as $Y_{(a+b)t}$. The corresponding length densities are $\eta_1(Y_{at}) = at$, $\eta_1(Y_{bt}) = bt$, and $\eta_1(Y_{at} \boxplus \underline{Y_{bt}}) = (a+b)t$ according to (16.11). To study the remaining I-segment in $Y_{at} \boxplus \underline{Y_{bt}}$, we consider the typical point of the 1-skeleton $\partial(Y_{at} \boxplus \underline{Y_{bt}})$ with respect to the Palm distribution induced by the length measure on the 1-skeleton of the iterated tessellation. Due to the length densities, this point lies with probability $\frac{a}{a+b}$ on the 1-skeleton of the frame Y_{at} and with probability $\frac{b}{a+b}$ on a segment of a tessellation nested inside a cell of the frame. Consequently, the distribution of the remaining I-segment in $Y_{at} \boxplus \underline{Y_{bt}}$ is a mixture with weights $\frac{a}{a+b}$ and $\frac{b}{a+b}$ of the distribution of the remaining I-segment in the frame and the distribution of the remaining I-segment of the nested tessellations.

If the typical point x lies on a segment of the frame tessellation, then the remaining I-segment $\mathrm{I}_R(x)$ is a remaining I-segment of Y_{at}, which starts from this point, for example, x_1 in Figure 16.5. If the typical point lies on a segment of a nested tessellation, then the corresponding remaining I-segment is a remaining I-segment of Y_{bt}; see, for example, x_2 in Figure 16.5, or it can be truncated by the frame; see x_3 in Figure 16.5. In this case, it has a $t\chi$-length, which is the minimum of the $t\chi$-length of the remaining I-segment in Y_{bt} and the $t\chi$-distance from the starting point to the frame in the direction of that segment. As mentioned above, this $t\chi$-distance is exponentially distributed with parameter a from the frame tessellation, using the scaling property $Y_{at} \overset{\mathrm{D}}{=} \frac{1}{a}Y_t$ from (6.2).

In formula (16.38) for $G_{\mathrm{I}_R}(A,r)$ the survival function for the $t\chi$-length is used. The survival function of the minimum of the $t\chi$-length and the $t\chi$-distance is the product of the survival functions, because the frame and the nested tessellations are independent.

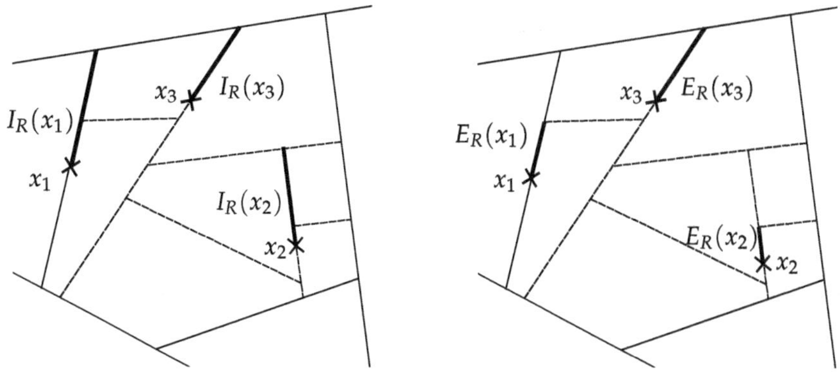

Figure 16.5: A cell of the frame tessellation together with a nested tessellation and three cases x_1, x_2, x_3 for the typical point of the 1-skeleton of the iterated tessellation. Left: The remaining I-segment $I_R(x_i)$ related to the typical point x_i. Right: The remaining edge $E_R(x_i)$ related to x_i, $i \in \{1, 2, 3\}$.

Using that $\mathcal{L}_{t\chi}(S) > r$ if and only if $\mathcal{L}_{at\chi}(S) > ar$ for a line segment S and $a > 0$, we thus get by (16.35) and (16.38) that

$$
\begin{aligned}
&G_{\mathrm{I}_R(Y_{(a+b)t})}(A, (a+b)r)\\
&= \frac{1}{(a+b)t} \mathbb{E} \int\limits_{\partial Y_{(a+b)t}} \mathbb{1}\{x \in [0,1]^2\}\, \mathbb{1}\{\mathrm{aff}_o \mathrm{I}_R(x) \in A\}\\
&\quad \times \mathbb{1}\{\mathcal{L}_{(a+b)t\chi}(\mathrm{I}_R(x)) > (a+b)r\}\, \mathscr{H}^1(\mathrm{d}x)\\
&= \frac{1}{(a+b)t} \Bigg[\mathbb{E} \int\limits_{\partial Y_{at}} \mathbb{1}\{x \in [0,1]^2\}\, \mathbb{1}\{\mathrm{aff}_o \mathrm{I}_R(x) \in A\}\\
&\quad \times \mathbb{1}\{\mathcal{L}_{at\chi}(\mathrm{I}_R(x)) > ar\}\, \mathscr{H}^1(\mathrm{d}x)\\
&\quad + \mathbb{E} \int\limits_{\partial Y_{bt}} \mathbb{1}\{x \in [0,1]^2\}\, \mathbb{1}\{\mathrm{aff}_o \mathrm{I}_R(x) \in A\}\\
&\quad \times \mathbb{1}\{\mathcal{L}_{bt\chi}(\mathrm{I}_R(x)) > br\}\, \mathrm{e}^{-ar}\, \mathscr{H}^1(\mathrm{d}x) \Bigg]
\end{aligned}
$$

for $A \in \mathcal{B}(\mathcal{H}_o)$ and $r, a, b > 0$. This yields the following functional equation for G_{I_R}, using the independence of t for G_{I_R}:

$$
G_{\mathrm{I}_R}(A, (a+b)r) = \frac{a}{a+b} G_{\mathrm{I}_R}(A, ar) + \frac{b}{a+b} G_{\mathrm{I}_R}(A, br)\, \mathrm{e}^{-ar}. \tag{16.39}
$$

To solve the equation, we first fix $A \in \mathcal{B}(\mathcal{H}_o)$ and use the function $U(r) := r\, G_{\mathrm{I}_R}(A, r)$. With the abbreviations $c := ar$ and $d := br$, the functional equation (16.39) can be transformed to

$$
U(c+d) = U(c) + U(d)\, \mathrm{e}^{-c}
$$

for $0 < c, d < \infty$. Interchanging c, d yields

$$U(c+d) = U(d) + U(c)\,\mathrm{e}^{-d}$$

for $0 < c, d < \infty$, and combining the last two equations leads to

$$U(c) + U(d)\,\mathrm{e}^{-c} = U(d) + U(c)\,\mathrm{e}^{-d} \quad \text{or} \quad \frac{U(c)}{1-\mathrm{e}^{-c}} = \frac{U(d)}{1-\mathrm{e}^{-d}}\,.$$

We see that the expression $U(c)/(1-\mathrm{e}^{-c})$ does not depend on c, and therefore it must be a constant. Of course, it can depend on A. Hence $U(c) = \mathrm{const}(A)(1-\mathrm{e}^{-c})$, and therefore

$$G_{\mathrm{I}_R}(A, r) = \frac{U(r)}{r} = \mathrm{const}(A)\,\frac{1-\mathrm{e}^{-r}}{r}\,.$$

The definition of G_{I_R} implies that as $r \to 0$, the expression $G_{\mathrm{I}_R}(A, r)$ converges to the probability that the direction of the remaining I-segment is in A, which is $\chi(A)$, that is,

$$\lim_{r\to 0} G_{\mathrm{I}_R}(A, r) = \mathrm{const}(A) = \chi(A)\,.$$

Hence the unique solution of the functional equation (16.39) for $0 < r < \infty$ and $A \in \mathcal{B}(\mathcal{H}_o)$ is

$$G_{\mathrm{I}_R}(A, r) = \chi(A)\,\frac{1-\mathrm{e}^{-r}}{r}\,.$$

If for any direction $L \in \mathcal{H}_o$ the $t\chi$-length r is replaced by the usual Euclidean length $s_{t\chi}(L)r$, then we obtain $(1-\mathrm{e}^{-s_{t\chi}(L)r})/s_{t\chi}(L)\,r$ for the survival function of the Euclidean length of the remaining I-segment, given its direction L. The derivative

$$\frac{\mathrm{d}}{\mathrm{d}r}\left(1 - \frac{1-\mathrm{e}^{-s_{t\chi}(L)r}}{s_{t\chi}(L)\,r}\right) = \frac{1}{s_{t\chi}(L)} \int\limits_0^{s_{t\chi}(L)} a\,\mathrm{e}^{-ar}\,\mathrm{d}a$$

yields (16.36), which completes the proof. □

Furthermore, the following relationship between the joint distributions $\mathbb{Q}_{\mathrm{I}_R:h,\ell}$ and $\mathbb{Q}_{\mathrm{I}:h,\ell}$ of direction and length for the remaining I-segment and the typical I-segment of Y_t, respectively, will be used in the proof of Theorem 16.10.

Lemma 16.13. *For a STIT tessellation Y_t with an arbitrary directional distribution and for any nonnegative measurable function $f : \mathcal{H}_o \times (0,\infty) \to \mathbb{R}$,*

$$\int\limits_{\mathcal{H}_o\times(0,\infty)} f(L,r)\,\mathbb{Q}_{\mathrm{I}_R:h,\ell}(\mathrm{d}(L,r)) = \frac{\zeta_{t\chi}}{2t} \int\limits_{\mathcal{H}_o\times(0,\infty)} \int\limits_{(0,r)} f(L,a)\,\lambda_1(\mathrm{d}a)\,\mathbb{Q}_{\mathrm{I}:h,\ell}(\mathrm{d}(L,r))\,,$$

where the constant $\zeta_{t\chi}$ is given in (16.18).

Proof. By (16.35) the left-hand side of the equation can be rewritten as follows:

$$\begin{aligned}
&\int\limits_{\mathcal{H}_o\times(0,\infty)} f(L,r)\,\mathbb{Q}_{\mathrm{I}_R:h,\ell}(\mathrm{d}(L,r))\\
&= \frac{1}{t}\,\mathbb{E}\int\limits_{\partial Y_t} \mathbb{1}\{x\in[0,1]^2\} f(\operatorname{aff}_o I_R(x), V_1(I_R(x)))\,\mathscr{H}^1(\mathrm{d}x)\\
&= \frac{1}{t}\,\mathbb{E}\sum_{S\in\mathrm{I}(Y_t)}\int\limits_{\mathbb{R}^2} \mathbb{1}\{x\in S\cap[0,1]^2\} f(\operatorname{aff}_o S, V_1(I_R(x)))\,\mathscr{H}^1(\mathrm{d}x)\\
&= \frac{1}{t}\,\mathbb{E}\sum_{S\in\mathrm{I}(Y_t)}\int\limits_{(-\frac{1}{2}V_1(S),\frac{1}{2}V_1(S))} \mathbb{1}\{c(S)+au\in[0,1]^2\} f\Big(\operatorname{aff}_o S, \frac{V_1(S)}{2}-a\Big)\,\lambda_1(\mathrm{d}a),
\end{aligned}$$

where u is a unit vector with direction $\operatorname{aff}_o S$, and the center function $c:\mathcal{P}_1\to\mathbb{R}^2$ assigns to each line segment S its midpoint. Obviously, for $\mathscr{H}^1$-almost all $x\in S$, both segments S and $I_R(x)$ have the same direction. Relation (2.27) yields

$$\begin{aligned}
&\frac{1}{t}\,\mathbb{E}\sum_{S\in\mathrm{I}(Y_t)}\int\limits_{(-\frac{1}{2}V_1(S),\frac{1}{2}V_1(S))} \mathbb{1}\{c(S)+au\in[0,1]^2\} f\Big(\operatorname{aff}_o S, \frac{V_1(S)}{2}-a\Big)\,\lambda_1(\mathrm{d}a)\\
&= \frac{1}{t}\,\gamma_{\mathrm{I}}\int\limits_{\mathcal{P}_{1,c}}\int\limits_{(-\frac{1}{2}V_1(S),\frac{1}{2}V_1(S))}\int\limits_{\mathbb{R}^2} \mathbb{1}\{x+au\in[0,1]^2\} f\Big(\operatorname{aff}_o S, \frac{V_1(S)}{2}-a\Big)\\
&\qquad \lambda_2(\mathrm{d}x)\,\lambda_1(\mathrm{d}a)\,\mathbb{Q}_{\mathrm{I}}(\mathrm{d}S)\\
&= \frac{1}{t}\,\gamma_{\mathrm{I}}\int\limits_{\mathcal{P}_{1,c}}\int\limits_{(-\frac{1}{2}V_1(S),\frac{1}{2}V_1(S))} f\Big(\operatorname{aff}_o S, \frac{V_1(S)}{2}-a\Big)\,\lambda_1(\mathrm{d}a)\,\mathbb{Q}_{\mathrm{I}}(\mathrm{d}S).
\end{aligned}$$

The substitution $a' = \frac{V_1(S)}{2} - a$ and the relation $\gamma_{\mathrm{I}} = \frac{1}{2}t^2\zeta_\chi = \frac{1}{2}\zeta_{t\chi}$, given in Lemma 16.6, result in

$$\begin{aligned}
&\frac{1}{t}\,\gamma_{\mathrm{I}}\int\limits_{\mathcal{P}_{1,c}}\int\limits_{(0,V_1(S))} f(\operatorname{aff}_o S, a')\,\lambda_1(\mathrm{d}a')\,\mathbb{Q}_{\mathrm{I}}(\mathrm{d}S)\\
&\quad= \frac{\zeta_{t\chi}}{2t}\int\limits_{\mathcal{H}_o\times(0,\infty)}\int\limits_{(0,r)} f(L,a)\,\lambda_1(\mathrm{d}a)\,\mathbb{Q}_{\mathrm{I}:h,\ell}(\mathrm{d}(L,r)).
\end{aligned}$$

This completes the proof. □

Note that the prefactor of the right-hand side of the relation in Lemma 16.13 is equal to the reciprocal of the mean length of the typical I-segment, which follows from Lemma 16.6 with $\zeta_{t\chi} = t^2\zeta_\chi$. Based on these results, Theorem 16.10 can now be proven.

Proof of Theorem 16.10. By the properties for the remaining I-segment given in Lemmas 16.12 and 16.13 we obtain that for all $A \in \mathcal{B}(\mathcal{H}_o)$ and $s > 0$,

$$
\begin{aligned}
&\frac{\zeta_{t\chi}}{2t} \int_{\mathcal{H}_o\times(0,\infty)} \int_0^r \mathbb{1}\{L \in A\}\, \mathbb{1}\{b \le s\}\, db\, \mathbb{Q}_{\mathrm{I}:h,\ell}(d(L,r)) \\
&= \int_{\mathcal{H}_o} \frac{1}{s_{t\chi}(L)} \int_0^\infty \mathbb{1}\{L \in A\}\, \mathbb{1}\{r \le s\} \int_0^{s_{t\chi}(L)} a\, e^{-ar}\, da\, dr\, \chi(dL) \\
&= \int_{\mathcal{H}_o} \frac{1}{s_{t\chi}(L)} \int_0^\infty \int_0^r \mathbb{1}\{L \in A\}\, \mathbb{1}\{b \le s\}\, db \int_0^{s_{t\chi}(L)} a^2\, e^{-ar}\, da\, dr\, \chi(dL)\,,
\end{aligned}
$$

where, in the final step, integration by parts is used for the integral $\int_0^\infty \mathbb{1}\{L \in A\}\, \mathbb{1}\{r \le s\} \int_0^{s_{t\chi}(L)} a\, e^{-ar}\, da\, dr$ with

$$
\frac{d}{dr} \int_0^s a\, e^{-ar}\, da = -\int_0^s a^2\, e^{-ar}\, da \tag{16.40}
$$

for $s > 0$. Hence $\mathbb{Q}_{\mathrm{I}:h,\ell}$ has the pdf

$$
(L,r) \mapsto \frac{1}{s_{t\chi}(L)} \frac{2t}{\zeta_{t\chi}} \int_0^{s_{t\chi}(L)} a^2\, e^{-ar}\, da
$$

with respect to $\chi \otimes \lambda_+$. This is the result for $q_{\mathrm{I}:h,\ell}$ given in (16.31). □

By Theorem 16.10 and relation (16.32) we can conclude that for $\mathbb{Q}_{\mathrm{I}:h}$-almost all $L \in \mathcal{H}_o$, the conditional distribution $\mathbb{Q}_{\mathrm{I}:\ell|h=L}$ of the length of the typical I-segment, given its direction $L \in \mathcal{H}_o$, has a pdf with respect to λ_+ with the following integral representation:

$$
q_{\mathrm{I}:\ell|h=L}(r) = \frac{2}{s_{t\chi}^2(L)} \int_0^{s_{t\chi}(L)} a^2\, e^{-ar}\, da\,, \quad r > 0\,. \tag{16.41}
$$

Consequently, for the typical I-segment, the conditional length distribution $\mathbb{Q}_{\mathrm{I}:\ell|h=L}$ is a mixture of exponential distributions, where the mixture distribution for the parameter of the exponential distribution is a probability measure on $(0, s_{t\chi}(L))$ with pdf $a \mapsto 2a/s_{t\chi}^2(H)$, that is,

$$
\mathbb{Q}_{\mathrm{I}:\ell|h=L} = \frac{2}{s_{t\chi}^2(L)} \int_0^{s_{t\chi}(L)} a\, \mathsf{Expo}(a)\, da\,,
$$

or in other words, for a Borel set $B \in \mathcal{B}(\mathbb{R}^+)$,

$$\mathbb{Q}_{\mathrm{I}:\ell|h=L}(B) = \frac{2}{s_{t\chi}^2(L)} \int_0^{s_{t\chi}(L)} a\,\mathbb{P}(\mathrm{E}_a \in B)\,\mathrm{d}a\,, \tag{16.42}$$

where E_a is an exponentially distributed random variable with parameter $a > 0$.

By Lemma 16.12 we obtain that for $\mathbb{Q}_{\mathrm{I}_R:h}$-almost all $L \in \mathcal{H}_o$, the conditional distribution of the length of the remaining I-segment given its direction $L \in \mathcal{H}_o$ is also a mixture of exponential distributions. Here the mixture distribution for the parameter of the exponential distribution is the uniform distribution on $(0, s_{t\chi}(L))$, that is, for $B \in \mathcal{B}(\mathbb{R}^+)$,

$$\mathbb{Q}_{\mathrm{I}_R:\ell|h=L}(B) = \frac{1}{s_{t\chi}(L)} \int_0^{s_{t\chi}(L)} \mathbb{P}(\mathrm{E}_a \in B)\,\mathrm{d}a\,. \tag{16.43}$$

Some of the integrals in this section are related to the *lower incomplete gamma function*

$$\Gamma(\upsilon+1, w) := \int_0^w a^\upsilon\,\mathrm{e}^{-a}\,\mathrm{d}a\,, \quad \upsilon, w > 0\,. \tag{16.44}$$

With $\int_0^w a^\upsilon\,\mathrm{e}^{-ar}\,\mathrm{d}a = \Gamma(\upsilon+1, rw)/r^{\upsilon+1}$, according to Theorem 16.10, for the pdf of the joint distribution of length and direction for the typical I-segment, we obtain

$$q_{\mathrm{I}:h,\ell}(L,r) = \frac{2t}{\zeta_{t\chi}\,s_{t\chi}(L)}\,\frac{1}{r^3}\,\Gamma(3, r\,s_{t\chi}(L))\,, \quad (L,r) \in \mathcal{H}_o \times (0,\infty)\,.$$

Furthermore, relations (16.36) and (16.41) result in

$$q_{\mathrm{I}_R:h,\ell}(L,r) = \frac{1}{s_{t\chi}(L)}\,\frac{1}{r^2}\,\Gamma(2, r\,s_{t\chi}(L))\,, \quad (L,r) \in \mathcal{H}_o \times (0,\infty)\,,$$

$$q_{\mathrm{I}:\ell|h=L}(r) = \frac{2}{s_{t\chi}^2(L)}\,\frac{1}{r^3}\,\Gamma(3, r\,s_{t\chi}(L))\,, \quad r > 0\,,$$

for $\mathbb{Q}_{\mathrm{I}:h}$-almost all $L \in \mathcal{H}_o$.

16.3.2 Edges

In (16.26) the directional distribution $\mathbb{Q}_{\mathrm{E}:h}$ of the typical edge is given for a planar STIT tessellation, and in Corollary 16.4, it was shown that $\mathbb{Q}_{\mathrm{E}:h} = \mathbb{Q}_{\mathrm{E}:h}^{\mathrm{PHT}}$.

To establish the result in Theorem 16.15 for the joint distribution $\mathbb{Q}_{\mathrm{E}:h,\ell}$ of direction and length of the typical edge, the same technique can be used as in the previous section for the I-segments. Therefore we will focus mainly on the differences between I-segments and edges. The definition of the **remaining edge** related to the typical point of the 1-skeleton ∂Y_t of a tessellation Y_t, abbreviated the remaining edge, is analogous to that for the remaining I-segment given in (16.34). By $\mathbb{Q}_{\mathrm{E}_R}$ the distribution of the remaining edge is denoted, similarly to (16.35).

Lemma 16.14. *For a planar STIT tessellation Y_t with directional distribution χ in the typical point of ∂Y_t, given in* (16.8)*, and with rose of intersections $s_{t\chi}$ (see* (16.17)*), the joint distribution $\mathbb{Q}_{\mathrm{E}_R:h,\ell}$ of direction and length of the remaining edge has the pdf $q_{\mathrm{E}_R:h,\ell}$ on $\mathcal{H}_o \times (0,\infty)$, that is,*

$$q_{\mathrm{E}_R:h,\ell}(L,r) = \frac{1}{s_{t\chi}(L)} \int_{s_{t\chi}(L)}^{2s_{t\chi}(L)} a\,\mathrm{e}^{-ar}\,\mathrm{d}a\,, \quad (L,r) \in \mathcal{H}_o \times (0,\infty)\,, \tag{16.45}$$

with respect to $\chi \otimes \lambda_+$.

Proof. Similarly to (16.38), for $A \in \mathcal{B}(\mathcal{H}_o)$ and $r > 0$, we define

$$G_{\mathrm{E}_R}(A,r) := \int_{\mathcal{P}_{1,R}} \mathbb{1}\{\mathrm{aff}_o\, S \in A\}\, \mathbb{1}\{\mathcal{L}_{t\chi}(S) \ge r\}\, \mathbb{Q}_{\mathrm{E}_R}(\mathrm{d}S)\,,$$

the probability that the remaining edge of Y_t has a direction in A and its $t\chi$-length $\mathcal{L}_{t\chi}$ is greater than r. The definition of the $t\chi$-length was given in (16.37). Similarly to G_{I_R}, the probability $G_{\mathrm{E}_R}(A,r)$ is independent of t. We consider the remaining edge in the iterated tessellation $Y_{at} \boxplus \underline{Y_{bt}}$. Recall that the typical point on the 1-skeleton of $Y_{at} \boxplus \underline{Y_{bt}}$ lies on the frame with probability $\frac{a}{a+b}$ and on a segment of a tessellation nested inside the frame cells with probability $\frac{b}{a+b}$.

In contrast to the behavior of the remaining I-segment, we now have the following situation. If the typical point on $\partial(Y_{at} \boxplus \underline{Y_{bt}})$ lies on the frame Y_{at}, then the remaining edge can be truncated by a segment that belongs to a tessellation iterated in one of the two adjacent frame-cells; see, for example, x_1 in the right-hand panel of Figure 16.5. These two tessellations will be nested into the neighboring cells independently. Each of them induces a Poisson point process with $t\chi$-intensity b on the line containing the remaining edge. The superposition of them is again a Poisson point process, now with $t\chi$-intensity $2b$. If the typical point lies on a segment of a nested tessellation, then we have the same situation for the remaining edge as for the remaining I-segment, for example, x_2, x_3 in Figure 16.5. Thus the functional equation for the remaining edge is

$$G_{\mathrm{E}_R}(A,(a+b)r) = \frac{a}{a+b}\, G_{\mathrm{E}_R}(A,ar)\,\mathrm{e}^{-2br} + \frac{b}{a+b}\, G_{\mathrm{E}_R}(A,br)\,\mathrm{e}^{-ar}$$

with the unique solution

$$G_{\mathrm{E}_R}(A,r) = \chi(A)\,\frac{1-\mathrm{e}^{-r}}{r}\,\mathrm{e}^{-r}$$

for $0 < r < \infty$ and $A \in \mathcal{B}(\mathcal{H}_o)$. Hence the distribution of the direction of the remaining edge coincides with the directional distribution χ of the tessellation.

The survival function of the $t\chi$-length of the remaining edge is given by

$$r \mapsto \frac{1 - \mathrm{e}^{-r}}{r}\,\mathrm{e}^{-r}.$$

As in the proof of Lemma 16.12, the $t\chi$-length r is now replaced by the usual Euclidean length $s_{t\chi}(L)r$ for any direction $L \in \mathcal{H}_o$, and the derivative

$$\frac{\mathrm{d}}{\mathrm{d}r}\left(1 - \frac{1 - \mathrm{e}^{-s_{t\chi}(L)r}}{s_{t\chi}(L)r}\,\mathrm{e}^{-s_{t\chi}(L)r}\right) = \frac{1}{s_{t\chi}(L)} \int\limits_{s_{t\chi}(L)}^{2s_{t\chi}(L)} a\,\mathrm{e}^{-ar}\,\mathrm{d}a$$

yields (16.45). This completes the proof. □

Theorem 16.15. *Let Y_t be a planar STIT tessellation with directional distribution χ in the typical point of ∂Y_t, given in* (16.8), *with rose of intersections $s_{t\chi}$ (see* (16.17)), *and with the constant $\zeta_{t\chi}$ given in* (16.18). *The joint distribution $\mathbb{Q}_{\mathrm{E}:h,\ell}$ of direction and length of the typical edge of Y_t has the pdf $q_{\mathrm{E}:h,\ell}$ on $\mathcal{H}_o \times (0,\infty)$,*

$$q_{\mathrm{E}:h,\ell}(L,r) = \frac{2t}{3\,\zeta_{t\chi}\,s_{t\chi}(L)} \int\limits_{s_{t\chi}(L)}^{2s_{t\chi}(L)} a^2\,\mathrm{e}^{-ar}\,\mathrm{d}a\,, \quad (L,r) \in \mathcal{H}_o \times (0,\infty)\,, \tag{16.46}$$

with respect to the product measure $\chi \otimes \lambda_+$.

Proof. Analogously to Lemma 16.13, the joint distributions of direction and length of the typical edge and of the remaining edge are related as follows:

$$\int\limits_{\mathcal{H}_o \times (0,\infty)} f(L,r)\,\mathbb{Q}_{\mathrm{E}_R:h,\ell}(\mathrm{d}(L,r)) = \frac{3\,\zeta_{t\chi}}{2\,t} \int\limits_{\mathcal{H}_o \times (0,\infty)} \int\limits_{(0,r)} f(L,a)\,\lambda_1(\mathrm{d}a)\,\mathbb{Q}_{\mathrm{E}:h,\ell}(\mathrm{d}(L,r))$$

for nonnegative measurable functions $f : \mathcal{H}_o \times (0,\infty) \to \mathbb{R}$. Here the prefactor of the right-hand side is the reciprocal of the mean length of the typical edge, as given in Lemma 16.6. By (16.45) and (16.40) and similar calculations as in proof of Theorem 16.10 we obtain that for all $A \in \mathcal{B}(\mathcal{H}_o)$ and $s > 0$,

$$\frac{3\,\zeta_{t\chi}}{2\,t} \int\limits_{\mathcal{H}_o \times (0,\infty)} \int\limits_0^r \mathbb{1}\{L \in A\}\,\mathbb{1}\{b \le s\}\,\mathrm{d}b\,\mathbb{Q}_{\mathrm{E}:h,\ell}(\mathrm{d}(L,r))$$

$$= \int\limits_{\mathcal{H}_o} \frac{1}{s_{t\chi}(L)} \int\limits_0^\infty \int\limits_0^r \mathbb{1}\{L \in A\}\,\mathbb{1}\{b \le s\}\,\mathrm{d}b \int\limits_{s_{t\chi}(L)}^{2s_{t\chi}(L)} a^2\,\mathrm{e}^{-ar}\,\mathrm{d}a\,\mathrm{d}r\,\chi(\mathrm{d}L)\,.$$

Thus $\mathbb{Q}_{\mathrm{E}:h,\ell}$ has the pdf

$$(L,r) \mapsto \frac{1}{s_{t\chi}(L)} \frac{2t}{3\zeta_{t\chi}} \int\limits_{s_{t\chi}(L)}^{2s_{t\chi}(L)} a^2 \,\mathrm{e}^{-ar} \,\mathrm{d}a$$

with respect to $\chi \otimes \lambda_+$, which completes the proof. □

Integration with respect to r on the right-hand side of (16.46) confirms the result for $\mathbb{Q}_{\mathrm{E}:h}$ given in (16.26).

As for the typical I-segment, we obtain the following representations for the conditional length distributions of the typical edge and of the remaining edge. From Theorem 16.15 and relation (16.26) we can conclude that for $\mathbb{Q}_{\mathrm{E}:h}$-almost all $L \in \mathcal{H}_o$, the conditional pdf of the length of the typical edge, given its direction $L \in \mathcal{H}_o$, has the integral representation

$$q_{\mathrm{E}:\ell|h=L}(r) = \frac{2}{3\,s_{t\chi}^2(L)} \int\limits_{s_{t\chi}(L)}^{2s_{t\chi}(L)} a^2 \,\mathrm{e}^{-ar} \,\mathrm{d}a \tag{16.47}$$

with respect to λ_+. Thus the conditional length distribution $\mathbb{Q}_{\mathrm{E}:\ell|h=L}$ is a mixture of exponential distributions, where the mixture distribution for the parameter of the exponential distribution is a probability measure on the interval $(s_{t\chi}(L), 2s_{t\chi}(L))$ with pdf $a \mapsto 2a/3s_{t\chi}^2(L)$. That is, for $B \in \mathcal{B}(\mathbb{R}^+)$,

$$\mathbb{Q}_{\mathrm{E}:\ell|h=L}(B) = \frac{2}{3\,s_{t\chi}^2(L)} \int\limits_{s_{t\chi}(L)}^{2s_{t\chi}(L)} a\,\mathbb{P}(\mathrm{E}_a \in B)\,\mathrm{d}a\,, \tag{16.48}$$

where E_a is an exponentially distributed random variable with parameter $a > 0$.

By (16.45), for $\mathbb{Q}_{\mathrm{E}_R:h}$-almost all $L \in \mathcal{H}_o$, the conditional distribution of the length of the remaining edge given its direction $L \in \mathcal{H}_o$ is a mixture of exponential distributions, where the mixture distribution for the parameter of the exponential distribution is the uniform distribution on the interval $(s_{t\chi}(L), 2s_{t\chi}(L))$. This means that for $B \in \mathcal{B}(\mathbb{R}^+)$,

$$\mathbb{Q}_{\mathrm{E}_R:\ell|h=L}(B) = \frac{1}{s_{t\chi}(L)} \int\limits_{s_{t\chi}(L)}^{2s_{t\chi}(L)} \mathbb{P}(\mathrm{E}_a \in B)\,\mathrm{d}a\,. \tag{16.49}$$

Similarly to Section 16.3.1, the integrals occurring here are related again to the lower incomplete gamma function defined in (16.44). With $\int_0^w a^\upsilon \,\mathrm{e}^{-ar}\,\mathrm{d}a = \Gamma(\upsilon+1, rw)/r^{\upsilon+1}$, relations (16.46), (16.45), and (16.47) result in

$$q_{\mathsf{E}:h,\ell}(L,r) = \frac{2t}{3\,\zeta_{t\chi}\,s_{t\chi}(L)}\,\frac{1}{r^3}\,(\Gamma(3,2r\,s_{t\chi}(L)) - \Gamma(3,r\,s_{t\chi}(L)))\,,$$

$$q_{\mathsf{E}_R:h,\ell}(L,r) = \frac{1}{s_{t\chi}(L)}\,\frac{1}{r^2}\,(\Gamma(2,2r\,s_{t\chi}(L)) - \Gamma(2,r\,s_{t\chi}(L)))$$

for $(L,r) \in \mathcal{H}_o \times (0,\infty)$, and

$$q_{\mathsf{E}:\ell|h=L}(r) = \frac{2}{3\,s_{t\chi}^2(L)}\,\frac{1}{r^3}\,(\Gamma(3,2r\,s_{t\chi}(L)) - \Gamma(3,r\,s_{t\chi}(L)))\,, \quad r>0\,,$$

for $\mathbb{Q}_{\mathsf{E}:h}$-almost all $L \in \mathcal{H}_o$.

16.3.3 J-segments

Much simpler is the situation in the case of the typical J-segment. Recall that the J-segments are defined in (16.28) as 1-faces of the tessellation cells, marked with their multiplicity. For the following result, the same center function $c : \mathcal{P}_1 \to \mathbb{R}^2$ is used for the J-segments in the STIT tessellation and also for the edges in the Poisson line tessellation.

Lemma 16.16. *The distribution $\mathbb{Q}_{\mathsf{J}}$ of the typical J-segment in a planar STIT tessellation Y_t and the distribution $\mathbb{Q}_{\mathsf{E}}^{\mathrm{PHT}}$ of the typical edge in a stationary Poisson line tessellation X_t, both tessellations with the directional distribution χ, are the same,*

$$\mathbb{Q}_{\mathsf{J}} = \mathbb{Q}_{\mathsf{E}}^{\mathrm{PHT}}\,.$$

Proof. By (16.29), for $A \in \mathcal{B}(\mathcal{P}_{1,c})$, we obtain

$$\begin{aligned}
\gamma_{\mathsf{J}}\,\mathbb{Q}_{\mathsf{J}}(A) &= \mathbb{E} \sum_{(S,m(S))\in\mathsf{J}(Y_t)} \mathbb{1}\{c(S) \in [0,1]^2\}\,\mathbb{1}\{S - c(S) \in A\}\,m(S) \\
&= \mathbb{E} \sum_{S\in\mathcal{F}_1(Y_t)} \mathbb{1}\{c(S) \in [0,1]^2\}\,\mathbb{1}\{S - c(S) \in A\} \sum_{z\in\mathsf{Z}(Y_t)} \mathbb{1}\{S \in \mathcal{F}_1(z)\} \\
&= \mathbb{E} \sum_{z\in\mathsf{Z}(Y_t)} \mathbb{1}\{c(z) \in [0,1]^2\} \sum_{S\in\mathcal{F}_1(Y_t)} \mathbb{1}\{S \in \mathcal{F}_1(z)\}\,\mathbb{1}\{S - c(S) \in A\}\,,
\end{aligned}$$

using the exchange formula, given in (2.61), for two sets $\mathsf{Z}(Y_t)$ and $\mathcal{F}_1(Y_t)$ of components with the functions $f(S,Y_t) = \mathbb{1}\{S - c(S) \in A\}$ and $h(z,Y_t) = 1$. Furthermore, for a cell $z \in \mathsf{Z}(Y_t)$ and a 1-face $S \in \mathcal{F}_1(Y_t)$, we apply $\mathbb{1}\{z \sim_{Y_t} S\} = \mathbb{1}\{S \in \mathcal{F}_1(z)\}$. By (6.4) in Theorem 6.2 it follows that the last mean value in the equation above is the same for Y_t and X_t, because Y_t has a Poisson typical cell. In a Poisson line tessellation, all elements of $\mathcal{F}_1(z)$ are edges. Using this and applying again the exchange formula, now for $\mathsf{Z}(X_t)$ and $\mathsf{E}(X_t)$ with an analogous relation as above, we can continue as follows:

$$\begin{aligned}
&\mathbb{E} \sum_{z\in Z(X_t)} \mathbb{1}\{c(z) \in [0,1]^2\} \sum_{e\in E(X_t)} \mathbb{1}\{e \in \mathcal{F}_1(z)\}\, \mathbb{1}\{e - c(e) \in A\} \\
&= \mathbb{E} \sum_{e\in E(X_t)} \mathbb{1}\{c(e) \in [0,1]^2\}\, \mathbb{1}\{e - c(e) \in A\} \sum_{z\in Z(X_t)} \mathbb{1}\{e \in \mathcal{F}_1(z)\} \\
&= 2\,\gamma_{\mathsf{E}}^{\mathrm{PHT}}\, \mathbb{Q}_{\mathsf{E}}^{\mathrm{PHT}}(A)\,,
\end{aligned}$$

because $\sum_{z\in Z(X_t)} \mathbb{1}\{e \in \mathcal{F}_1(z)\} = 2$ for any edge e in X_t. The relation $\gamma_{\mathsf{J}} = 2\gamma_{\mathsf{E}}^{\mathrm{PHT}}$ for the intensities, given in (16.30), completes the proof. □

The equality of the distributions in the lemma and the relation $\gamma_{\mathsf{J}} = 2\gamma_{\mathsf{E}}^{\mathrm{PHT}}$ are special cases of a result for arbitrary dimensions, which is discussed in Remark 17.4.

Using Corollary 16.4 and relation (16.33), we thus obtain that in a planar STIT tessellation Y_t the directional distributions of the typical segments of the three different types are the same,

$$\mathbb{Q}_{\mathsf{J}:h} = \mathbb{Q}_{\mathsf{E}:h} = \mathbb{Q}_{\mathsf{I}:h} \tag{16.50}$$

with pdf $s_\chi/\zeta_\chi(\cdot)$ with respect to χ.

By Lemma 16.16 it follows that $\mathbb{Q}_{\mathsf{I}:\ell|h=L} = \mathbb{Q}^{\mathrm{PHT}}_{\mathsf{E}:\ell|h=L}$. By Corollary 2.39 and (16.20) we can conclude that for $\mathbb{Q}_{\mathsf{J}:h}$-almost all $L \in \mathcal{H}_o$, the conditional distribution of the length of the typical J-segment with direction $L \in \mathcal{H}_o$ is the exponential distribution with parameter $s_{t\chi}(L)$. Combining this with (16.50) yields the following result.

Corollary 16.17. *For a planar STIT tessellation Y_t with directional distribution χ, the joint distribution $\mathbb{Q}_{\mathsf{J}:h,\ell}$ of direction and length of the typical J-segment has pdf $q_{\mathsf{J}:h,\ell}$ on $\mathcal{H}_o \times (0,\infty)$ given by*

$$q_{\mathsf{J}:h,\ell}(L,r) = \frac{t\, s_{t\chi}^2(L)}{\zeta_{t\chi}}\, \mathrm{e}^{-s_{t\chi}(L)\, r}\,, \quad (L,r) \in \mathcal{H}_o \times (0,\infty)\,, \tag{16.51}$$

with respect to the product measure $\chi \otimes \lambda_+$.

16.3.4 The isotropic and Mondrian cases

For a stationary and isotropic planar STIT tessellation Y_t with edge length intensity t, the rose of intersections does not depend on the direction L, that is, $s_{t\chi}(L) = \frac{2}{\pi}t$ and $\zeta_{t\chi} = \frac{2}{\pi}t^2$, given in (16.22). Define by $\mathbb{F}_{\mathsf{A}:\ell}(r) := \mathbb{Q}_{\mathsf{A}:\ell}((0,r])$, $r > 0$, the distribution function for the length of the typical A-segment, $\mathsf{A} \in \{\mathsf{I},\mathsf{E},\mathsf{J}\}$. Recall that J-segments are not components in the strict sense of Definition 2.9. However, according to Remark 16.9, all three segment types can be treated in a similar manner if the multiplicity of a J-segment is included in the calculations.

According to relations (16.31), (16.46), and (16.51), for a stationary and isotropic planar STIT tessellation, the distribution functions for the length of the typical A-segment,

$A \in \{I, E, J\}$, are given by

$$\mathbb{F}_{I:\ell}(r) = 1 - \frac{\pi^2}{2t^2}\frac{1}{r^2}\left[1 - \left(\frac{2t}{\pi}r + 1\right)e^{-\frac{2t}{\pi}r}\right], \tag{16.52}$$

$$\mathbb{F}_{E:\ell}(r) = 1 - \frac{\pi^2}{6t^2}\frac{1}{r^2}\left[\left(\frac{2t}{\pi}r + 1\right)e^{-\frac{2t}{\pi}r} - \left(\frac{4t}{\pi}r + 1\right)e^{-\frac{4t}{\pi}r}\right], \tag{16.53}$$

$$\mathbb{F}_{J:\ell}(r) = 1 - e^{-\frac{2t}{\pi}r},$$

and the corresponding pdfs are

$$q_{I:\ell}(r) = \frac{\pi^2}{t^2}\frac{1}{r^3}\left[1 - \left(\frac{2t^2}{\pi^2}r^2 + \frac{2t}{\pi}r + 1\right)e^{-\frac{2t}{\pi}r}\right], \tag{16.54}$$

$$q_{E:\ell}(r) = \frac{\pi^2}{3t^2}\frac{1}{r^3}\left[\left(\frac{2t^2}{\pi^2}r^2 + \frac{2t}{\pi}r + 1\right)e^{-\frac{2t}{\pi}r} - \left(\frac{8t^2}{\pi^2}r^2 + \frac{4t}{\pi}r + 1\right)e^{-\frac{4t}{\pi}r}\right],$$

$$q_{J:\ell}(r) = \frac{2t}{\pi}e^{-\frac{2t}{\pi}r} \tag{16.55}$$

for $r > 0$. These densities are shown in Figure 16.6. Note that in Corollary 11.19 the result for $q_{I:\ell}$ is given for general dimension.

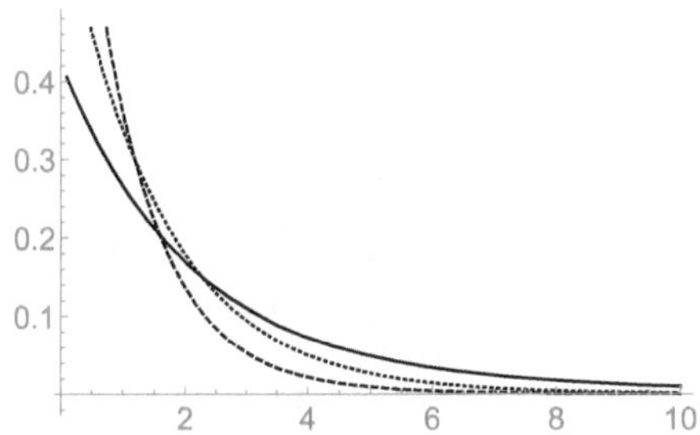

Figure 16.6: Densities $q_{I:\ell}(r)$ (solid), $q_{E:\ell}(r)$ (dashed) and $q_{J:\ell}(r)$ (dotted) in the isotropic case with $t = 1$.

Another particular case is the Mondrian case for which the definition in (4.4) and (16.23) is given, and a simulation can be seen in Figure 1.3. For $p_1 = p_2 = \frac{1}{2}$, the directional distribution χ in the typical point of the 1-skeleton is

$$\chi = \frac{1}{2}\,\delta_{L(e_1)} + \frac{1}{2}\,\delta_{L(e_2)},$$

where e_1 and e_2 are orthogonal unit vectors. Of course, the directional distribution of the typical A-segment, $A \in \{I, E, J\}$, in such a Mondrian STIT tessellation is concentrated on $L(e_1)$ and $L(e_2)$. According to (16.24), we obtain $s_{t\chi}(H) = \frac{1}{2}t$ for $H \in \{L(e_1), L(e_2)\}$ and $\zeta_{t\chi} = \frac{1}{2}t^2$. This results in the following densities of the length distribution:

$$q_{I:\ell}(r) = \frac{2}{t^2r^3}\left[8 - (t^2r^2 + 4tr + 8)\,e^{-\frac{1}{2}tr}\right],$$

$$q_{\mathrm{E}:\ell}(r) = \frac{2}{3t^2r^3}\left[(t^2r^2 + 4tr + 8)\,\mathrm{e}^{-\frac{1}{2}tr} - (4t^2r^2 + 8tr + 8)\,\mathrm{e}^{-tr}\right],$$
$$q_{\mathrm{J}:\ell}(r) = \frac{1}{2}t\,\mathrm{e}^{-\frac{1}{2}tr}$$

for $r > 0$, which again follows from (16.31), (16.46), and (16.51). These densities are illustrated in Figure 16.7.

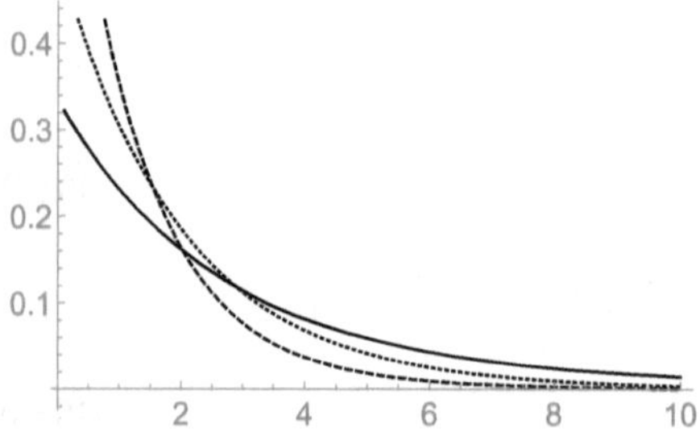

Figure 16.7: Densities $q_{\mathrm{I}:\ell}(r)$ (solid), $q_{\mathrm{E}:\ell}(r)$ (dashed), and $q_{\mathrm{J}:\ell}(r)$ (dotted) in the Mondrian case with $t = 1$.

The result for $q_{\mathrm{I}:\ell}$ in the Mondrian case agrees with the result given in Corollary 11.20 for $d = 2$.

16.3.5 Probabilistic representations

According to (16.52) and (16.53), in a stationary and isotropic planar STIT tessellation, both distribution functions for the length of the typical I-segment and for the length of the typical edge do not belong to one of the standard types in probability theory. We will now derive some alternative representations of these distributions. To simplify the constants, in this section, we choose $t = \frac{\pi}{2}$. Hence the rose of intersections $s_{t\chi}(\cdot)$ is equal to 1. Similarly to (16.42) and (16.48), we will find relationships to known distributions of probability theory for these length distributions. Recall that $\mathbb{F}_{\mathrm{A}:\ell}(r) = \mathbb{Q}_{\mathrm{A}:\ell}((0,r]),\ r > 0$, denotes the distribution function for the length of the typical A-segment.

Lemma 16.18. *Let* U *and* E *be independent random variables,* U *uniformly distributed on* $(0,1)$, *and* E *exponentially distributed with parameter 1. Then for a stationary and isotropic STIT tessellation* Y_t *with* $t = \frac{\pi}{2}$ *and for* $r > 0$,

$$\mathbb{F}_{\mathrm{I}_R:\ell}(r) = \mathbb{P}\left(\frac{\mathrm{E}}{\mathrm{U}} \le r\right), \qquad \mathbb{F}_{\mathrm{I}:\ell}(r) = \mathbb{P}\left(\frac{\mathrm{E}}{\sqrt{\mathrm{U}}} \le r\right),$$
$$\mathbb{F}_{\mathrm{E}_R:\ell}(r) = \mathbb{P}\left(\frac{\mathrm{E}}{1+\mathrm{U}} \le r\right), \quad \mathbb{F}_{\mathrm{E}:\ell}(r) = \mathbb{P}\left(\frac{\mathrm{E}}{\sqrt{1+3\mathrm{U}}} \le r\right).$$

The distribution functions for the length of the segments of the respective types satisfy the following chain of inequalities:

$$\mathbb{F}_{I_R:\ell} \leq \mathbb{F}_{I:\ell} \leq \mathbb{F}_{J:\ell} \leq \mathbb{F}_{E_R:\ell} \leq \mathbb{F}_{E:\ell} .$$

Proof. For the random variables U and E, we have

$$\mathbb{P}\left(\frac{\mathrm{E}}{\mathrm{U}} \leq r\right) = \int_0^1 \mathbb{P}(\mathrm{E} \leq ar)\, da = \int_0^1 (1 - \mathrm{e}^{-ar})\, da = 1 - \frac{1}{r}(1 - \mathrm{e}^{-r}) .$$

It follows from (16.36) that in the isotropic case with $t = \frac{\pi}{2}$ the pdf for the length of the remaining I-segment is $q_{I_R:\ell}(r) = \frac{1}{r^2}(1 - (r+1)\mathrm{e}^{-r})$. This is the derivative of $1 - \frac{1}{r}(1 - \mathrm{e}^{-r})$.

We obtain the result for the typical I-segment using (16.52):

$$\mathbb{P}\left(\frac{\mathrm{E}}{\sqrt{\mathrm{U}}} \leq r\right) = \int_0^1 (1 - \mathrm{e}^{-\sqrt{a}r})\, da = 2\int_0^1 a(1 - \mathrm{e}^{-ar})\, da = 1 - \frac{2}{r^2}(1 - (r+1)\,\mathrm{e}^{-r}) .$$

Analogously, the results for the remaining and typical edges follow by (16.45) and (16.53) and the integral representations

$$\mathbb{P}\left(\frac{\mathrm{E}}{1+\mathrm{U}} \leq r\right) = \int_1^2 (1 - \mathrm{e}^{-ar})\, da \quad \text{and} \quad \mathbb{P}\left(\frac{\mathrm{E}}{\sqrt{1+3\mathrm{U}}} \leq r\right) = \frac{2}{3}\int_1^2 a(1 - \mathrm{e}^{-ar})\, da .$$

According to Lemma 16.16, the relation $\mathbb{F}_{J:\ell}(r) = \mathbb{P}(\mathrm{E} \leq r)$ applies for the typical J-segment.

The inequalities $0 \leq \mathrm{U} \leq 1$ yield

$$\mathrm{U}^2 \leq \mathrm{U} \leq 1 \leq 1 + 2\mathrm{U} + \mathrm{U}^2 \leq 1 + 3\mathrm{U} \quad \text{or} \quad \mathrm{U} \leq \sqrt{\mathrm{U}} \leq 1 \leq 1 + \mathrm{U} \leq \sqrt{1 + 3\mathrm{U}} .$$

Finally, the chain

$$\frac{\mathrm{E}}{\mathrm{U}} \geq \frac{\mathrm{E}}{\sqrt{\mathrm{U}}} \geq \mathrm{E} \geq \frac{\mathrm{E}}{1+\mathrm{U}} \geq \frac{\mathrm{E}}{\sqrt{1+3\mathrm{U}}}$$

completes the proof. □

The relations $\mathbb{F}_{I_R:\ell} \leq \mathbb{F}_{I:\ell}$ and $\mathbb{F}_{E_R:\ell} \leq \mathbb{F}_{E:\ell}$ may seem surprising at first sight. However, note that the remaining segments I_R and E_R are determined by the typical point on the 1-skeleton of the tessellation, which is a *length*-weighted selection. In contrast, the typical I-segment and the typical edge are defined using the centroids of the corresponding segment type, which is a *number*-weighted selection.

The integral representations of the length distributions given in the above proof show that they can be expressed as mixtures of exponential distributions:

$$\mathbb{Q}_{I_R:\ell} = \int_0^1 \mathsf{Expo}(a)\, da , \quad \mathbb{Q}_{I:\ell} = 2\int_0^1 a\, \mathsf{Expo}(a)\, da , \tag{16.56}$$

$$\mathbb{Q}_{E_R:\ell} = \int_1^2 \mathsf{Expo}(a)\,da\,, \quad \mathbb{Q}_{E:\ell} = \frac{2}{3}\int_1^2 a\,\mathsf{Expo}(a)\,da\,,$$

where $\mathsf{Expo}(a)$ is the exponential distribution with parameter $a > 0$. These results for a stationary and isotropic STIT tessellation are special cases of (16.43), (16.42), (16.49), and (16.48).

16.3.6 Moments of the length

The mean lengths ℓ_I, ℓ_E, ℓ_J of the typical segments of the respective types were given in Corollary 16.5 and Lemma 16.6. We denote by $\ell_{I|h=L}$, $\ell_{E|h=L}$, $\ell_{J|h=L}$ the mean lengths of such a typical segment with respect to the conditional distribution for a given direction $L \in \mathcal{H}_o$. For $k \in \mathbb{R}$, the k-moments of the lengths are denoted by $\ell^{(k)}_{I|h=L}$, $\ell^{(k)}_{E|h=L}$, $\ell^{(k)}_{J|h=L}$. For example, for the typical I-segment for $\mathbb{Q}_{I:h}$-almost all $L \in \mathcal{H}_o$,

$$\ell^{(k)}_{I|h=L} = \int_0^\infty r^k\, q_{I:\ell|h=L}(r)\,dr\,,$$

where the pdf $q_{I:\ell|h=L}(r)$ is given in (16.41). Furthermore, $\ell^{(1)}_{I|h=L} = \ell_{I|h=L}$.

Lemma 16.19. *Let Y_t be a planar STIT tessellation with directional distribution χ in the typical point of ∂Y_t, given in (16.8). For the length of the typical I-segment of Y_t with respect to the conditional distribution for a given direction $L \in \mathcal{H}_o$, the k-moments, $k \in \mathbb{R}$, are given by*

$$\ell^{(k)}_{I|h=L} = \begin{cases} \frac{2}{2-k}\,\frac{\Gamma(k+1)}{s^k_{t\chi}(L)}\,, & k < 2\,,\\ +\infty\,, & k \ge 2\,,\end{cases}$$

for $\mathbb{Q}_{I:h}$-almost all $L \in \mathcal{H}_o$, where Γ is the gamma function.

Proof. By (16.41), for $\mathbb{Q}_{I:h}$-almost all $L \in \mathcal{H}_o$, we obtain

$$\ell^{(k)}_{I|h=L} = \int_0^\infty r^k \left(\frac{2}{s^2_{t\chi}(L)} \int_0^{s_{t\chi}(L)} a^2\,e^{-ar}\,da\right) dr\,.$$

Using that $\int_0^\infty r^k\,e^{-ar}\,dr = a^{-k-1}\Gamma(k+1)$, Fubini's theorem yields

$$\ell^{(k)}_{I|h=L} = \frac{2}{s^2_{t\chi}(L)}\,\Gamma(k+1) \int_0^{s_{t\chi}(L)} a^2\,a^{-k-1}\,da\,.$$

The integral $\int_0^{s_{t\chi}(L)} a^{1-k}\,\mathrm{d}a$ is infinite for $k \geq 2$. Integration for $k < 2$ completes the proof. □

Analogous considerations for the typical edge result in

$$\ell^{(k)}_{\mathrm{E}|h=L} = \begin{cases} \frac{2}{3\,s^k_{t\chi}(L)}\,\Gamma(k+1)\,\frac{2^k-4}{2^k(k-2)}\,, & k \neq 2\,,\\ \frac{4}{3\,s^2_{t\chi}(L)}\,\ln 2\,, & k = 2\,, \end{cases}$$

for $\mathbb{Q}_{\mathrm{E}:h}$-almost all $L \in \mathcal{H}_o$. In contrast to the I-segment, all moments of the length of the typical edge with given direction L are finite.

Because the length of the typical J-segment with direction L is exponentially distributed with parameter $s_{t\chi}(L)$, we obtain

$$\ell^{(k)}_{\mathrm{J}|h=L} = \frac{1}{s^k_{t\chi}(L)}\,\Gamma(k+1)$$

for $\mathbb{Q}_{\mathrm{J}:h}$-almost all $L \in \mathcal{H}_o$.

For $k = 1$, the mean lengths for a given direction L are

$$\ell_{\mathrm{I}|h=L} = \frac{2}{s_{t\chi}(L)}\,, \qquad \ell_{\mathrm{E}|h=L} = \frac{2}{3\,s_{t\chi}(L)}\,, \qquad \ell_{\mathrm{J}|h=L} = \frac{1}{s_{t\chi}(L)}$$

for $\mathbb{Q}_{\mathrm{A}:h}$-almost all $L \in \mathcal{H}_o$, $\mathrm{A} \in \{\mathrm{I},\mathrm{E},\mathrm{J}\}$.

By applying (16.50) and (16.32) we know that the directional distribution of the typical segment of type $\mathrm{A} \in \{\mathrm{I},\mathrm{E},\mathrm{J}\}$ has the pdf $s_\chi/\zeta_\chi(\cdot)$ with respect to χ. With

$$\ell_{\mathrm{A}} = \int_{\mathcal{H}_o} \ell_{\mathrm{A}|h=L}\,\frac{t\,s_{t\chi}(L)}{\zeta_{t\chi}}\,\chi(\mathrm{d}L)\,, \qquad \mathrm{A} \in \{\mathrm{I},\mathrm{E},\mathrm{J}\}\,,$$

the results in Corollary 16.5 and Lemma 16.6 for the mean length of the typical segments of the respective types are confirmed.

Next, we will analyze the variances of the length. Let $\mathrm{var}(V_1(\mathrm{A}))$ be the variance of the length of the typical segment of type $\mathrm{A} \in \{\mathrm{I},\mathrm{E},\mathrm{J}\}$. Furthermore, denote by $\mathrm{var}(V_1(\mathrm{A})|h = L)$ the conditional variance of the length of the typical segment of type A for $\mathbb{Q}_{\mathrm{A}:h}$-almost all $L \in \mathcal{H}_o$. Using the relation $\mathrm{var}(V_1(\mathrm{A})|h = L) = \ell^{(2)}_{\mathrm{A}|h=L} - (\ell_{\mathrm{A}|h=L})^2$, the moments of the length are given by

$$\begin{aligned} \mathrm{var}(V_1(\mathrm{I})|h = L) &= +\infty\,,\\ \mathrm{var}(V_1(\mathrm{E})|h = L) &= \frac{4}{3\,s^2_{t\chi}(L)}\left(\ln 2 - \frac{1}{3}\right),\\ \mathrm{var}(V_1(\mathrm{J})|h = L) &= \frac{1}{s^2_{t\chi}(L)}\,. \end{aligned}$$

For a stationary and isotropic STIT tessellation Y_t with $s_{t\chi}(\cdot) = \frac{2}{\pi}t$, the mean values and variances for the length of the typical A-segment, $\mathrm{A} \in \{\mathrm{I}, \mathrm{E}, \mathrm{J}\}$, are

$$\ell_{\mathrm{I}} = \frac{\pi}{t}, \quad \ell_{\mathrm{E}} = \frac{\pi}{3t}, \quad \ell_{\mathrm{J}} = \frac{\pi}{2t}$$

and

$$\operatorname{var}(V_1(\mathrm{I})) = +\infty, \quad \operatorname{var}(V_1(\mathrm{E})) = \frac{\pi^2}{3t^2}\left(\ln 2 - \frac{1}{3}\right), \quad \operatorname{var}(V_1(\mathrm{J})) = \frac{\pi^2}{4t^2}. \tag{16.57}$$

For the Mondrian case defined in (4.4), with directional distribution $\chi = \frac{1}{2}\delta_{L(e_x)} + \frac{1}{2}\delta_{L(e_y)}$ in the typical point of the 1-skeleton, we have $s_{t\chi}(L) = \frac{1}{2}t, L \in \{L(e_x), L(e_y)\}$. In this case the mean values and variances for the length of the typical A-segment, $\mathrm{A} \in \{\mathrm{I}, \mathrm{E}, \mathrm{J}\}$, are

$$\ell_{\mathrm{I}} = \frac{4}{t}, \quad \ell_{\mathrm{E}} = \frac{4}{3t}, \quad \ell_{\mathrm{J}} = \frac{2}{t}$$

and

$$\operatorname{var}(V_1(\mathrm{I})) = +\infty, \quad \operatorname{var}(V_1(\mathrm{E})) = \frac{16}{3t^2}\left(\ln 2 - \frac{1}{3}\right), \quad \operatorname{var}(V_1(\mathrm{J})) = \frac{4}{t^2}.$$

16.4 Typical I-segment in the STIT tessellation process

16.4.1 Birth time distribution

This section deals with the birth times of the I-segments. Distributions and also conditional distributions for the typical I segment are examined, which relate to the birth times. In addition, internal vertices of I-segments, conditioned on the birth time and the length of the segment, are considered. For this purpose, let $\mathbf{Y} = (Y_t, t > 0)$ be the planar STIT tessellation process driven by a translation-invariant line measure Θ given in (4.1), satisfying (4.3) in Convention 4.1. As shown in Lemma 6.8, a STIT tessellation Y_t of the considered process is stationary. Furthermore, as stated in (16.15) and (16.16), it has the length density $\eta_1 = t$ and directional distribution χ in the typical point of the 1-skeleton ∂Y_t.

To explain the birth time of an I-segment, we use the division events defined in Section 4.1. A division event (P, H, β) in the planar case describes that the cell P is divided by a line $H \in [\operatorname{int} P]$ at time $\beta > 0$. The intersection $P \cap H$ is an I-segment S, which has the **birth time** $\beta(S) = \beta$. Since each I-segment in the STIT tessellation process $\mathbf{Y} = (Y_t, t > 0)$ is created by such a division, to each I-segment a birth time can be assigned.

So far we have focussed on the length and direction of an I-segment. Now we are also interested in its birth time and introduce this as a mark as follows: For $t > 0$,

$$\mathrm{I}^*(Y_t) := \{(S, \beta(S)) \,:\, S \in \mathrm{I}_t(Y_t)\} = \mathcal{F}_1^{\max *}(Y_t)\,,$$

where $\mathrm{I}(Y_t) := \mathcal{F}_1^{\max}(Y_t)$ is the set of all I-segments of Y_t, defined in (8.9) for general dimension. In addition, $\beta(S) < t$ applies to all $S \in \mathrm{I}(Y_t)$. The "*" in the notation $\mathrm{I}^*(Y_t)$ indicates that birth time marked I-segments are considered.

In this section, various distributions and conditional distributions for the typical I-segment related to the birth time are considered. The existence of the conditional distributions follows by (2.35) in Corollary 2.22.

We start our investigations with the distribution of the birth time of the typical I-segment.

Theorem 16.20. *For $t > 0$ and a planar STIT tessellation Y_t with directional distribution χ in the typical point of ∂Y_t, given in (16.8), the distribution $\mathbb{Q}_{\mathrm{I}_t^*:\beta}$ of the birth time of the typical I-segment of Y_t has the pdf given by*

$$q_{\mathrm{I}_t^*:\beta}(s) = \frac{2s}{t^2}\,, \quad s \in (0, t)\,, \tag{16.58}$$

with respect to λ_+, which does not depend on the directional distribution χ.

Proof. The STIT tessellation process has the property

$$\mathrm{I}^*(Y_s) \subseteq \mathrm{I}^*(Y_t)\,, \quad 0 < s < t\,,$$

because an I-segment in a tessellation of the process persists to be an I-segment at any time after its birth. Let $\mathbb{F}_{\mathrm{I}_t^*:\beta}(s) := \mathbb{Q}_{\mathrm{I}_t^*:\beta}((0, s])$, $s > 0$, be the distribution function for the birth time of the typical I-segment in Y_t. According to (16.2), for $0 < s < t$, we have that

$$\begin{aligned} \gamma_{\mathrm{I}_t^*}\, \mathbb{F}_{\mathrm{I}_t^*:\beta}(s) &= \mathbb{E} \sum_{S \in \mathrm{I}^*(Y_t)} \mathbb{1}\{c(S) \in [0,1]^2\}\, \mathbb{1}\{\beta(S) < s\} \\ &= \mathbb{E} \sum_{S \in \mathrm{I}^*(Y_s)} \mathbb{1}\{c(S) \in [0,1]^2\} = \gamma_{\mathrm{I}_s^*}\,. \end{aligned}$$

Using that $\gamma_{\mathrm{I}_t^*} = \gamma_{\mathrm{I}_t} = \frac{1}{2} t^2 \zeta_\chi$, given in Lemma 16.6, we obtain $\mathbb{F}_{\mathrm{I}_t^*:\beta}(s) = s^2/t^2$, which has the stated pdf. □

The result corresponds to that in Corollary 11.25 for $d = 2$. This is a special case of Corollary 14.13(a) for $d = 2$, $k = 1$, and $j = 0$.

From now on, in this section we examine stationary STIT tessellations that are additionally isotropic.

Theorem 16.21. *Let $\mathbf{Y} = (Y_t,\, t > 0)$ be a process of stationary and isotropic STIT tessellations. Then the joint distribution function $\mathbb{F}_{\mathrm{I}_t^*:\beta,\ell}$ of the birth time and length of the typical I-segment of Y_t and the corresponding pdf $q_{\mathrm{I}_t^*:\beta,\ell}$ with respect to $\lambda_+ \otimes \lambda_+$ are given by*

$$\mathbb{F}_{I_t^*:\beta,\ell}(s,r) = \frac{s^2}{t^2} - \frac{\pi^2}{2t^2}\frac{1}{r^2}\left[1 - \left(\frac{2s}{\pi}r + 1\right)e^{-\frac{2s}{\pi}r}\right], \tag{16.59}$$

$$q_{I_t^*:\beta,\ell}(s,r) = \frac{4s^2}{\pi t^2}\,e^{-\frac{2s}{\pi}r}, \quad (s,r) \in (0,t) \times (0,\infty). \tag{16.60}$$

Proof. As in the proof of Theorem 16.20, we can conclude that

$$\begin{aligned}\gamma_{I_t^*}\,\mathbb{F}_{I_t^*:\beta,\ell}(s,r) &= \mathbb{E}\sum_{S\in I^*(Y_t)} \mathbb{1}\{c(S) \in [0,1]^2\}\,\mathbb{1}\{\beta(S) < s\}\,\mathbb{1}\{V_1(S) < r\} \\ &= \mathbb{E}\sum_{S\in I^*(Y_s)} \mathbb{1}\{c(S) \in [0,1]^2\}\,\mathbb{1}\{V_1(S) < r\} \\ &= \gamma_{I_s^*}\,\mathbb{F}_{I_s^*:\ell}(r)\end{aligned}$$

for $0 < s < t$ and $r > 0$. With $\gamma_{I_t^*} = \frac{t^2}{\pi}$ from (16.22) in the isotropic case, applying relation (16.52) for $t = s$ yields (16.59). The distribution function $\mathbb{F}_{I_t^*:\beta,\ell}$ is differentiable. Thus the pdf in (16.60) is obtained as $q_{I_t^*:\beta,\ell}(s,r) = \frac{\partial^2}{\partial s\,\partial r}\mathbb{F}_{I_t^*:\beta,\ell}(s,r)$. □

Relation (16.60) is a special case of Corollary 11.26.

Obviously, the marginal density function for the birth time coincides with $q_{I_t^*:\beta}$ in (16.58), and the marginal density function for the length of the typical I-segment agrees with the result for $q_{I:\ell}$ given in (16.54) for fixed $t > 0$ in the isotropic case.

In the isotropic case, the conditional density function of the length of the typical I-segment of Y_t under the condition that it has the birth time $\beta = s$ is

$$q_{I_t^*:\ell|\beta=s}(r) = \frac{2s}{\pi}\,e^{-\frac{2s}{\pi}r}, \quad r > 0, \quad 0 < s < t, \tag{16.61}$$

which is the density of the exponential distribution with parameter $\frac{2s}{\pi}$. This result is in line with Corollary 11.26 for the planar case.

The result can be explained as follows. An I-segment that is born at time s is a random chord in a cell of Y_s. Intuitively, it has the same length distribution as the typical interval of the point process generated when the STIT tessellation Y_s is intersected by a line. According to Lemma 6.13 with $c_{0,2} = \frac{2}{\pi}$, the section of a stationary and isotropic STIT tessellation Y_s with a fixed line results in a stationary Poisson point process on this line with intensity $\frac{2s}{\pi}$. Thus the length of the typical interval is exponentially distributed with parameter $\frac{2s}{\pi}$.

The combination of (16.61) with Theorem 16.20 provides a further interpretation of the length distribution.

Corollary 16.22. *The length distribution of the typical I-segment of a stationary and isotropic STIT tessellation Y_t is a mixture of exponential distributions*

$$q_{I_t:\ell}(r) = \frac{2}{t^2}\int_0^t s\,\frac{2s}{\pi}\,e^{-\frac{2s}{\pi}r}\,ds, \quad r > 0.$$

This equation is consistent with (16.54) and likewise with (16.56) for the special case $t = \frac{\pi}{2}$ and with the result given in Corollary 11.19 for $d = 2$.

With Corollary 11.24, the statement that the length distribution of the typical I-segment is a mixture of exponential distributions also applies to tessellations with arbitrary directional distributions, without requiring isotropy. In this case, we follow Corollary 2.39 and use that the conditional length distribution $\mathbb{Q}^{\mathrm{PHT}}_{1,t|h=L}$ of the typical edge of a Poisson line tessellation for $\mathbb{Q}^{\mathrm{PHT}}_{1,t:h}$-almost all $L \in \mathcal{H}_o$ is the exponential distribution with parameter $t\, V_1(\Pi_\Theta|L) = t\, s_\chi(L)$.

16.4.2 Internal vertices in the isotropic case

In contrast to the edges, an I-segment S in a STIT tessellation Y_t can have vertices in its relative interior. For short, we call them **internal vertices**. In the planar case, an I-segment is always born without internal vertices. They are generated by later born I-segments that have one endpoint on S. These I-segments appear in the construction in the time interval (s,t), where s is the birth time of S. Let ν_{I} denote the number of internal vertices of the typical I-segment of Y_t. In Lemma 16.7, it is shown that the mean number of internal vertices in the typical I-segment of a planar STIT tessellation is 2. In Theorem 14.18 the distribution of ν_{I} (which is $\nu_1^{(0)}$ in Corollary 14.19) is given for STIT tessellations in $\mathbb{R}^d$ as an application of the Mecke-type equation in Theorem 14.1. Note that the isotropy of Y_t is not required in this theorem. Applying $\mathbb{P}(\nu_{\mathrm{I}} = m) = \mathrm{p}_{1,0}(m)$ for $m \in \mathbb{N}_0$ and using the substitution $w = 1 - s$, we get that for $d = 2$,

$$\begin{aligned} \mathbb{P}(\nu_{\mathrm{I}} = m) &= 2^{m+1} \int_0^1 (1-w)^2 \frac{w^m}{(1+w)^{m+1}}\, dw \\ &= \frac{2^{m+2}}{(m+1)(m+2)(m+3)}\, {}_2F_1(m+1, m+1; m+4; -1)\,, \end{aligned} \tag{16.62}$$

where ${}_2F_1$ is the Gaussian hypergeometric function, which has the following integral representation, the so-called Euler integral:

$${}_2F_1(u, v; w; z) = \frac{\Gamma(w)}{\Gamma(v)\,\Gamma(w-v)} \int_0^1 \frac{1}{a^{1-v}\,(1-a)^{1-w+v}\,(1-za)^u}\, da \tag{16.63}$$

for $u, v, w > 0$, $z \in \mathbb{R}$.

The probabilities $\mathbb{P}(\nu_{\mathrm{I}} = m)$ are shown in Figure 16.8 for $m \in \{0, \dots, 10\}$ and also in Figure 14.1, where the notation $\mathbb{P}(\nu_{\mathrm{I}} = m) = p_{1,0}(m)$ is used. There they are given in comparison with the probabilities $p_{1,1}(m)$ that the typical *length-weighted* I-segment has m internal vertices.

The result (16.62) indicates that $\mathbb{P}(\nu_{\mathrm{I}} = m)$ does not depend on t. Hence the distribution of the number of internal vertices of the typical I-segment of Y_t is identical for all

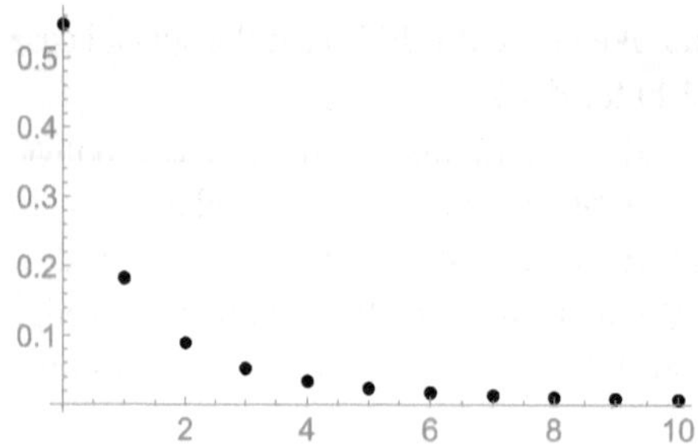

Figure 16.8: The probabilities $\mathbb{P}(\nu_{\mathrm{I}} = m)$ for $m \in \{0, \ldots, 10\}$.

$t > 0$. This can also be concluded from the scaling property $t\,Y_t \overset{\mathrm{D}}{=} Y_1$ for all $t > 0$, given in (6.2), since the number of internal vertices of the typical I-segment is not altered by a rescaling $Y_t \to t\,Y_t$.

Furthermore, the distribution of the number of vertices in the relative interior of the typical I-segment does not depend on the directional distribution χ. At first sight, this might seem surprising. To illustrate it, we consider for simplification a planar STIT tessellation driven by the Mondrian hyperplane measure, given in (4.4). A simulation of a planar Mondrian STIT tessellation can be seen in Figure 1.3. The directional distribution χ in the typical point of the 1-skeleton is $\chi = p_1\,\delta_{L(e_1)} + p_2\,\delta_{L(e_2)}$, where $p_1, p_2 \in (0,1)$ with $p_1 + p_2 = 1$. However, intuitively, we can argue as follows: For small p_1, the I-segments with direction e_1 are rare and short. They are hit by I-segments with direction e_2 with high intensity. On the other hand, I-segments with direction e_2 appear more often and are comparatively long. They are hit by I-segments with direction e_1 with low intensity. Surprisingly, these two effects are balancing and lead to a distribution of ν_{I}, which is independent from the direction of the segment and here, in this special case, independent from p_1, p_2.

We now proceed with special investigations for stationary and isotropic planar STIT tessellations to obtain further results regarding internal vertices.

The internal vertices of an I-segment S in a planar STIT tessellation can be partitioned into **"upper"** and **"lower"** vertices. These two types of vertices are shown in Figure 16.9. The line through the I-segment S generates an upper and a lower half-plane. An upper vertex arises by an I-segment located in the upper half-plane with a 0-face lying in the relative interior of S. If the I-segment S is vertical, then it is the left half-plane. Anal-

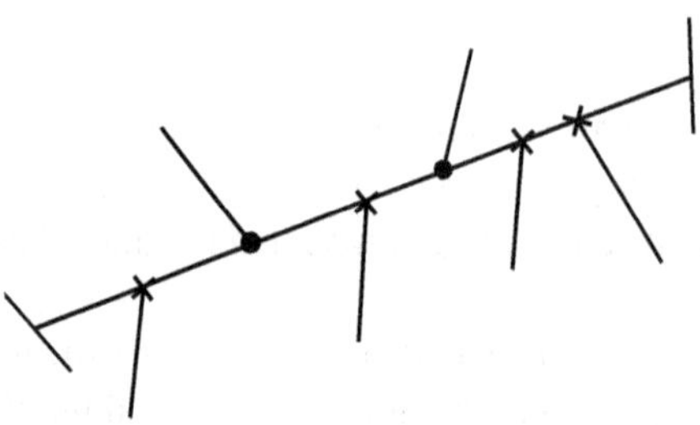

Figure 16.9: An I-segment with two upper vertices and four lower vertices.

ogously, the lower vertices are generated by I-segments located in the lower (or right) half-plane. Hence $\nu_{\mathrm{I}} = \nu_{\mathrm{I}}^{+} + \nu_{\mathrm{I}}^{-}$, where ν_{I}^{+} and ν_{I}^{-} are the numbers of upper and lower internal vertices of the typical I-segment, respectively.

The random variables ν_{I}, ν_{I}^{+}, ν_{I}^{-} are discrete with values in $\mathbb{N}_0$. At first, we are interested in some conditional distributions of ν_{I}, ν_{I}^{+}, and ν_{I}^{-}. We consider the typical I-segment of Y_t under the condition that it has birth time $s \in (0, t)$ and length $r > 0$ and the conditional distribution $\mathbb{Q}_{\mathrm{I}_t^*:\nu_{\mathrm{I}}|\beta=s,\ell=r}$ of the number of internal vertices. We denote by

$$\mathbb{P}(\nu_{\mathrm{I}} = m|\beta = s, \ell = r) = \mathbb{Q}_{\mathrm{I}_t^*:\nu_{\mathrm{I}}|\beta=s,\ell=r}(\{m\})$$

the probability that the typical I-segment has $m \in \mathbb{N}_0$ internal vertices under the given conditions with $(s, r) \in (0, t) \times (0, \infty)$.

Theorem 16.23. *For a stationary and isotropic planar STIT tessellation Y_t, the conditional distribution $\mathbb{Q}_{\mathrm{I}_t^*:\nu_{\mathrm{I}}|\beta=s,\ell=r}$ of the number of internal vertices of the typical I-segment, conditioned to its birth time and length, is given by the probabilities*

$$\mathbb{P}(\nu_{\mathrm{I}} = m|\beta = s, \ell = r) = \frac{(\frac{4}{\pi}r(t-s))^m}{m!}\, \mathrm{e}^{-\frac{4}{\pi}r(t-s)}, \quad m \in \mathbb{N}_0\,, \tag{16.64}$$

for $(s, r) \in (0, t) \times (0, \infty)$. This is the Poisson distribution with parameter $\frac{4}{\pi}r(t-s)$.

Proof. In a first step, we show that the conditional distribution of the number of upper vertices ν_{I}^{+} in the relative interior of the typical I-segment has the distribution

$$\mathbb{P}(\nu_{\mathrm{I}}^{+} = k|\beta = s, \ell = r) = \frac{(\frac{2}{\pi}r(t-s))^k}{k!}\, \mathrm{e}^{-\frac{2}{\pi}r(t-s)}, \quad k \in \mathbb{N}_0\,.$$

The STIT construction yields that in the planar case an I-segment is born without any vertex in its relative interior. For an I-segment S with $\beta(S) = s$, such internal vertices can be generated in the time interval (s, t). We know that the construction in this time interval yields the same distribution as the iteration with a STIT tessellation with length density $\eta_1 = t - s$, because $Y_t \overset{\mathrm{D}}{=} Y_s \boxplus \underline{Y_{t-s}}$, which is given in (6.1). If we consider vertices generated from only one side of an I-segment – here the upper side – then the effect of an iteration is the same as that of a section with a STIT tessellation Y_{t-s}. According to Lemma 6.13, any linear section $\partial Y_{t-s} \cap L$ of a stationary and isotropic STIT tessellation Y_{t-s} results in a Poisson point process on the line L with intensity $\frac{2}{\pi}(t-s)$. Thus the number of upper vertices in the relative interior of an I-segment with length r and birth time s has a Poisson distribution with parameter $\frac{2}{\pi}r(t-s)$ in the isotropic case.

It is obvious that the conditional distribution of ν_{I}^{-} is the same as that of ν_{I}^{+}. It is now straightforward to determine the conditional distribution of $\nu_{\mathrm{I}} = \nu_{\mathrm{I}}^{+} + \nu_{\mathrm{I}}^{-}$ given in (16.64). The sum of two independent random variables with Poisson distribution is also Poisson distributed, and the parameter is the sum of the parameters of the addends. □

By the construction, ν_I^+ and ν_I^- are conditionally independent, conditioned on $\beta = s$, $\ell = r$. This also results in the following joint conditional distribution of (ν_I^+, ν_I^-):

$$\begin{aligned} \mathbb{P}(\nu_I^+ = k, \nu_I^- = j | \beta = s, \ell = r) &= \mathbb{P}(\nu_I^+ = k | \beta = s, \ell = r)\, \mathbb{P}(\nu_I^- = j | \beta = s, \ell = r) \\ &= \frac{(\frac{2}{\pi} r(t-s))^{k+j}}{k!\, j!}\, \mathrm{e}^{-\frac{4}{\pi} r(t-s)}, \quad k, j \in \mathbb{N}_0 ; \end{aligned}$$

see also the superposition theorem, Theorem 19.13. The pdf $q_{I_t^*:\beta,\ell}$ of the joint distribution of birth time and length of the typical I-segment, given in (16.60), is continuous. Therefore we obtain the joint pdf $q_{I_t^*:\nu_I,\beta,\ell}$ as the product of the conditional one with respect to the counting measure on $\mathbb{N}_0$ and the density belonging to the condition, and similarly for ν_I^+, ν_I^- instead of ν_I.

Corollary 16.24. *Let Y_t be a stationary and isotropic planar STIT tessellation, and let $j, k, m \in \mathbb{N}_0$, $s \in (0, t)$, $r > 0$. Then*

$$\begin{aligned} q_{I_t^*:\nu_I^+,\beta,\ell}(k, s, r) &= \mathbb{P}(\nu_I^+ = k | \beta = s, \ell = r)\, q_{I_t^*:\beta,\ell}(s, r) \\ &= \frac{4s^2}{\pi t^2} \frac{(\frac{2}{\pi} r(t-s))^k}{k!}\, \mathrm{e}^{-\frac{2}{\pi} rt}, \\ q_{I_t^*:\nu_I^+,\nu_I^-,\beta,\ell}(k, j, s, r) &= \mathbb{P}(\nu_I^+ = k, \nu_I^- = j | \beta = s, \ell = r)\, q_{I_t^*:\beta,\ell}(s, r) \\ &= \frac{4s^2}{\pi t^2} \frac{(\frac{2}{\pi} r(t-s))^{k+j}}{k!\, j!}\, \mathrm{e}^{-\frac{2}{\pi} r(2t-s)}, \\ q_{I_t^*:\nu_I,\beta,\ell}(m, s, r) &= \mathbb{P}(\nu_I = m | \beta = s, \ell = r)\, q_{I_t^*:\beta,\ell}(s, r) \\ &= \frac{4s^2}{\pi t^2} \frac{(\frac{4}{\pi} r(t-s))^m}{m!}\, \mathrm{e}^{-\frac{2}{\pi} r(2t-s)} \end{aligned}$$

with respect to $(\sum_{n\in\mathbb{N}_0} \delta_n) \otimes \lambda_+ \otimes \lambda_+$ and $(\sum_{n\in\mathbb{N}_0} \delta_n) \otimes (\sum_{n\in\mathbb{N}_0} \delta_n) \otimes \lambda_+ \otimes \lambda_+$, respectively.

It is now straightforward to calculate the marginal distributions.

Lemma 16.25. *For the stationary and isotropic planar STIT tessellation Y_t, the distribution of the number of upper vertices ν_I^+ of the typical I-segment is*

$$\mathbb{P}(\nu_I^+ = k) = \frac{4}{(k+1)(k+2)(k+3)}, \quad k \in \mathbb{N}_0 . \tag{16.65}$$

Proof. Corollary 16.24 yields

$$\begin{aligned} \mathbb{P}(\nu_I^+ = k) &= \int_0^\infty \int_0^t q_{I_t^*:\nu_I^+,\beta,\ell}(k, s, r)\, \mathrm{d}s\, \mathrm{d}r \\ &= \frac{4}{\pi\, k!\, t^2} \left(\frac{2}{\pi}\right)^k \int_0^\infty r^k\, \mathrm{e}^{-\frac{2}{\pi} rt}\, \mathrm{d}r \int_0^t (t-s)^k s^2\, \mathrm{d}s . \end{aligned}$$

The last integral with respect to s can be calculated by a twofold integration by parts, which yields $2t^{k+3}/(k+1)(k+2)(k+3)$. The integral with respect to r is equal to $(\frac{\pi}{2t})^{k+1}\,\Gamma(k+1) = (\frac{\pi}{2t})^{k+1}\,k!$, using the substitution $v = \frac{2}{\pi}rt$. Combining these results, we obtain (16.65). □

Some particular values for the distribution of ν_I^+ are

$$\mathbb{P}(\nu_\mathrm{I}^+ = 0) = \frac{2}{3}, \quad \mathbb{P}(\nu_\mathrm{I}^+ = 1) = \frac{1}{6}, \quad \mathbb{P}(\nu_\mathrm{I}^+ = 2) = \frac{1}{15}.$$

The expectation for ν_I^+ can either be calculated directly from (16.65) or obtained from the known value $\mathbb{E}\,\nu_\mathrm{I} = 2$, given in Lemma 16.7. We obtain $\mathbb{E}\,\nu_\mathrm{I}^+ = 1$, and the second moment is $\mathbb{E}\,(\nu_\mathrm{I}^+)^2 = \infty$. This is not surprising because we have already shown in Section 16.3.6 that the second moment of the length of the typical I-segment is infinite.

To find a probabilistic interpretation for the distribution of ν_I^+, which is given in the above lemma, let us first consider the conditional distribution $\mathbb{P}(\nu_\mathrm{I}^+ = k|\beta = s)$. We obtain by (16.58) and Corollary 16.24 that

$$\mathbb{P}(\nu_\mathrm{I}^+ = k|\beta = s) = \frac{q_{\mathrm{I}_t^*:\nu_\mathrm{I}^+,\beta}(k,s)}{q_{\mathrm{I}_t^*:\beta}(s)} = \frac{\int_0^\infty q_{\mathrm{I}_t^*:\nu_\mathrm{I}^+,\beta,\ell}(k,s,r)\,\mathrm{d}r}{q_{\mathrm{I}_t^*:\beta}(s)} = \frac{s}{t}\left(1-\frac{s}{t}\right)^k \tag{16.66}$$

for $0 < s < t$ and $k \in \mathbb{N}_0$. This shows that the conditional distribution of the number of upper vertices of the typical I-segment of Y_t, conditioned on the birth time $s < t$, is a geometric distribution with parameter $\frac{s}{t}$.

The following relationship between the conditional distribution $\mathbb{P}(\nu_\mathrm{I}^+ = k|\beta = s)$ and exponential distributions can be derived. From (16.61) we know that the length distribution of the typical I-segment under the condition that it is born at time s is an exponential distribution with parameter $\frac{2s}{\pi}$. A segment between two adjacent upper vertices, generated on an I-segment between time s and t, is a J-segment with multiplicity 1. By (16.55) the length distribution of such a J-segment is also an exponential distribution, now with parameter $\frac{2(t-s)}{\pi}$. Let $\mathrm{E}, \mathrm{E}_1, \mathrm{E}_2, \ldots$ be independent exponentially distributed variables, where the distribution of E has the parameter $\frac{2s}{\pi}$, and the distribution of all the other variables $\mathrm{E}_1, \mathrm{E}_2, \ldots$ has the parameter $\frac{2(t-s)}{\pi}$. Since exponential distributions are memoryless, we can show that

$$\begin{aligned}\mathbb{P}(\mathrm{E}_1 + \cdots + \mathrm{E}_k \le \mathrm{E} \mid \mathrm{E}_1 + \cdots + \mathrm{E}_{k-1} \le \mathrm{E}) &= \mathbb{P}(\mathrm{E}_1 \le \mathrm{E}) \\ &= \int_0^\infty \left(1 - \mathrm{e}^{-\frac{2(t-s)}{\pi}x}\right)\frac{2s}{\pi}\,\mathrm{e}^{-\frac{2s}{\pi}x}\,\mathrm{d}x \\ &= 1 - \frac{s}{t} \end{aligned} \tag{16.67}$$

for all $k \ge 2$.

A $(k+1)$-fold application of (16.67) yields

$$\mathbb{P}(\mathrm{E}_1 + \cdots + \mathrm{E}_k \le \mathrm{E} < \mathrm{E}_1 + \cdots + \mathrm{E}_{k+1}) = \left(1 - \frac{s}{t}\right)^k \frac{s}{t} = \mathbb{P}(\nu_{\mathrm{I}}^+ = k \mid \beta = s)$$

for $k \in \mathbb{N}$.

Now let us turn to the distribution ν_{I}^+ of the number of upper vertices in the typical I-segment. By (16.58) and (16.66) we can conclude that it is a mixture of geometric distributions, because

$$\mathbb{P}(\nu_{\mathrm{I}}^+ = k) = \int_0^t \frac{s}{t}\left(1 - \frac{s}{t}\right)^k \frac{2s}{t^2}\, \mathrm{d}s = 2\int_0^1 (1-r)\, \mathbb{P}(\mathrm{G}_{1-r} = k)\, \mathrm{d}r\,,$$

using the substitution $r = 1 - \frac{s}{t}$. Here G_{1-r} is a geometrically distributed random variable with parameter $1 - r \in (0,1)$, that is, $\mathbb{P}(\mathrm{G}_{1-r} = k) = (1-r)\, r^k$, $k \in \mathbb{N}_0$.

We continue with the distribution of $(\nu_{\mathrm{I}}^+, \nu_{\mathrm{I}}^-)$.

Lemma 16.26. *For the stationary and isotropic planar STIT tessellation Y_t, the joint distribution of $(\nu_{\mathrm{I}}^+, \nu_{\mathrm{I}}^-)$ is given by*

$$\mathbb{P}(\nu_{\mathrm{I}}^+ = k, \nu_{\mathrm{I}}^- = j) = 2\binom{k+j}{j} \int_0^1 (1-w)^2 \frac{w^{k+j}}{(1+w)^{k+j+1}}\, \mathrm{d}w\,, \quad k,j \in \mathbb{N}_0\,. \tag{16.68}$$

Proof. Corollary 16.24 yields

$$\begin{aligned}
&\mathbb{P}(\nu_{\mathrm{I}}^+ = k, \nu_{\mathrm{I}}^- = j) \\
&= \int_0^\infty \int_0^t \mathbb{P}(\nu_{\mathrm{I}}^+ = k, \nu_{\mathrm{I}}^- = j \mid \beta = s, \ell = r)\, q_{\mathrm{I}_t^*:\beta,\ell}(s,r)\, \mathrm{d}s\, \mathrm{d}r \\
&= \frac{4}{\pi t^2} \frac{1}{k!\, j!} \left(\frac{2}{\pi}\right)^{k+j} \int_0^t \left(\int_0^\infty r^{k+j}\, \mathrm{e}^{-\frac{2}{\pi} r(2t-s)}\, \mathrm{d}r \right) s^2 (t-s)^{k+j}\, \mathrm{d}s\,, \quad k,j \in \mathbb{N}_0\,.
\end{aligned}$$

The inner integral with respect to r is equal to $(\frac{\pi}{2(2t-s)})^{k+j+1}(k+j)!$. By the substitution $s = (1-w)t$ we obtain (16.68). □

Recall that the random numbers ν_{I}^+ and ν_{I}^- are not stochastically independent, but conditionally independent, conditioned on $\beta = s$, $\ell = r$. The integral in (16.68) depends on the sum $k + j$ but not on the single summands. An easy calculation of the distribution of the sum $\nu_{\mathrm{I}} = \nu_{\mathrm{I}}^+ + \nu_{\mathrm{I}}^-$,

$$\mathbb{P}(\nu_{\mathrm{I}} = m) = \sum_{k=0}^m \mathbb{P}(\nu_{\mathrm{I}}^+ = k, \nu_{\mathrm{I}}^- = m-k)\,, \quad m \in \mathbb{N}_0\,,$$

results in

$$\mathbb{P}(\nu_{\mathrm{I}} = m) = 2^{m+1} \int_0^1 (1-w)^2 \frac{w^m}{(1+w)^{m+1}}\, \mathrm{d}w\,, \quad m \in \mathbb{N}_0\,. \tag{16.69}$$

This is consistent with Theorem 14.18 for $d = 2$, which is also given in (16.62). Note that the isotropy of Y_t is not required there.

The distribution of ν_{I}^+ and also the joint distribution of ν_{I}^+ and ν_{I}^- do not depend on t, like it was already observed for ν_{I}.

For further evaluation of the integral in (16.69), we consider

$$\frac{1}{2^{m+1}}\, \mathbb{P}(\nu_{\mathrm{I}} = m) = I(m,m) - 2I(m+1,m) + I(m+2,m)\,,$$

where

$$I(m,n) := \int_0^1 \frac{z^m}{(1+z)^{n+1}}\, \mathrm{d}z$$

for $m, n \in \mathbb{N}_0$. Substituting $1+z$ by v, after binomial expansion, we get

$$I(m,n) = \int_1^2 \frac{(v-1)^m}{v^{n+1}}\, \mathrm{d}v = \sum_{k=0}^{m} \binom{m}{k} (-1)^k \int_1^2 v^{m-k-n-1}\, \mathrm{d}v\,.$$

Hence

$$\begin{aligned}
&\frac{1}{2^{m+1}}\, \mathbb{P}(\nu_{\mathrm{I}} = m) \\
&= \sum_{k=0}^{m} \binom{m}{k} (-1)^k \int_1^2 v^{-k-1}\, \mathrm{d}v - 2 \sum_{k=0}^{m+1} \binom{m+1}{k} (-1)^k \int_1^2 v^{-k}\, \mathrm{d}v \\
&\quad + \sum_{k=0}^{m+2} \binom{m+2}{k} (-1)^k \int_1^2 v^{1-k}\, \mathrm{d}v \\
&= \sum_{k=1}^{m} \binom{m}{k} (-1)^k \frac{1-2^{-k}}{k} - 2 \sum_{k=0, k\neq 1}^{m+1} \binom{m+1}{k} (-1)^k \frac{1-2^{1-k}}{k-1} \\
&\quad + \sum_{k=0, k\neq 2}^{m+2} \binom{m+2}{k} (-1)^k \frac{1-2^{2-k}}{k-2} + \left(1 + \frac{1}{2}(m+1)(m+6)\right) \ln 2 \\
&= -m - \frac{5}{2} + \sum_{k=1}^{m} \frac{(-1)^k}{k} \left[\binom{m}{k} + 2\binom{m+1}{k+1} + \binom{m+2}{k+2} \right] (1 - 2^{-k}) \\
&\quad + \left(1 + \frac{1}{2}(m+1)(m+6)\right) \ln 2\,.
\end{aligned}$$

This shows that the probability that the typical I-segment in a stationary planar STIT tessellation has exactly m internal vertices can always be expressed as $q + p\ln 2$ with rational numbers p and q.

Some particular values for the number of vertices in the relative interior of the typical I-segment are

$$\begin{aligned} \mathbb{P}(\nu_{\mathrm{I}} = 0) &= 8\ln 2 - 5 \approx 0.545\,, \\ \mathbb{P}(\nu_{\mathrm{I}} = 1) &= 32\ln 2 - 22 \approx 0.181\,, \\ \mathbb{P}(\nu_{\mathrm{I}} = 2) &= 104\ln 2 - 72 \approx 0.087\,; \end{aligned} \tag{16.70}$$

see also Figure 16.8. Remember that according to Theorem 14.18, the isotropy of the tessellation is not required for these values.

Lemma 16.26 and (16.69) yield that the conditional distribution of ν_{I}^+, conditioned on $\nu_{\mathrm{I}} = \nu_{\mathrm{I}}^+ + \nu_{\mathrm{I}}^- = m$, is the binomial distribution with parameters m and $\frac{1}{2}$.

A probabilistic representation of the distribution of ν_{I} can be obtained analogously to that of ν_{I}^+. A straightforward calculation yields that the conditional distribution of ν_{I} under the condition that the birth time of the typical I-segment is s, $0 < s < t$, is the geometric distribution with parameter $\frac{s}{2t-s}$. By the substitution $w = 1 - \frac{s}{t}$ in (16.62) we obtain that for the distribution of ν_{I},

$$\begin{aligned} \mathbb{P}(\nu_{\mathrm{I}} = m) &= \int_0^t \frac{s}{2t-s}\left(1 - \frac{s}{2t-s}\right)^m \frac{2s}{t^2}\,\mathrm{d}s \\ &= 8\int_0^1 \frac{1-r}{(2-r)^3}\,\mathbb{P}(\mathrm{G}_{1-r} = m)\,\mathrm{d}r \\ &= 2\int_0^1 (1-w)\,\mathbb{P}(\mathrm{G}_{1-\frac{2w}{1+w}} = m)\,\mathrm{d}w\,. \end{aligned}$$

Recall that G_r denotes a geometrically distributed random variable with parameter $r \in (0,1)$, that is, $\mathbb{P}(\mathrm{G}_r = m) = r\,(1-r)^m$, $m \in \mathbb{N}_0$.

A probabilistic interrelation between the length and number of internal vertices of the typical I-segment is mentioned in note 14 of this chapter.

16.5 Classification of edges

In Section 16.2.2 a first classification of segments into edges (1-primitives), J-segments (1-faces marked with their multiplicity), and I-segments (maximal segments) is given. We have seen that such a classification enables a deeper study of the tessellation structure. We now consider a further classification, which refers to the edges and is based on the so-called equality relationship between edges and 1-faces of cells. The loss of equality of edges and 1-faces, but not the loss of their adjacency, occurs when a tessellation is not face-to-face. An edge e is always adjacent to two cells z_1, z_2. Let $s(z_1) \in \mathcal{F}_1(z_1)$ and $s(z_2) \in \mathcal{F}_1(z_2)$ be the 1-faces of the cells with $e \subseteq s(z_1)$ and $e \subseteq s(z_2)$ (equalities are possible). These two 1-faces can be equal. If this is the case, then they are identical with a J-segment with multiplicity 2 and $e = s(z_1) = s(z_2)$. Furthermore, in tessellations that are not face-to-face, an edge can also be equal to no J-segment or to a J-segment with multiplicity 1. Accordingly, for a planar tessellation T, the class $E(T)$ can be partitioned into three pairwise disjoint subclasses $E\langle J, i\rangle$, $i \in \{0, 1, 2\}$. We say that an edge $e \in E\langle J, i\rangle$ equals a 1-face of i cells. We abbreviate this by just saying that e is of type i and put

$$E\langle J, i\rangle := \{e \in E(T) : |\{z \in Z(T) : e \in \mathcal{F}_1(z)\}| = i\}.$$

These edge types are illustrated in Figure 16.10.

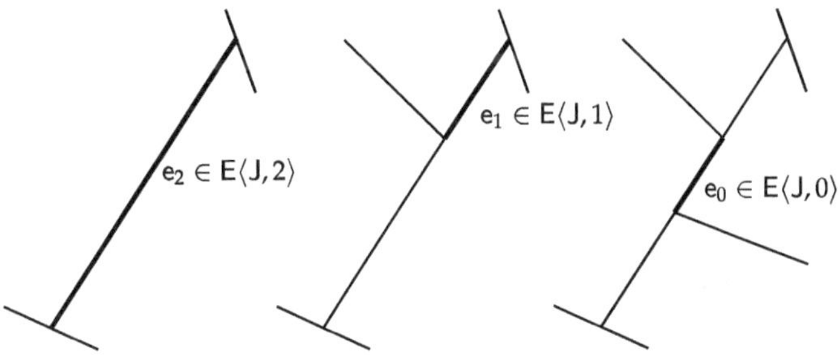

Figure 16.10: Classification of edges based on the equality relationship between edges and 1-faces of cells.

Obviously, we have that $E(T) = E\langle J, 0\rangle \cup E\langle J, 1\rangle \cup E\langle J, 2\rangle$. Following (16.1), the corresponding intensities are denoted by $\gamma_{E\langle J,i\rangle}$. We define

$$p_{E\langle J,i\rangle} := \frac{\gamma_{E\langle J,i\rangle}}{\gamma_E}, \quad i \in \{0, 1, 2\},$$

with $p_{E\langle J,0\rangle} + p_{E\langle J,1\rangle} + p_{E\langle J,2\rangle} = 1$. We can say that $p_{E\langle J,i\rangle}$ denotes the proportion of edges of type i in a random stationary tessellation. In other words, $p_{E\langle J,i\rangle}$ is the probability that the typical edge belongs to $E\langle J, i\rangle$.

For a STIT tessellation, let $\nu_{I:E\langle J,i\rangle}$ denote the number of edges of type i on the typical I-segment, $i \in \{0, 1, 2\}$.

Lemma 16.27. *For a planar STIT tessellation Y_t with directional distribution χ in the typical point of ∂Y_t, given in* (16.8), *the mean numbers of edges of type $i \in \{0, 1, 2\}$ on the typical I-segment are*

$$\mathbb{E}\,\nu_{I:E\langle J,0\rangle} = 2(2\ln 2 - 1), \quad \mathbb{E}\,\nu_{I:E\langle J,1\rangle} = 2(5 - 6\ln 2), \quad \mathbb{E}\,\nu_{I:E\langle J,2\rangle} = 8\ln 2 - 5.$$

The proportions of edges of type $i \in \{0,1,2\}$ in the tessellation are given by

$$p_{\mathrm{E}\langle \mathrm{J},0\rangle} = \frac{2}{3}(2\ln 2 - 1) \approx 0.257\,53\,,$$
$$p_{\mathrm{E}\langle \mathrm{J},1\rangle} = \frac{2}{3}(5 - 6\ln 2) \approx 0.560\,75\,,$$
$$p_{\mathrm{E}\langle \mathrm{J},2\rangle} = \frac{1}{3}(8\ln 2 - 5) \approx 0.181\,73\,.$$

Proof. An edge in a planar STIT tessellation equals a J-segment with multiplicity 2 if and only if this edge is an I-segment. So, if an I-segment comprises only one edge, then this edge is of type 2. According to (16.62), this is an event with probability $\mathbb{P}(\nu_{\mathrm{I}} = 0) = 8\ln 2 - 5$. Therefore $\mathbb{P}(\nu_{\mathrm{I}:\mathrm{E}\langle \mathrm{J},2\rangle} = 1) = \mathbb{P}(\nu_{\mathrm{I}} = 0)$, and $\mathbb{P}(\nu_{\mathrm{I}:\mathrm{E}\langle \mathrm{J},2\rangle} = 0) = 1 - \mathbb{P}(\nu_{\mathrm{I}} = 0)$. Consequently, $\mathbb{E}\,\nu_{\mathrm{I}:\mathrm{E}\langle \mathrm{J},2\rangle} = \mathbb{P}(\nu_{\mathrm{I}} = 0)$.

An I-segment with $m \geq 1$ internal vertices has $m+1$ edges. Two of these edges have a 0-face, which is equal to a 0-face of the I-segment. Thus these two edges must a. s. be of type 1. The remaining $m-1$ edges, also called internal edges, can be of type 0 or 1. In Section 16.4.2, it was shown that the internal vertices can be separated into “upper” and “lower” vertices; see also Figure 16.9. An internal edge is of type 0 if one of its 0-faces is an upper vertex and the other one is a lower vertex. If both 0-faces are of the same type, then the edge is of type 1. For each internal vertex of the I-segment, the choice of the “upper” or “lower” vertex is independent with probability $\frac{1}{2}$. This can be verified using the following consideration, which does not assume that the tessellation is isotropic.

Let $S \subset L \in \mathcal{H} = A(2,1)$ be an I-segment with birth time $\beta(S) = s \in (0,t)$ and direction $L_o \in \mathcal{H}_o$, parallel to L. In the planar case an I-segment is born without any vertex in its relative interior. Internal vertices on S arise in the time interval (s,t) by dividing the two adjacent cells. From (6.1) in Theorem 6.2 we know that the construction in this time interval yields the same distribution as the iteration with a STIT tessellation with length density $\eta_1 = t - s$, because $Y_t \overset{\mathrm{D}}{=} Y_s \boxplus \underline{Y_{t-s}}$. If we consider vertices generated from only one side of S – for example, the “upper” vertices – then the effect of an iteration is the same as that of a section with a STIT tessellation Y_{t-s}. Using Corollary 6.14, the “upper” vertices of S are therefore the restriction to S of a Poisson point process on the line L with intensity $s_{t\chi}(L_o)$. The same applies to the “lower” vertices. By the construction these two point processes are independent.

Let Φ_k be a stationary marked Poisson point process on L with mark space $\{1,2\}$ and intensity measure $s_{t\chi}(L_o)\,\lambda_1 \otimes \delta_k$, $k \in \{1,2\}$. A point $(x,1)$ of Φ_1 denotes an upper vertex, whereas a point $(x,2)$ of Φ_2 describes a lower vertex on L. Both point processes are independent. By Theorem 19.13 we obtain that the superposition $\Phi = \Phi_1 + \Phi_2$ is again a stationary marked Poisson point process on L with mark space $\{1,2\}$ and intensity measure $s_{t\chi}(L_o)\,\lambda_1 \otimes (\delta_1 + \delta_2)$. By Theorem 19.14, rewriting $s_{t\chi}(L_o)\,\lambda_1 \otimes (\delta_1 + \delta_2) = 2s_{t\chi}(L_o)\,\lambda_1 \otimes (\frac{1}{2}\delta_1 + \frac{1}{2}\delta_2)$, it follows that Φ is independently marked with mark distribution $\frac{1}{2}\delta_1 + \frac{1}{2}\delta_2$. This point process Φ restricted to the relative interior of the I-segment

$S \subset L$ represents the internal vertices of S. Hence, along the I-segment S, the vertex type of the internal vertices is determined by independent coin tosses with probabilities $\frac{1}{2}$.

Consequently, $\nu_{\mathrm{I:E}\langle \mathrm{J},0\rangle}$ and $\nu_{\mathrm{I:E}\langle \mathrm{J},1\rangle} - 2$ have a binomial distribution with parameter $(m-2, \frac{1}{2})$ conditioned on $\nu_{\mathrm{I}} = m$. Recall that in Section 16.4.2, it is shown in a different way with Lemma 16.26 that the conditional distribution of ν_{I}^{+}, conditioned on $\nu_{\mathrm{I}} = m$, is the binomial distribution with parameter $(m, \frac{1}{2})$.

Thus the joint distribution of $(\nu_{\mathrm{I:E}\langle \mathrm{J},0\rangle}, \nu_{\mathrm{I:E}\langle \mathrm{J},1\rangle})$ is

$$\mathbb{P}(\nu_{\mathrm{I:E}\langle \mathrm{J},0\rangle} = k, \nu_{\mathrm{I:E}\langle \mathrm{J},1\rangle} = j) = \begin{cases} \mathbb{P}(\nu_{\mathrm{I}} = 0)\,, & k = j = 0\,,\\ 0\,, & j = 1\,,\\ \mathbb{P}(\nu_{\mathrm{I}} = k+j-1)\binom{k+j-2}{k}(\frac{1}{2})^{k+j-2}\,, & \text{otherwise}\,, \end{cases}$$

for $k, j \in \{0, 1, 2, \ldots\}$. Hence the distribution of the number of edges of type 0 is given by the formulae

$$\mathbb{P}(\nu_{\mathrm{I:E}\langle \mathrm{J},0\rangle} = 0) = \mathbb{P}(\nu_{\mathrm{I}} = 0) + \sum_{j=2}^{\infty} \mathbb{P}(\nu_{\mathrm{I}} = j-1)\binom{j-2}{0}\left(\frac{1}{2}\right)^{j-2}$$

and

$$\mathbb{P}(\nu_{\mathrm{I:E}\langle \mathrm{J},0\rangle} = k) = \sum_{j=2}^{\infty} \mathbb{P}(\nu_{\mathrm{I}} = k+j-1)\binom{k+j-2}{k}\left(\frac{1}{2}\right)^{k+j-2}\,, \quad k \geq 1\,.$$

Using relation (16.62), we obtain that for $k \geq 1$,

$$\begin{aligned} \mathbb{P}(\nu_{\mathrm{I:E}\langle \mathrm{J},0\rangle} = k) &= \sum_{j=2}^{\infty} \binom{k+j-2}{k}\left(\frac{1}{2}\right)^{k+j-2} 2^{k+j} \int_0^1 (1-w)^2 \frac{w^{k+j-1}}{(1+w)^{k+j}}\, \mathrm{d}w \\ &= 4 \int_0^1 (1-w)^2 \left[\sum_{j=2}^{\infty} \binom{k+j-2}{k} \frac{w^{k+j-1}}{(1+w)^{k+j}}\right] \mathrm{d}w \\ &= 4 \int_0^1 (1-w)^2 \frac{w^{k-1}}{(1+w)^k} \left(\frac{w}{1+w}\right)^2 \frac{1}{(1-\frac{w}{1+w})^{k+1}}\, \mathrm{d}w \\ &= 4 \int_0^1 \frac{(1-w)^2\, w^{k+1}}{1+w}\, \mathrm{d}w \\ &= \frac{8}{(k+2)(k+3)(k+4)}\, {}_2F_1(1, k+2; k+5; -1)\,. \end{aligned}$$

Again, ${}_2F_1$ is the Gaussian hypergeometric function with integral representation (16.63). It is finite in the present context because ${}_2F_1(a, b; c; -1)$ is an absolutely convergent series when $c - a - b > 0$, as is the case here. The calculation also gives us

$$\begin{aligned}\mathbb{P}(\nu_{\mathrm{I:E}\langle\mathrm{J},0\rangle} = 0) &= \mathbb{P}(\nu_{\mathrm{I}} = 0) + 4\int_0^1 \frac{(1-w)^2\, w}{1+w}\, dw \\ &= 8\ln 2 - 5 + \frac{2}{3}(17 - 24\ln 2) \\ &\approx 0.7882\,.\end{aligned}$$

The expectation of $\nu_{\mathrm{I:E}\langle\mathrm{J},0\rangle}$ is

$$\mathbb{E}\,\nu_{\mathrm{I:E}\langle\mathrm{J},0\rangle} = 4\int_0^1 \frac{(1-w)^2}{1+w}\sum_{k=1}^{\infty} k\, w^{k+1}\, dw = 4\int_0^1 \frac{w^2}{1+w}\, dw = 2(2\ln 2 - 1)\,.$$

By similar calculations we obtain that for the edges of type 1 for $k \geq 2$,

$$\mathbb{P}(\nu_{\mathrm{I:E}\langle\mathrm{J},1\rangle} = k) = 4\int_0^1 \frac{(1-w)^2\, w^{k-1}}{1+w}\, dw = \frac{8}{k(k+1)(k+2)}\; {}_2F_1(1,k;k+3;-1)$$

and

$$\mathbb{E}\,\nu_{\mathrm{I:E}\langle\mathrm{J},1\rangle} = 4\int_0^1 \frac{(1-w)^2}{1+w}\sum_{k=2}^{\infty} k\, w^{k-1}\, dw = 4\int_0^1 \frac{w(2-w)}{1+w}\, dw = 2(5 - 6\ln 2)\,.$$

To calculate the probabilities $p_{\mathrm{E}\langle\mathrm{J},i\rangle}$, we use the exchange formula for the adjacencies between the edges of $\mathrm{E}\langle\mathrm{J},i\rangle$ and the I-segments, given in (16.7). We obtain $\gamma_{\mathrm{E}\langle\mathrm{J},i\rangle} = \gamma_{\mathrm{I}}\mathbb{E}\,\nu_{\mathrm{I:E}\langle\mathrm{J},i\rangle}$, because $\mu_{\mathrm{E}\langle\mathrm{J},i\rangle,\mathrm{I}} = 1$ and $\mu_{\mathrm{I},\mathrm{E}\langle\mathrm{J},i\rangle} = \mathbb{E}\,\nu_{\mathrm{I:E}\langle\mathrm{J},i\rangle}$. By the relation $\gamma_{\mathrm{I}} = \frac{1}{3}\gamma_{\mathrm{E}}$ from Section 16.2 we obtain

$$p_{\mathrm{E}\langle\mathrm{J},i\rangle} = \frac{1}{3}\mathbb{E}\,\nu_{\mathrm{I:E}\langle\mathrm{J},i\rangle}\,, \quad i \in \{0,1,2\}\,.$$

This completes the proof. □

16.6 Internal vertices of the typical J-segment

In this section, we focus on the J-segments. Recall that a J-segment of a planar tessellation is a 1-face marked with its multiplicity. The definition is given in Section 16.2.2. The multiplicity of a J-segment is equal to the number of tessellation cells that have this segment as a 1-face. Thus, for a random stationary tessellation X, the set of all J-segments $\mathrm{J}(X)$ can be understood as a point process on $\mathcal{P}_1$ that is not simple. The multiplicity m can be 1 or 2. Let $\mathrm{J}\langle m = i\rangle$ be the set of J-segments with multiplicity $i \in \{1,2\}$, that is,

$$\mathrm{J}\langle m = i\rangle := \{(S, m(S)) \in \mathrm{J}(X) \,:\, m(S) = i\}$$

with the corresponding intensities

$$\gamma_{\mathrm{J}\langle m=1\rangle} = \mathbb{E} \sum_{(S,m(S))\in \mathrm{J}\langle m=1\rangle} \mathbb{1}\{c(S) \in [0,1]^2\},$$

$$\gamma_{\mathrm{J}\langle m=2\rangle} = 2\,\mathbb{E} \sum_{(S,m(S))\in \mathrm{J}\langle m=2\rangle} \mathbb{1}\{c(S) \in [0,1]^2\}.$$

In a planar STIT tessellation, all vertices are T-vertices. If a J-segment has multiplicity 2, then it is equal to an I-segment. An I-segment is equal to a J-segment with multiplicity 2 if and only if it has no internal vertices. Hence

$$\gamma_{\mathrm{J}\langle m=2\rangle} = \gamma_{\mathrm{J}}\,\mathbb{P}(m_{\mathrm{J}} = 2) = 2\gamma_{\mathrm{I}}\,\mathbb{P}(\nu_{\mathrm{I}} = 0),$$

where m_{J} is the multiplicity of the typical J-segment. With $\gamma_{\mathrm{J}} = 4\gamma_{\mathrm{I}}$, given in Lemma 16.6, and $\mathbb{P}(\nu_{\mathrm{I}} = 0) = 8\ln 2 - 5 \approx 0.545$, from (16.70) we obtain the value

$$\mathbb{P}(m_{\mathrm{J}} = 2) = 4\ln 2 - \frac{5}{2} \approx 0.273$$

for the probability that the typical J-segment has multiplicity 2.

Although all J-segments with multiplicity 2 do not have internal vertices, a J-segment with multiplicity 1 can have internal vertices. Let ν_{J} be the random number of internal vertices of the typical J-segment of Y_t, that is,

$$\begin{aligned}\nu_{\mathrm{J}} &:= \frac{1}{\gamma_{\mathrm{J}}} \sum_{(S,m(S))\in \mathrm{J}(Y_t)} \mathbb{1}\{c(S) \in [0,1]^2\}\, m(S) \sum_{\mathrm{v}\in \mathrm{V}(Y_t)} \mathbb{1}\{\mathrm{v} \in \operatorname{relint} S\}\\ &= \frac{1}{\gamma_{\mathrm{J}}} \sum_{(S,m(S))\in \mathrm{J}(Y_t)} \mathbb{1}\{c(S) \in [0,1]^2\}\, \mathbb{1}\{m(S) = 1\} \sum_{\mathrm{v}\in \mathrm{V}(Y_t)} \mathbb{1}\{\mathrm{v} \in \operatorname{relint} S\}.\end{aligned}$$

From Lemma 16.7 we already know that

$$\mathbb{E}\,\nu_{\mathrm{J}} = \mu_{\operatorname{relint}\mathrm{J},\mathrm{V}} = \frac{1}{2}$$

applies for the corresponding mean value.

Theorem 16.28. *For a planar STIT tessellation Y_t with an arbitrary directional distribution, the distribution of the number ν_{J} of internal vertices of the typical J-segment is*

$$\mathbb{P}(\nu_{\mathrm{J}} = k) = \int_0^1 \frac{w^k}{(1+w)^{k+1}}\,\mathrm{d}w, \quad k \in \mathbb{N}_0.$$

Proof. For a J-segment $(S, m(S)) \in \mathrm{J}(Y_t)$, let $n_{\mathrm{V}}(S) \in \mathbb{N}_0$ be the number of internal vertices of S. Furthermore, the set $\mathrm{J}\langle k\rangle \subseteq \mathrm{J}(Y_t)$ of J-segments with $k \in \mathbb{N}_0$ internal vertices is defined by

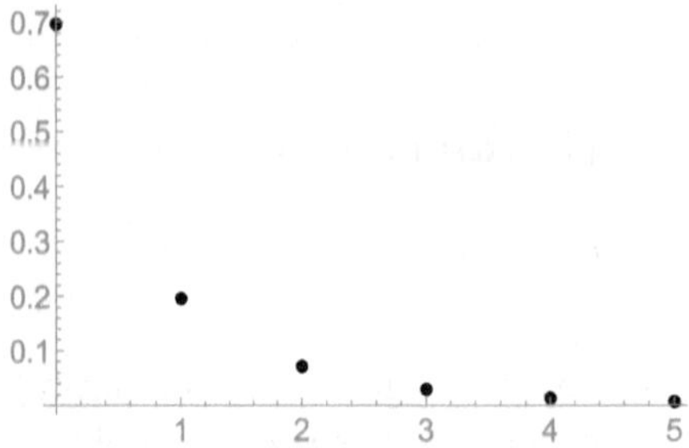

Figure 16.11: The probabilities $\mathbb{P}(\nu_{\mathrm{J}} = k)$ for $k \in \{0, \ldots, 5\}$.

$$\mathrm{J}\langle k\rangle := \{(S, m(S)) \in \mathrm{J}(Y_t) \, : \, n_{\mathrm{V}}(S) = k\} .$$

Recall that J-segments are not components in the strict sense of Definition 2.9; see Remark 16.9. However, they can be treated as 1-faces together with their multiplicities. Taking this into account, the exchange formula in Theorem 2.29 for the adjacencies between elements of $\mathrm{J}\langle k\rangle(Y_t)$ and $\mathrm{I}(Y_t)$ with a center function c results in

$$\begin{aligned} \gamma_{\mathrm{J}\langle k\rangle} &= \gamma_{\mathrm{J}} \, \mathbb{P}(\nu_{\mathrm{J}} = k) \\ &= \mathbb{E} \sum_{(S,m(S)) \in \mathrm{J}(Y_t)} \mathbb{1}\{c(S) \in [0,1]^2\} \, \mathbb{1}\{n_{\mathrm{V}}(S) - k\} \, m(S) \sum_{I \in \mathrm{I}(Y_t)} \mathbb{1}\{S \subset I\} \\ &= \mathbb{E} \sum_{I \in \mathrm{I}(Y_t)} \mathbb{1}\{c(I) \in [0,1]^2\} \sum_{(S,m(S)) \in \mathrm{J}(Y_t)} \mathbb{1}\{S \subseteq I\} \, \mathbb{1}\{n_{\mathrm{V}}(S) = k\} \, m(S) \\ &= \gamma_{\mathrm{I}} \, \mu_{\mathrm{I},\mathrm{J}\langle k\rangle} , \end{aligned}$$

where $\mu_{\mathrm{I},\mathrm{J}\langle k\rangle}$ denotes the mean number of J-segments, including the multiplicities, with k internal vertices on the typical I-segment. This relation is the same as $\gamma_{\mathrm{J}\langle k\rangle} \, \mu_{\mathrm{J}\langle k\rangle,\mathrm{I}} = \gamma_{\mathrm{I}} \, \mu_{\mathrm{I},\mathrm{J}\langle k\rangle}$, because any J-segment is adjacent to exactly one I-segment, and hence $\mu_{\mathrm{J}\langle k\rangle,\mathrm{I}} = 1$.

Let $\mu_{\mathrm{I},\mathrm{J}\langle k\rangle|\nu_{\mathrm{I}}=n}$ be the mean number of J-segments with k internal vertices on the typical I-segment under the condition that this I-segment has n internal vertices.

According to (16.6), this can be formalized as

$$\mu_{\mathrm{I},\mathrm{J}\langle k\rangle|\nu_{\mathrm{I}}=n} = \frac{1}{\mathbb{P}(\nu_{\mathrm{I}} = n)} \int_{\mathcal{T}} \sum_{S \in \mathrm{J}\langle k\rangle(T)} \mathbb{1}\{S \subseteq S^o\} \, m(S) \, \mathbb{1}\{n_{\mathrm{V}}(S^o) = n\} \, \mathbb{P}_{\mathrm{I}}^o(\mathrm{d}T) ,$$

where S^o is the I-segment of T with $c(S^o) = o$. Note that $\mathbb{P}_{\mathrm{I}}^o$ is concentrated on $\mathcal{T}^{\mathrm{I}} := \{T \in \mathcal{T} : \exists S \in \mathrm{I}(T) : c(S) = o\}$.

The relation $\mu_{\mathrm{I},\mathrm{J}\langle k\rangle} = \sum_{n=0}^{\infty} \mu_{\mathrm{I},\mathrm{J}\langle k\rangle|\nu_{\mathrm{I}}=n} \, \mathbb{P}(\nu_{\mathrm{I}} = n)$ yields

$$\mathbb{P}(\nu_{\mathrm{J}} = k) = \frac{\gamma_{\mathrm{I}}}{\gamma_{\mathrm{J}}} \sum_{n=0}^{\infty} \mu_{\mathrm{I},\mathrm{J}\langle k\rangle|\nu_{\mathrm{I}}=n} \, \mathbb{P}(\nu_{\mathrm{I}} = n) . \tag{16.71}$$

To calculate $\mu_{\mathrm{I},\mathrm{J}\langle k\rangle|\nu_{\mathrm{I}}=n}$, note that $\mu_{\mathrm{I},\mathrm{J}\langle k\rangle|\nu_{\mathrm{I}}=n} = 0$ if $k > n$. For $n \geq 1$, n vertices located in the relative interior of the I-segment are labeled from one endpoint to the other with

$1, 2, \ldots, n$. In addition, the 0-face of the I-segment next to the internal vertex with label 1 is labeled with 0. A J-segment on the I-segment is now labeled with the smaller of the two labels of its 0-faces. Note that two J-segments are given the label 0. One of these is a 1-face of a cell "above" the I-segment, and the other is a 1-face of a cell "below" it. For $n \geq k \geq 0$ and $n \geq 1$, we obtain that for the J-segments on the I-segment with n internal vertices,

$$\begin{aligned}\mu_{\mathrm{I,J}\langle k\rangle|\nu_{\mathrm{I}}=n} &= 2\,\mathbb{P}(\text{J-segment labelled 0 has } k \text{ internal vertices}|\nu_{\mathrm{I}} = n) \\ &\quad + \sum_{j=1}^{n-k} \mathbb{P}(\text{J-segment labelled } j \text{ has } k \text{ internal vertices}|\nu_{\mathrm{I}} = n)\,,\end{aligned}$$

where the sum is interpreted as 0 if $n = k$. The upper bound in the summation is $n - k$ because $\mathbb{P}(\text{J-segment labeled } j \text{ has } k \text{ internal vertices}|\nu_{\mathrm{I}} = n) = 0$ for $j > n - k$.

In the proof of Lemma 16.27, it was shown that for each internal vertex of the I-segment, the choice of the type "upper" or "lower" vertex is independent with probability $\frac{1}{2}$. Thus looking along the I-segment, the vertex type of the internal vertices is determined by independent coin tosses with probabilities $\frac{1}{2}$. All internal vertices of a J-segment must be of the same type, either "upper" or "lower" vertices. An "internal" J-segment is a segment with 0-faces that differ from the 0-faces of the I-segment. In this case, both 0-faces of the J-segment must have the same vertex type, either both "upper" or both "lower" vertices, and its internal vertices are of the opposite type. Thus

$$\mathbb{P}(\text{J-segment labeled 0 has } k \text{ internal vertices}|\nu_{\mathrm{I}} = n) = \begin{cases} (\frac{1}{2})^{k+1}, & n > k\,, \\ (\frac{1}{2})^{k}, & n = k\,. \end{cases}$$

For $1 \leq j \leq n - k$, we obtain

$$\mathbb{P}(\text{J-segment labeled } j \text{ has } k \text{ internal vertices}|\nu_{\mathrm{I}} = n) = \begin{cases} (\frac{1}{2})^{k+1}, & j < n - k\,, \\ (\frac{1}{2})^{k}, & j = n - k\,. \end{cases}$$

Therefore, for $k \leq n$,

$$\mu_{\mathrm{I,J}\langle k\rangle|\nu_{\mathrm{I}}=n} = \begin{cases} 2\,(\frac{1}{2})^{k} = (\frac{1}{2})^{k-1}, & n = k\,, \\ 2\,(\frac{1}{2})^{k+1} + \sum_{j=1}^{n-k-1} (\frac{1}{2})^{k+1} + (\frac{1}{2})^{k} = (n - k + 3)(\frac{1}{2})^{k+1}, & n > k\,. \end{cases}$$

By $\gamma_{\mathrm{J}} = 4\,\gamma_{\mathrm{I}}$, which is given in Lemma 16.6, using (16.62) for $\mathbb{P}(\nu_{\mathrm{I}} = n)$, relation (16.71) yields

$$\begin{aligned}&\mathbb{P}(\nu_{\mathrm{J}} = k) \\ &= \frac{1}{4}\left(\frac{\mathbb{P}(\nu_{\mathrm{I}} = k)}{2^{k-1}} + \sum_{n=k+1}^{\infty} \mathbb{P}(\nu_{\mathrm{I}} = n)\,\frac{n + 3 - k}{2^{k+1}} \right)\end{aligned}$$

$$
\begin{aligned}
&= \frac{1}{4}\left(4 \int_0^1 (1-w)^2 \frac{w^k}{(1+w)^{k+1}}\, \mathrm{d}w + \int_0^1 (1-w)^2 \sum_{n=k+1}^{\infty} 2^{n-k} \frac{(n+3-k)\, w^n}{(1+w)^{n+1}}\, \mathrm{d}w \right) \\
&= \frac{1}{4}\left(4 \int_0^1 (1-w)^2 \frac{w^k}{(1+w)^{k+1}}\, \mathrm{d}w + 4 \int_0^1 \frac{(2-w)\, w^{k+1}}{(1+w)^{k+1}}\, \mathrm{d}w \right) \\
&= \int_0^1 \frac{w^k}{(1+w)^{k+1}}\, \mathrm{d}w\,.
\end{aligned}
$$

This completes the argument. □

The expression for $\mathbb{P}(\nu_{\mathrm{J}} = k)$ is consistent with $\mathbb{E}\, \nu_{\mathrm{J}} = \frac{1}{2}$.

To calculate the integral $\int_0^1 \frac{w^k}{(1+w)^{k+1}}\, \mathrm{d}w$, we again use the method described in Section 16.4.2 to evaluate the integral in (16.69). This results in

$$
\mathbb{P}(\nu_{\mathrm{J}} = k) = \sum_{n=1}^{k} \binom{k}{n} (-1)^n \frac{1-2^{-n}}{n} + \ln 2\,.
$$

This shows that the probability that the typical J-segment in a stationary planar STIT tessellation has exactly k internal vertices can always be expressed as $q + \ln 2$ with a rational number q. Some particular values for the number of vertices in the relative interior of the typical J-segment are

$$
\begin{aligned}
\mathbb{P}(\nu_{\mathrm{J}} = 0) &= \ln 2 && \approx 0.693\,, \\
\mathbb{P}(\nu_{\mathrm{J}} = 1) &= \ln 2 - \frac{1}{2} && \approx 0.193\,, \\
\mathbb{P}(\nu_{\mathrm{J}} = 2) &= \ln 2 - \frac{5}{8} && \approx 0.068\,.
\end{aligned}
$$

Figure 16.11 shows the probabilities $\mathbb{P}(\nu_{\mathrm{J}} = k)$ for $k \in \{0, \ldots, 5\}$.

16.7 Comparison of isotropic STIT and Poisson line tessellations with Poisson–Voronoi tessellations

In this section, we consider three random tessellation models, which play an important role as reference models for tessellated structures: Poisson–Voronoi tessellations, abbreviated by PVTs, Poisson hyperplane tessellations (PHTs), and STIT tessellations. In Figure 1.1 simulations of these tessellations are shown. We present a systematic comparison of first- and second-order quantities of these models in $\mathbb{R}^2$ and later in Section 17.5 also in $\mathbb{R}^3$. This study aims to contribute to a deeper understanding of similarities and differences between the three model types and will therefore be valuable when choosing models for particular applications. A brief comparison of ergodic and mixing properties of these three tessellation models for arbitrary dimension d can be found at the end of this section.

A Poisson–Voronoi tessellation is generated by a stationary Poisson point process. According to Corollary 3.2.1 in [172], a stationary Poisson point process is also isotropic. Consequently, PVTs are also isotropic. Therefore we restrict the comparison for STIT tessellations and PHTs to the isotropic case as well. A brief overview of known results for PVTs in $\mathbb{R}^2$ and $\mathbb{R}^3$ is given in Section 9.7 in [33].

The results for the mean values of these three models indicate that with respect to the topological structure, STIT tessellations are closer to Poisson–Voronoi tessellations than to Poisson hyperplane tessellations. On the other hand, we know from (6.4) in Theorem 6.2 that for STIT tessellations and PHTs, the distributions of their typical cells coincide.

In the following comparisons, the values for the STIT tessellations are indicated by an upper index ^{STIT}, for PVTs by an upper index ^{PVT}, and for PHTs by ^{PHT}. Note that for $d = 2$, the hyperplanes are lines.

16.7.1 First-order comparison

We start with a comparison of the *mean values* of STIT tessellations, Poisson line tessellations, and Poisson–Voronoi tessellations in $\mathbb{R}^2$.

In the isotropic case, if the directional distribution χ in the typical point of the 1-skeleton is invariant under rotations on $\mathcal{H}_0$, then we obtain by (16.22) that $\zeta_\chi = \frac{2}{\pi}$. In a first comparison, we assume that all three tessellations have the same intensity of cells, that is, $\gamma_Z^{\mathrm{PHT}} = \gamma_Z^{\mathrm{STIT}} = \gamma_Z^{\mathrm{PVT}}$, which also implies that the mean areas of the typical cells are equal.

The mean values for a stationary and isotropic Poisson line tessellation, given in Section 9.5.1 in [33], as well as those for PVTs in Section 9.7.2 in [33] and equation (16.25) and Corollary 16.4 for STIT tessellations, yield

$$
\begin{aligned}
2\,\gamma_V^{\mathrm{PHT}} &= \gamma_V^{\mathrm{STIT}} = \gamma_V^{\mathrm{PVT}} && \text{(intensity of vertices)},\\
\frac{3}{2}\,\gamma_E^{\mathrm{PHT}} &= \gamma_E^{\mathrm{STIT}} = \gamma_E^{\mathrm{PVT}} && \text{(intensity of edges)},\\
\gamma_Z^{\mathrm{PHT}} &= \gamma_Z^{\mathrm{STIT}} = \gamma_Z^{\mathrm{PVT}} && \text{(intensity of cells)},\\
\frac{2}{3}\,\ell_E^{\mathrm{PHT}} &= \ell_E^{\mathrm{STIT}} = \frac{\sqrt{\pi}}{2}\,\ell_E^{\mathrm{PVT}} && \text{(mean length of the typical edge)},\\
\ell_Z^{\mathrm{PHT}} &= \ell_Z^{\mathrm{STIT}} = \frac{\sqrt{\pi}}{2}\,\ell_Z^{\mathrm{PVT}} && \text{(mean perimeter of the typical cell)},\\
a_Z^{\mathrm{PHT}} &= a_Z^{\mathrm{STIT}} = a_Z^{\mathrm{PVT}} && \text{(mean area of the typical cell)},\\
\eta_1^{\mathrm{PHT}} &= \eta_1^{\mathrm{STIT}} = \frac{\sqrt{\pi}}{2}\,\eta_1^{\mathrm{PVT}} && \text{(length density of the 1-skeleton)}.
\end{aligned}
$$

Furthermore, for the mean values for the adjacencies, defined in (16.4), and the mean number of k-faces of the typical cell $\nu_k(Z)$, $k \in \{0,1\}$, we obtain the following results.

	PHT	STIT	Voronoi
$\mu_{V,E} = \mu_{V,Z}$	4	3	3
$\mu_{Z,V} = \mu_{Z,E}$	4	6	6
$\nu_0(Z) = \nu_1(Z)$	4	4	6

It is shown in Theorem 6.2 that the distribution of the typical cell of a STIT tessellation Y_t is the same as that of a PHT X_t if both tessellations have the same directional distribution χ. This means that the respective mean values ℓ_Z and $\nu_0(Z)$ are equal. On the other hand, the intensities γ_V and γ_E and also the mean values for the adjacencies that express some features of the mutual arrangement of the cells are conform for STIT tessellations and PVTs. Only the mean length of the typical edge is an exception, because ℓ_E^{STIT} is not between ℓ_E^{PHT} and ℓ_E^{PVT}. This can be explained as follows. Unlike PVTs and PHTs, STIT tessellations are not face-to-face. Hence the edges, which are the 1-primitives, do not need to be 1-faces of the cells. A 1-face in a STIT tessellation can contain more than one edge. Thus the edges in a STIT tessellation are in the mean shorter than the edges in the tessellations of the other two models.

In a second comparison, we assume that the length densities for all three tessellations are the same, that is, $\eta_1^{\mathrm{PHT}} = \eta_1^{\mathrm{STIT}} = \eta_1^{\mathrm{PVT}}$. In this case, we obtain

$$
\begin{aligned}
2\,\gamma_V^{\mathrm{PHT}} &= \gamma_V^{\mathrm{STIT}} = \frac{4}{\pi}\,\gamma_V^{\mathrm{PVT}},\\
\frac{3}{2}\,\gamma_E^{\mathrm{PHT}} &= \gamma_E^{\mathrm{STIT}} = \frac{4}{\pi}\,\gamma_E^{\mathrm{PVT}},\\
\gamma_Z^{\mathrm{PHT}} &= \gamma_Z^{\mathrm{STIT}} = \frac{4}{\pi}\,\gamma_Z^{\mathrm{PVT}},\\
\frac{2}{3}\,\ell_E^{\mathrm{PHT}} &= \ell_E^{\mathrm{STIT}} = \frac{\pi}{4}\,\ell_E^{\mathrm{PVT}},\\
\ell_Z^{\mathrm{PHT}} &= \ell_Z^{\mathrm{STIT}} = \frac{\pi}{4}\,\ell_Z^{\mathrm{PVT}},\\
a_Z^{\mathrm{PHT}} &= a_Z^{\mathrm{STIT}} = \frac{\pi}{4}\,a_Z^{\mathrm{PVT}},\\
\eta_1^{\mathrm{PHT}} &= \eta_1^{\mathrm{STIT}} = \eta_1^{\mathrm{PVT}}.
\end{aligned}
$$

The relations in lines 1, 2, 4 are interesting in this comparison, because the characteristics for the STIT tessellation are not between those for PVT and PHT. Again, this is because STIT tessellations are not face-to-face. If the length densities of the 1-skeletons are equal, then the intensities of vertices and edges are larger for STIT tessellations than for

PHTs and PVTs. Consequently, the mean length of the typical edge for STIT tessellations is smaller than for PVTs and PHTs.

16.7.2 Second-order comparison

In this section, we present a systematic comparison of second-order properties for PHTs, STIT tessellations, and PVTs in $\mathbb{R}^2$. If available, we make use of analytic formulae for the considered quantities. In the case where explicit formulae are still unknown, the quantities were estimated by simulation. All tessellations in this section are isotropic. Furthermore, we assume that they have the same length density $\eta_1^{\mathrm{PHT}} = \eta_1^{\mathrm{STIT}} = \eta_1^{\mathrm{PVT}} = t$ of the 1-skeleton.

Chapter 12 contains a detailed second-order analysis of the total intrinsic volumes of the maximal $(d-1)$-polytopes for STIT tessellations in $\mathbb{R}^d$ and some special results for $d = 2$. An important tool for this study is the **pair correlation function** g_Φ of a random stationary and isotropic measure Φ on $\mathbb{R}^d$. It is defined in Section 12.7.1. For $d = 2$, this means

$$g_\Phi(r) = \frac{1}{2\pi r}\,\frac{\mathrm{d}K_\Phi(r)}{\mathrm{d}r}\,, \qquad r > 0\,, \tag{16.72}$$

where K_Φ stands for the K-function of Φ. Results for pair correlation functions of stationary and isotropic STIT tessellations were derived in Sections 12.7 and 12.8. Pair correlation functions for stationary and isotropic Poisson hyperplane processes were studied in Section 10.5 in [82].

For a random stationary and isotropic measure Φ on $\mathbb{R}^2$ and a window $W \in \mathcal{P}_2$, the **variance** $\operatorname{var}\Phi(W)$ can be calculated from the pair correlation function g_Φ via

$$\operatorname{var}\Phi(W) = \gamma_\Phi^2\, 2\pi \int_0^\infty \bar{c}_W(r)\, g_\Phi(r)\, r\, \mathrm{d}r - \left(\gamma_\Phi V_2(W)\right)^2 \tag{16.73}$$

with intensity $\gamma_\Phi = \mathbb{E}\,\Phi(B)$. This relation is the special case $d = 2$ of equation (7.25) in [33]. The isotropized set-covariogram function $\bar{c}_W(r)$ of W, as used here, is described in Definition 12.6. The quantities $g_\Phi(r)$ and $\operatorname{var}\Phi(W)$ for different measures Φ serve as the basis for our second-order comparison. In analysis of the variances $\operatorname{var}\Phi(W)$, we restrict ourselves to the *asymptotic* variances. Chapter 12 examines variances and asymptotic variances for stationary and isotropic STIT tessellations.

Vertices

Firstly, we deal with second-order quantities for the *set of vertices* $\mathsf{V} = \mathsf{V}(X_t)$ of a random stationary and isotropic tessellation X_t with length density $t > 0$. Sections 12.8 and 12.9 analyze the second-order structure of the vertex point process V for a stationary and isotropic planar STIT tessellation. From the first-order comparison in Section 16.7.1 we obtain with relation (16.25) and with $\zeta_\chi = \frac{2}{\pi}$ in the isotropic case, given in (16.22), that the vertex point process V for the tessellations considered here has the intensity

$$\gamma_{\mathrm{V}}^{\mathrm{PHT}} = \frac{1}{\pi} t^2, \quad \gamma_{\mathrm{V}}^{\mathrm{STIT}} = \frac{2}{\pi} t^2, \quad \gamma_{\mathrm{V}}^{\mathrm{PVT}} = \frac{1}{2} t^2,$$

respectively.

According to Remark 19.3, the vertex point process V can be considered as a random measure. Using $\Phi = \mathrm{V}$, we define as $g_{0,t} := g_{\mathrm{V}}$ the pair correlation function of V. In this case, $\lambda_{\mathrm{V}} K_{\mathrm{V}}(r)$ describes the mean number of vertices in a circle with radius $r > 0$ centered in the typical vertex; see (16.72).

We start with a comparison of the pair correlation functions $g_{0,t}$ for the three models. For PHTs and STIT tessellations, both isotropic and with length density $\eta_1^{\mathrm{PHT}} = \eta_1^{\mathrm{STIT}} = t > 0$, we have that

$$g_{0,t}^{\mathrm{PHT}}(r) = 1 + \frac{4}{t\pi r}, \tag{16.74}$$

$$g_{0,t}^{\mathrm{STIT}}(r) = 1 + \frac{2}{t^2 r^2} - \frac{\pi}{t^3 r^3} + \frac{\pi^2}{4t^4 r^4} - \left(\frac{1}{2t^2 r^2} - \frac{\pi}{2t^3 r^3} + \frac{\pi^2}{4t^4 r^4} \right) \mathrm{e}^{-\frac{2}{\pi} tr} \tag{16.75}$$

for $r > 0$. Relation (16.74) for stationary and isotropic Poisson line tessellations is given in equation (12.57), whereas the result (16.75) for STIT tessellations corresponds to Theorem 12.34. The exact pair correlation function $g_{0,t}^{\mathrm{PVT}}$ of the point process of vertices for Poisson–Voronoi tessellations can be found in Section 4 of [72]. It is a sum of several intricate integrals, which can only be analyzed numerically.

Plots of the three pair correlation functions for $t = 1$ are illustrated in Figure 16.12. It becomes clear that for the pair correlation function $g_{0,t}$ of the process of vertices, a STIT tessellation shows an intermediate behavior between PHT and PVT.

Furthermore, let $\Sigma_{0,t,W} := |\mathrm{V} \cap W|$ be the number of vertices of the tessellation inside a window $W \in \mathcal{P}_2$, and let $\operatorname{var} \Sigma_{0,t,W}$ denote the variance of this number. Note that this notation $\Sigma_{0,t,W}$ differs from $\Sigma_{V_j,t,W}$, which describes the total intrinsic volume V_j of the maximal $(d-1)$-polytopes in W, $j \in \{0, \ldots, d-1\}$, for STIT tessellations in $\mathbb{R}^d$. It was introduced in Section 10.4.

To obtain results for the asymptotic variances, we follow Section 12.5 and consider the tessellations in a family of growing windows aW with $W \in \mathcal{P}_2$ and $a \to \infty$. Thus, using the notation $\sim$ for asymptotic equivalence as introduced in Chapter 1, the asymptotic variances of the number of vertices $\operatorname{var} \Sigma_{0,t,aW}$ for the three tessellation models in the isotropic case are

$$\operatorname{var} \Sigma_{0,t,aW}^{\mathrm{PHT}} \sim \frac{4}{\pi^3} t^3 I_2(W) a^3, \tag{16.76}$$

$$\operatorname{var} \Sigma_{0,t,aW}^{\mathrm{STIT}} \sim \frac{16}{\pi} t^2 V_2(W) a^2 \log a, \tag{16.77}$$

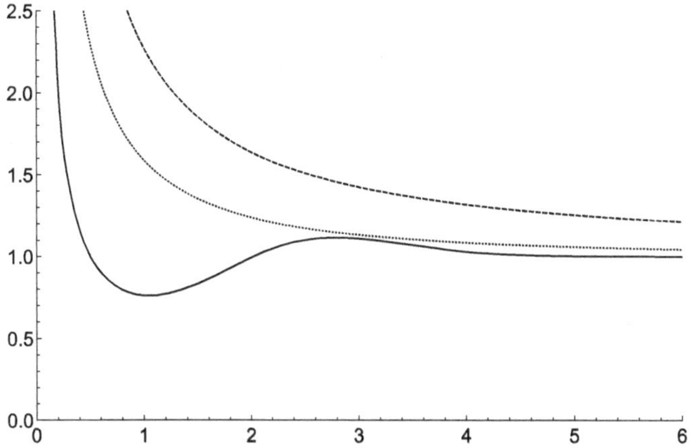

Figure 16.12: Plots for the pair correlation functions $g_{0,t}^{\mathrm{PHT}}(r)$ (dashed), $g_{0,t}^{\mathrm{STIT}}(r)$ (dotted) and $g_{0,t}^{\mathrm{PVT}}(r)$ (solid) for $t = 1$.

$$\operatorname{var} \Sigma_{0,t,aW}^{\mathrm{PVT}} \sim t^2 \, V_2(W) \, a^2 \tag{16.78}$$

as $a \to \infty$, where $I_2(W)$ is the chord power integral of W of order 2, defined in (12.26). Relation (16.76) was established in [70]. In another way, it can also be obtained directly from (16.73) together with some integral-geometric tools. The result for STIT tessellations given in (16.77) is in accordance with Theorem 12.18, where in equation (12.36) the variance $\operatorname{var} \Sigma_{V_0,t,W}$ of the total number of I-segments in W is considered. In a planar STIT tessellation, each vertex is a 0-face of precisely one I-segment, and each I-segment has two 0-faces. This explains the additional factor of 4 in (16.77) compared to (12.36). The result for Poisson–Voronoi tessellations (16.78) was shown in [72].

1-skeleton

In the next step, we focus on the *edges* of a tessellation. For the considered tessellations with length density $\eta_1^{\mathrm{PHT}} = \eta_1^{\mathrm{STIT}} = \eta_1^{\mathrm{PVT}} = t > 0$, the intensities of edges are

$$\gamma_{\mathrm{E}}^{\mathrm{PHT}} = \frac{2}{\pi} t^2, \quad \gamma_{\mathrm{E}}^{\mathrm{STIT}} = \frac{3}{\pi} t^2, \quad \gamma_{\mathrm{E}}^{\mathrm{PVT}} = \frac{3}{4} t^2,$$

respectively. This follows from the first-order comparison in Section 16.7.1 and relation (16.25) together with $\zeta_\chi = \frac{2}{\pi}$ in the isotropic case, given in (16.22).

The union of all edges of a tessellation X_t forms the 1-skeleton ∂X_t. For a planar STIT tessellation Y_t, the 1-skeleton ∂Y_t is as well the union of the I-segments. The random length measure on the 1-skeleton of X_t is $\mathscr{H}_t^1(\cdot) := \mathscr{H}^1(\cdot \cap \partial X_t)$, defined on $(\mathbb{R}^2, \mathcal{B}(\mathbb{R}^2))$, and $\mathscr{H}^1$ is the one-dimensional Hausdorff measure. With $\Phi = \mathscr{H}_t^1$ and (16.72), we obtain the pair correlation function $g_{1,t} := g_{\mathscr{H}_t^1}$ of the length measure on the 1-skeleton. The corresponding K-function is $K_{\mathscr{H}_t^1}$, and $\eta_1 K_{\mathscr{H}_t^1}$ describes the mean total edge length in a circle with radius $r > 0$ centered in the typical point of the 1-skeleton. For PHTs and STIT tessellations, which are both stationary and isotropic with edge length density $t > 0$, we have that

$$g_{1,t}^{\text{PHT}}(r) = 1 + \frac{1}{t\pi r}, \tag{16.79}$$

$$g_{1,t}^{\text{STIT}}(r) = 1 + \frac{1}{2t^2r^2}\left(1 - e^{-\frac{2}{\pi}tr}\right), \quad r > 0. \tag{16.80}$$

An analytic formula for the pair correlation function $g_{1,t}^{\text{PVT}}$ for a Poisson–Voronoi tessellation is not known until now. Equation (16.79) for PHT is given in (12.45), and that for a STIT tessellation (16.80) follows from Theorem 12.26 for $d = 2$.

Plots of the pair correlation functions $g_{1,t}$ are shown in Figure 16.13 for $t = 1$. The pair correlation function for the Poisson–Voronoi tessellation was estimated by simulations. Further details can be found in [165]. Again, the STIT tessellation shows an intermediate behavior between PHT and PVT.

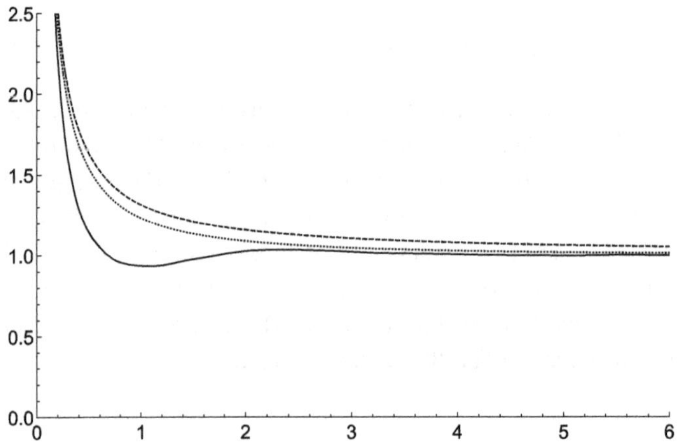

Figure 16.13: Plots for the pair correlation functions $g_{1,t}^{\text{PHT}}(r)$ (dashed), $g_{1,t}^{\text{STIT}}(r)$ (dotted) and $g_{1,t}^{\text{PVT}}(r)$ (solid, simulated) for $t = 1$.

Next, let us consider the total edge length of a tessellation X_t inside a window $W \in \mathcal{P}_2$, which is denoted by $\Sigma_{1,t,W}$, and its variance by $\operatorname{var} \Sigma_{1,t,W}$. For a planar STIT tessellation, we have that $\Sigma_{1,t,W}^{\text{STIT}} = \Sigma_{V_1,t,W}$, and for the definition of $\Sigma_{V_1,t,W}$, see (10.33). In Section 12.4 the variance of $\Sigma_{V_j,t,W}$, the sum of the jth intrinsic volumes of maximal $(d-1)$-polytopes of $Y_{t,W}$ was examined for stationary and isotropic STIT tessellations in $\mathbb{R}^d$, $j \in \{0, \dots, d-1\}$.

The variance of the total edge length inside the growing window aW behaves asymptotically, as $a \to \infty$, as

$$\operatorname{var} \Sigma_{1,t,aW}^{\text{PHT}} \sim \frac{1}{\pi}\, t\, I_2(W)\, a^3, \tag{16.81}$$

$$\operatorname{var} \Sigma_{1,t,aW}^{\mathrm{STIT}} \sim \pi V_2(W)\, a^2 \log a\,, \tag{16.82}$$

where $I_2(W)$ is the chord power integral of W of order 2, defined in (12.26). For (16.81), see [70] or use (16.73) to obtain an expression for the variance given in [165]. The result (16.82) follows from (12.27). For PVT, the variance of the total edge length in a square $a\,[0,1]^2$ with edge length $a > 0$ was computed in relation (8.14) in [20] as approximately

$$\operatorname{var} \Sigma_{1,t,a[0,1]^2}^{\mathrm{PVT}} \approx 1.0445\, a^2\,.$$

This is the result of the numerical evaluation of a rather intricate multiple integral.

Furthermore, in Chapter 12.8 the pair correlation functions $g_{0,t}^{\mathrm{STIT}}$ and $g_{1,t}^{\mathrm{STIT}}$ of a planar STIT tessellation are compared in (12.65) and (12.64) with the corresponding functions of a Boolean model of line segments. In Chapter 12.9, a cross-correlation function $g_{0,1,t}$ for planar STIT tessellations between the vertex point process V and the length measure $\mathscr{H}_t^1$ on the 1-skeleton is defined. An explicit expression is given in Theorem 12.36. This function is then also compared with the corresponding functions for a Poisson line tessellation and a Boolean model of line segments.

16.7.3 Length distributions of the maximal 1-polytopes

Intuitively, the length distribution of the maximal 1-polytopes in a planar tessellation is related to the range of spatial dependencies in this tessellation. In a planar STIT tessellation the I-segments are the maximal 1-polytopes. The distribution function $\mathbb{F}_{\mathrm{I}:\ell}^{\mathrm{STIT}}$ for the length of the typical I-segment in a stationary and isotropic planar STIT tessellation is given in (16.52). According to Lemma 16.6, the mean length of the typical I-segment is finite, but the variance of the length of the typical I-segment is infinite; see (16.57). A PVT is based on a Poisson point process and the tessellation is face-to-face. Therefore in a PVT the edges are also the I-segments in the sense of the maximum union of collinear and connected edges. This means that the dependencies in a PVT have a relatively short range. In contrast, PHTs in $\mathbb{R}^2$ are generated by lines, which causes dependencies of infinite range. In this respect, there are considerable differences between the three models.

To take a closer look at these differences between STIT tessellations and PVTs, we analyze the tail $\mathbb{Q}_{\mathrm{I}:\ell}^{\mathrm{STIT}}((u,\infty)) = 1 - \mathbb{F}_{\mathrm{I}:\ell}^{\mathrm{STIT}}(u)$ of the length distribution of the typical I-segment of a STIT tessellation and the tail $\mathbb{Q}_{\mathrm{E}:\ell}^{\mathrm{PVT}}((u,\infty)) = 1 - \mathbb{F}_{\mathrm{E}:\ell}^{\mathrm{PVT}}(u)$ of the length distribution of the typical edge of a PVT.

For arbitrary dimension d and a suitably chosen constant $c > 0$, which depends only on t and d, for a stationary and isotropic STIT tessellation Y_t, Corollary 11.19 gives

$$\mathbb{Q}_{\mathrm{I}:\ell}^{\mathrm{STIT}}((u,\infty)) \sim c\,\frac{1}{u^d} \quad \text{as } u \to \infty\,.$$

By (16.52) we obtain $c = \frac{\pi^2}{2t^2}$ for the constant in the planar case, and thus for $d = 2$,

$$\mathbb{Q}^{\mathrm{STIT}}_{\mathrm{I}:\ell}((u,\infty)) \sim \frac{\pi^2}{2t^2}\frac{1}{u^2} \quad \text{as } u \to \infty\,.$$

According to Remark 4 in [80], for a PVT X_t in $\mathbb{R}^d$, we obtain

$$\mathbb{Q}^{\mathrm{PVT}}_{\mathrm{E}:\ell}((u,\infty)) \le c_1\,\mathrm{e}^{-c_2 u^d}\,, \quad u > 0\,,$$

for appropriately selected constants $c_1, c_2 > 0$, which depend only on t and d.

16.7.4 Ergodic and mixing properties

For a planar random stationary tessellation, ergodic or mixing properties express the stochastic dependence of the restrictions of the tessellation to different regions in the plane, as the distance between these regions grows. In Chapter 15, mixing properties of STIT tessellations in arbitrary dimension are studied in detail.

Intuitively, comparing PHTs, STIT tessellations, and PVTs, already the results for the length distributions of the respective maximal segments, summarized in Section 16.7.3, suggest different ranges of dependencies for these models. The concepts of ergodicity, mixing, tail triviality, and β-mixing for random stationary tessellations are defined in Sections 15.1.1 and 15.1.2. The hierarchy between these properties is provided in Lemma 15.13.

With respect to these concepts, there are the following results for dimension $d \ge 2$.

For the *stationary and isotropic PHT* X_t, $t > 0$ (see (2.66)) with distribution $\mathbb{P}_{X_t}$, by Lemma 15.29

(a) X_t is mixing,
(b) the tail σ-algebra $\mathcal{B}_\infty(\mathcal{T})$ is not $\mathbb{P}_{X_t}$-trivial.

Thus by Lemma 15.13(a) the PHT is not β-mixing, and therefore it is not reasonable to consider the β-mixing coefficient as it is done below for the other two models

For the *stationary and isotropic STIT tessellation* Y_t, $t > 0$ (see Definition 6.1), with distribution $\mathbb{P}_{Y_t}$, Theorem 15.25 yields the following proposition.

For $0 < a < b$, denote $W_1 = [-a,a]^d$ and $W_2 = [-b,b]^d$. Then for the β-mixing coefficient $\beta(a,b)$, defined in (15.10), we have that

(a) the STIT tessellation Y_t is β-mixing in space in the sense of Definition 15.11(b), that is,

$$\lim_{b\to\infty} \beta(a,b) = 0$$

for all $a > 0$,

(b) for all $a > 0$ and $q \in (0,1)$, there exist constants $\rho(d,t,a,q) < \infty$ and $b(d,t,a,q) < \infty$ such that for $b > b(d,t,a,q)$,

$$\beta(a,b) \le \rho(d,t,a,q)\, b^{-q}\,. \tag{16.83}$$

The proof of Theorem 15.25 also shows that the asymptotic decay of the $\beta(a,b)$ is at least of polynomial order with degree $q \in (0,1)$ in $(b-a)$. By Lemma 15.13 this implies that also the tail σ-algebra $\mathcal{B}_\infty(\mathcal{T})$ is $\mathbb{P}_{Y_1}$-trivial.

Finally, consider the *stationary and isotropic PVT* with cell intensity γ_Z. By Theorem 2.2 of [69] there are positive constants c_0, c_1, c_2, depending on the dimension d, such that, with the same notation as used for the STIT tessellation,

$$\beta(a,b) \le \begin{cases} c_1\, d\,\left(\frac{b-a}{a}\right)^{d-1} \mathrm{e}^{-(2a)^{d-1}\gamma_Z\,(b-a)/24}\,, & b-a > c_0\, a\,,\\ c_2\, d\,\left(\frac{a}{b-a}\right)^{d-1} \mathrm{e}^{-\left(\frac{2}{c_0}\right)^{d-1}\gamma_Z\,(b-a)^d/24}\,, & b-a \le c_0\, a\,. \end{cases} \tag{16.84}$$

Even if it is not shown that for STIT tessellations, the upper bound (16.83) for $\beta(a,b)$ is sharp, this result suggests that for PVT, the decay of the β-mixing coefficient is considerably faster than for STIT tessellation.

16.8 Notes

1. Mecke's results in [125] form the basis for the relationships between the mean values of random stationary tessellations in $\mathbb{R}^2$. He has shown that from three suitably selected mean values, γ_V, γ_Z, η_1, a variety of other quantities of a planar tessellation can be calculated. This is used, for example, in Corollary 16.5. The proofs of the results from [125] can be found in [196], which are given there for tessellations with general nonconvex cell shapes. For tessellations that are not face-to-face, further mean values and their relations are studied in [193]. The book [33] is cited here as a reference for all these mean value relationships.
2. All vertices in a planar STIT tessellation are T-shaped vertices. Such a T-vertex with exactly three emanating edges is a special case of a π-vertex. Cowan introduced the term π-vertex. A π-vertex is a vertex with $k \in \{3,4,\ldots\}$ emanating edges, where one of the angles between neighboring edges is equal to π. The importance of these π-vertices in characterizing planar tessellations that are not face-to-face was first discussed in [35, 36] and then further explored in [193]. In [144], Lemma 3, it was shown that a necessary condition for a random stationary tessellation to be a STIT tessellation is that all vertices are T-shaped vertices, if a STIT tessellation exists. At that time the existence of STIT tessellations for dimension $d > 1$ was an open problem.
3. Lemma 16.2 can be found in [194], where the parameter triple γ_V, γ_Z, η_1 given in [125] was extended by two directional distributions, and the behavior of these five characteristics under iteration was investigated.
4. Most of the material presented in Section 16.2 was provided in [146].
5. The classification of line segments in Section 16.2.2 goes back to Miles. He introduced in [106] the so-called I-, J-, and K-segments for planar rectangular tessellations that are not face-to-face. In [136], this classification is used to study tessellations with typical cells distributed identically to those of Poisson line tessellations. In our notation the K-segments are the edges, which are the 1-primitives. In [44] the classification of segments was extended by the so-called M-segments, which play a role in tessellations that are not face-to-face and do not only contain π-vertices. An M-segment is a maximal connected component of dimension 1 of the 1-skeleton of a tessellation; see also (10.22) for general

dimension. The I-segments are special M-segments, and in a planar STIT-tessellation, all M-segments are I-segments.

6. Section 16.3 contains results from [128] for various distributions in the isotropic case, from [126] and [186] in the nonisotropic case and from [184] for the moments of the length.
In [128] the three different types of line segments, the edges, J-segments, and I-segments, and their length distributions are considered for stationary and isotropic planar STIT tessellations. For these investigations, the term "remaining I-segment" was introduced there for the first time. Later, the anisotropic case is treated, and the joint distribution of direction and length is given in [126] for the typical I-segment and in [186] for the typical edge and the typical J-segment, respectively.
Note that in these papers the results for the densities are sometimes given with respect to $t\chi \otimes \lambda_+$, whereas in Section 16.3, they are always given with respect to $\chi \otimes \lambda_+$.
7. The $t\chi$-length, used in the proofs of Lemmas 16.12 and 16.14, was defined in [126]. The idea for this came from Ambartzumian [3, 4], who introduced a metric $d_\kappa : \mathbb{R}^2 \times \mathbb{R}^2 \to [0,\infty)$, where the distance between points $x, y \in \mathbb{R}^2$ is given by
$$d_\kappa(x,y) = \frac{1}{2}\Lambda_\kappa\big(\{L \in \mathcal{H} : L \text{ separates } x,y\}\big)$$
for a translation-invariant line measure Λ_κ with directional measure κ on $\mathcal{H}_o$. Then the "length" of a segment $S \in \mathcal{P}_1$ with endpoints x, y measured in the d_κ-metric is
$$\mathcal{L}^*_\kappa(S) = d_\kappa(x,y) = \frac{1}{2}\Lambda_\kappa\big([S]\big) = \frac{1}{2}s_\kappa(\operatorname{aff}_o S)\, V_1(S)\,,$$
where s_κ is the rose of intersections defined in (16.17). Another normalization leads with $\Lambda_\kappa = \Theta$ and $\kappa = t\chi$ to the $t\chi$-length in (16.37).
8. Lemma 16.13 can be found in [126], where it is given without proof.
9. The results for the densities in the isotropic case in Section 16.3.4 are in accordance with those in [128] and [184]. However, it should be noted that the special case $t = \frac{\pi}{2}$ is considered there.
10. For the planar stationary and isotropic tessellation model considered in [136], which is not face-to-face and has the same vertex-type as STIT tessellations, it was shown that the length distributions of the typical edge, typical J-segment, and typical I-segment are all exponential distributions with various parameters. We have seen in Section 16.3.4 that this is not the case for segments in a planar STIT tessellations.
11. In Section 16.3.5 the probabilistic representations of some length distributions are given for the isotropic case. Further results can be found in Lemmas 9 and 14 in [128]. In addition, similar representations in the nonisotropic case are presented in Corollaries 2 and 6 in [126] for the joint distribution of direction and length of the typical I-segment and the remaining I-segment.
12. Section 16.3.6 is a generalized version of [184] with a correction of the formula for the variance of the length of the typical edge including the resulting consequences.
13. Most of the material of Section 16.4 was provided in [131] for stationary and isotropic planar STIT tessellations. Later, the result for the distribution of the internal vertices of the typical I-segment was also shown for the nonisotropic case, first, for $d = 2$ in [185] with a proof based on the martingale techniques from [174] and then for $d \geq 2$ in Theorem 14.18 as an application of the Mecke-type equation.
14. Section 16.4.2 contains results for distributions and conditional distributions of the random variables ν_{I} and ν_{I}^+, as well as some probabilistic representations of these distributions. Further representations of this type can be found in [131], where it is shown that the distribution $\mathbb{Q}_{\mathrm{I}:\ell}$ of the length of the typical I-segment in a stationary and isotropic STIT tessellation Y_t in $\mathbb{R}^2$ is a mixture of Erlang distributions $\mathsf{Erl}(j, \frac{2t}{\pi})$, that is,
$$\mathbb{Q}_{\mathrm{I}:\ell} = \sum_{j=1}^{\infty} \frac{4}{j(j+1)(j+2)}\,\mathsf{Erl}\Big(j, \frac{2t}{\pi}\Big).$$

In addition, the length $V_1(\mathrm{I})$ of the typical I-segment and the number ν_I^+ of upper vertices in its relative interior are related as follows:

$$V_1(\mathrm{I}) \stackrel{\mathrm{D}}{=} \sum_{i=1}^{\nu_\mathrm{I}^+ +1} \mathrm{E}_i\,,$$

where $\mathrm{E}_1, \mathrm{E}_2, \ldots$ are independent exponentially distributed random variables with parameter $\frac{2t}{\pi}$.

15. Sections 16.5 and 16.6 contain generalized material from [40] without the assumption of isotropy. The study of the structure of internal vertices of the typical I-segment leads to further results in [40]. For example, metric properties are investigated, so the mean lengths of the typical edge of type 0,1,2, respectively, are shown there to be $\ell_{\mathrm{I:E}\langle \mathrm{J},0\rangle} \approx 0.883\ell_\mathrm{E}$, $\ell_{\mathrm{I:E}\langle \mathrm{J},1\rangle} \approx 0.972\ell_\mathrm{E}$, and $\ell_{\mathrm{I:E}\langle \mathrm{J},2\rangle} \approx 1.251\ell_\mathrm{E}$. Hence in the mean 22.74 % of the 1-skeleton of a planar stationary and isotropic STIT tessellation is covered by edges of type 2 or, equivalently, by I-segments with no internal vertices. The same proportion of the 1-skeleton is covered by edges of type 0 and 54.52 % by edges of type 1.
    These quantities and associated adjacencies are also considered in [42], but there in the more general context of tessellations which are not face-to-face.
16. The first-order comparison in Section 16.7.1 follows [147], whereas the second-order comparison in Section 16.7.2 uses material from [165]. More detail on the simulations used in Section 16.7.2 can be found there. In addition, this paper contains results for the variances $\operatorname{var} \Sigma_{0,t,W}$ and $\operatorname{var} \Sigma_{1,t,W}$ of STIT tessellations, Poisson line tessellations, and Poisson–Voronoi tessellations.
    The result for the pair correlation function for the length measure of the STIT tessellation in (16.80) was first discovered in [195].
17. In [164] the arrangement of cells in planar STIT tessellations and Poisson line tessellations is studied using characteristics of the cells neighboring the typical cell. In this paper, also the pair correlation function and various mark correlation functions for the cell centroids are considered. The marks are the area, perimeter, and number of 0-faces of the cells.
18. Further investigations for planar STIT tessellations can be found in the PhD thesis [132]. Chapter 5 analyzes second-order properties of Mondrian STIT tessellation within an observation window, for example, the variance for the number of I-segments and the weighted total edge length and the corresponding covariance. Additionally, Mondrian analogues of the pair correlation functions for the length measure on the 1-skeleton and the vertex point process, as well as the cross-correlation function, are deduced. These results are also shown in [16].
    Chapter 6 of [132] deals with differently weighted I-segments and their density functions. Some of these results are used to refine geometrical and metrical properties proven in [40], now with birth time dependency. Furthermore, these considerations do not require isotropy.
19. The subject of [127] are inhomogeneous random planar tessellations generated by lines in a window, called mixed line-generated tessellations. It is not assumed that the line measure is invariant under translations. Therefore the tessellations do not have to be stationary. If the line measure is invariant under translations, then the analyzed mean values coincide with those for planar STIT tessellations. It remains open whether in this case the mixed line-generated tessellation has the same distribution as the corresponding random STIT tessellation in the window.
20. In [163], further second-order properties of the 1-skeleton of random tessellations are considered. In particular, the spectral density of the random length measure of the edges is analyzed. It turns out that the location of its first local maximum is proportional to the mean number of pores per inch in a foam, which is one of the most important geometric characteristic of foams. For this purpose, it is assumed that the core of a foam model is the 1-skeleton of a tessellation. In Example 3 of this paper, planar stationary and isotropic STIT tessellations are studied.

17 Spatial STIT tessellations

This chapter is devoted to tessellations in $\mathbb{R}^3$, which are also referred to as spatial tessellations. The structure of a spatial STIT tessellation is considerably more complex and intricate compared with its planar counterpart. This has various reasons. As in the planar case, the cells of a spatial STIT tessellation are not necessarily in a face-to-face position. This can be seen in Figures 17.1 and 17.2, where spatial STIT tessellations are shown for the isotropic and Mondrian cases.

On the boundary of a cell, additional vertices and edges can appear. As a consequence, both one- and the two-dimensional faces of the cells can have an internal structure. This leads to a variety of new effects. For example, a spatial STIT tessellation may contain vertices that are 0-primitives but are not a 0-face of any of their adjacent cells. Another difference to planar STIT tessellations is that a spatial STIT tessellation has two different types of vertices. Figure 17.3 shows the so-called T- and X-vertices. There it can be seen that an X-vertex is not a 0-face of a cell. This causes that in a detailed analysis, we have to take into account the additional effects generated by various types of vertices.

As in the planar case, we will consider k-faces with their multiplicities, namely 1-faces and 2-faces of the cells and in addition 1-faces of the 2-primitives, called plates.

Here, too, maximal polytopes will once again play an important role in our investigations. For a spatial STIT tessellation, these are the I-polygons, which are born during the cell division process. In addition, the 1-faces of the I-polygons are the maximal segments, called the I-segments. Another difference to the planar case is the fact that the I-segments in a spatial STIT tessellation can already have vertices in their relative interior at the time of their birth, as shown in Figure 17.7.

17.1 Basic combinatorial structure

As in the planar case, we start with the **primitives**, because they can be seen as building blocks of the tessellation structure. The set of k-primitives $\mathcal{F}_k^{\mathrm{pr}}(T)$ is defined in Section 2.1.2 for a tessellation T in $\mathbb{R}^d$, $k \in \{0, \dots, d\}$. To illustrate this notation, we can say that a k-primitive is a k-dimensional polytope that has no structure of the tessellation in its relative interior. The primitives of a spatial tessellation are called **vertices** ($k = 0$), **edges** ($k = 1$), **plates** ($k = 2$), and **cells** ($k = 3$). For the corresponding sets associated with a tessellation T, we use the following notation, similar to the planar case:

$$\mathsf{V}(T) := \mathcal{F}_0^{\mathrm{pr}}(T), \quad \mathsf{E}(T) := \mathcal{F}_1^{\mathrm{pr}}(T), \quad \mathsf{P}(T) := \mathcal{F}_2^{\mathrm{pr}}(T), \quad \mathsf{Z}(T) := \mathcal{F}_3^{\mathrm{pr}}(T) = T.$$

Although a primitive does not contain lower-dimensional primitives in its relative interior, the boundary structure of plates or cells can be rather intricate. A plate can have vertices on its boundary that are no 0-faces of it. A cell can have additional vertices and edges nested into its two-dimensional faces.

https://doi.org/10.1515/9783112222409-021

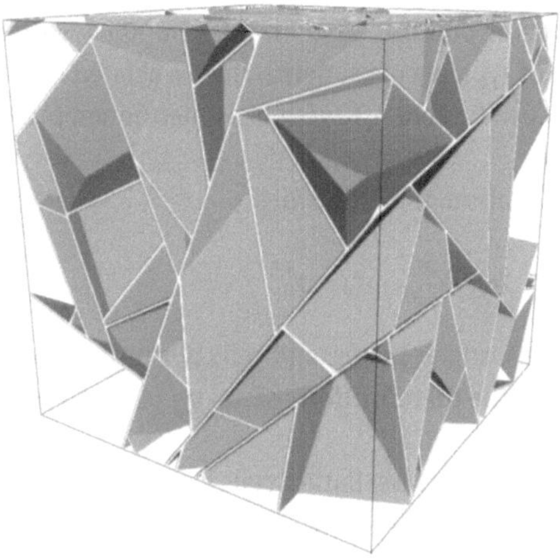

Figure 17.1: Simulation of an isotropic STIT tessellation in $\mathbb{R}^3$.

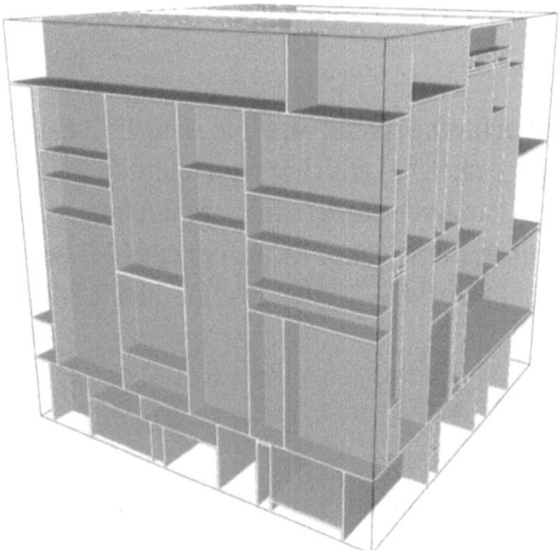

Figure 17.2: Simulation of a Mondrian STIT tessellation in $\mathbb{R}^3$.

Recall that A-**components** of a tessellation in $\mathbb{R}^d$ are k-dimensional polytopes, as, for example, k-primitives or k-faces of the tessellation cells, $k \in \{0, \dots, d\}$. Definition 2.9 introduced the notation $\mathsf{A}(T)$ for a set of A-components of a tessellation T in $\mathbb{R}^d$. For a set $\mathsf{A}(X)$ of k-dimensional components of a random stationary tessellation X, the **intensity of A-components** in (2.25) is defined as

$$\gamma_{\mathsf{A}} := \mathbb{E} \sum_{\mathsf{a} \in \mathsf{A}(X)} \mathbb{1}\{c(\mathsf{a}) \in [0,1]^d\}$$

for a given center function $c : \mathcal{P}_k \to \mathbb{R}^d$. With (2.26) the probability measure $\mathbb{Q}_{\mathsf{A}}$ on $\mathcal{P}_{k,c}$,

$$\mathbb{Q}_{\mathsf{A}}(A) := \frac{1}{\gamma_{\mathsf{A}}} \mathbb{E} \sum_{\mathsf{a} \in \mathsf{A}(X)} \mathbb{1}\{c(\mathsf{a}) \in [0,1]^d\} \, \mathbb{1}\{\mathsf{a} - c(\mathsf{a}) \in A\}, \quad A \in \mathcal{B}(\mathcal{P}_{k,c}), \tag{17.1}$$

denotes the **distribution of the typical** A-**component**. Since in this chapter we only consider translation-invariant functionals of typical components, the associated quantities are independent of the choice of the specific center function, as mentioned in Remark 2.18.

For the primitives in a spatial STIT tessellation, the intensities

$$\gamma_{\mathsf{V}} = \gamma_0^{\mathrm{pr}}, \quad \gamma_{\mathsf{E}} = \gamma_1^{\mathrm{pr}}, \quad \gamma_{\mathsf{P}} = \gamma_2^{\mathrm{pr}}, \quad \gamma_{\mathsf{Z}} = \gamma_3^{\mathrm{pr}}$$

are also called the **mean numbers** of vertices, edges, plates, and cells per unit volume, respectively. Furthermore, $\mathbb{Q}_{\mathsf{E}}$, $\mathbb{Q}_{\mathsf{P}}$, $\mathbb{Q}_{\mathsf{Z}}$ denote the **distributions** of the typical edge, typical plate, and typical cell, respectively.

In Section 2.1 the **skeletons** of the k-faces of the tessellation cells $\mathrm{skel}_k\, T$ and of the k-primitives $\mathrm{skel}_k^{\mathrm{pr}}\, T$, respectively, were defined in (2.2) and (2.15) as

$$\mathrm{skel}_k\, T = \bigcup_{F \in \mathcal{F}_k(T)} F \quad \text{and} \quad \mathrm{skel}_k^{\mathrm{pr}}\, T = \bigcup_{F \in \mathcal{F}_k^{\mathrm{pr}}(T)} F$$

for $T \in \mathcal{T}$ and $k \in \{0, \ldots, d-1\}$.

With the results of Corollary 2.8, we obtain for the spatial case that $\mathrm{skel}_1\, T = \mathrm{skel}_1^{\mathrm{pr}}\, T$ and $\mathrm{skel}_2\, T = \mathrm{skel}_2^{\mathrm{pr}}\, T = \partial T$. Note that $\mathrm{skel}_0 T \subseteq \mathrm{skel}_0^{\mathrm{pr}}\, T$, which is equivalent to $\mathcal{F}_0(T) \subseteq \mathcal{F}_0^{\mathrm{pr}}(T)$. This means that a spatial tessellation can have vertices that are not 0-faces of a tessellation cell. Figure 17.3 shows such vertices, the so-called X-vertices in a STIT tessellation T. An X-vertex is not contained in $\mathcal{F}_0(T)$. For $d = 3$, from Lemma 14.8 it follows that for the corresponding intensities,

$$\gamma_{\mathsf{V}} = \gamma_0^{\mathrm{pr}} = \frac{3}{2}\gamma_0^{\max}.$$

This means that in a spatial STIT tessellation, in the mean, one third of all vertices are not 0-faces of the cells; see also Lemma 17.5.

Following (2.47) and (2.49), the 1-skeleton of a random stationary spatial tessellation X has the density

$$\eta_1 = \mathbb{E} \sum_{\mathsf{e} \in \mathsf{E}(X)} V_1(\mathsf{e} \cap [0,1]^3).$$

We refer to η_1 as the **mean total length of tessellation edges per unit volume** or edge length density for short. Furthermore,

$$\eta_2 = \mathbb{E} \sum_{\mathsf{p} \in \mathsf{P}(X)} V_2(\mathsf{p} \cap [0,1]^3)$$

denotes the **mean total area of tessellation plates per unit volume** or surface density.

Furthermore, we recall the adjacency concept introduced in Definition 16.1. Let A and B be two sets of components in a tessellation. An element $\mathsf{a} \in \mathsf{A}$ is said to be **adjacent** to an element $\mathsf{b} \in \mathsf{B}$ if $\mathsf{a} \subseteq \mathsf{b}$ or $\mathsf{b} \subseteq \mathsf{a}$. For a random and stationary tessellation X, we denote by $\mu_{\mathsf{A},\mathsf{B}}$ the mean number of B-components adjacent to the typical A-component,

$$\mu_{\mathsf{A},\mathsf{B}} = \frac{1}{\gamma_{\mathsf{A}}} \mathbb{E} \sum_{\mathsf{a} \in \mathsf{A}(X)} \mathbb{1}\{c(\mathsf{a}) \in [0,1]^2\} \sum_{\mathsf{b} \in \mathsf{B}(X)} \mathbb{1}\{\mathsf{a} \subseteq \mathsf{b} \text{ or } \mathsf{b} \subseteq \mathsf{a}\}.$$

Another description for $\mu_{\mathsf{A},\mathsf{B}}$ is given in (16.6). For example, $\mu_{\mathsf{Z},\mathsf{V}}$, $\mu_{\mathsf{Z},\mathsf{E}}$, $\mu_{\mathsf{Z},\mathsf{P}}$ denote the mean numbers of vertices, edges, plates of the typical cell, respectively. In contrast, $\mu_{\mathsf{V},\mathsf{Z}}$ describes the mean number of cells adjacent to the typical vertex.

From the **exchange formula** in Theorem 2.29, for $f(\mathsf{a}, X) = h(\mathsf{b}, X) = 1$ and for the adjacency relation, it follows that

$$\gamma_{\mathsf{A}}\, \mu_{\mathsf{A},\mathsf{B}} = \gamma_{\mathsf{B}}\, \mu_{\mathsf{B},\mathsf{A}},$$

which is already mentioned in (16.7). This formula will be used frequently in our study of mean value relations.

As in Chapter 16, the direction of a line segment $S \in \mathcal{P}_1$ is given by the line $\operatorname{aff}_o S \in G(3,1)$ through the origin, which is parallel to S. In a spatial tessellation T for $\mathcal{H}^1$-almost all $x \in \operatorname{skel}_1 T$, there exists a unique tessellation edge $\mathsf{e}(x) \in \mathsf{E}(T) \subset \mathcal{P}_1$ with $x \in \mathsf{e}(x)$, and its direction is $\operatorname{aff}_o \mathsf{e}(x)$. We now follow (2.50). For a random stationary spatial tessellation X, the probability measure χ_1 on $G(3,1)$ given by

$$\begin{aligned} \chi_1(A) &:= \frac{1}{\eta_1} \mathbb{E} \int_{\mathbb{R}^3} \mathbb{1}\{x \in [0,1]^3 \cap \operatorname{skel}_1 X\}\, \mathbb{1}\{\operatorname{aff}_o \mathsf{e}(x) \in A\}\, \mathcal{H}^1(\mathrm{d}x) \\ &= \frac{1}{\eta_1} \mathbb{E} \sum_{\mathsf{e} \in \mathsf{E}(X)} \mathbb{1}\{\operatorname{aff}_o \mathsf{e} \in A\}\, V_1(\mathsf{e} \cap [0,1]^3), \quad A \in \mathcal{B}(G(3,1)), \end{aligned} \tag{17.2}$$

is the **directional distribution in the typical point of the 1-skeleton** $\operatorname{skel}_1 X$. It is also called the length-weighted directional distribution of $\operatorname{skel}_1 X$. The measure χ_1 coincides with the measure χ_1^{pr} in (2.50) for $d = 3$.

For a polygon $F \in \mathcal{P}_2$, the line through the origin orthogonal to the plane containing F is denoted by $F^\perp \in G(3,1)$. This line $F^\perp$ defines the direction of F more favorable for the investigations of the spatial case in this chapter. In a tessellation T in $\mathbb{R}^3$, for $\mathcal{H}^2$-almost all $x \in \partial T$, there exists a unique tessellation plate $\mathsf{p}(x) \in \mathsf{P}(T) \subset \mathcal{P}_2$ with $x \in \mathsf{p}(x)$ and direction $\mathsf{p}^\perp(x)$. Then the probability measure ψ_o on $G(3,1)$ defined by

$$\begin{aligned} \psi_o(A) &:= \frac{1}{\eta_2} \mathbb{E} \int_{\mathbb{R}^3} \mathbb{1}\{x \in [0,1]^3 \cap \partial X\}\, \mathbb{1}\{\mathsf{p}^\perp(x) \in A\}\, \mathcal{H}^2(\mathrm{d}x) \\ &= \frac{1}{\eta_2} \mathbb{E} \sum_{\mathsf{p} \in \mathsf{P}(X)} \mathbb{1}\{\mathsf{p}^\perp \in A\}\, V_2(\mathsf{p} \cap [0,1]^3), \quad A \in \mathcal{B}(G(3,1)), \end{aligned} \tag{17.3}$$

describes the **directional distribution in the typical point of the 2-skeleton** ∂X, also called the area-weighted directional distribution of ∂X. This corresponds to (2.51) for

$d = 3$. The measure χ_2^{pr} defined in (2.50) on $G(3,2)$ is related to ψ_o by $\chi_2^{\text{pr}}(A) = \psi_o(A^\perp)$ for $A \in \mathcal{B}(G(3,2))$. In (2.52) the distribution ψ is defined as an even probability measure on S^{d-1}. There is a one-to-one correspondence between ψ and ψ_o. Hence, in some notation, for example, $\zeta_{t\psi}^{(2)}$ and $\zeta_{t\psi}^{(3)}$, given in (17.9) and (17.10), for better readability, we use ψ as the lower index instead of ψ_o.

We now consider a spatial STIT tessellation Y_t based on the measure $t\,\Theta$, where the hyperplane measure Θ is given in (4.1) and satisfies Convention 4.1. This convention guarantees that the cells of the structure constructed with the measure Θ are bounded and that the structure is a. s. a random tessellation. By (6.6) and (6.8) in Theorem 6.9 the tessellation Y_t has the surface density

$$\eta_2 = t \tag{17.4}$$

for $d = 3$. In addition, according to (17.3), the directional distribution in the typical point of the 2-skeleton ∂Y_t is

$$\psi_o(A) = \varphi_o(A)\,, \quad A \in \mathcal{B}(G(3,1))\,, \tag{17.5}$$

where φ_o is the directional distribution characterizing the hyperplane measure Θ; see Section 4.1 and definition (4.2). Henceforth, in this chapter, for $t > 0$, we will refer to the STIT tessellation Y_t based on the measure $t\,\Theta$ as the STIT tessellation Y_t with directional distribution ψ_o in the typical point of ∂Y_t. The value $t\psi_o(A)$ can be interpreted as the mean total area per unit volume of all plates of Y_t having direction in $A \in \mathcal{B}(G(3,1))$. Theorem 7.9 implies that for any probability measure ψ_o on $G(3,1)$ that fulfils Convention 4.1 and for any $t > 0$, there exists a random spatial STIT tessellation with directional distribution ψ_o in the typical point of the 2-skeleton and with surface density t.

Let Y_t be a spatial STIT tessellation, and let X_t be a Poisson plane tessellation in $\mathbb{R}^3$, assuming that both tessellations are based on the same measure $t\,\Theta$. The distributions χ_1 and ψ_o of Y_t are defined in (17.2) and (17.3). The directional distribution χ_k^{PHT} in the typical point of the k-skeleton of X_t is given in Section 2.3.4 for general dimension d and $k \in \{0, \dots, d-1\}$ as a probability measure on $G(d,k)$.

Theorem 6.9 yields that for $d = 3$ and $A \in \mathcal{B}(G(3,1))$,

$$\begin{aligned}
&\eta_2(Y_t) = \eta_2(X_t) = t\,,\\
&\eta_1(Y_t) = 2\eta_1(X_t) = t^2 \int_{G(3,1)} \int_{G(3,1)} |\sin \sphericalangle(L_1, L_2)|\, \psi_o(\mathrm{d}L_1)\, \psi_o(\mathrm{d}L_2)\,,
\end{aligned} \tag{17.6}$$

$$\psi_o(A) = \chi_2^{\text{PHT}}(A^\perp) = \varphi_o(A)\,, \tag{17.7}$$

$$\begin{aligned}
\chi_1(A) &= \chi_1^{\text{PHT}}(A)\\
&= \frac{\int_{G(3,1)} \int_{G(3,1)} \mathbb{1}\{L_1^\perp \cap L_2^\perp \in A\}\, |\sin \sphericalangle(L_1, L_2)|\, \psi_o(\mathrm{d}L_1)\, \psi_o(\mathrm{d}L_2)}{\int_{G(3,1)} \int_{G(3,1)} |\sin \sphericalangle(L_1, L_2)|\, \psi_o(\mathrm{d}L_1)\, \psi_o(\mathrm{d}L_2)}\,.
\end{aligned} \tag{17.8}$$

For our further studies, it is useful to define two auxiliary constants depending on $t > 0$ and the probability measure ψ_o on $G(3,1)$:

$$\zeta_{t\psi}^{(2)} := t^2 \int_{G(3,1)} \int_{G(3,1)} \nabla_2(L_1, L_2)\, \psi_o(\mathrm{d}L_1)\, \psi_o(\mathrm{d}L_2) \tag{17.9}$$

$$= t^2 \int_{G(3,1)} \int_{G(3,1)} |\sin \sphericalangle(L_1, L_2)|\, \psi_o(\mathrm{d}L_1)\, \psi_o(\mathrm{d}L_2),$$

$$\zeta_{t\psi}^{(3)} := t^3 \int_{G(3,1)} \int_{G(3,1)} \int_{G(3,1)} \nabla_3(L_1, L_2, L_3)\, \psi_o(\mathrm{d}L_1)\, \psi_o(\mathrm{d}L_2)\, \psi_o(\mathrm{d}L_3). \tag{17.10}$$

The abbreviated notation $\nabla_2(L_1, L_2)$ and $\nabla_3(L_1, L_2, L_3)$ stand for $\nabla_2(u(L_1), u(L_2))$ and $\nabla_3(u(L_1), u(L_2), u(L_3))$, respectively, where $u(L)$ is the unit vector in the upper half-sphere with direction L. Then $\nabla_2(L_1, L_2) = |\sin \sphericalangle(L_1, L_2)|$, which is equal to the area of the parallelogram spanned by the unit vectors $u(L_1)$ and $u(L_2)$, and $\nabla_3(L_1, L_2, L_3)$ is equal to the volume of the parallelepiped spanned by $u(L_1)$, $u(L_2)$, and $u(L_3)$. Obviously,

$$\zeta_{t\psi}^{(2)} = t^2\, \zeta_{\psi}^{(2)} \quad \text{and} \quad \zeta_{t\psi}^{(3)} = t^3\, \zeta_{\psi}^{(3)}. \tag{17.11}$$

These relations are often used to emphasize the dependence on t. In addition, similarly to the planar case, the constants reflect the influence of the directional distribution ψ_o on several mean values and intensities of the random tessellation.

For the length density of a STIT tessellation Y_t, by (17.6) we obtain that

$$\eta_1 = \zeta_{t\psi}^{(2)}. \tag{17.12}$$

This is in line with relation (14.12), which is an application of Corollary 14.7.

Recall that $\Pi_{t\Theta}$ is the **associated zonoid** of the pair (t, Θ), which is introduced in (2.68). Section 2.3.4 deals with the **intersection process** $\widehat{X}_t^{(d-k)}$ of order $d - k$, $k \in \{0, \ldots, d-1\}$, for the Poisson hyperplane process $\widehat{X}_t$. For $d = 3$ and $k \in \{0, 1, 2\}$, the intersection process $\widehat{X}_t^{(3-k)}$ of order $3 - k$ is a process of k-dimensional affine subspaces generated by the intersection of $3 - k$ pairwise different hyperplanes of $\widehat{X}_t$. According to (2.81) and (2.82), the two constants $\zeta_{t\psi}^{(3-k)}$ are related to the associated zonoid $\Pi_{t\Theta}$ and to the intersection process $\widehat{X}_t^{(3-k)}$ of order $3 - k$ for $k \in \{0, 1\}$ as follows:

$$\frac{1}{(3-k)!}\, \zeta_{t\psi}^{(3-k)} = \eta_{k,t}(\widehat{X}_t^{(3-k)}) = V_{3-k}(\Pi_{t\Theta}).$$

In the *isotropic case*, if ψ_o is invariant under rotations on $G(3,1)$, then the associated zonoid $\Pi_{t\Theta_{\text{iso}}}$ is by (2.70) a ball with radius $\frac{1}{4}t$. Thus, for a stationary and isotropic spatial STIT tessellation Y_t, the two constants have the particular values

$$\zeta_{t\psi}^{(2)} = \frac{\pi}{4} t^2 \quad \text{and} \quad \zeta_{t\psi}^{(3)} = \frac{\pi}{8} t^3. \tag{17.13}$$

Another particular case is the *Mondrian case*. The directional distribution ψ_o in the typical point of the 2-skeleton of a spatial Mondrian STIT tessellation Y_t has the form

$$\psi_o = p_1\,\delta_{L(e_1)} + p_2\,\delta_{L(e_2)} + p_3\,\delta_{L(e_3)}\,, \tag{17.14}$$

where $\{e_1, e_2, e_3\}$ is the standard orthonormal basis in $\mathbb{R}^3$, $L(e) \in G(3,1)$ is the line that contains e, $\delta_{L(e)}$ is the Dirac measure at $L(e)$ on $G(3,1)$, and $p_1, p_2, p_3 \in (0,1)$ with $p_1 + p_2 + p_3 = 1$. For $d = 3$, this is in accordance with definition (4.4). According to (2.71), in the Mondrian case the associated zonoid Π_Θ is a cuboid with edges parallel to e_1, e_2, e_3, edge lengths p_1, p_2, p_3, volume $p_1p_2p_3$, and surface area $2(p_1p_2 + p_2p_3 + p_1p_3)$. Therefore in this case the constants $\zeta^{(2)}_{t\psi}$ and $\zeta^{(3)}_{t\psi}$ are

$$\zeta^{(2)}_{t\psi} = 2t^2\,(p_1p_2 + p_2p_3 + p_1p_3) \quad\text{and}\quad \zeta^{(3)}_{t\psi} = 6t^3\,p_1p_2p_3\,. \tag{17.15}$$

To start with, we consider the following mean values for the primitives.

	Mean...
ℓ_E	– length of the typical edge,
ℓ_P, a_P	– perimeter and area of the typical plate,
ℓ_Z, b_Z, a_Z, v_Z	– total edge length, mean width, surface area, volume of the typical cell,
$\mu_{V,E}, \mu_{V,P}, \mu_{V,Z}$	– number of edges, plates, cells adjacent to the typical vertex,
$\mu_{E,P} = \mu_{E,Z}$	– number of plates and cells adjacent to the typical edge,
$\mu_{P,V} = \mu_{P,E}$	– number of vertices and edges adjacent to the typical plate,
$\mu_{Z,V}, \mu_{Z,E}, \mu_{Z,P}$	– number of vertices, edges, plates adjacent to the typical cell,
$\nu_k(Z)$	– number of k-faces of the typical cell, $k \in \{0,1,2\}$.

Note that the mean total edge length ℓ_Z involves all edges adjacent to the typical cell. Because a STIT tessellation is not face-to-face, a cell may have edges on its boundary that are located in the relative interior of its 2-faces and not on its 1-faces.

The definitions for some of the mean values are listed below as examples. For a STIT tessellation Y_t and for a center function $c : \mathcal{P}_k \to \mathbb{R}^3$,

$$\ell_E = \int_{\mathcal{P}_{1,c}} V_1(S)\,\mathbb{Q}_E(dS) = \frac{1}{\gamma_E}\,\mathbb{E} \sum_{e\in E(Y_t)} V_1(e)\,\mathbb{1}\{c(e) \in [0,1]^3\}$$

is the mean length of the typical edge, whereas

$$\ell_P = \int_{\mathcal{P}_{2,c}} \sum_{F\in\mathcal{F}_1(P)} V_1(F)\,\mathbb{Q}_P(dP) = \frac{1}{\gamma_P}\,\mathbb{E} \sum_{p\in P(Y_t)} \sum_{S\in\mathcal{F}_1(p)} V_1(S)\,\mathbb{1}\{c(p) \in [0,1]^3\}$$

is the mean perimeter of the typical plate, and

$$\ell_Z = \int_{\mathcal{P}_{3,c}} \sum_{e \in E(Y_t)} V_1(e \cap P)\, \mathbb{Q}_Z(dP) = \frac{1}{\gamma_Z} \mathbb{E} \sum_{z \in Z(Y_t)} \sum_{e \in E(Y_t)} V_1(e \cap z)\, \mathbb{1}\{c(z) \in [0,1]^3\}$$

is the mean total length of edges adjacent to the typical cell. Again, we point out that all these mean values for a STIT tessellation Y_t may depend on t, but they are independent of the choice of the specific center function. As long as we consider Y_t for a fixed $t > 0$, we omit t in the notation for the mean values for better readability and to avoid an overloaded notation.

From (6.4) we know that a STIT tessellation has a Poisson typical cell. Applying equation (3.19) in [82] for $k = d = 3$, we obtain the following mean numbers of k-faces of the typical cell of a spatial STIT tessellation for $k \in \{0, 1, 2\}$:

$$\nu_0(Z) = 8\,, \quad \nu_1(Z) = 12\,, \quad \nu_2(Z) = 6\,. \tag{17.16}$$

To calculate the other mean values given above, we use that all these mean values for the primitives can be expressed by a system of seven characteristics, namely

$$\gamma_V\,, \ \gamma_Z\,, \ \gamma_{E+P} := \gamma_E + \gamma_P\,, \ \eta_1\,, \ \eta_2$$

and

$$m_V := \gamma_V\, \mu_{V,Z}\,, \quad m_E := \gamma_Z\, \ell_Z\,; \tag{17.17}$$

see Section 9.4 in [33]. The characteristic m_V can be interpreted as the mean *multiple* number of vertices per unit volume, and the characteristic m_E as the mean total *multiple* edge length per unit volume. The multiplicity of a vertex or an edge is defined here by the number of adjacent cells.

Lemma 17.1. *Let Y_t be a spatial STIT tessellation with directional distribution ψ_o in the typical point of ∂Y_t, given in* (17.5)*, and with two constants $\zeta_\psi^{(2)}$ and $\zeta_\psi^{(3)}$ (see* (17.9) *and* (17.10)*, respectively). Together with relation* (17.11)*, the seven characteristics in* (17.17) *are calculated for Y_t as follows:*

$$\gamma_V = t^3\, \zeta_\psi^{(3)} \quad \textit{(intensity of vertices)}, \tag{17.18}$$

$$\gamma_Z = \frac{1}{6}\, t^3\, \zeta_\psi^{(3)} \quad \textit{(intensity of cells)}, \tag{17.19}$$

$$\gamma_{E+P} = \frac{19}{6}\, t^3\, \zeta_\psi^{(3)} \quad \textit{(sum of the intensities of edges and plates)}, \tag{17.20}$$

$$\eta_1 = t^2\, \zeta_\psi^{(2)} \quad \textit{(length density)}, \tag{17.21}$$

$$\eta_2 = t \quad \textit{(surface density)}, \tag{17.22}$$

$$m_V = 4\, t^3\, \zeta_\psi^{(3)} \quad \textit{(multiple vertex intensity)}, \tag{17.23}$$

$$m_E = 3\, t^2\, \zeta_\psi^{(2)} \quad \textit{(multiple length density)}. \tag{17.24}$$

Proof. Equation (17.18) follows from (14.15) in Lemma 14.8, because $\gamma_V = \gamma^{pr}_{0,t}$. Relations (17.21) and (17.22) for the densities η_1 and η_2 have already been given in (17.12) and (17.4), respectively.

The following facts are used to prove the other formulae:

(i) According to (6.4), a STIT tessellation has a Poisson typical cell. Furthermore, $\chi_1 = \chi_1^{PHT}$, which is given by (6.10) or (17.8).

(ii) In a spatial STIT tessellation, each edge has exactly three adjacent tessellation cells, which follows from (8.13) for $d = 3$. This is also illustrated in Figures 17.1 and 17.2.

(iii) By (6.1) and (6.2) it follows that $Y_t \overset{D}{=} 2Y_t \boxplus \underline{2Y_t}$ for all $t > 0$.

(iv) Let $\gamma_A, \eta_1, \eta_2, \ell_A$ be the mean values for the spatial tessellation Y_t, and let $\gamma'_A, \eta'_1, \eta'_2, \ell'_A$ be the respective mean values for $2Y_t \overset{D}{=} Y_{\frac{t}{2}}$. They relate to each other as follows:

$$\begin{aligned} \gamma'_A &= \frac{1}{8}\,\gamma_A\,, \quad A \in \{V, E, P, Z\}\,,\\ \eta'_1 &= \frac{1}{4}\,\eta_1\,,\\ \eta'_2 &= \frac{1}{2}\,\eta_2\,,\\ \ell'_A &= 2\,\ell_A\,, \quad A \in \{E, P, Z\}\,. \end{aligned}$$

For the Poisson plane tessellation X_t, it is known that $\gamma_Z^{PHT} = \gamma_V^{PHT}$ and $\gamma_V^{PHT} = \frac{1}{6}\,t^3\zeta_\psi^{(3)}$, which is obtained from (2.81) with $d = 3$, $k = 0$, and $\gamma_V^{PHT} = \eta_{0,t}^{PHT}$. Due to (i), the mean volume of the typical cell is $v_Z = v_Z^{PHT}$. Since the relation $\gamma_Z = 1/v_Z$ generally applies to random tessellations, it follows that $\gamma_Z = \gamma_Z^{PHT}$, which leads to (17.19).

Now we consider the 1-skeleton of the iterated tessellation $2Y_t \boxplus \underline{2Y_t}$. Due to (iii), it has the length density η_1. The 1-skeleton is formed by the union of edges and contains three types of edges. Firstly, the edges from the frame tessellation $2Y_t$ with length density $\frac{1}{4}\eta_1$, according to (iv). Secondly, the edges on $\operatorname{skel}_1(2Y_{t,k}) \cap 2z_k$, $k \in \mathbb{N}_0$. They arise when $2Y_{t,k}$ is nested into the frame cell $2z_k$. These edges in $2Y_t \boxplus \underline{2Y_t}$ also have the length density $\frac{1}{4}\eta_1$. Also, there is a third type of edges located on the boundary of the frame cells. They are formed by intersections of plates of the frame cell $2z_k$ with plates of $2Y_{t,k}$, that is, $\partial(2Y_{t,k}) \cap \partial(2z_k)$, $k \in \mathbb{N}_0$. We will refer to those edges as "*new*" edges. They have the length density $\frac{1}{2}\eta_1$, since the sum of the length densities of the three edge types must be equal to η_1.

Each of these new edges is adjacent to exactly three cells. Furthermore, since $m_E = \gamma_Z\,\ell_Z$ for Y_t, the mean total multiple edge length per unit volume of $2Y_t$ is $\frac{1}{8}\,\gamma_Z\,2\,\ell_Z = \frac{1}{4}\,m_E$ by (iv). Thus, due to (iii), we obtain

$$m_E = \frac{1}{4}\,m_E + \frac{1}{4}\,m_E + 3\,\frac{1}{2}\,\eta_1\,,$$

which implies $m_E = 3\eta_1$. This yields (17.24).

We now consider the vertices of $2Y_t \boxplus \underline{2Y_t}$. Due to (iii), they have the intensity γ_V. Similarly to edges, we can also distinguish different types of vertices in an iterated tessellation. There are the vertices of the frame tessellation $2Y_t$ with intensity $\frac{1}{8}\gamma_V$, according to (iv). Furthermore, there are vertices of $\text{skel}_0^{\text{pr}}(2Y_{t,k}) \cap 2z_k$, $k \in \mathbb{N}_0$, which arise when $2Y_{t,k}$ is nested into the frame cell $2z_k$. Recall that $\text{skel}_0^{\text{pr}}(Y_t)$ is equal to the set of all vertices of Y_t. So these vertices are vertices of $2Y_{k,t}$ located in the interior of the frame cell $2z_k$. The vertices of this type in $2Y_t \boxplus \underline{2Y_t}$ again have the intensity $\frac{1}{8}\gamma_V$. The third type of vertices are the *"new"* vertices generated by iteration. The intensity of these new vertices is $\frac{3}{4}\gamma_V$, since the sum of the intensities of the three vertex types must be equal to γ_V.

The new vertices can appear on the 1-faces of the frame cells. This is the case where a plate of the nested tessellation intersects such a 1-face of the corresponding frame cell. Such vertices are referred to as T-vertices in Section 17.2.1. Or new vertices can be located in the relative interior of the 2-faces of the frame cells. They are created either by an intersection of a plate of the nesting tessellation with an already existing edge in a 2-face of the frame cell or by an intersection of two plates of nested tessellations in the two neighboring frame cells. Later in Section 17.2.1, we will refer to such vertices as X-vertices. In any case a new vertex in the iteration $2Y_t \boxplus \underline{2Y_t}$ has exactly four adjacent cells and exactly four adjacent edges.

From $m_V = \gamma_V \mu_{V,Z}$ we conclude by (iii) and (iv) that

$$m_V = \frac{1}{8}\, m_V + \frac{1}{8}\, m_V + 4\,\frac{3}{4}\,\gamma_V\,.$$

Thus relation (17.18) yields (17.23).

For the mean number of adjacent edges of the typical vertex $\mu_{V,E}$, we obtain

$$\gamma_V\,\mu_{V,E} = \frac{1}{8}\,\gamma_V\,\mu_{V,E} + \frac{1}{8}\,\gamma_V\,\mu_{V,E} + 3\,\gamma_V\,,$$

and thus $\mu_{V,E} = 4$. By the mean value relation $\gamma_V\,\mu_{V,E} = 2\,\gamma_E$ this yields $\gamma_E = 2\gamma_V$. Using this and the relation $\gamma_V - \gamma_E + \gamma_P - \gamma_Z = 0$, given in (9.51) in [33], we obtain $\gamma_P = \gamma_V + \gamma_Z$ and $\gamma_{E+P} = 3\gamma_V + \gamma_Z$. Relation (17.20) therefore results from (17.18) and (17.19). □

Results for the other intensities and mean values of a spatial STIT tessellation can now be derived directly from the known mean value formulae for spatial tessellations given in Section 9.4 of [33].

Corollary 17.2. *Let Y_t be a spatial STIT tessellation with directional distribution ψ_o in the typical point of ∂Y_t, given in* (17.5), *and with two constants $\zeta_\psi^{(2)}$ and $\zeta_\psi^{(3)}$ (see* (17.9) *and* (17.10)*, respectively). Then*

$$\gamma_E = 2\,t^3\,\zeta_\psi^{(3)}\,,\quad \gamma_P = \frac{7}{6}\,t^3\,\zeta_\psi^{(3)} \qquad \textit{(intensities of edges and plates)},$$

$$\ell_E = \frac{1}{2t}\,\frac{\zeta_\psi^{(2)}}{\zeta_\psi^{(3)}} \qquad \textit{(mean length of the typical edge)},$$

$$\ell_{\mathrm{P}} = \frac{18}{7t}\frac{\zeta_\psi^{(2)}}{\zeta_\psi^{(3)}}, \quad a_{\mathrm{P}} = \frac{6}{7t^2}\frac{1}{\zeta_\psi^{(3)}} \quad \text{(mean perimeter and mean area of the typical plate)},$$

$$\ell_{\mathrm{Z}} = \frac{18}{t}\frac{\zeta_\psi^{(2)}}{\zeta_\psi^{(3)}}, \quad b_{\mathrm{Z}} = \frac{3}{2t}\frac{\zeta_\psi^{(2)}}{\zeta_\psi^{(3)}} \quad \text{(mean total edge length and mean mean width of the typical cell)},$$

$$a_{\mathrm{Z}} = \frac{12}{t^2}\frac{1}{\zeta_\psi^{(3)}}, \quad v_{\mathrm{Z}} = \frac{6}{t^3}\frac{1}{\zeta_\psi^{(3)}} \quad \text{(mean surface area and mean volume of the typical cell)}.$$

The mean adjacency numbers are

$$\begin{aligned}
&\mu_{\mathrm{V,E}} = 4, && \mu_{\mathrm{V,P}} = 6, \quad \mu_{\mathrm{V,Z}} = 4,\\
&\mu_{\mathrm{E,P}} = \mu_{\mathrm{E,Z}} = 3, &&\\
&\mu_{\mathrm{P,V}} = \mu_{\mathrm{P,E}} = \frac{36}{7}, &&\\
&\mu_{\mathrm{Z,V}} = 24, && \mu_{\mathrm{Z,E}} = 36, \quad \mu_{\mathrm{Z,P}} = 14.
\end{aligned}$$

All mean values in Lemma 17.1 and Corollary 17.2 can be calculated explicitly for a stationary and *isotropic* STIT tessellation with the particular values $\zeta_\psi^{(2)} = \frac{\pi}{4}$ and $\zeta_\psi^{(3)} = \frac{\pi}{8}$ given in (17.13). With $\frac{\zeta_\psi^{(2)}}{\zeta_\psi^{(3)}} = 2$, we obtain

$$\eta_1 = \frac{\pi}{4}t^2, \quad \eta_2 = t, \quad m_{\mathrm{V}} = \frac{\pi}{2}t^3, \quad m_{\mathrm{E}} = \frac{3\pi}{4}t^2$$

and

A	γ_{A}	ℓ_{A}	a_{A}	b_{A}	v_{A}
V	$\frac{\pi}{8}t^3$				
E	$\frac{\pi}{4}t^3$	$\frac{1}{t}$			
P	$\frac{7\pi}{48}t^3$	$\frac{36}{7}\frac{1}{t}$	$\frac{48}{7\pi}\frac{1}{t^2}$		
Z	$\frac{\pi}{48}t^3$	$36\frac{1}{t}$	$\frac{96}{\pi}\frac{1}{t^2}$	$3\frac{1}{t}$	$\frac{48}{\pi}\frac{1}{t^3}$

The *Mondrian case* is defined in (17.14), and the values for $\zeta_\psi^{(2)}$ and $\zeta_\psi^{(3)}$ are given in (17.15). A simulation of a Mondrian STIT tessellation is shown in Figure 17.2. For $p_1 = p_2 = p_3 = \frac{1}{3}$, we obtain $\zeta_\psi^{(2)} = \frac{2}{3}$, $\zeta_\psi^{(3)} = \frac{2}{9}$, and $\frac{\zeta_\psi^{(2)}}{\zeta_\psi^{(3)}} = 3$. This yields

$$\eta_1 = \frac{2}{3}t^2, \quad \eta_2 = t, \quad m_{\mathrm{V}} = \frac{8}{9}t^3, \quad m_{\mathrm{E}} = 2t^2$$

and

A	γ_{A}	ℓ_{A}	a_{A}	b_{A}	v_{A}
V	$\frac{2}{9}t^3$				
E	$\frac{4}{9}t^3$	$\frac{3}{2}\frac{1}{t}$			
P	$\frac{7}{27}t^3$	$\frac{54}{7}\frac{1}{t}$	$\frac{27}{7}\frac{1}{t^2}$		
Z	$\frac{1}{27}t^3$	$54\,\frac{1}{t}$	$54\,\frac{1}{t^2}$	$\frac{9}{2}\frac{1}{t}$	$27\,\frac{1}{t^3}$

17.2 Refined combinatorial structure

Apart from the primitives, the *maximal polytopes* are important for the study of the topological and combinatorial structure of a STIT tessellation.

For spatial STIT tessellations, these maximal polytopes are I-segments and I-polygons. They can be characterized as follows. The **I-polygons** are the cell-dividing polygons. When a cell z is divided by a plane $H \in [\mathrm{int}\, \mathrm{z}]$, exactly one I-polygon $\mathrm{z} \cap H$ is born, whereas at least three I-segments are born simultaneously, because the **I-segments** are the 1-faces of the I-polygon. Hence with (8.10), for the spatial STIT tessellation Y_t, we define

$$\mathrm{I}_{\mathsf{P}}(Y_t) := \mathcal{F}_2^{\max}(Y_t) \quad - \quad \text{set of I-polygons,} \tag{17.25}$$

$$\mathrm{I}_{\mathsf{S}}(Y_t) := \mathcal{F}_1^{\max}(Y_t) \quad - \quad \text{set of I-segments.} \tag{17.26}$$

For the study of the I-polygons and I-segments, it is not sufficient to consider only the primitives of the STIT tessellation. To obtain results for their intensities and mean values and to investigate their internal structure, a finer classification of the vertices, segments, and polygons in the tessellation is required, which we now present.

17.2.1 Classification of vertices, segments, and polygons

Types of vertices

Recall that planar STIT tessellations have only T-shaped vertices. This means that any vertex has three outgoing edges and two of them are collinear. From the construction of a STIT tessellation in a window, which is the subject of Section 4.1, or likewise from the global construction given in Chapter 8 it follows that a spatial STIT tessellation has two different types of vertices. We call them **T-vertices** and **X-vertices**. These designations are based on the polygons that create the vertices.

Figure 17.3 shows the two possible situations for the vertex types. A T-vertex appears on a 1-face of a cell when this cell is intersected by a plane. Due to the cell division, a new I-polygon is born, and this T-vertex is a 0-face of that I-polygon. Hence a T-vertex is always a maximal polytope of dimension 0. Recall that in general dimension d the maximal k-polytopes are the k-faces of the maximal $(d-1)$-polytopes, $k \in \{0, \dots, d-2\}$. An X-vertex, on the other hand, is created as the intersection of two I-polygons in neighboring cells within the interior of the common plate. An X-vertex is therefore not a 0-face of any of these two I-polygons and thus not a maximal polytope of dimension 0.

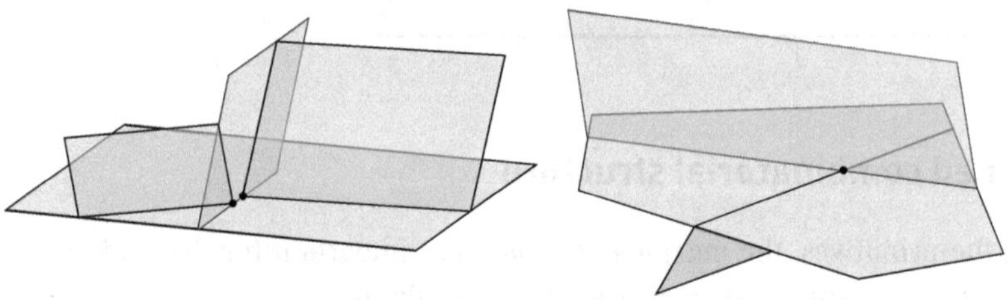

Figure 17.3: Two types of vertices in a spatial STIT tessellation. Left: Two T-vertices. Right: An X-vertex.

The classes of T- and X-vertices are denoted by $\mathsf{V}\langle\mathsf{T}\rangle$ and $\mathsf{V}\langle\mathsf{X}\rangle$, and the corresponding intensities are $\gamma_{\mathsf{V}\langle\mathsf{T}\rangle}$ and $\gamma_{\mathsf{V}\langle\mathsf{X}\rangle}$, respectively. Obviously, $\gamma_{\mathsf{V}} = \gamma_{\mathsf{V}\langle\mathsf{T}\rangle} + \gamma_{\mathsf{V}\langle\mathsf{X}\rangle}$. In addition, the following applies to a spatial STIT tessellation Y_t:

$$\mathcal{F}_0 = \mathcal{F}_0^{\max} = \mathsf{V}\langle\mathsf{T}\rangle \subseteq \mathcal{F}_0^{\mathrm{pr}} = \mathsf{V},$$

and

$$\mathsf{V}\langle\mathsf{X}\rangle = \mathcal{F}_0^{\mathrm{pr}} \setminus \mathcal{F}_0^{\max}.$$

Two types of vertices that occur in a spatial STIT tessellation and the differences between them are essential for our further considerations.

Types of segments

In Section 16.2.2, three different types of segments in a planar STIT tessellation were considered: edges as the primitives, I-segments as the maximal segments, and the J-segments, which are the 1-faces of the cells marked with their multiplicities. A similar concept will be used now for a classification of segments contained in the 1-skeleton of a spatial STIT tessellation.

- The **edges** are the 1-primitives of the tessellation. Hence an edge is a line segment in the 1-skeleton between two vertices with no further vertex in its relative interior. The set of all edges is denoted by E.

- The **I-segments** were defined in (17.26). They are the 1-faces of the maximal polygons born during the construction of the tessellation. Hence an I-segment is a maximal union of connected and collinear tessellation edges. It cannot be enlarged by another collinear edge, which is a consequence of Lemma 8.8(b). The class of all I-segments is denoted by I_S.
- A $\mathbf{J^{(3)}}$**-segment** is a 1-face of a **cell**, marked with its multiplicity as follows. For a tessellation T, the class of all $\mathrm{J}^{(3)}$-segments is defined by

$$\mathrm{J}_\mathrm{S}^{(3)}(T) := \{(S, m(S)) \,:\, S \in \mathcal{F}_1(T),\, m(S) = |\{\mathrm{z} \in \mathrm{Z}(T) \,:\, S \in \mathcal{F}_1(\mathrm{z})\}|\},$$

where $\mathcal{F}_1(T)$ is the set of all 1-faces of cells of T, given in (2.1). Thus the multiplicity $m(S)$ is the number of *cells* that have S as a 1-face. For a spatial STIT tessellation Y_t, the multiplicity of a $\mathrm{J}^{(3)}$-segment is 1 or 2. Consequently, $\mathrm{J}_\mathrm{S}^{(3)}(Y_t)$ can be understood as a marked point process on $\mathcal{P}_1$ with mark space $M = \{1, 2\}$.
- A $\mathbf{J^{(2)}}$**-segment** is a 1-face of a **plate**, marked with its multiplicity as follows. For a tessellation T, the class of all $\mathrm{J}^{(2)}$-segments is defined by

$$\mathrm{J}_\mathrm{S}^{(2)}(T) := \left\{(S, m(S)) \,:\, S \in \bigcup_{\mathrm{p}\in\mathrm{P}(T)} \mathcal{F}_1(\mathrm{p}),\, m(S) = |\{\mathrm{p} \in \mathrm{P}(T) \,:\, S \in \mathcal{F}_1(\mathrm{p})\}|\right\}.$$

Here the multiplicity $m(S)$ is the number of *plates* that have S as a 1-face. For a $\mathrm{J}^{(2)}$-segment S in a spatial STIT tessellation Y_t, we obtain that $m(S) \in \{1, 2, 3\}$. Therefore $\mathrm{J}_\mathrm{S}^{(2)}(Y_t)$ is a marked point process on $\mathcal{P}_1$ with mark space $M = \{1, 2, 3\}$.

In Figure 17.4 the different segment types are illustrated. In Figure 17.5, it is shown that the classes $\mathrm{J}_\mathrm{S}^{(2)}(T)$ and $\mathrm{J}_\mathrm{S}^{(3)}(T)$ do not coincide. Furthermore, they are not necessarily disjoint. For example, in Figure 17.4 the vertical cyan segment is a 1-face of exactly one plate and also a 1-face of exactly one cell. It is therefore a $\mathrm{J}^{(2)}$-segment with multiplicity 1 and a $\mathrm{J}^{(3)}$-segment, also with multiplicity 1. In addition, there may be segments that are in-

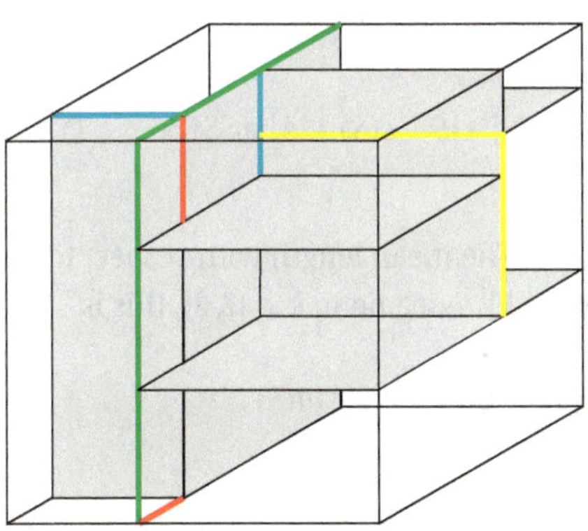

Figure 17.4: Two edges (red), two $\mathrm{J}^{(2)}$-segments (cyan), two $\mathrm{J}^{(3)}$-segments (yellow), and two I-segments (green).

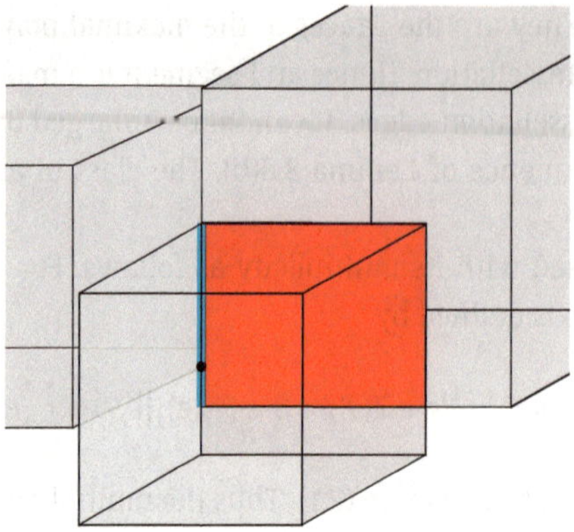

Figure 17.5: A plate (red) with a $J^{(2)}$-segment (cyan) on its boundary. The $J^{(2)}$-segment has multiplicity one, and it is not a $J^{(3)}$-segment.

cluded in both sets but have different multiplicities. The horizontal yellow segment in Figure 17.4 is a 1-face of two cells and also a 1-face of three plates.

Further illustrations for different segments in a spatial STIT tessellation can be seen in Figure 17.10.

Similarly to the planar case, we would like to point out that $J^{(2)}$- and $J^{(3)}$-segments are not components in the strict sense of Definition 2.9. However, according to Remark 16.9, the four different segment types can be treated in a similar manner if the multiplicity of a $J^{(2)}$- or $J^{(3)}$-segments is included in the calculations. The notation of the form $\mathsf{A} \in \{\mathsf{E}, \mathsf{I}_\mathsf{S}, \mathsf{J}_\mathsf{S}^{(2)}, \mathsf{J}_\mathsf{S}^{(3)}\}$ is based on this.

For a STIT tessellation Y_t, the intensity of segments of type $\mathsf{A} \in \{\mathsf{E}, \mathsf{I}_\mathsf{S}, \mathsf{J}_\mathsf{S}^{(2)}, \mathsf{J}_\mathsf{S}^{(3)}\}$ is denoted by γ_A and defined in (2.25). For $\mathsf{A} \in \{\mathsf{E}, \mathsf{I}_\mathsf{S}\}$, this is equal to the intensity of the stationary point process of centroids of A-components. For $\mathsf{A} \in \{\mathsf{J}_\mathsf{S}^{(2)}, \mathsf{J}_\mathsf{S}^{(3)}\}$, this is the intensity of the point process with multiplicities, which means that for $k \in \{2, 3\}$,

$$\gamma_{\mathsf{J}_\mathsf{S}^{(k)}} = \mathbb{E} \sum_{(S,m(S)) \in \mathsf{J}_\mathsf{S}^{(k)}(Y_t)} \mathbb{1}\{c(S) \in [0,1]^3\}\, m(S)\,. \tag{17.27}$$

Recall that the intensity γ_A does not depend on the choice of the center function c. The distribution $\mathbb{Q}_{\mathsf{J}_\mathsf{S}^{(k)}}$ of the typical $J^{(k)}$-segment, $k \in \{2, 3\}$, is defined by

$$\mathbb{Q}_{\mathsf{J}_\mathsf{S}^{(k)}}(A) := \frac{1}{\gamma_{\mathsf{J}_\mathsf{S}^{(k)}}} \mathbb{E} \sum_{(S,m(S)) \in \mathsf{J}_\mathsf{S}^{(k)}(Y_t)} \mathbb{1}\{c(S) \in [0,1]^3\}\, \mathbb{1}\{S - c(S) \in A\}\, m(S) \tag{17.28}$$

for $A \in \mathcal{B}(\mathcal{P}_{1,c})$ with respect to a center function c. The mean length with respect to $\mathbb{Q}_\mathsf{A}$ of the typical segment of type A is ℓ_A. For the typical $J^{(k)}$-segment, $k \in \{2, 3\}$, this is

$$\ell_{\mathsf{J}_\mathsf{S}^{(k)}} := \frac{1}{\gamma_{\mathsf{J}_\mathsf{S}^{(k)}}} \mathbb{E} \sum_{(S,m(S)) \in \mathsf{J}_\mathsf{S}^{(k)}(Y_t)} \mathbb{1}\{c(S) \in [0,1]^3\}\, V_1(S)\, m(S)\,,$$

which is independent of the choice of c; see also Remark 2.18.

The mean number of vertices in the relative interior of the typical A-segment can be denoted by $\mu_{\mathrm{relint\,A,V}}$ using the concept of adjacent components. The relative interior of a segment is the segment without its two 0-faces. This leads to the following obvious result for the typical segment of type $\mathsf{A} \in \{\mathsf{E}, \mathrm{I_S}, \mathrm{J_S^{(2)}}, \mathrm{J_S^{(3)}}\}$:

$$\mu_{\mathsf{A,V}} = \mu_{\mathrm{relint\,A,V}} + 2 = \mu_{\mathsf{A,E}} + 1 . \tag{17.29}$$

Using this formula, the treatment of objects with multiplicities, as described in Remark 16.9, will be formally demonstrated for the $\mathrm{J_S^{(k)}}$-segments, $k \in \{2,3\}$:

$$\begin{aligned}
\mu_{\mathrm{J_S^{(k)}},\mathsf{V}} &= \frac{1}{\gamma_{\mathrm{J_S^{(k)}}}} \mathbb{E} \sum_{(S,m(S))\in \mathrm{J_S^{(k)}}(Y_t)} \mathbb{1}\{c(S) \in [0,1]^2\}\, m(S) \sum_{\mathsf{v}\in \mathsf{V}(Y_t)} \mathbb{1}\{\mathsf{v} \subset S\} \\
&= \frac{1}{\gamma_{\mathrm{J_S^{(k)}}}} \mathbb{E} \sum_{(S,m(S))\in \mathrm{J_S^{(k)}}(Y_t)} \mathbb{1}\{c(S) \in [0,1]^2\}\, m(S) \Big(\sum_{\mathsf{v}\in \mathsf{V}(Y_t)} \mathbb{1}\{\mathsf{v} \subset \operatorname{relint} S\} + 2 \Big) \\
&= \frac{1}{\gamma_{\mathrm{J_S^{(k)}}}} \mathbb{E} \sum_{(S,m(S))\in \mathrm{J_S^{(k)}}(Y_t)} \mathbb{1}\{c(S) \in [0,1]^2\}\, m(S) \sum_{\mathsf{v}\in \mathsf{V}(Y_t)} \mathbb{1}\{\mathsf{v} \subset \operatorname{relint} S\} \\
&\quad + \frac{2}{\gamma_{\mathrm{J_S^{(k)}}}} \mathbb{E} \sum_{(S,m(S))\in \mathrm{J_S^{(k)}}(Y_t)} \mathbb{1}\{c(S) \in [0,1]^2\}\, m(S) \\
&= \mu_{\mathrm{relint\,J_S^{(k)}},\mathsf{V}} + 2 .
\end{aligned}$$

Types of polygons

Similarly to the different types of segments, the following classification of polygons can be given.

- The **plates** are the 2-primitives of the tessellation. Hence a plate is a two-dimensional convex polygon in the 2-skeleton, bounded by edges of the 1-skeleton and without edges and vertices in its relative interior. The set of all plates is denoted by P.
- The **I-polygons** were defined in (17.25). They arise when a cell is divided. Thus they are the maximal polygons in a spatial STIT tessellation. From Lemma 8.8(b) it follows that an I-polygon is the maximal union of coplanar and connected plates and cannot be enlarged by another coplanar plate. The set of all I-polygons is denoted by $\mathrm{I_P}$.
- A **J-polygon** is a 2-face of a **cell**, marked with its multiplicity as follows. For a tessellation T, the class of all J-polygons is defined by

$$\mathrm{J_P}(T) := \{(P, m(P)) \;:\; P \in \mathcal{F}_2(T),\, m(P) = |\{\mathsf{z} \in \mathsf{Z}(T) \;:\; P \in \mathcal{F}_2(\mathsf{z})\}|\} .$$

The multiplicity $m(P)$ is the number of cells that have P as a 2-face. For a spatial STIT tessellation Y_t, the multiplicity of a J-polygon is 1 or 2, where $m(P) = 2$ means that the two neighboring cells with the common 2-face P are in a face-to-face position.

Similarly to the $J^{(k)}$-segments, the random set $J_P(Y_t)$ can be understood as a point process on $\mathcal{P}_2$ with mark space $M = \{1, 2\}$.

In Figure 17.6 the different types of polygons are shown.

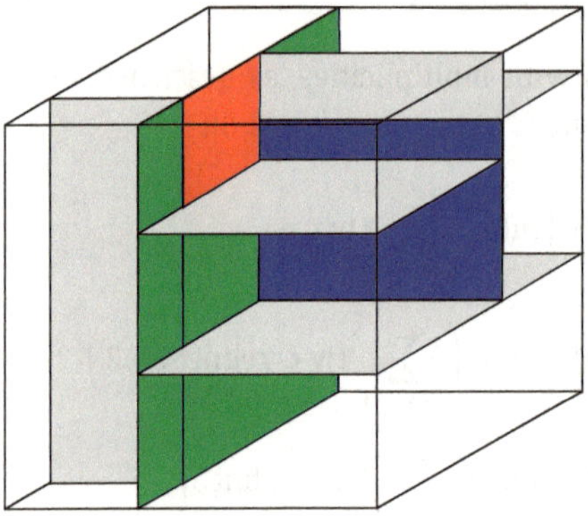

Figure 17.6: A plate (red), a J-polygon (blue) with multiplicity one, and an I-polygon (green) in a spatial STIT tessellation.

Note that we use the subscript $_P$ to distinguish the classifications for polygons from the classification for segments, where we have used the subscript $_S$.

According to the notational style of this chapter, the following notation will be used for polygons of type $A \in \{P, J_P, I_P\}$:

γ_A – intensity of A-polygons,
ℓ_A, a_A – mean perimeter and mean area of the typical A-polygon,
$\mu_{\text{relint}\,A,V}$ – mean number of vertices in the relative interior of the typical A-polygon,
$\mu_{\partial A,V}$ – mean number of vertices on the boundary of the typical A-polygon,
$\mu_{\text{relint}\,A,E}$ – mean number of edges in the relative interior of the typical A-polygon,
$\mu_{\partial A,E}$ – mean number of edges on the boundary of the typical A-polygon.

An edge $e \in E$ is located in the relative interior of a polygon $a \in A$ if $\text{relint}\, e \subseteq \text{relint}\, a$. Furthermore,

$\mu_{J_P,P}$ – mean number of plates adjacent to the typical J-polygon,
$\mu_{I_P,P}$ – mean number of plates adjacent to the typical I-polygon.

Like the $J^{(k)}$-segments, the J-polygons are polygons marked with their multiplicities. Hence their intensity is given by

$$\gamma_{J_P} = \mathbb{E} \sum_{(P,m(P)) \in J_P(Y_t)} \mathbb{1}\{c(P) \in [0,1]^3\}\, m(P)\,.$$

The multiplicities are also important for the definition of mean values of the typical J-polygon. For example, the mean area and the mean number of plates adjacent to the typical J-polygon are

$$a_{\mathrm{J_P}} = \frac{1}{\gamma_{\mathrm{J_P}}} \mathbb{E} \sum_{(P,m(P))\in \mathrm{J_P}(Y_t)} \mathbb{1}\{c(P) \in [0,1]^3\}\, V_2(P)\, m(P)\,,$$

$$\mu_{\mathrm{J_P,P}} = \frac{1}{\gamma_{\mathrm{J_P}}} \mathbb{E} \sum_{(P,m(P))\in \mathrm{J_P}(Y_t)} \mathbb{1}\{c(P) \in [0,1]^3\}\, m(P) \sum_{\mathrm{p}\in \mathrm{P}(Y_t)} \mathbb{1}\{\mathrm{p} \subseteq P\}\,.$$

Thus $\mu_{\mathrm{J_P,P}}$ is the mean of the number of plates that partition the typical J-polygon.

The following mean value relations for the typical polygon of type $\mathrm{A} \in \{\mathrm{P}, \mathrm{J_P}, \mathrm{I_P}\}$ can be easily derived:

$$\mu_{\mathrm{A,V}} = \mu_{\mathrm{relint\,A,V}} + \mu_{\partial\mathrm{A,V}}\,, \quad \mu_{\mathrm{A,E}} = \mu_{\mathrm{relint\,A,E}} + \mu_{\partial\mathrm{A,E}}\,, \qquad (17.30)$$
$$\mu_{\partial\mathrm{A,V}} = \mu_{\partial\mathrm{A,E}}\,.$$

Note that the relative interior of a polygon is that polygon without its boundary.

We will now investigate which relationships exist between edges and plates in a Poisson plane tessellation and segments and polygons of a suitably selected type in spatial STIT tessellations. In the planar case, we have seen in Lemma 16.16 that the typical J-segment of a planar STIT tessellation Y_t has the same distribution as the typical edge of a Poisson line tessellation X_t. We will obtain similar results for the distribution of the typical $\mathrm{J}^{(3)}$-segment and the typical J-polygon in a spatial STIT tessellation. Recall that a Poisson plane tessellation is face-to-face. That is why the k-faces there are also the k-primitives, $k \in \{0,1,2,3\}$. A 1-face is an edge adjacent to four plates and to four cells, and a 2-face is a plate.

From (6.4) in Theorem 6.2 we know that a spatial STIT tessellation Y_t has a Poisson typical cell. This means that the typical cell of Y_t has the same distribution as the typical cell of a Poisson plane tessellation X_t if both tessellations X_t and Y_t have the same directional distribution ψ_o. In addition, relation (6.6) shows that both have the same surface density $\eta_2^{\mathrm{STIT}} = \eta_2^{\mathrm{PHT}} = t$. The distribution $\mathbb{Q}_{\mathrm{J_S^{(3)}}}$ for the typical $\mathrm{J}^{(3)}$-segment is given in (17.28), and $\mathbb{Q}_{\mathrm{J_P}}$ is obtained in a similar way. The distributions of the typical edge $\mathbb{Q}_{\mathrm{E}}^{\mathrm{PHT}}$ and of the typical plate $\mathbb{Q}_{\mathrm{P}}^{\mathrm{PHT}}$ of a Poisson line tessellation X_t result from (17.1) with $\mathrm{A} = \mathrm{E}$ and $\mathrm{A} = \mathrm{P}$, respectively. With regard to a center function $c : \mathcal{P}_k \to \mathbb{R}^3$, $k \in \{1,2\}$, the distributions $\mathbb{Q}_{\mathrm{J_S^{(3)}}}$ and $\mathbb{Q}_{\mathrm{E}}^{\mathrm{PHT}}$ are probability measures on $\mathcal{P}_{1,c}$, and analogously $\mathbb{Q}_{\mathrm{J_P}}$ and $\mathbb{Q}_{\mathrm{P}}^{\mathrm{PHT}}$ are probability measures on $\mathcal{P}_{2,c}$.

Lemma 17.3. *Let Y_t be a spatial STIT tessellation with distributions $\mathbb{Q}_{\mathrm{J_S^{(3)}}}$ and $\mathbb{Q}_{\mathrm{J_P}}$ and intensities $\gamma_{\mathrm{J_S^{(3)}}}$ and $\gamma_{\mathrm{J_P}}$. Furthermore, let X_t be a Poisson plane tessellation with $\mathbb{Q}_{\mathrm{E}}^{\mathrm{PHT}}$, $\mathbb{Q}_{\mathrm{P}}^{\mathrm{PHT}}$ and $\gamma_{\mathrm{E}}^{\mathrm{PHT}}$, $\gamma_{\mathrm{P}}^{\mathrm{PHT}}$. Assume that both tessellations have the same directional distribution ψ_o. Then for the distributions, we obtain*

$$\mathbb{Q}_{\mathsf{J}_\mathsf{S}^{(3)}} = \mathbb{Q}_\mathsf{E}^{\mathrm{PHT}} \quad \textit{and} \quad \mathbb{Q}_{\mathsf{J}_\mathsf{P}} = \mathbb{Q}_\mathsf{P}^{\mathrm{PHT}},$$

and for the intensities,

$$\gamma_{\mathsf{J}_\mathsf{S}^{(3)}} = 4\gamma_\mathsf{E}^{\mathrm{PHT}} \quad \textit{and} \quad \gamma_{\mathsf{J}_\mathsf{P}} = 2\gamma_\mathsf{P}^{\mathrm{PHT}}.$$

Proof. Recall that for a $\mathsf{J}^{(3)}$-segment $(S, m(S)) \in \mathsf{J}_\mathsf{S}^{(3)}(Y_t)$, the following applies: $S \in \mathcal{F}_1(Y_t)$ and $m(S) = \sum_{\mathsf{z}\in\mathsf{Z}(Y_t)} \mathbb{1}\{S \in \mathcal{F}_1(\mathsf{z})\}$. For a J-polygon $(P, m(P)) \in \mathsf{J}_\mathsf{P}(Y_t)$, we have that $P \in \mathcal{F}_2(Y_t)$ and $m(P) = \sum_{\mathsf{z}\in\mathsf{Z}(Y_t)} \mathbb{1}\{P \in \mathcal{F}_2(\mathsf{z})\}$.

Furthermore, in the Poisson plane tessellation X_t, all 1-faces of cells are edges of X_t. Hence $\mathcal{F}_1(X_t) = \mathsf{E}(X_t)$, and all 2-faces of cells are plates, $\mathcal{F}_2(X_t) = \mathsf{P}(X_t)$.

Any edge $\mathsf{e} \in \mathsf{E}(X_t)$ in a Poisson plane tessellation is adjacent to exactly four cells, $\sum_{\mathsf{z}\in\mathsf{Z}(X_t)} \mathbb{1}\{\mathsf{e} \in \mathcal{F}_1(\mathsf{z})\} = 4$, and any plate $\mathsf{p} \in \mathsf{P}(X_t)$ is adjacent to exactly two cells, $\sum_{\mathsf{z}\in\mathsf{Z}(X_t)} \mathbb{1}\{\mathsf{p} \in \mathcal{F}_2(\mathsf{z})\} = 2$.

From now on, the proof is similar to that of Lemma 16.16 in the planar case.

For a cell $\mathsf{z} \in \mathsf{Z}(Y_t)$ and a 1-face $S \in \mathcal{F}_1(Y_t)$, the relation $\mathsf{z} \sim_{Y_t} S \iff S \in \mathcal{F}_1(\mathsf{z})$ is used as well as $\mathsf{z} \sim_{Y_t} P \iff P \in \mathcal{F}_2(\mathsf{z})$ for a plate $P \in \mathcal{F}_2(Y_t)$.

The exchange formula given in (2.61), applied to these relations, results in

$$\gamma_{\mathsf{J}_\mathsf{S}^{(3)}}\, \mathbb{Q}_{\mathsf{J}_\mathsf{S}^{(3)}} = 4\gamma_\mathsf{E}^{\mathrm{PHT}}\, \mathbb{Q}_\mathsf{E}^{\mathrm{PHT}} \quad \text{and} \quad \gamma_{\mathsf{J}_\mathsf{P}}\, \mathbb{Q}_{\mathsf{J}_\mathsf{P}} = 2\gamma_\mathsf{P}^{\mathrm{PHT}}\, \mathbb{Q}_\mathsf{P}^{\mathrm{PHT}},$$

which completes the proof. □

Remark 17.4. *Lemmas* 16.16 *and* 17.3 *are special cases of the following relation between STIT tessellations and Poisson hyperplane tessellations for arbitrary dimension d.*

For a tessellation T in $\mathbb{R}^d$, we define the class of k-faces of cells, $k \in \{0, \dots, d\}$, marked with their multiplicities

$$\mathcal{F}_k^m(T) := \{(F, m(F)) : F \in \mathcal{F}_k(T),\ m(F) = |\{\mathsf{z} \in \mathsf{Z}(T) : F \in \mathcal{F}_k(\mathsf{z})\}|\}.$$

Here the multiplicity $m(F) \in \mathbb{N}$ of F is equal to the number of cells that have F as a k-face. In particular, $\mathcal{F}_1^m(T) = \mathsf{J}(T)$ for $d = 2$, and $\mathcal{F}_1^m(T) = \mathsf{J}_\mathsf{S}^{(3)}(T)$, $\mathcal{F}_2^m(T) = \mathsf{J}_\mathsf{P}(T)$ for $d = 3$. The corresponding intensity $\gamma_{\mathcal{F}_k^m}$ and the distribution $\mathbb{Q}_{\mathcal{F}_k^m}$ of the typical multiple k-face in a random tessellation are defined in the same way as in (17.27) *and* (17.28).

Furthermore, the intensity γ_k^{pr} of the k-primitives is defined in (2.42), *and the distribution $\mathbb{Q}_k^{\mathrm{pr}}$ of the typical k-primitive is defined in* (2.44), *$k \in \{0, \dots, d\}$.*

A Poisson hyperplane tessellation is face-to-face; consequently, the k-primitives are also the k-faces of the cells. This means that $\gamma_k^{\mathrm{pr}} = \gamma_{\mathcal{F}_k}$ and $\mathbb{Q}_k^{\mathrm{pr}} = \mathbb{Q}_{\mathcal{F}_k}$, where $\gamma_{\mathcal{F}_k}$ denotes the intensity of k-faces, and $\mathbb{Q}_{\mathcal{F}_k}$ denotes the distribution of the typical k-face. A k-face of a PHT lies in the intersection of $(d-k)$ pairwise different hyperplanes. Hence a k-face of a PHT is adjacent to 2^{d-k} cells. For a STIT tessellation Y_t and a PHT X_t, both with measure Θ, it results for $k \in \{0, \dots, d\}$ and the center function $c : \mathcal{P}_k \to \mathbb{R}^d$ in

$$\mathbb{Q}^{\mathrm{STIT}}_{\mathcal{F}^m_k} = \mathbb{Q}^{\mathrm{PHT}}_{\mathcal{F}_k} \quad \textit{and} \quad \gamma^{\mathrm{STIT}}_{\mathcal{F}^m_k} = 2^{d-k}\, \gamma^{\mathrm{PHT}}_{\mathcal{F}_k}\,,$$

where both distributions $\mathbb{Q}^{\mathrm{STIT}}_{\mathcal{F}^m_k}$ *and* $\mathbb{Q}^{\mathrm{PHT}}_{\mathcal{F}_k}$ *are probability measures on* $\mathcal{P}_{k,c}$. *Also, for dimension d, the proof is analogous to that of Lemma* 16.16 *and to* (16.30) *in the planar case.*

17.2.2 Topology of vertices

Two different types of vertices and moreover their topology, which are shown in Figure 17.3, play a crucial role in the further investigations of the tessellation structure. We start the considerations with the T-vertices and their adjacencies in a spatial STIT tessellation Y_t with distribution $\mathbb{P}_{Y_t}$. The following relationships are valid $\mathbb{P}_{Y_t}$-a. s. The results are obtained from the two variants of intersections of three I-polygons, which generate either a T-vertex or an X-vertex. Figure 17.3 can be helpful to verify these results.

– Any T-vertex $\mathsf{v} \in \mathsf{V}\langle\mathsf{T}\rangle$ of Y_t is adjacent to exactly 4 *edges*; more precisely,

$$\sum_{\mathsf{e}\in\mathsf{E}(Y_t)} \mathbb{1}\{\mathsf{v} \in \mathcal{F}_0(\mathsf{e})\} = 4\,.$$

– Any T-vertex $\mathsf{v} \in \mathsf{V}\langle\mathsf{T}\rangle$ of Y_t is adjacent to exactly 6 *plates*; more precisely,

$$\sum_{\mathsf{p}\in\mathsf{P}(Y_t)} \mathbb{1}\{\mathsf{v} \in \mathcal{F}_0(\mathsf{p})\} = 5\,, \qquad \sum_{\mathsf{p}\in\mathsf{P}(Y_t)} \mathbb{1}\{\mathsf{v} \subset \partial\mathsf{p},\ \mathsf{v} \notin \mathcal{F}_0(\mathsf{p})\} = 1\,.$$

– Any T-vertex $\mathsf{v} \in \mathsf{V}\langle\mathsf{T}\rangle$ of Y_t is adjacent to exactly 4 *cells*; more precisely,

$$\sum_{\mathsf{z}\in\mathsf{Z}(Y_t)} \mathbb{1}\{\mathsf{v} \in \mathcal{F}_0(\mathsf{z})\} = 2\,,$$

$$\sum_{\mathsf{z}\in\mathsf{Z}(Y_t)} \sum_{S\in\mathcal{F}_1(\mathsf{z})} \mathbb{1}\{\mathsf{v} \subset \operatorname{relint} S\} = 1\,,$$

$$\sum_{\mathsf{z}\in\mathsf{Z}(Y_t)} \sum_{P\in\mathcal{F}_2(\mathsf{z})} \mathbb{1}\{\mathsf{v} \subset \operatorname{relint} P\} = 1\,.$$

– Any T-vertex $\mathsf{v} \in \mathsf{V}\langle\mathsf{T}\rangle$ of Y_t is adjacent to exactly 3 *I-segments*; more precisely,

$$\sum_{S\in\mathsf{I}_\mathsf{S}(Y_t)} \mathbb{1}\{\mathsf{v} \in \mathcal{F}_0(S)\} = 2\,, \qquad \sum_{S\in\mathsf{I}_\mathsf{S}(Y_t)} \mathbb{1}\{\mathsf{v} \subset \operatorname{relint} S\} = 1\,.$$

– Any T-vertex $\mathsf{v} \in \mathsf{V}\langle\mathsf{T}\rangle$ of Y_t is adjacent to exactly 3 *I-polygons*; more precisely,

$$\sum_{P\in\mathsf{I}_\mathsf{P}(Y_t)} \mathbb{1}\{\mathsf{v} \in \mathcal{F}_0(P)\} = 1\,,$$

$$\sum_{P\in\mathsf{I}_\mathsf{P}(Y_t)} \mathbb{1}\{\mathsf{v} \subset \partial P,\ \mathsf{v} \notin \mathcal{F}_0(P)\} = 1\,,$$

$$\sum_{P \in \mathrm{I}_{\mathrm{P}}(Y_t)} \mathbb{1}\{\mathrm{v} \subset \mathrm{relint}\, P\} = 1\,.$$

- Any T-vertex $\mathrm{v} \in \mathrm{V}\langle\mathrm{T}\rangle$ of Y_t is adjacent to $J^{(2)}$*-segments* whose total sum of multiplicities equals 11; more precisely,

$$\sum_{(S,m(S)) \in \mathrm{J}_{\mathrm{S}}^{(2)}(Y_t)} \mathbb{1}\{\mathrm{v} \in \mathcal{F}_0(S)\}\, m(S) = 10\,,$$

$$\sum_{(S,m(S)) \in \mathrm{J}_{\mathrm{S}}^{(2)}(Y_t)} \mathbb{1}\{\mathrm{v} \subset \mathrm{relint}\, S\}\, m(S) = 1\,.$$

- Any T-vertex $\mathrm{v} \in \mathrm{V}\langle\mathrm{T}\rangle$ of Y_t is adjacent to $J^{(3)}$*-segments* whose total sum of multiplicities equals 7; more precisely,

$$\sum_{(S,m(S)) \in \mathrm{J}_{\mathrm{S}}^{(3)}(Y_t)} \mathbb{1}\{\mathrm{v} \in \mathcal{F}_0(S)\}\, m(S) = 6\,,$$

$$\sum_{(S,m(S)) \in \mathrm{J}_{\mathrm{S}}^{(3)}(Y_t)} \mathbb{1}\{\mathrm{v} \subset \mathrm{relint}\, S\}\, m(S) = 1\,.$$

- Any T-vertex $\mathrm{v} \in \mathrm{V}\langle\mathrm{T}\rangle$ of Y_t is adjacent to *J-polygons* whose total sum of multiplicities equals 9; more precisely,

$$\sum_{(P,m(P)) \in \mathrm{J}_{\mathrm{P}}(Y_t)} \mathbb{1}\{\mathrm{v} \in \mathcal{F}_0(P)\}\, m(P) = 6\,,$$

$$\sum_{(P,m(P)) \in \mathrm{J}_{\mathrm{P}}(Y_t)} \mathbb{1}\{\mathrm{v} \subset \partial P,\ \mathrm{v} \notin \mathcal{F}_0(P)\}\, m(P) = 2\,,$$

$$\sum_{(P,m(P)) \in \mathrm{J}_{\mathrm{P}}(Y_t)} \mathbb{1}\{\mathrm{v} \subset \mathrm{relint}\, P\}\, m(P) = 1\,.$$

We now turn to the adjacencies for the X-vertices in a spatial STIT tessellation Y_t with distribution $\mathbb{P}_{Y_t}$. Again, all relationships are valid for $\mathbb{P}_{Y_t}$-a. s.

- Any X-vertex $\mathrm{v} \in \mathrm{V}\langle\mathrm{X}\rangle$ of Y_t is adjacent to exactly 4 *edges*; more precisely,

$$\sum_{\mathrm{e} \in \mathrm{E}(Y_t)} \mathbb{1}\{\mathrm{v} \in \mathcal{F}_0(\mathrm{e})\} = 4\,.$$

- Any X-vertex $\mathrm{v} \in \mathrm{V}\langle\mathrm{X}\rangle$ of Y_t is adjacent to exactly 6 *plates*; more precisely,

$$\sum_{\mathrm{p} \in \mathrm{P}(Y_t)} \mathbb{1}\{\mathrm{v} \in \mathcal{F}_0(\mathrm{p})\} = 4\,, \qquad \sum_{\mathrm{p} \in \mathrm{P}(Y_t)} \mathbb{1}\{\mathrm{v} \subset \partial \mathrm{p},\ \mathrm{v} \notin \mathcal{F}_0(\mathrm{p})\} = 2\,.$$

- Any X-vertex $\mathrm{v} \in \mathrm{V}\langle\mathrm{X}\rangle$ of Y_t is adjacent to exactly 4 *cells*; more precisely,

$$\sum_{\mathrm{z} \in \mathrm{Z}(Y_t)} \sum_{S \in \mathcal{F}_1(\mathrm{z})} \mathbb{1}\{\mathrm{v} \subset \mathrm{relint}\, S\} = 4\,.$$

- Any X-vertex $\mathsf{v} \in \mathsf{V}\langle\mathsf{X}\rangle$ of Y_t is adjacent to exactly 2 *I-segments*; more precisely,

$$\sum_{S \in \mathrm{I}_\mathsf{S}(Y_t)} \mathbb{1}\{\mathsf{v} \subset \operatorname{relint} S\} = 2\,.$$

- Any X-vertex $\mathsf{v} \in \mathsf{V}\langle\mathsf{X}\rangle$ of Y_t is adjacent to exactly 3 *I-polygons*; more precisely,

$$\sum_{P \in \mathrm{I}_\mathsf{P}(Y_t)} \mathbb{1}\{\mathsf{v} \subset \partial P,\ \mathsf{v} \notin \mathcal{F}_0(P)\} = 2\,, \qquad \sum_{P \in \mathrm{I}_\mathsf{P}(Y_t)} \mathbb{1}\{\mathsf{v} \subset \operatorname{relint} P\} = 1\,.$$

- Any X-vertex $\mathsf{v} \in \mathsf{V}\langle\mathsf{X}\rangle$ of Y_t is adjacent to $J^{(2)}$*-segments* whose total sum of multiplicities equals 10; more precisely,

$$\sum_{(S,m(S)) \in \mathrm{J}^{(2)}_\mathsf{S}(Y_t)} \mathbb{1}\{\mathsf{v} \in \mathcal{F}_0(S)\}\, m(S) = 8\,,$$
$$\sum_{(S,m(S)) \in \mathrm{J}^{(2)}_\mathsf{S}(Y_t)} \mathbb{1}\{\mathsf{v} \subset \operatorname{relint} S\}\, m(S) = 2\,.$$

- Any X-vertex $\mathsf{v} \in \mathsf{V}\langle\mathsf{X}\rangle$ of Y_t is adjacent to $J^{(3)}$*-segments* whose total sum of multiplicities equals 4; more precisely,

$$\sum_{(S,m(S)) \in \mathrm{J}^{(3)}_\mathsf{S}(Y_t)} \mathbb{1}\{\mathsf{v} \subset \operatorname{relint} S\}\, m(S) = 4\,.$$

- Any X-vertex $\mathsf{v} \in \mathsf{V}\langle\mathsf{X}\rangle$ of Y_t is adjacent to *J-polygons* whose total sum of multiplicities equals 8; more precisely,

$$\sum_{(P,m(P)) \in \mathrm{J}_\mathsf{P}(Y_t)} \mathbb{1}\{\mathsf{v} \subset \partial P,\ \mathsf{v} \notin \mathcal{F}_0(P)\}\, m(P) = 8\,.$$

From now on, we will frequently use these topological relations in our considerations.

17.2.3 Mean values

In this section, we present the intensities and mean values for various types of vertices, segments, and polygons in a spatial STIT tessellation based on the given refined classification. In the proofs, we will repeatedly use the special case (16.7) of the exchange formula

$$\gamma_\mathsf{A}\, \mu_{\mathsf{A},\mathsf{B}} = \gamma_\mathsf{B}\, \mu_{\mathsf{B},\mathsf{A}}$$

for the adjacency relation for two sets A, B of components.

Lemma 17.5. *Let Y_t be a spatial STIT tessellation with directional distribution ψ_o in the typical point of ∂Y_t, given in* (17.5), *and with two constants $\zeta_\psi^{(2)}$ and $\zeta_\psi^{(3)}$ (see* (17.9) *and* (17.10), *respectively). The intensities of the two types of vertices are*

$$\gamma_{V\langle T\rangle} = \frac{2}{3}\,t^3\,\zeta_\psi^{(3)} \quad \text{and} \quad \gamma_{V\langle X\rangle} = \frac{1}{3}\,t^3\,\zeta_\psi^{(3)}.$$

This means that these two types of vertices have the proportion $\gamma_{V\langle T\rangle} : \gamma_{V\langle X\rangle} = 2 : 1$.

The mean values for the number of primitives adjacent to the typical T*-vertex and the typical* X*-vertex, respectively, are*

$$\mu_{V\langle T\rangle,E} = \mu_{V\langle X\rangle,E} = 4\,, \quad \mu_{V\langle T\rangle,P} = \mu_{V\langle X\rangle,P} = 6\,, \quad \mu_{V\langle T\rangle,Z} = \mu_{V\langle X\rangle,Z} = 4\,.$$

Proof. Relation (17.18) states that the intensity of vertices γ_V in a spatial STIT tessellation is equal to $t^3\,\zeta_\psi^{(3)}$. Hence $\gamma_{V\langle T\rangle} + \gamma_{V\langle X\rangle} = t^3\,\zeta_\psi^{(3)}$. In Section 17.2.2, we pointed out that each T-vertex is a 0-face of exactly two cells. Furthermore, all 0-faces of a cell are T-vertices, because an X-vertex cannot be a 0-face of any cell. With $\nu_0(Z) = 8$ given in (17.16), using the exchange formula (16.7), we obtain $8\,\gamma_Z = 2\,\gamma_{V\langle T\rangle}$. The equality $\gamma_Z = \frac{1}{6}\,t^3\,\zeta_\psi^{(3)}$ in relation (17.19) implies

$$\gamma_{V\langle T\rangle} = 4\,\gamma_Z = \frac{2}{3}\,t^3\,\zeta_\psi^{(3)} \quad \text{and} \quad \gamma_{V\langle X\rangle} = \gamma_V - \gamma_{V\langle T\rangle} = \frac{1}{3}\,t^3\,\zeta_\psi^{(3)}.$$

The results for the mean number of adjacent primitives follow immediately from the topology of vertices, which is studied in Section 17.2.2 and illustrated in Figure 17.3. □

Any T-vertex in a spatial STIT tessellation is a 0-face of one I-polygon. Hence a T-vertex is a maximal 0-polytope. Because the X-vertices are not maximal 0-polytopes, we have $V\langle T\rangle(Y_t) = \mathcal{F}_0^{\max}(Y_t)$. Thus the result for $\gamma_{V\langle T\rangle}$ in the lemma above is in line with the result for $\gamma_{0,t}^{\max}$ in Lemma 14.8 for $d = 3$.

Investigations for the number of T- and X-vertices in the relative interior of the typical I-segment follow in Section 17.3.2. In Theorem 17.15 the probability that the typical I-segment has exactly m vertices of type T and k vertices of type X in its relative interior is given. Furthermore the neighborhood of the typical vertex and likewise the typical T-vertex and the typical X-vertex, respectively, will be considered in Theorem 17.27.

Our aim is now to calculate the intensities of the four different types of segments $E, J_S^{(2)}, J_S^{(3)}, I_S$ and furthermore the mean length and the mean number of vertices in the relative interior of the typical segment of the respective type.

Lemma 17.6. *Let Y_t be a spatial STIT tessellation with directional distribution ψ_o in the typical point of ∂Y_t, given in* (17.5), *and with two constants $\zeta_\psi^{(2)}$ and $\zeta_\psi^{(3)}$ (see* (17.9) *and* (17.10), *respectively). The mean values for the four different segment types $A \in \{E, J_S^{(2)}, J_S^{(3)}, I_S\}$ are as follows.*

A	E	$J_S^{(2)}$	$J_S^{(3)}$	I_S
γ_A	$2\,t^3\,\zeta_\psi^{(3)}$	$\frac{14}{3}\,t^3\,\zeta_\psi^{(3)}$	$2\,t^3\,\zeta_\psi^{(3)}$	$\frac{2}{3}\,t^3\,\zeta_\psi^{(3)}$
ℓ_A	$\frac{1}{2t}\,\frac{\zeta_\psi^{(2)}}{\zeta_\psi^{(3)}}$	$\frac{9}{14t}\,\frac{\zeta_\psi^{(2)}}{\zeta_\psi^{(3)}}$	$\frac{1}{t}\,\frac{\zeta_\psi^{(2)}}{\zeta_\psi^{(3)}}$	$\frac{3}{2t}\,\frac{\zeta_\psi^{(2)}}{\zeta_\psi^{(3)}}$
$\mu_{\mathrm{relint}\,A,V}$	0	$\frac{2}{7}$	1	2
$\mu_{A,V}$	2	$\frac{16}{7}$	3	4
$\mu_{\mathrm{relint}\,A,E}$	1	$\frac{9}{7}$	2	3

Proof. The results in the first column of the table, concerning edges, follow from known mean values, given in Corollary 17.2. Moreover, an edge is a 1-primitive, and hence

$$\mu_{\mathrm{relint}\,E,V} = 0\,.$$

The results in the last two rows of the table follow immediately by (17.29) from the respective values in the third row.

We now turn to the results for the different segment types. To show these, various adjacencies are required. Therefore we begin with considerations for T- and X-vertices. From Section 17.2.2 we know that any T-vertex is a 0-face of exactly 5 plates and any X-vertex is a 0-face of exactly 4 plates. Together with the exchange formula (16.7), for the mean number of 0-faces $\nu_0(P)$ of the typical plate, we obtain

$$\gamma_P\,\nu_0(P) = 5\,\gamma_{V\langle T\rangle} + 4\,\gamma_{V\langle X\rangle} = \frac{14}{3}\,t^3\zeta_\psi^{(3)}.$$

Hence, with intensity $\gamma_P = \frac{7}{6}\,t^3\zeta_\psi^{(3)}$, given in Corollary 17.2, it follows that the mean number of 0-faces of the typical plate is

$$\nu_0(P) = 4\,. \tag{17.31}$$

Recall that the J-segments are 1-faces of plates and cells, respectively, marked with their multiplicities. Furthermore, any edge in a spatial STIT tessellation is contained in the boundary of exactly 3 plates, that is, $\mu_{E,P} = 3$. With $\gamma_P\,\nu_0(P) = \gamma_P\,\nu_1(P) = \gamma_{J_S^{(2)}}$ and $\gamma_{J_S^{(2)}}\,\ell_{J_S^{(2)}} = 3\eta_1 = 3\,t^2\zeta_\psi^{(2)}$, given in (17.12) for the length density, we obtain

$$\gamma_{J_S^{(2)}} = \frac{14}{3}\,t^3\zeta_\psi^{(3)} \quad\text{and}\quad \ell_{J_S^{(2)}} = \frac{9}{14t}\,\frac{\zeta_\psi^{(2)}}{\zeta_\psi^{(3)}}\,.$$

The relation $\mu_{P,V} = \frac{36}{7}$ in Corollary 17.2 states that the typical plate has in the mean $\frac{36}{7}$ vertices on its boundary. By (17.31), in the mean 4 of them are 0-faces of the plate. Using the exchange formula (16.7), we get $\gamma_{J_S^{(2)}}\,\mu_{\mathrm{relint}\,J_S^{(2)},\mathrm{V}} = \gamma_{\mathrm{P}}\,(\mu_{\mathrm{P,V}} - 4)$, because

$$\gamma_{J_S^{(2)}}\,\mu_{\mathrm{relint}\,J_S^{(2)},\mathrm{V}} = \mathbb{E} \sum_{(S,m(S))\in J_S^{(2)}(Y_t)} \mathbb{1}\{c(S)\in[0,1]^3\} \sum_{\mathrm{v}\in \mathrm{V}(Y_t)} \mathbb{1}\{\mathrm{v}\subset \mathrm{relint}\, S\}\, m(S)$$
$$= \mathbb{E} \sum_{\mathrm{v}\in \mathrm{V}(Y_t)} \mathbb{1}\{\mathrm{v}\in[0,1]^3\} \sum_{(S,m(S))\in J_S^{(2)}(Y_t)} \mathbb{1}\{\mathrm{v}\subset \mathrm{relint}\, S\}\, m(S)\,.$$

With

$$\sum_{(S,m(S))\in J_S^{(2)}(Y_t)} \mathbb{1}\{\mathrm{v}\subset \mathrm{relint}\, S\}\, m(S) = \sum_{\mathrm{p}\in \mathrm{P}(Y_t)} \left(\mathbb{1}\{\mathrm{v}\subset \mathrm{p}\} - \mathbb{1}\{\mathrm{v}\in \mathcal{F}_0(\mathrm{p})\}\right)$$

for vertices $\mathrm{v}\in \mathrm{V}(Y_t)$, we can continue as follows:

$$\mathbb{E} \sum_{\mathrm{v}\in \mathrm{V}(Y_t)} \mathbb{1}\{\mathrm{v}\in[0,1]^3\} \sum_{\mathrm{p}\in \mathrm{P}(Y_t)} \left(\mathbb{1}\{\mathrm{v}\subset \mathrm{p}\} - \mathbb{1}\{\mathrm{v}\in \mathcal{F}_0(\mathrm{p})\}\right)$$
$$= \mathbb{E} \sum_{\mathrm{p}\in \mathrm{P}(Y_t)} \mathbb{1}\{c(\mathrm{p})\in[0,1]^3\} \left(\sum_{\mathrm{v}\in \mathrm{V}(Y_t)} \mathbb{1}\{\mathrm{v}\subset \mathrm{p}\} - \sum_{\mathrm{v}\in \mathrm{V}(Y_t)} \mathbb{1}\{\mathrm{v}\in \mathcal{F}_0(\mathrm{p})\} \right)$$
$$= \gamma_{\mathrm{P}}\,(\mu_{\mathrm{P,V}} - \nu_0(\mathrm{P}))\,.$$

This yields

$$\mu_{\mathrm{relint}\,J_S^{(2)},\mathrm{V}} = \frac{1}{4}\left(\frac{36}{7} - 4\right) = \frac{2}{7}\,.$$

Hence the mean number of vertices in the relative interior of the typical $\mathrm{J}^{(2)}$-segment is $\frac{2}{7}$. This completes the proof of the results for the $\mathrm{J}^{(2)}$-segments.

Lemma 17.3 yields $\ell_{J_S^{(3)}} = \ell_{\mathrm{E}}^{\mathrm{PHT}}$ for a Poisson plane tessellation X_t with the same directional distribution ψ_o as Y_t. The vertex intensity $\gamma_{\mathrm{V}}^{\mathrm{PHT}}$ of X_t is equal to the intensity of the intersection process of order 3, that is, $\gamma_{\mathrm{V}}^{\mathrm{PHT}} = \frac{1}{6}t^3\zeta_\psi^{(3)}$, which follows from (2.81) for $d = 3$, $k = 0$, and $\gamma_{\mathrm{V}}^{\mathrm{PHT}} = \eta_{0,t}^{\mathrm{PHT}}$. To calculate $\ell_{\mathrm{E}}^{\mathrm{PHT}}$, we use the mean value relations $\gamma_{\mathrm{E}}\,\ell_{\mathrm{E}} = \eta_1$ and $\gamma_{\mathrm{V}}\,\mu_{\mathrm{V,E}} = 2\,\gamma_{\mathrm{E}}$. In a Poisson plane tessellation, any vertex has 6 emanating edges, that is, $\mu_{\mathrm{V,E}}^{\mathrm{PHT}} = 6$. Again equation (2.81), now for $k = 1$, yields $\eta_1(X_t) = \frac{1}{2}t^2\zeta_\psi^{(2)}$. Hence

$$\ell_{\mathrm{E}}^{\mathrm{PHT}} = \ell_{J_S^{(3)}} = \frac{1}{t}\,\frac{\zeta_\psi^{(2)}}{\zeta_\psi^{(3)}}\,.$$

The fact that in a spatial STIT tessellation, each edge is adjacent to 1-faces of two cells implies

$$\gamma_{J_S^{(3)}} = \frac{2\,\eta_1}{\ell_{J_S^{(3)}}} = 2\,t^3\zeta_\psi^{(3)}\,.$$

To obtain the mean number of vertices in the relative interior of the typical $\mathrm{J}^{(3)}$-segment $\mu_{\mathrm{relint}\,\mathrm{J}_\mathrm{S}^{(3)},\mathrm{V}}$, we observe that any T-vertex is located in the relative interior of one $\mathrm{J}^{(3)}$-segment (with multiplicity 1) and any X-vertex in the relative interior of $\mathrm{J}^{(3)}$-segments whose multiplicities add up to 4. Lemma 17.5 for the intensities of the two vertex types, together with the exchange formula (16.7), yields that

$$\gamma_{\mathrm{J}_\mathrm{S}^{(3)}}\,\mu_{\mathrm{relint}\,\mathrm{J}_\mathrm{S}^{(3)},\mathrm{V}} = \gamma_{\mathrm{V}\langle\mathrm{T}\rangle} + 4\,\gamma_{\mathrm{V}\langle\mathrm{X}\rangle} = 2\,t^3\zeta_\psi^{(3)}\,,$$

and hence $\mu_{\mathrm{relint}\,\mathrm{J}_\mathrm{S}^{(3)},\mathrm{V}} = 1$.

The last column of the table contains the mean values for the I-segments, the maximal segments. To calculate their intensity $\gamma_{\mathrm{I}_\mathrm{S}}$, we notice that any T-vertex is the endpoint of exactly two I-segments and any I-segment has two endpoints, and they are T-vertices. Thus $2\,\gamma_{\mathrm{I}_\mathrm{S}} = 2\,\gamma_{\mathrm{V}\langle\mathrm{T}\rangle}$, which leads to

$$\gamma_{\mathrm{I}_\mathrm{S}} = \frac{2}{3}\,t^3\zeta_\psi^{(3)}\,.$$

For $k = 1$, the relation $\gamma_{\mathrm{I}_\mathrm{S}}\,\ell_{\mathrm{I}_\mathrm{S}} = \eta_1$ results from (8.17). This gives us

$$\ell_{\mathrm{I}_\mathrm{S}} = \frac{\eta_1}{\gamma_{\mathrm{I}_\mathrm{S}}} = \frac{3}{2t}\,\frac{\zeta_\psi^{(2)}}{\zeta_\psi^{(3)}}\,.$$

Moreover, each T-vertex is located in the relative interior of exactly one I-segment and each X-vertex in the relative interior of exactly two. Taking into account the intensities of T- and X-vertices and using the exchange formula (16.7), we arrive at

$$\mu_{\mathrm{relint}\,\mathrm{I}_\mathrm{S},\mathrm{V}} = \frac{1}{\gamma_{\mathrm{I}_\mathrm{S}}}\left(\gamma_{\mathrm{V}\langle\mathrm{T}\rangle} + 2\,\gamma_{\mathrm{V}\langle\mathrm{X}\rangle}\right) = \frac{1}{\gamma_{\mathrm{I}_\mathrm{S}}}\,\frac{4}{3}\,t^3\zeta_\psi^{(3)} = 2\,,$$

meaning that the mean number of vertices in the relative interior of the typical I-segment is equal to 2. This is the special case $d = 3$ of the result in Corollary 14.19 for $\mathbb{E}\nu_1^{(0)}$ and general dimension d. □

The result for the intensity of I-segments $\gamma_{\mathrm{I}_\mathrm{S}} = \frac{2}{3}\,t^3\zeta_\psi^{(3)}$ corresponds to relation $\gamma_{\mathrm{I}_\mathrm{S}} = \gamma_{1,t}^{\max} = \frac{4}{3}\gamma_{1,t}^{\mathrm{PHT}}$ in Theorem 11.8. To see this, use $\gamma_{1,t}^{\mathrm{PHT}} = 3\gamma_{0,t}^{\mathrm{PHT}} = \frac{1}{2}\,t^3\zeta_\psi^{(3)}$, which follows from (2.81) for $d = 3$ and $k = 0$.

By Lemma 17.6 for the mean length of the typical segments of the four respective types, the following mean value relation arises:

$$\ell_\mathrm{A} = \mu_{\mathrm{relint}\,\mathrm{A},\mathrm{E}}\,\ell_\mathrm{E}\,,\quad \mathrm{A} \in \{\mathrm{E}, \mathrm{J}_\mathrm{S}^{(2)}, \mathrm{J}_\mathrm{S}^{(3)}, \mathrm{I}_\mathrm{S}\}\,.$$

We now consider three different types of polygons, the set P of plates, which are the 2-primitives, the set J_P of J-polygons, which are the 2-faces of the cells marked with their multiplicities, and the set I_P of the maximal I-polygons.

Lemma 17.7. *Let Y_t be a spatial STIT tessellation with directional distribution ψ_o in the typical point of ∂Y_t, given in (17.5), and with two constants $\zeta_\psi^{(2)}$ and $\zeta_\psi^{(3)}$ (see (17.9) and (17.10), respectively). The mean values for the three different polygon types are as follows.*

A	P	J_P	I_P
γ_A	$\frac{7}{6}t^3\zeta_\psi^{(3)}$	$t^3\zeta_\psi^{(3)}$	$\frac{1}{6}t^3\zeta_\psi^{(3)}$
a_A	$\frac{6}{7t^2}\frac{1}{\zeta_\psi^{(3)}}$	$\frac{2}{t^2}\frac{1}{\zeta_\psi^{(3)}}$	$\frac{6}{t^2}\frac{1}{\zeta_\psi^{(3)}}$
ℓ_A	$\frac{18}{7t}\frac{\zeta_\psi^{(2)}}{\zeta_\psi^{(3)}}$	$\frac{4}{t}\frac{\zeta_\psi^{(2)}}{\zeta_\psi^{(3)}}$	$\frac{6}{t}\frac{\zeta_\psi^{(2)}}{\zeta_\psi^{(3)}}$
$\nu_0(A)$	4	4	4
$\mu_{\mathrm{relint}\,A,V}$	0	$\frac{2}{3}$	6
$\mu_{\partial A,V}$	$\frac{36}{7}$	8	12
$\mu_{A,V}$	$\frac{36}{7}$	$\frac{26}{3}$	18
$\mu_{\mathrm{relint}\,A,E}$	0	2	12
$\mu_{\partial A,E}$	$\frac{36}{7}$	8	12
$\mu_{A,E}$	$\frac{36}{7}$	10	24
$\mu_{A,P}$	1	$\frac{7}{3}$	7

Proof. Before we start the calculations, for certain results, the following relations between selected rows in the table should be clarified. For $A \in \{P, J_P, I_P\}$, it is easy to deduce that

$$\mu_{\mathrm{relint}\,A,V} + \mu_{\partial A,V} = \mu_{A,V}\,,$$
$$\mu_{\mathrm{relint}\,A,E} + \mu_{\partial A,E} = \mu_{A,E}\,,$$
$$\mu_{A,V} - \mu_{A,E} + \mu_{A,P} = 1\,;$$

see also (17.30). In the proof the topological results for the vertices given in Section 17.2.2 and the exchange formula (16.7) for the adjacency relation will be used repeatedly.

The first column of the table contains the results for the plates, which are the 2-primitives of the tessellation. Corollary 17.2 already presents the results for γ_P, a_P, ℓ_P and $\mu_{P,V}$, $\mu_{P,E}$.

In the proof of Lemma 17.6, we have shown in (17.31) that $\nu_0(P) = 4$. Moreover, from the definition of plates as the 2-primitives it follows that

$$\mu_{\mathrm{relint}\,P,V} = \mu_{\mathrm{relint}\,P,E} = 0 \quad \text{and} \quad \mu_{P,P} = 1\,.$$

Let us now look at the results for the typical J-polygon, listed in the second column of the table. It should be remembered that a J-polygon is a 2-face of a cell, marked with its multiplicity. This is the number of cells containing this polygon as a 2-face.

Lemma 17.3 yields that $\nu_0(\mathsf{J_P}) = 4$ and, furthermore, that the mean area and the mean perimeter of the typical J-polygon of Y_t are the same as the mean area and the mean perimeter of the 2-face of the typical cell of the Poisson plane tessellation X_t. The intensity of vertices $\gamma_{\mathsf{V}}^{\mathrm{PHT}} = \frac{1}{6}t^3\zeta_\psi^{(3)}$ of X_t is given in (2.81). Together with $\mu_{\mathsf{V,P}}^{\mathrm{PHT}} = 12$ and $\mu_{\mathsf{P,V}}^{\mathrm{PHT}} = 4$ for X_t (see Section 9.5.2 in [33]), the mean value relation $\gamma_{\mathsf{V}}\,\mu_{\mathsf{V,P}} = \gamma_{\mathsf{P}}\,\mu_{\mathsf{P,V}}$ yields $\gamma_{\mathsf{P}}^{\mathrm{PHT}} = \frac{1}{2}t^3\zeta_\psi^{(3)}$. With $\gamma_{\mathsf{P}}^{\mathrm{PHT}}\,a_{\mathsf{P}}^{\mathrm{PHT}} = \eta_2(X_t) = t$ and $\gamma_{\mathsf{P}}^{\mathrm{PHT}}\,\ell_{\mathsf{P}}^{\mathrm{PHT}} = 4\,\gamma_{\mathsf{E}}^{\mathrm{PHT}}\,\ell_{\mathsf{E}}^{\mathrm{PHT}} = 4\,\eta_1(X_t) = 2\,t^2\zeta_\psi^{(2)}$, because any edge in X_t is adjacent to exactly 4 plates, we obtain

$$a_{\mathsf{J_P}} = a_{\mathsf{P}}^{\mathrm{PHT}} = \frac{2}{t^2}\,\frac{1}{\zeta_\psi^{(3)}}\,, \qquad \ell_{\mathsf{J_P}} = \ell_{\mathsf{P}}^{\mathrm{PHT}} = \frac{4}{t}\,\frac{\zeta_\psi^{(2)}}{\zeta_\psi^{(3)}}\,.$$

The intensity $\gamma_{\mathsf{J_P}}$ can be calculated from the relation $\gamma_{\mathsf{J_P}}\,a_{\mathsf{J_P}} = 2t$, where the factor 2 is caused by the multiplicities of the J-polygons. This yields

$$\gamma_{\mathsf{J_P}} = t^3\,\zeta_\psi^{(3)}\,.$$

To consider the mean number of edges in the relative interior of the typical J-polygon, we use the following topological property of STIT tessellations: Any edge of the tessellation is adjacent to exactly three cells. In one of them, this edge is contained in the relative interior of a 2-face of this cell. In the other two adjacent cells the edge is contained in a 1-face, or it is equal to a 1-face of one or both of these cells. Hence $\gamma_{\mathsf{J_P}}\,\mu_{\mathrm{relint}\,\mathsf{J_P,E}} = \gamma_{\mathsf{E}}$, which yields

$$\mu_{\mathrm{relint}\,\mathsf{J_P,E}} = \frac{\gamma_{\mathsf{E}}}{\gamma_{\mathsf{J_P}}} = 2\,.$$

Next, we remark that any T-vertex is contained in the relative interior of exactly one J-polygon (with multiplicity 1), whereas an X-vertex is not in the relative interior of any J-polygon. This implies

$$\mu_{\mathrm{relint}\,\mathsf{J_P,V}} = \frac{\gamma_{\mathsf{V\langle T\rangle}}}{\gamma_{\mathsf{J_P}}} = \frac{2}{3}\,.$$

As shown in Section 17.2.2, when considering the topology of the vertices, each vertex of a STIT tessellation, regardless of its type, is located on the boundary of J-polygons whose multiplicities add up to 8. This gives us

$$\mu_{\partial\mathsf{J_P,V}} = \mu_{\partial\mathsf{J_P,E}} = \frac{8\gamma_{\mathsf{V}}}{\gamma_{\mathsf{J_P}}} = 8\,.$$

Thus $\mu_{J_P,V} = \frac{26}{3}$ and $\mu_{J_P,E} = 10$. The mean number of plates adjacent to the typical J-polygon is given by

$$\mu_{J_P,P} = \frac{2\gamma_P}{\gamma_{J_P}} = \frac{7}{3},$$

because each plate is a part of the J-polygons of neighboring cells. This agrees with $\mu_{J_P,P} = 1 - \mu_{J_P,V} + \mu_{J_P,E}$.

We now consider the typical I-polygon, the typical maximal polygon in a spatial STIT tessellation. The results can be found in the third column of the table. Following Section 11.2, Theorem 11.8, the intensity of I-polygons for $d = 3$ is given by

$$\gamma_{I_P} = \gamma_{2,t}^{\max} = \frac{1}{3}\gamma_{2,t}^{\text{PHT}} = \frac{1}{6}t^3\,\zeta_\psi^{(3)},$$

where $\gamma_{2,t}^{\text{PHT}}$ is the intensity of plates in a Poisson plane tessellation X_t with directional distribution ψ_o.

The relation $\gamma_{I_P}\,a_{I_P} = t$ implies

$$a_{I_P} = \frac{t}{\gamma_{I_P}} = \frac{6}{t^2}\frac{1}{\zeta_\psi^{(3)}}.$$

From $\eta_1 = \gamma_{I_P}\,\ell_{I_P}$ we can calculate the mean perimeter of the typical I-polygon, namely

$$\ell_{I_P} = \frac{\eta_1}{\gamma_{I_P}} = \frac{6}{t}\frac{\zeta_\psi^{(2)}}{\zeta_\psi^{(3)}}.$$

These results for the mean area and the mean perimeter of the typical I-polygon are in line with the results in Corollary 11.12 related to the associated zonoid.

According to the topology of vertices given in Section 17.2.2, in the next step, we consider the vertices together with their three adjacent I-polygons. Any T-vertex is a 0-face of exactly one I-polygon. It is also located on the boundary of a second I-polygon but is not a 0-face of this polygon, and it is contained in the relative interior of a third one. On the other hand, any X-vertex is located on the boundary of two I-polygons, and it is not a 0-face of these polygons. Furthermore, it is contained in the relative interior of a third I-polygon. These considerations lead to

$$\begin{aligned}
\mu_{\text{relint}\, I_P,V} &= \frac{\gamma_V}{\gamma_{I_P}} &&= 6\,,\\
\mu_{\partial I_P,V} &= \frac{2\gamma_V}{\gamma_{I_P}} &&= 12\,,\\
\mu_{I_P,V} &= \mu_{\text{relint}\, I_P,V} + \mu_{\partial I_P,V} &&= 18\,,\\
\nu_0(I_P) &= \frac{\gamma_{V\langle T\rangle}}{\gamma_{I_P}} &&= 4\,.
\end{aligned}$$

For any edge of a STIT tessellation, its relative interior is located in the relative interior of exactly one I-polygon, so the mean number of edges in the relative interior of the typical I-polygon can be calculated from

$$\mu_{\mathrm{relint}\,\mathrm{I_p},\mathrm{E}} = \frac{\gamma_{\mathrm{E}}}{\gamma_{\mathrm{I_p}}} = 12\,.$$

Because any edge is adjacent to exactly one I-segment, it is located on the boundary of exactly one I-polygon, which yields

$$\mu_{\partial \mathrm{I_p},\mathrm{E}} = \frac{\gamma_{\mathrm{E}}}{\gamma_{\mathrm{I_p}}} = 12\,.$$

Thus $\mu_{\mathrm{I_p},\mathrm{E}} = 24$. For the mean number of plates adjacent to the typical I-polygon, we obtain

$$\mu_{\mathrm{I_p},\mathrm{P}} = \frac{\gamma_{\mathrm{P}}}{\gamma_{\mathrm{I_p}}} = 7\,.$$

This agrees with $\mu_{\mathrm{I_p},\mathrm{P}} = 1 - \mu_{\mathrm{I_p},\mathrm{V}} + \mu_{\mathrm{I_p},\mathrm{E}}$. □

With the mean values, given in the table in Lemma 17.7, we find the following mean value relation for the typical J-polygon in a spatial STIT tessellation:

$$a_{\mathrm{J_p}} = \mu_{\mathrm{J_p},\mathrm{P}}\, a_{\mathrm{P}}\,.$$

Analogously, for the typical I-polygon, we get that

$$a_{\mathrm{I_p}} = \mu_{\mathrm{I_p},\mathrm{P}}\, a_{\mathrm{P}}\,.$$

Furthermore, in the mean, the typical I-polygon is tessellated into $\mu_{\mathrm{I_p},\mathrm{P}} = 7$ plates and it is adjacent to $\mu_{\mathrm{I_p},\mathrm{V}} = 18$ vertices and $\mu_{\mathrm{I_p},\mathrm{E}} = 24$ edges.

We conclude this section with a few more mean values for the typical cell. Recall that a STIT tessellation has a Poisson typical cell. According to (17.16), the mean numbers of j-faces, $j \in \{0,1,2\}$, of the typical cell are therefore $\nu_0(\mathrm{Z}) = 8$, $\nu_1(\mathrm{Z}) = 12$, $\nu_2(\mathrm{Z}) = 6$, respectively. Furthermore, in Corollary 17.2 the mean number of primitives adjacent to the typical cell are given as

$$\mu_{\mathrm{Z},\mathrm{V}} = 24\,, \quad \mu_{\mathrm{Z},\mathrm{E}} = 36\,, \quad \mu_{\mathrm{Z},\mathrm{P}} = 14\,.$$

Another feature of a cell z is its 1-skeleton $\mathrm{skel}_1\,\mathrm{z}$, which is defined as the union of its 1-faces

$$\mathrm{skel}_1\,\mathrm{z} := \bigcup_{S \in \mathcal{F}_1(\mathrm{z})} S\,.$$

Lemma 17.8. *Let Y_t be a spatial STIT tessellation with directional distribution ψ_o in the typical point of ∂Y_t, given in* (17.5)*, and with two constants $\zeta_\psi^{(2)}$ and $\zeta_\psi^{(3)}$ (see* (17.9) *and* (17.10)*, respectively). The mean total length of the 1-skeleton of the typical cell of Y_t is*

$$\ell_{\mathrm{skel}_1\,Z} = \frac{12}{t}\,\frac{\zeta_\psi^{(2)}}{\zeta_\psi^{(3)}}, \tag{17.32}$$

and the mean numbers of vertices and edges adjacent to the 1-skeleton are

$$\mu_{\mathrm{skel}_1\,Z,\mathrm{V}} = 20 \quad \textit{and} \quad \mu_{\mathrm{skel}_1\,Z,\mathrm{E}} = 24\,.$$

Proof. In this proof, we use the results from Lemma 17.1 for η_1, γ_Z, Corollary 17.2 for γ_E, and Lemma 17.5 for $\gamma_{V\langle T\rangle}, \gamma_{V\langle X\rangle}$.

Any edge of a STIT tessellation is adjacent to the 1-skeletons of two cells. Applying the exchange formula (16.7) for the adjacency relation leads to

$$\ell_{\mathrm{skel}_1\,Z} = \frac{2\,\eta_1}{\gamma_Z} = \frac{12}{t}\,\frac{\zeta_\psi^{(2)}}{\zeta_\psi^{(3)}}$$

and

$$\mu_{\mathrm{skel}_1\,Z,\mathrm{E}} = \frac{2\,\gamma_E}{\gamma_Z} = 24\,.$$

Furthermore, in Section 17.2.2, it was shown that each T-vertex is located on the 1-skeletons of three cells, whereas an X-vertex is adjacent to the 1-skeletons of four cells. This means that

$$\mu_{\mathrm{skel}_1\,Z,\mathrm{V}} = \frac{3\,\gamma_{V\langle T\rangle}}{\gamma_Z} + \frac{4\,\gamma_{V\langle X\rangle}}{\gamma_Z} = 20\,.$$

This completes the argument. □

With (6.4), the result in (17.32) is consistent with (11.26) and the subsequent considerations for the typical cell of a Poisson hyperplane tessellation in Section 11.3 of [82].

The following can be summarized for the typical cell of a spatial STIT tessellation:

The mean number of 0-faces is 8, the mean number of 1-faces is 12, and the mean number of 2-faces is 6. This is consistent with (17.16). On the other hand, from Corollary 17.2 we know for the primitives that the mean number of vertices is 24, the mean number of edges is 36, and the mean number of plates is 14. Due to Lemma 17.8, on the 1-skeleton of the typical cell, in the mean, there are 20 vertices and 24 edges. Consequently, in the relative interior of the 2-faces of the typical cell, the mean number of vertices is 4, and the mean number of edges is 12.

All mean values investigated in this section can be specialized for a stationary and isotropic STIT tessellation with the particular values for $\zeta_{\psi}^{(2)}$ and $\zeta_{\psi}^{(3)}$ given in (17.13) and for a Mondrian STIT tessellation using (17.15).

17.3 Typical I-segment

17.3.1 Birth times and types of vertices

The I-segments and I-polygons are the maximal 1- and 2-polytopes in a spatial STIT tessellation Y_t. Let $S \in \mathrm{I_S}(Y_t)$ be an I-segment, and let $P(S) \in \mathrm{I_P}(Y_t)$ be the I-polygon that has S as 1-face. Regarding the construction of the STIT tessellation process, the I-segment S is born when a plane $H \in \mathcal{H}$ generating $P(S)$ intersects an already existing I-polygon, shown in Figure 17.7. This polygon is called the **carrying I-polygon** $P_{\mathrm{carr}}(S) \in \mathrm{I_P}(Y_t)$ of the I-segment S. Then $\operatorname{relint} S \subset \operatorname{relint} P_{\mathrm{carr}}(S)$, because only the 0-faces of S can be located on the boundary of the carrying I-polygon $P_{\mathrm{carr}}(S)$. Let $\beta(S)$ be the **birth time** of the I-segment S and also of the I-polygon $P(S)$, and let $\beta_{\mathrm{carr}}(S)$ be the birth time of the carrying I-polygon. For these birth times, we have $0 < \beta_{\mathrm{carr}}(S) < \beta(S) < t$.

We now analyze the vertices adjacent to an I-segment of the STIT tessellation Y_t. Both 0-faces of an I-segment are T-vertices. The vertices in the relative interior, also called internal vertices, can be of both types, T or X.

We start our investigations for the internal vertices with X-vertices. In contrast to the planar case, an I-segment in a spatial STIT tessellation can have vertices in its relative interior at the time when it is born. These nodes are generated by already existing I-segments "on the backside" of the carrying I-polygon, the light-colored lines in the dark grey polygon in Figure 17.7. All these vertices are X-vertices. After the birth of S, the ongoing subdivision of the cells "on the backside" of the carrying I-polygon generates further

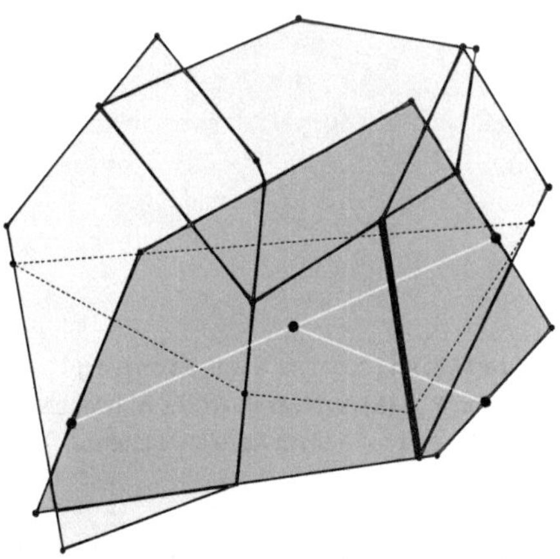

Figure 17.7: An I-segment (thick line) together with its carrying I-polygon (dark grey region) and a further subdivision (dashed lines) after the birth of the I-segment. At the time of its birth, the I-segment has two vertices in its relative interior, which are formed by the intersection with already existing edges (light lines).

X-vertices in the relative interior of S. Hence all X-vertices in the relative interior of an I-segment are born in the time interval $(\beta_{carr}(S), t)$.

From Corollary 6.14 and relation (11.13) it follows that for a STIT tessellation Y_t with measure $t\,\Theta$, the section $\partial Y_t \cap L$ generates a stationary Poisson point process on $L \in G(3,1)$ with intensity $t\,\Theta([e(L)])$. Here $e(L)$ is a segment of unit length on L, and $[e(L)]$ denotes the set of hyperplanes intersecting $e(L)$. From (6.1) in Theorem 6.2 and the Mecke-type equation in Theorem 14.1 we can conclude that the X-vertices in the relative interior of the typical I-segment of Y_t given its direction $L \in G(3,1)$ and the birth times $\beta = s \in (0,t)$ and $\beta_{carr} = r \in (0,s)$ are the restriction of a stationary Poisson point process with intensity $(t-r)\,\Theta([e(L)])$ to the typical I-segment. Note that this intensity depends on $\beta_{carr} = r$ and not on the birth time β of the typical I-segment.

We now turn to the T-vertices. Any T-vertex in the relative interior of an I-segment S has exactly 4 adjacent edges. Two of them are collinear and contained in S, whereas a third one is located in $P(S)$. The fourth edge lies in $P_{carr}(S)$. The location of this edge leads to a classification of the T-vertices, which is used for further results in Section 17.4. A plane $H \in \mathcal{H}$ creates two half-spaces H^+ and H^-, as defined in Section 1.3. Let H now be the plane that generates $P(S)$. The polygon $P_{carr}(S)$ gets divided by H into two sub-polygons $P^+_{carr} := P_{carr} \cap H^+$ and $P^-_{carr} := P_{carr} \cap H^-$. At birth time s, two tessellation cells exist with $P(S)$ as their common 2-face and with another 2-face contained in P^+_{carr} or P^-_{carr}, respectively. By construction of the STIT tessellation process the number of further I-polygons born in the time interval (s,t) in each of these two cells and intersecting the I-segment S are independent and identically distributed. We attribute the label "+" to these T-vertices in the relative interior of the I-segment S for which the fourth emanating edge is contained in P^+_{carr}, and the label "–" otherwise, that is, if the fourth emanating edge is contained in P^-_{carr}. The two types of T-vertices are shown in Figure 17.8.

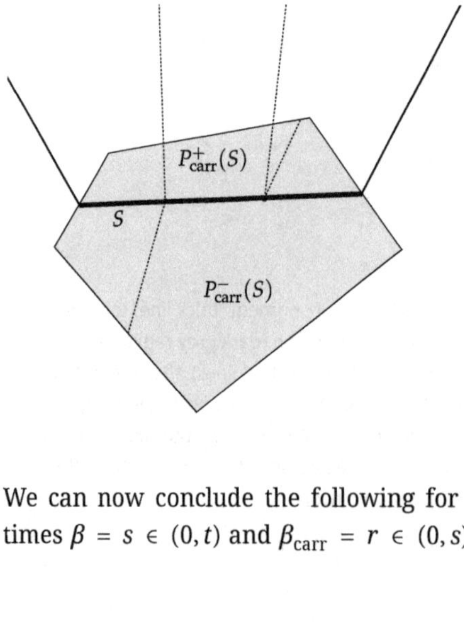

Figure 17.8: An I-segment S with its carrying I-polygon $P_{carr}(S)$ and two T-vertices with labels "+" (right) and "–" (left) in its relative interior.

We can now conclude the following for the typical I-segment of Y_t given the birth times $\beta = s \in (0,t)$ and $\beta_{carr} = r \in (0,s)$ and the direction $L \in G(3,1)$. Relation (6.1) in Theorem 6.2 and Corollary 6.14, together with (11.13), yield that both T-vertices with labels "+" and "–" in the relative interior of the typical I-segment are the restriction to the typical I-segment of a stationary Poisson point process on L, each with intensity $(t - s)\,\Theta([e(L)])$. In contrast to the X-vertices, these intensities depend on the birth time β of the I-segment, not on β_{carr}.

Furthermore, the construction of the STIT tessellation process yields the following. For $L \in G(3,1)$ and $k \in \{+,-,\mathrm{X}\}$, let Φ_k be a stationary Poisson point process on L with mark space $\{+,-,\mathrm{X}\}$ and intensity measure $(t - s)\,\Theta([e(L)])\,\lambda_1 \otimes \delta_k$ for $k \in \{+,-\}$ and $(t-r)\,\Theta([e(L)])\,\lambda_1 \otimes \delta_{\mathrm{X}}$ for $k = \mathrm{X}$. A point (x,k) of Φ_k denotes a T-vertex on L with label "+" for $k = +$ and label "–" for $k = -$ and an X-vertex for $k = \mathrm{X}$. The three point processes are conditionally independent given L and the birth times β and β_{carr}. By Theorems 19.13 and 19.14 it follows that the superposition $\Phi = \Phi_+ + \Phi_- + \Phi_{\mathrm{X}}$ is a stationary and independently marked Poisson point process on L with mark space $\{+,-,\mathrm{X}\}$ and intensity $(3t - 2s - r)\,\Theta([e(L)])$. The mark distribution is

$$\frac{(t-s)\,\Theta([e(L)])}{(3t-2s-r)\,\Theta([e(L)])}\,\delta_+ + \frac{(t-s)\,\Theta([e(L)])}{(3t-2s-r)\,\Theta([e(L)])}\,\delta_- + \frac{(t-r)\,\Theta([e(L)])}{(3t-2s-r)\,\Theta([e(L)])}\,\delta_{\mathrm{X}}\,.$$

This point process Φ restricted to the relative interior of the typical I-segment of Y_t with direction L and birth times $\beta_{\text{carr}} = r$ and $\beta = s$ represents its internal vertex structure.

The following lemma summarizes these considerations.

Lemma 17.9. *Let Y_t be a spatial STIT tessellation based on the measure $t\,\Theta$. The following applies to the vertices in the relative interior of the typical I-segment of Y_t given its direction $L \in G(3,1)$, its birth time $\beta = s \in (0,t)$, and the birth time of the carrying I-polygon $\beta_{\text{carr}} = r \in (0,s)$:*

(a) *The X-vertices in the relative interior of the typical I-segment are the restriction to this typical I-segment of a stationary Poisson point process on L with intensity $(t-r)\,\Theta([e(L)])$.*
(b) *Both T-vertices with labels "+" and "–" in the relative interior of the typical I-segment are the restriction to this typical I-segment of a stationary Poisson point process on L, each with intensity $(t-s)\,\Theta([e(L)])$.*
(c) *The restrictions to the typical I-segment of the three point processes for the vertex types are conditionally independent, given length, direction L, and the birth times β and β_{carr} of the typical I-segment.*
(d) *The internal vertices of the typical I-segment are represented by the restriction of a stationary Poisson point process on L with intensity $(3t-2s-r)\,\Theta([e(L)])$.*

17.3.2 Distributional results

After the detailed examination of the mean values in Sections 17.1 and 17.2, we are now interested in distributional results.

In Section 17.3.1, we have explained that for investigations concerning the vertices in the relative interior of an I-segment S, the birth times $\beta_{\text{carr}}(S)$ and $\beta(S)$ play an important role. Hence we study from now on the **marked process of I-segments of Y_t** with two marks $(\beta_{\text{carr}}, \beta)$:

$$\mathrm{I}_S^*(Y_t) := \{(S, \beta_{\text{carr}}(S), \beta(S)) \,:\, S \in \mathrm{I}_S(Y_t)\},$$

where $\mathrm{I}_S(Y_t)$ is the set of all I-segments of Y_t. The point process has the mark space $M = (0,t) \times (0,t)$, where $0 < \beta_{\text{carr}}(S) < \beta(S) < t$ applies to the marks. The process $I_S^*(Y_t)$ is a stationary marked segment process for any $t > 0$. As before, for considerations with a fixed time t, we drop Y_t in the notation in most cases.

We will use the style of notation for distributions and probability density functions (pdf) of the typical I-segment of $\mathrm{I}_S^*(Y_t)$ as already introduced in Section 16.1. For example, $\mathbb{Q}_{\mathrm{I}_S^*}$ is the distribution of the typical I-segment and $q_{\mathrm{I}_S^*}$ is the corresponding pdf. Furthermore, $\mathbb{Q}_{\mathrm{I}_S^*:\ell,\beta_{\text{carr}},\beta,h}$ is the joint distribution of length, birth time of the carrying I-polygon, birth time, and direction of the typical I-segment with corresponding pdf $q_{\mathrm{I}_S^*:\ell,\beta_{\text{carr}},\beta,h}$.

Recall that for the spatial STIT tessellation Y_t based on the measure $t\,\Theta$, the directional distribution ψ_o in the typical point of the 2-skeleton of Y_t is given in (17.3), and the directional distribution χ_1 in the typical point of the 1-skeleton is given in (17.2). Note that both distributions ψ_o and χ_1 are defined as probability measures on $G(3,1)$. For the corresponding Poisson plane tessellation X_t based on the same measure $t\,\Theta$, we obtain by (17.7) and (17.8) the relations $\chi_2^{\text{PHT}} = \psi_o$ and $\chi_1^{\text{PHT}} = \chi_1$.

A Poisson plane tessellation is in contrast to a STIT tessellation face-to-face. Consequently, in a Poisson plane tessellation the 1-faces are also the edges. The directional distribution $\mathbb{Q}_{\mathrm{E}:h}^{\text{PHT}}$ of the typical edge of the Poisson plane tessellation X_t with measure $t\,\Theta$ is given in (11.12) for all $A \in \mathcal{B}(G(3,1))$ by

$$\mathbb{Q}_{\mathrm{E}:h}^{\text{PHT}}(A) = \mathbb{Q}_{1,t:h}^{\text{PHT}}(A) = \frac{V_2(\Pi_\Theta)}{3\,V_3(\Pi_\Theta)} \int_A V_1(\Pi_\Theta|L)\,\chi_1^{\text{PHT}}(\mathrm{d}L),$$

where Π_Θ is the associated zonoid related to the measure Θ, introduced in (2.68). From (2.82) it follows that $\frac{1}{k!}\,\zeta_\psi^{(k)} = V_k(\Pi_\Theta)$ for $k \in \{2,3\}$. With $V_1(\Pi_\Theta|L) = \Theta([e(L)])$, given in (11.13), this distribution can also be written in the following form:

$$\mathbb{Q}_{\mathrm{E}:h}^{\text{PHT}}(A) = \frac{\zeta_\psi^{(2)}}{\zeta_\psi^{(3)}} \int_A \Theta([e(L)])\,\chi_1^{\text{PHT}}(\mathrm{d}L), \quad A \in \mathcal{B}(G(3,1)). \tag{17.33}$$

According to Corollary 2.39, the conditional length distribution $\mathbb{Q}_{\mathrm{E},\ell|h=L}^{\text{PHT}}$ of the typical edge of the Poisson plane tessellation X_t, given its direction $L \in G(3,1)$ is the exponential distribution with parameter $t\,\Theta([e(L)])$.

We start our considerations with some results for distributions associated with the typical I-segment, which follow from results in the previous parts of this book for general dimension.

Theorem 17.10. *Let Y_t be a spatial STIT tessellation with directional distribution ψ_o in the typical point of ∂Y_t, given in* (17.5).

(a) *The distribution of the direction of the typical I-segment $\mathbb{Q}_{I_S^*:h}$ of Y_t is equal to the distribution of the direction of the typical edge of the corresponding Poisson plane tessellation X_t,*

$$\mathbb{Q}_{I_S^*:h} = \mathbb{Q}_{E:h}^{PHT} .$$

(b) *The joint distribution $\mathbb{Q}_{I_S^*:\beta_{carr},\beta}$ of the birth times β_{carr} and β of the typical I-segment of Y_t has the probability density function (pdf)*

$$q_{I_S^*:\beta_{carr},\beta}(r,s) = \frac{3\,s}{t^3}\,\mathbb{1}\{0 < r < s < t\} \tag{17.34}$$

with respect to the Lebesgue measure on $\mathbb{R}^2$.

(c) *For $\mathbb{Q}_{I_S^*:h}$-almost all directions $L \in G(3,1)$, the joint conditional pdf of the length and birth times of the typical I-segment, given its direction $L \in G(3,1)$, is*

$$q_{I_S^*:\ell,\beta_{carr},\beta|h=L}(a,r,s) = \frac{3\,\Theta([e(L)])s^2}{t^3}\,\mathrm{e}^{-\Theta([e(L)])sa}\,\mathbb{1}\{0 < r < s < t\} \tag{17.35}$$

for $a > 0$ with respect to the Lebesgue measure on $\mathbb{R}^3$.

Proof. Part (a) of the theorem follows from Corollary 11.10 with $d = 3$ and $k = 1$.

The birth times of the two I-polygons, which form the typical I-segment, are β and β_{carr}. Hence, part (b) is the particular case of Theorem 14.12 with $d = 3$, $k = 1$, and $j = 0$.

The proof of part (c) begins with the following considerations. Based on Theorem 2.21, the disintegration of the joint distribution $\mathbb{Q}_{I_S^*:\ell,\beta_{carr},\beta,h}$ with respect to $\mathbb{Q}_{I_S^*:\ell,\beta_{carr},\beta|h}$ yields

$$\begin{aligned}&\int_{(0,\infty)^3\times G(3,1)} g(u,s_1,s_2,L)\,\mathbb{Q}_{I_S^*:\ell,\beta_{carr},\beta,h}(\mathrm{d}(u,s_1,s_2,L))\\ &= \int_{G(3,1)}\int_{(0,\infty)^3} g(u,s_1,s_2,L)\,\mathbb{Q}_{I_S^*:\ell,\beta_{carr},\beta|h=L}(\mathrm{d}(u,s_1,s_2))\,\mathbb{Q}_{I_S^*:h}(\mathrm{d}L)\end{aligned}$$

for all nonnegative measurable functions $g : (0,\infty)^3 \times G(3,1) \to \mathbb{R}$.

Now we apply Corollary 14.17 for $d = 3$, $k = 1$, and $j = 0$. Note that $\overline{V}_{1,0,t}^{\max} = 1$, which follows from definition (14.16). Disintegration of $\mathbb{Q}_{E,s_2:\ell,h}^{PHT}$ with respect to $\mathbb{Q}_{E,s_2:\ell|h}^{PHT}$ results in

$$\int_{(0,\infty)^3\times G(3,1)} g(u,s_1,s_2,L)\,\mathbb{Q}_{I_S^*:\ell,\beta_{carr},\beta,h}(\mathrm{d}(u,s_1,s_2,L))$$

$$= \frac{3}{t^3} \int\limits_{(0,\infty)^2} \int\limits_{(0,\infty)\times G(3,1)} g(u, s_1, s_2, L)\, \mathbb{Q}^{\mathrm{PHT}}_{\mathrm{E},s_2:\ell,h}(\mathrm{d}(u, L))$$

$$\times s_2\, \mathbb{1}\{0 < s_1 < s_2 < t\}\, \lambda_2(\mathrm{d}(s_1, s_2))$$

$$= \frac{3}{t^3} \int\limits_{(0,\infty)^2} \int\limits_{G(3,1)} \int\limits_{(0,\infty)} g(u, s_1, s_2, L)\, \mathbb{Q}^{\mathrm{PHT}}_{\mathrm{E},s_2:\ell|h=L}(\mathrm{d}u)\, \mathbb{Q}^{\mathrm{PHT}}_{\mathrm{E},s_2:h}(\mathrm{d}L)$$

$$\times s_2\, \mathbb{1}\{0 < s_1 < s_2 < t\}\, \lambda_2(\mathrm{d}(s_1, s_2)),$$

where $\mathbb{Q}^{\mathrm{PHT}}_{\mathrm{E},s_2:\ell,h}$ denotes the joint distribution of length and direction of the typical edge in the Poisson plane tessellation X_{s_2} with surface density s_2 and the same directional distribution as Y_t. From (17.33) it follows that the directional distribution of the typical edge in a Poisson plane tessellation X_t does not depend on t. Using this, for $a > 0$, $0 < r < s < t$, $A \in \mathcal{B}(G(3,1))$, and the special function

$$g(u, s_1, s_2, L) = \mathbb{1}\{u \le a\}\, \mathbb{1}\{s_1 \le r\}\, \mathbb{1}\{s_2 \le s\}\, \mathbb{1}\{L \in A\},$$

we obtain that

$$\int\limits_{G(3,1)} \int\limits_{(0,\infty)^3} \mathbb{1}\{u \le a\}\, \mathbb{1}\{s_1 \le r\}\, \mathbb{1}\{s_2 \le s\}\, \mathbb{Q}_{\mathrm{I}_5^*:\ell,\beta_{\mathrm{carr}},\beta|h=L}(\mathrm{d}(u, s_1, s_2))$$

$$\times \mathbb{1}\{L \in A\}\, \mathbb{Q}_{\mathrm{I}_5^*:h}(\mathrm{d}L)$$

$$= \frac{3}{t^3} \int\limits_{G(3,1)} \int\limits_{(0,\infty)^2} \int\limits_{(0,\infty)} \mathbb{1}\{u \le a\}\, \mathbb{Q}^{\mathrm{PHT}}_{\mathrm{E},s_2:\ell|h=L}(\mathrm{d}u)\, s_2\, \mathbb{1}\{0 < s_1 < s_2\}$$

$$\times \mathbb{1}\{s_1 \le r\}\, \mathbb{1}\{s_2 \le s\}\, \lambda_2(\mathrm{d}(s_1, s_2))\, \mathbb{1}\{L \in A\}\, \mathbb{Q}^{\mathrm{PHT}}_{\mathrm{E}:h}(\mathrm{d}L). \tag{17.36}$$

For $\mathbb{Q}^{\mathrm{PHT}}_{\mathrm{E}:h}$-almost all directions $L \in G(3,1)$, the conditional length distribution of the typical edge in a Poisson plane tessellation X_t, given its direction $L \in G(3,1)$, is the exponential distribution with parameter $t\,\Theta([e(L)])$. Hence

$$\int\limits_{(0,\infty)} \mathbb{1}\{u \le a\}\, \mathbb{Q}^{\mathrm{PHT}}_{\mathrm{E},s_2:\ell|h=L}(\mathrm{d}u) = 1 - \mathrm{e}^{-\Theta([e(L)])s_2 a}$$

for $a > 0$. The integral equation (17.36) holds for all $A \in \mathcal{B}(G(3,1))$. By part (a) of Theorem 17.10 this results in

$$\int\limits_{(0,\infty)^3} \mathbb{1}\{u \le a\}\, \mathbb{1}\{s_1 \le r\}\, \mathbb{1}\{s_2 \le s\}\, \mathbb{Q}_{\mathrm{I}_5^*:\ell,\beta_{\mathrm{carr}},\beta|h=L}(\mathrm{d}(u, s_1, s_2))$$

$$= \frac{3}{t^3} \int\limits_{(0,\infty)^2} (1 - \mathrm{e}^{-\Theta([e(L)])s_2 a})\, s_2\, \mathbb{1}\{0 < s_1 < s_2\}\, \mathbb{1}\{s_1 \le r\}\, \mathbb{1}\{s_2 \le s\}\, \lambda_2(\mathrm{d}(s_1, s_2))$$

$$= \frac{3}{t^3} \int_0^r \int_{s_1}^s (1 - \mathrm{e}^{-\Theta([e(L)])s_2 a})\, s_2\, \mathrm{d}s_2\, \mathrm{d}s_1$$

$$= \frac{3}{t^3}\left[\frac{1}{2}\left(s^2 r - \frac{r^3}{3}\right) + \frac{r}{(\Theta([e(L)])a)^2}(\Theta([e(L)])\,sa+1)\,\mathrm{e}^{-\Theta([e(L)])\,sa}\right.$$
$$\left. + \frac{1}{(\Theta([e(L)])a)^3}(\Theta([e(L)])\,ra+2)\,\mathrm{e}^{-\Theta([e(L)])\,ra} - \frac{2}{(\Theta([e(L)])a)^3}\right]$$

for $a > 0$ and $0 < r < s < t$.

The third-order partial derivative with respect to a, s, r yields the joint conditional pdf of the length and birth times β_{carr} and β of the typical I-segment given its direction. This derivative is equal to $\frac{3\,\Theta([e(L)])s^2}{t^3}\,\mathrm{e}^{-\Theta([e(L)])sa}$. This completes the proof. □

Notice that the joint distribution of the birth time vector $(\beta_{\mathrm{carr}}, \beta)$ in part (b) of Theorem 17.10 is independent of Θ and therefore also of ψ_o.

We will now draw some consequences of the previous result. First, the pdf of the birth time of the typical I-segment $q_{\mathrm{I}_S^*:\beta}$ is calculated, as well as the pdf of the birth time of the carrying I-polygon of the typical I-segment $q_{\mathrm{I}_S^*:\beta_{\mathrm{carr}}}$. Moreover, the corresponding conditional pdf of the birth times $q_{\mathrm{I}_S^*:\beta|\beta_{\mathrm{carr}}=r}$ and $q_{\mathrm{I}_S^*:\beta_{\mathrm{carr}}|\beta=s}$ are given. All the results for these pdf follow immediately from (17.34) by integration.

Corollary 17.11. *For a spatial STIT tessellation Y_t, we have*

$$q_{\mathrm{I}_S^*:\beta_{\mathrm{carr}}}(r) = \frac{3}{2}\frac{t^2 - r^2}{t^3}, \quad 0 < r < t,$$
$$q_{\mathrm{I}_S^*:\beta}(s) = \frac{3\,s^2}{t^3}, \quad 0 < s < t. \tag{17.37}$$

Moreover,

$$q_{\mathrm{I}_S^*:\beta|\beta_{\mathrm{carr}}=r}(s) = \frac{2\,s}{t^2 - r^2}, \quad r < s < t,$$
$$q_{\mathrm{I}_S^*:\beta_{\mathrm{carr}}|\beta=s}(r) = \frac{1}{s}, \quad 0 < r < s,$$

with respect to the Lebesgue measure on $\mathbb{R}$. All these functions are independent of Θ.

These densities are illustrated in Figure 17.9. Note that the conditional birth time distribution of the carrying I-polygon of the typical I-segment, given that the I-segment

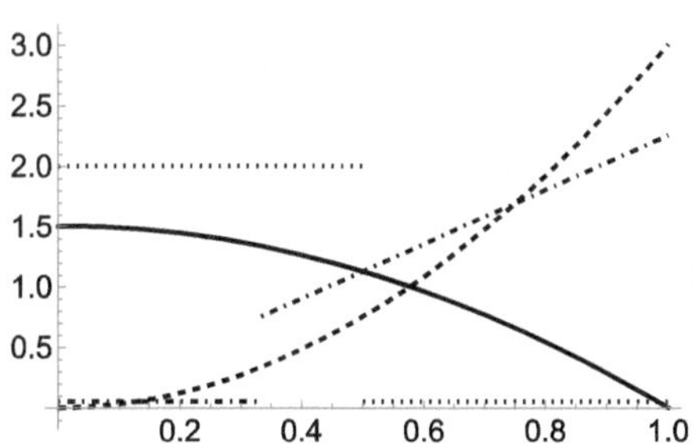

Figure 17.9: The densities $q_{\mathrm{I}_S^*:\beta_{\mathrm{carr}}}(r)$ (solid), $q_{\mathrm{I}_S^*:\beta}(s)$ (dashed) for $t = 1$, $q_{\mathrm{I}_S^*:\beta|\beta_{\mathrm{carr}}=r}$ (dash – dot) for $r = \frac{1}{3}$ and $t = 1$, and $q_{\mathrm{I}_S^*:\beta_{\mathrm{carr}}|\beta=s}(r)$ (dotted) for $s = \frac{1}{2}$.

is born at time $0 < s < t$, is the uniform distribution on $(0, s)$. This is a particular case of Corollary 14.13(b) for $d = 3$ and $j = 0$, and relation (17.37) is in line with Corollary 11.25.

In equation (17.35), integration with respect to u yields

$$q_{\mathrm{I}_S^*:\beta_{\mathrm{carr}},\beta|h=L}(r, s) = \frac{3\,s}{t^3}\,\mathbb{1}\{0 < r < s < t\}$$

for $\mathbb{Q}^{\mathrm{PHT}}_{\mathrm{I}_S^*:h}$-almost all directions $L \in G(3, 1)$.

This is in line with (17.34) and shows the independence of the direction h and birth times $(\beta_{\mathrm{carr}}, \beta)$. Furthermore, by (17.35) in Theorem 17.10 we obtain that the conditional pdf of the length of the typical I-segment of the STIT tessellation Y_t, given $(h, \beta_{\mathrm{carr}}, \beta) = (L, r, s) \in G(3, 1) \times (0, t)^2$ with $0 < r < s < t$, is

$$q_{\mathrm{I}_S^*:\ell|h=L,\beta_{\mathrm{carr}}=r,\beta=s}(a) = \Theta([e(L)])s\,\mathrm{e}^{-\Theta([e(L)])sa}$$

with respect to the Lebesgue measure on $\mathbb{R}$ for $a > 0$ and $\mathbb{Q}^{\mathrm{PHT}}_{\mathrm{I}_S^*:h}$-almost all directions $L \in G(3, 1)$.

Integration of (17.35), together with $\mathbb{Q}_{\mathrm{I}_S^*:h} = \mathbb{Q}^{\mathrm{PHT}}_{\mathrm{E}:h}$, which is given in Theorem 17.10, yields the marginal pdf for the length of the typical I-segment of a spatial STIT tessellation.

Corollary 17.12. *The distribution of the length of the typical I-segment of Y_t has the following pdf on $(0, \infty)$:*

$$q_{\mathrm{I}_S^*:\ell}(a) = \int\limits_{G(3,1)} \frac{3}{\Theta([e(L)])^3 t^3 a^4}\,\Big(6 - \big(6 + 6\,\Theta([e(L)])ta + 3\,\Theta([e(L)])^2 t^2 a^2$$
$$+\,\Theta([e(L)])^3 t^3 a^3\big)\,\mathrm{e}^{-\Theta([e(L)])ta}\Big)\,\mathbb{Q}^{\mathrm{PHT}}_{\mathrm{E}:h}(\mathrm{d}L)\,.$$

Note that $\mathbb{Q}^{\mathrm{PHT}}_{\mathrm{E}:h}$ of X_t and χ_1 of Y_t are related by (17.33), using $\chi_1^{\mathrm{PHT}} = \chi_1^{\mathrm{STIT}}$. In the isotropic case, $\Theta = \Theta_{\mathrm{iso}}$, this reduces to

$$q_{\mathrm{I}_S^*:\ell}(a) = \frac{3}{t^3 a^4}\big(48 - (48 + 24ta + 6t^2a^2 + t^3a^3)\,\mathrm{e}^{-\frac{1}{2}ta}\big), \quad a > 0\,.$$

This density coincides with the result in Theorem 11.18 for $\mathbb{R}^d$ and Corollary 11.19 for the isotropic case.

In the Mondrian case given in (17.14) with $p_1 = p_2 = p_3 = 1/3$, we obtain

$$q_{\mathrm{I}_S^*:\ell}(a) = \frac{3}{t^3 a^4}\big(162 - (162 + 54ta + 9t^2a^2 + t^3a^3)\,\mathrm{e}^{-\frac{1}{3}ta}\big), \quad a > 0\,,$$

which is in line with Corollary 11.20 for $d = 3$.

17.3.3 Internal vertices

In this section, we consider the vertices in the relative interior of the typical I-segment, also called the internal vertices. First, we are interested in the distribution of their number. This is followed by investigations for various vertex types, the X- and T-vertices.

The **number of vertices in the relative interior of the typical I-segment** is denoted by ν_{I} as in the planar case. Theorem 17.10 provides the basis for the following result, which for $d = 3$ corresponds to the result in Theorem 14.18. It is proven there as an application of the Mecke-type equation in Theorem 14.1. The proof given here applies specifically to $d = 3$.

Theorem 17.13. *Let Y_t be a spatial STIT tessellation with directional distribution ψ_o in the typical point of ∂Y_t, given in* (17.5). *The probability $\mathbb{P}(\nu_{\mathrm{I}} = n)$, $n \in \mathbb{N}_0$, that the typical I-segment of Y_t has exactly n vertices in its relative interior is given by*

$$\mathbb{P}(\nu_{\mathrm{I}} = n) = 3\int_0^1\int_0^1 (1-v)^3 \frac{(3-(1-v)(3-w))^n}{(3-(1-v)(2-w))^{n+1}}\, \mathrm{d}w\, \mathrm{d}v\,. \tag{17.38}$$

Proof. The first step is to verify the following expression for $\mathbb{P}(\nu_{\mathrm{I}} = n)$:

$$\begin{aligned}\mathbb{P}(\nu_{\mathrm{I}} = n) = &\int_0^t\int_0^s \int_{G(3,1)} \int_0^\infty \frac{3\,\Theta([e(L)])s^2}{t^3}\, \mathrm{e}^{-\Theta([e(L)])sa}\, \frac{(\Theta([e(L)])a(3t-2s-r))^n}{n!}\\ &\times \mathrm{e}^{-\Theta([e(L)])a(3t-2s-r)}\, \mathrm{d}a\, \mathbb{Q}^{\mathrm{PHT}}_{\mathrm{E}:h}(\mathrm{d}L)\, \mathrm{d}r\, \mathrm{d}s\,.\end{aligned} \tag{17.39}$$

Recall that an I-segment in a spatial STIT tessellation arises as the intersection of two I-polygons. One of them is the carrying I-polygon with birth time β_{carr}. The other polygon contains the I-segment as a 1-face on its boundary. That polygon and the I-segment have the same birth time β with the pdf given in (17.37).

In Section 17.3.1, we have classified the internal vertices of an I-segment and described the corresponding Poisson point processes. To show (17.39), we need the following result from Lemma 17.9. For the typical I-segment of Y_t with measure $t\,\Theta$, given its direction $L \in G(3,1)$, its birth time $\beta = s \in (0,t)$, and the birth time $\beta_{\mathrm{carr}} = r \in (0,s)$ of its carrying I-polygon, the internal vertices form a Poisson point process with intensity $\Theta([e(L)])(3t-2s-r)$. Of course, the number of internal vertices depends on the length of the typical I-segment. Given its length $a \in (0,\infty)$, the number is Poisson distributed with parameter $\Theta([e(L)])a(3t-2s-r)$.

The conditional pdf of the length and the birth time vector for a fixed direction is given in (17.35) of Theorem 17.10. Thus mixing the Poisson distributed number of vertices with respect to this density and using that $\mathbb{Q}_{\mathrm{I}_{\mathrm{S}}^*:h} = \mathbb{Q}^{\mathrm{PHT}}_{\mathrm{E}:h}$, we arrive at (17.39).

We now show that (17.39) is equivalent to (17.38). By

$$\int_0^\infty (\Theta([e(L)])a)^n \, e^{-\Theta([e(L)])sa} \, e^{-\Theta([e(L)])a(3t-2s-r)} \, da = \frac{n!}{\Theta([e(L)])(3t-s-r)^{n+1}}$$

we see that (17.39) can be transformed into

$$\begin{aligned}
\mathbb{P}(\nu_{\mathrm{I}} = n) &= \int_0^t \int_0^s \int_{G(3,1)} \frac{3s^2}{t^3} \frac{(3t-2s-r)^n}{(3t-s-r)^{n+1}} \, \mathbb{Q}_{\mathrm{E}:h}^{\mathrm{PHT}}(\mathrm{d}L) \, \mathrm{d}r \, \mathrm{d}s \\
&= \int_0^t \int_0^s \frac{3s^2}{t^3} \frac{(3t-2s-r)^n}{(3t-s-r)^{n+1}} \, \mathrm{d}r \, \mathrm{d}s \, .
\end{aligned}$$

Substituting $w = 1 - r/s$ into the inner integral yields

$$\mathbb{P}(\nu_{\mathrm{I}} = n) = \int_0^t \int_0^1 \frac{3s^3}{t^3} \frac{(3t - s(3-w))^n}{(3t - s(2-w))^{n+1}} \, \mathrm{d}w \, \mathrm{d}s \, .$$

Now the substitution $v = 1 - s/t$ leads to (17.38). This completes the proof. □

It is interesting to note that the distribution in Theorem 17.13 does not depend on the measure Θ or, equivalently, on the directional distribution ψ_o. It is also independent of the time parameter t. The latter can be explained as follows. The number of vertices in the relative interior of the typical I-segment does not change when the tessellation is scaled in space. However, by the scaling property (6.2) of STIT tessellations such a scaling is equivalent to a rescaling of time.

Some particular values for $\mathbb{P}(\nu_{\mathrm{I}} = n)$ are

$$\mathbb{P}(\nu_{\mathrm{I}} = 0) = \frac{189}{8} \ln 3 - 26 \ln 2 - \frac{15}{2} \approx 0.43289 \, , \tag{17.40}$$

$$\mathbb{P}(\nu_{\mathrm{I}} = 1) = \frac{1593}{16} \ln 3 - 107 \ln 2 - 35 \approx 0.21384 \, , \tag{17.41}$$

$$\mathbb{P}(\nu_{\mathrm{I}} = 2) = \frac{5319}{16} \ln 3 - 350 \ln 2 - \frac{245}{2} \approx 0.11841 \, ,$$

$$\mathbb{P}(\nu_{\mathrm{I}} = 3) = \frac{31617}{32} \ln 3 - 1025 \ln 2 - \frac{4499}{12} \approx 0.07075 \, .$$

These probabilities were already given in Section 14.6, where the notation $\mathbb{P}(\nu_{\mathrm{I}} = m) = p_{1,0}(m)$ is used. There they are shown in comparison with the probabilities $p_{1,1}(m)$ that the typical *length-weighted* I-segment has m internal vertices.

Formula (17.38) in Theorem 17.13 now gives the following result.

Corollary 17.14. *The mean number of vertices in the relative interior of the typical I-segment is equal to 2. Moreover, the variance equals* 59/3, *and all higher moments of that random variable are infinite, that is,*

$$\mathbb{E}\,\nu_{\mathrm{I}} = 2\,, \quad \operatorname{var}\nu_{\mathrm{I}} = \frac{59}{3}\,, \quad \mathbb{E}\,\nu_{\mathrm{I}}^{k} = \infty\,, \quad k > 2\,.$$

Proof. The numbers can immediately be calculated from the explicit formula in Theorem 17.13. To see this, write

$$\begin{aligned}\sum_{n=0}^{\infty} n\,\mathbb{P}(\nu_{\mathrm{I}} = n) &= 3\int_0^1\int_0^1 (1-v)^3 \sum_{n=0}^{\infty} n\,\frac{(3-(1-v)(3-w))^n}{(3-(1-v)(2-w))^{n+1}}\,\mathrm{d}w\,\mathrm{d}v\\ &= 3\int_0^1\int_0^1 (1-v)\big(v(3-w)+w\big)\,\mathrm{d}w\,\mathrm{d}v\\ &= 2\end{aligned}$$

and, similarly,

$$\begin{aligned}&\sum_{n=0}^{\infty} n^2\,\mathbb{P}(\nu_{\mathrm{I}} = n)\\ &= 3\int_0^1\int_0^1 (1-v)^3 \sum_{n=0}^{\infty} n^2\,\frac{(3-(1-v)(3-w))^n}{(3-(1-v)(2-w))^{n+1}}\,\mathrm{d}w\,\mathrm{d}v\\ &= 3\int_0^1\int_0^1 \big[v^2(2w^2-11w+15) - v(4w^2-10w-3) + w(1+2w)\big]\,\mathrm{d}w\,\mathrm{d}v\\ &= \frac{71}{3}\,,\end{aligned}$$

which yields the variance value $\frac{71}{3} - 2^2 = \frac{59}{3}$.

In general, for $k \ge 1$, we have that

$$\begin{aligned}\mathbb{E}\,\nu_{\mathrm{I}}^{k} &= \sum_{n=0}^{\infty} n^k\,\mathbb{P}(\nu_{\mathrm{I}} = n)\\ &= 3\int_0^1\int_0^1 (1-v)^3 \sum_{n=0}^{\infty} n^k\,\frac{(3-(1-v)(3-w))^n}{(3-(1-v)(2-w))^{n+1}}\,\mathrm{d}w\,\mathrm{d}v\\ &= 3\int_0^1\int_0^1 \frac{1}{(1-v)^{k-2}}\,(3-(1-v)(3-w))(3-(1-v)(2-w))^{k-1}\\ &\quad\times A_k\!\left(\frac{3-(1-v)(3-w)}{3-(1-v)(2-w)}\right)\mathrm{d}w\,\mathrm{d}v\end{aligned}$$

using the relation $\sum_{n=0}^{\infty} n^k x^n = x \frac{A_k(x)}{(1-x)^{k+1}}$, where $A_k(x)$ is an Eulerian polynomial $A_k(x) = \sum_{m=0}^{k-1} A_{k,m} x^m$ with nonnegative Eulerian numbers $A_{k,m}$ as coefficients. With $A_1(x) = 1$ and $A_2(x) = 1 + x$, we obtain the above results for $k = 1$ and $k = 2$. From $\frac{3-(1-v)(3-w)}{3-(1-v)(2-w)} < 1$ we get $A_k\left(\frac{3-(1-v)(3-w)}{3-(1-v)(2-w)}\right) > \left(\frac{3-(1-v)(3-w)}{3-(1-v)(2-w)}\right)^{k-1} \sum_{m=0}^{k-1} A_{k,m}$. Now using $3 - (1-v)(3-w) > 3 - (3-w) = w$, we obtain that for a fixed $w \in (0,1)$,

$$\begin{aligned}&\frac{1}{(1-v)^{k-2}}\,(3-(1-v)(3-w))(3-(1-v)(2-w))^{k-1} A_k\left(\frac{3-(1-v)(3-w)}{3-(1-v)(2-w)}\right)\\&\geq \frac{1}{(1-v)^{k-2}}\,w^k \sum_{m=0}^{k-1} A_{k,m}\,.\end{aligned}$$

When integrating with respect to v, the factor $1/(1-v)^{k-2}$ in the case of $k > 2$ causes the higher moments $\mathbb{E}\, \nu_{\mathrm{I}}^k$ to be infinite. □

Note that the mean value is in accordance with the result in Lemma 17.6 and also with a result given in Corollary 14.19. Furthermore, it is not surprising that the higher moments of ν_{I} are infinite, because the moments of order greater than 2 of the length of the typical I-segment of Y_t are also infinite; see (11.14) for general dimension.

The next step is to distinguish the internal vertices according to their type. A spatial STIT tessellation has two different types of vertices, T-vertices and X-vertices. Illustrations of these two types of vertices are shown in Figure 17.3. Given a carrying I-polygon, a T-vertex is generated in its relative interior if two further I-polygons intersect in the *same* half-space determined by the carrying I-polygon. An X-vertex is generated by an intersection of two further I-polygons in two *different* half-spaces specified by the carrying I-polygon, which is illustrated in Figure 17.7. The number ν_{I} of vertices in the relative interior of the typical I-segment can be partitioned as $\nu_{\mathrm{I}} = \nu_{\mathrm{I}}^{\mathrm{T}} + \nu_{\mathrm{I}}^{\mathrm{X}}$, where $\nu_{\mathrm{I}}^{\mathrm{T}}$ and $\nu_{\mathrm{I}}^{\mathrm{X}}$ are the numbers of T- and X-vertices, respectively, in the relative interior.

Theorem 17.15. *For a spatial STIT tessellation Y_t with directional distribution ψ_o in the typical point of ∂Y_t, given in* (17.5), *the probability that the typical I-segment has exactly m vertices of type* T *and k vertices of type* X *in its relative interior is given by*

$$\begin{aligned}&\mathbb{P}(\nu_{\mathrm{I}}^{\mathrm{T}} = m, \nu_{\mathrm{I}}^{\mathrm{X}} = k)\\&= 3 \cdot 2^m \binom{m+k}{m} \int_0^1 \int_0^1 (1-v)^3\, v^m \frac{(1-(1-v)(1-w))^k}{(3-(1-v)(2-w))^{m+k+1}}\, dw\, dv \qquad (17.42)\end{aligned}$$

for $m, k \in \mathbb{N}_0$.

Proof. We consider the typical I-segment of Y_t given its birth time $\beta = s \in (0,t)$ and the birth time $\beta_{\mathrm{carr}} = r \in (0,s)$ of its carrying I-polygon. In Section 17.3.1, we explained that the number of X-vertices in the relative interior of the typical I-segment is Poisson distributed and depends on β_{carr}. The number of T-vertices is also Poisson distributed but depends on β. According to Lemma 17.9, for given direction $L \in G(3,1)$ and length $a \in (0,\infty)$ of the typical I-segment, the parameters of these Poisson distributions are $\Theta([e(L)])a(t-r)$ for the X-vertices and $\Theta([e(L)])a(2t-2s)$ for the T-vertices. From here on, the proof readily follows the lines of the proof of Theorem 17.13. □

From (17.42) the following can be concluded.

Corollary 17.16. *The mean numbers of* T*-vertices and of* X*-vertices in the relative interior of the typical I-segment are both equal to* 1,

$$\mathbb{E}\,\nu_{\mathrm{I}}^{\mathsf{T}} = \mathbb{E}\,\nu_{\mathrm{I}}^{\mathsf{X}} = 1\,.$$

Furthermore, the variance of the number of T*-vertices is* 8*, and that of the number of* X*-vertices is* 11/3*. All higher moments are infinite. In addition, the covariance of the number of* T*- and* X*-vertices equals* 4*,*

$$\begin{aligned}\operatorname{var}\nu_{\mathrm{I}}^{\mathsf{T}} = 8\,, \quad \operatorname{var}\nu_{\mathrm{I}}^{\mathsf{X}} = \frac{11}{3}\,, \quad \operatorname{cov}(\nu_{\mathrm{I}}^{\mathsf{T}}, \nu_{\mathrm{I}}^{\mathsf{X}}) = 4\,,\\ \mathbb{E}\,(\nu_{\mathrm{I}}^{\mathsf{T}})^k = \mathbb{E}\,(\nu_{\mathrm{I}}^{\mathsf{X}})^k = \infty\,, \quad k > 2\,.\end{aligned}$$

Proof. All these results follow from Theorem 17.15 by direct calculations. For example, the mean number of T-type vertices in the relative interior of the typical I-segment can be calculated as follows:

$$\begin{aligned}\mathbb{E}\,\nu_{\mathrm{I}}^{\mathsf{T}} = \sum_{m=0}^{\infty} m \sum_{k=0}^{\infty} \mathbb{P}(\nu_{\mathrm{I}}^{\mathsf{T}} = m, \nu_{\mathrm{I}}^{\mathsf{X}} = k) &= 3 \sum_{m=0}^{\infty} m \int_0^1\!\!\int_0^1 \frac{(1-v)^3}{1+v}\left(\frac{2v}{1+v}\right)^m \mathrm{d}w\,\mathrm{d}v\\ &= 6 \int_0^1 v(1-v)\,\mathrm{d}v \int_0^1 \mathrm{d}w\\ &= 1\,.\end{aligned}$$

For the second moment, we find

$$\begin{aligned}\mathbb{E}\,(\nu_{\mathrm{I}}^{\mathsf{T}})^2 &= \sum_{m=0}^{\infty} m^2 \sum_{k=0}^{\infty} \mathbb{P}(\nu_{\mathrm{I}}^{\mathsf{T}} = m, \nu_{\mathrm{I}}^{\mathsf{X}} = k)\\ &= 3 \sum_{m=0}^{\infty} m^2 \int_0^1\!\!\int_0^1 \frac{(1-v)^3}{1+v}\left(\frac{2v}{1+v}\right)^m \mathrm{d}w\,\mathrm{d}v\\ &= 6 \int_0^1 v(1+3v)\,\mathrm{d}v \int_0^1 \mathrm{d}w\\ &= 9\,,\end{aligned}$$

so that the variance equals $9 - 1^2 = 8$. Similar calculations yield the results for the X-vertices. Furthermore, the product of the number of T- and X-vertices in the relative interior of the typical I-segment has the expectation

$$\mathbb{E}\,(\nu_{\mathrm{I}}^{\mathrm{T}}\nu_{\mathrm{I}}^{\mathrm{X}}) = \sum_{m=0}^{\infty} m \sum_{k=0}^{\infty} k\,\mathbb{P}(\nu_{\mathrm{I}}^{\mathrm{T}} = m, \nu_{\mathrm{I}}^{\mathrm{X}} = k) = 12 \int_0^1 \int_0^1 v(v + w - vw)\,\mathrm{d}w\,\mathrm{d}v = 5\,.$$

Consequently, the covariance equals 4, which completes the proof. □

In Lemma 17.5, it is shown that the proportion of the intensities of T- and X-vertices in a spatial STIT tessellation is 2 : 1. Whereas any T-vertex is located in the relative interior of exactly one I-segment, an X-vertex is in the relative interior of two I-segments. This confirms the first statement of Corollary 17.16.

From now on, we will only consider I-segments that have internal vertices. For this purpose, let $\mathrm{I_S}\langle \mathrm{relint\,V}\rangle \subseteq \mathrm{I_S}(Y_t)$ be the subset that contains all I-segments of Y_t with vertices in its relative interior. These internal vertices can be of type T or X. The set $\mathrm{I_S}\langle \mathrm{relint\,V}\rangle$ is a set of components of the tessellation as introduced in Definition 2.9. We call its typical element the **typical I-segment *with* internal vertices**. We are interested in the probabilities p_{T} and p_{X} that a randomly chosen vertex in the relative interior of the typical I-segment with internal vertices is of one of these types. To formalize this, recall that we denote by $\mathbb{P}(\nu_{\mathrm{I}}^{\mathrm{T}} = m, \nu_{\mathrm{I}}^{\mathrm{X}} = k)$ the probability that the typical I-segment contains exactly m internal vertices from $\mathrm{V}\langle\mathrm{T}\rangle$ and k internal vertices from $\mathrm{V}\langle\mathrm{X}\rangle$, $m, k \in \mathbb{N}_0$. Moreover, $1 - \mathbb{P}(\nu_{\mathrm{I}} = 0)$ is the probability that the typical I-segment has internal vertices. Thus, given the typical I-segment contains exactly $n \geq 1$ internal vertices and given that $m \in \{0, \dots, n\}$ of them are of type T, the probability for choosing a T-vertex is m/n. Averaging over all possible numbers m and n, we define

$$p_{\mathrm{T}} := \frac{1}{1 - \mathbb{P}(\nu_{\mathrm{I}} = 0)} \sum_{n=1}^{\infty} \sum_{m=0}^{n} \frac{m}{n}\, \mathbb{P}(\nu_{\mathrm{I}}^{\mathrm{T}} = m, \nu_{\mathrm{I}}^{\mathrm{X}} = n - m)\,, \tag{17.43}$$

where the factor $1/(1 - \mathbb{P}(\nu_{\mathrm{I}} = 0))$ comes from conditioning on the event that the typical I-segment has internal vertices as explained above. Furthermore, denote $p_{\mathrm{X}} := 1 - p_{\mathrm{T}}$.

Lemma 17.17. *For a spatial STIT tessellation Y_t with an arbitrary directional distribution, a randomly chosen vertex in the relative interior of the typical I-segment with internal vertices is a* T*-vertex with probability*

$$p_{\mathrm{T}} = \frac{4(21 + 56\ln 2 - 54\ln 3)}{68 + 208\ln 2 - 189\ln 3} \approx 0.433053$$

and an X*-vertex with probability*

$$p_{\mathrm{X}} = \frac{27\ln 3 - 16\ln 2 - 16}{68 + 208\ln 2 - 189\ln 3} \approx 0.566947\,.$$

Proof. Combining (17.43) with (17.42), we calculate

$$
\begin{aligned}
p_{\mathrm{T}} &= \frac{1}{1-\mathbb{P}(\nu_{\mathrm{I}}=0)} \sum_{n=1}^{\infty}\sum_{m=0}^{n} \frac{m}{n}\,\mathbb{P}(\nu_{\mathrm{I}}^{\mathrm{T}}=m, \nu_{\mathrm{I}}^{\mathrm{X}}=n-m) \\
&= \frac{3}{1-\mathbb{P}(\nu_{\mathrm{I}}=0)} \int_0^1\int_0^1 \sum_{n=1}^{\infty}\sum_{m=0}^{n} \frac{m}{n}\binom{n}{m} \\
&\quad \times (2v)^m(1-v)^3 \frac{(1-(1-v)(1-w))^{n-m}}{(3-(1-v)(2-w))^{n+1}}\,dw\,dv \\
&= \frac{3}{1-\mathbb{P}(\nu_{\mathrm{I}}=0)} \int_0^1\int_0^1 \frac{2v(1-v)^2}{1+2v+w-vw}\,dw\,dv \\
&= \frac{3}{1-\mathbb{P}(\nu_{\mathrm{I}}=0)} \left(\frac{7}{2}+\frac{28}{3}\ln 2 - 9\ln 3\right).
\end{aligned}
$$

The explicit value for p_{T} is obtained by taking into account that $\mathbb{P}(\nu_{\mathrm{I}}=0)=\frac{189}{8}\ln 3-26\ln 2-\frac{15}{2}$, as given in (17.40). In addition, the relation $p_{\mathrm{T}}+p_{\mathrm{X}}=1$ implies the value for p_{X}, which completes our argument. □

Lemma 17.17 deals with the typical I-segment that has *at least* one internal vertex. Interestingly, if we condition the typical I-segment on having *exactly* $n \geq 1$ internal vertices, then the probability for a randomly chosen vertex being of T- or X-type changes with n. We denote these probabilities by $p_{\mathrm{T}|\nu_{\mathrm{I}}=n}$ and $p_{\mathrm{X}|\nu_{\mathrm{I}}=n}$, respectively, $n \geq 1$. Formally, they can be introduced as follows:

$$
\begin{aligned}
p_{\mathrm{T}|\nu_{\mathrm{I}}=n} &:= \frac{1}{\mathbb{P}(\nu_{\mathrm{I}}=n)} \sum_{m=0}^{n} \frac{m}{n}\,\mathbb{P}(\nu_{\mathrm{I}}^{\mathrm{T}}=m, \nu_{\mathrm{I}}^{\mathrm{X}}=n-m), \\
p_{\mathrm{X}|\nu_{\mathrm{I}}=n} &:= 1-p_{\mathrm{T}|\nu_{\mathrm{I}}=n}\,.
\end{aligned}
\tag{17.44}
$$

By Theorems 17.13 and 17.15 some particular values are $p_{\mathrm{T}|\nu_{\mathrm{I}}=1}\approx 0.3854$, $p_{\mathrm{T}|\nu_{\mathrm{I}}=2}\approx 0.4114$ or $p_{\mathrm{T}|\nu_{\mathrm{I}}=20}\approx 0.6058$. This suggests that in I-segments with a small number of internal vertices, the probability of selecting a vertex of type X is greater than in I-segments with a large number of vertices in their relative interior. We will see in the next section that the probabilities $p_{\mathrm{T}|\nu_{\mathrm{I}}=n}$ and $\mathbb{P}(\nu_{\mathrm{I}}=n)$ play an important role in our refined combinatorial analysis.

In Section 17.3.1, we have shown that due to the construction of the STIT tessellation process, the types of two neighboring internal vertices are independent. This means that looking along a typical I-segment with n internal vertices, the vertex type T or X is determined by independent coin tosses with probabilities $p_{\mathrm{T}|\nu_{\mathrm{I}}=n}$ and $p_{\mathrm{X}|\nu_{\mathrm{I}}=n}=1-p_{\mathrm{T}|\nu_{\mathrm{I}}=n}$. As a reminder, in Section 16.4.2, we considered "upper" and "lower" vertices in the relative interior of the typical I-segment for the planar case. In the proof of Lemma 16.27, it was shown that looking along a typical I-segment in a planar STIT tessellation, the type "upper" or "lower" vertex is also determined by independent coin tosses, but in this case with the probabilities $\frac{1}{2}$ and $\frac{1}{2}$ and independent of the number of internal vertices. In Lemma 16.26 the distribution of $(\nu_I^{\mathrm{l}}, \nu_I)$ is given.

The last step in our investigations for the internal vertices is the classification of the T-vertices with the labels "+" and "−", which was already introduced in Section 17.3.1 and shown in Figure 17.8. Any T-vertex is located in the relative interior of exactly one I-segment. This way, each vertex of type T receives one of the labels "+" and "−". This leads to a partition of the class $\mathsf{V}\langle \mathsf{T}\rangle$ of all T-vertices of the tessellation into $\mathsf{V}\langle \mathsf{T+}\rangle$ and $\mathsf{V}\langle \mathsf{T-}\rangle$, respectively, with $\mathsf{V}\langle \mathsf{T}\rangle = \mathsf{V}\langle \mathsf{T+}\rangle \cup \mathsf{V}\langle \mathsf{T-}\rangle$. Note that $\mathsf{V}\langle \mathsf{T+}\rangle$ and $\mathsf{V}\langle \mathsf{T-}\rangle$ are stationary point processes in $\mathbb{R}^3$. By the construction of the STIT tessellation process for the typical vertex $\mathsf{v} \in \mathsf{V}$ of the tessellation we have that

$$\mathbb{P}(\mathsf{v} \in \mathsf{V}\langle \mathsf{T+}\rangle) = \mathbb{P}(\mathsf{v} \in \mathsf{V}\langle \mathsf{T-}\rangle) = \frac{1}{2}\,\mathbb{P}(\mathsf{v} \in \mathsf{V}\langle \mathsf{T}\rangle)\,.$$

For the typical I-segment with exactly $n \geq 1$ internal vertices, we denote by $p_{\mathsf{T+}|\nu_I=n}$ and $p_{\mathsf{T-}|\nu_I=n}$ the probabilities that a uniformly selected internal vertex of it belongs to $\mathsf{V}\langle \mathsf{T+}\rangle$ or $\mathsf{V}\langle \mathsf{T-}\rangle$, respectively. Then, with the same arguments as above,

$$p_{\mathsf{T+}|\nu_I=n} = p_{\mathsf{T-}|\nu_I=n} = \frac{1}{2}\,p_{\mathsf{T}|\nu_I=n}\,. \tag{17.45}$$

This will be used in our proofs below.

17.4 Further classifications

17.4.1 Classification of edges

In this section, we will further investigate the combinatorial structure of spatial STIT tessellations. Our refined analysis is based on the distributional properties of the typical I-segment given in the previous section.

In Section 17.2.1, we give a first classification of segments in a spatial STIT tessellation. There we study edges, $\mathsf{J}^{(2)}$-segments, $\mathsf{J}^{(3)}$-segments, and I-segments. We now refine the analysis by partitioning the class $\mathsf{E}(T)$ of edges into subclasses, whereby we examine several cases. The first one is based on a classification according to the type of their endpoints, the 0-faces of an edge. They can be T- or X-vertices. The other two classifications focus on an equality relation. In a tessellation that is not face-to-face, an edge is not necessarily a 1-face of a plate or a 1-face of a cell. However, the adjacency between edges and 1-faces of plates or cells still remains, and equality is a special case. For example, any edge in a spatial STIT tessellation is adjacent to exactly three plates. The edge is *equal* to a 1-face of these plates for one or two or three of them. In Figure 17.10, different edge types on an I-segment are illustrated. In Section 16.5 a similar study has been carried out for the planar STIT tessellation. Nevertheless, in planar STIT tessellations, all vertices are T-vertices. The situation with T- and X-vertices in the spatial case is much more involved, produces a lot of new effects, and allows several different classifications.

Remark 17.18 (Explanations to Figure 17.10). *This figure illustrates different edge types together with their adjacent I-segment. This is the I-segment, which contains the edge.*

First, we consider an edge according to the types of its endpoints, which are the 0-faces of the edge. These 0-faces can be T- *or* X-*vertices. If the 0-face is an* X-*vertex, then it must be located in the relative interior of the I-segment adjacent to the edge. If it is a* T-*vertex, then it can also be a 0-face of this adjacent I-segment. This results in seven different cases in which the edges should now be classified according to their equality with 1-faces of plates and with 1-faces of cells.*

First row, left side: *This edge is a 1-face of three plates and a 1-face of two cells. Thus the edge is of type* $\mathsf{E}\langle\mathsf{TT}\rangle$, $\mathsf{E}\langle J^{(2)},3\rangle$, $\mathsf{E}\langle J^{(3)},2\rangle$.

First row, right side: *This edge is a 1-face of two plates and a 1-face of one cell. Thus the edge is of type* $\mathsf{E}\langle\mathsf{TT}\rangle$, $\mathsf{E}\langle J^{(2)},2\rangle$, $\mathsf{E}\langle J^{(3)},1\rangle$.

Second row, left and right sides: *These edges are 1-faces of two plates, and they are not a 1-face of any cell. Thus the edges are of type* $\mathsf{E}\langle\mathsf{TX}\rangle$ *(left) or* $\mathsf{E}\langle\mathsf{XX}\rangle$ *(right) and* $\mathsf{E}\langle J^{(2)},2\rangle$, $\mathsf{E}\langle J^{(3)},0\rangle$.

Third row, left side: *This edge is a 1-face of one plate and is not a 1-face of any cell. Thus the edge is of type* $\mathsf{E}\langle\mathsf{TX}\rangle$, $\mathsf{E}\langle J^{(2)},1\rangle$, $\mathsf{E}\langle J^{(3)},0\rangle$.

Third row, right side: *This edge is a 1-face of two plates and a 1-face of one cell. Thus the edge is of type* $\mathsf{E}\langle\mathsf{TT}\rangle$, $\mathsf{E}\langle J^{(2)},2\rangle$, $\mathsf{E}\langle J^{(3)},1\rangle$.

Fourth row: *This edge is a 1-face of one plate and is not a 1-face of a cell. Thus the edge is of type* $\mathsf{E}\langle\mathsf{TT}\rangle$, $\mathsf{E}\langle J^{(2)},1\rangle$, $\mathsf{E}\langle J^{(3)},0\rangle$.

Classification of edges according to the types of their endpoints

The first classification of an edge is based on the types of its two endpoints, the 0-faces of the edge. Recall that a spatial STIT tessellation has T- and X-type vertices. These two types are considered in Section 17.2.1 and illustrated in Figure 17.3.

Thus, for a spatial STIT tessellation Y_t, the class of vertices $\mathsf{V}(Y_t)$ can be split into two subclasses $\mathsf{V}\langle\mathsf{T}\rangle$ and $\mathsf{V}\langle\mathsf{X}\rangle$ with $\mathsf{V}(Y_t) = \mathsf{V}\langle\mathsf{T}\rangle \cup \mathsf{V}\langle\mathsf{X}\rangle$. An edge can have two T-vertices or two X-vertices or one T- and one X-vertex as its endpoints, respectively. This way, the class $\mathsf{E}(Y_t)$ of edges can be split into three subclasses $\mathsf{E}\langle\mathsf{TT}\rangle$, $\mathsf{E}\langle\mathsf{XX}\rangle$, and $\mathsf{E}\langle\mathsf{TX}\rangle$, whose elements are referred to as TT-, XX-, and TX-edges, respectively. For these subclasses, $\mathsf{E}(Y_t) = \mathsf{E}\langle\mathsf{TT}\rangle \cup \mathsf{E}\langle\mathsf{XX}\rangle \cup \mathsf{E}\langle\mathsf{TX}\rangle$. The subclasses are sets of components in the sense of Definition 2.9. The corresponding intensities are $\gamma_{\mathsf{E}\langle\cdot\rangle}$, where $\mathsf{E}\langle\cdot\rangle$ stands for one of the subclasses $\mathsf{E}\langle\mathsf{TT}\rangle$, $\mathsf{E}\langle\mathsf{XX}\rangle$ and $\mathsf{E}\langle\mathsf{TX}\rangle$. We are now interested in the probabilities that the typical edge belongs to one of these classes. Formally, we define

$$p_{\mathsf{E}\langle\cdot\rangle} := \frac{\gamma_{\mathsf{E}\langle\cdot\rangle}}{\gamma_{\mathsf{E}}}. \tag{17.46}$$

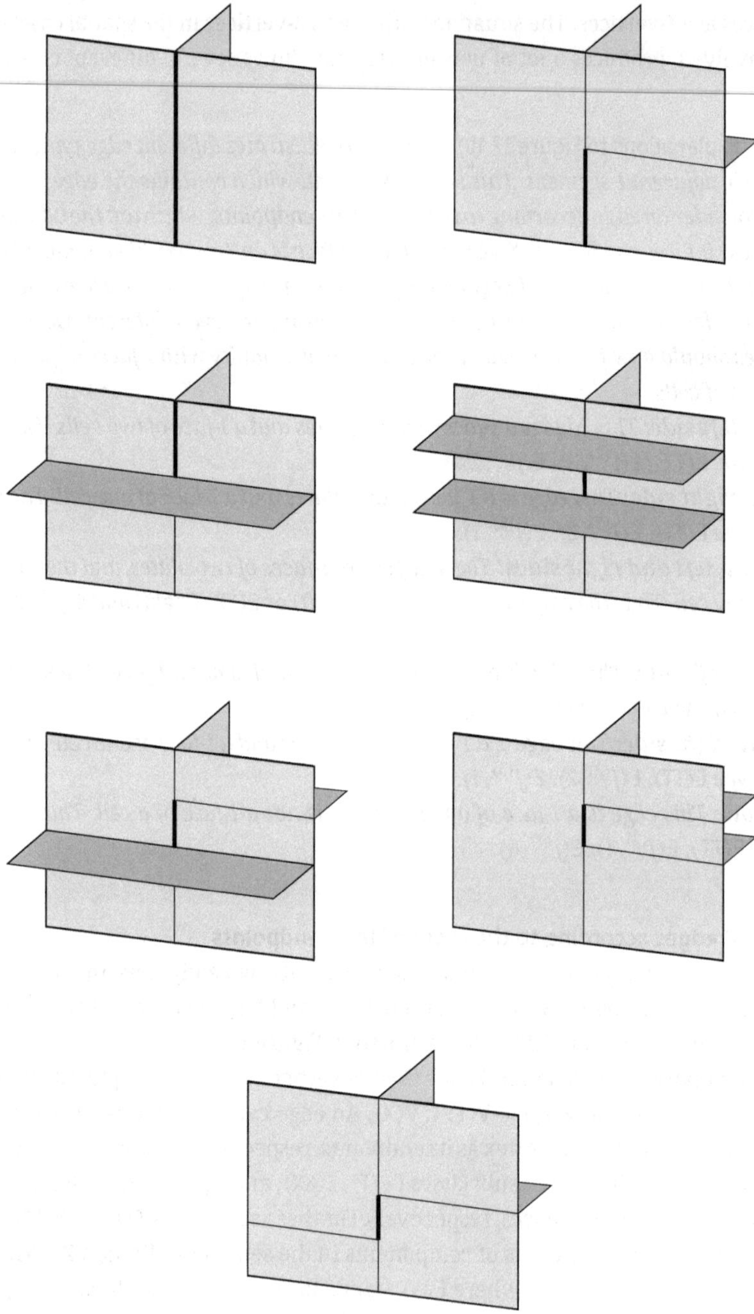

Figure 17.10: Illustration of different edge types on an I-segment of a spatial STIT tessellation. Further explanations for this figure are given in Remark 17.18.

We can say, for example, that $p_{\mathsf{E}\langle\mathsf{TT}\rangle}$ is the proportion of TT-edges of a STIT tessellation or, in other words, the probability that the typical edge belongs to the subclass $\mathsf{E}\langle\mathsf{TT}\rangle$.

Theorem 17.19. *For any spatial STIT tessellation Y_t with an arbitrary directional distribution, we have that*

$$\begin{aligned} p_{\mathsf{E}\langle\mathsf{TT}\rangle} &= \frac{1}{3}\,\mathbb{P}(\nu_\mathrm{I}=0) + \frac{2}{3}\,p_{\mathsf{T}|\nu_\mathrm{I}=1}\,\mathbb{P}(\nu_\mathrm{I}=1) \\ &\quad + \frac{1}{3}\sum_{n=2}^{\infty}(2p_{\mathsf{T}|\nu_\mathrm{I}=n} + (n-1)p^2_{\mathsf{T}|\nu_\mathrm{I}=n})\mathbb{P}(\nu_\mathrm{I}=n) \approx 0.442796\,, \\ p_{\mathsf{E}\langle\mathsf{XX}\rangle} &= p_{\mathsf{E}\langle\mathsf{TT}\rangle} - \frac{1}{3} \approx 0.109463\,, \\ p_{\mathsf{E}\langle\mathsf{TX}\rangle} &= \frac{4}{3} - 2p_{\mathsf{E}\langle\mathsf{TT}\rangle} \approx 0.447741\,, \end{aligned}$$

where $\mathbb{P}(\nu_\mathrm{I}=n)$, $n\in\mathbb{N}_0$, is given by (17.38), *and $p_{\mathsf{T}|\nu_\mathrm{I}=n}$, $n\in\mathbb{N}$, by* (17.44).

Proof. We start by noting that any edge is located on exactly one I-segment and that the two endpoints of an I-segment are T-vertices. Now let us look at the different types of edges on the typical I-segment created by the different types of internal vertices. If the typical I-segment has no internal vertices, $\nu_\mathrm{I}=0$, then it is equal to an edge of type TT. If $\nu_\mathrm{I}=1$, then the typical I-segment comprises two TT-edges if the internal vertex is of type T, an event with probability $p_{\mathsf{T}|\nu_\mathrm{I}=1}$, or we have two TX-edges if the internal vertex is an X-vertex with probability $p_{\mathsf{X}|\nu_\mathrm{I}=1}$. For $\nu_\mathrm{I}\geq 2$, we first consider edges whose endpoints both are internal vertices of the I-segment. For $\nu_\mathrm{I}=n$, there are $n-1$ of these edges on the typical I-segment. If we uniformly select one of them, then it is a TT-edge with probability $p^2_{\mathsf{T}|\nu_\mathrm{I}=n}$, thanks to the independence of the vertex type of two neighboring vertices on an I-segment as given in Lemma 17.9. In the same way, it is an XX-edge with probability $p^2_{\mathsf{X}|\nu_\mathrm{I}=n}$ and a TX-edge with probability $2p_{\mathsf{T}|\nu_\mathrm{I}=n}p_{\mathsf{X}|\nu_\mathrm{I}=n}$. Furthermore, there are two "boundary" edges, where one endpoint is also an endpoint of the I-segment and therefore a T-vertex. These edges are of type TT with probability $p_{\mathsf{T}|\nu_\mathrm{I}=n}$, of type TX with probability $p_{\mathsf{X}|\nu_\mathrm{I}=n}$, and of type XX with probability 0.

For this reason, the mean number $\mu_{\mathrm{I_S},\mathsf{E}\langle\mathsf{TT}\rangle}$ of edges of type TT that are located on the typical I-segment of the tessellation can be calculated by

$$\begin{aligned} \mu_{\mathrm{I_S},\mathsf{E}\langle\mathsf{TT}\rangle} &= \mathbb{P}(\nu_\mathrm{I}=0) + 2p_{\mathsf{T}|\nu_\mathrm{I}=1}\,\mathbb{P}(\nu_\mathrm{I}=1) \\ &\quad + \sum_{n=2}^{\infty}(2p_{\mathsf{T}|\nu_\mathrm{I}=n} + (n-1)p^2_{\mathsf{T}|\nu_\mathrm{I}=n})\mathbb{P}(\nu_\mathrm{I}=n)\,. \end{aligned} \tag{17.47}$$

Using the exchange formula (16.7) together with $\mu_{\mathsf{E}\langle\mathsf{TT}\rangle,\mathrm{I_S}}=1$, the intensity $\gamma_{\mathsf{E}\langle\mathsf{TT}\rangle}$ of TT-edges equals $\gamma_{\mathsf{E}\langle\mathsf{TT}\rangle}=\gamma_{\mathrm{I_S}}\,\mu_{\mathrm{I_S},\mathsf{E}\langle\mathsf{TT}\rangle}$.

Furthermore, by Lemma 17.6 we have the relation $\gamma_\mathsf{E}=3\gamma_{\mathrm{I_S}}$. Thus the probability $p_{\mathsf{E}\langle\mathsf{TT}\rangle}$ is given by

$$p_{E\langle TT\rangle} = \frac{\gamma_{E\langle TT\rangle}}{\gamma_E} = \frac{1}{3}\mu_{I_S,E\langle TT\rangle}\,,$$

which, together with (17.47), proves our first statement.

To calculate the intensities $\gamma_{E\langle XX\rangle}$ and $\gamma_{E\langle TX\rangle}$ of XX- and TX-edges, we observe that any vertex (regardless of its type) has exactly 4 emanating edges. The exchange formula (16.7) yields

$$\begin{aligned}2\gamma_{E\langle XX\rangle} &= \mathbb{E}\sum_{e\in E\langle XX\rangle} \mathbb{1}\{c(e)\in[0,1]^3\}\sum_{v\in V\langle X\rangle}\mathbb{1}\{v\subset e\}\\ &= \mathbb{E}\sum_{v\in V\langle X\rangle}\mathbb{1}\{c(v)\in[0,1]^3\}\sum_{e\in E\langle XX\rangle}\mathbb{1}\{v\subset e\}\end{aligned}$$

and, analogously,

$$\gamma_{E\langle XT\rangle} = \mathbb{E}\sum_{v\in V\langle X\rangle}\mathbb{1}\{c(v)\in[0,1]^3\}\sum_{e\in E\langle XT\rangle}\mathbb{1}\{v\subset e\}\,.$$

Consequently, we obtain

$$\begin{aligned}2\gamma_{E\langle XX\rangle} + \gamma_{E\langle TX\rangle} &= \mathbb{E}\sum_{v\in V\langle X\rangle}\mathbb{1}\{c(v)\in[0,1]^3\}\Big(\sum_{e\in E\langle XX\rangle}\mathbb{1}\{v\subset e\} + \sum_{e\in E\langle XT\rangle}\mathbb{1}\{v\subset e\}\Big)\\ &= 4\gamma_{V\langle X\rangle}\,,\end{aligned}$$

and in the same way,

$$2\gamma_{E\langle TT\rangle} + \gamma_{E\langle TX\rangle} = 4\gamma_{V\langle T\rangle}\,.$$

Thus

$$p_{E\langle TX\rangle} = \frac{\gamma_{E\langle TX\rangle}}{\gamma_E} = \frac{4\gamma_{V\langle T\rangle} - 2\gamma_{E\langle TT\rangle}}{\gamma_E} = 4\frac{\gamma_{V\langle T\rangle}}{\gamma_E} - 2p_{E\langle TT\rangle} = \frac{4}{3} - 2p_{E\langle TT\rangle}\,,$$

and similarly $p_{E\langle XX\rangle} = p_{E\langle TT\rangle} - \frac{1}{3}$, which completes the proof. □

Classification of edges according to equality with 1-faces of plates

In Section 17.2.1 a $J^{(2)}$-segment was introduced as a one-dimensional face of a plate marked with its multiplicity, which is the number of plates having this $J^{(2)}$-segment as a 1-face. For a $J^{(2)}$-segment in a spatial STIT tessellation, the multiplicity can be 1, 2, or 3. In comparison, a $J^{(3)}$-segment is a 1-face of a cell marked with its multiplicity, which is the number of cells having this segment as a 1-face. The upper index 2 or 3 in the notation stands for the dimension of the "owner" of this 1-face.

According to $\mu_{E,P} = 3$ in Corollary 17.2, any edge of a STIT tessellation is adjacent to exactly three plates. The next classification of edges is based on the following observations. For any edge, there exists at least one of the adjacent plates with a 1-face, which is *equal* to this edge. This 1-face is a $\mathrm{J}^{(2)}$-segment. It can have multiplicity 1, 2, or 3; see Figure 17.10, which illustrates the different types of edges on an I-segment and Remark 17.18 for explanations of this figure. In the classification according to equality with 1-faces of plates, we partition for a STIT tessellation Y_t the class $\mathsf{E}(Y_t)$ of edges into the subclasses $\mathsf{E}\langle J^{(2)},1\rangle$, $\mathsf{E}\langle J^{(2)},2\rangle$, and $\mathsf{E}\langle J^{(2)},3\rangle$, where $\mathsf{E}\langle J^{(2)},i\rangle$ is the set of all edges that are equal to a $\mathrm{J}^{(2)}$-segment with multiplicity $i \in \{1,2,3\}$. This is formally

$$\mathsf{E}\langle J^{(2)},i\rangle := \{\mathsf{e} \in \mathsf{E}(Y_t) : \mathsf{e} = S \text{ and } (S,i) \in J^{(2)}(Y_t)\}, \quad i \in \{1,2,3\}.$$

We obtain $\mathsf{E}(Y_t) = \mathsf{E}\langle J^{(2)},1\rangle \cup \mathsf{E}\langle J^{(2)},2\rangle \cup \mathsf{E}\langle J^{(2)},3\rangle$. Note that edges are segments without multiplicities. Therefore, the elements of $\mathsf{E}\langle J^{(2)},i\rangle$ are also segments without multiplicities. Define

$$p_{\mathsf{E}\langle J^{(2)},i\rangle} := \frac{\gamma_{\mathsf{E}\langle J^{(2)},i\rangle}}{\gamma_{\mathsf{E}}}, \quad i \in \{1,2,3\}, \tag{17.48}$$

as the probability that the typical edge belongs to class $\mathsf{E}\langle J^{(2)},i\rangle$. A set $\mathsf{E}\langle J^{(2)},i\rangle$ is a set of components in the sense of Definition 2.9, and thus the intensity $\gamma_{\mathsf{E}\langle J^{(2)},i\rangle}$ of $\mathsf{E}\langle J^{(2)},i\rangle$ is defined in (2.25). In other words, $p_{\mathsf{E}\langle J^{(2)},i\rangle}$ describes the proportion of edges of the STIT tessellation that are equal to a $\mathrm{J}^{(2)}$-segment with multiplicity i.

Theorem 17.20. *A spatial STIT tessellation Y_t with an arbitrary directional distribution satisfies*

$$\begin{aligned}
p_{\mathsf{E}\langle J^{(2)},1\rangle} &= \frac{1}{3}\sum_{n=2}^{\infty}\left(\frac{1}{2}(n-1)p^2_{\mathsf{T}|\nu_{\mathrm{I}}=n} + 2(n-1)p_{\mathsf{T}|\nu_{\mathrm{I}}=n}\,p_{\mathsf{X}|\nu_{\mathrm{I}}=n}\right)\mathbb{P}(\nu_{\mathrm{I}}=n)\\
&\approx 0.300533,\\
p_{\mathsf{E}\langle J^{(2)},2\rangle} &= \frac{2}{3}\mathbb{P}(\nu_{\mathrm{I}}=1) + \frac{1}{3}\sum_{n=2}^{\infty}\left(2+\frac{1}{2}(n-1)p^2_{\mathsf{T}|\nu_{\mathrm{I}}=n} + (n-1)p^2_{\mathsf{X}|\nu_{\mathrm{I}}=n}\right)\mathbb{P}(\nu_{\mathrm{I}}=n)\\
&\approx 0.554985,\\
p_{\mathsf{E}\langle J^{(2)},3\rangle} &= \frac{1}{3}\mathbb{P}(\nu_{\mathrm{I}}=0) = \frac{68}{8}\ln 3 - \frac{26}{3}\ln 2 - \frac{5}{2} \approx 0.144296
\end{aligned}$$

with $\mathbb{P}(\nu_{\mathrm{I}}=n)$, $n \in \mathbb{N}_0$, given by (17.38) and $p_{\mathsf{T}|\nu_{\mathrm{I}}=n}$ and $p_{\mathsf{X}|\nu_{\mathrm{I}}=n}$, $n \in \mathbb{N}$, by (17.44).

Proof. The values of the probabilities $p_{\mathsf{E}\langle J^{(2)},i\rangle}$ can be calculated with the same technique that was already used to prove Theorem 17.19. We refer again to Figure 17.10 and Remark 17.18, which may be helpful in reviewing some of the following considerations.

Again with $\gamma_{\mathsf{E}} = 3\gamma_{\mathrm{I_S}}$ from Lemma 17.6, we obtain

$$p_{\mathsf{E}\langle J^{(2)},i\rangle} = \frac{\gamma_{\mathsf{E}\langle J^{(2)},i\rangle}}{\gamma_{\mathsf{E}}} = \frac{1}{3}\mu_{\mathrm{I_S},\mathsf{E}\langle J^{(2)},i\rangle}, \quad i \in \{1,2,3\},$$

for the mean number $\mu_{\mathrm{I_S},\mathsf{E}\langle J^{(2)},i\rangle}$ of edges from the subclass $\mathsf{E}\langle J^{(2)}, i\rangle$ that are located in the relative interior of the typical I-segment.

However, an edge belongs to class $\mathsf{E}\langle J^{(2)}, 3\rangle$ if and only if it is also an I-segment. Hence $\mu_{\mathrm{I_S},\langle J^{(2)},3\rangle} = \mathbb{P}(\nu_{\mathrm{I}} = 0)$, which can be evaluated using relation (17.38). If $\nu_{\mathrm{I}} = 1$, then both edges on the I-segment belong to $\mathsf{E}\langle J^{(2)}, 2\rangle$.

For $\nu_{\mathrm{I}} \geq 2$, the two "boundary" edges, where one endpoint is also an endpoint of the I-segment, are again in $\mathsf{E}\langle J^{(2)}, 2\rangle$. An edge whose both endpoints are internal vertices of the I-segment is an element of subclass $\mathsf{E}\langle J^{(2)}, 2\rangle$ if both endpoints are X-vertices. If both endpoints of type T, then it belongs to $\mathsf{E}\langle J^{(2)}, 2\rangle$ only if both endpoints have the label "+" or both have label "–"; see Figure 17.8, Figure 17.10, and Remark 17.18. An edge with two internal vertices as endpoints belongs to $\mathsf{E}\langle J^{(2)}, 1\rangle$ if it has one endpoint of type T and the other one of type X or one of type T+ and the other one of type T–, respectively. These two cases are illustrated in Figure 17.10, third row left and last row. So, using $p_{\mathsf{T}+|\nu_{\mathrm{I}}=n} = p_{\mathsf{T}-|\nu_{\mathrm{I}}=n} = \frac{1}{2}p_{\mathsf{T}|\nu_{\mathrm{I}}=n}$, as given in (17.45), we find

$$\mu_{\mathrm{I_S},\mathsf{E}\langle J^{(2)},2\rangle} = 2\,\mathbb{P}(\nu_{\mathrm{I}} = 1) + \sum_{n=2}^{\infty}\left(2 + \frac{1}{2}(n-1)p_{\mathsf{T}|\nu_{\mathrm{I}}=n}^{2} + (n-1)p_{\mathsf{X}|\nu_{\mathrm{I}}=n}^{2}\right)\mathbb{P}(\nu_{\mathrm{I}} = n)$$

and similarly

$$\mu_{\mathrm{I_S},\mathsf{E}\langle J^{(2)},1\rangle} = \sum_{n=2}^{\infty}\left(\frac{1}{2}(n-1)p_{\mathsf{T}|\nu_{\mathrm{I}}=n}^{2} + 2(n-1)p_{\mathsf{T}|\nu_{\mathrm{I}}=n}p_{\mathsf{X}|\nu_{\mathrm{I}}=n}\right)\mathbb{P}(\nu_{\mathrm{I}} = n)\,, \tag{17.49}$$

which completes the proof. □

Classification of edges according to equality with 1-faces of cells

Another partition of $\mathsf{E}(Y_t)$ is obtained by taking the equality of an edge with a $\mathrm{J}^{(3)}$-segment instead a $\mathrm{J}^{(2)}$-segment as the basis. A $\mathrm{J}^{(3)}$-segment in a spatial STIT tessellation is a 1-face of a cell marked with its multiplicity, which can be 1 or 2. In this context, we observe that an edge of a spatial STIT tessellation is always adjacent to three cells, and only for two of them, the edge is adjacent to a 1-face of this cell. Hence it can be equal to a $\mathrm{J}^{(3)}$-segment with multiplicity 1 or 2, or the edge is not equal to any $\mathrm{J}^{(3)}$-segment. This is illustrated in Figure 17.10, and some explanations are given in Remark 17.18. Thus we can partition $\mathsf{E}(Y_t)$ into $\mathsf{E}\langle J^{(3)}, 0\rangle \cup \mathsf{E}\langle J^{(3)}, 1\rangle \cup \mathsf{E}\langle J^{(3)}, 2\rangle$ and define

$$\begin{aligned}
\mathsf{E}\langle J^{(3)}, 0\rangle &:= \mathsf{E}(Y_t) \setminus \mathcal{F}_1(Y_t)\,,\\
\mathsf{E}\langle J^{(i)}, i\rangle &:= \{\mathsf{e} \in \mathsf{E}(Y_t) : \mathsf{e} = S \text{ and } (S, i) \in J^{(3)}(Y_t)\}\,, \quad i \in \{1, 2\}\,,\\
p_{\mathsf{E}\langle J^{(3)},i\rangle} &:= \frac{\gamma_{\mathsf{E}\langle J^{(3)},i\rangle}}{\gamma_{\mathsf{E}}}\,, \quad i \in \{0, 1, 2\}\,,
\end{aligned} \tag{17.50}$$

where $\mathsf{E}\langle J^{(3)}, i\rangle$ is the class of tessellation edges that are equal to a $\mathrm{J}^{(3)}$-segment with multiplicity $i \in \{1, 2\}$, and $\mathsf{E}\langle J^{(3)}, 0\rangle$ are edges that are not equal to any 1-face of the cells.

Theorem 17.21. *For a spatial STIT tessellation Y_t with an arbitrary directional distribution, the values of $p_{\mathsf{E}\langle J^{(3)},i\rangle}$, $i \in \{0,1,2\}$, are given by*

$$\begin{aligned} p_{\mathsf{E}\langle J^{(3)},0\rangle} &= \frac{2}{3}\,\mathbb{P}(\nu_{\mathrm{I}}=1)p_{\mathsf{X}|\nu_{\mathrm{I}}=1} + \frac{1}{3}\sum_{n=2}^{\infty}\Big(2p_{\mathsf{X}|\nu_{\mathrm{I}}=n} + (n-1)p^2_{\mathsf{X}|\nu_{\mathrm{I}}=n} \\ &\quad + 2(n-1)p_{\mathsf{T}|\nu_{\mathrm{I}}=n}\,p_{\mathsf{X}|\nu_{\mathrm{I}}=n} + \frac{1}{2}(n-1)p^2_{\mathsf{T}|\nu_{\mathrm{I}}=n}\Big)\mathbb{P}(\nu_{\mathrm{I}}=n) \\ &\approx 0.624406\,, \\ p_{\mathsf{E}\langle J^{(3)},1\rangle} &= \frac{2}{3}\,\mathbb{P}(\nu_{\mathrm{I}}=1)p_{\mathsf{T}|\nu_{\mathrm{I}}=1} + \frac{1}{3}\sum_{n=2}^{\infty}\Big(2p_{\mathsf{T}|\nu_{\mathrm{I}}=n} + \frac{1}{2}(n-1)p^2_{\mathsf{T}|\nu_{\mathrm{I}}=n}\Big)\mathbb{P}(\nu_{\mathrm{I}}=n) \\ &\approx 0.231113\,, \\ p_{\mathsf{E}\langle J^{(3)},2\rangle} &= \frac{1}{3}\,\mathbb{P}(\nu_{\mathrm{I}}=0) = \frac{68}{8}\ln 3 - \frac{26}{3}\ln 2 - \frac{5}{2} \approx 0.144296\,, \end{aligned}$$

where $\mathbb{P}(\nu_{\mathrm{I}}=n)$, $n \in \mathbb{N}_0$, and $p_{\mathsf{T}|\nu_{\mathrm{I}}=n}$, $n \in \mathbb{N}$, are as in (17.38) *and* (17.44).

Proof. Because this follows once again by similar arguments used to prove Theorems 17.19 and 17.20, we restrict ourselves to a rough sketch. In this classification, we also have

$$p_{\mathsf{E}\langle J^{(3)},i\rangle} = \frac{\gamma_{\mathsf{E}\langle J^{(3)},i\rangle}}{\gamma_{\mathsf{E}}} = \frac{1}{3}\,\mu_{\mathrm{I}_S,\mathsf{E}\langle J^{(3)},i\rangle}\,, \qquad i \in \{0,1,2\}\,,$$

where $\mu_{\mathrm{I}_S,\mathsf{E}\langle J^{(3)},i\rangle}$ is the mean number of edges from $\mathsf{E}\langle J^{(3)},i\rangle$ in the relative interior of the typical I-segment. Hence it remains to determine $\mu_{\mathrm{I}_S,\mathsf{E}\langle J^{(3)},i\rangle}$.

As in the proof of Theorem 17.20, $\mu_{\mathrm{I}_S,\mathsf{E}\langle J^{(3)},2\rangle} = \mathbb{P}(\nu_{\mathrm{I}}=0)$, because an edge belongs to $\mathsf{E}\langle J^{(3)},2\rangle$ if and only if it is also an I-segment. Furthermore,

$$\mu_{\mathrm{I}_S,\mathsf{E}\langle J^{(3)},1\rangle} = 2p_{\mathsf{T}|\nu_{\mathrm{I}}=1}\mathbb{P}(\nu_{\mathrm{I}}=1) + \sum_{n=2}^{\infty}\Big(2p_{\mathsf{T}|\nu_{\mathrm{I}}=n} + \frac{1}{2}(n-1)p^2_{\mathsf{T}|\nu_{\mathrm{I}}=n}\Big)\mathbb{P}(\nu_{\mathrm{I}}=n),$$

and

$$\begin{aligned} \mu_{\mathrm{I}_S,\mathsf{E}\langle J^{(3)},0\rangle} &= 2p_{\mathsf{X}|\nu_{\mathrm{I}}=1}\mathbb{P}(\nu_{\mathrm{I}}=1) + \sum_{n=2}^{\infty}\Big(2p_{\mathsf{X}|\nu_{\mathrm{I}}=n} + (n-1)p^2_{\mathsf{X}|\nu_{\mathrm{I}}=n} \\ &\quad + 2(n-1)p_{\mathsf{T}|\nu_{\mathrm{I}}=n}\,p_{\mathsf{X}|\nu_{\mathrm{I}}=n} + \frac{1}{2}(n-1)p^2_{\mathsf{T}|\nu_{\mathrm{I}}=n}\Big)\mathbb{P}(\nu_{\mathrm{I}}=n)\,. \end{aligned}$$

This is achieved by first looking at $\nu_{\mathrm{I}} = 1$. If the internal vertex is a T-vertex, then both edges belong to $\mathsf{E}\langle J^{(3)},1\rangle$ and for an X-vertex to $\mathsf{E}\langle J^{(3)},0\rangle$. For $\nu_{\mathrm{I}} \geq 2$, in addition to the “boundary” edges, the various cases for the “internal” edges must be analyzed based on the possible types of their endpoints. All these cases are shown in Figure 17.10. □

Any of the probabilities calculated above carries additional information about the structure of a spatial STIT tessellation. In Section 17.4.2, this will be used to obtain new geometric mean values that arise in a refined analysis of the combinatorial structure of STIT tessellations.

Remark 17.22. *It is worth noticing that with the values from Theorems* 17.20 *and* 17.21 *and from known mean value relations, further probabilities can be calculated. For example, for the probabilities* $p_{J_S^{(2)}\langle E,1\rangle}$ *and* $p_{J_S^{(3)}\langle E,1\rangle}$ *that the typical* $J^{(2)}$*-segment and typical* $J^{(3)}$*-segment, respectively, is equal to one edge, we obtain,*

$$p_{J_S^{(2)}\langle E,1\rangle} = \frac{\gamma_E}{\gamma_{J_S^{(2)}}}(p_{E\langle J^{(2)},1\rangle} + 2p_{E\langle J^{(2)},2\rangle} + 3p_{E\langle J^{(2)},3\rangle}) \approx 0.790025$$

and

$$p_{J_S^{(3)}\langle E,1\rangle} = \frac{\gamma_E}{\gamma_{J_S^{(3)}}}(p_{E\langle J^{(3)},1\rangle} + 2p_{E\langle J^{(3)},2\rangle}) \approx 0.519705\,.$$

These values can be interpreted as follows:

$$p_{J_S^{(2)}\langle E,1\rangle} = \mathbb{P}(\textit{the typical } J^{(2)}\textit{-segment has no internal vertices})\,,$$
$$p_{J_S^{(3)}\langle E,1\rangle} = \mathbb{P}(\textit{the typical } J^{(3)}\textit{-segment has no internal vertices})\,.$$

17.4.2 Mean values for adjacencies

In this section, we extend the studies initiated in Section 17.1 and continued in Section 17.2 to calculate further mean values for adjacencies in a spatial STIT tessellation with respect to the classifications of edges just considered.

Vertex-edge adjacencies

In this paragraph, we deal with certain vertex-edge adjacencies. These adjacencies will be important to derive Theorem 17.27 at the end of this section. In this theorem the neighboring vertices of the typical vertex are considered, whereby these neighbors are distinguished according to their vertex type. Recall the adjacency notation; for example, $\mu_{V,E\langle TT\rangle}$ denotes the mean number of TT-edges emanating from the typical vertex. A TT-edge is an edge, where both endpoint are T-vertices. We start with mean values for the adjacencies for the typical vertex.

Lemma 17.23. *For the typical vertex in a spatial STIT tessellation* Y_t *with an arbitrary directional distribution, we have*

$$\mu_{V,E\langle TT\rangle} = 4\,p_{E\langle TT\rangle}\,, \qquad \mu_{V,E\langle TX\rangle} = \frac{16}{3} - 8\,p_{E\langle TT\rangle}\,,$$
$$\mu_{V,E\langle XX\rangle} = 4\,p_{E\langle TT\rangle} - \frac{4}{3}\,,$$
$$\mu_{V,E\langle J^{(2)},i\rangle} = 4\,p_{E\langle J^{(2)},i\rangle}\,, \qquad \mu_{V,E\langle J^{(3)},j\rangle} = 4\,p_{E\langle J^{(3)},j\rangle}$$

for $i \in \{1,2,3\}$ *and* $j \in \{0,1,2\}$.

We now refine these results. Firstly, we present the mean values for the vertex-edge adjacencies for the typical T-vertex.

Lemma 17.24. *For the typical* T*-vertex in a spatial STIT tessellation* Y_t *with an arbitrary directional distribution, we have that*

$$\mu_{V\langle T\rangle,E\langle TT\rangle} = 6\,p_{E\langle TT\rangle}\,, \quad \mu_{V\langle T\rangle,E\langle TX\rangle} = 4 - 6\,p_{E\langle TT\rangle}\,, \quad \mu_{V\langle T\rangle,E\langle XX\rangle} = 0\,,$$
$$\mu_{V\langle T\rangle,E\langle J^{(2)},1\rangle} = \sum_{n=2}^{\infty}(n-1)(p^2_{T|\nu_I=n} + 2p_{T|\nu_I=n}p_{X|\nu_I=n})\mathbb{P}(\nu_I = n)\,,$$
$$\mu_{V\langle T\rangle,E\langle J^{(2)},2\rangle} = 4 - \sum_{n=2}^{\infty}(n-1)(p^2_{T|\nu_I=n} + 2p_{T|\nu_I=n}p_{X|\nu_I=n})\mathbb{P}(\nu_I = n) - 6\,p_{E\langle J^{(2)},3\rangle}\,,$$
$$\mu_{V\langle T\rangle,E\langle J^{(2)},3\rangle} = 6\,p_{E\langle J^{(2)},3\rangle}\,,$$
$$\mu_{V\langle T\rangle,E\langle J^{(3)},0\rangle} = 4 - 6(p_{E\langle J^{(3)},1\rangle} + p_{E\langle J^{(3)},2\rangle})\,, \quad \mu_{V\langle T\rangle,E\langle J^{(3)},j\rangle} = 6\,p_{E\langle J^{(3)},j\rangle}\,, \quad j = \{1,2\}\,.$$

Finally, we summarize these mean values for the typical X-vertex.

Lemma 17.25. *The following adjacency relations apply to the typical* X*-vertex in a spatial STIT tessellation* Y_t *with an arbitrary directional distribution:*

$$\mu_{V\langle X\rangle,E\langle TT\rangle} = 0\,, \quad \mu_{V\langle X\rangle,E\langle TX\rangle} = 8 - 12\,p_{E\langle TT\rangle}\,, \quad \mu_{V\langle X\rangle,E\langle XX\rangle} = 12\,p_{E\langle TT\rangle} - 4\,,$$
$$\mu_{V\langle X\rangle,E\langle J^{(2)},1\rangle} = 4\sum_{n=2}^{\infty}(n-1)p_{T|\nu_I=n}p_{X|\nu_I=n}\mathbb{P}(\nu_I = n) = 4 - \mu_{V\langle X\rangle,E\langle J^{(2)},2\rangle}\,,$$
$$\mu_{V\langle X\rangle,E\langle J^{(2)},3\rangle} = 0\,,$$
$$\mu_{V\langle X\rangle,E\langle J^{(3)},0\rangle} = 4\,, \quad \mu_{V\langle X\rangle,E\langle J^{(3)},j\rangle} = 0\,, \quad j \in \{1,2\}\,.$$

The numerical values for these vertex-edge adjacencies are given in Tables 17.1 and 17.2.

Table 17.1: Numerical values for the vertex-edge adjacencies (part 1).

	TT	XX	TX
$\mu_{V,E\langle\cdot\rangle}$	1.77118	0.43785	1.79097
$\mu_{V\langle T\rangle,E\langle\cdot\rangle}$	2.65678	0	1.34322
$\mu_{V\langle X\rangle,E\langle\cdot\rangle}$	0	1.31355	2.68645

Table 17.2: Numerical values for the vertex-edge adjacencies (part 2).

	$E\langle J^{(2)},1\rangle$	$E\langle J^{(2)},2\rangle$	$E\langle J^{(2)},3\rangle$	$E\langle J^{(3)},0\rangle$	$E\langle J^{(3)},1\rangle$	$E\langle J^{(3)},2\rangle$
$\mu_{V,\cdot}$	1.20231	2.21994	0.57718	2.49762	0.92445	0.57718
$\mu_{V\langle T\rangle,\cdot}$	1.10376	2.03046	0.86578	1.74755	1.38668	0.86578
$\mu_{V\langle X\rangle,\cdot}$	1.39888	2.60112	0	4	0	0

The following relationships apply to the numerical values in the rows in Table 17.1:

$$\mu_{A,E\langle TT\rangle}+\mu_{A,E\langle XX\rangle}+\mu_{A,E\langle TX\rangle}=4 \quad \text{for } A\in\{V,V\langle T\rangle,V\langle X\rangle\},$$

because the typical vertex of each type has 4 emanating edges. Furthermore, in the columns,

$$\mu_{V,B}=\frac{2}{3}\mu_{V\langle T\rangle,B}+\frac{1}{3}\mu_{V\langle X\rangle,B} \quad \text{for } B\in\{E\langle TT\rangle,E\langle XX\rangle,E\langle TX\rangle\}.$$

The reason for this relation is the proportion $\gamma_{V\langle T\rangle}:\gamma_{V\langle X\rangle}=2:1$ for the two vertex types given in Lemma 17.5. Similar relationships exist for the values in Table 17.2.

Proof of Lemmas 17.23–17.25. (a) We start with the mean values $\mu_{V,E\langle TT\rangle}$, $\mu_{V,E\langle XX\rangle}$, and $\mu_{V,E\langle TX\rangle}$ and the analogous values for the typical T- and X-vertex.

With $\gamma_{V\langle T\rangle}\,\mu_{V\langle T\rangle,E\langle TT\rangle}=2\gamma_{E\langle TT\rangle}$, which follows from the exchange formula (16.7) and the relation $\gamma_{V\langle T\rangle}=\frac{2}{3}\gamma_V=\frac{1}{3}\gamma_E$, given in Section 17.1 together with Lemma 17.5, we obtain

$$\mu_{V\langle T\rangle,E\langle TT\rangle}=6\,\frac{\gamma_{E\langle TT\rangle}}{\gamma_E}=6\,p_{E\langle TT\rangle}$$

for the mean number of TT-edges emanating from the typical T-vertex. For the other edge types, we easily see that

$$\mu_{V\langle T\rangle,E\langle XX\rangle}=0 \quad \text{and} \quad \mu_{V\langle T\rangle,E\langle TX\rangle}=4-6\,p_{E\langle TT\rangle},$$

because $\mu_{V\langle T\rangle,E}=4$. Similar considerations for the typical X-vertex yield

$$\mu_{V\langle X\rangle,E\langle XX\rangle}=12\,\frac{\gamma_{E\langle XX\rangle}}{\gamma_E}=12\,p_{E\langle XX\rangle}=12\,p_{E\langle TT\rangle}-4$$

and

$$\mu_{V\langle X\rangle,E\langle TT\rangle}=0\,,\quad \mu_{V\langle X\rangle,E\langle TX\rangle}=8-12\,p_{E\langle TT\rangle}\,.$$

Using Lemma 17.5, for the typical vertex, we obtain that

$$\mu_{V,E\langle TT\rangle}=4\,p_{E\langle TT\rangle}\,,$$

which follows from $\gamma_{\mathrm{V}}\,\mu_{\mathrm{V,E}\langle\mathrm{TT}\rangle} = \gamma_{\mathrm{V}\langle\mathrm{T}\rangle}\,\mu_{\mathrm{V}\langle\mathrm{T}\rangle,\mathrm{E}\langle\mathrm{TT}\rangle}$, and

$$\mu_{\mathrm{V,E}\langle\mathrm{XX}\rangle} = 4\,p_{\mathrm{E}\langle\mathrm{TT}\rangle} - \frac{4}{3}\,,$$

which follows from $\gamma_{\mathrm{V}}\,\mu_{\mathrm{V,E}\langle\mathrm{XX}\rangle} = \gamma_{\mathrm{V}\langle\mathrm{X}\rangle}\,\mu_{\mathrm{V}\langle\mathrm{X}\rangle,\mathrm{E}\langle\mathrm{XX}\rangle}$.

Finally,

$$\mu_{\mathrm{V,E}\langle\mathrm{TX}\rangle} = \frac{16}{3} - 8\,p_{\mathrm{E}\langle\mathrm{TT}\rangle}\,.$$

(b) Let us now look at $\mu_{\mathrm{V,E}\langle J^{(2)},i\rangle}$, $i \in \{1,2,3\}$, and $\mu_{\mathrm{V,E}\langle J^{(3)},j\rangle}$, $j \in \{0,1,2\}$, for the typical vertex.

With $\gamma_{\mathrm{V}}\,\mu_{\mathrm{V,E}\langle J^{(2)},i\rangle} = 2\gamma_{\mathrm{E}\langle J^{(2)},i\rangle}$ and a similar relation for the edges from $\mathrm{E}\langle J^{(3)},j\rangle$, we obtain with $\gamma_{\mathrm{E}} = 2\gamma_{\mathrm{V}}$ that

$$\mu_{\mathrm{V,E}\langle J^{(2)},i\rangle} = 4\,p_{\mathrm{E}\langle J^{(2)},i\rangle}\,,\quad i \in \{1,2,3\}\,,\quad \text{and}\quad \mu_{\mathrm{V,E}\langle J^{(3)},j\rangle} = 4\,p_{\mathrm{E}\langle J^{(3)},j\rangle}\,,\quad j \in \{0,1,2\}\,.$$

(c) The adjacencies for the typical T-vertex and also for the typical X-vertex, respectively, and the edges from the classes $\mathrm{E}\langle J^{(2)},i\rangle$ and $\mathrm{E}\langle J^{(3)},j\rangle$ are more complicated.

All edges in the subclasses $\mathrm{E}\langle J^{(2)},3\rangle$, $\mathrm{E}\langle J^{(3)},1\rangle$, and $\mathrm{E}\langle J^{(3)},2\rangle$ are TT-edges. Hence

$$\mu_{\mathrm{V}\langle\mathrm{X}\rangle,\mathrm{E}\langle J^{(2)},3\rangle} = \mu_{\mathrm{V}\langle\mathrm{X}\rangle,\mathrm{E}\langle J^{(3)},1\rangle} = \mu_{\mathrm{V}\langle\mathrm{X}\rangle,\mathrm{E}\langle J^{(3)},2\rangle} = 0\,.$$

This means that $\mu_{\mathrm{V}\langle\mathrm{X}\rangle,\mathrm{E}\langle J^{(3)},0\rangle} = 4$ for the typical X-vertex. This result can be explained by the X-vertex geometry as follows. An X-vertex is an internal vertex of 1-faces of all four adjacent cells. Therefore none of the emanating edges from this vertex is a 1-face of a cell, as shown in Figure 17.3.

Using the mean value relations

$$\gamma_{\mathrm{V}\langle\mathrm{T}\rangle}\,\mu_{\mathrm{V}\langle\mathrm{T}\rangle,\mathrm{E}\langle J^{(2)},3\rangle} = 2\gamma_{\mathrm{E}\langle J^{(2)},3\rangle}\quad \text{and}\quad \gamma_{\mathrm{E}\langle J^{(2)},3\rangle} = \gamma_{\mathrm{E}}\,p_{\mathrm{E}\langle J^{(2)},3\rangle}$$

and the two intensity relations $\gamma_{\mathrm{E}} = 2\gamma_{\mathrm{V}}$ and $\gamma_{\mathrm{V}\langle\mathrm{T}\rangle} = \frac{2}{3}\gamma_{\mathrm{V}}$ given in Lemma 17.5, we obtain the result for $\mu_{\mathrm{V}\langle\mathrm{T}\rangle,\mathrm{E}\langle J^{(2)},3\rangle}$. Analogous considerations yield the relations for $\mu_{\mathrm{V}\langle\mathrm{T}\rangle,\mathrm{E}\langle J^{(3)},j\rangle}$, $j = 1,2$, for the typical T-vertex. From

$$\mu_{\mathrm{V}\langle\mathrm{T}\rangle,\mathrm{E}\langle J^{(3)},0\rangle} + \mu_{\mathrm{V}\langle\mathrm{T}\rangle,\mathrm{E}\langle J^{(3)},1\rangle} + \mu_{\mathrm{V}\langle\mathrm{T}\rangle,\mathrm{E}\langle J^{(3)},2\rangle} = \mu_{\mathrm{V}\langle\mathrm{T}\rangle,\mathrm{E}} = 4$$

we conclude that

$$\mu_{\mathrm{V}\langle\mathrm{T}\rangle,\mathrm{E}\langle J^{(3)},0\rangle} = 4 - 6(p_{\mathrm{E}\langle J^{(3)},1\rangle} + p_{\mathrm{E}\langle J^{(3)},2\rangle})\,.$$

(d) In a final step, we consider $\mu_{\mathrm{V}\langle\cdot\rangle,\mathrm{E}\langle J^{(2)},1\rangle}$ and $\mu_{\mathrm{V}\langle\cdot\rangle,\mathrm{E}\langle J^{(2)},2\rangle}$ for the typical T- and X-vertices, respectively.

To find their relations, we can use the same technique as for the proofs of Theorems 17.19–17.21.

To establish the relations for $\mu_{\mathsf{V}\langle\cdot\rangle,\mathsf{E}\langle J^{(2)},1\rangle}$, we first look at the edges in the subclass $\mathsf{E}\langle J^{(2)},1\rangle$. Such an edge lies on an I-segment, and both endpoints of the edge are internal vertices of that I-segment. Moreover, one of these endpoints must be of type T and the other one of type X, or one is of type T+ and the other one of type T–. This is shown in Figure 17.10 and explained in Remark 17.18. To make the following step easier to understand, we divide the set $\mathsf{E}\langle J^{(2)},1\rangle$ into $\mathsf{E}\langle J^{(2)},1,\mathsf{TX}\rangle$ and $\mathsf{E}\langle J^{(2)},1,\mathsf{T}+\mathsf{T}-\rangle$ according to the endpoints of the edges in these subclasses. The exchange formula (16.7) now leads to

$$\begin{aligned}\gamma_{\mathsf{V}\langle\mathsf{T}\rangle}\,\mu_{\mathsf{V}\langle\mathsf{T}\rangle,\mathsf{E}\langle J^{(2)},1\rangle} &= \gamma_{\mathsf{E}\langle J^{(2)},1\rangle}\,\mu_{\mathsf{E}\langle J^{(2)},1\rangle,\mathsf{V}\langle\mathsf{T}\rangle}\\ &= \gamma_{\mathsf{E}\langle J^{(2)},1,\mathsf{TX}\rangle}\,\mu_{\mathsf{E}\langle J^{(2)},1,\mathsf{TX}\rangle,\mathsf{V}\langle\mathsf{T}\rangle} + \gamma_{\mathsf{E}\langle J^{(2)},1,\mathsf{T+T-}\rangle}\,\mu_{\mathsf{E}\langle J^{(2)},1,\mathsf{T+T-}\rangle,\mathsf{V}\langle\mathsf{T}\rangle}\\ &= \gamma_{\mathsf{E}\langle J^{(2)},1,\mathsf{TX}\rangle} + 2\gamma_{\mathsf{E}\langle J^{(2)},1,\mathsf{T+T-}\rangle}\\ &= \gamma_{\mathrm{I_S}}\,\mu_{\mathrm{I_S},\mathsf{E}\langle J^{(2)},1,\mathsf{TX}\rangle} + 2\gamma_{\mathrm{I_S}}\,\mu_{\mathrm{I_S},\mathsf{E}\langle J^{(2)},1,\mathsf{T+T-}\rangle}\,.\end{aligned}$$

Considerations similar to those leading to relation (17.49) yield

$$\mu_{\mathsf{V}\langle\mathsf{T}\rangle,\mathsf{E}\langle J^{(2)},1\rangle} = \sum_{n=2}^{\infty} 2(n-1)p_{\mathsf{T}|\nu_{\mathrm{I}}=n}\,p_{\mathsf{X}|\nu_{\mathrm{I}}=n}\,\mathbb{P}(\nu_{\mathrm{I}}=n) + \sum_{n=2}^{\infty}(n-1)p^2_{\mathsf{T}|\nu_{\mathrm{I}}=n}\,\mathbb{P}(\nu_{\mathrm{I}}=n)\,,$$

using $\gamma_{\mathrm{I_S}} = \gamma_{\mathsf{V}\langle\mathsf{T}\rangle} = \frac{2}{3}\gamma_{\mathsf{V}}$ from Section 17.2.3.

Analogously, for the typical X-vertex, we obtain

$$\gamma_{\mathsf{V}\langle\mathsf{X}\rangle}\,\mu_{\mathsf{V}\langle\mathsf{X}\rangle,\mathsf{E}\langle J^{(2)},1\rangle} = \gamma_{\mathrm{I_S}}\sum_{n=2}^{\infty} 2(n-1)p_{\mathsf{T}|\nu_{\mathrm{I}}=n}\,p_{\mathsf{X}|\nu_{\mathrm{I}}=n}\,\mathbb{P}(\nu_{\mathrm{I}}=n)\,.$$

With $\gamma_{\mathrm{I_S}} = \frac{2}{3}\gamma_{\mathsf{V}}$ and $\gamma_{\mathsf{V}\langle\mathsf{X}\rangle} = \frac{1}{3}\gamma_{\mathsf{V}}$ from Section 17.2.3, this yields

$$\mu_{\mathsf{V}\langle\mathsf{X}\rangle,\mathsf{E}\langle J^{(2)},1\rangle} = 4\sum_{n=2}^{\infty}(n-1)p_{\mathsf{T}|\nu_{\mathrm{I}}=n}\,p_{\mathsf{X}|\nu_{\mathrm{I}}=n}\,\mathbb{P}(\nu_{\mathrm{I}}=n)\,.$$

The relations for $\mu_{\mathsf{V}\langle\cdot\rangle,\mathsf{E}\langle J^{(2)},2\rangle}$ for both types of typical vertices follow from the identity $\mu_{\mathsf{V}\langle\cdot\rangle,\mathsf{E}\langle J^{(2)},1\rangle} + \mu_{\mathsf{V}\langle\cdot\rangle,\mathsf{E}\langle J^{(2)},2\rangle} + \mu_{\mathsf{V}\langle\cdot\rangle,\mathsf{E}\langle J^{(2)},3\rangle} = 4.$ □

Adjacencies for edges

We are now interested in adjacencies and their mean values for edges in a spatial STIT tessellation Y_t with distribution $\mathbb{P}_{Y_t}$. The topological relationships listed below are valid $\mathbb{P}_{Y_t}$-a. s. for all edges $\mathsf{e} \in \mathsf{E}(Y_t)$.

We start with the adjacencies between edges and the other primitives. Any edge is adjacent to exactly 3 plates and to exactly 3 cells of the STIT tessellation. This is $\mu_{\mathrm{E,P}} = \mu_{\mathrm{E,Z}} = 3$, given in Corollary 17.2. In addition, $\mu_{\mathrm{E,V}} = 2$.

Next, we look at the adjacencies of edges with I-polygons and J-polygons. We refer again to Figure 17.10 and Remark 17.18, which can be helpful to verify the following results.

Any edge $\mathrm{e} \in \mathrm{E}(Y_t)$ is adjacent to exactly 2 I-polygons; more precisely,

$$\sum_{P\in \mathrm{I_P}(Y_t)} \mathbb{1}\{\mathrm{e} \subset \partial P\} = 1 \quad \text{and} \quad \sum_{P\in \mathrm{I_P}(Y_t)} \mathbb{1}\{\operatorname{relint} \mathrm{e} \subset \operatorname{relint} P\} = 1\,.$$

Furthermore, any edge $\mathrm{e} \in \mathrm{E}(Y_t)$ is adjacent to J-polygons whose total sum of multiplicities equals 5. More precisely,

$$\sum_{(P,m(P))\in \mathrm{J_P}(Y_t)} \mathbb{1}\{\mathrm{e} \subset \partial P\}\, m(P) = 4\,,$$

and

$$\sum_{(P,m(P))\in \mathrm{J_P}(Y_t)} \mathbb{1}\{\operatorname{relint} \mathrm{e} \subset \operatorname{relint} P\}\, m(P) = 1\,.$$

The same holds for the edges of all the considered subclasses $\mathrm{E}\langle\cdot\rangle$ of the three classifications introduced in Section 17.4.1.

The exchange formula $\gamma_{\mathrm{A}}\,\mu_{\mathrm{A,B}} = \gamma_{\mathrm{B}}\,\mu_{\mathrm{B,A}}$ from (16.7) applies to all sets considered here. In this context, we recall that according to Remark 16.9, objects with multiplicities can also be treated in a similar way if the multiplicity is included in the calculations. This gives the following result.

Lemma 17.26. *For a spatial STIT tessellation Y_t with an arbitrary directional distribution and for the three edge classifications considered in Section* 17.4.1 *with subclasses* $\mathrm{E}\langle\mathrm{TT}\rangle$, $\mathrm{E}\langle\mathrm{XX}\rangle$, $\mathrm{E}\langle\mathrm{TX}\rangle$, *and* $\mathrm{E}\langle J^{(2)},1\rangle$, $\mathrm{E}\langle J^{(2)},2\rangle$, $\mathrm{E}\langle J^{(2)},3\rangle$, *and* $\mathrm{E}\langle J^{(3)},0\rangle$, $\mathrm{E}\langle J^{(3)},1\rangle$, $\mathrm{E}\langle J^{(3)},2\rangle$, *we have*

$$\mu_{\mathrm{P,E}\langle\cdot\rangle} = \frac{36}{7}\,p_{\mathrm{E}\langle\cdot\rangle}\,, \quad \mu_{\mathrm{Z,E}\langle\cdot\rangle} = 36\,p_{\mathrm{E}\langle\cdot\rangle}\,,$$
$$\mu_{\mathrm{J_P,E}\langle\cdot\rangle} = 10\,p_{\mathrm{E}\langle\cdot\rangle}\,, \quad \mu_{\operatorname{relint}\mathrm{J_P,E}\langle\cdot\rangle} = 2\,p_{\mathrm{E}\langle\cdot\rangle}\,, \quad \mu_{\partial\mathrm{J_P,E}\langle\cdot\rangle} = 8\,p_{\mathrm{E}\langle\cdot\rangle}\,,$$
$$\mu_{\mathrm{I_S,E}\langle\cdot\rangle} = 24\,p_{\mathrm{E}\langle\cdot\rangle}\,, \quad \mu_{\operatorname{relint}\mathrm{I_S,E}\langle\cdot\rangle} = 12\,p_{\mathrm{E}\langle\cdot\rangle}\,, \quad \mu_{\partial\mathrm{I_S,E}\langle\cdot\rangle} = 12\,p_{\mathrm{E}\langle\cdot\rangle}\,,$$

where $p_{\mathrm{E}\langle\cdot\rangle}$ are the probabilities introduced in (17.46), (17.48), *and* (17.50).

Proof. The calculation is shown for one of the equations above. Using the exchange formula given in (16.7),

$$\gamma_{\mathrm{P}}\,\mu_{\mathrm{P,E}\langle\cdot\rangle} = \gamma_{\mathrm{E}\langle\cdot\rangle}\,\mu_{\mathrm{E}\langle\cdot\rangle,\mathrm{P}}\,,$$

with $\mu_{E\langle\cdot\rangle,P} = 3$, we obtain

$$\mu_{P,E\langle\cdot\rangle} = 3\frac{\gamma_{F\langle\cdot\rangle}}{\gamma_P}.$$

The relation $\gamma_P = \frac{7}{12}\gamma_E$ from Corollary 17.2 completes the calculation. □

One motivation for investigating this refined combinatorial structure is the following. Each vertex $v \in V(Y_t)$ in a spatial STIT tessellation Y_t has 4 emanating edges. The 0-faces of these edges, different from v, are called the neighboring vertices of v. Thus each vertex has 4 neighboring vertices of type T or X. We are now interested in the mean number $\nu_{V,V\langle T\rangle}$ of neighboring T-vertices and the mean number $\nu_{V,V\langle X\rangle}$ of neighboring X-vertices of the typical vertex. Furthermore, we want to compare these means with those in which the typical vertex is replaced by the typical T- or X-vertex. The corresponding mean values are denoted by $\nu_{V\langle T\rangle,V\langle T\rangle}$, $\nu_{V\langle T\rangle,V\langle X\rangle}$ and $\nu_{V\langle X\rangle,V\langle T\rangle}$, $\nu_{V\langle X\rangle,V\langle X\rangle}$.

Theorem 17.27. *For a spatial STIT tessellation Y_t with an arbitrary directional distribution, we have that*

$$\nu_{V,V\langle T\rangle} = \frac{8}{3} \quad \text{and} \quad \nu_{V,V\langle X\rangle} = \frac{4}{3}.$$

Moreover, for the typical T- *and* X-*vertices, the mean values are*

$$\nu_{V\langle T\rangle,V\langle T\rangle} = 6\,p_{E\langle TT\rangle} \approx 2.65678, \quad \nu_{V\langle T\rangle,V\langle X\rangle} = 4 - 6\,p_{E\langle TT\rangle} \approx 1.34322,$$
$$\nu_{V\langle X\rangle,V\langle T\rangle} = 8 - 12\,p_{E\langle TT\rangle} \approx 2.68645, \quad \nu_{V\langle X\rangle,V\langle X\rangle} = 12\,p_{E\langle TT\rangle} - 4 \approx 1.31355.$$

Proof. By definition we have that

$$\nu_{V\langle T\rangle,V\langle\cdot\rangle} = \mu_{V\langle T\rangle,E\langle T\cdot\rangle}, \quad \nu_{V\langle X\rangle,V\langle\cdot\rangle} = \mu_{V\langle X\rangle,E\langle X\cdot\rangle},$$

and similarly

$$\nu_{V,V\langle\cdot\rangle} = p_{V\langle T\rangle}\,\mu_{V\langle T\rangle,E\langle T\cdot\rangle} + p_{V\langle X\rangle}\,\mu_{V\langle X\rangle,E\langle X\cdot\rangle}.$$

The results from Lemmas 17.23–17.25 now complete the proof. □

Theorem 17.27 formalizes, for example, that in a spatial STIT tessellation the typical X-vertex favors a slightly higher number of T-neighbors compared with the typical T-vertex or the typical tessellation vertex.

Let us finally remark that for some quantities considered in this section, we can derive the whole distribution. However, these distributions appear to be complicated in general, and therefore we have restricted our attention to mean values. To give an example, let us consider the number $\nu_{E\langle XX\rangle}$ of XX-edges located on the typical I-segment.

We have that

$$\mathbb{P}(\nu_{\mathrm{E}\langle\mathrm{XX}\rangle} = 0) = \mathbb{P}(\nu_{\mathrm{I}} = 0) + \mathbb{P}(\nu_{\mathrm{I}} = 1) + \sum_{n=2}^{\infty}(1 - p^2_{\mathrm{X}|\nu_{\mathrm{I}}=n})^{n-1}\mathbb{P}(\nu_{\mathrm{I}} = n)$$

and

$$\mathbb{P}(\nu_{\mathrm{E}\langle\mathrm{XX}\rangle} = k) = \sum_{n=k+1}^{\infty} \binom{n-1}{k} p^{2k}_{\mathrm{X}|\nu_{\mathrm{I}}=n}(1 - p^2_{\mathrm{X}|\nu_{\mathrm{I}}=n})^{n-1-k}, \quad k \geq 1,$$

with $\mathbb{P}(\nu_{\mathrm{I}} = n)$ and $p_{\mathrm{X}|\nu_{\mathrm{I}}=n} = 1 - p_{\mathrm{T}|\nu_{\mathrm{I}}=n}$ given by (17.38) and (17.44), respectively. For small k, the numerical values are

$$\mathbb{P}(\nu_{\mathrm{E}\langle\mathrm{XX}\rangle} = 0) \approx 0.795161, \quad \mathbb{P}(\nu_{\mathrm{E}\langle\mathrm{XX}\rangle} = 1) \approx 0.131115,$$
$$\mathbb{P}(\nu_{\mathrm{E}\langle\mathrm{XX}\rangle} = 2) \approx 0.046467, \quad \mathbb{P}(\nu_{\mathrm{E}\langle\mathrm{XX}\rangle} = 3) \approx 0.016359.$$

17.5 Comparison of isotropic STIT and Poisson plane tessellations with Poisson–Voronoi tessellations

In this section, we consider Poisson–Voronoi tessellations (PVTs), Poisson plane tessellations (PHTs), and STIT tessellations in the three-dimensional space. Figure 1.1 shows simulations of the corresponding planar tessellations. As in Section 16.7, we compare first- and second-order quantities to identify similarities and differences between these three model types. The results can be helpful when selecting models for certain applications. The values for STIT tessellations are specified with an upper index ^{STIT}, for PVTs with ^{PVT}, and for PHTs with ^{PHT}.

Again, we restrict the comparison to the isotropic case, because a PVT, which is generated by a stationary Poisson point process, is isotropic. This results from Corollary 3.2.1 in [172].

In addition to this first- and second-order comparison, Sections 16.7.3 and 16.7.4 describe ergodic and mixing properties for the considered tessellation types for arbitrary dimensions.

17.5.1 First-order comparison

In analogy to the planar case, we assume for this first-order comparison that the considered tessellations have the same intensity of cells, $\gamma_{\mathrm{Z}}^{\mathrm{PHT}} = \gamma_{\mathrm{Z}}^{\mathrm{STIT}} = \gamma_{\mathrm{Z}}^{\mathrm{PVT}}$. This implies that the mean volumes of the typical cells of all three tessellations are the same.

Two constants $\zeta_{\psi}^{(2)}$ and $\zeta_{\psi}^{(3)}$ are given in (17.9) and (17.10) for the directional distribution ψ_o on $G(3,1)$. In the isotropic case, when ψ_o is invariant under rotations on $G(3,1)$, these two constants have the particular values $\zeta_{\psi}^{(2)} = \frac{\pi}{4}$ and $\zeta_{\psi}^{(3)} = \frac{\pi}{8}$; see (17.13).

Mean values for the seven characteristics for a spatial STIT tessellation are given in Lemma 17.1, and further mean values are given in Corollary 17.2. Together with the mean values for a stationary and isotropic PHT in Section 9.5.2 in [33] and for a PVT in Section 9.7.3 in [33], we obtain the following.

Intensities for primitives:

$$
\begin{aligned}
6\,\gamma_{\mathrm{V}}^{\mathrm{PHT}} &= \gamma_{\mathrm{V}}^{\mathrm{STIT}} \approx 0.887\,\gamma_{\mathrm{V}}^{\mathrm{PVT}},\\
4\,\gamma_{\mathrm{E}}^{\mathrm{PHT}} &= \gamma_{\mathrm{E}}^{\mathrm{STIT}} \approx 0.887\,\gamma_{\mathrm{E}}^{\mathrm{PVT}},\\
\frac{7}{3}\,\gamma_{\mathrm{P}}^{\mathrm{PHT}} &= \gamma_{\mathrm{P}}^{\mathrm{STIT}} \approx 0.901\,\gamma_{\mathrm{P}}^{\mathrm{PVT}},\\
\gamma_{\mathrm{Z}}^{\mathrm{PHT}} &= \gamma_{\mathrm{Z}}^{\mathrm{STIT}} = \gamma_{\mathrm{Z}}^{\mathrm{PVT}}.
\end{aligned}
$$

Typical edge:

$$
\frac{1}{2}\,\ell_{\mathrm{E}}^{\mathrm{PHT}} = \ell_{\mathrm{E}}^{\mathrm{STIT}} \approx 0.935\,\ell_{\mathrm{E}}^{\mathrm{PVT}}.
$$

Typical plate:

$$
\begin{aligned}
\frac{9}{14}\,\ell_{\mathrm{P}}^{\mathrm{PHT}} &= \ell_{\mathrm{P}}^{\mathrm{STIT}} \approx 0.920\,\ell_{\mathrm{P}}^{\mathrm{PVT}},\\
\frac{3}{7}\,a_{\mathrm{P}}^{\mathrm{PHT}} &= a_{\mathrm{P}}^{\mathrm{STIT}} \approx 0.946\,a_{\mathrm{P}}^{\mathrm{PVT}}.
\end{aligned}
$$

Typical cell:

$$
\begin{aligned}
\frac{3}{2}\,\ell_{\mathrm{Z}}^{\mathrm{PHT}} &= \ell_{\mathrm{Z}}^{\mathrm{STIT}} \approx 0.829\,\ell_{\mathrm{Z}}^{\mathrm{PVT}},\\
b_{\mathrm{Z}}^{\mathrm{PHT}} &= b_{\mathrm{Z}}^{\mathrm{STIT}} \approx 0.829\,b_{\mathrm{Z}}^{\mathrm{PVT}},\\
a_{\mathrm{Z}}^{\mathrm{PHT}} &= a_{\mathrm{Z}}^{\mathrm{STIT}} \approx 0.853\,a_{\mathrm{Z}}^{\mathrm{PVT}},\\
v_{\mathrm{Z}}^{\mathrm{PHT}} &= v_{\mathrm{Z}}^{\mathrm{STIT}} = v_{\mathrm{Z}}^{\mathrm{PVT}}.
\end{aligned}
$$

Densities:

$$
\begin{aligned}
2\,\eta_1^{\mathrm{PHT}} &= \eta_1^{\mathrm{STIT}} \approx 0.829\,\eta_1^{\mathrm{PVT}},\\
\eta_2^{\mathrm{PHT}} &= \eta_2^{\mathrm{STIT}} \approx 0.853\,\eta_2^{\mathrm{PVT}}.
\end{aligned}
$$

The exact values for a PVT can be found in Section 9.7.3 in [33]. The approximate values are sufficient for the comparison here. Furthermore, for the mean values for the adjacencies defined in (16.4) and for the mean number of k-faces of the typical cell $\nu_k(\mathrm{Z})$, $k \in \{0, 1, 2\}$, we obtain the following results.

	PHT	STIT	PVT
$\mu_{V,E}$	6	4	4
$\mu_{V,P}$	12	6	6
$\mu_{V,Z}$	8	4	4
$\mu_{E,P} = \mu_{E,Z}$	4	3	3
$\mu_{P,V} = \mu_{P,E}$	4	$\frac{36}{7} \approx 5.143$	≈ 5.228
$\mu_{Z,V}$	8	24	≈ 27.071
$\mu_{Z,E}$	12	36	≈ 40.606
$\mu_{Z,P}$	6	14	≈ 15.535
$\nu_0(Z)$	8	8	≈ 27.071
$\nu_1(Z)$	12	12	≈ 40.606
$\nu_2(Z)$	6	6	≈ 15.535

With the normalization $\gamma_Z^{\text{PHT}} = \gamma_Z^{\text{STIT}} = \gamma_Z^{\text{PVT}}$, it is obvious that the mean values for the typical cell b_Z, a_Z, v_Z, and $\nu_k(Z)$, $k \in \{0,1,2\}$, for STIT tessellations and PHTs coincide. Only ℓ_Z is different for these tessellations, because STIT tessellations are not face-to-face. Therefore a cell in a STIT tessellation may have edges on its boundary that are located in the relative interior of its 2-faces and not on its 1-faces. For most of the quantities, the values for a STIT tessellation lie between those for PHT and PVT. Exceptions are the mean value ℓ_E for the typical edge and the mean values ℓ_P and a_P for the typical plate. Once again, this can be explained by the fact that STIT tessellations are not face-to-face. An edge, that is a 1-primitive, does not need to be a 1-face of a cell, and plates can differ from the 2-faces of the cells. Thus edges in a STIT tessellation are in the mean shorter, and plates are in the mean smaller than the comparable corresponding components in PHT and PVT.

The mean values for the adjacencies, which provide some information about the mutual arrangement of the cells within a tessellation, clearly show that stationary and isotropic STIT tessellations and PVTs are quite close to each other. There are significant differences in these values between stationary and isotropic STIT tessellations and PHTs.

17.5.2 Second-order comparison

We will now analyze the second-order properties and compare them for the three models in the three-dimensional space. If available, we use analytic formulae for the considered quantities. In case where explicit formulae are still unknown, the quantities were studied by simulation. All tessellations in this section are stationary and isotropic. Similarly to the planar case, we now assume that the tessellations have the same surface density $\eta_2^{\text{PHT}} = \eta_2^{\text{STIT}} = \eta_2^{\text{PVT}} = t > 0$. Thus the considered tessellations have the following intensities of primitives:

$$\gamma_V^{PHT} = \frac{\pi}{48} t^3, \quad \gamma_V^{STIT} = \frac{\pi}{8} t^3, \quad \gamma_V^{PVT} \approx 0.275\, t^3,$$
$$\gamma_E^{PHT} = \frac{\pi}{16} t^3, \quad \gamma_E^{STIT} = \frac{16}{\pi} t^3, \quad \gamma_E^{PVT} \approx 0.526\, t^3,$$
$$\gamma_P^{PHT} = \frac{\pi}{16} t^3, \quad \gamma_P^{STIT} = \frac{7\pi}{48} t^3, \quad \gamma_P^{PVT} \approx 0.315\, t^3,$$

and edge length densities

$$\eta_1^{PHT} = \frac{\pi}{8} t^2, \quad \eta_1^{STIT} = \frac{\pi}{4} t^2, \quad \eta_1^{PVT} \approx 0.688\, t^2.$$

This follows for STIT tessellations from (17.18), (17.21), and Corollary 17.2, together with (17.13) in the isotropic case. The results for PHT and PVT can be found in [33], Sections 9.5.2 and 9.7.3. The exact values for PVT can also be found there.

For the second-order analysis, we will use as in Section 16.7.2 the *pair correlation function* g_Φ of a stationary and isotropic random measure Φ on $\mathbb{R}^d$, defined in Section 12.7.1. For $d = 3$, it is

$$g_\Phi(r) = \frac{1}{4\pi^2 r} \frac{\mathrm{d}K_\Phi(r)}{\mathrm{d}r}, \quad r > 0, \tag{17.51}$$

where K_Φ stands for the K-function of Φ.

Furthermore, for a random stationary and isotropic measure Φ on $\mathbb{R}^3$ with intensity $\gamma_\Phi = \mathbb{E}\,\Phi([0,1]^3)$ and window $W \in \mathcal{P}_3$, the **variance** $\operatorname{var}\Phi(W)$ can be calculated from the pair correlation function g_Φ via

$$\operatorname{var}\Phi(W) = \gamma_\Phi^2\, 4\pi \int_0^\infty \bar{c}_W(r)\, g_\Phi(r)\, r^2\, \mathrm{d}r - (\gamma_\Phi V_3(W))^2. \tag{17.52}$$

This relation is the special case $d = 3$ of equation (7.25) in [33]. The isotropized set-covariogram function $\bar{c}_W(r)$ of W is given in Definition 12.6.

Firstly, we consider second-order quantities for the *set of vertices* $\mathsf{V} = \mathsf{V}(X_t)$ of a random stationary and isotropic tessellation X_t with surface density $t > 0$. According to Remark 19.3, the vertex point process V can be considered as a random measure. Using $\Phi = \mathsf{V}$, we define that $g_{0,t}(r) := g_\mathsf{V}(r)$ is the pair correlation function of V. In this case, $\lambda_\mathsf{V} K_\mathsf{V}(r)$ describes the mean number of vertices at a given distance $r > 0$ from the typical vertex; see (17.51).

In addition, let $\Sigma_{0,t,W}$ be the number of vertices of the tessellation inside a window W, analogous to the planar case in Section 16.7.2, and let $\operatorname{var}\Sigma_{0,t,W}$ be the variance of this number.

Next, we are interested in the pair correlation function related to the *edges* and in that related to the *plates* of the tessellations. For this reason, we consider the random length measure concentrated on the 1-skeleton $\operatorname{skel}_1 X_t$ and the random area measure concentrated on the 2-skeleton $\operatorname{skel}_2 X_t$, respectively. For a spatial STIT tessellation Y_t, it follows by Corollary 2.8 that

$$\mathrm{skel}_1\, Y_t = \mathrm{skel}_1^{\mathrm{pr}}\, Y_t \quad \text{and} \quad \mathrm{skel}_2\, Y_t = \mathrm{skel}_2^{\mathrm{pr}}\, Y_t = \partial Y_t\,.$$

This is obvious for PHT and PVT because these tessellations are face-to-face.

The random length measure on the i-skeleton of X_t defined on $(\mathbb{R}^3, \mathcal{B}(\mathbb{R}^3))$ is $\mathcal{H}_t^i(\cdot) := \mathcal{H}^i(\cdot \cap \mathrm{skel}_i\, X_t)$, where $\mathcal{H}^i$ is the i-dimensional Hausdorff measure, $i \in \{1, 2\}$. With $\Phi = \mathcal{H}_t^i$ and (17.51), we obtain the pair correlation function $g_{i,t} := g_{\mathcal{H}_t^i}$ of the length ($i = 1$) or the area ($i = 2$) measure on the i-skeleton, respectively. Using the corresponding K-function $K_{\mathcal{H}_t^1}$, we can say that $\eta_1 K_{\mathcal{H}_t^1}(r)$ describes the mean total edge length in a ball with radius $r > 0$ and center in a typical point of the 1-skeleton. Similarly, $\eta_2 K_{\mathcal{H}_t^2}(r)$ is the mean total area of plates in a ball with radius $r > 0$ and center in a typical point of the 2-skeleton.

Furthermore, for the tessellation X_t inside a polygonal window $W \in \mathcal{P}_3$, we consider the total edge length of X_t in W, which is denoted by $\Sigma_{1,t,W}$ and its variance by $\mathrm{var}\, \Sigma_{1,t,W}$. In addition, $\Sigma_{2,t,W}$ is the total surface area of X_t in W, and $\mathrm{var}\, \Sigma_{2,t,W}$ is its variance. Note that this notation $\Sigma_{j,t,W}$ differs from $\Sigma_{V_j,t,W}$, which was introduced in Section 10.4 for STIT tessellations in $\mathbb{R}^d$. It describes the total intrinsic volume V_j of the maximal $(d-1)$-polytopes in $W, j \in \{0, \ldots, d-1\}$.

We now compare these second-order quantities $g_{i,t}$ and $\mathrm{var}\, \Sigma_{i,t,W}$, $i \in \{0, 1, 2\}$, for the three tessellation models. As in the planar case, here we restrict our analysis of the variances to the asymptotic variances.

Variances and asymptotic variances for stationary and isotropic STIT tessellations are considered in Chapter 12. Results for pair correlation functions of stationary and isotropic STIT tessellations are given in Sections 12.7 and 12.8. For stationary and isotropic Poisson hyperplane processes, pair correlation functions are studied in Section 10.5 of [82].

It appears that in the spatial case, much less is known analytically. Many of the characteristics can only be approximated by simulation. Details of these simulations can be found in [165].

Vertices

We start the comparison of the three models with $g_{0,t}$ and $\mathrm{var}\, \Sigma_{0,t,W}$ based on the vertex point process V. Note that the three tessellations are assumed to be stationary and isotropic with the same surface density $\eta_2^{\mathrm{PHT}} = \eta_2^{\mathrm{STIT}} = \eta_2^{\mathrm{PVT}} = t$.

The pair correlation function $g_{0,t}^{\mathrm{PHT}}$ equals

$$g_{0,t}^{\mathrm{PHT}}(r) = 1 + \frac{9}{2\,t\,r} + \frac{36}{t^2\,\pi^2\,r^2}\,, \qquad r > 0\,.$$

This results from the much more general case in (12.68) with $K = 0$, $i = j = 0$, and $d = 3$ or can be obtained directly from the Mecke equation, given in (19.5).

For STIT tessellations, in contrast to the planar case, no second-order characteristics for the point process of vertices are currently known. They have to be estimated by simulation.

For a PVT in $\mathbb{R}^3$, an analytic formula for the pair correlation function $g_{0,t}^{\mathrm{PVT}}$ was proven in [72]. As in the planar case, it consists of a sum of rather complicated integrals, which have to be evaluated numerically. A table of values can also be found in [72].

Plots of the three pair correlation functions for $t = 1$ are shown in Figure 17.11. We can see that the STIT tessellation shows an intermediate behavior between PHT and PVT, as in the planar case.

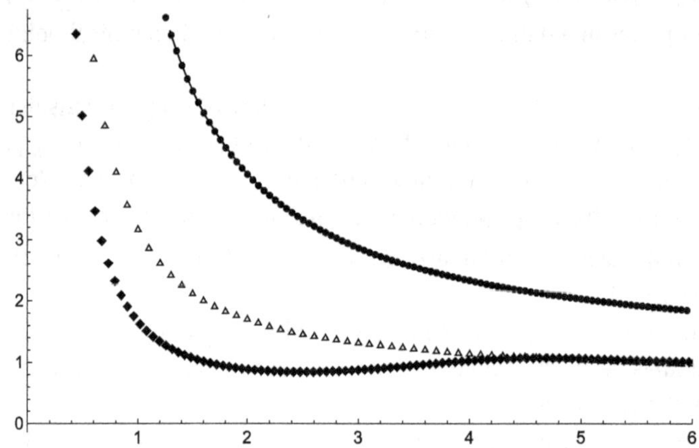

Figure 17.11: Plots for the pair correlation functions $g_{0,t}^{\mathrm{PHT}}(r)$ (dots), $g_{0,t}^{\mathrm{STIT}}(r)$ (triangles, simulated), and $g_{0,t}^{\mathrm{PVT}}(r)$ (squares) for $t = 1$.

Using the notation ~ for asymptotic equivalence as introduced in Chapter 1, the asymptotic variances for PHT and PVT in the isotropic case for a window W, as $a \to \infty$, are

$$\begin{aligned} \mathrm{var}\, \Sigma_{0,t,aW}^{\mathrm{PHT}} &\sim \frac{\pi^2}{1536} t^5 I_3(W)\, a^5 , \\ \mathrm{var}\, \Sigma_{0,t,aW}^{\mathrm{PVT}} &\sim 1.9517\, t^3 V_3(W)\, a^3 . \end{aligned}$$

Here $I_3(W)$ is the chord power integral of W of order 3, defined in (12.26). The first result can be found in [165]. Furthermore, a formula for the asymptotic variance of a stationary and isotropic PHT in each dimension is given there, which can be derived from the results shown in [122]. The second one was proven in [72] and [73]. For STIT tessellations, the asymptotic variance is not yet known.

1-skeleton

We now consider the pair correlation function $g_{1,t}$ of the random length measure concentrated on the 1-skeleton and the variance $\operatorname{var}\Sigma_{1,t,W}$ of the total edge length inside a window W.

For a stationary and isotropic PHT in $\mathbb{R}^3$ with surface density $t > 0$, the pair correlation function $g_{1,t}^{\mathrm{PHT}}$ is given by

$$g_{1,t}^{\mathrm{PHT}}(r) = 1 + \frac{2}{t\,r} + \frac{4}{t^2\,\pi^2\,r^2}\,, \qquad r > 0\,.$$

This again follows from the much more general case in (12.68), here with $K = 0$, $i = j = 1$, and $d = 3$.

With this pair correlation function $g_{1,t}^{\mathrm{PHT}}$ and (17.52), a formula for the variance $\operatorname{var}\Sigma_{1,t,W}^{\mathrm{PHT}}$ can be given. For a window W, as $a \to \infty$, this results in

$$\operatorname{var}\Sigma_{1,t,aW}^{\mathrm{PHT}} \sim \frac{\pi^2}{96}\,t^3\,I_3(W)\,a^5$$

with the chord power integral $I_3(W)$ of W of order 3, defined in (12.26); see also [165]. The second-order characteristics for the 1-skeleton are known neither for STIT tessellations nor for PVTs. They have to be obtained by simulation. In Figure 17.12, plots of the pair correlation functions for $t = 1$ are given. As in the planar case, a STIT tessellation shows an intermediate behavior between PHT and PVT.

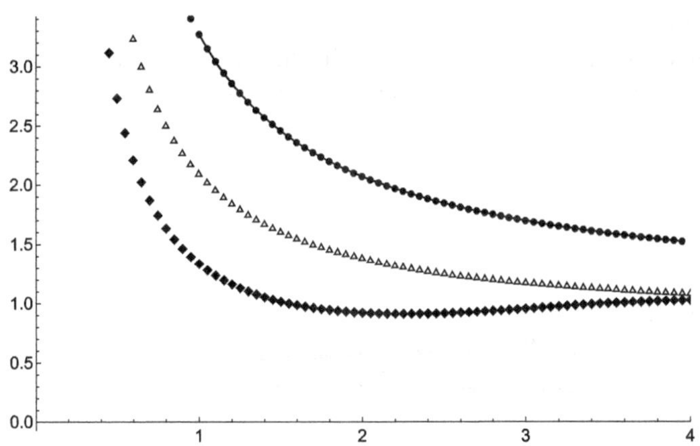

Figure 17.12: Plots for the pair correlation functions $g_{1,t}^{\mathrm{PHT}}(r)$ (dots), $g_{1,t}^{\mathrm{STIT}}(r)$ (triangles, simulated), and $g_{1,t}^{\mathrm{PVT}}(r)$ (squares, simulated) for $t = 1$.

2-skeleton

Finally, we compare for the three models the pair correlation functions $g_{2,t}$ of the random area measure concentrated on the 2-skeleton and the asymptotic variances $\operatorname{var}\Sigma_{2,t,W}$ of the total surface area inside a window W. The pair correlation functions for $r, t > 0$ are

$$g_{2,t}^{\mathrm{PHT}}(r) = 1 + \frac{1}{2\,t\,r}\,,$$
$$g_{2,t}^{\mathrm{STIT}}(r) = 1 + \frac{1}{t^2 r^2}\left(1 - \mathrm{e}^{-\frac{1}{2}tr}\right).$$

The relation for PHT follows from (12.45) or from (10.22) in [82]. The pair correlation function for STIT tessellations is given in Theorem 12.26.

Again, with the pair correlation function $g_{2,t}$ and (17.52), a formula for the variance $\operatorname{var}\Sigma_{2,t,W}$ can be given for PHT and STIT tessellations. This yields the asymptotic variances

$$\operatorname{var}\Sigma_{2,t,aW}^{\mathrm{PHT}} \sim \frac{1}{6}\,t\,I_3(W)\,a^5\,,$$
$$\operatorname{var}\Sigma_{2,t,aW}^{\mathrm{STIT}} \sim I_2(W)\,a^4$$

for a window W as $a \to \infty$, with the chord power integral $I_k(W)$ of W of order $k \in \{2,3\}$, defined in (12.26). Note that the result for STIT tessellations, which follows from (12.28) in Theorem 12.13 for $d = 3$ with $\Sigma_{V_2,t,aW} = \Sigma_{2,t,aW}^{\mathrm{STIT}}$, is independent from the surface density $\eta_2^{\mathrm{STIT}} = t$. Again, it appears that no second-order characteristics are known for PVT. They are only accessible by simulation.

Plots of the pair correlation functions for $t = 1$ are shown in Figure 17.13. Again, the STIT tessellation shows an intermediate behavior between PHT and PVT.

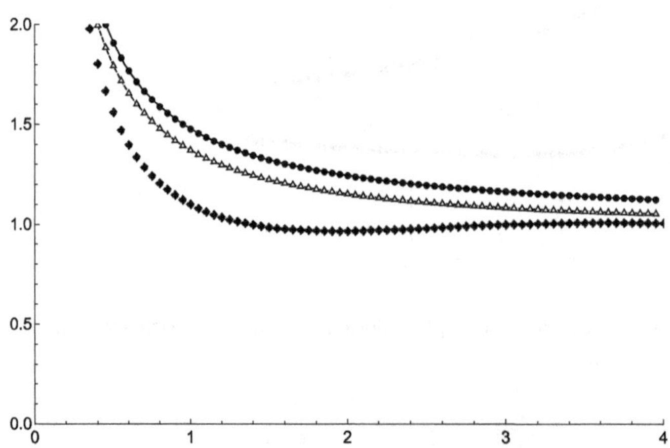

Figure 17.13: Plots for the pair correlation functions $g_{2,t}^{\mathrm{PHT}}(r)$ (dots), $g_{2,t}^{\mathrm{STIT}}(r)$ (triangles), and $g_{2,t}^{\mathrm{PVT}}(r)$ (squares, simulated) for $t = 1$.

17.6 Notes

1. Mecke [125] showed for spatial tessellations that a large variety of mean values can be calculated from seven suitably selected mean values. The proofs for these results can be found in [196]. They are given there for tessellations with general cell shapes; convexity is not required. The parameters are calculated for a spatial STIT tessellation in Lemma 17.1, and these mean value relations are used, for example, in Corollary 17.2. For tessellations that are not face-to-face, further mean values are analyzed in [193]. The book [33] is cited here as a reference for all these mean value relationships.
2. Mean values for spatial STIT tessellations were analyzed for the first time in [147]. Lemma 17.1 and Corollary 17.2 can be found in this paper.
3. The finer classification of the vertices, segments, and polygons in a spatial STIT tessellation from Section 17.2 goes back to [187]. We have already used the notion of I-segments in Section 16.2.2 for planar tessellations. An I-segment can be defined as a maximal union of collinear and connected line segments that appear in the 1-skeleton of a tessellation. Obviously, this definition does not depend on the dimension d of the ambient space and will henceforth also be used for the spatial case $d = 3$. Generalizing this concept, an I-polygon is a maximal union of coplanar and convex planar polygons from the 2-skeleton of the tessellation, which itself is convex. Here "maximal" is understood in the sense that there is no extension to a larger coplanar and convex set in the 2-skeleton.
   The results for the mean values given in Lemmas 17.5, 17.6, and 17.7 are contained in [187]. Furthermore, in this paper, several extremal problems for these mean values are solved using the associated zonoid. By applying classical inequalities for intrinsic volumes of convex bodies from convex geometry it can be deduced that some of the mean values achieve their minimal values if and only if the associated zonoid is a ball. This is the case when the STIT tessellation is additionally isotropic. For the maximal polytopes, which in the spatial case are the I-segments and the I-polygons, these results agree with Corollary 11.16.
4. In Remark 17.4, it is shown that the typical multiple k-face of a STIT tessellation Y_t in $\mathbb{R}^d$ has the same distribution as the typical k-face of the PHT X_t if both tessellations have intensity measure $t\,\Theta$, $k \in \{0, \dots, d\}$. In Theorem 11.8 a relation for the distribution of the typical maximal k-polygon of the STIT tessellation Y_t and the distribution of typical k-face of the PHT X_t is given. It is still an open question whether there are similar relations for the typical k-face of a STIT tessellation (without multiplicities) and the typical k-face of a PHT.
5. Sections 17.3.1 and 17.3.2 follow [189]. The main result in this paper is an explicit formula for the distribution of the number of vertices in the relative interior of the typical I-segment, which is given in Theorem 17.13. For these investigations, the term "carrying I-polygon" of an I-segment was introduced there for the first time. The paper contains further results for the length-weighted typical I-segment.
   Note that it would also be interesting to calculate the distribution of the number of vertices in the relative interior of the typical I-polygon. However, this quantity is currently not accessible, because to this end, the area distribution of a typical Poisson polygon is needed. Determining this distribution is a long-standing open problem in stochastic geometry.
6. In Theorem 17.15 the probability $\mathbb{P}(\nu_{\mathrm{I}}^{\mathrm{T}} = m, \nu_{\mathrm{I}}^{\mathrm{X}} = k)$ that the typical I-segment has exactly m vertices of type T and k vertices of type X in its relative interior is calculated for $m, k \in \mathbb{N}_0$. In addition, the probability $\mathbb{P}(\nu_{\mathrm{I}}^{\mathrm{T}+} = m, \nu_{\mathrm{I}}^{\mathrm{T}-} = k)$ that the typical I-segment contains exactly m T-vertices with label "+" and k with label "–" is given in [188] by

$$\mathbb{P}\big(\nu_{\mathrm{I}}^{\mathrm{T}+} = m, \nu_{\mathrm{I}}^{\mathrm{T}-} = k\big) = 3\binom{m+k}{m} \int_0^1 (1-w)^3 \frac{w^{m+k}}{(1+w)^{m+k+1}}\, \mathrm{d}w$$

   for $m, k \in \mathbb{N}_0$.

7. Most of the material in Sections 17.3.3 and 17.4 was provided in [188]. Our presentation also includes a correction to Theorem 3, Propositions 3 and 4 in [188], and an improvement in the calculation of numerical values.
   The refined combinatorial structures for spatial STIT tessellations considered in Section 17.4 are based only on edges and not on a similar classification of plates, which are the two-dimensional primitives. Clearly, a classification by the type of vertices is not very promising, because a plate can have arbitrarily many 0-faces or vertices on its boundary. However, a classification according to the equality relationship for plates with the 2-faces of the cells would be possible. This would lead to a partition of P into the classes $\mathsf{P}\langle J_P, i\rangle$, $i \in \{0,1,2\}$. For any plate $\mathsf{p} \in \mathsf{P}\langle J_P, i\rangle$, there exist i cells with a 2-face equal to this plate p. In other words, p is equal to a J-polygon with multiplicity i. The related probabilities that the typical plate belongs to $\mathsf{P}\langle J_P, i\rangle$ are $p_{\mathsf{P}\langle J_P,i\rangle}$ for $i \in \{0,1,2\}$. To calculate these probabilities, we would have to investigate the geometry of I-polygons (instead of I-segments). From (11.3) it is known that the distribution of the typical I-polygon of Y_t is a mixture of distributions of the typical 2-face of Poisson plane tessellations X_s, $0 < s < t$. By the Mecke equation (19.4) these distributions are related to the distribution of the typical cell in a Poisson line tessellation. Thus, to calculate $p_{\mathsf{P}\langle J_P,i\rangle}$, we would have to know the area distribution of a typical cell in a Poisson line tessellation. However, this is one of the open problems in stochastic geometry; see also Note 5. This way, a calculation of the probabilities $p_{\mathsf{P}\langle J_P,i\rangle}$ is currently out of reach.
8. The classification of edges in Section 17.4 leads to new geometric mean values, for example, for the neighborhood of the typical vertex, which are given in Theorem 17.27. To our best knowledge, for other tessellation models, there do not exist results of this nature dealing with the neighborhood of the typical vertex.
9. The first-order comparison of mean values and densities of spatial STIT tessellations with PVT and PHT in Section 17.5.1 follows [147]. The second-order comparison in Section 17.5.2 uses material from [165]. In cases where explicit formulae are still unknown, the characteristics are studied by simulation. More detail on these simulations can be found there. In addition, this paper contains results for the variances $\operatorname{var}\Sigma_{i,t,W}$, $i \in \{0,1,2\}$, of STIT tessellations, Poisson line tessellations, and Poisson–Voronoi tessellations.
10. In [97] a wide variety of three-dimensional tessellation models, including STIT tessellations, is considered in a simulation study. The problem is addressed which properties of a random stationary tessellation can be captured by geometric shape indices of the typical cell.
11. The paper [151] deals with the so-called column tessellations. The model described there is also based on a special form of random divisions. To define the spatial construction, we start with a stationary planar tessellation X. From each cell of X we form an infinite column perpendicular to the plane in which X lies. To create the spatial tessellation, each column is intersected independently of the neighboring columns and according to a predefined division rule. This is comparable to a one-dimensional fragmentation process described in Section 9.1. Various mean values and densities of the spatial tessellation can be calculated from suitably chosen parameters of the planar tessellation.
12. A crucial feature of a STIT tessellation Y_t is that sections of Y_t with affine subspaces inherit the STIT property; see Lemma 6.10. This is used in [142], where stationary and isotropic spatial STIT tessellations and their planar sections are considered, and stereological relations are given. Various mean values and densities of Y_t can be calculated by the corresponding quantities of the planar section of Y_t.

Part V: **Appendix**

18 Spaces of sets and of tessellations

In this chapter, we first present fundamental facts on the topology of closed subsets in locally compact spaces and then about spaces of tessellations. We also establish measurability properties for maps on tessellations, which play a central role throughout this book.

18.1 The space of closed sets

Let E be a locally compact topological space that by convention satisfies the Hausdorff separation property and has a countable base. Denote by $\mathcal{F}$, $\mathcal{G}$, and $\mathcal{C}$ the sets of all nonempty closed, nonempty open, and nonempty compact subsets of E, respectively.

Most of the propositions in this section are drawn from [172]. However, in that reference, the statements are formulated for the class $\mathcal{F} \cup \{\emptyset\}$, which includes the empty set, and for the classes of open and compact sets. This distinction is not problematic, as the necessary results can be adapted by restricting to nonempty sets. The only significant difference is that although $\mathcal{F} \cup \{\emptyset\}$ with the Fell topology forms a compact space, $\mathcal{F}$ with restricted Fell topology is merely locally compact.

The topology of closed convergence, also known as the Fell topology, on $\mathcal{F}$ is generated by the following system of open sets:

$$\big\{\{F \in \mathcal{F} : F \cap C = \emptyset\} : C \in \mathcal{C} \cup \{\emptyset\}\big\} \cup \big\{\{F \in \mathcal{F} : F \cap G \neq \emptyset\} : G \in \mathcal{G}\big\}. \tag{18.1}$$

This makes $\mathcal{F}$ a locally compact topological space with the Hausdorff separation property and a countable base; see Theorem 12.2.1 and Remark (b) on p. 565 of [172]. According to Lemma 2.1.1 in [172], the canonical Borel σ-algebra $\mathcal{B}(\mathcal{F})$ on $\mathcal{F}$ is generated by the system $\{\{F \in \mathcal{F} : F \cap C = \emptyset\} : C \in \mathcal{C} \cup \{\emptyset\}\}$, that is,

$$\mathcal{B}(\mathcal{F}) = \sigma\big(\{\{F \in \mathcal{F} : F \cap C = \emptyset\} : C \in \mathcal{C} \cup \{\emptyset\}\}\big). \tag{18.2}$$

Furthermore,

$$\mathcal{C} \in \mathcal{B}(\mathcal{F});$$

see Lemma 2.1.2 of [172]. For $\mathcal{A} \in \mathcal{B}(\mathcal{F})$, we denote the restriction of the σ-algebra by $\mathcal{B}(\mathcal{A}) := \mathcal{B}(\mathcal{F}) \llcorner \mathcal{A}$.

In the special case $E = \mathbb{R}^d$, we consider additional classes of sets. The set of all nonempty compact convex subsets of $\mathbb{R}^d$, known as convex bodies, is denoted by $\mathcal{K}$. The convex ring consisting of all finite unions of elements of $\mathcal{K}$ is denoted by $\mathcal{R}$. A polytope in $\mathbb{R}^d$ is the convex hull of a nonempty finite set of points. We denote by $\mathcal{P}_k$ the set of all k-dimensional polytopes in $\mathbb{R}^d$ for $k \in \{0, \dots, d\}$ and write

https://doi.org/10.1515/9783112222409-023

$$\mathcal{P} := \bigcup_{k=0}^{d} \mathcal{P}_k .$$

Furthermore, for $k \in \{0, \dots, d\}$, the Grassmannian of all k-dimensional linear subspaces is denoted by $G(d, k)$, and the set of all k-dimensional affine subspaces in $\mathbb{R}^d$ by $A(d, k)$. Both are equipped with the Borel σ-algebras induced by $\mathcal{B}(\mathcal{F})$ on their respective sets. We also write

$$\mathcal{H} := A(d, d-1)$$

for the set of all hyperplanes in $\mathbb{R}^d$ and

$$\mathcal{H}_o := G(d, d-1)$$

for the set of all hyperplanes passing through the origin o.

We have

$$\{\mathcal{K}, \mathcal{R}, \mathcal{P}, \mathcal{P}_d, \mathcal{P}_k, G(d,k), A(d,k) : k \in \{0, \dots, d-1\}\} \subset \mathcal{B}(\mathcal{F});$$

see Theorem 2.4.2, p. 448, and p. 39 of [172]. Note that

$$\mathcal{P}_d = \mathcal{P} \cap \{F \in \mathcal{F} : \operatorname{int} F \neq \emptyset\} \in \mathcal{B}(\mathcal{F}) .$$

Furthermore, for $k \in \{0, \dots, d-1\}$, we have

$$\mathcal{P}_k = \bigcup_{r \in \mathbb{N}} \{K \in \mathcal{K} : \dim K = k, K \subseteq rB^d\} \cap \mathcal{P} \in \mathcal{B}(\mathcal{F});$$

see the proof of Lemma 10.1.2 of [172].

For $E = \mathbb{R}^d$, the **Hausdorff metric** on $\mathcal{C}$ is defined by

$$\delta^{\text{Hausd}}(C, C') := \min\{\varepsilon \geq 0 : C \subseteq C' + \varepsilon B^d \text{ and } C' \subseteq C + \varepsilon B^d\} .$$

The restrictions $\mathcal{B}(\mathcal{F}) \llcorner \mathcal{C}$ and $\mathcal{B}(\mathcal{F}) \llcorner \mathcal{P}$ coincide with the Borel σ-algebras when $\mathcal{C}$ and $\mathcal{P}$ are equipped with the Hausdorff metric and the corresponding topology; see Theorem 2.4.1 and Theorem 12.3.4 of [172].

For a general locally compact topological space E with countable base and $\mathcal{F}$ equipped with the topology of closed convergence, denote by $\overline{\mathcal{F}}(\mathcal{F})$ the set of all closed subsets of $\mathcal{F}$, including $\emptyset$. The topology of closed convergence is also defined on $\overline{\mathcal{F}}(\mathcal{F})$, making it a compact topological space that satisfies the Hausdorff separation property and has a countable base; see Theorem 12.2.1 of [172]. Furthermore, let $\overline{\mathcal{F}}_{\ell f}(\mathcal{F})$ denote the set of all locally finite subsets of $\mathcal{F}$. This means that $\mathcal{A} \subseteq \mathcal{F}$ belongs to $\overline{\mathcal{F}}_{\ell f}(\mathcal{F})$ if and only if the number

$$|\{A \in \mathcal{A} : A \cap C \neq \emptyset\}|$$

is finite for all compact sets $C \in \mathcal{C}$. Note that

$$\overline{\mathcal{F}}_{ef}(\mathcal{F}) \in \mathcal{B}(\overline{\mathcal{F}}(\mathcal{F}))$$

according to Lemma 3.1.4 of [172].

Lemma 18.1. *The topological spaces $\mathcal{F}$ and $\overline{\mathcal{F}}(\mathcal{F})$ are Polish spaces, which means that they are metrizable so that they become separable complete metric spaces. This implies that they are also Borel spaces.*

Proof. By Satz 10.15 of [162] a locally compact topological space with countable base and thus, in particular, $\mathcal{F}$ and $\overline{\mathcal{F}}(\mathcal{F})$ are metrizable and σ-compact. Hence Satz 13.17 ibidem can be applied, which says that every locally compact, metrizable, and σ-compact space is a Polish space. Finally, by Theorem 1.8 of [89] every Polish space is a Borel space. □

18.2 Spaces of tessellations

In this section, we consider the set $\mathcal{T}$ of all tessellations in $\mathbb{R}^d$, $d \geq 1$, as introduced in Definition 2.1. For $W \in \mathcal{P}_d$, the set $\mathcal{T}_W$ of tessellations in W is defined in Definition 2.10.

Lemma 18.2. *We have that $\mathcal{T} \in \mathcal{B}(\overline{\mathcal{F}}_{ef}(\mathcal{F}))$ and $\mathcal{T}_W \in \mathcal{B}(\overline{\mathcal{F}}_{ef}(\mathcal{F}))$ for $W \in \mathcal{P}_d$. In particular, $\mathcal{T}$ and $\mathcal{T}_W$ are Borel spaces.*

Proof. By definition, $\mathcal{T} \subseteq \overline{\mathcal{F}}_{ef}(\mathcal{F})$; see Definition 2.1(c). Furthermore, $\mathcal{T} \in \mathcal{B}(\overline{\mathcal{F}}(\mathcal{F}))$ by Lemma 10.1.2 of [172], and hence $\mathcal{T} \in \mathcal{B}(\overline{\mathcal{F}}_{ef}(\mathcal{F}))$. As a measurable subset of the Polish space $\overline{\mathcal{F}}(\mathcal{F})$, the set $\mathcal{T}$ is a Borel space; see Theorem 1.8 of [89].

The proof of Lemma 10.1.2 of [172] can analogously be applied to $\mathcal{T}_W$. This can be seen from the observation that

$$\mathcal{T}_W = \left\{T \in \overline{\mathcal{F}}_{ef}(\mathcal{F}) \cap \overline{\mathcal{F}}(\mathcal{P}_d) : \bigcup_{P \in T} = W\right\}$$
$$\cap \left\{T \in \overline{\mathcal{F}}_{ef}(\mathcal{F}) \cap \overline{\mathcal{F}}(\mathcal{P}_d) : \sum_{P \in T} V_d(P) = V_d(W)\right\}.$$

This yields that $\mathcal{T}_W$ is a Borel subset of $\overline{\mathcal{F}}(\mathcal{F})$. Hence also $\mathcal{T}_W \in \mathcal{B}(\overline{\mathcal{F}}_{ef}(\mathcal{F}))$. Again, as a measurable subset of the Polish space, $\mathcal{T}_W$ is a Borel space by Theorem 1.8 of [89]. □

It follows that

$$(\mathcal{T}, \mathcal{B}(\mathcal{T})) = (\mathcal{T}, \mathcal{B}(\overline{\mathcal{F}}_{ef}(\mathcal{F})) \llcorner \mathcal{T}) \tag{18.3}$$

is the standard measurable space of tessellations in $\mathbb{R}^d$. A generating system of the σ-algebra is provided in the following lemma. A bijection between two measurable spaces is called **bimeasurable** if both the map and its inverse are measurable.

Lemma 18.3. $\mathcal{B}(\mathcal{T}) = \sigma(\{\{T \in \mathcal{T} : |T \cap A| = k\} : A \in \mathcal{B}(\mathcal{P}_d), k \in \mathbb{N}_0\})$.

Proof. The mapping $i_\varsigma : \sum_{P \in T} \delta_P \mapsto T$, which maps the sum of Dirac measures on $\mathcal{P}_d$ to its support, is a bimeasurable bijection; see Lemma 19.1. Furthermore, $\mathcal{P}_d \in \mathcal{B}(\mathcal{F})$, and thus Lemma 19.1, together with (19.1), yields the statement. □

We now address measurability issues concerning the skeletons of tessellations and the sets of faces of cells.

Lemma 18.4. *For $W \in \mathcal{P}_d$, $\partial\mathcal{T}_W := \{\partial T : T \in \mathcal{T}_W\} \in \mathcal{B}(\mathcal{F})$, and $\partial\mathcal{T}_W$ is a Borel space. Furthermore, $\partial\mathcal{T} := \{\partial T : T \in \mathcal{T}\} \in \mathcal{B}(\mathcal{F})$, and $\partial\mathcal{T}$ is a Borel space.*

Proof. For $W \in \mathcal{P}_d$, define

$$\partial\mathcal{T}_W(k,n) := \left\{\partial T : T \in \mathcal{T}_W, |T| = k, \text{inradius}(P) \geq \frac{1}{n} \text{ for all } P \in T\right\}, \quad k, n \in \mathbb{N}.$$

Let $(\partial T_j, j \in \mathbb{N})$ be a sequence of elements $\partial T_j \in \partial\mathcal{T}_W(k,n)$ that has a limit $F \in \mathcal{F}$. Because all these sets are compact, F is the limit with respect to the Hausdorff metric, that is, for all $\varepsilon > 0$, there exists $j(\varepsilon) \in \mathbb{N}$ such that $\partial T_j \subseteq F + \varepsilon B^d$ and $F \subseteq \partial T_j + \varepsilon B^d$ for all $j \geq j(\varepsilon)$. Since all ∂T_j arise from tessellations of W that have exactly k cells, each with an inradius of at least $\frac{1}{n}$, it follows that the limit F also lies in $\partial\mathcal{T}_W(k,n)$. Hence $\partial\mathcal{T}_W(k,n)$ is a closed subset of $\mathcal{F}$. This yields $\partial\mathcal{T}_W = \bigcup_{k\in\mathbb{N}} \bigcup_{n\in\mathbb{N}} \partial\mathcal{T}_W(k,n) \in \mathcal{B}(\mathcal{F})$. It follows that $\partial\mathcal{T}_W$, as a measurable subset of a Polish space, is a Borel space by Theorem 1.8 of [89].

Next, we put $W_i := [-i, i]^d$, $i \in \mathbb{N}$, and for ℓ, k, $n \in \mathbb{N}$ with $\ell > i$,

$$\begin{aligned}\partial\mathcal{T}(i,\ell,k,n) := \{F \in \mathcal{F} : \exists T \in \mathcal{T} : (F \cap W_\ell) \cup \partial W_\ell = \partial(T \wedge W_\ell),\\ \partial(T \wedge W_\ell) \in \partial\mathcal{T}_{W_\ell}(k,n), \forall P \in T : P \cap \operatorname{int} W_i \neq \emptyset \Rightarrow P \subseteq W_\ell\}.\end{aligned}$$

Let $(F_j, j \in \mathbb{N})$ be a sequence of elements of $\partial\mathcal{T}(i,\ell,k,n)$ that has a limit $F \in \mathcal{F}$. Because $F_j \to F$ if and only if $F_j \cap W_\ell \to F \cap W_\ell$ and $F_j \cap \text{cl}(W_\ell^c) \to F \cap \text{cl}(W_\ell^c)$, by analogous arguments as for $\partial\mathcal{T}_W(k,n)$ above we can show that the limit F satisfies $F \in \partial\mathcal{T}(i,\ell,k,n)$, and hence $\partial\mathcal{T}(i,\ell,k,n)$ is a closed subset of $\mathcal{F}$. This yields $\partial\mathcal{T} = \bigcap_{i\in\mathbb{N}} \bigcup_{\ell>i} \bigcup_{k\in\mathbb{N}} \bigcup_{n\in\mathbb{N}} \partial\mathcal{T}(i,\ell,k,n) \in \mathcal{B}(\mathcal{F})$. Hence $\partial\mathcal{T}$ is a Borel space with the same argument as before. □

In (2.1) and (2.4) the sets $\mathcal{F}_k(T)$ of k-faces and $\mathcal{F}_k^{\text{pr}}(T)$ of k-primitives of a tessellation T are defined. Here we establish the measurability of the corresponding maps.

Lemma 18.5. *For $k \in \{0, \dots, d-1\}$, the following maps are measurable:*

(a) $\mathcal{F}_k(\cdot) : \mathcal{T} \to \boldsymbol{\mathcal{F}}_{\ell f}(\mathcal{F})$,

(b) $\mathcal{F}_k^{\text{pr}}(\cdot) : \mathcal{T} \to \boldsymbol{\mathcal{F}}_{\ell f}(\mathcal{F})$.

For $k = d - 1$, both maps are injective.

Proof. (a) Measurability: The proof of the corresponding part of Lemma 10.1.2 of [172] can be applied also to tessellations that are not necessarily face-to-face. At the same time, we use that $\overline{\mathcal{F}}_{ef}(\mathcal{F}) \in \mathcal{B}(\overline{\mathcal{F}}(\mathcal{F}))$.

For $k = d-1$, the injectivity of the map follows from (2.2), (2.1), and the observation that $\mathcal{F}_{d-1}^{-1}(D) = \{P \in \mathcal{P}_d : \operatorname{int} P \text{ is a connected component of } (\bigcup_{F\in D} F)^c\}$ if $D = \mathcal{F}_{d-1}(T)$ for some $T \in \mathcal{T}$.

(b) Measurability: For $T \in \mathcal{T}$ and $n \in \mathbb{N}$, denote $A(T,n) := \{\bigcap_{P\in B} P : B \subset T, |B| = n\}$. The mapping $T \mapsto \sum_{F\in A(T,n)} \delta_F = \frac{1}{n!}\sum^{\neq}_{P_1\ldots P_n\in T^n} \delta_{P_1\cap\cdots\cap P_n}$, where finite intersections of pairwise different polytopes are applied, is measurable by Theorem 12.2.6(a) of [172] and the measurability statement in Theorem 3.1.3 of [172]. Note that $A(T,n) \in \overline{\mathcal{F}}_{ef}(\mathcal{F})$. Hence, with the notation $A_k(T)$ introduced in Definition 2.2, also $T \mapsto \sum_{F\in A_k(T)} \delta_F = \sum_{n\in\mathbb{N}} \sum_{F\in A(T,n)} \mathbb{1}\{F \in \mathcal{P}_k\}\, \delta_F$ is measurable by Theorem 12.2.7 of [172].

By Definition 2.2, $F \in A_k(T)$ is an element of $\mathcal{F}_k^{\mathrm{pr}}(T)$ if and only if for all $F' \in A_k(T)$, either $F \cap F' \notin A_k(T)$ or $F \cap F' = F$. Thus the sum of Dirac measures with the support set $\mathcal{F}_k^{\mathrm{pr}}(T)$ is described by

$$\sum_{F\in\mathcal{F}_k^{\mathrm{pr}}(T)} \delta_F = \sum_{F\in A_k(T)} \prod_{F'\in A_k(T)} \left(\mathbb{1}\{F\cap F' \notin A_k(T)\} + \mathbb{1}\{F\cap F' \in \{F\}\}\right)\delta_F .$$

The injectivity for $k = d-1$ follows by (2.17) and the fact that $(\mathcal{F}_{d-1}^{\mathrm{pr}})^{-1}(D) = \{P \in \mathcal{P}_d : \operatorname{int} P \text{ is a connected component of } (\bigcup_{F\in D} F)^c\}$ if $D = \mathcal{F}_{d-1}^{\mathrm{pr}}(T)$ for some $T \in \mathcal{T}$. □

The k-skeleton $\operatorname{skel}_k T$ and the skeleton of k-primitives $\operatorname{skel}_k^{\mathrm{pr}} T$ of a tessellation T are introduced in (2.2) and (2.15) as the union of all k-faces of cells and the union of all k-primitives of T, respectively.

Lemma 18.6. *For $k \in \{0,\ldots d-1\}$, the maps $\operatorname{skel}_k : \mathcal{T} \to \mathcal{F}$ and $\operatorname{skel}_k^{\mathrm{pr}} : \mathcal{T} \to \mathcal{F}$ are measurable.*

Proof. The statement can be shown using Lemma 18.5 and then applying analogous arguments as in the proofs of Lemmas 18.4 and 18.5. To start with, we restrict ourselves to a bounded window and to finite unions of k-faces of cells or k-primitives of T, respectively. Then extend this to whole $\mathbb{R}^d$ and countable unions of the respective polytopes. □

An essential tool in proofs involving STIT tessellations is the capacity functional, defined by (2.23), of the $(d-1)$-skeleton. We note that by Lemma 18.4, with the notation $\partial T := \operatorname{skel}_{d-1} T$,

$$\partial\mathcal{T} := \{\partial T : T \in \mathcal{T}\} \in \mathcal{B}(\mathcal{F}),$$

which implies that $\partial\mathcal{T}$ is a Borel space.

Lemma 18.7. *The map $\partial : \mathcal{T} \to \partial\mathcal{T}$ is a bimeasurable bijection.*

Proof. For $F \in \partial\mathcal{T}$, the corresponding tessellation

$$\partial^{-1}(F) = \{P \subset \mathcal{P}_d : P \text{ is the closure of a connected component of } F^c\}$$

is uniquely determined, and hence the map ∂ is injective. Furthermore, the map ∂ is measurable by Lemma 18.6. Because $\mathcal{T}$ and $\partial\mathcal{T}$ are Borel spaces according to Lemmas 18.2 and 18.4, the measurability and injectivity of ∂ yield the measurability of ∂^{-1}; see Theorem A1.1 of [89]. □

Now the σ-algebra $\mathcal{B}(\mathcal{T})$ given in (18.3) and Lemma 18.3 can also be generated in the following way.

Corollary 18.8. *We have*

$$\mathcal{B}(\mathcal{T}) = \sigma(\{\{T \in \mathcal{T} : \partial T \cap C = \emptyset\} : C \in \mathcal{C}\}). \tag{18.4}$$

Proof. By (18.2) and Lemma 18.4 the Borel σ-algebra on $\partial\mathcal{T}$ is given by $\mathcal{B}(\partial\mathcal{T}) = \sigma(\{\{\partial T \in \partial\mathcal{T} : \partial T \cap C = \emptyset\} : C \in \mathcal{C}\})$. Hence Lemma 18.7 yields that $\sigma(\{\{T \in \mathcal{T} : \partial T \cap C = \emptyset\} : C \in \mathcal{C}\}) \subseteq \mathcal{B}(\mathcal{T})$. On the other hand, to show equality, note that for all $A \in \mathcal{B}(\mathcal{T})$, by bimeasurability we have $\partial A \in \mathcal{B}(\partial\mathcal{T})$. Thus

$$\begin{aligned} A \in \partial^{-1}\mathcal{B}(\partial\mathcal{T}) &= \partial^{-1}\sigma(\{\{\partial T \in \partial\mathcal{T} : \partial T \cap C = \emptyset\} : C \in \mathcal{C}\}) \\ &= \sigma(\partial^{-1}\{\{\partial T \in \partial\mathcal{T} : \partial T \cap C = \emptyset\} : C \in \mathcal{C}\}) \\ &= \sigma(\{\{T \in \mathcal{T} : \partial T \cap C = \emptyset\} : C \in \mathcal{C}\}). \end{aligned}$$

Here we used that for a generating system $\mathcal{E}$ of $\mathcal{B}(\partial\mathcal{T})$ and an injection $f : \mathcal{T} \to \partial\mathcal{T}$, by the definition of the generated σ-algebra, $f^{-1}(\sigma(\mathcal{E})) = \sigma(f^{-1}(\mathcal{E}))$. □

As the generating system of $\mathcal{B}(\mathcal{T})$ given in (18.4) is $\cap$-stable, the preceding lemma shows that the distribution of a random tessellation is determined by the capacity functional for its $(d-1)$-skeleton.

Using Lemma 18.2, the proofs of Lemmas 18.5, 18.6, and 18.7 can be analogously applied to $\mathcal{T}_W$. Recall the notation $\partial(\mathcal{T}_W) := \{\partial T : T \in \mathcal{T}_W\}$ introduced in Lemma 18.4. This yields the following statement.

Lemma 18.9. *For $W \in \mathcal{P}_d$ and $k \in \{0, \dots d-1\}$, the maps $\mathrm{skel}_k : \mathcal{T}_W \to \mathcal{F}$ and $\mathrm{skel}_k^{\mathrm{pr}} : \mathcal{T}_W \to \mathcal{F}$ are measurable. The map $\partial : \mathcal{T}_W \to \partial(\mathcal{T}_W)$ is a bimeasurable bijection.*

The next propositions establish the measurability of the restriction of a tessellation to a window $W \in \mathcal{P}_d$ and of its section with an affine subspace A of $\mathbb{R}^d$.

Lemma 18.10. *For any fixed $W \in \mathcal{P}_d$, the mapping $T \mapsto T \wedge W$ from $\mathcal{T}$ to $\mathcal{T}_W$, introduced in Definition 2.11, is measurable.*

Proof. For fixed $W \in \mathcal{P}_d$, consider the composition of measurable mappings

$$T \mapsto \partial T \mapsto \partial T \cap W \mapsto (\partial T \cap W) \cup \partial W \mapsto T \wedge W\,.$$

The statement follows now from Lemma 18.7, Theorems 12.2.6(a) and 12.2.3 of [172], and Lemma 18.9. □

For $k \in \{1, \dots d-1\}$ and $A \in A(d,k)$, the set $\mathcal{T}_A$ of tessellations in A is introduced in Definition 2.12.

Lemma 18.11. *Let $k \in \{1, \dots d-1\}$ and $A \in A(d,k)$. Then $\mathcal{T}_A \in \mathcal{B}(\overline{\mathcal{F}}_{\ell f}(\mathcal{F}))$, and $\mathcal{T}_A$ is a Borel space.*

Proof. By definition, $\mathcal{T}_A \subseteq \overline{\mathcal{F}}_{\ell f}(\mathcal{F})$. Denote by $\overline{\mathcal{F}}(\mathcal{K})$ the set of all closed subsets of $\mathcal{K}$, including $\emptyset$. Furthermore, because $\{A\} \in \mathcal{B}(\mathcal{F})$, the proof of Lemma 10.1.2 of [172] can be adapted to $\mathcal{T}_A$ using the representation

$$\begin{aligned}\mathcal{T}_A = &\left\{T \in \overline{\mathcal{F}}_{\ell f}(\mathcal{F}) \cap \overline{\mathcal{F}}(\mathcal{K}) : \bigcup_{K \in T} K = A\right\} \\ &\cap \bigcap_{r \in \mathbb{N}} \left\{T \in \overline{\mathcal{F}}_{\ell f}(\mathcal{F}) \cap \overline{\mathcal{F}}(\mathcal{K}) : T \cap \mathcal{K}_r^{k-1} = \emptyset,\ \sum_{K \in T} \lambda_k(K \cap rB^d) = \lambda_k(A \cap rB^d)\right\}\end{aligned}$$

with $\mathcal{K}_r^{k-1} := \{K \in \mathcal{K} : \dim K \le k-1,\ K \subset rB^d\}$. This yields that $\mathcal{T}_A \in \mathcal{B}(\overline{\mathcal{F}}(\mathcal{F}))$. Hence also $\mathcal{T}_A \in \mathcal{B}(\overline{\mathcal{F}}_{\ell f}(\mathcal{F}))$. Thus $\mathcal{T}_A$, as a measurable subset of a Polish space, is a Borel space by Theorem 1.8 of [89]. □

The set of tessellations $\mathcal{T}(A)$ is defined in (2.21).

Lemma 18.12. *Fix $k \in \{1, \dots, d-1\}$ and $A \in A(d,k)$. Then $\mathcal{T}(A) \in \mathcal{B}(\mathcal{T})$, and the map $\mathcal{T}(A) \to \mathcal{T}_A$ with $T \mapsto T \cap A$ is Borel measurable.*

Proof. For fixed $A \in A(d,k)$, the mapping $P \mapsto P \cap A$, $P \in \mathcal{P}_d$, is measurable by Theorem 12.2.6(a) of [172]. Hence also $T \mapsto \sum_{P \in T} \delta_{P \cap A}$, $T \in \mathcal{T}$, is measurable; see the measurability statement in Theorem 3.1.3 of [172]. Thus an application of Lemma 3.1.4 of [172] yields that $T \mapsto \{P \cap A : P \in T\}$ is a $\mathcal{B}(\mathcal{T}) - \overline{\mathcal{F}}_{\ell f}(\mathcal{F})$-measurable mapping. Adapting an idea in the proof of Lemma 10.1.2 of [172], we use the set $\mathcal{K}_r^{k-1}$ introduced in the proof of Lemma 18.11. For all $r \in \mathbb{N}$, this set is closed in $\mathcal{F}$. Then

$$\mathcal{T}(A) = \bigcap_{r \in \mathbb{N}} \{T \in \mathcal{T} : \{P \cap A : P \in T\} \cap \mathcal{K}_r^{k-1} = \emptyset\}\,,$$

which shows that $\mathcal{T}(A) \in \mathcal{B}(\mathcal{T})$. Hence also the restriction of the mapping $T \mapsto \{P \cap A : P \in T\}$ to $\mathcal{T}(A)$ is Borel measurable. □

For tessellations generated by the division of cells, we will use the following statement.

Lemma 18.13. *The map* $\oslash : \{(P,H,T) \in \mathcal{P}_d \times \mathcal{H} \times \mathcal{T} : P \in T, H \in [\mathrm{int}\, P]\} \to \mathcal{T}$*, defined in* (4.6)*, is measurable.*

Proof. We use Theorem 12.2.7 of [172], which implies that the map $E \times (\mathcal{F} \cup \{\emptyset\}) \to \mathbb{R}$ with $(x,F) \mapsto \mathbb{1}\{x \in F\}$ is measurable if E is a locally compact topological space with countable base. Choosing $E = \mathcal{F}$ and observing that $\mathcal{T} \in \mathcal{B}(\overline{\mathcal{F}}(\mathcal{F}))$ and $\mathcal{P}_d \in \mathcal{B}(\mathcal{F})$ yield that the set $\{(P,T) \in \mathcal{P}_d \times \mathcal{T} : P \in T\}$ is measurable, and hence also $\{(P,H,T) \in \mathcal{P}_d \times \mathcal{H} \times \mathcal{T} : P \in T\}$ is measurable.

Furthermore, choosing $E = \mathcal{F}$ again and noting that for all $P \in \mathcal{P}_d$, we have $\{F \in \mathcal{F} : F \cap \mathrm{int}\, P \neq \emptyset\} \in \mathcal{B}(\overline{\mathcal{F}}(\mathcal{F}))$ and $\mathcal{P}_d \in \mathcal{B}(\mathcal{F})$, and $\mathcal{H} \in \mathcal{B}(\mathcal{F})$ yields that the set $\{(P,H) \in \mathcal{P}_d \times \mathcal{H} : H \in [\mathrm{int}\, P]\}$ is measurable, and hence also $\{(P,H,T) \in \mathcal{P}_d \times \mathcal{H} \times \mathcal{T} : H \in [\mathrm{int}\, P]\}$ is measurable. As a consequence, the set $\{(P,H,T) \in \mathcal{P}_d \times \mathcal{H} \times \mathcal{T} : P \in T, H \in [\mathrm{int}\, P]\}$ is measurable.

To show the measurability of the map $\oslash$, first note that on $\mathcal{H}$ the mappings $H \mapsto H^+$ and $H \mapsto H^-$ are continuous, and hence by Theorem 12.2.6(a) of [172] the mapping $(P,H) \mapsto (P \cap H^+, P \cap H^-)$ is measurable. Now relying on Lemma 19.1, we can switch to the counting measures corresponding to these sets. According to (2.37), we write $\delta(T,\cdot) = \sum_{P \in T} \delta_P(\cdot)$. It remains to observe that on the set

$$\{(\delta_P, \delta_H, \delta(T,\cdot)) \in \mathsf{N}_s(\mathcal{P}_d) \times \mathsf{N}_s(\mathcal{H}) \times \mathsf{N}_s(\mathcal{P}_d) : \delta(T,\{P\}) > 0, H \in [\mathrm{int}\, P]\}$$

the mapping

$$(\delta_P, \delta_H, \delta(T,\cdot)) \mapsto \delta(T,\cdot) - \delta_P + \delta_{P\cap H^+} + \delta_{P \cap H^-}$$

is measurable. □

Finally, it is shown that a completed probability space on the space of tessellations becomes a Lebesgue probability space. For a probability space $(\Omega, \mathbf{A}, \mathbb{P})$, we denote by $(\Omega, \mathbf{A}_{\mathbb{P}}, \mathbb{P})$ the completed probability space, meaning that $\mathbf{A}_{\mathbb{P}}$ also includes all subsets of sets $A \in \mathbf{A}$ with $\mathbb{P}(A) = 0$.

Definition 18.14. *A* ***Lebesgue probability space*** *(or a Lebesgue–Ro(k)hlin space, or a standard probability space) is a complete probability space isomorphic to a complete probability space* $([0,1], \mathcal{B}([0,1])_{\mathbb{P}'}, \mathbb{P}')$ *with a probability measure* $\mathbb{P}'$ *of the form* $\mathbb{P}' = \alpha \lambda_1 \llcorner [0,1] + (1-\alpha) \sum_{i\in\mathbb{N}} p_i \delta_{x_i}$ *with* $\alpha \in [0,1]$, $p_i \geq 0$, $x_i \in [0,1]$, $i \in \mathbb{N}$, $\sum_{i\in\mathbb{N}} p_i = 1$*, that is,* $\mathbb{P}'$ *is a convex combination of the restricted Lebesgue measure and a pure atomic measure.*

Lemma 18.15. *Let* $\mathbb{P}$ *be a probability measure on* $(\mathcal{T}, \mathcal{B}(\mathcal{T}))$*. Then the completed probability space* $(\mathcal{T}, \mathcal{B}(\mathcal{T})_{\mathbb{P}}, \mathbb{P})$ *is a Lebesgue probability space. Let* $W \in \mathcal{P}_d$*, and let* $\mathbb{P}$ *be a probability measure on* $(\mathcal{T}_W, \mathcal{B}(\mathcal{T}_W))$*. Then the completed probability space* $(\mathcal{T}_W, \mathcal{B}(\mathcal{T}_W)_{\mathbb{P}}, \mathbb{P})$ *is a Lebesgue probability space.*

Proof. If $(\Omega, \mathbf{A})$ is a Polish space endowed with its Borel σ-algebra and $\mathbb{P}$ is a probability measure on it, then the completed probability space $(\Omega, \mathbf{A}_{\mathbb{P}}, \mathbb{P})$ is a Lebesgue probability space by Theorem 2–3 of [48] or Theorem 2.4.1 of [84]. Furthermore, if $(\Omega, \mathbf{A}_{\mathbb{P}}, \mathbb{P})$ is a Lebesgue probability space and $A \in \mathbf{A}$ with $\mathbb{P}(A) > 0$, then also $(A, \mathbf{A}_{\mathbb{P}} \llcorner A, \mathbb{P} \llcorner A)$ is a Lebesgue probability space; see Proposition 5 of [66]. Hence, if $A' \in \mathcal{B}(A)$ is a measurable set of a Polish space and $\mathbb{P}'$ is a probability measure on $(A', \mathcal{B}(A'))$, then the completed probability space $(A', \mathcal{B}(A')_{\mathbb{P}'}, \mathbb{P}')$ is Lebesgue. Thus the corresponding arguments in the proof of Lemma 18.2 yield the statement. □

18.3 Notes

1. The proof of Lemma 18.4 was provided by Daniel Hug in a personal communication.
2. The measurability of the map $\oslash$ in Lemma 18.13 was alternatively shown for the so-called splitting tessellations on spheres in Lemma 2.6 of [83].

19 Facts from point process theory

In this chapter, we collect the fundamental definitions and propositions concerning point processes that are frequently used throughout the book. Our primary reference is Chapter 3 of [172], which deals extensively with marked point processes and related measurability considerations. For Poisson point processes, the principal references are [101] and [94].

Throughout this chapter, let E be a locally compact topological space – including, by convention, the Hausdorff separation property – with a countable base. Denote by $\mathcal{B}(E)$ the Borel σ-algebra on E.

19.1 Point processes

A measure μ on $(E, \mathcal{B}(E))$ is a **locally finite measure** if $\mu(C) < \infty$ for all compact sets $C \in \mathcal{C}$. A measure μ on E is a **counting measure** if μ is locally finite and $\mu(A) \in \mathbb{N}_0 \cup \{\infty\}$ for all $A \in \mathcal{B}(E)$, and it is a **simple counting measure** if it is a counting measure with $\mu(\{x\}) \leq 1$ for all $x \in E$. Let $\mathsf{N}(E)$ and $\mathsf{N}_s(E)$ denote the sets of all counting measures and all simple counting measures on E, respectively. On $\mathsf{N}(E)$ the σ-algebra

$$\mathcal{N}(E) := \sigma(\{\{\mu \in \mathsf{N}(E) : \mu(A) = k\} : A \in \mathcal{B}(E),\ k \in \mathbb{N}_0\}) \tag{19.1}$$

is defined. Correspondingly, the restriction $\mathcal{N}_s(E) := \mathcal{N}(E) \llcorner \mathsf{N}_s(E)$ is the σ-algebra on $\mathsf{N}_s(E)$ which will be used. The **support set** of a counting measure $\mu \in \mathsf{N}(E)$ can be described as

$$\operatorname{supp} \mu := \{x \in E : \mu(\{x\}) > 0\}.$$

A subset $A \subseteq E$ is **locally finite** if $A \cap C$ is finite for all $C \in \mathcal{C}$. Denote by $\overline{\mathcal{F}}_{ef}$ the set of all locally finite subsets of E. Recall definition (18.2) of $\mathcal{B}(\mathcal{F})$, the σ-algebra on the space of nonempty closed subsets of E.

Lemma 19.1. *The set of all locally finite subsets of E is measurable, that is, $\overline{\mathcal{F}}_{ef} \in \mathcal{B}(\mathcal{F})$. Furthermore, the mapping $i_s : \mathsf{N}_s(E) \to \overline{\mathcal{F}}_{ef}$ with $i_s : \mu \mapsto \operatorname{supp} \mu$ is bijective and bimeasurable, which means that i_s and i_s^{-1} are measurable. The trace σ-algebra $\mathcal{B}(\overline{\mathcal{F}}_{ef}) := \mathcal{B}(\mathcal{F}) \llcorner \overline{\mathcal{F}}_{ef}$ satisfies $\mathcal{N}_s(E) = i_s^{-1}(\mathcal{B}(\overline{\mathcal{F}}_{ef}))$ and $\mathcal{B}(\overline{\mathcal{F}}_{ef}) = i_s(\mathcal{N}_s(E))$.*

Proof. This is Lemma 3.1.4 of [172]. □

Lemma 19.2. *For all $\mu \in \mathsf{N}(E)$, the support set $\operatorname{supp} \mu$ is finite or countable, and there exist measurable mappings $\mathsf{m}_i : \mathsf{N}(E) \to E$, $i \in \{0, \ldots, \mu(E) - 1\}$, such that*

$$\mu = \sum_{i=0}^{\mu(E)-1} \delta_{\mathsf{m}_i(\mu)} \tag{19.2}$$

with the convention that $\mu(E) - 1 := \infty$ if $\mu(E) = \infty$.

https://doi.org/10.1515/9783112222409-024

Proof. This is Lemma 3.1.3 of [172]. □

Note that (19.2) can also be applied to counting measures $\mu \in \mathsf{N}(E)$ that are not in $\mathsf{N}_s(E)$, and thus their support set contains points $x \in E$ with multiplicities $\mu(\{x\}) > 1$. Then there is an index set $\{i_1, \dots, i_{\mu(\{x\})}\} \subseteq \{0, \dots, \mu(E) - 1\}$ with $m_{i_j}(\mu) = x$.

A **point process on** E is a random variable on a probability space $(\Omega, \mathbf{A}, \mathbb{P})$ with values in $\mathsf{N}(E)$, and it is a **simple point process on** E if its values are in $\mathsf{N}_s(E)$ a. s.

Remark 19.3. *According to Lemma* 19.1*, we can describe a simple point process alternatively as a random measure or as a random subset of E, and we will switch between these options arbitrarily. Thus for a simple point process* Φ*, we will shortly write* $x \in \Phi$ *if* $x \in \operatorname{supp} \Phi$.

Furthermore, by Lemma 19.2 the elements of the support set of Φ can be numbered. This is applied in Chapter 3 in the definition of the iteration of tessellations.

The **intensity measure** of a point process Φ is the measure Λ on E defined by

$$\Lambda(A) := \mathbb{E}\Phi(A), \quad A \in \mathcal{B}(E).$$

Theorem 19.4 (Campbell theorem). *Let* Φ *be a point process on E with intensity measure* Λ*, and let* $f : E \to \mathbb{R}$ *be a nonnegative measurable function. Then*

$$\mathbb{E} \int_E f(x)\, \Phi(\mathrm{d}x) = \int_E f(x)\, \Lambda(\mathrm{d}x).$$

Proof. This is Theorem 3.1.2 of [172]. □

In the space $E = \mathbb{R}^d$, the translation by $x \in \mathbb{R}^d$ is defined as $z \mapsto z + x$ for all $z \in \mathbb{R}^d$. The image measure of a counting measure μ under such a translation is denoted $\mu + x$ and correspondingly, $\Phi + x$ for a point process Φ. Analogously, the notation $\vartheta\,\Phi$ is used for $\vartheta \in SO_d$, where SO_d is the rotation group on $\mathbb{R}^d$. A point process Φ on $\mathbb{R}^d$ is called **stationary** if $\Phi + x \overset{\mathrm{D}}{=} \Phi$ for all $x \in \mathbb{R}^d$, and it is called **isotropic** if $\vartheta\,\Phi \overset{\mathrm{D}}{=} \Phi$ for all $\vartheta \in SO_d$.

If Φ is a stationary point process on $\mathbb{R}^d$ with intensity measure $\Lambda \neq 0$ and $\Lambda(C) < \infty$ for all $C \in \mathcal{C}$, then there is a number $\gamma \in \mathbb{R}^+$, called the **intensity**, such that $\Lambda = \gamma\,\lambda_d$, that is, a multiple of the Lebesgue measure.

Let Φ be a stationary point process with intensity $\gamma \in \mathbb{R}^+$. Then its **Palm distribution** $\mathbb{P}^o$ is defined by

$$\gamma\,\mathbb{P}^o(A) := \mathbb{E} \int_{\mathbb{R}^d} \mathbb{1}_B(x)\, \mathbb{1}_A(\Phi - x)\, \Phi(\mathrm{d}x)$$

for all Borel sets $B \subset \mathbb{R}^d$ with $\lambda_d(B) = 1$ and $A \in \mathcal{N}(\mathbb{R}^d)$. The Palm distribution can intuitively be understood as a conditional distribution of Φ, given that a “typical point” of Φ is located at the origin $o \in \mathbb{R}^d$.

Theorem 19.5 (Refined Campbell theorem). *Let Φ be a stationary point process on $\mathbb{R}^d$ with intensity $\gamma \in \mathbb{R}^+$. Then for all nonnegative measurable functions $f : \mathbb{R}^d \times \mathsf{N}(\mathbb{R}^d) \to \mathbb{R}$,*

$$\mathbb{E} \int_{\mathbb{R}^d} f(x, \Phi)\, \Phi(\mathrm{d}x) = \gamma \int_{\mathbb{R}^d} \int_{\mathsf{N}(\mathbb{R}^d)} f(x, \mu + x)\, \mathbb{P}^o(\mathrm{d}\mu)\, \lambda_d(\mathrm{d}x)\,.$$

Proof. This is Theorem 3.3.3 of [172]. □

In stochastic geometry, it is often convenient to consider marked point processes, where the marks take values in a locally compact topological space M with countable base.

Definition 19.6. *A **marked point process in** $\mathbb{R}^d$ with **mark space** M is a simple point process Φ on $\mathbb{R}^d \times M$ with intensity measure Λ satisfying*

$$\Lambda(C \times M) < \infty \quad \text{for all } C \in \mathcal{C}\,. \tag{19.3}$$

In this context the translation by $x \in \mathbb{R}^d$ is defined by $(z, m) \mapsto (z + x, m)$ for all $(z, m) \in \mathbb{R}^d \times M$, that is, the marks remain unchanged. Analogously, $\mu + x$ denotes the image measure of a counting measure μ under this translation and correspondingly, $\Phi + x$ for a marked point process to Φ. A marked point process Φ is **stationary** if $\Phi + x \overset{\mathrm{D}}{=} \Phi$ for all $x \in \mathbb{R}^d$.

Theorem 19.7. *If Φ is a stationary marked point process in $\mathbb{R}^d$ with mark space M and intensity measure $\Lambda \neq 0$, then there are a number $\gamma \in \mathbb{R}^+$, called the **intensity** of Φ, and a uniquely determined probability measure $\mathbb{Q}$ on M, called the **mark distribution**, such that*

$$\Lambda = \gamma \lambda_d \otimes \mathbb{Q}\,.$$

Proof. This is Theorem 3.5.1 of [172]. □

Let Φ be a stationary marked point process with intensity $\gamma \in \mathbb{R}^+$ and mark distribution $\mathbb{Q}$. Then its **Palm distribution** $\mathbb{P}^o$ is defined by

$$\gamma\, \mathbb{P}^o(A') := \mathbb{E} \sum_{(x,m) \in \Phi} \mathbb{1}_B(x)\, \mathbb{1}_{A'}(m, \Phi - x)$$

for all Borel sets $B \subset \mathbb{R}^d$ with $\lambda_d(B) = 1$ and $A' \in \mathcal{B}(M) \otimes \mathcal{N}_s(\mathbb{R}^d \times M)$. Furthermore, there exists a family $(\mathbb{P}^{o,m},\ m \in M)$ of probability measures on $\mathsf{N}_s(\mathbb{R}^d \times M)$ such that

$$\mathbb{P}^o(A \times N) = \int_A \mathbb{P}^{o,m}(N)\, \mathbb{Q}(\mathrm{d}m)$$

for all $A \in \mathcal{B}(M)$ and $N \in \mathcal{N}_s(\mathbb{R}^d \times M)$. The family $(\mathbb{P}^{o,m},\ m \in M)$ is uniquely determined up to a set of $\mathbb{Q}$-measure zero. The probability measures $\mathbb{P}^{o,m}$ are also called Palm distributions, and they can intuitively be understood as the distribution of Φ under the condition that a point of Φ with mark m is located at the origin $o \in \mathbb{R}^d$.

Theorem 19.8 (Refined Campbell theorem for marked point processes). *Let Φ be a stationary marked point process in $\mathbb{R}^d$ with mark space M, intensity $\gamma \in \mathbb{R}^+$, and mark distribution $\mathbb{Q}$. Then for all nonnegative measurable functions $f : \mathbb{R}^d \times M \times \mathsf{N}_s(\mathbb{R}^d \times M) \to \mathbb{R}$,*

$$\mathbb{E} \sum_{(x,m)\in\Phi} f(x,m,\Phi) = \gamma \int_{\mathbb{R}^d} \int_M \int_{\mathsf{N}_s(\mathbb{R}^d\times M)} f(x,m,\mu + x)\, \mathbb{P}^{o,m}(\mathrm{d}\mu)\, \mathbb{Q}(\mathrm{d}m)\, \lambda_d(\mathrm{d}x)\,.$$

Proof. This is Theorem 3.5.5 of [172]. □

Definition 19.9. *Let Φ be a marked point process in $\mathbb{R}^d$ with mark space M. It is called **independently marked** if it has a representation*

$$\Phi = \sum_{i=0}^{\Phi(\mathbb{R}^d\times M)-1} \delta_{(x_i,m_i)}$$

according to (19.2)*, where the marks $m_0, m_1, m_2, \ldots$ are i. i. d. and independent of the value $\Phi(\mathbb{R}^d \times M)$ and of the points $x_0, x_1, x_2, \ldots$*

19.2 Poisson point processes

In this section, Λ denotes a nonzero locally finite measure on E. It is assumed that Λ has no atoms, that is, $\Lambda(\{x\}) = 0$ for all $x \in E$.

Definition 19.10. *A **Poisson point process on** E **with intensity measure** Λ is a point process Φ with two properties:*
(a) *For every $A \in \mathcal{B}(E)$ with $0 < \Lambda(A) < \infty$, the random number $\Phi(A)$ of points of Φ in A is Poisson distributed with parameter $\Lambda(A)$.*
(b) *For every $m \in \mathbb{N}$ and all pairwise disjoint sets $A_1, \ldots, A_m \in \mathcal{B}(E)$, the random variables $\Phi(A_1), \ldots, \Phi(A_m)$ are independent.*

By Theorem 3.2.1 of [172], for each atomless locally finite measure Λ on E, there exists a Poisson point process on E with intensity measure Λ, and its distribution is uniquely determined. Furthermore, by Corollary 3.2.2 of [172] condition (a) of Definition 19.10 implies (b).

Moreover, if E is a Borel space and Λ is a σ-finite atomless measure, then conditions (a) and (b) of the definitions determine each other. This means that (a) implies (b), and if (b) holds, then there exists a measure Λ such that (a) holds; for details, see Chapter 6 of [101].

Theorem 19.11 (Multivariate Mecke equation). *Let Φ be a Poisson point process on E with intensity measure Λ. Then for all $n \in \mathbb{N}$ and all nonnegative measurable functions $f : \mathsf{N}(E) \times E^n \to \mathbb{R}$,*

$$\mathbb{E} \sum_{(x_1,\dots,x_n)\in\Phi^n_{\neq}} f(\Phi, x_1, \dots, x_n) = \int_E \dots \int_E \mathbb{E} f\left(\Phi + \sum_{i=1}^{n} \delta_{x_i}, x_1, \dots, x_n\right) \Lambda(\mathrm{d}x_1) \dots \Lambda(\mathrm{d}x_n), \tag{19.4}$$

where $\Phi^n_{\neq}$ denotes the set of all n-tuples of pairwise different points of Φ.

Proof. This is Corollary 3.2.3 of [172]. □

For $n = 1$, (19.4) is referred to as the Mecke equation. This equation can be applied to characterize Poisson point processes in the following manner.

Theorem 19.12 (Mecke–Slivnyak theorem). *Let Φ be a point process on E with atomless intensity measure Λ. Then Φ is a Poisson point process if and only if*

$$\mathbb{E} \sum_{x\in\Phi} f(\Phi, x) = \int_E \mathbb{E} f(\Phi + \delta_x, x)\, \Lambda(\mathrm{d}x) \tag{19.5}$$

for all nonnegative measurable functions $f : \mathsf{N}(E) \times E \to \mathbb{R}$.

Proof. This is Theorem 3.2.5 of [172]. □

A further useful proposition concerns the superposition of independent Poisson point processes.

Theorem 19.13 (Superposition theorem). *Let Φ_i be independent Poisson point processes on E with intensity measures Λ_i, $i \in I \subseteq \mathbb{N}$. Then the* ***superposition*** *$\Phi := \sum_{i\in I} \Phi_i$ is a Poisson point process on E with intensity measure $\Lambda := \sum_{i\in I} \Lambda_i$ if Λ is locally finite.*

Proof. This is Theorem 3.3 of [101]. □

A Poisson point process on a product space can also be regarded as a marked point process. The following theorem states the independence of marking.

Theorem 19.14. *Let M be a locally compact topological space with countable base, the mark space, and let Φ be a stationary Poisson point process on $\mathbb{R}^d \times M$ with intensity measure $\Lambda = \gamma \lambda_d \otimes \mathbb{Q}$ as in Theorem 19.7. Then Φ is an independently marked point process.*

Proof. This is Theorem 3.5.8 of [172]. □

19.3 Notes

1. In Section 19.2, we recalled propositions for Poisson point processes as they are shown in [172]. This is sufficient for the purposes of the present volume. Note that in [101] the respective results are shown in considerably more general settings.

Bibliography

[1] K. Adamczyk-Chauvat, M. Kassa, K. Kiêu, J. Papaïx, H. Monod, and R. S. Stoica. Gibbsian T-tessellation model for agricultural landscape characterization. *arXiv:2007.16094*, pages 1–14, 2020.

[2] K. Adamczyk-Chauvat, M. Kassa, J. Papaïx, K. Kiêu, and R. S. Stoica. Statistical inference for random T-tessellations models. Application to agricultural landscape modeling. *Ann. Inst. Stat. Math.*, 76(3):447–479, 2024.

[3] R. V. Ambartzumian. A note on pseudo-metrics on the plane. *Z. Wahrscheinlichk.theor. Verw. Geb.*, 37(2):145–155, 1976/77.

[4] R. V. Ambartzumian. *Combinatorial Integral Geometry*. Wiley Series in Probability and Mathematical Statistics: Tracts on Probability and Statistics. John Wiley & Sons, Inc., New York, 1982. With applications to mathematical stereology, Edited and with an appendix by Adrian Baddeley.

[5] R. V. Ambartzumian. *Factorization Calculus and Geometric Probability*, volume 33 of *Encyclopedia of Mathematics and its Applications*. Cambridge University Press, Cambridge, 1990.

[6] T. Arak, P. Clifford, and D. Surgailis. Point-based polygonal models for random graphs. *Adv. Appl. Probab.*, 25(2):348–372, 1993.

[7] T. Arak and D. Surgailis. Markov fields with polygonal realizations. *Probab. Theory Relat. Fields*, 80(4):543–579, 1989.

[8] T. Arak and D. Surgailis. Consistent polygonal fields. *Probab. Theory Relat. Fields*, 89(3):319–346, 1991.

[9] S. Axler, P. Bourdon, and W. Ramey. *Harmonic Function Theory*, volume 137 of *Graduate Texts in Mathematics*. Springer-Verlag, New York, second edition, 2001.

[10] K. Ball. An elementary introduction to modern convex geometry. In *Flavors of Geometry*, volume 31 of *Mathematical Sciences Research Institute Publications*, pages 1–58. Cambridge Univ. Press, Cambridge, 1997.

[11] R. Baptista, E. O'Reilly, and Y. Xie. TrIM: Transformed iterative Mondrian forests for gradient-based dimension reduction and high-dimensional regression. *arXiv:2407.09964*, pages 1–39, 2024.

[12] M. Beil, S. Eckel, F. Fleischer, H. Schmidt, V. Schmidt, and P. Walther. Fitting of random tessellation models to keratin filament networks. *J. Theor. Biol.*, 241(1):62–72, 2006.

[13] V. Beneš and J. Rataj. *Stochastic Geometry: Selected Topics*. Kluwer Academic Publishers, Boston, MA, 2004.

[14] J. Bertoin. *Random Fragmentation and Coagulation Processes*, volume 102 of *Cambridge Studies in Advanced Mathematics*. Cambridge University Press, Cambridge, 2006.

[15] C. Betken, D. Hug, and C. Thäle. Intersections of Poisson k-flats in constant curvature spaces. *Stoch. Process. Appl.*, 165:96–129, 2023.

[16] C. Betken, T. Kaufmann, K. Meier, and C. Thäle. Second-order properties for planar Mondrian tessellations. *Methodol. Comput. Appl. Probab.*, 25(2):Paper No. 47, 28, 2023.

[17] J. D. Biggins. Martingale convergence in the branching random walk. *J. Appl. Probab.*, 14(1):25–37, 1977.

[18] W. Blaschke. *Vorlesungen über Integralgeometrie*. Deutscher Verlag der Wissenschaften, Berlin, 3te Aufl. 1955.

[19] R. C. Bradley. *Introduction to Strong Mixing Conditions. Vol. 1*. Kendrick Press, Heber City, UT, 2007.

[20] K. A. Brakke. Statistics of random plane Voronoi tessellations. Technical report, Susquehanna University Selinsgrove, Pennsylvania, 2005.

[21] L. Breiman. *Probability*, volume 7 of *Classics in Applied Mathematics*. Society for Industrial and Applied Mathematics (SIAM), Philadelphia, PA, 1992. Corrected reprint of the 1968 original.

[22] A. Burchard and H. Hajaiej. Rearrangement inequalities for functionals with monotone integrands. *J. Funct. Anal.*, 233(2):561–582, 2006.

[23] A. Burchard and M. Schmuckenschläger. Comparison theorems for exit times. *Geom. Funct. Anal.*, 11(4):651–692, 2001.

https://doi.org/10.1515/9783112222409-025

[24] J. Burridge and R. Cowan. Planar tessellations that have the half-Gilbert structure. *Adv. Appl. Probab.*, 48(2):574–584, 2016.

[25] J. Burridge, R. Cowan, and I. Ma. Full- and half-Gilbert tessellations with rectangular cells. *Adv. Appl. Probab.*, 45(1):1–19, 2013.

[26] P. Calka and N. Chenavier. Extreme values for characteristic radii of a Poisson-Voronoi tessellation. *Extremes*, 17(3):359–385, 2014.

[27] A. Callegaro and M. I. Roberts. A spatially-dependent fragmentation process. *Probab. Theory Relat. Fields*, 192:163–266, 2025.

[28] P. Cesana and B. M. Hambly. A probabilistic model for interfaces in a martensitic phase transition. *J. Appl. Probab.*, 59(4):1081–1105, 2022.

[29] N. Chenavier. A general study of extremes of stationary tessellations with examples. *Stoch. Process. Appl.*, 124(9):2917–2953, 2014.

[30] N. Chenavier and R. Hemsley. Extremes for the inradius in the Poisson line tessellation. *Adv. Appl. Probab.*, 48(2):544–573, 2016.

[31] N. Chenavier and W. Nagel. The largest order statistics for the inradius in an isotropic STIT tessellation. *Extremes*, 22(4):571–598, 2019.

[32] S. N. Chiu. Mean-value formulae for the neighbourhood of the typical cell of a random tessellation. *Adv. Appl. Probab.*, 26(3):565–576, 1994.

[33] S. N. Chiu, D. Stoyan, W. S. Kendall, and J. Mecke. *Stochastic Geometry and its Applications*. Wiley Series in Probability and Statistics. John Wiley & Sons, Ltd., Chichester, third edition, 2013.

[34] T. Courtat, L. Decreusefond, and P. Martins. Stochastic simulation of urban environments. Application to path-loss in wireless systems. *arXiv:1604.00688*, pages 1–22, 2016.

[35] R. Cowan. The use of the ergodic theorems in random geometry. *Adv. Appl. Probab.*, (suppl):47–57, 1978. Spatial patterns and processes (Proc. Conf., Canberra, 1977).

[36] R. Cowan. Properties of ergodic random mosaic processes. *Math. Nachr.*, 97:89–102, 1980.

[37] R. Cowan. A collection of problems in random geometry. In *Stochastic Geometry, Geometric Statistics, Stereology (Oberwolfach, 1983)*, volume 65 of *Teubner-Texte zur Mathematik*, pages 64–68. Teubner, Leipzig, 1984.

[38] R. Cowan. A mosaic of triangular cells formed with sequential splitting rules. *J. Appl. Probab.*, 41(A):3–15, 2004.

[39] R. Cowan. New classes of random tessellations arising from iterative division of cells. *Adv. Appl. Probab.*, 42(1):26–47, 2010.

[40] R. Cowan. Line segments in the isotropic planar STIT tessellation. *Adv. Appl. Probab.*, 45(2):295–311, 2013.

[41] R. Cowan and S. Chen. The random division of faces in a planar graph. *Adv. Appl. Probab.*, 28(2):377–383, 1996.

[42] R. Cowan and C. Thäle. The character of planar tessellations which are not side-to-side. *Image Anal. Stereol.*, 33(1):39–54, 2014.

[43] R. Cowan and V. Weiss. Constraints on the fundamental topological parameters of spatial tessellations. *Math. Nachr.*, 288(5–6):540–565, 2015.

[44] R. Cowan and V. Weiss. Line segments which are unions of tessellation edges. *Image Anal. Stereol.*, 37(1):83–98, 2018.

[45] D. J. Daley and D. Vere-Jones. *An Introduction to the Theory of Point Processes. Elementary Theory and Methods. Vol. I.* Probability and its Applications (New York). Springer-Verlag, New York, second edition, 2003.

[46] D. J. Daley and D. Vere-Jones. *An Introduction to the Theory of Point Processes. General Theory and Structure. Vol. II.* Probability and its Applications (New York). Springer, New York, second edition, 2008.

[47] M. H. A. Davis. *Markov Models and Optimization*, volume 49 of *Monographs on Statistics and Applied Probability*. Chapman & Hall, London, 1993.

[48] T. de la Rue. Espaces de Lebesgue. In *Séminaire de Probabilités, XXVII*, volume 1557 of *Lecture Notes in Mathematics*, pages 15–21. Springer, Berlin, 1993.
[49] C. Deuß. *Zufällige Mosaike und das Kovariogramm*. PhD Thesis. Ruhr University Bochum, 2016.
[50] C. Deuss, J. Hörrmann, and C. Thäle. A random cell splitting scheme on the sphere. *Stoch. Process. Appl.*, 127(5):1544–1564, 2017.
[51] C. Deuß and C. Thäle. The mixing property of STIT tessellations revisited. *North-West. Eur. J. Math.*, 2:1–15, 2016.
[52] P. Embrechts, C. Klüppelberg, and T. Mikosch. *Modelling Extremal Events*, volume 33 of *Applications of Mathematics (New York)*. Springer-Verlag, Berlin, 1997.
[53] S. N. Ethier and T. G. Kurtz. *Markov Processes. Characterization and Convergence*. Wiley Series in Probability and Mathematical Statistics: Probability and Mathematical Statistics. John Wiley & Sons, Inc., New York, 1986.
[54] X. Fan, B. Li, and S. Sisson. The binary space partitioning-tree process. In *Proceedings of the Twenty-First International Conference on Artificial Intelligence and Statistics. Proceedings of Machine Learning Research*, volume 84, pages 1859–1867, 2018.
[55] N. A. Friedman and D. S. Ornstein. On isomorphism of weak Bernoulli transformations. *Adv. Math.*, 5:365–394, 1970.
[56] S. Ge, S. Wang, and L. Elliott. Shape modeling with spline partitions. *Stat. Comput.*, 33:13:1–12, 2023.
[57] S. Ge, S. Wang, Y. W. Teh, L. Wang, and L. Elliott. Random tessellation forests. *Adv. Neural Inf. Process. Syst.*, 32:9571–9581, 2019.
[58] H.-O. Georgii, T. Schreiber, and C. Thäle. Branching random tessellations with interaction: A thermodynamic view. *Ann. Probab.*, 43(4):1892–1943, 2015.
[59] E. N. Gilbert. Random plane networks and needle-shaped crystals. In *Applications of Undergraduate Mathematics in Engineering*. Macmillan, New York, 1967.
[60] B. V. Gnedenko and A. N. Kolmogorov. *Limit Distributions for Sums of Independent Random Variables*. Addison-Wesley Publishing Co., Reading, Mass.-London-Don Mills, Ont., revised edition, 1968. Translated from the Russian, annotated, and revised by K. L. Chung. With appendices by J. L. Doob and P. L. Hsu.
[61] I. S. Gradshteyn and I. M. Ryzhik. *Table of Integrals, Series, and Products*. Elsevier/Academic Press, Amsterdam, eighth edition, 2015. Translated from the Russian. Translation edited and with a preface by Daniel Zwillinger and Victor Moll.
[62] D. E. Grady. Dynamic fragmentation of solids. In *Shock Waves Science and Technology Reference Library, Vol. 3*. Springer-Verlag, Berlin, Heidelberg, 2009.
[63] D. E. Grady and M. E. Kipp. Geometric statistics and dynamic fragmentation. *J. Appl. Phys.*, 58:1210–1222, 1985.
[64] A. Gusakova, Z. Kabluchko, and C. Thäle. The β-Delaunay tessellation IV: Mixing properties and central limit theorems. *Stoch. Dyn.*, 23(3):Paper No. 2350021, 38, 2023.
[65] H. Hadwiger. Das Wills'sche Funktional. *Monatshefte Math.*, 79:213–221, 1975.
[66] J. Haezendonck. Abstract Lebesgue-Rohlin spaces. *Bull. Soc. Math. Belg.*, 25:243–258, 1973.
[67] A. Hafver, E. Jettestuen, M. Kobchenko, D. K. Dysthe, P. Meakin, and A. Malthe-Sørenssen. Classification of fracture patterns by heterogeneity and topology. *Europhys. Lett.*, 105:56004-p1–56004-p6, 2014.
[68] E. F. Harding and D. G. Kendall, editors. *Stochastic Geometry*. Wiley Series in Probability and Mathematical Statistics. John Wiley & Sons, London-New York-Sydney, 1974. A tribute to the memory of Rollo Davidson.
[69] L. Heinrich. Normal approximation for some mean-value estimates of absolutely regular tessellations. *Math. Methods Stat.*, 3(1):1–24, 1994.
[70] L. Heinrich. Central limit theorems for motion-invariant Poisson hyperplanes in expanding convex bodies. *Rend. Circ. Mat. Palermo (2) Suppl.*, (81):187–212, 2009.

[71] L. Heinrich and I. S. Molchanov. Central limit theorem for a class of random measures associated with germ-grain models. *Adv. Appl. Probab.*, 31(2):283–314, 1999.

[72] L. Heinrich and L. Muche. Second-order properties of the point process of nodes in a stationary Voronoi tessellation. *Math. Nachr.*, 281(3):350–375, 2008.

[73] L. Heinrich and L. Muche. Erratum: Second-order properties of the point process of nodes in a stationary Voronoi tessellation. *Math. Nachr.*, 283(11):1674–1676, 2010.

[74] L. Heinrich, H. Schmidt, and V. Schmidt. Limit theorems for stationary tessellations with random inner cell structures. *Adv. Appl. Probab.*, 37(1):25–47, 2005.

[75] L. Heinrich, H. Schmidt, and V. Schmidt. Limit theorems for functionals on the facets of stationary random tessellations. *Bernoulli*, 13(3):868–891, 2007.

[76] I. S. Helland. Central limit theorems for martingales with discrete or continuous time. *Scand. J. Stat.*, 9(2):79–94, 1982.

[77] G. Hernandez, R. Leon, L. Salinas, and E. Dimnet. A fragmentation model with neighborhood interaction. *Appl. Math. Model.*, 36:1694–1702, 2012.

[78] F. Herold. *Random Mosaics in Hyperbolic Space*. PhD Thesis. Karlsruher Institut für Technologie (KIT), 2020.

[79] F. Herold, D. Hug, and C. Thäle. Does a central limit theorem hold for the k-skeleton of Poisson hyperplanes in hyperbolic space? *Probab. Theory Relat. Fields*, 179(3–4):889–968, 2021.

[80] D. Hug and R. Schneider. Faces of Poisson-Voronoi mosaics. *Probab. Theory Relat. Fields*, 151(1–2):125–151, 2011.

[81] D. Hug and R. Schneider. Faces with given directions in anisotropic Poisson hyperplane mosaics. *Adv. Appl. Probab.*, 43(2):308–321, 2011.

[82] D. Hug and R. Schneider. *Poisson Hyperplane Tessellations*. Springer Monographs in Mathematics. Springer Nature Switzerland, 2024.

[83] D. Hug and C. Thäle. Splitting tessellations in spherical spaces. *Electron. J. Probab.*, 24:Paper No. 24, 2019.

[84] K. Itô. *Introduction to Probability Theory*. Cambridge University Press, Cambridge, 1984.

[85] K. Iyer, B. Jamtveit, J. Mathiesen, A. Malthe-Sørenssen, and J. Feder. Reaction-assisted hierarchical fracturing during serpentinization. *Earth Planet. Sci. Lett.*, 267:503–516, 2008.

[86] J. Jacod and A. N. Shiryaev. *Limit Theorems for Stochastic Processes*, volume 288 of *Grundlehren der mathematischen Wissenschaften [Fundamental Principles of Mathematical Sciences]*. Springer-Verlag, Berlin, second edition, 2010.

[87] A. Jakubowski. A non-Skorohod topology on the Skorohod space. *Electron. J. Probab.*, 2:No. 4, 21 pp. 1997.

[88] J. Kahn. How many T-tessellations on k lines? Existence of associated Gibbs measures on bounded convex domains. *Random Struct. Algorithms*, 47(3):561–587, 2015.

[89] O. Kallenberg. *Foundations of Modern Probability*, volume 99 of *Probability Theory and Stochastic Modelling*. Springer Nature Switzerland, third edition, 2021.

[90] I. Karatzas and S. E. Shreve. *Brownian Motion and Stochastic Calculus*, volume 113 of *Graduate Texts in Mathematics*. Springer-Verlag, New York, second edition, 1991.

[91] W. S. Kendall and J. Mecke. The range of the mean-value quantities of planar tessellations. *J. Appl. Probab.*, 24(2):411–421, 1987.

[92] J. Kern. Skorokhod topologies: What they are and why we should care. *Math. Semesterber.*, 71(1):1–18, 2024.

[93] H. Kesten and B. P. Stigum. A limit theorem for multidimensional Galton-Watson processes. *Ann. Math. Stat.*, 37:1211–1223, 1966.

[94] J. F. C. Kingman. *Poisson Processes*, volume 3 of *Oxford Studies in Probability*. The Clarendon Press, Oxford University Press, New York, 1993.

[95] K. Kiêu, K. Adamczyk-Chauvat, H. Monod, and R. S. Stoica. A completely random T-tessellation model and Gibbsian extensions. *Spat. Stat.*, 6:118–138, 2013.

[96] D. A. Klain and G.-C. Rota. *Introduction to Geometric Probability*. Lezioni Lincee [Lincei Lectures]. Cambridge University Press, Cambridge, 1997.
[97] M. A. Klatt, G. Last, K. Mecke, C. Redenbach, F. M. Schaller, and G. E. Schröder-Turk. Cell shape analysis of random tessellations based on Minkowski tensors. In *Tensor Valuations and Their Applications in Stochastic Geometry and Imaging*, volume 2177 of *Lecture Notes in Mathematics*, pages 385–421. Springer, 2017.
[98] F. C. Klebaner. *Introduction to Stochastic Calculus with Applications*. Imperial College Press, London, second edition, 2005.
[99] T. G. Kurtz. Random time changes and convergence in distribution under the Meyer-Zheng conditions. *Ann. Probab.*, 19(3):1010–1034, 1991.
[100] R. Lachièze-Rey. Mixing properties for STIT tessellations. *Adv. Appl. Probab.*, 43(1):40–48, 2011.
[101] G. Last and M. D. Penrose. *Lectures on the Poisson Process*, volume 7 of *Institute of Mathematical Statistics Textbooks*. Cambridge University Press, Cambridge, 2018.
[102] R. León, E. Montero, and W. Nagel. Parameter optimization on a tessellation model for crack pattern simulation. *IEEE Access*, 125421–125438, 2023.
[103] R. León, W. Nagel, J. Ohser, and S. Arscott. Modeling crack patterns by modified STIT tessellations. *Image Anal. Stereol.*, 39(1):33–46, 2020.
[104] E. H. Lieb. Sharp constants in the Hardy-Littlewood-Sobolev and related inequalities. *Ann. Math. (2)*, 118(2):349–374, 1983.
[105] M. Lotz, M. B. McCoy, I. Nourdin, G. Peccati, and J. A. Tropp. Concentration of the intrinsic volumes of a convex body. In *Geometric Aspects of Functional Analysis. Vol. II*, volume 2266 of *Lecture Notes in Mathematics*, pages 139–167. Springer, Cham, 2020.
[106] M. S. Mackisack and R. E. Miles. Homogeneous rectangular tessellations. *Adv. Appl. Probab.*, 28(4):993–1013, 1996.
[107] R. Maier. *Iterated Random Tessellations with Applications to Spatial Modelling of Telecommunication Networks*. PhD thesis, University of Ulm, 2003.
[108] R. Maier, J. Mayer, and V. Schmidt. Distributional properties of the typical cell of stationary iterated tessellations. *Math. Methods Oper. Res.*, 59(2):287–302, 2004.
[109] R. Maier and V. Schmidt. Stationary iterated tessellations. Technical report, University of Ulm, 2002.
[110] R. Maier and V. Schmidt. Stationary iterated tessellations. *Adv. Appl. Probab.*, 35(2):337–353, 2003.
[111] V. S. Mandrekar. *Weak Convergence of Stochastic Processes. With Applications to Statistical Limit Theorems*. De Gruyter Graduate. De Gruyter, Berlin, 2016.
[112] N. F. G. Martin. The Rosenblatt mixing condition and Bernoulli shifts. *Ann. Probab.*, 2:333–338, 1974.
[113] S. Martínez. STIT tessellations are Bernoulli and standard. *Ergod. Theory Dyn. Syst.*, 34(3):876–892, 2014.
[114] S. Martínez and W. Nagel. Ergodic description of STIT tessellations. *Stochastics*, 84(1):113–134, 2012.
[115] S. Martínez and W. Nagel. STIT tessellations have trivial tail σ-algebra. *Adv. Appl. Probab.*, 46(3):643–660, 2014.
[116] S. Martínez and W. Nagel. STIT process and trees. *Markov Process. Relat. Fields*, 21(4):847–868, 2015.
[117] S. Martínez and W. Nagel. The β-mixing rate of STIT tessellations. *Stochastics*, 88(3):396–414, 2016.
[118] S. Martínez and W. Nagel. STIT tessellations – ergodic limit theorems and bounds for the speed of convergence. *arXiv:1609.01138*, pages 1–14, 2016.
[119] S. Martínez and W. Nagel. Regenerative processes for Poisson zero polytopes. *Adv. Appl. Probab.*, 50(4):1217–1226, 2018.
[120] S. Martínez and W. Nagel. Tessellation-valued processes that are generated by cell division. *J. Appl. Probab.*, 61(2):697–715, 2024.
[121] G. Matheron. Hyperplans Poissoniens et compacts de Steiner. *Adv. Appl. Probab.*, 6:563–579, 1974.
[122] G. Matheron. *Random Sets and Integral Geometry*. John Wiley & Sons, New York-London-Sydney, 1975.
[123] P. Mattila. *Geometry of Sets and Measures in Euclidean Spaces*, volume 44 of *Cambridge Studies in Advanced Mathematics*. Cambridge University Press, Cambridge, 1995. Fractals and rectifiability.

[124] J. Mecke. Palm methods for stationary random mosaics. In *Combinatorial Principles in Stochastic Geometry*, pages 124–132. Armenian Academy of Sciences Publ., Erevan, 1980.

[125] J. Mecke. Parametric representation of mean values for stationary random mosaics. *Math. Oper.forsch. Stat., Ser. Stat.*, 15(3):437–442, 1984.

[126] J. Mecke. Joint distribution of direction and length of the typical I-segment in a homogeneous random planar tessellation stable under iteration. *Izv. Nats. Akad. Nauk Armen. Mat.*, 44(1):73–84, 2009.

[127] J. Mecke. Inhomogeneous random planar tessellations generated by lines. *Izv. Nats. Akad. Nauk Armen. Mat.*, 45(6):63–76, 2010.

[128] J. Mecke, W. Nagel, and V. Weiss. Length distributions of edges in planar stationary and isotropic STIT tessellations. *Izv. Nats. Akad. Nauk Armen. Mat.*, 42(1):39–60, 2007.

[129] J. Mecke, W. Nagel, and V. Weiss. A global construction of homogeneous random planar tessellations that are stable under iteration. *Stochastics*, 80(1):51–61, 2008.

[130] J. Mecke, W. Nagel, and V. Weiß. The iteration of random tessellations and a construction of a homogeneous process of cell divisions. *Adv. Appl. Probab.*, 40:49–59, 2008.

[131] J. Mecke, W. Nagel, and V. Weiß. Some distributions for I-segments of planar random homogeneous STIT tessellations. *Math. Nachr.*, 284(11–12):1483–1495, 2011.

[132] K. Meier. *Probabilistic Analysis of Coxeter Statistics and STIT Tessellations*. PhD thesis, Ruhr University Bochum, 2023.

[133] R. E. Miles. On the repeated random splitting of a planar domain. In *Stochastic Geometry, Geometric Statistics, Stereology (Oberwolfach, 1983)*, volume 65 of *Teubner-Texte zur Mathematik*, pages 110–123. Teubner, Leipzig, 1984.

[134] R. E. Miles. A large class of random tessellations with the classic Poisson polygon distributions (abstract). *Adv. Appl. Probab.*, 30:285, 1998.

[135] R. E. Miles and M. S. Mackisack. Further random tessellations with the classic Poisson polygon distributions (abstract). *Adv. Appl. Probab.*, 28:338–339, 1996.

[136] R. E. Miles and M. S. Mackisack. A large class of random tessellations with the classic Poisson polygon distributions. *Forma*, 17(1):1–17, 2002.

[137] I. Molchanov. *Theory of Random Sets*. Springer-Verlag, London, second edition, 2017.

[138] J. Møller. Random tessellations in $\mathbf{R}^d$. *Adv. Appl. Probab.*, 21(1):37–73, 1989.

[139] L. J. Mosser and S. K. Matthäi. Tessellations stable under iteration. Evaluation of application as an improved stochastic discrete fracture modeling algorithm. In *International Discrete Fracture Network Engineering Conference*, DFNE 2014 – 163, 2014.

[140] W. Nagel and E. Biehler. Consistency of constructions for cell division processes. *Adv. Appl. Probab.*, 47(3):640–651, 2015.

[141] W. Nagel, J. Mecke, J. Ohser, and V. Weiß. A tessellation model for crack patterns on surfaces. *Image Anal. Stereol.*, 27(2):73–78, 2008.

[142] W. Nagel, J. Mecke, and V. Weiß. Homogeneous STIT tessellations – a stereological aspect. In V. Capasso, G. Aletti, and A. Micheletti, editors, *Stereology and Image Analysis. ECS10 The 10th European Congress of ISS*, pages 231–234, Bologna, 2009.

[143] W. Nagel, N. L. Nguyen, C. Thäle, and V. Weiß. A Mecke-type formula and Markov properties for STIT tessellation processes. *ALEA Lat. Am. J. Probab. Math. Stat.*, 14(2):691–718, 2017.

[144] W. Nagel and V. Weiss. Limits of sequences of stationary planar tessellations. *Adv. Appl. Probab.*, 35(1):123–138, 2003. In honor of Joseph Mecke.

[145] W. Nagel and V. Weiß. Crack STIT tessellations: Characterization of stationary random tessellations stable with respect to iteration. *Adv. Appl. Probab.*, 37:859–883, 2005.

[146] W. Nagel and V. Weiss. STIT tessellations in the plane. *Rend. Circ. Mat. Palermo (2) Suppl.*, (77):441–458, 2006.

[147] W. Nagel and V. Weiß. Mean values for homogeneous STIT tessellations in 3D. *Image Anal. Stereol.*, 27:29–37, 2008.

[148] G. Nassif, C. Gloaguen, and P. Martins. A combined stochastic and physical framework for modeling indoor 5G millimeter wave propagation. *IEEE Trans. Antennas Propag.*, 70(6):4712–4727, 2022.

[149] D. Neuhäuser, C. Hirsch, C. Gloaguen, and V. Schmidt. Ratio limits and simulation algorithms for the Palm version of stationary iterated tessellations. *J. Stat. Comput. Simul.*, 84(7):1486–1504, 2014.

[150] D. Neuhäuser, C. Hirsch, C. Gloaguen, and V. Schmidt. A stochastic model for multi-hierarchical networks. *Methodol. Comput. Appl. Probab.*, 18(4):1129–1151, 2016.

[151] N. L. Nguyen, V. Weiss, and R. Cowan. Column tessellations. *Image Anal. Stereol.*, 34(2):87–100, 2015.

[152] F. W. J. Olver, D. W. Lozier, R. F. Boisvert, and C. W. Clark, editors. *NIST Handbook of Mathematical Functions*. U.S. Department of Commerce, National Institute of Standards and Technology, Washington, DC; Cambridge University Press, Cambridge, 2010. With 1 CD-ROM (Windows, Macintosh and UNIX).

[153] E. O'Reilly. Statistical advantages of oblique randomized decision trees and forests. *arXiv:2407.02458*, pages 1–43, 2024.

[154] E. O'Reilly and N. M. Tran. Stochastic geometry to generalize the Mondrian process. *SIAM J. Math. Data Sci.*, 4(2):531–552, 2022.

[155] E. O'Reilly and N. M. Tran. Minimax rates for high-dimensional random tessellation forests. *J. Mach. Learn. Res.*, 25:1–32, 2024.

[156] D. S. Ornstein. *Ergodic Theory, Randomness, and Dynamical Systems*. Yale University Press, New Haven, Conn.-London, 1974. James K. Whittemore Lectures in Mathematics given at Yale University, Yale Mathematical Monographs, No. 5.

[157] D. S. Ornstein and B. Weiss. Entropy and isomorphism theorems for actions of amenable groups. *J. Anal. Math.*, 48:1–141, 1987.

[158] C. Osborne and E. O'Reilly. The uniformly rotated Mondrian kernel. In *Proceedings of the 28th International Conference on Artificial Intelligence and Statistics*, pages 4663–4671, 2025.

[159] M. Otto. Extremal behavior of large cells in the Poisson hyperplane mosaic. *Electron. J. Probab.*, 28(162):1–29, 2023.

[160] Z. Pawlas. Self-crossing points of a line segment process. *Methodol. Comput. Appl. Probab.*, 16(2):295–309, 2014.

[161] W. H. Press, S. A. Teukolsky, W. T. Vetterling, and B. P. Flannery. *Numerical Recipes. The Art of Scientific Computing*. Cambridge University Press, third edition, 2007.

[162] B. v. Querenburg. *Mengentheoretische Topologie*. Springer-Verlag, Berlin-New York, third edition, 2001.

[163] C. Redenbach, J. Ohser, and A. Moghiseh. Second-order characteristics of the edge system of random tessellations and the PPI value of foams. *Methodol. Comput. Appl. Probab.*, 18:59–79, 2016.

[164] C. Redenbach and C. Thäle. On the arrangement of cells in planar STIT and Poisson line tessellations. *Methodol. Comput. Appl. Probab.*, 15(3):643–654, 2013.

[165] C. Redenbach and C. Thäle. Second-order comparison of three fundamental tessellation models. *Statistics*, 47(2):237–257, 2013.

[166] D. Revuz and M. Yor. *Continuous Martingales and Brownian Motion*, volume 293 of *Grundlehren der mathematischen Wissenschaften [Fundamental Principles of Mathematical Sciences]*. Springer-Verlag, Berlin, third edition, 1999.

[167] S. M. Ross. *Stochastic Processes*. Wiley Series in Probability and Statistics: Probability and Statistics. John Wiley & Sons, Inc., New York, second edition, 1996.

[168] D. M. Roy and Y. W. Teh. The Mondrian process. In *Proceedings of the 21st International Conference on Neural Information Processing Systems*, pages 1377–1384, 2008.

[169] L. A. Santaló. *Integral Geometry and Geometric Probability*. Cambridge Mathematical Library. Cambridge University Press, Cambridge, second edition, 2004.

[170] L. A. Santaló. Mixed random mosaics. *Math. Nachr.*, 117(1):129–133, 1984.

[171] R. Schneider. *Convex Bodies: The Brunn-Minkowski Theory*. Cambridge University Press, second edition, 2013.

[172] R. Schneider and W. Weil. *Stochastic and Integral Geometry*. Probability and its Applications (New York). Springer-Verlag, Berlin, 2008.

[173] T. Schreiber and N. Soja. Limit theory for planar Gilbert tessellations. *Probab. Math. Stat.*, 31(1):149–160, 2011.

[174] T. Schreiber and C. Thäle. Second-order properties and central limit theory for the vertex process of iteration infinitely divisible and iteration stable random tessellations in the plane. *Adv. Appl. Probab.*, 42(4):913–935, 2010.

[175] T. Schreiber and C. Thäle. Intrinsic volumes of the maximal polytope process in higher dimensional STIT tessellations. *Stoch. Process. Appl.*, 121(5):989–1012, 2011.

[176] T. Schreiber and C. Thäle. Second-order theory for iteration stable tessellations. *Probab. Math. Stat.*, 32(2):281–300, 2012.

[177] T. Schreiber and C. Thäle. Geometry of iteration stable tessellations: Connection with Poisson hyperplanes. *Bernoulli*, 19(5A):1637–1654, 2013.

[178] T. Schreiber and C. Thäle. Limit theorems for iteration stable tessellations. *Ann. Probab.*, 41(3B):2261–2278, 2013.

[179] T. Schreiber and C. Thäle. Shape-driven nested Markov tessellations. *Stochastics*, 85:510–531, 2013.

[180] J. Serra. *Image Analysis and Mathematical Morphology*. Academic Press, Inc. [Harcourt Brace Jovanovich, Publishers], London, 1984. English version revised by Noel Cressie.

[181] S. Sheffield. Gaussian free fields for mathematicians. *Probab. Theory Relat. Fields*, 139(3–4):521–541, 2007.

[182] D. Stoyan. Inequalities and bounds for variances of point processes and fibre processes. *Math. Oper.forsch. Stat., Ser. Stat.*, 14(3):409–419, 1983.

[183] D. Stoyan and J. Ohser. Cross-correlation measures of weighted random measures and their estimation. *Teor. Veroyatn. Primen.*, 29(2):338–347, 1984.

[184] C. Thäle. Moments of the length of line segments in homogeneous planar STIT tessellations. *Image Anal. Stereol.*, 28(2):69–76, 2009.

[185] C. Thäle. The distribution of the number of nodes in the relative interior of the typical I-segment in homogeneous planar anisotropic STIT tessellations. *Comment. Math. Univ. Carol.*, 51(3):503–512, 2010.

[186] C. Thäle. Supplementary results for length distributions in planar STIT tessellations. *Forma*, 26(1):1–6, 2011.

[187] C. Thäle and V. Weiß. New mean values for homogeneous spatial tessellations that are stable under iteration. *Image Anal. Stereol.*, 29(3):143–157, 2010.

[188] C. Thäle and V. Weiß. The combinatorial structure of spatial STIT tessellations. *Discrete Comput. Geom.*, 50(3):649–672, 2013.

[189] C. Thäle, V. Weiß, and W. Nagel. Spatial STIT tessellations: Distributional results for I-segments. *Adv. Appl. Probab.*, 44(3):635–654, 2012.

[190] C. Thomas. Extremum properties of the intersection densities of stationary Poisson hyperplane processes. *Math. Oper.forsch. Stat., Ser. Stat.*, 15(3):443–449, 1984.

[191] J. A. van Casteren. *Markov Processes, Feller Semigroups and Evolution Equations*, volume 12 of *Series on Concrete and Applicable Mathematics*. World Scientific Publishing Co. Pte. Ltd., Hackensack, NJ, 2011.

[192] A. M. Vershik. Theory of decreasing sequences of measurable partitions. *Algebra Anal.*, 6(4):1–68, 1994.

[193] V. Weiss and R. Cowan. Topological relationships in spatial tessellations. *Adv. Appl. Probab.*, 43(4):963–984, 2011.

[194] V. Weiss and W. Nagel. Interdependences of directional quantities of planar tessellations. *Adv. Appl. Probab.*, 31(3):664–678, 1999.

[195] V. Weiss, J. Ohser, and W. Nagel. Second moment measure and K-function for planar STIT tessellations. *Image Anal. Stereol.*, 29(2):121–131, 2010.

[196] V. Weiss and M. Zähle. Geometric measures for random curved mosaics of R^d. *Math. Nachr.*, 138:313–326, 1988.

[197] W. Whitt. *Stochastic-Process Limits*. Springer Series in Operations Research. Springer-Verlag, New York, 2002.

[198] W. Whitt. Proofs of the martingale FCLT. *Probab. Surv.*, 4:268–302, 2007.

[199] J. A. Wieacker. Intersections of random hypersurfaces and visibility. *Probab. Theory Relat. Fields*, 71(3):405–433, 1986.

[200] J. A. Wieacker. Geometric inequalities for random surfaces. *Math. Nachr.*, 142:73–106, 1989.

[201] W. Xu. Entropy of chord distribution of convex bodies. *Proc. Am. Math. Soc.*, 147(7):3131–3141, 2019.

[202] K.-i. Yoshihara. Limiting behavior of U-statistics for stationary, absolutely regular processes. *Z. Wahrscheinlichk.theor. Verw. Geb.*, 35(3):237–252, 1976.

[203] X. Yu, T. Courtat, P. Martins, L. Decreusefond, and J.-M. Kelif. Crack STIT tessellations for city modeling and impact of terrain topology on wireless propagation. In *12th International Symposium on Modeling and Optimization in Mobile, Ad Hoc, and Wireless Networks (WiOpt)*, pages 22–29, 2014.

[204] M. Zähle. Random processes of Hausdorff rectifiable closed sets. *Math. Nachr.*, 108:49–72, 1982.

[205] G. Zhang. Isoperimetric inequalities for integral geometric invariants of random lines. *Acta Math. Sci. Ser. B Engl. Ed.*, 45(1):189–199, 2025.

Author index

https://doi.org/10.1515/9783112222409-026

Subject index

https://doi.org/10.1515/9783112222409-027

Notation index

https://doi.org/10.1515/9783112222409-028

De Gruyter Studies in Mathematics

Volume 104
Maria Fragoulopoulou
Banach and C*-Algebras, 2026
ISBN 978-3-11-114988-2, e-ISBN 978-3-11-115055-0, e-ISBN (ePUB) 978-3-11-115078-9

Volume 91
Johnny Henderson, Rodica Luca
Boundary Value Problems for Second-Order Finite Difference Equations and Systems, 2025
ISBN 978-3-11-222482-3, e-ISBN 978-3-11-222483-0, e-ISBN (ePUB) 978-3-11-222484-7

Volume 103
Nick Gurski, Juan Orendain, David Yetter
Compact Closed 2-Categories. Duality, Enrichment, and Strictification, 2025
ISBN 978-3-11-914815-3, e-ISBN 978-3-11-218854-5, e-ISBN (ePUB) 978-3-11-220458-0

Volume 102
Changxing Miao, Bo Zhang, Jiqiang Zheng
Harmonic Analysis Methods in Partial Differential Equations, 2025
ISBN 978-3-11-138451-1, e-ISBN 978-3-11-138472-6, e-ISBN (ePUB) 978-3-11-138486-3

Volume 101
Marko Kostić
Almost Periodic Type Solutions. to Integro-Differential-Difference Equations, 2025
ISBN 978-3-11-168728-5, e-ISBN 978-3-11-168974-6, e-ISBN (ePUB) 978-3-11-169021-6

Volume 64
Dorina Mitrea, Irina Mitrea, Marius Mitrea, Michael Taylor
The Hodge–Laplacian. Boundary Value Problems on Riemannian Manifolds, 2nd Edition, 2025
ISBN 978-3-11-148098-5, e-ISBN 978-3-11-148140-1, e-ISBN (ePUB) 978-3-11-148389-4

Volume 100
Changxing Miao, Ruipeng Shen
Regularity and Scattering of Dispersive Wave Equation. Multiplier Method and Morawetz Estimate, 2025
ISBN 978-3-11-148754-0, e-ISBN 978-3-11-148835-6, e-ISBN (ePUB) 978-3-11-148940-7

Volume 99
Marcus Laurel, Marius Mitrea
Weighted Morrey Spaces. Calderón-Zygmund Theory and Boundary Problems, 2024
ISBN 978-3-11-145816-8, e-ISBN 978-3-11-145827-4, e-ISBN (ePUB) 978-3-11-146145-8

www.degruyterbrill.com